Solidification and Crystallization Processing in Metals and Alloys

Solidification and Crystallization Processing in Metals and Alloys

Hasse Fredriksson
KTH, Royal Institute of Technology, Stockholm, Sweden

and

Ulla Åkerlind
University of Stockholm, Stockholm, Sweden

A John Wiley & Sons, Ltd., Publication

This edition first published 2012

Registered office
John Wiley & Sons Ltd, The Atrium, Southern Gate, Chichester, West Sussex, PO19 8SQ, United Kingdom

For details of our global editorial offices, for customer services and for information about how to apply for permission to reuse the copyright material in this book please see our website at www.wiley.com.

Library of Congress Cataloging-in-Publication Data

Fredriksson, Hasse.
Solidification and crystallization processing in metals and alloys / Hasse Fredriksson and Ulla Åkerlind.
p. cm.
Includes bibliographical references and index.
ISBN 978-1-119-99305-6 (cloth)
1. Metal crystals–Growth. I. Åkerlind, Ulla. II. Title.
TN690.F73 2012
669′.94–dc23

2011014030

A catalogue record for this book is available from the British Library.

ISBN: 978-1-119-99305-6
Set in 10/12pt, Times by Thomson Digital, Noida
Printed in [Country] by [Printer]

Contents

Preface

The present book is the third and last book in a series of three:

1. Materials Processing during Casting, Wiley 2006
2. Physics of Functional Materials, Wiley 2008
3. Solidification and Crystallization Processing in Metals and Alloys, Wiley 2012

Solidification and Crystallization Processing in Metals and Alloys represents a deeper interpretation of the solidification and crystallization processes than that treated in the book *Materials Processing during Casting*, written for the undergraduate level. The aim of the present book is to analyze the solidification and crystallization processes from a general point of view and in accordance with generally accepted results and experimental evidence of modern research in the field. Hence, the book does not treat applications on casting other than as occasional examples.

The book may be useful and suitable as a text book on courses at the Master and PhD level. The mathematical level is not discouragingly high. Ordinary basic courses in Mathematics at university level are enough. On the other hand, genuine knowledge of Physics is often required. The second book, *Physics of Functional Materials*, or any other Physics book with any other equivalent content, will cover this want for those who need it. Numerous citations to the second book are given in the present book.

Solidification and Crystallization Processing in Metals and Alloys starts with a chapter of basic thermodynamics. Chapter 1 is a review of the thermodynamics that later will be applied on metals and alloys. It may be a tough and abstract introduction. Alternatively, it can be studied in connection with later applications. Energy conditions play an important role for understanding the driving forces of solidification processes in metals and alloys. These topics are treated in Chapter 2. The structure and properties of interfaces between two phases and the nucleation of embryos and forming of stable nuclei are closely related to crystallization processes. The basic outlines of these fields are given in Chapters 3 and 4, respectively.

After these four basic and general chapters, Chapters 5 and 6 follow, where the mechanisms of the solidification and crystallization processes in vapours and liquids are extensively discussed. Heat transport during solidification processes is treated in Chapter 7, which also includes an orientation about modern methods of thermal analysis. Chapter 8 deals with crystal growth controlled by heat and mass transport.

The rest of the book is devoted to the structures of the solid phases that form during different types of solidification processes, i.e. faceted and dendritic structures (Chapter 9), eutectic structures (Chapter 10), peritectic structures (Chapter 11) and structure of Metallic Glasses (Chapter 12).

Solidification and Crystallization Processing in Metals and Alloys contains many solved examples in the text and exercises for students at the end of each chapter. Answers to all the exercises are given at the end of the book. In a 'Guide to Exercises' full solutions to all the exercises are given on the Internet at http://www.wiley.com/go/fredriksson3

Acknowledgements

This book is based on lectures and exercises presented to PhD and Master students over the years. We want to express our sincere thanks to them for their engagement and for the fruitful discussions we have had with them about different scientific topics in the field. The communication with them helped us to identify the items that were difficult for them to understand. In this way, we could try to explain better and improve the book.

We are most grateful to MSc Per Olov Nilsson for many fruitful mathematical discussions and general support through the years. We also wish to express our sincere thanks to Dr Jonas Åberg, Thomas Bergström (Casting of Metals, KTH, Stockholm), Dr Gunnar Edvinsson (University of Stockholm), Dr Bengt Örjan Jonsson (KTH), Dr Hani Nasser and Dr Sathees Ranganathan (Casting of Metals, KTH) for their valuable support concerning practical matters such as annoying computer problems and application of some special computer programs. We also thank Dr Lars Åkerlind warmly for patient assistance with checking parts of the manuscript.

We are very grateful for financial support from the Iron Masters Association in Sweden and the Foundation of Sven and Astrid Toresson, which made it possible for us to fulfill the last part of this project.

In particular, we want to express our sincere gratitude to Karin Fredriksson and Lars Åkerlind. Without their constant support and great patience through the years this book would never have been written.

Finally, we want to thank each other for a more than 10 years long and pleasant cooperation. It has been the perfect symbiosis. Neither of us could have written this trilogy without the other. We complemented each other well. One of us (guess who) contributed many years of research in the field and an ever-lasting enthusiasm to his devoted subject. The other one contributed with some Mathematics and Physics together with many years experience of book production and teaching Physics at university level and also plenty of *time* for the extensive project.

Hasse Fredriksson
Ulla Åkerlind
Stockholm, May 2010

1

Thermodynamic Concepts and Relationships

Solidification and Crystallization Processing in Metals and Alloys, First Edition. Hasse Fredriksson and Ulla Åkerlind.

1.1 Introduction

Solidification or crystallization is a process where the atoms are transferred from the disordered liquid state to the more ordered solid state. The rate of the crystallization process is described and controlled by kinetic laws. These laws give information of the movements of the atoms during the rearrangement. In most cases a driving force is involved that makes it possible to derive the rate of the solidification process.

The aim of this book is to study the solidification processes in metals and alloys. The laws of thermodynamics and other fundamental physical laws, which control the solidification, rule these processes.

In this preparatory chapter the basic concepts and laws of thermodynamics are introduced. They will be the tools in following chapters. In particular, this is true for the second chapter, where the driving forces of solidification for pure metals and binary alloys are derived and the relationship between energy curves of solid metals and metal melts and their phase diagrams is emphasized.

1.2 Thermodynamic Concepts and Relationships

1.2.1 First Law of Thermodynamics. Principle of Conservation of Energy

One of the most fundamental laws of physics is the law of conservation of energy. So far, it is known to be valid without any exceptions. When applied in thermodynamics it is called the first law of thermodynamics and is written

$$Q = U + W \tag{1.1}$$

where

Q = heat energy added to a closed system
U = the internal energy of the system or the sum of the kinetic and potential energies of the atoms
W = work done by the system.

Differentiating Equation (1.1) gives the relationship

$$\mathrm{d}Q = \mathrm{d}U + \mathrm{d}W = \mathrm{d}U + p\mathrm{d}V \tag{1.2}$$

The added heat $\mathrm{d}Q$ is used for the increase $\mathrm{d}U$ of the internal energy of the system and for the external work $p\mathrm{d}V$ against the surrounding pressure p when the volume of the system increases by the amount $\mathrm{d}V$.

1.2.2 Enthalpy

The enthalpy of a closed system is defined as

$$H = U + pV \tag{1.3}$$

The enthalpy H can be described as the 'heat content' of the system. In the absence of phase transformations we have

$$H = \int_0^T nC_\mathrm{p}\mathrm{d}T \tag{1.4}$$

where
H = enthalpy of the system
n = number of kmol
C_p = molar heat capacity of the system (J/kmol).

When the heat content of a system is changed the enthalpy changes by the amount ΔH. When the system *absorbs* heat from the surroundings the heat content of the system is *increased* and ΔH *is positive*.

When the system *emits* heat to the surroundings and *reduces* its heat content, ΔH is *negative*. The amount of heat absorbed by the surroundings is $(-\Delta H)$, which is a positive quantity.

Differentiating Equation (1.3) gives

$$dH = dU + pdV + Vdp \tag{1.5}$$

By use of Equation (1.2) we obtain

$$dH = dQ + Vdp$$

At constant pressure, the heat absorbed by a system equals its enthalpy increase:

$$(dH)_p = (dQ)_p \tag{1.6}$$

1.2.3 *Second Law of Thermodynamics. Entropy*

Second Law of Thermodynamics

By experience it is known that heat always is transferred spontaneously from a warmer to a colder body, never the contrary.

It is possible to transfer heat into mechanical work but never to 100%. Consider a closed system (no energy exchange with the surroundings) which consists of an engine in contact with two heat reservoirs. If an amount Q_1 of heat is emitted from reservoir 1 at a temperature T_1 to the engine and an amount Q_2 of heat is absorbed by reservoir 2 at temperature T_2, the energy difference $Q_1 - Q_2$ is transferred into mechanical work in the ideal case.

The reversed process is shown in Figure 1.1. It can be described as follows.

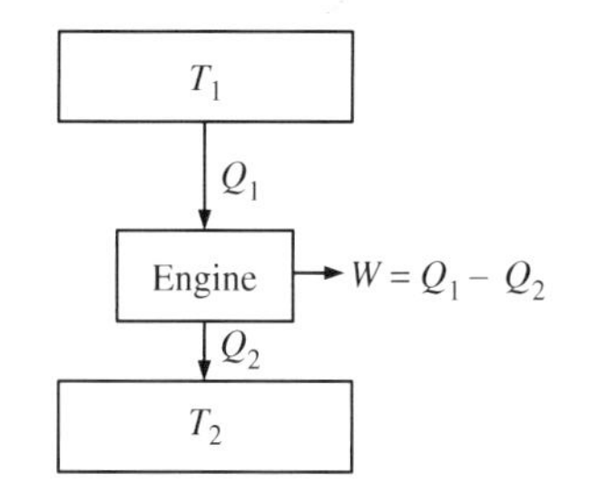

Figure 1.1 Engine in contact with two heat reservoirs. $T_1 > T_2$. The energy law gives $Q_1 = Q_2 + W$.

- Heat can be transferred from a colder body to a warmer body only if work or energy is supplied.

This is one of many ways to express the second law of thermodynamics.

In practice, heat is transferred into mechanical work in heat engines. The process in such a machine can theoretically be described by the Carnot cycle (Figure 1.2a). Fuel is burned and the combustion gas runs through the following cycle in the engine

1. The gas absorbs the amount Q_1 of heat and expands isothermally at temperature T_1.
2. The gas expands adiabatically. The temperature decreases from T_1 to T_2.
3. The gas is compressed isothermally at temperature T_2 and emits the amount Q_2 of heat.
4. The gas is compressed adiabatically. The temperature changes from T_2 back to T_1.

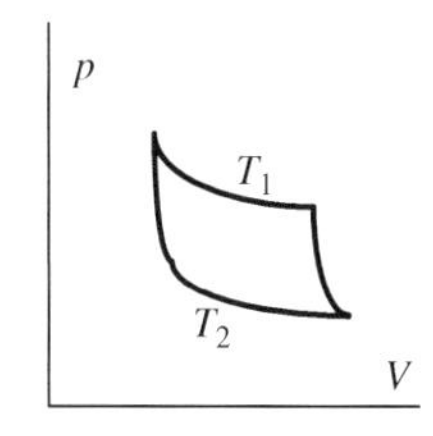

Figure 1.2a The Carnot cycle. The enclosed area represents the work done by the heat engine during a cycle.

Calculations of the expansion and compression works show that

$$\frac{Q_1}{Q_2} = \frac{T_1}{T_2} \tag{1.7}$$

In the ideal case, the efficiency η of the Carnot cycle is

$$\eta = \frac{Q_1 - Q_2}{Q_1} = \frac{T_1 - T_2}{T_1} \tag{1.8}$$

Entropy

A very useful and important quantity in thermodynamics is the entropy S. Entropy is defined by the relationship

$$dS = \frac{dQ}{T} \tag{1.9}$$

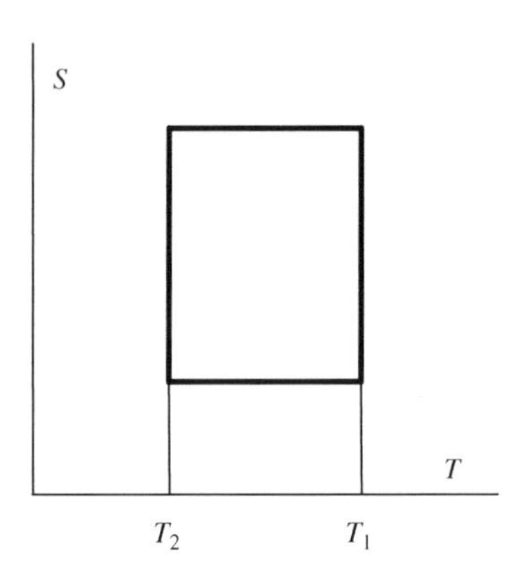

Figure 1.2b The Carnot cycle. The enclosed area represents the work done by the heat engine during a cycle.

As an example we represent the Carnot cycle in a TS diagram (Figure 1.2b). Horizontal lines represent the isothermal expansion and compression. The vertical lines illustrate the adiabatic steps of the cycle. The area of the rectangle equals the work done by the gas. A reversible adiabatic process is also called *isentropic*.

Entropy Change at an Isothermal Irreversible Expansion of a Gas
In many solidification processes the entropy change is of great interest. As a first example we will consider the isothermal, irreversible expansion of an ideal gas and calculate the entropy change.

Example 1.1

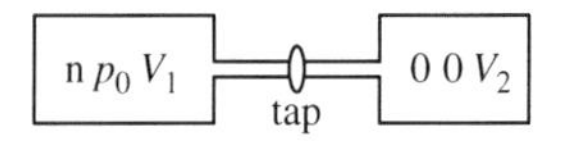

Calculate the change in entropy when n kmol of an ideal gas with the pressure p_0 and the volume V_1 expands irreversibly to the volume $V_1 + V_2$ in the way shown in the figure.

Solution:

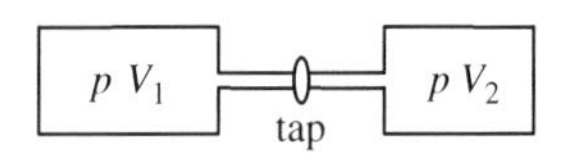

There are no forces between the molecules in the gas and therefore there is no change of internal energy when the gas expands. No change of internal energy means no temperature change.

When the tap is opened, the gas expands isothermally from volume V_1 to $V_1 + V_2$. The first law of thermodynamics and the definition of entropy give

$$\Delta S = \int \frac{dq}{T} = \int_{V_1}^{V_1+V_2} \frac{p\,dV}{T} \tag{1'}$$

Using the general gas law $pV = nRT$ to eliminate T we obtain

$$\Delta S = \int_{V_1}^{V_1+V_2} \frac{nR\,dV}{V} = nR \ln \frac{V_1 + V_2}{V_1} \tag{2'}$$

Answer: The entropy increases by the amount $nR \ln \dfrac{V_1 + V_2}{V_1}$.

The final state in Example 1.1 is far more likely than the initial state. When the tap is opened the molecules move into the empty container until the pressures in the two containers are equal rather than that no change at all occurs. Hence, the system changes spontaneously from one state to another, more likely state and the entropy increases.

All experience shows that the result in Example 1.1 can be generally applied. The following statements are generally valid:

- If a process is reversible the entropy change is zero.

$$\Delta S = 0.$$

- The entropy increases in all irreversible processes.

$$\Delta S > 0.$$

Entropy of Mixtures

Entropy Change at Mixing Two Gases

As a second example of deriving entropy changes we will calculate the entropy change when two gases mix.

Example 1.2

A short tube and a closed tap connect two gases A and B of equal pressures, each in a separate closed container. When the tap is opened the two gases mix irreversibly. No changes in pressure and temperature are observed.

Calculate the total change of entropy as a function of the initial pressure p and the final partial pressures when the two gases mix. The data of the gases are given in the figure.

$n_A\ p\ V_1$ tap $n_B\ p\ V_2$

n = number of kmol
p = pressure
V = volume
The temperature T is constant.

Solution:

When the tap is opened the two gases mix by diffusion. The diffusion goes on until the composition of the gas is homogeneous. It is far more likely that the gases mix by diffusion than that the gases remain separate. Hence, the total entropy change is expected to be positive.

In a gas, the distances between the molecules are large and the interaction between them can be neglected. The diffusion of each gas is independent of the other. The total entropy change can be regarded as the sum of the entropy change of each gas after its separate diffusion from one container into the other.

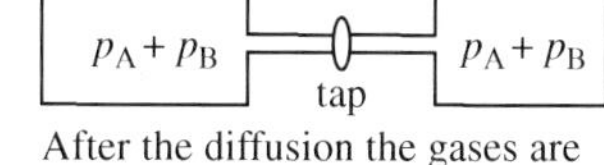

After the diffusion the gases are mixed and the pressure is equal in the two containers. The temperature T is constant.

$$\Delta S^{mix} = \Delta S_A^{mix} + \Delta S_B^{mix} \qquad (1')$$

n_A kmol of gas A change their pressure from p to p_A where p_A is its final partial pressure. In the same way n_B kmol of gas B change their pressure from p to p_B.

The initial pressure and the final total pressure are equal as no pressure change is observed.

$$p = p_A + p_B \qquad (2')$$

Using the result of Example 1.1 we obtain

$$\Delta S_A^{mix} = n_A R \ln \frac{V_1 + V_2}{V_1} \qquad (3')$$

and

$$\Delta S_B^{mix} = n_B R \ln \frac{V_1 + V_2}{V_2} \qquad (4')$$

The partial pressures can be calculated with the aid of Boyle's law applied on each gas:

$$pV_1 = p_A(V_1 + V_2) \qquad (5')$$

and

$$pV_2 = p_B(V_1 + V_2) \qquad (6')$$

The ratio of the volumes from Equations (5′) and (6′) are inserted into Equations (3′) and (4′) and we obtain

$$\Delta S_A^{mix} = n_A R \ln \frac{p}{p_A} \tag{7'}$$

and

$$\Delta S_B^{mix} = n_B R \ln \frac{p}{p_B} \tag{8'}$$

The total entropy change is

$$S^{mix} = \Delta S_A^{mix} + \Delta S_B^{mix} = n_A R \ln \frac{p}{p_A} + n_B R \ln \frac{p}{p_B} \tag{9'}$$

The ratio of the pressures is >1 and the entropy change is therefore positive, as predicted above.

Answer: The total entropy change when the gases mix is

$$n_A R \ln \frac{p}{p_A} + n_B R \ln \frac{p}{p_B}, \quad \text{where} \quad p = p_A + p_B.$$

Entropy Change at Mixing Two Liquids or Solids

Diffusion occurs not only in gases but also in liquids and solids. The entropy change ΔS^{mix} or simply S^{mix}, owing to mixing of two compounds in a melt or a solid can be calculated if we make a minor modification of Equation (9′) in Example 1.2.

Instead of the partial pressures of the two gases we introduce the mole fractions x_A and x_B:

$$x_A = \frac{p_A}{p_A + p_B} \quad \text{and} \quad x_B = \frac{p_A}{p_A + p_B}$$

Equation (9′) can then be written as

$$S^{mix} = -n_A R \ln x_A - n_B R \ln x_B \tag{1.10}$$

By use of the relationship $n = n_A + n_B$ where n is the total number of kmole we obtain

$$S^{mix} = -nR(x_A \ln x_A + x_B \ln x_B) \tag{1.11}$$

The molar entropy of mixing will then be

$$S_m^{mix} = -R(x_A \ln x_A + x_B \ln x_B) \tag{1.12}$$

Equations (1.11) and (1.12) are directly applicable on mixtures of gases but also on liquids and solids. These applications will be discussed later.

Entropy and Probability

The two examples given above indicate that entropy in some way is related to probability. The probability function can be found by the following arguments.

Consider N molecules in a container with the volume V (Figure 1.3). The molecules do not interact at all. Each molecule is free to move within the volume V. The probability of finding it within a unit volume is the same everywhere. Hence the probability of finding a molecule within a volume V_1 is V_1/V. The probability of finding two molecules within the same volume V_1 equals the product of their probabilities $(V_1/V)^2$. The probability of finding N molecules within a particular volume V_1 is $(V_1/V)^N$.

Equation (2′) in Example 1.1 on page 4 and Equation (9′) on page 6 give us the clue to relating entropy and probability. We have seen above that the overall probability is the *product* of the probabilities of independent events. We also know that the entropy of two systems is the *sum* of their entropies. It is striking that the logarithmic function converts the multiplicative property of probability to the additive property of entropy.

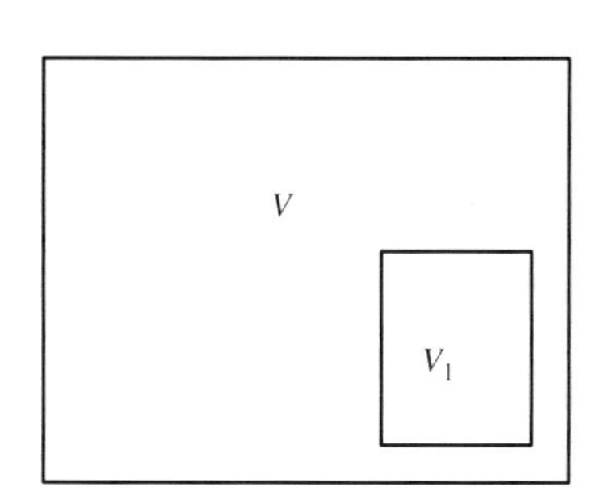

Figure 1.3 Volume element.

These arguments led the physicists Maxwell and Boltzmann in 1877 to suggest a relationship between entropy and probability. Boltzmann interpreted entropy as a measure of the order, or rather disorder, of a system. The more probable a state of a system is and the greater its disorder is, the higher will be its entropy.

Boltzmann derived an alternative expression of entropy by using statistical thermodynamics.

$$S = k_B \ln P \tag{1.13}$$

where
k_B = Boltzmann's constant
P = probability of the state.

Equation (1.13) is the fundamental relationship between entropy and probability. P is called the statistical weight of the state, i.e. its configuration.

Consider a binary system, a gas or a liquid, of two components B and C. N_B atoms of B and N_C atoms of C are arranged at random among $N = N_B + N_C$ sites. This can be done in many different ways and is equivalent to mixing the two components. The statistical weight P is the number of alternative ways to arrange the B and C atoms among the N sites. Statistical considerations give the result

$$P = \frac{N!}{N_B! N_C!} = \frac{N!}{N_B!(N - N_B)!} \tag{1.14}$$

We use Stirling's formula

$$\lim N! \to \sqrt{2\pi}\, N^{N+1/2} e^{-N} \quad \text{when} \quad n \to \infty$$

for the very high numbers N, N_B and $N_C = N - N_B$ and obtain

$$P = \frac{N^{N+1/2}}{N_B^{N_B+1/2}(N - N_B)^{N - N_B + 1/2}} \frac{e^{-N}}{e^{-N_B} e^{-(N - N_B)}} \tag{1.15}$$

The last factor in Equation (1.15) is equal to 1. The term 1/2 in the exponents can be neglected in comparison with N and N_B. Taking the logarithm of both sides of Equation (1.15) we obtain

$$\ln P = N \ln N - N_B \ln N_B - (N - N_B)\ln(N - N_B) \tag{1.16}$$

If we introduce the mole fractions

$$x_B = \frac{N_B}{N} \quad \text{and} \quad x_C = \frac{N - N_B}{N} \quad \text{and the relationship} \quad x_B + x_C = 1$$

Equation (1.16) can, after some calculations, be transformed into

$$\ln P = N(-x_B \ln x_B - x_C \ln x_C) \tag{1.17}$$

Instead of N we introduce n, the number of kmol with the aid of the relationship:

$$n = \frac{N}{N_A} \tag{1.18}$$

where N_A is Avogadro's number, i.e. the number of molecules of 1 kmol. Hence, we obtain

$$\ln P = nN_A(-x_B \ln x_B - x_C \ln x_C) \tag{1.19}$$

Instead of N_A we introduce R/k_B into Equation (1.18), where k_B is Boltzmann's constant.

$$k_B \ln P = -nR(x_B \ln x_B + x_C \ln x_C) \tag{1.20}$$

According to Equation (1.13) $k_B \ln P$ equals the entropy change S^{mix} when the two components B and C mix. Hence, we obtain

$$S^{mix} = -nR(x_B \ln x_B + x_C \ln x_C) \tag{1.21}$$

Equation (1.21) is valid for homogeneously mixed solids and liquids.

Equation (1.21) is identical with Equation (1.11) on page 6 derived in an entirely different way. The conclusion is that the concept of entropy as a function of probability is in complete agreement with the classical theory and does not lead to any contradictions.

Entropy Change during Solidification

Freezing and boiling causes changes in the molecular order and are therefore expected to lead to changes in entropy.

At freezing and melting the system is in equilibrium and the temperature is constant. Energy is transferred reversibly and isothermally in the shape of heat between the system and its surroundings.

At constant pressure the heat *absorbed* by a system equals its enthalpy *increase*:

$$(\Delta Q_{tr})_p = (\Delta H_{tr})_p \tag{1.22}$$

and the entropy change will be

$$\Delta S = \frac{\Delta H_{tr}}{T_{tr}} \tag{1.23}$$

Subscript 'tr' stands for transformation.

At *solidification* the system *emits* heat and the phase transition is exothermic. In this case the enthalpy change is negative and *the entropy change is negative.* The system changes from a disordered state (liquid) to a more ordered state (solid). The disorder decreases.

In the case of melting the opposite is valid. The phase transition is endothermic and the entropy change is positive. Heat has to be added to the system and its disorder increases.

1.2.4 Gibbs' Free Energy

Consider a system in thermal equilibrium with its surroundings at the temperature T. A spontaneous change of state of the system always leads to an increase of the entropy of the system. This can be expressed mathematically by Clausius' inequality

$$dS - \frac{dQ_{reversible}}{T} \geq 0 \tag{1.24}$$

If the process is reversible the sign of equality is valid. In the case of an irreversible process the upper sign has to be used. Heat transfer at constant pressure is of special interest in chemistry and metallurgy. If there is no other work than expansion work, $dQ_{rev} = dH$ and Equation (1.24) can be written as

$$TdS - dH \geq 0 \tag{1.25}$$

Equation (1.25) is valid if either of the following two conditions is fulfilled

1. The enthalpy of the system is constant.

$$\mathrm{d}S_{\mathrm{H,p}} \geq 0 \tag{1.26}$$

2. The entropy of the system is constant.

$$\mathrm{d}H_{\mathrm{S,p}} \leq 0 \tag{1.27}$$

The conditions 1 and 2 can be expressed in a simpler and understandable way by introduction of another thermodynamic function, the *Gibbs' free energy* or *Gibbs' function*

$$G = H - TS \tag{1.28}$$

When the state of the system changes at constant temperature the change of G equals

$$\mathrm{d}G = \mathrm{d}H - T\mathrm{d}S \tag{1.29}$$

Equation (1.25) can be expressed in terms of the Gibbs' function as

$$\mathrm{d}G_{\mathrm{T,p}} \leq 0 \tag{1.30}$$

- At constant temperature and pressure spontaneous processes always occur in the direction of decreasing G.

The deviation of the thermodynamic function G of a system from its equilibrium value can be regarded as the driving force in chemical and metallurgical reactions. At equilibrium

$$\mathrm{d}G_{\mathrm{T,p}} = 0 \tag{1.31}$$

and the driving force is zero. This condition will be applied later in this chapter.

Gibbs' free energy is a most useful instrument for studying the driving forces of various processes, for example chemical reactions, solution processes, melting/solidification and evaporation/condensation processes.

1.2.5 Intensive and Extensive Thermodynamic Quantities

Consider a single-component system, for example a pure metal, which consists of a certain amount of the pure substance. The system can be described by a number of quantities such as volume V, pressure p, absolute temperature T, number of kmol n of the pure element, internal energy U and entropy S.

The quantities of the system can be divided into two groups

- *intensive* quantities, which are independent of the amount of matter;
- *extensive* quantities, which are proportional to the amount of matter.

The pressure p, the temperature T are intensive quantities while volume, energy and entropy of the system are examples of extensive quantities.

A single-component system may consist of several phases, e.g. solid and liquid phases. A single-component system can be regarded as a special case of a multiple-component system. Phases in single- and multiple-component systems will be discussed in Section 1.5.1.

Extensive quantities, which refer to 1 kmol, will normally be denoted by capital letters with no index, otherwise with an index or by the product if the capital letter and the number of kmol.

In metallurgy an index 'm' for molar (1 kmol) is often used. We will also use this principle of designation. As we strictly use the SI system the concept *molar* is used in the sense of 1 kmol.

Gibbs' free energy G is an extensive thermodynamic quantity that will be frequently used in this and the following chapters, especially in Chapter 2.

1.3 Thermodynamics of Single-Component Systems

A single-component system may consist of only one phase, for example a solid, a liquid or a gas depending on the temperature. At certain temperatures discontinuous phase transformations occur when the system changes from one phase to another.

Examples of such transformations are melting of solid metals and evaporation of liquids. In this section we will study this type of processes in single-component systems. The results are general but will entirely be applied on the special single-component systems, which consist of *pure metals*.

At the transformation temperatures two phases are in equilibrium with each other. The conditions for equilibrium will be set up in terms of thermodynamics and the influence of some parameters on the transformation temperatures will be studied.

1.3.1 *Clausius–Clapeyron's Law*

The general expression for equilibrium between two arbitrary phases α and β in contact with each other is

$$G^{\alpha} = G^{\beta} \tag{1.32}$$

The Gibbs' free energies can be regarded as functions of the independent variables pressure p and temperature T.

A change of one of the independent variables results in changes of G^{α} and G^{β}. If the equilibrium is to be maintained a simultaneous and corresponding change of the other independent variable must occur. There must be a relationship between Δp and ΔT. In order to find it we use the condition for maintaining the equilibrium of the system:

$$\mathrm{d}G^{\alpha} = \mathrm{d}G^{\beta} \tag{1.33}$$

Gibbs' free energy is defined as

$$G = H - TS = U + pV - TS \tag{1.34}$$

The total differential of G is

$$\mathrm{d}G = \mathrm{d}U + p\mathrm{d}V + V\mathrm{d}p - T\mathrm{d}S - S\mathrm{d}T \tag{1.35}$$

By use of the first law of thermodynamics, $\mathrm{d}Q = \mathrm{d}U + p\mathrm{d}V$ and the definition of entropy, $\mathrm{d}S = \mathrm{d}Q/T = (\mathrm{d}U + p\mathrm{d}V)/T$, Equation (1.35) can be reduced to

$$\mathrm{d}G = V\mathrm{d}p - S\mathrm{d}T \tag{1.36}$$

Equation (1.36) is applied on the two phases α and β and the results are introduced into Equation (1.33).

$$V^{\alpha}\mathrm{d}p - S^{\alpha}\mathrm{d}T = V^{\beta}\mathrm{d}p - S^{\beta}\mathrm{d}T \tag{1.37}$$

or

$$\frac{dT}{dp} = \frac{V^{\beta} - V^{\alpha}}{S^{\beta} - S^{\alpha}} = \frac{\Delta V^{\alpha \to \beta}}{\Delta S^{\alpha \to \beta}} \quad \text{Clausius–Clapeyron's law} \tag{1.38}$$

where
$\Delta V^{\alpha \to \beta}$ = difference between the volumes of the β and α phases, i.e. $V^{\beta} - V^{\alpha}$
$\Delta S^{\alpha \to \beta}$ = difference between the entropies of the β and α phases, i.e. $S^{\beta} - S^{\alpha}$.

Equation (1.38) is the *Clausius–Clapeyron's law*, which is the desired relationship between the changes in pressure and temperature in a two-phase system at equilibrium. It will be applied on the melting and boiling processes below.

1.3.2 Equilibrium between Liquid and Solid Phases. Influence of Pressure and Crystal Curvature on Melting Point

Influence of Pressure on Melting Point

Consider a solid metal at the melting point in equilibrium with the corresponding metal melt. If the pressure p_0 is increased by an amount dp to $p = p_0 + \Delta p$, a new equilibrium is developed and the temperature is changed by the amount dT.

The temperature change or the change of melting point as a function of the pressure change can be found if we apply Clausius–Clapeyron's law on the liquid–solid system. Solidification means that a liquid phase α is transformed into a solid phase β at the melting-point temperature.

If we apply Equation (1.38) on 1 kmol of the liquid, which solidifies, we obtain

$$\frac{dT}{dp} = \frac{\Delta V_{m}^{L \to s}}{\Delta S_{m}^{L \to s}} \tag{1.39}$$

Subscript 'm' stands for molar (1 kmol).

If $\Delta S_{m}^{L \to s}$ is substituted by $\Delta H_{m}^{L \to s}/T_M$ (compare Equation (1.23) on page 8) in Equation (1.39) we obtain

$$\frac{dT}{dp} = \frac{T_M \Delta V_{m}^{L \to s}}{\Delta H_{m}^{L \to s}} = \frac{T_M \left(V_{m}^{s} - V_{m}^{L}\right)}{-H_{m}^{\text{fusion}}} \tag{1.40}$$

or

$$\Delta T_M = T_M(p) - T_M(p_0) = \frac{T_M \left(V_{m}^{L} - V_{m}^{s}\right)}{H_{m}^{\text{fusion}}} \Delta p \tag{1.41}$$

where
ΔT_M = change of melting point, i.e. $T_M(p) - T_M(p_0)$
p_0 = normal pressure, usually 1 atm
p = pressure p equal to $p_0 + \Delta p$
$T_M(p)$ = melting-point temperature at pressure p
$\Delta V_{m}^{L \to s}$ = difference in molar solid and liquid volumes at temperature T_M, i.e. $V^{s} - V^{L}$
$-\Delta H_{m}^{L \to s}$ = molar heat of fusion $H_{m}^{\text{fusion}} = -(H^{s} - H^{L})$.

As the molar volume of a metal melt normally is larger than that of the solid metal and the heat of fusion is positive, an *increase* of the pressure leads to an *increase* of the melting point.

The pressure dependence on the melting-point temperature in metal melts is very small. Moderate changes of the pressure during the casting and solidification processes hardly influence the liquidus temperature at all. The effect can normally be neglected except for pressures of the magnitude $\geq 10^2$ atm.

Influence of Crystal Curvature on Melting Point

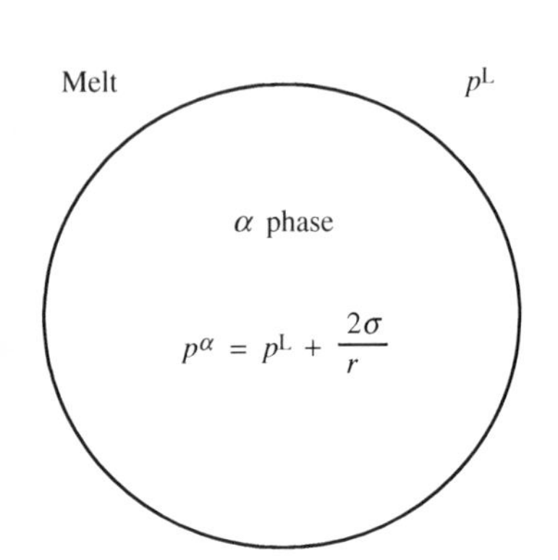

Figure 1.4a Spherical particle in a melt. Curvature K equals $1/r$.

It is a common practical situation that one phase has a much smaller volume than the other phase while the phases still are at equilibrium with each other. An example of such a case is shown in Figure 1.4a.

A small spherical α phase particle or a gas pore with radius r is floating freely in a melt of pressure p^L. The surface tension between particle and melt causes the pressure inside the particle to be higher than outside in the melt. The pressure inside the melt is obtained by a mechanical equilibrium study, which gives the result

$$p^\alpha = p^L + \frac{2\sigma}{r} \tag{1.42}$$

where σ is the surface tension between the melt and the α phase.

Equation (1.42) is a special case of Laplace's formula that states that there will be a pressure difference, owing to surface tension at a point on a curved surface between two phases

$$p^\alpha = p^L + \sigma\left(\frac{1}{r_{max}} + \frac{1}{r_{min}}\right) \tag{1.43}$$

where r_{max} and r_{min} are the maximum and minimum radii of curvature of the particle that is no longer necessarily spherical. We introduce the concept *curvature*, which is denoted by K and defined by the relationship

$$K = \frac{1}{2}\left(\frac{1}{r_{max}} + \frac{1}{r_{min}}\right) \tag{1.44}$$

Equation (1.43) can then be written as

$$p_K^\alpha = p^L + 2\sigma K \tag{1.45}$$

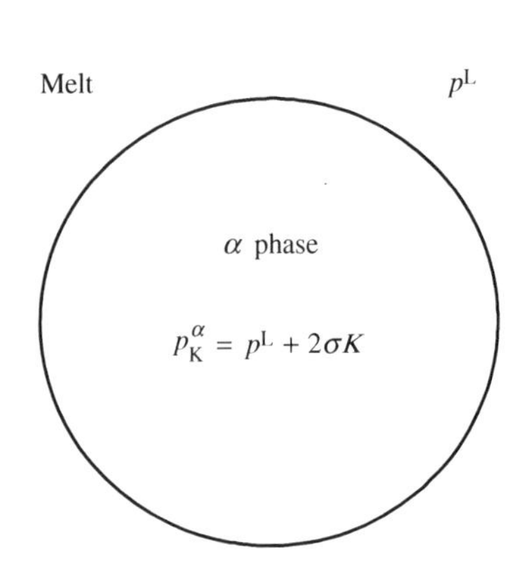

Figure 1.4b Spherical particle in a melt. Curvature K equals $1/r$.

Equation (1.45) is introduced into Figure 1.4b. At a planar surface K is zero and $p_K^\alpha = p^L$.

The pressure p_K^α will therefore change when the solid surface is curved instead of planar. This pressure change $p_K^\alpha - p_0^\alpha$ is accompanied by a corresponding change of the melting point of the solid. The difference in melting-point temperature can be calculated with the aid of Equation (1.41) while the following conditions are assumed to be valid:

1. the α phase and the melt are in equilibrium with each other
2. the melt maintains a constant pressure p^L independent of the particle curvature.

According to Equation (1.45) the pressure p_0^α at a planar interface equals the constant liquid pressure p^L. It is also identical with the pressure p_0 in Equation (1.41). Hence, we obtain

$$T_M(p_K^\alpha) - T_M(p_0^\alpha) = \frac{T_M\left(V_m^\alpha - V_m^L\right)}{H_m^{fusion}}\left[\left(p^L + 2\sigma K\right) - p^L\right]$$

or

$$T_M\left(p_K^\alpha\right) - T_M\left(p_0^\alpha\right) = \frac{T_M\left(V_m^\alpha - V_m^L\right)}{H_m^{fusion}} 2\sigma K \tag{1.46}$$

where
$T_M(p_K^\alpha)$ = melting point at curvature K of the particle and constant liquid pressure p^L
$T_M(p_0^\alpha)$ = melting point at curvature 0 and constant liquid pressure p^L
σ = surface tension
K = average curvature of the particle
V_m^α = molar volume of the α phase at the melting point
V_m^L = molar volume of the melt at the melting point.

K is positive when the curvature is convex. This is the case for a growing dendrite tip in a metal melt. As the volume change in Equation (1.46) is negative, there will be a *decrease* in melting-point temperature in the neighbourhood of a growing dendrite tip. The more curved the dendrite tip is, the lower will be the melting point.

1.3.3 Equilibrium between Liquid and Gaseous Phases. Influence of Pressure on Boiling Point. Bubble Formation in Melts

Influence of Pressure on Boiling Point

Clausius–Clapeyron's law (Equation (1.38) on page 11) is also valid for evaporation and condensation. It can be used for calculation of the change in boiling-point temperature of a liquid and its vapour in equilibrium with each other, caused by a change of pressure of the two phases. Boiling means that a liquid phase L is transformed into a gaseous phase g at the boiling-point temperature.

If we apply Equation (1.38) on 1 kmol of the liquid, which evaporates, we obtain

$$\frac{dT}{dp} = \frac{\Delta V_m^{L \to g}}{\Delta S_m^{L \to g}} \tag{1.47}$$

If $\Delta S_m^{L \to g}$ is substituted by $\Delta H_m^{L \to g}/T_B$ in Equation (1.47) we obtain

$$\frac{dT}{dp} = \frac{T_B \Delta V_m^{L \to g}}{\Delta H_m^{L \to g}} = \frac{T_B\left(V_m^g - V_m^L\right)}{H_m^{evap}} \tag{1.48}$$

or

$$\Delta T_B = T_B(p) - T_B(p_0) = \frac{T_B\left(V_m^g - V_m^L\right)}{H_m^{evap}} \Delta p \tag{1.49}$$

where
ΔT_B = change of boiling-point temperature
p_0 = normal pressure, usually 1 atm
p = pressure equal to $p_0 + \Delta p$
$\Delta V_m^{L \to g}$ = difference in molar gaseous and liquid volumes at boiling-point temperature T_B
$\Delta H_m^{L \to g}$ = molar heat of evaporation H_m^{evap}.

As the molar volume of the vapour is very much larger than that of the liquid, an *increased* pressure corresponds to an *increase* of the boiling-point temperature and vice versa.

The boiling point change is of the magnitude 10^4 times larger than the melting point change for the same change of pressure and can not be neglected.

Equation (1.49) can be simplified by neglecting the molar volume of the liquid in comparison with that of the vapour. If we replace the quantities ΔT and Δp by the corresponding differentials and assume, somewhat doubtfully, that the gas behaves like an ideal gas, we obtain for a temperature T close to the boiling point

$$dT = \frac{T}{H_m^{evap}} V_m^g dp = \frac{T}{H_m^{evap}} \frac{RT}{p} dp \tag{1.50}$$

or

$$\frac{dp}{p} \approx \frac{H_m^{evap}}{R} \frac{dT}{T^2} \tag{1.51}$$

It is reasonable to assume that the heat of evaporation is approximately independent of pressure and temperature.

$$\int_{p_0}^{p} \frac{dp}{p} \approx \frac{H_m^{evap}}{R} \int_{T_B}^{T} \frac{dT}{T^2} \tag{1.52a}$$

or

$$\ln \frac{p}{p_0} = \frac{H_m^{evap}}{R} \left(\frac{1}{T_B} - \frac{1}{T} \right) \tag{1.52b}$$

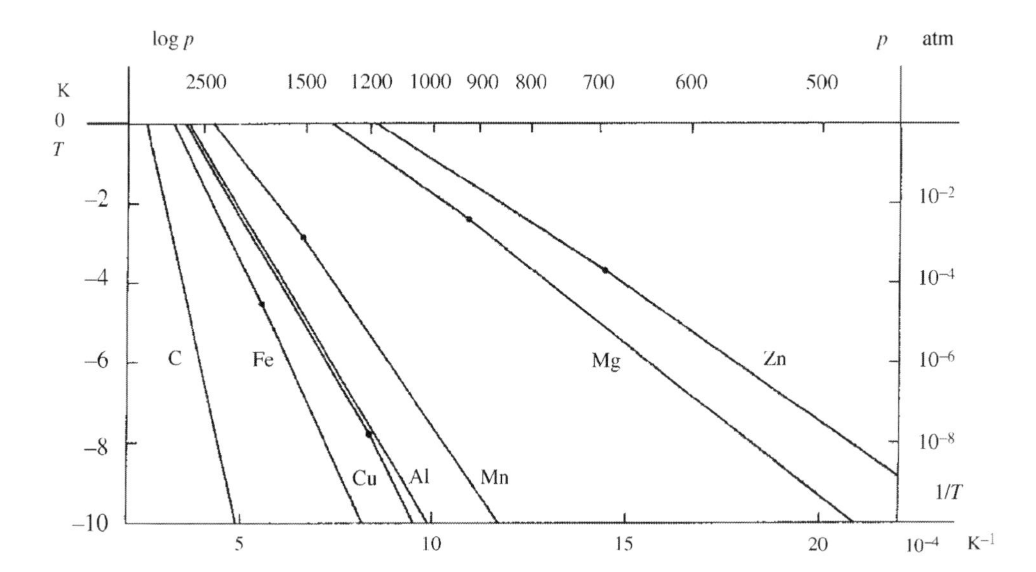

Figure 1.5 Vapour pressure as a function of $1/T$ for some common elements.

The melting points of the metals are marked by small black dots.

The rectangle represents four different axes. The functions and the corresponding scales are found in the corners. Reproduced from Elsevier © 1983.

Figure 1.5 shows $\ln p$ as a function of $1/T$ for some common elements. The straight lines in the figure shows that the approximations we made above seem to be justified and acceptable.

Equation (1.52b) can alternatively be written as

$$\Delta T_B = T_B(p) - T_B(p_0) = \frac{RT_B(p)T_B(p_0)}{H_m^{evap}} \ln \frac{p}{p_0} \tag{1.53}$$

which gives the change of boiling-point temperature as a function of pressure. The molar heat of evaporation is positive. At $p = p_0$ the boiling point is $T_B(p_0)$. An increase of the pressure leads to a positive value of ΔT, i.e. the boiling point increases. A decrease of the pressure results in a decrease of the boiling point.

Bubble Formation in Melts

The condition for bubble formation in a liquid is that the pressure inside the bubble $\geq$ the pressure of the surrounding liquid.

To treat bubble formation we substitute the α phase in Figure 1.4a by a gas phase of pressure $p = p_K^g$ (Figure 1.6a). The pressure inside a bubble with the curvature K (page 12) is derived in the same way as on page 12. In analogy with Equation (1.45) we obtain the total pressure inside the bubble equals

$$p_K^g = p^L + 2\sigma K \tag{1.54}$$

where $p^L = p_0 + \rho g h$.

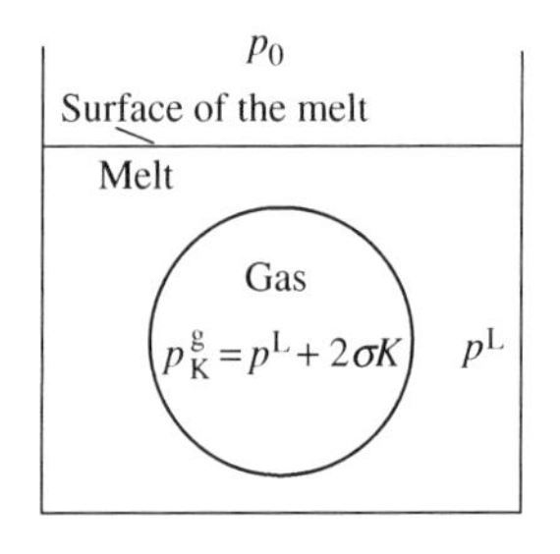

Figure 1.6a Pressure inside a spherical gas pore in a melt. Curvature K equals $1/r$ where r is the radius of the gas pore. p_0 is the pressure above the surface.

The smaller the bubble is when it is formed, the larger will be p_K^g, which facilitates bubble formation. If we replace p by p_K^g in Equation (1.53) we obtain

$$\Delta T_B = T_B(p_K^g) - T_B(p_0) = \frac{RT_B(p_K^g)T_B(p_0)}{H_m^{evap}} \ln \frac{p^L + 2\sigma K}{p_0} \tag{1.55}$$

The third term in Equation (1.55) is always positive. This leads to the obvious conclusion that a gas bubble with a higher internal pressure than the surroundings can only be formed *above* the normal boiling temperature T_B (p_0) of the melt. Energy for evaporation must be supplied, or else the boiling will stop.

The equilibrium between phases will be discussed more generally in the following sections.

If the pressure p_0 is kept low by pumping away the gas above the surface of the melt (Figure 1.6b), the pressure p^L, which equals $p_0 + \rho g h$, will also go down and the boiling-point temperature becomes strongly reduced. The consequence is that the bubble-formation condition is easier to fulfil and the boiling starts at a much lower temperature than normally.

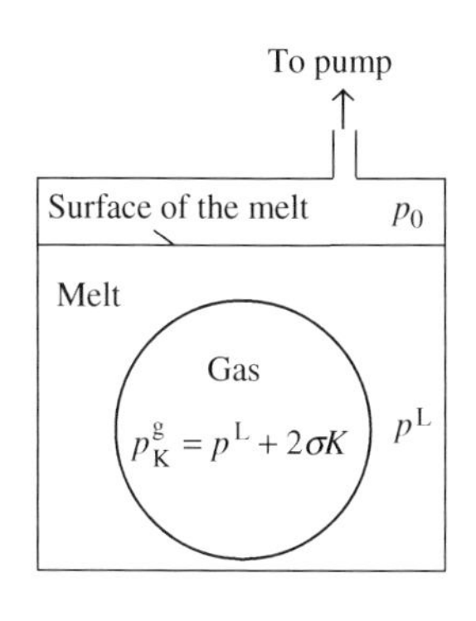

Figure 1.6b Pressure inside a spherical gas pore in a melt. Curvature K equals $1/r$ where r is the radius of the gas pore. h is the depth of the bubble below the surface.

1.3.4 Molar Gibbs' Free Energy of a Pure Metal

The Gibbs' free energy of a single phase, for example a pure metal, is defined by Equation (1.28) on page 9. If we apply this definition to the Gibbs' free energy of 1 kmol of the pure metal and use an index m, which stands for molar, we obtain

$$G_m = H_m - TS_m \tag{1.56}$$

Each metal has its own characteristic values of the thermodynamic quantities. The values of the molar enthalpy and the molar entropy of the solid metal are different from those of the liquid metal. The values of H_m are fairly constant but S_m varies with the temperature for each phase.

Figures 1.7 and 1.8, which show the functions

$$G_m^s = H_m^s - TS_m^s \tag{1.57a}$$

$$G_m^L = H_m^L - TS_m^L \tag{1.57b}$$

confirm the statements above. We can conclude that the thermodynamic quantities vary with temperature, as the curves are non-linear.

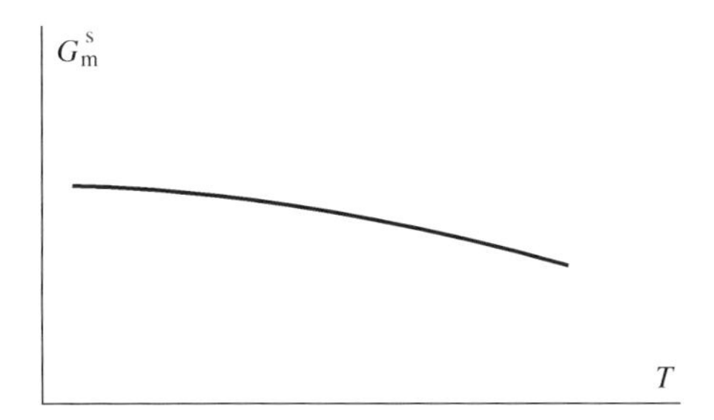

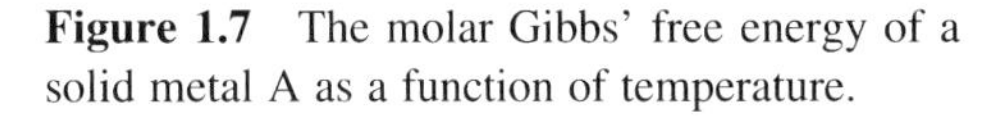

Figure 1.7 The molar Gibbs' free energy of a solid metal A as a function of temperature.

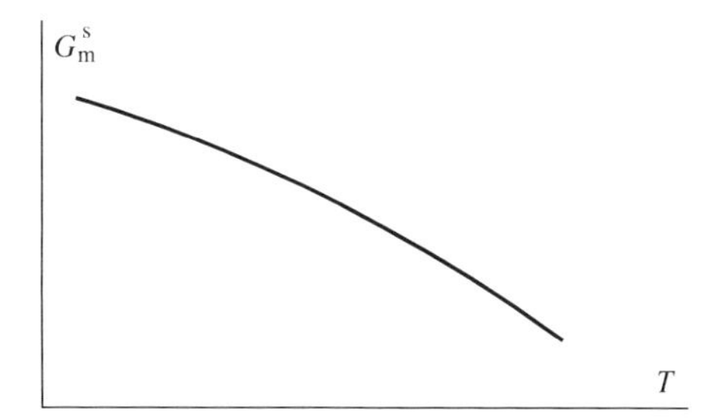

Figure 1.8 The molar Gibbs' free energy of a liquid metal A as a function of temperature.

Often, a reference temperature T_{ref} is used instead of zero and relative values ΔH and ΔS are introduced:

$$\Delta H_m^i = \int_{T_{ref}}^{T} C_p(T)\mathrm{d}T \quad \text{and} \quad \Delta S_m^i = \int_{T_{ref}}^{T} \frac{C_p(T)}{T}\mathrm{d}T \tag{1.58a+b}$$

Calculation of Molar Enthalpy and Entropy of Solids and Liquids

The curves in Figures 1.7 and 1.8 are calculated from Equations (1.57a) and (1.57b) in combination with the relationships

$$H_{\mathrm{m}}^{\mathrm{i}} = \int_0^T C_{\mathrm{p}}(T)\mathrm{d}T \quad \text{and} \quad S_{\mathrm{m}}^{\mathrm{i}} = \int_0^T \frac{C_{\mathrm{p}}(T)}{T}\mathrm{d}T \tag{1.59a+b}$$

where i = s and L, respectively.

The entropy of the liquid phase is normally larger than that of the solid. For this reason the curve in Figure 1.8 is steeper than that in Figure 1.7.

1.4 Thermodynamics of Multiple-Component Systems

Both single- and multiple-component systems may be *homogeneous*, i.e. appear as a single phase with uniform composition and physical properties, or *heterogeneous*, i.e. consist of several phases with different physical properties. At the phase boundaries these properties change abruptly. In multiple-component systems the composition of each phase must be specified.

Below, we will introduce a number of new thermodynamic concepts and relationships between them, which will be used on binary alloys in the remaining sections in this chapter and in Chapter 2. They are generally valid for many quantities in multiple-component systems, for example volume V, enthalpy H, entropy S and Gibbs' free energy G and also for several other thermodynamic functions.

To make the description below as short and surveyable as possible and avoid boring repetitions we use the volume V as a principle example throughout this section, when new concepts and relationships are introduced. However, we want to point out strongly that the concepts and relationships are also valid for other thermodynamic quantities.

- In all definitions, concepts and relationships the enthalpy H, the entropy S or the Gibb's free energy G can replace the volume V.

Independent Thermodynamic Variables of Multiple-Component Systems

As a multiple-component system we will consider an alloy to make the description below realistic. Alloys can be described by the following independent variables

- temperature;
- pressure;
- composition.

The thermodynamic quantities of a multiple-component system, for example an alloy, changes very little when the pressure is changed and the influence of pressure can be neglected in most cases.

The influence of temperature on thermodynamic variables, for example the Gibbs' free energy, is often strong and must definitely be taken into consideration.

The composition of a system is expressed as concentration of its components. Usual measures are number of kmol, mole fraction and atomic per cent (at%).

1.4.1 *Partial Molar Thermodynamic Quantities*

Consider a system, which consists of components A, B ... c where c corresponds to the number of components. Index i refers to component i of the system where i = A, B, ... c. The number of kmol of component i in the alloy is n_{i}.

Each system has its own characteristic thermodynamic quantities such as volume V and entropy S. In addition, it is convenient and useful to define so-called *partial molar thermodynamic quantities* or shortly *molar quantities* for multiple-component systems. A partial molar quantity is the corresponding molar quantity of each component instead the molar quantity of the total system. As an example we use the volume V.

$$\text{Partial molar quantity } \overline{V}_i = \frac{\partial V}{\partial n_i} \tag{1.60}$$

The definition is general and can be applied on *any* thermodynamic quantity. It is applied in Table 1.1 below on four of the most important and frequent basic thermodynamic quantities.

By use of basic mathematics we will derive some important and useful relationships. They are derived below for the volume but are valid for all extensive thermodynamic quantities. Examples are given in Table 1.1.

Table 1.1 Some partial molar thermodynamic quantities (V, S, H and G)

Partial molar volume	Partial molar entropy	Partial molar enthalpy	Partial molar Gibbs' free energy
$\overline{V}_i = \frac{\partial V}{\partial n_i}$	$\overline{S}_i = \frac{\partial S}{\partial n_i}$	$\overline{H}_i = \frac{\partial H}{\partial n_i}$	$\overline{G}_i = \frac{\partial G}{\partial n_i}$

When the numbers of kmol of the components are changed by dn_A, $dn_B \ldots dn_c$, the volume dV of the system changes by the amount

$$dV = \frac{\partial V}{\partial n_A}dn_A + \frac{\partial V}{\partial n_B}dn_B + \cdots + \frac{\partial V}{\partial n_c}dn_c. \tag{1.61}$$

If we introduce the partial molar volume we obtain

$$dV = \overline{V}_A dn_A + \overline{V}_B dn_B + \cdots + \overline{V}_c dn_c \cdots \tag{1.62}$$

Systems with Constant Composition

At constant pressure and temperature the intensive quantities $\overline{V}_i$ remain constant. The total volume of a system, which consists of $n_A, n_B, \ldots n_c$ kmol of the components A, B ... c, is found by integrating Equation (1.62).

$$V = n_A\overline{V}_A + n_B\overline{V}_B + \cdots + n_c\overline{V}_c \tag{1.63}$$

If Equation (1.63) is divided by the total number of kmol $(n_A + n_B + \cdots n_c)$ the molar volume V_m of the alloy (subscript m stands for molar) as a function of the mol fractions is obtained:

$$V_m = x_A\overline{V}_A + x_B\overline{V}_B + \cdots + x_c\overline{V}_c \tag{1.64}$$

Systems with Variable Composition

If we differentiate Equation (1.63) for the case that the composition of the system may change we obtain

$$\begin{aligned} dV = {} & \left(n_A d\overline{V}_A + n_B d\overline{V}_B + \cdots + n_c d\overline{V}_c\right) \\ & + \left(\overline{V}_A dn_A + \overline{V}_B dn_B + \cdots + \overline{V}_c dn_c\right) \end{aligned} \tag{1.65}$$

From a comparison between Equations (1.62) and (1.65) we can conclude that the first bracket in Equation (1.65) must be zero. If we divide it with $(n_A + n_B + \cdots n_c)$ we obtain

$$x_A d\overline{V}_A + x_B d\overline{V}_B + \cdots + x_c d\overline{V}_c = 0 \tag{1.66}$$

where $d\overline{V}_A, d\overline{V}_B, \ldots, d\overline{V}_c$ are the changes in volume in the individual partial molar quantities when p and T are constant.

Graphical Construction of Partial Molar Quantities for Binary Systems

If we apply Equation (1.66) on a binary system, where x_B is regarded as the independent composition variable, we can derive some useful relationships. From Equation (1.66) it follows that

$$x_A \frac{\partial \overline{V}_A}{\partial x_B} + x_B \frac{\partial \overline{V}_B}{\partial x_B} = 0. \tag{1.67}$$

This relationship was first derived by Duhem. If we take the derivative of the binary version of Equation (1.64) with respect to x_B we obtain

$$\frac{\partial V_m}{\partial x_B} = \frac{\partial x_A}{\partial x_B}\overline{V}_A + x_A \frac{\partial \overline{V}_A}{\partial x_B} + \overline{V}_B + x_B \frac{\partial \overline{V}_B}{\partial x_B}$$

Since $x_A + x_B = 1$ and $\partial x_A/\partial x_B = -1$ and Equation (1.67) is valid we obtain

$$\frac{\partial V_m}{\partial x_B} = \overline{V}_B - \overline{V}_A \tag{1.68}$$

If we combine Equations (1.64) and (1.68) to an equation system we can solve the partial molar volumes as functions of $x_A, x_B, \overline{V}_m$ and its partial derivatives with respect to x_A and x_B. The solution is

$$\overline{V}_A = V_m - x_B \frac{\partial V_m}{\partial x_B} = V_m + (1 - x_A)\frac{\partial V_m}{\partial x_A} \tag{1.69}$$

$$\overline{V}_B = V_m + (1 - x_B)\frac{\partial V_m}{\partial x_B} \tag{1.70}$$

where $V_m = x_A\overline{V}_A + x_B\overline{V}_B$ according to Equation (1.64).

Equations (1.69) and (1.70) are called *Gibbs–Duhem's equations.*

Both V_m, $\overline{V}_A$ and $\overline{V}_B$ obviously varies with x_B. This is shown in Figure 1.9, which also shows a useful graphical way to construct the partial molar volumes $\overline{V}_A$ and $\overline{V}_B$ when the $x_B V_m$ curve is known. The construction is based on Equations (1.69) and (1.70).

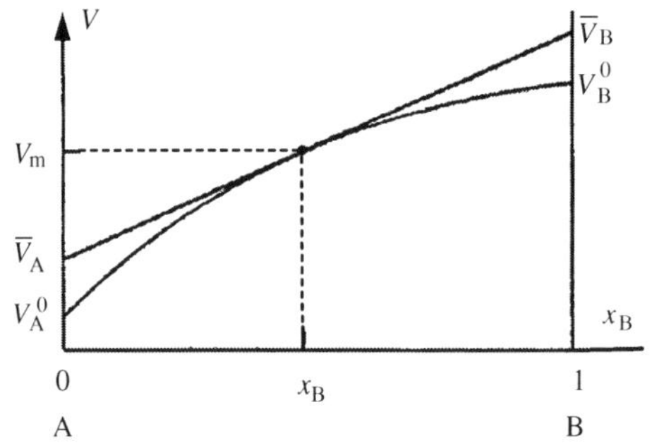

Figure 1.9 Molar volume V_m of a binary alloy as a function of the mole fraction x_B.

$$\overline{V}_A = V_m - x_B \frac{\partial V_m}{\partial x_B}$$

$$\overline{V}_B = V_m + (1 - x_B)\frac{\partial V_m}{\partial x_B}$$

The intersections V_A^0, respectively V_B^0, between the vertical axes and the V_m curve represent the molar volumes of pure elements A and B, respectively, at a given temperature.

As we have emphasized before, the theory of partial molar quantities given above and completely analogous formulas are valid for all thermodynamic functions.

1.4.2 Relative Thermodynamic Quantities and Reference States. Relative Partial Molar Thermodynamic Quantities or Partial Molar Quantities of Mixing

In the case of potential energy it is well known that a zero level has to be defined. This is also true for thermodynamic quantities. An example is the entropy S. Its zero level appears as an integration constant and is fixed by the third law of thermodynamics, which says that $S=0$ for a pure crystalline substance at $T=0\,\mathrm{K}$.

For the thermodynamic quantities specific reference states or *standard states* are defined. With the aid of these reference states *relative partial molar quantities* can be defined. Normally, the state of the pure substance is used as a reference state.

Lewis and Randall suggested that

A relative partial molar quantity = the difference between a partial molar quantity of a component in a solution and the molar quantity of the pure substance.

The relative partial molar volume is also called the *partial molar volume of mixing* of substance i and is designated by V_i^{mix}.

$$\overline{V}_i^{\mathrm{mix}} = \overline{V}_i - V_i^0 \tag{1.71}$$

where

$\overline{V}_i$ = partial molar volume of component i in a solution, i.e. the volume change of an infinitely large quantity of a solution of a given composition when 1 kmol of pure substance i is added at specified pressure and temperature

V_i^0 = molar volume of the pure substance i in its standard state, i.e. at specified values of p and T

$\overline{V}_i^{\mathrm{mix}}$ = the difference between the partial molar volume of component i in a solution and the molar volume of pure substance i at fixed values of p and T.

The relative partial molar thermodynamic quantities are marked by a bar, a component subscript and the designation 'mix'.

Relationships between Partial Molar Quantities of Mixing in a Binary System

The relationships (1.61)–(1.70), which were derived for the partial molar volume V, are also valid for the relative molar thermodynamic quantities or simply for quantities with subscript 'mix'.

In analogy with Equation (1.67) on page 18 there are partial differential equations that relate the relative partial molar quantities of the components in a multiple component solution:

$$x_A \frac{\partial \overline{V}_A^{\mathrm{mix}}}{\partial x_B} + x_B \frac{\partial \overline{V}_B^{\mathrm{mix}}}{\partial x_B} + \cdots + x_c \frac{\partial \overline{V}_c^{\mathrm{mix}}}{\partial x_c} = 0 \tag{1.72}$$

If we apply Equation (1.72) on a binary system and use the relationship $x_A = 1 - x_B$ we obtain $\overline{V}_A^{\mathrm{mix}}$ as a function of $\overline{V}_B^{\mathrm{mix}}$ and x_B:

$$\overline{V}_A^{\mathrm{mix}} = -\int_0^{x_B} \frac{x_B}{1-x_B} \frac{\partial \overline{V}_B^{\mathrm{mix}}}{\partial x_B} \mathrm{d}x_B \tag{1.73}$$

Partial integration of Equation (1.73) gives

$$\overline{V}_A^{\mathrm{mix}} = \int_0^{x_B} \frac{\overline{V}_B^{\mathrm{mix}}}{(1-x_B)^2} \mathrm{d}x_B - \frac{x_B \overline{V}_B^{\mathrm{mix}}}{1-x_B} \tag{1.74}$$

1.4.3 *Relative Integral Molar Thermodynamic Quantities or Integral Molar Quantities of Mixing*

In addition to the relative partial molar quantities we have to define *relative integral molar quantities.* As an example we choose the relative integral molar volume V_m^{mix}.

V_m^{mix} = the difference between the volume of 1 kmol of a solution V_m and the sum of the volumes of the corresponding pure component substances

$$V_m^{mix} = V_m - (x_A V_A^0 + x_B V_B^0 + \cdots + x_c V_c^0) \tag{1.75}$$

With the aid of Equation (1.64) on page 18 and Equation (1.71) on page 19, Equation (1.75) can be transformed into

$$V_m^{mix} = x_A \overline{V}_A^{mix} + x_B \overline{V}_B^{mix} + \cdots + x_c \overline{V}_c^{mix} \tag{1.76}$$

The relative integral molar volume is the volume of mixing or the volume of 1 kmol of the alloy formed by x_A kmol of pure substance A + x_B kmol of pure substance B + ... + x_C kmol of pure substance C.

For a binary system the relative integral molar quantities appear in a set of formulas, which are analogous with Equations (1.69) and (1.70) on page 18. These equations give the partial molar quantities of mixing as a function of the relative integral molar quantities.

The partial molar volumes of mixing for a binary system are:

$$\overline{V}_A^{mix} = V_m^{mix} + (1 - x_A)\frac{\partial V_m^{mix}}{\partial x_A} \quad \text{Gibbs–Duhem's} \tag{1.77}$$

$$\overline{V}_B^{mix} = V_m^{mix} + (1 - x_B)\frac{\partial V_m^{mix}}{\partial x_B} \quad \text{equations} \tag{1.78}$$

The integral molar volume of mixing of a binary alloy can be obtained as a function of the partial molar free energy of mixing of component B by inserting the expression in Equation (1.74) for $\overline{V}_A^{mix}$ into Equation (1.76). The result is

$$V_m^{mix} = (1 - x_B)\int_0^{x_B} \frac{\overline{V}_B^{mix}}{(1 - x_B)^2} dx_B \tag{1.79}$$

Graphical Construction of Partial and Integral Molar Quantities of Mixing for a Binary Alloy

Equations (1.77) and (1.78) are called *Gibbs–Duhem's equations.* V^{mix}, $\overline{V}_A^{mix}$ and $\overline{V}_B^{mix}$ can be derived graphically if the molar volume V_m of the binary alloy is known as a function of composition. The principle is shown in Figure 1.10. The construction is based on Equations (1.71), (1.75), (1.77) and (1.78).

As an example of a thermodynamic quantity other than the volume, Figure 1.11 shows the molar entropy of mixing of a binary system as a function of composition. Entropy can be regarded as a measure of the disorder of a system. Mixing corresponds to disorder. Hence the entropy of mixing is always positive, i.e. the curve is convex.

Molar enthalpy of mixing will be discussed in Section 1.5 in connection with ideal and non-ideal solutions. The Gibbs' free energy is discussed in Sections 1.6 and 1.7 and applied in Chapter 2.

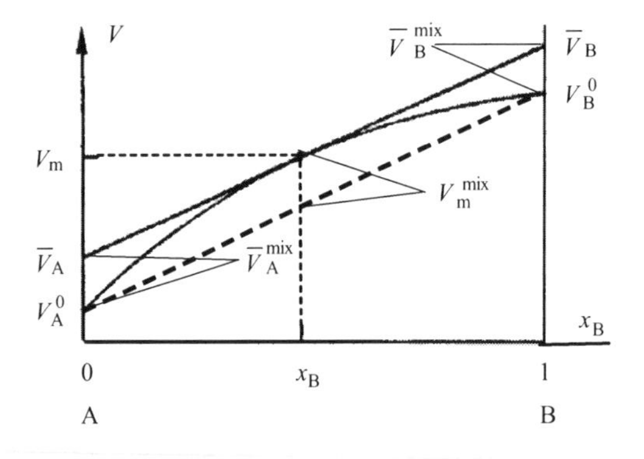

Figure 1.10 The molar volume V_m of a system as a function of the mole fraction x_B.

The partial and integral molar volumes of mixing, $\overline{V}_A^{mix}$, V_B^{mix} and V_m^{mix} are marked in the figure.

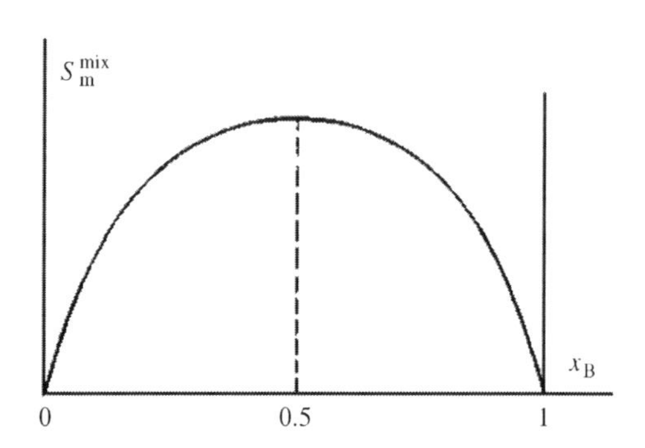

Figure 1.11 Molar entropy of mixing of a binary system as a function of composition.

1.4.4 Other Thermodynamic Functions and Relationships

The theory of thermodynamics of multiple-component systems is not complete without the theory of *ideal* and *non-ideal solutions*.

These concepts, together with the concepts of *activity* and *excess quantities,* belong to the general theory of thermodynamics but they will be introduced and penetrated in connection with the thermodynamics of alloys in this book. The concepts are of utmost importance for the study of the Gibbs' free energy of alloys, which is the aim of Chapter 2.

1.5 Thermodynamics of Alloys

An *alloy* is a multiple-component metallic system. Depending on the pressure, temperature and composition of the alloy it can either be *homogeneous*, i.e. consist of a single phase, or *heterogeneous*, i.e. consist of several phases.

When a second metallic component (solute) is added to a pure liquid metal (solvent) a solution or a single phase with a given composition is normally formed. When the solution solidifies a homogeneous solid solution is produced. If the metals can be mixed in any proportions, the miscibility is said to be complete and either of the metals can be considered as solvent.

If the miscibility is *not* complete, the alloy becomes heterogeneous at certain temperatures and compositions and a new phase is precipitated. The two immiscible phases have different compositions, which depend on temperature. In a temperature–composition diagram there exists a region with two phases.

The *phase diagram* of a metallic system describes the phases of the alloy and their compositions as a function of temperature. These topics will be discussed in Chapter 2.

The so-called *intermediate phases* are a special type of phases. Their compositions correspond to chemical compounds with integer proportions between the two kinds of metal atoms.

Classification of Phases

Phases of alloys may be classified according to their degree of atomic order.

- *Liquid phase*: A multiple-component phase with no long-range order.
- *Disordered substitutional solid solution*: Random distribution of the solute atoms on the lattice sites of the solvent.
- *Interstitial solid solution*: Random distribution of interstitial solute atoms in a solvent lattice.
- *Intermediate phase or chemical compound*: Solvent and solute atoms, respectively, form two independent but interpenetrating lattices.

In this chapter only binary alloys will be treated. In Chapter 2 a brief discussion of ternary alloys is included in addition to the dominating binary alloys.

Binary Alloys

In the introduction to this chapter and this book the close connection between the driving force of solidification and the solidification process in metals and alloys was pointed out. The driving forces of solidification will be defined and calculated for pure metals and alloys in Chapter 2. They are closely related to the Gibbs' free energy of the solid and liquid phases.

It is fairly easy to find the driving force of solidification for pure metals. A binary system is far more complicated than a single-component system. It is necessary to start with a study of the thermodynamics and stability of ideal and non-ideal solutions to find the molar Gibbs' free energy as a function of composition and temperature. When these functions are known, an expression of the driving force of solidification of binary alloys can be derived.

In Section 1.4 the thermodynamics of multiple-component systems has been treated. By deriving expressions of enthalpy and entropy of ideal and non-ideal solutions we obtain the contributions required to find the molar Gibbs' free energies for the liquid and solid phases of binary alloys as functions of composition and temperature.

These functions are the basis of the Gibbs' molar free-energy curves, which are drawn for the liquid and solid phases in the next chapter.

1.5.1 Heat of Mixing

The enthalpy of mixing or, simpler, the *heat of mixing* is entirely a function of the forces between the atoms in a metal or an alloy. These forces depend on the type of atoms involved and on the interatomic distances. The temperature dependence of the heat of mixing is therefore weak. On the other hand the heat of mixing of an alloy is strongly influenced by its composition.

Consider one kmol of a binary alloy, which contains N_{AB} randomly distributed A–B atoms pairs and an excess of either A–A or B–B atom pairs. The alloy has been formed by mixing $\frac{1}{2}$ N_{AB} A–A atom pairs with $\frac{1}{2}$ N_{AB} B–B atom pairs. Figure 1.12 shows the energy changes of the system when the alloy is formed, i.e. when the original A–A and B–B bonds are broken and pairs of A–B bonds are formed instead. There is no energy change of the excess A–A or B–B atom pairs before and after the mixing, as their bonds remain unchanged.

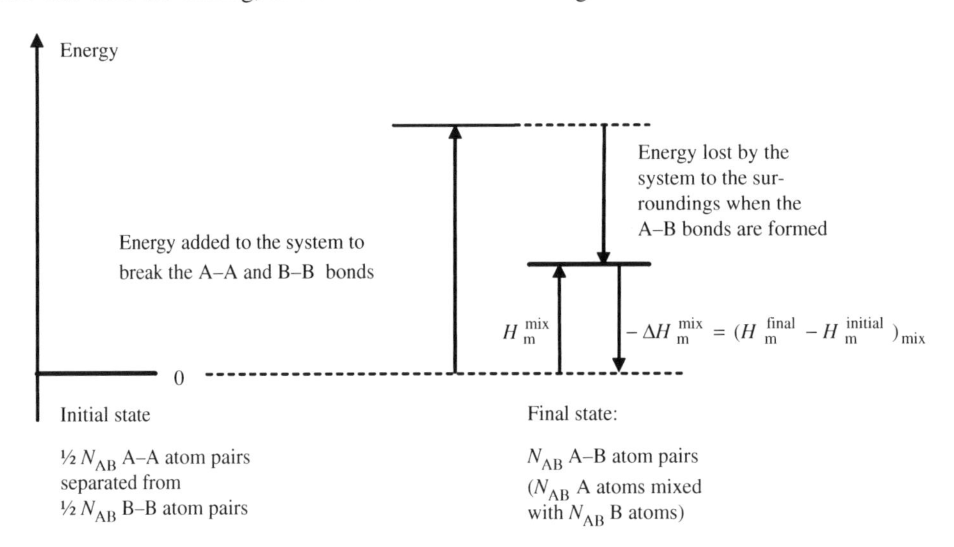

Figure 1.12a Enthalpy change and heat of mixing when $\frac{1}{2}$ N_{AB} A–A atom pairs mix with $\frac{1}{2}$ N_{AB} B–B atom pairs.

The *molar heat of mixing* is defined as

H_m^{mix} = the net energy, which has to be added to the system to mix N_{AB} A atoms and N_{AB} B atoms.

Figure 1.12a shows that the heat of mixing has the *same magnitude* as the enthalpy change of mixing but *opposite sign.*

$$H_m^{mix} = -(\Delta H_m)^{mix} = N_{AB}\left[E_{AB} - \frac{1}{2}(E_{AA} + E_{BB})\right] \tag{1.80}$$

where

N_{AB} = number of atom pairs A–B per kmol of the alloy
$-(\Delta H_m)^{mix}$ = molar enthalpy change at mixing
E_{ij} = binding energy between specified types of atoms.

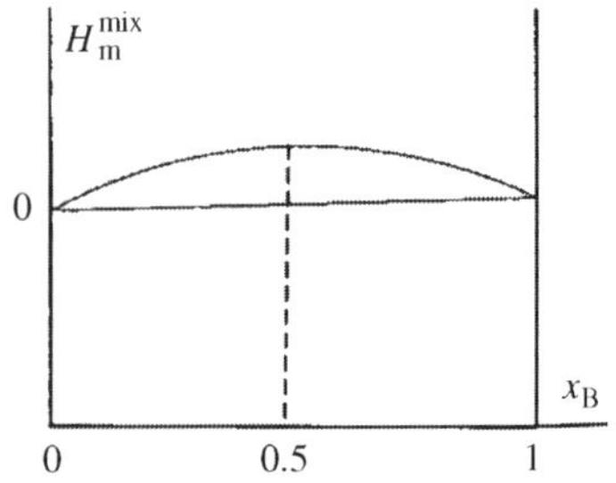

Figure 1.12b. Molar heat of mixing of a binary alloy as a function of composition. H_m^{mix} is *positive* if the forces between *unlike* atoms are *weaker* than the forces between *like* atoms.

When the forces between the A and B atoms are *weaker* than the forces between A–A atoms and B–B atoms, the heat of mixing is *positive*. This is the situation in Figures 1.12a and b.

The solution process to split up strong A–A and B–B bonds and replace them with weaker A–B bonds requires additional energy. Heat is *absorbed* by the system when the two components mix. Primarily, the temperature of the alloy decreases and heat is successively transferred from the surroundings.

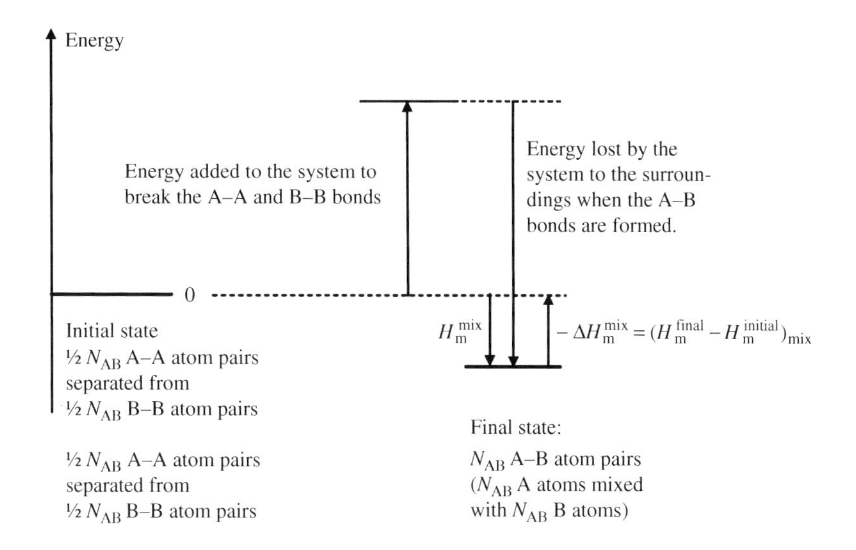

Figure 1.12c Enthalpy change and heat of mixing when ½ N_{AB} A–A atom pairs mix with ½ N_{AB} B–B atom pairs.

When the forces between the A and B atoms are *stronger* than the forces between A–A atoms and B–B atoms, the heat of mixing is *negative*. This is the situation in Figures 1.12c and d.

The solution process to split up weak A–A and B–B bonds and replace them with stronger A–B bonds *releases* energy. Heat is *emitted* by the system when the two components mix (Figure 1.12d).

Primarily, the temperature of the alloy increases and heat is gradually transferred to the surroundings.

The solution process is spontaneous and a stable alloy is formed. Apart from initial activation energy the solution process does not require energy addition from outside.

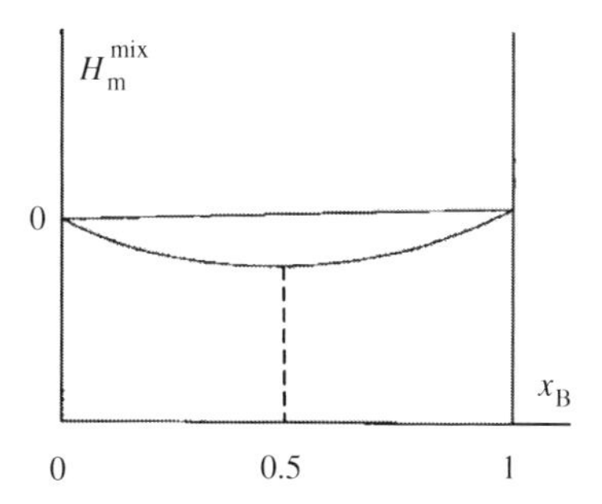

Figure 1.12d Molar heat of mixing of a binary alloy as a function of composition. H_m^{mix} is *negative* if the forces between *unlike* atoms are *stronger* than the forces between *like* atoms.

Theoretical Model of Heat of Mixing

The number N_{AB} of mixed atom pairs is proportional to the presence of the two types of atoms, i.e. the concentrations of A and B. The heat of mixing H_m^{mix}, based on the forces between neighbouring atoms, can then be written

$$H_m^{mix} = -(\Delta H_m)^{mix} = Lx_Ax_B \tag{1.81}$$

where L is a constant, which can be derived by identification of Equations (1.80) and (1.81). L includes both the proportionality constant and the bracket factor, as the binding energies of the three molecule types are constant.

For ideal solutions $L = 0$ and the heat of mixing is zero. The solubility of an ideal solution is total at all proportions.

Depending on the interatomic forces and the sign of the bracket factor, the constant L and therefore the molar heat of mixing H_m^{mix} can be either positive or negative for non-ideal solutions (Figures 1.12). Equation (1.81) will be used in next chapter in connection with a simple solution model.

1.5.2 Ideal and Non-Ideal Solutions

An *ideal solution* is defined as a solution in which

1. The homogeneous attractive forces (A–A and B–B) and the heterogeneous attractive forces (A–B) are equal. From condition 1 it follows that
2. The heat of mixing is zero.
3. The solubility of the components is complete, i.e. the components dissolve at all proportions.

Solutions that do not fulfil the conditions 1–3, are *non-ideal*. The thermodynamics of non-ideal solutions will be discussed in Section 1.7.

Survey of Concepts and Relationships of Gibbs' Free Energy for Binary Solutions

Partial Molar Gibbs' Free Energy

$$\overline{G}_{\mathrm{i}} = \frac{\partial G}{\partial n_{\mathrm{i}}} \tag{1'}$$

$$G_{\mathrm{m}} = x_{\mathrm{A}}\overline{G}_{\mathrm{A}} + x_{\mathrm{B}}\overline{G}_{\mathrm{B}} \tag{2'}$$

Solutions with variable composition:

$$x_{\mathrm{A}}\mathrm{d}\overline{G}_{\mathrm{A}} + x_{\mathrm{B}}\mathrm{d}\overline{G}_{\mathrm{B}} = 0 \tag{3'}$$

Gibbs–Duhem's equations:

$$\overline{G}_{\mathrm{A}} = G_{\mathrm{m}} + (1 - x_{\mathrm{A}})\frac{\partial G_{\mathrm{m}}}{\partial x_{\mathrm{A}}} \tag{4'}$$

$$\overline{G}_{\mathrm{B}} = G_{\mathrm{m}} + (1 - x_{\mathrm{B}})\frac{\partial G_{\mathrm{m}}}{\partial x_{\mathrm{B}}} \tag{5'}$$

Relative Partial Molar Gibbs' Free Energy or Partial Molar Gibbs' Free Energy of Mixing

$$\overline{G}_{\mathrm{i}}^{\mathrm{mix}} = \overline{G}_{\mathrm{i}} - G_{\mathrm{i}}^{0} \tag{6'}$$

$$\overline{G}_{\mathrm{i}}^{\mathrm{mix}} = RT\ln\frac{p_{\mathrm{i}}}{p_{\mathrm{i}}^{0}} \tag{7'}$$

Duhem's relationship:

$$x_{\mathrm{A}}\frac{\partial \overline{G}_{\mathrm{A}}^{\mathrm{mix}}}{\partial x_{\mathrm{B}}} + x_{\mathrm{B}}\frac{\partial \overline{G}_{\mathrm{B}}^{\mathrm{mix}}}{\partial x_{\mathrm{B}}} = 0 \tag{8'}$$

Relationship between Gibbs' free energy of mixing and the partial Gibbs' free energy of mixing

$$\overline{G}_{\mathrm{A}}^{\mathrm{mix}} = -\int_{0}^{x_{\mathrm{B}}}\frac{x_{\mathrm{B}}}{1 - x_{\mathrm{B}}}\frac{\partial \overline{G}_{\mathrm{B}}^{\mathrm{mix}}}{\partial x_{\mathrm{B}}}\mathrm{d}x_{\mathrm{B}} \tag{9'}$$

and

$$\overline{G}_{\mathrm{A}}^{\mathrm{mix}} = \int_{0}^{x_{\mathrm{B}}}\frac{\overline{G}_{\mathrm{B}}^{\mathrm{mix}}}{(1 - x_{\mathrm{B}})^{2}}\mathrm{d}x_{\mathrm{B}} - \frac{x_{\mathrm{B}}\overline{G}_{\mathrm{B}}^{\mathrm{mix}}}{1 - x_{\mathrm{B}}} \tag{10'}$$

Relative Integral Molar Gibbs' Free Energy or Integral Molar Gibbs' Free Energy of Mixing

$$G_{\mathrm{m}}^{\mathrm{mix}} = G_{\mathrm{m}} - \left(x_{\mathrm{A}}V_{\mathrm{A}}^{0} + x_{\mathrm{B}}V_{\mathrm{B}}^{0}\right) \tag{11'}$$

$$G_{\mathrm{m}}^{\mathrm{mix}} = x_{\mathrm{A}}\overline{G}_{\mathrm{A}}^{\mathrm{mix}} + x_{\mathrm{B}}\overline{G}_{\mathrm{B}}^{\mathrm{mix}} \tag{12'}$$

Gibbs–Duhem's equations:

$$\overline{G}_{\mathrm{A}}^{\mathrm{mix}} = G_{\mathrm{m}}^{\mathrm{mix}} + (1 - x_{\mathrm{A}})\frac{\partial G_{\mathrm{m}}^{\mathrm{mix}}}{\partial x_{\mathrm{A}}} \tag{13'}$$

$$\overline{G}_{\mathrm{B}}^{\mathrm{mix}} = G_{\mathrm{m}}^{\mathrm{mix}} + (1 - x_{\mathrm{B}})\frac{\partial G_{\mathrm{m}}^{\mathrm{mix}}}{\partial x_{\mathrm{B}}} \tag{14'}$$

Relationship between the integral molar Gibbs' free energy of mixing and the partial molar Gibbs' free energy of mixing:

$$G_{\mathrm{m}}^{\mathrm{mix}} = (1 - x_{\mathrm{B}})\int_{0}^{x_{\mathrm{B}}}\frac{\overline{G}_{\mathrm{B}}^{\mathrm{mix}}}{(1 - x_{\mathrm{B}})^{2}}\mathrm{d}x_{\mathrm{B}} \tag{15'}$$

1.6 Thermodynamics of Ideal Binary Solutions

Molar Enthalpy of Mixing

In an ideal solution of A and B atoms the A–B forces equals the A–A and B–B forces. From Equation (1.80) on page 22 we can conclude that the *molar enthalpy of mixing* must be zero when two pure substances A and B mix. Hence, we have

$$H_{\rm m}^{\rm mix} = 0 \tag{1.82}$$

Molar Entropy of Mixing

In Section 1.2.3 we derived the desired expression. According to Equation (1.12) on page 6 the molar entropy of mixing

$$S_{\rm m}^{\rm mix} = -R(x_{\rm A}\ln x_{\rm A} + x_{\rm B}\ln x_{\rm B}) \tag{1.12}$$

Molar Gibbs' Free Energy of Mixing

The molar Gibbs' free energy of mixing can easily be derived when $H_{\rm m}^{\rm mix}$ and $S_{\rm m}^{\rm mix}$ of an ideal solution are given.

$$G_{\rm m}^{\rm mix} = H_{\rm m}^{\rm mix} - TS_{\rm m}^{\rm mix} \tag{1.83}$$

As $H_{\rm m}^{\rm mix}$ is zero we obtain

$$G_{\rm m}^{\rm mix} = RT(x_{\rm A}\ln x_{\rm A} + x_{\rm B}\ln x_{\rm B}) \tag{1.84}$$

1.6.1 Molar Gibbs' Free Energy of Ideal Binary Solutions

Mechanical Mixture of Two Components

In a *mechanical mixture* of two pure elements there is no interaction between the two phases. In such cases, overall thermodynamic quantities can be calculated as the sum of corresponding extensive thermodynamic quantities of the individual components. The latter are often known and can be found in tables.

Molar Gibbs' Free Energy of an Ideal Binary Solution

If the two components form a *solution* instead of a mechanical mixture the interaction between the atoms and the energy of mixing must be taken into consideration.

Consider a binary solution of components A and B. The molar Gibbs' free energy $G_{\rm m}$ of the solution is equal to the sum of the molar Gibbs' free energies of the components plus the molar Gibbs' free energy of mixing:

$$G_{\rm m} = x_{\rm A}G_{\rm A}^0 + x_{\rm B}G_{\rm B}^0 + G_{\rm m}^{\rm mix} \tag{1.85}$$

where

$G_{\rm m}$ = molar Gibbs' free energy of the solution
$G_{\rm i}^0$ = molar Gibbs' free energy of the pure components (i = A, B)
$x_{\rm i}$ = number of kmole of the components (i = A, B)
$G_{\rm m}^{\rm mix}$ = molar Gibbs' free energy of mixing.

If the expression of G_m^{mix} in Equation (1.84) is introduced in Equation (1.85) we obtain

$$G_m = x_A G_A^0 + x_B G_B^0 + RT(x_A \ln x_A + x_B \ln x_B) \tag{1.86}$$

As $x_A = 1 - x_B$ we obtain alternatively

$$G_m = (1 - x_B)G_A^0 + x_B G_B^0 + RT[(1 - x_B)\ln(1 - x_B) + x_B \ln x_B] \tag{1.87}$$

Equation (1.86) is the desired expression for the Gibbs' molar free energy G_m of an ideal binary solution as a function of composition and temperature.

Equation (1.87) has been plotted in the Figures 1.13 for a given temperature. The G_m curve will be frequently used in Chapter 2.

The dotted line in Figure 1.13a represents the two first terms in Equation (1.87). The last term is negative, as both mole fractions in Equation (1.87) are < 1 As the entropy of mixing S_m^{mix} is always positive (Equation (1.12) on page 6 and Figure 1.11 on page 20), the term $-TS_m^{mix}$ in Equation (1.83) is always negative. Hence, the G_m curve must be convex downwards and lie below the dotted line.

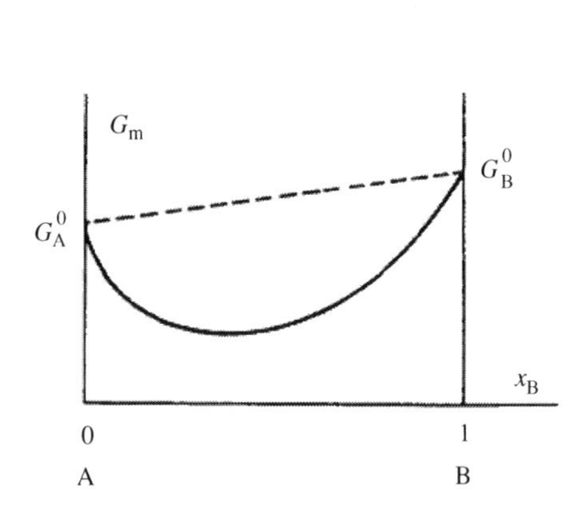

Figure 1.13a The molar Gibbs' free energy of ideal binary solutions as a function of composition x_B.

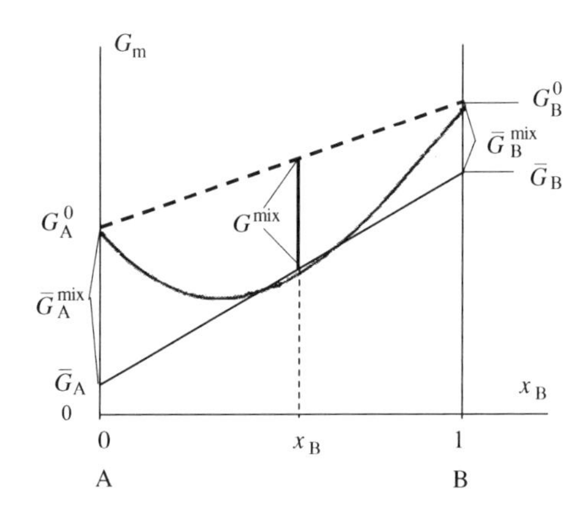

Figure 1.13b The molar Gibbs' free energy of ideal binary solutions as a function of composition. The partial and integral molar quantities are marked in the figure.

1.7 Thermodynamics of Non-Ideal Binary Solutions

Before it makes sense to discuss the corresponding quantities of non-ideal solutions we will introduce some new concepts, which are closely related to non-ideal solutions.

1.7.1 Activities of Non-Ideal Solutions Raoult's and Henry's Laws

Fugacity

Consider a *pure liquid* or *solid substance* in equilibrium with its vapour phase. The condition for equilibrium is that the Gibbs' free energies of the two phases are equal.

$$G_i(\text{l or c}) = G_i(\text{g}) \tag{1.88}$$

We neglect the pressure dependence of $G_i(\text{l or c})$ and assume that $G_i(\text{l or c}) \approx G_i^0(\text{l or c})$ where $G_i^0(\text{l or c})$ is the Gibbs' free energy of the liquid or solid at standard pressure 1 atm and standard temperature 273 K. This is most reasonable at ordinary pressures.

The Gibbs' free energy of the gas $G_i(g)$ depends on the pressure of the gas. We assume that the vapour behaves like an ideal gas. On page 10 we derived the relationship

$$dG = Vdp - SdT \tag{1.36}$$

For an ideal gas at constant temperature $V = RT/p$ and $dT = 0$ These expressions are introduced into Equation (1.34) on page 10. After integration we obtain

$$\int_{G_i^0(g)}^{\overline{G}_i(g)} dG = \int_{p_0}^{p_i} \frac{RT dp}{p}$$

which gives

$$\overline{G}_i(g) = G_i^0(g) + RT \ln p_i/p_i^0 \tag{1.89}$$

where

$\overline{G}_i(g)$ = Gibbs' free energy of the ideal gas phase i

G_i^0 = standard Gibbs' free energy, i.e. the Gibbs' free energy at STP (273 K, $p_i^0 = 1$ atm).

Next we consider a *liquid* or *solid solution* in equilibrium with its gas phase. The pressure of the gas is composed of the partial pressures of the component gases in equilibrium with the components of the solution. The components of the liquid or solid solution are *not* assumed to be ideal. Equation (1.89) is only valid for component i in a non-ideal liquid or solid solution if the pressure p_i is replaced by a corrected pressure or *fugacity* f_i

$$\overline{G}_i(\text{l or c}) = G_i^0(g) + RT \ln f_i/f_i^0 \tag{1.90}$$

where

$\overline{G}_i(\text{l or c})$ = partial molar Gibbs' free energy of component i in the solid or liquid solution

$G_i^0(g)$ = standard Gibbs' free energy of pure component i in the gas phase (273 K, 1 atm).

f_i = fugacity or corrected partial pressure of component i in the gas phase over the solution

f_i^0 = fugacity of component i over the pure condensed phase.

If the solution is replaced by a pure condensed phase of component i the equilibrium condition will be

$$\overline{G}_i(\text{l or c}) \approx G_i^0(\text{l or c}) = G_i(\text{g}) = G_i^0(\text{g}) + RT \ln f_i^0 \tag{1.91}$$

where the pressure of the gas has been replaced by the fugacity f_i^0 over the pure condensed phase.

If we subtract Equation (1.91) from Equation (1.90) we obtain by definition the partial molar Gibbs' free energy of mixing of component i in the solution

$$\overline{G}_i(\text{l or c}) = G_i^0(\text{l or c}) + RT \ln f_i/f_i^0$$

or

$$\overline{G}_i^{\text{mix}}(\text{l or c}) = RT \ln f_i/f_i^0 \tag{1.92}$$

Analogously, Equation (1.89) can be written

$$\overline{G}_i^{\text{mix}}(\text{g}) = RT \ln p_i/p_i^0 \tag{1.93}$$

As the gas and the solution are in equilibrium with each other, we can conclude that $\overline{G}_i^{mix}(\text{l or c}) = \overline{G}_i^{mix}(\text{g})$ and therefore

$$\frac{f_i}{f_i^0} = \frac{p_i}{p_i^0} \tag{1.94}$$

Definitions of Activity and Activity Coefficient

Activity

The activity of component i in a solution is defined as the ratio of the fugacity f_i of component i over the solution to the fugacity of the standard state equal to the pure stable condensed phase at the pressure 1 atm and temperature 273 K.

$$a_i = f_i/f_i^0 \tag{1.95}$$

$$\overline{G}_i(\text{l or c}) = G_i^0(\text{l or c}) + RT\ln a_i \tag{1.96}$$

The activity always has the value 1 at the standard state.

Activity Coefficient

The activity coefficient γ_i or f_i is defined by the relationship

$$a_i = \gamma_i x_i \quad \text{or} \quad a_i = f_i x_i \tag{1.97}$$

where x_i is the mole fraction of component i in the solution. At low pressures the fugacities can be replaced by the pressures

$$a_i \equiv \gamma_i x_i \approx p_i/p_i^0 \quad \text{or} \quad a_i \equiv f_i x_i \approx p_i/p_i^0 \tag{1.98}$$

Raoult's Law

Raoult's law is the statement that the activity and the mole fraction are equal, e.g. the activity coefficient equals 1.

If we designate the activity coefficient f_i, Raoult's law is a special case of Equation (1.97) $a = f_i x_i$ and can be written

$$a_i = x_i \quad \text{or} \quad f_i = 1 \tag{1.99}$$

It has been observed that Raoult's law is obeyed in the limiting case when x_i approaches unity. This is equivalent to the statement that the *solvent* in a dilute solution obeys Raoult's law when the concentration of the solute approaches zero.

$$\lim_{x_i \to 1} a_i^{solvent} = x_i \quad \text{or} \quad \lim_{x_i \to 1} f_i = 1 \tag{1.100}$$

which gives

$$\overline{G}_i = G_i^0 + RT\ln x_i \tag{1.101}$$

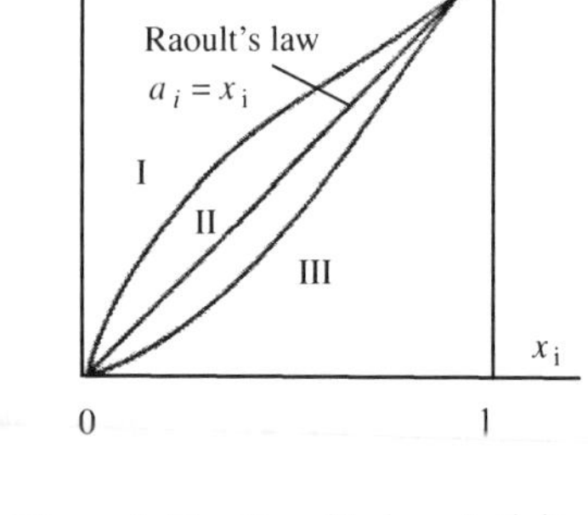

Figure 1.14 Raoult's law. Activity of solute component as a function of mole fraction.

The deviation from Raoult's law for real solutions (I and III) (Figure 1.14) can be either positive or negative. Case II corresponds to an ideal solution. Equation (1.101) is only valid for the solvent in dilute solutions. For a binary solution index i equals A.

Henry's Law

The activity $a_i^{solute} = \gamma_i x_i$ of the *solute* is proportional, but not equal, to the mole fraction when x_i approaches zero, i.e. in dilute solutions. The designation γ_i is used here for the activity of the solute to make it possible to distinguish between the activities of the solvent and the solute. In a binary solution index i = B

$$\lim_{x_i \to 0} a_i^{solute} = \lim_{x_i \to 0} \gamma_i x_i = \gamma_i^0 x_i \quad \text{or} \quad \lim_{x_i \to 0} \gamma_i = \gamma_i^0 \tag{1.102}$$

where γ_i^0 is a constant.

Hence, at low pressures

$$a_i^{solute} \equiv \frac{f_i}{f_i^0} \approx \gamma_i^0 x_i \quad \text{and as} \quad \frac{f_i}{f_i^0} \approx \frac{p_i}{p_i^0}$$

we obtain

$$p_i = p_i^0 \gamma_i^0 x_i = const \times x_i \tag{1.103}$$

- The partial pressure p over a dilute solution of component i is proportional to its mole fraction in the solution.

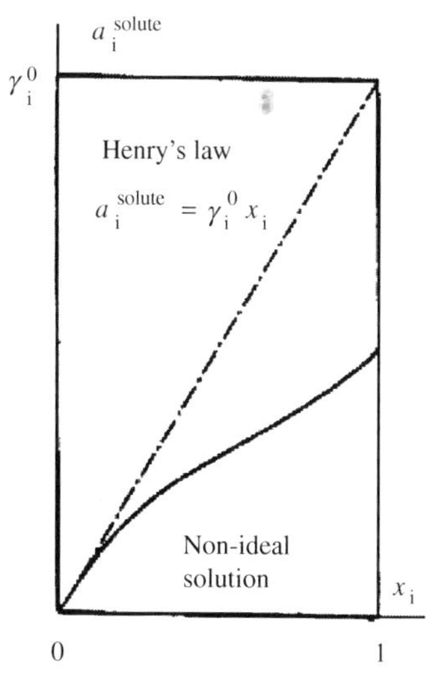

Figure 1.15 Henry's law. For a binary solution $i = B$.

The deviation from Henry's law for real solutions can be either positive or negative.

This is Henry's law, which has been known for more than 200 years. As Figure 1.15 shows Henry's law is not valid for large values of x_i. The activity γ_i of the solute is generally described by the constant γ_i^0 and a *relative activity coefficient*, defined by the relationship

$$\gamma_i^{rel} = \frac{\gamma_i}{\gamma_i^0} \tag{1.104}$$

When Henry's law is valid the value of γ_i^{rel} is 1. When Henry's law is not valid for solute component i its activity can generally be written

$$a_i^{solute} = \gamma_i x_i \quad \text{or} \quad a_i^{solute} = \gamma_i^{rel} \gamma_i^0 x_i \tag{1.105}$$

By use of Equations (1.96) and (1.105) the Gibbs' free energy of component i in the solution can be written as

$$\overline{G}_i = G_i^0 + RT \ln \gamma_i^0 + RT \ln \gamma_i^{rel} x_i \tag{1.106}$$

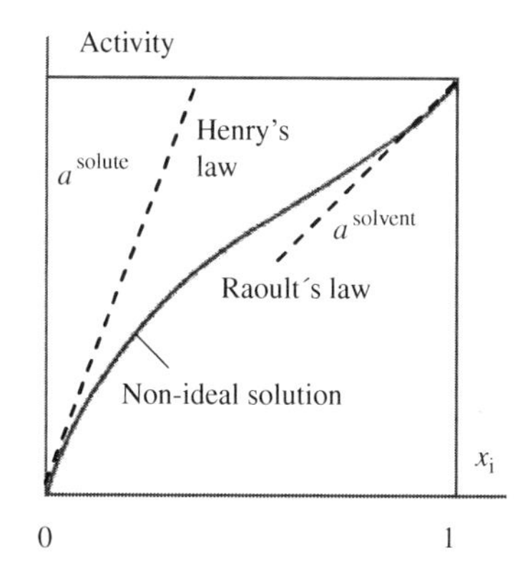

Figure 1.16 Activity as a function of component *i* in a solution.

It can be seen from Figure 1.16 that a positive deviation from Raoult's law corresponds to a negative deviation from Henry's law and vice versa.

It should be noticed that the activities and activity coefficients are not the same in Raoult's and Henry's laws. They differ by a factor γ_i^0. This is shown in Table 1.2.

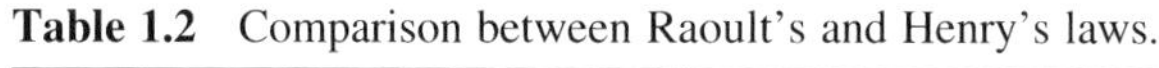

Table 1.2 Comparison between Raoult's and Henry's laws.

Law	Activity	Gibb's free energy
General law	$a_i = f_i x_i$	$\overline{G}_i = G_i^0 + RT \ln a_i$
Raoult's law, valid for solvents in dilute solutions	$a_i = x_i$ and $f_i = 1$ valid when $x_i \to 1$	$\overline{G}_i = G_i^0 + RT \ln x_i$
General laws	$a_i = \gamma_i x_i = \gamma_i^{rel} \gamma_i^0 x_i$	$\overline{G}_i = G_i^0 + RT \ln \gamma_i^0 + RT \ln \gamma_i^{rel} x_i$
Henry's law, valid for solutes in dilute solutions	$a_i = \gamma_i^0 x_i$ and $\gamma_i^{rel} = 1$ valid when $x_i \to 0$	$\overline{G}_i = G_i^0 + RT \ln \gamma_i^0 + RT \ln x_i$

Calculation of Activity Coefficients in Binary Solutions

With the aid of Equation (1.98) on page 28 Equation (1.93) on page 27 can be written as

$$\overline{G}_{\mathrm{i}}^{\mathrm{mix}} = RT \ln \frac{p_{\mathrm{i}}}{p_{\mathrm{i}}^{0}} = RT \ln \gamma_{\mathrm{i}} x_{\mathrm{i}} \tag{1.107}$$

These expressions are introduced into Equation (8′) on page 24.

After division with RT we obtain:

$$x_{\mathrm{A}} \frac{\partial \ln \gamma_{\mathrm{A}} x_{\mathrm{A}}}{\partial x_{\mathrm{B}}} + x_{\mathrm{B}} \frac{\partial \ln \gamma_{\mathrm{B}} x_{\mathrm{B}}}{\partial x_{\mathrm{B}}} = 0 \tag{1.108}$$

or

$$x_{\mathrm{A}} \frac{\partial (\ln \gamma_{\mathrm{A}} + \ln x_{\mathrm{A}})}{\partial x_{\mathrm{B}}} + x_{\mathrm{B}} \frac{\partial (\ln \gamma_{\mathrm{B}} + \ln x_{\mathrm{B}})}{\partial x_{\mathrm{B}}} = 0$$

Partial derivation in combination with the conditions $x_{\mathrm{A}} = 1 - x_{\mathrm{B}}$ and $\partial x_{\mathrm{A}} / \partial x_{\mathrm{B}} = -1$ gives

$$x_{\mathrm{A}} \frac{\partial \ln \gamma_{\mathrm{A}}}{\partial x_{\mathrm{B}}} - 1 + x_{\mathrm{B}} \frac{\partial \ln \gamma_{\mathrm{B}}}{\partial x_{\mathrm{B}}} + 1 = 0$$

or

$$x_{\mathrm{A}} \frac{\partial \ln \gamma_{\mathrm{A}}}{\partial x_{\mathrm{B}}} + x_{\mathrm{B}} \frac{\partial \ln \gamma_{\mathrm{B}}}{\partial x_{\mathrm{B}}} = 0 \tag{1.109}$$

If we replace x_{A} by $(1 - x_{\mathrm{B}})$ and integrate Equation (1.109) we obtain

$$\ln \gamma_{\mathrm{A}} = - \int_{0}^{x_{\mathrm{B}}} \frac{x_{\mathrm{B}}}{1 - x_{\mathrm{B}}} \frac{\partial \ln \gamma_{\mathrm{B}}}{\partial x_{\mathrm{B}}} \partial x_{\mathrm{B}} \tag{1.110}$$

By partial integration we obtain

$$\ln \gamma_{\mathrm{A}} = - \frac{x_{\mathrm{B}}}{(1 - x_{\mathrm{B}})^2} + \int_{0}^{x_{\mathrm{B}}} \frac{\ln \gamma_{\mathrm{B}}}{(1 - x_{\mathrm{B}})^2} \partial x_{\mathrm{B}} \tag{1.111}$$

If γ_{B} is known as a function of x_{B} it is possible to calculate γ_{A}. The accuracy may be poor if x_{B} is close to unity. In this case both $\ln \gamma_{\mathrm{B}}$ and $(1 - x_{\mathrm{B}})$ approach zero.

1.7.2 Excess Quantities of Non-Ideal Solutions

The concept excess quantity of a non-ideal solution is closely related to activity and activity coefficient. The definition of an excess quantity X^{Ex} is

$$(X)_{\text{non-ideal solution}} = (X)_{\text{ideal solution}} + X^{\mathrm{Ex}} \tag{1.112}$$

- Excess quantity = a quantity of a non-ideal solution minus the corresponding quantity of an ideal solution.

The quantity X can for example be H, S or G or any other thermodynamic functions of these quantities. By definition all excess quantities are zero for ideal solutions. The Gibbs–Duhem's equations and the other relationships given in Section 1.4 are also valid for excess quantities.

As examples of excess quantities of non-ideal solutions we choose the excess partial molar quantities and the excess integral molar quantities of enthalpy, entropy and Gibbs' free energy to illustrate the definition.

Excess Enthalpy

As $H_i^{mix} = 0$ for an ideal solution, the partial molar enthalpy of mixing $\overline{H}_i^{mix}$ of component i of a non-ideal solution is simply equal to the excess enthalpy

As $(\overline{H}_i^{mix})_{non\ ideal} = 0 + \overline{H}_i^{Ex}$ we obtain

$$\overline{H}_i^{Ex} = \overline{H}_i^{mix} \tag{1.113}$$

The Equation (1.112) on page 30 is valid both for partial and integral molar quantities. Hence we have for a non-ideal solution

$$H_m^{Ex} = \left(H_m^{mix}\right)_{non\text{-}ideal} \tag{1.114}$$

Excess Entropy

According to Equation (1.12) on page 6 the integral molar entropy of mixing or simply the *molar entropy of mixing* can be written as

$$(S_m^{mix})_{ideal} = -R(x_A \ln x_A + x_B \ln x_B + \cdots) \tag{1.12}$$

According to Equation (1.112) on page 30 the molar entropy of mixing of a non-ideal solution can therefore be written as

$$\left(S_m^{mix}\right)_{non\ ideal} = -R(x_A \ln x_A + x_B \ln x_B + \cdots) + S_m^{Ex} \tag{1.115}$$

or

$$S_m^{Ex} = \left(S_m^{mix}\right)_{non\ ideal} + R(x_A \ln x_A + x_B \ln x_B + \cdots) \tag{1.116}$$

Analogously to Equation (1.64) on page 17 we obtain

$$S_m^{Ex} = x_A \overline{S}_A^{Ex} + x_B \overline{S}_B^{Ex} + \cdots \tag{1.117}$$

$$S_m^{mix} = x_A \overline{S}_A^{mix} + x_B \overline{S}_B^{mix} + \cdots \tag{1.118}$$

The *excess partial molar entropy* of component i of a non-ideal solution will be

$$\overline{S}_i^{Ex} = \left(\overline{S}_i^{mix}\right)_{non\text{-}ideal} + R \ln x_i \tag{1.119}$$

Excess Gibbs' Free Energy

The excess partial molar Gibbs' free energy of component i is found with the aid of the basic relationship $G = H - TS$ (Equation (1.28) on page 9):

$$\overline{G}_i^{Ex} = \overline{H}_i^{Ex} - T\overline{S}_i^{Ex} \tag{1.120}$$

With the aid of Equations (1.113) and (1.120) we obtain

$$\overline{G}_i^{Ex} = \overline{H}_i^{mix} - T(\overline{S}_i^{mix} + R \ln x_i)$$

or

$$\overline{G}_i^{Ex} = \overline{G}_i^{mix} - RT \ln x_i \tag{1.121}$$

By use of the relationship

$$G_m^{Ex} = x_A \overline{G}_A^{Ex} + x_B \overline{G}_B^{Ex} + \cdots + x_c \overline{G}_c^{Ex} \tag{1.122}$$

and Equation (1.121) we obtain the *excess molar Gibbs' free energy*:

$$G_{\mathrm{m}}^{\mathrm{Ex}} = \left(G_{\mathrm{m}}^{\mathrm{mix}}\right)_{\text{non ideal}} - RT(x_{\mathrm{A}}\ln x_{\mathrm{A}} + x_{\mathrm{B}}\ln x_{\mathrm{B}} + \cdots) \tag{1.123}$$

Gibbs–Duhem's Equations for a Binary Non-Ideal Solution

Gibbs–Duhem's equations, analogous to Equations (13′) and (14′) on page 24 are also valid for the excess partial molar Gibbs' free energies and the excess integral molar Gibbs' free energy:

$$\overline{G}_{\mathrm{A}}^{\mathrm{Ex}} = G_{\mathrm{m}}^{\mathrm{Ex}} + (1 - x_{\mathrm{A}})\frac{\partial G_{\mathrm{m}}^{\mathrm{Ex}}}{\partial x_{\mathrm{A}}} \tag{1.124}$$

$$\overline{G}_{\mathrm{B}}^{\mathrm{Ex}} = G_{\mathrm{m}}^{\mathrm{Ex}} + (1 - x_{\mathrm{B}})\frac{\partial G_{\mathrm{m}}^{\mathrm{Ex}}}{\partial x_{\mathrm{B}}} \tag{1.125}$$

The excess molar Gibbs' free energy of a binary solution as a function of the excess partial molar Gibbs' free energy of component B is, in analogy with Equation (15′) on page 25,

$$G_{\mathrm{m}}^{\mathrm{Ex}} = (1 - x_{\mathrm{B}})\int_{0}^{x_{\mathrm{B}}} \frac{\overline{G}_{\mathrm{B}}^{\mathrm{Ex}}}{(1 - x_{\mathrm{B}})^{2}}\,\mathrm{d}x_{\mathrm{B}} \tag{1.126}$$

The excess Gibbs' free energy

$$\overline{G}_{\mathrm{i}}^{\mathrm{Ex}} = \overline{H}_{\mathrm{i}}^{\mathrm{Ex}} - T\overline{S}_{\mathrm{i}}^{\mathrm{Ex}} \tag{1.120}$$

is a very important quantity. It is used for accurate calculations of the molar Gibbs' free energies of non-ideal liquid and solid binary alloys as functions of temperature.

Relationship between the Excess Molar Gibbs' Free Energy and Activity

With the aid of Equation (1.121) and Equation (1.107) on page 31 we obtain

$$\overline{G}_{\mathrm{i}}^{\mathrm{Ex}} = RT\ln \gamma_{\mathrm{i}}x_{\mathrm{i}} - RT\ln x_{\mathrm{i}} = RT\ln \gamma_{\mathrm{i}} \tag{1.127}$$

With the aid of Equations (1.122) and (1.127) we obtain

$$G_{\mathrm{m}}^{\mathrm{Ex}} = RT(x_{\mathrm{A}}\ln \gamma_{\mathrm{A}} + x_{\mathrm{B}}\ln \gamma_{\mathrm{B}} + \cdots) \tag{1.128}$$

1.7.3 Molar Gibbs' Free Energies of Non-Ideal Binary Solutions

Calculation of accurate functions, which describe the molar Gibbs' free energy of non-ideal liquid and solid metals, is very important. These functions are the basis of the phase diagrams of alloy systems.

The Gibbs' molar free energy of a non-ideal solution is the sum of that of an ideal solution plus the excess Gibbs' molar free energy. With the aid of Equation (1.86) on page 27 and Equation (1.112) on page 31 we obtain

$$(G_{\mathrm{m}})_{\text{non-ideal}} = x_{\mathrm{A}}G_{\mathrm{A}}^{0} + x_{\mathrm{B}}G_{\mathrm{B}}^{0} + RT(x_{\mathrm{A}}\ln x_{\mathrm{A}} + x_{\mathrm{B}}\ln x_{\mathrm{B}}) + G_{\mathrm{m}}^{\mathrm{Ex}} \tag{1.129}$$

or

$$(G_{\mathrm{m}})_{\text{non-ideal}} = x_{\mathrm{A}}G_{\mathrm{A}}^{0} + x_{\mathrm{B}}G_{\mathrm{B}}^{0} + RT(x_{\mathrm{A}}\ln x_{\mathrm{A}} + x_{\mathrm{B}}\ln x_{\mathrm{B}}) + H_{\mathrm{m}}^{\mathrm{Ex}} - TS_{\mathrm{m}}^{\mathrm{Ex}} \tag{1.130}$$

According to Equation (1.114) on page 31 we obtain

$$(G_m)_{non\text{-}ideal} = x_A G_A^0 + x_B G_B^0 + RT(x_A \ln x_A + x_B \ln x_B) + \left(H_m^{mix}\right)_{non\text{-}ideal} - TS_m^{Ex} \qquad (1.131)$$

Equation (1.131) is the desired expression for Gibbs' molar free energy of a non-ideal solution as a function of composition and temperature.

Equation (1.131) is not as easy to calculate and show graphically as Equation (1.86) on page 20 because of the $(H_m^{mix})_{non\text{-}ideal}$ and S_m^{Ex} terms.

On page 24 we discussed a theoretical model for the molar heat of mixing. It has been checked experimentally because the $(H_m^{mix})_{non\text{-}ideal}$ term can be determined experimentally, as we will see in next section. The molar heat of mixing is therefore comparatively easy to determine.

The difficulties to make accurate calculations of $(G_m)_{non\text{-}ideal}$ for non-ideal metal solutions emanate from the S_m^{Ex} term in Equation (1.131). There exists no satisfactory theoretical model of the excess molar entropy and it cannot be measured directly by experiments.

The entropy of an ideal solid solution consists of two terms, one due to lattice vibrations and one originating from the binding energies of the component atoms.

The entropy of a pure metal is essentially due to lattice vibrations. The vibrational entropy of an ideal solution of metals is approximately the same as that of pure metals and does not contribute to the excess entropy.

No satisfactory model for the exchange energy and therefore also for the excess entropy has been found so far. However, there is very convincing experimental evidence, described in Figure 1.17, that the molar heat of mixing and the excess entropy of non-ideal solutions are proportional. This seems reasonable as both $(H_m^{mix})_{non\text{-}ideal}$ and S_m^{Ex} depend on the binding energy of the component atoms.

$$S_m^{Ex} = A\left(H_m^{mix}\right)_{non\text{-}ideal} \qquad (1.132)$$

The proportionality constant A is obtained as the slope of the line in Figure 1.17. Then S_m^{Ex} can be calculated if $(H_m^{mix})_{non\text{-}ideal}$ is known.

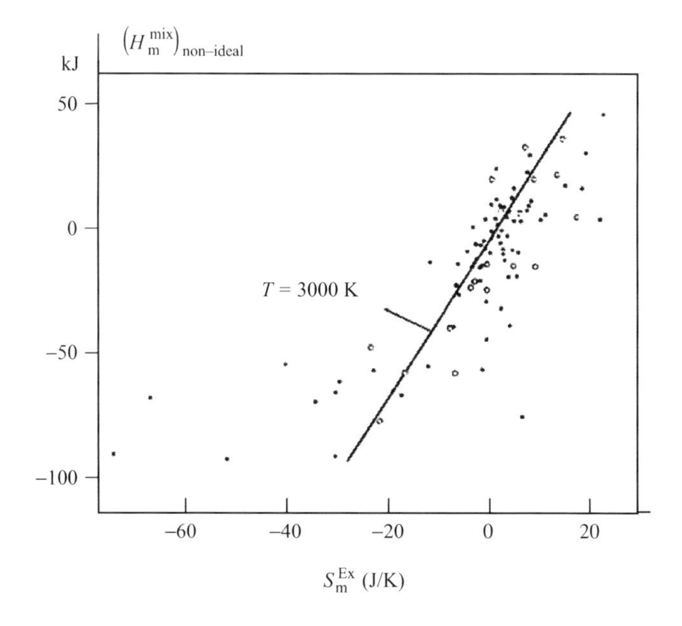

Figure 1.17 Molar heat of mixing as a function of excess molar entropy of liquid and solid alloys. Reproduced from Elsevier © 1983.

1.8 Experimental Determination of Thermodynamic Quantities of Binary Alloys

In Section 1.4.2 theoretical formulas of the relative thermodynamic quantities *S*, *H* and *G* have been derived for alloys and graphical methods to construct partial molar and relative partial molar quantities have been indicated (pages 19 and 20).

Only two of the relative thermodynamic quantities can be measured experimentally

- molar heat of mixing H_m^{mix}
- partial molar Gibbs' free energy of mixing $\overline{G}_i^{mix}$.

From experimental values of these quantities it is possible to calculate other thermodynamic quantities with the aid of for example Equation (1.132) on page 33 and equations and relationships given in Section 1.4 and other parts of the text. An example is given on page 35.

1.8.1 Determination of Molar Heat of Mixing of Binary Alloys

The best way to measure enthalpy quantities is direct measurement of heat, i.e. calorimetric measurements. As most metals have high melting points, measurements of heat when two metal melts are mixed, require high-temperature calorimetry. Such measurements are hard to perform and require accurate knowledge of heat capacities of the metals, good insulation arrangements and vessels with low heat capacities.

Determination of the Molar Heat of Mixing

Method 1

A crucible with a given amount of solute metal A is placed in a calorimeter at room temperature. A known quantity of molten solvent metal B is poured into the crucible. The alloy is formed and allowed to cool to room temperature and the total heat is measured by the calorimeter. A blind experiment is then performed. The same experiment is carried out without metal A. From the difference in total heat and the known weights of the metal samples H_m^{mix} of the alloy can be calculated.

The method is not very accurate, owing to variation of the heat losses between the two experiments and the risk of contamination, but it is fairly rapid. The H_m^{mix} values of many liquid iron–silicon alloys have been determined by this method.

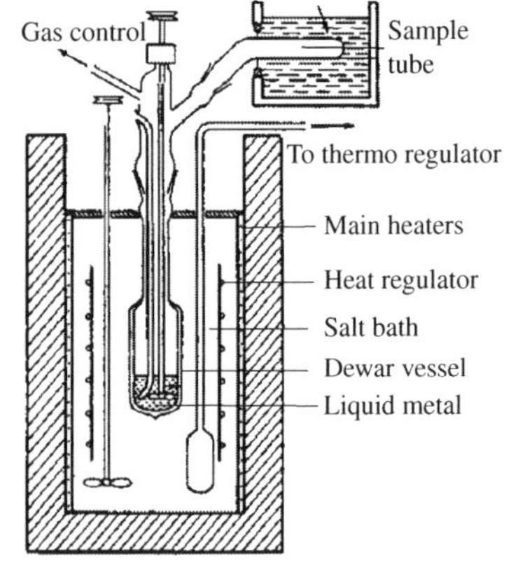

Figure 1.18 Liquid-metal calorimeter. Reproduced from Elsevier © 1979.

Method 2

A flask, which is surrounded by a salt bath, contains a given amount of metal A, which is molten by electrical heating until a steady temperature is achieved (Figure 1.18). A sample of the solute metal B is placed in the sample tube and cooled by a water-ice bath to 0 °C. The sample tube is rotated into a vertical position and the sample is dropped into the molten metal A. The alloy is formed and heated until the molten alloy reaches the initial steady temperature. The heat of mixing can be calculated when the masses of tin and sample metal, temperature, supplied amount of heat and material constants are known.

The accuracy of the method has been successively improved and is of the magnitude 5%. In recent years differential and much more accurate thermal methods have been developed. They will be discussed in Chapter 7 (thermal analysis).

Derivation of the Partial Molar Heat of Mixing

H_m^{mix} is determined as a function of composition by experiments on a series of alloys with different x_B values. An $x_B H_m^{mix}$ curve is drawn and the partial molar heat of mixing for each component can be calculated from Equations (1.86) and (1.87) on page 27 or derived graphically from the diagram in the way shown in Figure 1.9 on page 19.

1.8.2 Determination of Partial Molar Gibbs' Free Energy of Mixing of Binary Alloys

Method 1
The partial molar Gibbs' free energies $\overline{G}_i^{mix}$ can be determined experimentally by measuring the partial vapour pressure of component i in equilibrium with the solution and the vapour pressure in equilibrium with the corresponding pure element under equivalent conditions.

If the pressure is so low that the ideal gas laws are valid and if there is no change in the chemical composition of the vapour, $\overline{G}_i^{mix}$ can be calculated from the ratio of the measured pressures with the aid of Equation (1.93) on page 28

$$\overline{G}_i^{mix} = RT \ln \frac{p_i}{p_i^0} \tag{1.93}$$

This is the case for most metallic elements, for example Al, In, Pb, Cr, Mn, Ni, Cu, Ag, Au and Zn, which have monoatomic vapours. This is not the case for semiconductor elements, covalent elements and most metals in group III and V in the periodic table, for example C, Si, Ge, Sn, Sb, Bi, P, As and S. In these cases polyatomic molecules are formed in the vapour.

Method 2
Alternatively $\overline{G}_i^{mix}$ can be determined by use of a galvanic cell. The positive pole of the cell consists of an electrode made of the alloy with component i. The negative pole consists of pure element i. The electrolyte is an ionic conductor with ions of metal i. It is illustrated by the scheme

	−	ionic conductor	+
− − −	pure metal i	with ions of metal i	alloy with component i

The work that is required to transport 1 kmol of ions reversibly from one electrode to the other when the cell is closed by an outer conductor equals

$$\overline{G}_i^{mix} = Fz_i\varepsilon \tag{1.133}$$

where
F = Faraday's constant, 96 500 × 10^3 Coulomb/kmol
z_i = valence of the charge carrier in the electrolyte
ε = reversible emf of the cell.

Calculation of the Molar Gibbs' Free Energy of Mixing

In the case of a binary solution the partial molar Gibbs' free energy of mixing $\overline{G}_B^{mix}$ is determined for a series of alloys of various compositions by method 1 or 2 above. For each composition x_B the value of G_m^{mix} can be calculated with the aid of Equation (15') on page 24.

$$G_m^{mix} = (1 - x_B) \int_0^{x_B} \frac{\overline{G}_B^{mix}}{(1 - x_B)^2} dx_B \tag{1.134}$$

Determination of Entropy Quantities for Binary Alloys

Entropy cannot be measured directly and has to be calculated indirectly from other measurable quantities.

If H_m^{mix} and G_m^{mix} are known the entropy of mixing S_m^{mix} can be calculated from the relationship

$$G_m^{mix} = H_m^{mix} - TS_m^{mix} \tag{1.135}$$

If the $\overline{G_i}^{mix}$ values are measured at various temperatures it is possible to calculate G_m^{mix} as a function of temperature. Then S_m^{mix} can be calculated from the relationship

$$S_m^{mix} = -\frac{\partial G_m^{mix}}{\partial T} \tag{1.136}$$

Today, it is also possible to determine G_m^{mix} from measurements of the compositions of different phases in equilibrium with each other. The equilibrium is achieved by heat treatment of the specimen.

A method to determine the excess molar Gibbs' free energy G_m^{Ex} is described on page 36.

Summary

■ Thermodynamic Concepts and Relationships

$Q = U + W$ U = Internal energy

$H = U + pV$ $(\mathrm{d}H)_p = (\mathrm{d}Q)_p$ Enthalpy

$\mathrm{d}S = \frac{\mathrm{d}Q}{T}$ $S = k_B \ln P$ Entropy

Molar entropy of mixing (solids, liquids and gases):

$$S_m^{mix} = -R(x_A \ln x_A + x_B \ln x_B)$$

$G = H - TS$ Gibbs' free energy

At constant temperature and pressure spontaneous processes always occur in the direction of decreasing G.

The deviation of the thermodynamic function G of a system from its equilibrium value can be regarded as the driving force in chemical and metallurgical reactions.

■ Thermodynamics of Single-Component Systems

Equilibrium Condition between two Phases α and β

$$G^\alpha = G^\beta$$

Change of Transformation Temperature as a Function of Change of Pressure

$$\frac{\mathrm{d}T}{\mathrm{d}p} = \frac{V^\beta - V^\alpha}{S^\beta - S^\alpha} = \frac{\Delta V^{\alpha\to\beta}}{\Delta S^{\alpha\to\beta}}$$ Clausius–Clapeyron's law

This law can be applied on melting and vaporization processes.

Change of Melting Point owing to Pressure

$$\Delta T_M = T_M(p) - T_M(p_0) = \frac{T_M\left(V_m^L - V_m^s\right)}{H_m^{fusion}}\Delta p$$

The melting points of metals increase with increasing pressure.

Change of Melting Point owing to Crystal Curvature

$$\Delta T_M = T_M(p_K^\alpha) - T_M(p_0^\alpha) = \frac{T_M\left(V_m^\alpha - V_m^L\right)}{H_m^{fusion}} 2\sigma K$$

$$K = \frac{1}{2}\left(\frac{1}{r_{max}} + \frac{1}{r_{min}}\right) \qquad \Delta p = 2\sigma K$$

An application is the decrease in temperature, which occurs in the neighbourhood of a growing dendrite tip.

Change of Boiling Point owing to Pressure

$$\Delta T_B = T_B(p) - T_B(p_0) = \frac{T_B\left(V_m^g - V_m^L\right)}{H_m^{evap}} \Delta p$$

or

$$\Delta T_B = T_B(p) - T_B(p_0) = \frac{RT_B(p)T_B(p_0)}{H_m^{evap}} \ln \frac{p}{p_0}$$

Bubble Formation in Melts

$$\Delta T_B = T_B(p_K^g) - T_B(p_0) = \frac{RT_B(p_K^g)T_B(p_0)}{H_m^{evap}} \ln\left(\frac{p^L + 2\sigma K}{p_0}\right)$$

■ Gibbs' Free Energy of a Pure Metal

$$G_m^s = H_m^s - TS_m^s$$
$$G_m^L = H_m^L - TS_m^L$$

Consider Figures 1.7 and 1.8 on page 15.
The entropy of the liquid phase is normally larger than that of the solid. For this reason the liquid curve is steeper than the solid curve.

■ Thermodynamics of Multiple-Component Systems

Designations of Components:

Multiple-Component Systems:	Binary Systems:
$i = A, B, \ldots$	$i = A, B$

Concepts and Relationships

The concepts and relationships below are valid for many thermodynamic quantities, among them

Volume V	
Enthalpy H	*Common designation*:
Entropy S	
Gibbs' free energy G	$X = V, H, S, G$

Partial Molar Quantities (X = V, H, S and G)

$$X = n_A\overline{X}_A + n_B\overline{X}_B + \cdots$$

$$dX = \overline{X}_A dn_A + \overline{X}_B dn_B + \cdots$$

Definition: $\overline{X}_i = \frac{\partial X}{\partial n_i}$

Systems with Constant Composition:

$$X_m = x_A\overline{X}_A + x_B\overline{X}_B + \cdots \quad (1\ \text{kmol})$$

Systems with Variable Composition:

$$x_A d\overline{X}_A + x_B d\overline{X}_B + \cdots = 0$$

Graphical Construction of the Partial Molar Quantities of a Binary System
Consider Figure 1.9 on page 19.
The construction is based on the relationships

$$x_A + x_B = 1$$

$$X_m = x_A\overline{X}_A + x_B\overline{X}_B$$

$$\overline{X}_A = X_m + (1 - x_A)\frac{\partial X_m}{\partial x_A} \quad \text{Gibbs–Duhem's}$$

$$\overline{X}_B = X_m + (1 - x_B)\frac{\partial X_m}{\partial x_B} \quad \text{equations}$$

X is plotted as a function of x_B. At point (x_B, X) a tangent to the curve is drawn. The intersections between the tangent and the lines $x_B = 0$ and $x_B = 1$ give $\overline{X}_A$ and $\overline{X}_B$, respectively.

Relative Partial Molar Quantities (X = V, H, S and G)
Partial Molar Quantities of Mixing

Definition:

A relative partial molar quantity = the difference between a partial molar quantity of a component in a solution and the molar quantity of the pure substance.
Relative partial molar quantities are also called *partial molar quantities of mixing* of substance i and are designated by X_i^{mix}.

$$\overline{X}_i^{mix} = \overline{X}_i - X_i^0$$

Relationships between Partial Molar Quantities of Mixing in a Binary System

$$\overline{X}_A^{mix} = -\int_0^{x_B} \frac{x_B}{1 - x_B}\frac{\partial \overline{X}_B^{mix}}{\partial x_B}\,dx_B \quad \text{or} \quad \overline{X}_A^{mix} = \int_0^{x_B} \frac{\overline{X}_B^{mix}}{(1 - x_B)^2}\,dx_B - \frac{x_B\overline{X}_B^{mix}}{1 - x_B}$$

Relative Integral Molar Quantities (X = V, H, S and G)
Integral Molar Quantities of Mixing

Definition:

X_m^{mix} = difference between a molar quantity X_m of a solution and the sum of the quantities of the corresponding pure component substances.

$$X_m^{mix} = X_m - (x_A X_A^0 + x_B X_B^0 + \cdots + x_c X_c^0)$$

Relationships between Partial and Integral Molar Quantities of Mixing

$$X_m^{mix} = x_A\overline{X}_A^{mix} + x_B\overline{X}_B^{mix} + \cdots + x_c\overline{X}_c^{mix}$$

$$\overline{X}_A^{mix} = X_m^{mix} + (1 - x_A)\frac{\partial X_m^{mix}}{\partial x_A} \quad \text{Gibbs–Duhem's}$$

$$\overline{X}_B^{mix} = X_m^{mix} + (1 - x_B)\frac{\partial X_m^{mix}}{\partial x_B} \quad \text{equations}$$

$$X_m^{mix} = (1 - x_B)\int_0^{x_B} \frac{\overline{X}_B^{mix}}{(1 - x_B)^2}\,dx_B$$

Graphical Construction of Partial and Integral Molar Quantities of Mixing

X is plotted as a function of x_B.
Graphical construction of the above quantities is described in the text.
Study Figure 1.10 on page 21 carefully.

■ Thermodynamics of Alloys

An *alloy* is a multiple-component metallic system. An alloy of a given pressure, temperature and composition can either be *homogeneous*, e.g. consist of a single phase, or *heterogeneous*, e.g. consist of several phases.

Heat of Mixing

$$H_m^{mix} = -(\Delta H_m)^{mix} = N_{AB}\left[W_{AB} - \frac{1}{2}(W_{AA} + W_{BB})\right]$$

Theoretical Model of Heat of Mixing

$$H_m^{mix} = -(\Delta H_m)^{mix} = Lx_A x_B$$

If $L > 0$ $\quad H_m^{mix} = -\Delta H_m^{mix}$ is positive
If $L < 0$ $\quad H_m^{mix} = -\Delta H_m^{mix}$ is negative.

Ideal and Non-Ideal Solutions of Two Components

- A–A and B–B and A–B forces are equal.
- The heat of mixing is zero.
- The solubility is complete.

All other solutions are non-ideal.

■ Thermodynamics of Binary Ideal Solutions

Enthalpy of Mixing: $H_m^{mix} = 0$
Entropy of Mixing: $S_m^{mix} = -R(x_A \ln x_A + x_B \ln x_B)$
Gibbs' Free Energy of Mixing:

$$G_m^{mix} = H_m^{mix} - TS_m^{mix} = RT(x_A \ln x_A + x_B \ln x_B)$$

Gibbs' Free Energy of a Binary Ideal Solution

$$G_m = x_A G_A^0 + x_B G_B^0 + G_m^{mix}$$

or

$$G_m = x_A G_A^0 + x_B G_B^0 + RT(x_A \ln x_A + x_B \ln x_B)$$

G_m is plotted as a function of x_B.
Graphical construction of the quantities in the equations above is given in the text. See Figure 1.13b on page 27.

■ Thermodynamics of Non-Ideal Solutions

Activities and Activity Coefficients

The *activity of component* i *in a solution* is defined as the ratio of the fugacity of component i over the solution to the fugacity of the standard state equal to the pure stable condensed phase at the pressure 1 atm and temperature 273 K.

$$a_i = f_i / f_i^0$$

$$\overline{G}_i(\text{l or c}) = G_i^0(\text{l or c}) + RT \ln a_i$$

Raoult's law: $a_i = x_i$ or $f_i = 1$

Henry's law: $\lim_{x_i \to 0} a_i^{\text{solute}} = \lim_{x_i \to 0} \gamma_i x_i = \gamma_i^0 x_i$ or $\lim_{x_i \to 0} \gamma_i = \gamma_i^0$

See Figure 1.14 on page 29 and Figures 1.15 and 1.16 on page 29.

Law	Activity	Gibbs' free energy
General law	$a_i = f_i x_i$	$\overline{G}_i = G_i^0 + RT \ln a_i$
Raoult's law, valid for solvents in dilute solutions	$a_i = x_i$ and $f_i = 1$ valid when $x_i \to 1$	$\overline{G}_i = G_i^0 + RT \ln x_i$
General law	$a_i = \gamma_i x_i = \gamma_i^{\text{rel}} \gamma_i^0 x_i$	$\overline{G}_i = G_i^0 + RT \ln \gamma_i^0 + RT \ln \gamma_i^{\text{rel}} x_i$
Henry's law, valid for solutes in dilute solutions	$a_i = \gamma_i^0 x_i$ and $\gamma_i^{\text{rel}} = 1$ valid when $x_i \to 0$	$\overline{G}_i = G_i^0 + RT \ln \gamma_i^0 + RT \ln x_i$

■ Excess Quantities of Non-Ideal Solutions

Definition:

Excess quantity = relative partial molar quantity of a non-ideal solution minus that of an ideal solution.

$$(X)_{\text{non-ideal solution}} = (X)_{\text{ideal solution}} + X^{\text{Ex}}$$

Relationships between Partial and Molar Excess Quantities and Thermodynamic Quantities of Mixing

$$\overline{H}_i^{\text{Ex}} = \overline{H}_i^{\text{mix}} \qquad H_m^{\text{Ex}} = \left(H_m^{\text{mix}}\right)_{\text{non-ideal}}$$

$$\overline{S}_i^{\text{Ex}} = \overline{S}_i^{\text{mix}} - R \ln x_i \qquad S_m^{\text{Ex}} = \left(S_m^{\text{mix}}\right)_{\text{non-ideal}} + R(x_A \ln x_A + x_B \ln x_B + \cdots)$$

$$\overline{G}_i^{\text{Ex}} = \overline{G}_i^{\text{mix}} - RT \ln x_i \qquad G_m^{\text{Ex}} = \left(G_m^{\text{mix}}\right)_{\text{non-ideal}} - RT(x_A \ln x_A + x_B \ln x_B + \cdots$$

For ideal solutions the excess quantities are zero.

The excess Gibbs' free energy

$$\overline{G}_i^{\text{Ex}} = \overline{H}_i^{\text{Ex}} - T\overline{S}_i^{\text{Ex}}$$

is a very important quantity. It is used for accurate calculations of the molar Gibbs' free energies of non-ideal liquid and solid binary alloys as functions of temperature.

■ Gibbs' Free Energy of a Non-Ideal Solution

Gibbs–Duhem's equations, analogous to Equations (13′) and (14′) on page 25 are also valid for the excess partial molar Gibbs' free energies and the excess integral molar Gibbs' free energy:

$$\overline{G}_A^{Ex} = G_m^{Ex} + (1 - x_A)\frac{\partial G_m^{Ex}}{\partial x_A}$$

$$\overline{G}_B^{Ex} = G_m^{Ex} + (1 - x_B)\frac{\partial G_m^{Ex}}{\partial x_B}$$

$$G_m^{Ex} = (1 - x_B)\int_0^{x_B} \frac{\overline{G}_B^{Ex}}{(1 - x_B)^2}\,dx_B$$

Relationships between Partial Excess Gibbs' Free Energy and Activity Coefficients:

$$\overline{G}_i^{Ex} = RT\ln\gamma_i x_i - RT\ln x_i = RT\ln\gamma_i$$

$$G_m^{Ex} = RT(x_A\ln\gamma_A + x_B\ln\gamma_B + \cdots + x_c\ln\gamma_c)$$

Molar Gibbs' Free Energies of Non-Ideal Binary Solutions

Calculation of accurate functions, which describe the molar Gibbs' free energy of non-ideal liquid and solid metals, is very important. These functions are the basis of the phase diagrams of alloy systems.

$$(G_m)_{non\text{-}ideal} = x_A G_A^0 + x_B G_B^0 + RT(x_A\ln x_A + x_B\ln x_B) + G_m^{Ex}$$

$$(G_m)_{non\text{-}ideal} = x_A G_A^0 + x_B G_B^0 + RT(x_A\ln x_A + x_B\ln x_B) + H_m^{Ex} - TS_m^{Ex}$$

$$(G_m)_{non\text{-}ideal} = x_A G_A^0 + x_B G_B^0 + RT(x_A\ln x_A + x_B\ln x_B) + \left(H_m^{mix}\right)_{non\text{-}ideal} - TS_m^{Ex}$$

The only measurable thermodynamic quantities are

- molar heat of mixing H_m^{mix}
- partial molar Gibbs' free energy of mixing $\overline{G}_i^{mix}$

However, as both $\left(H_m^{mix}\right)_{non\text{-}ideal}$ and S_m^{Ex} depend on the binding energy of the component atoms, it is reasonable to assume that

$$S_m^{Ex} = A\left(H_m^{mix}\right)_{non\text{-}ideal}$$

Calculation of other thermodynamic quantities are described briefly in the text.

Further Reading

A.H. Carter, Classical and Statistical Thermodynamics, *Upper Saddle River, New Jersey*, US, Prentice Hall, 2001.

L. Darken, R.W. Gurry, Physical Chemistry of Metals, New York, McGraw-Hill Book Company, 1953.

D. R. Gaskell, Introduction to Metallurgical Thermodynamics, *Bristol*, US, Taylor & Francis, 1981.

G. Grimvall, Thermophysical Properties of Materials, Amsterdam, Elsevier Science B.V. 1999.

M. Kaufman, Principles of Thermodynamics, New York, Marcel Dekker Inc, 2002.

2

Thermodynamic Analysis of Solidification Processes in Metals and Alloys

2.1 Introduction

Solidification processes in metals and alloys are controlled by kinetic laws and depend on the driving force of the process. In Chapter 1 on page 9 it is mentioned that the change of Gibbs' free energy of a system can be regarded as the driving force of chemical and metallurgical reactions. The driving force

Solidification and Crystallization Processing in Metals and Alloys, First Edition. Hasse Fredriksson and Ulla Åkerlind.

of solidification equals the change in Gibbs' free energy when the system is transferred from a liquid to a solid state.

To be able to analyze the solidification process we must primarily find the Gibbs' free energy of liquid and solid metals and alloys as a function of composition and temperature. The tools we need for this purpose are found in Chapter 1.

In this chapter the driving force of solidification of pure metals is derived. A binary system is far more complicated than a single-component system. After an extensive study of the thermodynamics and stability of ideal and non-ideal solutions the driving force of solidification of binary alloys is derived.

Special care has been devoted to the fundamental condition of equilibrium between phases and the relationships between the Gibbs' free energies of liquid and solid binary alloys and the corresponding phase diagrams of the alloys. Ternary alloys are discussed shortly. The chapter ends with a short section on the thermodynamics of lattice defects in metals.

This theoretical basis together with the laws of physics will be applied in the following chapters where different aspects of the solidification of metals and alloys will be studied and analyzed.

2.2 Thermodynamics of Pure Metals

2.2.1 *Driving Force of Solidification*

The concept of Gibbs' free energy can be used to understand the behaviour of a metal close to its melting point T_M.

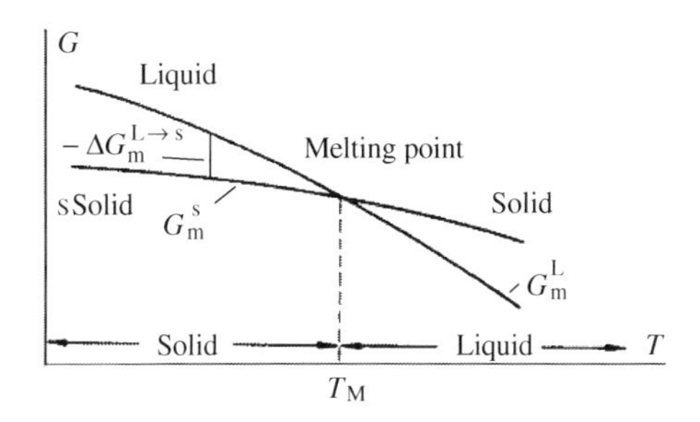

Figure 2.1 The molar Gibbs' free energy of a metal as a liquid and a solid as a function of temperature. The stable state is the one with lowest possible free energy.

Figure 2.1 is a combination of Figures 1.7 and 1.8 on page 16 in Chapter 1. It shows how the molar Gibbs' free energies G_m of a pure liquid and of the corresponding solid metal varies as a function of temperature.

At the melting point the molar Gibbs' free energies of the melt and the solid are equal. Below the melting point the solid has a lower free energy than the melt and is therefore the stable phase. Above the melting point the reverse is true and the liquid is more stable than the solid.

The difference in molar Gibbs' free energy between the liquid and the solid at the same temperature below the melting point acts as a driving force for solidification of the undercooled liquid. The greater the driving force is, the stronger will be the tendency of crystallization.

During a spontaneous solidification process the Gibbs' free energy change $\Delta G_m^{L\to s} = G_m^s - G_m^L$ is *negative* because heat is lost to the surroundings. It is desirable that the driving force is positive for a spontaneous process and for this reason it is defined as

Driving force = the negative change of the Gibbs' free energy at solidification.

$$-\Delta G_m^{L\to s} = G_m^L - G_m^s \quad (2.1)$$

In order to derive a useful expression of the driving force of solidification we apply Equation (1.29) on page 9 in Chapter 1 on 1 kmol of the liquid and solid phases and take the differences between these equations. The result is

$$-\Delta G_m^{L\to s} = -\left(\Delta H_m^{L\to s} - T\,\Delta S_m^{L\to s}\right) \quad (2.2)$$

At the melting point $T = T_M$ the liquid and solid phases are in equilibrium with each other, which can be expressed as $\Delta G_m^{L \to s} = 0$. Inserting these values into Equation (2.2) we obtain

$$\Delta S_m^{L \to s} = \frac{\Delta H_m^{L \to s}}{T_M} \tag{2.3}$$

This expression of $\Delta S_m^{L \to s}$ is introduced into Equation (2.2):

$$-\Delta G_m^{L \to s} = -\left(\Delta H_m^{L \to s} - T\frac{\Delta H_m^{L \to s}}{T_M}\right) \tag{2.4}$$

After rearrangement of Equation (2.4) the driving force can be written

$$-\Delta G_m^{L \to s} = \frac{T_M - T}{T_M}(-\Delta H_m^{L \to s}) = \frac{\Delta T}{T_M} H_m^{fusion} \tag{2.5}$$

where
$-\Delta G_m^{L \to s}$ = driving force of solidification
T_M = melting-point temperature
ΔT = undercooling ($T_M - T$)
$-\Delta H_m^{L \to s}$ = molar heat of fusion H_m^{fusion}.

The driving force of solidification close to the melting point is proportional to the undercooling and the heat of fusion.

2.3 Thermodynamics of Binary Alloys

The molar Gibbs' free energies of ideal and non-ideal binary solutions as functions of their compositions have been treated briefly in Sections 1.6 and 1.7, respectively, in Chapter 1. The topic will be further penetrated here and the functions will be shown graphically.

In Chapter 1 the molar Gibbs' free energies of ideal and non-ideal binary solutions as functions of composition with the temperature as a parameter were derived. Most of the graphical illustrations of the molar Gibbs' free energy of ideal and non-ideal binary solutions in the following sections are based on these functions, i.e. on Equations (1.86) on page 26 and (1.130) on page 32 in Chapter 1.

Molar Gibbs' free energy of an ideal binary solution

$$G_m = x_A G_A^0 + x_B G_B^0 + RT\,(x_A \ln x_A + x_B \ln x_B) \tag{2.6}$$

Molar Gibbs' free energy of a non-ideal binary solution

$$G_m = x_A G_A^0 + x_B G_B^0 + RT\,(x_A \ln x_A + x_B \ln x_B) + H_m^{mix} - TS_m^{Ex} \tag{2.7}$$

The different terms and the possibilities to measure or calculate the excess terms have been discussed in Sections 1.7 and 1.8 in Chapter 1.

2.3.1 *Gibbs' Free Energy of Ideal Binary Solutions*

The molar Gibbs' free energy of an ideal binary solution is shown in Figure 2.2. It is identical to Figure 1.13a on page 26 in Chapter 1. The dotted line represents the first two terms in Equation (2.6). The third term in Equation (2.6) is negative and equal to ($-TS_m^{mix}$) (Chapter 1 on page 25). It causes a negative deviation from the dotted line in Figure 2.2. No excess-energy term is present. Equation (2.6) and Figure 2.2 may refer either to a liquid or to a solid solution.

The composition of the solution is given in mole fraction of the solute B in the figure. This is the common unit when Gibbs' free energy is concerned.

The molar Gibbs' free energies G_A^0 of pure component A and G_B^0 of pure component B are given by the intersections between the curve and the vertical lines $x_B = 0$ and $x_B = 1$. Superscript "0" indicates that the elements A and B are at their standard states. G_A^0 and G_B^0 *vary with temperature* according to Equations (1.58) and (1′) and Figures 1.7 and 1.8 are four on page 15 in Chapter 1. The standard state values are constant *at a fixed temperature.*

The whole G_m curve lies below the dashed line, which means that a solution of *any* composition of components A and B is stable, in agreement with the definition of an ideal solution. Every solution, independent of composition, has lower free energy than the unmixed components. Hence, a spontaneous mixing will occur when the components are brought together.

The dissolving process can be followed stepwise in Figure 2.3. Initially, we have a mixture of two components A and B in the proportions $x_A = 1 - x$ and $x_B = x$. The initial molar free energy is G_m^0, found at the intersection between the dashed line and the vertical line $x_B = x$.

When the components start to dissolve, the concentrations near the components A and B changes gradually. This is described by two sliding points on the curve, moving from P_0 to P_1 to P_2, respectively, Q_0 to Q_1 to Q_2. Simultaneously, the molar free energy decreases from G_m^0 to G_1 to G_2 and finally to its lowest value G_3.

The final points P_3 and Q_3 coincide and correspond to a single homogeneous stable phase. The line through the points P_3 and Q_3 is the tangent to the curve at the point $P_3 = Q_3$. The molar Gibbs' free energy at point P_3 is G_3. Other ways along the curve can occur, depending on how the dissolution process proceeds in detail, but the end will always be the same ($P_3 = Q_3$ and G_3).

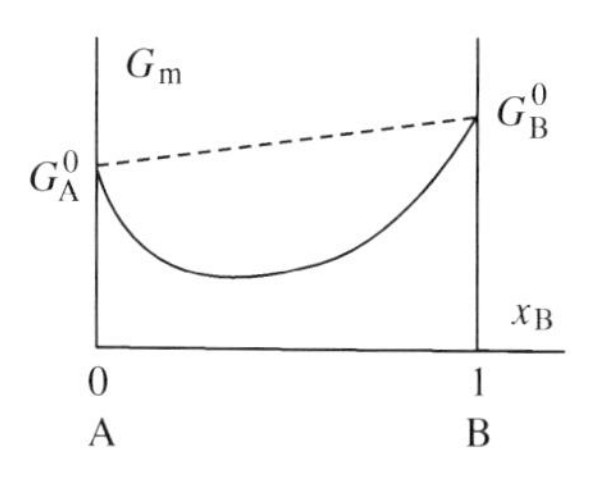

Figure 2.2 The molar Gibbs' free energy of an ideal binary solution as a function of composition.

The composition is given in mole fraction x in the figure. This is the usual unit when Gibbs' free energy is concerned.

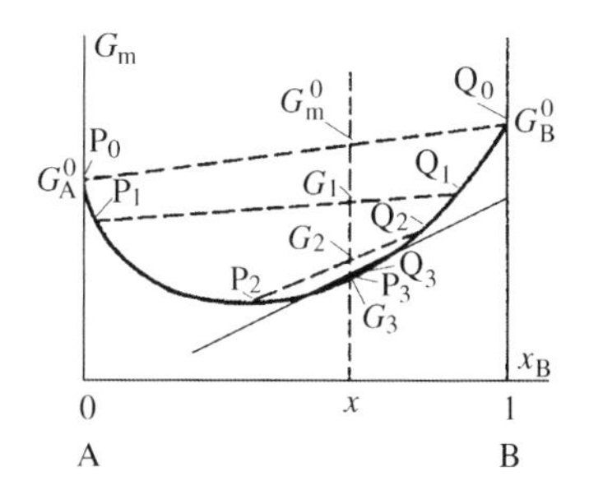

Figure 2.3 Dissolution process resulting in a homogeneous single phase.

The molar Gibbs' free energy as a function of composition expressed as molar fraction.

2.3.2 *Gibbs' Free Energy of Non-Ideal Solutions*

The molar excess Gibbs' free energy is the difference between the molar Gibbs' free energy of the real non-ideal solution and the one of an imaginary ideal solution of the same set of components.

$$G_m^{Ex} = G_m^{non\ ideal} - G_m^{ideal} \tag{2.8}$$

Equation (2.8) can be written with the aid of Equation (2.6)

$$G_m = x_A G_A^0 + x_B G_B^0 + RT\,(x_A \ln x_A + x_B \ln x_B) + G_m^{Ex} \tag{2.9}$$

Excess quantities of non-ideal solutions were discussed in Chapter 1 (Section 1.7.2 on page 30). The term G_m^{Ex} in Equation (2.9) has to be replaced by quantities, which can be calculated, or measured. Using the basic relationship $G = H - TS$ on the excess functions we obtain

$$G_m^{Ex} = H_m^{Ex} - TS_m^{Ex} \tag{2.10}$$

According to Equation (1.114) on page 31 in Chapter 1 we have

$$H_m^{Ex} = H_m^{mix} \tag{2.11}$$

The remaining problem is to find the excess entropy. As was mentioned on page 33 in Chapter 1 it is difficult to calculate the molar excess entropy of a non-ideal binary solution in the absence of a convenient model. A feasible way is to use the very convincing experimental evidence, reported on pages 33 in Chapter 1, that the heat of mixing and the excess entropy of non-ideal solutions are proportional.

$$S_m^{Ex} = AH_m^{mix} \tag{2.12}$$

where the constant A has the dimension K^{-1}.

If we combine Equations (2.10), (2.11) and (2.12) we obtain

$$G_m^{Ex} = H_m^{mix}(1 - AT) \tag{2.13}$$

Hence, the excess Gibbs' molar free energy of a non-ideal solution can be calculated if the heat of mixing is known. In Section 1.8 in Chapter 1 it was mentioned that H_m^{mix} is one of the few thermodynamic quantities that can be measured experimentally. Hence, it is possible to calculate G_m^{Ex} and consequently also the molar Gibbs' free energy of the non-ideal solution.

An alternative is to use a theoretical model to calculate H_m^{mix}.

2.3.3 The Regular Solution Model. Miscibility Gap in a Regular Solution

Hildebrand proposed the so-called *regular-solution model* in 1929 [1]. It can be regarded as an intermediate step between an ideal solution and a real non-ideal solution. It is not an adequate model of a real non-ideal solution but it can be used to illustrate the changes of molar Gibbs' free energy with temperature and the origin of observed miscibility gaps in real non-ideal solutions.

The Regular Solution Model

The molar Gibbs' free energy of a non-ideal binary solution can be written (Equation (2.7) on page 44)

$$G_m = x_A G_A^0 + x_B G_B^0 + RT\,(x_A \ln x_A + x_B \ln x_B) + H_m^{mix} - TS_m^{Ex} \tag{2.7}$$

where the excess molar Gibbs' free energy of mixing is the sum of the last two terms.

The regular-solution model is a hybrid between an ideal and a non-ideal solution. It is based on the following two assumptions

1. The *molar enthalpy of mixing or* simply *the molar heat of mixing* is assumed to be equal to the expression in Equation (1.81) on page 23 in Chapter 1

$$H_m^{mix} = (-\Delta H_m)^{mix} = Lx_A x_B \tag{2.14}$$

The physical background of this model has been discussed on page 23 in Chapter 1. The model takes the difference in strength of the forces between equal and unequal atoms, respectively, into consideration.

2. The molar entropy of mixing is assumed to be the same as the one for an ideal solution, i.e.

$$S_m^{Ex} = 0 \tag{2.15}$$

This assumption is doubtful, as the temperature dependence is strong. The advantage is its simplicity.

If the expressions in Equations (2.14) and (2.15) are introduced into Equation (2.7) we obtain

$$G_m = x_A G_A^0 + x_B G_B^0 + Lx_A x_B + RT\,(x_A \ln x_A + x_B \ln x_B) \tag{2.16}$$

or

$$G_m = x_A G_A^0 + x_B G_B^0 + G_m^{mix} \tag{2.17}$$

where

$$G_m^{mix} = Lx_A x_B + RT(x_A \ln x_A + x_B \ln x_B) \tag{2.18}$$

L is called the regular solution parameter.

The relative partial molar Gibbs' free energy of mixing can be found by introducing the expression in Equation (2.18) of the molar Gibbs' free energy of mixing into Equations (1.13') and (1.14') on page 24 in Chapter 1. The result is two simple equations

$$\overline{G}_{\mathrm{A}}^{\mathrm{mix}} = L\,(1 - x_{\mathrm{A}})^2 + RT \ln x_{\mathrm{A}} \tag{2.19}$$

$$\overline{G}_{\mathrm{B}}^{\mathrm{mix}} = L(1 - x_{\mathrm{B}})^2 + RT \ln x_{\mathrm{B}} \tag{2.20}$$

Miscibility Gap in a Regular Solution

Figures 2.4a–f show the molar enthalpy $H_{\mathrm{m}}^{\mathrm{mix}} = L\,x_{\mathrm{A}}x_{\mathrm{B}}$, the term $-TS_{\mathrm{m}}^{\mathrm{mix}} = RT\,(x_{\mathrm{A}} \ln x_{\mathrm{A}} + x_{\mathrm{B}} \ln x_{\mathrm{B}})$ and the molar Gibbs' free energy $G_{\mathrm{m}}^{\mathrm{mix}} = H_{\mathrm{m}}^{\mathrm{mix}} - TS_{\mathrm{m}}^{\mathrm{mix}}$ for a regular solution with $L = 1.2 \times 10^5$ J/kmol as functions of composition at six different temperatures.

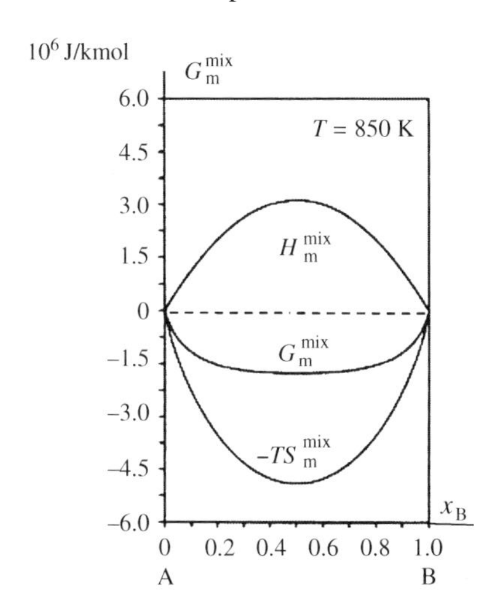

Figure 2.4a $G_{\mathrm{m}}^{\mathrm{mix}} = H_{\mathrm{m}}^{\mathrm{mix}} - TS_{\mathrm{m}}^{\mathrm{mix}}$ as a function of x_{B}. $T = 850$ K.

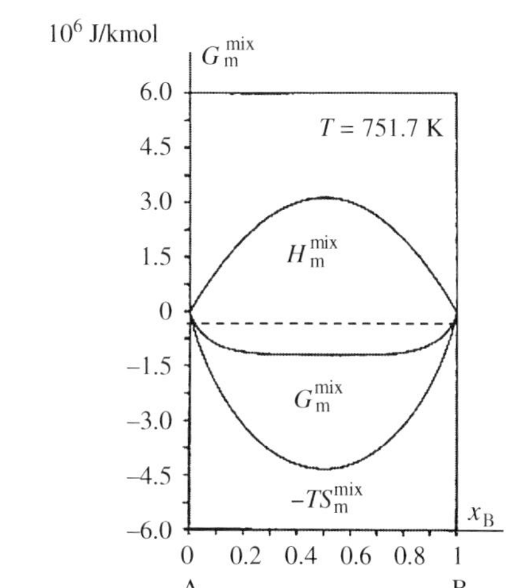

Figure 2.4b $G_{\mathrm{m}}^{\mathrm{mix}} = H_{\mathrm{m}}^{\mathrm{mix}} - TS_{\mathrm{m}}^{\mathrm{mix}}$ as a function of x_{B}. $T = 751.7$ K.

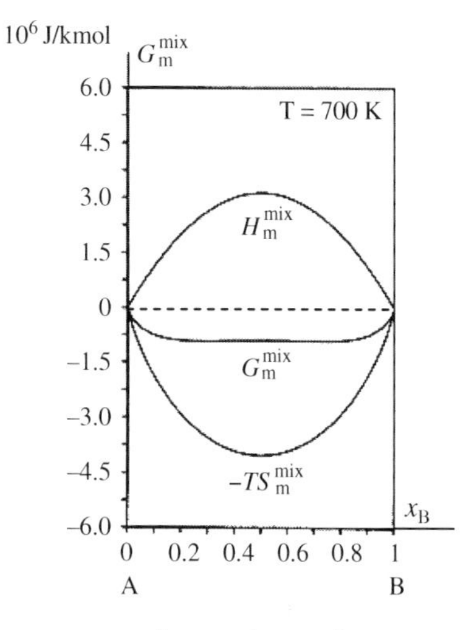

Figure 2.4c $G_{\mathrm{m}}^{\mathrm{mix}} = H_{\mathrm{m}}^{\mathrm{mix}} - TS_{\mathrm{m}}^{\mathrm{mix}}$ as a function of x_{B}. $T = 700$ K.

At high temperatures, $T > 751.7$ K, Figures 2.4a and b, the $x_{\mathrm{B}}G_{\mathrm{m}}^{\mathrm{mix}}$ curve is of the same type as the G_{m} curve in Figures 2.2 and 2.3. The curve is concave (seen from above) over the whole interval $x_{\mathrm{B}} = 0$ to $x_{\mathrm{B}} = 1$. The regular solution appears as a single phase, which is stable at any composition.

The following $G_{\mathrm{m}}^{\mathrm{mix}}$ curves at lower temperatures, Figures 2.4d–f show an increasing tendency to form a central maximum surrounded by two minima. At the same time the $G_{\mathrm{m}}^{\mathrm{mix}}$ curves are

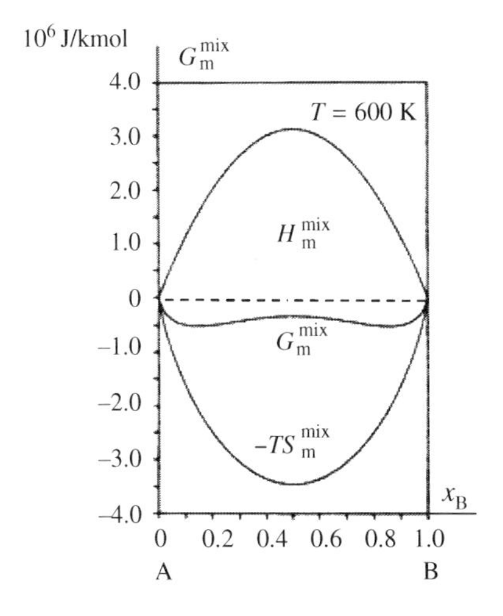

Figure 2.4d $G_{\mathrm{m}}^{\mathrm{mix}} = H_{\mathrm{m}}^{\mathrm{mix}} - TS_{\mathrm{m}}^{\mathrm{mix}}$ as a function of $\mathrm{x_B}$. $T = 600$ K.

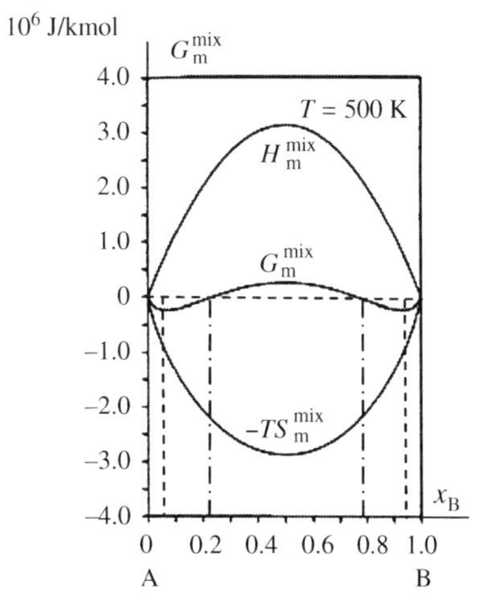

Figure 2.4e $G_{\mathrm{m}}^{\mathrm{mix}} = H_{\mathrm{m}}^{\mathrm{mix}} - TS_{\mathrm{m}}^{\mathrm{mix}}$ as a function of $\mathrm{x_B}$. $T = 500$ K.

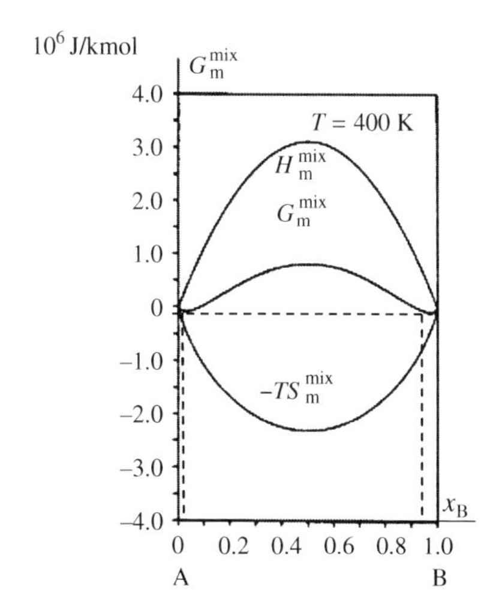

Figure 2.4f $G_{\mathrm{m}}^{\mathrm{mix}} = H_{\mathrm{m}}^{\mathrm{mix}} - TS_{\mathrm{m}}^{\mathrm{mix}}$ as a function of $\mathrm{x_B}$. $T = 400$ K.

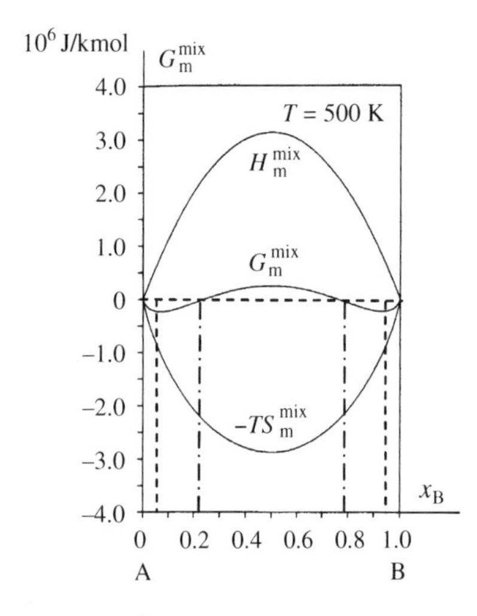

Figure 2.4e G_m^{mix} as a function of mole fraction x_B at temperature 500 K. The mole fraction values, which correspond to the minimum values and the inflection point values of the G_m^{mix} curve in Figure 2.4e, and the temperature 500 K are plotted in the x_B–T diagram in Figure 2.5a.

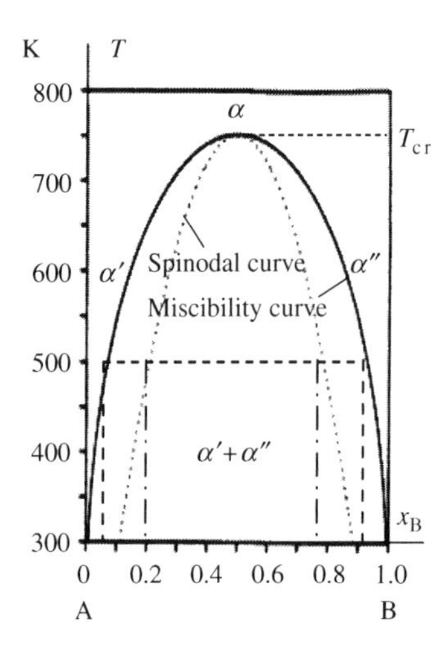

Figure 2.5a The construction, described in the text below Figure 2.4 e, gives two points on the miscibility curve and two points on the spinodal curve. By the same construction for other values of the temperature T the whole curves can be derived.

moved upwards. The latter effect is easy to explain. The figures show that the H_m^{mix} curves are independent of temperature. The reason is that the forces between the atoms do not depend on temperature. When H_m^{mix} is a constant and the temperature is lowered, $|TS_m^{mix}|$ is reduced and G_m^{mix} increases.

The tendency to form two minima and a maximum appears at a specific critical temperature T_{cr}. By analyzing the composition dependence of G, dG/dx, d^2G/dx^2 and d^3G/dx^3, it can be shown that the critical temperature occurs at $x = 0.5$ and has the value

$$T_{cr} = \frac{L}{2R} \tag{2.21}$$

Below the critical temperature the single phase is no longer stable at all compositions. At a central interval around the maximum composition the single phase is unstable, decomposes and forms two stable phases with different compositions.

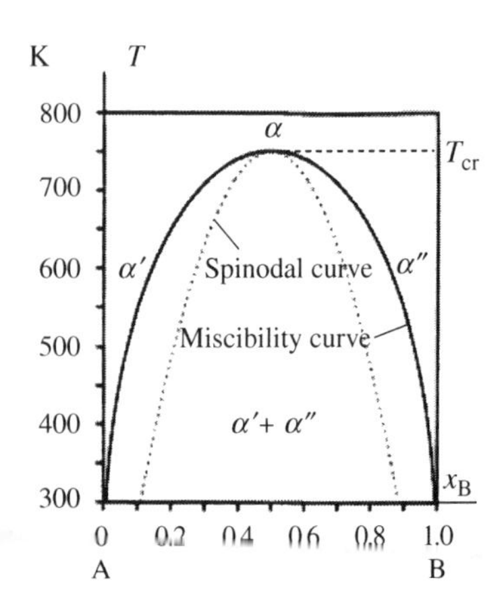

Figure 2.5b Miscibility and spinodal curves of a binary alloy.

It should be emphasized that this occurs only for positive values of L. If L is negative G_m decreases monotonously with T and no minima are ever formed.

The compositions of the stable phases correspond to the lowest possible total Gibbs' free energy of the system. The compositions of the stable phases at a given temperature are the x_B values that correspond to the two minimum values of the G_m^{mix} curve. They are indicated in Figures 2.4d–f.

The composition of the two phases varies with temperature. A comparison between the Figures 2.4d–f shows that the higher the temperature is, the closer will be the composition of the two phases and vice versa.

If we plot the compositions of the phases as a function of the temperature T we obtain the continuous curve in Figure 2.5a, which is a simple phase diagram. The dotted curves are the so-called *spinodal curves*, which correspond to the inflection points between the maximum and the minimum (Figure 2.4e is chosen as an example) The construction of the curves is described in the text below the Figures 2.4e and 2.5a.

Figure 2.5b, together with Figures 2.6 and 2.7 below, describe the formation and presence of stable phases of a binary alloy at different temperatures and compositions.

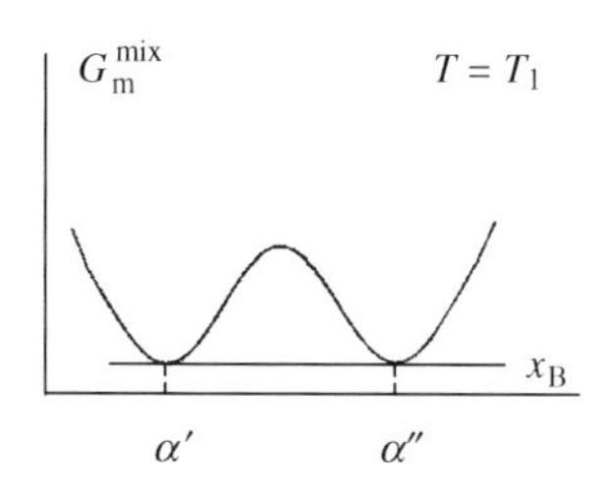

Figure 2.6 The compositions of the stable phases can be found by drawing the common tangent to the two parts of the G_m^{mix} curve.

Identification of Stable Phases with No Attention to the Spinodal Curve

At a high temperature, for example 800 K (Figure 2.5b), the alloy consists of a single phase independent of composition. Below the critical temperature two phases appear in the composition region $\alpha' < x_B < \alpha''$ where the compositions of the two phases correspond to the intersection between the line T = constant and the miscibility curve. At compositions $x_B > \alpha''$ and $x_B < \alpha'$ the alloy consists of a single phase.

The compositions of the two phases in the two-phase region are plotted in Figure 2.6 as the coordinates of the two minima, which correspond to the lowest possible total Gibbs' free energy of the system. Hence, the miscibility curve represents the borderline between the single-phase and the two-phase region.

Identification of Stable Phases with Attention to the Spinodal Curve

Further study of the G_m curves is required to show the sense and influence of the spinodal curve on the stable phases at different temperatures and compositions of a binary alloy.

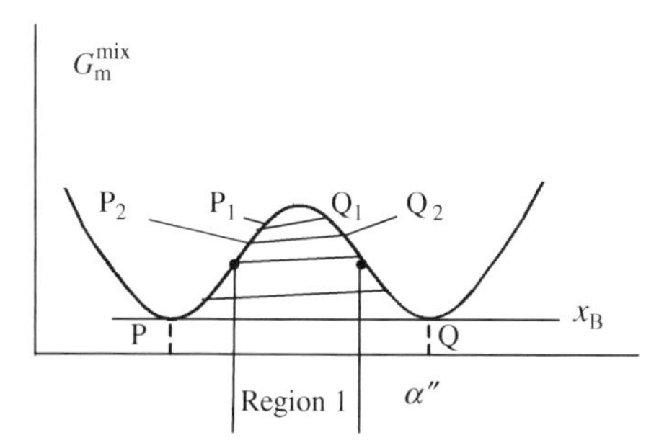

Figure 2.7a G_m^{mix} curve of a binary alloy at temperatures and compositions, which correspond to the region inside the spinodal curve.

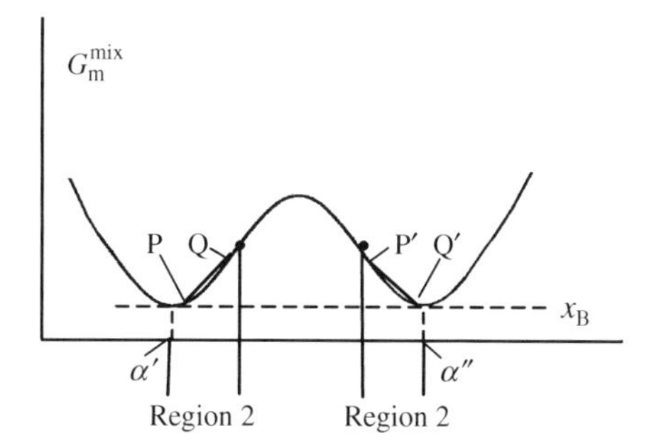

Figure 2.7b G_m^{mix} curve of a binary alloy at temperatures and compositions, which correspond to the region between the miscibility and spinodal curves.

In the areas outside the miscibility curve (Figure 2.5b) the alloy exists as a single phase. The region inside the miscibility curve has to be regarded as two separate regions 1 and 2. The G_m^{mix} curves for each of them at a given temperature below the critical temperature are seen in Figures 2.7a and b, respectively.

The points P_1 and Q_1 within region 1 in Figure 2.7a describe a binary alloy, which consists of two phases. Their compositions correspond to the region inside the spinodal curve. The phases are *not* in equilibrium with each other and a successive transformation of the phases occurs. The line $P_2 Q_2$ and others, seen in the figure, describe this process. The points P_i and Q_i slide downwards all the time, which agrees with the condition for a spontaneous process $dG_m^{mix} < 0$. The final stable compositions of the two phases correspond to the double tangent PQ. Hence, the stable state of the alloy is the two phases with compositions α' and α'', which correspond to the intersection between the miscibility curve and the common tangent.

The phases of the alloy, which correspond to the points P and Q in Figure 2.7b, do *not* correspond to energetically stable phases. P and Q are situated between the minimum point and the inflection point, i.e. in region 2 between the miscibility and the spinodal curves. Owing to the lack of enough energy the point Q cannot move uphill and across the maximum and therefore a metastable single homogeneous phase is formed. It corresponds to the tangent in the region, which leads to the lowest possible total Gibbs' free energy. If energy is supplied in one way or other, the point Q slips over the obstruction and the single phase splits up into two stable phases with compositions α' and α'' (dotted line in Figure 2.7b). The same arguments are valid for the point P in the region on the other side of the maximum.

Hence, the sense of the spinodal curve is that

- Alloys with compositions that correspond to regions *outside* of the miscibility curve (both sides), consist of a *stable single* phase.
- Alloys with compositions within each of the regions between the miscibility and spinodal curves, consist of a *metastable single* phase.
- Alloys with compositions that correspond to regions *inside* the spinodal curve, consist of *two stable* phases.

2.4 Equilibrium between Phases in Binary Solutions. Phase Diagrams of Binary Alloys

2.4.1 Gibbs' Phase Rule

Consider a multiple-component system with N_c components and N_{ph} phases in equilibrium with each other.

Each component has its own concentration in each phase. Hence, we have $(N_c N_{ph})$ variables. Common variables are the pressure p and the temperature T. Hence, the total number of variables is $(N_c N_{ph} + 2)$.

The equilibrium conditions give a number of relationships between the variables. If we can estimate the number of equilibrium conditions we can find the number of independent variables.

We know that

1. The general expression for equilibrium between two arbitrary phases α and β in contact with each other is

$$\overline{G_A^\alpha} = \overline{G_A^\beta}$$

2. Gibbs' free energy G is a function of concentration.

For each of the N_c components A, B, ... we have an equilibrium condition of the type $\overline{G_A^I} = \overline{G_A^{II}} = \ldots = \overline{G_A^{N_{ph}}}$ e.g. $(N_{ph} - 1)$ equations.

We also have one relationship of the type $x_A + x_B + \ldots + x_{N_c} = 1$ for each phase. That makes another N_{ph} equations.

Hence, there are $[N_c (N_{ph} - 1) + N_{ph}]$ relationships between the $(N_c N_{ph} + 2)$ variables. The number of independent variables, which we call degrees of freedom, will then be

$$N_{free} = (N_c N_{ph} + 2) - [N_c (N_{ph} - 1) + N_{ph}] = N_c - N_{ph} + 2$$

where

N_{free} = number of degrees of freedom
N_c = number of components of the system
N_{ph} = number of phases.

After reduction we obtain

$$N_{ph} + N_{free} = N_c + 2 \tag{2.22}$$

- In a system at equilibrium the number of phases and degrees of freedom equals the number of components added by 2.

This is *Gibbs' phase rule*, which is a very useful relationship when dealing with phase diagrams of systems in equilibrium.

In a binary system at equilibrium at constant pressure *three* phases can exist simultaneously within a temperature interval and *two* phases at constant pressure and temperature according to Gibbs' phase rule. This will be shown below.

The significance of Gibbs' phase rule will be illustrated by the system ice + water + steam Figure 2.8

At point Q1 there is only one phase, water. Hence, the number of degrees of freedom is 2. Both pressure and temperature can be varied independently of each other.

At point Q2 there are two phases, water and steam in equilibrium with each other. The number of degrees of freedom is 1. Pressure *or* temperature can be varied. The remaining variable, temperature or pressure, is then fully determined and cannot be varied independently.

At the triple point Q3 there are three phases, ice, water and steam in equilibrium with each other. The number of degrees of freedom is zero. Neither temperature nor pressure can be varied. Table 2.1 gives a review of the discussion given above.

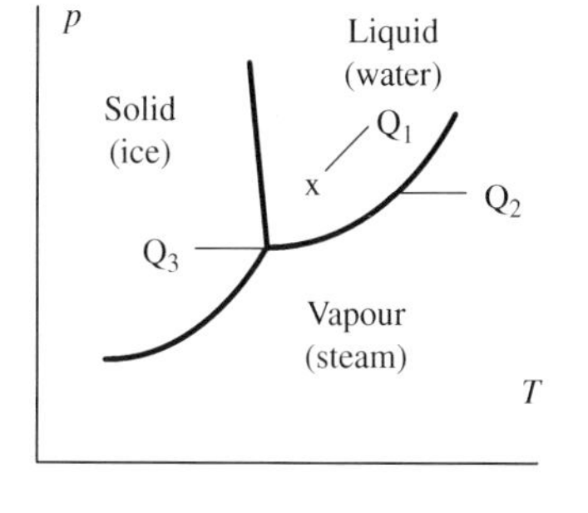

Figure 2.8. Schematic phase diagram of the H_2O system.

Table 2.1 Application of Gibbś phase rule on the system ice-water-steam

Point in the phase diagram	Number of phases	Number of components	Number of degrees of freedom according to Gibbś phase rule
Q_1	1	1	2
Q_2	2	1	1
Q_3	3	1	0

2.4.2 Gibbs' Free Energy Curves for Solid and Liquid Binary Solutions

With Chapter 1 and Sections 2.2 and 2.3 we have at our disposal the tools required to calculate and show graphically the Gibbs' free energy of solid and liquid phases of ideal and non-ideal solutions of binary alloys. A complete survey is given below.

The shapes of the Gibbs' free-energy curves depend on H_m^{mix}, which is of special importance. This quantity has been discussed extensively in Section 1.5.1 on pages 22–23 in Chapter 1. It can be either positive, zero or negative, depending on the interatomic forces.

The thermodynamic quantities of mixing and the Gibbs' free energy of ideal solutions have been treated on page 25 in Chapter 1. The Gibbs' free energy of non-ideal solutions is found in Section 1.7.3 on page 32 in Chapter 1.

All Gibbs' free-energy curves depend strongly on temperature.

Molar Gibbs' Free Energy Curves of Solid and Liquid Ideal Binary Solutions

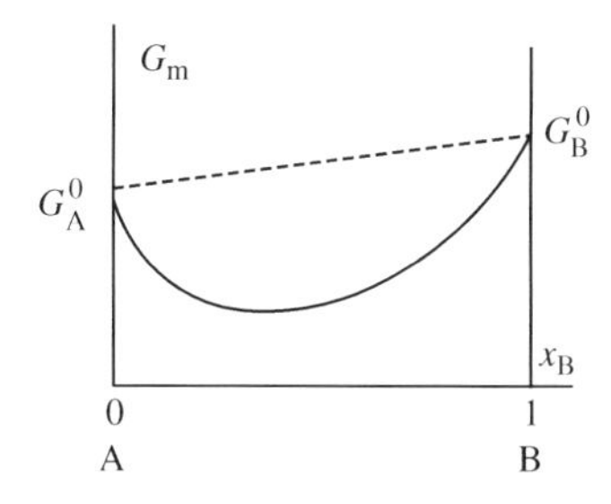

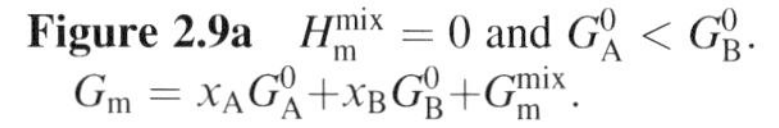

Figure 2.9a $H_m^{mix} = 0$ and $G_A^0 < G_B^0$.
$G_m = x_A G_A^0 + x_B G_B^0 + G_m^{mix}$.

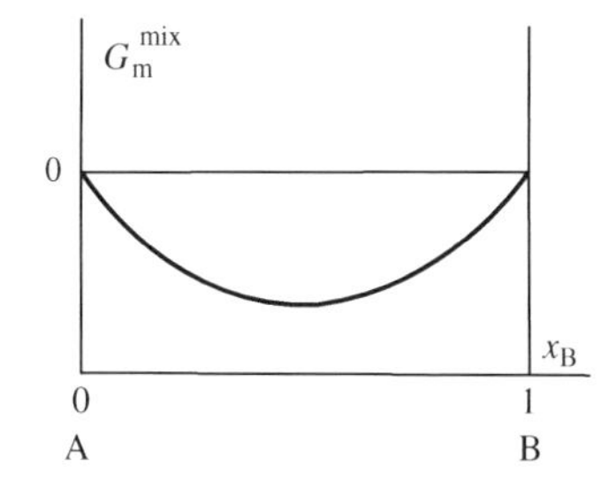

Figure 2.9b $G_m^{mix} = RT(x_A \ln x_A + x_B \ln x_B)$.
G_m^{mix} is < 0 as both x_A and x_B < 1.

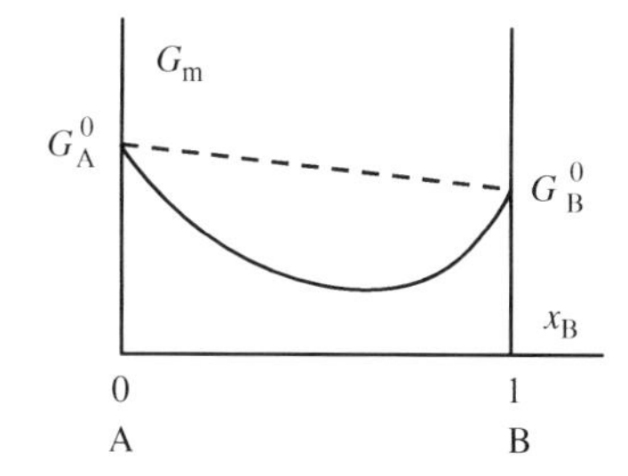

Figure 2.9c $H_m^{mix} = 0$ and $G_A^0 > G_B^0$.
$G_m = x_A G_A^0 + x_B G_B^0 + G_m^{mix}$.

The heat of mixing $H_m^{mix} = 0$ for all ideal solutions.

$$G_m^{ideal} = x_A G_A^0 + x_B G_B^0 + G_m^{mix} \tag{2.23}$$

where

$$G_m^{mix} = RT\,(x_A \ln x_A + x_B \ln x_B) \qquad (2.24)$$

Gibbs' free-energy curves of ideal binary solutions depend on temperature and material constants. The appearance of three such curves are shown above.

The curves of the corresponding solid and the liquid solutions are of the same type but have not identical shapes.

Molar Gibbs' Free Energy Curves of Solid and Liquid Non-Ideal Binary Solutions

The Gibbs' free energy curves of non-ideal binary solutions are more complicated than those of ideal solutions. The shapes of the non-ideal G_m curves depend on temperature and material constants and on the heat of mixing. In the case of solid solutions the shape of the curves also depend on the crystal structures of the two components A and B. The heat of mixing can either be positive or negative.

Theoretical basis:

$$G_m^{non\text{-}ideal} = G_m^{ideal} + G_m^{Ex} \qquad (2.25)$$

$$G_m^{Ex} = H_m^{Ex} - TS_m^{Ex} = H_m^{mix} - TS_m^{Ex} \approx H_m^{mix}(1 - AT) \qquad (2.26)$$

$$G_m = x_A G_A^0 + x_B G_B^0 + RT\,(x_A \ln x_A + x_B \ln x_B) + H_m^{mix} - TS_m^{Ex} \qquad (2.7)$$

Gibbs' Free Energy Curves of Solid Non-Ideal Solutions when the two Components have Equal Crystal Structures

In a mechanical mixture of elements A and B each component has its own G_m curve. If A and B have equal structures the two curves join into one when a solution is formed. The shape of the G_m curve of the solution depends on the heat of mixing and the temperature. There are three different cases.

If H_m^{mix} is *positive* and the temperature $T < T_{cr}$ the G_m curve has two minimum (compare the regular-solution model on page 46). The contributions of the last two terms in Equation (2.7) have opposite signs. This case is shown in Figure 2.10a.

If H_m^{mix} is *positive* and the temperature $T > T_{cr}$ the G_m curve has only one minimum and is similar to that of an ideal solution (Figure 2.9a). The minimum in Figure 2.10b is deeper than those in Figure 2.10a because the last term in Equation (2.7) becomes more negative with increasing temperature and lowers the G_m curve in the middle figure.

If H_m^{mix} is *negative*, the G_m curve of the binary solution is a single curve (Figure 2.10c) with a shape such as that in Figure 2.10b but with a deeper minimum than the corresponding curve for positive H_m^{mix} values at the same temperature. The reason is that the contributions to the G_m curve of the last two terms in Equation (2.7) are both negative for negative values of H_m^{mix}.

Gibbs' Free-Energy Curves of Liquid Non-Ideal Binary Solutions

Liquid metals and alloys have no stationary structure. Non-ideal liquid solutions can be said to have equal 'crystal structures' of the two components.

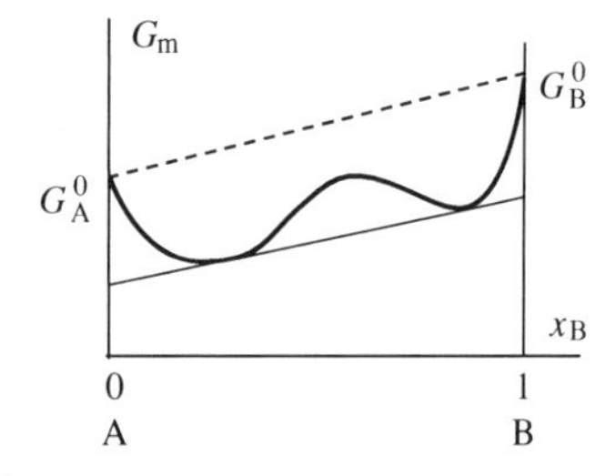

Figure 2.10a $H_m^{mix} > 0$ and $T < T_{cr}$. $G_A^0 < G_B^0$. $G_m = (G_m)_{ideal} + H_m^{mix} - TS_m^{Ex}$.

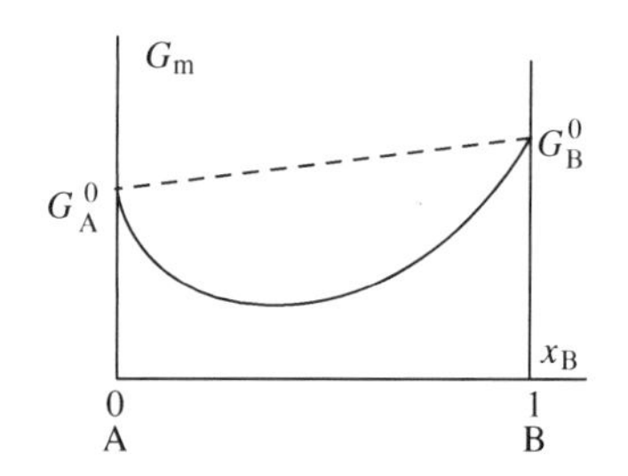

Figure 2.10b $H_m^{mix} > 0$ and $T > T_{cr}$. $G_A^0 < G_B^0$. $G_m^{mix} = (G_m^{mix})_{ideal} + H_m^{mix} - TS_m^{Ex}$.

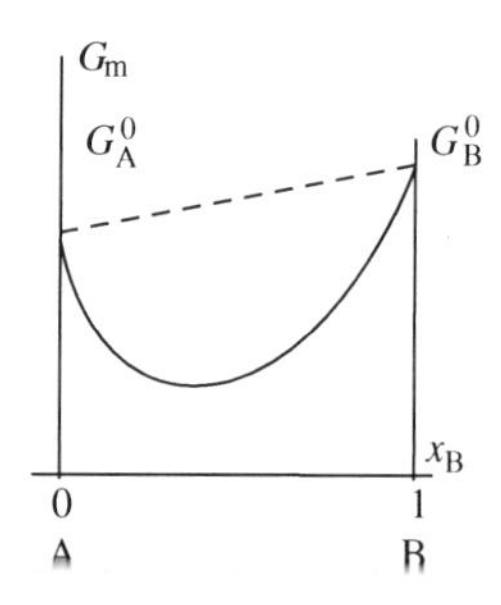

Figure 2.10c $H_m^{mix} < 0$ and $T > T_{cr}$. $G_A^0 < G_B^0$. $G_m = (G_m)_{ideal} + H_m^{mix} - TS_m^{Ex}$.

The G_m curves of non-ideal *liquid* binary solutions have the same principal shapes as those of non-ideal *solid* solutions when the pure components have *equal* crystal structures. Such curves for solid solutions are shown in Figures 2.10a–c.

Gibbs' Free Energy Curves of Solid Non-Ideal Solutions when the two Components have Unequal Crystal Structures

If the components A and B have unequal crystal structures, each component maintains a separate G_m curve even when a solution is formed. In these cases, the G_m curve of the solution consists of *two* curves such as those in Figure 2.11, independent of the sign of H_m^{mix}. Equation (2.7) is valid for each curve.

The G_m curves with negative H_m^{mix} values have deeper minima than those with positive values of heat of mixing.

It should be emphasized that liquid solutions of components with unequal crystal structures in the solid state have G_m curves of the types shown above in Figures 2.10a or 2.10c.

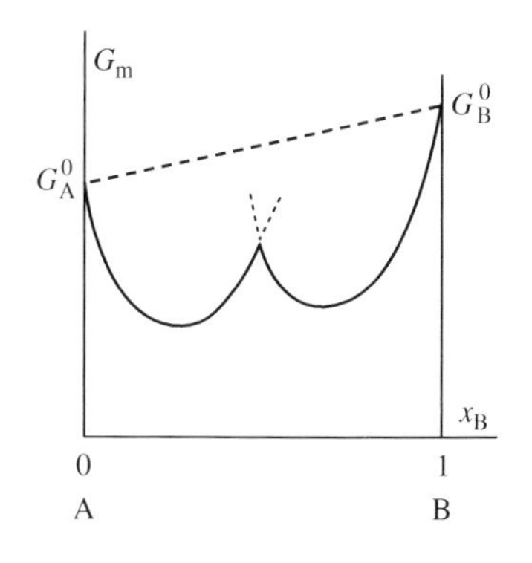

Figure 2.11 G_m curve of a solid solution of two components with unequal crystal structures.

Temperature Dependence of Molar Gibbs' Free Energy Curves

On page 52 it was pointed out that the molar Gibbs' free energy curves depend strongly on temperature. The temperature dependence of the terms in Equations (2.7) on page 44 and (2.9) on page 45 will be briefly discussed below.

$$G_m = x_A G_A^0 + x_B G_B^0 + RT\,(x_A \ln x_A + x_B \ln x_B) + G_m^{Ex} \tag{2.9}$$

or

$$G_m = x_A G_A^0 + x_B G_B^0 + RT\,(x_A \ln x_A + x_B \ln x_B) + H_m^{mix} - TS_m^{Ex} \tag{2.7}$$

G_A^0 and G_B^0

G_A^0 and G_B^0 represent molar Gibbs' free energy of the pure metals A and B. The superscript means "pure" (metal) and not "constant" in this case. The terms depend on temperature, which can be seen from Figure 2.12, which is analogous to Figure 2.1 on page 43. It shows the molar Gibbs' free energy values of the liquid and solid states of a pure element as functions of temperature. The figure also shows that the slope of the liquid curve is steeper than that of the solid curve.

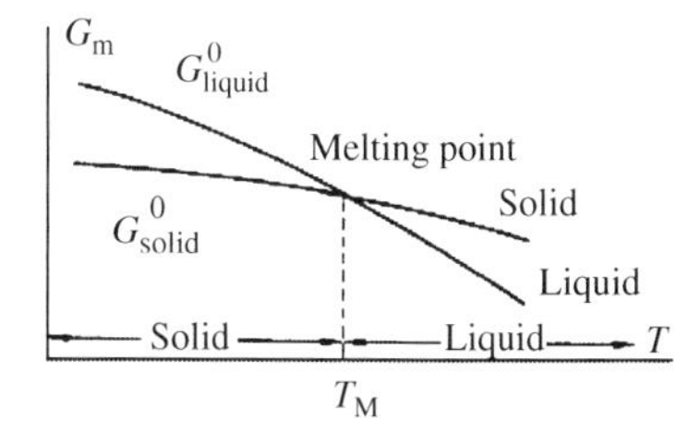

Figure 2.12 Gibbs' free energies of the liquid and solid states of a pure metal as functions of temperature.

Each metal has two free energy values, one for the liquid state and one for the solid state. When the temperature changes, the vertical positions of both the liquid and solid values change. Depending on the material constants of the liquid and solid states, the *liquid* value G_A^0 of a metal A rises more than its solid value when the temperature *decreases*. When the temperature increases the liquid value sinks more than the solid value. This is true of most metals.

The temperature dependence of G_A^0 and G_B^0 means that the *liquid* molar Gibbs' free energy curve G_m^L of a metal normally moves *upwards*, relative to its corresponding solid curve G_m^s, when the temperature *decreases* and vice versa.

Third Term in Equation (2.7)

The temperature dependence of the third term is obvious. The factor $R(x_A \ln x_A + x_B \ln x_B)$ is independent of temperature. Therefore, this term varies linearly with T.

Excess Terms

The temperature dependence of the fourth term in Equation (2.7) is weak as the forces between the atoms are little influenced by the temperature.

The temperature dependence of the fifth term in Equation (2.7) is hard to find as there is no good model of the excess entropy so far.

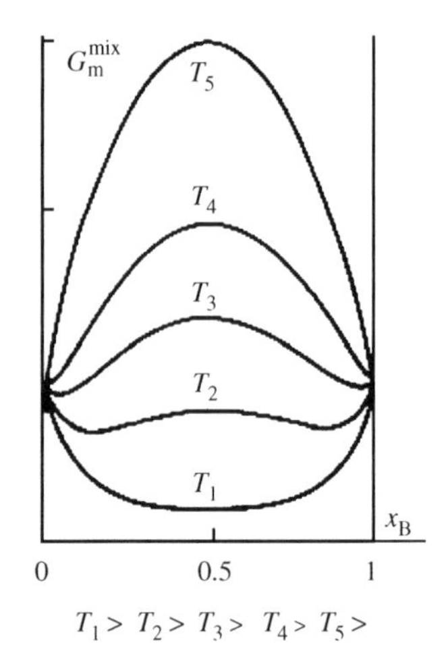

Figure 2.13 G_m^{mix} of a binary solution as a function of concentration with temperature as a parameter. $H_m^{mix} > 0$.

A rough approximation is to neglect the excess entropy. Then its temperature dependence is zero and the temperature dependence of G_m^{mix} can be studied in terms of the regular-solution model [Equation (2.17) on page 46 and Equation (2.18) on page 46]

$$G_m = x_A G_A^0 + x_B G_B^0 + G_m^{mix} \tag{2.17}$$

$$G_m^{mix} = L x_A x_B + RT\,(x_A \ln x_A + x_B \ln x_B) \tag{2.18}$$

where Lx_Ax_B is the heat of mixing H_m^{mix}. Equation (2.18) shows that the Gibbs' free energy of mixing is a symmetric function of composition. It is shown for a number of different temperatures in Figure 2.13, provided that the heat of mixing is positive. If H_m^{mix} is negative ($L < 0$) G_m^{mix} decreases monotonously with temperature. Compare also Figures 2.4a–f (pages 47–48).

Alternatively, the experimental relationship between H_m^{mix} and S_m^{Ex} on [Equation (1.131) on page 33 in Chapter 1] can be used to estimate the temperature dependence of the excess entropy

$$G_m^{Ex} = H_m^{mix}(1 - AT) \tag{2.27}$$

to analyze the temperature dependence of the fifth term in Equation (2.9) on page 45.

Temperature Dependence of Molar Gibbs' Free Energy Curves

The expressions (2.25) and (2.27) are introduced into Equation (2.9):

$$G_m = x_A G_A^0 + x_B G_B^0 + RT\,(x_A \ln x_A + x_B \ln x_B) + H_m^{mix}(1 - AT) \tag{2.28}$$

The first two terms determine the *position* of the G_m curve and the last two terms determine its *shape*. The same is valid for temperature variations in position and shape.

2.4.3 The Tangent to Tangent Method to Predict Phases in Binary Solutions

Below, we introduce a general and very useful method to analyze and interpret Gibbs' free energy curves and find the compositions and number of phases in binary solutions.

Phase Formation in Binary Solutions

The two types of binary systems, formed by components with the same or different structures, form the same types of phases. As an example we will discuss two cases of phase formation more in detail.

The molar Gibbs' free energy curve of a solid non-ideal binary solution with components of the same crystal structure and a positive heat of mixing is given in Figure 2.14 on page 55. It has one maximum and two minima, just like the Gibbs' free energy curves of the regular solution model (page 45). We will study the solution process in the same way as we did for ideal solutions (Figure 2.2 on page 45).

The lateral parts of the free-energy curve in Figure 2.14 offer no problems. They have the same shape as the Gibbs' free energy curve of an ideal solution. All solutions with a composition $\leq x^\alpha$ form stable homogeneous solutions. A acts as a solvent for B. Similarly, all solutions with compositions $\geq x^\beta$ form homogeneous solutions of A solved in B.

The middle part of the Gibbs' free energy curve shows a maximum, which indicates an unstable state. An alloy with composition x has the Gibbs' free energy G_m when homogeneous. Owing to instability it splits up into two phases corresponding to points P_1 and Q_1 and the Gibbs' free energy decreases to G_1. The free energy continues to fall and the points slide on the curve until G_m via G_2 has reached its lowest possible value G_3. This corresponds to the stable state, which is a mixture of two separate phases with the compositions x^α and x^β (compare Figure 2.7a on page 49).

The line P_3Q_3 is the common tangent to the curve in the double points P_3 and Q_3. In the composition interval x^α to x^β the alloy consists of a mixture of an A-rich phase α with composition x^α and a B-rich phase β with composition x^β.

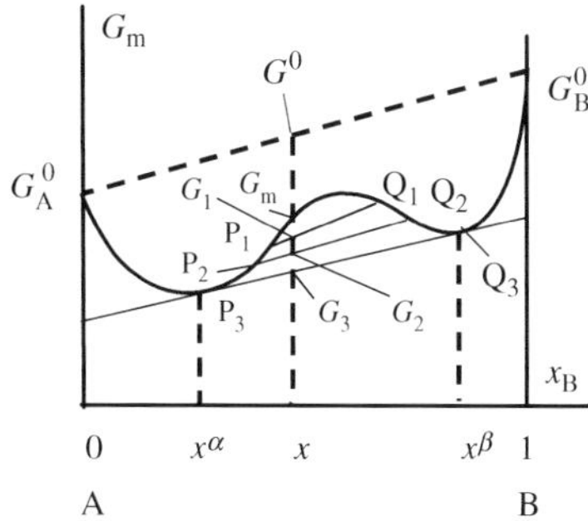

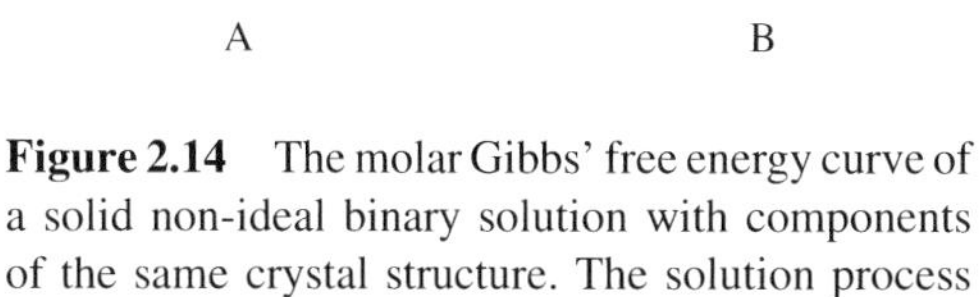

Figure 2.14 The molar Gibbs' free energy curve of a solid non-ideal binary solution with components of the same crystal structure. The solution process results in two different phases.

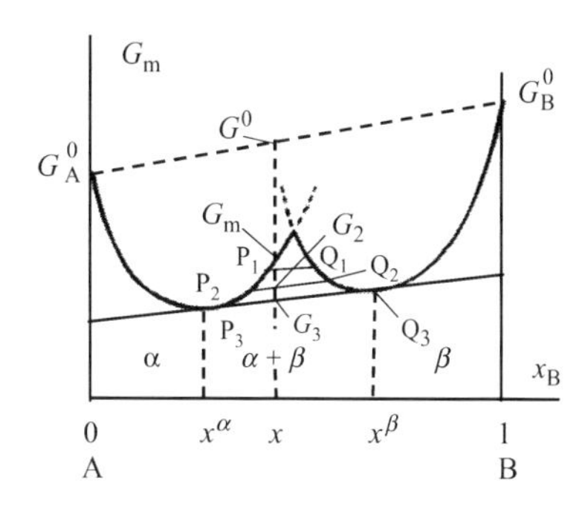

Figure 2.15 The molar Gibbs' free energy curve of a solid non-ideal binary solution with components of different crystal structures. The solution process results in two different phases.

Figure 2.15 shows an analogous process for a non-ideal binary solution with components of different crystal structures. The mechanical mixture of components A and B has initially the Gibbs' free energy G^0. When an unstable solution has been formed the Gibbs' free energy is G_m. The systems changes to more likely states and lowers its total Gibbs' free energy gradually. In the end, a stable state has been reached with the lowest possible energy G_3. The tangent points define the compositions of the two phases x^α and x^β (compare Figure 2.7b on page 49.)

The proportions of the two phases are given by the lever rule

$$\text{fraction of } \alpha = \frac{x^\beta - x}{x^\beta - x^\alpha} \qquad \text{fraction of } \beta = \frac{x - x^\alpha}{x^\beta - x^\alpha} \tag{2.29}$$

The Tangent to Tangent Method

The common tangent construction is fundamental in all interpretation of free energy curves. It is used above to find the stable state of a non-ideal solution, i.e. the number of phases and their free energies, compositions and fractions. The method and the main properties of the phases are summarized below.

1. A common tangent is drawn between two free energy curves.
2. The stable state is a mechanical mixture of two phases.
3. The points where the common tangent is in contact with the free energy curves, give the compositions of the two phases.
4. The point on the tangent, which corresponds to the given overall composition, gives the free energy of the system at equilibrium.
5. The fractions of the two phases vary with the overall composition in accordance with the lever rule.

These rules will be frequently used in the following sections. They will be verified theoretically in Section 2.4.4 in connection with chemical potentials.

If the described tangent to tangent method is used at all relevant temperatures in the AB system, a complete phase diagram can be derived. Phase diagrams will be discussed in Section 2.4.5.

2.4.4 Calculation of Chemical Potentials from Gibbs' Free Energy Diagrams

Definitions

The *chemical potential of a pure element* is defined as the Gibb's free energy of the pure element. For an element A the definition can be written as

$$\mu_A^0 = G_A^0 \tag{2.30}$$

If the element is a component in a solution a more general definition is required.

The *chemical potential* μ_A of an element A is by definition equal to the partial molar Gibbs' free energy of A

$$\mu_A = \overline{G}_A \tag{2.31}$$

The chemical potential is an important quantity as

- Equilibrium between phases means that their chemical potentials are equal.

The tangent to tangent method is based on Gibbs–Duhem's law, which relates the molar Gibbs' free energy G_m and the chemical potential μ of each component. This law has appeared several times. It is valid for $\overline{G}_A$ [Equation (1.4') on page 24 in Chapter 1] as well as for $\overline{G}_A^{mix}$ (Equations (1.13') on page 24 in Chapter 1) and $\overline{G}_A^{Ex}$ [Equation (1.123) on page 32 in Chapter 1). By combining the expression for $\overline{G}_A$, analogous to Equations (1.69) and (1.70) on page 19 in Chapter 1, with Equation (2.31) above we obtain

$$\mu_A = \overline{G}_A = G_m + (1 - x_A)\left(\frac{\partial G_m}{\partial x_A}\right)_T \tag{2.32}$$

$$\mu_B = \overline{G}_B = G_m + (1 - x_B)\left(\frac{\partial G_m}{\partial x_B}\right)_T \tag{2.33}$$

Graphical Construction of Chemical Potentials from the Molar Gibbs' Free-Energy Curve of an Ideal Binary Liquid Solution

The free energy G_m of an ideal solution of two components is a function of its composition, i.e. of the mole fraction x_B. Each component has a chemical potential in the solution.

Equations (2.32) and (2.33) are linear relationships between G_m and x_B. They can both be interpreted as the *equation of the tangent* to the molar Gibbs' free-energy curve. We apply the equations for a liquid solution and add a superscript L.

If $x_B^L = 0$ is inserted into Equation (2.32) we obtain $G_m^L = \mu_A^L < \overline{G}_A^{0L}$

If $x_B^L = 1$ is inserted into Equation (2.33) we obtain $G_m^L = \mu_B^L < \overline{G}_B^{0L}$

At a given composition the chemical potentials of the components A and B can be found in the free energy diagram as the intercepts of the tangent and the axes given by $x_B^L = 0$ and $x_B^L = 1$.

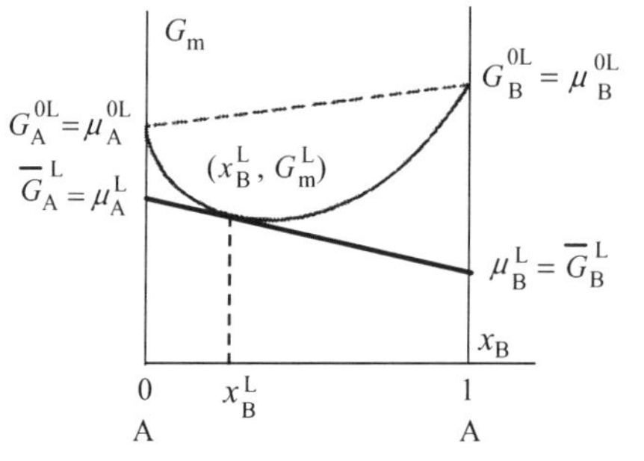

Figure 2.16 Free energy curve of an ideal liquid solution at a given temperature.

The tangent intersects the line $x_B^L = 0$ at $\mu_A^L = \overline{G}_A^L$ and the line $x_B^L = 1$ at $\mu_B^L = \overline{G}_B^L$.

By use of the relationship $x_A^L + x_B^L = 1$ and its derivative, Equation (2.32) with the superscript L can be transformed into

$$\mu_A^L = \overline{G}_A^L = G_m^L - x_B^L\left(\frac{\partial G_m^L}{\partial x_B^L}\right)_T \tag{2.34}$$

The partial derivative is eliminated between Equation (2.33) with the superscript L and Equation (2.34) and the new equation is solved for G_m^L, which gives

$$G_m^L = \left(1 - x_B^L\right)\mu_A^L + x_B^L\,\mu_B^L \tag{2.35}$$

Equation (2.35) is another way to express the equation of the tangent. This equation can be used to find the chemical potentials of the two components graphically if the G_m^L curve is known.

The procedure is illustrated graphically for a liquid solution in Figure 2.16. The tangent to the G_m^L curve for the composition x_B^L is drawn and the chemical potentials of A and B are found as the intersections with two vertical axes.

The construction is the same for solid solutions. In this case superscript s replaces superscript L.

Graphical Construction of Chemical Potentials from the Molar Gibbs' Free-Energy Curves of a Non-Ideal Solution

Consider a non-ideal solution of components A and B with a liquid and a solid phase. The chemical potentials μ_A and μ_B are functions of the compositions of the liquid phase (superscript L) and the solid phase (superscript s). As usual, the compositions are defined by the mole fractions x_B^L and x_B^s. When the compositions of the two phases are known the four chemical potentials can be calculated from the relationships

$$\mu_A^L = \mu_A^{0L} + RT \ln\left(1 - x_B^L\right) + \overline{G}_A^{Ex\,L} \tag{2.36}$$

$$\mu_A^s = \mu_A^{0s} + RT \ln\left(1 - x_B^s\right) + \overline{G}_A^{Ex\,s} \tag{2.37}$$

$$\mu_B^L = \mu_B^{0L} + RT \ln x_B^L + \overline{G}_B^{Ex\,L} \tag{2.38}$$

$$\mu_B^s = \mu_B^{0s} + RT \ln x_B^s + \overline{G}_B^{Ex\,s} \tag{2.39}$$

where

T = absolute temperature

μ_j^i = chemical potential of component j (A or B) in phase i (L or s) at the given temperature T

$\mu_j^{0\,i}$ = chemical potential of pure element j (A or B) in phase i (L or s) at the given temperature T

x_B^i = mole fraction of component B in phase i (L or s)

$\overline{G}_j^{Ex\,i}$ = partial molar excess Gibbs' free energy of component j (A or B) in phase i (L or s).

By inserting the expressions for μ_A^s, μ_B^s and μ_A^L, μ_B^L, respectively, into Equation (2.35) we obtain the molar Gibbs' free energy of the different phases G_m^s and G_m^L as functions of the mole fraction x_B at all temperatures.

In Figure 2.17 (page 58) G_m^s and the corresponding G_m^L functions are plotted as functions of x_B at a given temperature. The shapes and positions of the curves depend on specific material constants and the excess functions G^{Ex}.

On pages 54–23 phase formation in binary solutions has been discussed. From the general principle that the lowest possible total energy of a system corresponds to equilibrium, the compositions of the two stable phases could be determined by simply drawing the common tangent of the two molar Gibbs' free energy curves.

This construction has been used in Figure 2.17, which shows the molar Gibbs' free energy curves of the liquid and solid phases. At the given temperature the initial unstable solution is a liquid and has the composition x_B^L. It decomposes into two stable phases with compositions $x_B^{\alpha/L}$ (solid phase) and $x_B^{L/\alpha}$ (liquid phase). The solid phase is frequently called α instead of s, which we do here.

In this case we have two tangent points. If we use the 'solid' tangent point the intersections with the axes or chemical potentials will be

$$\mu_A^{\alpha/L} = \overline{G}_A^{\alpha/L} \quad \text{and} \quad \mu_B^{\alpha/L} = \overline{G}_B^{\alpha/L} \tag{2.40}$$

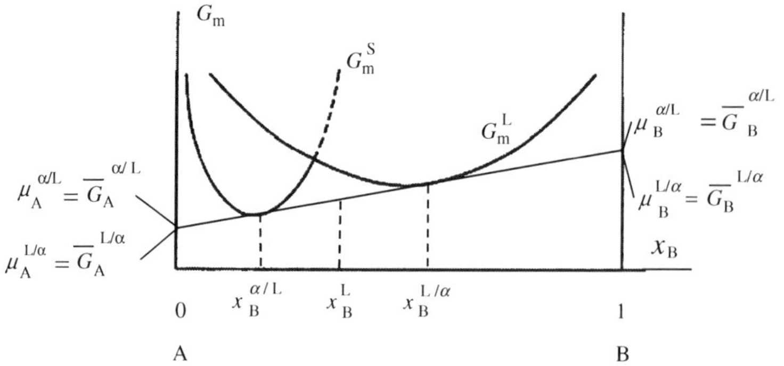

Figure 2.17 The molar Gibbs' free energies of the solid and liquid phases of a non-ideal binary solution as functions of composition at a given temperature.

On the other hand, if we consider the 'liquid' tangent point we obtain the following relationships

$$\mu_A^{L/\alpha} = \overline{G}_A^{L/\alpha} \quad \text{and} \quad \mu_B^{L/\alpha} = \overline{G}_B^{L/\alpha} \tag{2.41}$$

The different results give no contradiction as the condition for equilibrium is that the free energy or chemical potential of each component is the same in all phases of the system (compare the derivation of Gibbs' phase rule on page 50). In this case we obtain

$$\mu_A^{\alpha/L} = \mu_A^{L/\alpha} \quad \text{and} \quad \mu_B^{\alpha/L} = \mu_B^{L/\alpha} \tag{2.42}$$

Example 2.1

Consider a binary, non-ideal solution, which can be described with the aid of the regular solution model. Find expressions for the partial excess Gibbs' free energies of the solution in terms of the constant L and the mole fractions of the components A and B.

Solution:

According to Equation (1.114) on page 31 in Chapter 1 we have $H_m^{Ex} = H_m^{mix}$.
The regular-solution model is characterized by $H_m^{mix} = Lx_Ax_B$ [Equation (1.81) on page 24 in Chapter 1]. and $S_m^{Ex} = 0$.
By use of the definition of excess quantities and the equations above we obtain

$$G_m^{Ex} = H_m^{Ex} - TS_m^{Ex} = Lx_Ax_B - T \times 0 = Lx_Ax_B$$

The desired expressions are obtained by application of Gibbs–'Duhem's equations [Equations (1.123) and (1.124) on page 32 in Chapter 1]. For component A we obtain

$$\overline{G}_A^{Ex} = G_m^{Ex} + (1 - x_A)\frac{\partial G_m^{Ex}}{\partial x_A} = Lx_A(1 - x_A) + (1 - x_A).\ \frac{\partial[Lx_A(1 - x_A)]}{\partial x_A}, \text{ which gives}$$

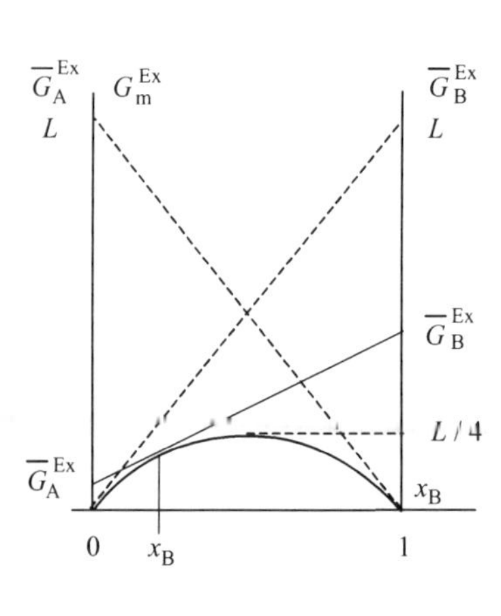

$$\overline{G}_A^{Ex} = Lx_A(1 - x_A) + (1 - x_A) \times L \times [1 - 2x_A] = L(1 - x_A)(x_A + 1 - 2x_A) = Lx_B^2$$

For component B we obtain analogously

$$\overline{G}_B^{Ex} = G_m^{Ex} + (1 - x_B)\frac{\partial G_m^{Ex}}{\partial x_B} = Lx_B(1 - x_B) + (1 - x_B).\frac{\partial[Lx_B(1 - x_B)]}{\partial x_B}$$

which gives $\overline{G}_B^{Ex} = Lx_B(1 - x_B) + (1 - x_B) \times L \times [1 - 2x_B] = L(1 - x_B)(x_B + 1 - 2x_B) = Lx_A^2$

Answer:

$\overline{G}_A^{Ex} = Lx_B^2$ and $\overline{G}_B^{Ex} = Lx_A^2$

2.4.5 *Phase Diagrams of Binary Alloys*

If the molar Gibbs' free-energy curves are drawn for the liquid and solid phases of a binary alloy, their relationships to the phase diagram for the binary system can be studied.

A binary phase diagram contains compressed information of all the alloys, which can be formed from two metals A and B. The phase diagram of the binary system AB is a diagram, which shows the compositions of the phases of the system as a function of temperature. Areas in the diagram represent the phases.

The equilibrium conditions define the boundary lines between the phases. Phase diagrams are always drawn with the temperature T on the vertical axis and the composition of the alloy as the horizontal axis. An example is given in Figure 2.18.

Phase diagrams are widely used in chemistry and metallurgy because they contain much information of the phases of an alloy in a surveyable way, for example the number and compositions of the phases at a given temperature, the temperatures at which phase transitions occur.

Phase diagrams are known and published in the scientific literature for all alloys of importance and collected in special publications and data bases.

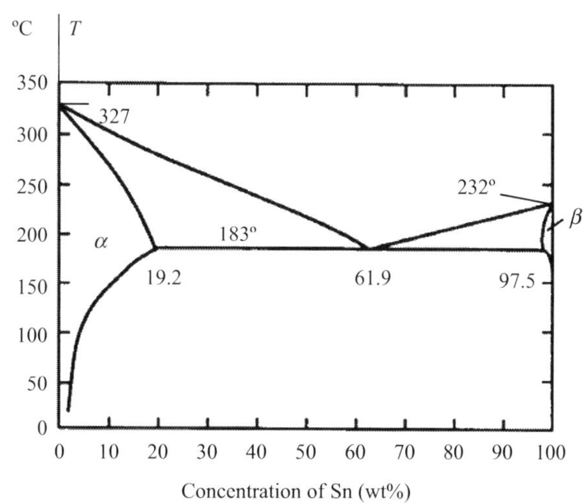

Figure 2.18 Phase diagram of the system PbSn. From Binary Alloy Phase Diagrams Vol. 2, Ed. Thaddeus B. Massalski, American Society for Metals 1986. Reprinted with permission of ASM International. All rights reserved. www.asminternational.org.

Phase Reactions or Phase Transformations as Functions of Temperature

A phase diagram shows the phases in equilibrium with each other at various temperatures. The temperature is a very important variable. A slight temperature change may create a new phase or lead to the disappearance of an existing phase. Such processes are called *phase reactions* or *phase transformations.*

The four most common types of reactions that involve liquids and their specific designations are listed in Table 2.2. They will frequently be used in the discussions of phase diagrams below.

Table 2.2 Four common tectic type reactions

Appearance in the phase diagram	Reaction when the alloy system is cooled	Name of reaction
	$L \rightarrow \alpha_{solid} + \gamma_{solid}$	Eutectic
	$L_1 \rightarrow L_2 + \gamma_{solid}$	Monotectic
	$\beta_{solid} \rightarrow \alpha_{solid} + L$	Metatectic
	$\alpha_{solid} + L \rightarrow \beta_{solid}$	Peritectic

The names of the solid phases are not fixed but vary from alloy to alloy. The general suffix 'tectic' is used for reactions, which involve one or two liquid phases. The upper three reactions belong to the particular tectic types, which involve only one reactant. When there are two reactants the reactions are said to be of the *peritectic* type.

2.4.6 Relationship Between Molar Gibbs' Free Energy Curves and Phase Diagrams. Construction of Phase Diagrams

There is a close relationship between the molar Gibbs' free energy curves and the phase diagram of a binary system. If the solid and liquid curves are known for various temperatures, the phase diagram can be constructed with the aid of the tangent to tangent method. This method gives a good understanding of the connection between free-energy curves and the corresponding phase diagram.

The appearance of the phase diagram depends entirely on the shapes and positions of the free-energy curves at various temperatures. The pressure has little influence for both liquids and solids and is supposed to be constant and equal to 1 atm. The heat of mixing influences the shape of the phase diagram very much.

There is a manifold of possible shapes of the molar Gibbs' free energy curves of liquids and solids, owing to different material constants and different forces between the atoms in pure elements and solutions. This also influences the appearances of the phase diagrams. We will restrict the discussion below to some common types of molar Gibbs' free energy curves and generation of a few principal types of phase diagrams from these curves.

Case I concerns two metals, which form ideal solutions at all compositions, both as liquids and solids. This results in simple molar Gibbs' free-energy curves and they generate the simplest possible type of binary phase diagrams.

Cases II, III and IV concern solid and liquid solutions, which are non-ideal, will be treated next. The theoretical background and the shapes of the molar Gibbs' free energy curves of solid and liquid solutions have been discussed on pages 46–50.

If the two solid components have *equal* structures (cases II and III), only one molar Gibbs' free energy curve of the solid solution is formed. In the case of *unequal* structures of the two solid components (case IV), there are two separate molar Gibbs' free energy curves of the solid solution. The influence of the solid structures and the heat of mixing on the phase diagrams will be studied for some different cases of non-ideal solutions.

Each molar Gibbs' free energy curve in the following sections is based on one of the three alternative models below.

Model for Ideal Binary Solutions

$$G_m = x_A G_A^0 + x_B G_B^0 + RT(x_A \ln x_A + x_B \ln x_B) \quad (2.6)$$

General Model for Non-Ideal Binary Solutions

$$G_m = x_A G_A^0 + x_B G_B^0 + RT(x_A \ln x_A + x_B \ln x_B) + G_m^{Ex} \quad (2.9)$$

$$G_m^{Ex} = H_m^{mix}(1 - AT) \quad (2.13)$$

Regular-Solution Model for Non-Ideal Binary Solutions

$$G_m = x_A G_A^0 + x_B G_B^0 + L x_A x_B + RT(x_A \ln x_A + x_B \ln x_B) \quad (2.16)$$

$H_m^{mix} = L x_A x_B$ and $S_m^{Ex} = 0$ in the regular solution model.

All terms in the three equations, except H_m^{mix} in most cases, depend on temperature. The first two terms in all three equations determine the position of the curves and the rest determine their shapes.

Case I. Ideal Solution of Two Metals $H_{\mathrm{m}}^{\mathrm{mix}} = 0$

In this case, the heat of mixing is zero for both the liquid and the solid alloy. The Gibbs' free-energy curves $G_{\mathrm{m}}^{\mathrm{L}}$ and $G_{\mathrm{m}}^{\mathrm{s}}$ will be of the same type as the G_{m} curve in Figure 2.2 on page 45.

The Gibbs' free-energy curves for the solid and the liquid states at temperature T_1 are calculated from Equation (2.6) on page 44 and plotted in an $x_{\mathrm{B}}G_{\mathrm{m}}$ diagram. The common tangent to the curves is drawn and the mole fractions x_{s} and x_{L} at the tangent points are noted and plotted in an $x_{\mathrm{B}}T$ diagram for the given temperature T_1. The procedure is repeated for T_2 and other temperatures. New pairs of mole fractions are found and plotted in the $x_{\mathrm{B}}T$ diagram. In this way, the phase diagram is found, which is characteristic of alloys with miscible components at all proportions.

Figures 2.19a–f illustrate how the phase diagram of a binary system forming ideal solutions can be derived by studying the relative position of the free-energy curves for the liquid and the solid at various temperatures. The construction is described step by step, partly in the text below the figures and partly in the current text. The most low-lying free-energy curve always represents the stable state.

The initial temperature T_1 is higher than the melting points of both metals. At decreasing temperatures both G_{m} curves rise, but the liquid curve rises more than the solid one (compare the discussion on page 53), crosses and passes the solid curve.

At temperature T_3 alloys with compositions between pure A and x^{s} exist as homogeneous solid solutions. Alloys with compositions $\geq x^{\mathrm{L}}$ remain as homogeneous liquid solutions. Alloys with compositions between x^{s} and x^{L} exist as a mixture of a solid phase of composition x^{s} and a liquid phase of composition x^{L}. (x^{s}, T_3) lies on the solidus line, (x^{L}, T_3) on the liquidus line. The relative proportions of solid and liquid phases are determined by the lever rule (Equations (2.29) on page 55).

The phase diagram in Figure 2.19f is typical for binary systems that form ideal solutions.

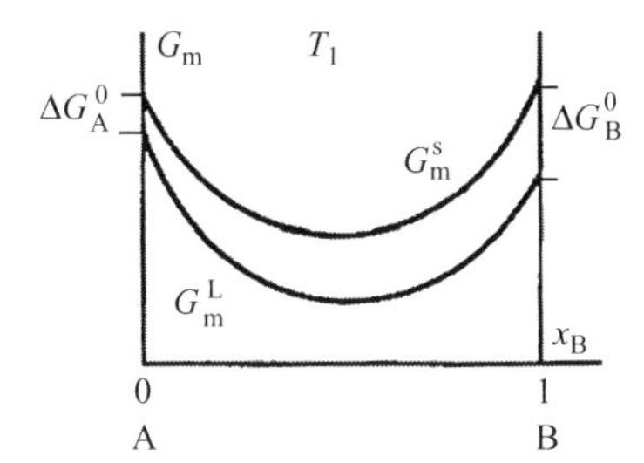

Figure 2.19a At temperature T_1 the free energy curve $G_{\mathrm{m}}^{\mathrm{L}}$ of the liquid lies entirely below the curve $G_{\mathrm{m}}^{\mathrm{s}}$ of the solid. Hence, the stable phase is a homogeneous liquid solution.

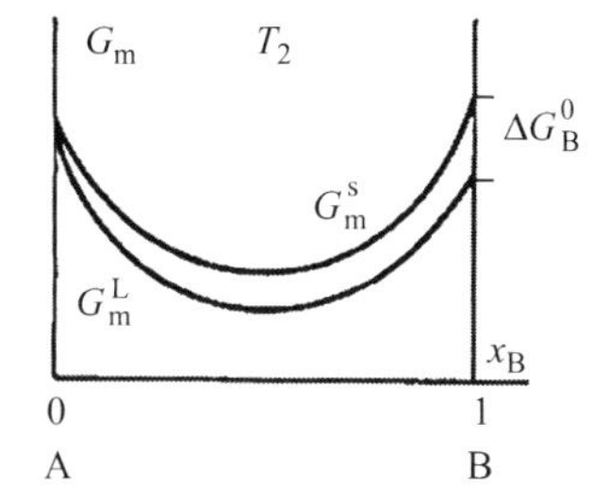

Figure 2.19b At temperature T_2 the $G_{\mathrm{m}}^{\mathrm{L}}$ and $G_{\mathrm{m}}^{\mathrm{s}}$ curves are in contact with each other at the axis $x_{\mathrm{B}} = 0$. Pure A can solidify. This corresponds to the point (0, T_2) in the phase diagram (Figure 2.19f).

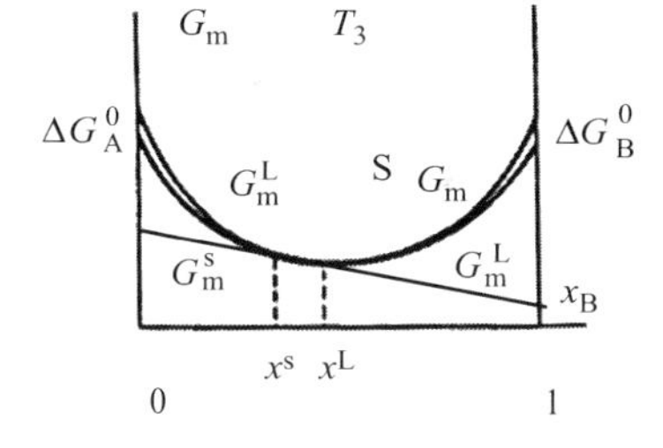

Figure 2.19c At temperature T_3 the $G_{\mathrm{m}}^{\mathrm{L}}$ and $G_{\mathrm{m}}^{\mathrm{s}}$ curves cross. The common tangent of the two curves is drawn. x^{s} and x^{L} are the points where the tangent touches the curves.

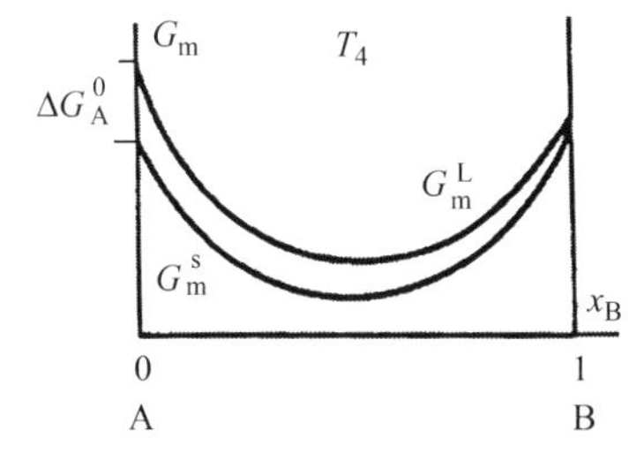

Figure 2.19d At temperature T_4 the Gibbs' free energy curves are in contact with each other at the axis $x_{\mathrm{B}} = 1$. Only pure B can exist as a liquid. This gives the point (1, T_4) in the phase diagram.

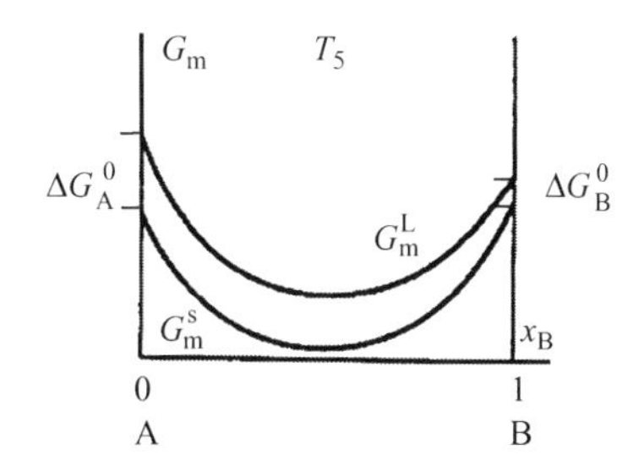

Figure 2.19e At temperature T_5 the entire $G_{\mathrm{m}}^{\mathrm{s}}$ curve lies below the $G_{\mathrm{m}}^{\mathrm{L}}$ curve. The stable phase is a homogeneous solid solution.

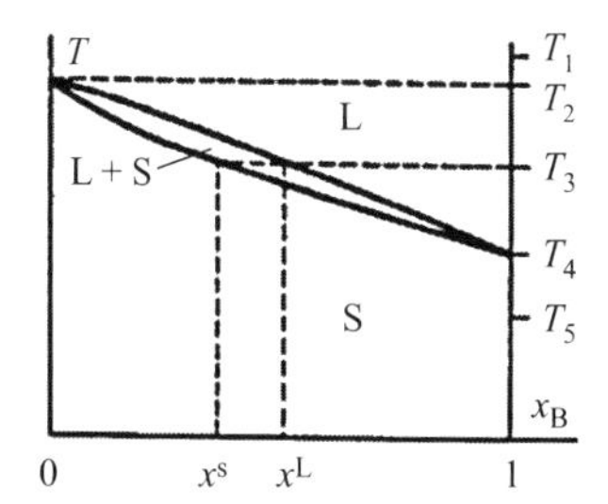

Figure 2.19f Phase diagram of a binary system with components, which form an ideal solution.

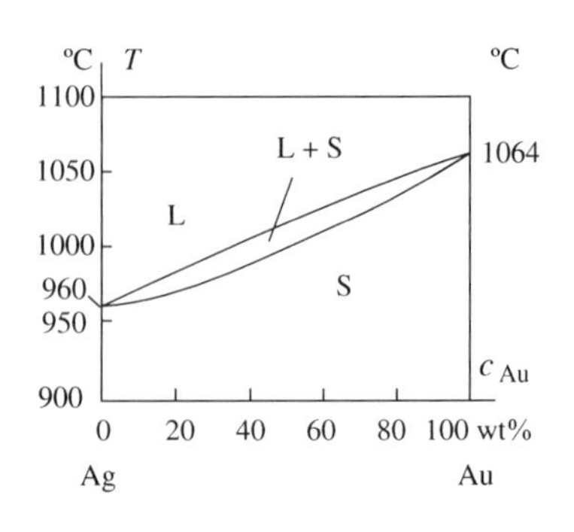

Figure 2.20 Phase diagram of the Ag–Au system. From Binary Alloy Phase Diagrams Vol. 2, Ed. Thaddeus B. Massalski, American Society for Metals 1986. Reprinted with permission of ASM International. www.asminternational.org.

The end points of the boundary lines correspond to the melting points of the pure metals. The 'direction' of the two-phase area might be 'tilted' the other way depending on the melting points of the pure components.

Figure 2.20 shows the phase diagram of the Ag–Au system as an example of such a system. It is nearly ideal.

Case II. Non-Ideal Solutions. Equal Structure of Solid Components

Positive H_m^{mix} of Solid. Ideal Liquid

The heat of mixing is a quantity of great importance for the shape of the Gibbs' free-energy curves. The topic has been discussed in Section 1.5.1 on pages 23-24 in Chapter 1 and particularly in Section 2.4.2.

The special conditions, which control the appearance of the G_m curves, have been extensively discussed on pages 51-53 in this chapter. A necessary but not sufficient condition for a G_m curve with one maximum and two minima, like those in Figures 2.21, is for example that H_m^{mix} is positive.

The phase diagram will be derived from known G_m curves for some cases with positive heat of mixing. In the first example we concentrate on construction of the phase diagram. In the second example the influence of the strength of the interatomic bonds is discussed. The third example shows how the shape of the phase diagram changes when the interatomic bonds are very strong.

Construction of the Phase Diagram

A simple example is shown in Figures 2.21a–f. At the initial temperature T_1, above the melting points of both metals, the Gibbs' free-energy curves do not intersect. When the temperature decreases the liquid curve raises more than the solid one.

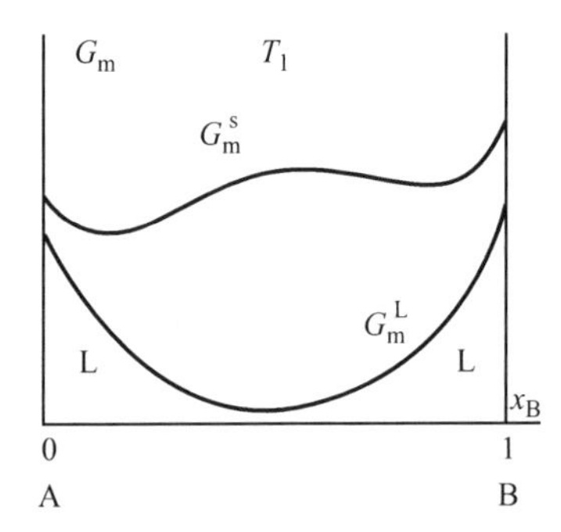

Figure 2.21a At temperature T_1 the stable phase is liquid at all compositions.

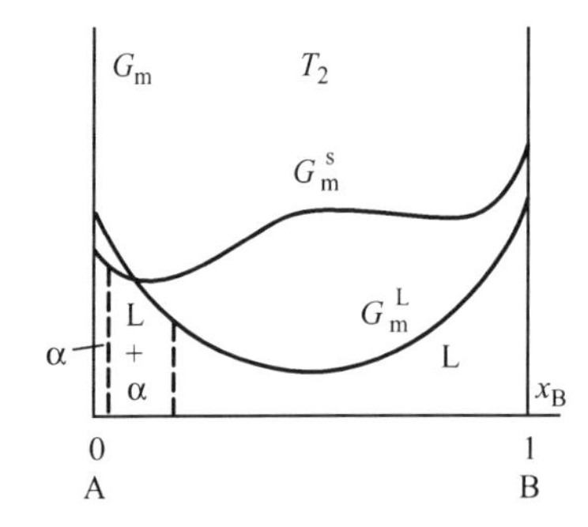

Figure 2.21b At temperature T_2 the tangent construction predicts an A-rich two-phase region.

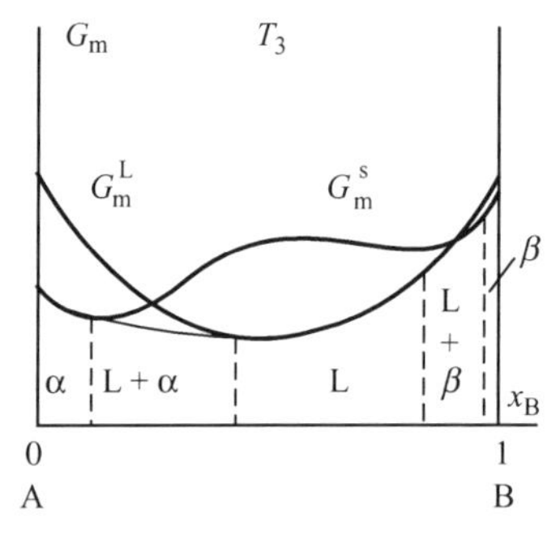

Figure 2.21c At temperature T_3 an A-rich and a B-rich two-phase region are found by the tangent construction.

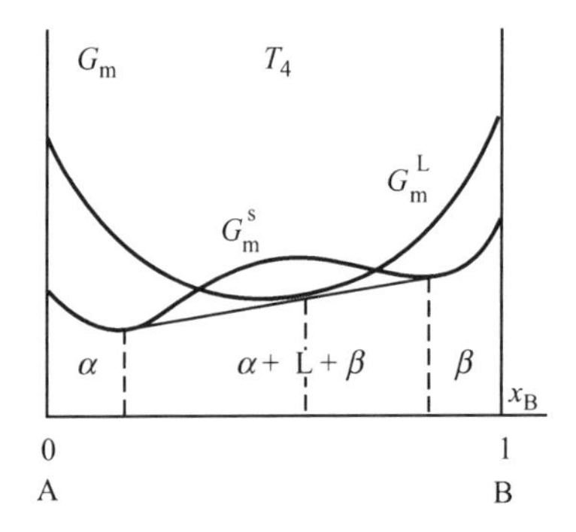

Figure 2.21d At temperature T_4 the tangent is in contact with three phases, a liquid phase and two solid phases. This corresponds to a triple point, the eutectic point, in the phase diagram.

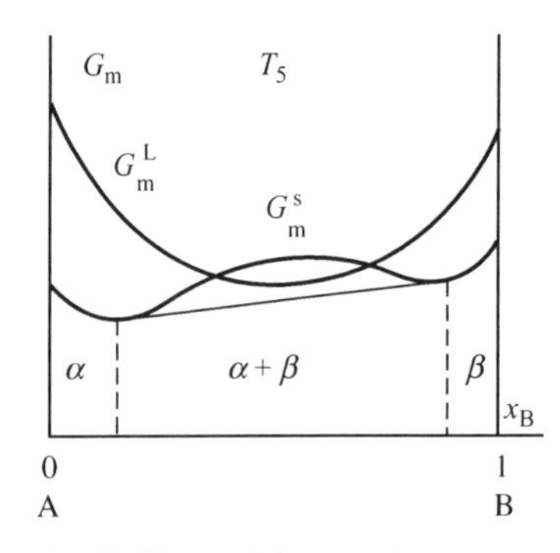

Figure 2.21e At temperature T_5 no liquid phase exists. The tangent construction predicts a mixture of two solid phases in the central region.

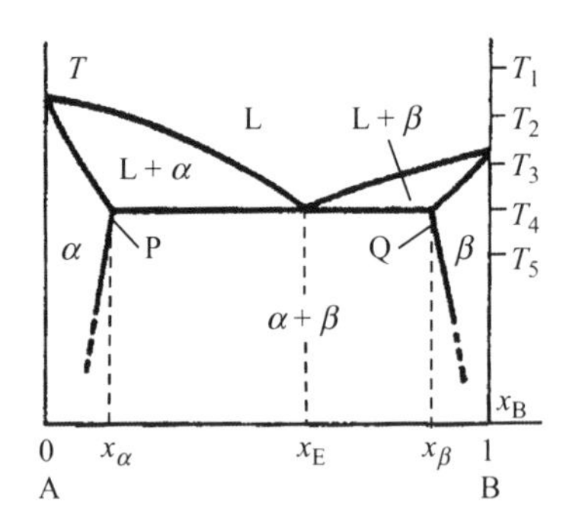

Figure 2.21f Phase diagram of a binary system with components, which are only partially miscible. The system is eutectic (see page 60.)

The construction of the phase diagram is obviously done in the same way as for the ideal solutions above (page 61).

The type of phase diagram, shown in Figure 2.21f, is very common.

Interatomic Bonds and Melting Points of Alloys with $H_m^{mix} > 0$

For positive values of H_m^{mix} the bonds between equal atoms are stronger than those between unequal atoms. Energy is required to break the A–A and B–B bonds and less energy is released when the weaker A–B bonds are formed (pages 23-24 in Chapter 1). Hence, the crystal lattice of the alloy is weaker than those of the pure metals. Consequently, the melting points of all solid alloys of this type are lower than the "melting-point line", which corresponds to a mechanical mixture of the two components. (The line is not drawn in Figure 2.22e below).

The corresponding phase diagram is constructed in the same way as above. At a temperature T_1, higher than the melting points of both metals, the Gibbs' free-energy curve of the solid solution lies above that of the liquid solution (Figure 2.22a) and the curves do not intersect.

When the temperature decreases the Gibbs' free-energy curve of the liquid solution raises more than that of the solid solution. The curves intersect, for example at the temperature T_2 and two tangents can be drawn, which give four points in the phase diagram (Figure 2.22b). At the special temperature, when the solidus and liquidus curves have a common minimum, all four tangent points coincide (Figure 2.22c). This point corresponds to the lowest melting point of all the alloys.

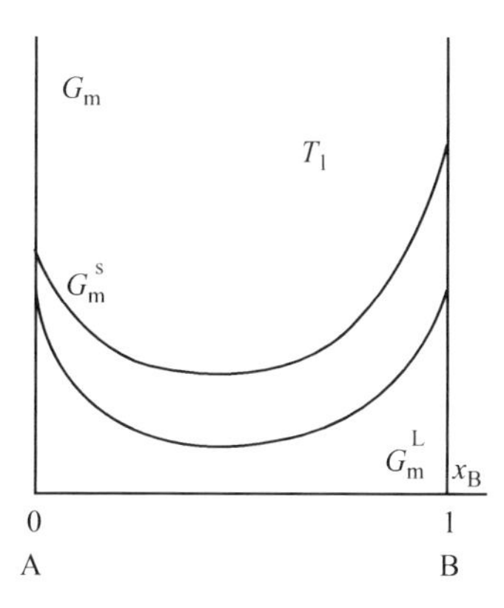

Figure 2.22a Molar Gibbs' free energy curves at temperature T_1 for a non-ideal solid solution of two metals A and B.

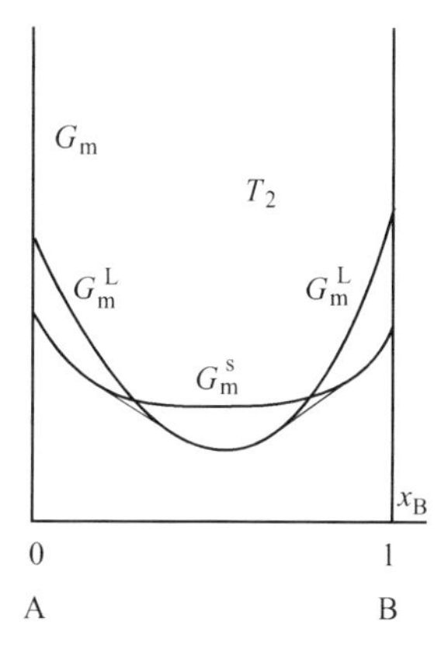

Figure 2.22b Molar Gibbs' free energy curves at temperature T_2 for the non-ideal solid solutions in Figure 2.22 a.

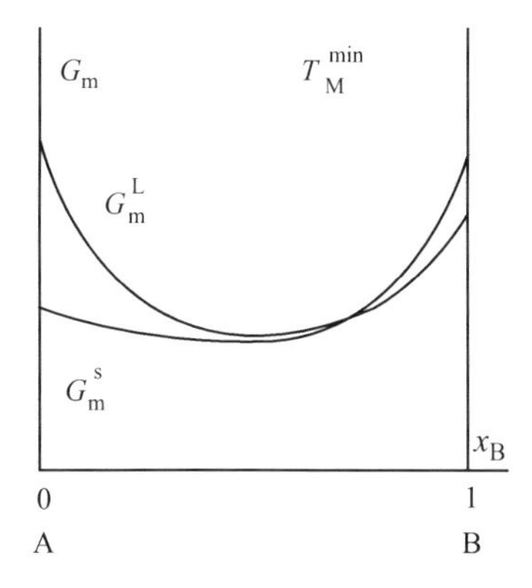

Figure 2.22c Molar Gibbs' free energy curves at temperature T_M^{min} for the non-ideal solid solutions in Figure 2.22 a.

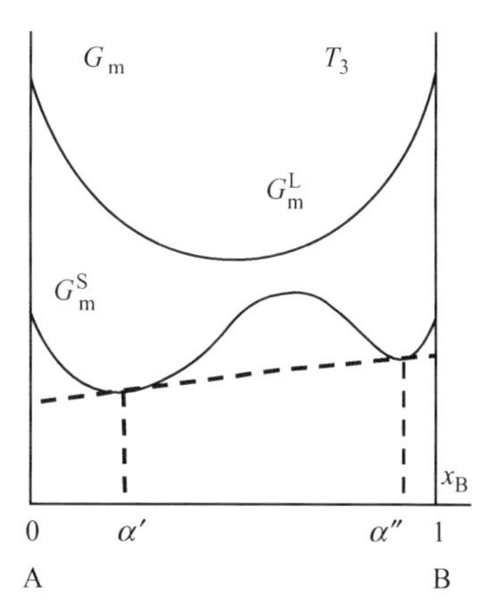

Figure 2.22d Molar Gibbs' free energy curves at low temperature T_3 for the non-ideal solid solutions in Figure 2.22 a.

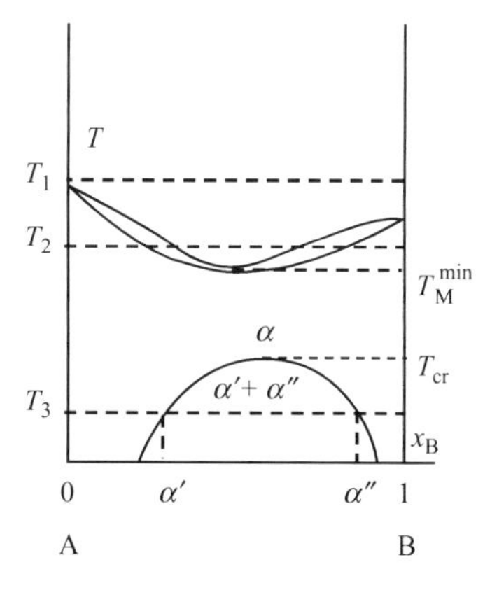

Figure 2.22e Phase diagram of the system AB when the heat of mixing is positive. The phase diagram includes a miscibility gap in the solid solution.

The solid solution is non-ideal. The two components of the alloy are not miscible at all proportions and a miscibility gap appears at low temperatures for the same reason as in a regular solution (page 46-47). The G_m curve of the solid solution has one maximum and two minima just like a regular solution if the temperature is below the critical temperature. The height of the maximum depends on the alloy and becomes more pronounced the lower the temperature is. The miscibility gap of the two phases can be constructed gradually from the G_m curves in the way illustrated in Figure 2.22d for $T = T_3$. The corresponding part of the phase diagram is shown in Figure 2.22e.

Strongly Positive H_m^{mix} of Solid

For alloys with a strongly positive H_m^{mix} value the melting point minimum in the phase diagram will be lower and the critical temperature of the miscibility gap will be higher than in Figure 2.22e above. If the H_m^{mix} value is large enough, the boundary curves get in touch with each other, which leads to a *eutectic* phase diagram (Figure 2.23b). Such phase diagrams are very common in binary alloy systems (see also the Figures 2.19).

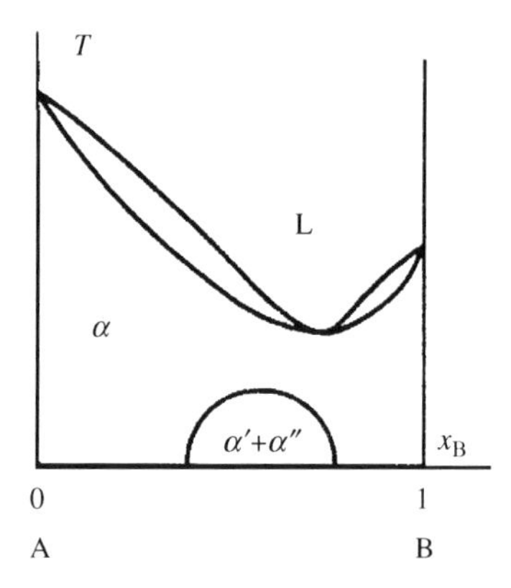

Figure 2.23a Phase diagram for the non-ideal solid solutions of components A and B. H_m^{mix} is positive.

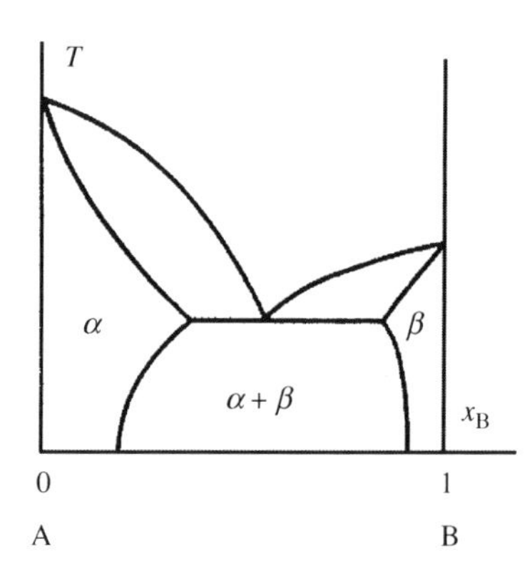

Figure 2.23b Eutectic phase diagram of the system AB. H_m^{mix} in the solid is strongly positive.

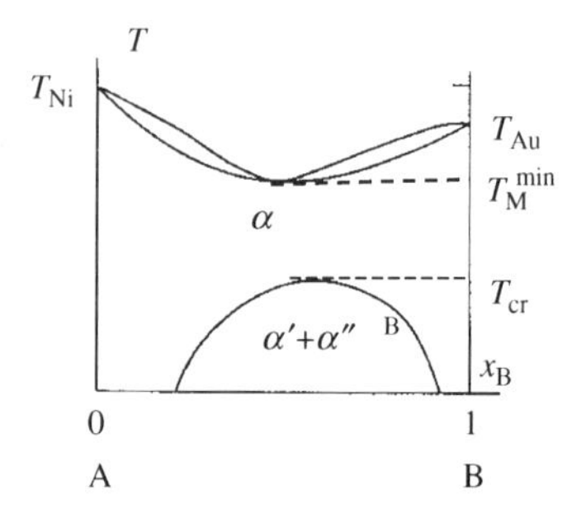

Phase diagram of the system AuNi.

Example 2.2

The phase diagram of Ni–Au is similar to the phase diagram shown in Figure 2.22e. The melting point minimum of the alloys is 950 °C, which correspond to the composition $x_{Au} = x_{Ni} = 0.5$. The melting points of Ni and Au are 1450 °C and 1063 °C, respectively. The molar heats of fusion for Ni and Au are 17×10^6 J/kmol and 12×10^6 J/kmol, respectively.

a) Calculate the molar heat of mixing of the system AuNi. The regular solution model is assumed to be valid for the solid Ni and Au phases.
b) The critical temperature of the solid two-phase region at low temperature is 750 °C. Use this information for an alternative calculation of the heat of mixing of the system Au–Ni.

Solution:

a) The regular solution model is an acceptable model for the molar Gibbs' free energy of the solid non-ideal solution. The liquid solution is assumed to be ideal.

With the usual designations the molar Gibbs' free energy of the solid, respectively, liquid: can be written

$$G_m^s = x_A^s\, G_A^{0s} + x_B^s\, G_B^{0\,s} + L x_A^s x_B^s + RT\left(x_A^s \ln x_A^s + x_B^s \ln x_B^s\right) \tag{1'}$$

$$G_m^L = x_A^L\, G_A^{0\,L} + x_B^L\, G_B^{0\,L} + RT\left(x_A^L \ln x_A^L + x_B^L \ln x_B^L\right) \tag{2'}$$

The two molar Gibbs' free-energy curves touch each other at

$$x_A^s = x_A^L = x_B^s = x_B^L = 0.5 \tag{3'}$$

when the four tangent points coincide. The corresponding point in the phase diagram is the minimum melting point T_M^{min} of the alloy with composition 0.5. The condition

$$G_m^L = G_m^s$$

in combination with Equations (1') and (2') gives

$$x_A^s\, G_A^{0s} + x_B^s\, G_B^{0\,s} + L x_A^s x_B^s = x_A^L\, G_A^{0\,L} + x_B^L\, G_B^{0\,L}$$

or, in combination with Equation (3')

$$L x_A^s x_B^s = x_A^s\left(G_A^{0\,L} - G_A^{0\,s}\right) + x_B^s\left(G_B^{0L} - G_B^{0s}\right) \tag{4'}$$

The difference in molar Gibbs' free energy between liquid and solid element A is related to the molar heat of melting. If we apply Equation (2.5) on page 44 and obtain

$$-\Delta G_m^{L\to s} = \frac{T_M - T}{T_M}\left(-\Delta H_m^{L\to s}\right) = \frac{\Delta T}{T_M} H_m^{fusion}$$

and insert $T_M = T_M^{min}$, $T = T_M^A$ and $T = T_M^B$ we obtain

$$G_A^{0L} - G_A^{0s} = \frac{T_M^{Ni} - T_M^{min}}{T_M^{min}} H_m^{fusion\ Ni} = \frac{1723 - 1223}{1223} \times 17 \times 10^6 = 6.9 \times 10^6 \text{J/kmol}$$

and

$$G_B^{0L} - G_B^{0s} = \frac{T_M^{Au} - T_M^{min}}{T_M^{min}} H_m^{fusion\ Au} = \frac{1336 - 1223}{1223} \times 12 \times 10^6 = 1.1 \times 10^6 \text{ J/kmol}$$

The desired heat of mixing is obtained from Equation (4′):

$$H_m^{mix} = x_A^s\left(G_A^{0L} - G_A^{0\,s}\right) + x_B^s\left(G_B^{0\,L} - G_B^{0\,s}\right)$$

or

$$H_m^{mix} = 0.5 \times 6.9 \times 10^6 + 0.5 \times 1.1 \times 10^6 = 4.0 \times 10^6 \text{ J/kmol}$$

b) By use of Equation (2.21) on page X10 we obtain

$$L = 2RT_{cr} = 2 \times 8.3 \times 10^3 \times (750 + 273)\ \text{J/kmol} = 17 \times 10^6\ \text{J/kmol}$$

which gives

$H_m^{mix} = L \times x_A \times x_B = 17 \times 10^6 \times 0.5 \times 0.5\ \text{J/kmol} = 4.2 \times 10^6\ \text{J/kmol}$ in good agreement with the value, obtained in a).

Answer:

The molar heat of mixing of the Ni–Au system is 4.1×10^6 J/kmol.

Case III. Non-Ideal Solutions. Equal Structure of Solid Components

Negative H_m^{mix} of Solid. Ideal Liquid

The liquid solution is assumed to be ideal (Equation (2.6) on page 44) and the solid solution is approximated by the regular-solution model (Equation (2.18) on page 48).

Interatomic Bonds and Melting Points of Alloys with $H_m^{mix} < 0$

When H_m^{mix} is negative instead of positive the conditions are reversed, compared to the discussion on page 63. According to the discussion on pages 23-24 in Chapter 1 negative heat of mixing means that the bonds between unequal atoms are stronger than the bonds between equal atoms. Less energy is required to break the A–A and B–B bonds than the energy, which is released when the strong A–B bonds of the alloy are formed. Hence, the crystal lattice of the alloy is stronger than the lattices of the pure metals.

Consequently, the melting points of all solid alloys of this type are higher than the melting points of the corresponding mechanical mixture of the components. The latter lie on the "melting-point line" of the pure components (not drawn in Figure 2.24c below).

If the initial temperature is low the equilibrium state of the alloy is solid (Figure 2.24a). When the temperature increases to T_2 both G_m curves 'sink'. The liquid curve sinks more than the solid one (page 53). The curves intersect and the two common tangents are drawn (Figure 2.24b). The tangent points x_1, x_2, x_3 and x_4 are plotted in an x_BT diagram.

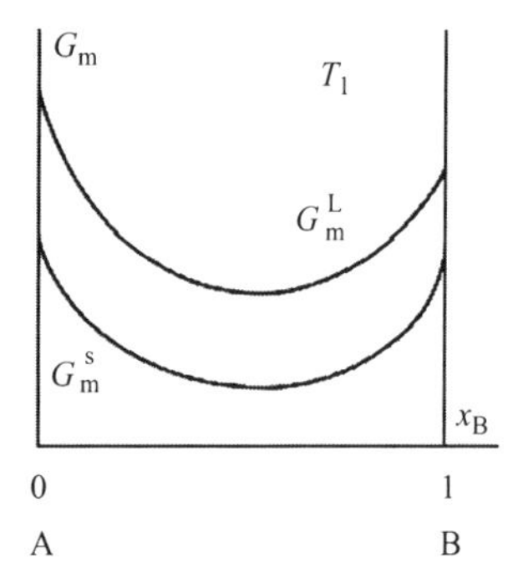

Figure 2.24a Molar Gibbs' free energy curves at temperature T_1 (Figure 2.24c) for the non-ideal solid solutions of metals A and B. H_m^{mix} is negative.

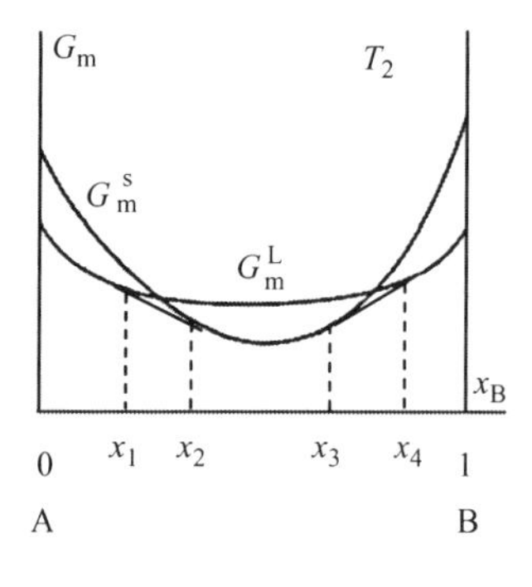

Figure 2.24b Molar Gibbs' free-energy curves at temperature T_2 for the non-ideal solid solutions in Figure 2.24a.

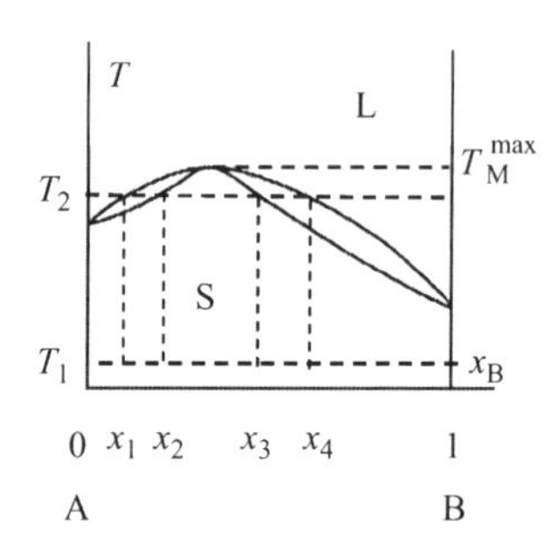

Figure 2.24c Phase diagram of the system AB when the heat of mixing is negative.

The tangent points define four intervals. Within the interval $0 - x_1$ the liquid phase is the stable state because the liquid free-energy curve is lower than that of the solid solution (Figure 2.24b). Within the interval $x_1 - x_2$ liquid and solid phases coexist. Within the interval $x_2 - x_3$ the solid phase is stable because the solid curve is everywhere lower than the liquid curve. Within the interval $x_3 - x_4$ liquid and solid phases coexist. In the last interval $x_4 - 1$ the liquid phase alone is stable.

These conclusions agree with the appearance of the phase diagram in Figure 2.24c. At the special temperature, when the solidus and liquidus curves have a common maximum, all four tangent points coincide. This point corresponds to the highest melting point of all the alloys.

Figure 2.24c does not represent the whole phase diagram. The alloy represents a non-ideal solution. The two components of the alloy are not miscible at all proportions. Phase diagrams with miscibility gaps also occur for positive H_m^{mix} values. Examples of complete phase diagram of this type are shown in Figures 2.22e and 2.24d.

Strongly Negative H_m^{mix} of Solid

In alloys with strongly negative H_m^{mix} values a stable so-called *intermediate phase*, here called the γ phase, with a structure different from those of the parent phases, may be formed.

The more negative the H_m^{mix} value is the higher will be the maximum temperature T_{max} of the γ phase. The increase of the critical temperature may even result in a change of the phase diagram from the type, shown in Figure 2.24d, to the type, drawn in Figure 2.25a, where the two parts join.

The more stable the γ phase is the higher will be the melting point of the alloy. As the intermediate phase has a different structure, compared to the parent phases, it has its own molar Gibbs' free-energy curve. It is shown in Figure 2.25b.

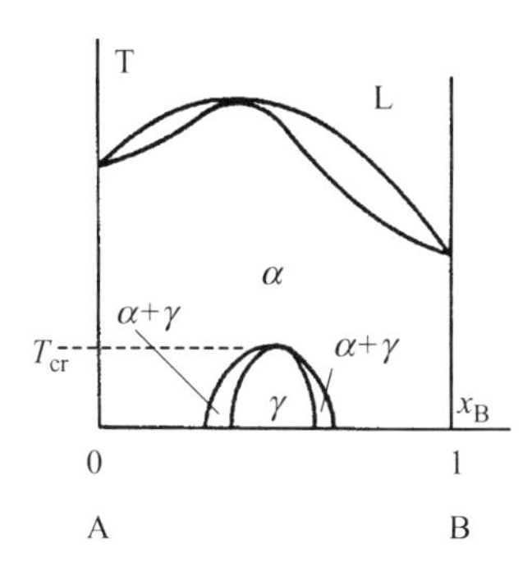

Figure 2.24d Complete phase diagram of the system AB when the heat of mixing is negative.

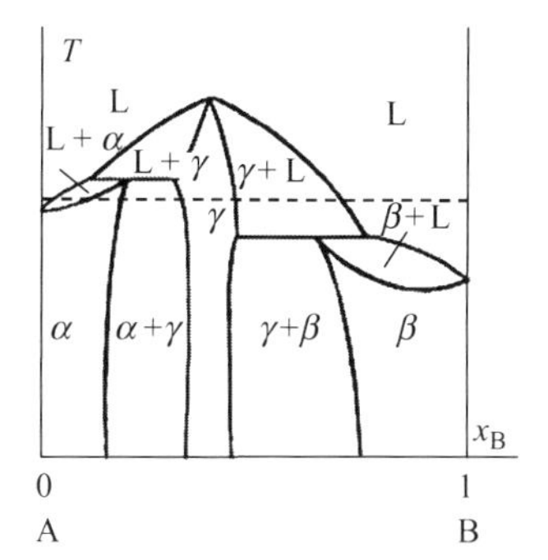

Figure 2.25a Complete phase diagram of the system AB when H_m^{mix} is strongly negative. The system is peritectic.

The six intersections between the dotted horizontal line and the boundary lines in Figure 2.25a correspond to the six tangent points in Figure 2.25b.

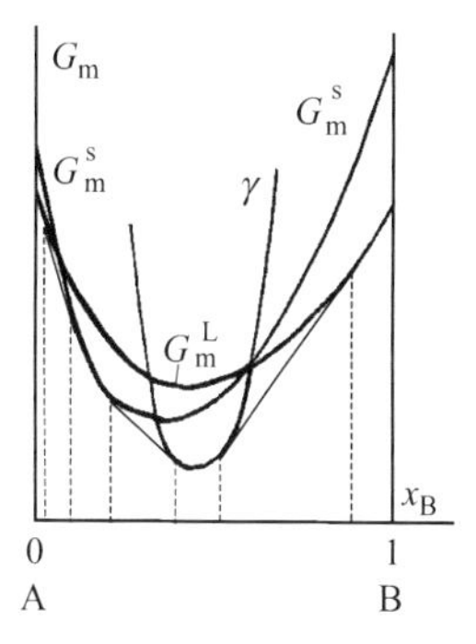

Figure 2.25b Molar Gibbs' free-energy curves of the solid, liquid and intermediate phases at the 'dotted' temperature in Figure 2.25 a. Three common tangents can be drawn at this temperature, which results in six tangent points.

The phase diagram of the system in Figure 2.25a is *peritectic* and contains one intermediate phase. Phase diagrams with intermediate phases will be further discussed later in this chapter.

Case IV. Non-Ideal Solutions. Unequal Structures of Solid Components

If the two components of an alloy have unequal structures they have separate G_m curves. As in the case of non-ideal solutions with equal solid structures the heat of mixing is an important parameter. The value and sign of H_m^{mix} has a greater influence on the shape of the phase diagram than the question whether the structures of the solid metals are equal or unequal.

Below, we will give two examples of the relationship between the Gibbs' free-energy curves and the corresponding phase diagram in the case of unequal structures.

Positive H_m^{mix} of Solid and Liquid

A eutectic phase diagram is obtained in most cases, when H_m^{mix} is positive for both solid and liquid. However, if the melting points of the two components differ very much a peritectic phase diagram may be obtained instead of a eutectic diagram. Such a case is shown below.

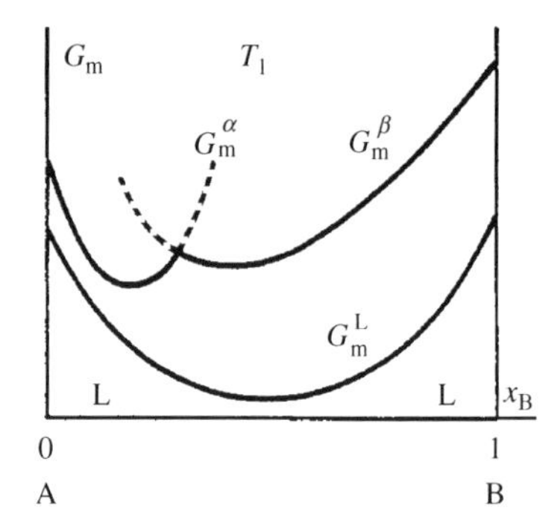

Figure 2.26a At temperature T_1 the stable phase is liquid at all compositions.

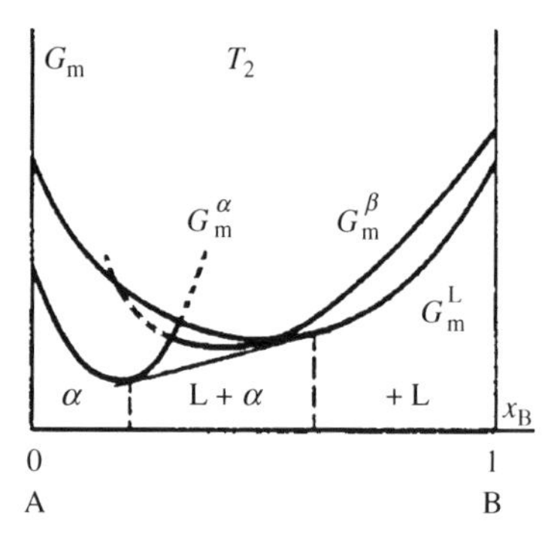

Figure 2.26b At temperature T_2 no liquid phase exists. The tangent construction predicts a mixture of two phases in the central region.

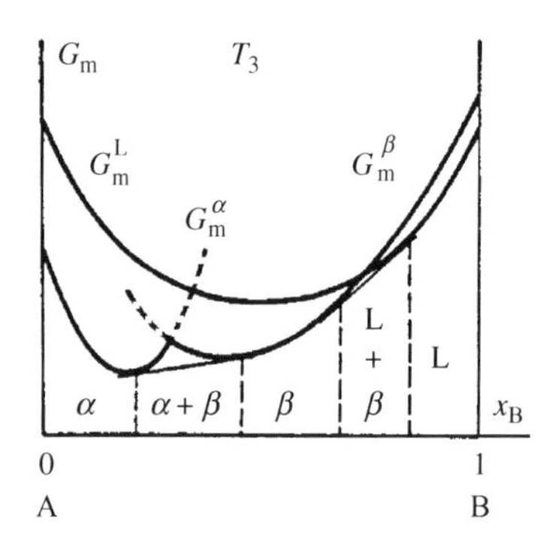

Figure 2.26c At temperature T_3 an A-rich and a B-rich two-phase regions are found by the tangent construction.

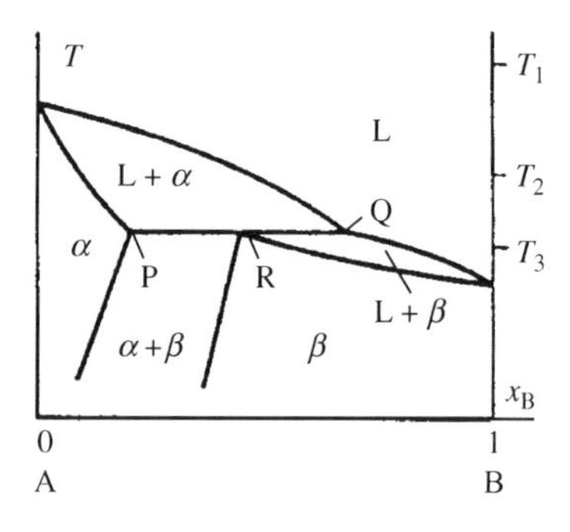

Figure 2.26d The free-energy curves lead to a peritectic phase diagram.

Negative H_m^{mix} of Solid and Liquid

Figure 2.25a on page 67 shows a phase diagram that corresponds to equal solid structures and a strongly negative value of H_m^{mix}. It can be seen from the sequence in Figures 2.27a–f that a phase diagram with an intermediate phase may arise even when the structures of the components of the binary alloy are unequal. In Figure 2.27 below we have assumed that the liquid also has a strongly negative value of H_m^{mix}. In Figure 2.25a the liquid was assumed to be ideal.

The phase diagram in Figure 2.27f is of the eutectic type and contains one intermediate phase. In more complicated cases several intermediate phases may be formed.

Alternatively a peritectic phase diagram may be formed instead. Peritectic reactions cause intermediate phases in many alloys.

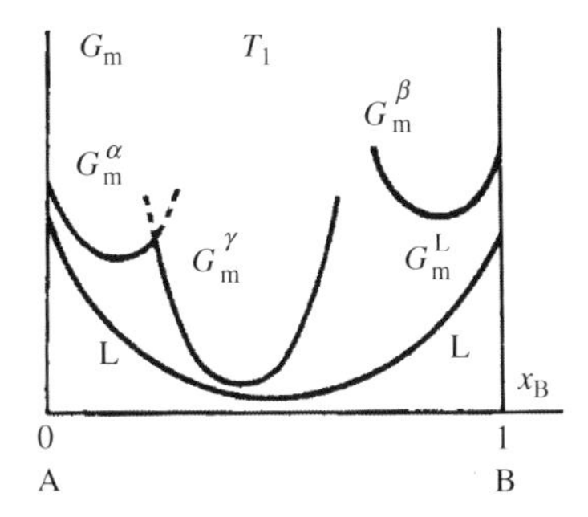

Figure 2.27a At temperature T_1 the stable phase is liquid at all compositions.

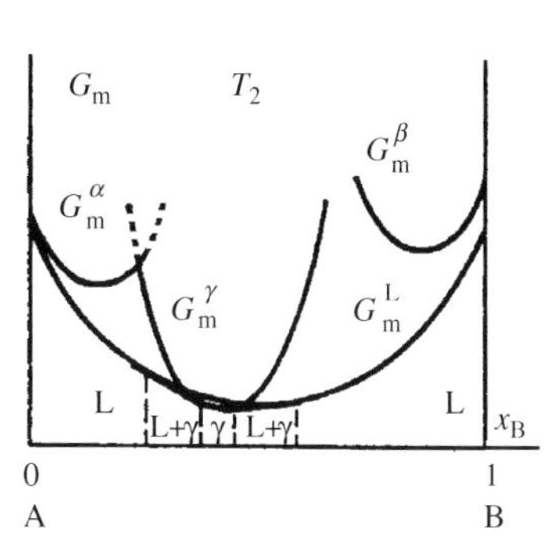

Figure 2.27b At temperature T_2 the G_m^L curve intersects the intermediate phase γ and two two-phase regions are formed.

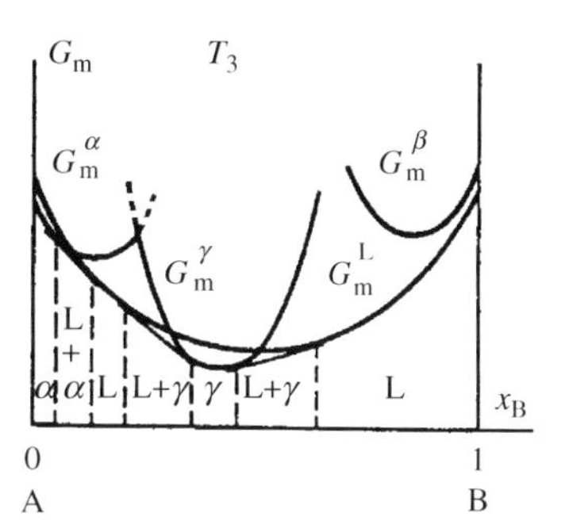

Figure 2.27c At temperature T_3 the tangent construction predicts three two-phase regions.

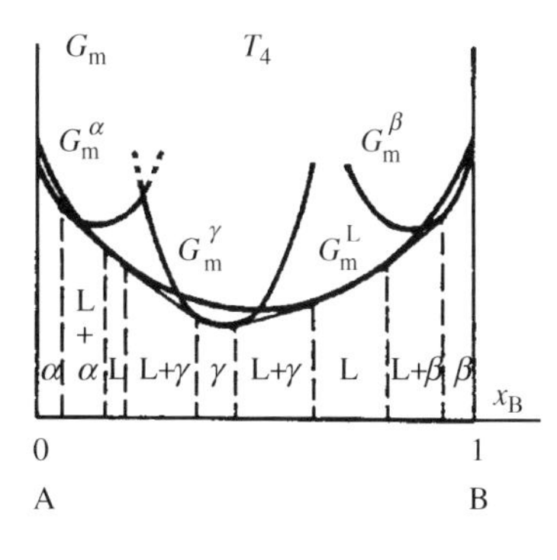

Figure 2.27d At temperature T_4 the tangent construction predicts four two-phase regions.

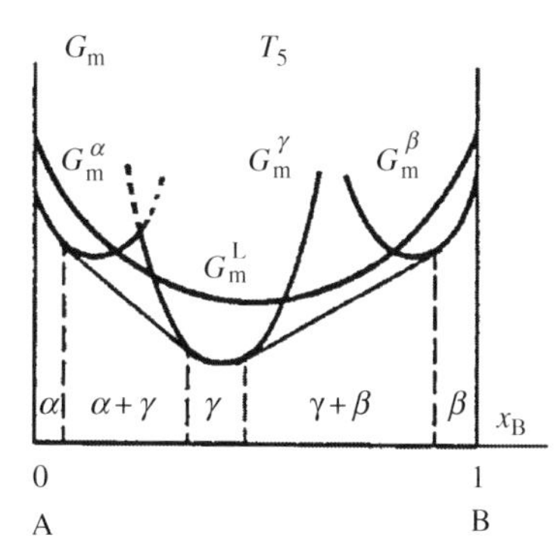

Figure 2.27e At temperature T_5 no liquid phase exists. The tangent construction predicts two two-phase regions.

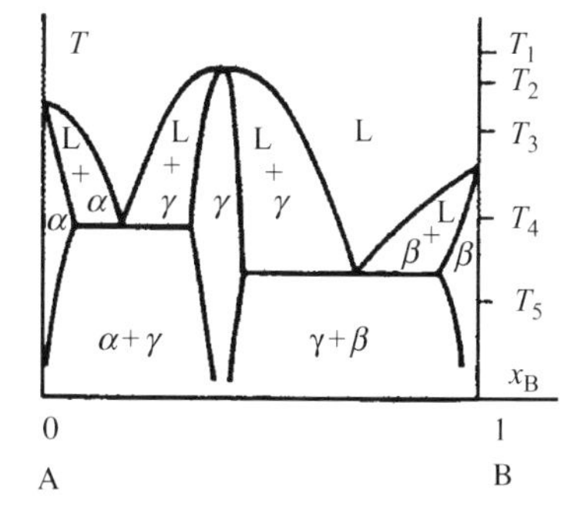

Figure 2.27f The resulting phase diagram. The system is eutectic.

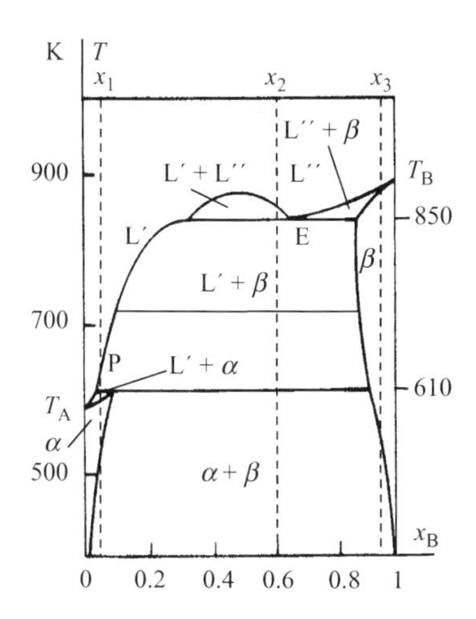

Example 2.3

The phase diagram, given in the margin, shows the equilibrium between phases of a binary system AB.

a) Describe the phase diagram briefly, i.e. the significance of the horizontal lines and the type of each phase. Name the phases.
b) How many phase reactions are there in the phase diagram? List the "reaction formulae" for them all.
c) Describe what happens to three molten alloys 1, 2 and 3 of the compositions x_1, x_2 and x_3 when they cool to room temperature. Equilibrium is reached in each case.

Solution and Answer:

a) The intersections with the lines $x_B = 0$ and $x_B = 1$ are the melting-points of the components A and B.

The horizontal lines represent transformations of some kind. The temperature remains constant until the transformation is completely finished.

The shape of the 'top hill' indicates that it is a miscibility gap. It is surrounded by liquid on both sides and the two-phase area must be a mixture of two liquids.

The conventional names of the phases have been introduced into the figure on the next page.

b) There are two reactions involved in the phase diagram:
A monotectic reaction of the eutectic type at point E:

$$L'' \rightarrow L' + \beta_{\text{solid}}$$

A peritectic reaction at point P:

$$\text{L}' + \beta_{\text{solid}} \rightarrow \alpha_{\text{solid}}$$

c) Alloy 1 remains as a homogeneous liquid until the temperature is about 620 K when a solid β phase starts to precipitate. The proportions between the β and L′ phases are determined by the lever rule at equilibrium conditions.

$$\text{Fraction of L}' = \frac{x^{\beta} - x_1}{x^{\beta} - x^{\text{L}'}} \qquad \text{Fraction of } \beta = \frac{x_1 - x^{\text{L}'}}{x^{\beta} - x^{\text{L}'}}$$

where the x^{β} and x^{L} are the intersections between the boundaries and a horizontal line, which defines the temperature, or simply the end points of the horizontal line.

At $T = 610$ K the alloy consists of a mechanical mixture of L′ and β. The β phase is transformed into α phase at constant temperature by a peritectic reaction $\text{L}' + \beta \rightarrow \alpha$. When the transformation is complete the alloy starts to cool again. It consists of liquid L′ and solid α. The proportions between L′ and α are determined by the lever rule. When all liquid has solidified the alloy consists of pure α phase with composition x_1.

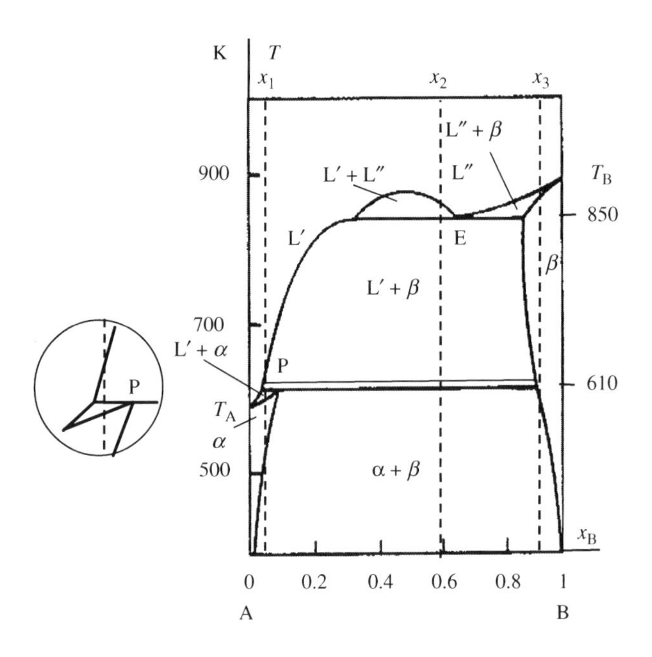

At about 550 K some of the α phase starts to be transformed into β phase. The alloy consists of a mechanical mixture of solid α and β phases, which cools to room temperature. The fractions of the α and β phases are determined by the lever rule at each temperature.

The average composition of the alloy is constant at all temperatures and in this case equal to x_1. The cooling processes can briefly be described by the schemes:

Alloy 1
L′ → mixture of $\text{L}' + \beta$ → peritectic reaction $\text{L}' + \beta \rightarrow \alpha$ at constant temperature (610 K) → mixture of $\text{L}' + \alpha$ → L′ gives α until all L′ is finished → pure α → mixture of $\alpha + \beta$.

Alloy 2

Analogous processes occur for the other alloys. The proportions of the two-phase areas are always determined by the lever rule.

The average composition of the alloy is constant at all temperatures and equal to x_2.

The cooling process can be described briefly by the scheme

L $\rightarrow$ mixture of $L' + L''$ $\rightarrow$ monotectic reaction $L'' \rightarrow L' + \beta$ at constant temperature (850 K) $\rightarrow$ mixture of $L' + \beta$ $\rightarrow$ peritectic reaction $L' + \beta \rightarrow \alpha$ at constant temperature (610 K) $\rightarrow$ mixture of $\alpha + \beta$.

Alloy 3

The average composition of the alloy is constant at all temperatures and equal to x_3.

The cooling process can be described briefly by the scheme

L solidifies and gives β (870 K) $\rightarrow$ mixture of $L + \beta$ $\rightarrow$ pure β $\rightarrow$ mixture of $\alpha + \beta$ (below ~590 K).

2.4.7 *Influence of Parameters on Phase Diagrams*

In the preceding section we found that the free energies of phases are closely related to the phase diagram of an alloy. The phase diagram plays an important role when quantitative calculations on different processes, are performed.

In the present section the influence of various parameters on the phase diagram, particularly on the so-called *partition coefficient* (see below), will be discussed.

Definition of Partition Coefficient

Consider a binary alloy at such a temperature that there are two phases, a melt and a solid, in equilibrium with each other. The two metal components A and B occur in both phases. As usual B is the solute component and A the solvent component. The composition of the alloy is expressed as the concentration of component B, e.g. the mole fraction of component B.

The *partition coefficient* or *distribution coefficient* is defined as the ratio of the concentrations of component B in the liquid and in the solid phase in equilibrium with each other.

$$k_B = \frac{x_B^{s/L}}{x_B^{L/s}} \tag{2.43}$$

where

k_B = partition coefficient of element B (*not* Boltzmann's constant here)

$x_B^{s/L}$ = mole fraction of element B in the solid phase close to the interface

$x_B^{L/s}$ = mole fraction of element B in the liquid phase close to the interface.

If the liquid and solid phases are homogeneous the definition (2.43) can be replaced by the simpler expression

$$k_B = \frac{x_B^s}{x_B^L} \tag{2.44}$$

The partition coefficient is an important quantity in solidification processes. In order to find a useful expression of the partition constant k_B we will study conditions for equilibrium between the two phases. They depend on both A and B.

Calculation of the Partition Coefficient and the Slope of the Liquidus Line in an Ideal Binary Solution

Equilibrium of Component A

At equilibrium, the chemical potentials of the two phases in a binary system are equal. Hence, we obtain for component A

$$\mu_A^L = \mu_A^s \tag{2.45}$$

To derive an expression of the partition coefficient k_B as a function of temperature we introduce the expressions in Equations (2.36) and (2.37) on page 57 for the chemical potentials into Equation (2.45). These expressions, with the excess quantities = 0, are valid for ideal solutions.

$$\mu_A^{0L} + RT\ln(1 - x_B^L) = \mu_A^{0s} + RT\ln(1 - x_B^s) \tag{2.46}$$

or

$$\mu_A^{0L} - \mu_A^{0s} = RT\ln\frac{1 - x_B^s}{1 - x_B^L} \tag{2.47}$$

The difference of the Gibbs' free energy of pure A, close to its melting point, can be written [Equation (2.5) on page 44 with the present nomenclature] as

$$-\Delta G_A^{L\to s} = -\left(\mu_A^{0L} - \mu_A^{0s}\right) = \frac{T_M^A - T}{T_M^A} H_m^{\text{fusion A}} \tag{2.48}$$

The melting-point temperature T_M^A of pure A is used as the standard state. Combination of Equations (2.47) and (2.48) gives

$$\frac{T_M^A - T}{T_M^A} H_m^{\text{fusion A}} = RT\ln\left(1 + \frac{x_B^L - x_B^s}{1 - x_B^L}\right) \tag{2.49}$$

or, after series expansion and simplification ($x_B^L \ll 1$),

$$\frac{T_M^A - T}{T_M^A} H_m^{\text{fusion A}} = RT\left(x_B^L - x_B^s\right) \tag{2.50}$$

After division with x_B^L Equation (2.50) can be solved for k_B

$$k_B = 1 - \frac{T_M^A - T}{x_B^L R T_M^A T} H_m^{\text{fusion A}} \tag{2.51}$$

where

T_M^A = melting-point temperature of pure element A
x_B^L = mole fraction of component B in the melt
$H_m^{\text{fusion A}}$ = molar heat of fusion of pure element A.

Equilibrium of Component B

Equation (2.51) contains two unknown quantities, x_B^L and T. Hence, it is necessary to find a second relationship between them in order to eliminate x_B^L and calculate k_B as a function of T.

This relationship is found by an analogous study of the equilibrium for component B.

$$\mu_B^L = \mu_B^s \tag{2.52}$$

The expressions for the chemical potentials [Equations (2.38) and (2.39) on page 57] are introduced into Equation (2.52)

$$\mu_B^{0L} + RT \ln x_B^L = \mu_B^{0s} + RT \ln x_B^s \tag{2.53}$$

and we obtain

$$\mu_B^{0L} - \mu_B^{0s} = RT \ln \frac{x_B^s}{x_B^L} \tag{2.54}$$

The driving force of solidification of pure B close to its melting point can be written as

$$-\Delta G_B^{L\rightarrow s} = \mu_B^{0s} - \mu_B^{0L} = -RT \ln k_B = \frac{T_M^B - T}{T_M^B} H_m^{\text{fusion B}} \tag{2.55}$$

We solve Equation (2.55) for k_B and obtain

$$k_B = e^{-\frac{1}{RT}\frac{T_M^B - T}{T_M^B} H_m^{\text{fusion B}}} \tag{2.56}$$

where

T_M^B = melting-point temperature of pure element B

$H_m^{\text{fusion B}}$ = molar heat of fusion of pure element B.

Equation (2.56) is the desired expression for the partition coefficient.

van't Hoff's Equation. Slope of the Liquidus Line

Equation (2.56) gives the partition coefficient as a function of temperature. Another method is to use the slope of the liquidus line in the phase diagram (Figure 2.28) to calculate k_B.

Equation (2.51) is solved for x_B^L and then derived with respect to T:

$$\frac{dx_B^L}{dT} = -\frac{H_m^{\text{fusion A}}}{RT^2(1 - k_B)} \tag{2.57}$$

This relationship is called van't Hoff's equation. The derivative equals the inverted slope m_L of the liquidus line and can be obtained from the phase diagram.

When m_L is known k_B can be calculated.

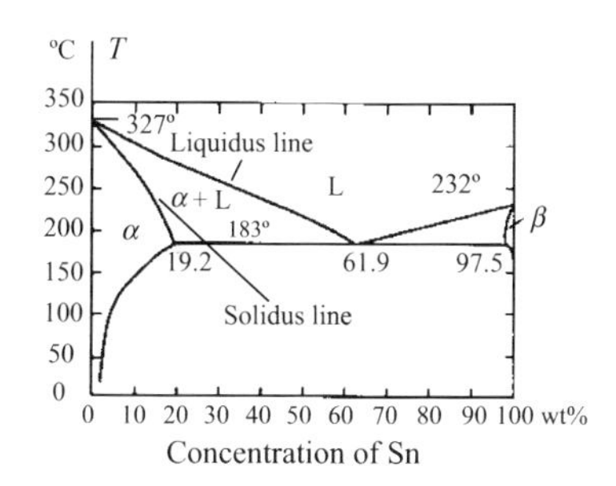

Figure 2.28 Phase diagram of the system PbSn. The liquidus and solidus lines are marked in the figure. From Binary Alloy Phase Diagrams Vol. 2, Ed. Thaddeus B. Massalski, American Society for Metals 1986. Reprinted with permission of ASM International. All rights reserved. www.asminternational.org.

Calculation of the Partition Coefficient as a Function of Temperature in a Non-Ideal Binary Solution

Consider a non-ideal binary solution and assume that the regular-solution model is valid. In this case the equilibrium Equations (2.46) on page 72 and (2.53) for an ideal solution have to be replaced by

$$\mu_A^{0L} + RT \ln(1 - x_B^L) + L^L (x_B^L)^2 = \mu_A^{0s} + RT \ln(1 - x_B^s) + L^s (x_B^s)^2 \tag{2.58}$$

$$\mu_B^{0L} + RT \ln x_B^L + L^L (1 - x_B^L)^2 = \mu_B^{0s} + RT \ln x_B^s + L^s (1 - x_B^s)^2 \tag{2.59}$$

The last terms in Equations (2.58) and (2.59) are associated with the regular-solution parameters for the liquid and the solid phases, originate from Equations (2.19) and (2.20) on page 47.

In a dilute solution ($x_B^L \ll 1$) Equation (2.58) can be approximated by Equation (2.46) on page 72 and Equation (2.59) on page 73 can be written as

$$\mu_B^{0L} + RT\ln x_B^L + L^L = \mu_B^{0s} + RT\ln x_B^s + L^s \tag{2.60}$$

Further calculations show that the expression in Equation (2.56) for k_B of an ideal solution will be replaced by Equation (2.61) for a non-ideal solution

$$k_B = e^{-\frac{1}{RT}\left[\frac{T_M^B - T}{T_M^B} H_m^{\text{fusion B}} + \left(L^s - L^L\right)\right]} \tag{2.61}$$

Influence of the Pressure on the Partition Coefficient

Equations (2.36)–(2.39) on page 57 are valid only if the pressure of the system is constant. If the pressure varies a term has to be added to each equation. If we assume that the excess free energy is zero, Equations (2.38) and (2.39) change into

$$\mu_B^L = \mu_B^{0L} + RT\ln x_B^L + \overline{V}_B^L \Delta p \tag{2.62}$$

$$\mu_B^s = \mu_B^{0s} + RT\ln x_B^s + \overline{V}_B^s \Delta p \tag{2.63}$$

where $\overline{V}_B^L$ and $\overline{V}_B^s$ are the partial molar volumes of B in the liquid and the solid. We also assume that $T \approx T_M^A$.

By combining Equations (2.62), (2.63), (2.44) on page 73 and (2.53) and series expansion of the logarithm functions we can derive the following expression for the partition coefficient as a function of the pressure difference

$$k'_B = k_B - \frac{\overline{V}_B^s - \overline{V}_B^L}{RT_M^A} \Delta p \tag{2.64}$$

where k'_B is the partition coefficient at a pressure change of Δp.

Influence of the Pressure on the Slope of the Liquidus Line

The slope m'_L of the liquidus line can be calculated with the aid of the inverse of van't Hoff's equation [Equation (2.57)]. We assume that $T \approx T_M^A$.

$$m'_L = -\frac{dT}{dx_B^L} = -\frac{R\left(T_M^A\right)^2\left(1 - k'_B\right)}{H_m^{\text{fusion A}}} \tag{2.65}$$

where k'_B, which is the partition coefficient at a pressure change of Δp, can be calculated from Equation (2.64).

Figure 2.29 shows the change of the liquidus and solidus lines, owing to the external pressure change, in a binary A–B system. The melting point change, owing to change of pressure, is calculated

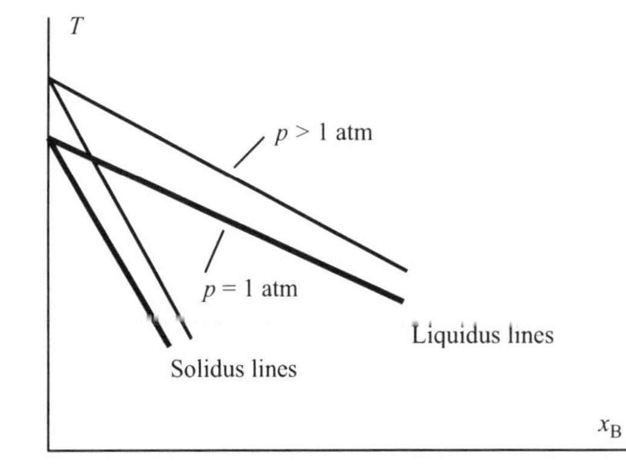

Figure 2.29 Solidus and liquidus lines in a binary system with and without an external pressure. The effect is exaggerated in the figure. Merton C. Flemings © McGraw Hill 1974.

from Equation (1.41) on page12 in Chapter 1, based on Clausius–Clapeyron's Equation (1.38) on page 11 in Chapter 1.

If the pressure difference Δp does not exceed 10^2 atm it is reasonable to assume that $k_B{}' = k_B$. If $k_B^{'} \approx k_B$ we obtain $m_L^{'} \approx m_L$.

Influence of the Phase-Boundary Curvature on the Partition Coefficient

If we take the curvature of the phase boundary into account, a new term has to be added to each of the Equations (2.36)–(2.39) on page 57. Equations (2.38) and (2.39) have to be replaced by

$$\mu_B^L = \mu_B^{0L} + RT \ln x_B^L \tag{2.66}$$

$$\mu_B^s = \mu_B^{0s} + RT \ln x_B^s + 2\overline{V}_B^s \sigma K \tag{2.67}$$

where
$\overline{V}_B^{s,L}$ = partial molar volumes of B in solid and the liquid, respectively
σ = surface tension
K = curvature (page 12 in Chapter 1).

At equilibrium, the chemical potentials μ_B^s and μ_B^L are equal. If we put the expressions in Equations (2.66) and (2.67) equal, use Equations (2.44) (page 73) and (2.53) and assume that $T \approx T_M^A$ we obtain the relationship

$$RT \ln k_B = RT \ln k_B^{''} + 2\overline{V}_B^s \sigma K \tag{2.68}$$

where $k_B^{''}$ and k_B are the partition coefficients with and without curvature K, respectively. By series expansion of the logarithm functions the expression for the partition coefficient can approximately be written as

$$k_B^{''} = k_B - \frac{2\overline{V}_B^s \sigma K}{RT_M^A} \tag{2.69}$$

The partition coefficient decreases with increasing curvature. Equation (2.69) is valid for all dilute solutions of B. k''_B does not deviate noticeably from k_B when the phase boundary curvature is lower than $10^{-8}\,m^{-1}$. A normal value of the curvature is around $10^{-6}\,m^{-1}$ and a reasonable approximation is $k''_B = k_B$.

Influence of the Phase-Boundary Curvature on the Melting Point of Solvent A

This topic has been treated for gases in Chapter 1 and is included more generally here for the sake of completeness. With the present nomenclature the resulting Equation (1.46) on page 13 in Chapter 1 can be written

$$T_M^A(K) = T_M^A - \frac{T_M^A\left(\overline{V}_A^L - \overline{V}_A^s\right)}{H_m^{\text{fusion A}}} 2\sigma K \tag{2.70}$$

where
$T_M^A(K)$ = melting-point temperature at curvature K
T_M^A = melting-point temperature at curvature 0
$\overline{V}_A^s$ = partial molar volume of A in the solid phase
$\overline{V}_A^L$ = partial molar volume of A in the liquid phase.

Influence of the Phase-Boundary Curvature on the Slope of the Liquidus Line

The decrease of the melting point of A as a consequence of the phase-boundary curvature causes a displacement of the solidus and liquidus lines of A. In analogy with Equation (2.65) we obtain the slope of the liquidus line

$$m''_{\mathrm{L}} = -\frac{\mathrm{d}T}{\mathrm{d}x_{\mathrm{B}}^{\mathrm{L}}} = -\frac{R\left(T_{\mathrm{M}}^{\mathrm{A}}\right)^2 (1 - k''_{\mathrm{B}})}{H_{\mathrm{m}}^{\mathrm{fusion\ A}}} \tag{2.71}$$

where m_{L}'' is the slope of the liquidus line at the curvature K. For normal curvatures (page 75) a good approximation is $k''_{\mathrm{B}} \approx k_{\mathrm{B}}$ and then we obtain $m''_{\mathrm{L}} \approx m_{\mathrm{L}}$.

Owing to the effect of the phase-boundary curvature, the liquidus and solidus lines will be transferred downwards compared to their normal positions. This is shown in Figure 2.30.

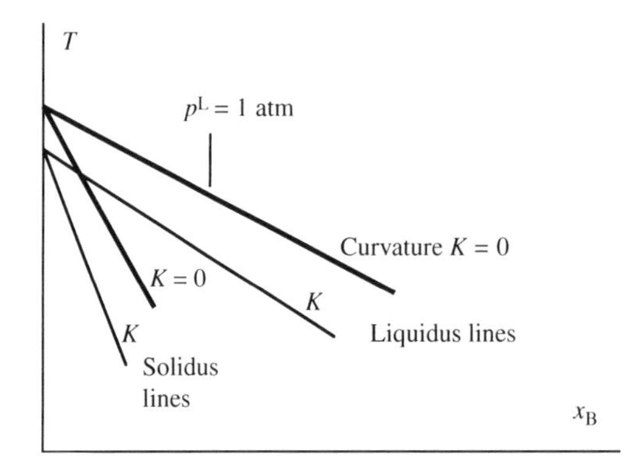

Figure 2.30 Influence of phase-boundary curvature on the positions of the liquidus and solidus lines in a phase diagram of a binary system. The effect is exaggerated. Merton C. Flemings © McGraw Hill 1974.

Influence of the Compressibility of a Gas on the Chemical Potential

In the derivations above we have assumed incompressibility of both the solid and the liquid phase in a binary system. This is not correct if we have a precipitated gas pore in a metal melt. Examples of such pores in metals are carbon monoxide in cast iron and hydrogen in various metals.

It is necessary to take into account that the gas phase is compressible. The chemical potential of B in the gas phase is

$$\mu_{\mathrm{B}}^{\mathrm{gas}} = \mu_{\mathrm{B}}^{0\ \mathrm{gas}} + RT \ln \frac{p_{\mathrm{pore}}}{p_0} \tag{2.72}$$

where

$p_0 =$ the standard pressure, normally 1 atm and

$$p_{\mathrm{pore}} = p_0 + 2\sigma K \tag{2.73}$$

At equilibrium, the chemical potentials of the melt and the gas pore must be equal.

$$\mu_{\mathrm{B}}^{0\mathrm{L}} + RT \ln x_{\mathrm{B}}^{\mathrm{L}} = \mu_{\mathrm{B}}^{0\ \mathrm{gas}} + RT \ln \frac{p_0 + 2\sigma K}{p_0} \tag{2.74}$$

2.5 Driving Force of Solidification in Binary Alloys

In Section 2.2.1 the driving force of solidification in a single-component system was derived (Equation (2.5) on page 44). The aim of this section is to derive an analogous expression for binary systems.

Graphical Construction of the Driving Force of Solidification

Free energy diagrams can be used to find out

- the composition of the solid that is formed at solidification of a liquid with given composition
- the magnitude of the driving force of solidification.

As an example, we examine the two cases shown in Figures 2.31 and 2.32.

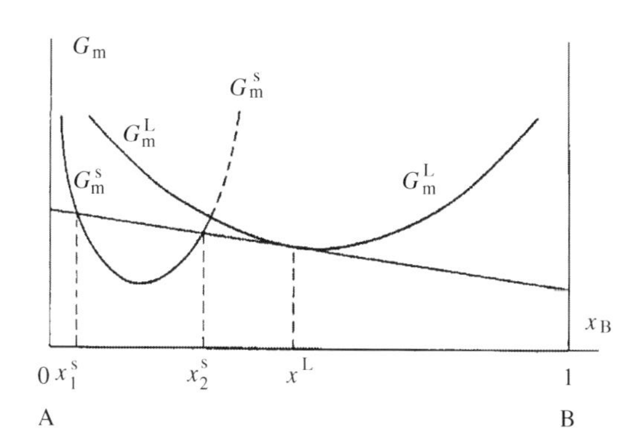

Figure 2.31 Partial molar free energy of components in an alloy as a function of composition of the alloy.

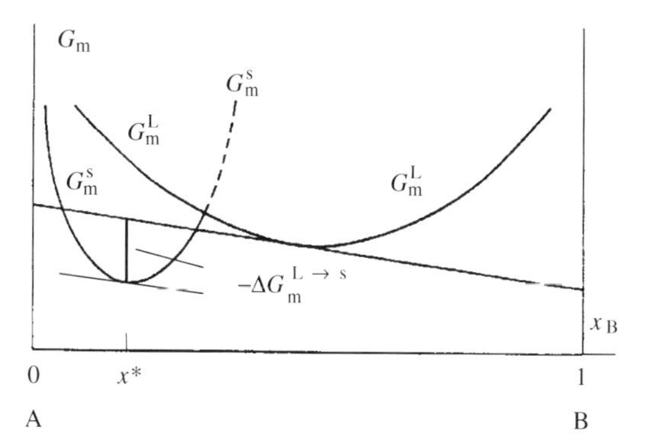

Figure 2.32 Construction in Figure 2.31 to find the driving force of solidification.

Consider a liquid of composition x^L. Its molar free energy curve is shown in Figure 2.31. A tangent is drawn to the free-energy curve in the point on the curve with composition x^L. All possible compositions of a tentative crystallized solid phase, in equilibrium with the liquid phase of composition x^L, falls along the tangent.

This line intersects the free energy curve of the solid at two points with compositions x_1^s and x_2^s.

For energy reasons the liquid can only form stable solid solutions within the composition interval x_1^s to x_2^s within which the free energy of the solid falls below the tangent.

The driving force $-\Delta G_m^{L\rightarrow s}$ of the solidification process is the difference in free energy between the line and the free-energy curve of the solid. The most likely solid to be formed is the one that has the largest driving force, i.e. gives the largest decrease in free energy.

The composition of this solid phase is found by drawing a tangent to the free energy curve of the solid, parallel to the primarily drawn tangent to the free energy curve of the liquid. This composition has been marked as x^* in Figure 2.32. However, the equilibrium value $x^{s/L}$ between the solid and liquid phases often replaces x^* for the sake of simplicity. It will be used below.

Analytical Expression of the Driving Force of Solidification

To find an analytical expression of the driving force of solidification in alloys, analogous to Equation (2.5) on page 44 for a single-component system, we will use the lower part of Figure 2.33.

The driving force $-\Delta G_m^{L\rightarrow s}$ for precipitation of an α phase with composition $x_B^{\alpha/L}$ from a liquid with composition x_0^L can be derived with the aid of the lower part of Figure 2.33.

The driving force is obtained as the difference PR – QR. The distances PR and QR are calculated using two top triangles having a common horizontal height equal to 1. Their vertical bases are functions of the chemical potentials of the elements involved.

With the aid of lower Figure 2.33 we obtain

$$-\Delta G^{L\rightarrow s} = PR - QR = \left(\mu_A^L - \mu_A^{L/\alpha}\right)\frac{1 - x_B^{\alpha/L}}{1} - \left(\mu_B^{L/\alpha} - \mu_B^L\right)\frac{x_B^{\alpha/L}}{1} \qquad (2.75)$$

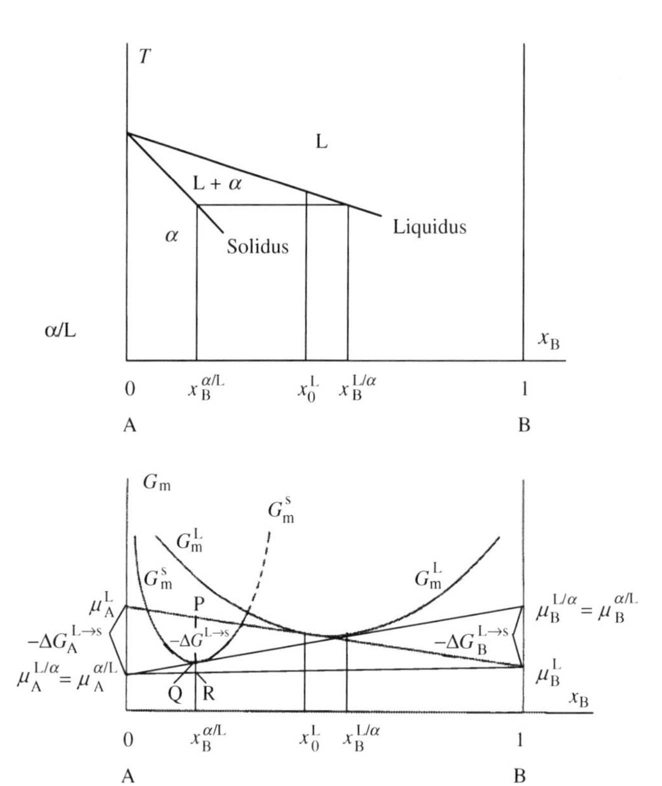

Figure 2.33 Phase diagram and the corresponding free energy diagram of a binary alloy AB.

which gives the desired expression of the driving force of binary alloys:

$$-\Delta G^{L\to s} = \left(1 - x_B^{\alpha/L}\right)\left(\mu_A^L - \mu_A^{L/\alpha}\right) + x_B^{\alpha/L}\left(\mu_B^L - \mu_B^{L/\alpha}\right) \tag{2.76}$$

where

$-\Delta G_m^{L\to s}$ = driving force of solidification of the alloy

$x_B^{\alpha/L}$ = composition of the solid precipitated phase

$\mu_i^{L/\alpha}$ = chemical potential of solid element i (i = A, B) in equilibrium with liquid (i = A, B)

μ_i^L = chemical potential of element i in the original liquid (i = A, B).

μ_A^L and μ_B^L are functions of the composition of the melt x_0^L.

The chemical potentials are functions of temperature and composition of the alloy or, rather, of the activity a_i of the alloying element.

$$\mu_i = \mu_i^0 + RT\ \ln a_i \approx \mu_i^0 + RT\ \ln x_i + \overline{G}_i^{Ex} \tag{2.77}$$

where

μ_i = chemical potential of element i

μ_i^0 = chemical potential of the standard state of element i (i = A, B)

a_i = activity of element i (i = A, B)

$\overline{G}_i^{Ex}$ = molar partial excess Gibbs' free energy of element i.

2.6 Thermodynamics of Ternary Alloys

The general thermodynamics of multiple-component systems has been discussed in Section 1.4 in Chapter 1. In Sections 1.5, 1.6 and 1.7, all in Chapter 1, and in Section 2.3, 2.4 and 2.5 it has been applied on binary alloys.

A similar extensive analysis of multiple component alloys is far beyond the scope of this book. We will only discuss ternary alloys briefly and restrict the theoretical treatment to some outlines of their thermodynamics. The elementary introduction to graphical representation of Gibbs' free energy and phase diagrams of ternary alloys that is given, is based on a comparison with binary alloys.

Thermodynamics

In a ternary alloy the number of components is three. Two independent variables are necessary to describe the composition of a ternary alloy. Normally, the mole fractions x_A, x_B and x_C of the components A, B and C are used together with the relationship

$$x_A + x_B + x_C = 1 \tag{2.78}$$

By use of Gibbs' phase rule (Equation (2.22) on page 52)

$$N_{ph} + N_{free} = N_c + 2 \tag{2.22}$$

we can calculate the number of independent variables, as the number of phases is three.

$$N_{free} = 3 + 2 - N_{ph}$$

If the number of phases is two (a solid and a liquid phase) the number of independent variables becomes three, temperature and two molar fractions.

When the number of independent variables is zero, the maximum number of phases is five. In reality, the number of phases does not exceed four as experience shows that the pressure has little influence on thermodynamic quantities and can be disregarded as an independent variable.

Gibbs' Free Energy

Analytical expressions for the Gibbs' free energy and other thermodynamic quantities can be derived for ternary alloys, analogous to those of binary alloys.

Molar thermodynamic functions, partial molar and relative molar quantities can be set up for ternary alloys. A simple example is the molar Gibbs' free energy for a non-ideal ternary solution

$$G_m = x_A^0 G_A^0 + x_B^0 G_B^0 + x_C^0 G_C^0 + RT\,(x_A \ln x_A + x_B \ln x_B + x_C \ln x_C) + G_m^{Ex}(x_A,\, x_B,\, x_C) \tag{2.79}$$

Excess Quantities

$G_m^{Ex} = 0$ in the case of ideal solid and liquid solutions. The most difficult problem of non-ideal ternary solutions is to find an adequate model of G_m^{Ex}. The simplest model is the regular solution model, which is an extension of that of binary alloys (page 46).

$$G_m^{Ex} = H_m^{mix} = L_{AB}\, x_A\, x_B + L_{AC}\, x_A\, x_C + L_{BC}\, x_B\, x_C \tag{2.80}$$

where $L_{ij}x_ix_j$ represents the binding energy between the i and j atoms. Alternatively G_m^{Ex} can be expressed in terms of activity coefficients (compare Equation (1.127) for binary alloys on page 32 in Chapter 1)

$$G_m^{Ex} = RT(x_A \ln \gamma_A + x_B \ln \gamma_B + x_C \ln \gamma_C) \tag{2.81}$$

A MacLaurin expansion is often used to describe the interaction between the different alloying elements. In this case, we obtain for example

$$\ln \gamma_B = \ln \gamma_B^0 + x_B \varepsilon_B^B + x_C \varepsilon_B^C \tag{2.82}$$

where

γ_B^0 = activity coefficient of B in a dilute solution

ε_B^i = interaction coefficient between components B and i (B and C).

If the molar excess Gibbs' free energy function is known, the partial molar excess Gibbs' free energy functions can be derived from the general relationship (Chapter 1)

$$\overline{G}_i^{Ex} = G_m^{Ex} + (1 - x_i)\frac{\partial G_m^{Ex}}{\partial x_i}. \tag{2.83}$$

where i = A, B, C.

Graphical Representation of Ternary Systems

Gibbs' Concentration Triangle

The composition of a ternary alloy with the components A, B and C is best described by an equilateral triangle ABC (Figure 2.34). Its properties are listed below.

- The point P inside the triangle determines the composition of the ternary alloy. The distances PP_A, PP_B, respectively, PP_C are proportional to the mole fractions x_A, x_B and x_C, respectively, which are marked on the triangle edges, identical with the mole fraction axes.
- The lines parallel to the side BC represent compositions at constant distances from BC, i.e. they represent constant x_A values. The other sets of parallel lines represent constant x_B and x_C values, respectively.
- the triangle has the property that the sum $PP_A + PP_B + PP_C$ is equal to the height of the triangle.

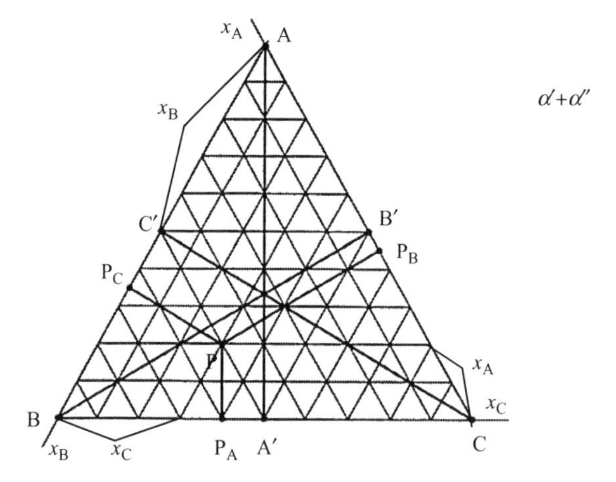

Figure 2.34 Gibbs' concentration triangle. The point P describes the composition of a ternary alloy. The edge of the equilateral triangle is chosen as 1 unit length. Reproduced from Elsevier © 1983.

As the triangle side is chosen as 1 unit length the mole fractions x_A, x_B and x_C can directly be read on the mole fraction axes, as the triangle edges are divided into ten equal parts. The mole fractions that corresponds to point P in Figure 2.34 are $x_A = 0.2$, $x_B = 0.5$ and $x_C = 0.3$.

The sum of the mole fractions equals 1 and is equal to the length of the triangle edge. This is why the triangle is called Gibbs' concentration triangle.

The corner points represent the pure elements A, B and C, respectively. The triangle edges correspond to binary alloys of all compositions depending on the position of points P_A, P_B and P_C, respectively, along the edges. The edge AB corresponds to the binary alloy AB as $PP_C = 0$ and $x_C = 0$. $PP_B = 0$ ($x_B = 0$) and $PP_A = 0$ ($x_A = 0$) correspond to the binary alloys AC and BC, respectively.

Graphical Representation of Gibbs' Free Energy

The Gibbs' free energy G_m is a symmetric function of x_A, x_B and x_C. The function in Equation (2.79) on page 79 is plotted in a three-dimensional diagram with the concentration triangle as a horizontal bottom and a vertical G_m axis. The coordinate system has the shape of a prism.

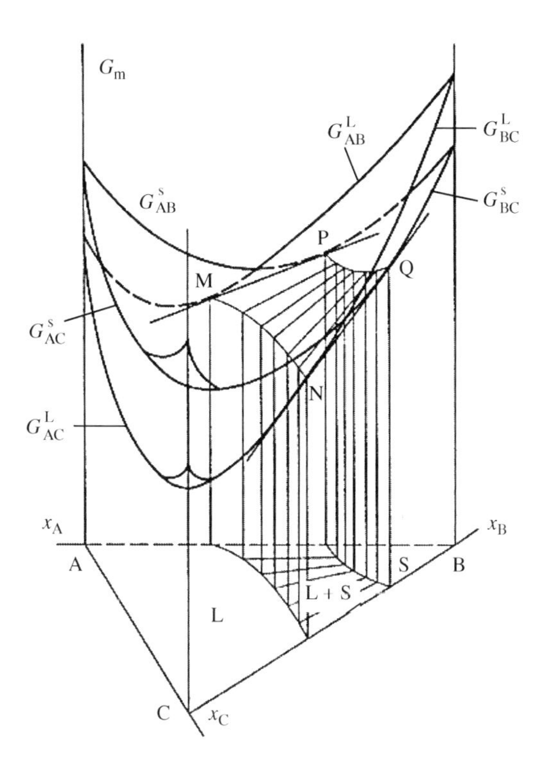

Figure 2.35 Gibbs' free energy surfaces for the solid and liquid phases of an ideal ternary alloy. Reproduced from Elsevier © 1983.

Figure 2.35 shows a simple ternary G_m diagram where both the solid and liquid solutions are ideal. It is hard to represent a three-dimensional diagram on a flat paper page but in this case the comparison with the G_m curves of binary alloys facilitates the understanding.

Figure 2.35 shows two asymmetric bowl-shaped surfaces, which correspond to the Gibbs' free energy curves of the solid and liquid phases of a binary alloy. When the two Gibbs' free energy surfaces and the side planes intersect we obtain the free energy curves of the binary alloys AB, AC and BC, respectively. Figure 2.35 shows these six curves.

The solid and liquid surfaces move upwards when the temperature decreases, just like the curves of a binary alloy. The liquid surface moves faster than the solid curve. The diagram is drawn for a given temperature *T*.

The diagram can be used for construction of the phase diagram of the ternary alloy in analogy with the binary case. Tangent planes are drawn between the two bowl-shaped surfaces and the coordinates of the tangent points are projected on the concentration triangle at the bottom. In the three-dimensional case we do not obtain only one pair of tangent points but two tangent point lines MN and PQ.

The vertical tangent point surfaces, which are marked in the figure, define the various phases. The prism space is divided in a liquid region L, a two-phase region L + S and a solid region S.

Graphical Representation of Phase Diagrams

For a ternary system a three-dimensional $x_B x_C T$ phase diagram replaces the two-dimensional $x_B T$ phase diagram of a binary system. In ternary phase diagrams equilibrium lines correspond to equilibrium points in binary phase diagrams. Similarly, boundary surfaces in ternary diagrams replace boundary lines in binary diagrams. A ternary phase diagram of an ideal solution contains an upper convex liquidus surface and a lower concave solidus surface.

Figure 2.36 shows the phase diagram of the ternary alloy in Figure 2.35. The intersections between a ternary phase diagram and the planes $x_C = 0$ or $x_B = 0$ are identical with the phase diagrams of the binary AB and AC systems, respectively. The three ideal phase diagrams of the typical 'lens' type are shown in Figure 2.36.

The construction of a ternary phase diagram is done by the tangent plane to tangent plane method. The tangent point lines constructed in Figure 2.35 are transferred to Figure 2.36 and define the phase regions in Figure 2.36. They are the same as those in Figure 2.35.

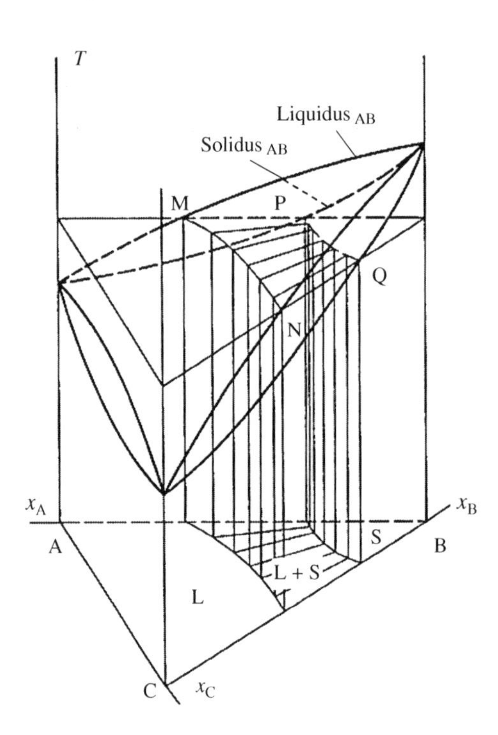

Figure 2.36 Phase diagram of a ternary alloy. Reproduced from Elsevier © 1983.

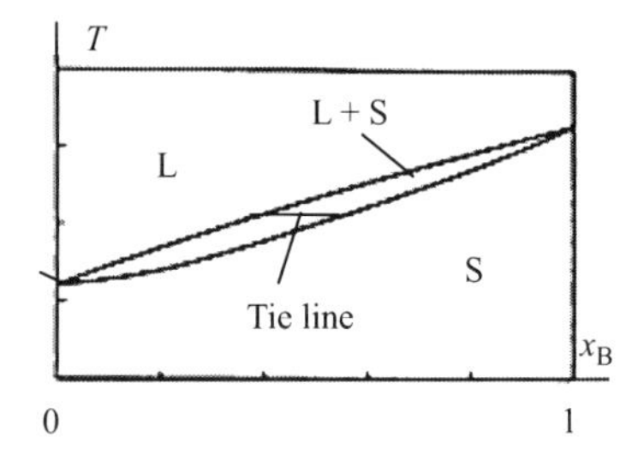

Figure 2.37 Phase diagram of an ideal solution binary alloy.

Figure 2.37 shows part of a binary phase diagram of an ideal solution alloy. A line that connects the two tangent points such as that in upper Figure 2.33 for a binary alloy is called a *tie line*. Such tie lines are drawn in the three-dimensional case, too. A whole set of tie lines, which tie the pairs of tangent points together, are shown both in Figures 2.35 and 2.36.

The tie lines in ternary phase diagrams are all horizontal, which corresponds to $T = \text{constant}$, and end in coordinate pairs of the type (x_B^s, x_C^s) and (x_B^L, x_C^L). All the former ones lie on the solidus surface and the latter ones on the liquidus surface.

We will save the reader from more complicated examples of ternary phase diagrams, for example a three-dimensional diagram with intermediate phases (compare for example Figure 2.27f on page 69). Instead, we will once more demonstrate the close relationships of binary and ternary systems. Figure 2.38 shows a miscibility gap in a ternary system.

Figure 2.38 shows that the miscibility gap is a three-dimensional version of the miscibility gap we discussed for binary alloys. For each value of temperature T there is a critical point. The joined critical points form the critical line.

The two-phase region is found inside the miscibility surface, indicated by the black lines in Figure 2.38. It is surrounded by a single-phase region. Miscibility gaps occur both in solid and liquid phases. The spinodal surface lies inside the miscibility surface.

The intersection of the ternary miscibility surface and the vertical plane AB results in the miscibility curve of the binary alloy AB.

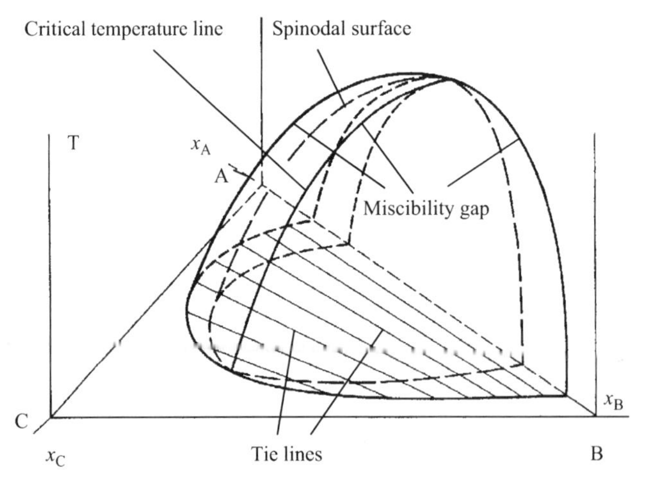

Figure 2.38 Miscibility gap of a ternary system. Reproduced from Elsevier © 1983.

2.7 Thermodynamics of Vacancies in Pure Metals and Alloys

Vacancies, dislocations and other stacking faults are common structural defects in materials. Those types of defects are often formed during the solidification processes. The number of vacancies influences a number of properties of the solid, such as melting point, diffusion rate and creep properties, as well as the formation of pores. Vacancies always exist in the structure, which can be brought to equilibrium by heat treatment.

2.7.1 Vacancies in Pure Metals

So far, thermodynamics has only been applied on alloying elements in alloys. It can also be applied on vacancy formation in metals and alloys during solidification.

This was done in 1997 by Emi and Fredriksson [2], who studied the effects of vacancies and alloying order in metals and binary alloys during cooling and solidification. It would require too much space to include their many thermodynamic equations here. For this reason we restrict the presentation below to a brief discussion of their results.

Vacancies at Random Distribution in Pure Metals

Many experimental investigations have shown that a large fraction of vacancies are created and frozen in during solidification. The concentration of frozen vacancies exceeds the equilibrium concentration at the temperature of the cooled metal or alloy.

The fraction of frozen vacancies varies with the cooling rate of the specimen. At *high* cooling rates a totally disordered state is formed with a large supersaturation of vacancies. At *medium* cooling rates some of the frozen vacancies form divacancies. At *low* cooling rates the vacancies form larger clusters, dislocations or move to grain boundaries.

Vacancy Concentration

In this section we will use a slightly different definition of the vacancy concentration. The concept mole fraction in binary system A + B is related to the share of the total number of atoms A + B. In this case B represents vacancies. The mole fraction of vacancies in a metal A can be written

$$x_{vac} = \frac{N_{vac}}{N_A + N_{vac}} \tag{2.84}$$

where

x_{vac} = mole fraction of vacancies
N_A = number of metal atoms A (not Avogadro's number here)
N_{vac} = number of vacancies
$N_A + N_{vac}$ = number of lattice sites.

Instead of x_{vac} we will use y_{vac}, which is defined by the relationship

$$y_{vac} = \frac{N_{vac}}{N_A} \tag{2.85}$$

or

$$y_{vac} = \frac{\dfrac{N_{vac}}{N_{vac} + N_A}}{\dfrac{N_A}{N_{vac} + N_A}} = \frac{x_{vac}}{x_A} \tag{2.86}$$

where we will call y_{vac} the *relative mole fraction of vacancies*. When the fraction of vacancies is small the difference between x_{vac} and y_{vac} is negligible.

x_{vac} is related to the *total number of lattice sites*, while y_{vac} refers to the *number of atoms A* of the specimen. y_{vac} is strictly proportional to the number of vacancies. Thermodynamical calculations are often based on 1 kmol of A atoms. In this case N_A equals Avogadro's number = 6.02×10^{26} atom/kmol.

2.7.2 *Calculation of the Relative Mole Fraction of Vacancies in Pure Metals*

Vacancies can be regarded as a special type of alloying element in a metal and the theory of binary alloys, which has been discussed earlier in this chapter, can therefore be applied on vacancies. The binary system consists of A atoms and vacancies. The number of vacancies present at equilibrium in a metal A can be calculated in the following way:

Consider a crystal that contains N_A A atoms (1 kmol) and N_{vac} vacancies. We want to set up an expression for Gibbs' free energy of the crystal expressed in terms of the relative mole fraction y_{vac}. This is done in the box below and the result is

$$G = G_A^0 + y_{vac}G_{vac} + RT\,[y_{vac}\ln y_{vac} - (1 + y_{vac})\ln(1 + y_{vac})] + G^{Ex} \tag{2.87}$$

Calculation of the Gibbs' Free Energy of a Crystal as a Function of the Relative Mole Fraction of Vacancies

Consider a crystal, which contains N_A A atoms (1 kmol) and N_{vac} vacancies ($N_{vac} \ll N_A$) where N_A = Avogadro's number = 6.02×10^{26} atoms/kmol.

The Gibbs' free energy of the system can be written as

$$G = x_A G_A^0 + x_{vac}G_{vac} - TS + G_m^{Ex} \tag{1'}$$

We apply the definitions (2.85) and (2.86) on page 83

$$y_A = \frac{N_A}{N_A} = 1 \quad \text{and} \quad y_{vac} = \frac{N_{vac}}{N_A} \tag{2'}$$

x_A and x_{vac} can be expressed in terms of y_{vac} by means of the relationships (2′):

$$x_A = \frac{N_A}{N_A + N_{vac}} = \frac{1}{1 + y_{vac}} \quad \text{and} \quad x_{vac} = \frac{N_{vac}}{N_A + N_{vac}} = \frac{y_{vac}}{1 + y_{vac}} \tag{3'}$$

With the aid of the statistical definition of entropy ($S = k_B \ln P$) and Stirling's formula ($\ln x! \approx x\ln x - x$) the entropy term in Equation (1′) can be written as $-TS = -k_B T\ln P$ or

$$\begin{aligned} -TS &= -k_B T\ln\frac{(N_A + N_{vac})!}{N_A!\,N_{vac}!} \approx k_B T\,[N_A \ln N_A! + N_{vac}\ln N_{vac}! - \ln(N_A + N_{vac})!] \\ &= -k_B T[N_A \ln N_A + N_{vac}\ln N_{vac} - (N_A + N_{vac})\ln(N_A + N_{vac})] \end{aligned} \tag{4'}$$

as the first-order terms in Sterling's formula cancel. Equation (4′) can be transformed into

$$\begin{aligned} -TS &= k_B T\,N_A\left[\frac{N_A}{N_A}\ln\frac{N_A}{N_A + N_{vac}} + \frac{N_{vac}}{N_A}\ln\frac{N_{vac}}{N_A + N_{vac}}\right] \\ &= RT\left[\ln\frac{1}{1 + y_{vac}} + y_{vac}\ln\frac{y_{vac}}{1 + y_{vac}}\right] = RT[y_{vac}\ln y_{vac} - (1 + y_{vac})\ln(1 + y_{vac})] \end{aligned} \tag{5'}$$

The relationships (2′), (3′) and (5′) are introduced into Equation (1′):

$$G = \frac{G_A^0}{1 + y_{vac}} + \frac{y_{vac}G_{vac}}{1 + y_{vac}} + RT[y_{vac}\ln y_{vac} - (1 + y_{vac})\ln(1 + y_{vac})] + G_m^{Ex}$$

As $y_{vac} \ll 1$ we can replace $(1 + y_{vac})$ in the denominators by 1. This approximation gives

$$G = G_A^0 + y_{vac}G_{vac} + RT[y_{vac}\ln y_{vac} - (1 + y_{vac})\ln(1 + y_{vac})] + G^{Ex} \tag{6'}$$

where
G = total Gibbs' free energy of the crystal
G_A^0 = Gibbs' free energy of 1 kmol of A in its standard state
G_{vac} = Gibbs' free energy of formation of 1 kmol of vacancies
G^{Ex} = excess Gibbs' free energy of the crystal
y_A = relative mole fraction of A atoms in the crystal
y_{vac} = relative mole fraction of vacancies in the crystal
k_B = Boltzmann's constant
T = temperature.

The relative equilibrium mole fraction of vacancies can be calculated from the condition that G has a minimum at equilibrium, i.e. the derivative of G with respect to y_{vac} must be zero.

$$\frac{\partial G}{\partial y_{vac}} = G_{vac} + \frac{\partial G^{Ex}}{\partial y_{vac}} + RT\left[y_{vac}\frac{1}{y_{vac}} + \ln y_{vac}\right] - RT\left[(1+y_{vac})\frac{1}{1+y_{vac}} + \ln(1+y_{vac})\right] = 0 \tag{2.88}$$

The relative mole fraction of vacancies y_{vac} is normally very small. It is reasonable to neglect the term $\ln(1+y_{vac}) \approx \ln 1 = 0$ in Equation (2.88). After approximation and reduction of Equation (2.88) we obtain

$$\frac{\partial G}{\partial y_{vac}} = G_{vac} + \frac{\partial G^{Ex}}{\partial y_{vac}} + RT \ln y_{vac} = 0 \tag{2.89}$$

Special Case

Equation (2.89) can be further developed if we apply the *regular-solution model* to obtain an explicit expression of G^{Ex} and assume that $x_A \approx 1$ and $x_{vac} \approx y_{vac}$.

$$G^{Ex} = H^{Ex} - TS^{Ex} = L_{A\ vac} x_A\ x_{vac} - 0 \approx L_{A\ vac}\ y_{vac} \tag{2.90}$$

The derivative of this expression with respect to y_{vac} is introduced into Equation (2.89)

$$\frac{\partial G}{\partial y_{vac}} = G_{vac} + L_{A\ vac} + RT \ln y_{vac}^{eq} = 0$$

which gives the relative equilibrium fraction of vacancies

$$y_{vac}^{eq} = e^{-\frac{G_{vac}+L_{A\ vac}}{RT}} \tag{2.91}$$

where the superscript 'eq' stands for equilibrium.

The interaction coefficient $L_{A\ vac}$ is normally included in G_{vac}. By use of the general relationship

$$G_{vac} = H_{vac} - TS_{vac}$$

we obtain

$$y_{vac}^{eq} = e^{\frac{S_{vac}}{R}}\ e^{-\frac{H_{vac}}{RT}} \tag{2.92}$$

where
y_{vac}^{eq} = relative equilibrium fraction of vacancies at temperature T
S_{vac} = molar entropy of formation of vacancies
H_{vac} = molar enthalpy of formation of vacancies.

Formation and Absorption of Excess Vacancies

At most solidification processes a lot of vacancies are formed in the structure. Hence, the number of vacancies will be larger than the equilibrium value given above, owing to kinetic effects.

The Gibbs' free energy of a solid is increased when the number of vacancies exceeds the equilibrium number. This results in a decrease of the melting-point.

The equilibrium concentration of vacancies in the matrix can be estimated by use of Equation (2.91) or Equation (2.92). When the alloy is cooled rapidly from a high temperature there will be no time for the new equilibrium concentration of vacancies to be established and the high vacancy concentration will initially be maintained in the solid.

The vacancies have a tendency to attract each other and form clusters. Some clusters collapse into dislocation loops, which can grow by absorbing more vacancies. Other consumers of excess vacancies are grain boundaries and other interfaces within the specimen. Because vacancies have high diffusivities (diffusion coefficients) it is difficult to avoid losing vacancies in the vicinity of the grain boundaries and interfaces.

2.7.3 Decrease of Melting Point and Heat of Fusion as Functions of the Relative Mole Fraction of Vacancies in Pure Metals

Decrease of Melting Point owing to Vacancies at Random Distribution in Metals

Increase of the number of lattice faults above the equilibrium value by rapid solidification causes the free energy of the solid phase to increase. As a result of this the melting point of the alloy decreases.

The increasing number of lattice faults and the decrease of the melting point depend on the cooling rate and the undercooling during the solidification process. This dependence can be calculated by using the thermodynamic relationship, which states that the molar Gibbs' free energies of the liquid and the solid phases are equal at equilibrium.

$$G_m^L = G_m^s \tag{2.93}$$

Small Relative Mole Fraction of Vacancies

The molar Gibbs' free energy of the solid is a function of the mole fraction of vacancies. For small values of the relative mole fraction of vacancies the molar Gibbs' free energy of the solid can be written

$$G_m^s = G_m^{s\,eq} + y_{vac}RT\ln\frac{y_{vac}}{y_{vac}^{eq}} \tag{2.94}$$

where

y_{vac} = relative mole fraction of vacancies

y_{vac}^{eq} = relative equilibrium mole fraction of vacancies

G_m^s = molar Gibbs' free energy at the relative mole fraction of vacancies y_{vac}

$G_m^{s\,eq}$ = equilibrium molar Gibbs' free energy at the relative equilibrium mole fraction of vacancies y_{vac}^{eq}.

To find the decrease of the melting-point temperature as a function of the relative mole fraction of vacancies we use the equilibrium condition between a solid with a supersaturation of vacancies and a liquid $G_m^L = G_m^s$. With the aid of Equation (2.94) we obtain

$$G_m^L - G_m^{s\,eq} = y_{vac}RT\ln\frac{y_{vac}}{y_{vac}^{eq}} \tag{2.95}$$

The difference $G_m^L - G_m^{s\,eq}$ is described by Equation (2.5) on page 46, i.e. by the expression for the driving force of solidification

$$-\Delta G_{m}^{L \to s} = G_{m}^{L} - G_{m}^{s\ eq} = \frac{(T_{M}^{eq} - T)}{T_{M}^{eq}} H_{m}^{fusion} \tag{2.96}$$

where
H_{m}^{fusion} = molar heat of fusion
T_{M}^{eq} = melting-point temperature at relative equilibrium mole fraction of vacancies y_{vac}^{eq}
T_{M} = melting-point temperature at the relative mole fraction of vacancies y_{vac}.

The driving force is zero at $T = T_{M}^{eq}$, which is the equilibrium melting-point temperature. At this temperature the relative equilibrium mole fraction of vacancies is y_{vac}^{eq}. With the aid of Equation (2.95) Equation (2.96) can be written for $T = T_{M}$:

$$\frac{(T_{M}^{eq} - T_{M})}{T_{M}^{eq}} H_{m}^{fusion} \approx y_{vac} RT_{M} \ln \frac{y_{vac}}{y_{vac}^{eq}} \tag{2.97}$$

As $T_{M}^{eq} \approx T_{M}$ Equation (2.97) can be written as

$$T_{M}^{eq} - T_{M} = \frac{R(T_{M}^{eq})^{2}}{H_{m}^{fusion}} y_{vac} \ln \frac{y_{vac}}{y_{vac}^{eq}} \tag{2.98}$$

Equation (2.98) gives the melting-point decrease as function of the relative mole fraction of vacancies. It is only valid for *small* values of y_{vac}, which normally is the case.

Large Relative Mole Fraction of Vacancies

Emi and Fredriksson [2] have derived an expression of the decrease of melting point or the undercooling, which is valid also for *large* values of the relative mole fraction of vacancies.

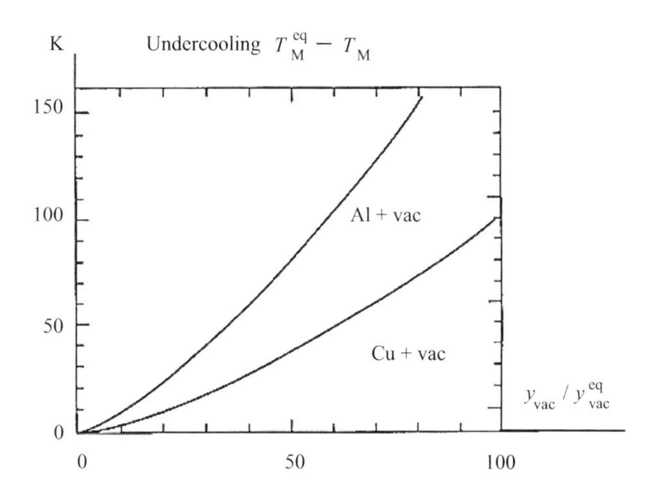

Figure 2.39 Decrease of melting point as a function of the ratio of relative mole fractions of vacancies for Al and Cu. H. Fredriksson and T. Emi, in Materials Transactions © The Japan Institute of Metals 1998.

The formula is neither given nor derived here. However, as an illustration ΔT is calculated for pure Al and pure Cu as functions of the ratio of the relative mole fraction of vacancies. The result is shown in Figure 2.39.

$$\Delta T = T_{M}^{eq} - T_{M} \tag{2.99}$$

Decrease of Heat of Fusion owing to Vacancies at Random Distribution in Pure Metals

The molar heat of fusion of a pure metal as a function of fraction of vacancies can be obtained from the relationship

$$H_m^{fusion} = H_{tab}^{fusion} - \left(y_{vac} - y_{vac}^{eq}\right) H_{vac}^{formation} \tag{2.100}$$

where

H_m^{fusion} = molar heat of fusion owing to excess vacancies
H_{tab}^{fusion} = tabulated value of the molar heat of fusion of the pure metal containing the relative fraction y_{vac}^{eq} of vacancies
$H_{vac}^{formation}$ = molar heat of formation of vacancies in the pure metal
y_{vac} = true relative fraction of frozen vacancies in the pure metal
y_{vac}^{eq} = equilibrium relative fraction of vacancies in the pure metal at the melting point.

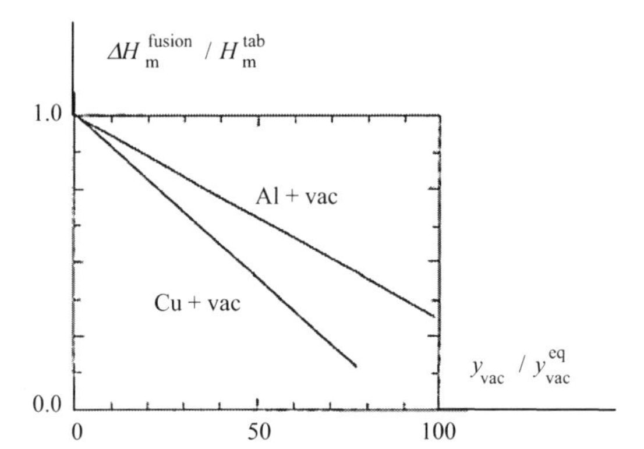

Figure 2.40 Relative heat of fusion as a function of the ratio of the relative fractions of vacancies for Al and Cu. H. Fredriksson and T. Emi, in Materials Transactions © The Japan Institute of Metals 1998.

If the ratio of the heat of fusion to the tabulated heat of fusion is plotted versus the relative vacancy concentrations for Al and Cu the curves in Figure 2.40 are obtained.

Decrease of Melting Point and Heat of Fusion owing to Vacancies and Divacancies at Random Distribution in Metals

The heat of formation of vacancies in a metal is of the order of 10^5 J/kmol, which is a very high value. The formation energy can also be considered as a heat of mixing. A large value of the heat of mixing is often found in alloys where the difference in atomic volume between the two mixed species is very large. The value of the excess entropy is also very large, which indicates that the atomic structure is partly ordered.

These statements agree well with the fact that a fraction of the monovacancies spontaneously and easily forms divacancies. The reaction can be written as

$$\text{vac} + \text{vac} \rightarrow \text{divac} \tag{2.101}$$

Emi and Fredriksson calculated the Gibbs' free energy G_{divac} of formation of 1 kmol of divacancies. They set up an expression G_m for the Gibbs' free energy of a system, which consists of 1 kmol of the metal, the fraction y_{vac} monovacancies and the fraction y_{divac} divacancies. The equilibrium value of y_{divac} was found from the condition $dG_m = 0$.

Two vacancies are required to form one divacancy. It should be noticed that the total number of vacancies is *reduced* when a number of vacancies form divacancies. For this reason the undercooling and the reduction of the heat of fusion probably are likely to be *smaller* for mono- and divacancies than for monovacancies alone. Fredriksson and Emi performed detailed calculations for Al. The two revised curves are shown in Figures 2.41 and 2.42.

A comparison between the curves with and without divacancies shows that the formation of divacancies strongly influences the decrease of melting point and the heat of fusion of the metal. As expected, both effects are considerably reduced by the presence of divacances, especially at high monovacancy concentrations.

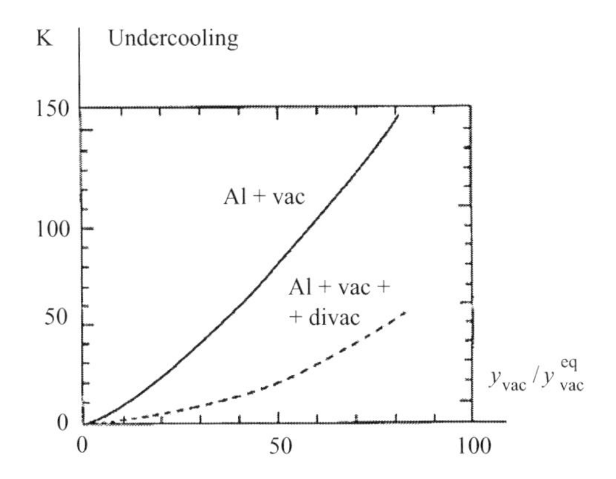

Figure 2.41 Decrease of the melting-point as a function of the ratio of relative fractions of vacancies for pure Al with monovacancies and with mono- and divacancies, respectively. H. Fredriksson and T. Emi, in Materials Transactions © The Japan Institute of Metals 1998.

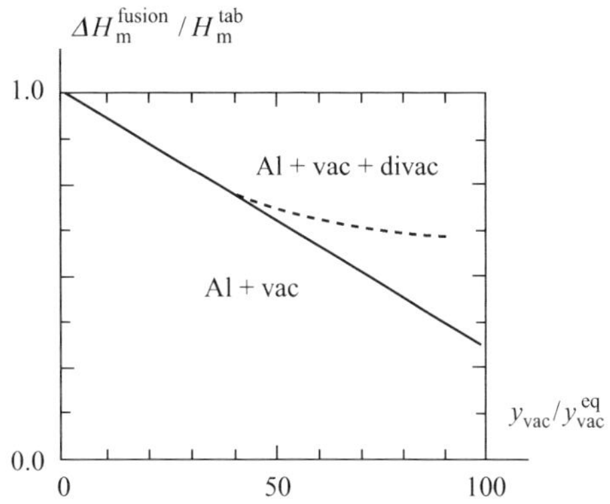

Figure 2.42 Relative heat of fusion as a function of the ratio of the relative fractions of vacancies for pure Al with monovacancies and with mono- and divacancies, respectively. H. Fredriksson and T. Emi, in Materials Transactions © The Japan Institute of Metals 1998.

2.7.4 Vacancies in Binary Alloys

Gibbs' Free Energy

In binary alloys the number of vacancies will affect the excess Gibbs' free energies and the activities γ_i^s of the elements involved. The activities are described by the interaction coefficients in the regular-solution model. The activity coefficients can be calculated by assuming equilibrium between solid and liquid. The binary alloy can be regarded as a ternary alloy with vacancies as the third component.

Consider a ternary alloy with the relative mole fraction y_A lattice sites, occupied by A atoms, the relative mole fraction y_B lattice sites, occupied by B atoms, and the relative fraction *yvac* vacant sites. The Gibbs' free energy of this alloy can be written as

$$G = y_A G_A^0 + y_B G_B^0 + y_{vac} G_{vac} + G^{Ex} \\ + RT\left[y_A \ln y_A + y_B \ln y_B + y_{vac} \ln y_{vac} - (1 + y_{vac})\ln(1 + y_{vac})\right] \tag{2.102}$$

where
G_A^0 = Gibbs' free energy of 1 kmol of A
G_B^0 = Gibbs' free energy of 1 kmol of alloying element B
G_m^{Ex} = molar excess Gibbs' free energy.

The equilibrium between two phases is normally described thermodynamically by assuming that the regular-solution model is valid. In this case the excess entropy is assumed to be zero and the excess Gibbs' free energy of a ternary alloy equals the excess enthalpy or the heat of mixing.

$$G_m^{Ex} = L_{AB}\, y_A y_B + L_{A\,vac}\, y_A y_{vac} + L_{B\,vac}\, y_B y_{vac} \tag{2.103}$$

where
L_{AB} = interaction coefficient between A and B
L_{Avac} = interaction coefficient between A and a vacant site
$L_{B\,vac}$ = interaction coefficient between B and a vacant site.

Thermodynamics of Vacancies in Binary Alloys

This chapter illustrates what a fundamental quantity Gibbs' free energy is and how useful it is for theoretical thermodynamic calculations of various properties of metals and alloys. It is closely related to the chemical potential, which is of utmost importance in theoretical chemistry. Good examples are the phase diagrams of binary alloys, which are of great importance both in chemistry and metallurgy.

The general condition for equilibrium between phases is expressed in terms of Gibbs' free energy. This relationship is frequently the fundamental basis of thermodynamic calculations. Examples are calculations of change of melting points, activity, partition coefficients in solutions, driving-force calculations and solidification rates in casting processes.

Partition Coefficient as a Function of Vacancy Concentration

As a last example, we will briefly describe the calculation of the partition coefficient as a function of the vacancy concentration in a binary alloy. The binary alloy is treated as a ternary alloy with the vacancies as the third component.

Starting with the general expression for the molar Gibbs' free energy the partial molar Gibbs' free energies of the components in the solid and liquid solutions are derived. They represent the chemical potentials. With the aid of them the equilibrium conditions are set up for the components in the liquid and solid phases. From these equations it is possible to calculate among other things the influence of vacancies on the partition coefficient. The result of extensive calculations is

$$k_{vac} = \frac{y_B^s}{y_B^L} = k_{vac}^{eq}\, e^{-\frac{\left(y_{vac} - y_{vac}^{eq}\right)\left(G_{vac} + L_{B\ vac}\right)}{RT}} \tag{2.104}$$

where

B = alloying element

k_{vac} = partition coefficient at the relative mole fraction of vacancies y_{vac}

k_{vac}^{eq} = partition coefficient with no excess vacancies, i.e. equilibrium concentration of vacancies

y_{vac} = relative mole fraction of vacancies in the alloy

G_{vac} = Gibbs' free energy of formation of 1 kmol of vacancies

$L_{B\ vac}$ = interaction coefficient between vacancies and atoms of the alloying element B

T = temperature.

Summary

■ Thermodynamics of Pure Metals

Driving Force of Solidification

$$-\Delta G_m^{L\to s} = \frac{T_M - T}{T_M}\left(-\Delta H_m^{L\to s}\right) = \frac{\Delta T}{T_M} H_m^{fusion}$$

■ Thermodynamics of Binary Alloys

Gibbs' Free Energy of Ideal Binary Solutions

$$G_m = x_A G_A^0 + x_B G_B^0 + RT\left(x_A \ln x_A + x_B \ln x_B\right)$$

■ Gibbs' Free Energy of Non-Ideal Binary Solutions

Some Thermodynamic Relationships

$$G_m^{Ex} = G_m^{non\text{-}ideal} - G_m^{ideal}$$

$$G_m^{Ex} = H_m^{Ex} - TS_m^{Ex}$$

$$H_m^{Ex} = H_m^{mix}$$

Gibbs' Free Energy of Non-Ideal Solutions

$$G_m = x_A G_A^0 + x_B G_B^0 + RT(x_A \ln x_A + x_B \ln x_B) + G_m^{Ex}$$

or

$$G_m = x_A G_A^0 + x_B G_B^0 + RT(x_A \ln x_A + x_B \ln x_B) + H_m^{mix} - TS_m^{Ex}$$

In non-ideal solutions the excess Gibbs' free energy is approximately proportional to the heat of mixing.

$$G_m^{Ex} = H_m^{mix}(1 - AT)$$

The Regular-Solution Model

Assumptions:

$$H_m^{mix} = L\,x_A x_B \quad \text{and} \quad S_m^{Ex} = 0$$

$$G_m^{mix} = Lx_A x_B + RT\,(x_A \ln x_A + x_B \ln x_B)$$

Miscibility Gap in a Regular Solution

Depending on composition and temperature a hypothetical regular solution appears in a diagram (page 48) as a

- single phase solution (areas α', α and α'')
- metastable single phase solution (areas between the miscibility and spinodal curves)
- two-phase solution (area inside the spinodal curve).

■ Gibbs' Phase Rule

$$N_{phases} + N_{degrees\ of\ freedom} = N_{components} + 2$$

■ Equilibrium between Phases in Binary Solutions

Gibbs' Free Energy Curves for Liquid and Solid Phases of Binary Alloys

Gibbs' free energy G_m for different types of binary solid solutions is plotted as a function the solute concentration within the interval $0 \le x_B \le 1$.

– Ideal solution: $H_m^{mix} = 0$. The curve 2.9a (page 52) has a minimum.

– Non-ideal solution: Components with equal structures. $H_m^{mix} > 0$ and $T < T_{cr}$.

The curve 2.10a (page 52) has two minima and a maximum in between.

–Non-ideal solution: Components with equal structures. $H_m^{mix} > 0$ and $T < T_{cr}$ or $H_m^{mix} < 0$.

The curve 2.10b (page 52) has a deeper minimum than that of an ideal solution.

–Non-ideal solution: Components with unequal structures. $H_m^{mix} > 0$ or $H_m^{mix} < 0$.

Two separate curves (page 53).

The corresponding curve for the liquid alloy has the similar shape as a solid solution in the case of components with the same structure.

■ The Tangent to Tangent Method to Predict Phases in Binary Solutions

- A common tangent is drawn between two Gibbs' free energy curves or a free energy curve with two minima (page 55).
- The stable state is a mechanical mixture of two phases.
- The point on the tangent, corresponding to the given overall composition, gives the free energy of the system in equilibrium.
- The points where the common tangent is in contact with the frée-energy curves/curve, give the compositions of the two phases with compositions x^α and x^β.
- The fractions of the two phases vary with the overall composition in accordance with the lever rule.

$$\text{Fraction of } \alpha = \frac{x^\beta - x}{x^\beta - x^\alpha} \qquad \text{Fraction of } \beta = \frac{x - x^\alpha}{x^\beta - x^\alpha}$$

■ Calculation of Chemical Potentials from Free-Energy Diagrams

Equilibrium between phases means that their chemical potentials are equal.
The *chemical potential* μ_A of a pure metal A is, by definition, equal to the molar free Gibbs' energy of A

$$\mu_A^0 = G_A^0$$

The *chemical potential* μ_A of a component A in an alloy is by definition equal to the partial molar Gibbs' free energy of A

$$\mu_A = \overline{G}_A$$

The partial molar Gibbs' free energies of the components and the molar Gibbs' free energy of the alloy are related by Gibbs–'Duhem's relationships

$$\mu_A = \overline{G}_A = G_m + (1 - x_A)\left(\frac{\partial G_m}{\partial x_A}\right)_T$$

$$\mu_B = \overline{G}_B = G_m + (1 - x_B)\left(\frac{\partial G_m}{\partial x_B}\right)_T$$

The tangent to tangent method is based on these relationships

Chemical Potentials of Liquids and Solids in Binary Alloys

$$\mu_A^L = \mu_A^{0L} + RT\ln(1 - x_B^L) + \overline{G}_A^{Ex\,L}$$

$$\mu_A^s = \mu_A^{0s} + RT\ln(1 - x_B^s) + \overline{G}_A^{Ex\,s}$$

$$\mu_B^L = \mu_B^{0L} + RT\ln x_B^L + \overline{G}_B^{Ex\,L}$$

$$\mu_B^s = \mu_B^{0s} + RT\ln x_B^s + \overline{G}_B^{Ex\,s}$$

The chemical potentials are found as intersections between the vertical axes and a tangent to the G_m curve (page 58).

The Molar Gibbs' Free Energy of a Binary Alloy

$$G_m = (1 - x_B)\mu_A + x_B\mu_B$$

A simple case is shown on page 56.

■ Tectic Phase Reactions

See page Table 2.2 on page 60.

■ Relationship between Gibbs' Molar Free Energy Curves and Phase Diagrams

The phase diagrams of binary alloys are functions of the shapes and positions of the corresponding Gibbs' free-energy curves of the solid and liquid phases at various temperatures.

Ideal solutions have simple free energy curves corresponding to simple lens-shaped phase diagrams

The complex Gibbs' free energy curves of non-ideal solutions are the basis of all other types of phase diagrams, for example eutectic systems, peritectic systems and systems with intermediate phases.

■ Influence of Parameters on Phase Diagrams

Parameters such as external pressure and phase boundary curvature influence the partition coefficient, the slope of the liquidus line and the melting point of alloys.

If gas pores are involved in a melt, care has to be taken to the influence of the compressibility of the gas on the chemical potential of the elements.

Mathematical expressions are given in the text.

■ Driving Force of Solidification in Binary Alloys

$$-\Delta G^{L\rightarrow s} = \left(1 - x_B^{\alpha/L}\right)\ \left(\mu_A^L - \mu_A^{L/\alpha}\right) + x_B^{\alpha/L}\left(\mu_B^L - \mu_B^{L/\alpha}\right)$$

Graphical construction of the driving force and various thermodynamical quantities are described in the text (page 77).

■ Thermodynamics of Ternary Alloys

Some Thermodynamic Relationships

$$G_m = x_A^0 G_A^0 + x_B^0 G_B^0 + x_C^0 G_C^0 + RT(x_A \ln x_A + x_B \ln x_B + x_C \ln x_C) + G_m^{Ex}(x_A,\ x_B,\ x_C)$$

$$G_m^{Ex} = H_m^{mix} = L_{AB}x_Ax_B + L_{AC}x_Ax_C + L_{BC}x_Bx_C$$

$$G_m^{Ex} = RT\ (x_A \ln \gamma_A + x_B \ln \gamma_B + x_C \ln \gamma_C)$$

$$\ln \gamma_B = \ln \gamma_B^0 + x_B\varepsilon_B^B + x_C\varepsilon_B^C$$

$$\overline{G}_i^{Ex} = G_m^{Ex} + .(1 - x_i)\ \frac{\partial G_m^{Ex}}{\partial x_i}$$

Gibbs' Concentration Triangle

The composition of a ternary alloy with the components A, B and C is best described by an equilateral triangle ABC. Its properties are listed in the text.

Graphical Representation of Gibbs' Free Energy. Ternary Phase Diagrams

Gibbs' free energy surfaces for the solid and liquid phases of an ideal ternary alloy and the corresponding phase diagram can be shown in a three-dimensional diagram (page 80).

■ Thermodynamics of Vacancies in Metals and Alloys

Relative Fraction of Vacancies

$$y_{vac} = \frac{N_{vac}}{N_A} = \frac{x_{vac}}{x_A}$$

The fraction of vacancies in a material varies strongly with temperature.

Relative Equilibrium Mole Fraction of Vacancies

$$y_{vac}^{eq} = e^{-\frac{G_{vac} + L_{A\ vac}}{RT}} \quad \text{(Regular solution model)}$$

Decrease of Melting Point Temperature

Vacancies, like other 'particles', cause a melting point decrease, which is proportional to the mole fraction of vacancies, in a pure metal.

The effect is reduced by divacancy formation.

Decrease of Heat of Fusion Owing to Vacancies in Pure Metals

$$H_m^{fusion} = H_{tab}^{fusion} - \left(y_{vac} - y_{vac}^{eq}\right) H_{vac}^{formation}$$

The effect is reduced by divacancy formation.

Vacancies in Binary Alloys

A binary alloy can be treated as a ternary alloy with vacancies as the third component.

Partition Coefficient as a Function of Vacancy Concentration in Binary Alloys

$$k_{vac} = \frac{y_B^s}{y_B^L} = k_{vac}^{eq}\, e^{-\frac{\left(y_{vac} - y_{vac}^{eq}\right)\left(G_{vac} + L_{B\ vac}\right)}{RT}}$$

Exercises

2.1 The Gibbs' free energy functions of a solid metal and its melt can be used to calculate the driving force of solidification at an arbitrary temperature and the corresponding undercooling.

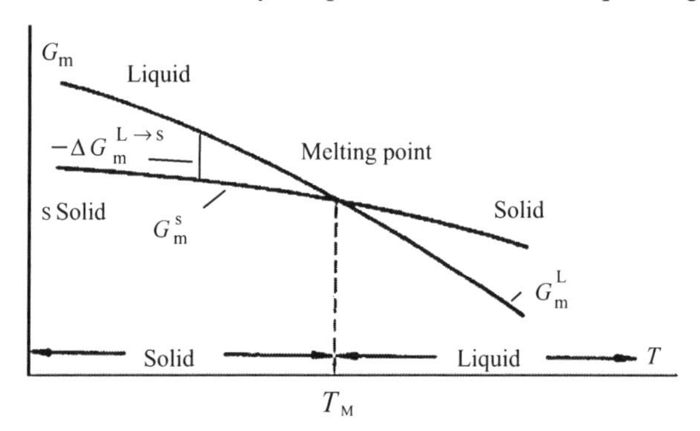

Figure 2.1–1

a) Use the figure above (Figure 2.1 on page 43 in Chapter 2) to predict qualitatively the change of the solidification temperature for the case that the Gibbs' free energy of the solid is increased (for example by an increased number of dislocations or other lattice faults).

b) When a melt is stirred its Gibbs' free energy increases. The energy added from outside is used to increase the distances between the atoms in the melt. Predict qualitatively the change of the solidification temperature when the Gibbs' free energy of the melt is increased by stirring during the solidification process.

c) The equilibrium melting point of pure Ni is 1726 K at standard pressure 1 atm. Calculate the molar driving force of solidification of nickel at the maximum undercooling 319 K.
d) Compare the molar driving force in c) to the molar driving force of vapour deposition on a substrate at the same temperature and pressure. The heat of fusion of Ni is 17.2×10^3 J/kmol.

2.2 The equilibrium melting point at zero curvature (flat surface) is 1726 K. Calculate the temperature decrease of the melting point when the solid nickel has the shape of a sphere with the radius

a) 1.0 cm
b) 1.0 μm
c) 0.010 μm.

Material constants of nickel:

$M = 58.7$ kg/kmol
$\rho_s = 8.9 \times 10^3$ kg/m^3 at a temperature close to the melting point 1726 K
$\rho_L = 8.5 \times 10^3$ kg/m^3 at a temperature close to the melting point 1726 K
$\sigma_{L/s} = 0.255$ J/m^2
$-\Delta H_m^{fusion} = 17.2 \times 10^3$ J/kmol.

2.3 The figure below shows the phase diagram and the Gibbs' free-energy curves at a given temperature for a binary alloy A–B. Consider a dilute alloy with the composition x_0^L at temperature T. The compositions of the solid and liquid phases at the interphase are marked in the figure.

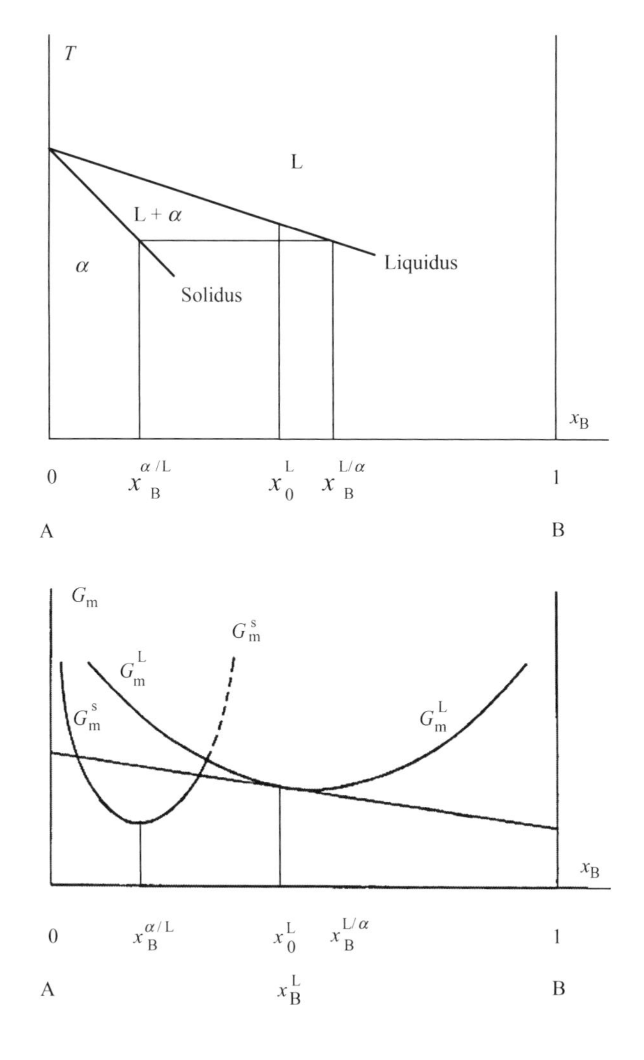

Figure 2.3–1

a) Construct the driving force of this alloy in a free-energy diagram and express it in terms of $-\Delta G_A^{L\to s}$ and $-\Delta G_B^{L\to s}$. Justify your answer.
b) The growth rate of a spherical particle of the solid phase with a radius r, growing in the melt, can be written as:

$$\frac{dr}{dt} = D\,\frac{x_B^{L/\alpha} - x_0^L}{x_B^{L/\alpha} - x_B^{\alpha/L}}\,\frac{V_m^\alpha}{V_m^L}\,\frac{1}{r}$$

where
D = diffusion coefficient of the alloying element B in the melt
V_m^α, V_m^L = the molar volumes of solid and liquid phases, respectively.

Derive a relationship between the driving force and the growth rate of the particle.

2.4 The phase diagram of the Al Pb system contains a nearly symmetric immiscibility gap. with a maximum temperature in the liquid state of ≈ 1670 K shown in Figure 1 below. The monotectic region is shown in the enlarged Figure 2.

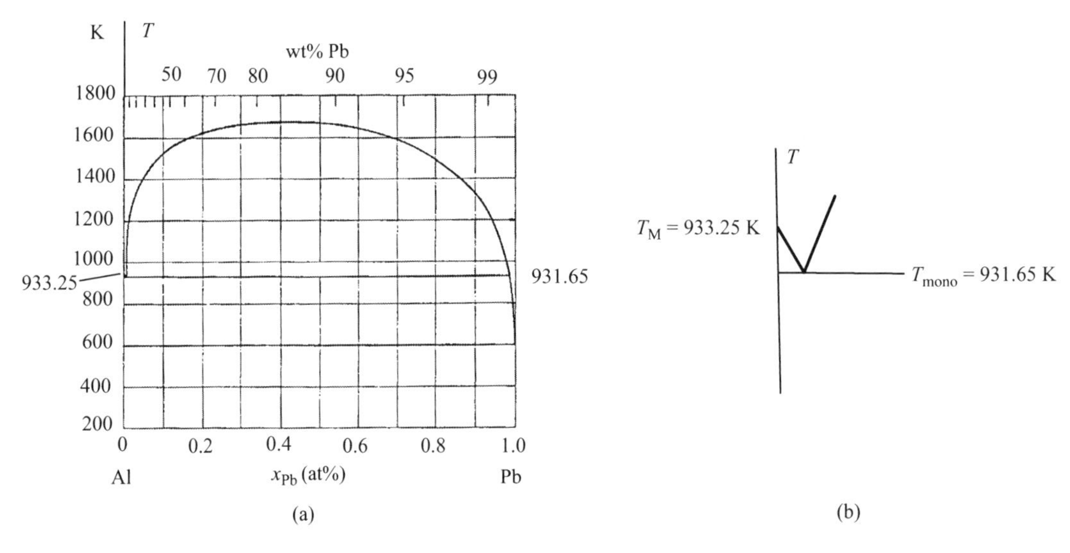

Figure 2.4–1 **Figure 2.4–2**

a) Read the maximum temperature of the miscibility gap and the temperature difference between the melting-point temperature of Al and the monotectic temperature from the figure.
b) Calculate the value of the regular solution parameter for use in c.
c) Calculate the composition of the melt at the monotectic point. Lead has practically no solubility in the solid state.

2.5 Consider a hypothetic binary alloy A B that forms an ideal liquid solution at all temperatures and a non-ideal solid solution. Assume that the heat of mixing of the solid phase is positive and relatively low.

a) Construct the expected shape of the phase diagram for such a binary system. Justify your answer by drawing the free energy curves of the liquid and solid phases at different temperatures.
b) The phase diagram of the binary system Cs Rb is shown below. This binary system follows the characteristics described above. The regular solution model is valid for the solid phase.

Calculate the heat of mixing of the system at the minimum liquidus temperature and explain the absence of a miscibility gap in the phase diagram.

Hint: Calculate the critical temperature of the system.

The minimum liquidus temperature $T_M^{min} = 9.7°C$ obtained at $x_{Rb} = 0.47$. The heats of fusion of Cs and Rb are $H_{m\,Cs}^{fusion} = 2.09 \times 10^3$ kJ/kmol and $H_{m\,Rb}^{fusion} = 2.20 \times 10^3$ kJ/kmol, respectively.

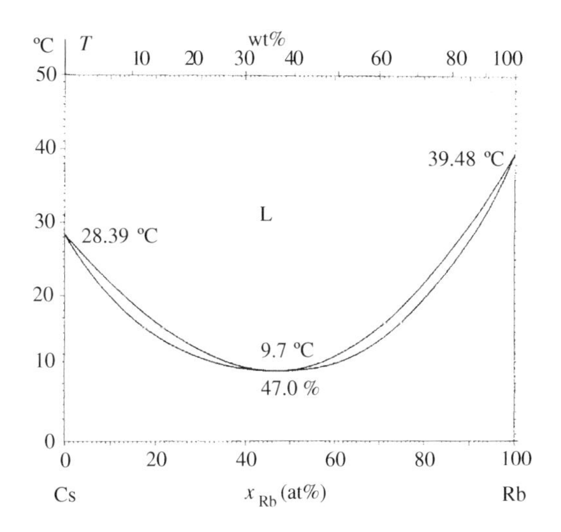

Figure 2.5–1

2.6 When metals solidify during ordinary conditions they contain an excess of vacancies and become supersaturated. The solubility of vacancies in the metal at a given temperature T can be calculated when the melting point, the heat of fusion and the equilibrium mol fraction of vacancies y_{vac}^{eq} at the given temperature are known.

Use the data for Al and other information given below to calculate the decrease of the melting point for pure Al as a function of $n \times \ln n$ where $n = y_{vac}/y_{vac}^{eq}$, i.e. the relative mol fraction of supersaturated vacancies at $T_M = 900$ K. Plot the decrease of the melting point as a function of n for $n = 1$–20 in a diagram.

Material constants for Al are:

Molar weight 27.0 kg/kmol

Heat of fusion 398 kJ/kg

Melting point at equilibrium conditions 933 K.

At $T_M = 900$ K the mol fraction of vacancies at equilibrium is 3.0×10^{-3}.

2.7 During the solidification of an aluminium alloy melt containing 2% Cu. The Cu concentration in the solid phase was found to be twice the equilibrium concentration expected from the phase diagram. One explanation could be that the fraction of vacancies in the solid exceeds the equilibrium value and that the interaction between the vacancies and the Cu atoms enhance the solubility of Cu in the melt. The regular solution model is supposed to be valid.

Estimate the regular solution parameter L between vacancies and the Cu atoms in the matrix with the aid of the following data.

The liquidus temperature of Al is 933 K. The solidification occurred at an undercooling of 3.0 K. The equilibrium concentration of vacancies in the alloy at the temperature 930 K was 3.2×10^{-3} (mol fraction). The heat of fusion of the alloy is $(-\Delta H_m^{fusion}) = 10.7 \times 10^3$ kJ/kmol. The free energy of vacancies $G_{vac} = 75.3 \times 10^3$ kJ/kmol.

References

1. J. H. Hildebrand, *Physical Review*, **34**, (1929), 984.
2. T. Emi and H. Fredriksson, *Materials Transaction, JIM*, **39**, (1998).

Further Reading

R.T. de Hoff, *Thermodynamics in Materials Science*, New York, McGraw-Hill Inc, 1993.
N. A, Goksen, *Statistical Thermodynamics of Alloys*, New York, Plenum Press, 1986.
E. A Guggenheim, *Thermodynamics*, New York, Amsterdam, North Holland, 1967.
M. Hillert. *Phase Equilibria*, Cambridge, Cambridge University Press, 1967.
C. H. P. Lupis, *Chemical Thermodynamics of Materials*, New York, Elsevier Science Publisher, 1983.
C. Wagner, *Thermodynamics of Alloys*, London, Addison-Wesley Publishing Company, 1952.

3

Properties of Interfaces

3.1 Introduction

In the book 'Physics of Functional Materials' [1] the structure of crystals and the properties of melts and solids, such as viscosity, diffusion, heat capacity and thermal conduction have been treated. All these properties of the matter in bulk play important roles in crystallization processes.

The properties of the interface between the crystal and the melt or between the crystal and its surrounding gas phase are of great importance in crystal-growth processes. Hence, it is critical to

Solidification and Crystallization Processing in Metals and Alloys, First Edition. Hasse Fredriksson and Ulla Åkerlind.

examine the conditions concerning energy, structure and other properties of interfaces and discuss their influence on various processes. This has been done in the present chapter.

Throughout this chapter we will use the notation *interface energy* for the surface energy between two phases. Both interface energy and surface energy occur in the literature for this type of energy. Since we will work with crystallization processes in future chapters, the notation interface energy is a better choice than surface energy. The well-known concept of surface tension will also be used and the relationship between interface energy and surface tension will be discussed.

3.2 Classical Theory of Interface Energy and Surface Tension

3.2.1 *Basic Concepts and Definitions of Interface Energy and Surface Tension*

To make it possible to treat the theory of interfaces properly, we have to define some basic concepts associated with interface energy and connect the new concepts to thermodynamics.

- The *interface energy* σ = the work that has to be done to form one unit area of the interface. In the literature, the designation γ is often used instead of σ for interface energy. In this book we will use the letter σ for both interface energy and surface tension except in the few cases when a distinction between the two quantities is necessary to avoid confusion. See Sections 3.2.2 and 3.2.3 below.
- The *free energy* of the interface is defined as Gibbs' free energy. The definition equation is

$$G^{\mathrm{I}} = H^{\mathrm{I}} - TS^{\mathrm{I}} \tag{3.1}$$

where the superscript I stands for interface and the free energy G^{I}, the enthalpy H^{I} and the entropy S^{I} refer to one kmol.

The relationship between the interface energy, measured in J/m^2, and the free energy of a solid/liquid interface, measured in J/kmol, is given by the equation

$$\Delta G^{\mathrm{I}} = G^{\mathrm{I}} - G^{\mathrm{L}} = \sigma \frac{V_{\mathrm{m}}}{d} \tag{3.2}$$

where
σ = interface energy per unit area (J/m^2)
V_{m} = the molar volume
d = the interface thickness.

The factor $V_{\mathrm{m}}^{\mathrm{I}}/d$ relates the thermodynamic quantities and the surface quantities.

3.2.2 *Interface Energy and Surface Tension*

It is well known that the forces between the atoms in a gas or a vapour are much weaker than those in a melt. Hence, the attraction of neighbouring atoms in the liquid on an atom in the surface (for example the black one in Figure 3.1) gives a net force directed vertically towards the liquid. Energy is required to move a surface atom from the liquid phase to the gas phase.

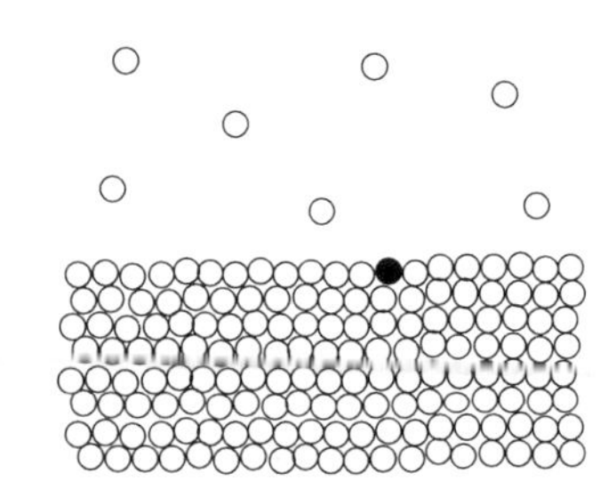

Figure 3.1 A vapour/liquid interface.

The net force on an atom in bulk is zero for symmetry reasons. The atom is easily moved within the liquid. Hence, the energy of an atom at the surface is higher than that of an atom in bulk. The energy per unit area is called *the interface energy* or *surface energy* and is denoted by γ or σ. It is measured in J/m^2. A sphere has the smallest surface/volume ratio of all possible shapes for a given volume. At equilibrium, a given volume of the liquid tends to form a sphere, which corresponds to the energy minimum.

A liquid surface can be regarded as an elastic skin. If we make a virtual cut in the surface, net forces perpendicular to the edges will act along them as shown in Figure 3.2. These net forces try to

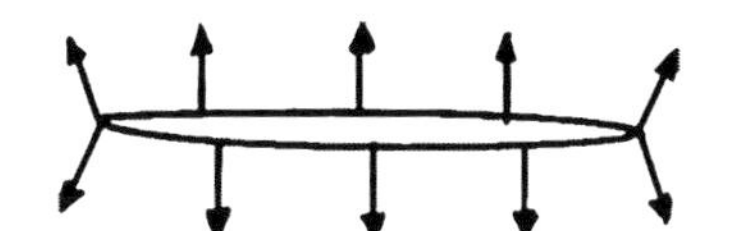

Figure 3.2 Surface tension forces acting on two sides of a cut in a liquid surface.

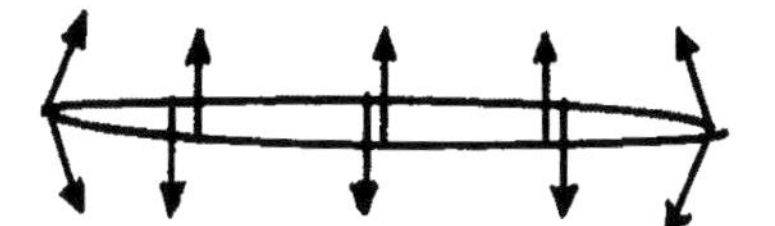

Figure 3.3 Outer reaction forces acting on the edges of the cut to keep them together.

minimize the surface to reach energy minimum. The force per unit length is called *the surface tension* γ and is measured in N/m. To keep the surface intact, outer reaction forces have to be applied along the edges (Figure 3.3) to keep the edges together.

3.2.3 *Surface Tension as a Function of Interface Energy*

Relationship between Surface Tension and Interface Energy under Isothermal Conditions

There is a simple relationship between interface energy and surface tension that can be derived in the following way, using the definitions of the two quantities in Sections 3.2.1 and 3.2.2.

Consider a liquid film of length l and width x (Figure 3.4). Its area can be varied reversibly by moving the bar AB gently. The surface tension forces are overcome by adapting an opposite total force F

$$F = \gamma \times 2l \tag{3.3}$$

The work to move the bar from position A_0, B_0 to AB can be obtained in two ways. The force times the distance is equal to the change in surface energy of the double surface at constant temperature.

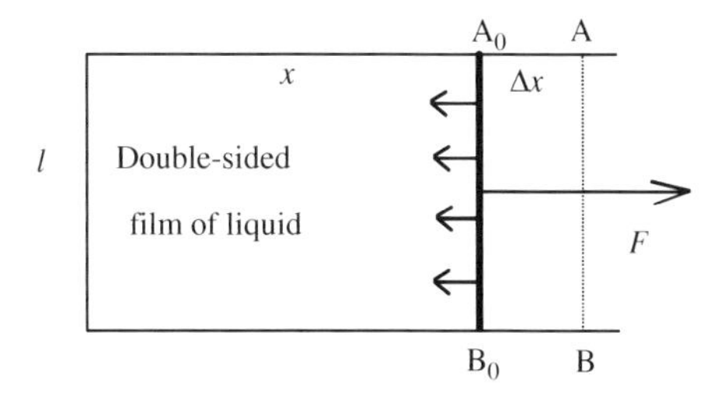

Figure 3.4 Extension of a double-sided liquid film.

$$F\Delta x = \gamma \times 2l\Delta x \tag{3.4}$$

The work equals the increase of the interface energy:

$$\gamma 2l \times \Delta x = \sigma \times 2l\Delta x$$

or

$$\gamma = \sigma \tag{3.5}$$

- The surface tension is equivalent to the interface energy under isothermal conditions in materials that are incapable of resisting shear.

This is the reason why the same letter can be used for both interface energy and surface tension. No distinction will be made between interface energy and surface tension, both of which are denoted by σ when no confusion is possible.

The shear condition will be discussed below. Liquid/liquid interfaces and liquid/vapour interfaces belong to this case. Hence, Equation (3.5) is valid for metal melts, which are of special interest in this book.

Two Ways to Enlarge a Surface

The relationship between surface tension and interface energy differs from that in Equation (3.5) if the conditions given above are not valid.

The surface in Figure 3.5a can be enlarged in two ways. One way is to stretch the surface (Figure 3.5b). This can only be done in materials that are capable of resisting shear. Solid materials, for example crystals, resist shear. The other way is to transfer atoms from the interior to the surface and prolong rows of atoms (Figure 3.5c). This case is the one treated above.

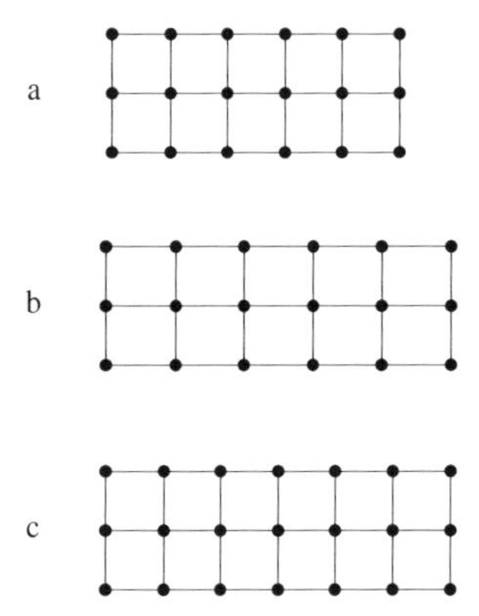

Figure 3.5 a, b and c The surface in Figure 3.5 a can be enlarged either by b) stretching the surface; c) bringing additional atoms to the surface from the interior.

In Figure 3.5b the surface is intact in the sense that the surface is *stretched but not broken.* The number of atoms and their relative positions remain unchanged. Instead, the interatomic distances increase and the number of atoms per unit area goes down. The interface energy increases as work has to be done to increase the distances between the atoms against the attraction forces between the atoms.

Using tensor formalism a general relationship between surface tension and interface energy can be derived. It is performed below.

General Relationship between Surface Tension and Interface Energy

Consider a flat crystal surface with the area A (Figure 3.6). Tangential forces act on the surface and cause a stretching or contraction dA. The surface deformation can be described by a strain tensor ε_{ij}

$$dA = \sum_{ij} A d\varepsilon_{ij}\delta_{ij} \quad (i, j = x, y) \tag{3.6}$$

where δ_{ij} is Kronecker's delta function:

$$\delta_{ij} = 1 \quad \text{for} \quad i = j \quad \text{and} \quad \delta_{ij} = 0 \quad \text{for} \quad i \neq j$$

The work to enlarge the surface area equals the work done against the surface-tension forces or the surface stress g_{ij} corresponding to an infinitesimal strain $d\varepsilon_{ij}$

$$dW = \sum_{ij} A g_{ij} d\varepsilon_{ij} \tag{3.7}$$

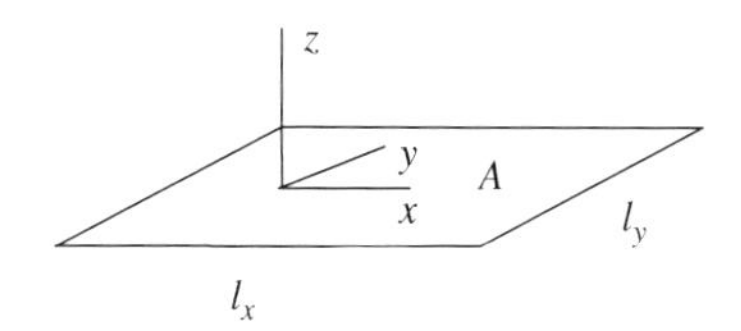

Figure 3.6 Surface A with area *A* in a coordinate system *xyz*.

This work is also equal to the increase in surface free energy γ

$$dW = d(\gamma A) = \gamma dA + A d\gamma = \sum_{ij} \gamma A \delta_{ij} d\varepsilon_{ij} + \sum_{ij} A \frac{\partial \gamma}{\partial \varepsilon_{ij}} d\varepsilon_{ij} \tag{3.8}$$

Identification of Equations (3.7) and (3.8) gives a relationship between the surface stress g_{ij} per unit length and the elastic strain ε_{ij}, induced by the stress

$$\sum_{ij} g_{ij} = \sum_{ij} \gamma \delta_{ij} + \sum_{ij} \frac{\partial \gamma}{\partial \varepsilon_{ij}} \tag{3.9}$$

Equation (3.9) is the general relationship between the surface energy σ and the surface stress g_{ij} for an anisotropic solid.

The surface stress is measured in J/m^2 or N/m.

The surface tension σ in an anisotropic solid can be defined as the average value of the surface stress in two perpendicular directions in the surface

$$\sigma = \frac{g_{xx} + g_{yy}}{2} \tag{3.10}$$

The stress g_{zz} is zero as there is no stress perpendicular to the surface.

Isotropic Surfaces

If the general formula (3.9) is applied on an *isotropic solid,* we obtain a simpler expression for the surface stress tensor

$$g_{xx} = g_{yy} = g \tag{3.11}$$

and the only components of the elastic strain tensor will be (Figure 3.6)

$$\varepsilon_{xx} = \frac{dl_x}{l_x} \quad \text{and} \quad \varepsilon_{yy} = \frac{dl_y}{l_y} \tag{3.12}$$

Equations (3.10) and (3.11) gives for the left-hand side of Equation (3.9)

$$\sigma = g \tag{3.13}$$

The right-hand side of Equation (3.9) can be transformed as follows

$$\begin{aligned}
\sum_{ij} \gamma \delta_{ij} + \sum_{ij} \frac{\partial \gamma}{\partial \varepsilon_{ij}} &= \gamma + \frac{\partial \gamma}{\partial (\varepsilon_{xx} + \varepsilon_{yy})} = \gamma + \frac{\partial \gamma}{\partial \left(\frac{dl_x}{l_x} + \frac{dl_y}{l_y} \right)} \\
\gamma + \frac{\partial \gamma}{\partial \left(\frac{dl_x}{l_x} + \frac{dl_y}{l_y} \right)} &= \gamma + \frac{l_x l_y \partial \gamma}{\partial \left(l_y dl_x + l_x dl_y \right)} = \gamma + A \frac{d\gamma}{dA}
\end{aligned} \tag{3.14}$$

Combining Equations (3.9), (3.10), (3.11), (3.12) and (3.14) we obtain

$$g = \gamma + A \frac{d\gamma}{dA} \tag{3.15}$$

Introducing the surface tension instead of g gives, according to Equation (3.13),

$$\sigma = \gamma + A\frac{d\gamma}{dA} \tag{3.16}$$

In the special case when *the surface cannot resist shear*, stretching causes the surface area to increase by transfer of atoms from the interior to the surface. The interatomic distances and the number of atoms per unit area remain unchanged as shown in Figure 3.5c on page 102. This means that $d\gamma/dA = 0$ in Equation (3.15) and we obtain

$$\sigma = \gamma$$

in agreement with Equation (3.5) on page 101.

Elastic strain energy will be discussed later (Section 3.4.7, page 142–143).

3.2.4 Interface Energy as a Function of Temperature and Impurity Concentration

Temperature Dependence

It is normally easy to measure the interface energy of liquids and solids in a gas phase, as will be discussed in Section 3.2.6 on page 107. In order to illustrate the effect of temperature on the interface energy the results of some measurements are shown in Figure 3.7.

Figure 3.7 shows the variation of the interface energy of pure copper surrounded by a gas atmosphere as a function of temperature. Obviously, the interface energies of both solid and liquid copper *decrease* with temperature. The temperature dependence of interface energy can be described by the slope of the line. The slope is identical with the molar heat capacity of the interface

$$C_p^I = \left(\frac{\partial H}{\partial T}\right)_p \tag{3.17}$$

which can be transformed into heat capacity per unit area

$$c_p^I = \frac{V_m}{d}\left(\frac{\partial \sigma}{\partial T}\right)_p \tag{3.18}$$

C_p^I is measured in J/kmol and c_p^I in J/m^2.

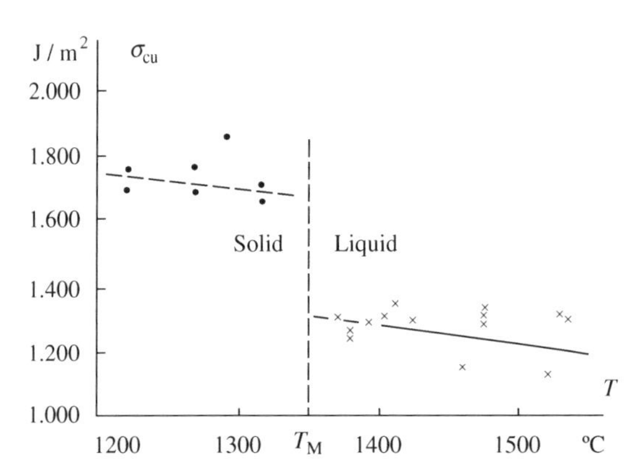

Figure 3.7 The interface energy of Cu as a function of temperature.

The sudden change in surface tension at the melting point is associated with the heat of fusion ($-\Delta H$).

Figure 3.7 also shows that the liquid has considerably lower interface energy than the solid. This is always the case as the forces in a solid are stronger than those in a liquid. In general, the interface energies of solids are 1.1 to 1.3 times larger than those of the corresponding liquids.

Impurity Dependence in Metal Melts

The qualities of pure metals change considerably by adding even small amounts of non-metallic elements. These may be present in the ore and may be difficult to remove during the production process. Other elements are added deliberately to give the alloy better and special qualities. They also change the interface energy.

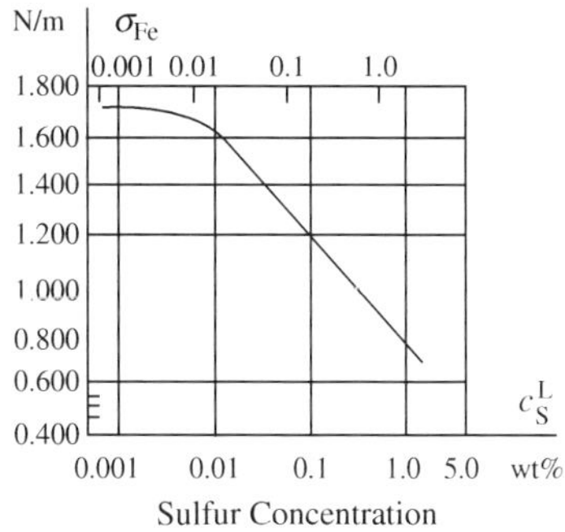

Figure 3.8 The surface tension of liquid iron at 1570 °C as a function of sulfur concentration. Reproduced from Elsevier © 1963.

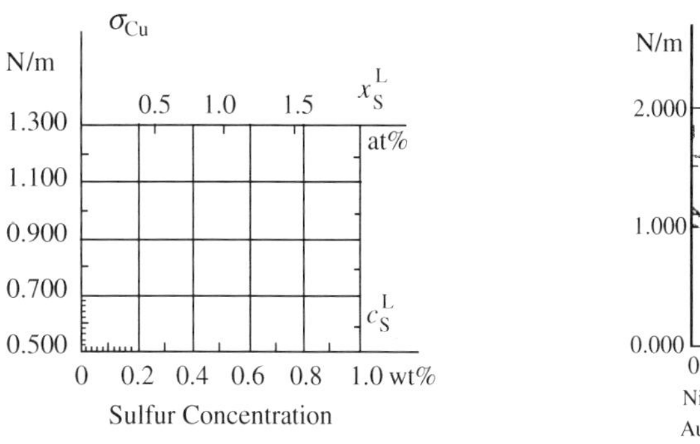

Figure 3.9 The surface tension of liquid copper at 1120 °C as a function of sulfur concentration. Reproduced from Springer © 1953.

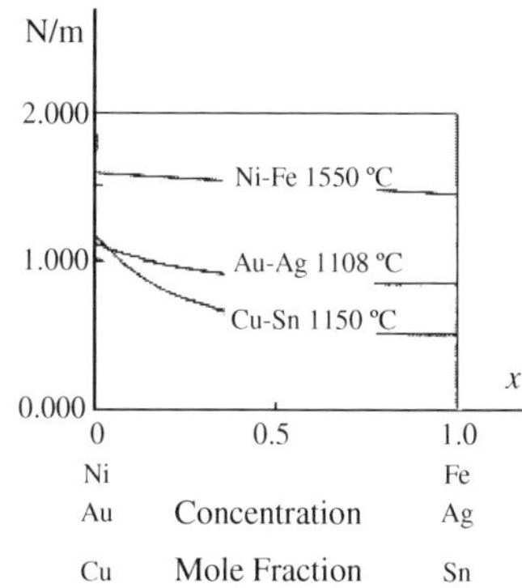

Figure 3.10 The surface tension of some binary liquid alloys as a function of composition.

In both cases the surface tension will be affected by the composition of the melt. Figures 3.8, 3.9 and 3.10 show that the interface energy decreases with increasing concentration of impurities or alloying elements.

3.2.5 *Wetting*

The shape and extension of a liquid droplet on a solid surface depends on the balance of three surface-tension forces between solid–liquid, solid–vapour and liquid–vapour, respectively. The directions of the tension forces are given by the condition that they will minimize the area of the surface.

Derivation of the General Wetting Equation

In the general case the three surface-tension forces have different directions and different sizes as in Figure 3.11. The general wetting equation is derived in the box on next page.

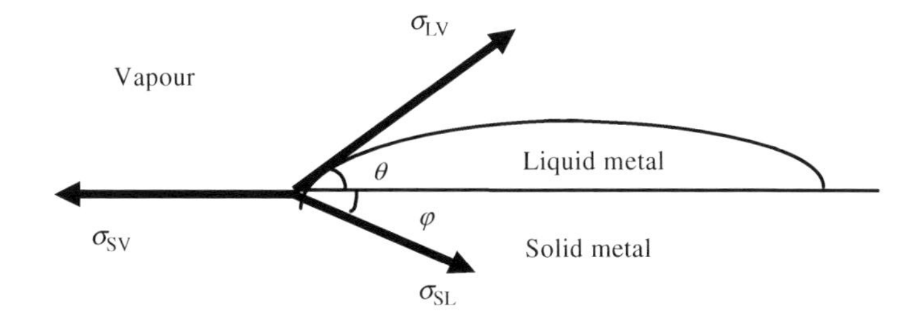

Figure 3.11 Surface-tension forces on a liquid droplet on a planar solid surface.

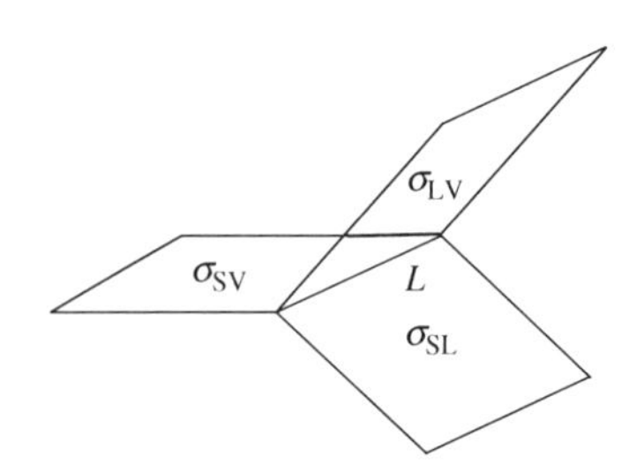

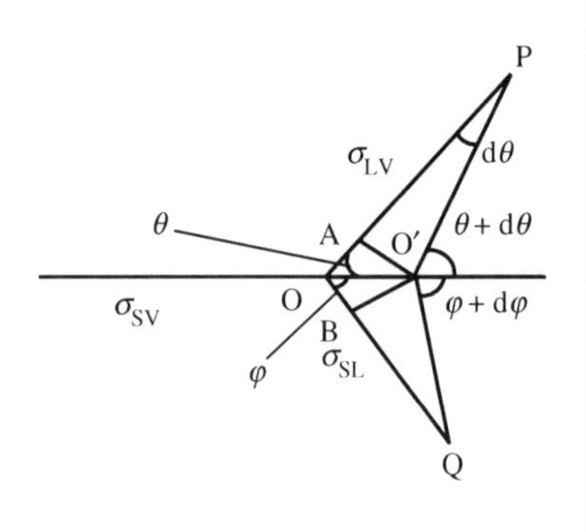

Derivation of the General Wetting Equation

Consider three planar surfaces that are the boundaries between a vapour phase, a liquid phase and a solid phase. The interface energies are indicated and named in the upper box figure.

To find the equilibrium condition we make a virtual infinitesimal displacement of the line L parallel to itself with the rest intact and calculate the corresponding energy change.

A cross-section of the system is shown in the lower box figure. When the line is moved from position O to O' the interface areas and the angles θ and φ change.

Total Surface Energy Change

The area changes cause the following total surface energy change:

$$\sigma_{SV} L \times OO' - \sigma_{LV} L \times OO' \times \cos\theta - \sigma_{SL} L \times OO' \times \cos\varphi \qquad (1')$$

where

σ_{SV} = surface tension between the solid and the vapour
σ_{SL} = surface tension between the solid and the liquid
σ_{LV} = surface tension between the liquid and the vapour
$\theta,\ \varphi$ = contact angles.

From the lower box figure we can conclude that

$$\cos\theta = \frac{OA}{OO'} \quad \text{and} \quad \cos\varphi = \frac{OB}{OO'}$$

The change of the surface energy owing to a change $d\theta$ of the angle θ cannot be shown in the lower box figure but can, for mathematical reasons be written:

$$LPO' \frac{\partial \sigma_{LV}}{\partial \theta} d\theta$$

or, for geometrical reasons ($PO'd\theta = OO'\sin\theta$):

$$LOO'\sin\theta \frac{\partial \sigma_{LV}}{\partial \theta} \qquad (2')$$

Similarly, we obtain the

$$L \times OO' \times \sin\varphi \frac{\partial \sigma_{SL}}{\partial \varphi} \qquad (3')$$

σ_{SV} is independent of the angles and gives therefore no 'angular' contribution to the energy change.

Condition for Equilibrium

Hence, the sum of the total surface energy changes will be

$$\sigma_{SV} L \times OO' - \sigma_{LV} L \times OO'\cos\theta - \sigma_{SL} L \times OO'\cos\varphi + LOO'\sin\theta \frac{\partial \sigma_{LV}}{\partial \theta} + L \times OO'\sin\varphi \frac{\partial \sigma_{SL}}{\partial \varphi}$$

If the total surface energy change is zero the system will be in equilibrium. This condition gives

$$\sigma_{SV} = \sigma_{LV}\cos\theta + \sigma_{SL}\cos\varphi - \sin\theta \frac{\partial \sigma_{LV}}{\partial \theta} - \sin\varphi \frac{\partial \sigma_{SL}}{\partial \varphi} \qquad (4')$$

The equilibrium is independent of L and is valid for interfaces of any shape.

The last equation in the box is the *general wetting equation.*

$$\sigma_{SV} = \sigma_{LV}\cos\theta + \sigma_{SL}\cos\varphi - \sin\theta\frac{\partial\sigma_{LV}}{\partial\theta} - \sin\varphi\frac{\partial\sigma_{SL}}{\partial\varphi} \tag{3.19}$$

Simplified Wetting Equation

In many cases it is reasonable to simplify Equation (3.19). If the contact angle φ is small and σ_{LV} and σ_{SL} are supposed to be independent of the contact angle, Equation (3.19) can be written (Figure 3.12):

$$\sigma_{SV} = \sigma_{SL} + \sigma_{LV}\cos\theta \tag{3.20}$$

Equation (3.20) is valid at constant temperature and applicable for the case when the liquid is insoluble in the solid. If it is solved for cos θ we obtain

$$\cos\theta = \frac{\sigma_{SV} - \sigma_{SL}}{\sigma_{LV}} \tag{3.21}$$

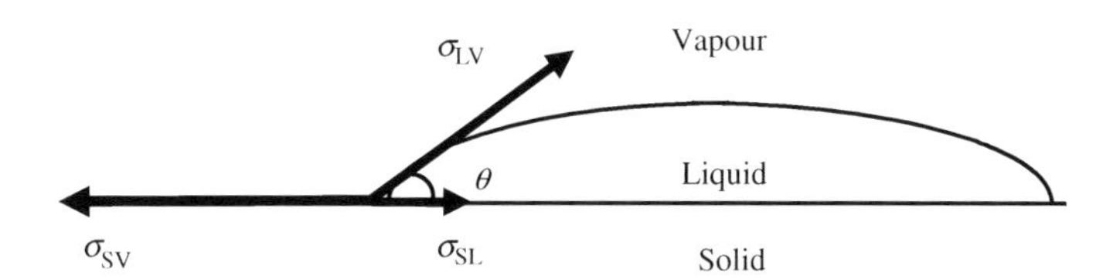

Figure 3.12 The shape and extension of the liquid droplet on a solid surface depends of the balance between the three surface tension forces.

$\theta > 0$ if $\sigma_{SV} \geq \sigma_{SL}$. Complete wetting corresponds to $\theta = 0$ or $\sigma_{SV} - \sigma_{SL} \geq \sigma_{LV}$. It means that the liquid spreads as a thin layer all over the solid phase. Normally $\sigma_{SV} > \sigma_{SL} > \sigma_{LV}$.

Wetting is promoted by decreasing θ. The wetting angle also decreases if

- σ_{SV} is increased
- σ_{SL} is decreased
- σ_{LV} is decreased.

σ_{SV} is maximized by cleaning the surface of the solid. The term σ_{SL} is constant at a fixed temperature and set of materials. σ_{SL} is highly temperature dependent and decreases usually rapidly with increasing temperature. Spreading can therefore be controlled by changing the temperature.

3.2.6 Measurements of Interface Energies

It is comparatively easy to measure the interface energy of liquids. We will discuss some methods to measure interface energies in order to obtain a possibility to compare experimental results with the theoretical models discussed later on.

Experimental

Liquid/Vapour Interface Energies

Most methods of measuring interface energies are based on the balance between tension forces and gravity forces. The most common methods used for metals are

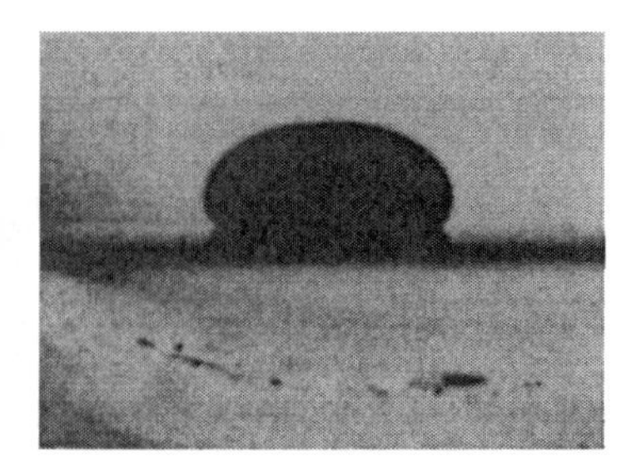

Figure 3.13 Photograph of a Hg sessile drop at 25 °C.

1. study of a sessile drop
2. study of pendant drops
3. drop weight method
4. maximum bubble-pressure method.

Methods 1–4 are used for liquid/vapour interfaces. The measurements on metals require high temperatures because most metals have high melting points. Methods 2 and 3 are most accurate. The different methods will be briefly described below.

A small amount of the metal in a tube, filled with a purified inert atmosphere, is placed on a polished ceramic planar surface in a closed tube. It is heated by a high-frequency generator to the wanted temperature above the melting point of the metal. A *sessile drop* is formed, which is illuminated and photographed. An example is shown in Figure 3.13.

The liquid/vapour interface energy can be derived from the shape of the drop. Care has to be taken to avoid impurity contamination.

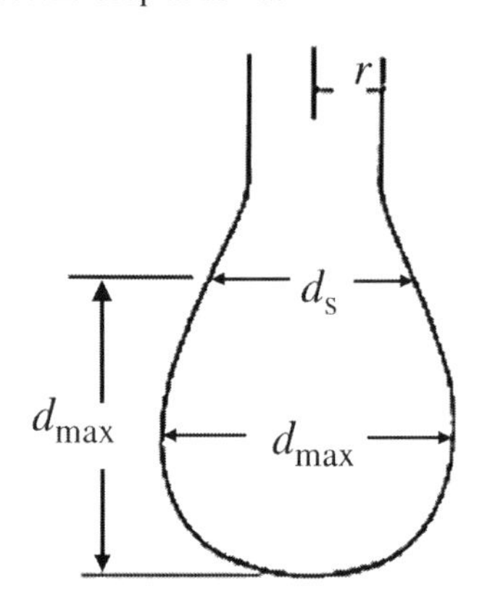

Figure 3.14 Pendant drop before falling.

The balance between the surface-tension forces and the gravitation forces determines the shape of a *pendant drop*. The drop is photographed and the geometric distances, marked in Figure 3.14, are measured. A metal rod is heated by electron bombardment to create the drop. The liquid/vapour interface energy can then be calculated from the equation

$$\sigma_{LV} = Jg\rho_L d_{max}^2 \tag{3.22}$$

J is the shape factor, a tabulated function of the ratio d_s/d_{max}.

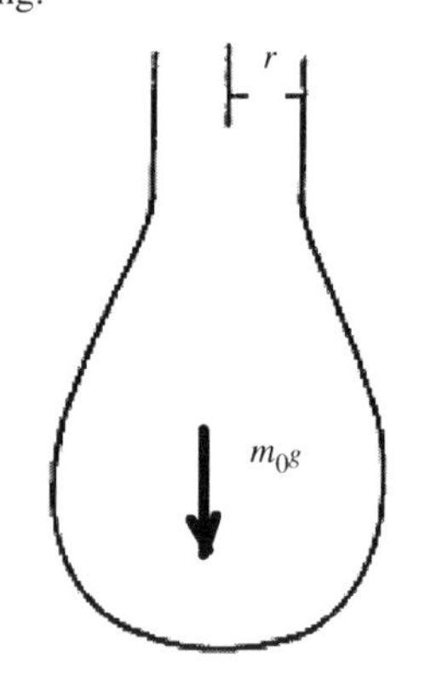

Figure 3.15 Pendant drop after falling.

Alternatively, the pendant drop described above is allowed to fall (Figure 3.15). It is reasonable to assume that the tension forces balance the gravitation force of the drop just before it falls. The interface energy can be calculated from

$$\gamma_{LV} = C\frac{m_0 g}{r} \tag{3.23}$$

where C is a tabulated function of the ratio V/r^3 where V is the volume of the fallen drop. A number of drops are collected in a container with high walls and weighed. Hence, the drop weight m_0 can be calculated, r can be measured and the interface energy can be derived.

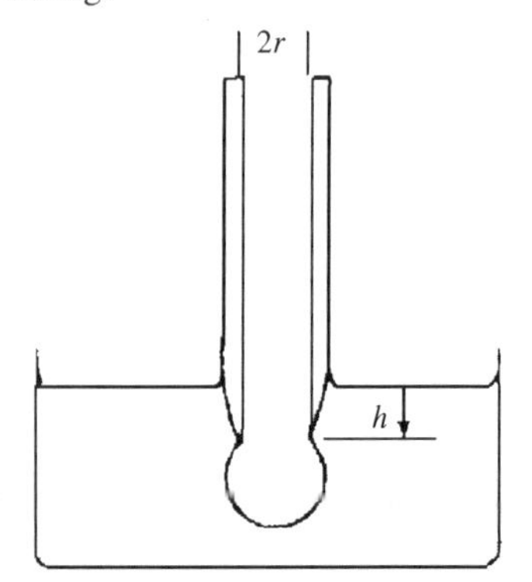

Figure 3.16 Equipment for measuring liquid/vapour interface energies.

The equipment of the *maximum bubble-pressure method* is shown in Figure 3.16.

A gas bubble is pressed through a capillary tube immersed in a molten metal (Figure 3.16). The maximum gas pressure that can be applied without loss of gas to the environment depends on the surface tension of the melt. If the density of the gas is neglected in comparison with the density of the metal, the interface energy is given approximately by

$$\gamma_{LV} = \frac{rp_{max}}{2}\left[1 - \frac{2}{3}\frac{g\rho_L r}{p_{max}} - \frac{1}{6}\left(\frac{g\rho_L r}{p_{max}}\right)^2\right] \tag{3.24}$$

where ρ_L is the density of the melt.

Solid/Vapour Interface Energies

Measurements of solid/vapour interface energies of non-metallic, brittle materials can be performed by using a controlled cleavage technique. The energy required to form a new surface by extension of a crack is equal to the interface energy of the new surface. Determination of solid/vapour interface energies can also be performed using heat of solution calorimetry.

Liquid/Liquid Interface Energies

Method 4 described above can also be used for liquid/liquid interfaces, provided that the two liquids are immiscible.

If the gas is replaced by another liquid metal of different density an equilibrium is developed (Figure 3.17). The maximum pressure is measured. The interface energy can be calculated from Equation (3.24) if ρ_L is replaced by $\rho_1 - \rho_2$, i.e. by the difference of the two liquid densities.

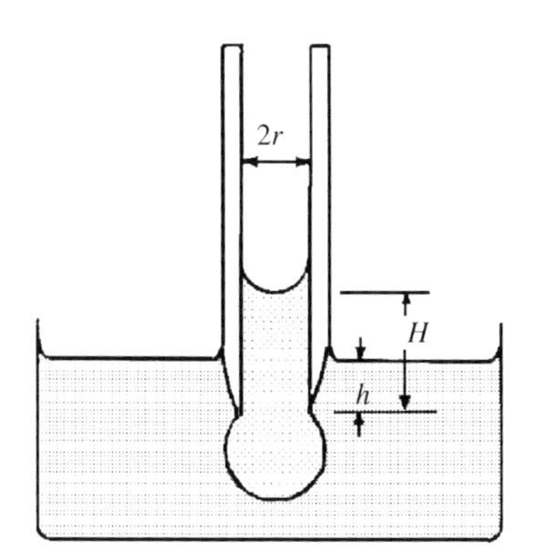

Figure 3.17 Equipment for measuring liquid/liquid interface energies.

This capillary method is the only one for direct measurement of interface energies. Unfortunately, it is neither very accurate nor convenient to use at temperatures. Measurements have been performed on metals and alloys with low melting points such as Hg, Na, Bi, Sn and related alloys.

The most common method to measure liquid/liquid interface energies or rather the tension forces is to study the shape of the boundaries between the different liquid phases. Its application on metal melts will be described below.

Consider two immiscible liquids 1 and 3. Another immiscible liquid 2 is added. When liquid 2 is entirely surrounded by liquid 1 (Figure 3.18a), its shape will be spherical corresponding to the minimum interface energy at a given volume.

When liquid 2 reaches liquid 3 a new equilibrium, shown in Figure 3.18b, is developed. The tension forces balance each other (Figure 3.18c) and at isothermal conditions we obtain with the aid of the sinus theorem in a triangle with the sides σ_{12}, σ_{23} and σ_{13}:

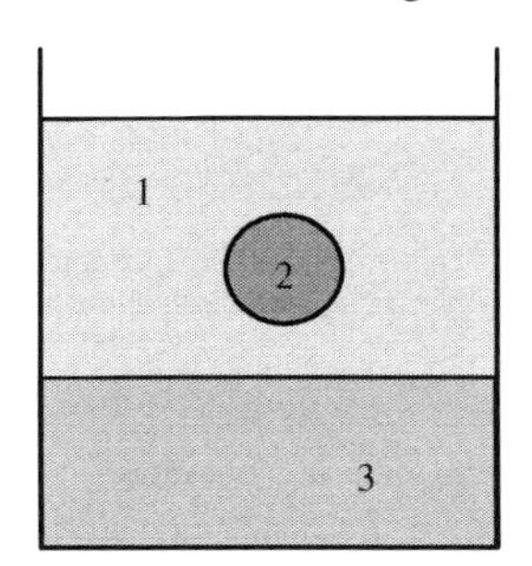

Figure 3.18a Spherical droplet.

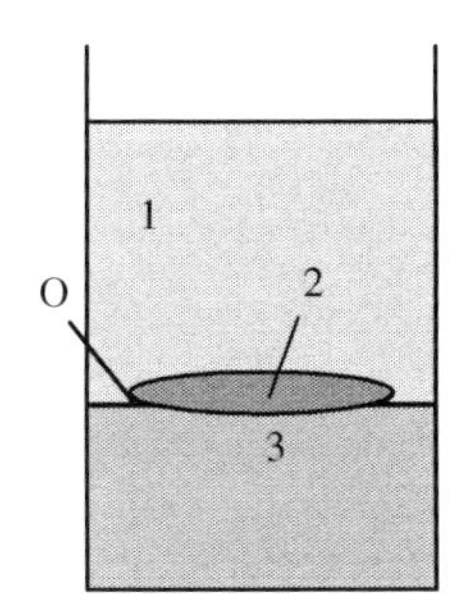

Figure 3.18b Lens-shaped droplet.

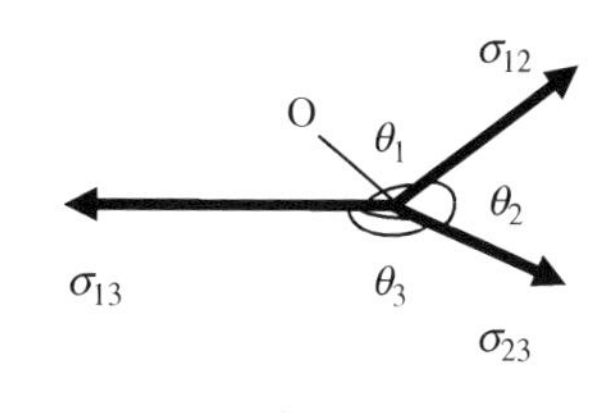

Figure 3.18c Surface-tension forces, acting on the droplet in Figure 3.18 b, in equilibrium.

$$\frac{\sigma_{23}}{\sin\theta_1} = \frac{\sigma_{13}}{\sin\theta_2} = \frac{\sigma_{12}}{\sin\theta_3} \tag{3.25}$$

Equation (3.25) gives the relative sizes of the interface energies if the angles are known.

The angles are measured experimentally by quenching the system and deriving the angles by examining a cross-section of the solid body.

Solid/Liquid Interface Energies

If liquid 3 in Figure 3.18b is replaced by a planar solid phase (Figure 3.19) the equilibrium condition will be

$$\sigma_{13} - \sigma_{23} = \sigma_{12}\cos\theta_2 \tag{3.26}$$

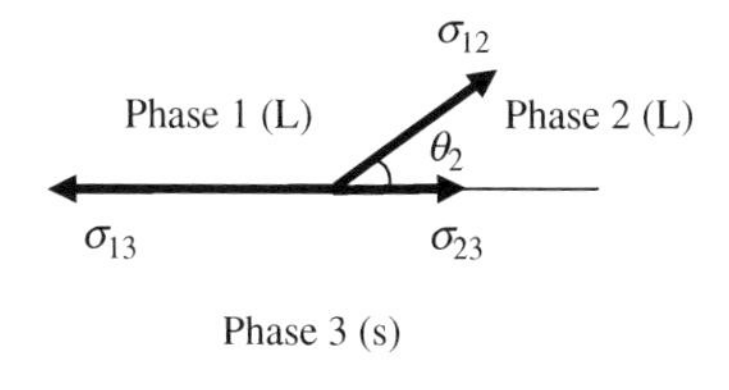

Figure 3.19 Surface-tension forces in equilibrium.

If $\sigma_{13} - \sigma_{23} > \sigma_{12}$ Equation (3.26) cannot be satisfied by any value of θ_2. This situation corresponds to complete wetting. An example is copper spreading over a steel surface. Equation (3.26) is also the reason why a metal melt stays in a sand form while water just runs through.

Measurements of the angle θ_2 is performed in the same way as described above. Figure 3.20 shows the surface tnesion balance between two different phases under equilibrium. Consider an alloy at the end of the solidification (Figure 3.21). Solid grains have been formed and the remaining liquid phase (phase 2) occurs as small volumes among the grains. The qualities of the alloy depend very much on the interface energies between the three phases. The angle θ is determined by the equilibrium condition

$$\sigma_{11} = 2\sigma_{12}\cos\frac{\theta}{2} \tag{3.27}$$

θ is called the dihedral angle. It represents the angle between two adjacent interphase boundaries at a junction point in the microstructure at equilibrium.

Figure 3.20 At equilibrium the three forces are in balance with each other.

Figure 3.21 The force balance between two grains and the remaining liquid in a solidifying alloy. GB = subscript for the grain-boundary interface. SL = subscript for the solid/liquid interface.

If θ is known the ratio of the interface energies can be obtained. The method to measure θ is to anneal the alloy at constant temperature for many hours to make sure that equilibrium is achieved. Then, the alloy is quenched to solidify the liquid phase and cut into planar sections and analyzed. The dihedral angle is measured. This procedure can be repeated at a range of different temperatures.

The result varies very much depending on the ratio between the interface energies. Two examples will illustrate this.

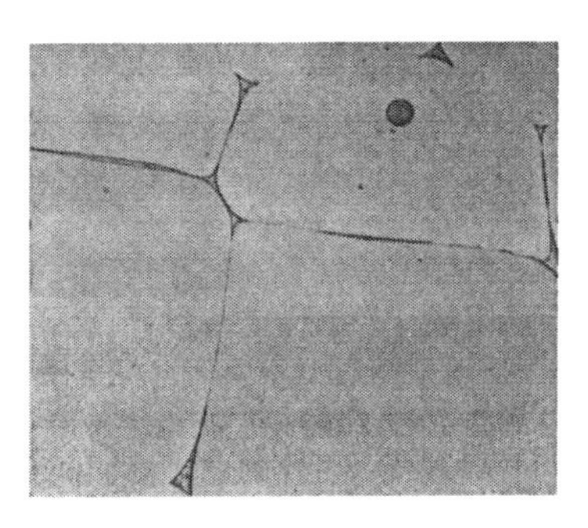

Figure 3.22 Cu-grains+liquid Bi. The alloy was annealed at 750 °C for 16 hours, quenched and dichromate etched. Reproduced from Springer © 1948.

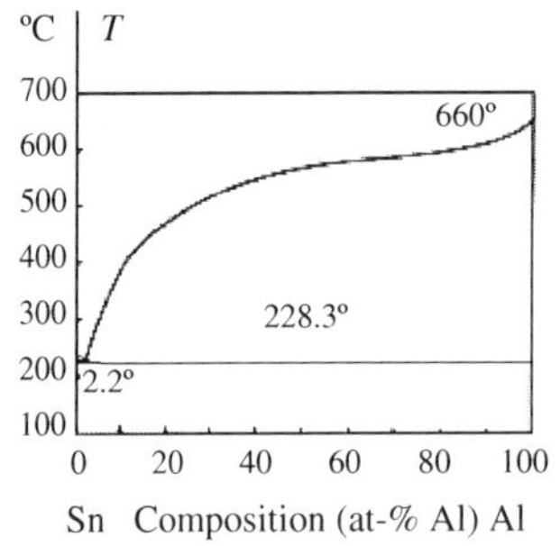

Figure 3.23 Phase diagram of the binary system Sn–Al. The temperature determines the composition of the liquid. Max Hansen © McGraw Hill 1958.

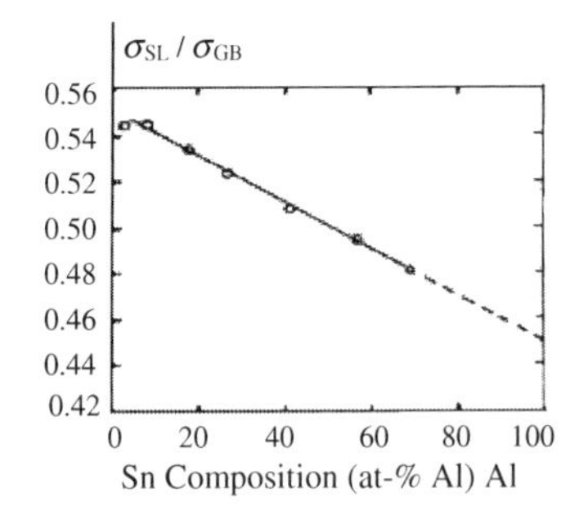

Figure 3.24 The relative interface energy of Sn–Al alloys as a function of the Al concentration. From Solidification, American Society for Metals 1971. Reprinted with permission of ASM International. All rights reserved. www.asminternational.org.

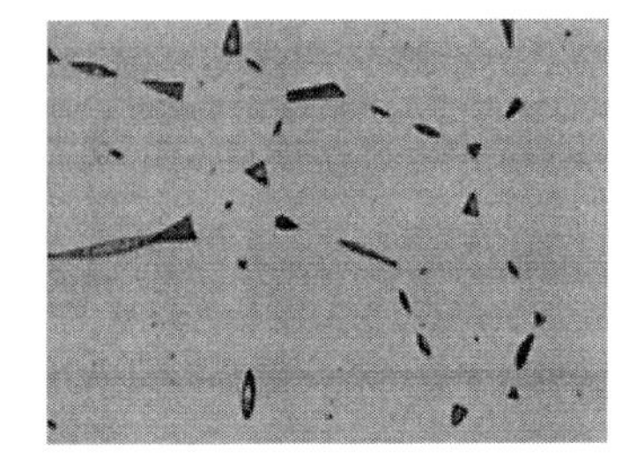

Figure 3.25 Structure of an alloy Al–Sn 10% annealed at 475 °C for 24 h and then quenched in water. Courtesy of R. T. Southin, Solidification, American Society for Metals 1971. Reprinted with permission of ASM International. All rights reserved. www.asminternational.org.

The first example is copper alloyed with small amounts of bismuth (Figure 3.22). In this case the dihedral angle is zero (complete wetting) and the liquid spreads over all the boundary surfaces as a thin film before solidification. This makes the alloy extremely brittle.

The second example is Sn–Al alloys. The dihedral angles have been measured for a number of alloys of different liquid composition in the range from eutectic composition (Figure 3.23) up to the composition which gives $\theta = 0$. Figure 3.24 shows that the ratio of the interface energies is proportional to the atomic per cent of Al dissolved in the liquid. The structure of one of the alloys is seen in Figure 3.25.

Solid/Solid Interface Energies

Consider a case where three solid phases (grains) according to Figure 3.26 meet in a junction point. No immediate change will occur to fulfill the condition of equilibrium as in the case when a liquid is involved. The grains are not completely rigid and can, at sufficiently high temperatures, slowly change their shapes by diffusion or plastic deformation until equilibrium is achieved and the relative interface energies and the dihedral angle fulfill Equation (3.28).

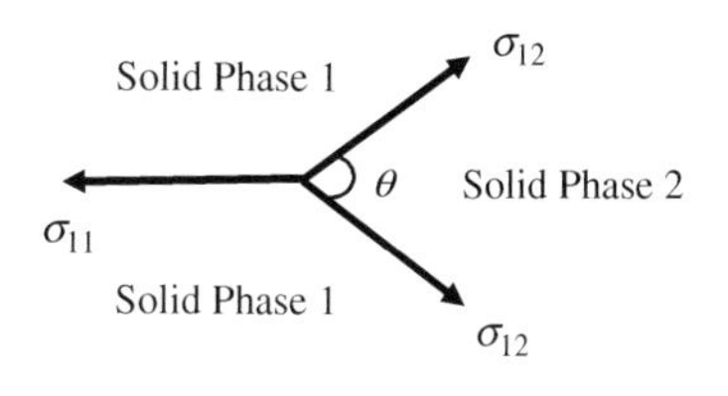

Figure 3.26 Equilibrium between three solid grains in a junction point.

$$\sigma_{11} = 2\sigma_{12} \cos\frac{\theta}{2} \tag{3.28}$$

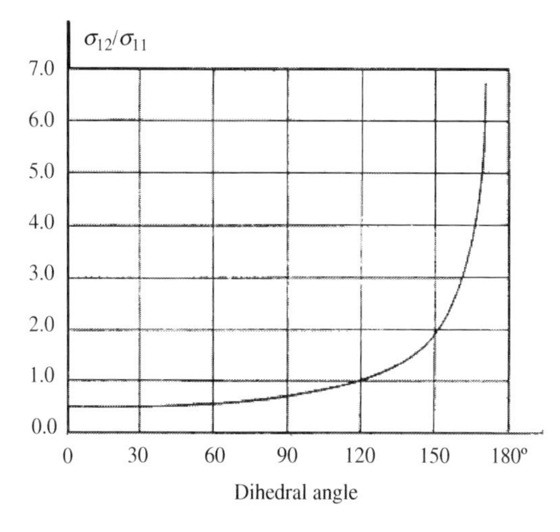

Figure 3.27 Ratio of interface boundary energy σ_{12} and grain-boundary energy σ_{11} as a function of the dihedral angle θ of phase 2 in the θ interval 0–180°. Reproduced from Springer © 1948.

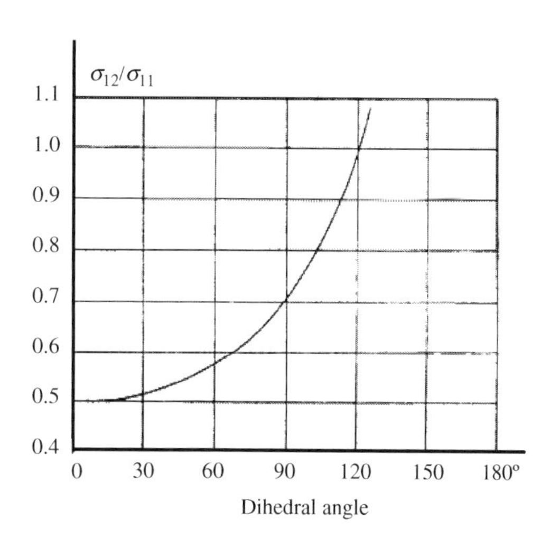

Figure 3.28 Enlargement of the first part of the curve in Figure 3.27. Ratio of interface-boundary energy σ_{12} and grain-boundary energy σ_{11} as a function of the dihedral angle θ of phase 2 in the θ interval 0–120°. Reproduced from Springer © 1948.

Figures 3.27 and 3.28 show the relative interface energies as a function of the dihedral angle. The equilibrium shape of the second solid phase is shown in Figure 3.29.

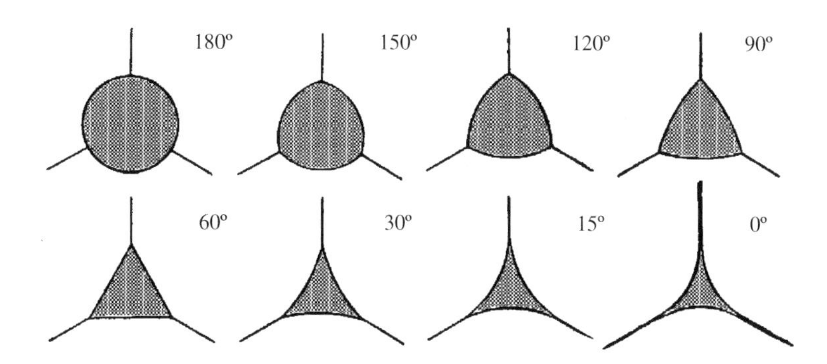

Figure 3.29 The influence of the dihedral angle on the shape of phase 2 at equilibrium in a three-grain junction. Reproduced from Springer © 1948.

Results

In Tables 3.1 and 3.2 some values of interface energies are listed.

Table 3.1 Interface energies of some metals and semiconductor elements

Metal	Liquid–Vapour		Solid–Vapour		Solid–Liquid	
	Interface energy J/m^2	Temperature °C	Interface energy J/m^2	Temperature °C	Interface energy J/m^2	Temperature °C
Al	0.866	658	0.980	450	0.093	658
Au	1.140	1063	1.400	1000	0.132	1063
Bi			0.500	240	0.061	271
Cu	1.300	1083	1.520	1000	0.177	1083
Fe BCC	1.880	1535	1.930	1475	0.204	1535
Fe FCC	1.880	1535	2.100	1350	0.204	1535
Ni	1.780	1455	1.940	1400	0.255	1455
Pt	1.800	1773	1.950	1700	0.240	1773
Sn	0.550	232	0.670	200	0.055	232
Ge	0.610	960				
In	0.560	156	0.630	140		
Sb	0.380	640				
Si	0.730	1410				

Table 3.2 Interface energies of some molten salts and inorganic materials

Salts or oxides	Liquid–Vapour		Non metallic compounds (structure)	Solid–Vapour	
	Interface energy J/m^2	Temperature °C		Interface energy J/m^2	Temperapture °C
Al_2O_3	0.700	2353	Al_2O_3	0.905	1850
Ag Cl	0.179	728	$CaCO_3$ (1010)	0.230	25
B_2O_3	0.080	1173	CaF_2 (111)	0.450	25
$CaBr_2$	0.120	1015	KCl	0.110	25
CuCl	0.092	703	MgO	1.000	25
FeO	0.585	1693	NiO	1.100	25
LiCl	0.196	883	TiC	1.190	1100
LiF	0.236	121	LiF (100)	0.340	25
NaCl	0.114	1074	NaCl (100)	0.300	25
NaF	0.186	1269			
$NaNO_3$	0.120	583			
Na_2SO_4	0.195	1157			
$SnCl_2$	0.104	518			

Conclusions

From the data in Tables 3.1 and 3.2 we can conclude that

1. The liquid/vapour interface energies of salts and oxides are generally lower than those of metals.
2. The solid/vapour interface energies of salts and oxides are generally lower than those of metals.
3. The solid/liquid interface energies of metals are roughly ten times smaller than their solid/vapour interface energies.
4. The solid/liquid interface energies are generally smaller than the difference between their solid/vapour and liquid/vapour interface energies.

Condition for Complete Wetting

We reconsider the concept of wetting introduced in Section 3.2.5 on page 105. Wetting means that the liquid covers the solid substrate. This occurs when the contact angle is zero. The equilibrium condition for the tension forces is

$$\sigma_{SV} = \sigma_{SL} + \sigma_{LV} \cos\theta \tag{3.20}$$

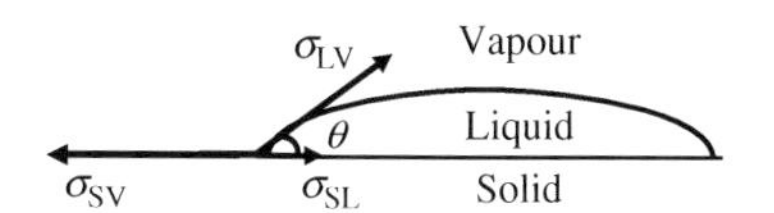

Figure 3.12 The shape and extension of the liquid droplet on a solid surface depends of the balance between the three surface tension forces.

At compete wetting, $\theta = 0$ and the factor cos $\theta = 1$, which is its maximum value. Hence, the condition for complete wetting is

$$\sigma_{SV} = \sigma_{SL} + \sigma_{LV} \tag{3.29}$$

If the wetting is *not* complete (Figure 3.13 on page 108) we have

$$\sigma_{SV} \leq \sigma_{SL} + \sigma_{LV} \tag{3.30}$$

If the condition

$$\sigma_{SV} \geq \sigma_{SL} + \sigma_{LV} \tag{3.31}$$

is valid, the wetting is certainly complete.

Solid/Vapour and Liquid/Vapour Interface Energies

The condition for complete wetting, Equation (3.29), can be simplified if we neglect the solid/liquid interface energy, which is known to be low (point 3 on page 112).

Complete wetting occurs if the following simple relationship holds

$$\sigma_{SV} \geq \sigma_{LV} \tag{3.32}$$

This is the case for all the metals in Table 3.1. A comparison between Tables 3.1 and 3.2 shows that the relationship (3.32) is fulfilled when a liquid layer of an oxide is in contact with the corresponding metal or other solid metals. This is valid for most metals.

- Liquid metals wet their own solids.
- Oxides normally wet metals.

This is of great importance in the technologies of brazing and soldering. A liquid metal or alloy serves as the necessary 'glue' to join two solid metal pieces rigidly. We will come back to this in a later chapter.

Another very important application is the joining of integrated circuit chips to their packages. Electrical connections to hundreds of identical microscopic circuits on a chip are made simultaneously. Each molten drop of solder is prevented from spreading beyond its connector area by oxides separating the gold-plated connectors. The thin gold layer prevents undesired oxidation of the connector surfaces.

Solid/Liquid Interface Energies

As mentioned above the solid/liquid interface energy is much smaller than the surface free energies.

Rough experimental rules to calculate the molar solid/liquid interface energy are
for *Metals*:

$$\text{The molar solid/liquid interface energy} \approx \sigma_{SL}\frac{V_m}{d} = 0.5(-\Delta H) \tag{3.33a}$$

where $(-\Delta H)$ = the molar heat of fusion.

$$\text{Another rough expression for the molar solid/liquid interface energy} \approx N_A^{1/3}\, V_m^{2/3}\, \sigma_{SL} \tag{3.33b}$$

where N_A is Avogadro's number.

for *Semi-conductors and Organic Compounds*:

$$\text{The molar solid/liquid interface energy} \approx \sigma_{SL}\frac{V_m}{d} = 0.5(-\Delta H) \tag{3.33c}$$

where $(-\Delta H)$ = the molar heat of fusion.

3.3 Thermodynamics of Interphases

3.3.1 Two Thermodynamic Models of Interfaces

There are two radically different approaches to thermodynamics of interfaces, 'Gibbs' classical model and Guggenheim's model, developed at the end of the 1920s.

According to *'Gibbs' model* (Figure 3.30a) the interface is treated as an infinitely thin surface, called the dividing surface, which separates two phases of different structure. The advantage of this model is that a thermodynamic description of the interface is possible without detailed knowledge of the structure of the intermediate region.

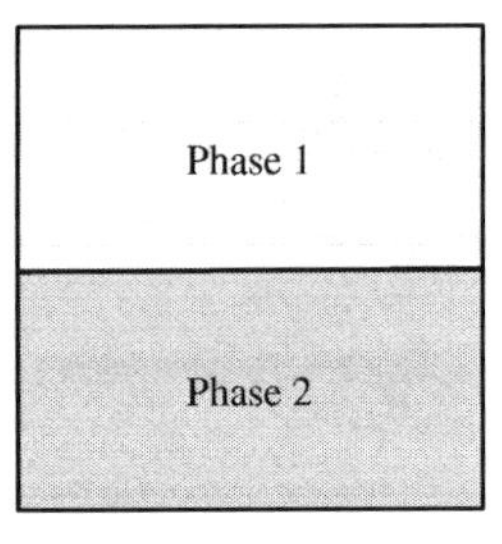

Figure 3.30a Gibbs' thermodynamic model of an interface is a dividing plane of zero thickness.

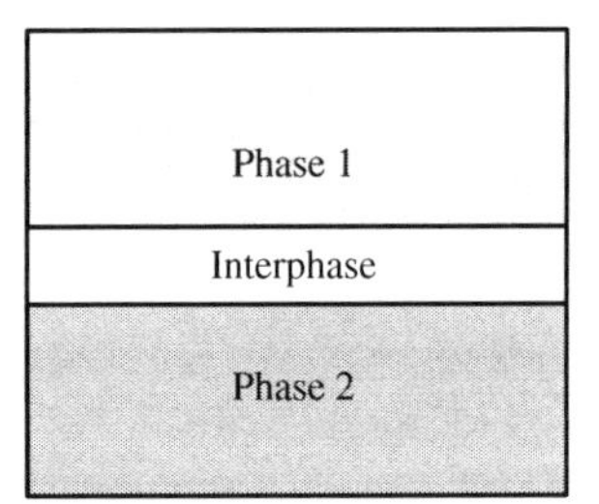

Figure 3.30b Guggenheim's thermodynamical model of an interphase is a layer of finite and uniform thickness.

According to *Guggenheim's pseudo-thermodynamic model* (Figure 3.30b) the interface is regarded as a separate phase, the interface phase or shorter *interphase*. It has a finite thickness and is treated thermodynamically as a third phase between two neighbouring homogeneous phases.

The aim of theoretical models is to give a fruitful and good description of the experimental reality. A demand and a test of good models is that they can successfully predict new facts as a guide to further experimental research. There is no contradiction in using different models on various occasions.

Guggenheim's model has proved to be successful so far when dealing with recrystallization in metals and adsorption of solutes in the grain boundaries. We will use both models in this chapter but preferably Guggenheim's model when dealing with adsorption.

3.3.2 Gibbs' Model of Interphases

Figure 3.31a shows a body of infinite size ($R = \infty$). The surface area per unit volume is zero as the ratio area to volume is zero. Let the Gibbs' free energy per kmol of the infinite body be G^0. In this body we select a small volume containing n kmol without separating it from the rest of the body (Figure 3.31a). The Gibbs' free energy of the n kmol is nG^0.

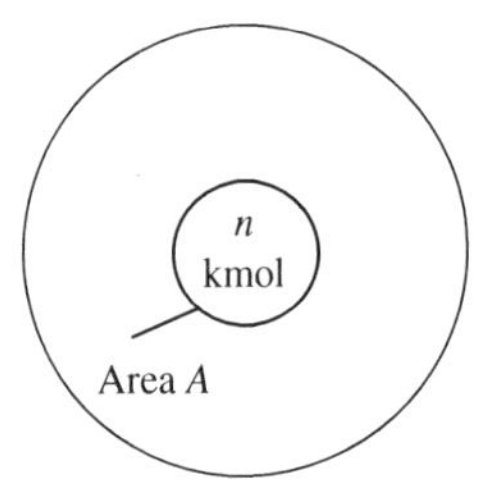

Figure 3.31a A body of infinite size. A selected volume of material contains n kmol of matter in bulk.

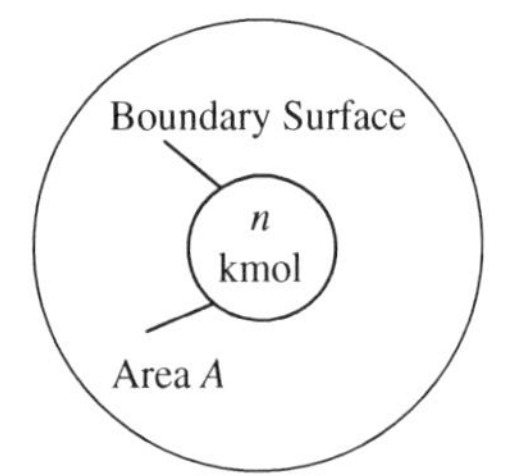

Figure 3.31b Selected volume of material containing n kmol separated from the rest of the infinite body.

When the selected volume is separated from the infinite body, it becomes associated with an interface of area A. The ratio area to volume is no longer zero. The atoms within the selected volume become differently arranged, especially those near the surface.

The solute atoms are redistributed to fulfil the condition of a constant chemical potential independent of position. The result is that the Gibbs' free energy per kmol changes from G^0 to G_A within the selected body. The total Gibbs' free energy of n kmol of material, included within the area A, equals nG_A and exceeds nG^0.

The excess Gibbs' free energy G^{Ex} is defined as

$$G^{Ex} = \frac{nG_A - nG^0}{A} \tag{3.34}$$

Instead of developing expressions for G^{Ex} we will use Gibbs' general definition of the free energy of the interface given in Section 3.2.1 on page 100.

$$G^I = H^I - TS^I \tag{3.1}$$

where the superscript I stands for interface and the free energy G^I, the enthalpy H^I and the entropy S^I refers to one kmol. The thermodynamic quantities are always measured *per kmol.*

Consider for example a solid metal, surrounded by its melt. The thermodynamic quantities of the solid, its Gibbs' interface and the melt are shown in Figure 3.32.

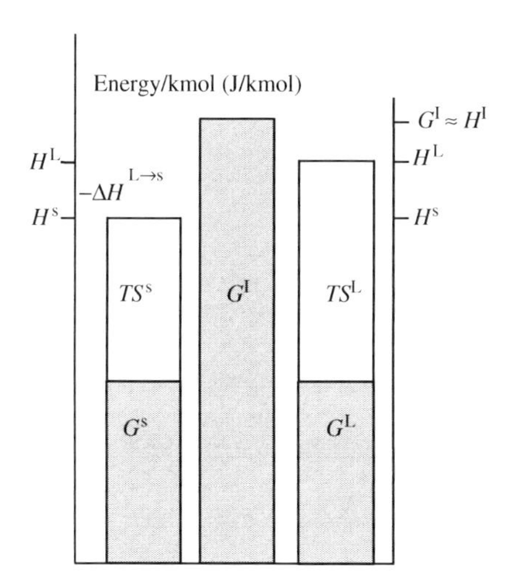

Figure 3.32 Solid/liquid interface. Thermodynamic quantities of the solid phase, the Gibbs' interface and the liquid phase.

The Gibbs' interface has zero thickness and its entropy is very low. It is customary to set S^I equal to zero and consequently $H^I \approx G^I$. The interface region has a high enthalpy (surface energy or surface tension). H^I is always larger than H^s and either smaller or larger than H^L. The latter alternative is shown in Figure 3.32.

Equation (3.2) is not simply applicable on a Gibbs' interface as the thickness of the interface is zero. Instead, the enthalpies of the surfaces are studied in terms of the energy of broken bonds between the atoms.

Gibbs' model of interfaces is applicable to crystallographic surfaces with low Miller indices (page 16 in chapter 1 in [1]) in solids, to surfaces that form low angles with such surfaces and to planar interfaces between solid–vapour and solid–liquid phases.

3.3.3 Guggenheim's Model of Interphase

So far, we have treated interfaces as 'mathematical' surfaces without thickness. Guggenheim developed his theory based on the assumption that an interface, which we simply call *interphase,*

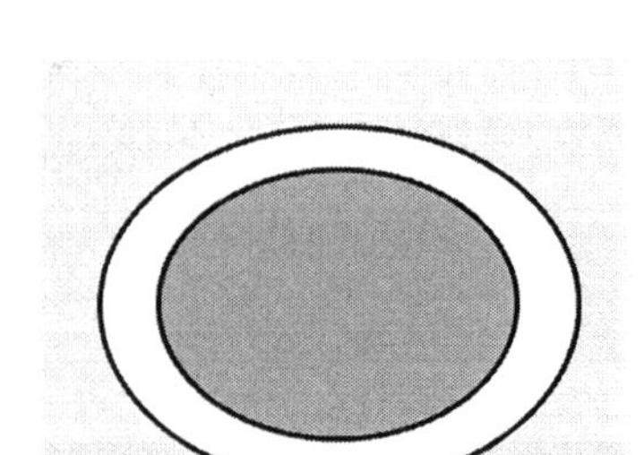

Figure 3.33 A solid grain in a liquid surrounded by a boundary phase (interphase) of finite thickness.

can be described as a separate phase with homogeneous properties and a finite thickness (Figure 3.33). The model is used for its simplicity. It describes surfaces with the aid of the usual thermodynamic quantities and relationships. The thermodynamic quantities are expressed per kmol instead of per unit area.

A great advantage of the model is that it is possible to find a relationship between thermodynamic quantities and surface tension or interphase energy, which is expressed per unit area. One has to use either Equation (3.34) on page 115 or Equation (3.2) on page 100 to transfer the interface energy per unit area into molar quantities.

We will apply the general model on solid/liquid interphases and demonstrate the close relationship between the interphase energy and the thermodynamic quantities. The discussion can be applied on any type of interphase between two phases.

Next, we will apply the model on a more specific problem and use it to analyze the segregation of alloying elements to grain boundaries, liquid/vapour and liquid/solid interphases.

Gibbs' Free Energies of Solid/Liquid and Solid/Vapour Interphases

Solid/Liquid Interphases

According to the general condition of equilibrium the Gibbs' free energies of a solid and a liquid in contact with each other are equal.

$$G^{s} = H^{s} - TS^{s} = H^{L} - TS^{L} = G^{L} \tag{3.35}$$

A principle sketch of the energy a liquid/solid interphase is shown in Figure 3.34.

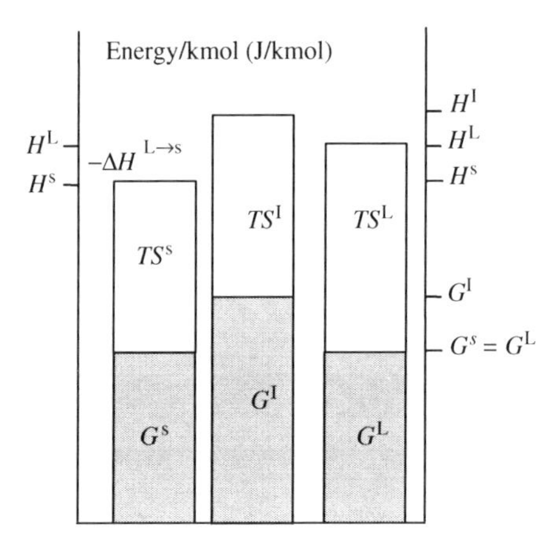

Figure 3.34 Solid/liquid interphase. Thermodynamic quantities of the solid phase, the interphase and the liquid phase. $-\Delta H^{L\to s}$ is the heat of fusion.

It is well known that the *enthalpies* of the two phases differ by the amount of heat of fusion. Heat has to be added to the solid to melt it. Therefore, $H^{L} > H^{s}$ and we have

$$-\Delta H^{L\to s} = H^{L} - H^{s} \tag{3.36}$$

The entropy cannot be neglected for Guggenheim interphases. As the free Gibbs' energies of the two phases are equal, the *entropy* of the liquid must be higher than that of the solid. The liquid has a more disordered structure than the solid and consequently a higher entropy. The entropy of the solid and the liquid are both positive and the Gibbs' free energies G^{s} and G^{L} consequently *smaller* than the corresponding enthalpies.

At the melting point the liquid and solid phases are in equilibrium with each other, which can be expressed as

$$G^{s} = G^{L} \tag{3.37}$$

The Gibbs' free energy of the *interphase* can be described in the same way as for liquids and solids. In analogy with Equation (3.35) we obtain

$$G^{\mathrm{I}} = H^{\mathrm{I}} - TS^{\mathrm{I}} \tag{3.1}$$

The interphase is not in equilibrium with the solid and liquid phases and is assumed to consist of a mixture of 'solid' atoms and 'liquid atoms' with slightly different properties. The mixing of the atoms result in mixing entropy, which has to be considered.

In the case of a liquid/solid interphase the enthalpy of the interphase is larger than that of the solid but smaller than that of the liquid. The following relationships are valid for the interphase:

$$-\Delta H^{\mathrm{I}\rightarrow\mathrm{s}} = H^{\mathrm{I}} - H^{\mathrm{s}} \tag{3.38}$$

$$-\Delta H^{\mathrm{L}\rightarrow\mathrm{I}} = H^{\mathrm{L}} - H^{\mathrm{I}} \tag{3.39}$$

($-\Delta H^{\mathrm{I}\rightarrow\mathrm{s}}$) equals the surface tension of the solid, i.e.

$$-\Delta H^{\mathrm{I}\rightarrow\mathrm{s}} = \frac{\sigma V_{\mathrm{m}}}{d} \tag{3.40}$$

where
V_{m} = molar volume ($\mathrm{m^3/kmol}$)
d = thickness of the interphase.

The factor V_{m}/d is a unit factor that transfers the surface energy unit $\mathrm{J/m^2}$ into the thermodynamic energy unit J/kmol.

Solid/Vapour Interphases
Figure 3.35 shows a principle sketch of a vapour/solid interphase. It can be compared with the corresponding liquid/solid interphase in Figure 3.34.

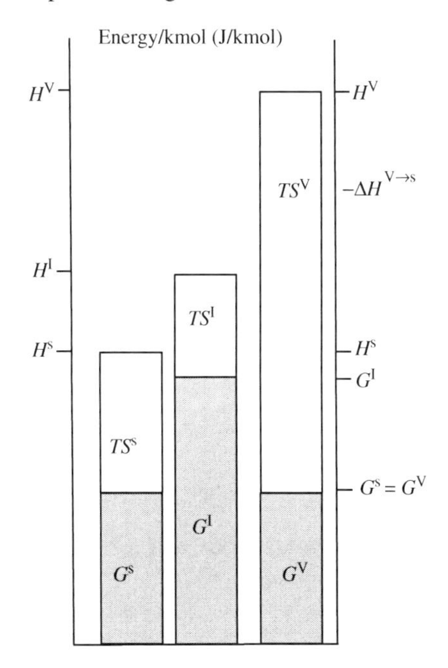

Figure 3.35 Solid/vapour interphase. Thermodynamic quantities of the solid phase, the interphase and the vapour phase. $-\Delta H^{\mathrm{V}\rightarrow\mathrm{s}}$ is the evaporation energy.

The vapour and solid phases are in equilibrium with each other, which means that $G^{\mathrm{V}} = G^{\mathrm{s}}$.

The heat of evaporation is always larger, often very much larger, than the heat of fusion. Hence, $H^{\mathrm{V}} \gg H^{\mathrm{s}}$, which is indicated in Figure 3.35.

The entropy of the vapour is very much larger than that of the solid and also much larger than that of the liquid. The reason is the degree of disorder in the three phases.

The entropy of the interphase depends of the structure of the rough surface and cannot be neglected. The structure of the interphase is much more like the solid structure than that of the vapour. Therefore, the enthalpy H^I is located much closer to H^s than to H^V.

Segregation of Alloying Elements in Binary Alloys to Grain Boundaries or Liquid/Vapour Interphases

Guggenheim's model is very useful for analyzing segregation of alloying elements to grain boundaries or liquid/vapour interphases. It can be used to describe the drastic decrease of surface tension and interphase energy shown in Figures 3.8, 3.9 and 3.10 on page 105 when small amounts of sulfur or oxygen are added to iron and copper.

These phenomena will be discussed in terms of Gibbs' free energy of the type introduced into Chapter 2. Consider as a basic example an alloy melt and its interphase surrounded by a vapour (Figure 3.36a). Figure 3.36b shows the free-energy curves of the liquid phase and the interphase. It can be used to find the relationship between compositions of the two phases graphically.

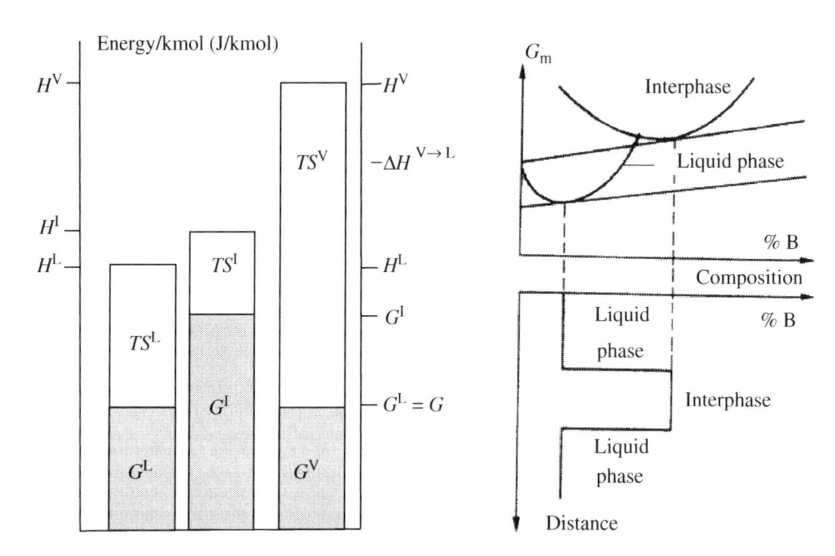

Figure 3.36a Liquid/vapour interphase of an alloy. Thermodynamic quantities of the liquid phase, the Guggenheim interphase and the vapour phase. $-\Delta H^{V \to L}$ is the heat of evaporation.

Figure 3.36b Guggenheim interphase and liquid phase of the alloy melt in Figure 3.36a. Upper figure: Gibbs' free-energy curves of the liquid phase and the Guggenheim interphase.

Lower figure: The concentration of solute B in the two phases as functions of position.

The Gibbs' free-energy curves in the upper Figure 3.36b can be used to find the composition of interphase. The compositions of the liquid phase and the interphase are found by drawing parallel tangents to the two Gibbs' free energy curves.

The lower tangent to the liquid curve is drawn through the given concentration of element B in the liquid. The intersection point between the interphase curve and the tangent corresponds to the concentration of element B in the interphase. The theoretical basis for this procedure is given in Chapter 2 on page 77.

The lower Figure 3.36b shows that the concentration of element B is higher in the interphase than that in the liquid. Element B has been segregated in the interphase.

Guggenheim's theory is used as a basis for the present thermodynamic treatment of interphases and especially adsorption of elements on surfaces.

3.3.4 *Partition of Alloying Element between Liquid and Interphase in Binary Alloys in Equilibrium with a Gas or a Solid*

Consider a liquid in contact with an interphase (Figure 3.37). A pure solvent (subscript 1) contains a solute (subscript 2). We will examine the concentration of the solute in the interphase (superscript I) and compare it to the concentration of the solute in the liquid phase (superscript L). The comparison will be done by calculation of the *distribution coefficient* or *partition coefficient*.

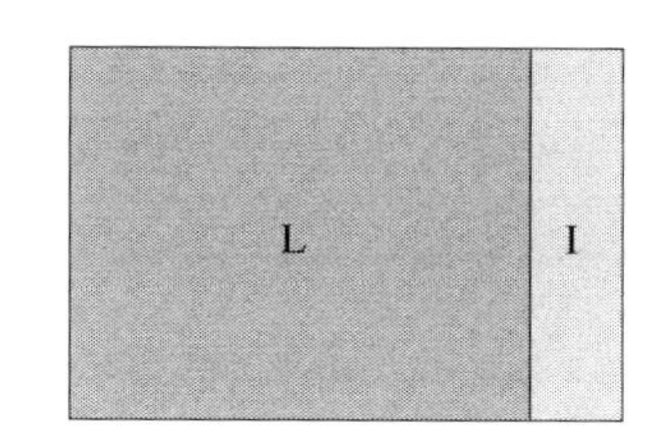

Figure 3.37 A liquid in contact with an interphase.

The interphase I is obviously a Guggenheim interphase. The definition of the interface Gibbs' free energy can be applied on the present interphase. The Gibbs' free energy G is expressed in the unit J/kmol. The surface energy can alternatively be regarded as the surface energy σ of the interphase, which is expressed in J/m^2 and a factor V_m^I/d has to be introduced in analogy with Equation (3.2) on page 100 and Equation (3.40) on pages 117.

$$G_1^{0I} - G_1^{0L} = \frac{\sigma_1^0 V_m^I}{d} \tag{3.41}$$

$$G_2^{0I} - G_2^{0L} = \frac{\sigma_2^0 V_m^I}{d} \tag{3.42}$$

where
G_i^{0I} = Gibbs' free energy per kmol of pure component i (i = 1 or 2) in the interphase
G_i^{0L} = Gibbs' free energy per kmol of pure component I (i = 1 or 2) in the liquid phase
σ_i^0 = surface tension of pure component i (i = 1 or 2)
V_m^I = molar volume of the interphase.

Since the Gibbs' free energy of the interphase is an *increase* of the total Gibbs' free energy of the system, the normal thermodynamic relationships for equilibrium (equal chemical potentials in the two phases of components 1 and 2, respectively) can *not* be used. Instead, we will use the fact that *the free energy increase is a minimum at equilibrium*. This is achieved when the tangents to the two Gibbs' free energy curves are parallel, according to the discussion on pages 117–118. This condition can be expressed in thermodynamic terms as

$$G_1^I - G_1^L = G_2^I - G_2^L \tag{3.43}$$

The total Gibbs' free energy of each component can be written

$$G_i = G_i^0 + RT\ln x_i + G_i^{Ex} \tag{3.44}$$

Equation (3.44) is applied and introduced into Equation (3.43) for each component and each phase and we obtain for the interphase

$$G_2^I - G_1^I = G_2^{0I} + RT\ln x_2^I + G_2^{Ex\,I} - \left(G_1^{0I} + RT\ln x_1^I + G_1^{Ex\,I}\right) \tag{3.45}$$

and for the liquid phase

$$G_2^L - G_1^L = G_2^{0L} + RT\ln x_2^L + G_2^{Ex\,L} - \left(G_1^{0L} + RT\ln x_1^L + G_1^{Ex\,L}\right) \tag{3.46}$$

where
$G_1^{Ex\,I}$ = excess Gibbs' free energy of solvent in the interphase
$G_1^{Ex\,L}$ = excess Gibbs' free energy of solvent in the liquid
$G_2^{Ex\,I}$ = excess Gibbs' free energy of solute in the interphase
$G_2^{Ex\,L}$ = excess Gibbs' free energy of solute in the liquid
x_i^I = concentration of component i in the interphase
x_i^L = concentration of component i in the liquid phase.

The expressions on the right-hand sides of Equations (3.45) and (3.46) are equal according to Equation (3.43). After rearranging the terms we obtain

$$\frac{x_2^{\mathrm{I}}}{x_1^{\mathrm{I}}} = \frac{x_2^{\mathrm{L}}}{x_1^{\mathrm{L}}} \mathrm{e}^{-\frac{\left(G_2^{0\mathrm{I}} - G_2^{0\mathrm{L}}\right) - \left(G_1^{0\mathrm{I}} - G_1^{0\mathrm{L}}\right) + \left(G_2^{\mathrm{Ex\ I}} - G_2^{\mathrm{Ex\ L}}\right) - \left(G_1^{\mathrm{Ex\ I}} - G_1^{\mathrm{Ex\ L}}\right)}{RT}}$$

or

$$\frac{x_2^{\mathrm{I}}}{x_2^{\mathrm{L}}} = \frac{x_1^{\mathrm{I}}}{x_1^{\mathrm{L}}} \mathrm{e}^{-\frac{\sigma_2^0 - \sigma_1^0}{RT} \frac{V_{\mathrm{m}}^{\mathrm{I}}}{d}} \mathrm{e}^{-\frac{\Delta G_2^{\mathrm{Ex\ I}} - \Delta G_1^{\mathrm{Ex\ I}}}{RT}} \tag{3.47}$$

where

$\Delta G_2^{\mathrm{Ex\ I}} = G_2^{\mathrm{Ex\ I}} - G_2^{\mathrm{Ex\ L}}$

$\Delta G_1^{\mathrm{Ex\ I}} = G_1^{\mathrm{Ex\ I}} - G_1^{\mathrm{Ex\ L}}$

Equation (3.47) is the desired expression of the partition coefficient of the solute in the two phases.

A high partition coefficient or concentration of the solute to the interphase is favoured if the second factor on the right-hand side of Equation (3.47) is as large as possible.

Equation (3.47) shows that this condition is fulfilled if

$$G_1^{\mathrm{Ex\ I}} \approx G_1^{\mathrm{Ex\ L}} \quad \text{and} \quad G_2^{\mathrm{Ex\ L}} \gg G_2^{\mathrm{Ex\ I}}.$$

The differences in excess Gibbs' free energies of the solute can be written

$$G_2^{\mathrm{Ex\ L}} - G_2^{\mathrm{Ex\ I}} = -RT \ln \frac{\gamma_2^{\mathrm{L}}}{\gamma_2^{\mathrm{I}}} \tag{3.48}$$

$$G_1^{\mathrm{Ex\ L}} - G_1^{\mathrm{Ex\ I}} = -RT \ln \frac{\gamma_1^{\mathrm{L}}}{\gamma_1^{\mathrm{I}}} \tag{3.49}$$

where the γ quantities are the activity coefficients of the solvent and the solute in the liquid phase and the interphase, respectively. If we introduce them instead of the excess quantities into Equation (3.47) we obtain the *partition coefficient* or *segregation ratio* β

$$\beta = \frac{x_2^{\mathrm{I}}}{x_2^{\mathrm{L}}} = \frac{x_1^{\mathrm{I}}}{x_1^{\mathrm{L}}} \frac{\gamma_1^{\mathrm{I}}}{\gamma_1^{\mathrm{L}}} \frac{\gamma_2^{\mathrm{L}}}{\gamma_2^{\mathrm{I}}} \mathrm{e}^{-\frac{\Delta G^{\mathrm{I}}}{RT}} \tag{3.50}$$

where

$$\Delta G^{\mathrm{I}} = \left(\sigma_2^0 - \sigma_1^0\right) \frac{V_{\mathrm{m}}^{\mathrm{I}}}{d}$$

If we consider dilute solutions and 'Raoult's law (Section 1.7.1 page 26–29 in Chapter 1) is valid, then x_1^{I}, x_1^{L}, γ_1^{L} and γ_1^{I} in Equation (3.50) are all approximately equal to 1 and we obtain

$$\beta = \frac{x_2^{\mathrm{I}}}{x_2^{\mathrm{L}}} = \frac{\gamma_2^{\mathrm{L}}}{\gamma_2^{\mathrm{I}}} \mathrm{e}^{-\frac{\sigma_2^0 - \sigma_1^0}{RT} \frac{V_{\mathrm{m}}^{\mathrm{I}}}{d}} \tag{3.51}$$

Figure 3.38 shows some experimental values of the segregation ratio for adsorption at grain boundaries as a function of the solid solubility x_2^* of the solute in metallic hosts. The figure shows that Equation (3.51) seems to be valid also for complete solid solubility ($x_2^* = 1$). The Equation (3.51) is general and not specific to metals. Hence, it can be applied not only to metals but also to ceramics.

As we have found on page 104–105 the presence of impurities in a metal has a great effect on its surface tension. The surface tension decreases dramatically even at small amounts of the impurity. As we can conclude from Equation (3.48) the surface tension strongly influences the segregation ratio of the impurity in the two phases.

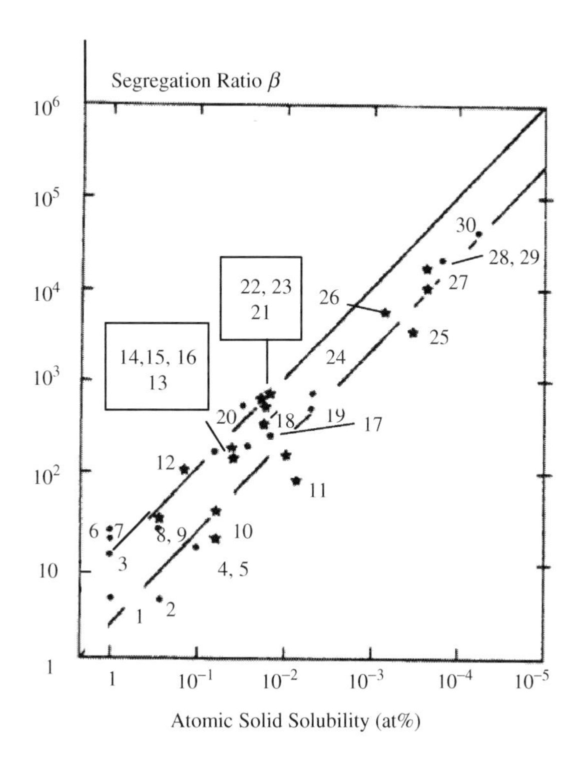

Figure 3.38 Measured grain-boundary segregation ratios as a function of atomic solid solubility.

Figure explanation: 1 Cu–Fe, 2 Fe–Si, 3 γFe–Ni, 4 δFe–Si, 5 Cu–Sb, 6 γFe–Mn, 7 γFe–Cr, 8 γFe–Si, 9 αFe–Al, 10 αFe–Ni, 11 αFe–Cu, 12 αFe–Zn, 13 αFe–P, 14 αFe–Sn, 15 αFe–As, 16 δFe–Sn, 17 γFe–P, 18 αFe–P, 19 δFe–P, 20 αFe–N, 21 αFe–P, 22 αFe–Sn, 23 αFe–Sb, 24 Cu–Sb, 25 αFe–B, 26 Ni–B, 27 αFe–C, 28 Cu–B, 29 αFe–S, 30 Cu–B. Reproduced from Elsevier © 1983.

Example 3.1

A copper melt contains 1.2 at% sulfur. The temperature of the melt is 1120 °C. The surface tension as a function of the composition of the melt at this temperature is known and given in the diagram.

Calculate the partition coefficient of sulfur between the melt and its interphase at the given temperature.

The thickness of the interphase is estimated to 10^{-9} m.

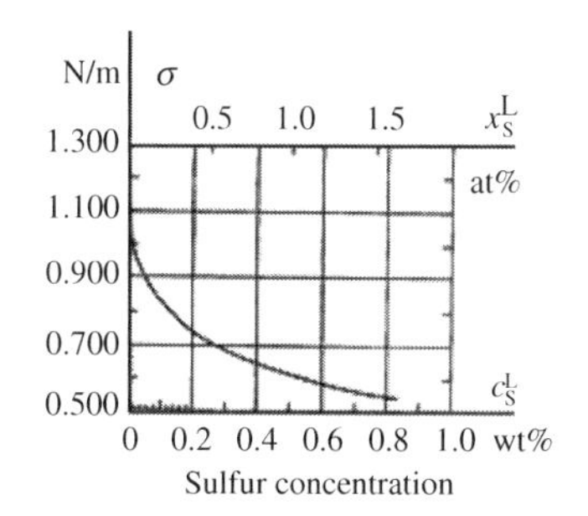

Solution:

Equation (3.51) cannot simply be applied as the surface tension of pure sulfur cannot be obtained from the diagram. Instead, we study the Gibbs' free energy of the solvent Cu in the interphase and in the solution. Equations (3.44) and (3.49) give:

$$G_{Cu}^{I} - G_{Cu}^{L} = G_{Cu}^{0I} + RT \ln x_{Cu}^{I} \gamma_{Cu}^{I} - \left(G_{Cu}^{0L} + RT \ln x_{Cu}^{L} \gamma_{Cu}^{L}\right) \quad (1')$$

The sulfur solution can be regarded to be dilute, which gives for the solvent $\gamma_{Cu}^{I} = \gamma_{Cu}^{L} = 1$ and we obtain

$$G_{Cu}^{I} - G_{Cu}^{L} = G_{Cu}^{0I} - G_{Cu}^{0L} + RT\left(\ln x_{Cu}^{I} - \ln x_{Cu}^{L}\right) \quad (2')$$

The differences in Gibbs' free energy are related to the interphase energy. Equation (3.38) gives in this case for pure copper

$$G_{\text{Cu}}^{0\text{I}} - G_{\text{Cu}}^{0\text{L}} = \frac{\sigma^0 V_{\text{m}}^{\text{I}}}{d} \qquad (3')$$

Analogously, we have for the solution

$$G_{\text{Cu}}^{\text{I}} - G_{\text{Cu}}^{\text{L}} = \frac{\sigma V_{\text{m}}^{\text{I}}}{d} \qquad (4')$$

We substitute the Gibbs' free energy differences in Equations (3′) and (4′) into Equation (2′) and obtain

$$(\sigma - \sigma^0)\frac{V_{\text{m}}}{d} = RT\left(\ln x_{\text{Cu}}^{\text{I}} - \ln x_{\text{Cu}}^{\text{L}}\right) \qquad (5')$$

We want to calculate the segregation ratio of sulfur in the two phases. For this reason, we introduce the atomic fraction of sulfur instead of copper:

$$x_{\text{Cu}}^{\text{I}} = 1 - x_{\text{S}}^{\text{I}} \quad \text{and} \quad x_{\text{Cu}}^{\text{L}} = 1 - x_{\text{S}}^{\text{L}}$$

These expressions are introduced into Equation (5′):

$$(\sigma - \sigma^0)\frac{V_{\text{m}}}{d} = RT\left[\ln\left(1 - x_{\text{S}}^{\text{I}}\right) - \ln\left(1 - x_{\text{S}}^{\text{L}}\right)\right]$$

As the x_{S}^{I} and x_{S}^{L} are small compared to 1, the ln functions can be expanded in series $[\ln(1 - x) \approx -x]$ and we obtain

$$(\sigma - \sigma^0)\frac{V_{\text{m}}}{d} = RT\left(x_{\text{S}}^{\text{L}} - x_{\text{S}}^{\text{I}}\right) = RT\,x_{\text{S}}^{\text{L}}\left(1 - \frac{x_{\text{S}}^{\text{I}}}{x_{\text{S}}^{\text{L}}}\right) = RT\,x_{\text{S}}^{\text{L}}(1 - \beta) \qquad (6')$$

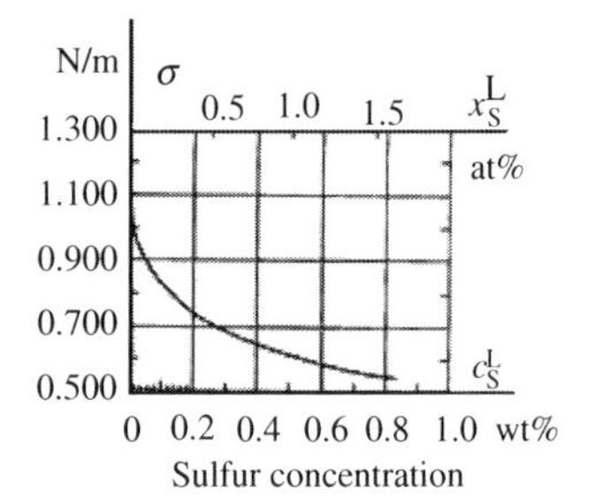

Equation (6′) is solved for the desired partition coefficient β:

$$\beta = 1 + \frac{(\sigma^0 - \sigma)}{RT\,x_{\text{S}}^{\text{L}}}\frac{V_{\text{m}}}{d} \qquad (7')$$

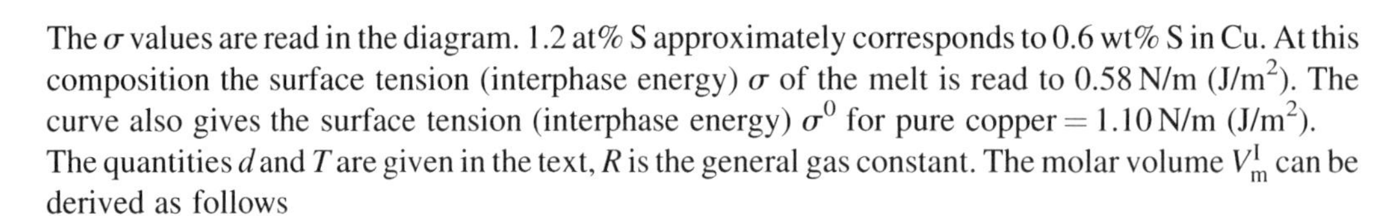

The σ values are read in the diagram. 1.2 at% S approximately corresponds to 0.6 wt% S in Cu. At this composition the surface tension (interphase energy) σ of the melt is read to 0.58 N/m (J/m^2). The curve also gives the surface tension (interphase energy) σ^0 for pure copper = 1.10 N/m (J/m^2). The quantities d and T are given in the text, R is the general gas constant. The molar volume V_{m}^{I} can be derived as follows

$$V_{\text{m}}^{\text{I}} = \frac{M_{\text{Cu}}}{\rho_{\text{Cu}}} = \frac{63.57\ \text{kg/kmol}}{8.93 \times 10^3\ \text{kg/m}^3} = 7.1 \times 10^{-3}\ \text{m}^3\ /\text{kmol}$$

Inserting the values given above we obtain

$$\beta = 1 + \frac{(\sigma^0 - \sigma)V_{\text{m}}}{RT\,x_{\text{S}}^{\text{L}}d} = 1 + \frac{(1.10 - 0.58) \times 7.1 \times 10^{-3}}{8.314 \times 10^3(1120 + 273) \times 0.012 \times 10^{-9}} \approx 28$$

Answer:
The partition coefficient $\beta = x_S^I/x_S^L \approx 30$. Sulfur is concentrated in the interphase.

3.4 Structures of Interfaces

So far, we have not discussed the various physical factors that determine the interface energies and the structures of different interfaces. Very little experimental work has been done in this field owing to practical difficulties. The existing theoretical models are also very rough.

According to a simple analysis there are five contributions to the interface energy per unit area:

1. A chemical interaction between the atoms resulting in an interface energy σ_{ch}, i.e. the energy required to form 1 unit area of an interface by breaking bonds in 'interior' atoms to transform them into 'surface atoms'.
2. An electrochemical potential at the surface, caused by a change of the number of free[1] electrons at the surface interphase, resulting in an interphase energy σ_e.
3. An entropy term σ_S associated with the mixing of different species or atoms at the interphase.
4. A strain (deformation) energy term σ_d associated with straining of the atoms at the interphase (see pages 104–106).
5. A chemical adsorption term σ_a (see pages 116–117).

Hence, the total interphase energy σ per unit area will be

$$\sigma = \sigma_{ch} + \sigma_e + \sigma_S + \sigma_d + \sigma_a \tag{3.52}$$

Types of Interphases

The chemical interaction term in Equation (3.52) is the dominating contribution to the surface energy. It will be discussed most carefully of the five terms. For this purpose we will introduce some definitions and concepts that will be used later.

Figure 3.39 shows two Gibbs' interfaces 1 and 3. Other names used for this type of extremely thin interfaces are *singular* or *planar* interfaces. Solid/vapour or liquid/vapour interfaces are only half-filled and contain atoms merely from the solid or liquid phase.

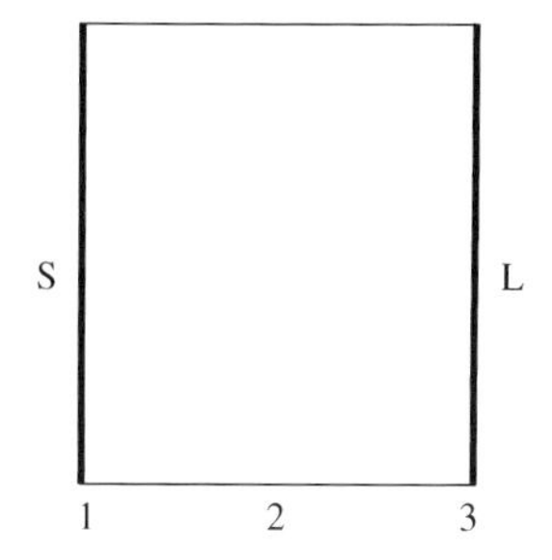

Figure 3.39 Guggenheim interphase (2) surrounded by two Gibbs' interfaces (1 and 3).

A Guggenheim interphase, marked by 2 in Figure 3.39, has a thickness of one or several atomic layers.

A complete Guggenheim interphase can be described as consisting of one interphase and two Gibbs' interfaces. Such an interphase is often called a *rough interphase*. Planar and rough interfaces will be carefully studied in this chapter.

We will discuss the five terms in Equation (3.52) for different types of interphases: liquid/vapour, solid/vapour, liquid/liquid, liquid/solid and solid/solid interfaces. In some cases only the most important one or two of the five terms will be discussed. The liquid/solid interphase, which is of special importance in this book, has been treated more extensively than the others.

3.4.1 Liquid/Vapour Interphases

The simplest and most direct way of calculating the interphase energy is to count the number of broken bonds per unit area when 'liquid' atoms are transformed into 'vapour' atoms. This can be done in the following way for metal melts.

The heat of evaporation per atom is the change of enthalpy required to transfer one atom from the interior of the melt to the gas phase. If the molar heat of evaporation is denoted $-\Delta H_v$ (or L_v) the heat of evaporation per atom will be $-\Delta H_v/N_A$, where N_A is Avogadro's number.

Each atom in the interior of the melt has on average 10 nearest neighbours. Three or four bonds per atom are broken when an 'interior atom' or 'bulk' atom is moved to the surface of the melt.

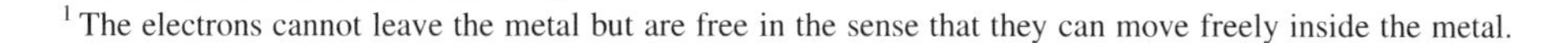

[1] The electrons cannot leave the metal but are free in the sense that they can move freely inside the metal.

Therefore, each atom at the surface is most likely to be surrounded by six or seven 'liquid' neighbours. Hence, the first term σ_{ch} of the interphase energy in Equation (3.52) can be estimated to

$$\frac{-\Delta H_v}{10} \times 3 < \sigma_{ch} \frac{V_m^l}{d} < \frac{-\Delta H_v}{10} \times 4 \tag{3.53}$$

The calculated values of σ_{ch} according to Equation (3.53) are about twice as high as the experimental values of the liquid/vapour interphase energies of metals in Table 3.1 on page 112 in this chapter. The reason for these discrepancies is the drastic change of the electronic structure of the electron cloud around the 'surface' atoms compared to the 'liquid' atoms. This is not included in the theoretical calculations. The change of electronic structure of the atoms changes the interatomic distances and also the surface energy.

The interatomic distances in the interphase are larger than those in the melt and lead to strain of the interphase. This circumstance contributes to the rough structure of the interface. Our assumption that the interphase is of the Guggenheim type seems to be reasonable.

3.4.2 *Solid/Vapour Interphases*

The interphase energy σ_{ch} between a solid and a vapour phase can be calculated in the same way as the interphase energy between a liquid and a vapour phase. However, when a solid is involved a new factor must be taken into consideration: the crystal structure of the solid. Crystal surfaces are smooth planar surfaces. Hence, Gibb's interface model is most convenient in this case. As an introductory example we regard a metal with FCC structure.

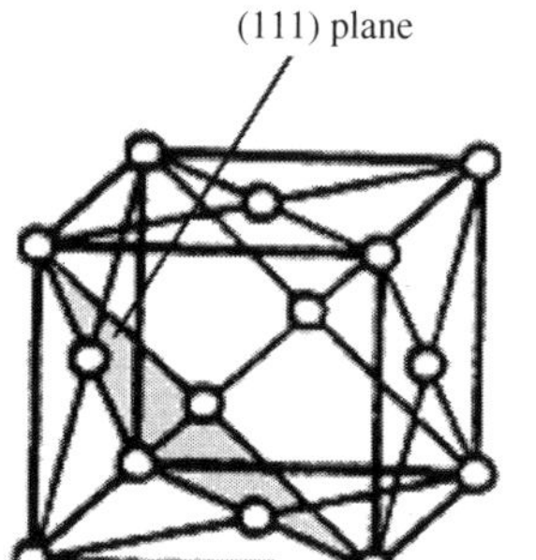

Example 3.2

Consider an FCC metal in contact with a gas.

Calculate the interface energy per atom in two crystal faces, cut along

a) the (111) plane of the FCC metal.
b) the (100) plane of the FCC metal.
c) Calculate the ratio of the interface energies in cases a) and b).

The (100) plane has fewer atoms per unit area than the (111) plane. The ratio is $n_{100}/n_{111} = 1.00/1.15$.

The interfaces are supposed to be planar. The molar heat of evaporation of the metal is L_v. The coordination number of FCC crystals is 12. Formation of an (100) interface requires 4 broken bonds per surface atom. To form a (111) interface it is necessary to break 3 bonds per surface atom.

Solution:

The enthalpy change, associated with formation of the crystal/vapour interface, is equal to the interface energy. We count the number of nearest neighbours in the lattice and assume that the binding energies between each of them and the reference atom are equal.

a) Consider a closely packed (111) plane of an FCC metal (upper figure in the text). Each interface atom in the plane has 9 solid neighbours and 3 broken bonds (upper figure in the text). The enthalpy change is equal to the surface energy. Hence, the interface energy per atom equals 3/12 times L_v/N_A, where N_A is Avogadro's number
b) In the less closely packed (100) plane of an FCC crystal, each interface atom has 8 solid neighbours and 4 broken bonds (lower figure in the text). Hence, the interface energy per atom equals 4/12 times L_v/N_A.

c) The energy required to form 1 unit area of the (111) interface is:

$$\sigma_{111} = n_{111} \times \frac{3}{12}\frac{L_v}{N_A} \qquad (1')$$

$$\sigma_{100} = n_{100} \times \frac{4}{12}\frac{L_v}{N_A} \qquad (2')$$

The ratio of the Gibbs' free interface energies per atom of the (111) and (100) planes is obtained by division of Equations (1′) and (2′).

$$\frac{\sigma_{111}}{\sigma_{100}} = \frac{n_{111}}{n_{100}} \times \frac{1/4}{1/3} = 1.15 \times \frac{3}{4} = 0.87$$

Answer:

a) The interface energy per atom of the (111) plane is $L_v/4N_A$.
b) The interface energy per atom of the (100) plane is $L_v/3N_A$.
c) The ratio of the interface energy per atom of the (111) and the (100) planes is 0.87.

Like the values calculated by Equation (3.53) on page 124, the solid/vapour interface energies calculated above are large compared to the measured values, presented in Tables 3.1 on page 112. The reason is the same as in the last section: the structure of the electron cloud around the atom nucleus is changed and the distances between the atoms become increased.

Mobile Defects in Pure Metal Surfaces

There are several types of lattice defects in crystals (see for example Chapter 1 in [1]). The thermodynamics of lattice defects (vacancies) has been treated briefly in Chapter 2.

A crystal lattice in thermal equilibrium contains a number of vacancies per unit volume, which is a function of temperature.

$$n_{vac} = n_0\, e^{-\frac{U_{vac}}{k_B T}} \; n_{vac} = n_0\, e^{-\frac{G_{a\,act}^{vac}}{k_B T}} \qquad (3.54)$$

where
n_{vac} = the number of vacancies per unit volume at the temperature T
n_0 = the number of lattice points per unit volume
$G_{a\,act}^{vac}$ = the activation energy or the energy required to form a vacancy.

On a planar (singular) surface adsorbed atoms and surface vacancies appear spontaneously. In exactly the same way the statistical number of surface vacancies and adsorbed atoms, respectively, on a pure metal surface can be calculated in analogy with the three-dimensional case:

$$n_{vac\ surface} = n_{0\ surface}\, e^{-\frac{U_{vac\ surface}}{k_B T}} \; n_S^{vac} = n_{0S}\, e^{-\frac{G_{a\,act}^{vac}}{k_B T}} \qquad (3.55)$$

and

$$n_{\text{a surface}} = n_{0\text{ surface}} e^{-\frac{U_{\text{a surface}}}{k_B T}} \quad n_S^{\text{ad}} = n_{0S} e^{-\frac{G_{\text{a act}}^{\text{ad}}}{k_B T}} \tag{3.56}$$

where
n_{0S} = the number of surface lattice positions per unit area of the surface
n_S^{vac} = the number of surface vacancies per unit area of the surface
n_S^{ad} = the number of adsorbed atoms per unit area of the surface
$G_{\text{a act}}^{\text{vac}}$ = the energy required to form a surface vacancy
$G_{\text{a act}}^{\text{ad}}$ = the energy required to adsorb an atom on the surface.

The exponential factors of Equations (3.54)–(3.56) show that all the point defects increase rapidly with temperature. The relative values of the formation energies lead to the conclusions that

- the relative occurrence of point defects will be larger on the surface than in the volume;
- the concentration of surface vacancies is larger than that of adsorbed atoms.

Fluctuations of the thermal vibrations cause the defects to jump to a neighbouring site or to leave the surface.

The mobile point defects on crystal surfaces play an important role in transport processes, especially in crystal growth from a vapour.

3.4.3 *Liquid/Liquid Interfaces*

Liquid/liquid interfaces occur in systems with an immiscibility gap (Figure 3.40). The free energy of this type of system consisting of two components A and B is described by the relationship

$$G_m = x_A G_A^0 + x_B G_B^0 + RT(x_A \ln x_A + x_B \ln x_B) + \Omega x_A x_B \tag{3.57}$$

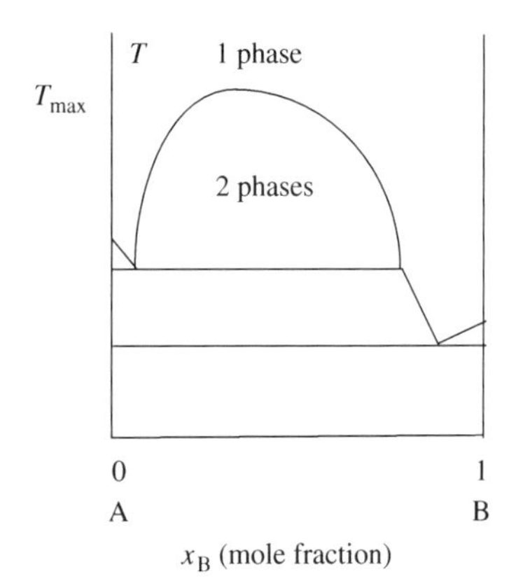

Figure 3.40 Phase diagram of two partially miscible liquids.

where
G_m = the total Gibbs' free energy per kmol of the interface
$G_{A,B}^0$ = the molar Gibbs' free energy of each component
$x_{A,B}$ = the molar fraction of each components A and B, respectively
Ω = a constant.

The constant Ω is related to the maximum point of the immiscibility gap by the relationship

$$\Omega = 2RT_{max} \tag{3.58}$$

where T_{max} is the maximum temperature of the immiscibility gap (Figure 3.40). Ω is related to the binding energy between the atoms by the relationship

$$\Omega = ZE_{AB}N_A \tag{3.59}$$

where
Z = the number of nearest neighbours of the atom
E_{AB} = the binding energy between the two kinds of atoms
N_A = Avogadro's number.

Becker [2] used this model in the middle of the 1930s to calculate the free energy of the interface between two phases. His calculations are shown in Example 3.3 below.

Example 3.3

Calculate the interface energy between two immiscible liquids L_1 and L_2 formed by the exchange process shown in the figure. The answer shall be given as a function of binding energies and concentrations in at%.

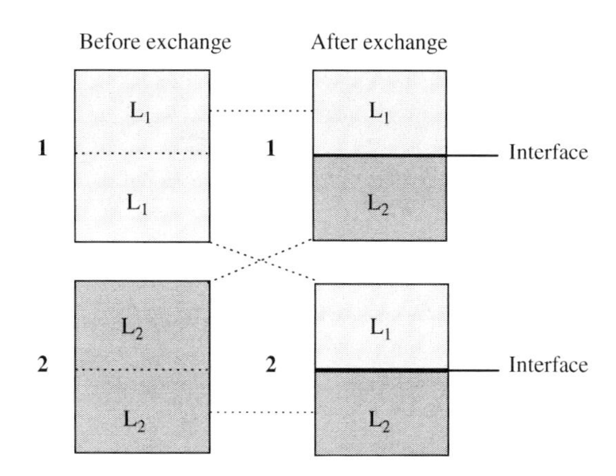

Solution:

To be able to calculate the interface energy we start with two equal containers. Liquid L_1 in container 1 consists of a mixture of atoms A and B. Liquid L_2 in container 2 consists of another mixture of atoms A and B. Suppose that half the volume in container 1 is transferred to container 2 and half the volume in container 2 is poured into container 1. The *interfaces* in both containers change. Energy has to be supplied to form the new interface. When the old interfaces disappear energy is gained. Our aim is to calculate the change of interface energy.

As indicated in the text the calculations will be based on binding energies between pairs of atoms. We will express the interface energy as a function of the binding energies of A–A, B–B and A–B atoms and the known concentrations of the A and B atoms in the two containers before and after the exchange.

The concentrations of A and B atoms in container 1 before the exchange are indicated by the symbol. "The concentrations of A and B atoms in container 2 before the exchange are indicated by the symbol".

Part 1. Calculation of Change of the Number of Bonds

We make the general and basic assumption that the number of XY molecules is proportional to the concentrations of X atoms and Y atoms.

Before the exchange of volumes:

$$\text{Number of A}-\text{A bonds} \quad \propto \quad {x'_A}^2 \tag{1'}$$

$$\text{Number of B}-\text{B bonds} \quad \propto \quad {x'_B}^2 \tag{2'}$$

$$\text{Number of A}-\text{B bonds} \quad \propto \quad x'_A x'_B + x'_B x'_A = 2x'_A x'_B \tag{3'}$$

Before the exchange of volumes:

$$\text{Number of A}-\text{A bonds} \quad \propto \quad {x''_A}^2 \tag{4'}$$

$$\text{Number of B}-\text{B bonds} \quad \propto \quad {x''_B}^2 \tag{5'}$$

$$\text{Number of A}-\text{B bonds} \quad \propto \quad x''_A x''_B + x''_B x''_A = 2x''_A x''_B \tag{6'}$$

After the exchange of volumes:

$$\text{Number of A}-\text{A bonds} \quad \propto \quad x'_A x''_A \tag{7'}$$

$$\text{Number of B}-\text{B bonds} \quad \propto \quad x'_B x''_B \tag{8'}$$

$$\text{Number of A}-\text{B bonds} \quad \propto \quad x'_A x''_B + x'_B x''_A \tag{9'}$$

After the exchange of volumes:

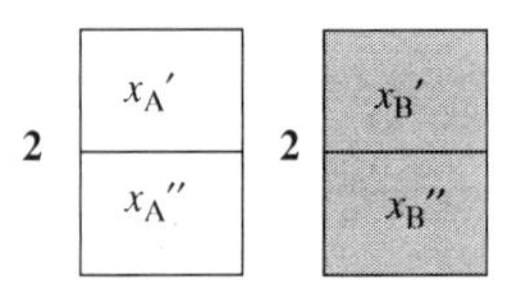

$$\text{Number of A} - \text{A bonds} \propto x''_A x'_A \tag{10'}$$

$$\text{Number of B} - \text{B bonds} \propto x''_B x'_B \tag{11'}$$

$$\text{Number of A} - \text{B bonds} \propto x''_A x'_B + x'_A x''_B \tag{12'}$$

The total change of the number of A–A bonds is obtained from

$$(7') + (10') - (1') - (4')$$

After some calculations we obtain $\propto -(x'_A - x''_A)^2$

The total change of the number of B–B bonds is, after reduction,

$$(8') + (11') - (2') - (5') \propto -(x'_B - x''_B)^2$$

The total change of the number of A–B bonds is

$$(9') + (12') - (3') - (6') \propto -2(x'_A - x''_A)(x'_B - x''_B)$$

Using the relationships

$$x'_A + x'_B = 1 \text{ and } x''_A + x''_B = 1 \text{ we obtain}$$

The total change of the number of A–B bonds $\propto 2(x'_A - x''_A)^2$

Part 2. Calculation of the Energy Change Associated with the Change of the Number of Bonds
The change of the number of A–A bonds and B–B bonds are negative while the number of A–B bonds has increased. A–A and B–B bonds are broken and new A–B bonds have been formed instead.

The net energy change associated with the change of the number of bonds of the three kinds can be calculated if the binding energies are known. We introduce the designations

$$E_{aXY} = \text{the binding energy of an X–Y bond.}$$

The binding energy depends on the environment of the atom and is therefore a function of concentration. Hence, we must introduce and distinguish between $E'_{a\,AB}$ and $E''_{a\,AB}$.

The net energy required to form the interface equals the net energy associated with the created and disappeared bonds. If the net number of bonds crossing the interface per unit area is called n we obtain the following expression for the surface energy per unit area

$$\sigma = \frac{n}{2}\left[-E_{aAA}(x'_A - x''_A)^2 - E_{aBB}(x'_B - x''_B)^2 - 2E'_{a\,AB}x'_A x'_B\right]$$
$$[-2E''_{a\,AB}x''_A x''_B + E'_{a\,AB}(x'_A x''_B + x''_A x'_B) + E''_{a\,AB}(x'_A x''_B + x''_A x'_B)] \tag{13'}$$

The factor $^1/_2$ appears in Equation (13′) because the binding energy of each bond is shared between the two atoms involved.
Equation (13′) will be simplified if we introduce a quantity that is defined as

$$\zeta = E_{a\,AB} - \frac{1}{2}(E_{aAA} - E_{a\,BB}) \tag{14'}$$

where ζ represents the difference in binding energy between an A–B bond and the average of the A–A and B–B bonds. The expressions

$$E_{\text{a AB}} = \zeta + \frac{1}{2}(E_{\text{a AA}} - E_{\text{a BB}}) \quad (15')$$

$$E'_{\text{a AB}} = \zeta' + \frac{1}{2}(E_{\text{a AA}} - E_{\text{a BB}}) \quad (16')$$

$$E''_{\text{a AB}} = \zeta'' + \frac{1}{2}(E_{\text{a AA}} - E_{\text{a BB}}) \quad (17')$$

are introduced in Equation (13′) and we obtain the answer below, after reduction.

Answer: The interface energy is

$$\sigma = \frac{n}{2}[(\zeta' + \zeta'')(x'_A x''_B + x''_A x'_B) - 2x'_A x'_B \zeta' - 2x''_A x''_B \zeta''] \quad (18')$$

where n is the net number of bonds crossing the interface per unit area.

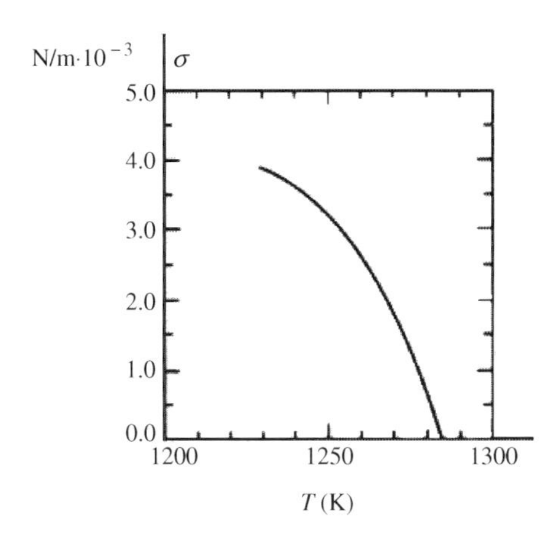

Figure 3.41 Surface tension (interface energy) of Cu–Pb alloys as a function of temperature.

Bergman Fredriksson and Shahani [3] used Becker's model to calculate the liquid/liquid interface energy in the Cu–Pb system. The result is shown in Figure 3.41.

Figure 3.41 shows that the interface energy decreases with increasing temperature and becomes zero at the maximum temperature. An extrapolation towards lower temperatures leads to a constant value of the interface energy that is independent of temperature.

3.4.4 *Liquid/Solid Interphases*

The liquid/solid interfaces are of special importance as the most common process of solidification is the solidification of a melt. Two radically different types of solid/liquid interfaces have been found.

The *first type of interface* has a narrow transition zone and is extended over just a few atom layers. These interfaces are atomically flat and stepped as is shown schematically in Figures 3.42 and 3.43. The energies of such sharp and smooth interfaces are *anisotropic*, e.g. they depend on orientation. Gibbs' interface model is to be preferred in this case.

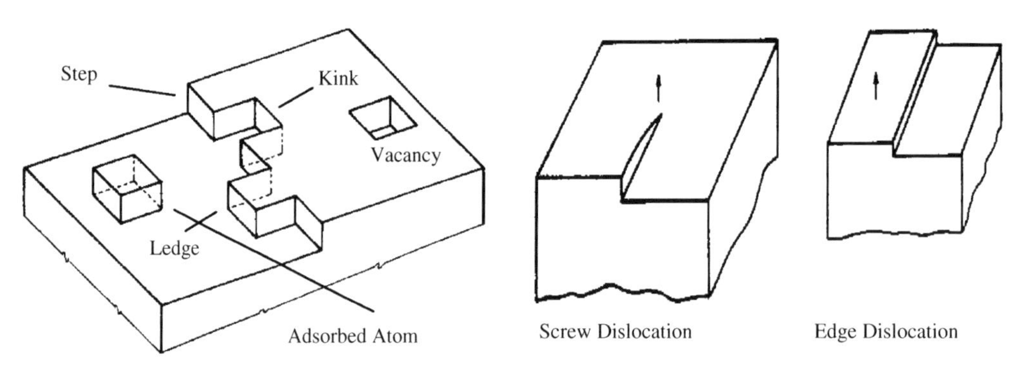

Figure 3.42 Model of an atomically flat solid/liquid interface.

Figure 3.43 A screw dislocation forms a ramp at the surface. An edge dislocation forms a step on the surface.

Interfaces of the second type, called *diffuse* or *rough interphases*, are spread over several atomic layers with a gradual change of order within the transition zone of the interphase (Figures 3.44 and 3.45). The atomic disorder close to the liquid phase changes gradually to the regular lattice site order characteristic of a solid. Guggenheim's interphase model is the best one for this type of rough interface.

Energies of *diffuse* interphases are *isotropic*, e.g. independent of orientation. A diffuse interphase is *rough* and not smooth like the atomically flat interfaces. For this reason diffuse interphases are often called *rough interphases*.

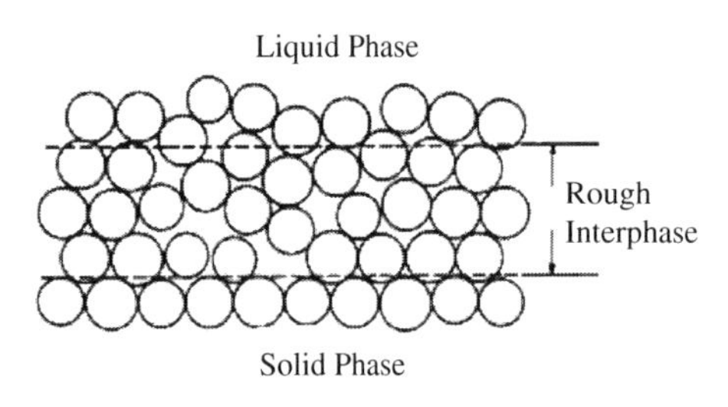

Figure 3.44 Positions of atoms in a liquid/solid interphase of the diffuse or rough interphase type.

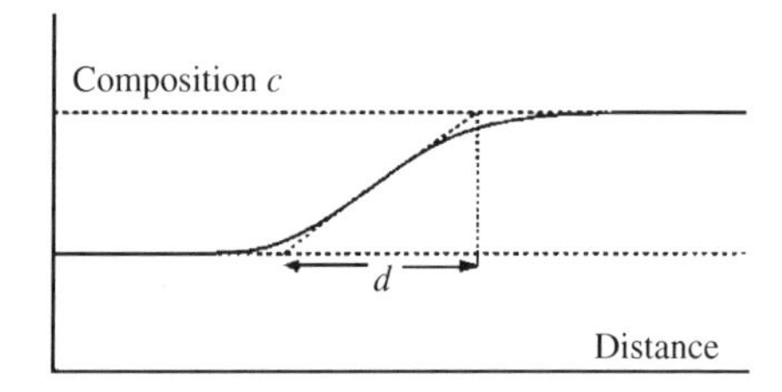

Figure 3.45 Composition in a diffuse interphase between two phases with the same crystal structure as a function of distance.

To describe the relationship between the structure of the interface and the kinetic solidification process at the interface, two theoretical models for the interface energy and its relationship to the atomic structure will be introduced, one for planar and one for rough solid/liquid interfaces. The two models will be presented below supported by mathematical calculations.

The models are based on the following assumptions:

1. The atoms are hard spheres with the diameter d.
2. The forces between the atoms are located to the centres of the atoms.
3. The energy of a Gibbs' interface is a function of the forces between the atoms in one atomic plane to the atoms in the neighbouring atomic plane. Only nearest-neighbour interaction is taken into account.
4. The energy of a Guggenheim interphase is a function of the forces between liquid and solid atoms within one plane, i.e. only lateral forces (bonds between nearest neighbours in the atomic plane).

5. A complete Guggenheim interphase consists of one or several Guggenheim layers plus two neighbouring Gibbs interfaces, one on each side of the Guggenheim interphase.
6. Normally, we express the energy as surface tension (J/m^2) instead of the Gibbs' free energy ΔG (J/kmol). The relationship between these units is a function of the molar volume V_m and the diameter d of the atoms.

$$\begin{array}{cccc} \Delta G & = \sigma & \times & \dfrac{V_m}{d} \\ \text{J/kmol} & \text{J/m}^2 & & \text{area/kmol} = \text{m}^3\text{/kmol} \times 1\text{/m} \end{array}$$

7. The energy calculations are coarse. We express the energy in the shape of the number of broken bonds between the atoms (S–S, S–L and L–L). The heat of fusion of the metal is called L_f.

3.4.5 *Atomically Planar Solid/Liquid Interfaces*

As a first attempt we will use a very simple model to calculate the interface energy between a solid and a liquid. We consider the interface as atomically planar and disregard the entropy of the interface. Hence, Gibbs' model of an interface is the natural choice.

The interface energy can then be calculated by adding the enthalpy of the bonds between the 'solid' and the 'liquid' atoms over the interface to the free energy of the solid and the liquid. The interface energy is also in this case a function of the crystal structure. The procedure is exactly the same as the one we used in Example 3.2 on page 124 in the case of a metal with FCC structure, surrounded by a gas phase.

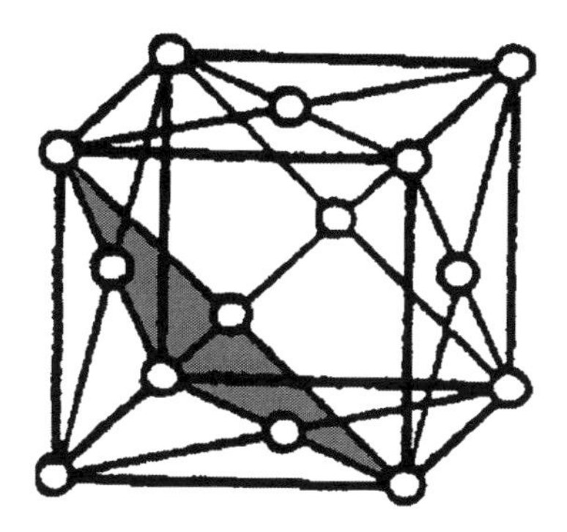

Figure 3.46a The (111) plane in an FCC metal.

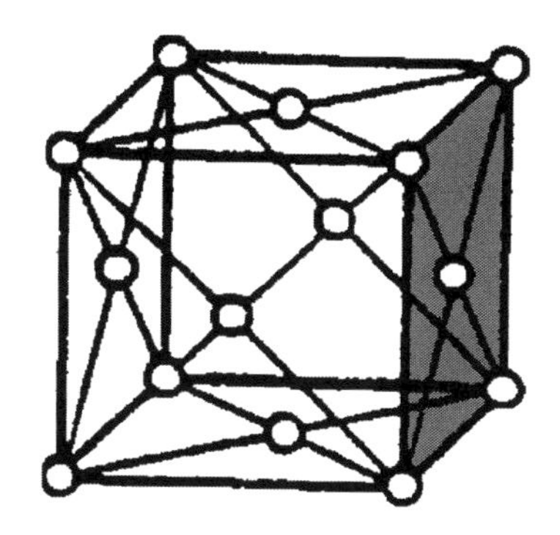

Figure 3.46b The (100) plane in an FCC metal.

Example 3.2 on page 124 shows that the closely packed (111) plane (Figure 3.46a) has a lower interface Gibbs' free energy than the (100) plane (Figure 3.46b). This is a general conclusion. It can be shown in the same way for other crystal structures that

- The closely packed planes have a lower interface energy per unit area than all other planes.

Calculation of the Enthalpy and the Surface Energy of an Atomically Planar Solid/Liquid Interface

The enthalpy per kmol and the surface energy can be described by the following general expression, based on the number of broken bonds per surface atom relative to bulk atoms:

$$-\Delta H_m^I = n \frac{Z_b}{Z_{coord}} \frac{L_f}{N_A} \frac{V_m^I}{d} \qquad (3.60)$$

The order of a crystal is high and consequently the entropy is small and therefore $-\Delta H_{\mathrm{m}}^{\mathrm{I}} \approx -\Delta G_{\mathrm{m}}^{\mathrm{I}} = \frac{\sigma V_{\mathrm{m}}^{\mathrm{I}}}{d}$.

Hence,

$$\sigma = n \frac{Z_{\mathrm{b}}}{Z_{\mathrm{coord}}} \frac{L_{\mathrm{f}}}{N_{\mathrm{A}}} \tag{3.61}$$

where

σ = surface energy per unit area or surface tension
$-\Delta H_{\mathrm{m}}^{\mathrm{I}}$ = molar interface enthalpy of the solid/liquid interface
$V_{\mathrm{m}}^{\mathrm{I}}$ = molar volume of the interphase
d = thickness of the interphase
n = number of possible adsorption sites per unit area of the interphase (number of atoms/m^2)
Z_{b} = number of broken solid–liquid atomic bonds of a solid surface atom
Z_{coord} = total number of atomic bonds to nearest neighbours or the coordination number of an atom in bulk
L_{f} = molar heat of fusion (J/kmol) and the energy required to break all the bonds to nearest neighbours
N_{A} = Avogadro's number (number of atoms/kmol).

The surface tensions, calculated with the aid of Equation (3.61), are larger than the corresponding experimental values in analogy with the discussion on page 123–124.

3.4.6 *Atomically Rough Liquid/Solid Interphases*

In an atomically rough interphase there are a number of incompletely filled atomic planes (Figure 3.47). For an elementary theoretical treatment with the aid of Guggenheim's model it is necessary to make some approximations.

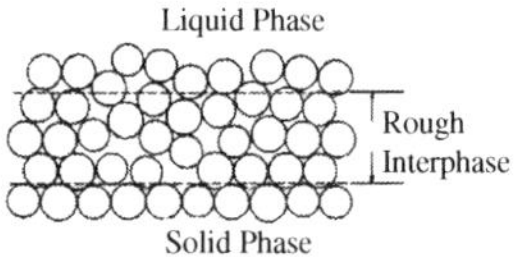

Figure 3.47 Atomically rough interphase.

Jackson's Model

In 1958 Jackson [4] suggested a two-level model with an inner smooth filled layer and a partially filled atomic layer outside. The number of possible adsorption sites per unit area in the filled layer is n. We assume that n_{a} atoms per unit area are adsorbed at random on the surface at temperature T. They form the partially filled outer layer. This layer is filled to the fraction n_{a}/n.

Every surface atom has Z_{I} lateral bonds to its neighbours. The number depends on the crystal structure and which crystal plane the atom layers represent. In an FCC crystal for example, $Z_{\mathrm{I}} = 4$ for the (100) plane and $Z_{\mathrm{I}} = 6$ for the (111) plane.

We want to find the interphase energy per unit area from an initial to a final state as a function of the fraction n_{a}/n. The initial state is the smooth atomic layer with no adsorbed atoms. In the final state n_{a} atoms per unit area have been adsorbed.

If the atoms are randomly distributed and no clustering is present, it is reasonable to assume that *each* atom *on the average* will have $Z_{\mathrm{I}} n_{\mathrm{a}}/n$ saturated bonds per unit area and $Z_{\mathrm{I}}(1 - n_{\mathrm{a}}/n)$ unsaturated bonds per unit area. The total number of unsaturated bonds of the occupied sites per unit area is then n_{a} times the number of unsaturated bonds per adsorbed atom.

Jackson considered the interface as a mixture of 'solid' atoms and 'liquid' atoms. He defined the fraction x^{s} of 'solid' atoms as the fraction of adsorbed atoms and the rest as 'liquid' atoms:

$$x^{\mathrm{s}} = \frac{n_{\mathrm{a}}}{n} \tag{3.62}$$

$$x^{\mathrm{L}} = 1 - \frac{n_{\mathrm{a}}}{n} \tag{3.63}$$

Monolayer of Mixed 'Liquid' and 'Solid' Atoms

Jackson applied the regular theory of solutions (page 46 in Chapter 2). See also Section 5.2.6 in Chapter 5 in [1] concerning the mixture of 'liquid' and 'solid' atoms at the interface. He derived his model for a monoatomic interface.

Here, we will apply Jackson's theory on a Guggenheim interphase and start with a monolayer of mixed 'solid' and 'liquid' atoms with the thickness d.

The Guggenheim layer consists of n atoms/m^2, all in the same plane. The fraction x^{L} consists of 'liquid' atoms and the rest $(1 - x^{\mathrm{L}})$ are 'solid' atoms. They are mixed at random.

We refer to the mixing theory on pages 126–129 and count the broken bonds to nearest neighbours. The number of broken bonds per atom pair is Z_1. In analogy with Equation (3.60) on page 131 we obtain the enthalpy

$$-\Delta H^{\mathrm{I}}_{\mathrm{mix}} = \frac{Z_1}{Z_{\mathrm{coord}}}\frac{L_{\mathrm{f}}}{N_{\mathrm{A}}} n x^{\mathrm{L}}(1 - x^{\mathrm{L}}) \times \frac{V_{\mathrm{m}}}{d} \tag{3.64}$$

where

Z_1 = number of lateral bonds between an atom and its nearest neighbours in the surface plane

n = number of surface atoms per unit area

d = the atomic diameter and the distance between the centres of two atoms in touch with each other

V_{m} = the molar volume of the metal.

An entropy term has to be added, owing to mixing that causes an increased disorder.

$$\Delta S^{\mathrm{I}}_{\mathrm{mix}} = -nk_{\mathrm{B}}\left[x^{\mathrm{L}}\ln x^{\mathrm{L}} + (1 - x^{\mathrm{L}})\ln(1 - x^{\mathrm{L}})\right] \times \frac{V_{\mathrm{m}}}{d} \tag{3.65}$$

The Gibbs' free energy ($\Delta G^{\mathrm{I}}_{\mathrm{mix}} = \Delta H^{\mathrm{I}}_{\mathrm{mix}} - T\Delta S^{\mathrm{I}}_{\mathrm{mix}}$) will be

$$-\Delta G^{\mathrm{I}}_{\mathrm{mix}} = \frac{Z_1}{Z_{\mathrm{coord}}}\frac{L_{\mathrm{f}}}{N_{\mathrm{A}}} n x^{\mathrm{L}}(1 - x^{\mathrm{L}}) + nk_{\mathrm{B}}T\left[x^{\mathrm{L}}\ln x^{\mathrm{L}} + (1 - x^{\mathrm{L}})\ln(1 - x^{\mathrm{L}})\right] \times \frac{V_{\mathrm{m}}}{d}$$

If we introduce the abbreviation $\alpha = \dfrac{Z_1}{Z_{\mathrm{coord}}}\dfrac{L_{\mathrm{f}}}{N_{\mathrm{A}}k_{\mathrm{B}}T}$ that will often be used below, the molar free energy for a mixed Guggenheim surface can be written

$$-\Delta G^{\mathrm{I}}_{\mathrm{mix}} = nk_{\mathrm{B}}T\left[\alpha x^{\mathrm{L}}(1 - x^{\mathrm{L}}) + x^{\mathrm{L}}\ln x^{\mathrm{L}} + (1 - x^{\mathrm{L}})\ln(1 - x^{\mathrm{L}})\right] \times \frac{V_{\mathrm{m}}}{d} \tag{3.66}$$

If we drop the last factor on the right-hand side of Equation (3.66) we obtain the surface energy (surface tension):

$$\sigma_{\mathrm{mix}} = nk_{\mathrm{B}}T\left[\alpha x^{\mathrm{L}}(1 - x^{\mathrm{L}}) + x^{\mathrm{L}}\ln x^{\mathrm{L}} + (1 - x^{\mathrm{L}})\ln(1 - x^{\mathrm{L}})\right] \tag{3.67}$$

The function is shown in Figure 3.48 on next page.

As Figure 3.48 shows, all the curves are symmetrical around the line $x^{\mathrm{L}} = 0.5$ and have either a maximum with two minima or only a minimum for $x^{\mathrm{L}} = 0.5$.

If we take the derivative of the Equation (3.67) the second derivative of $\Delta\sigma_{\mathrm{mix}}$ is found to be

$$\frac{\mathrm{d}^2}{\mathrm{d}x^{L^2}}\left(\frac{\sigma_{\mathrm{mix}}}{nk_{\mathrm{B}}T}\right) = -2\alpha + \frac{1}{x^{\mathrm{L}}} + \frac{1}{1 - x^{\mathrm{L}}} \tag{3.68}$$

For $x^{\mathrm{L}} = 0.5$ we obtain

$$\frac{\mathrm{d}^2}{\mathrm{d}x^{\mathrm{L}^2}}\left(\frac{\sigma_{\mathrm{mix}}}{nk_{\mathrm{B}}T}\right)_{x^{\mathrm{L}} = 0.5} = (4 - 2\alpha)\frac{V^{\mathrm{I}}_{\mathrm{m}}}{d} \tag{3.69}$$

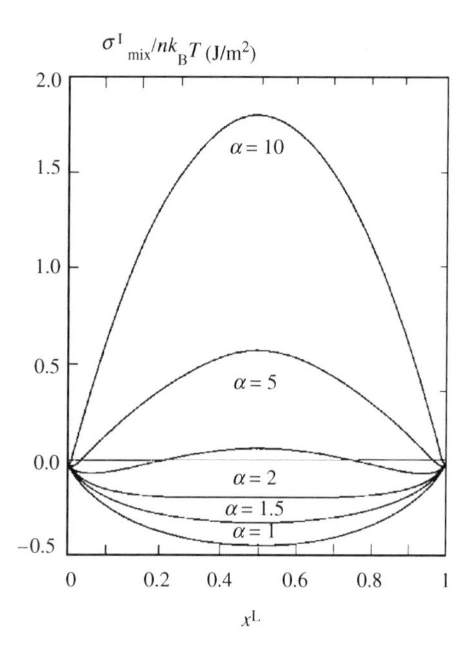

Figure 3.48 $\sigma^{\mathrm{I}}_{\mathrm{mix}}/nk_{\mathrm{B}}T$ as a function of the fraction liquid atoms x^{L} (compare Equation (3.67) with α as a parameter. The temperature is an indirect parameter. The thickness of the Guggenheim interphase $= d$. From Liquid Metals and Solidification, K. A. Jackson, American Society for Metals 1958. Reprinted with permission of ASM International. All rights reserved. www.asminternational.org.

Principally, there are two types of curves in Figure 3.48:

- $\alpha < 2$

The second derivative is positive and the curves show minima at $x^{\mathrm{L}} = 0.5$. If the atomic layer is half-filled, the rough interface is in a stable state.

For metals α is always less than 2 and rough interfaces are the stable state.

- $\alpha > 2$

The second derivative is negative and the curves show maxima at $x^{\mathrm{L}} = 0.5$. The functions have shallow minima for x^{L} values very close to 1 and 0. Accordingly, the equilibrium state is a filled atomic layer with either a small deficiency of atoms or a small surplus. An approximately planar interface is in both these cases the stable state.

It is much easier for an atom from outside to enter a rough interphase than an atomically flat surface. The number of possible sites inside a rough interphase is much larger than at a planar surface. This is normally expressed as a high probability for a 'liquid' atom to find a site in the solid.

Below, we will use Jacksons's model in combination with the energy law to find the total energy, i.e. the surface tension of an interphase between a planar crystallographic metal surface and its corresponding melt at the melting-point temperature.

We will also make an attempt to find the surface tension of a multilayer solid/liquid interphase of the Jackson model for the metal.

Total Surface Energy of a Monolayer Guggenheim Interphase

Assumptions and Simplifications

We will use the same method to estimate energies as Jackson, i.e. count the number of broken bonds. In addition, it is necessary to introduce some rough simplifications:

1. Bonds between solid–solid atoms are assumed to be equally strong as bonds between solid–liquid atoms.
2. Bonds between liquid–liquid atoms are weak and can be neglected in comparison with any solid–liquid bonds.
3. Only interaction between nearest neighbours is taken into account.

The same designations as above of the quantities involved will be used, for example Z_{coord}, Z_b, Z_I and α.

Z_{coord}, and Z_I are material constants and determine the structure of the metal. For symmetrical reasons the number of bonds between an atom and its nearest neighbours in an upper and lower parallel crystal plane, respectively, must be equal. In this case, the relationship

$$Z_{coord} = Z_I + 2Z_b \tag{3.70}$$

is valid.

Calculation of the Total Surface Energy of a Monolayer Guggenheim Interphase

A real interphase consists of a Guggenheim monolayer connected to a solid and a liquid surface, i.e. of a Guggenheim layer surrounded by two Gibbs interfaces.

To be able to calculate the total surface energy per unit area we will build up 1 m^2 of the surface in three steps (Figure 3.49).

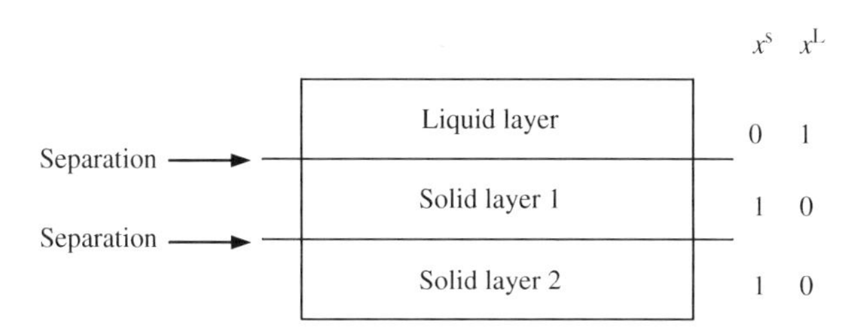

Figure 3.49 A crystallographic planar surface of a solid metal in connection with the corresponding metal melt.

1. Separation of the liquid layer and solid layer 1 and Separation of solid layers 1 and 2.
   It is necessary to supply energy to break the bonds between the atoms. In the first case nZ_b liquid–solid bonds/m^2 have to be broken. In the second case nZ_b S–S bonds/m^2 have to be broken. According to Equation (3.60) the energy/m^2 required for these operations are equal to $n\dfrac{Z_b}{Z_{coord}}\dfrac{L_f}{N_A} + n\dfrac{Z_b}{Z_{coord}}\dfrac{L_f}{N_A}$.
   If we introduce α (page 133), the energy/m^2 can be written as on page 131
$$nk_BT\alpha\frac{2Z_b}{Z_I}.$$
2. Transformation of the insulated solid layer 1 into a Guggenheim monolayer.
   The Guggenheim layer contains nx^S 'solid' atoms and nx^L 'liquid' atoms. They are distributed over the surface at random.
   The lateral bonds between the surface atoms have to be broken and 'liquid' and 'solid' atoms have to be mixed at random.
   According Equation (3.67) on page 133 the required separation energy/m^2 plus the following mixing energy is
$$\sigma^{I}_{mix} = nk_BT\left[\alpha x^L(1-x^L) + x^L\ln x^L + (1-x^L)\ln(1-x^L)\right] \tag{3.67}$$
3. Connection of the Guggenheim layer to the solid layer (crystal surface) and liquid.
   The formation of stable bonds *releases* energy, which is *negative* as it is removed from the system. This energy is transformed into heat and is not involved in the calculation of the surface tension.

Now, we have a complete interphase of a monolayer. The energy per unit area (J/m^2) or the surface tension (N/m), that is required to form it is

$$\sigma^{mono}_{total} = nk_BT\alpha\frac{2Z_b}{Z_I} + nk_BT\left[\alpha x^L(1-x^L) + x^L\ln x^L + (1-x^L)\ln(1-x^L)\right] \tag{3.71}$$

The last two terms will be negative because x^L and $(1 - x^L)$ are both < 1 and consequently $\ln x^L$ and $\ln(1 - x^L)$ are negative.

Total Surface Energy of a Multilayer Guggenheim Interphase

Consider a multilayer that consists of p monolayers with the same composition of mixed 'solid' and 'liquid' atoms as the monolayer (Figure 3.50). We will calculate the energy/m^2 or the surface tension (N/m) with the same methods and assumptions as above.

Number of atomic layer		Fraction of 'liquid' atoms	Fraction of 'solid' atoms
	Liquid	1	0
p		x^L	$x^S = 1 - x^L$
p − 1		x^L	$x^S = 1 - x^L$
⋮		⋮	⋮
i		x^L	$x^S = 1 - x^L$
⋮		⋮	⋮
2		x^L	$x^S = 1 - x^L$
1		x^L	$x^S = 1 - x^L$
0	Solid	0	1

Figure 3.50 Multilayer of the Jackson model. The fractions of liquid and solid atoms for each layer are indicated in the figure.

Calculation of the Total Surface Energy of a Multilayer Guggenheim Interphase
Consider 1 m^2 of a solid metal surface that contains n atoms/m^2 in connection with the corresponding melt at the melting-point temperature. The multilayer will be built up in three steps, layer after layer, and we will list the total energy of each layer.

Total Energy of Interphase Layer p
The initial layer is layer p between the liquid and the solid. The total energy required to form this layer is equal to that of the monolayer in Equation (3.71).

$$\sigma^{p} = nk_{B}T\alpha\frac{2Z_{b}}{Z_{I}} + nk_{B}T\left[\alpha x^{L}(1 - x^{L}) + x^{L}\ln x^{L} + (1 - x^{L})\ln(1 - x^{L})\right]$$

Total Energy of Interphase Layer $p - 1$
Interphase layer $p - 1$ is formed by separation of solid layer 2 from interphase layer p and solid layer 3 (Figure 3.51), transformation of solid layer 2 into a Guggenheim monolayer, (interphase layer $p - 1$) and connection of this layer to interphase layer p and solid layer 3.

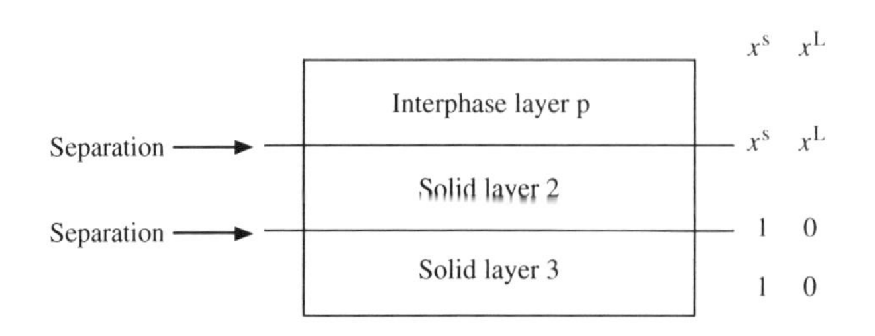

Figure 3.51 A crystallographic planar surface of a solid metal in connection with the corresponding metal melt.

Separation of solid layer 2 and interphase layer 1 requires $nx^{s}Z_{b}$ broken solid–solid bonds $+nx^{L}Z_{b}$ broken L–S bonds or nZ_{b} bonds. nZ_{b} S–S bonds/m^2 have to be broken to separate the solid layers 2 and 3. Altogether, 2 nZ_{b} bonds must be broken.

The formation of the Guggenheim layer will be the same as for a monolayer and connecting the new interphase layer to interphase layer 1 and solid layer 3 requires no energy.

The total energy of interphase layer 2 will therefore be the same as that of interphase layer 1.

$$\sigma^{p-1} = nk_{B}T\alpha\frac{2Z_{b}}{Z_{I}} + nk_{B}T\left[\alpha x^{L}(1-x^{L}) + x^{L}\ln x^{L} + (1-x^{L})\ln(1-x^{L})\right]$$

Total Energy of Interphase Layers p – 2, p – 3,... Down to Layer 1

Corresponding calculations for the following interphase layers will be exactly the same as those for interphase layer $p-1$. Hence, every interphase layer has the same energy/m^2 or surface tension as a monolayer.

Total Energy of the Guggenheim Multilayer Interphase

Therefore, the total energy or surface tension of the multilayer interphase will be

$$\sigma_{total}^{Jackson} = p \times \sigma_{total}^{mono} \tag{3.72}$$

or

$$\sigma_{total}^{Jackson} = p \times \left[nk_{B}T\,\alpha\frac{2Z_{b}}{Z_{I}} + nk_{B}T\left[\alpha x^{L}(1-x^{L}) + x^{L}\ln x^{L} + (1-x^{L})\ln(1-x^{L})\right]\right] \tag{3.73}$$

Below the theory is applied on the (111) plane of copper.

Example 3.4

Consider a Cu crystal in a Cu melt at melting-point temperature. We assume that the interphase between the crystal and the melt consists of

a) a monolayer interphase of a (111) facet.
b) three interphase layers of a (111) facets.

Each layer consists of 50% 'liquid atoms' and 50% 'solid' atoms. Calculate the surface tension of the interphase in a) and b) according to Jackson's model.

Cu has FCC structure and its lattice constant equals 0.362 nm. Assume that on the average 3 bonds are broken when a melt layer and a solid (111) layer are separated.

Some material constants are given in the margin. Other constants are found in standard tables.

Material constants of Cu:

T_{M} = 1083 °C
ρ_{Cu}^{s} = 8.24 × 10^3 kg/m^3 at T_M
Z_{b} = 3
Z_{111} = 6
Z_{coord} = 12
n_{111} = 1.77 × 10^{19} atoms/m^2
M = 63.5 kg/kmol
L_{f} = 206 × 10^3 J/kg.

Solution:

We will use Equations (3.71) and (3.72). In addition to the material constants given in the margin of the text we need the following constants:

N_{A} = 6.02 × 10^{26} atoms/kmol
R = 8.31 × 10^3 J/kmol K
k_{B} = 1.38 × 10^{-23} J/K
T_{M} = 1083 + 273 = 1356 K
L_{f} = 63.5 × 206 × 10^3 = 1.31 × 10^{7}J/kmol.

It simplifies the calculations if we calculate the numerical values of some quantities in advance.

$$\alpha = \frac{Z_{I}}{Z_{coord}}\frac{L_{f}}{RT} = \frac{6}{12} \times \frac{1.31 \times 10^{7}}{8.31 \times 10^{3} \times 1356} = 0.581$$

With the aid of $a = 0.362$ nm the surface density has been calculated as $n_{111} = 1.77 \times 10^{19}$ atoms/m^2.

$$n_{111}k_BT = 1.77 \times 10^{19} \times 1.38 \times 10^{-23} \times 1356 = 0.331 \text{ J/m}^2$$

Calculations

a)

$$\sigma_{\text{total}}^{\text{mono}} = nk_\text{B}T\left[\alpha\frac{2Z_\text{b}}{Z_\text{I}} + \alpha x^\text{L}(1 - x^\text{L}) + x^\text{L}\ln x^\text{L} + (1 - x^\text{L})\ln(1 - x^\text{L})\right]$$

Inserting the numerical values we obtain

$$\sigma_{\text{total}}^{\text{mono}} = 0.331\left[0.581 \times \frac{2 \times 3}{6} + 0.581 \times 0.5 \times 0.5 + 0.5\ln 0.5 + 0.5\ln 0.5\right]$$

or

$$\sigma_{\text{total}}^{\text{mono}} = 0.331[0.581 \times 1.25 - \ln 2] = 0.331[0.7262 - 0.6932] = 0.011 \text{ J/m}^2$$

b)
We use Equation (3.72):

$$\sigma_{\text{total}}^{\text{Jackson}} = 3\sigma_{\text{total}}^{\text{momo}} = 3 \times 0.011 \text{ J/m}^2$$

Answer:
The surface tension of the interface between a Cu crystal and its melt is, according to Jackson

a) 0.01 J/m^2 for the monolayer interphase.
b) 0.03 J/m^2 for the multilayer of Jackson's model.

Comments on the result are given on page 141.

Further Development of Jackson's Model
The Jackson model deals only with first-neighbour interaction and can be applied successfully only at temperatures close to the equilibrium temperature. Temkin [5] has extended Jackson's model to multilayers in the 1960s. In 1986 Chen, et al. [6] generalized Jackson's model by including many-body interaction with more distant neighbours. The Temkin model will be discussed below.

Temkin's Interphase Model

The Jackson model, developed from and inspired by Guggenheim's ideas, may in some respects appear too simple. Temkin assumed that an interphase between a solid and a liquid has a large fraction of 'liquid' atoms close to the liquid and a large fraction of 'solid' atoms close to the solid and gradually changing proportions in the atomic layers between the layers at the ends of the interphase.

Temkin's description of the interphase is sketched in Figure 3.32. It consists of a number of atomic layers, i.e. Guggenheim's layers. The fractions of 'liquid' and 'solid' atoms are listed close to the corresponding layer. They are no longer equal in all the layers, as in Jackson's model, because x^L and x^s varies gradually from layer to layer.

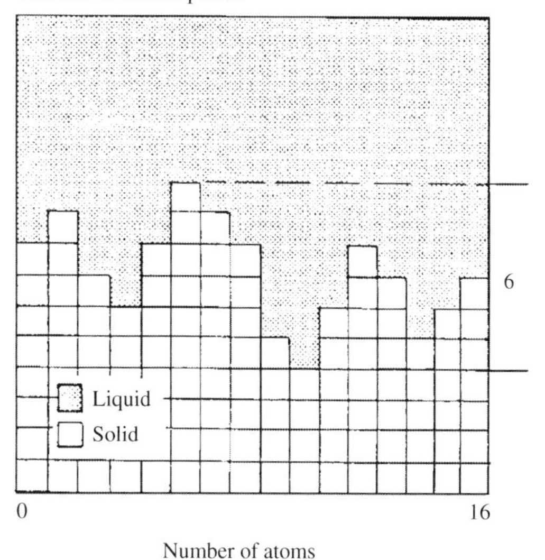

Figure 3.52 Sketch of Temkin's model of a liquid/solid interphase with $p = 6$ atomic layers and $n = 16$ adsorption sites per unit area, that accommodates a gradually changed number of 'liquid' and 'solid' atoms in the layers.

In real cases n is a very high number. Reproduced from Solid State Transformation, ed by N. N. Siroto, F. K. Gorskii and V. M. Varikash, Crystallisation Processes by D. E. Temkin, Copyright (1964) Springer.

Total Surface Energy of a Temkin Interphase

Characteristic of Temkin's model of an interphase is that the compositions of the layers vary gradually. Close to the solid layer the interphase layer contains more 'solid' atoms than 'liquid' atoms and vice versa close to the liquid layer. The simplest model is based on a linear change of the fractions over the interphase (Figure 3.53).

Calculation of the Surface Energy of a Temkin Interphase between a Solid and a Liquid

We use the same procedure as for Jackson's model to calculate the surface energy of a Temkin surface where the fractions of 'solid' and 'liquid' atoms vary gradually from 0 to 1 for the various layers. The designations are the same as above.

Consider 1 m^2 of a Temkin interphase with p atomic layers. If we assume that concentrations of 'solid' atoms and 'liquid' atoms vary linearly over the interphase, the concentration of 'liquid' atoms in the layers can be expressed in terms of index p, the number of atomic layers of the interphase, and the index i. This has been done in Figure 3.53.

Number of atomic layer		Fraction of 'liquid' atoms	Fraction of 'solid' atoms
p + 1	Liquid	1	0
p		p/(p + 1)	1 − p/(p + 1)
p−1		(p − 1)/(p + 1)	1 − (p − 1)/(p + 1)
⋮		⋮	⋮
i		i/(p + 1)	1 − i/(p + 1)
⋮		⋮	⋮
2		2/(p + 1)	1 − 2/(p + 1)
1		1/(p + 1)	1 − 1/(p + 1)
0	Solid	0	1

Figure 3.53 Multilayer of the Temkin model. The fractions of liquid and solid atoms for each layer are indicated in the figure.

The energy required to separate each solid layer is the same as for Jackson's model and equal to the first term in Equation (3.71). The energy required for the transformation of the solid layer into a Guggenheim interphase varies with the composition of the different layers and has to be summarized over the interphase.

Hence, the total interphase energy of the Temkin interphase is a modification of Equation (3.73) and can be written

$$\sigma_{\text{total}}^{\text{Temkin}} = nk_{\text{B}}T\left[p \times \alpha\frac{2Z_{\text{b}}}{Z_{\text{I}}} + \sum_{\text{i}=1}^{\text{p}}\left[\alpha x_{\text{i}}^{\text{L}}(1 - x_{\text{i}}^{\text{L}}) + x_{\text{i}}^{\text{L}}\ln x_{\text{i}}^{\text{L}} + (1 - x_{\text{i}}^{\text{L}})\ln(1 - x_{\text{i}}^{\text{L}})\right]\right] \tag{3.74}$$

Equation (3.74) represents the general expression for the Temkin model of an interphase.

The fractions of x_{i}^{L} and x_{i}^{s}, expressed as functions of indices p and i can be introduced into Equation (3.74) and used for numerical calculation. This has not been done here. The purpose is to avoid an unsurveyable formula.

We will calculate the surface tension of Cu with Temkin's model for comparison with Jackson's model.

Example 3.5

Consider a Cu crystal in a Cu melt at melting-point temperature. We assume that the interphase between the crystal and the melt consists of 3 layers of (111) facets. The compositions of the layers are:

- layer close to the crystal: 25% 'liquid' atoms +75% 'solid' atoms;
- middle layer: 50% 'liquid' atoms +50% 'solid' atoms;
- layer close to the melt: 75% 'liquid' atoms +25% 'solid' atoms.

Calculate the surface tension of the interphase according to Temkin's model.
All constants are the same as in Example 3.4.

Solution:

Fractions of 'solid' and 'liquid' atoms in the three Guggenheim layers

Layer	x_{p}^{s}	x_{p}^{L}
Liquid	0	1
Guggenheim 3	0.25	0.75
Guggenheim 2	0.50	0.50
Guggenheim 1	0.75	0.25
Solid	1	0

We start with a table of the fractions of the 'liquid' and 'solid' atoms in the three layers.

We use Equation (3.74) to find the surface tension.

$$\sigma_{\text{total}}^{\text{Temkin}} = nk_{\text{B}}T\left[\underbrace{p \times \alpha\frac{2Z_{\text{b}}}{Z_{\text{I}}}}_{\text{Term 1}} + \sum_{\text{i}=1}^{\text{p}}\left[\underbrace{\alpha x_{\text{i}}^{\text{L}}(1 - x_{\text{i}}^{\text{L}})}_{\text{Term 2}} + \underbrace{x_{\text{i}}^{\text{L}}\ln x_{\text{i}}^{\text{L}} + (1 - x_{\text{i}}^{\text{L}})\ln(1 - x_{\text{i}}^{\text{L}})}_{\text{Term 3}}\right]\right]$$

Inserting the numerical values into the three terms gives

$$\text{Term 1} = p \times \alpha\frac{2Z_{\text{b}}}{Z_{\text{I}}} = 3 \times 0.581 \times \frac{2 \times 3}{6} = 1.743$$

and

$$\begin{aligned}\text{Term 2} &= \sum_{\text{i}=1}^{3}\left[\alpha x_{\text{i}}^{\text{L}}(1 - x_{\text{i}}^{\text{L}})\right] \\ &= 0.581 \times (0.25 \times 0.75 + 0.50 \times 0.50 + 0.75 \times 0.25) \\ &= 0.581 \times 0.625 = 0.363\end{aligned}$$

and

$$\begin{aligned}\text{Term 3} &= \sum_{\text{i}=1}^{3}\left[x_{\text{i}}^{\text{L}}\ln x_{\text{i}}^{\text{L}} + (1 - x_{\text{i}}^{\text{L}})\ln(1 - x_{\text{i}}^{\text{L}})\right] \\ &= (0.25 \times \ln 0.25 + 0.75 \times \ln 0.75) + (0.50 \times \ln 0.50 + 0.50 \times \ln 0.50) \\ &\quad +(0.75 \times \ln 0.75 + 0.25 \times \ln 0.25) \\ &= 2 \times (0.25 \times \ln 0.25 + 0.50 \times \ln 0.50 + 0.75 \times \ln 0.75) = 2 \times (-0.9090) = -1.818\end{aligned}$$

Hence, we obtain the total energy/m²

$$\sigma_{\text{total}}^{\text{Temkin}} = 0.331 \times (1.743 + 0.363 - 1.818) = 0.096\ \text{J/m}^2$$

Answer:
The surface tension of a Cu crystal and its melt is $\sim 0.1\ \text{J/m}^2$, according to Temkin.

The agreement between the calculated values of the surface tension in Examples 3.4 and 3.5 and experimental values is poor but of the proper magnitude. The reasons for the discrepancy are many.

The assumption that the liquid–solid bonds and the solid–solid bonds are equally strong can certainly be questioned. We have neglected the liquid–liquid bonds between the atoms. Only the influence of nearest neighbours has been included in the calculations. The calculations show that the surface tension is derived as the difference between two much larger quantities, which reduces the accuracy considerably.

Experimental measurements at high temperature of the surface tension of metals are hard to perform. Therefore, the uncertainty of the experimental values may be considerable.

The fictive multilayer calculations of σ according to the Jackson and Temkin models show diverging results. The influence of the mixing profile of 'liquid' and 'solid' atoms in the interphase is obviously strong.

3.4.7 *Solid/Solid Interfaces*

Types of Solid/Solid Interfaces

The interface between two solid phases is most important in the case of chemical vapour deposition or heterogeneous nucleation of crystals (Chapter 4). Here we will briefly discuss the theoretical basis for calculation of the interface energy of solid/solid interfaces as a background for future use. Normal grain boundaries will not be discussed at all.

The solid/solid interfaces are classified as three different groups of interfaces

- incoherent interfaces
- coherent interfaces
- semicoherent interfaces.

Some of the different types of solid/solid interfaces are shown in Figures 3.54 and 3.55.

Incoherent interfaces show no similarities at all concerning the crystal structures and lattice constants of the two solid phases.

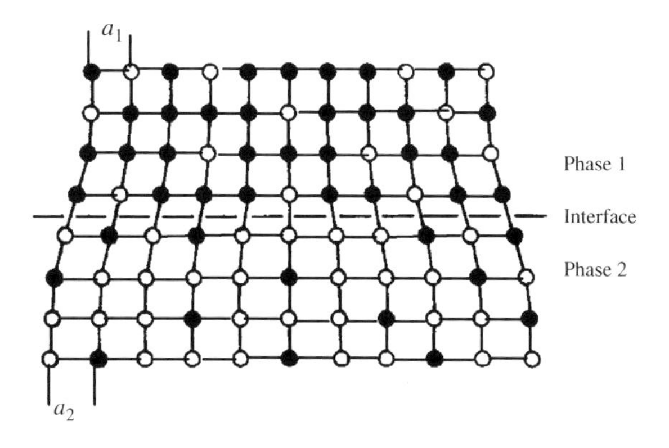

Figure 3.54 Nearly coherent solid/solid interface. The slight lack of matching causes lattice distortion and strain. © Pearson PLC 1992.

Coherent interfaces show a perfect match between the crystal lattices of the two solid phases. Perfect coherence never occurs in practice. Slight deviations from perfect coherence are possible and cause deformation of the lattice and lattice strain (Figure 3.54).

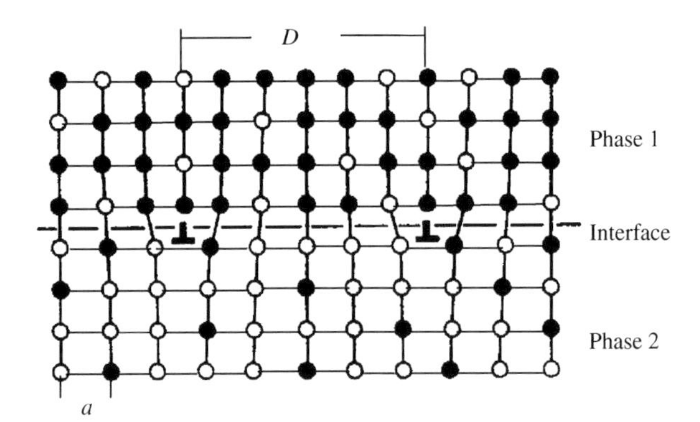

Figure 3.55 Semi-coherent solid/solid interface. Strain and dislocations overcome together the lack of agreement between the two crystal lattices. © Pearson PLC 1992.

Semi-coherent interfaces (Figure 3.55) can be regarded as a subgroup of coherent interfaces or an intermediate group between coherent and incoherent interfaces. There is no perfect match between the two crystal lattices but this can partly be overcome by *strain*, which changes the positions of the surface atoms in the direction of better agreement and by *dislocations* between the coherent parts, which regularly and discontinously correct the mismatch.

The coherent and semicoherent interfaces are of most interest and the further discussion will be restricted to them.

Mismatch or Disregistry

Definition

The degree of deviation of a partially coherent interface (between the foreign particle and the solidified melt) from a perfectly coherent interphase can be quantified by introducing the concept of *mismatch* or *disregistry* δ

$$\delta = \frac{|a_1 - a_2|}{a_2} \tag{3.75a}$$

where a_1 and a_2 are the lengths of the unit cells in the two lattices.

The smaller the disregistry is, the better will do the crystallographic planes of the substrate (foreign particle) and the nucleated phases (solidified melt) agree the more coherent will be the interface.

Special investigations have been performed to study whether good matching can be established between the crystalline structures of the foreign particle and the solidified melt. As shown in Figure 3.55 good matching can be obtained in a thin layer of solidified substance by deformation if the atomic spacings of the two structures are not too different.

Calculation of the Total Mismatching of a Semi-coherent Interface

A separate mismatch can be defined for the elastic strain and the dislocation strain, respectively:

$$\delta_{el} = \frac{\overline{\Delta a}}{a} \tag{3.75b}$$

and

$$\delta_{\perp} = \frac{1}{N_d} \tag{3.76}$$

where N_d is the number of atom rows between two adjacent dislocations.

The total mismatch is constant and can be written as

$$\delta_{total} = \delta_{el} + \delta_{\perp} \tag{3.77}$$

Elastic Strain Energy

For immediate use below an expression for *elastic strain energy* will be derived.

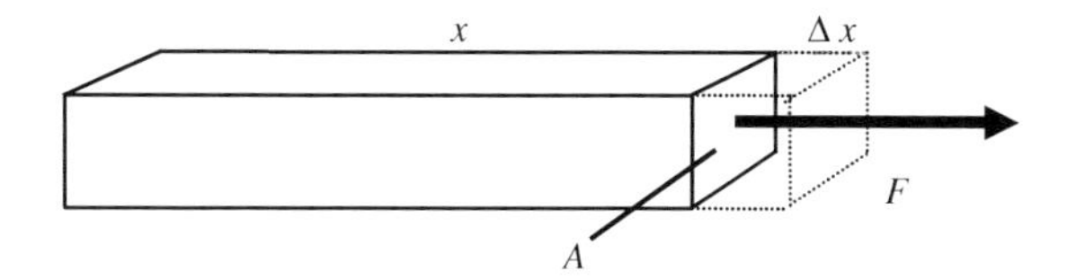

Figure 3.56 Extended bar.

A force F acts on a bar of length x and cross-section A and prolongs it a distance Δx (Figure 3.56). Basic theory of solid mechanics gives, if we assume that Hooke's law is valid

$$\sigma_{\text{tension}} = Y\,\varepsilon \tag{3.78}$$

where

σ_{tension} = the tensile stress F/A
Y = modulus of elasticity of the solid (Young's modulus)
ε = the relative stretch length.

$$\varepsilon = \Delta x/x \tag{3.79}$$

An infinitesimal change $\mathrm{d}x$ of x results in a change $\mathrm{d}\varepsilon$ of the relative stretch length

$$\mathrm{d}\varepsilon \approx \mathrm{d}x/x \tag{3.80}$$

Using Equations (3.78), (3.79) and (3.80) we obtain an expression for the work W required to extend the bar a distance Δx

$$W = \int \mathrm{d}W = \int_0^{\Delta x} A\sigma_{\text{tension}}\mathrm{d}x = \int_0^{\varepsilon} A(Y\varepsilon)(x\mathrm{d}\varepsilon) = \int_0^{\varepsilon} Ax\,Y\varepsilon\,\mathrm{d}\varepsilon$$

Ax is the volume of the bar. The volume is nearly constant and we obtain the *strain energy w per unit volume*

$$w = \frac{W}{V} = \int_0^{\varepsilon} Y\varepsilon\,\mathrm{d}\varepsilon$$

or

$$w = \frac{Y\varepsilon^2}{2} \tag{3.81}$$

which is a general expression, independent of the shape of the solid body.

Calculation of Solid/Solid Interphase Energies

The energy of a coherent solid/solid interface is given by the interaction between the atoms of the two solid phases. The interface energy can be calculated by Becker's exchange model, presented and applied on liquid/liquid interphase in Example 3.3 on page 126.

Figure 3.55 shows that, in addition to the same interaction between the atoms as mentioned above for coherent interphase, one has to add the 'elastic' strain energy σ_{el} caused by increasing or decreasing distances between the atoms and the strain energy $\sigma_{\perp}$ owing to edge dislocations.

Hence, the interface energy per unit area of a semicoherent interface will be the sum of three different terms

$$\sigma = \sigma_{ch} + \sigma_{el} + \sigma_{\perp} \tag{3.82}$$

where
σ_{ch} = surface energy per unit area owing to interaction between the atoms
σ_{el} = elastic strain energy per unit area
$\sigma_{\perp}$ = strain energy per unit area caused by edge dislocations.

If we replace ε in Equation (3.82) by the mismatching δ_{el} the elastic strain energy per unit area of the interface (area stretch energy = *twice* the linear stretch energy) can be written as

$$\sigma_{el} = Y\delta_{el}^2 d \times 1 \tag{3.83}$$

where d the thickness of the interface.

The strain energy $\sigma_{\perp}$ per unit area owing to edge dislocations (area strain energy = *twice* the linear strain energy), is given by

$$\sigma_{\perp} = \frac{2w_{\perp}}{D} \tag{3.84}$$

where
$w_{\perp}$ = energy per unit length required to form dislocations
D = distance between two adjacent dislocations.

Hence, the total solid/solid interface energy σ in Equation (3.82) is the sum of the two expressions in Equations (3.83) and (3.84) and the answer in Example 3.3 (page 126).

The two strain energy terms are related to each other, which is shown in Example 3.6 below.

Example 3.6

Consider a semicoherent interface between two solid phases. The lattice constants of the two crystal structures are a_1 and a_2, both $\approx a$. The dislocation energy per unit length is $w_{\perp}$. The elasticity modulus of the materials is Y. The thickness of the interface is d. The number of atom rows between two adjacent dislocations is N_d.

Calculate the elastic mismatch δ_{el} as a function of known quantities.

Solution:

The total mismatch can be written (Equations (3.75a and b) on page 142)

$$\delta_{total} = \delta_{el} + \delta_{\perp} \tag{1'}$$

or

$$\delta_{\perp} = \delta_{total} - \delta_{el} \tag{2'}$$

The total interface energy is

$$\sigma = \sigma_c + \sigma_{el} + \sigma_{\perp} \tag{3'}$$

Using Equations (3′), (3.76) and (3.77) on page 142 we obtain

$$\sigma = \sigma_c + Y\delta_{el}^2 d + \frac{2w_{\perp}}{D} \tag{4'}$$

We also know (page 142) that

$$D = N_{\perp} a = \frac{1}{\delta_{\perp}} a \tag{5'}$$

The expression for D in Equation (5′) is inserted into Equation (4′):

$$\sigma = \sigma_c + Y\delta_{el}^2 d + \frac{2w_{\perp}}{a}\delta_{\perp} \tag{6'}$$

or

$$\sigma = \sigma_c + Y\delta_{el}^2 d + \frac{2w_{\perp}}{a}(\delta_{total} - \delta_{el}) \tag{7'}$$

The general condition for equilibrium is that the free energy of the system has a minimum. Hence, equilibrium corresponds to a *minimum value* of σ. This σ value is found by taking the derivative of Equation (7′) with respect to δ_{el}.

$$\frac{d\sigma}{d\delta_{\perp}} = 2Y\delta_{el}d - \frac{2w_{\perp}}{a} \tag{8'}$$

The derivative equal to zero gives the equilibrium value of δ_{el}

$$\delta_{el} = \frac{w_{\perp}}{Yda} \tag{9'}$$

Equation (8′) shows that the second derivative is positive, which corresponds to a minimum.

Answer:
The desired relationship is $\delta_{el} = w_{\perp}/Yda$.

Magnitudes of Surface Energies of Solid/Solid Interfaces

The magnitudes of experimental values of the surface energies of coherent and incoherent solid/solid interfaces are listed in Table 3.3.

Table 3.3 Measured interface energies of coherent, semicoherent and incoherent interfaces

Type of solid-solid interface	Interface energy
Coherent boundaries	0.025 – 0.2 J/m^2
Semicoherent boundaries	0.2 – 0.5 J/m^2
Competely incoherent boundaries	0.5 – 1 J/m^2

Obviously, the interface energy is much smaller for coherent than for incoherent interfaces. We will come back to this in Chapter 6.

3.5 Equilibrium Shapes of Crystals

3.5.1 Gibbs–Curie–Wulff's Theorem

To find the equilibrium of a crystal (Figure 3.57) with its ambient phase we have to consider its total free energy. The thermodynamic condition of stability is that *the free energy of the crystal has a minimum value.*

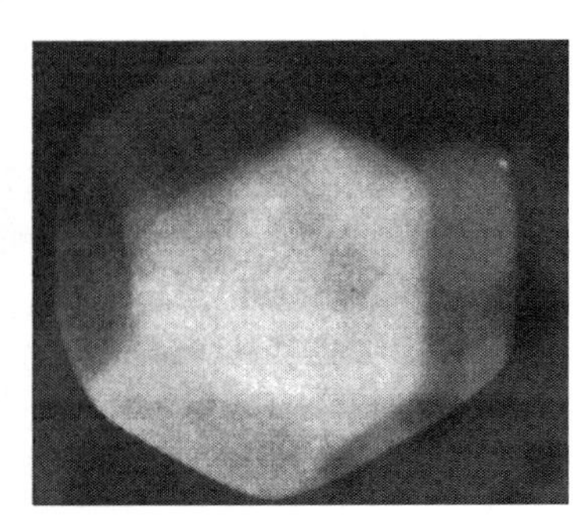

Figure 3.57 Typical FCC crystal of equilibrium shape. Ni crystal showing hexagonal and square surfaces corresponding to the (111) and (100) planes, respectively.

Consider a system of a crystal surrounded by either a vapour or a liquid phase. The total free energy of the crystal containing n_c kmol of material can be written as

$$G_c^{total} = n_c G + n_c V_m w_{strain} + \Sigma A_i \sigma_i \tag{3.85}$$

where

n_c = number of kmol in the crystal
G_c^{total} = total Gibbs' free energy of the crystal
G = Gibbs' free energy per kmol of the material
w_{strain} = strain energy per unit volume
A_i = area of the ith interface
σ_i = interface free energy per unit area of the interface
d = thickness of interface.

Some of the interfaces may also be stretched and possess strain energy. This energy is represented by the second term in Equation (3.85). The strain energy is often small except at low temperatures and can in most cases be neglected.

The total Gibbs' free energy is obviously a function of both the *shape* and *size* of the crystal. We assume that the molar free energy G and the molar volume V_m are constant and neglect the strain energy. Therefore, a crystal with a given volume the total Gibbs' free energy of the crystal only varies with the shape of the crystal.

$$G_c^{total} \approx n_c G + \Sigma A_i \sigma_i \tag{3.86}$$

The first term in Equation (3.86) represents the free energy of the crystal volume, which is constant and independent of shape. It is expressed as a function of the number of kmol in Equation (3.86) but can alternatively be given as a function of the volume of the crystal ($n_c = V_c/V_m$).

The interphase energy in the second term in Equation (3.81) can be regarded as the work to form one area unit of the surface. It depends on the structure of the crystal and the orientation of the crystal faces relative to the crystal planes. No general formula can be given. The term has to be found in each case.

The general condition for equilibrium is that the Gibbs' free energy of the system has a minimum.

- The equilibrium shape of a crystal of constant volume is the one at which the Gibbs' free energy of the crystal has a minimum.

To show that the minimum surface energy condition really determines the shape of the crystal we choose a simple fictive example.

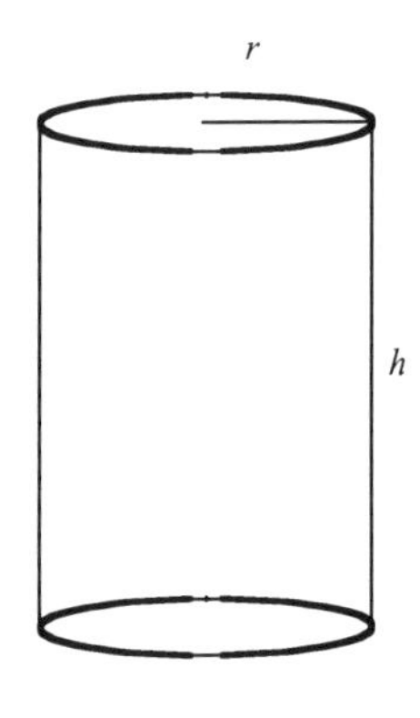

Example 3.7

A crystal solidifies in a melt in the shape of a cylinder. The solid/liquid interface energy of the base planes is σ_b and the cylinder surface has the solid/liquid interface energy σ_c.

The shape of the crystal is given by the ratio r/h. Find the equilibrium shape of the crystal as a function of σ_b and σ_c at constant volume of the crystal.

Solution:

The volume of the crystal is constant = V.

$$V = \pi r^2 h \tag{1'}$$

The total Gibbs' free energy of the crystal can be written

$$G_c^{total} = n_c G + \left(2\pi r^2 \sigma_b + 2\pi r h \sigma_c\right) = n_c G + \left(2\pi r^2 \sigma_b + 2\pi r \frac{V}{\pi r^2} \sigma_c\right)$$

where the volume term is constant.

The equilibrium condition is then

$$\frac{\partial G_c^{\text{total}}}{\partial r} \approx \left[4\pi\sigma_b r + 2V\sigma_c\left(-\frac{1}{r^2}\right)\right] = 0 \qquad (2')$$

which gives

$$r^3 = \frac{V}{2\pi}\frac{\sigma_c}{\sigma_b} \qquad (3')$$

or

$$r^3 = \frac{\pi r^2 h}{2\pi}\frac{\sigma_c}{\sigma_b}, \quad \text{which can be reduced to } \frac{h}{r} = 2\frac{\sigma_b}{\sigma_c}$$

Answer:
The shape of the cylindrical crystal is given by the condition $\frac{h}{r} = \frac{2\sigma_b}{\sigma_c}$

In real crystals the shape is more complicated to calculate as it depends on the orientation of the crystal faces relative to the crystal planes. Examples of a true crystal shape are given in Figures 3.57 and 3.58. The equilibrium shape is faceted but with rounded corners and edges.

However, if the interface energies of the various crystal surface planes are known, a very simple and useful geometrical construction of the crystal shape can be used. It is described below.

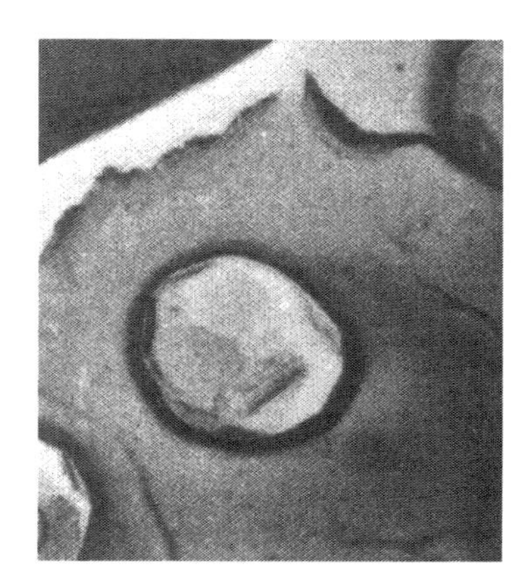

Figure 3.58 Replica of small FCC Ni crystals annealed in dry hydrogen at 1000 °C for 100 h.

Gibbs–Curie–Wulff's Theorem

On the basis of the work of Gibbs and Pierre Curie at the end of the nineteenth century an expression for the equilibrium shape of a crystal can be derived. A simplified derivation of this expression is given in the box below.

Derivation of Gibbs–Curie–Wulff's Theorem

Consider a crystal (subscript c, volume V_c) surrounded by a vapour or a liquid (index v, volume V_v). The crystal has the shape of a polyhedron confined by a number of crystal facets $A_1, A_2, \ldots A_n$.The crystal facets have the surface energies $\sigma_1, \sigma_2, \ldots \sigma_n$, respectively. The total volume V of the system crystal + vapour or liquid is constant.

$$V = V_c + V_v \qquad (1')$$

and

$$dV = dV_c + dV_v = 0 \qquad (2')$$

To find the equilibrium shape of the crystal we change its shape slightly. The surfaces are changed by the amounts dA_I (i = 1, 2 . . . n). The corresponding volume change is dV_c.

The associated change of the total Gibbs' free energy of the crystal consists of two terms, a volume term and an area term.

The *volume term* equals the work to enlarge the crystal against the pressure forces

$$dW = -p_c dV_c - p_v dV_v \qquad (3')$$

or in combination with Equation (2′)

$$dW = -(p_c - p_v)dV_c \quad (4')$$

$p_c > p_v$ and the work is therefore negative. This is in agreement with the fact that the crystal growth process is spontaneous.

The volume V_c of the crystal is the sum of n pyramids with a common apex in an arbitrary point within the crystal.

$$V_c = \frac{1}{3}\sum_n h_n A_n \quad (5')$$

where $h_1 h_2 \ldots h_n$ are the heights of the pyramids.

Differentiating Equation (5′) we obtain

$$dV_c = \frac{1}{3}\sum_n (h_n dA_n + A_n dh_n) \quad (6')$$

If every height h_n is increased by an amount dh_n the total volume change can alternatively be written as

$$dV_c = \sum_n A_n dh_n \quad (7')$$

The expressions in Equations (6′) and (7′) are equal, which gives

$$dV_c = \sum_n A_n dh_n = \frac{1}{2}\sum_n h_n dA_n \quad (8')$$

The equilibrium condition is

$$dG_c^{total} = -(p_c - p_v)dV_c + \sum_n \sigma_n dA_n = 0 \quad (9')$$

where the *area term* represents the change in surface energy.

The expression in Equation (8′) is introduced into Equation (9′), which gives

$$\sum_n \left[-(p_c - p_v)\frac{h_n}{2} + \sigma_n\right] dA_n = 0 \quad (10')$$

As every area change dA_n is independent of the others, all the coefficients must be zero to fulfill the condition (10′)

$$\frac{\sigma_n}{h_n} = \frac{1}{2}(p_c - p_v) \quad (11')$$

The pressure difference on the right-hand side is independent of the crystallographic orientation, which gives

$$\frac{\sigma_n}{h_n} = const \quad (12')$$

Equation (12′) in the box is called the *'Gibbs–Curie–Wulff's theorem* or *Wulff's rule*. It can be written as

$$\frac{\sigma_1}{h_1} = \frac{\sigma_2}{h_2} = \frac{\sigma_3}{h_3} = \ldots . = \frac{\sigma_n}{h_n} \tag{3.87}$$

It delivers the practical procedure to construct the equilibrium shape of the crystal, when the interphase energies $\sigma_1 \ldots \sigma_n$ are known:

1. Vectors normal to all the crystal surfaces from a common point (Wulff's point) are drawn within the crystal.
2. Distances proportional to the interface energies are marked on corresponding vectors.
3. Planes normal to the vectors are constructed through the marks.

A polyhedron shape is obtained. It is often convenient to choose the centre of the crystal as Wulff' s point, as is shown in Example 3.8.

Example 3.8

Solve Example 3.7 by applying Wulff's rule.

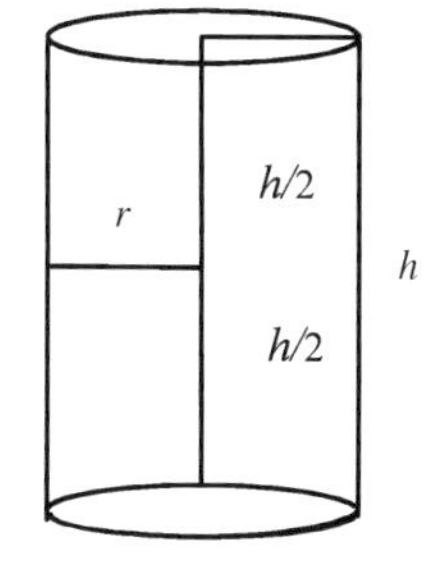

Solution and Answer:

Choose the centre of the cylinder as Wulff's point.

$$\frac{\sigma_b}{h/2} = \frac{\sigma_c}{r}, \quad \text{which gives} \quad \frac{h}{r} = \frac{2\sigma_b}{\sigma_c}$$

Constructions of crystal shapes based on Wulff's rule will be given in next section.

3.5.2 Interface Energy Dependence on Orientation

So far, we have assumed that the interface energy σ_n is constant over the whole surface area A_n. Here, we will discuss the interphase energy as a function of orientation in the crystal.

Consider a crystal surface that deviates slightly from the direction of one of the faces of the crystal with small Miller indices. Such a face is called *vicinal* and consists of terraces and steps. For simplicity we assume the steps to be monoatomic and equidistant.

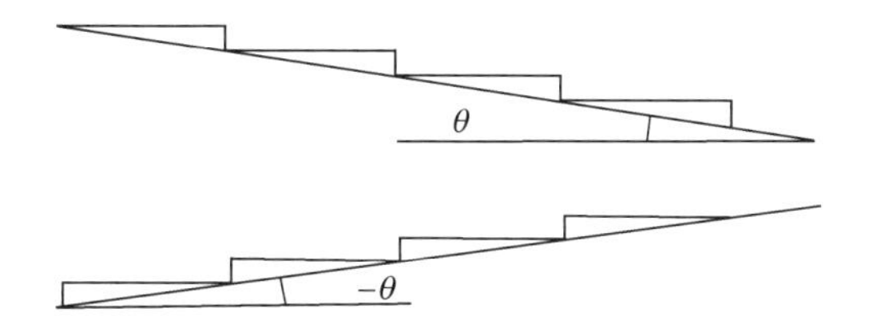

Figure 3.59 Vicinal surface, tilted an angle $\pm\theta$ relative to a plane with small Miller indices.

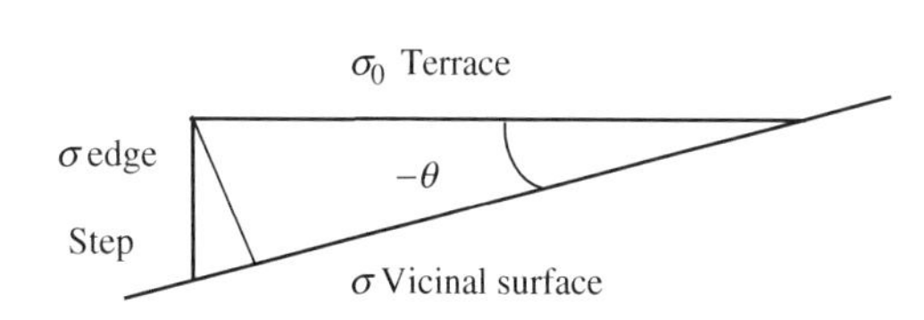

Figure 3.60 Enlargement of a step in Figure 3.59.

We want to find the average surface energy of the vicinal surface when the terrace surface energy σ_0 per unit area and the edge energy σ_{edge} per unit area are known. The vicinal surface energy is a function of the surface energy along a terrace and the surface energy along a step. It can be derived with the aid of the same method as we used on pages 123–124, i.e. counting the number of broken bonds per unit length in combination with the energy law. If the interaction between the steps is neglected the surface energy per vicinal unit area is found to be (Landau 1969).

$$\sigma = \frac{\sigma_{edge}}{a}\sin\theta + \sigma_0\cos\theta \quad (\text{positive } \theta) \tag{3.88a}$$

and (Figure 3.59)

$$\sigma = -\frac{\sigma_{edge}}{a}\sin\theta + \sigma_0\cos\theta \quad (\text{negative } \theta) \tag{3.88b}$$

where a = a dimensionless constant.

The equations can be summarized as

$$\sigma = \pm\frac{\sigma_{edge}}{a}\sin\theta + \sigma_0\cos\theta \tag{3.89}$$

Equation (3.89) represents the interface energy as a function of orientation. The function has been represented graphically in Figures 3.61a, b and c in polar coordinates, orthogonal coordinates and spherical coordinates, respectively, for arbitrary values of θ.

The slope of the tangent to the curve in Figure 3.61b is obtained if we take the derivative of Equation (3.89) with respect to the angle θ.

$$\frac{d\sigma}{d\theta} = \pm\frac{\sigma_{edge}}{a}\cos\theta - \sigma_0\sin\theta \tag{3.90}$$

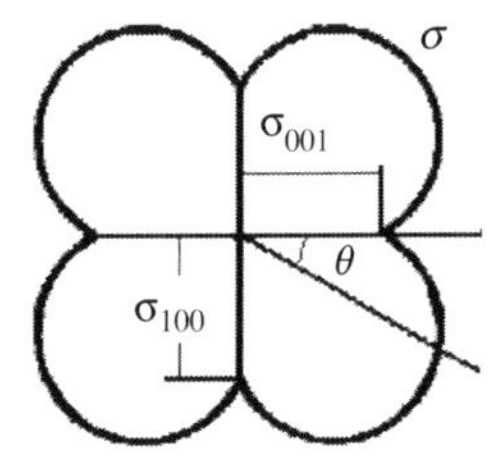

Figure 3.61a Surface energy as a function of orientation in polar coordinates. Reproduced from World Scientific Publishing Co © 1992.

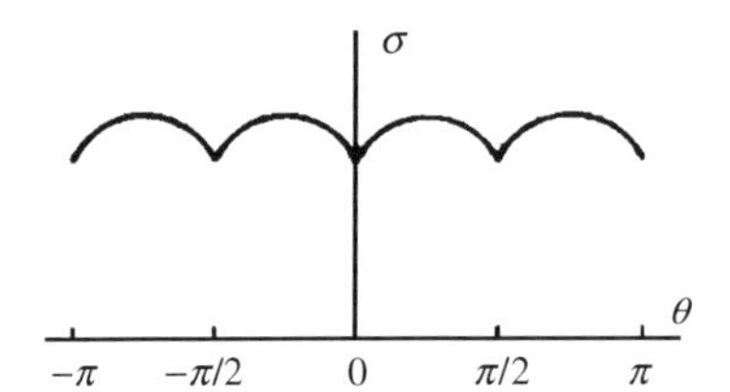

Figure 3.61b Surface energy as a function of orientation in orthogonal coordinates. Reproduced from World Scientific Publishing Co © 1992.

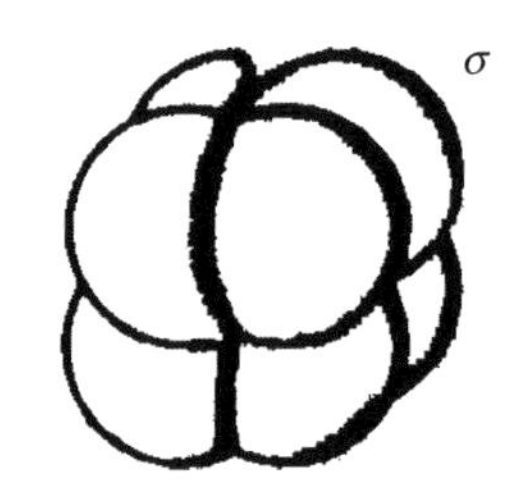

Figure 3.61c Surface energy as a function of orientation in spherical coordinates. Reproduced from World Scientific Publishing Co © 1992.

The function $\sigma(\theta)$ is continuous for all values of θ. Its derivative is also continuous except for $\theta = 0$ and all positive and negative multiples of $\pi/2$ where the derivative is discontinuous and switches from $+\sigma_{edge}/a$ to $-\sigma_{edge}/a$.

The curves in the Figures 3.61 are related to interface energies calculated from the first nearest neighbours only (Example 3.2 on page 124). Such a curve, named σ_1, is shown in Figure 3.62. The corresponding curve based on the second nearest neighbours only is the one called σ_2. Its orientation is changed by an angle $\pi/4$ compared to the curve σ_1 because the direction of the bonds differ by this amount, as is shown in Example 3.2.

The outer curve denoted σ_{1+2} represents the interface energy when the calculations of the broken bonds involve the interactions of both the first and the second nearest neighbours.

$$\sigma_{1+2} = \sigma_1 + \sigma_2 \tag{3.91}$$

Figure 3.62 also shows a polygon drawn through the points of the curve σ_{1+2} where the derivative is discontinuous. As we will see on page 150 it represents the equilibrium shape of the crystal obtained with the aid of Wulff's rule (pages 147–148).

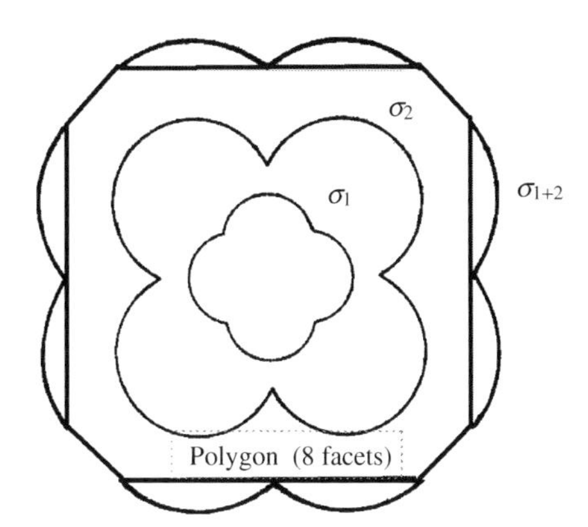

Figure 3.62 Polar diagram of interface energies of a crystal, involving the interaction of the first, the second and the first + second nearest-neighbour interactions.

The polygon represents the equilibrium shape of the crystal obtained with the aid of Wulff's rule. Reproduced from World Scientific Publishing Co © 1992.

Herring's Formula. Surface Stiffness

In Section 3.5.1 we derived the shape of a crystal based on the condition that the interface energy of the crystal has a minimum value at equilibrium. Earlier in this section we found that the interface energy depends on the orientation of the crystal surfaces (terraces and steps) when the crystal face deviates from crystal planes with small Miller indices. The theory in Section 3.5.1 has to be revised to account for the *influence of orientation* on interface energy.

This problem has been treated by Burton, et al. [7] in 1951, by Herring [8] in 1951. The extensive general theory will not be reported here, only the results will be presented.

- The calculations are made in spherical coordinates r, φ, θ.
- σ_n is replaced by σ, which is a function of both φ and θ.
- h_n is replaced by two principle radii of curvature $R_1$and R_2.

- The generalized Gibbs–Curie–Wulff theorem can be written as

$$p_c - p_v = \left[\frac{\sigma + \dfrac{\partial^2\sigma}{\partial\varphi^2}}{R_1} + \frac{\sigma + \dfrac{\partial^2\sigma}{\partial\theta^2}}{R_2}\right] = \left(\frac{\sigma_1^*}{R_1} + \frac{\sigma_2^*}{R_2}\right) \tag{3.92}$$

The generalized Gibbs–Curie–Wulff's theorem:

$$p_c - p_v = \frac{2\sigma_n}{h_n}$$

where the expressions

$$\sigma_1^* = \sigma + \frac{\partial^2\sigma}{\partial\varphi^2} \quad \text{and} \quad \sigma_2^* = \sigma + \frac{\partial^2\sigma}{\partial\theta^2} \tag{3.93}$$

are called *surface stiffnesses*.

The surface stiffness is a measure of the resistivity of the crystal surfaces against bending or roughening when a pressure is applied on them. As we will see in Section 3.5.3 surface stiffness is a most essential quantity for the stability of crystal surfaces.

The construction of the equilibrium shape of the crystal is based on Equation (3.92). It is done as is shown in Figure 3.63 on page 154:

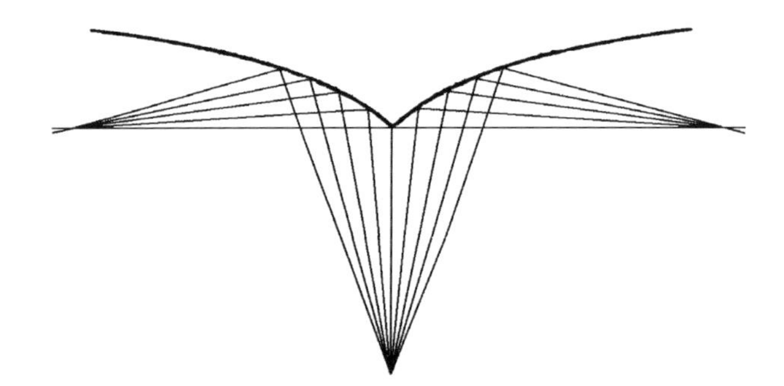

Figure 3.63 Construction of the equilibrium shape of a crystal from the polar diagram (see Figure 3.55 on page 142).

Figure 3.64 The polygon represents the equilibrium shape of the crystal.

1. A radius vector in any direction is drawn from the centre. At the point where it intersects the polar diagram surface, a plane perpendicular to the radius vector is drawn.
2. The procedure is repeated for many other directions until the polar diagram area is properly covered.
3. The inner envelope to all the planes is the equilibrium shape of the crystal.

The planes drawn through the 'discontinuity points' are of special interest. The first derivative in these points is discontinuous, which means that the second derivatives and the surface stiffnesses are infinitely large in these points. The pressure difference and the difference in chemical potentials are finite. Consequently, the radii of curvature must also be infinitely large and the corresponding crystal faces must be flat, as is shown in Figure 3.64.

3.5.3 *Stability of Crystal Surfaces*

In the last section we have seen how the equilibrium shape of a crystal can be constructed. The next step is to examine the stability of crystal surfaces and the necessary condition for maintaining them. It can be done by a study of the energy of the crystal surface at equilibrium and finding the condition for the equilibrium to be stable.

The energy of a crystal surface depends on geometrical factors like terrace length and step height but also on adsorption of impurity atoms on the surface. The impurity atoms saturate the unsaturated bonds on the crystal surface and cause a decrease of the surface energy. The decrease is proportional to the concentration of the impurity atoms.

The matter of stability of crystal surfaces was primarily studied by Chernov [9] in 1961 and further by Cabrera and Coleman [10] in 1963. The ideas of the latter scientists are presented below.

Derivation of the Stability Condition of Crystal Surfaces

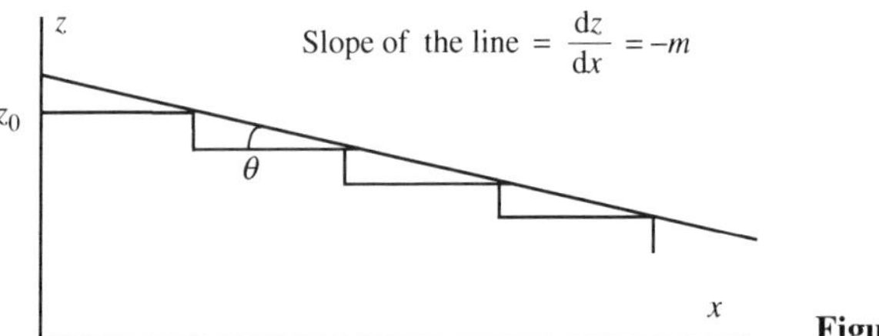

Figure 1′.

Consider a vicinal surface that forms an angle θ with the nearest crystal plane with low Miller indices. It consists of terraces and monoatomic steps. The crystal has a constant volume.

A coordinate system is chosen as is shown in Figure 1′. The crystal plane is represented by $z = 0$ and the vicinal surface by $z = z_0 - mx$ where $m = \tan\theta$.

Let the area of the vicinal crystal surface be A_0 (Figure 2′). Its projection on the crystal plane with low Miller indices has the area $A = A_0 \cos\theta$.

The step surface energy is neglected The total surface energy Σ_0 of the surface A_0 can be written as

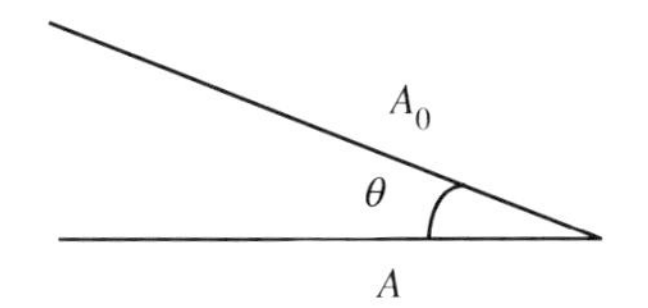

Figure 2′.

$$\Sigma_0 = \sigma(\theta)A_0 = \frac{\sigma(\theta)A}{\cos\theta} \tag{1′}$$

The interface energy Σ_0 is a function of θ or, better expressed, of $\tan\theta = m$ as we want to transform Σ_0 into a function of m. We have σ as a function of θ and want to express σ as a function of m, which is equal to $\tan\theta$. Equation (1′) can be written as

$$\Sigma_0 = \frac{\sigma(\theta)A}{\cos\theta} = \sigma(\theta)\sqrt{1 + (\tan\theta)^2}A = \sigma(m)A \tag{2′}$$

where

$$\sigma(m) = \sigma(\theta)\sqrt{1 + m^2} \tag{3′}$$

With the aid of Equations (2′) and (3′) we obtain

$$\Sigma_0 = \frac{\sigma(\theta)A}{\cos\theta} = \sigma(m)\sqrt{1 + m^2}A \tag{4′}$$

If we introduce the function

$$\xi(p) = \sigma(p)\sqrt{1 + p^2} \tag{5′}$$

we obtain

$$\Sigma_0 = \xi(m)A \tag{6′}$$

Stability Condition

The surface A_0 is replaced by two new surfaces with slightly different orientations Figure 3′:

$$m_1 = m + \delta m_1 \quad \text{and} \quad m_2 = m + \delta m_2 \tag{7′}$$

Together, the new surfaces 1 and 2 have the same projected area as the original surface. Projection on the plane $z = 0$ gives

$$A = A_1 + A_2 \tag{8}$$

Projection of the plane $x = 0$ gives ($m_i = \tan\theta_i$)

$$mA = m_1A_1 + m_2A_2 \tag{9′}$$

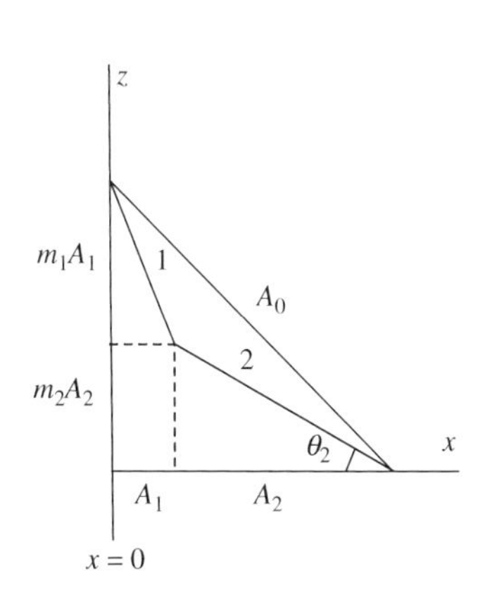

Figure 3′.

Substituting the two Equations (7′) and Equation (8′) into Equation (9′) we obtain

$$A_1\delta m_1 + A_2\delta m_2 = 0 \tag{10′}$$

The total interface energy Σ_{1+2} of the new surfaces can with the aid of Equation (6′) be written as

$$\Sigma_{1+2} = \xi(m_1)A_1 + \xi(m_2)A_2 \tag{11′}$$

The *condition for the initial surface to be the stable state* is that its interface energy Σ_0 is *lower* than the total energy Σ_{1+2} of the two surfaces.

$$\Delta\Sigma = \Sigma_{1+2} - \Sigma_0 > 0 \tag{12′}$$

which can be written as

$$\Delta\Sigma = \xi(m_1)A_1 + \xi(m_2)A_2 - \xi(m)(A_1 + A_2) > 0 \tag{13′}$$

The functions $\xi(m + \delta m_i)$ can be developed in a Taylor series

$$\xi(m + \delta m_i) = \xi(m) + \xi'(m)\delta m_i + \frac{\xi''(m)}{2}(\delta m_i)^2 \tag{14′}$$

The condition (12′) can then be reduced to

$$\Delta\Sigma = \xi'(m)(A_1\delta m_1 + A_2\delta m_2) + \frac{\xi''(m)}{2}\left[A_1(\delta m_1)^2 + A_2(\delta m_2)^2\right] > 0 \tag{15′}$$

The first term in Equation (15′) is zero because of Equation (10′). The second factor in the second term is always positive. Hence, the stability condition reduces to

$$\xi''(m) > 0 \tag{16′}$$

which is the desired stability condition.

Equation (15′) can be developed further. The calculation of $\xi''(m) = \mathrm{d}^2\xi/\mathrm{d}m^2$ has been indicated in the box below.

Calculation of $\xi''(m)$

Before taking the derivative of the function $\xi(p) = \sigma(p)\sqrt{1 + p^2}$ we will list a few mathematical relationships for later use.

$$\frac{\mathrm{d}\xi}{\mathrm{d}m} = \frac{\mathrm{d}\xi}{\mathrm{d}\theta}\frac{\mathrm{d}\theta}{\mathrm{d}m} \qquad \theta = \arctan m \qquad \frac{\mathrm{d}\theta}{\mathrm{d}m} = \frac{1}{1 + m^2}$$

ξ can be regarded as a function of m or θ.

Taking the derivative of $\xi = \sigma\sqrt{(1 + m^2)}$ twice with respect to m gives, after use of the relationships above, and some reduction

$$\xi' = \frac{\mathrm{d}\xi}{\mathrm{d}m} = \frac{\upsilon + \frac{\mathrm{d}\sigma}{\mathrm{d}\theta}}{\sqrt{1 + m^2}} \quad \text{and} \quad \xi'' = \frac{\mathrm{d}^2\xi}{\mathrm{d}m^2} = \frac{\sigma + \frac{\mathrm{d}^2\sigma}{\mathrm{d}\theta^2}}{\sqrt{1 + m^2}^3} \tag{17′}$$

As the denominator of the stability condition in the box for ξ'', i.e. Equation (17′), is always positive the stability condition can be written as

$$\sigma^* = \sigma + \frac{\mathrm{d}^2\sigma}{\mathrm{d}\theta^2} > 0 \tag{3.88}$$

Hence, the stability of the crystal surface depends entirely on the sign of the surface stiffness σ^* that was introduced on page 151 in this chapter:

- If $\sigma^* > 0$ the crystal surface is stable.
- If $\sigma^* < 0$ the crystal surface will break down into facets with positive surface stiffness.

Impurities lower the surface energy and may cause negative stiffness values. Carbon is a very dangerous impurity in the semiconductor industry. Metals also cause a decrease of the surface stiffness with far-reaching consequences.

Summary

■ Basic Concepts and Fundamental Relationships

Interface energy = the work that has to be done to form one unit area of the interface.
Surface tension = the force per unit length in the surface acting to minimize its area.

Under isothermal conditions the interface energy and the surface tension are equal. The letter σ is used for both interface energy and surface tension except in the few cases when a distinction between the two quantities is necessary to avoid confusion.

The molar Gibbs' free energy of an interface (for example solid/liquid) is defined as $G^{\mathrm{I}} = H^{\mathrm{I}} - TS^{\mathrm{I}}$

At equilibrium between a solid and a liquid the Gibbs' free energies of the solid and liquid are equal. $G^{\mathrm{s}} = G^{\mathrm{L}}$.

Relationship between the Molar Gibbs' Free Energy and the Surface Energy
The Gibbs' free energy of a solid/liquid interphase is always larger than that of the solid and the liquid. The difference is equal to the interface energy.

$$\Delta G^{\mathrm{I}} = G^{\mathrm{I}} - G^{\mathrm{L}} = \sigma \frac{V_{\mathrm{m}}^{\mathrm{I}}}{d}$$

The factor $V_{\mathrm{m}}^{\mathrm{I}}/d$ relates the thermodynamic quantities (J/kmol) and the surface quantities (J/m^2).

■ Interface Energy as a Function of Temperature and Impurity Concentration

Liquids in a gas atmosphere have lower interface energies than do equivalent solids.

The change in enthalpy at the melting point:

$-\Delta H$ = the molar heat of fusion = L_{f}.

Influence of Temperature
The interface energies decrease with increasing temperature for both liquids and solids.

Influence of Impurities
The surface tension and interface energy decreases strongly with the impurity concentration.

■ Wetting

Simplified Wetting Equation for Solid/Liquid/Vapour Interfaces:

$\sigma_{SV} = \sigma_{SL} + \sigma_{LV}\cos\theta$ (Page 107)

At compete wetting $\theta = 0$ and the factor $\cos\theta = 1$, which is its maximum value.

Liquid/Liquid Interfaces

$$\frac{\sigma_{23}}{\sin\theta_1} = \frac{\sigma_{13}}{\sin\theta_2} = \frac{\sigma_{12}}{\sin\theta_3} \quad \text{(Page 109)}$$

Solid/Liquid Interfaces

$\sigma_{13} - \sigma_{23} = \sigma_{12}\cos\theta_2$ (Page 109)

Condition for complete wetting:

$\sigma_{SV} \geq \sigma_{SL} + \sigma_{LV}$ (Page 113)

Solid/Solid Interfaces
Three solid phases (grains) that meet at a junction point.

$$\sigma_{11} = 2\sigma_{12}\cos\frac{\theta}{2} \quad \text{(Page 110)}$$

$\theta =$ dihedral angle.

Experimental Evidence

- The liquid/vapour interface energies of salts and oxides are generally lower than those of metals.
- The solid/vapour interface energies of salts and oxides are generally lower than those of metals.
- The solid/liquid interface energies of metals are roughly ten times smaller than their solid/vapour interface energies.
- The solid/liquid interface energies are generally smaller than the difference between their solid/vapour and liquid/vapour interface energies.

■ Thermodynamics of Interfaces

Models
Gibbs' Model:
A dividing plane of zero thickness between two phases.

Guggenheim's Model:
A layer of finite and uniform thickness between two neighbouring homogeneous phases. It is treated as a separate phase, the interface phase or shortly the interphase.

Graphical examples of the thermodynamic quantities G and H for solid/liquid, solid/vapour and liquid/vapour interphases are given in the text.

Segregation of Alloying Elements in Binary Alloys to Grain Boundaries or Liquid/Vapour Interfaces
If the free-energy curves of the liquid phase and interface phase are known, the concentrations of the alloying element in the liquid phase and the interface phase can be constructed graphically. Two parallel tangents are drawn (page 118).

The partition coefficient of the alloying element can be derived by thermodynamic calculations. The calculations are given in Section 3.3.4

The equations

$$G_1^{0\mathrm{I}} - G_1^{0\mathrm{L}} = \frac{\sigma_1^0 V_\mathrm{m}^\mathrm{I}}{d}$$

$$G_2^{0\mathrm{I}} - G_2^{0\mathrm{L}} = \frac{\sigma_2^0 V_\mathrm{m}^\mathrm{I}}{d}$$

$$G_1^\mathrm{I} - G_1^\mathrm{L} = G_2^\mathrm{I} - G_2^\mathrm{L}$$

$$G_\mathrm{i} = G_\mathrm{i}^0 + RT\ln x_\mathrm{i} + G_\mathrm{i}^\mathrm{Ex}$$

give the partition coefficient

$$\frac{x_2^\mathrm{I}}{x_2^\mathrm{L}} = \frac{x_1^\mathrm{I}}{x_1^\mathrm{L}} \mathrm{e}^{-\frac{\sigma_2^0 V_\mathrm{m}^\mathrm{I} - \sigma_1^0 V_\mathrm{m}^\mathrm{I}}{d\,RT} + \frac{\Delta G_2^{\mathrm{Ex\,I}} - \Delta G_1^{\mathrm{Ex\,I}}}{RT}}$$

The activity coefficients γ are introduced instead of G^Ex

$$G_2^{\mathrm{Ex\,L}} - G_2^{\mathrm{Ex\,I}} = -RT\ln\frac{\gamma_2^\mathrm{L}}{\gamma_2^\mathrm{I}}$$

and

$$G_1^{\mathrm{Ex\,L}} - G_1^{\mathrm{Ex\,I}} = -RT\ln\frac{\gamma_1^\mathrm{L}}{\gamma_1^\mathrm{I}}$$

Segregation ratio:

$$\beta = \frac{x_2^\mathrm{I}}{x_2^\mathrm{L}} = \frac{x_1^\mathrm{I}}{x_1^\mathrm{L}} \frac{\gamma_1^\mathrm{I}}{\gamma_1^\mathrm{L}} \frac{\gamma_2^\mathrm{L}}{\gamma_2^\mathrm{I}} \mathrm{e}^{-\frac{\Delta G^\mathrm{I}}{RT}}$$

where $\Delta G^\mathrm{I} = \left(\sigma_2^0 \frac{V_\mathrm{m}^\mathrm{I}}{d} - \sigma_1^0 \frac{V_\mathrm{m}^\mathrm{I}}{d}\right)$.

If the solution is dilute and Raoult's law is valid:

$$\beta = \frac{x_2^\mathrm{I}}{x_2^\mathrm{L}} = \frac{\gamma_2^\mathrm{L}}{\gamma_2^\mathrm{I}} \mathrm{e}^{-\frac{\Delta G^\mathrm{I}}{RT}}$$

■ Structures of Interfaces

There are five contributions to the total interface energy σ

$$\sigma = \sigma_\mathrm{ch} + \sigma_\mathrm{e} + \sigma_\mathrm{S} + \sigma_\mathrm{d} + \sigma_\mathrm{a}$$

- A chemical interaction between the atoms resulting in an interface energy σ_ch.
- A electrochemical potential at the surface, caused by a change of the number of free electrons at the surface interface, resulting in an interface energy σ_e.

- An entropy term σ_S associated with the mixing of different species or atoms at the interface.
- A strain energy term σ_d associated with straining of the atoms at the interface.
- A chemical adsorption term σ_a.

Liquid/Vapour Interfaces

Liquid/vapour interface energies can be calculated by counting the number of broken bonds per unit area when 'liquid' atoms are transformed into 'vapour' atoms.

The change of enthalpy to transfer one atom from the interior of the melt to a gas phase equals the heat of evaporation L_v divided by Avogadro's number N_A. Each atom in the melt has on average 10 nearest neighbours. At the surface each atom is most likely surrounded by 6 or 7 neighbours. Three or four bonds per atom are broken.

$$\frac{L_v}{10} \times 3 < \sigma_c \frac{V_m^I}{d} < \frac{L_v}{10} \times 4$$

Solid/Vapour Interfaces

The interface energy σ_{ch} between a solid and a vapour phase can be calculated in the same way as the interface energy between a liquid and a vapour phase. Care must also be taken as to the influence of the structure of the solid and the number of vacancies in the material.

Closely packed planes have a lower interface free energy per unit area than all other planes.

Liquid/Liquid Interfaces

The free energy of the interface between two immiscible liquids can be calculated by Becker's method of breaking old bonds and forming new ones when the liquids in two containers are partially transferred into each other.

Becker's method is demonstrated in Example 3.3 on page 126.

Solid/Liquid Interfaces

The liquid/solid interfaces are of special importance as the most common process of solidification is the solidification of a melt. Two radically different models of solid/liquid interfaces have been proposed.

Atomically Planar Interfaces (Gibbs' Interface)

Atomically flat and stepped interfaces have narrow transition zones. A 'Gibbs' interface is a plane between two atomic layers. The energies of such sharp and smooth interfaces are *anisotropic*, i.e. they depend on orientation.

If the interface is atomically planar and the entropy of the interface is disregarded, the interface energy can be calculated by adding the enthalpy of the bonds between the 'solid' and the 'liquid' atoms over the interface to the Gibbs' free energy of the solid and the liquid

$$-\Delta G_m^I \approx -\Delta H_m^I = n \frac{Z_b}{Z_{coord}} \frac{L_f}{N_A} \frac{V_m^I}{d}$$

The interface energy is also a function of the crystal structure. The procedure is analogous to that used in the case of a solid surrounded by a vapour.

The closely packed planes have a lower interface free energy per unit area than all other planes.

Diffuse or Rough Interphases (Guggenheim Interphase)

Diffuse interphases are spread over several atomic layers with a gradual change of order within the transition zone of the interphases. The atomic disorder close to the liquid phase changes gradually to the regular lattice site order characteristic of a solid.

Energies of diffuse interphases are *isotropic*, e.g. independent of orientation. A diffuse interphases is rough and not smooth as the atomically flat interfaces. For this reason diffuse interphases are often called rough interphases.

Jackson's Model

Jackson's two-level model offers an elementary theoretical treatment of the interphase energy of a rough interphase. He used the formalism of a solution to describe the outer partially filled level

$$x^{s} = \frac{n_{a}}{n} \quad \text{and} \quad x^{L} = 1 - \frac{n_{a}}{n}$$

and considered the broken lateral bonds of the atoms to their neighbour atoms.

$$-\Delta H^{I}_{mix} = \frac{Z_{l}}{Z_{coord}} \frac{L_{f}}{N_{A}} n x^{L}(1 - x^{L}) \frac{V^{I}_{m}}{d}$$

The entropy cannot be neglected for a rough interphase

$$\Delta S^{I} = -nk_{B}\left[x^{L}\ln x^{L} + (1 - x^{L})\ln(1 - x^{L})\right] \times \frac{V^{I}_{m}}{d}$$

The relationship $\Delta G^{I} = \Delta H^{I} - T\Delta S^{I}$ gives

$$\Delta G^{I}_{mix} = nk_{B}T\left[\alpha x^{L}(1 - x^{L}) + x^{L}\ln x^{L} + (1 - x^{L})\ln(1 - x^{L})\right] \times \frac{V^{I}_{m}}{d}$$

or

$$\sigma^{I}_{mix} = nk_{B}T\left[\alpha x^{L}(1 - x^{L}) + x^{L}\ln x^{L} + (1 - x^{L})\ln(1 - x^{L})\right]$$

where $\alpha = \frac{Z_{l}}{Z_{coord}} \frac{L_{f}}{N_{A}k_{B}T}$

- If $\alpha < 2$ a rough interface is the stable state. Metals belong to this case.
- If $\alpha > 2$ an atomically planar interface is the stable state.

A monolayer interphase consists of a Guggenheim interphase surrounded by two Gibbs interfaces.
Total surface tension of a monolayer interphase:

$$\sigma^{mono}_{total} = nk_{B}T\alpha\frac{2Z_{b}}{Z_{I}} + nk_{B}T\left[\alpha x^{L}(1 - x^{L}) + x^{L}\ln x^{L} + (1 - x^{L})\ln(1 - x^{L})\right]$$

A multilayer interphase of Jackson's type consisting of p monolayers has the total surface tension:

$$\sigma^{Jackson}_{total} = p \times \sigma^{mono}_{total}$$

If the composition of the multilayer varies gradually from the solid layer to the liquid layer the total surface tension is

$$\sigma^{Temkin}_{total} = nk_{B}T\left[p \times \alpha\frac{2Z_{b}}{Z_{I}} \sum_{i=1}^{p}\left[\alpha x^{L}_{i}(1 - x^{L}_{i}) + x^{L}_{i}\ln x^{L}_{i} + (1 - x^{L}_{i})\ln(1 - x^{L}_{i})\right]\right]$$

Solid/Solid Interfaces

There are three types of solid/solid interfaces

- coherent interfaces
- semicoherent interfaces
- incoherent interfaces.

Mismatch or Disregistry between Two Lattices

$$\delta = \frac{|a_1 - a_2|}{a_2}$$

The interface energy of a semicoherent interface consists of three terms owing to interaction between the atoms, elastic strain and dislocations

$$\sigma = \sigma_c + \sigma_{el} + \sigma_\perp$$

where

$$\sigma_{el} = Y\delta^2 d \quad \text{and} \quad \sigma_\perp = \frac{2w_\perp}{D}$$

The total mismatch of a semicoherent interface:

$$\delta_{total} = \delta_{el} + \delta_\perp$$

■ Equilibrium Shapes of Crystals

Total Gibbs' free energy of a crystal:

$$G_c^{total} \approx n_c G + \Sigma A_i \sigma_i$$

where $n_c G = (p_c - p_v)V_c$ and strain is neglected.

The equilibrium shape of a crystal of constant volume is the one at which the free energy of the crystal has a minimum.

Gibbs–Curie–Wulff's theorem or Wulff's rule (Pages 147–149).

$$\frac{\sigma_1}{h_1} = \frac{\sigma_2}{h_2} = \frac{\sigma_3}{h_3} = \ldots\ldots = \frac{\sigma_n}{h_n} = const$$

can be used to construct the equilibrium shape of a crystal.

Construction of the Equilibrium Shape of a Crystal

1. Vectors normal to all the crystal surfaces from a common point (Wulff's point) are drawn within the crystal.
2. Distances proportional to the interface energies are marked on corresponding vectors.
3. Planes normal to the vectors are constructed through the marks. They enclose the crystal (page 149).

Interface Energy as a Function of Orientation

Interface energy depends on the orientation of the crystal.

$$\sigma = \pm\frac{\sigma_{edge}}{a}\sin\theta + \sigma_0\cos\theta$$

■ Stability of Crystal Surfaces

Surface Stiffness

$$\sigma_1^* = \sigma + \frac{\partial^2 \sigma}{\partial \varphi^2} \quad \text{and} \quad \sigma_2^* = \sigma + \frac{\partial^2 \sigma}{\partial \theta^2}$$

where the angles φ and θ are spherical coordinates.

Stability Condition of a Crystal

$$\sigma^* = \sigma + \frac{d^2 \sigma}{d\theta^2} > 0$$

- If $\sigma^* > 0$ the crystal surface is stable.
- If $\sigma^* < 0$ the crystal surface will break down into facets with positive surface stiffnesses.

Exercises

3.1 An experimental method to determine the surface free energy between a liquid and a solid is to perform measurements on a surface tension balance in a three-phase junction, i.e. a triple point where three phases meet and the surface-tension forces balance each other at equilibrium.

The result of such measurements is shown in the figure below.

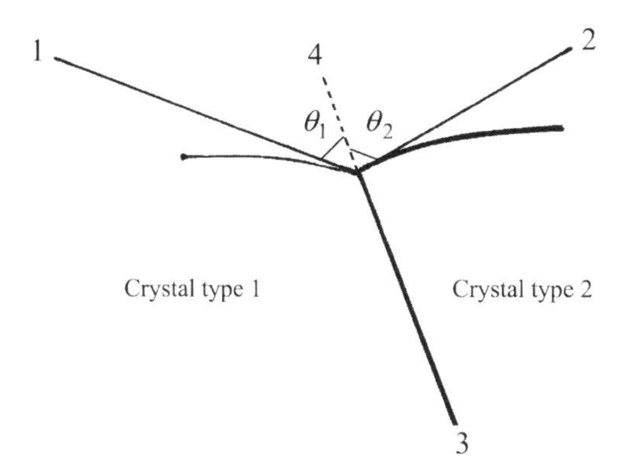

Figure 3.1–1

The angles θ_1 and θ_2 are 70° and 80°, respectively. The surface energy between crystals of types 1 and 2 (direction 3) is 0.40 J/m^2 (compare Table 3.3 on page 145 in Chapter 3).

Calculate the interface Gibbs' free energy between the liquid and crystals of type 1 (direction 1) and ditto between the liquid and crystals of type 2 (direction 2). What is the reason for the discrepancy between the two values?

3.2 Porous heat resistant materials such as rock wool and other porous fibres are filled with a metal melt. The melt fills the cavities in the materials and so-called metal matrix composites are formed. They have good mechanical properties and considerably lower density than pure metals and alloys and are used for a variety of industrial purposes, such as electronic components and rotating components in electrical machines.

The most common way to form such composites is to fill the cavities between the fibres of a so-called preform with a metal melt. The preform consists of a great number of parallel, more or less straight channels.

Consider a preform made of a great number of circular parallel fibres, which touch each other and have an average radius of 10 μm and are assumed to be straight (Figure 1). One end of a

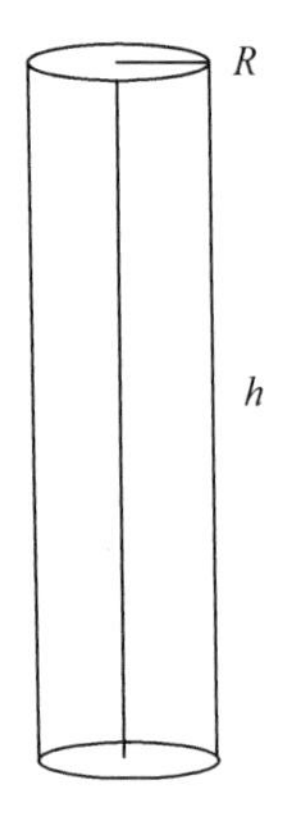

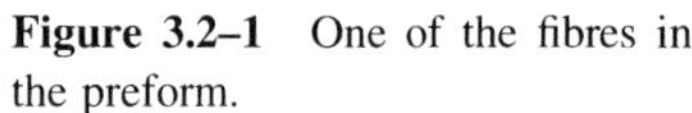

Figure 3.2–1 One of the fibres in the preform.

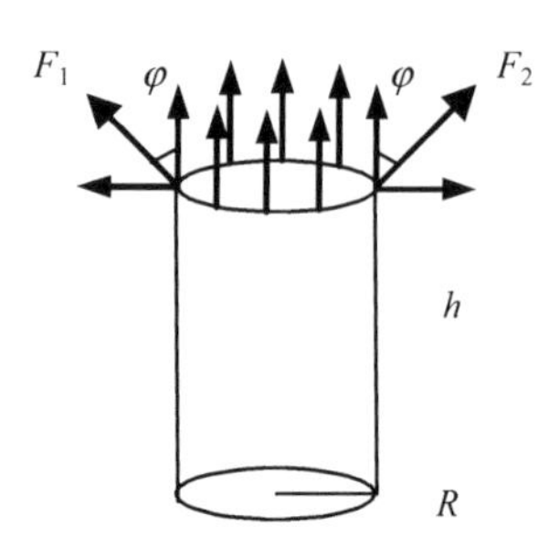

Figure 3.2–2 Shape of the cross-section of a channel.

preform with vertical fibres is dipped into an Al melt. The preform and the melt are kept at a temperature above the melting point 658 °C of Al during the whole process.

The melt is soaked into the cavities owing to the capillary forces. No external pressure is applied to fill the preform. The wetting between the melt and the fibre walls of the channels is assumed to be complete, i.e. $\varphi = 0$ in Figure 3.2–2.

The height h of the preform is 2.25 m. The density of Al is 2.7×10^3 kg/m^3. Is it possible to fill all the cavities in the channels up to the top by simply let the melt get in touch with the Al melt if the surface tension between the melt and the channel walls is 2.0 N/m?

3.3 Surface tension is discussed in Chapter 3. Surface-tension values for pure elements are listed in Table 3.1 on page 112. When alloying a metal the segregation of elements to the surface drastically changes the surface tension, which is of utmost importance in many applications.

Calculate the solid/liquid surface tension for the αFe-P alloy with the aid of Table 3.1 and Figure 3.38 on page 121 in Chapter 3. The thickness of the interface is assumed to be 1.0×10^{-9} m.

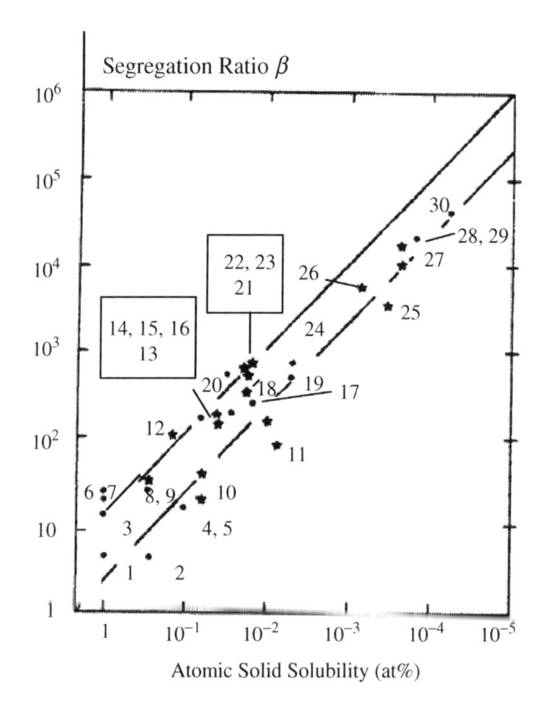

Figure 3.3–1 Interface energies of some metals and semiconductor elements. Measured grain-boundary segregation ratios as a function of atomic solid solubility. Figure explanation: 1 Cu–Fe, 2 Fe–Si, 3 γFe–Ni, 4 δFe–Si, 5 Cu–Sb, 6 γFe–Mn, 7 γFe–Cr, 8 γFe–Si, 9 αFe–Al, 10 αFe–Ni, 11 αFe–Cu, 12 αFe–Zn, 13 αFe–P, 14 αFe–Sn, 15 αFe–As, 16 δFe–Sn, 17 γFe–P, 18 αFe–P, 19 δFe–P, 20 αFe–N, 21 αFe–P, 22 αFe–Sn, 23 αFe–Sb, 24 Cu–Sb, 25 αFe–B, 26 Ni–B, 27 αFe–C, 28 Cu–B, 29 αFe–S, 30 Cu–B.

Material constants for α-Fe (BCC):

Melting point	1535 °C
Molar volume	7.9×10^{-3} m^3/kmol.

3.4 Immiscible alloys such as Al–Pb alloys are used as bearing materials. During the solidification of such alloys the movements of Pb droplets, precipitated in the liquid as a consequence of the surface tension, are important.

The relationship between the surface tension and the concentrations of the components A and B can be written as (Answer of Example 3.3 on page 126 in Chapter 3)

$$\sigma = \frac{n}{2}\left[(\zeta' + \zeta'')\left(x'_A x''_B + x''_A x'_B\right) - 2x'_A x'_B \zeta' - 2x''_A x''_B \zeta''\right]$$

Use the phase diagram for the system Al–Pb and the regular solution model and calculate

a) the excess interface energy ζ per bond
b) the excess Gibbs' free energy per unit area of the interphase
c) the surface tension at the monotectic temperature 659 °C.

Assume that the immiscibility gap has its maximum point at $x_{Pb} = x_{Al} = 0.50$ and that the gap is symmetrical around this composition. ($\zeta' = \zeta'' = \zeta$). The coordination numbers of liquid Al and Pb are 11. The coordination number of the interface is 3. The average radius of the Al and Pb atoms is 0.165 nm.

Hint: It is worthwhile to study Example 3.3 on pages 126–129 in Chapter 3.

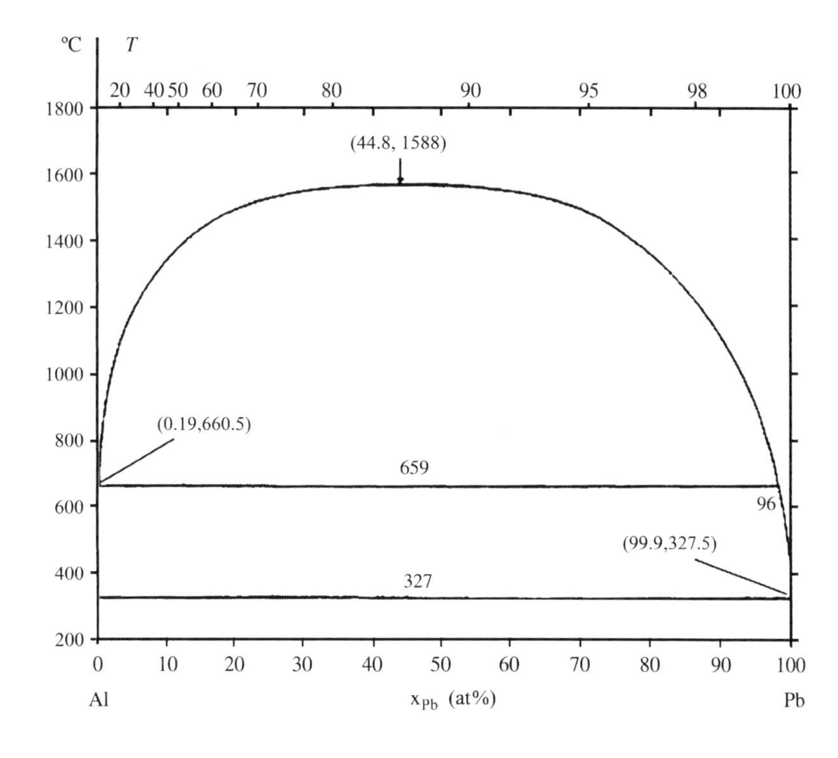

Figure 3.4–1

3.5 The structures of the phases of metals and alloys are of vital importance for the mechanical properties of the materials. An example of this is cast iron, which appears in several different morphologies. One of them is graphite, which shows two extremes in shape: flakes and nodules with spherical shape.

Graphite has a hexagonal structure (Figure 3.5–1). One specific feature of graphite is that elements such as oxygen and sulfur preferably adhere to the basal planes, while the prismatic planes remain unaffected.

Consider a cast iron melt that contains hexagonal crystals of graphite and traces of sulfur with the concentration x_0^{L}. At the vertical prismatic facets the surface tension is σ^0. The surface tension between the melt and the basal upper and lower planes depends on the sulfur concentration. Provided that the sulfur concentration is small the surface tension can be written as (see Example 3.1 on pages 121–122 in Chapter 3)

$$\left(\sigma - \sigma^0\right)\frac{V_{\mathrm{m}}}{d} = RT\left(x_0^{\mathrm{L}} - x_0^{\mathrm{I}}\right) = RT\,x_0^{\mathrm{L}}\left(1 - \frac{x_0^{\mathrm{I}}}{x_0^{\mathrm{L}}}\right) = RT\,x_0^{\mathrm{L}}(1 - \beta)$$

where β is the segregation ratio $x_0^{\mathrm{I}}/x_0^{\mathrm{L}}$ between sulfur concentrations in the interphase I and the melt L.

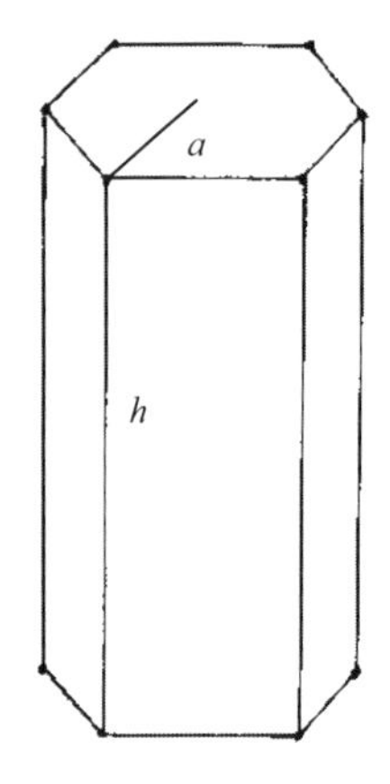

Figure 3.5–1

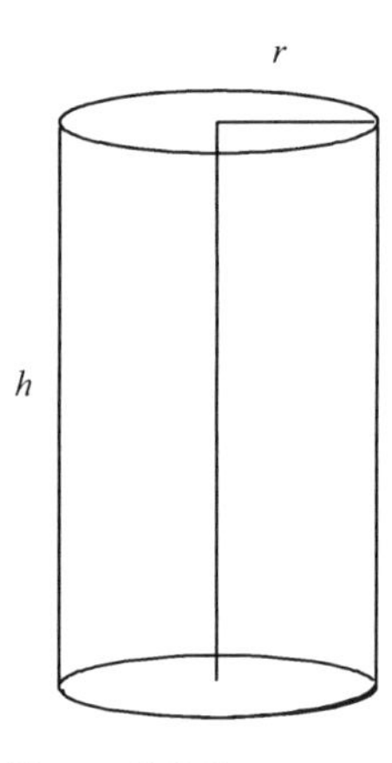

Figure 3.5–2

Approximate the hexagonal graphite crystals with straight cylinders with the height h and radius r and use Gibbs–Curie–Wulff's theorem, to calculate the shape h/r of hexagonal graphite crystals in the cast iron melt as a function of x_0^{L} prolong the line so the center of the bottom circle in Figure 2.

Plot h/r in a diagram as a function of x_0^{L} and describe the shape of the hexagonal graphite crystals for various values of x_0^{L} up to a maximum sulfur mole fraction of 1.8×10^{-4}.

Material constants

$\sigma_0 = 0.20\ \mathrm{J/m^2}$

$d = 1 \times 10^{-9}\ \mathrm{m}$

$V_{\mathrm{m}} = 7.1 \times 10^{-3}\ \mathrm{m^3/kmol}$ (graphite)

$T_{\mathrm{melt}} = 1200\ ^\circ\mathrm{C}$

$\beta = 230.$

3.6 Numerous thin and small plates of an Au alloy have been prepared for a thermal analysis experiment in order to study the melting-point temperature as a function of the grain diameter. The Au plates were produced by cold-deformation to different thicknesses and then reheated at various temperatures during various times. This procedure gave homogeneous samples with a wide variation in grain diameters.

With the aid of thermal analysis experiments (Chapter 7) the melting point temperatures of the samples were measured and studied as a function of the grain diameters. The resulting melting-point temperatures were plotted in the figure as a function of the diameter of the grains in a plate. Each plate corresponds to a point. The figure below shows that the interfacial energies (surface tensions) of the plates vary with the grain diameter. The asymptotic melting point of the alloys in the figure (infinite grain size) is 1306 K. The heats of fusion of the alloys can be set equal to that of pure gold in the calculations.

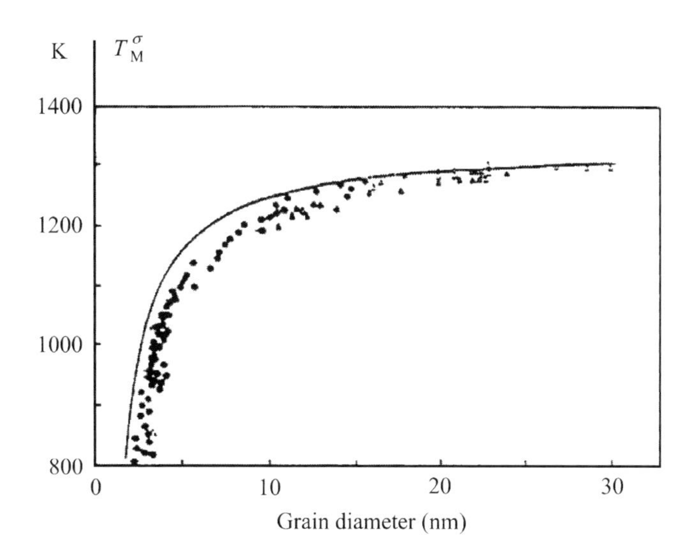

Figure 3.6–1 Melting point temperature of the grains as a function of the grain diameter.

The full line curve represents the melting point calculated as a function of the grain size and a constant surface tension. The measured values plotted as points in the figure shows that the melting point varies with the surface tension.

Calculate the surface tensions of the grains from some of the measured values of their melting points. Present the σ values and the corresponding grain diameters in a table and comment on the result. Give an explanation for the deviation of the melting points of the grains from the full line.

Material constants

$$T_M^{Au} = 1306\,\text{K}$$
$$-\Delta H = 480\,\text{kJ/kg}$$
$$M_{Au} = 197$$
$$\rho_{Au} = 19.3 \times 10^3\,\text{kg/m}^3.$$

Hint: Set up an expression for the Gibbs' free energy balance at the interface and assume that the phases are in equilibrium with each other. Perform the calculations for a number of grain diameters, for example $2r = 30$, 25, 20, 15, 10 and 5 nm.

3.7 a) Describe shortly the Gibbs' and Guggenheim's models of interfaces.

b) Derive the relationship between the Gibbs' free energy of an interface and the surface tension. Justify your answer.

Hint: Discuss the interface energy of 1 kmol of an interface and 1 kmol of an interphase.

c) Discuss shortly the features of Jackson's and Temkin's models.

References

1. H. Fredriksson and U. Åkerlind, *Physics of Functional Materials*, Chichester, England, John Wiley & Son, 2008.
2. R. von Becker, *Zeitschrift für Metallkunde*, **8**, 1937, 245.
3. A. Bergman, H Fredriksson, H Shahani, *Journal of Materials Science*, **23**, 1988, 1573–1579.
4. K. A. Jackson, *Liquid Metals and Solidification*, Cleveland, ASM, 1958.
5. D. E. Temkin, *Crystallization Processes, English Translation by Consultants Bureau*, New York, 1966.
6. J.-S. Chen, N.-B. Ming, F. Rosenberger, *Journal of Chemical Physics*, **84**, 1986, 2365.
7. W. K. Burton, N. Cabrera, F. C. Frank, *Philosophical Transaction*, **A243**, 1950, 299.
8. C. Herring,in: *Structure and Properties of Solid Surfaces*, Chicago, Chicago Press, 1953, 1.
9. A. A. Chernov, *Soviet Physics Uspekhi*, **4**, 1961, 116.
10. N. Cabrera R.V. Coleman, Theory of crystal growth from Vapor, in: *The Art and Science of Growing Crystals*, New York, John Wiley & Sons, 1963.

4

Nucleation

4.1 Introduction

The structures of cast materials are very much influenced by the number of crystals growing in the melt. The formation of crystals, generally called *nucleation*, is very important. Normally there are three different mechanisms:

1. homogeneous nucleation
2. heterogeneous nucleation
3. crystal multiplication.

The classical theory of *homogeneous nucleation* states that a very high supersaturation is needed before any crystals form. In pure metals it is extremely unlikely that homogeneous nucleation will occur. However, it is much easier to achieve this very high supersaturation for intermediate phases.

Solidification and Crystallization Processing in Metals and Alloys, First Edition. Hasse Fredriksson and Ulla Åkerlind.

In the following section we intend to analyze the theory of homogeneous nucleation of crystals and, in particular, the nucleation of intermediate phases.

The theory of *heterogeneous nucleation* is very closely related to homogeneous nucleation. Crystals are formed on heterogeneities or other small crystals in the melt. These heterogeneities can be formed by addition of certain elements to the melt. This process is known as *inoculation* of the melt. The theory of inoculation will also be discussed.

The third way in which crystals are formed is by splitting of existing crystals. This process is known as *crystal multiplication*. The methods used to achieve this will be discussed briefly at the end of this chapter.

We will mainly consider nucleation of a solid from a liquid or a gas phase. At the end of the chapter the nucleation of gas pores in metal melts is discussed.

4.2 Homogeneous Nucleation

In the general discussion of nucleation of a new phase, for example in a molten material, it is assumed that the new phase is formed as clusters. The clusters vary in size owing to statistical fluctuations. A cluster that reaches a minimum critical size can continue to grow. The rate of occurrence of continuously growing clusters is often designated as the *nucleation rate*.

Clusters in a new phase, which exceed the critical size required for continuous growth, are usually called *nuclei*. Clusters of subcritical size are often called *embryos* in order to distinguish them from the nuclei. The size of an embryo changes because of thermal fluctuations. A nucleus is unlikely to arise owing to an occasional thermal fluctuation. It is rather formed by an embryo, which gradually increases its size by assimilation of atoms from its surroundings. Alternatively, an embryo can lose atoms to its surroundings and disappear.

4.2.1 *Theory of Homogeneous Nucleation of Solid Crystals from Liquids*

Consider a metal melt at constant pressure not far from its melting point. Clusters of atoms form at random in the melt owing to thermal agitation.

These occasional solid conglomerates are called *embryos* (Figure 4.1). Their shapes, compositions, structures and sizes vary. In order to analyze the complicated process of nucleation we have to make some simplifications. We have seen in Chapter 3 on page 109 that the equilibrium shape of a solid body with an isotropic surface tension, surrounded by a liquid, is spherical. We assume that the compositions and structures of all the embryos are equal. The only remaining variable is the size of the particle or the embryo radius.

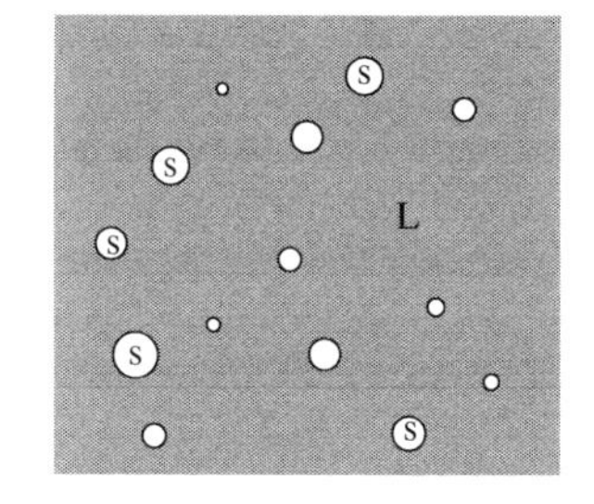

Figure 4.1 Embryos and clusters in a metal melt.

To study the nucleation process theoretically we have to consider the Gibbs' free energy of the system. The Gibbs' free energies of the liquid and the solid phases of a single component system have been discussed in Chapter 2 on pages 43–44.

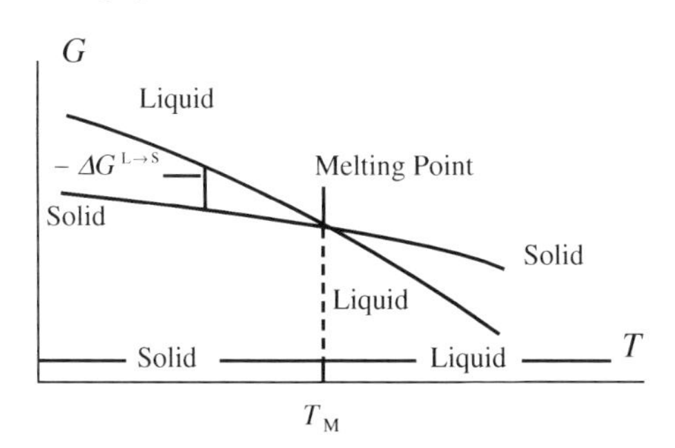

Figure 4.2 The molar Gibbs' free energy of a metal as a liquid and a solid, respectively, versus temperature.

Figure 4.2 shows that the melt L is more stable than the solid embryos at temperatures above the melting point. Hence, the embryos have a very uncertain existence. The majority of them decay shortly after their formation. Some of them survive, though, and increase in size. The larger the embryo radius and the difference in Gibbs' free energy between the melt and the embryo are the more unlikely will be the survival of the embryo. The opposite is true below the melting point. A stable statistical equilibrium distribution is developed although the individual embryos incessantly grow or disappear. It can be shown that the distribution is the well known Boltzmann distribution

$$n_r = ne^{-\frac{\Delta G_r}{k_B T}} \tag{4.1}$$

ΔG_r is the Gibbs' free energy of formation of an embryo with radius r. It is also the difference in Gibbs' free energy between the embryo and the melt. It consists of two parts, a *volume* part corresponding to the Gibbs' free energy of the embryo 'in bulk' and an *area* part owing to the interface energy between the melt and the embryo.

$$\Delta G_r = -\frac{\frac{4\pi r^3}{3}}{V_m}\left[-\Delta G^{L \to s})\right] + 4\pi r^2 \sigma^{L/s} \tag{4.2}$$

where

r = the radius of the embryo
n_r = the number of embryos with radius r per unit volume of the melt
n = the total number of sites where embryos can form per unit volume in the melt
ΔG_r = the Gibbs' free energy of formation of an embryo with radius r
G^L = the molar Gibbs' free energy of the melt
G^s = the molar Gibbs' free energy of the embryo
$-\Delta G^{L \to s}$ = $-(G^s - G^L)$ = the driving force of solidification
V_m = the molar volume of the material
$\sigma^{L/s}$ = σ = the interface energy of the interface melt/embryo.

Figure 4.2 shows that the driving force of solidification $-(G^s - G^L)$ depends strongly on the temperature. This is also the case for the formation energy of an embryo ΔG_r. Figure 4.3 shows ΔG_r as a function of r with temperature as a parameter. At temperatures *above* the melting point the liquid phase is more stable than the solid phase, and $-(G^s - G^L)$ is *negative*. Hence, the first term in Equation (4.2) is *positive* and the formation energy ΔG_r increases rapidly with r. It is therefore unlikely that big embryos can form and grow.

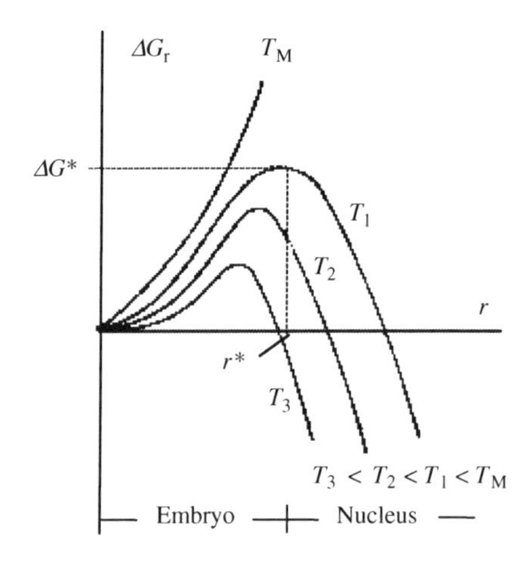

Figure 4.3 The Gibbs' free energy of the embryo ΔG_r as a function of particle size with the temperature as a parameter. If $T < T_m$ the driving force of solidification $-\Delta G_r^{L \to s} = -(G^s - G^L)$ is positive.

At temperatures *lower* than the melting point the opposite is true and the first term in Equation (4.2) is *negative*. At low r values the second term dominates and ΔG_r is positive. At large r values the first term dominates and ΔG_r becomes negative. Every curve has a maximum in between, at a critical radius corresponding to a critical size of the embryo. When an embryo of this size grows further it is transformed into a stable growing nucleus. r^* is called the *critical nucleus size.*

The values of the critical nucleus radii at various temperatures, all below the melting point, are easily found in the normal way by taking the derivative of ΔG_r with respect to r. The derivative equal to zero gives the value of the critical nucleus size and the corresponding value of ΔG_r, denoted ΔG^*, can be found by inserting the r^* value into Equation (4.2). The result is

$$r^* = \frac{2\sigma V_m}{-\Delta G^{L \to s}} \tag{4.3}$$

and

$$\Delta G^* = \frac{16\pi}{3} \frac{\sigma^3 {V_m}^2}{(-\Delta G^{L \to s})^2} \tag{4.4}$$

$\Delta G_r^{L \to s}$ is negative and the driving force of solidification is positive. The surface tension σ is identical with the surface tension $\sigma^{L/S}$ between the liquid and the solid nucleus.

By inserting the expression in Equation (4.4) into Equation (4.1) the number of embryos of critical size can be calculated. However, Equation (4.1) does not indicate whether the embryos will decay or continue to grow and form a stable crystal. The last alternative is described by the *nucleation rate*. This concept is introduced below.

4.2.2 Nucleation Rate

To find an expression for the nucleation rate we will consider what happens to embryos with radii over a certain critical value r^*. Figure 4.3 indicates that such an embryo may increase its size spontaneously without supply of energy and be transformed into a stable growing nucleus.

A necessary condition for this process is, however, that the embryo incorporates atoms from the liquid. These atoms must overcome the Gibbs' free-energy barrier or the *activation energy* U^I to reach the surface of the embryo. The number of atoms per unit volume that have energy enough to fulfil this condition is proportional to $Pe^{-\frac{U^I}{k_B T}}$, where P is the probability related to the entropy ($S = k_B \ln P$).

The equilibrium number of embryos per unit volume of the liquid is given by Equation (4.1)

$$n^* = ne^{-\frac{\Delta G^*}{k_B T}} \tag{4.5}$$

where

n^* = equilibrium number of embryos per unit volume of the liquid
n = total number of sites where embryos can form per unit volume in the melt.

The nucleation rate J can then be described by the equation

$$J = const \times n^* Pe^{-\frac{U^I}{k_B T}} \tag{4.6}$$

The expression of n^* in Equation (4.5) is introduced into Equation (4.6) and we obtain

$$J = Const \times Pe^{-\frac{\Delta G^* + U^I}{k_B T}} \tag{4.7}$$

where

J = the number of embryos per unit volume and unit time that are transformed into growing nuclei
ΔG^* = the Gibbs' free energy of formation of an embryo of critical size
U^I = the activation energy = the energy barrier
k_B = Boltzmann's constant
T = temperature
P = probability related to entropy.

4.2.3 Homogeneous Nucleation as a Function of Undercooling. Nucleation Temperature

The formation energy of an embryo is given by Equation (4.2) on page 168. The volume term is proportional to the driving force of solidification $-\Delta G^{L \to s} = -(G^s - G^L)$. As Figure 4.2 on page 167 shows, the curves can be approximated by straight lines near the melting point.

$$-\Delta G^{L\rightarrow s} = -(G^s - G^L) \approx const \times (T_M - T) \quad (4.8)$$

Basic thermodynamics gives the same result:

$$\Delta G^{L\rightarrow s} = \Delta H - T\Delta S \quad (4.9)$$

Near the melting point $(-\Delta H)$ is approximately equal to the heat of fusion and $\Delta S = -\Delta H/T_M$. If we assume that the heat of fusion and the entropy are independent of the temperature we obtain

$$-\Delta G^{L\rightarrow s} = -\Delta H - T\frac{-\Delta H}{T_M} = \frac{-\Delta H}{T_M}(T_M - T) \quad (4.10)$$

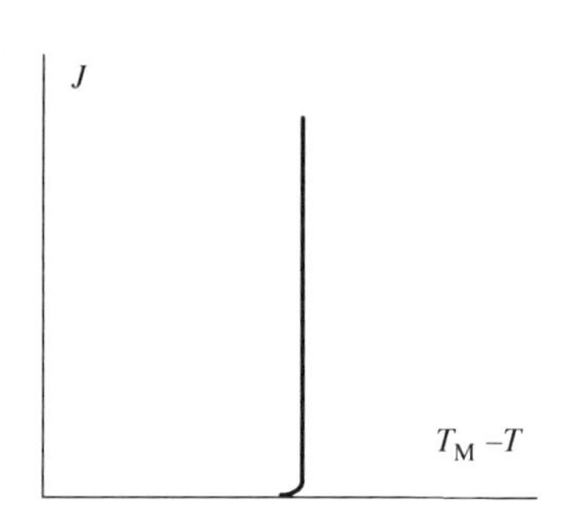

Figure 4.4 Nucleation rate as a function of undercooling.

$T_M - T$ is called the *undercooling*. It is evident from Equation (4.4) on page 169 that ΔG^* is a function of the undercooling. The nucleation rate is also a function of the undercooling according to Equation (4.7). This fact can be used to find the temperature at which nucleation suddenly occurs, the *nucleation temperature* T. The nucleation temperature can easily be found if the nucleation rate is plotted as a function of the undercooling, as is shown in Figure 4.4.

If the fraction of embryos of critical size is known, a relationship between ΔG^* and the nucleation temperature can easily be derived from Equation (4.5) on page 169, which can be written as

$$\frac{\Delta G^*}{k_B T} = \ln\frac{n}{n^*} \quad (4.11)$$

Example 4.1

Consider a metal melt with one embryo per mol, or 10^3 embryos per kmol, that has increased its energy by the critical amount ΔG^*. The embryos then become nuclei and are responsible for the homogeneous nucleation in the melt. Calculate the relationship between ΔG^* and the nucleation temperature T.

Solution:

Consider the volume V_m of 1 kmol of the melt.
$n^* = 10^3$ and $n = N_A = 6.023 \times 10^{26}$ are inserted into Equation (4.11) and we obtain

$$\frac{\Delta G^*}{k_B T} = \ln\frac{n}{n^*} = \ln\frac{6.023 \times 10^{26}/V_m}{10^3/V_m} \approx 55$$

Answer: The required relationship is $\Delta G^* = 55\, k_B T$.

Example 4.2

Use the classical theory for homogeneous nucleation and make a rough estimation of

a) the undercooling necessary for homogeneous nucleation in copper
b) the radius of the embryos of critical size.

The magnitude of the surface energy between a pure solid close-packed metal and its melt is normally of the magnitude 0.2 J/m^2. It is reasonable to assume that the fraction of embryos of critical size is one per mol. Material constants for copper are found in standard tables.

Solution:

a)
We assume that the nucleation temperature $\approx$ the melting point. According to Example 4.1 on page 170 and Equation (4.4) on page 169 we have

$$55\,k_B T_M = \Delta G^* = \frac{16\pi}{3}\frac{\sigma^3 V_m{}^2}{(\Delta G^{L\rightarrow s})^2} \tag{1'}$$

We also use Equation (4.10) on page 170

$$-\Delta G^{L\rightarrow s} = \frac{-\Delta H}{T_M}(T_M - T) \tag{2'}$$

Equation (2′) is introduced into Equation (1′). The undercooling is solved from the new equation:

$$T_M - T = \sqrt{\frac{16\pi\sigma T_M}{3 \times 55 k_B}}\left(\frac{\sigma V_m}{-\Delta H}\right) \tag{3'}$$

Inserting the values of material constants for copper and $-\Delta H = 343 \times 10^3$ J/kg $\times$ 63.6 kg/kmol $= 2.18 \times 10^7$ J/kmol and

$$V_m = \frac{63.6\,\text{kg/kmol}}{8.93 \times 10^3\text{kg/m}^3} = 7.12 \times 10^{-3}\text{m}^3/\text{kmol}$$

into Equation (3′) we obtain

$$T_M - T = \sqrt{\frac{16\pi \times 0.2 \times (1083 + 273)}{3 \times 55 \times 1.38 \times 10^{-23}}}\frac{0.2 \times 7.1 \times 10^{-3}}{2.2 \times 10^7} \approx 160\text{ K}$$

b)
The radius can be calculated from Equation (4.4) on page 169:

$$r^* = \frac{2\sigma V_m}{-\Delta G^{L\rightarrow s}} = \frac{2\sigma V_m}{\frac{-\Delta H}{T_M} \times (T_M - T)} = \frac{2\sigma V_m T_M}{-\Delta H(T_M - T)}$$

$$r^* = \frac{2 \times 0.2 \times 7.1 \times 10^{-3} \times (1083 + 273)}{2.2 \times 10^7 \times 160} = 1.1 \times 10^{-9}\text{m}$$

Answer:

a) An undercooling of the magnitude 200 K is required.
b) The corresponding radius of embryos of critical size is of the magnitude 10^{-9} m.

For practical reasons it is not possible to undercool a metal melt more than about 5 K. Similar calculations on other metals give results in agreement with copper. Therefore, homogeneous nucleation is highly improbable in pure metals.

In practice, it is often observed that nuclei are formed at a significantly lower supercooling than given by the theory of homogeneous nucleation. The reason for this is that there are other solid particles in the melt than the embryos that promote nucleation. These particles are called *active heterogeneities.*

In many cases it is believed that the active heterogeneities themselves are formed in the melt during cooling. It is puzzling why nucleation occurs more easily with active heterogeneities than with the parent metal embryos. This will be discussed later in this chapter (page 179).

4.2.4 *Homogeneous Nucleation as a Function of Concentration in Binary Alloys*

In the preceding section we have studied homogeneous nucleation as a function of undercooling. Here, we will discuss the phenomenon as a function of supersaturation. There is really not much difference between the two approaches. An undercooled liquid is supersaturated. The theory is basically the same. The only difference is that we obtain the driving force of solidification $-\Delta G^{L \to s}$ as a function of concentration instead of undercooling.

In order to demonstrate this we will consider a binary alloy AB. Its phase diagram is shown in the upper part of Figure 4.5.

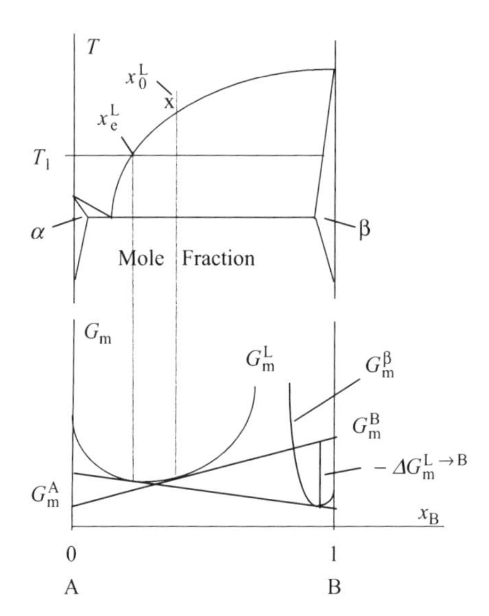

Figure 4.5 Phase diagram and Gibbs' free energy diagram of a binary alloy as a function of composition (mole fraction of component B).

Let a binary alloy melt of composition x_0^L be cooled to temperature T_0. The driving force $-\Delta G^{L \to s}$ for precipitation of solid B is shown in the Gibbs' free energy diagram in the lower part of Figure 4.5. The driving force, which has been discussed in Section 2.2.1 on page 43 in Chapter 2, is in this case given by the equation

$$-\Delta G^{L \to s} = RT \ln \frac{x^L}{x_{eq}^L} \tag{4.12}$$

where
x^L = the concentration of B atoms in the melt
x_{eq}^L = the equilibrium concentration of B atoms in the melt.

Equation (4.12), which is derived under the assumption that Henry's law is valid, often gives values of the driving force of homogenous nucleation in alloys, which are one order of magnitude higher than the results given by Equation (4.10) on page 170.

Example 4.3

The melting point of a solid/liquid system is 1000 K. The liquid phase with a surface energy of $0.2\ \text{J/m}^2$ requires a driving force of solidification of approximately 10^6 J/kmol according to the classical theory of homogeneous nucleation. Calculate the corresponding degree of supersaturation of the melt, i.e. the ratio of the true concentration x^{L} and the equilibrium concentration $x^{\mathrm{L}}_{\mathrm{eq}}$, necessary to achieve the same driving force.

Solution:

We use Equation (4.12)

$$RT \ln \frac{x^{\mathrm{L}}}{x^{\mathrm{L}}_{\mathrm{eq}}} = -\Delta G^{\mathrm{L} \rightarrow \mathrm{s}} \tag{1'}$$

to find the required ratio of concentrations.

$$\ln \frac{x^{\mathrm{L}}}{x^{\mathrm{L}}_{\mathrm{e}}} = \frac{-\Delta G^{\mathrm{L} \rightarrow \mathrm{s}}}{RT} = \frac{10^6}{8.315 \times 10^3 \times 1000} \approx 0.12 \tag{2'}$$

which gives

$$\frac{x^{\mathrm{L}}}{x^{\mathrm{L}}_{\mathrm{eq}}} = e^{\frac{-\Delta G_m}{RT}} = \mathrm{e}^{0.12} \approx 1.1 \tag{3'}$$

Answer:
The degree of supersaturation of the solution corresponding to the given driving force is 1.1.

A supersaturation such as that in Example 4.3 can easily be obtained, particularly if the heat of solution of the secondary phase in the melt is high.

To show the influence of the surface tension and the melt composition on nucleation we choose a concrete example.

Consider the precipitation of Al_2O_3 in a steel melt. The driving force of this reaction can be written as

$$-\Delta G^{\mathrm{L} \rightarrow \mathrm{s}} = RT \ln \frac{\left(x^{\mathrm{L}}_{\mathrm{Al}}\right)^2 \left(x^{\mathrm{L}}_{0}\right)^3}{\left(x^{\mathrm{eq}}_{\mathrm{Al}}\right)^2 \left(x^{\mathrm{eq}}_{0}\right)^3} \tag{4.13}$$

where
$x^{\mathrm{L}}_{\mathrm{Al}}$ = concentration of Al in the melt
x^{L}_{0} = concentration of oxygen in the melt
$x^{\mathrm{eq}}_{\mathrm{Al}}$ = equilibrium concentration of Al in the melt
x^{eq}_{0} = equilibrium concentration of oxygen in the melt.

By combining Equations (4.13), (4.4) on page 169 and (4.11) on page 170 the necessary concentrations of Al and O for homogeneous nucleation can be calculated. This has been done for different values of the surface tension and the result is shown graphically in Figure 4.6.

A realistic value of the solid–liquid surface tension is 1.5 J/m^2. In Figure 4.6 the curve denoted $\sigma = 0$ corresponds to the equilibrium concentrations.

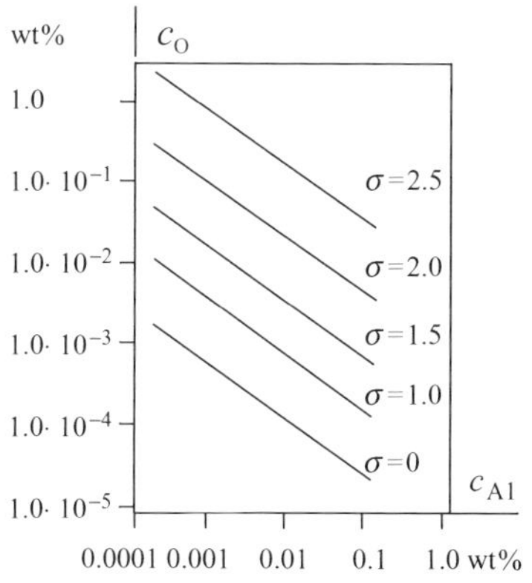

Figure 4.6 Influence of surface tension and melt composition on the nucleation of Al_2O_3 in a steel melt. Reproduced from Springer © 1980.

4.2.5 *Influence of Variable Surface Energy on Homogeneous Nucleation*

The theory of homogeneous nucleation, given in the preceding sections deals with the growth of small embryos and their possible transformations into growing nuclei. The basic equation is Equation (4.2) on page 168:

$$\Delta G_r = -\frac{\frac{4\pi r^3}{3}}{V_m}\left[-(G^s - G^L)\right] + 4\pi r^2 \sigma^{L/s} \tag{4.2}$$

Gibbs has stated that the surface tension is a function of curvature, owing to the influence of the thickness of the interphase between the solid and liquid phases. Larsson and Garside [1] and others, among them Tolman [2] and Ono and Kondo [3], have developed this theory further and applied it to the process of homogeneous nucleation.

The interphase region of a cluster or an embryo is defined by two border surfaces where the density is equal to the neighbouring phase (Figure 4.7). The solid line between the interphase is the equimolar surface, i.e. the numbers of kmol outside and inside the surface are equal. Tolman showed that

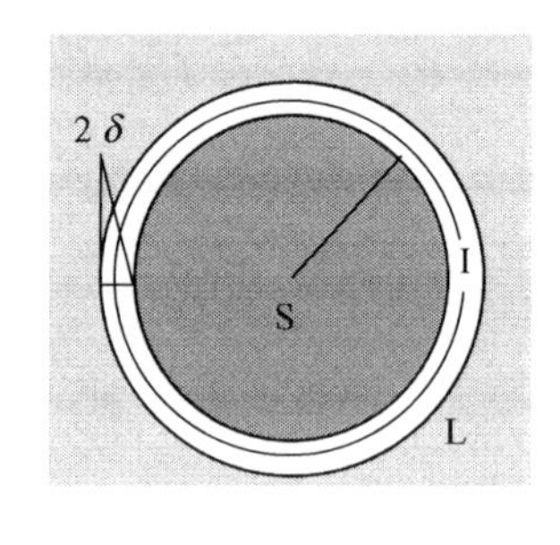

Figure 4.7 Spherical embryo in a melt surrounded by an interphase.

$$\delta = \frac{\rho_{surface}}{\rho_s - \rho_L} \tag{4.14}$$

where

δ = half the interphase thickness
$\rho_{surface}$ = surface density of the interphase
ρ_L = density of the melt
ρ_s = the density of the embryo.

If $\delta/r \ll 1$, where r the is the radius of the embryo, the resultant approximate formula, derived by Tolman, in the interval is

$$\frac{\sigma}{\sigma_\infty} \approx e^{-\frac{2\delta}{r}} \tag{4.15}$$

If $\delta/r > 0.1$ the approximate expression in the interval is

$$\frac{\sigma}{\sigma_\infty} \approx e^{-\frac{1.3\delta}{r}} \tag{4.16}$$

For small embryos the change in surface tension is considerable. The resulting Gibbs' free energy required to form an embryo with the radius r is

$$\Delta G_r = -\frac{\frac{4\pi r^3}{3}}{V_m}\left[-(G^s - G^L)\right] + 4\pi r^2 \sigma_\infty^{L/s} e^{-\frac{a\delta}{r}} \tag{4.17}$$

where

a = a constant
$\sigma_\infty^{L/s}$ = the surface tension of a flat liquid/solid interface.

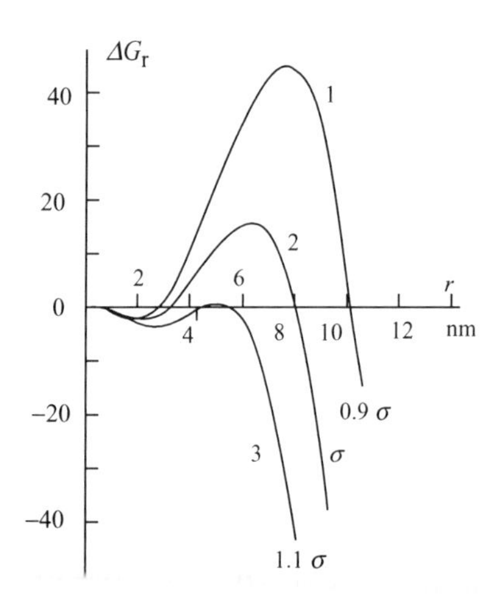

Figure 4.8 The formation Gibbs' free energy ΔG_r of an embryo as a function of particle size with surface tension as a parameter. Reproduced from Elsevier © 1986.

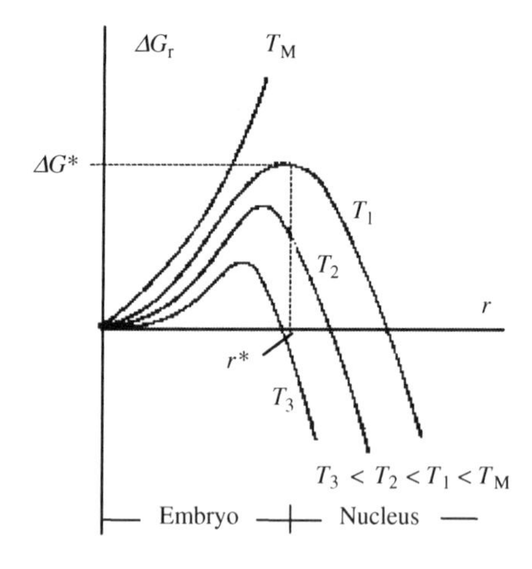

Figure 4.3 The formation Gibbs' free energy ΔG_r of an embryo as a funcion of particle size with temperature as a parameter.

The influence of the decrease of the surface tension for small embryos is shown in Figure 4.8. It shows that the formation Gibbs' free energy of the embryo as a function of its radius for three different σ values. As a comparison, a corresponding curve with an uncorrected σ value is given (Figure 4.3).

Figure 4.8 shows that small embryos can form spontaneously in the shallow minima but they do not lead to embryos of critical size unless energy is supplied. The smaller the surface tension the smaller will be the barrier ΔG^*.

Example 4.4

Suppose that numerous vacancies are formed during solidification to such an extent that the densities of the solid and the melt become equal.

Analyze the effects on nucleation of embryos of this special situation theoretically and practically.

Solution and **Answer**:

The denominator of the exponent in Equation (4.14) would be zero. Hence, the ratio $(-2\delta/r)$ in Equation (4.15) would be infinite and negative. This means that the exponential factor in Equation (4.17) would be zero and the whole second term would vanish. Only the first term, which is negative, would remain.

The practical consequence of this would be that the ΔG_r–r curves have no maxima as shown in the figure. There would be no barrier for small embryos. All of them can be transformed into growing nuclei.

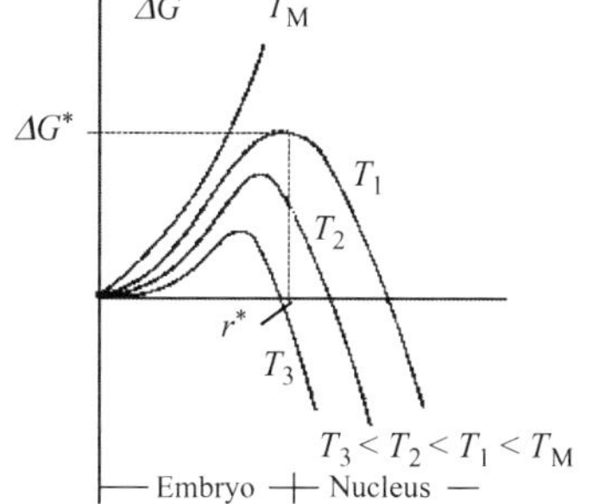

4.2.6 *Homogeneous Nucleation of Non-Spherical Embryos. Nucleation of Faceted Crystals*

In the preceding sections we have made the simplified assumption that the embryos formed on homogeneous nucleation are spherical. Spherical embryos are formed only if the surface energy is isotropic. This is seldom the case. In many cases the embryos are asymmetrical owing to the anisotropy of the surface tension.

Nucleation of Faceted Crystals

Asymmetrical embryos that reach the critical size and are transformed into growing nuclei form faceted crystals.

The shape of crystals has been studied in Section 3.5 in Chapter 3 (pages145–149). The result can shortly be summarized as follows

1. The equilibrium shape of a crystal of a given volume is found by minimizing the total Gibbs' free surface energy.
2. The resulting equilibrium condition is given by Gibbs–Curie–Wulff's theorem

$$\frac{\sigma_1}{h_1} = \frac{\sigma_2}{h_2} = \frac{\sigma_3}{h_3} = \ldots\ldots = \frac{\sigma_n}{h_n} = const \tag{4.18}$$

3. Close-packed crystal faces have lower total Gibbs' free energies than other crystal faces.

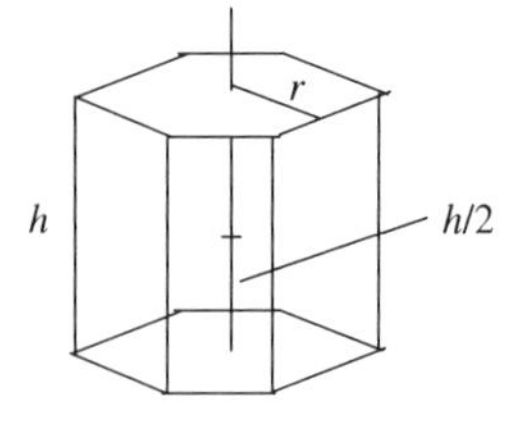

Example 4.5

Derive the equilibrium shape of a graphite nucleus in a cast iron melt as a function of the surface tension in the base plane divided by the surface tension in the prism planes.

Solution and **Answer:**

Graphite has a hexagonal structure as sketched in the figure. We use Wulff's rule and choose the centre of the nucleus as Wulff's point.

$$\frac{\sigma_b}{\frac{h}{2}} = \frac{\sigma_p}{r} \quad \Rightarrow \quad \frac{h}{r} = 2\frac{\sigma_b}{\sigma_p}$$

The Total Gibbs' Free Energy of an Embryo of Critical Size

In Chapter 3 Section 3.5 the condition of equilibrium shape of a crystal was derived in terms of Gibbs – Curie – Wulff's theorem. These results can be used to calculate the total Gibbs' free energy of a nucleus or an embryo. This is done in the box below.

Derivation of the Total Gibbs' Free Energy of an Embryo of Equilibrium Shape

As a basis for the calculations we use Equation (9′) on page 148 in Chapter 3. The total Gibbs' free energy of an embryo or crystal relative to the surrounding liquid can be written as

$$\Delta G_c^{total} = -(p_c - p_L)V_c + \sum_n \sigma_n A_n \qquad (1')$$

The volume of the embryo is the sum of n tetrahedrons with the faces A_n as bases and h_n as heights

$$V_c = \frac{1}{3}\sum_n h_n A_n \qquad (2')$$

This expression of V_c is introduced into Equation (1′)

$$\Delta G_c^{total} = -(p_c - p_L) \times \frac{1}{3} \times \sum_n h_n A_n + \sum_n \sigma_n A_n \qquad (3')$$

The condition of equilibrium shape of a crystal [Equation (11′) on page 148 in Chapter 3] in a melt can be written

$$\frac{\sigma_i}{h_i} = \frac{(p_c - p_L)}{2} \qquad i = 1, 2, 3 \ldots n \qquad (4')$$

Equations (4′) can be written as $h_i = \frac{2\sigma_i}{(p_c - p_L)}$.

These h_i values are substituted into Equation (3′) and we obtain

$$\Delta G_c^{total} = -(p_c - p_L) \times \frac{1}{3} \times \sum_n \frac{2\sigma_n A_n}{(p_c - p_L)} + \sum_n \sigma_n A_n$$

or

$$\Delta G_c^{total} = -\frac{2}{3}\sum_n \sigma_n A_n + \sum_n \sigma_n A_n \qquad (5')$$

which gives

$$\Delta G_c^{total} = \frac{1}{3}\sum_n \sigma_n A_n \qquad (6')$$

Equation (6′) in the box is independent of the symmetry of the crystal. It can be used to calculate the energy of formation of embryos or nuclei. The total Gibbs' free energy of an embryo of critical size is

$$\Delta G_{c}^{total*} = \frac{1}{3}\sum_{n}\sigma_{n}A_{n} \tag{4.19}$$

Equation (4′) in the box can be given an alternative formulation in terms of molar Gibbs' free energies. Equation (1′) in the box can be compared with the basic Equation (3.85) on page 146 in Chapter 3. The Gibbs' free energy of the crystal relative to the Gibbs' free energy of the liquid is given in the equation below: The first term in Equation (4.20) is negative.

$$\Delta G_{c}^{total} = n_{c}\left[(G^{s} - G^{L})\right] + \sum_{n}\sigma_{n}A_{n} = \frac{V_{c}}{V_{m}}\left[(G^{s} - G^{L})\right] + \sum_{n}\sigma_{n}A_{n} \tag{4.20}$$

where n_c is the number of kmoles of the crystal and $-(G^{s} - G^{L})$ is the driving force of solidification. Equation (1′) in the box

$$\Delta G_{c}^{total} = -(p_{c} - p_{L})V_{c} + \sum_{n}\sigma_{n}A_{n} \tag{1′}$$

can be identified with Equation (4.20), which gives

$$p_{c} - p_{L} = \frac{-(G^{s} - G^{L})}{V_{m}} \tag{4.21}$$

This value of the pressure difference is introduced into Equation (4′) together with the definition of the driving force, i.e. $-\Delta G^{L\rightarrow s} = -(G^{s} - G^{L})$ and we obtain

$$\frac{\sigma_{i}}{h_{i}} = \frac{-\Delta G^{L\rightarrow s}}{2V_{m}} \tag{4.22}$$

where $-\Delta G^{L\rightarrow s}$ is the driving force of solidification.

Equilibrium Shapes of FCC Crystals

As an example of faceted crystals we will study the equilibrium shapes of FCC crystals.

Owing to condition 1 on page 175, the dominating crystal faces are those that have the lowest possible total Gibbs' free energies. In Chapter 3 on page 131 we found that close-packed crystal planes have lower surface energies than all other planes.

In a cubic FCC system these surfaces are the close-packed (100) and (111) faces. It is reasonable to assume that only these two types of crystal faces will occur in the equilibrium shape of FCC crystals.

Equation (4.22) is applied on the (100) and (111) faces.

$$h_{111}^{*} = \frac{2\sigma_{111}V_{m}}{-\Delta G^{L\rightarrow s}} \tag{4.23}$$

$$h_{100}^{*} = \frac{2\sigma_{100}V_{m}}{-\Delta G^{L\rightarrow s}} \tag{4.24}$$

where $-\Delta G^{L\rightarrow s}$ is the driving force for precipitation of a solid from a liquid phase.

The shape of the crystal depends entirely on the ratio $\sigma_{100}/\sigma_{111}$. If $\sigma_{100}/\sigma_{111} > 3$ an octahedron is formed. If the ratio $\sigma_{100}/\sigma_{111} > 1/3$ a cubic nucleus is formed. When the ratio falls within the interval 1/3 – 3, the nucleus will be bounded by both the (100) and (111) faces as shown in Figure 4.9. The intermediate stages between a cube and an octahedron are shown in Figure 4.10.

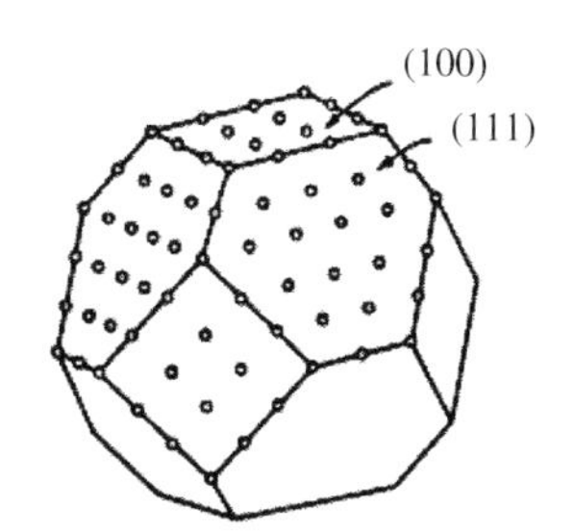

Figure 4.9 Possible structure of a critical nucleus. Bruce Chalmers, Courtesy of S. Chalmers.

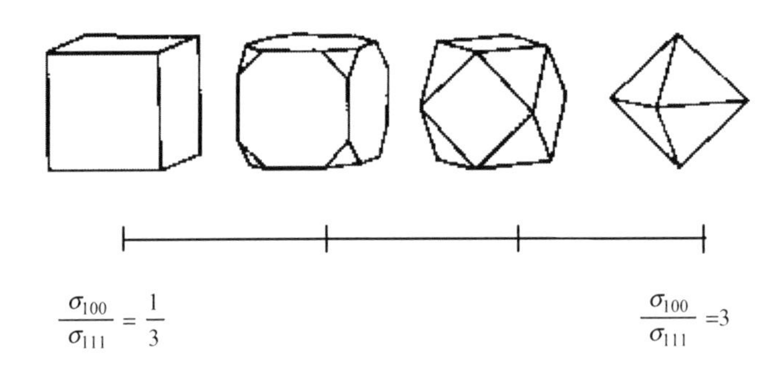

Figure 4.10 The value of the ratio $\sigma_{100}/\sigma_{111}$ determines the shape of the nucleus.

The value of the ratio is in each case determined by the minimum Gibbs' free energy condition of the crystal.

Gibbs' Free Energy of an FCC Crystal of Critical Size

As an application we will calculate the Gibbs' free energy of a crystal with FCC structure.

Example 4.6

An FCC embryo has the equilibrium shape of a truncated octahedron consisting of six square (100) faces and eight hexagonal (111) faces with equal edge length l^*, as sketched in the margin figure.

Calculate the ratio of the interface energies $\sigma_{100}/\sigma_{111}$ and the total surface free energy of the embryo as a function of l^* and the interface energy σ_{111}.

Solution:

G H A B F D C E

(a)

Truncated octahedron.

a)

σ_{100} is unknown and it is necessary to express it in terms of σ_{111}. The equilibrium shape condition, according to Wulff's rule, can be used. The centre of the embryo is used as Wulff's point.

$$\frac{\sigma_{100}}{\sigma_{111}} = \frac{h_{100}}{h_{111}} \qquad (1')$$

The ratio $h_{100}/h_{111}$is found from the geometry of the crystal. Triangle calculations give $h_{100} = l^*\sqrt{2}/2$ and $h_{111} = 3\,l^*/4$.

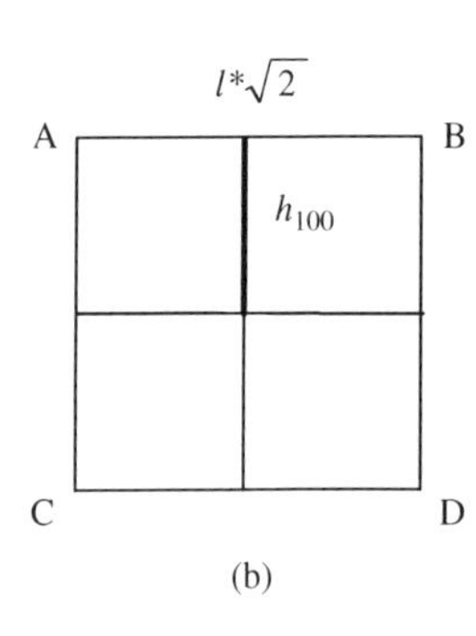

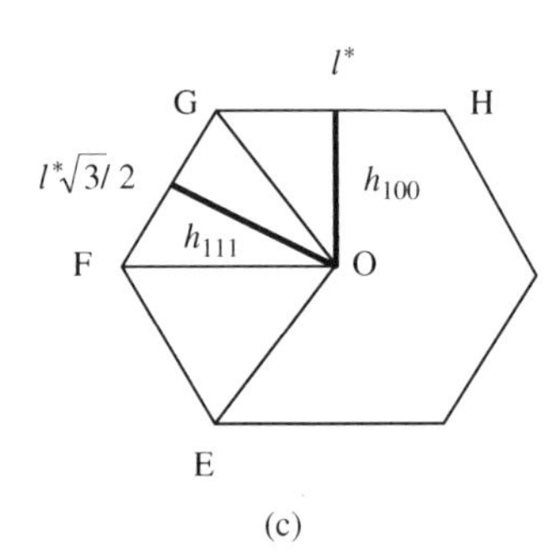

Lower left figure: Cross-section of the crystal in order to find h_{100}.

Lower right figure: Cross-section of the crystal in order to find h_{111}. The triangle FGO has equals has equal sides $= l^*\sqrt{3/2}$.

The figures above give $$\frac{h_{100}}{h_{111}} = \frac{l^*\sqrt{2}/2}{3l^*/4} = \frac{2\sqrt{2}}{3} \qquad (2')$$

b)
The total surface free energy of the embryo in relationship to the surface free energy of the melt can be written as

$$\Delta G^* = \frac{1}{3}\sum_{n} \sigma_n A_n = \frac{1}{3}\left(8\sigma_{111}\frac{l^{*2}}{4}\sqrt{3} + 6\sigma_{100}l^{*2}\right) \qquad (3')$$

The ratio $\sigma_{100}/\sigma_{111}$ from Equations (1′) and (2′) is inserted into Equation (3′):

$$\Delta G^* = \frac{\sigma_{111}}{3}\left(8\frac{l^{*2}}{4}\sqrt{3} + 6\frac{\sigma_{100}}{\sigma_{111}}l^{*2}\right) = \frac{\sigma_{111}l^{*2}}{3}\left(2\sqrt{3} + 6\frac{2\sqrt{2}}{3}\right)$$

Answer:

a) The ratio $\sigma_{100}/\sigma_{111}$ equals $2\sqrt{2}/3$.
b) The total surface free energy of the embryo is $\Delta G^* = \frac{\sigma_{111}l^{*2}}{3}(2\sqrt{3} + 4\sqrt{2})$.

4.3 Heterogeneous Nucleation. Inoculation

In Example 4.2 on page 170 we found that the undercooling necessary for homogeneous nucleation is of the magnitude 200 K, which is impossible to realize. Hence, homogeneous nucleation is highly improbable. The undercooling required for heterogeneous nucleation is of the order of a few degrees and is possible to realize in practice.

Heterogeneous nucleation is closely related to and promoted by inoculation.

4.3.1 Inoculation of Metal Melts

Inoculation of metallic melts is widely used in industry in order to improve the properties of cast materials. The purpose of inoculation in foundries is often to minimize the formation of undesirable textures and structures in castings and ingots. For instance, in foundry production of cast iron, inoculation is used to promote precipitation of graphite instead of cementite.

Experience shows that the process of dissolution of inoculants is of major importance for understanding and knowledge of the way an inoculant works. The process can be illustrated by the following basic model experiment.

An Fe–4%Ti melt was brought into contact with graphite and the course of events in the melt is studied by quenching and analyzing samples at various times. The result is shown in Figure 4.11. Part of the graphite was seen to the left in the figure. The dark particles are TiC crystals.

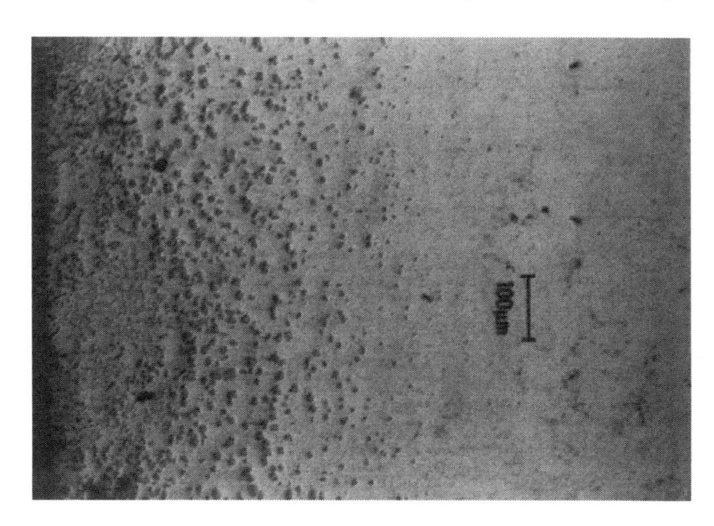

Figure 4.11 The dissolution process of graphite added to an Fe-4% Ti melt.

The graphite is the dark area to the left in the figure. The dark particles are TiC.

The temperature of the melt was 1570 °C. The sample was quenched 5 seconds after the graphite was added. Reproduced from Springer © 1997.

Soon after the contact between the melt and the graphite plate had been established small TiC particles were observed close to the graphite surface. The small TiC crystals were precipitated further away in the melt owing to diffusion of carbon into the titanium-rich liquid ahead of the growing crystals. The region containing small precipitated TiC particles increased in width with time. Simultaneously, Ti atoms started to diffuse through the melt towards the graphite surface. Because of this diffusion process, new TiC crystals formed continuously close to the graphite surface. The process went on until all graphite had dissolved or until a three-phase equilibrium between TiC, graphite and melt had been established throughout the system.

The diffusion process will be penetrated further on in connection with *in-situ* composites (Chapter 6).

When all the graphite had been dissolved, the TiC crystals started to dissolve to reach the equilibrium of the system. If the total carbon and titanium contents were high enough, a considerable amount of TiC crystals will remain in the melt when the equilibrium is reached. These crystals are effectively distributed into the melt by convection and acted as nucleants of solid phases during the solidification process.

It is reasonable to assume that other inoculants work in the same way as TiC. To understand the detailed process of heterogeneous nucleation we have to discuss the theory of inoculation and the solid/solid interface interaction. The study of crystal growth in a melt on the surface of a heterogeneous solid is essential for understanding heterogeneous nucleation and identifying the factors that influence the process.

4.3.2 Theory of Inoculation

The theory of the influence of an inoculant on heterogeneous nucleation is well known. In addition to the particle shape, the influence of surface tension are very important. The significant parameter is the wetting angle θ (Chapter 3 page 106–107).

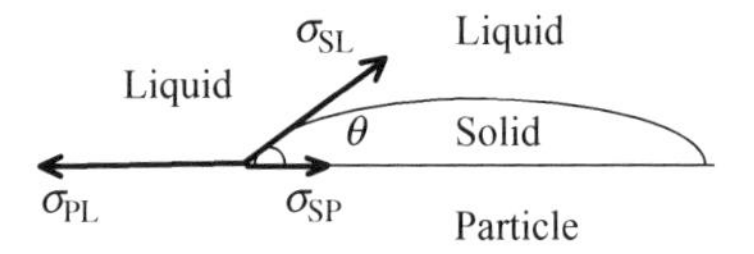

Figure 4.12 Formation of an embryo/crystal S on a foreign particle P.

The equilibrium condition is (see Figure 4.12)

$$\sigma_{PL} = \sigma_{SP} + \sigma_{SL} \cos\theta \tag{4.25}$$

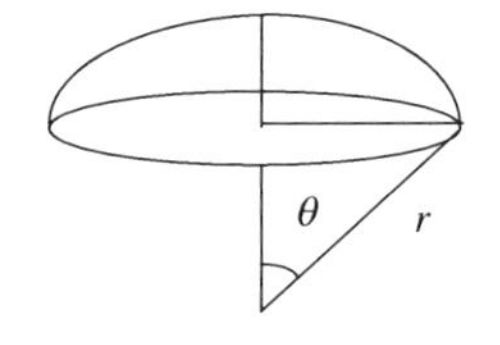

Figure 4.13 Spherical cap.

If the solid phase S (embryo/crystal) formed is a spherical cap, the change in Gibbs' free energy can be written, using 3-dimensional geometry and spherical coordinates

$$\Delta G_c = \frac{\pi r^3}{3}\left(2 - 3\cos\theta - \cos^3\theta\right)\frac{-(G^s - G^L)}{V_m} + \sigma_{SL}\left[2\pi r^2(1 - \cos\theta) - \pi r^2\left(1 - \cos^2\theta\right)\right] \tag{4.26}$$

where

ΔG_c = Gibbs' free energy of formation of the embryo or crystal
G^s = molar Gibbs' free energy of crystal
G^L = molar Gibbs' free energy of the liquid
$-(G^s - G^L)$ = driving force of solidification $= -\Delta G^{L\to s}$
σ_{SL} = surface tension between the embryo/crystal and the liquid
r = radius of the sphere
θ = coordinate angle.

ΔG_i is a function of the radius of the particle and the wetting angle θ. In analogy with the treatment of homogeneous nucleation (pages 167 and 169) we determine the maximum ΔG^* of ΔG_c and the corresponding r value at constant θ. Using the condition $d\Delta G_c/dr = 0$ we obtain

$$r^* = \frac{2\sigma_{SL} V_m \sin\theta}{-\Delta G^{L\to s}} \tag{4.27}$$

and

$$\Delta G_i^* = \frac{16}{3} \frac{\pi(\sigma_{SL})^3 V_m^2}{(-\Delta G^{L\to s})^2} \frac{(2+\cos\theta)(1-\cos\theta)^2}{4} \tag{4.28}$$

The expression of ΔG_i^* in Equation (4.4) on page 169 is called ΔG_{hom} and is introduced into Equation (4.28):

$$\Delta G_{het} = \Delta G_{hom} \frac{(2+\cos\theta)(1-\cos\theta)^2}{4} \tag{4.29}$$

Statistical thermodynamics provides an expression for the number of potentially growing crystals around a heterogeneous particle.

$$n^*(A_i) = n(A) e^{-\frac{\Delta G_i^*}{k_B T}} \tag{4.30}$$

where
$n(A)$ = the total number of atoms of the undercooled melt in contact with the heterogeneity
$n^*(A_i)$ = the number of critical nuclei on the surface A_i

Figure 4.14 shows the nucleation rate (page 169) as a function of the undercooling for various heterogeneities with different wetting angles.

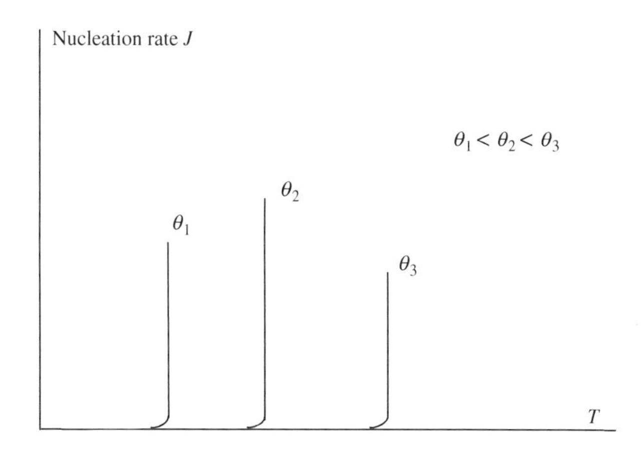

Figure 4.14 Nucleation rate J (the number of embryos per unit volume and unit time which are transformed into growing nuclei) as a function of undercooling for various heterogeneities. They have different wetting angles, which appear as a parameters in the curves.

Extensive studies have been carried out on the wetting angle of different materials with the aim of identifying active inoculants. However, these efforts have not been especially successful. The results are probably influenced negatively by impurities, adsorbed at the surfaces of the inoculants.

If a foreign particle with a planar surface is to act as an effective inoculant it is necessary that its wetting angle is very small, probably less than 10 degrees. In order to achieve this, the interface energy of the particle/melt interface must be high enough to exceed the interface energy of the particle/solid phase interface.

The wetting angle condition is not the only condition that has to be fulfilled for a good inoculant of a metal melt. The structures of the foreign particle and the solidified melt must also be taken into consideration. This topic will be discussed below in terms of energy.

4.3.3 *Solid/Solid Interface Energies*

Equations (4.27) and (4.28) show clearly that heterogeneous nucleation is influenced by the surface tension or interface energy between the embryo/crystal and the foreign particle, acting as an inoculant. The value of the interface energy between the embryo and the particle, depends on the crystal structures of the two solid–solid phases. These topics have been discussed in Chapter 3, Section 3.5.

An interface between two crystals is *coherent* if the planes of the two crystals forming the interface have the same atomic configuration and spacing.

Energies of Coherent Interfaces

Coherent interfaces may have slight deviations in spacing, which causes deformation of the lattice and lattice strain (Figure 4.15). This increases the interface energy by *elastic strain energy*.

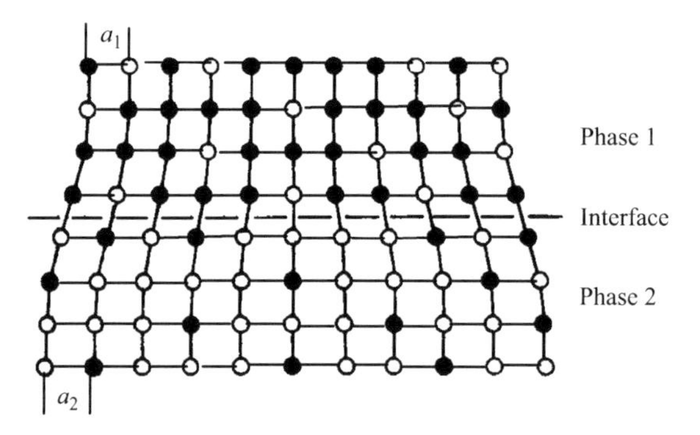

Figure 4.15 Slightly distorded but coherent interface between two crystals. © Pearson PLC 1992.

The elastic strain energy was derived in Chapter 3 on pages 142–144. The classical elastic deformation energy per unit volume, was found to be $w_{el} = Y\varepsilon^2/2$. When this formula is applied on a solid/solid interface with mismatching, ε has to be replaced by the mismatching δ_{el} and we obtain

$$\sigma_{el} = Y{\delta_{el}}^2 d \qquad (4.31)$$

where

σ_{el} = elastic strain energy per unit area
d = thickness of the interface
Y = modulus of elasticity of the material
δ_{el} = average mismatch between the two crystal lattices owing to the elastic strain.

The reason why the strain energy per unit area is twice the classical expression is that the surface is strained in *two* perpendicular directions while the expression $Y\delta_{el}^2/2$ refers to *linear* strain.

Energies of Semicoherent Interfaces

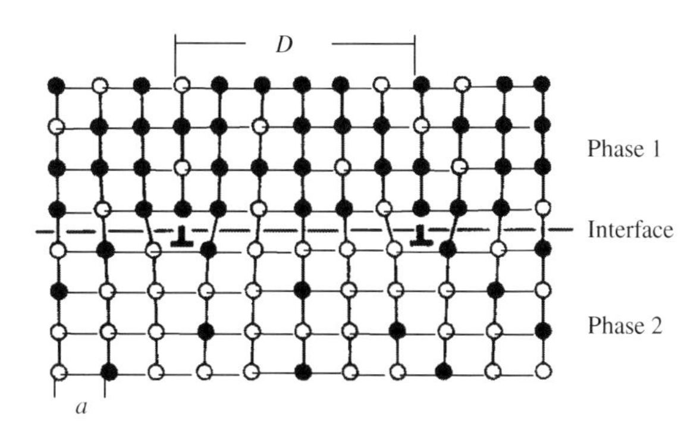

Figure 4.16 Semicoherent interface between two crystals. Coherent distorded areas are interrupted by dislocations. © Pearson PLC 1992.

If the deviations in spacing between two crystals are too large, the lattice strain becomes too strong and the structure in Figure 4.15 is not stable. Instead, *dislocations* are formed at regular intervals interrupted by more or less distorted areas. Such an interface is a *semicoherent* or *partly coherent interface* (Figure 4.16). This increases the interface energy. The total strain energy consists of *elastic strain energy* and *dislocation energy*.

The dislocation energy per unit area $\sigma_{\perp}$,owing to edge dislocations, is given by

$$\sigma_{\perp} = \frac{2w_{\perp}}{D} \tag{4.32}$$

where
$\sigma_{\perp}$ = energy per unit area required to form dislocations in two perpendicular directions
$w_{\perp}$ = energy per unit length required to form dislocations in a row (linear dislocation formation)
D = distance between two adjacent dislocations.

The total strain energy per unit area of a semicoherent interface can be written as

$$\sigma_{\text{strain}} = \sigma_{\text{el}} + \sigma_{\perp} \tag{4.33}$$

or

$$\sigma_{\text{strain}} = Y\delta_{\text{el}}^2 d + \frac{2w_{\perp}}{D} \tag{4.34}$$

Energies of Incoherent Interfaces

If there is no crystallographic matching between the atoms of the two crystals the interface is *incoherent*. The spacing and structure change rapidly from one crystal to the other over a distance of only a few atom layers.

No simple model has been developed for the energies of incoherent interfaces. The larger the mismatching is, the larger will be the contribution of strain energy to the total interface energy.

$$\sigma_{\text{incoherent}}^{\text{strain}} > \sigma_{\text{semicoherent}}^{\text{strain}} > \sigma_{\text{coherent}}^{\text{strain}}$$

Magnitudes of Interface Energies of Solid/Solid Interfaces

The magnitudes of experimental values of the surface energies of coherent and incoherent solid/solid interfaces are listed in Table 4.1.

Table 4.1 Measured interface energies of coherent, semi-coherent and incoherent interfaces

Type of interface	Interface energy J/m^2
Coherent	0.025 – 0.2
Semi-coherent	0.2 – 0.5
Incoherent	0.5 – 1

Experimental evidence obviously confirms that the interface energy is much smaller for coherent than for incoherent interfaces.

4.3.4 Influence of Strain on Coherent and Semi-coherent Nucleation

When a nucleus is formed on a substrate the strain energy must be taken into consideration. The strain energy has to be added to the total Gibbs' free energy of the nucleus. In addition, strain also occurs in both solid phases. Hence, the strain energy within the crystal must be taken into account.

Turnbull and Vonnegut [4] have suggested a theoretical model for the interface between a heterogeneous substrate and the nucleating phase of different lattice spacings.

$$\Delta G_{\mathrm{c}} = -\left(-\Delta G^{\mathrm{L}\to\mathrm{s}} + \Delta G_{\mathrm{strain}}\right)\frac{V}{V_{\mathrm{m}}} + \sum_{\mathrm{i}} A_{\mathrm{i}}\sigma_{\mathrm{i}} \tag{4.35}$$

where the driving force of solidification equals $-\Delta G^{\mathrm{L}\to\mathrm{s}} = -(G^{\mathrm{s}} - G^{\mathrm{L}})$.

The strain energy within the crystal can be regarded as a complement to the Gibbs' free energy of the nucleus, i.e. an increase of G^{s}. The strain energy depends on the type of interface.

Coherent Nucleation

In the case of coherent nucleation only elastic strain is present. The molar strain energy of the crystal is then increased by the amount

$$\Delta G_{\mathrm{strain}} = Y{\delta_{\mathrm{el}}}^2 V_{\mathrm{m}} \tag{4.36}$$

The total Gibbs' free energy of the nucleus/crystal can be written, using Equations (4.31) and (4.35)

$$\Delta G_{\mathrm{c}} = -\left[\left(-\Delta G^{\mathrm{L}\to\mathrm{s}}\right)\frac{V}{V_{\mathrm{m}}} + Y{\delta_{\mathrm{el}}}^2 V\right] + \sum_{\mathrm{i}} A_{\mathrm{i}}\sigma_{\mathrm{i}} + A\left(\sigma_{\mathrm{c}} + Y{\delta_{\mathrm{el}}}^2 d\right) \tag{4.37}$$

where A is the area of the solid/solid interface between the particle and the crystal with volume V. The summary term represents the interfaces between the other crystal surfaces and the melt.

Semi-coherent Nucleation

In the case of semi-coherent nucleation elastic strain and also dislocation strain are present. The edge dislocations occur only on the surface of the crystal and do not contribute to the bulk strain energy of the crystal. The molar strain energy is then the same as for coherent nucleation

$$\Delta G_{\mathrm{strain}} = Y{\delta_{\mathrm{el}}}^2 V_{\mathrm{m}} \tag{4.36}$$

The edge dislocations affect only the interface energy of the crystal and the total free energy of the nucleus can be written with the aid of Equations (4.34) and (4.35) in analogy with Equation (4.37)

$$\Delta G_{\mathrm{c}} = -(-\Delta G^{\mathrm{L}\rightarrow\mathrm{s}})\frac{V}{V_{\mathrm{m}}} + Y{\delta_{\mathrm{el}}}^{2}V + \sum_{\mathrm{i}} A_{\mathrm{i}}\sigma_{\mathrm{i}} + A(\sigma_{\mathrm{c}} + Y{\delta_{\mathrm{el}}}^{2}d + \frac{2w_{\perp}}{D}) \tag{4.38}$$

In Equations (4.37) and (4.38) we have disregarded the fact that the modulus of elasticity may differ in the interface and in bulk. Equation (4.38) can be used to calculate the critical size of a non-spherical embryo. The method is described below.

Calculation of the Critical Size of a Non-Spherical Embryo in the Case of Semi-coherent Nucleation

The solid crystals formed at solidification have in most cases an anisotropic surface tension. Instead of a spherical cap like the one shown in Figures 4.12 and 4.13 on page 180, it is more likely that a crystal of some other shape is formed. A simple example is shown in Figure 4.17.

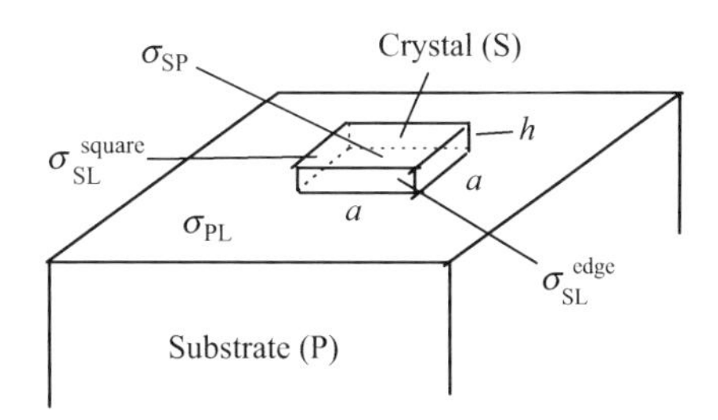

Figure 4.17 Formation of a square disk on a substrate.

Figure 4.17 illustrates the formation of a square disk formed on a substrate. The principles of the designations of the surface tensions/interface energies are the same as those given in Figure 4.12. As Figure 4.17 shows there are two different interface energies between the melt and the crystal, corresponding to the top and the edge faces, respectively, of the embryo/crystal, denoted by subscript S (solid).

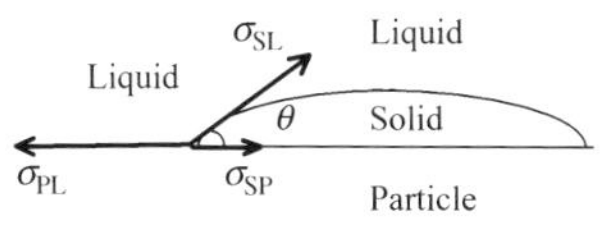

Equation (4.38) is applied to find the total free energy of the square disk. Three new interfaces between the melt and the crystal appear. The interface between the substrate, denoted by subscript P (particle), and the melt is replaced by a solid/solid interface between the substrate and the crystal.

Figure 4.12 Formation of a embryo/crystal on a foreign particle P. Figure 4.12 is only valid if the surface tension is isotrop.

$$\Delta G_{\mathrm{square}} = -\frac{ha^{2}}{V_{\mathrm{m}}}\left[-(G^{\mathrm{s}} - G^{\mathrm{L}})\right] + Y\delta_{\mathrm{el}}^{2}ha^{2} + 4ha\,\sigma_{\mathrm{SL}}^{\mathrm{edge}} + a^{2}\delta_{\mathrm{SL}}^{\mathrm{square}} + a^{2}(\sigma_{\mathrm{c}} + Y\delta_{\mathrm{el}}^{2}d + \frac{2w_{\perp}}{D} - \sigma_{\mathrm{PL}}) \tag{4.39}$$

where

$\Delta G_{\mathrm{square}}$ = total Gibbs' free energy of the square crystal
h, a = height and side of the square crystal
V_{m} = molar volume of the crystal
$\sigma_{\mathrm{SL}}^{\mathrm{edge}}$ = interface energy per unit area between the edge sides of the crystal and the melt
$\sigma_{\mathrm{SL}}^{\mathrm{square}}$ = interface energy per unit area between the square sides of the crystal and the melt
σ_{SP} = interface energy per unit area between the crystal and the substrate $= \sigma_{\mathrm{c}} + E\delta_{\mathrm{el}}^{2}d + 2w_{\perp}/D$
σ_{PL} = interface energy per unit area between the substrate and the melt.

Taking the derivative of Equation (4.2) with respect to r gives the critical radius of the spherical embryo. In this case, we have two variables h and a. To find the critical size of the square we have to take the partial derivatives with respect to both h and a under the assumption that the surface tensions are constant. The maximum of the ΔG function is identical with the Gibbs' free energy ΔG^* of the square of critical size.

$$\frac{\partial \Delta G_{\text{square}}}{\partial h} = 0 \tag{4.40}$$

$$\frac{\partial \Delta G_{\text{square}}}{\partial a} = 0 \tag{4.41}$$

Equations (4.40) and (4.41) form a linear equation system, which can be solved for h and a and gives the critical values h^* and a^*. These values are inserted into Equation (4.39) and we obtain

$$\begin{aligned}\Delta G^*_{\text{square}} = & -\frac{h^*a^{*2}}{V_{\text{m}}}\left[-(G^{\text{s}} - G^{\text{L}})\right] + Y\delta_{\text{el}}^2 h^*a^{*2} + 4h^*a^*\sigma_{\text{SL}}^{\text{edge}} \\ & + a^{*2}\sigma_{\text{SL}}^{\text{square}} + a^{*2}\left(\sigma_{\text{c}} + Y\delta_{\text{el}}^2 d + \frac{2w_{\perp}}{D} - \sigma_{\text{PL}}\right)\end{aligned} \tag{4.42}$$

The method of finding the critical size of a non-spherical embryo in the case of semi-coherent nucleation can be summarized as follows:

1. An expression of the total Gibbs' free energy of the embryo/crystal is set up including the interface energies and the strain energies. New interfaces are formed and one disappears in analogy with Equation (4.39).
2. The condition for minimum Gibbs' free energy is found by partial derivation of the Gibbs' free energy with respect to each of the independent variables. The derivatives are equal to zero and the obtained set of equations is solved for the independent variables. The solution represents the critical size and shape of the embryo.

It is obvious from the example with the square crystal above that the energy expressions depend on the material constants, for example the modulus of elasticity and the surface tensions. The equations have to be set up from case to case with the aid of the outlines given by Equations (4.37) and (4.38), respectively.

4.3.5 *Nucleating Agents of Iron Alloys*

Careful studies have been carried out of the influence of heterogeneous nucleation by various carbides and nitrides in a number of ferrous alloys that solidify as ferrite.

A sample of pure iron (99.95%) was heated to a temperature of at least 78 °C above the melting point in a vacuum-induction furnace to insure complete melting. The sample was then allowed to cool at a rate of about 100 °C per minute until complete solidification and the freezing temperature was registered. Several melting–cooling cycles were performed to give a reliable value of the undercooling for each nucleation agent. Then, the inoculant was added and a number of melting–cooling cycles were performed. The result of some typical experiments, performed by Bramfitt [5], is shown in Figures 4.18 and 4.19.

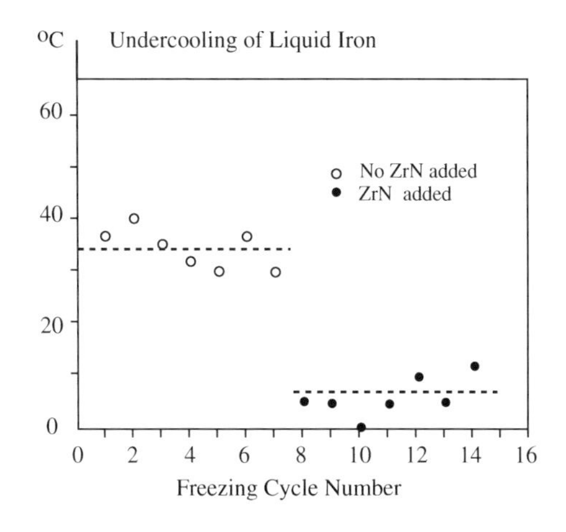

Figure 4.18 Undercooling of liquid iron, necessary for solidification, with and without addition of zirconium nitride ZrN according to Bramfitt. Reproduced from Springer © 1970.

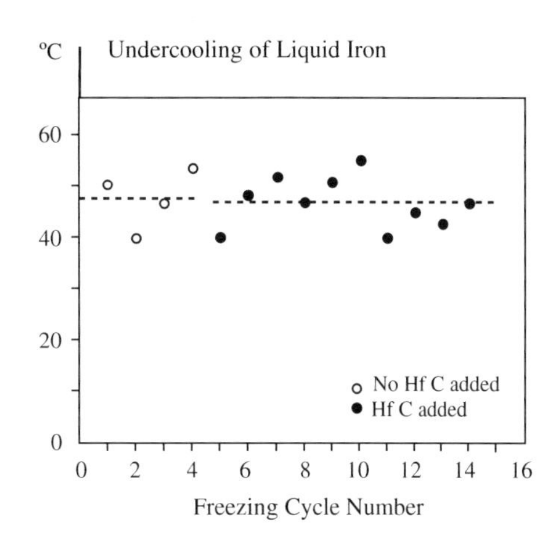

Figure 4.19 Undercooling of liquid iron, necessary for solidification, with and without addition of hafnium carbide HfC according to Bramfitt. Reproduced from Springer © 1970.

Figure 4.19 shows clearly that addition of hafnium carbide has no influence at all on the undercooling necessary for solidification. Zirconium nitride lowers the undercooling considerably (Figure 4.18) and is representative of good nucleating agents.

More than twenty nitrides and carbides were examined. Five of them were more or less successful nucleating agents for ferrite. They are listed in Table 4.2.

Table 4.2 Nucleating agents of liquid iron

Nucleation Agent	Characteristic undercooling °C	Mismatch relative ferrite %	Relative judgement
TiN	1.7	3.9	Very effective
TiC	1.8	5.9	Very effective
ZrN	7.0	11.2	Rather effective
ZrC	13.6	14.4	Least effective
WC	16.1	29.4	Least effective

Figure 4.20 on page 189 shows the measured undercooling of the five compounds as a function of mismatch δ in steel melts. The figure also includes the curve that represents the undercooling, calculated theoretically according to Turnbull and Vonnegut's theory. As we will see on page 188–189 the undercooling is a parabolic function of the mismatch and can be calculated from Equation (4.47).

Turnbull and Vonnegut [4] pointed out that the efficiency of an inoculant in promoting heterogeneous nucleation depends on the crystallographic mismatch δ between the substrate and the solidified melt. These topics have been discussed in Section 4.3.3 on page 182. The mismatch or disregistry is defined as

$$\delta = \frac{\overline{\Delta a}}{a} \tag{4.43}$$

where

a = the lattice constant of the nucleated phase

Δa = the average difference between the lattice constant of the substrate and the nucleated solid phase for a low-index plane.

For $\delta \leq 0.20$ the interface region can be explained by a simple distortion model, i.e. coherent nucleation with strain. The strain energy in this case is given by Equation (4.37) on page 186

$$\Delta G_c = -\left[(-\Delta G^{L\rightarrow s})\frac{V}{V_m} + Y\delta_{el}{}^2 V\right] + \sum_i A_i\sigma_i + A(\sigma_{ch} + Y\delta_{eli}{}^2 d) \qquad (4.37)$$

The stable state of the crystal is obtained when the total Gibbs' free energy of the crystal is a minimum. Turnbull and Vonnegut disregarded the interface energy in comparison with the energy of the crystal. Hence, the sum of the third and the fourth terms in Equation (4.37) is small in comparison with the sum of the first terms. Bramfitt assumed that the minimum value of the total Gibbs' free energy is zero as it cannot be negative.

$$\Delta G_c = -(-\Delta G^{L\rightarrow s})\frac{V}{V_m} + Y\delta_{el}{}^2 V = 0 \qquad (4.44)$$

for any value of V that implies that the factor in front of V must be zero:

$$\Delta G^{L\rightarrow s} + Y\delta_{el}{}^2 V_m = 0 \qquad (4.45)$$

Equation (4.45) is combined with Equation (4.10) on page 170:

$$-\Delta G^{L\rightarrow s} = \frac{-\Delta H}{T_M}(T_M - T) \qquad (4.10)$$

which gives

$$-\frac{-\Delta H}{T_M}(T_M - T) + Y\delta_{el}{}^2 V_m = 0 \qquad (4.46)$$

or, if we introduce $\delta_{el} = \dfrac{\Delta a}{a}$ and the undercooling $\Delta T = T_M - T$

$$\Delta T = \frac{T_M Y V_m}{-\Delta H}\left(\frac{\Delta a}{a}\right)^2 \qquad (4.47)$$

- The undercooling is proportional to the square of the mismatch in this case.
- Three conditions must be fulfilled for a compound to be a good nucleating agent

1. The compound must not dissolve in the melt.
2. The interface between low-index planes of the substrate and the nucleated solid must be coherent or nearly coherent.
3. The interfacial energy between the substrate and the melt must be higher than the interfacial energy between the nucleated solid and the melt.

Hafnium carbide is ruled out because it dissolves in the iron melt and does not provide nucleation sites for this reason.

TiN, TiC, ZrN and ZrC have a simple cubic structure. The mismatches relative to delta iron (ferrite) with BCC structure are listed in Table 4.2 on page 187. The smaller the mismatch is, the more coherent will be the interface and the better will be the nucleating power of the inoculant, Figure 4.20 on the next page.

VN and VC have δ values around 1% but are useless as nucleating agents as they dissolve in molten iron when the carbon or nitrogen contents are low enough for ferrite to be formed. Higher carbon and nitrogen values give precipitation of austenite. Even in this case VN and VC do not work as nucleation agents owing to poor matching between the inoculants and the solidifying melt.

The mismatch between delta iron (ferrite) and tungsten carbide is 29.4%. The coherency between the two phases is poor and WC cannot be used as a nucleating agent in an iron melt that solidifies as ferrite.

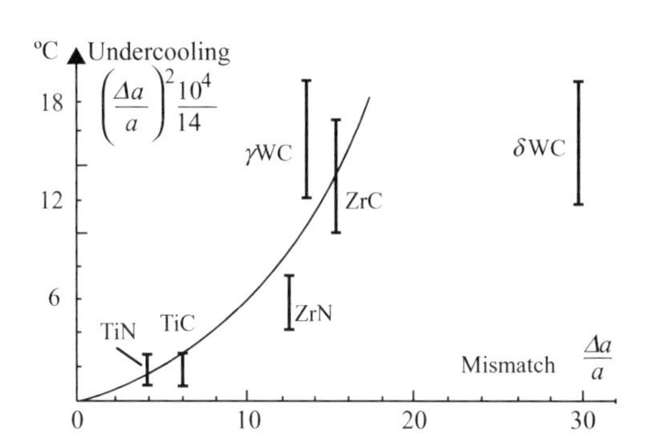

Figure 4.20 Undercooling, necessary for solidification, as a function of mismatch between ferrite and different substrates according to Bramfitt.

The value of γWC has been added by Fredriksson. Reproduced from Springer © 1970.

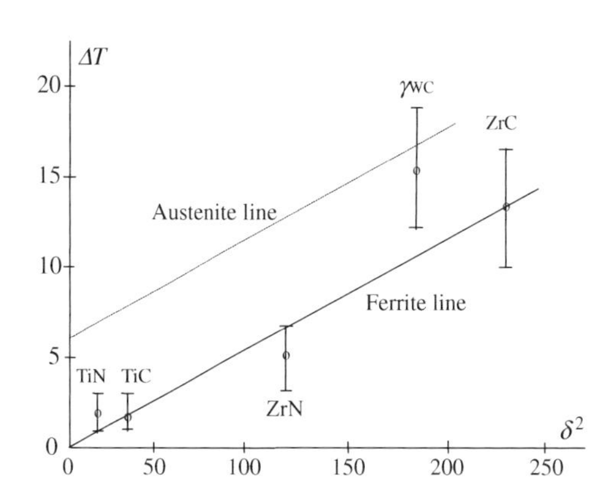

Figure 4.21 The undercooling, necessary for solidification, as a function of the square of the mismatch between the heterogeneity and the solidified melt.

The values of ΔT in Figure 4.20 have been plotted versus the corresponding δ^2 values.

As is seen from Figure 4.21, the values of ΔT and δ in Table 4.2 are, within the limits of error, in agreement with Equation (4.47) with the exception of δWC. It deviates strongly from the parabolic curve built on Equation (4.47). Obviously, this equation is not valid for δWC. For this reason it has been excluded from Figure 4.21.

If the iron melt solidifies as gamma iron (austenite) the result is entirely different. If the value for γWC is plotted in Figures 4.20 and 4.21, it can be seen that the value of γWC does not deviate considerably from those of the other nucleating agents. Hence, WC can be used as a nucleating agent if the iron melt solidifies as austenite.

The necessary undercooling temperature for solidification of austenite is only 6 °C below the melting point of ferrite. If a line, parallel to and located 6 °C below the 'ferrite line', is drawn we obtain the corresponding 'austenite line'. This has been done in Figure 4.21.

There is a simple explanation of the difference in behavior of δWC and γWC. To realize this we have to compare the structures of tungsten carbide, ferrite and austenite.

WC has a hexagonal structure. The number of nearest neighbours in the close-packed base plane (100) is 6.

The structure of ferrite is BCC (page 20 in Chapter 1 in [6]). The number of nearest neighbours in the close-packed plane (100) is only 4. The number of next-nearest neighbours is 8.

Austenite has FCC structure (page 20 in Chapter 1 in [6]). The number of nearest neighbours in the close-packed plane (111) is 6.

The atomic distances are much more similar and the general structures of WC and austenite resemble each other much more than those of WC and ferrite. This gives a much better coherency between WC and austenite than between WC and ferrite.

Consequently, the mismatch is much lower for γWC than for δWC. The mismatch given in Table 4.2 along the (110) plane of delta iron and the $(\bar{1}2\bar{1}0)$ plane of the tungsten carbide is as high as 29.4%, very much higher than the mismatch 13.5% for γWC.

4.4 Nucleation of Bubbles

4.4.1 *Homogeneous Nucleation of Spherical Pores*

In Section 4.2.1 we have derived expressions for the critical size of an embryo and the activation energy in a system at constant pressure. These expressions [Equations (4.3) and (4.4) on page 168–169] are only valid if the molar volume V_m is independent of the pressure.

This is not the case when the 'embryo' is a gas pore in a melt. Gibbs has treated bubble formation. He calculated the work to form a pore of critical size and assumed this bubble to be in chemical and mechanical equilibrium with its surroundings.

The pressure inside a spherical pore is not constant but depends on the radius [compare Equation (1.42) on page 12 in Chapter 1]

$$p_r = p_a + \frac{2\sigma}{r} \tag{4.48}$$

where

p_r = pressure inside the pore with radius r
p_a = ambient pressure
σ = interface energy between the gas pore and the melt
r = radius of the pore.

The work required to form a bubble consists of two terms, one for forming the 'embryo' and the other for forming the interface. It is equal to the total Gibbs' free energy of the bubble. According to Gibbs the formation energy of a pore of critical size is

$$\Delta G^* = -4\pi r^{*3}\frac{p^* - p_a}{3} + 4\pi\sigma r^{*2} \tag{4.49}$$

where

ΔG^* = Gibbs' free energy of a pore of critical size
r^* = radius of pore of critical size
p^* = pressure of a pore of critical size.

Elimination of r^* with the aid of Equations (4.48) and (4.49) gives

$$\Delta G^* = \frac{16\pi}{3}\frac{\sigma^3}{(p^* - p_a)^2} \tag{4.50}$$

In analogy with Equation (4.48), we obtain the relationship

$$p^* = p_a + \frac{2\sigma}{r^*} \tag{4.51}$$

Thermodynamics of Homogeneous Nucleation of Bubbles

The pressure difference $(p^* - p_a)$ in Equation (4.50) can be related to the Gibbs' free energy and the chemical potential of the system by relationships derived in Chapter 2. We will check it by applying it on growth of a solid embryo in a liquid.

According to Gibbs' theory chemical equilibrium exists between an embryo of critical size and the bulk liquid. As an introduction, we apply his theory on the case of a *solid embryo* and obtain the relationship

$$G^L = G^s + \int_{p_a}^{p^*} V_m \mathrm{d}p \tag{4.52}$$

In a solid, V_m is practically independent of the pressure and Equation (4.52) gives the driving force for embryo nucleation:

$$-\Delta G^{L\rightarrow s} = -(G^s - G^L) = V_m(p^* - p_a) \tag{4.53}$$

When this expression is introduced into Equation (4.50) we obtain Equation (4.4) on page 169, i.e. complete agreement with the theory of homogeneous nucleation given in Section 4.2.1.

Nucleation of a Gas Pore in a Metal Melt

Next, we apply Gibbs' theory on the formation of a gas pore of critical size in a melt. For a gas, the molar volume is related to the pressure by the general gas law. If we introduce $V_m = RT/p$ into Equation (4.52) we obtain

$$G^{pore} = G^0 + RT\int_{p_0}^{p_{pore}}\frac{dp}{p} \qquad \text{and} \qquad G^a = G^0 + RT\int_{p_0}^{p_a}\frac{dp}{p}$$

$$G^{pore} = G^0 + RT\ln\frac{p_{pore}}{p_0} \qquad \text{and} \qquad G^a = G^0 + RT\ln\frac{p_a}{p_0}$$

The combination of these equations gives

$$G^{pore} = G^a + RT\ln\frac{p_{pore}}{p_a} \tag{4.54}$$

where

G^0 = Gibbs' free energy at a standard pressure p_0

G^{pore} = Gibbs' free energy of the gas in a bubble of critical size with radius r^* in equilibrium with the liquid

G^a = Gibbs' free energy of the surroundings.

By inserting the relationship in Equation (4.54) into Equation (4.50) on page 190 we obtain the driving force for nucleation of bubbles:

$$\Delta G^* = \frac{16\pi}{3}\frac{\sigma^3}{\left(e^{\frac{G^{pore}-G^a}{RT}}\right)^2} \tag{4.55}$$

or

$$G^{pore} - G^a = \frac{RT}{2}\ln\left(\frac{16\pi}{3}\frac{\sigma^3}{\Delta G^*}\right) \tag{4.56}$$

We found in Example 4.1 on page 170 that if 10^3 nuclei per kmol of the melt are formed, ΔG^* equals $55k_BT$. If this expression in introduced into Equation (4.56) we obtain the special case

$$G^{pore} - G^a = \frac{RT}{2}\ln\left(\frac{16\pi}{3}\frac{\sigma^3}{55k_BT}\right) \tag{4.57}$$

It is also possible to express the driving force as a function of the concentration of the gas dissolved in the liquid. By defining a standard state μ^0 is determined at a standard pressure p_0, we obtain:

$$G^{pore} = \mu_L^0 + RT\ln x_{pore}^L = \mu_{gas}^0 + RT\ln p_{pore} \tag{4.58}$$

$$G^a = \mu_L^0 + RT\ln x_a^L = \mu_{gas}^0 + RT\ln p_a \tag{4.59}$$

By inserting the G values in Equations (4.58) and (4.59) into Equation (4.57) and replacing p_{pore} by p^* we obtain the special case

$$(p^* - p_a)^2 = \left(\frac{x^L_{pore}}{x^L_a}\right)^2 = \frac{16\pi}{3}\frac{\sigma^3}{55k_BT} \tag{4.60}$$

Equation (4.60) relates the required pressure for formation of a pore of critical size and the temperature of the melt.

As a concrete example we will discuss the formation of hydrogen bubbles in aluminium melts.

Example 4.7

Consider an Al melt with the surface tension σ at temperature T. Calculate the pressure of a tentative bubble of critical size that is relative to the ambient (hydrogen) pressure p_a required for homogeneous formation of bubbles in the Al melt. Assume that 10^3 nuclei per kmol of the melt are formed in the melt.

Present the result graphically, i.e. plot the ^{10}logarithm of the pressure difference $(p_{H_2} - p_{\underline{H}})$ in atm as a function of the temperature of the Al melt (900–1400 K) with the surface tension as a parameter ($\sigma = 1.0$, 0.10, 0.010 and 0.0010 N/m).

Solution:

If 10^3 nuclei per kmol of the melt are formed ΔG^* equals $55k_BT$ (Example 4.1 on page 170). We use Equation (4.60) and replace p^* by p_{H_2}, p_a by $p_{\underline{H}}$ and ΔG^* by $55k_BT$ in Equation (4.50) and obtain

$$p_{H_2} - p_{\underline{H}} = \sqrt{\frac{16\pi}{3}\frac{\sigma^3}{55k_BT}}$$

where

p_{H_2} = critical pressure in the bubble
$p_{\underline{H}}$ = pressure of solved H atoms in the melt.

The pressure difference $(p_{H_2} - p_{\underline{H}})$ is calculated as a function of the temperature for different values of the surface tension.

The result is given in the figure in the answer, where the ^{10}logarithm of the pressure difference $(p_{H_2} - p_{Al})$ is plotted as a function of the temperature T for each of the given values of the surface tension.

Answer:

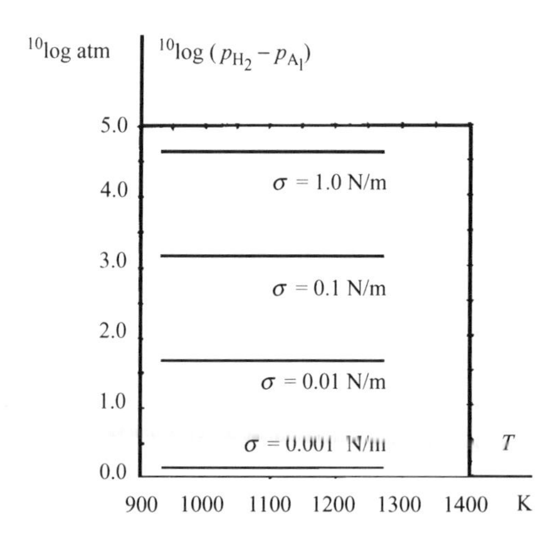

Required pressure of hydrogen for homogeneous pore nucleation in an aluminium melt as a function of the melt temperature.

A reasonable value of the surface tension of liquid aluminium is 0.9 N/m. Interpolation in the figure in Example 4.7 above shows that homogeneous nucleation would not occur unless the hydrogen pressure exceeds 10^4 atm. This is not a realistic value and therefore the hydrogen pores in Al melts are normally formed by heterogeneous nucleation.

4.4.2 *Heterogeneous Nucleation*

In a melt there are usually particles, such as oxides, which can cause heterogeneous nucleation of pores. The mechanism of this nucleation is normally treated by considering the gap, formed on a flat substrate as shown in Figure 4.22.

In a spherical substrate, the sizes of the particles and also the differences in surface tension, influence the spread of the gas (vapour) around the particle. For the sake of simplicity, we assume that the ratio

$$n_\sigma = \frac{\sigma_{PL}}{\sigma_{LV}}$$

is constant.

The designations stand for

σ_{PL} = surface tension at the particle/liquid interface
σ_{LV} = surface tension at the liquid/gas interface.

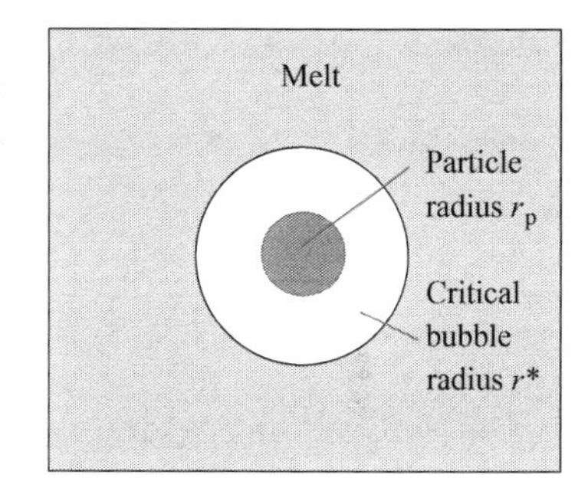

Figure 4.22 Heterogeneous nucleation on a spherical particle.

If $n_\sigma > 1$ (page 107 in Chapter 3) the bubble will spread around the particle and the conditions described in Example 4.7 (192) will be obtained. The formation energy of a pore of critical size can in this case be written as

$$W^* = \Delta G_r = 4\pi\left(r^{*2} - n_\sigma r_p^2\right)\sigma_{LV} - \frac{4\pi}{3}\left(r^{*3} - r_p^3\right)\left(p_{gas} - p_a\right) \tag{4.61}$$

where

W^* = formation energy of a pore of critical size
r^* = critical bubble radius
r_p = particle radius
n_σ = ratio between surface tensions (see above).

The formation energy of a pore of critical size equals 55 k_BT, which corresponds to 10^3 nucleus per kmol of the melt (see Example 4.1 on page 170). This value is inserted into Equation (4.61) and the new equation represents the pressure of hydrogen p_{H_2} required for heterogeneous nucleation of hydrogen gas in an Al melt as a function of the particle size r_p.

4.5 Crystal Multiplication

There are two methods of practical interest to stimulate nucleation in metal melts.

One is *inoculation*, which has been treated in Section 4.3.

The other is *crystal multiplication*, which implies that small pieces of the framework of growing dendrites (Figure 4.23) in a melt are broken off and moved to other parts of the melt where they act as nuclei for new crystals. The process of crystal multiplication then accelerates spontaneously.

Two different mechanisms for separation of dendrite tips have been suggested.

1. Small pieces of growing dendrite tips are torn away in an entirely mechanical way, for example by the natural convection in the melt or forced convection on purpose.
2. Secondary dendrite arms melt off at their roots, spontaneously or on purpose.

Figure 4.23 Growing dendrite with primary/secondary and tertiary arms.

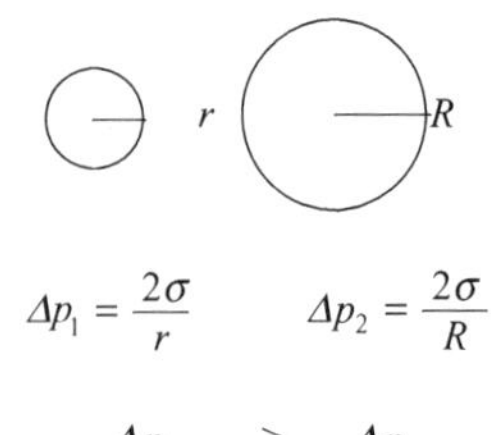

Figure 4.24 The excess pressure of the small drop is larger than that of the big drop.

The first mechanism can be used in many ways. Ultrasound in combination with strong forced convection is a successful method to obtain fine-grained steel and aluminium alloys. Another way (for steel melts) is to use an alternating and rotating magnetic field that causes inertial forces on the dendrite arms, strong enough to break them.

The second mechanism of crystal multiplication is based on the surface tension of the secondary dendrite arms. It is well known that the excess pressure in a sphere with the surface tension σ is

$$\Delta p = \frac{2\sigma}{r} \tag{4.62}$$

Figure 4.24 shows that the excess pressure is larger for small radii than for large ones. The excess pressure causes a decrease of the melting point, which is proportional to the pressure difference. Hence, the melting point is lowest at the narrow roots of the secondary dendrite arms (Figure 4.25) and they primarily melt off there.

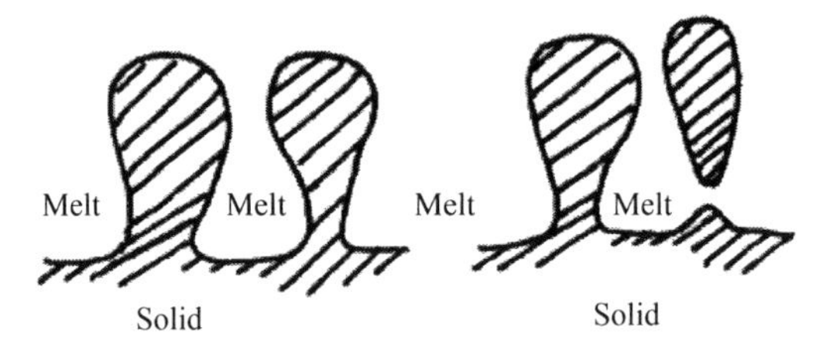

Figure 4.25 Release of dendrite tips by melting.

The simplest way to induce release of dendrite arms by melting is to add an adequate amount of hot melt to the solidifying melt.

Summary

■ Basic Concepts

Embryos and Nuclei

New phases are formed as *clusters*. The clusters vary in size by statistical fluctuations. Clusters or *embryos,* which reach a minimum critical size, can continue to grow and may be transformed into nuclei.

■ Homogeneous Nucleation

Stable Statistical Equilibrium Distribution of Embryos

$$n_r = n e^{-\frac{\Delta G_r}{k_B T}}$$

Gibbs' Free Energy of Formation of an Embryo with Radius r

$$\Delta G_r = -\frac{\frac{4\pi r^3}{3}}{V_m}\left[-(G^s - G^L)\right] + 4\pi r^2 \sigma^{L/s}$$

Gibbs' Free Energy of a Spherical Embryo of Critical Size

$$r^* = \frac{2\sigma V_m}{\Delta G^{L\to s}}$$

$$\Delta G^* = \frac{16\pi}{3}\frac{\sigma^3 {V_m}^2}{(\Delta G^{L\to s})^2}$$

Driving Force of Solidification:

$$-\Delta G^{\mathrm{L}\rightarrow \mathrm{s}} = -(G^{\mathrm{s}} - G^{\mathrm{L}})$$

Nucleation Rate

Nucleation rate = the number of embryos per unit volume and unit time that are transformed into growing nuclei.

$$J = Const \times P\mathrm{e}^{-\frac{\Delta G^{*}+U^{\mathrm{I}}}{k_{\mathrm{B}}T}} \qquad J = const \times n^{*}P\mathrm{e}^{-\frac{U^{\mathrm{I}}}{k_{\mathrm{B}}T}}$$

■ Homogeneous Nucleation as a Function of Undercooling. Nucleation Temperature

$$-\Delta G^{\mathrm{L}\rightarrow \mathrm{s}} = \frac{-\Delta H}{T_{\mathrm{M}}}(T_{\mathrm{M}} - T)$$

Undercooling = $T_{\mathrm{M}} - T$

Nucleation temperature = the temperature when nucleation suddenly occurs.

Relationship between ΔG* and Nucleation Temperature

$$\frac{\Delta G^{*}}{k_{\mathrm{B}}T} = \ln\frac{N}{n^{*}}$$

Homogeneous Nucleation as a Function of Concentration in Binary Alloys

$$-\Delta G^{\mathrm{L}\rightarrow \mathrm{s}} = RT\ln\frac{x^{\mathrm{L}}}{x^{\mathrm{L}}_{\mathrm{eq}}}$$

Influence of Variable Surface Energy on Homogeneous Nucleation

The smaller the surface tension is, the smaller will be the barrier.

$$\Delta G_{\mathrm{r}} = -\frac{\frac{4\pi r^{3}}{3}}{V_{\mathrm{m}}}\left[-(G^{\mathrm{s}} - G^{\mathrm{L}})\right] + 4\pi r^{2}\sigma_{\infty}^{\mathrm{L/s}}\mathrm{e}^{-\frac{a\delta}{r}}$$

Small embryos can form spontaneously in the shallow minima but they do not lead to embryos of critical size unless energy is supplied.

Homogeneous Nucleation of Non-Spherical Embryos. Nucleation of Faceted Crystals

- The equilibrium shape of a crystal of a given volume is found by minimizing the total Gibbs' free surface energy.
- Gibbs–Curie–Wulff's theorem

$$\frac{\sigma_1}{h_1} = \frac{\sigma_2}{h_2} = \frac{\sigma_3}{h_3} = \ldots\ldots = \frac{\sigma_n}{h_n} = const$$

- Close-packed crystal facets have lower total Gibbs' free energies than other crystal facets.

Total Gibbs' Free Energy of an Embryo of Critical Size

$$\Delta G_{\mathrm{c}}^{\mathrm{total}\,*} = \frac{1}{3}\sum_{n}\sigma_n A_n$$

■ Heterogeneous Nucleation. Inoculation

Size of Critical Nucleus

$$r^* = \frac{2\sigma_{SL} V_m \sin\theta}{-\Delta G^{L \to s}}$$

Gibbs' Free Energy of Critical Nucleus

$$\Delta G_i^* = \frac{16}{3} \frac{\pi(\sigma_{SL})^3 V_m^2}{(-\Delta G^{L \to s})^2} \frac{(2 + \cos\theta)(1 - \cos\theta)^2}{4}$$

or

$$\Delta G_{het} = \Delta G_{hom} \frac{(2 + \cos\theta)(1 - \cos\theta)^2}{4}$$

Number of Potentially Growing Crystals around a Heterogeneous Particle

$$n^*(A_i) = n(A) e^{-\frac{\Delta G_i^*}{k_B T}}$$

Influence of Strain on Coherent and Semi-coherent Nucleation

Strain energy consists of elastic strain energy and dislocation energy.

$$\sigma_{strain} = Y \delta_{el}^2 d + \frac{2 w_\perp}{D}$$

Coherent Nucleation

The total free energy of a nucleus with elastic strain energy:

$$\Delta G_c = -\left[(-\Delta G^{L \to s}) \frac{V}{V_m} + Y \delta_{el}^2 V \right] + \sum_i A_i \sigma_i + A(\sigma_c + Y \delta_{el}^2 d)$$

Semi-coherent Nucleation

The total Gibbs' free energy of a nucleus with elastic strain energy and dislocation energy:

$$\Delta G_c = -(-\Delta G^{L \to s}) \frac{V}{V_m} + Y \delta_{el}^2 V + \sum_i A_i \sigma_i + A(\sigma_c + Y \delta_{el}^2 d + \frac{2 w_\perp}{D})$$

Calculation of the Critical Size of a Non-Spherical Embryo in the Case of Semi-coherent Nucleation

1. The total free energy of a nucleus with strain energy and dislocation energy as a function of geometrical quantities is set up.
2. Partial derivation of the total free energy of the nucleus with respect to the geometrical variables is performed in order to find the minimum total energy.
3. These relationships form a linear equation system that is solved. The solution represents the critical size of the embryo.

■ Inoculation. Nucleating Agents

Inoculation means addition of small crystals to metal melts. This results in heterogeneous nucleation and a fine-grained solid product after solidification.

A compound is a good nucleating agent if

- The compound must not dissolve in the melt.
- The interface between low-index planes of the substrate and the nucleated solid must be coherent or near coherent.
- The interfacial energy between the substrate and the melt must be higher than the interfacial energy between the nucleated solid and the melt.

Undercooling as a Function of the Crystallographic Mismatch

$$\Delta T = \frac{T_M Y V_m}{-\Delta H}\left(\frac{\Delta a}{a}\right)^2$$

■ Nucleation of Bubbles

Homogeneous Nucleation

Formation energy of a pore of critical size:

$$\Delta G^* = \frac{16\pi}{3}\frac{\sigma^3}{(p^* - p_a)^2}, \quad \text{where} \quad p^* = p_a + \frac{2\sigma}{r^*}$$

Nucleation of a gas pore in a metal melt

$$G^{pore} = G^a + RT \ln \frac{p_{pore}}{p_a}$$

Pressure for formation of a pore of critical size:

$$(p^* - p_a)^2 = \left(\frac{x^L_{pore}}{x^L_a}\right)^2 = \frac{16\pi}{3}\frac{\sigma^3}{\Delta G^*}$$

Heterogeneous Nucleation

Activation energy of pore nucleation:

$$W^* = \Delta G_r = 4\pi\left(r^{*2} - n_\sigma r_p^2\right)\sigma_{LV} - \frac{4\pi}{3}\left(r^{*3} - r_p^3\right)(p_{gas} - p_a)$$

where $n_\sigma = \dfrac{\sigma_{PL}}{\sigma_{LV}}$

If $n_\sigma > 1$ the gas pore will spread around the nucleating particle.

■ Crystal Multiplication

Mechanisms:

- Small pieces of growing dendrite tips are torn away in an entirely mechanical way, for example by the natural convection in the melt or by forced convection on purpose.
- Secondary dendrite arms melt off at their roots, spontaneously or on purpose.

Exercises

4.1 Start with the Gibbs' free energy of an embryo (Equation (4.2) on page 168) and derive the critical radius r^* and the critical Gibbs' free energy ΔG^* of the nucleus that is able to grow (Equations (4.3) and (4.4) on pages 168–169 in Chapter 4).

4.2 The lowest possible undercooling of a pure nickel melt before the solidification starts has been measured to 319 K with the aid of accurate thermal measurements (thermal analysis). It is reasonable to assume that homogeneous nucleation occurs at 319 K below its melting point. Calculate the surface energy between the solid and liquid phases of pure nickel. You may assume that the number of nucleated embryos is 10^3 per kmol.

The melting point of Ni at STP is 1453 °C. The molar volume of Ni is $6.6 \times 10^{-3}\,\text{m}^3/\text{kmol}$ and its heat of fusion is 17.2 kJ/mol.

4.3 According to the classical theory of homogeneous nucleation, the surface tension is constant and independent of the particle size. However, it is well known that the surface tension is influenced by the size of the particles when they are small. This surface tension dependence can be described by the relationship

$$\sigma = \sigma_0 \left(1 - \frac{\delta}{r}\right)$$

where σ_0 is the surface tension for a flat surface, δ is the thickness of the particle/melt interphase and r is the particle radius.

Use this expression to show in a diagram how the activation energy for grain nucleation is changed for a pure metal, for example pure Fe, and how this activation energy varies with the undercooling. Discuss your results.

Material constants for pure Fe:

$T_M = 1811\,\text{K}$
$-\Delta H_m^{fusion} = 1.52 \times 10^7\,\text{J/kmol}$
$V_m = 7.60 \times 10^{-3}\,\text{m}^3/\text{kmol}$
$\sigma_0 = 0.010\,\text{J/m}^2$
$\delta = 2.5\,\text{nm}.$

4.4 An iron melt with the temperature of 1550 °C contains small concentrations of $\underline{\text{Al}}$ and $\underline{\text{O}}$ in equilibrium with solid Al_2O_3 and $FeOAl_2O_3$. Al_2O_3 is the stable phase at the concentrations in question.

During deoxidation of an iron melt at 1550 °C with 0.010 at% $\underline{\text{O}}$ and 0.040 at% $\underline{\text{Al}}$, precipitation of FeO Al_2O_3 has been observed instead of precipitation of Al_2O_3. Perform a theoretical analysis to explain why.

The solubility products at equilibrium of Al_2O_3 and $FeOAl_2O_3$ in the melt are known:

$$\left(x_{\underline{\text{Al}}}^{\text{L eq}}\right)^2_{\text{Al}_2\text{O}_3} \left(x_{\underline{\text{O}}}^{\text{L eq}}\right)^3_{\text{Al}_2\text{O}_3} = 2.50 \times 10^{-16} \quad (\text{at\%})^5$$

$$\left(x_{\underline{\text{Al}}}^{\text{L eq}}\right)^2_{\text{FeOAl}_2\text{O}_3} \left(x_{\underline{\text{O}}}^{\text{L eq}}\right)^4_{\text{FeOAl}_2\text{O}_3} = 2.56 \times 10^{-16} \quad (\text{at\%})^6$$

4.5 It is a-well known fact that cast iron can solidify as grey or white iron. The white solidification process is characterized by a high solidification rate. Certain researchers claim that it is easier to nucleate cementite than graphite.

Calculate the undercooling as a function of the carbon concentration for both cementite and graphite and plot the functions for cementite and graphite in the phase diagram of the system Fe–C below. Try to find some support for the mentioned hypothesis concerning the nucleation of cementite. You may assume that the number of nucleated embryos is 10^3 per kmol.

Material constants:

The surface tension between cementite and the liquid phase $\sigma^{\text{L/cementite}} = 0.20\,\text{J/m}^2$.
$\sigma^{\text{L/graphite}} = 0.20\,\text{J/m}^3$
$\sigma^{\text{L/cementite}} = 0.20\,\text{J/m}^3$
V_m graphite $= 5.4 \times 10^{-3}\,\text{m}^3/\text{kmol}$
V_m cementite $= 5.8 \times 10^{-3}\,\text{m}^3/\text{kmol}.$

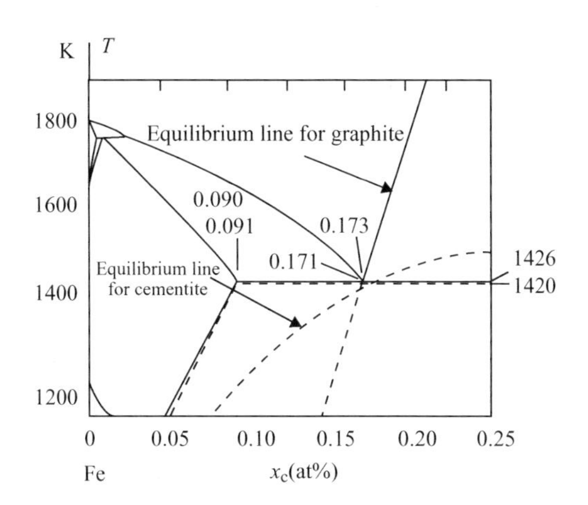

Figure 4.5–1

4.6 A Ni melt contains a large number of small particles. These particles have a wetting angle of 30° relative to solid nickel.

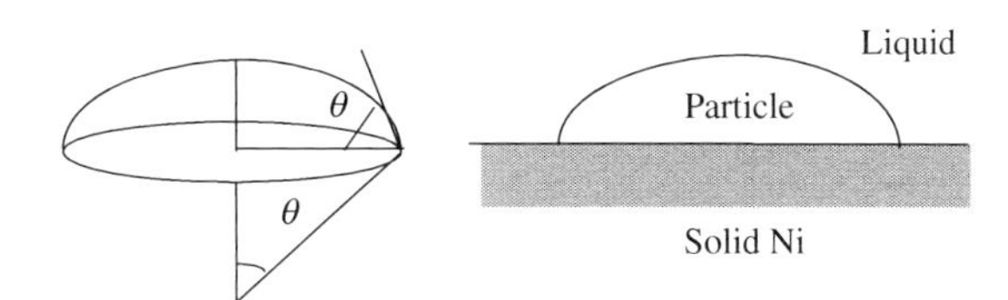

Figure 4.6–1

Calculate the maximum undercooling of the melt that can be achieved at continuous cooling. You may assume that the number of nucleated embryos is 10^{-3} per kmol.

Material constants for pure nickel:

M = 58.7 kg/kmol
ρ = 8.9×10^3 kg/m^3
T_M = 1453 °C = 1726 K
$\sigma_{L/s}$ = 0.255 J/m^2
$-\Delta H_m^{fusion}$ = 17.2×10^3 J/kmol.

4.7 The figure below shows the theoretical relationship between the undercooling of an Al melt and the mismatch δ of the old atomic layers and the new layers that form when the nucleated crystals grow. The figure is drawn for the case when four new atomic layers of atoms expand to the same interatomic distances as the old ones.

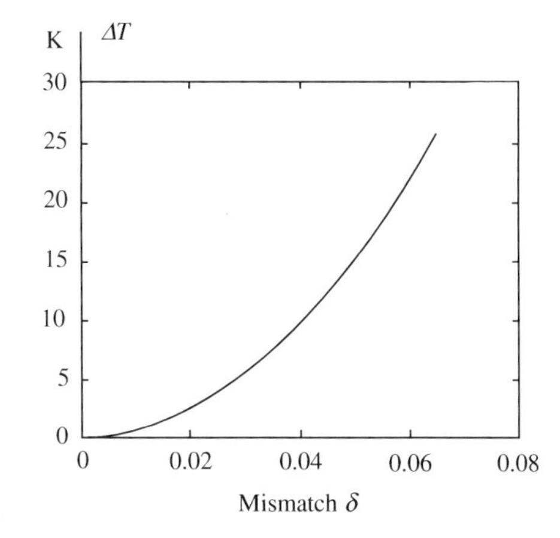

Figure 4.7–1

Derive and draw an analogous figure for the case when four outer atomic layers are expanded and the fifth layer has the normal interatomic distance of the bulk metal. The expansion decreases linearly from 100% down to 25%, distributed over four layers. Plot the average value of the undercooling as a function of the four strained atomic layers as a function of the mismatch.

Material constants of aluminium:

$T_M = 628\,°C = 931\,K$
$M = 27\,kg/kmol$
$\rho = 2.7 \times 10^3\,kg/m^3$
$Y = 7.0 \times 10^{10}\,N/m^2$
$-\Delta H^{fusion} = 10.7 \times 10^6\,J/kmol$

4.8 During casting of aluminium alloys the melt is usually inoculated with an Al 3wt% Ti alloy. A very fine-grained structure is obtained in this way. The reason for this effect is not well understood, but it has been suggested that the solidification process starts with a primary precipitation of Al_3Ti. Part of the phase diagram of the system Al–Ti is given below.

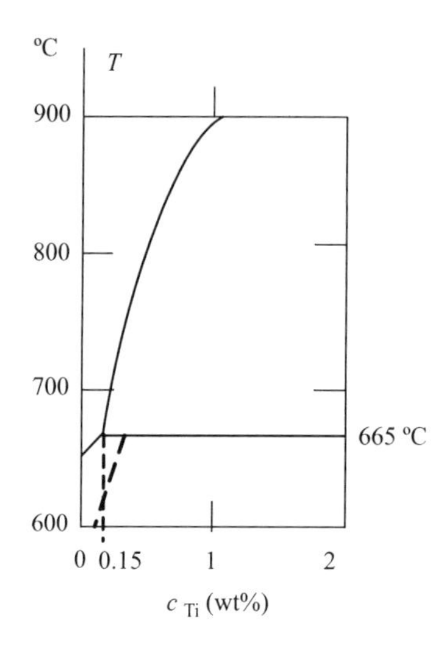

Figure 4.8–1

Examine this hypothesis with the aid of the classical homogeneous nucleation theory and calculate the minimum Ti concentration required for formation of Al_3Ti. Consider also the formation of Al_3Ti at the peritectic temperature 665 °C. You may assume that the number of nucleated embryos is 10^3 per kmol.

Material constants:
$M_{Al} = 27.0\,kg/kmol$
$M_{Ti} = 47.9\,kg/kmol.$

The interphase energy σ between the Al melt and the Al_3Ti grains is $0.090\,J/m^2$. The density of Al_3Ti is $3.70 \times 10^3\,kg/m^3$.

References

1. M. A. Larsson and J. Garside, Solute Clustering and Interfacial Tension, *Journal of Crystal Growth*, **76**, 1986, 88–92.
2. R. C. Tolman, *Journal of Chemical Physics*, **17**, 1949, 333.
3. S. Ono and S. Kondo, Molecular Theory of Surface Tension in Liquids, in *Handbuch der Physik, Band X*, Berlin, Springer, 1960, 134.
4. D. Turnbull and B. Vonnegut, *Industrial Engineering Chemistry*, **44**, 1952, 1292.
5. B. Bramfitt, The effect of carbide and nitride additions on the heterogeneous nucleation behavior of liquid iron, *Metallurgical Transactions* **1**, 1970, 1987–1995.
6. H. Fredriksson and U. Åkerlind, *'Physics of Functional Materials'* Chichester, England, John Wiley & Son, 2008.

5

Crystal Growth in Vapours

Solidification and Crystallization Processing in Metals and Alloys, First Edition. Hasse Fredriksson and Ulla Åkerlind.

5.1 Introduction

The phenomenon of crystal growth in a gas or a supersaturated vapour has been known and studied for more than a hundred years. The mechanisms and kinetics of such crystal growth are very complicated. The first successful attempts of explanation and understanding were made as late as the 1950s and 1960s.

The importance of the research field increased dramatically after the development of the semi-conductor industry. The production of semi-conductors is entirely based on the production of doped thin films of even thickness. A great number of practical methods for production of thin films have been published in recent years. There are also applications in metallurgy, for example production of wear-resistant coatings on cutting tools.

In this chapter we will discuss various types of methods to grow crystals in vapours and to produce continuous thin films (single crystals) or thin films of small crystals, so-called crystallites, together with the brief outlines of the associated present theoretical knowledge.

The crystal growth in vapours is influenced by many factors, such as for example temperature, supersaturation of the vapour, diffusion, crystal structure, occurrence of chemical reactions and impurity content.

At the end of the chapter the mechanical restrictions on thin films are mentioned briefly.

The basic terminology of crystal structure, i.e. nomenclature of crystallographic directions and planes in space, is treated in Chapter 1 in the book 'Physics of Functional Materials' [1]. A study of this section in [1] or any other equivalent literature is highly recommended.

5.2 Crystal Morphologies

Many different crystal morphologies (shape and structure) can be observed during crystal growth from a vapour phase. The most well known crystals are ice or snow crystals grown from water vapour. Figure 5.1 gives an example of snow crystals.

Ice crystals occur in many different morphologies other than the one shown in Figure 5.1. They have also been observed as hexagonal plates and hexagonal columns. Experiments show that the morphology is strongly related to the growth conditions, i.e. the supersaturation of the water vapour as well as the growth temperature. This is a general observation, valid for all sorts of crystals.

Volmer and Eastermann [2] primarily studied the growth of crystals from metal melts at the beginning of the twentieth century. They studied the growth of Hg crystals with an equipment like that in Figure 5.2.

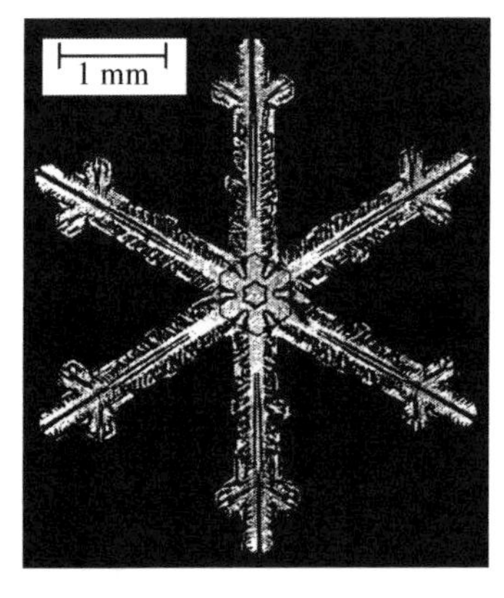

Figure 5.1 Snow crystal with dendritic structure photographed in reflected light. Courtesy of J. Gilman.

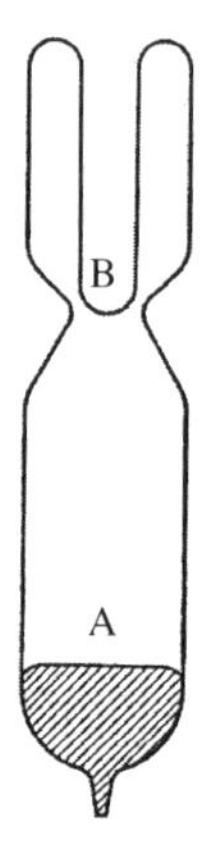

Figure 5.2 The Volmer-Estermann experiment. In an evacuated glass vessel Hg atoms evaporate from A (high temperature) and condense at B (low temperature). Figure provided by Dr Robert F. Mehl Copyright reserved.

During the 1930s Straumanis [3] studied the growth of Zn, Cd and Mg crystals by a method similar to that used by Volmer. He found that most crystals were faceted. Figure 5.3 shows pictures of three different Mg crystals with HCP structure. One can see that the crystals are bounded by (0001), $(10\bar{1}0)$ and $(10\bar{1}1)$ planes (pages 21–23 in Chapter 1 in [1]). He found that the growth temperature and the supersaturation of the Mg vapour determined the morphologies of the Mg crystals.

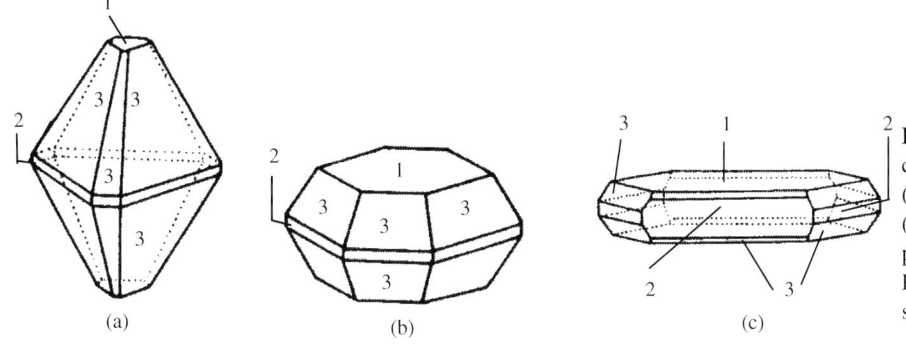

Figure 5.3 Shapes of magnesium crystals grown in water vapour (Straumanis 1931). (1) basal planes; (2) prism planes; (3) pyramidal planes. Straumanis, M. in Z. f. Physik. Chem. © Oldenbourg Wissenschaftsverlag GmbH 1931.

The anisotropy of the growth rate influences the formation of faceted crystals. This means that the growth rate of the crystal varies with the crystallographic direction. The crystallographic planes with the lowest growth rates are those that form the crystal surfaces. This is shown schematically in Figure 5.4 where the growth rate is faster in the [110] direction than in the [010] and [100] directions (page 21 in Chapter 1 in [1]).

The early experiments also showed that lateral growth of steps over the surfaces cause the crystal growth. The heights of these steps are frequently more than a hundred atomic distances. In some cases, columnar-like crystals, so-called whiskers, are found.

The mechanisms of various types of crystal growth will be discussed later in this chapter.

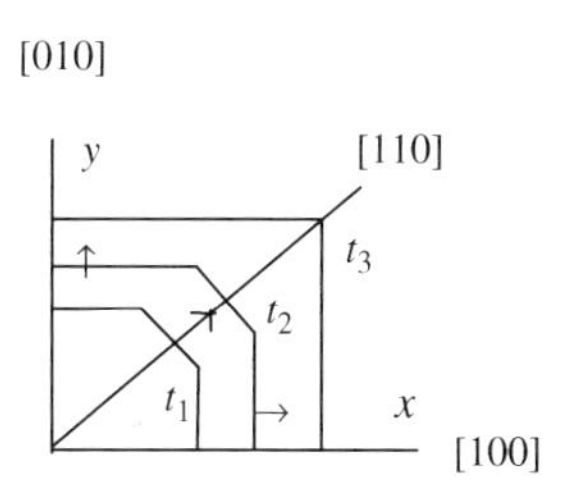

Figure 5.4 Crystal surfaces of part of a crystal at three different times. $t_1 < t_2 < t_3$. The surface with the fastest growth disappears.

5.3 Chemical Vapour Deposition

A large group of methods to produce crystals and thin films involve chemical reactions. The crystallization process is based on the chemical reactions in combination with a controlled temperature distribution of the apparatus. The reactions also supply material for the crystallization, which occurs separated from the source zone.

The chemical vapour deposition (CVD) methods have a number of great advantages:

- If a chemical reaction is involved the crystallization conditions become greatly facilitated. The chemical reaction is reversible and the crystallization process occurs near the equilibrium state. This provides a low supersaturation and a constant stoichiometric ratio of the reactants formed during the reaction.
- With CVD methods, a much larger variety of substances can be crystallized in a vapour phase than with the use of other methods. Methods that are more direct often require a high vapour pressure of the substance to induce crystallization. This condition would exclude the crystallization of a large number of substances, for example silicon and diamond, unless the chemical reactions are involved.

The CVD processes offer a highly efficient method to produce nearly perfect crystals and thin films owing to the qualities mentioned above. CVD processes are used in a wide range of applications both in metallurgical and in the electronics industry. In wear-resistant applications, many different carbides and nitride coatings, e.g. TiC, TiN and ZrC, are used. The temperature of the heated surface in the evacuated reactor is usually high, often close to the melting point of the growing crystals.

The large group of CVD methods can roughly be divided into three subgroups:

1. chemical transport methods;
2. vapour decomposition methods;
3. vapour synthesis methods.

The macroscopic characteristics of each subgroup method is briefly described below, with no attention to the mechanism of the CVD processes. The deposits grow by so-called *epitaxial growth*, which will be analyzed in Section 5.8.

5.3.1 Chemical Transport Methods

Consider a solid substance A with a low vapour pressure at the working temperature. It reacts chemically with a gas component B. The reaction products C and D are gases. They form when the reversible reaction proceeds from the left-hand to the right-hand side.

$$\mathrm{a\,A(s) + b\,B(g) \leftrightarrow c\,C(g) + d\,D(g)} \tag{5.1}$$

The reaction products form in the so-called source zone and become transferred to another zone, which has a different temperature. The chemical reaction now proceeds from the right to the left and the initial substances A and B form. Substance A crystallizes at a substrate. The result is that A has been transferred from the source zone and precipitated at a new place. The process is called *chemical transport*. It can be used in several different ways.

Substance B is called the transporting agent. Any substance that can react with substance A and give gaseous reaction products, can be used as a transporting agent. *Halogens*, usually I_2 but also Br_2 and Cl_2, *hydrogen halides*, usually HCl, *elements of group* 6 b *and* 5 b (S, Se, Te and As, P) and various chemical compounds, for example $SiCl_4$, are used.

Some typical examples of transport reactions used in crystal growth are

$$\mathrm{WCl_6 + 3\,H_2 \leftrightarrow 6\,HCl + W}$$

This reaction was used in the early days of the lamp industry to coat graphite fibres with tungsten.

This method is often used to produce metallic whiskers (Section 5.9) with remarkably good mechanical properties. For copper whiskers, for instance, the following reactions are used

$\mathrm{2Cu + 2\,HCl \leftrightarrow 2\,CuCl + H_2}$ at 430–800 °C
$\mathrm{2Cu + 2\,HBr \leftrightarrow 2\,CuBr + H_2}$ at 600 °C

There are a large number of equipments for chemical transport. Figure 5.2 on page 202 and Figure 5.5 show some examples of closed growth cells for condensation of metal vapours.

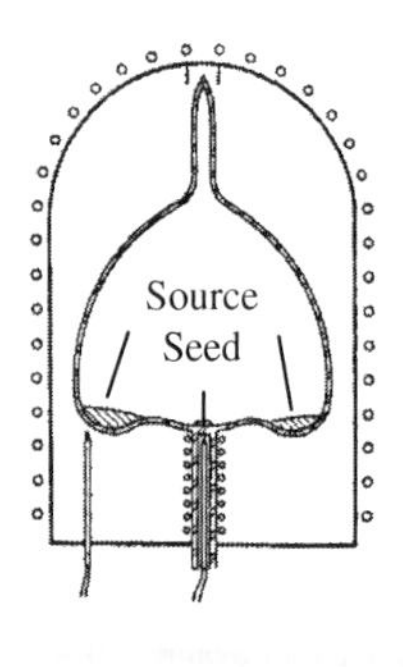

Figure 5.5 Crystallization ampoule with a radial gradient. Thermocouples control the temperature. Reproduced from Springer © 1984.

The simplest equipment for the chemical transport methods is a sealed ampoule with a certain amount of transporting agent included (for example Ar, H_2 or He). In Figure 5.5 the material to be transported is placed in the circular source zone and the deposition occurs in the central crystallization zone at lower temperature. The process is cyclic as the transporting agent, liberated at the reaction in the central zone, diffuses back to the source zone and the process starts again. The upper tip of the ampoule is cut off to introduce a seed into the crystallization zone. With this method single crystals of high purity are produced. Alternatively, open methods are also used. They are more flexible than the closed methods as they allow introduction of impurities (doping materials) and easier variation of the growth conditions. Figures 5.6a and b give some examples.

Diamond is one of the hardest materials known. Artificial diamond crystals are used extensively in industry. Silicon is essential for the semiconductor industry. For this reason, much research and great efforts have been devoted to finding good production methods of diamond and silicon crystals (pages 247–249). The most common methods are described in Figures 5.6a and b.

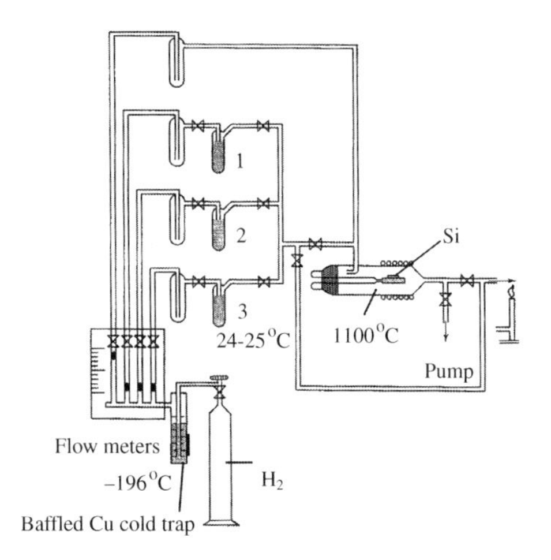

Figure 5.6a Equipment for chemical transport by open methods. Radio frequency-heated Si deposition system used for among other things teflon deposition. A gas mixture of H_2 and $SiHCl_3$ reacts in a chamber at 1100 °C. Courtesy of J. Gilman.

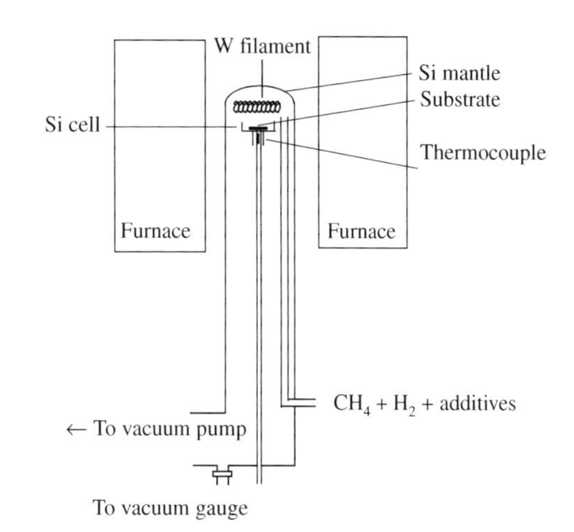

Figure 5.6b Equipment for production of diamond thin films. Near a hot tungsten filament, at a temperature of 2000 °C, a gas mixture of methane and dissociated hydrogen reacts and gives a deposition of diamond on a Si substrate. Reproduced from Elsevier © 1994.

5.3.2 *Vapour Decomposition Methods*

All vapour decomposition methods use a flow of a gas or a gas mixture. Examples are given in Figures 5.6a, 5.6b and 5.7. When the gas enters the high-temperature zone a chemical reaction occurs and the desired substance is one of the reaction products, while the others are carried away by the gas flow. The chemical reaction may be complex and often involves intermediate compounds.

The chemical reactions are of two kinds:

- reduction reactions
- thermal decomposition reactions.

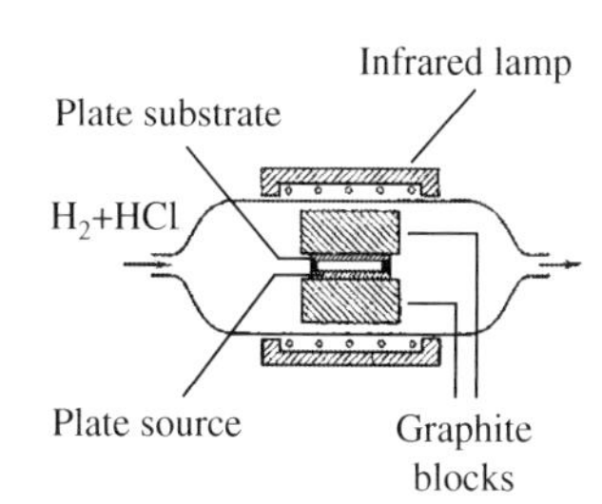

Figure 5.7 Crystallization with the aid of the sandwich process. Two parallel plates at a distance of some tenths of a mm from each other are kept at a temperature difference. The method has several advantages over the ampoule methods: higher transport rate and reduced risk of contamination. It is comparatively easy to control the process due to the low gas flow. Reproduced from Springer © 1984.

A typical example of the first type of reaction is reduction of silicon tetrachloride with hydrogen

$$SiCl_4 + 2\,H_2 \leftrightarrow Si + 4\,HCl$$

which is one way of producing thin Si films. Hydrogen serves as both the reducing agent and carrier gas. The deposition temperature is 1100 °C. n- and p-type films can be produced by adding vapours of doping element compounds such as $AsCl_3$ and BBr_3, respectively. To avoid undesired impurities it is advisable to use the lowest possible temperature.

Thermal decomposition is also called *pyrolysis*. Decomposition of silicon tetraiodide may serve as an example:

$$SiI_4 \leftrightarrow Si + 4\,I$$

A convenient deposition temperature for this reaction is 1000 °C.

5.3.3 *Vapour Synthesis Methods*

The vapour synthesis methods are used for production of *compound crystals*. A mixture of vapours of the elements involved or their compounds are carried to a high temperature zone where a chemical reaction occurs and the desired compound is formed.

The most important classes of compounds are carbides and semicompounds consisting of one element from group 3 and one element from group 5 in the periodic table.

A typical example of the first group is the production of silicon carbide from silicon tetrachloride and methane.

$$SiCl_4 + CH_4 \leftrightarrow SiC + 4\,HCl$$

Vapour synthesis methods are frequently used in the semiconductor industry. A very representative example is the production of gallium arsenide by the overall reaction

$$2\,Ga + 2\,AsCl_3 + 3\,H_2 \leftrightarrow 2\,GaAs + 6\,HCl$$

The reaction occurs often in a quartz tube placed in a furnace with three temperature zones.

5.4 Crystal Growth

The basic theory of transformation kinetics and chemical reactions is given in Chapter 5 in [1]. The chapter also includes a section on reaction rates and the kinetics of homogeneous reactions in gases. This domain constitutes useful reading as a complement to the text in this chapter and we will refer to it now and then below.

Growth rate has the dimension of velocity and is here defined as

- Growth rate = rate of thickness change of deposition layer.

Hitherto, we have given a review of the chemical vapour deposition methods of crystal production without discussing their theoretical background. To understand the mechanisms of crystal growth we have to consider the topic from an atomistic and thermodynamic point of view.

5.4.1 Simple Model for Growth Rate

All chemical vapour deposition methods are based on a process where the vapour is produced in a source zone at a given temperature, carried away to the deposition zone at another temperature where the crystallization occurs.

The overall growth rate of the complex crystallization process is a function of

- the mass transport rate, i.e. the rate of delivering the reactants to the growing crystal and the removal of the reaction products into the gas phase (*diffusion rate*)
- the rate of the processes on the growing surface, i.e. the adsorption of the reactants, their chemical interaction on the surface and the desorption of the reaction products (*kinetic rate*).

The mass transport occurs normally by diffusion of atoms to and from the interface. If the rate of mass transport (*diffusion rate*) is slow, the chemical reaction also slows down and therefore the overall rate of crystallization also will be low.

The rate of the surface processes is called the *kinetic rate* with a common name. The overall vapour deposition rate is a function of both the diffusion rate and the kinetic rate and the slowest one mainly determines the overall rate. The circumstances in each case determine which one of the two rates that dominates.

Diffusion Rate and Kinetic Rate of Crystal Growth. Driving Force of Surface Crystallization

The kinetic laws normally control the growth rate or the thickening of the deposition layer of a crystal. In simple cases this can be written as

$$\text{mass transport rate} = k(-\Delta G) \tag{5.2}$$

where

k = reaction rate coefficient

$-\Delta G$ = driving force of crystal growth.

The law is assumed to be valid for both the total growth process and for each substep, i.e. the diffusion process and the kinetic interface process. The latter two processes are said to be coupled in series.

In Chapter 2 on page 43–44 we found that the driving force of solidification is proportional to the undercooling ΔT. Analogously, it is natural to assume in this case that the driving force of crystal growth is proportional to the supersaturation of the vapour or to the concentration difference. All the way through this chapter volume concentrations will be expressed in the unit mole fraction x.

To express the driving forces as functions of concentrations in a binary vapour we make some rough assumptions as a first approximation (Figure 5.8).

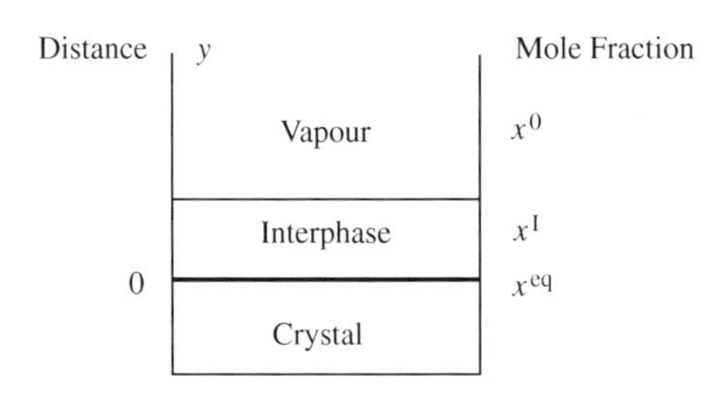

Figure 5.8 Interphase between crystal and vapour. The interphase is a Guggenheim surface.

- The mole fraction x^0 of the solute in the vapour is independent of the distance from the interphase.
- The mole fraction x^I of the solute within the interface is constant.
- The mole fraction of the solute at the surface of the crystal is equal to the equilibrium value x^{eq} obtained when the vapour and the crystal are in equilibrium with each other.

The driving force of the total growth process can then be written as

$$-\Delta G^{total} = const(x^0 - x^{eq}) \tag{5.3}$$

The mass transport or the diffusion to and from the interphase occurs between the vapour and the interphase. The driving force of the diffusion process can be written as

$$-\Delta G^{diff} = const(x^0 - x^I) \tag{5.4}$$

The kinetic interface processes occur at the surface of the crystal. The driving force of the kinetic processes can be written as

$$-\Delta G^{kin} = const(x^I - x^{eq}) \tag{5.5}$$

where
x^0 = bulk concentration of the solute in the vapour
x^{eq} = equilibrium concentration of the solute at the surface of the crystal
x^I = concentration of the solute within the interphase.

The total concentration difference can be separated into two parts corresponding to the two substeps:

$$x^0 - x^{eq} = \left(x^0 - x^I\right) + \left(x^I - x^{eq}\right) \tag{5.6}$$

Using Equations (5.3), (5.4) and (5.5) the concentration differences in Equation (5.6) can be replaced by the driving forces and we obtain

$$\left(-\Delta G^{total}\right) = \left(-\Delta G^{diff}\right) + \left(-\Delta G^{kin}\right) \tag{5.7}$$

In the steady state the transport rate must be equal to the kinetic rate of the crystallization process as material does not accumulate anywhere in the interface. This statement can be expressed in mathematical terms by applying Equation (5.2)

$$k_{total}\left(-\Delta G^{total}\right) = k_{diff}\left(-\Delta G^{diff}\right) + k_{kin}\left(-\Delta G^{kin}\right) \tag{5.8}$$

The Equations (5.8) are introduced into Equation (5.7), which gives

$$\frac{1}{k_{total}} = \frac{1}{k_{diff}} + \frac{1}{k_{kin}} \tag{5.9}$$

where
k_{total} = overall coefficient for crystallization
k_{diff} = diffusion coefficient for crystallization
k_{kin} = kinetic coefficient of interface reaction.

The quantity k_{diff} on the right-hand side in Equation (5.9) describes the material transport from the vapour to the interphase and vice versa. The dominating process is diffusion, but other mechanisms, such as convection and molecular flow, may also contribute to the transport and are included in the diffusion coefficient.

Both k_{kin} and k_{diff} depend on temperature. In Chapter 5 in [1] [Equation (5.82) on page 243] we found that the reaction rate coefficient depends exponentially on the temperature. The same temperature dependence is valid for k_{kin}.

$$k_{kin} = const\, e^{-\frac{G_{a\,act}}{k_B T}} = const\, e^{-\frac{G_{a\,act} N_A}{RT}} \tag{5.10a}$$

As $G = H - TS = U + pV - TS$ and the term pV is small for the crystal and at the interface it can be neglected. The exponential entropy term is often included in the constant, which will be the case in our further treatment of the interface kinetic reactions. Therefore $G_{a\,act}$ can be replaced by $U_{a\,act}$ in Equation (5.10a), which can be written as

$$k_{kin} = A\, e^{-\frac{U_{a\,act}}{k_B T}} = A\, e^{-\frac{U_{a\,act} N_A}{RT}} \tag{5.10b}$$

where subscript 'a' means 'per atom'
and
k_B = Boltzmann's constant
$G_{a\,act}$ or $U_{a\,act}$ = the activation energy per atom of the chemical reaction
N_A = Avogadro's number
R = the gas constant
T = the absolute temperature.

These arguments are valid here and in many of the equations in this chapter. They will be used without further explanation.

k_{diff} is associated with mass transport and diffusion. The diffusion coefficient D_v in a vapour is proportional to $T^{1/2}$. Therefore, k_{diff} grows slowly with increasing temperature while k_{kin} increases more rapidly. The magnitude of the activation energy indicates that in most cases a temperature increase of, for example, 100 K increases the reaction rate k_{kin} by a factor of 3 to 4, while the diffusion rate k_{diff} increases only slightly.

If $k_{kin} \ll k_{diff}$ at relatively low temperatures the crystallization process runs in the *kinetic mode* according to Equation (5.9) ($k_{total} \approx k_{kin}$). At higher temperatures k_{kin} increases strongly and dominates over k_{diff} and the crystallization process changes from the kinetic mode to the *diffusion mode* ($k_{total} \approx k_{kin}$).

It is always the slowest subprocess that controls the overall rate of a crystallization process. The overall rate can be varied by changing the three variables temperature, concentration and flow rate. The temperature variation is the most important instrument to control the crystallization process.

It is often desirable to keep the temperature as low as possible. The equipments become simpler and the fraction of defects in the crystals can be controlled better in this way. The solubilities of impurities decrease with temperature.

Determination of the Rate Constants

The driving forces of the diffusion and the kinetic subprocesses have been discussed briefly above.

The rate constant of the *diffusion* process has been analyzed in Section 5.5.4 on page 248 in Chapter 5 in [1].

The rate constant of the *kinetic* process at the interphase can be derived from the activated-complex theory (Section 5.5.5 on page 250 in Chapter 5 in [1]). This constant depends of the structure of the interface. This topic will be discussed in this chapter in Sections 5.5 and 5.6.

Relationship between Mass Transport Rate and Growth Rate

Mass transport rates are often expressed in terms of *the number of atoms transferred to the crystal per unit area and unit time*. In order to describe the growth rate as a change of thickness per unit time, the coefficients of surface crystallization introduced in the previous section have to be multiplied by the molar volume and divided by Avogadro's number. Hence, we obtain the following expression of the growth rate as a function of the coefficient of crystal growth and the driving force of crystal growth [Equation (5.2) on page 206]

$$V_{\text{growth}} = \frac{V_{\text{m}}}{N_{\text{A}}} k_{\text{total}} \left(- \Delta G^{\text{total}} \right) = a^3 k_{\text{total}} \left(- \Delta G^{\text{total}} \right) \tag{5.11}$$

where

V_{growth} = growth rate of solidification front or change of thickness per unit time
a = atomic distance (lattice constant)
k_{total} = overall coefficient of surface crystallization
N_{A} = Avogadro's number
V_{m} = molar volume of the crystal.

The constant $V_{\text{m}}/N_{\text{A}}$ is equal to average atomic volume a^3.

5.4.2 *Crystal Growth on Rough and Smooth Surfaces*

A crystal grows by addition of 'building blocks', normally atoms, from the ambient supersaturated liquid or vapour phase. The growth mechanism depends strongly on the structure of the crystal surface. The simple model presented in Section 5.4.1, pays no attention to this fact. It is desirable to analyze the effect of surface structure and find more satisfactory models for the crystallization process.

In Chapter 3 on page 114 we found that there are two different types of interfaces between the crystal and its ambient phase

- rough or diffuse interphases (Guggenhem interphases);
- smooth or singular interfaces (Gibbs interfaces).

In Chapter 3 the two types of interfaces have been described. In this chapter, we will distinguish between *rough*, *smooth* and *vicinal* (terraced) interfaces, respectively, in contact with a vapour phase. We will use the definition of a Gibbs interface on smooth interfaces and the definition of a Guggenheim interphase on rough interphases.

The rough interphases have an extension of one or several atomic layers and contain plenty of kinks (Figure 3.48 on page 134 in Chapter 3).

The smooth or vicinal interfaces (Section 3.5.2 on page 149 in Chapter 3) are, as the name indicates, atomically flat. The effect of the surface structure on crystal growth has been discussed since the 1920s. Today, the generally accepted view is that crystal growth is very much easier on rough interphases than on smooth interfaces.

In 1927 Kossel [4] pointed out the problems connected with crystal growth on smooth interfaces. One of his drawings of 'cubic' atoms on crystal interfaces is shown in Figure 5.9. He claimed that the

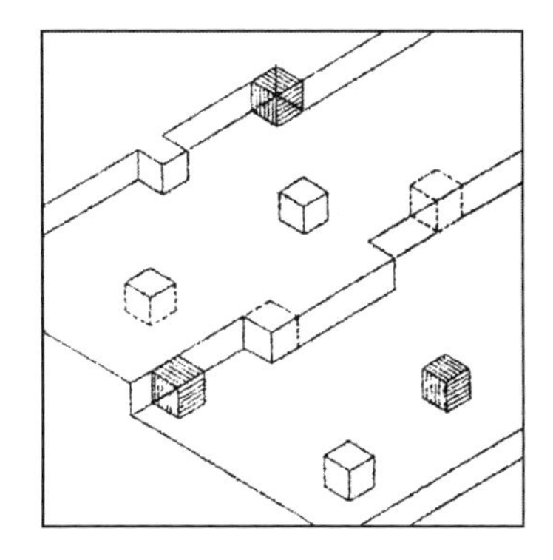

Figure 5.9 One of Kossel's original drawings of 'cubic' atoms. Kossel assigned different energies to the different parts of the crystal.

adsorbed atoms (called *ad-atoms*) on the crystal surface from the vapour phase form a step consisting of a monolayer of atoms and that the easiest way for a vapour atom to be added to a crystal is to be incorporated into the crystal by a kink. No extra energy was supposed to be required in that case.

Owing to the fundamental differences in structure between rough and smooth crystal surfaces the mechanisms of crystal growth are completely different in the two cases. They will be treated separately in Sections 5.5 and 5.6, respectively.

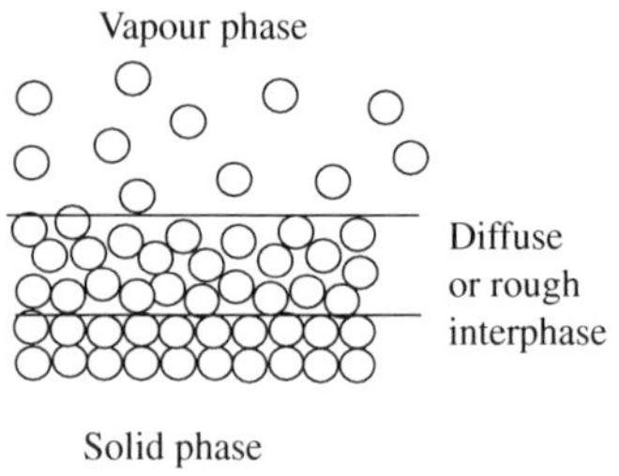

5.5 Normal Crystal Growth of Rough Surfaces in Vapours

Crystal growth of rough surfaces is often *dendritic growth* and is called *normal crystal growth* because it is the most common process.

A rough interphase extends over several atomic layers as shown in Figure 5.10. The surface is built up by layers with a large number of kinks.

It is easy to understand that growth on this type of interphases is much easier than on smooth ones.

Vapour phase

Diffuse or rough interphase

Solid phase

Figure 5.10 Rough interphase. © Pearson PLC 1993.

5.5.1 Driving Force and Activation Energy

By definition, an atom becomes part of the crystal if its Gibbs' free energy is equal to the Gibbs' free energy of the crystal. This is the case when an atom from the vapour phase occupies a kink (see Figure 5.9 on page 209) in a rough crystal surface. The bonds between the atom and the crystal lattice are strong. The crystallization process can be described graphically in a Gibbs' free energy diagram.

An atom in the vapour phase has a higher Gibbs' free energy (the level to the right in Figure 5.11) than the crystal (the level to the left). The difference in Gibbs' free energy between the supersaturated vapour phase and the crystal is the *driving force* ($-\Delta G^{\text{total}}$) of the crystallization process (page 240 in Chapter 5 in [1]). The larger the supersaturation is, the larger will be the driving force.

Before the vapour atom becomes incorporated in a kink position, i.e. in a vacancy at the crystal surface it has to overcome a barrier U_a. This barrier is the *activation energy* (page 239 in Chapter 5 in [1]) of the crystallization process. In the case of chemical vapour deposition it is caused by the chemical reaction that requires energy to start.

These concepts will be used below to derive the growth rate of a rough surface.

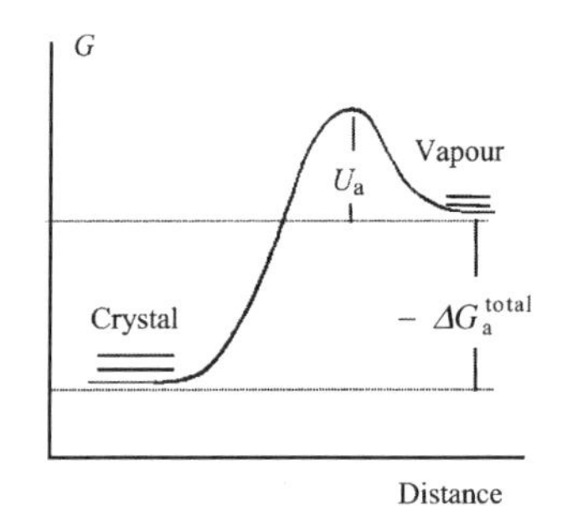

Figure 5.11 Gibbs' free energy levels of a crystal with a rough interphase and its vapour phase as a function of position.

As is indicated in the figures on page 240 in Chapter 5 in [1] all vapour molecules do not have the same energy. Their energies represent a Boltzmann distribution, which depends on the temperature. In such cases, the effect is that the activation energy corresponds to a narrow energy interval. For atoms, there is a single activation energy.

5.5.2 Rate of Normal Growth

Flux of Vapour Atoms

To derive a more accurate expression of the growth rate than in Section 5.4.1 on page 206 it is necessary to know the flux of vapour atoms towards the crystal surface and also the kinetics of the incorporation of atoms at the crystal surface. The growth rate can be derived by means of the kinetic theory of gases.

In the same way as in Section 5.4.1 the growth process can be regarded as two processes in series. The driving force consists of two parts given by Equation (5.7) on page 207

$$\left(-\Delta G^{\text{total}}\right) = \left(-\Delta G^{\text{diff}}\right) + \left(-\Delta G^{\text{kin}}\right) \tag{5.7}$$

In the analysis below we will assume that

- the transport of atoms from the vapour phase to the interphase is very fast
- the number of vapour atoms reaching the surface per unit area and unit time equals the collision frequency between the vapour atoms and the crystal surface.

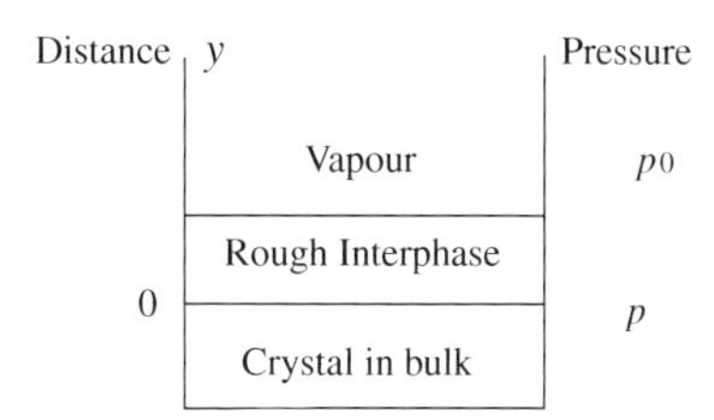

Figure 5.12 Vapour crystal interphase.

Consider a crystal with a rough interphase in a vapour phase (Figure 5.12). The flux of atoms is defined as

- Flux of atoms = the number of atoms that pass the unit area per unit time.

The number of atoms can be calculated by considering the volume of the vapour shown in Figure 5.13. The cross-sectional area of the box is 1 unit area and its length is the root mean square velocity v_{rms} of the vapour atoms multiplied by 1 unit of time.

According to the kinetic theory of gases (Equation (4.12) on page 173 in Chapter 4 in [1]) the root mean square velocity equals

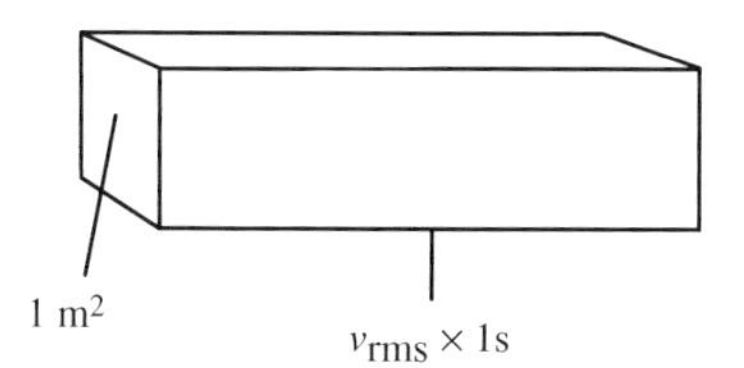

Figure 5.13 Vapour box.

$$v_{\text{rms}} = \sqrt{\frac{3RT}{M}} \tag{5.12}$$

In the absence of a concentration gradient towards the interphase ($\Delta G^{\text{diff}} = 0$) all directions are equivalent. It is reasonable to assume that one sixth of the vapour atoms move in a direction perpendicular to the cross-section towards the surface. Hence, the number of atoms per unit area and unit time, which hit the crystal surface, will be

$$\text{Incoming flux} = j_{\text{i}} = v_{\text{rms}} \frac{n}{6} N_{\text{A}} \tag{5.13}$$

where

j_{i} = number of atoms per unit area and unit time that hit the crystal
v_{rms} = root mean square velocity of vapour atoms
n = number of kmol per unit volume of vapour
N_{A} = Avogadro's number
nN_{A} = number of atoms per unit volume of vapour.

n can be expressed in terms of pressure p by means of the gas law, applied on $V = 1$ unit volume:

$$p \times 1 = nRT \tag{5.14}$$

where p is the pressure close to the crystal surface.

Equation (5.13) in combination with Equation (5.14) and use of the relationships $R = k_{\text{B}} N_{\text{A}}$ and $M = m\,N_{\text{A}}$ gives an approximate value of the total incoming flux that reaches the interphase

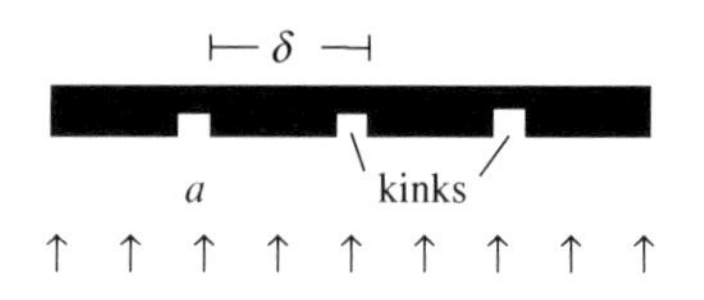

Figure 5.14 Principle sketch of the simplest possible rough surface (Guggenheim mono-atomic layer). The probability that a cubic ad-atom will be incorporated into the crystal is equal to $a/\delta \times a/\delta = a^2/\delta^2$ where; a = the length of the side of the cube; δ = average distance between the kinks. Figure 3.52 on page 139 in Chapter 3 shows a more general model of a rough surface ($a \approx \delta$).

$$j_i = \sqrt{\frac{3RT}{M}}\frac{p}{6RT}N_A = \frac{p}{\sqrt{12\, mk_BT}}$$

where

j_i = total incoming flux
p = pressure of the vapour close to the crystal surface
R = the gas constant
T = absolute temperature
M = molecular weight
k_B = Boltzmann's constant
m = mass of a vapour atom
a = atomic distance = lattice constant
a^2 = area of a kink.

The SI unit of the flux j is number/m^2 s. A more accurate derivation gives

$$j_i = \frac{p}{\sqrt{2\pi\, mk_BT}} \tag{5.15}$$

Of the total number of atoms, which move towards the surfaces only a certain fraction will be incorporated into the crystal. The reason for this is that all sites on the surface do not have the same capacity to incorporate atoms and that the vapour atoms have to adjust their sizes in the crystal and in the interphase. These effects are in most models described by a probability factor a^2/δ^2 (Figure 5.14). In addition, only the fraction $e^{-\frac{U_{a\,act}}{k_BT}}$ of the incoming atoms have energy enough to pass energy barrier (compare Equation (5.81) Section 5.4 on page 242 in [1]). Hence, the incoming flux j_+ that becomes incorporated into the crystal surface, can be written as

$$j_+ = \frac{p}{\sqrt{2\pi\, mk_BT}}\frac{a^2}{\delta^2}e^{-\frac{U_{a\,act}}{k_BT}} \tag{5.16}$$

The reverse flux j_-, i.e. the number of atoms from any site on the crystal surface that leave the unit area per unit time can be written as

$$j_- = const\, e^{-\frac{-\Delta G_a + U_{a\,act}}{k_BT}} \tag{5.17}$$

where $(-\Delta G_a + U_{a\,act})$ is the energy barrier (see Figures 5.15 and 5.11 on page 212). The value of the constant is determined by the equilibrium condition $j_+^{eq} = j_-^{eq}$ at $p = p_{eq}$, where p_{eq} is the vapour pressure in equilibrium with a crystal with an infinitely large surface.

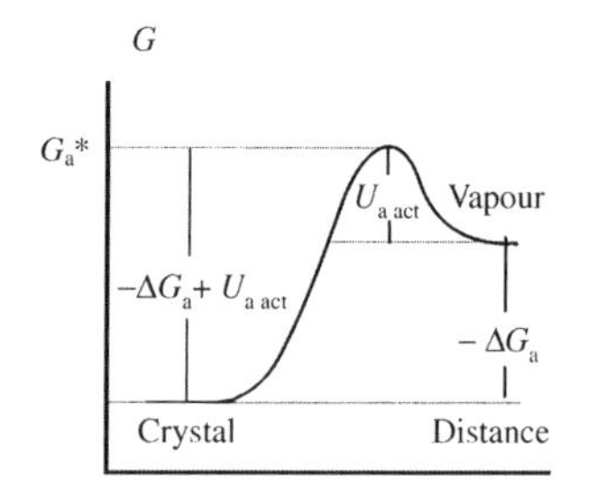

Figure 5.15 Gibbs' free energy levels of the system crystal/vapour.

$$\frac{p_{eq}}{\sqrt{2\pi\, mk_BT}}\frac{a^2}{\delta^2}e^{-\frac{U_{a\,act}}{k_BT}} = const\, e^{-\frac{-\Delta G_a + U_{a\,act}}{k_BT}} \tag{5.18}$$

which gives

$$j_- = \frac{p_e}{\sqrt{2\pi\, mk_BT}}\frac{a^2}{\delta^2}e^{-\frac{U_{a\,act}}{k_BT}} \tag{5.19}$$

Rate of Normal Growth

In order to calculate the rate of normal growth from the net flux of incorporated atoms we need a relationship between these quantities. It is derived in the box.

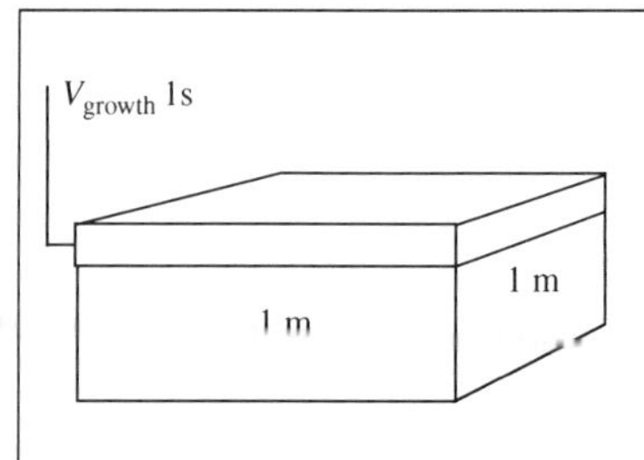

Relationship between j and V_{growth}

The net number of incorporated atoms (cubes with volume a^3 per atom) per second and m^2 is $(j_+ - j_-)$. These atoms form a layer that covers the whole area in question (1 m^2). If the average growth rate is V_{growth} then the layer grows a distance = $V_{growth} \times 1$ s in one second.

We have two expressions of the volume of the incorporated atoms per second:

$$(V_{growth} \times 1) \times 1^2 = a^3(j_+ - j_-)$$

which gives
$V_{growth} = a^3(j_+ - j_-)$.

The relationship between the total growth rate and the net flux of atoms, derived in the box, is

$$V_{\text{growth}} = a^3(j_+ - j_-) \tag{5.20}$$

According to Equations (5.16), (5.19) and (5.20) the total growth rate of the crystal at the pressure p is

$$V_{\text{growth}} = a^3(j_+ - j_-) = a^3\frac{a^2}{\delta^2}\left(\frac{p}{\sqrt{2\pi mk_\text{B}T}}\text{e}^{-\frac{U_{\text{a act}}}{k_\text{B}T}} - \frac{p_{\text{eq}}}{\sqrt{2\pi mk_\text{B}T}}\text{e}^{-\frac{U_{\text{a act}}}{k_\text{B}T}}\right) \tag{5.21}$$

where

V_{growth} = total growth rate of the crystal
$j_{+,-}$ = fluxes of attached and detached atoms
a = atomic distance
δ = average distance between kinks in the interphase
$\frac{a^2}{\delta^2}$ = probability of an incoming vapour atom to find a kink
p = vapour pressure
p_{eq} = vapour pressure in equilibrium with the crystal with an infinitely large surface.

For a *rough* surface $a \approx \delta$. This means that practically all incoming atoms will be incorporated and that the probability factor ≈ 1. For the sake of clarity, it is kept in all formulas.

According to Equation (5.21) the rate of normal crystal growth on a rough interphase can be written as

$$V_{\text{growth}} = a^3\frac{a^2}{\delta^2}\frac{1}{\sqrt{2\pi mk_\text{B}T}}\text{e}^{-\frac{U_{\text{a act}}}{k_\text{B}T}}(p - p_{\text{eq}}) \tag{5.22}$$

The activation energy can easily be determined by measuring the growth rate at several temperatures and plotting the logarithm of the growth rate as a function of $1/k_\text{B}T$. The activation energy is derived from the slope of the straight line (Figure 5.16).

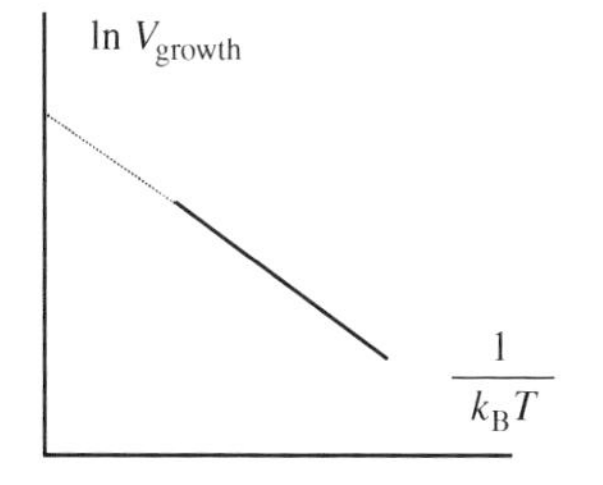

Figure 5.16 Graphical derivation of the activation energy.

Kinetic Coefficient of Growth Rate

The growth rate is proportional to the pressure difference.

$$V_{\text{growth}} = \beta_{\text{kin}}(p - p_{\text{eq}}) \tag{5.23}$$

The constant β_{vapour} is called the *kinetic growth coefficient in vapour*. Identification of Equations (5.22) and (5.23) gives

$$\beta_{\text{kin}} = a^3\frac{a^2}{\delta^2}\frac{1}{\sqrt{2\pi mk_\text{B}T}}\text{e}^{-\frac{U_{\text{a act}}}{k_\text{B}T}} \tag{5.24}$$

It can be seen from Equation (5.24) that the kinetic coefficient

- is proportional to the probability for a vapour atom to find a site on the rough surface, i.e. the probability of the atoms or building units to be incorporated into the crystal surface;
- is proportional to the exponential function of the activation energy for incorporation of a building unit into the crystal lattice.

5.6 Layer Crystal Growth on Smooth Surfaces in Vapours

In an ideal smooth crystal surface, all sites are occupied. Atoms from the ambient vapour phase can not be incorporated, only adsorbed at the surface. The only possibility of crystal growth is

- formation of *steps* or atom layers parallel to the smooth surface by two-dimensional nucleation, or dislocations reaching the surface;
- lateral growth of the steps at their rough edges.

Two factors determine the step growth: the diffusion of atoms towards the step edge and the kinetic incorporation of atoms into the crystal lattice at the step edge. The slowest of these processes determines the rate of step growth. In this section, we will analyze the step growth and set up models for determining the step length λ and the step height h of the steps. This knowledge is necessary for deriving the rate of crystal growth.

New steps form by nucleation on top of each other. Simultaneously, the steps grow by propagation of their edges. A pyramid of terraces is formed (Figure 5.17). The step height can be one, several or a great number of atom layers (macrosteps). The higher the step is, the lower will be the edge propagation. Hence, the velocities of macrostep edges may be very small.

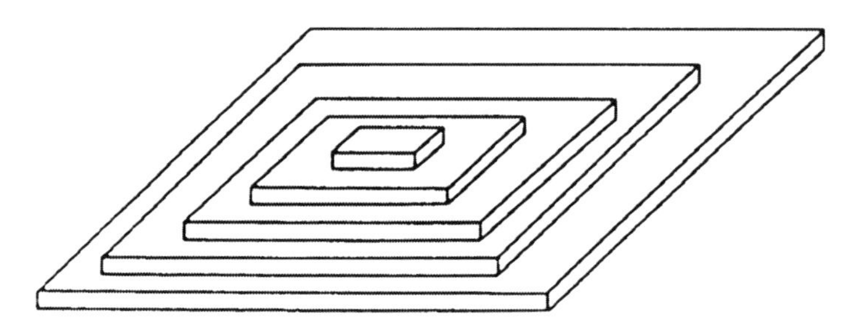

Figure 5.17 Pyramid of terraces. Courtesy of J. Gilman.

If screw dislocations are present in the crystal, they represent another source of steps. Hillocks of spiral steps are formed.

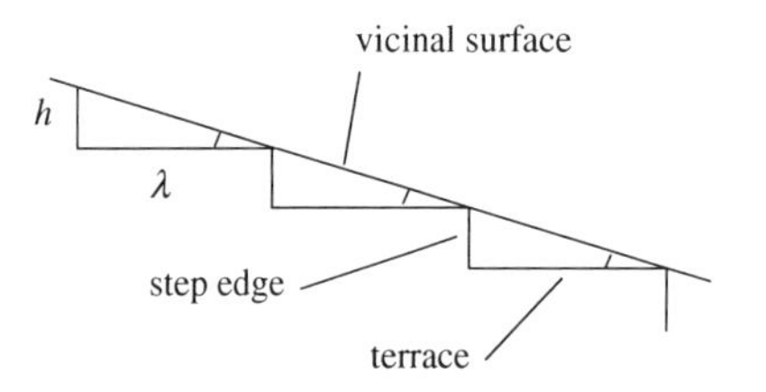

Figure 5.18a Terraces and steps on a growing smooth crystal surface.

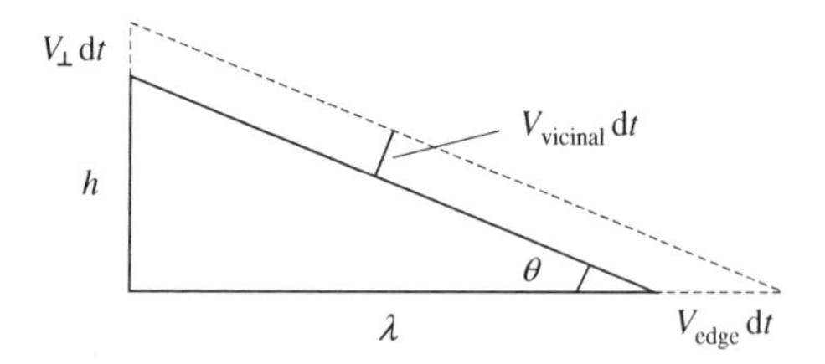

Figure 5.18b Cross-section of a step at $t = 0$ and $t = dt$. The position of the vicinal surface at the time $t = dt$ is marked by a dotted line in the figure.

The growth rate $V_\perp$ in a direction perpendicular to the singular surface and the terraces depends on the height h and width λ and the velocity V_{edge} of the edge movement. Conform triangles in Figures 5.18a and b gives

$$\frac{V_\perp}{h} = \frac{V_{edge}}{\lambda}$$

or

$$V_\perp = h\frac{V_{edge}}{\lambda} \tag{5.25}$$

The growth rate of the vicinal surface (see Figure 5.18b) perpendicular to itself is

$$V_{\text{vicinal}} = V_{\perp}\cos\theta \tag{5.26}$$

5.6.1 *Driving Forces and Activation Energies*

Consider a terraced vicinal crystal surface below its roughening temperature in contact with its supersaturated vapour (Figure 5.19). The system is in incessant movement. Vapour atoms bombard the terraces and steps. The flux of atoms going directly to the steps is much smaller than the flux of atoms to the terraces and the former can be neglected.

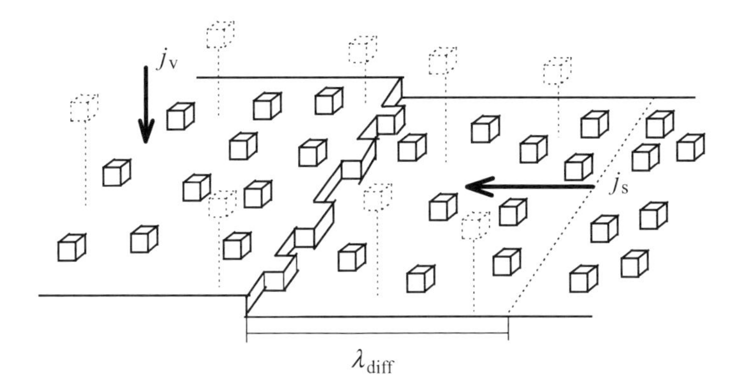

Figure 5.19 Monoatomic layer surrounded by terraces.
j_v = flux of (dotted) atoms from the vapour;
j_s = flux of adsorbed atoms diffusing towards the step;
λ_{diff} = mean distance travelled by the adsorbed atoms on diffusing towards the step during their mean lifetime;
δ = average spacing between the kinks (not indicated in the figure). Reproduced from World Scientific Publishing Co © 1992.

The following processes occur (Figure 5.19):

1. Some of the vapour atoms are adsorbed at the smooth terrace surface.
2. The adsorbed atoms, distributed all over the surface, start to diffuse towards to steps.
3. The atoms that reach a step become incorporated into kink sites.

Driving Force of Surface Diffusion

The random motion of the atoms causes the flux of atoms from the vapour to the surface. The adsorbed atoms become incorporated in kink sites at the steps by the mechanism described in Section 5.5. There remains to explain why the adsorbed atoms on the planar surface diffuse towards the steps.

When adsorbed atoms become incorporated into kink sites a lack of atoms arises close to the step. The consequence is that the surface concentration of adsorbed atoms is constantly low near the steps and a concentration gradient is developed. This concentration gradient is a result of the driving force between the adsorbed atoms on the interface and the step.

Activation Energies

The Gibbs' free energy diagram describing the growth process at a smooth surface differs in several respects from that of a rough surface:

- The interface is a Gibbs' interface (Chapter 3 page 114) and its width is zero.
- There are *three* Gibbs' free energy levels instead of two: one for the vapour, one for the adsorbed atoms and one for the crystal.
- The two activation energies of rough surfaces correspond to *four* activation energies for smooth surfaces (see Figure 5.20).

Two at the smooth interface:

- one for adsorption from and one for desorption to the vapour phase.

Two at the steps:

- one for the incorporation of adatoms into kinks and one for the reverse ejection process.

By definition, an atom becomes part of the crystal if its Gibbs' free energy is equal to the Gibbs' free energy of the crystal. This is the case when an atom from the vapour phase occupies a kink (see Figure 5.9 on page 209) on a rough crystal surface. The bonds between the atom and the crystal are strong.

An adsorbed atom on a smooth crystal surface does not have the same strong bonds to the crystal as an incorporated atom occupying a kink site in a rough surface. It *not* the same Gibbs' free energy and it is *not* a part of the crystal. Its Gibbs' free energy has a value between the Gibbs' free energies of the crystal and the vapour.

Figure 5.20 gives an example of a schematic diagram showing the Gibbs' free energy levels of a smooth surface and some of the designations of the quantities, which will be used in the next section on growth rates, are introduced.

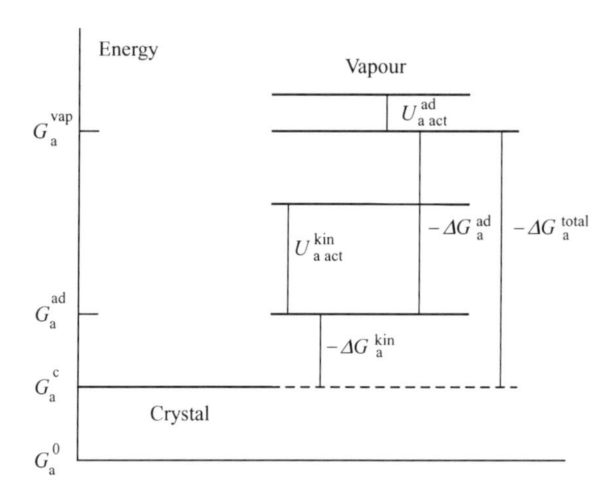

Figure 5.20 Gibbs' free energy levels of the system crystal/vapour during growth. The interface is a Gibbs interface and has the thickness zero.

5.6.2 *Two-Dimensional Nucleation*

Becker initially proposed the theory of two-dimensional nucleation in the early 1930s. We will apply his model for two-dimensional nucleation on smooth Zn interfaces in contact with Zn vapour.

Becker's model deals with the energy that is required to build up a layer of two-dimensional nuclei at the surface of the crystal. The theory of nucleation, given in Chapter 4, is valid and can be applied.

Consider a smooth crystal surface in contact with a vapour. As is shown in Figure 5.20 the incorporation of vapour atoms into the crystal lattice occurs in two steps. The first step is that vapour atoms become ad-atoms and occupy some of the available sites of the crystal surface. The higher the pressure of the vapour and its supersaturation are, the higher will be the number of adsorbed atoms per unit area of the crystal surface. The adatoms on the crystal surface cluster and planar embryos form. If the embryos reach a certain critical size, some of them change into what somewhat inadequately is called a *two-dimensional nucleus*, which continues to grow. The shape of the two-dimensional nuclei is an enlargement of the shape of the bottom plane of the individual adatoms. The growing nuclei cluster and a flat covering monolayer of adatoms forms.

The second step incorporation of the new adatom layer into the surface layer of the crystal, i.e. transfer of all the adatoms into atoms with the same radius and the same Gibbs' free energy as the lattice atoms.

Example 5.1

Zn crystals are growing in a supersaturated Zn vapour at a temperature of 400 °C. Flat two-dimensional nuclei of adatoms form, which grow continuously. The shape of such a nucleus is shown in the upper margin figure. It has the same shape as the bottom plane of a crystal with HCP structure. The height h is assumed to be equal to c, the height of an individual Zn crystal.

Material constants for Zn are given in the margin.

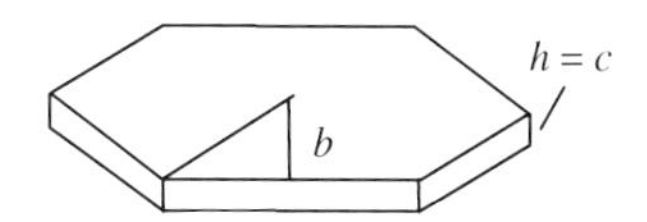

Material constants of Zn and some other constants.

$\sigma = 0.50\ \text{J/m}^2$
$c = 2.66 \times 10^{-10}\ \text{m}$
$M = 65.37$
$\rho = 7.1 \times 10^3\ \text{kg/m}^3$
$R = 8.31 \times 10^3\ \text{J/Kkmol}$
$k_B = 1.38 \times 10^{-23}\ \text{J/K}$

a) Derive the critical size of a two dimensional nucleus (the distance b in the figure) as a function of supersaturation p/p_{eq} The nuclei are formed on the basic plane of already existing Zn crystals.
b) Derive the necessary critical supersaturation p/p_{eq} for homogeneous nucleation of the layer discussed in a). Assume that the fraction of embryos of critical size is 1 embryo per mol (Example 4.1 on page 170 in Chapter 4).

Solution:

a)
The bottom and top areas of the nucleus are:

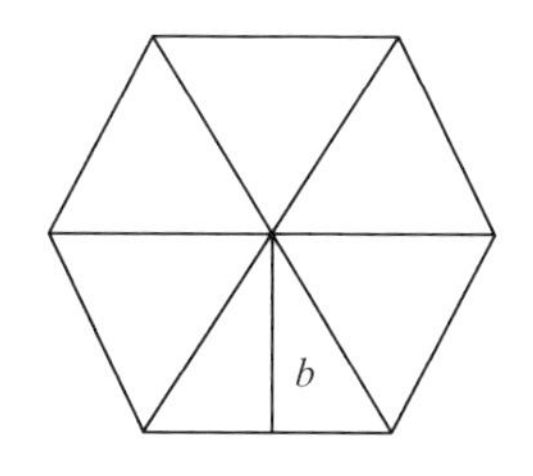

$$A = 6 \times \frac{b}{2} \times \frac{2b}{\sqrt{3}} = 2b^2\sqrt{3} \tag{1'}$$

The edge area is

$$A_{\text{edge}} = 6h \times \frac{2b}{\sqrt{3}} = 4\,hb\sqrt{3} \tag{2'}$$

The edge area can *not* be neglected in comparison with the hexagonal area of the nucleus.
The volume change from zero up to the volume of the flat nucleus is

$$\Delta V = V = Ah = 2\sqrt{3} \times b^2 h \tag{3'}$$

The crystal grows on a substrate of Zn crystals and the net increase of the area is the vertical envelope area

$$\Delta A = \lambda A_{\text{edge}} = 4\,hb\sqrt{3} \tag{4'}$$

because the hexagonal top area replaces the bottom area in contact with the substrate, which disappears. The net change is zero.

To find the critical size of the flat embryo we proceed in exactly the same way as on pages 168–169 in Chapter 4. We set up an expression of the total Gibbs' free energy of the embryo in analogy with Equation (4.2) on page 168 in Chapter 4. It consists of a volume part and an area part. In the present case we obtain

$$\Delta G_{\text{embryo}} = -\frac{\Delta V}{V_m}(-\Delta G_m) + \sigma\,\Delta A \tag{5'}$$

where σ is the interface energy/m^2 of the nucleus. In analogy with Equation (5.16) page 212 we have

$$-\Delta G_m = RT \ln \frac{p}{p_{eq}} \tag{6'}$$

Inserting (3′), (4′) and (6′) into Equation (5′) we obtain

$$\Delta G_{\text{embryo}} = -\frac{2\sqrt{3} \times b^2 h}{V_{\text{m}}} RT \ln \frac{p}{p_{\text{eq}}} + 4hb\sqrt{3} \times \sigma \tag{7'}$$

The critical size of the flat nucleus represents the maximum value of the Gibbs' free energy. We assume the height h of the nucleus is constant and equal to one atomic distance. To find the maximum $\Delta G^*_{\text{nucleus}}$ we use the standard method and take the derivative of the function in Equation (7′). The condition that the derivative is equal to zero gives the value of b^*_{cr}

$$\frac{d\Delta G_{\text{embryo}}}{db} = -\frac{4\sqrt{3} \times b_{\text{cr}} h}{V_{\text{m}}} \times RT \ln \frac{p}{p_{\text{eq}}} + 4h\sqrt{3} \times \sigma = 0 \tag{8'}$$

which gives

$$b^*_{\text{cr}} = \frac{\sigma V_{\text{m}}}{RT \ln \frac{p}{p_{\text{eq}}}} \tag{9'}$$

Inserting values of constants and material constants for Zn give

$$b^*_{\text{cr}} = \frac{0.50 \times (65.37) \times (7.1 \times 10^3)}{8.31 \times 10^3 (273 + 400) \times \ln \frac{p}{p_{\text{eq}}}} = \frac{0.83 \times 10^{-9}}{\ln \frac{p}{p_{\text{eq}}}} \tag{10'}$$

b)
To solve the problem we must find the Gibbs' free energy of a nucleus of critical size. The value of b^*_{cr} is inserted into Equation (7′). After some calculations we obtain

$$\Delta G^*_{\text{nucleus}} = 2\sqrt{3}\, \sigma\, h b^*_{\text{cr}} \tag{11'}$$

The Gibbs' free energy of a nucleus of critical size is related to the temperature by Equation (4.11) on page 170 in Chapter 4

$$\Delta G^*_{\text{nucleus}} = k_{\text{B}} T \ln \frac{n}{n^*} \tag{12'}$$

If we assume that 10^3 embryo per kmol has increased its energy by the critical amount $\Delta G^*_{\text{nucleus}}$ we obtain (Example 4.1 on page 170 in Chapter 4)

$$\Delta G^*_{\text{nucleus}} = 55\, k_{\text{B}} T \tag{13'}$$

Combination of Equations (9′), (10′), (11′) and (13′) and the condition $h = c$ gives

$$55\, k_{\text{B}} T = 2\sqrt{3} \times \sigma c \times \frac{0.83 \times 10^{-9}}{\ln \frac{p}{p_{\text{eq}}}}$$

UI

$$\ln \frac{p}{p_{\text{eq}}} = 2\sqrt{3} \times \sigma c \times \frac{0.83 \times 10^{-9}}{55\, k_{\text{B}} T} \tag{14'}$$

Inserting the numerical values gives

$$\ln \frac{p}{p_{eq}} = \frac{2\sqrt{3}}{55} \times 0.50 \times 2.66 \times 10^{-10} \times \frac{0.83 \times 10^{-9}}{1.38 \times 10^{-23} \times 673} = 0.75$$

Hence, $\frac{p}{p_{eq}} = e^{0.75} \approx 2.12$

Answer:

a) The desired function is

$$b_{cr} = \frac{\sigma V_m}{RT \ln \frac{p}{p_{eq}}} = \frac{0.83 \times 10^{-9}\text{m}}{\ln \frac{p}{p_{eq}}}$$

b) The critical supersaturation p/p_{eq} for homogeneous nucleation is equal to 2.

Surface Concentration of Ad-atoms on a Smooth Crystal Vapour Surface as a Function of the Vapour Pressure

All smooth interfaces have adatoms in equilibrium with the vapour phase. We want to find the relationship between the number of adatoms per unit area and the pressure of the vapour.

Relationship between the Equilibrium Surface Concentration of Ad-atoms on a Smooth Crystal Surface and the Vapour Pressure

When vapour atoms become adsorbed on the crystal surface they will be mixed at random among the available adsorption sites and the mixing energy has to be included in the Gibbs' free energy. The derivation of the Gibbs' free energy is analogous to that of vacancies in Section 2.7.2 on page 84 in Chapter 2 with the difference that we replace y_{vac} with the ratio of the surface concentration of the adatoms at equilibrium to the number of available adatoms sites per unit area n_S^{eq}/n_{0S}

$$G_{total}^{ad} = \frac{n_S^{eq}}{n_{0S}} G^{ad\ p_{eq}} + \left(RT \frac{n_S^{eq}}{n_{0S}} \ln \frac{n_S^{eq}}{n_{0S}} + RT \frac{n_{0S} - n_S^{eq}}{n_{0S}} \ln \frac{n_{0S} - n_S^{eq}}{n_{0S}} \right) \quad (1')$$

where the second and third terms represent the mixing energy term $-TS$ (compare page 132 in Chapter 3). As $n_S^{eq} \ll n_{0S}$ the last term is close to zero and can be neglected.

At equilibrium, G_{total}^{ad} has a minimum, which in this case becomes zero. Hence, we obtain

$$-\frac{n_S^{eq}}{n_{0S}} G^{ad\ p_{eq}} = RT \frac{n_S^{eq}}{n_{0S}} \ln \frac{n_S^{eq}}{n_{0S}} \quad (2')$$

or

$$n_S^{eq} = n_{0S}\, e^{\frac{-G^{ad\ p_{eq}}}{RT}} \approx n_{0S}\, e^{\frac{-U^{ad\ p_{eq}}}{RT}} e^{\frac{TS^{ad\ p_{eq}}}{RT}} = n_{0S} A\, e^{\frac{-U^{ad\ p_{eq}}}{RT}} \quad (3')$$

where A is a constant and equal to the entropy factor $e^{\frac{S^{ad\ p_{eq}}}{R}}$.

Equation (3′) in the box can be written as

$$n_{\mathrm{S}}^{\mathrm{eq}} = n_{0\mathrm{S}} A\,\mathrm{e}^{-\frac{U_{\mathrm{a\,act}}^{\mathrm{ad\,p_{eq}}}}{k_{\mathrm{B}}T}} \tag{5.27}$$

where

$n_{\mathrm{S}}^{\mathrm{eq}}$ = number of adsorbed atoms (adatoms) per unit area on the crystal surface at pressure p_{eq}

$p_{\mathrm{eq}} n_{0\mathrm{S}}$ = number of surface lattice positions per unit area of the crystal surface for adsorption of adatoms

$U_{\mathrm{a\,act}}^{\mathrm{ad\,p_{eq}}}$ = energy required to adsorb a vapour atom on the crystal surface at equilibrium pressure p_{eq}

A = a constant equal to the entropy factor in the box.

Subscript 'a' indicates here and generally in this book that one atom is involved instead of one kmol. Note that subscript or subscript 'm' (for example V_{m}) refers to 1 kmol.

Figure 5.21 describes the Gibbs' free energies of a smooth crystal surface in contact with a vapour. The Gibbs' free energies $G_{\mathrm{a}}^{\mathrm{p}}$ and $G_{\mathrm{a}}^{\mathrm{p_{eq}}}$, which correspond to the pressures p and p_{eq}, are marked in the figure. At the crystal surface there are $n_{0\mathrm{S}}$ available sites per unit area for adsorption of vapour atoms.

Figure 5.21 shows that if the pressure p of the vapour is higher than p_{eq} a lower activation energy than in Equation (5.27) is required for adsorption of vapour atoms.

$$n_{\mathrm{S}} = n_{0\mathrm{S}} A\mathrm{e}^{-\frac{-\Delta G_{\mathrm{a}}^{\mathrm{p}\rightarrow\mathrm{p_{eq}}}}{k_{\mathrm{B}}T}} \tag{5.28}$$

where

n_{S} = number of adsorbed atoms (ad-atoms) per unit area on the crystal surface at pressure p.

$\Delta G_{\mathrm{a}}^{\mathrm{p}\rightarrow\mathrm{p_{eq}}}$ = the activation energy difference $G_{\mathrm{a}}^{\mathrm{p}} - G_{\mathrm{a}}^{\mathrm{p_{eq}}}$

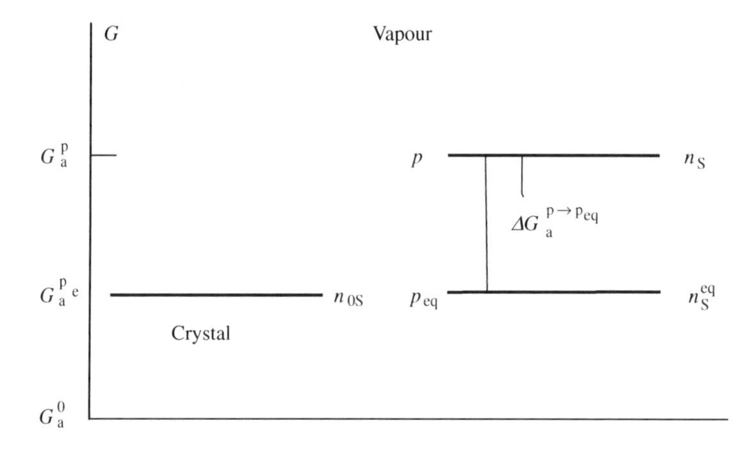

Figure 5.21 Gibbs' free energy of the crystal vapour system as a function of position, provided that the ad-atoms and the crystal surface are in equilibrium with each other.

Analogous to Equation (4.54) and other equations on page 191 in Chapter 4 the Gibbs' free energies can be written as

$$G_{\mathrm{a}}^{\mathrm{p}} - G_{\mathrm{a}}^{0} + k_{\mathrm{D}} T \ln p \tag{5.29}$$

$$G_{\mathrm{a}}^{\mathrm{p_{eq}}} = G_{\mathrm{a}}^{0} + k_{\mathrm{B}} T \ln p_{\mathrm{eq}} \tag{5.30}$$

Subtraction gives

$$\Delta G_{a}^{p \to p_{eq}} = G_{a}^{p} - G_{a}^{p_{eq}} = k_{B} T \ln \frac{p}{p_{eq}} \tag{5.31}$$

where

G_{a}^{0} = Gibbs' free energy of an ad-atom on the crystal interface when the surrounding pressure equals the standard pressure ($p_0 = 1$)

G_{a}^{p} = Gibbs' free energy of an ad-atom on the crystal interface when the surrounding vapour pressure is p

$G_{a}^{p_{eq}}$ = Gibbs' free energy of an ad-atom on the crystal surface when the surrounding vapour pressure is p_{eq}.

Combining Equations (5.28) and (5.31) we obtain

$$n_{S} = n_{0S} A \, e^{-\frac{U_{a\,act}^{ad\,p_{eq}} - k_{B} T \ln \frac{p}{p_{eq}}}{k_{B} T}} = n_{0S} A \frac{p}{p_{eq}} e^{-\frac{U_{a\,act}^{ad\,p_{eq}}}{k_{B} T}} \tag{5.32}$$

According to Equation (5.27) the product of the first and third factors on the right-hand side of Equation (5.32) is equal to n_{S}^{eq}. Hence, Equation (5.32) can be written as

$$\frac{n_{S}}{p} = \frac{n_{S}^{eq}}{p_{eq}} \tag{5.33}$$

- The number of ad-atoms per unit area on the crystal surface is proportional to the pressure of the vapour at equilibrium with the crystal.

Strain of Ad-atoms

In Example 5.1 we have assumed that the new layer of atoms is a monolayer. This is the most likely thickness if there are no other forces than the surface tension.

Strain (Chapter 3 page 142–143) will always arise at the incorporation of the vapour atoms into the crystal because the vapour atoms, the ad-atoms and the lattice atoms in the crystal differ in size. The main shrinkage of the atomic radius occurs at the incorporation of ad-atoms into the crystal. In addition, the radius of an atom normally decreases with decreasing coordination number.

The extension of the farthest electron shell around the nucleus of an atom determines its size. It takes a certain time, the so-called *relaxation time*, to contract the electron shell of a vapour atom to the proper size of an ad-atom or to adjust the size of an ad-atom to the size of an incorporated atom in the crystal lattice. The relaxation times are probably much longer than the time of forming a nucleus. During the time of lacking adaption of the atomic size to the new value we can apply the laws for nucleation on a substrate with a minor mismatch (Chapter 3 page 142) between the sizes of the ad-atoms and the atoms in the crystal lattice.

Example 5.2

Zn crystals are growing in a supersaturated Zn vapour at a temperature of 400 °C. Flat two-dimensional monolayer nuclei of adatoms of the critical size and the shape shown in the figure form close to the smooth crystal surface.

Calculate numerically the logarithm of the necessary supersaturation p/p_{eq} for homogeneous nucleation of the two-dimensional Zn nucleus in the margin figure as a function of the mismatch δ_{el} from zero up to 3.0% and show the function graphically.

It is necessary to pay attention to the fact that the nuclei consisting of ad-atoms are strained in comparison to the unstrained lattice atoms of the crystal surface.

The material constants given in Example 5.1 can also be used here. Young's modulus for zinc is 9.3×10^{10} N/m^2. Zn crystals have HCP structure. Assume that the fraction of embryos of critical size is 1 embryo per mol (Example 4.1 on page 170 in Chapter 4).

Solution:

Figure 5.21 on page 220 shows the Gibbs' free-energy levels of the system crystal–vapour and can be used to derive the Gibbs' free energy of the vapour as a function of the vapor pressure.
The Gibbs' free energy per kmol of the vapour at pressure p [Equation (5.29) on page 220] can be written as

$$G^{p} = G_c^0 + RT \ln p \tag{1'}$$

where G_c^0 is the Gibbs' free energy/kmol of the crystal surface.
The corresponding relationship for vapour at pressure p_{eq} [Equation (5.30) on page 220] is

$$G^{p_{eq}} = G_c^0 + RT \ln p_{eq} \tag{2'}$$

The driving force for growth of the nuclei formed of adatoms coming from the vapour phase will be

$$-\Delta G^{vap \to ad} = G^{p} - G^{p_{eq}} = \left(G_c^0 + RT \ln p\right) - \left(G_c^0 + RT \ln p_{eq}\right)$$

or

$$-\Delta G^{vap \to ad} = RT \ln \frac{p}{p_{eq}} \tag{3'}$$

If we take into consideration that the ad-atoms are strained, owing to the larger size than the lattice atoms in the crystal, we realize that the adatom layer has a *higher* energy than the crystal surface and that the driving force will be *reduced* by an amount equal to the strain energy/kmol.

$$-\Delta G_{strain}^{vap \to ad} = RT \ln \frac{p}{p_{eq}} - Y\delta_{el}^2 V_m \tag{4'}$$

where the last term is the strain energy and δ_{el} is the mismatch (Chapter 3 on page 142) owing to the strain.

Next, we consider a nucleus with the same dimensions as in Example 5.1.

V = the volume of the nucleus $= 2b^2h\sqrt{3}$
A = the bottom area $= 2b^2\sqrt{3}$
A_{edge} = the edge area $= 4bh\sqrt{3}$.

We set up the Gibbs' free energy of the strained embryo and proceed in the same way as in Example 5.1 to find the minimum critical size for a surviving nucleus. The difference is that strain energy must be included here.

In this case, we will apply an equation analogous to Equation (4.37) on page 184 in Chapter 4

$$\Delta G_c = -\left[\left(-\Delta G^{L \to s}\right)\frac{V}{V_m} + Y\delta_{el}^2 V\right] + \sum_i A_i\sigma_i + A\left(\sigma_c + Y\delta_{el}^2 h\right)$$

where we replace the expression within the brackets with the Gibbs' free energy of the nucleus including strain, e.g. with $-\Delta G_{\text{strain}}^{\text{vap}\to\text{ad}}$. The second term is the surface energy of the edge.

The last two terms correspond to the surface energy of the common bottom area between the crystal and the firmly adsorbed adatoms including the strain energy, owing to the mismatch. The first of these terms is close to zero as the adatoms and the layer atoms are of the same kind.

$-\Delta G_{\text{strain}}^{\text{vap}\to\text{ad}}$ is replaced with the right-hand side of Equation (4′) and we obtain

$$\Delta G_{\text{embryo}}^{\text{strain}} = -\left(RT\ln\frac{p}{p_{\text{eq}}} - Y\delta_{\text{el}}^2 V_{\text{m}}\right)\frac{V}{V_{\text{m}}} + A_{\text{edge}}\sigma_{\text{edge}} + \left(0 + Ah\,Y\delta_{\text{el}}^2\right) \tag{5′}$$

Inserting the expressions of V, A and A_{edge} and $h = c$ gives

$$\Delta G_{\text{embryo}}^{\text{strain}} = -\left(RT\ln\frac{p}{p_{\text{eq}}} - Y\delta_{\text{el}}^2 V_{\text{m}}\right)\frac{2b^2c\sqrt{3}}{V_{\text{m}}} + 4bc\sqrt{3}\sigma_{\text{edge}} + \left(0 + 2b^2c\sqrt{3}Y\delta_{\text{el}}^2\right)$$

or, after reduction,

$$\Delta G_{\text{embryo}}^{\text{strain}} = -\frac{2b^2c\sqrt{3}}{V_{\text{m}}}RT\ln\frac{p}{p_{\text{eq}}} + 4b^2c\sqrt{3}Y\delta_{\text{el}}^2 + 4bc\sqrt{3}\,\sigma_{\text{edge}} \tag{6′}$$

To find the nucleus of critical size b_{cr}^* we take the derivative of $\Delta G_{\text{embryo}}^{\text{strain}}$ with respect to b and set it equal to zero.

$$\frac{d\Delta G_{\text{embryo}}^{\text{strain}}}{\mathrm{d}b} = -\frac{4b_{\text{cr}}c\sqrt{3}}{V_{\text{m}}}RT\ln\frac{p}{p_{\text{eq}}} + 8b_{\text{cr}}c\sqrt{3}Y\delta_{\text{el}}^2 + 4c\sqrt{3}\,\sigma_{\text{edge}} = 0$$

which gives

$$b_{\text{cr}}^* = \frac{4c\sqrt{3}\sigma_{\text{edge}}}{\dfrac{4c\sqrt{3}}{V_{\text{m}}}RT\ln\dfrac{p}{p_{\text{eq}}} - 8c\sqrt{3}Y\delta_{\text{el}}^2}$$

or

$$b_{\text{cr}}^* = \frac{\sigma_{\text{edge}}V_{\text{m}}}{RT\ln\dfrac{p}{p_{\text{eq}}} - 2Y\delta_{\text{el}}^2 V_{\text{m}}} \tag{7′}$$

Inserting the value of b_{cr}^* into Equation (6′) gives the Gibbs' free energy of the nucleus of critical size.

$$\Delta G_{\text{nucleus}}^{\text{strain}}{}^* = -\frac{2b_{\text{cr}}^{*2}c\sqrt{3}}{V_{\text{m}}}RT\ \ln\frac{p}{p_{\text{eq}}} + 4b_{\text{cr}}^{*2}c\sqrt{3}Y\delta_{\text{el}}^2 + 4b_{\text{cr}}^*c\sqrt{3}\,\sigma_{\text{edge}} \tag{8′}$$

The Gibbs' free energy of the strained nucleus is also equal to $55\,k_{\text{B}}T$. If we insert this value into Equation (8′) we obtain

$$55k_{\text{B}}T = -\frac{2b_{\text{cr}}^{*2}c\sqrt{3}}{V_{\text{m}}}RT\ln\frac{p}{p_{\text{eq}}} + 4b_{\text{cr}}^{*2}c\sqrt{3}Y\delta_{\text{el}}^2 + 4b_{\text{cr}}^*c\sqrt{3}\,\sigma_{\text{edge}} \tag{9′}$$

where b_{cr}^* is equal to the expression in Equation (7′).

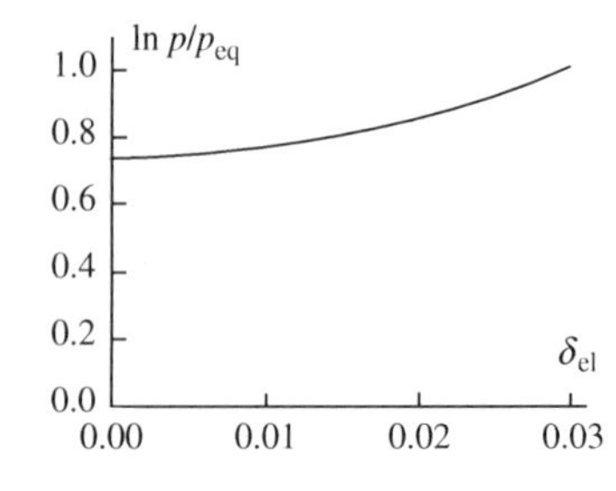

If we insert the expression (7′) into Equation (9′) we obtain a relationship between ln p/p_{eq} and δ_{el}, which is impossible to solve exactly. Instead we solve ln p/p_{eq} as a function of δ_{el} numerically and show the result graphically.

Answer:

The logarithm of the supersaturation p/p_{eq} as a function of the mismatch δ_{el} is shown in the margin figure.

The function is given in Equations (9′) and (7′) in the solution.

5.6.3 *Roughening Temperature. Influence of Pressure*

Jackson's Solution Model

Jackson's solution model [5] has been introduced on page 132 in Chapter 3 for liquid/solid interphase. In the case of vapour/solid interfaces the same theory is valid with the only difference that we have to replace the fusion energy L_f in Equation (3.64) on page 133–134 in Chapter 3 with the evaporation energy L_{ev}. As $L_{ev} \gg L_f$ the result is that the value of the parameter α will be larger for vapours than for liquids.

For liquid/solid interfaces the surface will be smooth if $\alpha > 2$. For vapour/metal interfaces α will be of the order 20, which means that metal/vapour surfaces remain smooth up to high temperatures.

Roughening Temperature

At sufficiently high temperature, equal to the so-called *roughening temperature*, smooth and vicinal interfaces transform from smooth to rough interphase at constant vapour pressure. Each metal has its own characteristic roughening temperature.

As we have shown on pages 219–221 the surface concentration of ad-atoms on a smooth crystal surface in contact with a vapour is proportional to the vapour pressure.

When the temperature increases, i.e. when the adatoms obtain increased average kinetic energy, the interaction between them increases gradually. The smooth surface collapses suddenly, i.e. the surface becomes rough when the interaction becomes high enough. This happens *either* at the special roughening temperature at constant pressure *or* at an increased supersaturation at constant temperature or at a combination of the two.

Influence of Pressure on the Transition from Smooth to Rough Metal Surfaces at Constant Temperature

When the pressure is increased above the equilibrium vapour pressure at constant temperature, the surface concentration of adatoms increases, the interatomic distances between them decrease and the interaction between the adatoms increases. At a certain pressure a sudden transition from smooth to rough surface occurs. The theory of two-dimensional nucleation offers a reasonable explanation of this phenomenon.

At normal growth the growth rate of a metal surface in contact with a supersaturated vapour with the pressure p is proportional to the excess pressure [Equation (5.23) on page 213].

$$V_{growth} = const \times (p - p_{eq}) \tag{5.34}$$

This is also true for layer growth. If we plot V_{growth} versus $(p - p_{eq})$ we obtain a straight line (Figure 5.22).

However, Examples 5.1 and 5.2 deal with the critical supersaturation at a given temperature, necessary for formation of embryos, which can grow to nuclei. At the given temperature plenty of new nuclei form when the pressure exceeds the critical supersaturation, which corresponds to a critical value of $(p - p_{eq})$. The critical supersaturation results in a drastic increase of the growth rate, as shown by the vertical line in Figure 5.22.

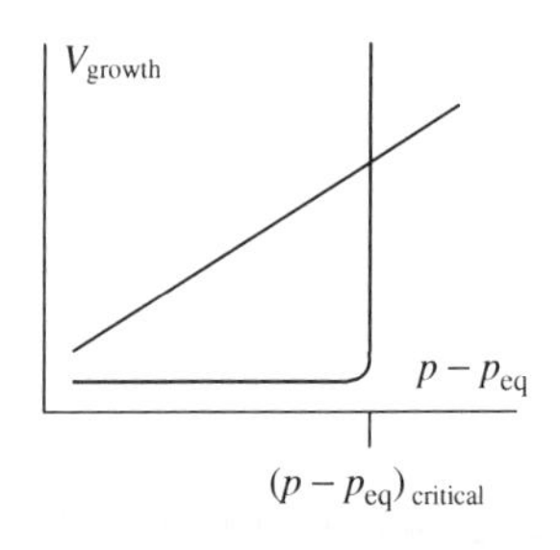

Figure 5.22 Transition from smooth to rough surface at the critical pressure for nucleation. The lowest value of V_{growth} is always valid.

The sudden increase of the number of nuclei causes the smooth crystal surface to break down and the surface becomes rough. The intersection point of the two curves in Figure 5.22 represents the transition from a smooth to a rough surface.

5.6.4 *Diffusion Process at the Step Terrace*

Subprocesses of Layer Growth

In analogy with normal growth, layer growth can be regarded as a series of consecutive processes. The three processes mentioned on page 215 are involved and the total driving force of layer growth can be written as

$$\left(-\Delta G^{\text{total}}\right) = \left(-\Delta G^{\text{ad}}\right) + \left(-\Delta G^{\text{diff}}\right) + \left(-\Delta G^{\text{kin}}\right) + \cdots \tag{5.35}$$

The first term on the right-hand side describes the adsorption of atoms on the planar interface. The second term describes the effect of diffusion along the smooth interface towards the step edge. The third term describes the incorporation and ejection of adatoms at the rough step edge.

A term describing the formation of new interfaces and a strain–stress term should also be added on the right-hand side of Equation (5.35) for completeness. This topic will be discussed in Sections 5.6.8 and 5.6.9.

We will start the analysis below by discussing the three terms on the right-hand side of Equation (5.35) in the order they are mentioned.

The rate-controlling subprocess is normally either the diffusion process along the smooth step terraces or the kinetic process at the step edge. Before the analysis of the diffusion process we will start with a short discussion of the adsorption process needed for the further discussion.

Rate of Adsorption Process. Mean Lifetime of Ad-atoms

In Equation (5.35) we have included the transport of atoms from the vapour to the interface. However, this is as a first approximation a very fast process with little influence on the rate of layer growth. The diffusion in the vapour phase is so rapid that there are no mass gradients. The surface concentration of adsorbed atoms on the planar terraces is given by Equation (5.32) on page 221.

Even if the influence of the adsorption process (Figure 5.23) on the rate of layer growth can be neglected, the mean lifetime of the adsorbed atoms on the smooth interface is of interest for the diffusion of ad-atoms towards the rough step edge and the following exchange of atoms between the interphase and the crystal.

As we have shown on page 221 the number of ad-atoms per unit area is proportional to the supersaturation. The vapour atoms incessantly arrive to and leave the interface. Hence, Equation (5.32) describes a 'kinetic' equilibrium of adatoms.

The mean lifetime τ_S of an ad-atom on the interface can be calculated by means of statistical mechanics. The minimum energy necessary for evaporation of an ad-atom on the interface is equal to the activation energy of desorption $-\Delta G_a^{\text{desorp}}$ (Figure 5.24).

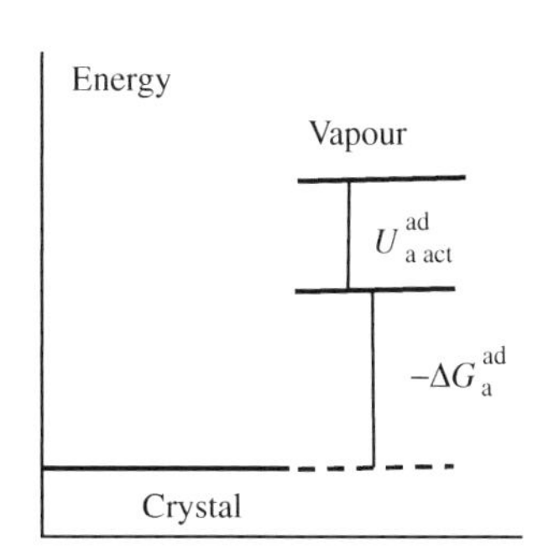

Figure 5.23 Free-energy levels of the system vapour–smooth crystal.

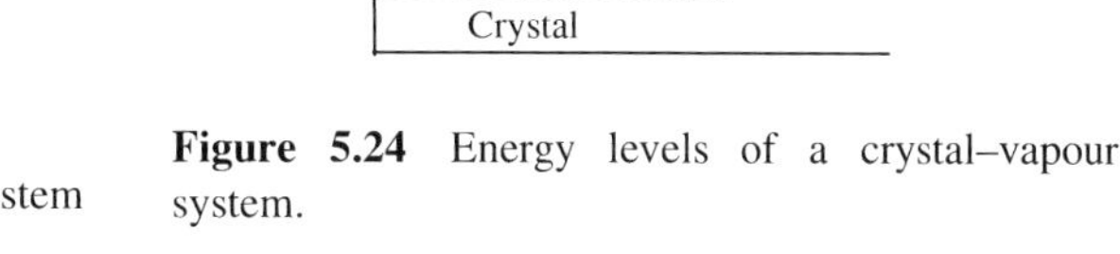

Figure 5.24 Energy levels of a crystal–vapour system.

The calculated mean lifetime is given by the relationship

$$\frac{1}{\tau_S} = \nu_{\perp} e^{-\frac{-\Delta G_a^{desorp}}{k_B T}} \tag{5.36}$$

where

τ_S = mean lifetime of adatoms on the smooth interface before desorption back to vapour
$\nu_{\perp}$ = vibration frequency of the adatoms perpendicular to the crystal surface
$-\Delta G_a^{desorp}$ = activation energy of desorption.

The first factor in Equation (5.36) represents the number of escape attempts of an ad-atom per unit time. The second factor is the Boltzmann factor, which represents the fraction of the total number of adatoms, which have energies enough to exceed the minimum energy of escape.

The more likely a desorption process is the shorter will be the mean lifetime. The mean lifetime of an adatom at the interface is independent of the concentration of ad-atoms.

The higher the activation energy for desorption of an adsorbed atom from the crystal surface is, the longer will be its mean lifetime on its way to the step edge. Equation (5.36) is applied on the adsorption–desorption process of vapour atoms (Figure 5.24) on the smooth interface and the mean lifetime can be written as

$$\tau_S = \frac{1}{\nu_{\perp}} e^{\frac{-\Delta G_a^{desorp}}{k_B T}} \tag{5.37}$$

Rate of Surface Diffusion of Ad-atoms towards the Step Edge

The basic equation of diffusion is

$$J = -D_S A \frac{dn_S}{dy} \tag{5.38}$$

This equation is applied on the surface diffusion of adatoms. We introduce the surface concentration instead of the volume concentration (n_S atoms/m^2 corresponds to the volume 1 m$^2 \times a$, which gives the volume concentration $n = n_S/a$)(Figure 5.25)

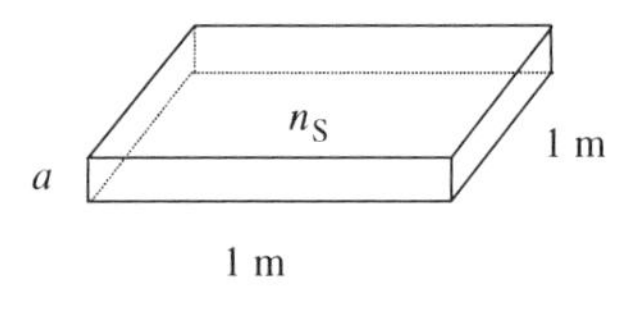

Figure 5.25 Volume element.

$$j_{diff} = -D_S \frac{d\frac{n_S}{a}}{dy} = \frac{D_S}{a} \frac{dn_S}{dy} \tag{5.39}$$

where

J = number of atoms per unit time and area A
j_{diff} = number of atoms per unit time and unit area
D_S = diffusion coefficient at surface diffusion
A = area of diffusion region perpendicular to the atom flow
a = atomic distance
n_S = surface concentration (number of atoms per unit area)
n = volume concentration (number of atoms per unit volume)
y = coordinate.

According to the theory of diffusion the diffusion coefficient D_S is strongly temperature dependent and can be written as

$$D_S = a^2 \nu_{=} e^{-\frac{\Delta U_{a\,act}^{diff}}{k_B T}} \tag{5.40}$$

where

$\nu_{=}$ = the vibration frequency of the adatoms parallel to the crystal surface
$-\Delta U_a^{diff}$ = activation energy of surface diffusion.

By studying the diffusion of the adatoms towards the step edge their mean distance travelled during their mean lifetime can be calculated and is found to be

$$\lambda_{\text{diff}} = \sqrt{D_S \tau_S} \tag{5.41}$$

where
λ_{diff} = the mean distance traveled along the surface by the adatoms during their mean lifetime.

If the expressions in Equations (5.37) and (5.40) are inserted into Equation (5.41) together with the assumption $\nu_{=} = \nu_{\perp} = \nu$, we obtain

$$\lambda_{\text{diff}} = a\,e^{\frac{\Delta G_a^{\text{desorp}} - \Delta U_{\text{a act}}^{\text{diff}}}{2k_B T}} \tag{5.42}$$

The rate of surface diffusion of the adatoms is defined as distance per unit time. Using the definition and Equations (5.41) and (5.42) we obtain

$$V_{\text{diff}} = \frac{\lambda_{\text{diff}}}{\tau_S} = \frac{D_S}{\lambda_{\text{diff}}} \tag{5.43}$$

Rate of Step Growth Controlled by the Diffusion Process at the Terraces. Diffusion Coefficient

At the step edge it is normally easy for the adatoms to be incorporated because the step interphase can be regarded as a rough interphase. The ad-atoms that arrive at the step edge, will partly be incorporated into the rough step edge and we can apply the theory for growth at rough interphases given on pages 212–213.

Figures 5.26 and 5.27 show the variation of the surface concentration of ad-atoms along the step. It is low near the step edge owing to the continuous flux of ad-atoms into the crystal surface.

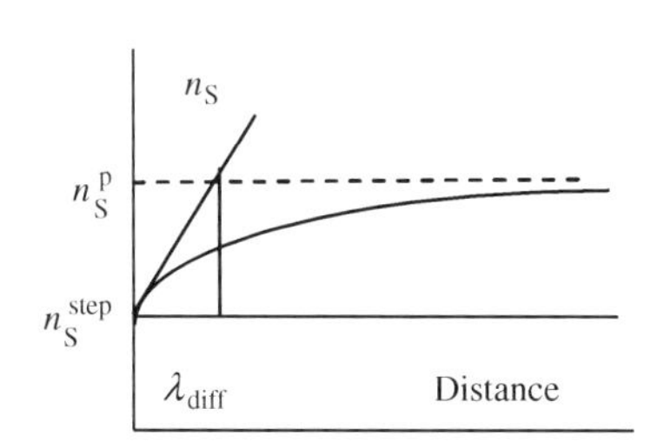

Figure 5.26 Approximate surface concentration of adatoms as a function of distance from the step edge.

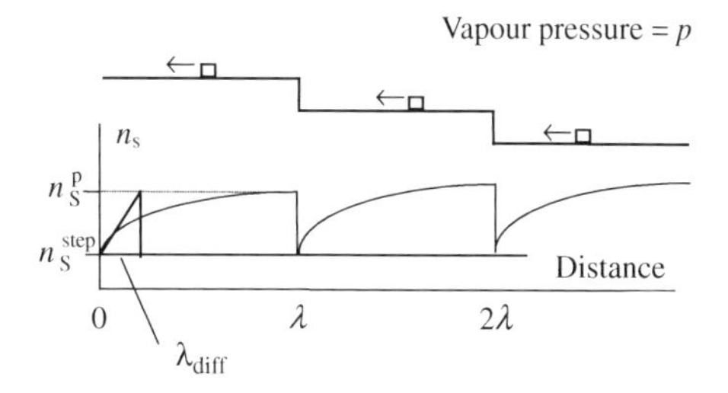

Figure 5.27 The surface concentration of ad-atoms as a function of the distance from the step edge on a smooth stepped surface. Compare Figure 5.20 on page 216.

Figure 5.14 on page 212. The probability that a cubic ad-atom will be incorporated into the crystal is equal to $a/\delta \times a/\delta = a^2/\delta^2$.

Equation (5.39)

$$j_{\text{diff}} = -D_S \frac{d\frac{n_S}{a}}{dy} = \frac{D_S}{a}\frac{dn_S}{dy} \tag{5.39}$$

relates the surface flux and the gradient of the surface concentration of adatoms. It can, in combination with Equation (5.44) be used to find the rate of edge step growth as a function of the surface concentrations.

Equation (5.39) in combination with Figure 5.26 gives approximately

$$j_{\text{diff}} = -\frac{D_S}{a}\frac{dn_S}{dy} = -\frac{D_S}{a}\frac{n_S^p - n_S^{\text{step}}}{\lambda_{\text{diff}}} \tag{5.44}$$

where

j_{diff} = flux of ad-atoms towards the step edge (numbers per unit area and unit time)
n_S^p = surface concentration of ad-atoms in equilibrium with vapour of pressure p
n_S^{step} = surface concentration of ad-atoms close to the step
λ_{diff} = diffusion distance (page 224).

In analogy with Equation (5.20) on page 213 Equation (5.44), gives the step-edge growth-controlled by diffusion as a function of the ad-atom concentration.

$$V_{\text{edge}} = -a^3 j_{\text{diff}} = -\left(-a^3 \frac{D_S}{a}\frac{n_S^p - n_S^{\text{step}}}{\lambda_{\text{diff}}}\right) \tag{5.45}$$

or

$$V_{\text{edge}} = a^2 \frac{D_S}{\lambda_{\text{diff}}}\left(n_S^p - n_S^{\text{step}}\right) \tag{5.46}$$

We define the *diffusion rate coefficient* β_{diff} with the aid of Equation

$$V_{\text{edge}} = \beta_{\text{diff}}\left(n_S^p - n_S^{\text{step}}\right) \tag{5.47}$$

Identification of Equations (5.46) and (5.47) gives the diffusion coefficient

$$\beta_{\text{diff}} = a^2 \frac{D_S}{\lambda_{\text{diff}}} \tag{5.48}$$

Equation (5.46) is only valid if the step length is larger than the diffusion distance, i.e. $\lambda > \lambda_{\text{diff}}$ (Figure 5.27). If this condition is fulfilled all the ad-atoms are likely to reach the step edge within their mean lifetime.

If $\lambda < \lambda_{\text{diff}}$ the surface concentration n_S^p will decrease because more ad-atoms diffuse to the step than are adsorbed from the vapour. This effect can be taken into consideration by multiplying n_S^p by the factor tanh $\lambda/\lambda_{\text{diff}}$. When this product decreases to a value close to n_S^{step} there is no longer any driving force for diffusion and the rate of step growth will be controlled by the adsorption of atoms from the vapour.

5.6.5 *Kinetic Process at the Step Edge*

The procedure to find the growth rate at the step edge and the kinetic coefficient in the present case is analogous to the one for rough surfaces. The only difference is that the incoming atoms are adatoms instead of vapour atoms, as in the case of normal growth.

The flux of adsorbed atoms towards the rough step edge and the flux of atoms leaving the kink sites of the rough step edge are considered and the flux is related to the adatom concentration near the step edge.

Surface Concentration of Ad-atoms at the Smooth Step Surface as a Function of Vapour Pressure

The ad-atoms on the smooth terrace surface are in equilibrium with both the vapour and the crystal at the same time. This statement will be used to calculate the equilibrium value of the ad-atom concentration n_S^{eq} near the step edge in terms of vapour pressure.

The step edge is a rough interphase and the theory of the balance between the interphase and the vapour, given in Section 5.5, is directly applicable. Equations (5.16) and (5.19) on page 212) are valid with the only difference that the atoms are *adsorbed* and not incorporated into kinks in the crystal lattice.

$$j_+ = \frac{p}{\sqrt{2\pi m k_B T}} e^{-\frac{U_{a\ act}^{ad}}{k_B T}} \tag{5.49}$$

$$j_- = \frac{p_{eq}}{\sqrt{2\pi m k_B T}} e^{-\frac{U_{a\ act}^{ad}}{k_B T}} \tag{5.50}$$

where

j_+ = flux of vapour atoms towards the terrace surface
j_- = flux of atoms leaving the surface and returning to the vapour
p = vapour pressure
p_{eq} = pressure at equilibrium between vapour, interface and crystal
$U_{a\ act}^{ad}$ = activation energy of adsorption.

The flux j_- of the ad-atoms, which evaporate from the interface and return to the vapour, can alternatively be written as the surface concentration of ad-atoms divided by mean lifetime τ_S of the atoms

$$j_- = \frac{n_S}{\tau_S} \tag{5.51}$$

At equilibrium, we obtain by use of Equations (5.49) and (5.50)

$$j_+^{eq} = j_-^{eq} = \frac{n_S^{eq}}{\tau_S} = \frac{p_{eq}}{\sqrt{2\pi m k_B T}} e^{-\frac{U_{a\ act}^{ad}}{k_B T}} \tag{5.52}$$

or

$$n_S^{eq} = \tau_S \frac{p_{eq}}{\sqrt{2\pi m k_B T}} e^{-\frac{U_{a\ act}^{ad}}{k_B T}} \tag{5.53}$$

If the vapour pressure is p we obtain analogously

$$n_S^{p} = \tau_S \frac{p}{\sqrt{2\pi m k_B T}} e^{-\frac{U_{a\ act}^{ad}}{k_B T}} \tag{5.54}$$

Net Flux of Ad-atoms at the Smooth Step Surface as a Function of the Ad-atom Concentration

The concentration of ad-atoms as a function of vapour pressure has been derived above. Here we will study the kinetic process at the step edge when adatoms become incorporated into kinks in the lattice of the crystal.

Figure 5.27 on page 227 shows that the ad-atom concentration is considerably lower close to the step edge than elsewhere. The reason for this is that ad-atoms continously become incorporated into the rough step edge.

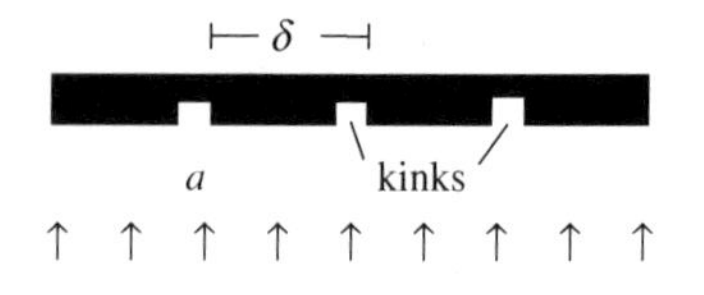

Figure 5.14 Principle sketch of the simplest possible rough surface (Guggenheim mono-atomic layer). The probability that a cubic adatom will be incorporated into the crystal is equal to; $a/\delta \times a/\delta = a^2/\delta^2$ where; a = the length of the side of the cube; δ = average distance between the kinks. Figure 3.52 on page 139 in Chapter 3 shows a more general model of a rough surface ($a \approx \delta$).

The basis of finding the net flow of adatoms into the rough step is the ad-atom concentration n_S^{step} near the step. Only the ad-atoms that have energies larger than the activation energy are able to overcome the potential barrier. The incoming flux is analogously decreased by the Boltzmann factor. As Figure 5.15 shows the flux is also reduced by the probability factor a/δ as only this fraction of adatoms will find kinks in the crystal lattice. Hence, we obtain the following expression for the flux of *attached* atoms at the rough step

$$j_+ = \frac{n_S^{step}}{\tau_S}\frac{a^2}{\delta^2} e^{-\frac{U_{a\,act}^{kin}}{k_B T}} \tag{5.55}$$

where

j_+ = incoming flux = number of attached adatoms per unit area and unit time that hit the crystal
n_S^{step} = surface concentration = number of atoms per unit area on the smooth terrace close to the step edge
τ_S = mean lifetime of ad-atoms
$U_{a\,act}^{kin}$ = activation energy to transfer an ad-atom from the smooth interface to a kink site in the rough step edge
a = atomic distance
δ = average distance between kinks
$\frac{a}{\delta}$ = probability of an incoming ad-atom at the step being incorporated into the crystal.

In analogy with the case of normal growth on pages 214–215 we obtain the following expression for the flux of *detached* atoms at the rough step (Figure 5.28)

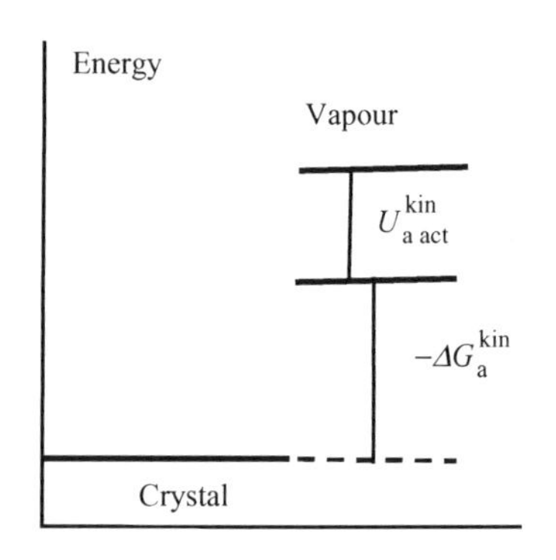

Figure 5.28 Gibbs' free energy levels of the system smooth interface/crystal.

$$j_- = const \frac{a^2}{\delta^2} e^{-\frac{-\Delta G_a^{kin} + U_{a\,act}^{kin}}{k_B T}} \tag{5.56}$$

At equilibrium, the two fluxes j_+ and j_- are equal and the concentration n_S^{step} of ad-atoms on the surface close to the step is equal to n_S^{eq}:

$$j_+^{eq} = \frac{n_S^{eq}}{\tau_S}\frac{a^2}{\delta^2} e^{-\frac{U_{a\,act}^{kin}}{k_B T}} = j_-^{eq} = const \frac{a^2}{\delta^2} e^{-\frac{-\Delta G_a^{kin} + U_{a\,act}^{kin}}{k_B T}} \tag{5.57}$$

A comparison between Equations (5.56) and (5.57) shows that Equation (5.56) can be written as

$$j_- = \frac{n_S^{eq}}{\tau_S}\frac{a^2}{\delta^2} e^{-\frac{U_{a\,act}^{kin}}{k_B T}} \tag{5.58}$$

where

n_S^{eq} = the surface concentration of ad-atoms close to the step at equilibrium with crystal and vapour of pressure p_{eq}.

By use of Equations (5.55) and (5.58) the net flux of ad-atoms at the step edge can be written as

$$j_+ - j_- = \frac{n_S^{step}}{\tau_S}\frac{a^2}{\delta^2} e^{-\frac{U_{a\,act}^{kin}}{k_B T}} - \frac{n_S^{eq}}{\tau_S}\frac{a^2}{\delta^2} e^{-\frac{U_{a\,act}^{kin}}{k_B T}} \tag{5.59}$$

or

$$j_+ - j_- = \frac{a^2}{\delta^2} e^{-\frac{U_{a\,act}^{kin}}{k_B T}} \frac{n_S^{step} - n_S^{eq}}{\tau_S} \tag{5.60}$$

Rate of Step Growth Controlled by the Kinetic Process at the Step Edge. Kinetic Growth Coefficient

In analogy with Equation (5.20) on page 213 the net flux ($j_+ - j_-$) towards the crystal is related to the rate of step-edge growth by

$$V_{\text{edge}} = a^3(j_+ - j_-) \tag{5.61}$$

where

V_{edge} = rate of the step-edge growth for an infinitely large crystal by incorporation of adatoms

j_+, j_- = fluxes of attached and detached adatoms at the crystal = number of atoms per unit area and unit time

a = atomic distance = lattice constant.

Equation (5.60) in combination with Equation (5.61) gives

$$V_{\text{edge}} = \frac{a^3}{\tau_S}\frac{a^2}{\delta^2} e^{-\frac{U_{\text{a act}}^{\text{kin}}}{k_B T}} \left(n_S^{\text{step}} - n_S^{\text{eq}}\right) \tag{5.62}$$

or

$$V_{\text{edge}} = \beta_{\text{kin}} \left(n_S^{\text{step}} - n_S^{\text{eq}}\right) \tag{5.63}$$

where β_{kin} is the *kinetic growth coefficient of the step* in analogy with the kinetic growth coefficient of the crystal surface in the case of rough surfaces (page 213). The concentration difference is the driving force of the growth of the step.

Identification of Equations (5.62) and (5.63) gives

$$\beta_{\text{kin}} = \frac{a^3}{\tau_S}\frac{a^2}{\delta^2} e^{-\frac{U_{\text{a act}}^{\text{kin}}}{k_B T}} \tag{5.64}$$

5.6.6 *Total Rate of Step Growth*

Total Rate of Step Growth as a Function of Vapour Pressure

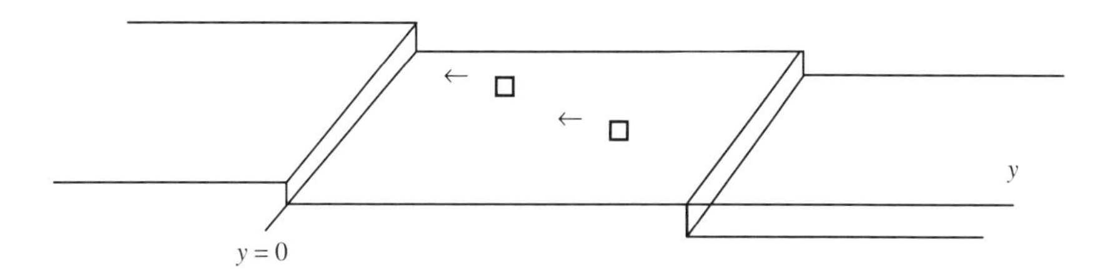

Figure 5.29 Step growth.

The total rate of step growth (Figure 5.29) is a function of the vapour pressure. The more the pressure deviates from the equilibrium pressure, the larger will be the total rate of crystallization. In analogy with Equation (5.60) we obtain

$$j_+ - j_- = \frac{a^2}{\delta^2} e^{-\frac{U_{\text{a act}}^{\text{kin}}}{k_B T}} \frac{n_S^{\text{p}} - n_S^{\text{eq}}}{\tau_S} \tag{5.65}$$

The equilibrium adatom concentration close to the step n_S^{eq} has been expressed in terms of equilibrium pressure in Equation (5.53) on page 229.

Similarly we have expressed the ad-atom concentration n_S^p as a function of the pressure p in Equation (5.54) on page 229.

If we introduce these expressions into Equation (5.65) we obtain

$$V_{edge}^{total} = \frac{a^3}{\sqrt{2\pi\, mk_B T}} \frac{a^2}{\delta^2} e^{-\frac{U_{a\,act}^{kin}}{k_B T}} \left(p - p_{eq}\right) \tag{5.66}$$

Diffusion and Kinetic Rates of Step Growth

Both in the case of normal growth and layer growth the adsorption process is very fast and the adsorption coefficient is very large compared to the diffusion coefficient and the kinetic coefficient. The adsorption process is *not* the rate determining process and can normally be disregarded in comparison with the surface diffusion and the kinetic incorporation of adatoms at the step.

An important question is whether the total growth rate of the step is determined by the transfer of atoms to the step (diffusion) or by the incorporation of adatoms at the rough step (kinetic growth).

The total driving force of the crystal layer growth is proportional to the vapour pressure difference $(p - p_{eq})$, which is proportional to the adatom concentration difference $\left(n_S^p - n_S^{eq}\right)$. To make a comparison between the subprocesses easier we divide the difference $\left(n_S^p - n_S^{eq}\right)$ into two parts:

$$n_S^p - n_S^{eq} = \left(n_S^p - n_S^{step}\right) + \left(n_S^{step} - n_S^{eq}\right) \tag{5.67}$$

The concentration differences can be replaced by the driving forces in analogy with the analysis of the simple growth model on pages 206–207. We introduce Equation (5.7) on page 207 here.

$$\left(-\Delta G^{total}\right) = \left(-\Delta G^{diff}\right) + \left(-\Delta G^{kin}\right) \tag{5.7}$$

In the steady state, the material transport must be the same during the adsorption from the vapour to the smooth surface and during the incorporation of adatoms into the crystal as material is not accumulated anywhere. Hence, we have the following expressions for the mass transport rates

$$k_{total}\left(-\Delta G^{total}\right) = k_{diff}\left(-\Delta G^{diff}\right) = k_{kin}\left(-\Delta G^{kin}\right) \tag{5.68}$$

These expressions are introduced into Equation (5.7) and we obtain

$$\frac{1}{k_{total}} = \frac{1}{k_{diff}} + \frac{1}{k_{kin}} \tag{5.69}$$

where

$k_{diff} = const \times \beta_{diff}$ (Equation (5.47) on page 228)

$k_{kin} = const \times \beta_{kin}$ (Equation (5.63) on page 231)

If $\beta_{diff} \gg \beta_{kin}$ the rate-controlling process is *incorporation of adsorbed atoms at the step* and Equation (5.63) on page 231 is valid

$$V_{edge} = \beta_{kin}\left(n_S^{step} - n_S^{eq}\right) \quad \text{Kinetic mode}$$

If $\beta_{diff} \ll \beta_{kin}$ the rate controlling process is *diffusion along the step terraces* and Equation (5.47) on page 228 is valid

$$V_{edge} = \beta_{diff}\left(n_S^p - n_S^{step}\right) \quad \text{Diffusion mode}$$

If we combine Equation (5.67) on page 232 with Equations (5.47) on page 228 and (5.63) on page 231 we obtain

$$n_S^p - n_S^{eq} = \frac{V_{edge}}{\beta_{diff}} + \frac{V_{edge}}{\beta_{kin}} \tag{5.70}$$

which gives

$$V_{edge} = \frac{\beta_{diff}\beta_{kin}}{\beta_{diff} + \beta_{kin}} (n_S^p - n_S^{eq}) \tag{5.71}$$

The growth rate can be related to the pressures p and p_{eq} with the aid of Equations (5.53) on page 229 and (5.54) on page 229.

5.6.7 *Spiral Steps and Screw Dislocations*

So far we have only discussed growth of planar steps, which is the most common case of layer growth. As Figure 5.30a shows steps can also form spirals.

The classical way to explain the formation of spiral steps is that they emerge from screw dislocations in the crystal. The formation of such a spiral step is described in Figure 5.30b. Screw dislocations have been discussed on pages 26–27 in Chapter 1 in [1].

It is seldom observed, however, that internal screw dislocations reach the surface and cause a series of steps. Spiral steps often occur more frequently than possibly can be explained by dislocations within the crystal. Hence, there must be some other cause and explanation of the appearance of frequent screw dislocations on the surfaces of crystals.

Figure 5.30a Microscope picture of a spiral step on the surface of a SiC crystal grown in vapour.

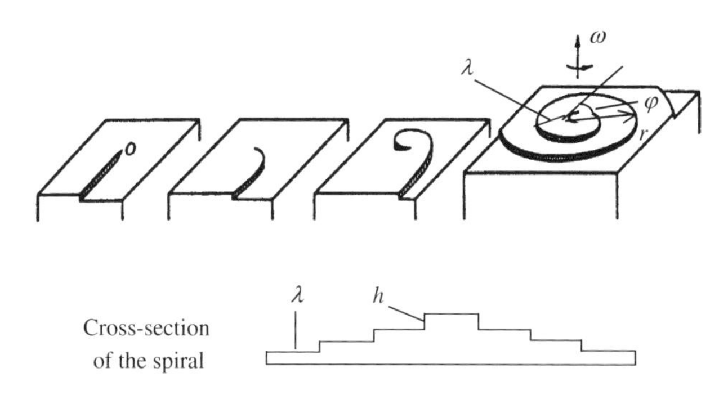

Figure 5.30b Formation of a spiral step around the origin of a screw dislocation on the crystal surface.

The lower figure shows the cross-section of the spiral step.

We assume that

- The origin of the screw dislocations is the difference in size between ad-atoms and freshly incorporated ad-atoms on one side and normal lattice atoms on the other.

The phenomenon has been discussed in Chapter 1 in [1]. The difference in atomic size causes strain and results in screw dislocations.

An Archimedes' spiral is used as a mathematical model of the spiral. It can be written as

$$r = A\varphi \tag{5.72}$$

where
r = distance from origin
A = a constant
φ = angle.

Rate of Spiral Step Growth. Layer Growth

The growth rate $V_{\perp}$ of a stepped crystal, perpendicular to its surface, depends on the step height h, the step length λ and the step growth rate V_{edge} according to Equation (5.25) on page 214:

$$V_{\perp} = h\frac{V_{edge}}{\lambda} \tag{5.25}$$

Each of these quantities will be discussed shortly.

Distance between Steps

We will calculate the distance λ between the steps in the present case by using the classical method described by Burton and Cabrera [6] in 1950 in connection with screw dislocations.

The growth of the spiral occurs *radially*. The part of the spiral close to the origin grows very slowly and the growth rate is zero at the origin. This is a mathematical quality of the spiral. The physical reality is that the radius of curvature of the spiral must exceed the critical radius r^* of a two-dimensional nucleus, otherwise the spiral cannot grow at all. The embryo/crystal has its centre at the origin and its radius of curvature at the origin is r^*. It can be shown mathematically that the constant A in Equation (5.72) on page 233 is $2r^*$ in this case

$$r = 2r^*\varphi \tag{5.73}$$

The radial distance λ between the steps (Figure 5.30a) is obtained when $\Delta\varphi = 2\pi$

$$\lambda = \Delta r = 2r^* \times 2\pi = 4\pi r^* \tag{5.74}$$

where r^* is the critical radius of a two-dimensional nucleus.

In analogy with Equation (4.3) on page 169 in Chapter 4 we have the following expression of the critical radius in terms of interface energy σ and driving force of nucleation

$$r^* = \frac{2\sigma V_m}{-\Delta G^{V \rightarrow s}} \tag{5.75}$$

where

σ = interface energy per unit area
V_m = molar volume of the crystal material
$-\Delta G^{V \rightarrow s}$ = driving force of solidification.

Rate of Step Growth

In the preceding section we have derived expressions for the rate of the step growth V_{edge} in the case of single steps. The same theory is valid also in this case.

The growth rate of the step depends of the growth mode. If the growth is controlled by the kinetic process at the steps then V_{edge} is given by Equation (5.63) on page 231.

If the diffusion along the smooth surface is the slowest process then V_{edge} is described by Equation (5.47) on page 228. This equation is only valid if the distance λ between the steps > the mean diffusion distance λ_s. This is the case when the supersaturation is very low.

In the opposite case, the surface concentration n_S^p will decrease because more ad-atoms diffuse to the step than are adsorbed from the vapour. This effect is taken into consideration by multiplying n_S^p by a factor tanh λ/λ_{diff} where $\lambda = 4\pi r^*$ (Equation (5.72) on page 233). When this product decreases to a value close to n_S^{step} then the whole process will be controlled by the adsorption of atoms from the vapour to the step.

Step Height
The rate of layer growth can be written as

$$V_{\perp} = h\frac{V_{edge}}{\lambda} = n_h a \frac{V_{edge}}{4\pi r^*} \qquad (5.76)$$

where
n_h = number of atomic layers
a = atomic distance.

The heights of the spiral are often found to be several tens of atomic layers.

Burton and Cabrera's method involves an expression for the step length but gives no information on the height of the steps. The calculation of the step height will be further discussed below.

5.6.8 Rate of Layer Growth with No Consideration to Strain

In Section 5.6.3 (page 224) the diffusion of ad-atoms along the smooth interface and in Section 5.6.4 (page 225) the incorporation of ad-atoms into the crystal lattice have been discussed and expressions of V_{edgee} have been derived. So far, no general determination of the height h and length λ of the steps has been performed. It is necessary to know these quantities if the rate $V_{\perp}$ of layer growth is to be calculated (Figure 5.18b).

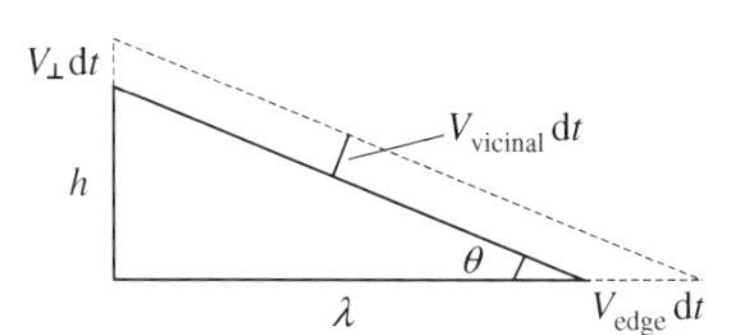

Figure 5.18b Cross-section of a step at $t = 0$ and $t = dt$. The position of the vicinal surface at the time $t = dt$ is marked by a dotted line in the figure.

Crystals are strained not only in the presence of spiral steps but *for any type of steps* because *the vapour atoms and the ad-atoms differ in size from the lattice atoms*. This fact can be used for calculation of the length and height of steps if the process is controlled by the adsorption of ad-atoms to the step.

Length and Height of Single Steps

Consider a step where the smooth interface, which normally is the most densely packed crystal plane, adsorbs vapour atoms. The atom radii in a gas or vapour are normally larger than those in a solid. This means that the outer parts of the smooth interface of the crystal are strained owing to the adatoms.

As long as the smooth interface is strained no adatoms can be incorporated into the crystal lattice. The growth will only occur by adsorption of vapour atoms into kinks at the rough step edge. During the growth, a *relaxation of the lattice* will occur simultaneously. The strain is reduced by formation of vacancies at the step edge. The vacancies are assumed to become frozen in the lattice during growth and annealed out of the lattice during relaxation. This process can occur in many alternative ways. One possibility is that the vacancies diffuse towards the interface, which acts as a sink, and annihilate there. The process is described in Figures 5.31a and b.

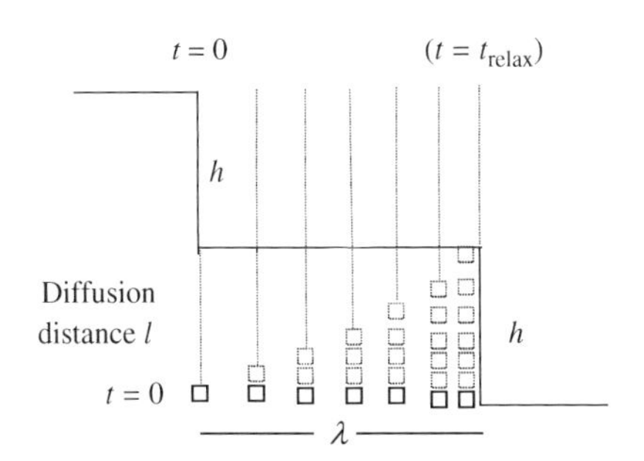

Figure 5.31a Crystal relaxation and simultaneous motion of step edge.

The figure describes the situation at $t = 0$. The number of vacancies close to the step edge is higher than that far from the edge.

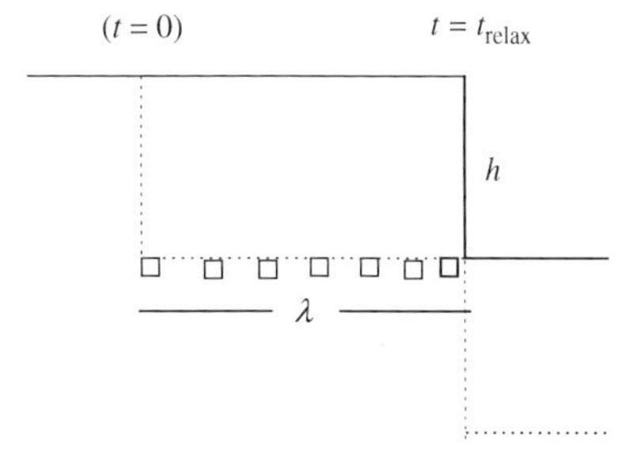

Figure 5.31b Crystal relaxation and simultaneous motion of step edge. The figure describes the situation at $t = t_{relax}$. All the vacancies in Figure 5.31a have arrived at the former terrace surface and annealed. The step edge has moved one step length, compared to Figure 5.31a, during the same time and a new step has been built up to the right.

Two motions, which occur simultaneously, are shown in Figures 5.31a and b. One is the horizontal step-edge motion with the constant velocity V_{edge} to the right. The other is the vertical motion of the vacancies towards the interface. The motions are coherent and can be described by the equations

$$y = V_{edge}\, t \tag{5.77}$$

$$l = \sqrt{D_{vac}\, t} \tag{5.78}$$

where
l = the diffusion distance of the vacancies in the crystal during time t.

When t is equal to the relaxation time t_{relax} (page 237) the edge has advanced a distance equal to the step distance λ and the vacancy in Figure 5.31a has just reached the interface after travelling a distance h:

$$\lambda = V_{edge}\, t_{relax} \tag{5.79}$$

$$h = \sqrt{D_{vac}\, t_{relax}} \tag{5.80}$$

where
h = height of a step
D_{vac} = diffusion coefficient of vacancies in the crystal
λ = length of a step
V_{edge} = velocity of edge movement
t = time
t_{relax} = time for vacancies to diffuse the distance h or relaxation time of the crystal lattice.

After relaxation, the smooth interface is free from strain and starts to adsorb new vapour atoms and a new step is formed.

Example 5.3

Calculate λ as a function of h with the growth rate of the step as a parameter.

Solution:

Consider Equations (5.77) and (5.78) at the time $t = t_{relax}$. The $y = h$ and $l = \lambda$. We square Equation (5.78) and eliminate t_{relax} between the new equation and Equation (5.77)

$$h^2 = D_{vac} t_{relax} \tag{1'}$$

$$\lambda = V_{edge} t_{relax} \tag{2'}$$

Elimination of t_{relax} results in the relationship

$$h^2 = \frac{D_{vac}\lambda}{V_{edge}} \tag{3'}$$

Answer:

The relationship is $\lambda = \dfrac{V_{edge}}{D_{vac}} h^2$, which is a parabola.

5.6.9 Rate of Layer Growth with Consideration to Strain

The rate of layer growth can be calculated from Equation (5.25) on page 214:

$$V_{\perp} = h\frac{V_{\text{edge}}}{\lambda} \tag{5.25}$$

The length and height of the steps are determined by the relaxation time by means of Equations (5.79) and (5.80) on page 236.

The rate of the step-edge growth is a function of surface concentration of adatoms close to the step and the activation energy of the process of incorporating adatoms into the crystal lattice. It is given by Equation (5.62) on page 231)

$$V_{\text{edge}} = \frac{a^3\,a^2}{\tau_{\text{s}}\,\delta^2}\mathrm{e}^{-\frac{U_{\text{a act}}^{\text{kin}}}{k_{\text{B}}T}}\left(n_{\text{S}}^{\text{step}} - n_{\text{S}}^{\text{eq}}\right) \tag{5.62}$$

Equation (5.62) is not valid here without a correction. The total driving force of crystal growth includes in this case a strain-energy term. A strained crystal has a higher Gibbs' free energy than a strain-free crystal, which is shown in Figure 5.32. The activation energy decreases by the amount $\Delta G_{\text{a}}^{\text{strain}}$and the proper expression for the rate of step-edge growth for a strained interface will be

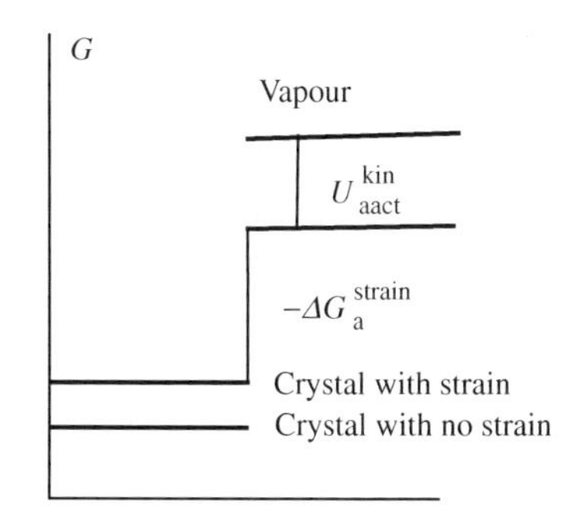

Figure 5.32 Influence of strain on the free energy of a crystal.

$$V_{\text{edge}}^{\text{strain}} = \frac{a^3\,a^2}{\tau_{\text{S}}\,\delta^2}\mathrm{e}^{-\frac{U_{\text{a act}}^{\text{kin}} - \Delta G_{\text{a}}^{\text{strain}}}{k_{\text{B}}T}}\left(n_{\text{S}}^{\text{step}} - n_{\text{S}}^{\text{eq}}\right) \tag{5.81}$$

Strain energy has been discussed in Sections 3.4.7 in Chapter 3 and Sections 4.3.3 and 4.3.4 in Chapter 4. $\Delta G_{\text{a}}^{\text{strain}}$ can be calculated by use of the theoretical analyses given there.

If the reaction is controlled by the kinetic process the layer growth rate will be given by

$$V_{\perp} = h\frac{V_{\text{edge}}^{\text{strain}}}{\lambda} \tag{5.82}$$

where $V_{\text{edge}}^{\text{strain}}$ is given by Equation (5.81). The surface concentrations are functions of the relaxation time t_{relax} (compare for example Equations (5.53) and (5.54) on page 229 when $t = t_{\text{relax}}$). If the relaxation time is known the rate of layer growth can be calculated.

Relaxation Time

Relaxation times occur in many types of processes and are determined experimentally. The basic formula is

$$\delta = \delta^0\mathrm{e}^{-\frac{t}{\tau}} \tag{5.83}$$

where

δ = the measured quantity that decreases with time
δ^0 = equlibrium value of the quantity measured at time $t = 0$
t = time
τ = relaxation time, characteristic time constant for the process.

The relaxation time is defined as the time when $\delta/\delta^0 = 1/\mathrm{e}$.

In this chapter, the concept has been applied on elastic strain, for example mismatch, and on vacancies. In the case of vacancies the relaxation time depends on the distance between the vacancy sinks. The density of sinks is given by the fraction of dislocations or other lattice defects in the crystals. The higher the density of sinks is, the lower will be the relaxation time.

When a crystal is strained, i.e. exposed to an elastic stress, which suddenly disappears, the distances between the atoms change back to their strain-free positions. This requires time and the process can be characterized by its relaxation time.

5.7 Influence of Impurities on Crystal Growth in Vapours

Impurities is a common name for desired or undesired minor amounts of foreign substances in a crystal. Impurities affect the appearance of the phase diagram and consequently the equilibrium and supersaturation or supercooling. Therefore, the whole growth process will be entirely changed. The description of thin-film production of silicon and diamond in Section 5.8 is an example of this.

5.7.1 Impurity Trapping

The impurities become trapped in the crystal during growth. If the impurities are of atomic size, i.e. atoms, ions or molecules, the trapping is *homogeneous*. If the trapped particles are much bigger than atomic size, e.g. colloid particles or macroscopic inclusions, the trapping is *heterogeneous*.

Homogeneous trapping results in a solid solution. If the impurity concentration of the crystal equals the thermodynamic equilibrium concentration at the growth temperature, the trapping is said to be *equilibrium trapping*, otherwise *non-equilibrium trapping*. Heterogeneous trapping is always non-equilibrium.

The impurity distribution is seldom constant within a crystal. There are three main types of non-uniformity

- sectorial non-uniformity;
- zonal non-uniformity;
- structural non-uniformity.

Sectorial non-uniformity implies that the concentration of impurities is different in different growth pyramids.

Under certain circumstances, which are poorly understood, the impurity concentration will vary from one crystal plane, parallel to the substrate, to another. Such non-uniformity is called *zonal.*

Structural non-uniformity means that there is a positive or negative deviation in impurity concentration at the boundaries of twins and grains, at dislocations or other structural defects of a crystal.

The mechanism of impurity trapping is very complex and not fully understood so far. The treatment will be simplified if we consider two limiting cases:

- The adsorbed impurities on the crystal interface are immobile. The impurity particles stay where they happen to be adsorbed.
- The adsorbed impurities on the crystal interface are completely mobile.

The first case is valid when *equilibrium* between the crystal and the vapour exists over the growing interface. It will be discussed in Section 5.7.2.

The second case is valid at *non-equilibrium*. It will be discussed in Section 5.7.3. In both cases the discussion will be restricted to homogeneous impurity trapping.

5.7.2 Equilibrium Trapping of Immobile Impurities

Equilibrium Partition of Immobile Impurities between Crystal and Vapour

Homogeneous trapping means that impurity particles such as atoms, molecules or ions, enter the crystal lattice according to the equilibrium conditions.

The chemical potential per kmol of the impurity atoms in the crystal can, by analogy with Equation (2.53) on page 73 in Chapter 2, be written as

$$\mu_{ic} = \mu_{ic}^0 + RT \ln \gamma_{ic} x_{ic} \tag{5.84}$$

If the concentration of impurity atoms in the vapour is expressed as pressure we obtain analogously

$$\mu_{iV} = \mu_{iV}^0 + RT \ln f_{iV} p_{iV} \tag{5.85}$$

where
μ_{ic} = chemical potential of an impurity atom in the crystal
μ_{iV} = chemical potential of an impurity atom in the vapour
μ_{ic}^0 = chemical potential of an impurity atom in the crystal in its standard state
μ_{iV}^0 = chemical potential of an impurity atom in the vapour in its standard state
γ_{ic} = activity coefficient of the impurity in the crystal
f_{iV} = activity coefficient of the impurity in the vapour
x_{ic} = molar fraction of the impurity in the crystal
p_{iV} = partial pressure of the impurity in the vapour phase
R = gas constant
T = absolute temperature.

The pressure p and the temperature T of crystallization of the pure main component determine the standard state of the vapour. The sizes of the atoms, their valences and the type of binding between the atoms in the lattice determine the chemical potentials. At equilibrium $\mu_{ic} = \mu_{iV}$. This relationship, in combination with Equations (5.84) and (5.85), gives an expression of the equilibrium concentration of the impurity in the solid crystal

$$x_{ic} = p_{iV} \frac{f_{iV}}{\gamma_{ic}} e^{\frac{\mu_{iV}^0 - \mu_{ic}^0}{RT}} \tag{5.86}$$

Equation (5.86) is fundamental and describes the equilibrium partition of the impurity atoms between the crystal and the vapour.

Effect of Crystal Curvature and Strain on the Equilibrium Partition of Immobile Impurities
The chemical potential of the solid crystal is influenced by many variables such as the curvature of the interface and the strain in the crystal. The influence on the partition of impurities between the crystal and the vapour is discussed below.

Effect of Crystal Curvature on Equilibrium Partition of Immobile Impurities
The curvature of a step increases the pressure p of the vapour in contact with the curved crystal surface. The new pressure will be

$$p_{eff} = p + \frac{2\sigma}{\rho_c} \tag{5.87}$$

where
σ = surface energy per unit area of the crystal surface relative to the vapour
ρ_c = radius of curvature of the crystal.

The changed pressure affects the chemical potential of the vapour and all other quantities, which depend on the pressure, for example Gibbs' free energy, the rate of normal growth and the step growth rate at layer growth.

Effect of Strain on Equilibrium Partition of Immobile Impurities
In Section 5.6.7 on page 233 we discussed the influence of strain in the crystal on the formation of spiral steps. Strain is likely to influence the chemical potential of the crystal strongly. If strain energy is taken into consideration Equation (3.80) on page 143 in Chapter 3, Equation (5.84) on page 143 will be (area strain = twice linear strain) replaced by

$$\mu_{ic} = \mu_{ic}^0 + RT \ln \gamma_{ic} x_{ic} + Y\varepsilon^2 V_m \tag{5.88}$$

and the equilibrium concentration of impurity atoms in the crystal will be (compare Equation (5.86) on page 239)

$$x_{ic} = p_{iV} \frac{f_{iV}}{\gamma_{ic}} e^{\dfrac{\mu_{iV}^0 - \mu_{ic}^0 - Y\varepsilon^2 V_m}{RT}} \tag{5.89}$$

where
Y = the modulus of elasticity of the crystal
ε = relative stretch length
V_m = molar volume of the crystal material.

Effect of Crystal Curvature on Rate of Step Growth

We have discussed the edge growth of a planar step in Section 5.6. The more the system deviates from equilibrium, the higher will be the growth rate.

$$V_{growth} = const \times \left(\mu_{vapour} - \mu_{crystal}\right) = const \times \Delta\mu \tag{5.90}$$

Equation (5.90) is valid when the step is *planar*, i.e. its radius of curvature is infinite. However, curved steps occur very often. The curvature changes the growth rate of the step considerably:

$$V_{\rho_c} = const \times \left(\mu_{vapour} - \mu_{\rho_c}\right) = const \times \Delta\mu\left(1 - \frac{\rho}{\rho_c}\right) \tag{5.91}$$

and according to the theory of two-dimensional nucleation

$$\rho_c = \frac{\sigma V_m}{-\Delta G_m} = \frac{\sigma V_m}{\Delta\mu} \tag{5.92}$$

where
V_{growth} = growth rate of a planar straight step
V_{ρ_c} = growth rate of a step with the radius of curvature ρ_c
μ_{vapour} = chemical potential of the vapour
μ_{ρ_c} = chemical potential of a step with the radius of curvature
$\Delta\mu$ = difference in chemical potential between of vapour and a flat crystal ρ_c
ρ = radius of curvature of the step
ρ_c = radius of curvature of the crystal
G_m = molar Gibbs' free energy of the crystal
$-G_m$ = $\Delta\mu = \mu_{vapour} - \mu_{\rho_c}$
σ = surface energy per unit area of the edge
V_m = molar volume of the crystal.

A rough estimate of the edge growth rate has been given by Cabrera and Vermilyea [7]:

$$V_{\rho_c} = V_{\text{growth}}\sqrt{\left(1 - \frac{2\rho_c}{d}\right)} \tag{5.93}$$

where d is the distance between two impurity particles. If $d < 2\rho_c$ Equation (5.93) gives an imaginary value of V_{ρ_c}. This means that the step cannot squeeze the impurity obstacles. The impurity particles act as stoppers and the further growth of the step is inhibited.

5.7.3 Non-Equilibrium Trapping of Impurities

During growth, two different processes will occur on the interface.

1. incorporation of adatoms into the crystal and rearranging them to fit into the ordered state of the crystal lattice
2. transport of impurity atoms over the interface in order to achieve equilibrium in the crystal.

The last process is a result of different composition in the interface and the crystal. It is normally much slower than the first process. At high growth rates the transport of impurities along the interface is not rapid enough to achieve and maintain equilibrium. Therefore, impurities become trapped and the crystal obtains a composition that deviates from equilibrium. The trapping is a function of growth rate.

Two examples of non-equilibrium trapping of impurities are of special interest and will be treated below: trapping of mobile impurities and vacancy trapping.

Trapping of Mobile Impurities

As was pointed out in Section 5.6, layer growth is based on diffusion of the adsorbed vapour atoms. If mobile impurity atoms also are present we obtain an extremely dynamic system in incessant motion and the impurity concentration along a step is in equilibrium with the impurity concentration in the vapour.

Consider a growing step on a smooth crystal surface with impurity atoms present (Figure 5.33). If the growth rate is high enough, impurity atoms become incorporated into the step. The number of foreign atoms is a function of the growth rate.

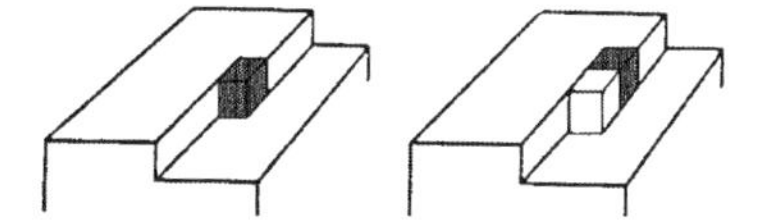

Figure 5.33 Incorporation of impurity particles into a crystal lattce. Reproduced from Springer © 1984.

If the growth of the step is not too high both the vapour atoms and the impurity atoms are adsorbed and desorbed at the crystal surface a tremendous number of times per unit time. Then, there is a good chance that the ratio of impurity atoms and vapour atoms in a row of the step and in the step layer as a whole corresponds to the equilibrium distribution when they become absorbed by kinks. The layer in surface equilibrium is covered by another layer during the further growth of the crystal and becomes part of the crystal in the bulk instead of a surface layer.

However, the surface equilibrium distribution of impurity particles is *not* the same as the corresponding distribution in the interior of the crystal. Hence, the normal situation is a state of non-equilibrium anyway. As is known from Section 5.6, the step can be considered as a rough interphase. Using this model the trapping mechanism can be described in the following way.

Trapping of impurity atoms into the crystal during growth is associated with the partition of the impurity atoms between the interphase and the vapour phase. The question is whether the impurity atoms become trapped into the solid, or if they concentrate in and follow the advancing interphase. The behaviour of the impurities is a function of

- the mobility of the impurity atoms;
- the velocity of the advancing interphase.

In Chapter 3 (page 169) Guggenheim's model of interphase was used to discuss the segregation of alloying elements in binary alloys to the liquid/vapour interphases and calculate the partition coefficient (Section 3.3.4 on pages 119–121 in Chapter 3). The results can be applied to the present analogous problem. If the liquid phase is replaced by a solid phase the Gibbs' free energy as a function of impurity concentration is essentially the same as that in Figure 3.36b on page 118 in Chapter 3. Upper Figure 5.34 can be used to construct the concentration of impurity atoms as a function of the distance from the surface, shown in lower Figure 5.34.

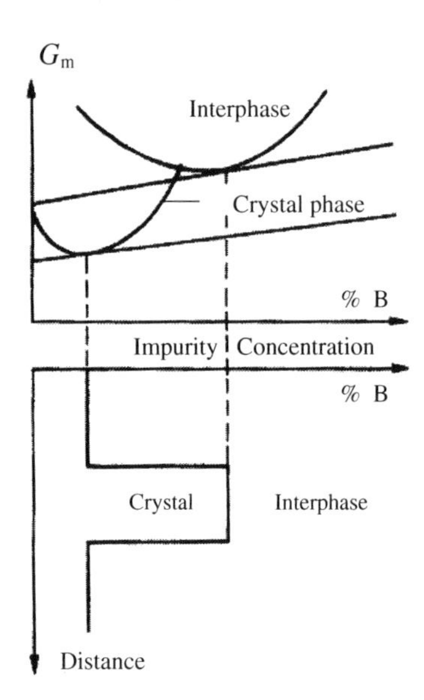

Figure 5.34 The interphase and the vapour phases are at equilibrium with each other.
Upper figure: Gibbs' free energy curves of the vapour phase and the interphase.
Lower figure: The concentration of the impurity B as a function of position.

Figure 5.34 is valid for equilibrium during growth. Two phenomena occur during growth. Firstly, a fraction of the impurities within the interphase become trapped inside the crystal. Secondly, the impurity atoms in the vapour phase do not arrive rapidly enough at the interphase to maintain its impurity concentration, that will decrease. Only at very high growth rates the composition of the crystal will be the same as that of the interphase.

Vacancy Trapping at Crystal Step Growth

The rough interphase of a step can be regarded as a volume of more loosely packed atoms than those in the interior of the crystal (Figure 5.10 on page 210). The interphase can be considered to be a normal crystal volume, which contains a number of vacancies.

Vacancy Concentration in a Vapour/Solid Interphase as a Function of the Rate of Step Growth
When the step front advances at solidification some of the vacancies will be trapped in the crystal. The concentration of vacancies varies with position and rate of step growth within the interphase.

The interphase in contact with the vapour has an equibrilium concentration of vacancies, which is called $n_{vac}^{0\,max}$ and is shown in Figure 5.35a. It is influenced by the structure of the interphase and the size of the vapour atoms absorbed inside the interphase. There is also an equibrilium between the concentration of vacancies n_{vac}^{c} in the crystal and in the interphase $n_{vac}^{0\,min}$, close to the crystal. The vacancy concentration $n_{vac}^{0\,min}$ inside the interphase, close to the crystal lattice, is lower than $n_{vac}^{0\,max}$.

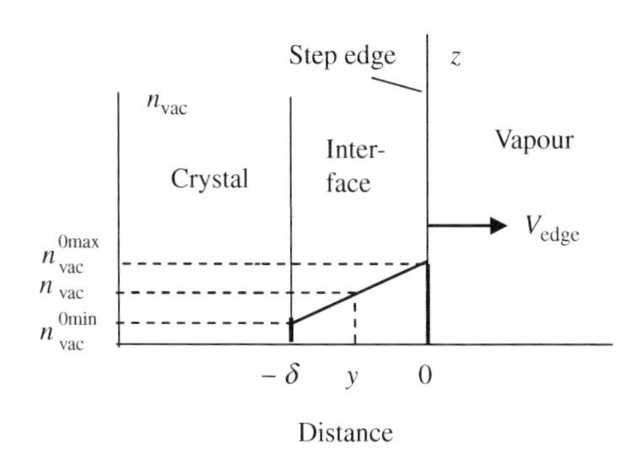

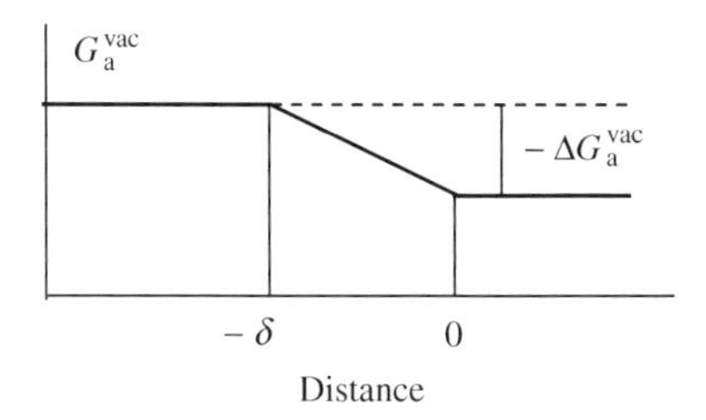

Figure 5.35a and b Schematic vacancy concentration as a function of position (distance from step edge). The step edge and the interphase moves forwards with the constant rate V_{edge}. Schematic Gibbs' free energy of vacancies as a function of position (distance from step edge).

Owing to the Gibbs' free energy difference (Figure 5.35b), caused by concentration differences of vacancies within the interphase, vacancies will diffuse from a lower to a higher vacancy concentration. The diffusion occurs by jumps. As shown on page 266 in Section 5.6 in Chapter 5 in [1], the total jump frequency of vacancies is proportional to the concentration of vacancies. It is also proportional to the vibration frequency of the atoms, i.e. to the number of attempts per unit time of an atom to overcome the potential barrier. Only the fraction of the vacancies that have a higher energy than the activation energy will be able to jump.

Some vacancies inside the interphase become trapped into the crystal during growth of the step. The Gibbs' free energy of the vacancies increases when the Gibbs' free-energy profile coupled to the interphase advances in the positive y direction, as is shown in Figure 5.35b. The trapped vacancies start to diffuse from higher to lower Gibbs' free energy, i.e. towards the interphase in spite of the low vacancy concentration in the crystal compared to that of the interphase (Figure 5.35a). This may seem to be a contradiction, but it is always the Gibbs' free energy condition that controls the behaviour of the vacancies and not the concentration gradient.

When the net flow of vacancies is calculated, care has to be taken of the different activation energies in the two directions (interphase to vapour and vapour to interphase). The net jump frequency of vacancies can be written as

$$\frac{dn_{vac}}{dt} = n_{vac}P\nu\, e^{-\frac{U_{a\,act}}{k_B T}} - n_{vac}P\nu\, e^{-\frac{U_{a\,act} - \Delta G_a^{vac}}{k_B T}} \tag{5.94}$$

where

n_{vac} = the number of vacancies per unit volume
t = time
P = probability factor (compare pages 237–241 in Chapter 5 in [1])
ν = vibration frequency of the atoms in the crystal lattice
$U_{a\,act}$ = activation energy of diffusion from interphase to vapour
$-\Delta G_a^{vac}$ = driving force of diffusion
k_B = Boltzmann's constant
T = temperature.

If ΔG_a^{vac} is small compared to $k_B T$ simple expansion of the e-function in series is possible and Equation (5.94) can approximately be written as

$$\frac{dn_{vac}}{dt} = n_{vac}P\nu\, e^{-\frac{U_{a\,act}}{k_B T}}\left(1 - e^{\frac{\Delta G_a^{vac}}{k_B T}}\right) = D_{vac}^{I}\, n_{vac}\frac{\Delta G_a^{vac}}{k_B T} \tag{5.95}$$

where $D_{vac}^{I} = P\nu\, e^{-\frac{U_{a\,act}}{k_B T}}$.

In Equation (5.95) we separate the variables n_{vac} and t, substitute dt by dy/V_{edge} and integrate the resulting equation.

$$\int_{n_{vac}^{0\,max}}^{n_{vac}} \frac{dn_{vac}}{n_{vac}} = \int D_{vac}^{I} \frac{\Delta G_{a}^{vac}}{k_B T} dt = \int_{0}^{y} D_{vac}^{I} \frac{\Delta G_{a}^{vac}}{k_B T} \frac{dy}{V_{edge}} \tag{5.96}$$

Integration of Equation (5.96) gives

$$\ln \frac{n_{vac}}{n_{vac}^{0\,max}} = D_{vac}^{I} \frac{\Delta G_{a}^{vac}}{k_B T} \frac{y}{V_{edge}}$$

or

$$n_{vac} = n_{vac}^{0\,max} e^{\frac{D_{vac}^{I} \frac{\Delta G_{a}^{vac}}{k_B T}}{V_{edge}} y} \qquad -\delta \leq y \leq 0 \tag{5.97}$$

where

$D_{vac}^{I} = P\nu e^{-\frac{U_{a\,act}}{k_B T}}$ = diffusion coefficient of vacancies in the interphase
n_{vac} = number of vacancies per unit volume at position y
$n_{vac}^{0\,max}$ = the number of vacancies per unit volume at $y = 0$
V_{edge} = rate of step growth
δ = thickness of the interphase.

Equation (5.97) represents the concentration of vacancies as a function of position (y is negative) and step growth rate V_{edge}.

For $y = -\delta$ we obtain $n_{vac}^{c} = n_{vac}^{0\,min}$. Hence, the number vacancies per unit volume in the interphase close to the crystal will be

$$n_{vac}^{c} = n_{vac}^{0\,min} = n_{vac}^{0\,max} e^{-\frac{D_{vac}^{I}}{V_{edge}} \frac{\Delta G_{a}^{vac}}{k_B T} \delta} \tag{5.98}$$

where δ is the thickness of the interphase (see Figure 5.35a).

The higher the step growth rate is, the higher will be the concentration of trapped vacancies in the crystal. If the step growth is rapid, there is little difference between $n_{vac}^{0\,min}$ and $n_{vac}^{0\,max}$.

5.8 Epitaxial Growth

The term *epitaxial growth* is a concept usually used for an oriented crystallization by vapour deposition of a thin film on a foreign substrate. The formation occurs by

1. cluster formation of adsorbed atoms on the surface of the substrate
2. nucleation of crystallites from the clusters
3. subsequent growth of the crystallites on the substrate surface.

If the vapour and the substrate are different the growth is *hetero-epitaxial*, otherwise *isoepitaxial*.

There are a large number of film deposition methods such as the use of an electron beam (Figure 5.36a), sputtering or chemical vapour deposition (CVD) (Figure 5.36b). The latter is described in Section 5.3 at the beginning of this chapter.

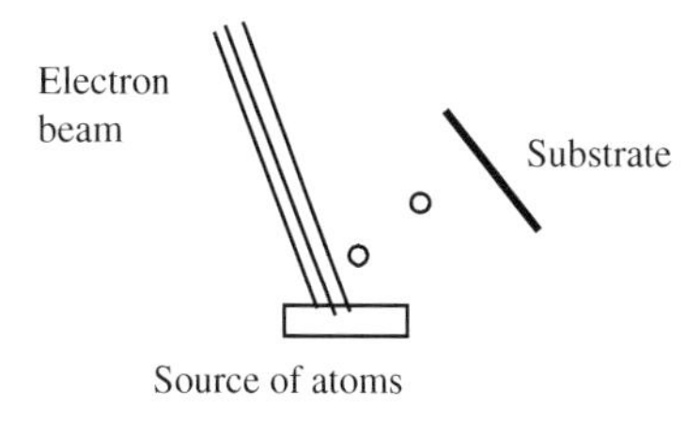

Figure 5.36a Electron beam method. An electron beam hits the source of atoms. The releasesd atoms strike the substrate.

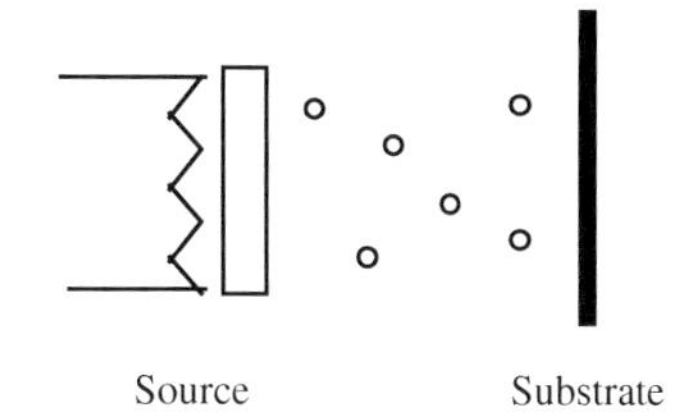

Figure 5.36b Sputtering. The source is heated directly or indirectly and the source emits atoms. The substrate serves as a target.

The treatment of epitaxial growth given below refers entirely to CVD processes and the formation of thin films directly from a vapour phase.

5.8.1 Epitaxial Growth of Thin Films

The main factors, which control the epitaxial growth of thin films, are

- nature of substrate
- temperature of substrate
- deposition rate
- influence of vapour impurities.

Nature of Substrate

As indicated by the definition of epitaxial growth the orientation of the thin film relative to the substrate is very important. Equilibrium shape and stability of crystals have been discussed in Chapter 3 in Section 3.5. Wulff's rule, which decides the equilibrium shape of a crystal, the discussion on crystal faces and their Gibbs' free energies, the conditions for stability of crystals and the concept of misfit in Section 3.4.7 are all highly relevant in connection with epitaxial growth.

It is inevitable that the crystallites formed on the substrate give a slight misfit that affects the epitaxial growth process. Monocrystalline substrates are necessary. To keep the number of defects down it is desirable to choose a smooth substrate surface. Cleaved surfaces of NaCl and mica are often used.

Metal substrates are used instead of ionic substrates in order to minimize twinning defects (page 28 in Chapter 1 in [1]). The surface must be kept free from oxide layers which otherwise may disturb the growth. All traces of oxygen and water vapour have to be carefully removed.

When metallic substrates are used for metallic depositions there is a risk for alloying at the interface. It is minimized by keeping the substrate temperature as low as possible.

Temperature of Substrate

The temperature is a very important parameter of epitaxial growth. The choice of temperature has three major effects. The temperature affects

1. the critical size and rate of formation of (100), (110), and (111) nuclei (Chapter 4, page 168)
2. the mobility of the adsorbed atoms
3. the annealing of defects in condensed films.

At very low temperatures the critical size of the nuclei is very small and the rate of nucleation high. The lateral rate of growth is also very low because the mobility of the adsorbed atoms is low and disoriented crystallites form easily. The resultant thin films become imperfect and polycrystalline (Figure 5.37a).

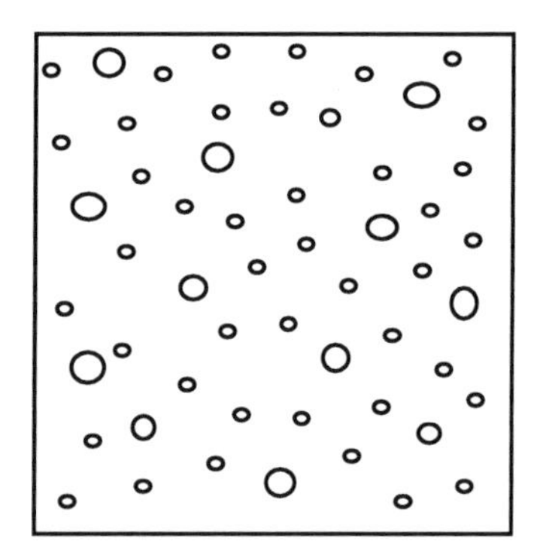

Figure 5.37a Low substrate temperatures give imperfect and polycrystalline thin films.

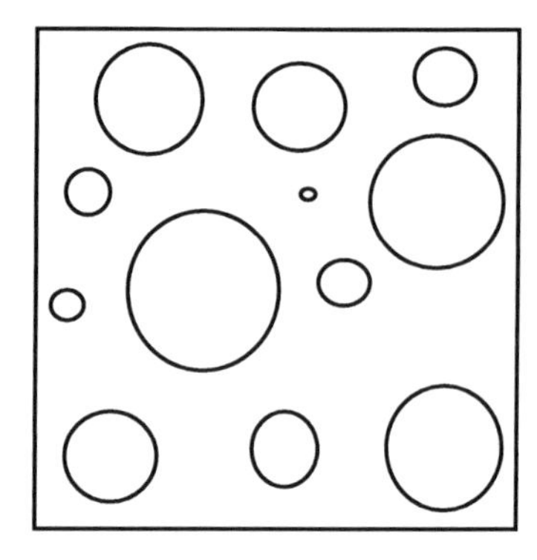

Figure 5.37b At higher temperature than in Figure 5.36a the rate of nucleation decreases and the surface mobility and the lateral growth rate increase. A low number of defects become incorporated. Fewer nuclei at larger distances form than in Figure 5.36a.

If the temperature is too high the crystallites will grow as rods or whiskers (see Section 5.9) and no uniform continuous film will be formed. Hence, there is an optimal range of epitaxial temperatures.

The optimal conditions are given in Table 5.1 for some common metals.

The substrate temperature also affects *annealing of defects, recrystallization* and *grain growth.* At annealing after deposition the crystallites reorient themselves relative to each other in order to minimize the total energy. This improves the quality of a film considerably.

Table 5.1 Optimal conditions for epitaxial growth of thin films for some metals

Metals	Crystal Orientation	Substructure	Temperature of substrate °C
Ag	(100)	NaCl	150
		CaF_2	> 510
	(111)	Mica	250–300
		$CaCO_3$	470
		Cu	250
Au	(100)	NaCl	400
	(110)	$CaCO_3$	510
	(111)	Mica	450
		$CaCO_3$	360
		CaF_2	380
		Cu	25
		Ag	270
Cu	(100)	NaCl	300
Pr	Variable	Ag	350
Ni	(100)	NaCl	370
	(100), (110), (111)	Polished NaCl	330
		Cu	330
Co	(100)	NaCl	> 540
	(100), (110), (111)	Polished NaCl	500
Al	(100)	NaCl	440
Cr	(100)	NaCl	> 540
Zr	(0001)	Cu	25

Deposition Rate

When the deposition rate is increased the quality of the film decreases. A likely explanation is that more disoriented nuclei form or that the rate of nucleation increases.

Influence of Vapour Impurities

Impurities in the vapour, especially strongly adsorbing gases such as oxygen, nitrogen and water vapour, have a strong influence on the deposition processes. A monolayer of adsorbed oxygen may even inhibit the formation of an oriented deposition of metals.

5.8.2 Epitaxial Growth of Silicon and Diamond

As concrete examples of epitaxial growth two most important examples will be chosen. The production of thin films of pure and doped silicon and of pure diamond will be discussed together with arguments, based on earlier knowledge of epitaxial growth, for choice of CVD methods (page 245).

A severe demand is highest possible cleanness and environment. Undesired contamination during the production processes must be avoided.

Silicon and other semiconductors are of a tremendous importance in the electronic industry, where the demands on quality of the products are extremely high. Thin films of diamond are also required by industry.

Silicon

Three types of chemical reactions have been developed and are used for thin film production of pure and doped silicon:

- pyrolytic or thermal decomposition reactions
- disproportionate reactions
- reduction reactions.

An example of a *pyrolytic decomposition reaction is*

$$SiH_4 \rightarrow Si + 2\,H_2$$

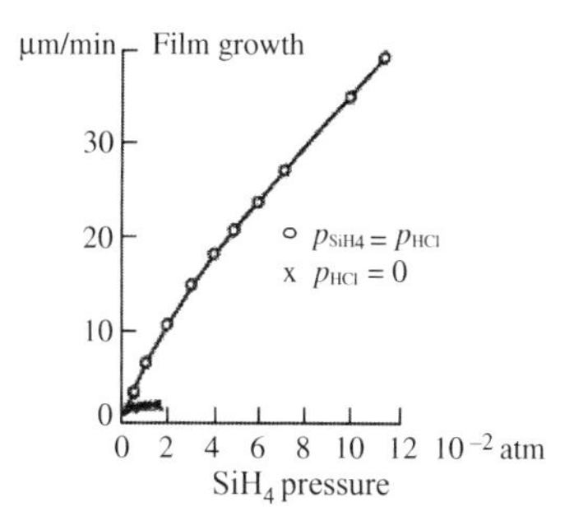

Figure 5.38 Film growth rate as a function of SiH_4 pressure for the reaction $SiH_4 \rightarrow Si + 2\,H_2$ with and without addition of HCl and a substrate temperature of 1150 °C.

The deposition rate is temperature dependent as is shown in the lower short curve in Figure 5.38. If HCl is added the rate increases strongly (upper curve in Figure 5.38). A much higher rate is obtained if some HCl is added. The reason for this is that clusters of Si atoms form in the vapour by homogeneous nucleation in the gas phase. This retards the rate of deposition of Si considerably. The added HCl gas destroys and dissolves the clusters by etching.

Pyrolytic decomposition reactions have the disadvantage of being performed in a sealed vessel where it is difficult to control the process and hence the quality of the product. The method is not convenient for industrial use.

A pressure and temperature controlled decomposition of a chemical compound that results in a similar compound is called a *disproportionate reaction*. An example of such a reaction is

$$2\,SiCl_2 \rightarrow Si + SiCl_4$$

$SiCl_2$, formed at a certain pressure and temperature, becomes instable when the temperature is changed. The reaction is performed in a sealed tube and the decomposition rate is low. The reaction cannot be used for industrial production.

The reduction of trichlorosilane with hydrogen (see Figure 5.6a on page 205) is a commercially useful modern method to produce semiconductor components.

$$SiHCl_3 + H_2 \leftrightarrow Si + 3\,HCl$$

It is an open, non-recycling method with a one way flow that allows control of the process, cleaning the reaction chamber and controlled addition of doping elements. The deposition rate is high. There is a great number of apparatus designs and methods for heating, for example electrical induction and microwave radiation. The temperature in the deposition chamber is measured by an optical pyrometer.

The growth rate varies with the pressure and with the crystallographic orientation as is shown from Figures 5.39 and 5.40.

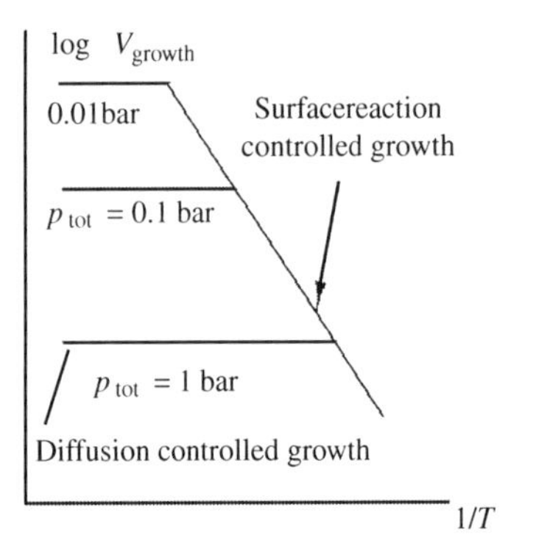

Figure 5.39 Pressure dependence of deposition rate for silicon. Reproduced from Cambridge University Press © 1991.

Figure 5.40 Orientation dependence of deposition rate for silicon. Reproduced from Cambridge University Press © 1991.

Table 5.2 Dissociation reactions of the Si-H-Cl system

$SiCl_4 + 2\,H_2 \leftrightarrow Si + 4\,HCl$
$SiHCl_3 + H_2 \leftrightarrow Si + 3\,HCl$
$SiH_2Cl_2 \leftrightarrow Si + 2\,HCl$
$SiH_3Cl \leftrightarrow Si + HCl + H_2$
$SiCl_2 + H_2 \leftrightarrow Si + 2\,HCl$
$SiCl + \frac{1}{2}\,H_2 \leftrightarrow Si + HCl$

The chemical reactions are complex, which is evident from Table 5.2.

Three conditions must be fulfilled for growing monocrystalline layers of silicon:

- the crystal structures, atomic dimensions and thermal expansion of the substrate and the deposition must be compatible
- the substrate surface must be clean and without serious surface damage
- the deposition temperature must be compatible with the deposition rate.

The first condition is automatically fulfilled when silicon disks are used as substrate.

Diamond

Great efforts have been spent to develop methods to produce thin films of diamond because of the excellent mechanical and chemical qualities of diamond. The produced films seem to have all the good qualities characteristic for natural diamonds. One important application is to use the films to cover cutting tools, which increases their efficiency and prolongs their lifetime considerably.

The task was not easy because the vapour pressure of carbon is very low and graphite is the thermodynamically most stable state of carbon. The phase transformation of graphite into diamond starts at the surface of graphite in bulk and depends strongly on the purity of the specimen and its crystallographic orientation.

The quality of the diamond surface can be examined with the aid of low energy electron diffraction (LEED) to check the structure. Evidence shows that heating in ultra high vacuum or treatment with hydrogen does not change the surface structure much at temperatures below 1300–1400 °C. Diamond is a metastable state of carbon.

Experiments confirm that treatment of a diamond surface with activated hydrogen, i.e. atomic hydrogen, extends the meta-stability considerably, up to 1600 °C.

Methods based on deposition of carbon by evaporation from carbon source are seldom used because of the low carbon pressure and low deposition rate. Other possible methods, which involve chemical reactions, have been tried, for example

- pyrolytic decomposition reactions
- reduction reactions.

Pyrolytic decomposition of methane $CH_4 \leftrightarrow C + 2\,H_2$ did not work because of the low deposition rate and the formation of graphite during deposition. Conventional reduction of methane with hydrogen, deposition of a mixture of methane and hydrogen gases did not work either because of low deposition rate.

The solution of the problem was

- observation of activation influence of additives on the stability of the diamond surface
- addition of small amounts of oxygen or water vapour to the activated gas mixture of methane and hydrogen.

The success was based on the action of the activated atomic hydrogen and the selected etching effect of oxygen or the OH^- radical on graphite, which is shown in Figure 5.41. The growth rate is nearly the same for diamond and graphite and the etching rate is much larger for graphite than for diamond. The net rate is positive for diamond and negative for graphite.

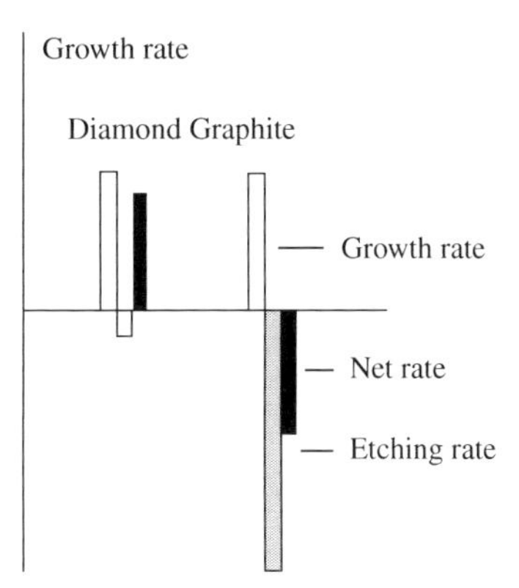

Figure 5.41 Growth and etching rates of diamond and graphite. The black piles represent the net rates.

The first epitaxially grown diamond thin films were grown in the middle of the 1960s. Partially dissociated hydrogen, heated to 1900 °C, was mixed with water-cooled pure methane in the ratio 95:5 and allowed to deposit at a temperature of 1020 – 1030 °C. A grey thin film, about 1 μm thick was formed in 1.5 hours. Examination with RHEED (reflexion high energy electron diffraction) showed all the characteristics of diamond and none of graphite.

Today there is a great variety of methods for production of artificial diamond thin films. Vapour activation is obtained by thermal heating, electrical discharges, radio frequency heating, microwave heating, chemical methods or photochemistry (laser). The most popular method for epitaxial and polycrystalline growth of diamond thin films is still the use of a hot filament of temperature 2000 °C and a gas mixture of methane and dissociated hydrogen with minor additives of oxygen and water vapour (see Figure 5.42 and Figure 5.6b on page 205).

The maximum thickness of the diamond film has increased from 1–5 μm in early experiments up to hundreds of μm or even mm size at the end of the 20th century.

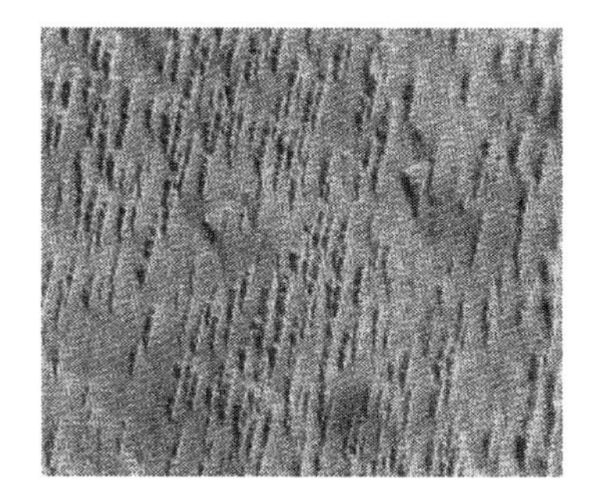

Figure 5.42 Epitaxial diamond film, grown on a polished natural diamond surface with a crystallographic orientation equal to 5° deviation from the (100) plane. Reproduced from Elsevier © 1994.

5.9 Whisker Growth

Whisker growth is a type of crystal growth, which is strongly anisotropic. The growth rate in one direction is much faster than in the lateral directions. The resultant crystals, so-called whiskers, have the shape of needles and are perpendicular to the substrate surface (Figures 5.43a and b).

Figure 5.43a Tungsten dendrite whiskers. Upper figure: [110] growth. Lower figure: [111] growth. Courtesy of J. Gilman.

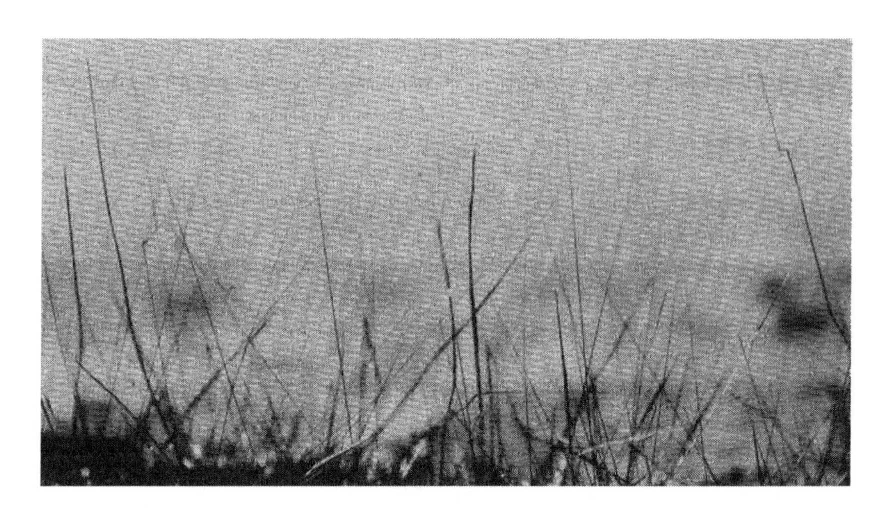

Figure 5.43b Iron whiskers grown from $FeBr_2$ at 700 °C. Courtesy of J. Gilman.

Thin films with whisker crystals offers for example an easy way to obtain plane polarized light (Chapter 7 in [1]). Al_2O_3 whiskers covered by Pt are used in the car industry for catalytic purification of combustion gases, owing to their large surface area.

Whisker crystals have unusual mechanical properties, better perfection (lower fraction of defects) and better crystal orientation than normal crystals. As an example, it can be mentioned that Zn whiskers have 100 –1000 times the strength of the bulk crystals of the metal. The disadvantage of whiskers is small size, which makes the crystals difficult to handle.

In thin film production in semiconductor industry, the formation of whiskers is highly inconvenient and must be carefully avoided because the coating of the substrate becomes very poor.

The whisker phenomenon is puzzling and no fully satisfactory and generally accepted explanation has been found so far. The growth methods are presented in Section 5.9.1. The prevailing explanations will be briefly discussed in Section 5.9.2.

5.9.1 Growth Methods

Volmer and Eastermann [2] primarily observed anisotropic growth in 1921 on mercury platelets, which grew at least 10^4 times faster in the length direction than in all perpendicular directions. Since then many experimental studies have been performed to investigate the conditions for whisker growth and to develop optimal methods for production of whiskers.

The main methods of whisker production are based on

- condensation of metal vapours in vacuum, sometimes in a hydrogen or inert gas atmosphere
- reduction of metal halides in hydrogen.

The experimental conditions determine the length and appearance of the crystals. The magnitude of the lengths varies between 0.1 mm and 20 mm.

Coleman and Sears [8] showed that addition of a pure inert gas increases the yield of Zn whiskers by condensation and prolongs the length of the crystals considerably.

The equipment for whisker growth is very simple. Figure 5.44 shows an example. A small open container filled with the solid metal halide is placed in a hot furnace, which quickly heats it up to the reaction temperature. The halide reacts with streaming hydrogen gas and the crystals nucleate on small metal or metal oxide crystals and grow in the cooling chamber.

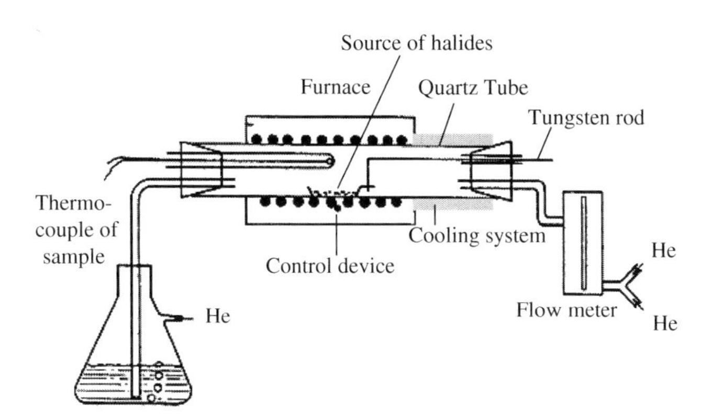

Figure 5.44 Equipment for production of whiskers from metal halides. Courtesy of J. Gilman.

5.9.2 Mechanisms of Whisker Growth

It has been suggested that whisker growth in highly anisotropic substances may be explained by the existence of a direction of extremely high growth rate. This does not seem to be the likely reason for two reasons: Whisker growth also occurs in many metals with symmetric structures, for example cubic structure. The ratio of other anisotropic properties does not show such extreme maxima as the whisker growth rate.

Whisker growth in vapours is assumed to occur in three steps. The first two steps have been verified experimentally.

1. Vapour atoms become adsorbed on the smooth walls of the whisker.
2. The adsorbed atoms diffuse towards the whisker tip.
3. The adsorbed atoms become incorporated into kink sites in a single axial screw dislocation at the tip of the whisker.

The main question to be explained is to tell why the growth rate parallel to the surface is low or zero. The most accepted theory of whisker growth is associated with the existence of impurities.

According to Section 5.7.2 on pages 238–241 impurity atoms influence the growth rate strongly. Immobile adsorbed particles on the smooth walls of a whisker may slow down and even stop the advancement of the steps away from the nucleated source. The impurity atoms at the tip become 'buried' in the growing crystal, which keeps the tip free from impurities and new layers can rapidly be generated.

5.10 Mechanical Restrictions on Thin Films

In the preceding sections various kinds of crystal growth and production of thin films of small crystals from vapours have been treated.

In the present section we will discuss the solid mechanics and stability of thin films and show that there are restrictions concerning their minimum thickness.

5.10.1 Stress in Thin Films

In Chapter 3 on pages 102–104 the elastic stress tensor ε_{ij} was introduced. This concept will be used to examine the stress of an epitaxially grown thin film attached to a substrate.

Consider a thin film on a substrate which is completely free from stress (Figure 5.45a). When the thin film is separated from the substrate it is still in a stress-free state (Figure 5.45b) and its lateral dimensions agree completely with those of the substrate.

If the dimensions of the film change, e.g. caused by atomic force relaxations, elastic strains and stresses will develop in the film. Suppose that the film experiences uniform volume shrinkage (Figure 5.45c). In this case of pure dilatational strain, the principal strain components are:

$$\varepsilon_{xx} = \varepsilon_{yy} = \varepsilon_{zz} = \frac{e_T}{3} \qquad (5.99)$$

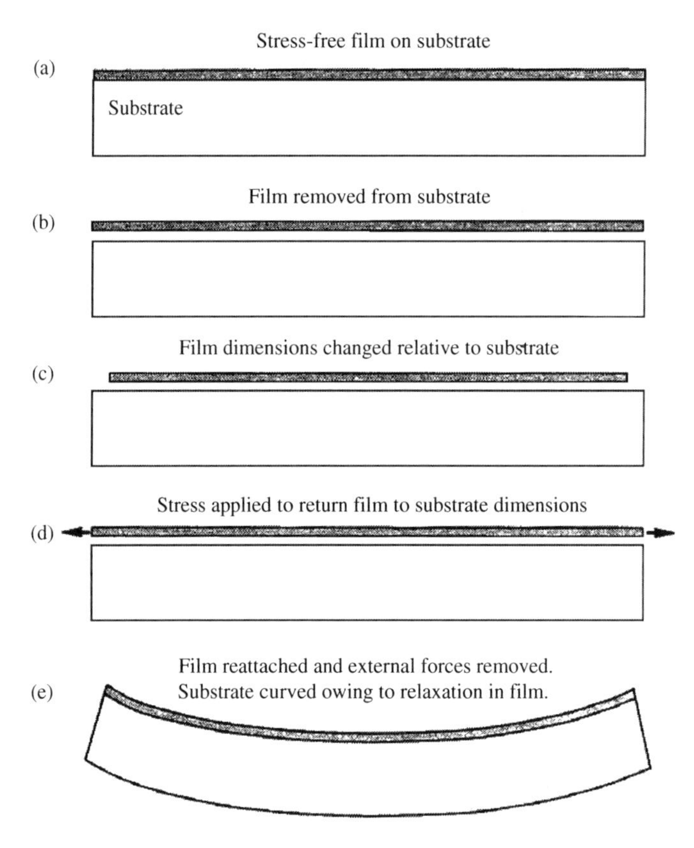

Figure 5.45 Biaxial stress development in a thin film and the associated bending of its attached substrate. Reproduced from Angelika Sperlich (angelika.sperlich@oldenbourg.de) © 1999.

where

$$\varepsilon_{xx} = \frac{dl_x}{l_x} \quad \varepsilon_{yy} = \frac{dl_y}{l_y} \quad \varepsilon_{zz} = \frac{dl_z}{l_z}$$

and

$\varepsilon_{xx}\varepsilon_{yy}\varepsilon_{zz}$ = principal strain components
$l_x l_y l_z$ = dimensions of the film
e_T = dilatation transformation strain.

Next, we reattach the film to the substrate. It does no longer fit and a biaxial stress must be exerted on the film to deform it elastically to bring the film back to the former substrate dimensions (Figure 5.45d). The stress required to do this creates elastic strain in the film, which exactly compensates the transformation strain:

$$\varepsilon = \varepsilon_{xx} = \varepsilon_{yy} = \varepsilon_{zz} = \frac{-e_T}{3} \tag{5.100}$$

According to Hooke's law the biaxial stress in the film will be

$$\sigma_f = \sigma_{xx} = \sigma_{yy} = M\varepsilon \tag{5.101}$$

where subscript 'f' means film and
σ_f = biaxial tensile stress in the film
M_f = biaxial elastic modulus of the film
ε_f = elastic strain in the film.

The biaxial modulus of the film in case of isotropic elasticity can be written as

$$M_f = \frac{Y}{1-\nu} \tag{5.102}$$

where
Y = Young's modulus of elasticity
ν = Poisson's ratio.

When the film is reattached on the substrate a couple of forces, reaction edge forces, associated with the stress in Figure 5.45d, will act on the film and the substrate, respectively. The edge forces, acting on the substrate, cause shear stresses, which result in an elastic bending of the substrate (Figure 5.45e). The radius of curvature of the bent substrate will be

$$\rho = \frac{M_s h_s}{6 \sigma_f h_f} \tag{5.103}$$

where
ρ = radius of curvature of the substrate
M_s = biaxial elastic modulus of the substrate
h_s = thickness of the substrate
σ_f = biaxial tensile stress in the film
h_f = thickness of the film.

The thicknesses of thin films are often of the magnitude 1–10 μm and the substrate thicknesses are normally ~0.5–1 mm. When several thin films are deposited on a much thicker substrate, each film causes a fixed bending determined by the stress and thickness of each film, independent of the others. The total substrate curvature will be the algebraic sum of the curvatures of all the films.

$$\frac{1}{\rho_{total}} = \frac{1}{\rho_1} + \frac{1}{\rho_2} + \cdots\cdots + \frac{1}{\rho_n} \tag{5.104}$$

5.10.2 *Mismatch Dislocation Formation in Epitaxial Films. Critical Film Thickness*

Semiconductor thin films shall preferably be grown on dislocation-free substrates by heteroepitaxial methods. In such cases, it is impossible to avoid mismatch between the crystals of the film and the substrate crystals. This mismatch causes a uniform strain in the film, which causes very large biaxial stresses in the film.

It would be natural to assume the stresses, caused by mismatch, to be relaxed by plastic flow in the film, dislocation nucleation and motion, independent of the film thickness.

However, if the energy relationships of the system are considered, the result is that no relaxation can occur in very thin films. In such films the dislocation energy, caused by relaxation processes, is larger than the corresponding reduction of the strain energy. There is a critical film thickness h_{cr} below which the film is thermodynamically stable or in other words if the film thickness $< h_{cr}$ no mismatch dislocations can be formed in heteroepitaxial structures.

Calculation of the Critical Film Thickness h_{cr}

The calculations given below offer a method, reported by Tiller [9], to calculate the smallest possible thickness of a film free from mismatch dislocations.

The excess energy per unit area of a very thin film is lowest when no dislocations are present. Such films are completely coherent with the substrate in their equilibrium states in the absence of dislocations. According to Equation (3.77) in Chapter 3 on page 142, in the absence of dislocations the energy per unit area is given by the relationship

$$w_{\mathrm{h}} = M_{\mathrm{s}} h\varepsilon^2 \tag{5.105}$$

where

w_{h} = the strain energy per unit area (biaxial energy is twice uniaxial energy)
M_{s} = biaxial elastic modulus of the substrate
h = film thickness
ε = the biaxial elastic strain which must be exerted on the film to bring the lattices of the film and the substrate to coincide.

In the presence of dislocations, the dislocation energy has to be added to w_{h} to find the total energy of the system. The dislocation energy per unit area is given by

$$w_{\mathrm{d}} = \frac{G(b^*)^2}{4\pi(1-\nu)}\frac{2}{\lambda}\ln\frac{\beta h}{b^*} \tag{5.106}$$

where

w_{d} = dislocation energy per unit area
G = shear modulus of the thin film and the substrate, which are assumed to be equal
b^* = Burgers vector [three-dimensional vector used in solid mechanics to describe the size and direction of the lattice defect in a dislocation (page 25 in Chapter 1 in [1])]
ν = Poisson's ratio
λ = distance between adjacent dislocations
β = a constant of magnitude 1.

The presence of dislocations reduces the strain energy some-what, which is shown in Equation (5.107) of the total energy.

Hence, the total energy per unit area of thin films containing mismatch dislocations can be written as

$$w = M_{\mathrm{f}}h\left(\varepsilon - \frac{b^*}{\lambda}\right)^2 + \frac{G(b^*)^2}{2\pi(1-\nu)}\frac{1}{\lambda}\ln\frac{\beta h}{b^*} \tag{5.107}$$

The three functions w, w_{h} and w_{d} are shown in Figure 5.46 as a function of b^*/λ for a film without and with mismatch dislocations. w_{h} varies linearly with h while w_{d} is a logarithmic function of h. $1/\lambda$ represents the number of mismatch dislocations per unit length.

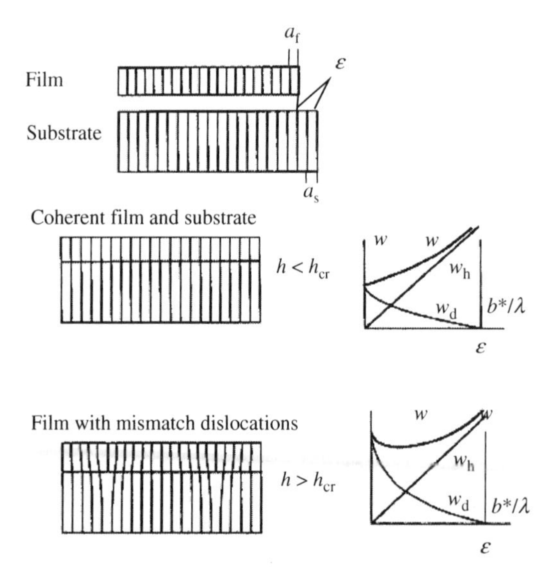

Figure 5.46 Upper figures: Non-coherent film and substrate.

Middle figures: w, w_{h} and w_{d} without mismatch dislocations as a function of b^*/λ.

Lower figures: w, w_{h} and w_{d} with mismatch dislocations as a function of b^*/λ. Reproduced from Cambridge University Press © 1991.

Figure 5.47 shows that only for h values above the critical thickness h_{cr} do the mismatch dislocations lead to a decrease of the energy of the system.

If we compare the slope of the tangent of the total energy curve at $b^*/\lambda = 0$ in the upper and lower diagrams we can conclude that

If $h < h_{cr}$ the derivative is > 0
If $h > h_{cr}$ the derivative is < 0
The conclusion is that
if $h = h_{cr}$ the derivative must be zero.

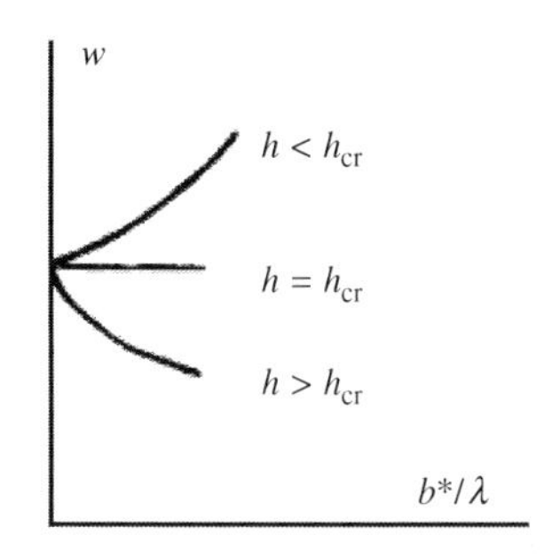

Figure 5.47 The total energy per unit area of thin films containing mismatch dislocations as a function of b^*/λ with h as a parameter.

This condition can be used to derive an expression of h_{cr}. We take the derivative of the total energy w in Equation (5.107) with respect to $1/\lambda$:

$$\frac{dw}{d\left(\frac{1}{\lambda}\right)} = 2M_f h\left(\varepsilon - \frac{b^*}{\lambda}\right)(-b^*) + \frac{G(b^*)^2}{2\pi(1-\nu)}\ln\frac{\beta h}{b^*} \tag{5.108}$$

The condition: derivative $= 0$ for $h = h_{cr}$ gives

$$h_{cr} = \ln\frac{\beta h_{cr}}{b^*}\frac{Gb^*}{4\pi(1-\nu)M_f\varepsilon} \tag{5.109}$$

The value of h_{cr} is not given explicitly in Equation (5.109) but can be calculated graphically or numerically.

Depending on the value of the film thickness there are two different cases:

$$h > h_{cr}$$

The equilibrium state includes mismatch dislocations which reduces the total energy of the film.

$$h < h_{cr}$$

The lowest possible energy corresponds to a film free from mismatch dislocations. Such an ideal completely coherent epitaxial film is thermodynamically stable.

For practical purposes it is advisable to let the film thickness always be lower than the critical value h_{cr} and to minimize the mismatch dislocations during the epitaxial growth of the film.

5.10.3 Instability of Thin Films

As shown in the preceding section epitaxial growth of thin films on heterogeneous substrates are always associated with strain owing to mismatch between the film and the substrate.

Relaxation of Thin Films

The classical way to achieve relaxation of strongly strained crystals is formation of dislocations when the strain energy is transformed into dislocation energy. However, this is not a realistic possibility in the case of thin films. As was shown on page 142 the activation energy for such a process is normally much higher than the available energy and the strained layer is therefore *metastable*, which prevents formation of dislocations.

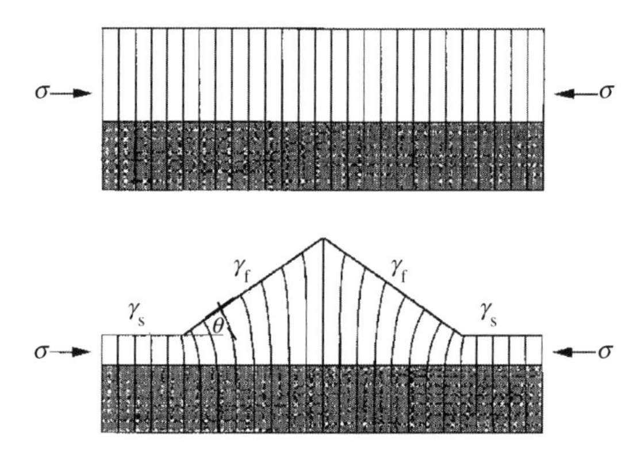

Figure 5.48 Upper figure: Cross-section of a coherently strained two-dimensional film.

Lower figure: Cross-section of a three-dimensional pyramidal island situated on top of a planar wetting layer.

σ = applied stress γ = surface energy per unit area.

At the end of the 20th century new ideas of the relaxation mechanism in strained epitaxial layers appeared in literature. Jesson [10] among others has reported convincing experimental evidence which shows that uniformly strained layers are instable, i.e. transform spontaneously from planar layers into dislocation-free strained islands of many different shapes. A simple example is given in Figure 5.48.

The uniform planar two-dimensional epitaxial film surface is spontaneously transformed into three-dimensional pyramidal islands, interrupted by pits and grooves. These provide points of great stress where the barrier to nucleation of dislocations is likely to be small.

Observations have shown that small islands have a compact and symmetrical shape. Above a certain critical size, the symmetry of the islands disappears. Large islands become spontaneously strikingly long and narrow. Such 'one-dimensional' thread-like islands are called 'quantum wires'.

It is of utmost interest and importance in the semiconductor industry to control and avoid the undesired phenomena described above.

Summary

■ Chemical Vapour Deposition Methods (CVD) to Produce Crystals and Thin Films

- Chemical transport methods
- Vapour decomposition methods
- Vapour synthesis methods.

The CVD methods require low supersaturation and provide a constant stoichiometric ratio of the reactants formed during the reactions. The methods can be used for a much larger variety of substances than other methods.

■ Simple Model of Crystal Growth

Growth rate:

V_{growth} = rate of thickness change of deposition layer.

Driving force = change in Gibbs' free energy $(-\Delta G)^{total}$ between the vapour and the crystal.

$$(-\Delta G^{total}) = (-\Delta G^{diff}) + (-\Delta G^{kin})$$

$$-\Delta G^{total} = const(x^{0} - x^{eq})$$

$$-\Delta G^{diff} = const(x^{0} - x^{I})$$

$$-\Delta G^{kin} = const(x^{I} - x^{eq})$$

Mass transport rate = number of atoms transferred to the crystal per unit area and unit time.

$$k_{\text{total}}(-\Delta G^{\text{total}}) = k_{\text{diff}}(-\Delta G^{\text{diff}}) = k_{\text{kin}}(-\Delta G^{\text{kin}})$$

Relationship between the reaction rate coefficients of solidification:

$$\frac{1}{k_{\text{total}}} = \frac{1}{k_{\text{diff}}} + \frac{1}{k_{\text{kin}}}$$

The slowest process controls the total growth rate.
Relationship between growth rate and mass transport rate:

$$V_{\text{growth}} = \frac{V_{\text{m}}}{N_{\text{A}}} k_{\text{total}}(-\Delta G^{\text{total}}) = a^3 k_{\text{total}}(-\Delta G^{\text{total}})$$

The crystal growth depends on the structure of the crystal surface:

- Normal growth on rough surfaces (Guggenheim interphase)
- Layer growth on smooth surfaces (Gibbs interface).

■ Normal Crystal Growth on Rough Crystal Surfaces in Vapours

Vapour atoms are

1. adsorbed on the interphase vapour/crystal
2. incorporated into kinks in the crystal.

The first process is very fast. Hence, the growth rate of normal growth is controlled by the second process.

Fluxes in and out of Kinks in the Crystal Surface:

$$j_+ = \frac{p}{\sqrt{2\pi m k_{\text{B}} T}} \frac{a^2}{\delta^2} e^{-\frac{U_{\text{a act}}}{k_{\text{B}} T}}$$

$$j_- = \frac{p_{\text{eq}}}{\sqrt{2\pi m k_{\text{B}} T}} \frac{a^2}{\delta^2} e^{-\frac{U_{\text{a act}}}{k_{\text{B}} T}}$$

Rate of normal growth:

$$V_{\text{growth}} = a^3 (j_+ - j_-)$$

or

$$V_{\text{growth}} = a^3 \frac{a^2}{\delta^2} \frac{1}{\sqrt{2\pi m k_{\text{B}} T}} e^{-\frac{U_{\text{a act}}}{k_{\text{B}} T}} (p - p_{\text{eq}})$$

Kinetic coefficient of normal growth rate:

$$V_{\text{growth}} = \beta_{\text{kin}} (p - p_{\text{eq}})$$

■ Layer Growth on Smooth Crystal Surfaces

In an ideal smooth crystal surface, all sites are occupied. Atoms from the ambient vapour phase can not be incorporated, only adsorbed at the surface. The only possibility of crystal growth is

- formation of *steps* or atom layers parallel to the smooth surface by two-dimensional nucleation
- lateral growth of the steps at their rough edges.

Two-dimensional Nucleation

Flat crystals are nucleated from ad-atoms close to the smooth crystal surface, grow, cluster and form a flat atomic layer. The new layer is strained and has a higher Gibbs' free energy than the crystal surface because the ad-atoms are larger than the lattice atoms.

Ad-atom Concentration at the Crystal Surface as a Function of Vapour Pressure

$$n_S = n_{0S} A\, e^{-\frac{U_{a\,act}^{ad\,p_{qe}} - \Delta G_a^{p \to p_{eq}}}{k_B T}}$$

where $\Delta G_a^{p \to p_{eq}} = G_a^p - G_a^{p_{eq}} = k_B T \ln \dfrac{p}{p_{eq}}$

or $\quad n_S = n_{0S} \dfrac{p}{p_{eq}} e^{-\frac{U_{a\,act}^{ad\,p_{eq}}}{k_B T}} \quad$ or $\quad \dfrac{n_S}{p} = \dfrac{n_S^{eq}}{p_{eq}}$

The number of ad-atoms per unit area on the crystal surface is proportional to the pressure of the vapour at equilibrium with the crystal.

Roughening Temperature. Influence of Pressure

At constant vapour pressure metal surfaces are normally smooth at low temperatures. Above the characteristic *roughening temperature* of the metal smooth and vicinal interfaces become rough.

At constant temperature a vapour pressure above the characteristic nucleation pressure also results in a transition from smooth to rough crystal surfaces.

Growth Process

Vapour atoms

1. are adsorbed on the smooth crystal surface but not incorporated into the crystal
2. diffuse along the smooth terraces of the step
3. are adsorbed by the interphase and incorporated into kinks in the crystal at the step edge.

The first process is very fast and does not influence the growth rate.

Driving force of layer growth:

$$(-\Delta G^{total}) = (-\Delta G^{ad}) + (-\Delta G^{diff}) + (-\Delta G^{kin})$$

The adsorption process is rapid and can be disregarded compared to the diffusion along the planar terrace step and the kinetic incorporation process.

Mechanism of Layer Growth

When the smooth surface is strained no vapour atoms can be adsorbed. The surface concentration close to the rough edge of the nearest step is permanently low close to the edge owing to incorporation of ad-atoms into the edge. For this reason the ad-atoms diffuse along the smooth surface towards the step edge.

Diffusion Process along the Terrace:

Mean lifetime of adatoms: $\tau_S = \frac{1}{\nu_\perp} e^{\frac{-\Delta G_a^{desorp}}{k_B T}}$

Basic equation of diffusion: $J = -D_S A \frac{dn_S}{dy}$

Diffusion coefficient: $D_S = a^2 \nu = e^{-\frac{\Delta U_{a\,act}^{diff}}{k_B T}}$

Mean distance traveled during the mean lifetime:

$$\lambda_{diff} = \sqrt{D_S \tau_S} \quad \text{or} \quad \lambda_{diff} = a\, e^{\frac{\Delta G_a^{desorp} - \Delta U_{a\,act}^{diff}}{2k_B T}}$$

Mean velocity of diffusing adatoms: $V_{diff} = \frac{\lambda_{diff}}{\tau_S} = \frac{D_S}{\lambda_{diff}}$

Diffusion flux: $j_{diff} = -\frac{D_S}{a}\frac{dn_S}{dy} = -\frac{D_S}{a}\frac{n_S^p - n_S^{step}}{\lambda_{diff}}$

Step-edge Growth controlled by Adatoms Diffusing along and from the Terrace

Velocity of step-edge growth, controlled by diffusion:

$$V_{edge} = -a^3 j_{diff} = -\left(-a^3 \frac{D_S}{a}\frac{n_S^p - n_S^{step}}{\lambda_{diff}}\right)$$

or

$$V_{edge} = a^2 \frac{D_S}{\lambda_{diff}}(n_S^p - n_S^{step})$$

Diffusion coefficient of step growth rate:

$$V_{edge} = \beta_{diff}\left(n_S^p - n_S^{step}\right)$$

where $\beta_{diff} = a^2 \frac{D_S}{\lambda_{diff}}$

Kinetic Process at the Step Edge

Fluxes to and from the step edge surface:

$$j_+ = \frac{p}{\sqrt{2\pi m k_B T}} e^{-\frac{U_{a\,act}^{ad}}{k_B T}} \qquad j_- = \frac{p_{eq}}{\sqrt{2\pi m k_B T}} e^{-\frac{U_{a\,act}^{ad}}{k_B T}}$$

Relationships between ad-atom surface concentrations and vapour pressure:

$$n_S^{eq} = \tau_S \frac{p_{eq}}{\sqrt{2\pi m k_B T}} e^{-\frac{U_{a\,act}^{ad}}{k_B T}} \qquad n_S^p = \tau_S \frac{p}{\sqrt{2\pi m k_B T}} e^{-\frac{U_{a\,act}^{ad}}{k_B T}}$$

Net flux of adatoms at the step edge:

$$j_+ - j_- = \frac{a^2}{\delta^2} e^{-\frac{U_{a\,act}^{kin}}{k_B T}} \frac{n_S^{step} - n_S^{eq}}{\tau_S}$$

Velocity of step growth owing to the incorporation of adatoms:

$$V_{\text{edge}} = \frac{a^3}{\tau_{\text{S}}}\frac{a^2}{\delta^2}\mathrm{e}^{-\frac{U_{\text{a act}}^{\text{kin}}}{k_{\text{B}}T}}\left(n_{\text{S}}^{\text{step}} - n_{\text{S}}^{\text{eq}}\right)$$

Kinetic growth coefficient of the step:

$$V_{\text{edge}} = \beta_{\text{kin}}\left(n_{\text{S}}^{\text{step}} - n_{\text{S}}^{\text{eq}}\right)$$

$$\text{where } \beta_{\text{kin}} = \frac{a^3}{\tau_{\text{S}}}\frac{a}{\delta}\mathrm{e}^{-\frac{U_{\text{a act}}^{\text{kin}}}{k_{\text{B}}T}}$$

Total Rate of Step Growth

The total rate of step growth is controlled by the slowest of the two subprocesses:

$$V_{\text{edge}} = \beta_{\text{diff}}\left(n_{\text{S}}^{\text{p}} - n_{\text{S}}^{\text{step}}\right) \quad V_{\text{edge}} = \beta_{\text{kin}}\left(n_{\text{S}}^{\text{step}} - n_{\text{S}}^{\text{eq}}\right)$$

The total rate of step growth:

$$V_{\text{edge}}^{\text{total}} = \frac{a^3}{\sqrt{2\pi m k_{\text{B}}T}}\frac{a^2}{\delta^2}\mathrm{e}^{-\frac{U_{\text{a act}}^{\text{kin}}}{k_{\text{B}}T}}\left(p - p_{\text{eq}}\right)$$

Mass transport rates:

$$\left(-\Delta G^{\text{total}}\right) = \left(-\Delta G^{\text{diff}}\right) + \left(-\Delta G^{\text{kin}}\right)$$

$$k_{\text{total}}\left(-\Delta G^{\text{total}}\right) = k_{\text{diff}}\left(-\Delta G^{\text{diff}}\right) = k_{\text{kin}}\left(-\Delta G^{\text{kin}}\right)$$

Relationship between the reaction rate coefficients:

$$\frac{1}{k_{\text{total}}} = \frac{1}{k_{\text{diff}}} + \frac{1}{k_{\text{kin}}}$$

Total growth rate when none of the subprocesses dominates over the other:

$$V_{\text{edge}} = \frac{\beta_{\text{diff}}\beta_{\text{kin}}}{\beta_{\text{diff}} + \beta_{\text{kin}}}\left(n_{\text{S}}^{\text{p}} - n_{\text{S}}^{\text{eq}}\right)$$

■ Spiral Steps and Screw Dislocations

The origin of spiral steps is screw dislocations. Most of the screw dislocations originate from difference in size between ad-atoms and the freshly incorporated ad-atoms on one side and the normal lattice atoms on the other.

Archimedes' spiral is used as a mathematical model of the spiral. It can be written as

$$r = A\varphi \quad \text{or} \quad r = 2r^*\varphi$$

where r^* is the critical size of the nucleus.

$$r^* = \frac{2\sigma V_m}{-\Delta G^{V \to s}}$$

The radial distance λ between the steps $= \Delta r = 4\pi r^*$

Rate of Layer Growth with no Consideration to Strain

$$V_{\perp} = h\frac{V_{edge}}{\lambda} = n_h a \frac{V_{edge}}{4\pi r^*}$$

Spiral step growth is a special case of layer growth. h and λ can not be determined separately.

Rate of Layer Growth with Consideration to Strain

Owing to difference in size between the ad-atoms and the lattice atoms the crystal surface is strained before relaxation. No ad-atoms can be incorporated at the smooth terrace surface, only at the rough step edge.

This fact can be used for determination of the height and length of steps and the rate of layer growth.

Determination of Height and Length of a Step

Two motions occur simultaneously. One is the horizontal step-edge growth with the constant velocity V_{edge}.

$$V_{edge} = \frac{a^3}{\tau_s}\frac{a^2}{\delta^2} e^{-\frac{U_a^{kin}}{k_B T}} \left(n_s^{step} - n_s^{eq}\right) = \beta_{kin}\left(n_s^{step} - n_s^{eq}\right)$$

The other one is the relaxation process when vacancies move vertically towards the interface. The two motions are coherent and can be described by the equations

$$y = V_{edge}\, t \quad \text{and} \quad l = \sqrt{D_{vac}\, t}$$

At $t = t_{relax}$ the surface is relaxed and the edge has moved a distance equal to the step length.

$$\lambda = V_{edge}\, t_{relax} \quad \text{and} \quad h = \sqrt{D_{vac}\, t_{relax}}$$

If t_{relax} is known h and λ can be calculated.

Rate of Layer Growth in Case of Strain

$$V_{edge}^{strain} = \frac{a^3}{\tau_S}\frac{a^2}{\delta^2} e^{-\frac{U_{a\,act}^{kin} - \Delta G_a^{strain}}{k_B T}} \left(n_S^{step} - n_S^{eq}\right)$$

where $-\Delta G_a^{strain} = G_a^{strain} - G_a^{no\ strain}$

If the reaction is controlled by the crystallization process the layer growth rate will be given by

$$V_{\perp} = h \frac{V_{\text{edge}}^{\text{strain}}}{\lambda}$$

■ Influence of Impurities on Crystal Growth in Vapours

The influence of impurities on crystal growth in vapours is large and complex. Two cases are treated in the text:

- Equilibrium trapping of immobile impurities.
- Non-equilibrium trapping of mobile impurities.

Vacancy Trapping
Vacancies can be regarded as a special sort of mobile impurities.

Like other mobile impurities vacancies become trapped in the crystal during step growth.

Vacancy Distribution inside the Interphase:

$$n_{\text{vac}} = n_{\text{vac}}^{0\,\text{max}} e^{\frac{D_{\text{vac}}^{\text{I}} \frac{\Delta G_{\text{a}}^{\text{vac}}}{k_{\text{B}}T}}{V_{\text{edge}}} y} \qquad -\delta \leq y \leq 0$$

Vacancy Distribution in the Crystal:
The vacancy concentration (number of vacancies per unit volume) in the crystal close to the interface as a function of the rate of step growth:

$$n_{\text{vac}}^{\text{eq}} = n_{\text{vac}}^{0\,\text{min}} = n_{\text{vac}}^{0\,\text{max}} e^{-\frac{D_{\text{vac}}^{\text{I}}}{V_{\text{edge}}} \frac{\Delta G_{\text{a}}^{\text{vac}}}{k_{\text{B}}T} \delta}$$

The more rapid the step growth rate is, the higher will be the concentration of trapped vacancies in the crystal.

■ Epitaxial Growth

Epitaxial growth = oriented crystallization by vapour deposition of a thin film on a substrate.

CVD methods are used for thin film production. Important examples are silicon and diamond films.

Formation process of a thin film:

1. clusters are formed on a substrate
2. clusters nucleate to crystallites
3. crystallites growth on the substrate.

Epitaxial growth is influenced by

- nature of substrates
- temperature of subtrate
- deposition rate
- influence of vapour impurities.

Temperature controls the critical size and the rate of formation of the nuclei, the mobility of the adsorbed atoms and the annealing of defects in the condensed films.
Crystal orientation and pressure also influence the growth rate

■ Whisker Growth

Whisker growth is strongly anisotropic. The growth rate in one direction is much faster than that in the lateral directions.

Whiskers have the shape of needles and are produced by condensation of metal vapours in vacuum or by reduction of metal halides in hydrogen.

Whisker growth is supposed to be associated with impurity atoms, which slow down or even stop the growth in the lateral directions.

Whiskers have good mechanical properties and low fraction of defects. They are used for catalytical purposes because of their large area. In thin film production, they are highly inconvenient and must be carefully avoided.

Exercises

5.1 Cu crystals have been epitaxially coated with a thin Zn layer. The Zn melt was held at a temperature of 500 °C. The temperature of the Cu crystals could be varied from 25 °C up to 400 °C. The Cu crystals exposed only {111} faces to the Zn vapour. The Zn layer formed on the Cu surface consisted of {I001} faces.

The equilibrium vapour pressure of Zn, measured in atm, at temperature T, is

$$p_{Zn}^{eq} = p_{ref} \exp\left(8.108 - \frac{6947}{T}\right)$$

where T is measured in K. The reference pressure $p_{ref} = 1$ atm.

Calculate the total driving force for precipitation of Zn on Cu as a function of the Cu temperature for two cases, one with no consideration to strain and the other with consideration to strain between the Zn atoms and the Cu atoms at the interface Zn/Cu. Plot the two functions in a diagram for the temperature interval 298–700 K and compare them.

To solve the task you must take epitaxial strain and thermal expansion effects into account.

Table 5.1–1

Material constants	Cu	Zn
Interatomic distance a	0.132 nm	0.128 nm
Young's modulus	12×10^{10} N/m^2	9.3×10^{10} N/m^2
Average linear thermal expansion coefficient α within the temperature interval 25–400 °C	1.6×10^{-5} K^{-1}	2.6×10^{-5} K^{-1}
Molar volume	7.1×10^{-3} m^3/kmol	9.3×10^{-3} m^3/kmol

5.2 At vapour deposition of thin films the epitaxial growth is influenced by the nature and temperature of the substrate and the interaction between the film atoms and the substrate atoms. Mismatch between these two types of atoms causes strain in the film. The strain reduces the driving force for epitaxial growth rate and consequently the deposition rate.

Derive the ratio between the deposition rate of the strained and unstrained films, respectively, as a function of the ratio of the vapour pressure in the presence of strain, and ditto at equilibrium without strain. Apply your relationship on gold deposition and plot $V_{growth}^{strain}/V_{growth}$ as a function of p/p_{eq} at room temperature 20 °C and a strain of 6%. Plot the curve for p/p_{eq} from 1.62 to 3.00.

Material constants for Au:

Young's modulus of elasticity $Y = 3.2 \times 10^{10}$ N/m^2

Strain $\varepsilon = 0.06$

Molar volume $V_m = 10.2 \times 10^{-3}$ m^3/kmol.

5.3 Vapour deposition of gold is a very common process in the jeweler branch. The process is shown in Figure 5.5 on page 204 in Chapter 5. Liquid gold is kept in a crucible at a constant temperature T_0. The temperature T_{comp} of the component to be depleted is kept on a lower temperature than T_0. The temperature T_0 controls the vapour pressure p_0 in the chamber. The vapour pressure p_{comp} of gold at temperature T_{comp} in the vicinity of the component is lower than p_0 because $T_{comp} < T_0$ and gold may be depleted on the component.

The temperature difference $(T_0 - T_{comp})$ controls the deposition rate. If the temperature T_{comp} is lower than a critical value, droplets of liquid gold will precipitate spontaneously in the vapour around the component and destroy the gentle deposition process of gold atoms on the surface of the component.

Derive an expression of T_0 as a function of T_{comp} and calculate the critical temperature T_{cr} for droplet precipitation. Plot T_0 and T_{comp} as functions of $x = \sqrt{T_{comp}}$ for the interval $33.4 \leq x \leq 36$ in a diagram. Use the diagram for choice of optimal values of T_0 and T_{comp} for gold coating. 1 nucleus per mol is assumed to be nucleated. See Example 4.1 on page 170 in Chapter 4.

The vapour pressure of gold is given by the relationship

$$\ln \frac{p}{p_{ref}} = 22.5 - \frac{40400}{T}$$

where p is expressed in atm. The temperature is expressed in K. The reference pressure p_{ref} is equal to 1 atm.

Material constants for gold:

Surface tension $\sigma = 1.14\ \text{J/m}^2$

Molar volume of gold $V_m = 10.2 \times 10^{-3}\ \text{m}^3/\text{kmol}$.

Hint: Set up an expression for the molar Gibbs' free energy for formation of gold droplets near the component.

5.4 Crystal growth in a vapour phase often occurs by layer growth of steps on a crystal interphase. The length and height of the steps are closely related to a relaxation process of the vapour atoms that occurs when they become incorporated into the crystal.

The relaxation process can alternatively be treated as a condensation of vacancies at the crystal interphase, which becomes part of the crystal lattice. Condensation of vacancies is in most cases related to formation of dislocations.

a) Assume primarily that the distances between vacancy sinks are constant. This condition gives a constant diffusion distance and a constant height of the steps. Predict the effect on the length of the steps.

b) Suppose that the number of dislocation sinks per unit volume is proportional to the pressure difference $(p - p_{eq})$ where p is the pressure of the vapour phase. The condensation process is rapid and does not occur at equilibrium between the vapour phase and the solid phase. p_{eq} is the equilibrium value of the pressure at equilibrium between the two phases.

Assume in this case that the height of the steps is inversely proportional to the pressure difference $(p - p_{eq})$. What will be the effect on the length of the steps?

c) Does the density of dislocations have any effect on the properties of the crystals?

5.5 The growth of an epitaxial film is thermally activated and collision-controlled. In the figure on next page the growth rate of a Si surface film is shown as a function of the growth temperature when the pressure difference is kept constant.

Calculate the activation energy per atom, expressed in eV, and the kintetic growth coefficient β_{kin} at 700 °C measured in SI units.

The diameter a of the diffusing Si atoms is 0.117 nm.

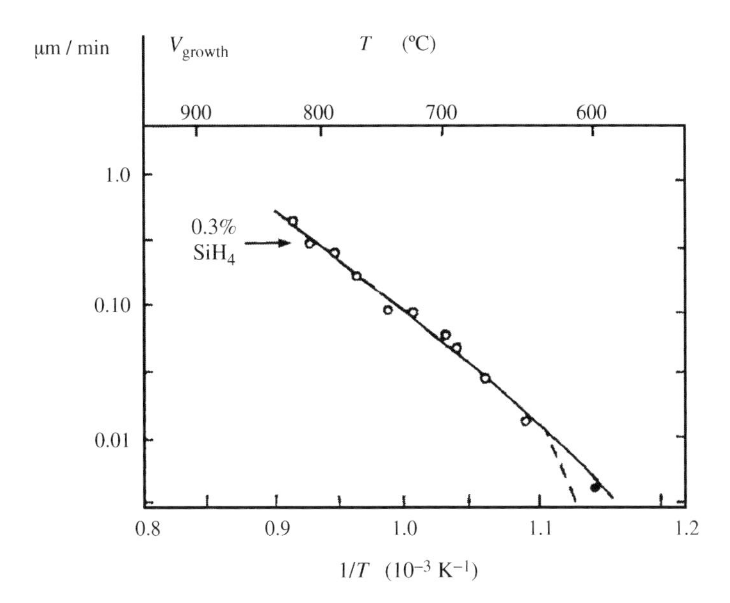

Figure 5.5–1

5.6 Diamond coatings can be performed in a *vacuum chamber* in an methan (CH_4) atmosphere. It is very urgent that the fitting between the atoms of the substrate to be coated and the atoms in the diamond film is good.

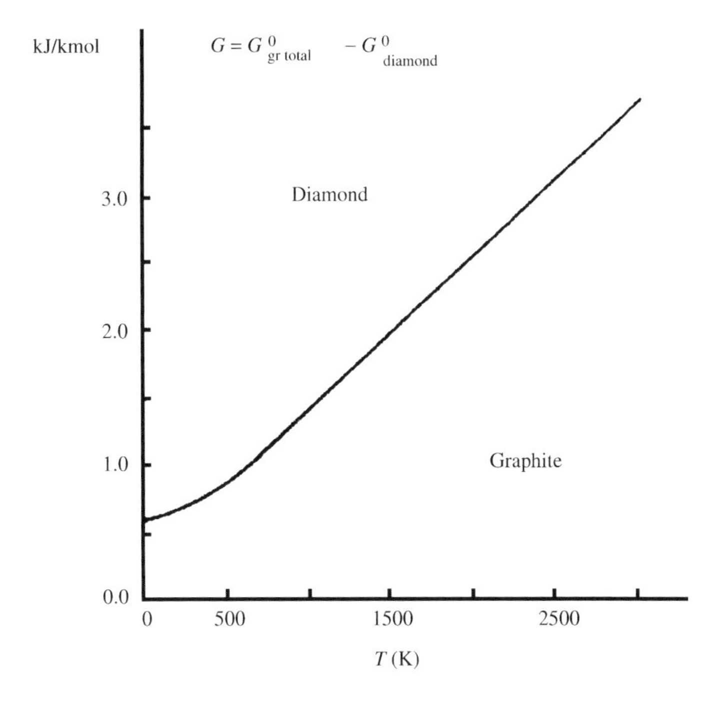

Figure 5.6–1

In a series of experiments where the temperature of the substrate was successively varied, a transformation of a diamond film into a graphite film was observed at 900 °C. Explain the reason for this transformation with the aid of the diagram above. The diagram shows the Gibbs' free energy change from diamond to graphite as a function of temperature.

References

1. H. Fredriksson, U. Åkerlind, *'Physics of Functional Materials'* Chichester, England, John Wiley & Son 2008.
2. M. Volmer, I. Eastermann, *Zeitshrift fur Physik*, **7**, 1921, 13.
3. M. Straumanis, *Zeitschrift für Physik und Chemie*, **B19**, 1934, 63.
4. W. Kossel, Nach, Ges. *Wissenschaft*, Gotlingen, **135**, 1927.
5. K. A. Jackson, *Liquid Metals and Solidification*, ASM Cleveland, 1958.
6. W. K. Burton, N. Cabrera, F. C. Frank, Philosophical Transactions, *Royal Society*, **243**, 1951, 209.
7. N. Cabrera, D. A. Vermilyea, *Growth and Perfection of Crystals*, John Wiley & Sons 1958, page 411.
8. R. V. Coleman, G. W. Sears, *Acta Metallurgica*, **5**, 1957, 131.
9. W. A. Tiller, *The Science of Crystallization, Macroscopic Phenomena and Defect Generation*, Cambridge University Press, Cambridge, 1991, page 433.
10. D. E. Jesson, K. M. Chen, S. J. Pennycook, T. Thundat, R. J. Wanack, Crack-Like Sources of Dislocation and Multiplication in Thin Films, *Science*, **268**, 1995, 1161–1163.

6

Crystal Growth in Liquids and Melts

Solidification and Crystallization Processing in Metals and Alloys, First Edition. Hasse Fredriksson and Ulla Åkerlind.

6.1 Introduction[1]

The properties of castings and single crystals grown from liquids are in most cases determined by the conditions, which control the solidification and crystallization processes. There are two kinds of conditions, macroscopic and atomic.

The macroscopic conditions, that control the rate of the solidification and crystallization processes, are related to the heat transport, i.e. the possibility of removing the released latent heat, and mass transport owing to convection. These topics will be treated in Chapters 7 and 8.

The atomic conditions are related to mass transport at the interface between the disordered liquid state and the ordered solid state, i.e. to the transport of atoms to, from and across the interface. These topics will be discussed in this chapter.

After an orientation about the structures of crystal structure and the structure of melts we give an introductory review of the solidification methods used in casting and the practical methods of single crystal production and refining. These sections are followed by a description of the processes at the interface for pure metals and later also for binary alloys. Growth at planar interfaces in pure metals and alloys are discussed and the growth of spheres in liquids. The influence of volume changes and relaxation on crystal growth are also considered.

6.2 Structures of Crystals and Melts

6.2.1 Experimental Evidence

The dominating source of information about the structures of crystals and melts is X-ray analysis.

Electrons are accelerated by a strong electric field in an evacuated tube towards a metal target. When they hit the anode they suddenly lose their kinetic energy in numerous successive collisions with the electrons around the target nuclei. These collisions can occur in many different ways and the kinetic energy of the accelerated electrons becomes transformed into electromagnetic radiation. A continuous X-ray radiation is emitted together with a few discrete X-ray wavelengths characteristic of the target atoms (Figure 6.1).

In an X-ray spectrometer of the Debye-Scherrer type, the crystal shown in Figure 6.2 is replaced by a powder of many small crystals with arbitrary orientations. In this way the scattered radiation in all possible directions can be registered simultaneously. Figure 6.3a shows that the scattered angle, which can be measured, equals 2θ. Figure 6.3b shows a typical Debye-Sherrer pattern.

[1] Consider some thermodynamic quantities, such as $-\Delta G_a$ and $-\Delta H_a$. The subscript 'a' means 'per atom'. Normally much larger quantities are used. There are three alternative units based on different amounts of material: 'per atom', 'per kmol' and 'per kg' (SI unit). There is no distinction (subscript) between the second and third alternatives.

If the numerator and denumerator of the following expression are multiplied by Avogadro's number N_A you obtain

$$\frac{-\Delta G_a N_A}{k_B T N_A} = \frac{-\Delta G}{RT}$$

There is also a link between alternatives two and three: 1 kmol = M kg where M is the molar weight.

The right-hand side expression above is valid for 1 kmol of the metal. The same distinction is valid for all material constants. You can use any set of units (the three alternatives) but be consequent and do not mix them!

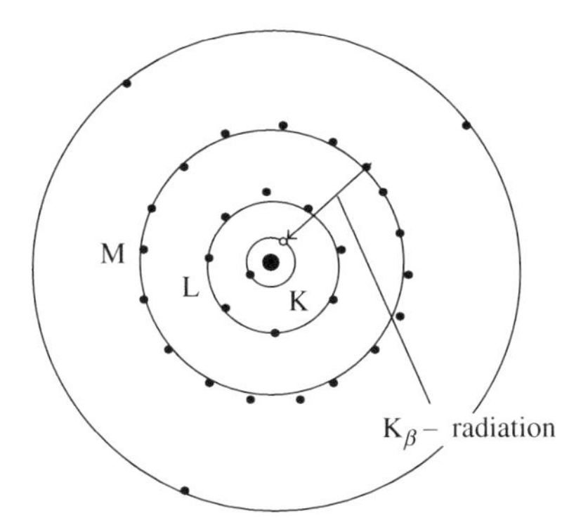

Figure 6.1 When a high-speed electron hits an electron in an inner shell of a target atom, both electrons may leave the atom and a vacancy is left in its inner shell. The vacancy is filled by an electron from an outer shell and a photon, in this case K_α radiation, is emitted. This is the origin of the characteristic X-ray spectrum of the target atoms.

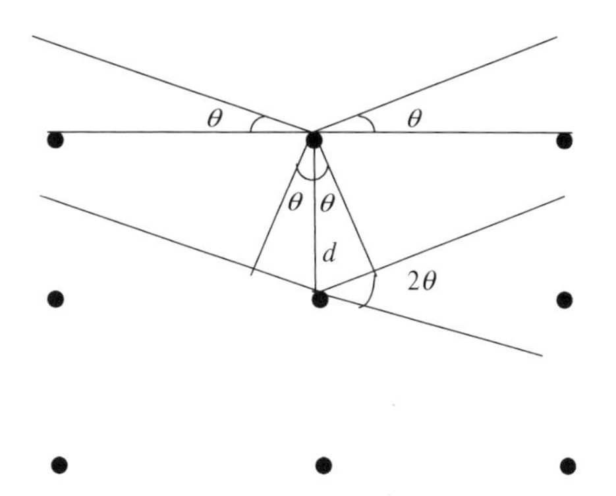

Figure 6.2 The wavelength of the characteristic X-ray lines can be determined by diffraction in a crystal with known lattice constant *d*. Bragg's law gives
$2d \sin\theta = n\lambda$,
where θ is the angle of the incident radiation and the crystal surface. *n* is the order of the diffraction. *n* and *d* are known, θ is measured and λ can be calculated.

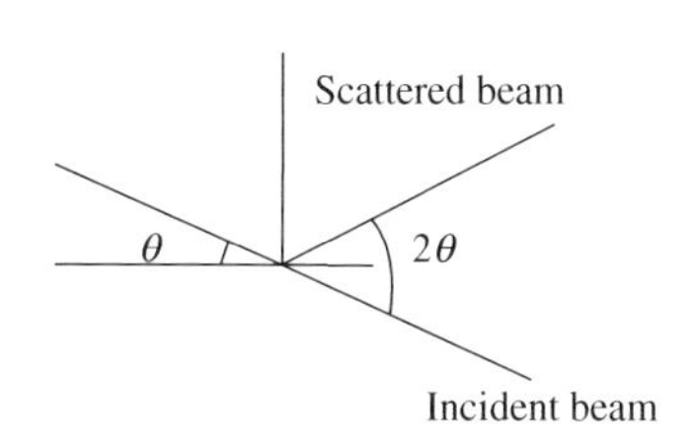

Figure 6.3a The angle 2θ of scattering is defined as the angle between the incident and scattered beam. Reproduced from Elsevier © 1979.

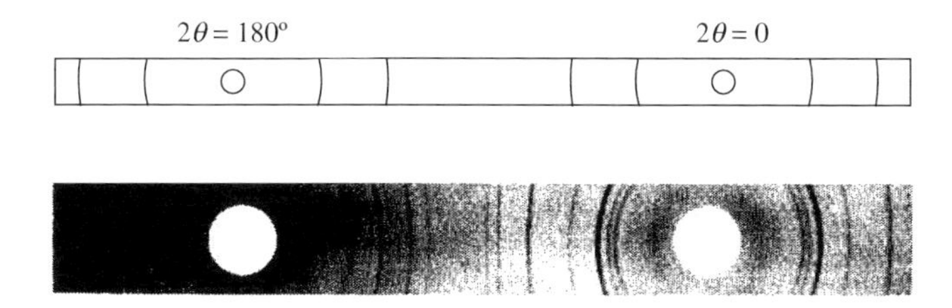

Figure 6.3b Debye–Scherrer pattern of zinc. The white circles represent the incident beam. © Pearson PLC 1956.

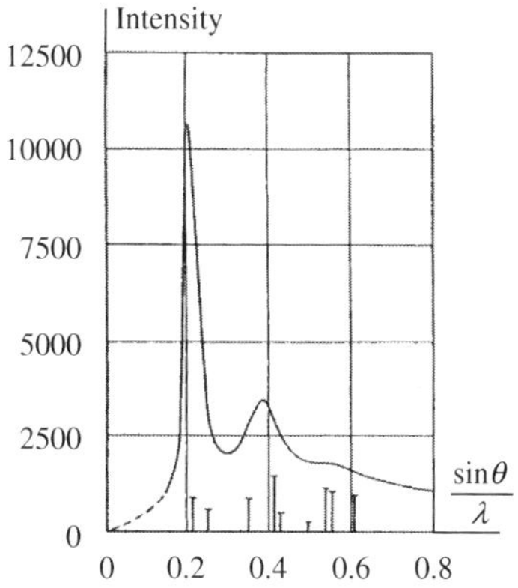

By replacing the powder compound by a metal melt at constant temperature the corresponding Debye-Sherrer measurements can be made on a liquid metal.

The results deviate very much from those obtained for crystalline powders. Instead of a few sharp well-defined X-ray lines, several wide and unsharp maxima and minima are obtained.

Figure 6.4 shows the intensity of the scattered X-ray radiation as a function of $\sin\theta/\lambda$ for liquid gold at 1100 °C. The vertical lines originate from scattering of crystalline gold.

The interpretation of the X-ray plot for liquid gold in Figure 6.4 will be further discussed below.

Intensity
12500
10000
7500
5000
2500
0
0 0.2 0.4 0.6 0.8
$\frac{\sin\theta}{\lambda}$

Figure 6.4 X-ray pattern of crystalline gold and liquid gold at 1100 °C.

6.2.2 *Structures of Crystals*

The explanation of the characteristic X-ray spectrum of solid metals has been given above. If the wavelength of the incident X-ray radiation is known, Bragg's law can be used for determination of the distance *d* between the parallel crystal planes which are the origin of the diffraction.

$$2d \sin\theta = n\lambda \tag{6.1}$$

where *n* is the order of the diffraction.

The analysis of the X-ray lines of a specimen results in a complete knowledge of the crystallographic structure of the metal, i.e. the distances and angles between all the reflecting planes. The structures of complicated nonmetallic compounds, such as cholesterol, insulin, penicillin and vitamin B12 have successively been solved with the aid of crystallography (Dorothy Crowfoot Hodkin, Nobel prize winner in Chemistry in 1964). The structures of even more complicated molecules have been solved, for examples the twinned double DNA spiral (Cricks and Watson, Nobel prize winners in Chemistry in 1962),

Volume and Surface Packing Fractions of Crystal Planes

The *volume packing fraction p* of a crystal is defined as

- p_V = the fraction of a crystal volume, which is occupied by atoms.

In analogy with the volume packing fraction we can define the concept *surface packing fraction* p_S for a crystal plane as

- p_S = the fraction of a crystal plane, which is occupied by atoms.

In Chapter 3 we have discussed the different surface energies of coherent and semicoherent solid/solid interfaces. The surface packing fraction is important for the solidification process at the solid/liquid interface. The closer the surface packing fractions for the crystal and the melt are, the easier will be the crystallization process. We will discuss this topic later in this chapter.

Most metals have a close packed crystal structure. As an example we will discuss surface packing fractions for FCC and BCC crystal structures (pages 21–22 in [1]).

Example 6.1

a) Calculate the surface packing fraction for {100} and {111} surfaces of a crystal with FCC structure.
b) Perform the same calculations for a crystal with BCC structure.

Solution:

a)
{100} *planes*

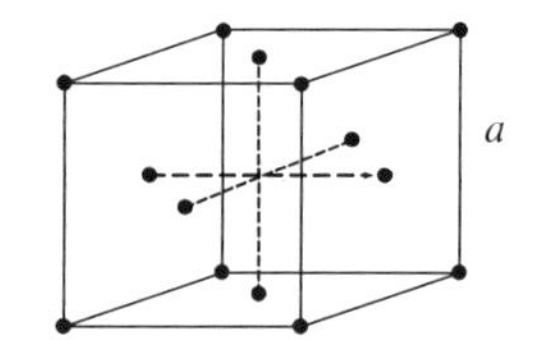

Figure 1′ Unit cell of the FCC structure. There is a central atom in each of the six lateral faces of the cube.

Figure 2′ Lateral face of a close-packed FCC structure. It represents a {100} crystal plane.

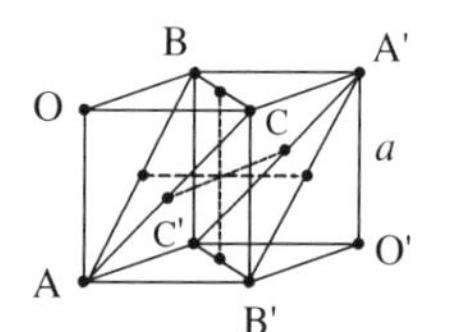

Figure 3′ The two triangles are included in two parallel {111} planes, perpendicular to the principal diagonal OO′.

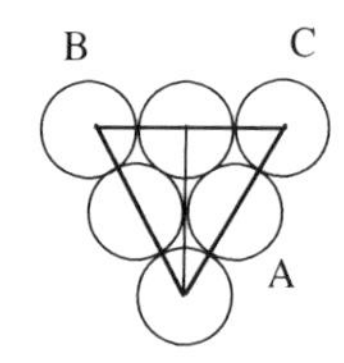

Figure 4′ Close-packed triangle in an FCC unit cell. It represents a {111} crystal plane.

We consider a unit cell of a crystal with FCC structure (Figure 1′).
To find the surface packing fraction we have to calculate the sum of the cross-section areas of the atoms inside one of the lateral faces, which represents an {100} plane. The latter is a square with the area a^2 (Figure 2′).

Figure 2′ gives two expressions of the diagonal in the square:

$$4R = a\sqrt{2} \qquad \Rightarrow \qquad R = \frac{a\sqrt{2}}{4} \tag{1'}$$

Hence, the cross-section area of an atom equals

$$\pi R^2 = \frac{2\pi a^2}{16} = \frac{\pi a^2}{8} \tag{2'}$$

Inside the square there are $(4 \times 1/4 + 1) = 2$ circles or 2 cross-section areas. Hence, the surface packing fraction will be

$$p_S^{100} = 2 \times \frac{\pi a^2}{8} / a^2 = \frac{\pi}{4} \qquad (3')$$

{111} *planes*

In the same way we can derive the surface packing fraction for the {111} planes. The equilateral triangles ABC and A′B′C′ in Figure 3′ are {111} planes.

Equations (1′) and (2′) are still valid. The area of the equilateral triangle is $(a\sqrt{2})^2 \times \sqrt{3/4} = a^2\sqrt{3}/2$.

Inside the triangle in Figure 4′ there are $(3 \times 1/6 + 3 \times 1/2 = 2$ circles or 2 cross-section areas. Hence, the surface packing $(a\sqrt{2})^2\sqrt{3}/4 = a^2\sqrt{3}/2$ fraction will be

$$p_S^{111} = \frac{2 \times \frac{\pi a^2}{8}}{\frac{a^2\sqrt{3}}{2}} = \frac{\pi\sqrt{3}}{6} \qquad (4')$$

b)

With the aid of Figures 5′–8′ we can derive the surface packing fractions of a BCC structure for the {100} and {111} planes.

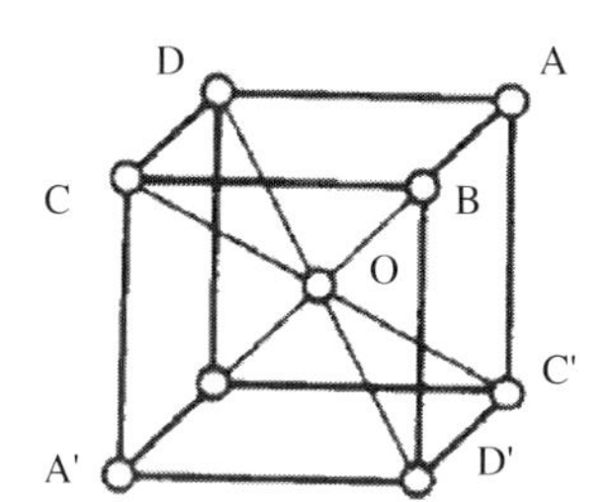

Figure 5′ Unit cell of the BCC structure.

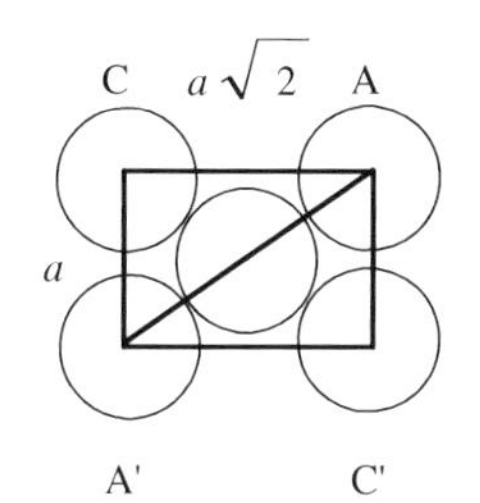

Figure 6′ Principal diagonal in a BCC unit cell.

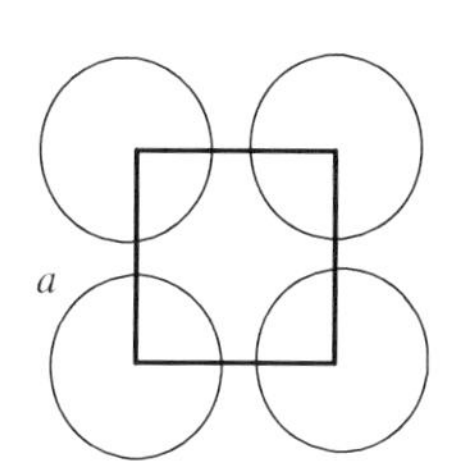

Figure 7′ Lateral face of a BCC unit cell. It represents a {100} crystal plane.

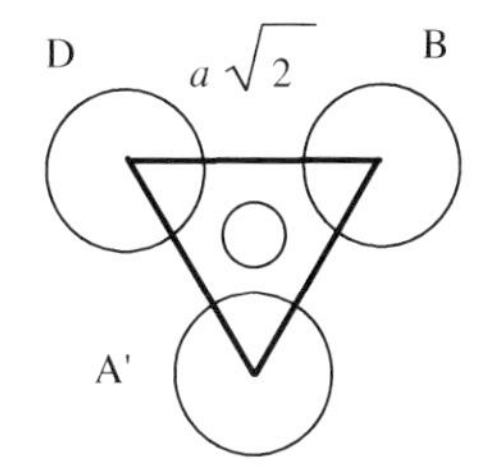

Figure 8′ Triangle in a BCC unit cell. It represents a {111} crystal plane. The small circle in the centre is the intersection area between the central atom O with radius R and the plane DBA′.

{100} *planes*

The length of the principal diagonal in the BCC unit cell can be written in two ways (Figure 6′)

$$4R = a\sqrt{3} \quad \Rightarrow \quad R = \frac{a\sqrt{3}}{4} \qquad (5')$$

Hence, each cross-section area is

$$\pi R^2 = \frac{\pi(3a^2)}{16} = \frac{3\pi a^2}{16} \qquad (6')$$

The square in Figure (7′) contains $(4 \times 1/4) = 1$ circle or 1 cross-section area. Consequently, its area equals that, given in Equation (6′) and the surface packing fraction will be

$$p_S^{100} = 1 \times \frac{3\pi a^2}{16} / a^2 = \frac{3\pi}{16} \qquad (7')$$

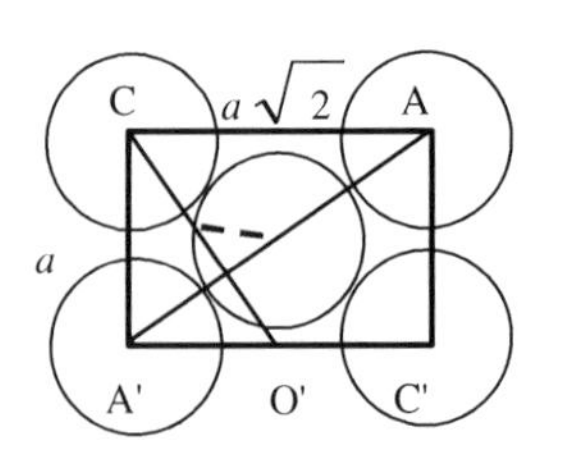

Figure 9′ Lateral face of a BCC unit cell. It represent the 111 crystal plane.

{111} *planes*

Equations (5′) and (6′) are still valid. The triangle in Figure (8′) is the total area of a {111} plane within the unit cell. The area occupied by atoms is (3 × 1/6) = ½ circle or ½ cross-section area plus the area of the small central circle. The latter can be calculated with the aid of the small triangle with one dashed side in Figure 9′. The area of the triangle A'CO' can be written as

$$\frac{1}{2}a \times \frac{a\sqrt{2}}{2} = \frac{1}{2}h \times \sqrt{a^2 + \left(\frac{a\sqrt{2}}{2}\right)^2} \tag{8'}$$

where h is the height of the triangle along the principle diagonal.

The dashed side equals the radius R of the atoms. We apply Pythagoras theorem on the triangle which gives

$$x^2 + \left(\frac{a\sqrt{3}}{2} - h\right)^2 = R^2 \tag{9'}$$

where x is the radius of the small circle in Figure 8′. The result of the calculations is

$$h = \frac{a}{\sqrt{3}} \quad \text{and} \quad x^2 = \frac{5a^2}{48}$$

Hence, the area of the small circle is $\pi x^2 = \pi \times \dfrac{5a^2}{48}$

The area of the triangle DBA' in Figure 8′ is $(a\sqrt{2})^2 \times \sqrt{3}/4 = a^2\sqrt{3}/2$ and the surface packing fraction can be written as

$$p_S^{111} = \frac{\frac{1}{2} \times \frac{3\pi a^2}{16} + \pi \times \frac{5a^2}{48}}{\frac{a^2\sqrt{3}}{2}} = \pi \times \frac{\frac{3}{16} + \frac{10}{48}}{\sqrt{3}} \tag{10'}$$

Answer:

The desired surface packing fractions are given in the table.

Crystal structure	p_S^{100}	p_S^{111}
FCC	$\frac{\pi}{4} \approx 0.78$	$\frac{\pi\sqrt{3}}{6} \approx 0.91$
BCC	$\frac{3\pi}{16} \approx 0.59$	$\pi \times \frac{\frac{3}{16} + \frac{10}{48}}{\sqrt{3}} \approx 0.72$

From the answer in Example 6.1 we can conclude that the surface packing fraction depends on

- the structure of the crystal
- the direction of the crystal plane
- the {111} planes are more close-packed than the {100} planes.

The first two conclusions are valid for all types of crystals. It is easy to understand that the surface packing fractions generally are higher for FCC than for BCC because FCC has 4 atoms per unit cell and BCC only 2 atoms per unit cell.

6.2.3 Structures of Metal Melts

Atomic Distribution Diagrams

The properties of liquids and melts are functions of the interatomic forces. It is extremely difficult to derive models of liquids and reliable expressions for the dynamic properties of liquids because the atomic motion in liquids can not be accurately described as a function of time.

The fundamental basis for knowledge of the structure of liquids and melts is their X-ray spectra. Figure 6.4 shows the intensity of the scattered radiation X-ray (Figure 6.3a) as a function of sin θ/λ where θ is half the scattering angle and λ is the wave-length of the radiation.

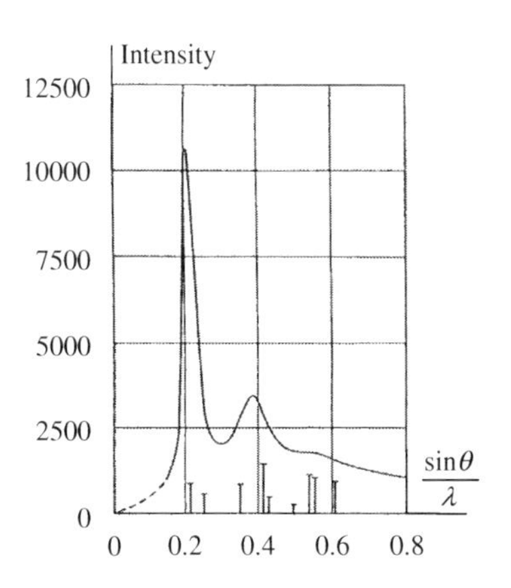

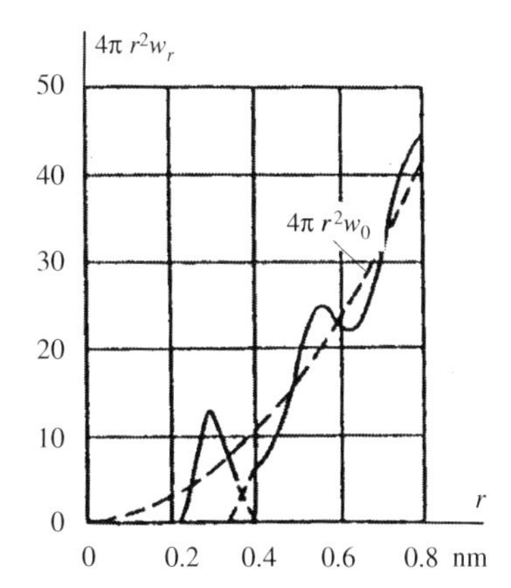

Figure 6.4 (left) X-ray pattern of crystalline gold and liquid gold at 1100 °C. Hendus, H. © Verlag der Zeitschrift für Naturforschung 1947.

Figure 6.5 (right) Atomic distribution diagram of liquid gold at 1100 °C. The probability of finding an atom within a spherical shell with the radius r as a function of r. Reproduced from Elsevier © 1976.

As is shown on pages 6–9 in Chapter 1 in [1] X-ray plots of the intensity of the X-ray radiation versus direction can be 'translated' into so-called *atomic distribution diagrams* where the function w_r is the probability of finding another atom in a volume element at distance r from an atom at the origin. This type of diagrams is much easier to interpret than the original experimental function. Figure 6.5 shows the atomic distribution diagram for liquid gold.

The probability of finding next atom within the volume element dV at the distance r from the origin atom equals the product of dV and wr. If we chose a spherical shell with the radius r and the thickness dr as volume element, the probability dWr of finding another atom within the spherical shell will be given by the expression

$$dW_r = w_r dV = w_r 4\pi r^2 dr \tag{6.2}$$

If the probability of finding another atom within the volume element were constant and equal to w_0 the dashed curve would be obtained. Figure 6.4 shows that this is not the case for the melt of liquid gold.

Interpretation of Atomic Distribution Diagrams

Figure 6.5 shows that the atoms around the 'origin atom' are *not* equally distributed. The wide maxima of the distribution curves can be interpreted as 'shells' where the probability to find atoms is higher than the average. Between the wide shells, the probability is lower than the average to find atoms at those distances.

This phenomenon can be interpreted as a *short range order* in the melt. The atoms are in incessant motion, collide with other atoms and change position all the time, but *on the average* each atom is surrounded by a *constant number* of nearest neighbours even if the surrounding atoms are not the same from time to time. Experimental evidence shows that if the temperature is increased, the maxima become wider and wider. Above a particular temperature, the shell structure disappears completely.

In a crystalline solid, the atoms are regularly distributed in permanent positions in atomic planes and are able to oscillate around their average positions. The result is a long range order and the characteristic sharp X-ray lines observed for solid metals.

The corresponding distribution of the atoms in a melt is totally different. Owing to the short range order the liquid consists of clusters of near neighbour atoms. The clusters interact with each other and with free atoms in the melt. Atoms leave or are incorporated into the clusters all the time but the *average* pattern is constant and is manifested in the shell structure of the atomic distribution diagrams of liquid metals.

Interatomic Distances and Coordination Numbers in Melts

It is possible to derive average interatomic distances and average coordination numbers from an atomic distribution diagram. As an example we use the diagram for liquid gold in Figure 6.5.

The *shortest possible distance between two atoms* in the melt is equal to the intersection point between the curve and the r axis. It can be read to 0.22 nm for liquid gold at 1100 °C.

The *average distance to the nearest atoms* is the r value, that corresponds to first shell, i.e. the first maximum of the curve. For liquid gold it is approximately 0.29 nm. The average distance to the next nearest atoms can be read from the second maximum of the curve.

The *average number of nearest neighbours*, i.e. the average number of atoms in the first shell, is approximately equals to the area under the first peak of the curve (black area in Figure 6.6).

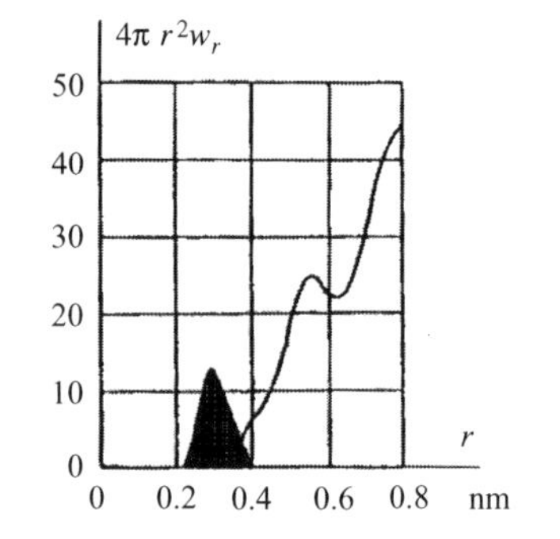

Figure 6.6 Derivation of the number of nearest neighbours.

$$\text{Number of atoms in coordination shell} = \int_{r_1}^{r_2} 4\pi r^2 w_r \mathrm{d}r \tag{6.3}$$

where r_1 and r_2 are the limiting r values of the peak.

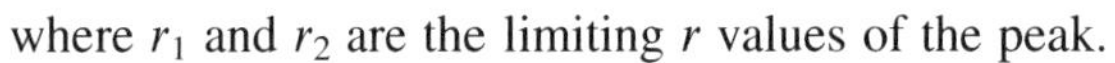

This number often deviates from integers for metal melts. Clusters may occasionally consist of 10, 11 or 12 atoms and the number varies with time. Hence, it is natural that the average number deviates from an integer.

The properties of liquids and melts are discussed in Chapter 8 in [1]. Tables 6.1 and 6.2 show the nearest neighbour distances and the coordination numbers for some pure solid metals and their corresponding melts. They are also included here.

Table 6.1 Nearest neighbour distances for some pure solid and liquid metals at their melting points

Metal	Position in the periodic table	Melting point (°C)	Crystal structure	Nearest neighbour distance (nm) for the solid metal	Nearest neighbour distance (nm) for the liquid metal
Na	1a	98	BCC	0.372	0.370
K	1a	64	BCC	0.452	0.470
Cu	1b	1083	FCC	0.256	0.257
Ag	1b	961	FCC	0.289	0.286
Au	1b	1063	FCC	0.288	0.285
Mg	2a	651	HCP	0.320	0.335
Hg	2b	−39	Rhombohedral	0.301 or 0.347*	0.307
Al	3a	660	FCC	0.286	0.296
Ge	4a	959	Diamond (two FCC)	0.245	0.270
Sn	4a	232	Tetrahedral	0.302 or 0.318*	0.327
Pb	4a	327	FCC	0.350	0.340
Sb	5a	631	Rhombohedral	0.291 or 0.335*	0.312
Bi	5a	271	Rhombohedral	0.309 or 0.353*	0.332

*Different directions. The unit cell is asymmetric.

Table 6.2 Coordination numbers for some pure solid and liquid metals at their melting points. Reproduced from Elsevier © 1974.

Metal	Position in the periodic table	Melting point (°C)	Crystal structure	Coordination number for the solid metal	Coordination number for the liquid metal
Na	1a	98	BCC	8	9.5
K	1a	64	BCC	8	9.5
Cu	1b	1083	FCC	12	11.5
Ag	1b	961	FCC	12	10.0
Au	1b	1063	FCC	12	8.5
Mg	2a	651	HCP	12	10.0
Hg	2b	−39	Rhombohedral	6 or 6*	10.0
Al	3a	660	FCC	12	10.6
Ge	4a	959	Diamond (two FCC)	4	8.0
Sn	4a	232	Tetrahedral	4 or 2*	8.5
Pb	4a	327	FCC	12	8.0
Sb	5a	631	Rhombohedral	3 or 3*	6.1
Bi	5a	271	Rhombohedral	3 or 3*	7–8

*Different directions. The unit cell is asymmetric.

6.3 Growth Methods

Crystal growth in solidifying melts is a process, which is the basis of all production of cast materials. The whole melt solidifies and the heat transport controls the solidification process. The cooling process determines the qualities of the product, for example its structure, grain size and microsegregation.

Another more selective way of using the process of crystal growth in melts is the production of single crystals.

Single crystals are mainly produced according to four basic principles:

1. growth in melts
2. growth in solutions
3. growth in solid phases
4. growth in gas phases.

The demand for purity of single crystals can be satisfied either during the production or afterwards, by purifying solid single crystals with the aid of zone refining.

6.3.1 Casting and Solidification Methods

Most casting processes start when the melt is brought into contact with a solid, which conducts heat away from the melt. The solidification process proceeds in one of the two alternative ways shown in Figure 6.7 and 6.8.

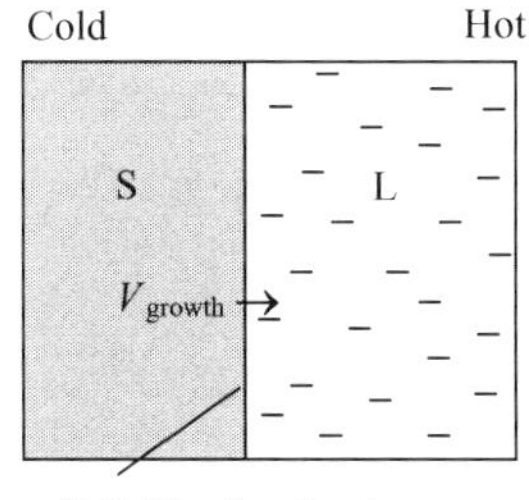

Figure 6.7 Solidification mode 1.

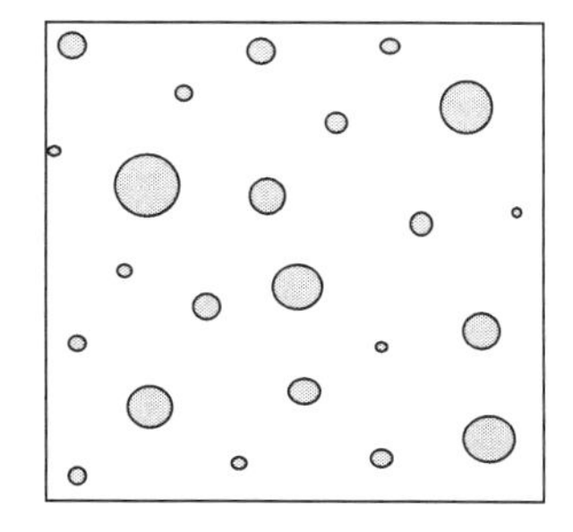

Figure 6.8 Solidification mode 2.

The figures show two different solidification modes. In Figure 6.7 the solidification front moves into the liquid. In Figure 6.8 numerous freely floating nuclei and crystals grow simultaneously at many different places in the melt. Intermediate solidification modes between these two extremes do also occur.

The type of solidification mode is determined by the cooling conditions, the mould material and by the properties of the element or alloy. These topics will be discussed in Chapter 7.

The cooling process has been refined over the years, which opens new possibilities to produce products with better properties than earlier. A typical example is the production of turbine blades, which will be discussed in Section 6.3.2. The method used in this case has many similarities with a common crystal growth method called the Bridgman method that is described below.

6.3.2 *Single Crystal Processes*

Crystal Growth in a Crucible

One of the oldest methods to produce crystals is to let the melt solidify in a crucible of adequate shape and size.

Advantages and Disadvantages of the Method

The advantages are simple equipment and a process with little need of control.

The disadvantages emanate from the interfaces, the one between the crucible and the solid phase and the boundary area between the solid phase and the melt. Undesirable nucleation and strain owing to differences in heat expansion appear very easily. Every irregularity on the surface of the crucible will affect the crystal growth. Such irregularities act as heterogeneous nucleii at the wall.

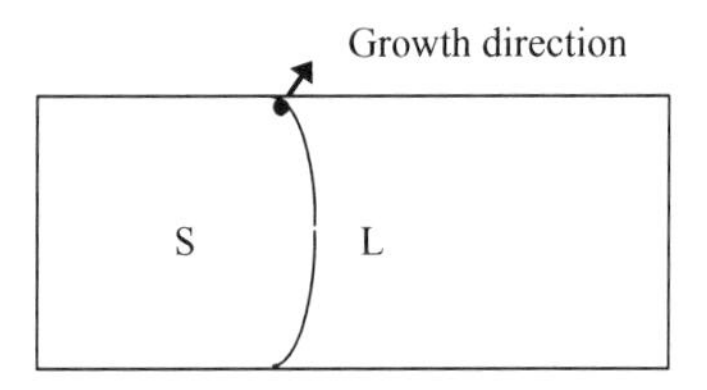

Figure 6.9 The wall prevents growth of the nuclei at the wall if the solidification front is convex.

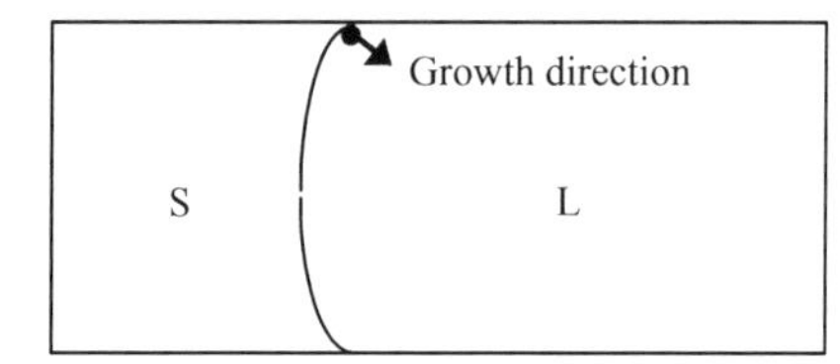

Figure 6.10 A concave solidification front allows growth of the nuclei at the wall.

The heat transport controls the shape of the solidification front. A slower heat transport at the wall than in the centre results in a convex solidification front surface (Figure 6.9). The opposite gives a concave solidification front surface (Figure 6.10). As is shown in the figures the growth conditions are different in the two cases.

Another disadvantage of the method is that the melt may react chemically with the crucible material.

Cooling of the Melt

The cooling of the melt can be done in various ways.

1. The crucible can be moved through a region with a temperature gradient.
2. The furnace and its temperature gradient can be moved in relation to a resting crucible.
3. The crucible and the furnace can both be at rest while the temperature decreases in such a way that the temperature gradient is kept unchanged.

An example of crystal growth in a crucible is the Bridgman technique shown in Figure 6.11. The crystals are produced according to method 1.

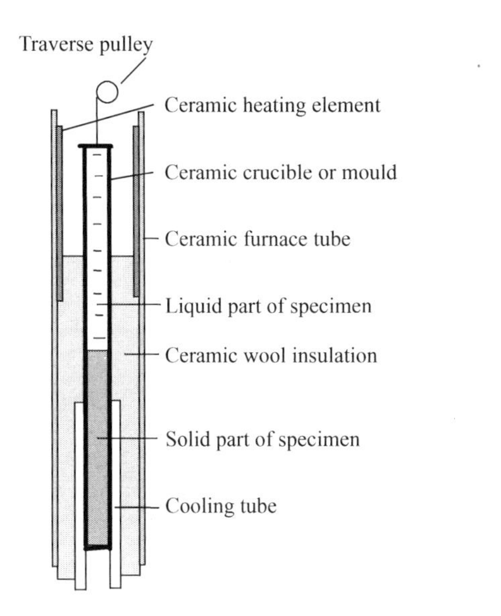

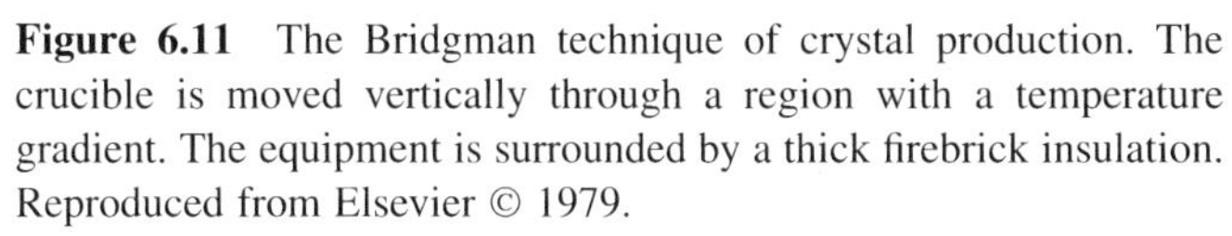
Figure 6.11 The Bridgman technique of crystal production. The crucible is moved vertically through a region with a temperature gradient. The equipment is surrounded by a thick firebrick insulation. Reproduced from Elsevier © 1979.

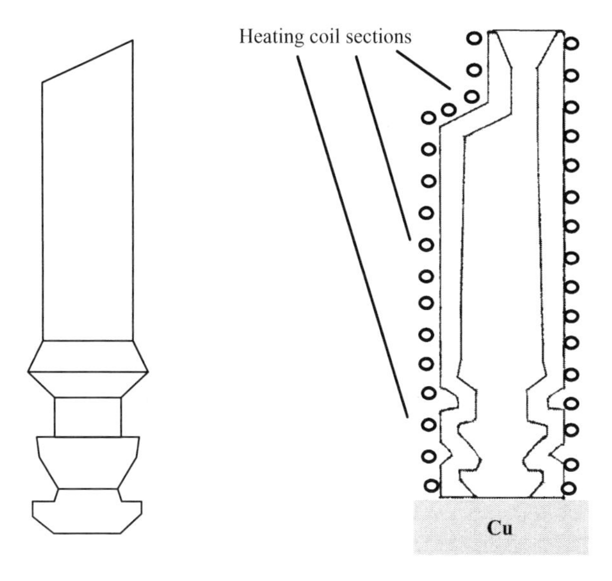

Figure 6.12 Single crystal production of turbine blades.
(a) Completed turbine blade.
(b) Turbine blade mould on a cooled Cu plate.
For more information see Chapter 8.

A concrete example is the production of turbine blades. When in use they will be exposed to severe mechanical strain at very high temperatures. Therefore, turbine blades must possess the best mechanical properties modern technique can offer. Today they are produced as single crystals by the method sketched in Figure 6.12.

A mould is placed on a water-cooled copper plate and the alloy melt is poured into the mould. Crystals are nucleated at the bottom of the mould close to the copper plate and grow upwards. Necks in the mould select one of the nucleated crystals, which is allowed to continue to grow. In order to prevent the liquid from cooling below the liquidus temperature, which would result in formation of new crystals ahead of the selected one, the mould and the melt are heated in a controlled way.

Crystal Growth by the Dragging Method

So far we have only described methods where the boundary between the solid and the liquid phases is kept in contact with the crucible. This can be avoided by using a method where the crystal grows at the surface of the melt and is dragged upwards slowly and simultaneously. The process is shown in Figure 6.13.

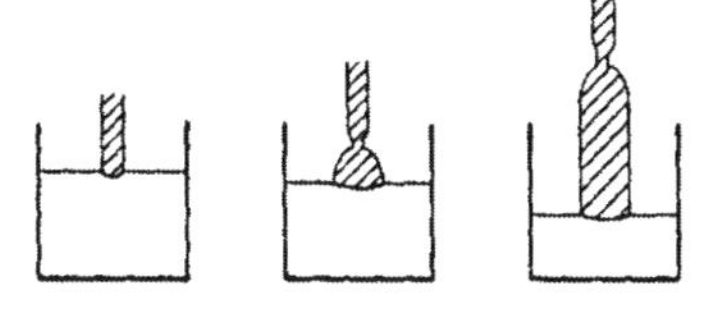
Figure 6.13 Dragging a crystal out of the melt.

The dragging method of producing single crystals has become very popular because of its speed. The process starts with a small seed crystal dipped in the melt. The crystal grows by solidification of melt and is slowly dragged upwards. The material must fulfil a number of requirements to force this process to work in a proper way. The material must have

1. a suitable melting point
2. high thermal conductivity
3. low vapour pressure
4. low viscosity
5. a suitable mechanism of growing
6. no phase changes between melting point temperature and room temperature
7. no pronounced cleavage.

It must be pointed out that the melt is kept in a crucible during the dragging process. For reactive materials it may be difficult to find a crucible, which will not react with the melt at the temperature of the growth process.

6.3.3 *Zone-Refining Methods*

Pfann invented the zone melting method at Bell Telephone Laboratories in USA in the early days of semiconductor industry at the end of the 1950s. Today it is one of the methods used for production of pure and doped semiconductor materials. It has also wide and vital applications in many other fields, for example metal refining.

Zone Melting in and without a Container

In a Bridgman process, all material is molten when the process starts. If *zone melting* is used, only a small fraction of the material is molten during the process (Figure 6.14). A hot zone with a temperature above the melting point of the material, is slowly moved along the solid specimen. The principle is that the molten zone solves impurities, that follow it, and hence leaves a recrystallized and purified material behind. The process will be analyzed some-what more in detail in Chapter 8.

The great advantage of zone melting is that the process can be repeated several times and give a more and more purified material. Depending on material constants and demand for purity of the final product, it may be necessary to repeat the zone melting process from a few to hundreds of times.

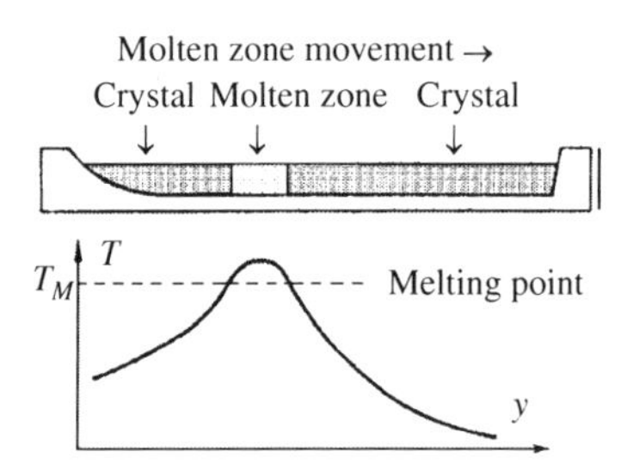

Figure 6.14 Principle of zone melting.

If any of the components of the zone-melted material is volatile, special care must be taken to prevent the component from extensive evaporation during the melting process. The process can for example be performed in an inert atmosphere or under high pressure. Corrosion and contamination between the material and the container is a problem at the high temperature of the molten zone.

This disadvantage can be eliminated simply by giving up the container and letting the melt be carried by its own solid phase. The melt is kept in proper position either by surface tension forces (Figure 6.15) or by electromagnetic forces in the case of magnetic materials.

Another group of zone refining methods is based on continuous zone melting. These methods have the advantage that the zone melting process can be repeated continuously and it is possible to achieve a high degree of purity. The method will be described in Chapter 8 in connection with the theory of zone refining.

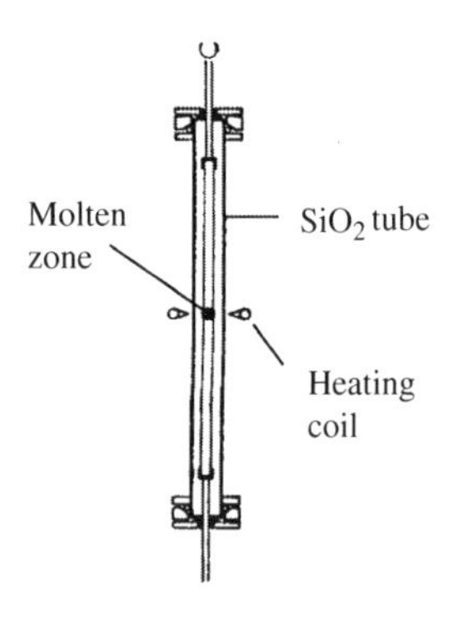

Figure 6.15 Zone refining without container. The molten zone moved slowly from bottom to the top of the tube-shaped specimen.

6.4 Crystal Growth

The structure of the interface between a liquid and a solid has been discussed extensively in the scientific literature. As in crystal growth in vapours, one normally distinguishes between rough and smooth interphases. At the end of the 1950s Jackson [2] made a simple theoretical analysis of the structure of an interphase (Chapter 3 page 132). Some years later Temkin [3] presented a more sophisticated treatment, which gave similar results.

Jackson's and Temkin's models show that diffuse or rough interphases are to be expected for most metals above certain temperatures. The most common mechanism of crystal growth in metal melts is normal growth on rough interphases, extended over several atomic layers. Normal growth is much easier than layer growth.

However, at temperatures below the roughening temperature (Chapter 5 page 224), lateral crystal growth of steps occurs at surface planes coinciding with crystal planes with low Miller indices or vicinal planes, in analogy with crystal growth in vapours.

In connection with their studies on the cellular growth phenomenon (Chapter 8) Elbaum and Chalmers [4] studied the topography of interfaces after rapid separation of the growing crystal from the melt. In spite of a thin film of melt left on the decanted surface they could show that the crystals grew by layer growth. An example is shown in Figure 6.16 where the terrace structure is clearly shown.

Elbaum made a number of experiments to analyze the formation of steps as a function of the crystallographic direction. He found that the surface was terraced if the surfaces deviated less than 10° from the < 100 > direction and less than 20° from the < 111 > direction for an FCC metal.

Normal growth and layer growth in liquids and melts will be treated in Sections 6.6 and 6.7, respectively. It is often observed that the growth surface is terraced during dendritic growth and that faceted crystals are formed. These topics will be discussed later in this chapter and in Chapter 9.

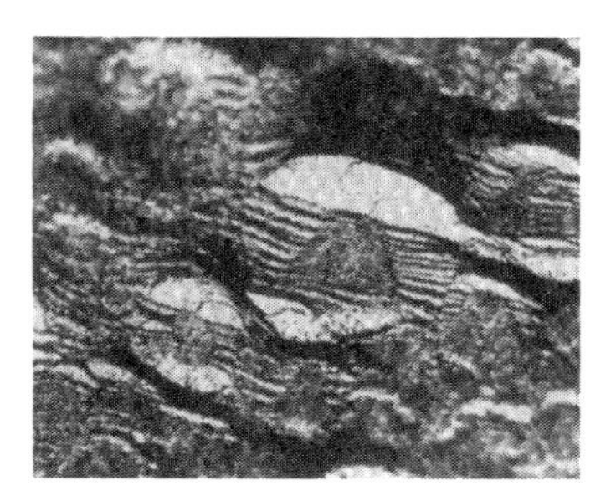

Figure 6.16 Microphotograph of a decanted cellular solidification front. Bruce Chalmers, courtesy of S. Chalmers.

6.4.1 Comparison between Crystal Growth in Vapours and Liquids

There are many similarities between crystal growth in vapours and in liquids. Concepts like driving force, activation energy, rough surfaces, smooth surfaces and roughening temperature, which we have introduced in Chapter 5, are the same and will also be used in this chapter on crystal growth in liquids.

The mechanisms of crystal growth in vapours and liquids both depend strongly on the structure of the interface. Therefore, we will distinguish between normal growth on rough interphases and layer growth on smooth interfaces for melts and also for vapours.

However, there is a large and important difference between crystal growth in vapours and in liquids. The contact between a solid and a melt is much closer than that of a solid and a vapour.

The contact between the solid and the liquid causes strain across and along the interface during the crystal growth. The stress forces, caused by the strain, vary with the crystallographic direction as they depend on the packing of the different crystal planes (Chapter 1 in [1]).

Strain occurs at solidification in both vapours and liquids. The solidification process consists of a kinetic part, when the atoms become incorporated into the crystal, and a so-called relaxation process. During solidification in vapours, the kinetic process is slow and determines the rate of solidification. For this reason, relaxation is not discussed in Chapter 5. During solidification in liquids, the relaxation process is often slower than the kinetic process and cannot be neglected. Relaxation will be treated extensively in this chapter.

6.5 Volume Changes and Relaxation Processes during Anelastic Crystal Growth in Metal Melts

6.5.1 Deformation Theory of Relaxation

In Chapter 3 (pages 142–143) and 5 (page 221) we have used simple theory of elasticity for describing deformation of solids under influence of stress. The fundamental law is Hooke's law, which is valid for small values of stress.

$$\sigma_{\text{tension}} = Y\varepsilon \tag{6.4}$$

where

σ_{tension} = the tensile stress F/A
Y = modulus of elasticity of the solid (Young's modulus)
ε = the relative stretch length.

The strain is

$$\varepsilon = \Delta l / l \tag{6.5}$$

Equation (6.4) is valid in case of ideal elasticity which is characterized by three principles:

1. Each value of the stress corresponds to a unique value of the strain.
2. The strain is proportional to the applied stress.
3. The applied stress gives instantly the corresponding strain.

No physical processes are independent of time. Elastic waves in crystals move with finite velocities. Even light transport through transparent crystals requires time even if it is so short that it can be neglected in most cases.

The three criteria 1–3 above can be used to describe the deformation processes in materials and their deformation properties. Some of them are given in Table 6.3.

Table 6.3 Some different types of deformation processes

Type of deformation process	1 Applied stress–unique strain	2 Stess and strain are proportion	3 Instantaneous strain as aresponse to stress	Characteristics of material
Ideal elasticity	yes	yes	yes	elastic
Anelasticity	yes	yes	yes	anelastic
Non-linear elasticity	yes	no	yes	non-linear elastic
Instantaneous plasticity	no	no	yes	instantaneous plastic
Linear visco-plasticity	no	yes	no	linear visco-plastic

The elastic and anelastic materials return to their original shape when the applied stress is removed. All elastic materials are in reality anelastic. If the time delay can be neglected they can be regarded as ideally elastic.

Plastic and linear visco-plastic materials do not return to their original shape when the stress is removed. There will be a permanent deformation.

A material is often elastic for small values of the stress and becomes plastic beyond a characteristic value of the stress, which is a material constant.

Figure 6.17 shows the strain as a function of time when a stress is applied up to a certain maximum value and shows also the strain when it decreases from the top value down to a new equilibrium value.

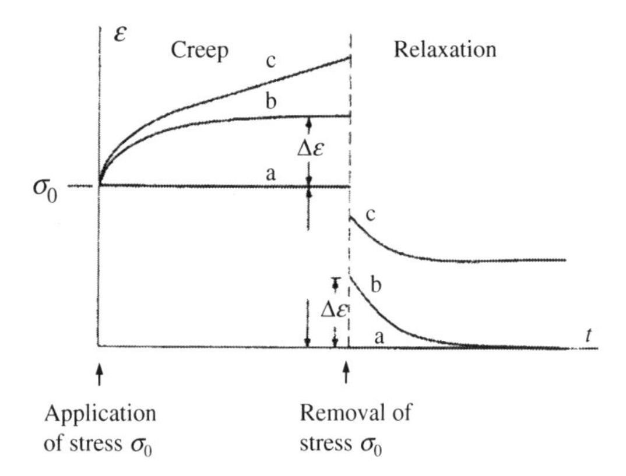

Figure 6.17 Strain as a function of time for some materials with different elastic properties when they are exposed to stress and when the stress is suddenly removed.
(a) ideally elastic material
(b) anelastic material
(c) linear viscoplastic material. Reproduced from Elsevier © 1972.

Figure 6.17 confirms that ideally elastic and anelastic materials go back to the same shape when the stress is removed. This process takes time for anelastic materials. It is spontaneous and is called *relaxation*. Relaxation processes also occur in plastic and visco-plastic materials.

6.5.2 *Volume Changes of Metals at Solidification*

A comparison between the densities of solid metals ρ_s and the corresponding liquids ρ_L shows that the total volume decrease at solidification of most common metals varies between 3% and 8%.

One reason is the different packing structures of melts and solid metals, which will be treated in Section 6.7.2. The forces between the liquid atoms are considerably weaker than those in crystalline solids.

The volume differences between the melt and the crystal cause primarily instability over the interface. Adaptation to a stable state occurs by *deformation*. There are three possible mechanisms of deformation:

- elastic deformation
- creep
- formation of vacancies.

Each of these alternatives will be examined below in terms of Gibbs' free energy. Two of them are discussed here. Vacancy formation is discussed in Section 6.7.5 on page 304. A comparison between the mechanisms is given on page 306. The calculations will indicate which one of the three mechanisms that will be the most likely one.

Elastic Deformation

Strain involves forces, which act on the atoms of the interface. If the atoms move under the influence of the strain forces from their initial positions, elastic forces arise which counteract the strain forces. The elastic forces increase until the net force is zero and a new equilibrium is established. This is elastic deformation and one way to overcome the strain of the interface.

The classical theory of elastic deformation is only valid for ideally elastic materials and is given in Chapter 3 on pages 142–144. If Hooke's law is valid, the linear elastic strain energy per unit volume of the interphase, is equal to the work required to achieve the relative stretch length ε. It can be written as

$$w_{\text{el}} = \frac{Y\varepsilon^2}{2} \tag{6.6}$$

The Gibbs' free energy change per kmol of the interface, owing to the elastic strain, is obtained if w_{el} is multiplied by the factor $2M/\rho$. where M is the atomic weight of the metal and ρ is the density of the interphase. The elastic strain is two-dimensional in the present case and the *area* stretch energy is equal to *twice* the *linear* stretch energy. The factor M/ρ represents the transition from J/m^3 to J/kmol. Hence, we obtain

$$-\Delta G_{\text{l}}^{\text{I}} = \frac{Y\varepsilon^2}{2}\frac{2M}{\rho}$$

or

$$-\Delta G_{\text{el}}^{\text{I}} = Y\varepsilon^2\frac{M}{\rho} \tag{6.7}$$

where

$-\Delta G_{\text{el}}^{\text{I}}$ = elastic strain energy per kmol of the interphase
Y = modulus of elasticity of the solid
ε = relative linear stretch length of the interphase caused by the strain of the interphase
M = atomic weight of the metal
ρ = density of the interphase.

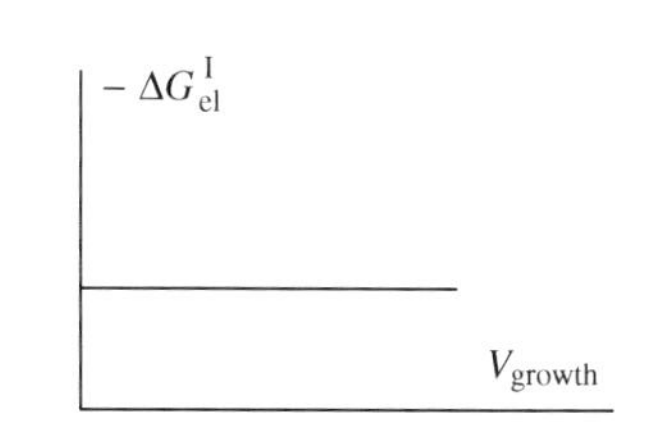

Figure 6.18 Elastic strain energy per kmol of the metal interphase as a function of the growth rate.

For ideally elastic materials the elastic strain energy is independent of the growth rate of the crystal (Figure 6.18). Hence, $-\Delta G_{\text{el}}^{\text{I}}$ appears as a horizontal line if it is plotted as a function of V_{growth}. The modulus of elasticity Y is an anisotropic quantity. Its value in the $< 111 >$ direction is much higher than in all other directions. The value given in tables is an average value over all directions. Y decreases with temperature.

Creep

Characteristic of an *ideally elastic deformation* is that it disappears when the outer force becomes zero. A *plastic deformation* remains even if the outer force disappears.

Creep is *a plastic deformation, which moves through a material.* A creep process is another way to overcome the strain of the interphase caused by volume changes during solidification of metals.

At high temperature, metals may be plastically deformed by a creep process, which sweeps over the interphase and removes the strain. This process can occur in many different ways. The two most common ones are

- creep by generation of vacancies followed by condensation of vacancies
- creep by rearrangement of atoms, where several atoms change their positions simultaneously.

The second type of process means that atoms gradually move during the creep process, which is similar to the flow of a liquid. This model will be used here. Before the visco-plastic model of creep is described, a box on the theory of plastic deformation is given below.

Plastic Deformation

An *elastic* deformation of a material is described by Hooke's law. When the strain is increased the so-called *yield point* is reached. Above this point the deformation becomes *plastic*, i.e. it remains after the strain on the material has been removed.

According to plasticity theory the relationship between stretch length rate, yield strain $\boldsymbol{\sigma}$ and temperature is given by the so-called Norton expression

$$\frac{\mathrm{d}\varepsilon}{\mathrm{d}t} = A\sigma^{1/\mathrm{m}}\mathrm{e}^{-Q/RT} \tag{1'}$$

where
A, m, Q = constants
R = gas constant
T = temperature.

We assume that $m = 1$ at visco-plastic processes. By introduction of another constant C the Norton expression can be transformed into

$$\sigma = \frac{1}{A}\mathrm{e}^{-CT}\frac{\mathrm{d}\varepsilon}{\mathrm{d}t} \tag{2'}$$

or

$$\sigma = \eta\frac{\mathrm{d}\varepsilon}{\mathrm{d}t} \tag{3'}$$

where $\eta = \frac{1}{A}\mathrm{e}^{-CT}$

η is strongly temperature dependent and decreases with increasing temperature.

The Visco-Plastic Model of Creep

We will start the analysis of the effect of volume changes on crystal growth by assuming that the creep process sweeps over the interphase. This creep process is complicated to describe mathematically. One practical way is to treat the interphase as a visco-plastic medium. If we use this model, the creep process can be described as a moving relative stretching of the atoms through the material.

The stretch can be described with the aid of the relative stretch length ε. The rate of the creep process is described by the time derivative of the relative stretch length, i.e. how fast the relative stretch moves. The motion resembles of the motion of a viscous liquid. The creep process can then be described by the following simple law, the correspondence to $\sigma = Y\varepsilon$ in the case of elastic deformation (Equation (3′) in the box on page 282)

$$\sigma = \eta \frac{d\varepsilon}{dt} \tag{6.8}$$

where

ε = strain, i.e. relative stretch length
σ = stress over the interphase caused by the strain ε
η = viscosity constant
t = time
$\frac{d\varepsilon}{dt}$ = the stretch length rate during the motion of the stretch through the material.

The work per unit volume, required to achieve the volume change which corresponds to the relative stretch length ε, is

$$w_{creep} = \sigma\varepsilon = \eta\varepsilon \frac{d\varepsilon}{dt} \tag{6.9}$$

It is related to the growth rate, by the following relationship

$$w_{creep} = \eta\varepsilon \frac{d\varepsilon}{dt} = \eta\varepsilon \frac{d\varepsilon}{dy}\frac{dy}{dt} = \eta\varepsilon \frac{d\varepsilon}{dy} V_{growth} \tag{6.10}$$

where y is a coordinate in the direction of the growth rate.

The driving force of the creep process equals the change of Gibbs' free energy per kmol of the interphase. It is also equal to the work per kmol required to achieve the relative volume change caused by the creep.

$$-\Delta G^{I}_{creep} = w_{creep} \frac{2M}{\rho} = 2\eta\varepsilon V_{growth} \frac{d\varepsilon}{dy} \frac{M}{\rho} \tag{6.11}$$

The factor 2 is introduced because *area* stretch energy is equal to *twice* the *linear* stretch energy. The factor M/ρ represents the transition from J/m^3 to J/kmol (compare page 281).

Equation (6.11) is integrated over the interphase

$$\int_0^d -\Delta G^{I}_{creep} dy = \int_0^{\varepsilon} 2\eta V_{growth} \frac{M}{\rho} \varepsilon d\varepsilon$$

or

$$-\Delta G^{I}_{creep} = \eta\varepsilon^2 \frac{V_{growth}}{d} \frac{M}{\rho} \tag{6.12}$$

where

$-\Delta G^{I}_{creep}$ = visco-plastic strain energy per kmol of the interphase
ε = relative stretch length
V_{growth} = growth rate of the crystal
d = thickness of interphase
M = atomic weight of the metal
ρ = density of the metal.

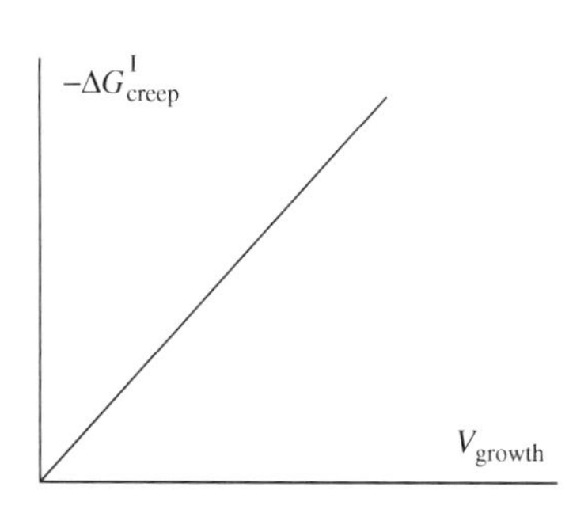

Figure 6.19 Gibbs' free-energy change per kmol of the metal interphase owing to creep as a function of the growth rate.

As an illustration of the theory given above, the Gibbs' free energy change per kmol of the metal interphase is plotted as a function of growth rate in Figure 6.19. According to Equation (6.12) on page 283 the strain energy per kmol, caused by creep, varies linearly with the growth rate.

6.5.3 *Relaxation Processes*

General Definition of Relaxation

Consider a system, which can be described by a variable as a function of one or several parameters, in equilibrium at given values of the parameters. At the time $t = 0$ one of the parameters is suddenly changed into a new value (see for example Figure 6.17 on page 280). The equilibrium is disturbed and the system tries to achieve a new equilibrium. This process, which requires time, is called *relaxation*. The rate of the change is followed by studying the variable as a function of time.

Relaxation processes have been discussed briefly earlier on page 237 in Chapter 5. Below relaxation will be treated more in detail with special accentuation on solidification processes, especially for anelastic materials.

Application on an Electrical Current. Relaxation Time

This case is discussed in Chapter 7 on pages 355–356 in [1].

A constant electrical current I runs in a resistor R. The carriers of the electrical charge are electrons, which are accelerated in the opposite direction of the current in an electrical field, caused by the voltage U (Figure 6.20). They frequently collide with other particles and change direction.

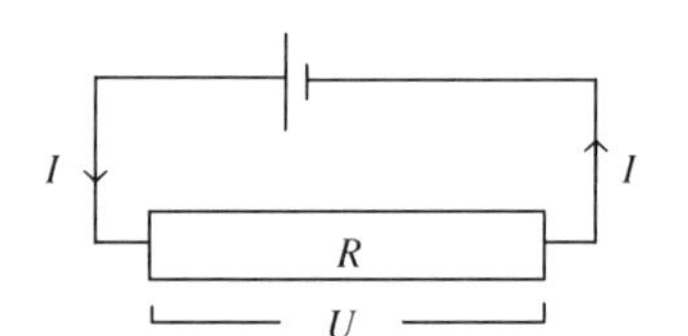

Figure 6.20 Simple electrical circuit.

The electrical field accelerates the conduction electrons between the collisions. The collisions can be regarded as a friction force. A force balance for the electron motion can be set up:

$$m\frac{\mathrm{d}v}{\mathrm{d}t} = eE - \frac{mv - 0}{\tau} \tag{6.13}$$

Resulting force	Electrical accelerating force	Retarding friction force

The time τ is the average time between two collisions.

Relaxation Time

The time τ has also another signification. If the electrical field is suddenly switched off [insert $E = 0$ into Equation (6.13)] the friction force gradually retards the electrons from their average velocity to zero.

Integration of equation (6.13) gives

$$\int_{v_e}^{v} \frac{\mathrm{d}v}{v} = \int_{0}^{t} -\frac{\mathrm{d}t}{\tau}$$

which can be written as

$$v = v_e e^{-\frac{t}{\tau}} \tag{6.14}$$

where

v = electron velocity at time t

v_e = drift velocity, i.e. velocity at time $t = 0$

τ = average time between two consecutive collisions of Equation (6.13) gives.

As the current I is proportional to the velocity v (page 354 in [1], Equation (6.14) can be written as

$$I = I_0 e^{-\frac{t}{\tau}} \tag{6.15}$$

Equation (6.15) is one example of the basic formula for relaxation. The general definition of relaxation time τ is:

$$\delta = \delta_0 e^{-\frac{t}{\tau}} \tag{6.16}$$

where
τ = the time required to reduce the initial value δ_0 to δ_0/e, where e is the base of natural logarithms.

In the present case, the relaxation time is equal to the average time between two consecutive collisions of the conduction electrons.

Other Examples of Relaxation

The concept relaxation can be applied on many phenomena, such as strain in crystals, vacancies in crystals, chemical reactions and solidification processes, which will be discussed in next section. The most common parameters in relaxation processes are temperature, pressure and electrical field.

The relaxation law is valid for strained crystals, When the stress (pressure) is suddenly released, the atoms go back to their strain-free positions, which requires time (Figure 6.17 on page 280). The process can be described by the relaxation time.

In the case of vacancies the relaxation time depends on the distance between the vacancy sinks. The density of sinks is given by the fraction of dislocations or other defects. The higher the density of sinks is, the shorter will be the relaxation time.

M Eigen introduced the study of rapid chemical reactions with the aid of relaxation methods in the middle of the 1950's. A sample of a system in equilibrium at an initial temperature T_0 is exposed to a rapid and short discharge. Hence, its temperature suddenly raises to a temperature T. The system starts to achieve a new equilibrium and the process is followed by continuous measurement of the concentration change Δx of some component. If the temperature change is not too large, the relaxation law is valid for the process

$$\Delta x = \Delta x_0 e^{-\frac{t}{\tau}} \tag{6.17}$$

τ is called the *chemical relaxation time* of the reaction. Depending on the type of reaction (first order, second order etc) a relationship between τ and the rate constants of the reaction can be derived in each case.

6.5.4 *Relaxation during Anelastic Crystal Growth in Metal Melts*

The Relaxation Law

The atoms have different potential energy in the liquid and in the solid. Figure 6.17 on page 280 shows the process of transferring the system back to equilibrium, i.e. the time required for the atoms to reach their lowest potential energy. The process consists of two steps: one in the interphase and one in the solid phase.

The interphase normally consists of both 'solid' and 'liquid' atoms and their positions are not stationary as in a solid. The change of the potential energy is much faster in the interphase than in the solid, owing to the presence of 'liquid' atoms.

To change the potential energy in the solid requires longer time than in the interphase because the potential energy of one atom depends on the potential energy of the surrounding atoms. The vibration of the atoms influences the required time for relaxation. The process can be described by the normal relaxation law, which in its most general form can be written as

$$\Delta E_{pot} = \Delta E_{pot}^0 \times Ce^{-\frac{t}{\tau}} \tag{6.18}$$

where
ΔE_{pot} = energy deviation from the equilibrium energy of the sample.
ΔE_{pot}^0 = initial energy deviation from the equilibrium energy of the sample.
C = dimensionless time-independent constant
t = time
τ = relaxation time.

The potential energy can be expressed in many different ways, for example as a change of latent heat or heat capacity. The change of potential energy is also related to a volume change of the material.

The relaxation time τ is related to the type of material, i.e. to the vibrational motion of the atoms which change their potential energies in order to reach the equilibrium positions.

The relaxation process is very similar to heat and mass transport processes in metals. In order to find reasonable values of the constants C and τ we will use relationships valid for well known diffusion laws below.

Relaxation of the Interphase

In all time-independent phase transformations Fick's first law is used to describe the velocity of the boundary. To find an expression for the mass transport to and from the phase boundary Fick's second law, a differential equation of second order, has to be solved. The constants in the solution are determined by given boundary conditions.

In the case of relaxation there is no mass transport. It concerns a change of the potential energy of the atoms. However, in spite of this we will apply the diffusion laws valid for heat and mass transport to describe the relaxation process.

The relaxation process can be regarded as coupled to the solidification process, i.e. the generation of heat that arises when the atoms change their relative position to each other. Each atomic layer generates a certain amount of heat. Integration over all the atom layers gives the total amount of heat, i.e. the heat of solidification.

In order to find the latent heat (heat of solidification) released over the interphase we have to find a solution of the differential equation

$$\frac{\partial(-\Delta H)}{\partial t} = D_{\text{relax}} \frac{\partial^2(-\Delta H)}{\partial y^2} \tag{6.19}$$

where

D_{relax} = material constant coupled to heat transport (m^2/s)
$(-\Delta H)$ = heat of solidification

which corresponds to Fick's second law. In our case we will use the exponential solution, which is described in the box and applied on concentration distribution of an alloying element at diffusion instead of generation of heat.

By replacing the concentration c by generation of heat per unit volume we obtain the solution of Equation (6.19)

$$-\Delta H = \frac{(-\Delta H_0) \times (1 \times \delta)}{\sqrt{\pi D_{\text{relax}}}} \frac{\exp(-y^2/4Dt)}{\sqrt{t}} \tag{6.20}$$

where

$-\Delta H$ = developed heat of solidification over the interphase at distance y and time t
$-\Delta H_0$ = total heat of solidification per unit volume
δ = thickness of interphase.

Concentration Distribution of the Diffusing Element as a Function of Time and Position at One-Dimensional Diffusion

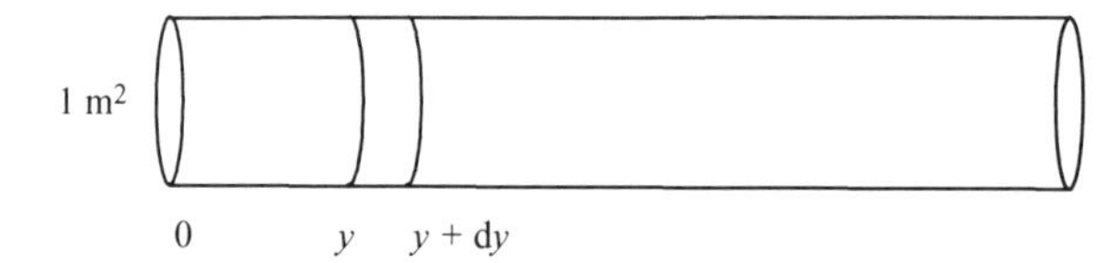

Consider a metal cylinder of infinite length with the cross-section area equal to 1 unit area. An amount m of an alloying element is located at $y = 0$ at time $t = 0$. The atoms start to diffuse in the direction of increasing x values. We want to find the concentration of the alloying element as a function of time and position. The diffusion coefficient is constant and equal to D.

We use Fick's second law

$$\frac{\partial c}{\partial t} = D\frac{\partial^2 c}{\partial y^2} \tag{1'}$$

It is easy to verify by partial differentiation that a solution of the differential equation can be written as

$$c = \frac{A}{\sqrt{t}}\exp(-y^2/4Dt) \tag{2'}$$

where A is a constant. To determine the value of A in the present case we set up an expression for the total amount of the alloying element which has passed the cylinder when the diffusion process is finished.

$$m = \int_0^\infty c \times 1 \times \mathrm{d}y \tag{3'}$$

We introduce the relationship

$$y^2/4Dt = \xi^2 \tag{4'}$$

and differentiate it

$$\mathrm{d}y = 2\sqrt{Dt}\,\mathrm{d}\,\xi \tag{5'}$$

Equations (2′), (4′) and (5′) are introduced into Equation (3′):

$$m = \int_0^\infty c\mathrm{d}y = \int_0^\infty \frac{A}{\sqrt{t}}\exp(-\xi^2) \times 2\sqrt{Dt}\,\mathrm{d}\,\xi = 2A\sqrt{D}\int_0^\infty \exp(-\xi^2)\mathrm{d}\xi \tag{6'}$$

The integral represents half the area under the error function and its value is equal to $\mathrm{d}y = \sqrt{\pi}/2$. We insert this value into Equation (6′) and use it for solving the constant A.

$$m = A\sqrt{\pi D} \qquad \Rightarrow \qquad A = m/\sqrt{\pi D} \tag{7'}$$

We insert this value of A into Equation (2′) and obtain the concentration distribution of the diffusing alloying element as a function of position and time and m, the total amount of the alloying element.

$$c = \frac{m}{\sqrt{\pi D}}\frac{\exp(-y^2/4Dt)}{\sqrt{t}} \tag{8'}$$

The function c/m as a function of position is shown in the figure below at three different times $t = 1$, 1/2 and 1/16 s.

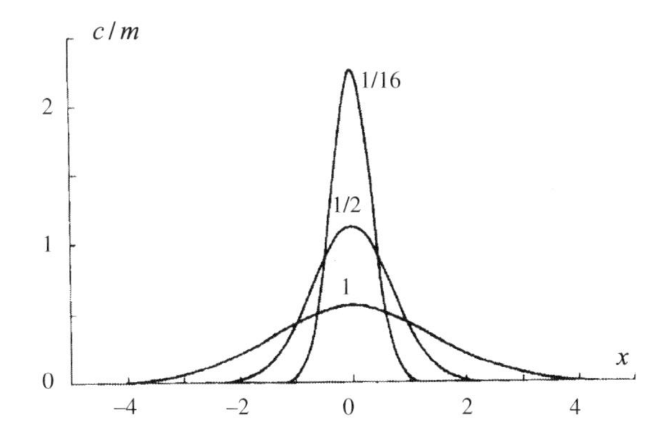

We want to find the released amount of heat per unit volume, i.e. how much the atoms have changed their potential energy over the interphase. The position of the interphase corresponds to $y = 0$. Hence, the exponential factor equals 1. The interphase moves with the growth rate V_{growth} during the solidification process and has a thickness δ. The relaxation time t of the interphase will be the thickness divided by the growth rate. When these conditions are considered, the solution, Equation (6.20) on page 286, can be written as

$$-\Delta H = -\Delta H_0 \frac{\delta}{\sqrt{\pi D_{relax}}} \frac{1}{\sqrt{\dfrac{\delta}{V_{growth}}}}$$

The conclusion from this equation is that the atoms in the solid phase has not yet reached their lowest potential energy. A certain amount of heat per unit volume is left in the solid after the relaxation time. This amount of heat per unit volume can be interpreted as *strain energy per unit volume*

$$-\Delta H_{strain} = -\Delta H_0 \sqrt{\frac{V_{growth}\delta}{\pi D_{relax}}} \tag{6.21}$$

where $-\Delta H_0$ is the total heat of solidification.

The strain energy will be released and transferred into heat during the continued relaxation process. This relaxation can occur either by an anelastic process (page 285) or by a creep process (page 282) related to formation of vacancies. Vacancies will be discussed on pages 304–306.

Relaxation Processes at Anelastic Solidification Processes in Metals

In Chapter 5 we discussed the effect of elastic deformation on the solidification process of a metal vapour to a crystalline solid provided that solid metal was ideally elastic.

This chapter deals with formation of a crystalline solid from a liquid metal. In this case we will consider the influence of time on the solidification process and consider the solid crystal as an anelastic material. As was mentioned on page 221 in Chapter 5 the surface of the crystal becomes strained owing to the difference in interatomic distances between the interior and the surface of the crystal.

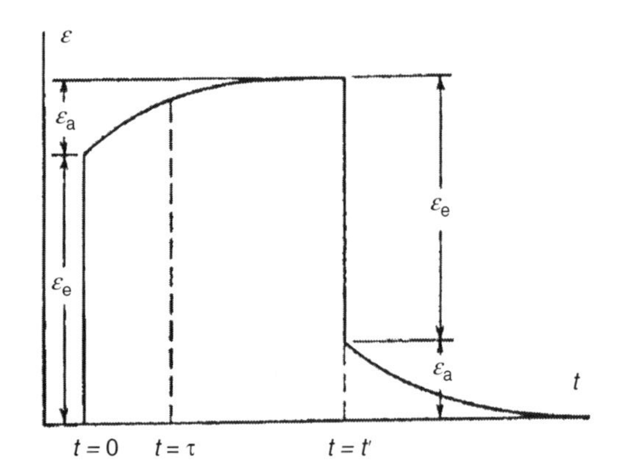

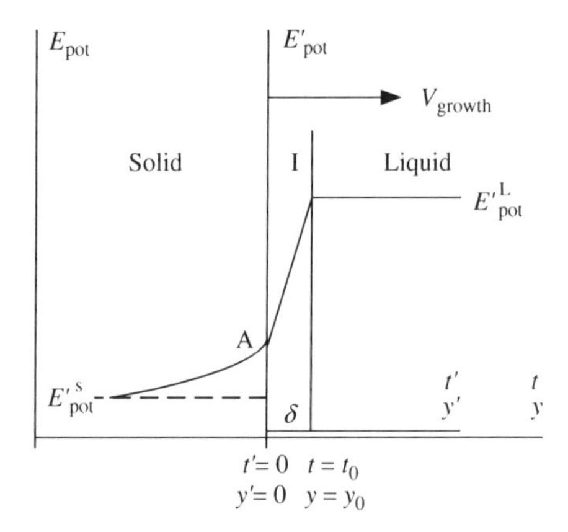

Figure 6.21a Strain as a function of time for an anelastic specimen when stress is applied and later removed. Reproduced from Elsevier © 1972.

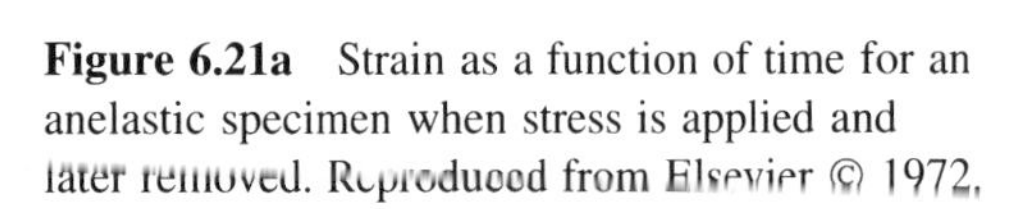

Figure 6.21b Potential energy of the atoms in the vicinity of the liquid/solid (crystal) interphase.

Figure 6.21a shows a sketch of the strain in a specimen as a function of time when it is exposed to stress. The specimen is assumed to be anelastic and has the properties described on pages 279–280.

When a constant stress is applied at $t = 0$, elastic strain is the immediate response followed by a slow increase of the strain owing to a relaxation process which lasts as long as the stress remain unchanged. It goes on until equilibrium has been achieved. When the stress is removed at time $t = t'$, the elastic part of the strain disappears rapidly and the rest declines exponentially by a slow relaxation process down to zero.

This strain process above can be used to describe the solidification process in a solidifying metal melt. The 'solid' atoms in the interior of a crystal are in equilibrium and have a constant average interatomic distance and a constant potential energy. In the melt the interatomic average distance is larger than that in the interior of the crystal. The larger distance corresponds to an anelastic strain and also to a higher potential energy in the liquid than in the interior of the crystal. The strain is related to the potential energy of the atoms.

Figure 6.21b describes the solidification process as the potential energy of the atoms is a function of time. Before the solidification a layer of 'liquid' atoms have a constant potential energy. When the solidification front reaches the layer of 'liquid' atoms they pass the interphase and their interatomic distances decreases rapidly, which corresponds to the elastic part of the strain. When the atoms reach the crystal surface and form a solid surface layer of the crystal, their interatomis distances are reduced but higher than that of the interior 'solid' atoms, owing to the remaining strain. The potential energy of the surface atoms are also smaller than those in the liquid but larger than those of the interior 'solid' atoms (point A in Figure 6.21b). A relaxation process starts immediately, which results in an exponential decrease of the interatomic distance and ditto potential energy down to the equilibrium state of the atoms in the interior of the crystal.

The interphase and the solid move into the liquid with the constant growth rate V_{growth}. In a mobile coordinate system, connected with them, the E'_{pot} profile will be constant. If we introduce $y = y_0$ at point A we can replace the time coordinate t by the position coordinate y on the horizontal axis in the stationary coordinate system. y is proportional to t, which can be written as $y = y_0 + V_{growth}\,(t - t_0)$.

6.6 Normal Crystal Growth in Pure Metal Melts

The mechanism of normal crystal growth is essentially the same in liquids and melts as in vapours. There is a rough interphase between a melt and the crystal surface with its regular structure and strong attraction forces between the atoms. The forces between the atoms in a melt are weaker than those in the solid and allow rather free relative motion of the atoms in the melt. However, the attraction forces between the atoms in a melt are strong enough to keep the melt together.

Like solid-vapour interphases a rough solid/liquid interphase is extended over several irregular atomic layers (Figure 6.22). The interphase contains a large number of kinks and atoms from the liquid, which can rather easily be incorporated into the crystal lattice.

Crystal growth from a melt on a rough interphase (Figure 6.22) is much easier than that on a smooth crystal surface.

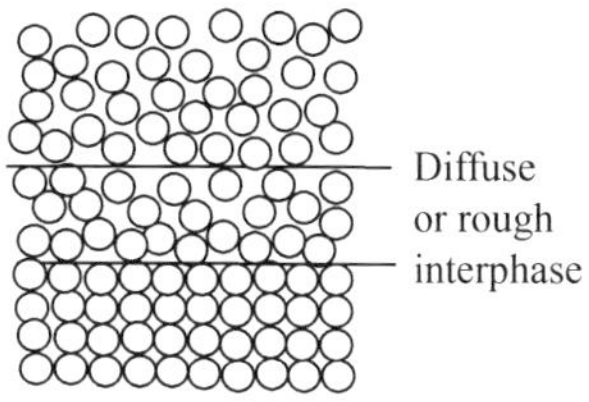

Liquid phase

Diffuse or rough interphase

Solid phase

Figure 6.22 Rough interphase between a liquid and a solid.

6.6.1 *Driving Force and Activation Energy with Consideration to Relaxation*

When an atom from a liquid or melt occupies a kink in a rough interphase (Figure 6.23) the bonds between the atom and the crystal lattice will be strong and the atom becomes part of the crystal. This means that its Gibbs' free energy will be equal to the Gibbs' free energy of the crystal.

An atom in the liquid phase has a higher Gibbs' free energy than the atoms in the crystal lattice. The difference in Gibbs' free energy between the melt and the crystal is the *driving force* of the crystallization process per atom (subscript 'a') (Figure 6.24).

$$-\Delta G_a^{L\rightarrow s} = -\left(G_a^s - G_a^L\right) = G_a^L - G_a^s = -\Delta G_a \tag{6.22}$$

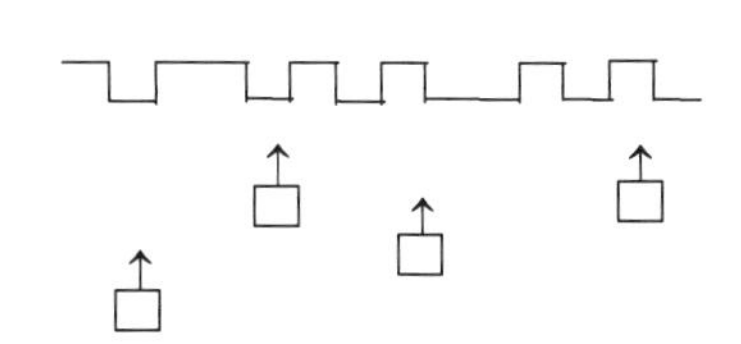

Figure 6.23 Adsorption of 'cubic' melt atoms with the side a in a rough interphase. The chance of adsorption is $a/\overline{\delta}$, where $\overline{\delta}$ is the average distance between the kinks.

In Chapter 2 (Section 2.2.1 pages 43–44) the driving force of solidification was found to be a function of the undercooling of the melt in front of the solidification front (Equation (2.5) on page 44 in Chapter 2). For small values of the undercooling $(T_M - T)$

- The driving force of solidification is proportional to the undercooling of the melt.

$$-\Delta G_{\mathrm{a}}^{\mathrm{total}} = (-\Delta H_{\mathrm{a}})\frac{T_{\mathrm{M}} - T}{T_{\mathrm{M}}} \tag{6.23}$$

where

$-\Delta G_{\mathrm{a}}^{\mathrm{total}}$ = total driving force of solidification per atom or = difference in Gibbs' free energy between an atom in the melt and an atom in the crystal = $-\Delta G_{\mathrm{a}}$

T_{M} = melting-point temperature

T = solidification temperature

$-\Delta H_{\mathrm{a}}$ = total heat of fusion (J/atom).

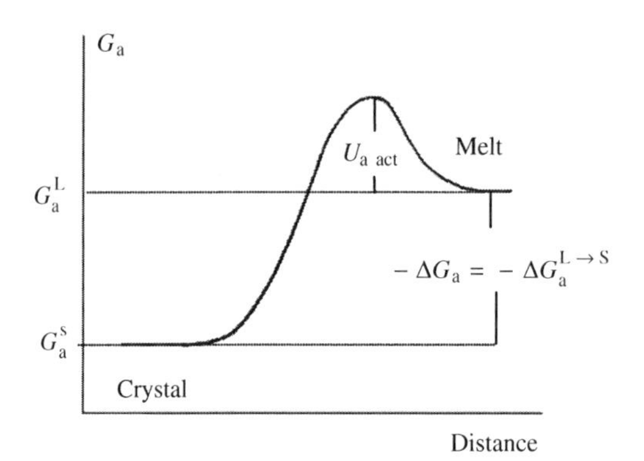

Figure 6.24 Gibbs' free energy of the interphase between a crystal with a rough interphase and its liquid phase as a function of position.

As we will see below the same theory is valid for liquids and melts as for vapours. Before an atom in the liquid can be incorporated into a kink it has to overcome the activation energy $U_{\mathrm{a\ act}}$ (Figure 6.24). The activation energy originates from a position change and a rotation of the atom to make it fit into the kink.

6.6.2 *Rate of Normal Growth in Pure Metals*

Flux of 'Liquid' Atoms

The total crystallization process consists of two subprocesses, diffusion of individual atoms in the melt to the interphase and incorporation of these atoms into the crystal lattice. These two processes are coupled in series and the total driving force can be divided into two parts, one related to the diffusion process and one related to the kinetic interphase process in analogy with the corresponding conditions in vapours. In this section we will discuss the ordering process. The diffusion process occurs in alloys and will be discussed in Section 6.8.

In analogy with crystal growth in vapours the growth rate is a function of the flux to and from the crystal surface. In Chapter 5 the flux of atoms was expressed in terms of the vapour pressure. The pressure is no relevant quantity in the case of liquids. Instead it is the *temperature* or *concentration differences* which *controls the solidification process*. However, the proportionality between the growth rate and the net flux of atoms that will be incorporated into the crystal [Equation (5.20) on page 213 in Chapter 5] holds for crystal growth in both vapours and liquids/melts.

$$V_{\mathrm{growth}} = a^3(j_+ - j_-) \tag{6.24}$$

where

a = side of a 'cubic' atom (page 212 in Chapter 5)

V_{growth} = growth rate (m/s)

j_+ = attached flux = number of incoming atoms which hit the crystal per unit area and unit time

j_- = detached flux = number of ejected atoms that leave the crystal per unit area and unit time.

On pages 289–290 we discussed the process when a 'liquid' atom occupies a kink and becomes incorporated into the crystal. In Chapter 5 on page 212 we described the probability as a factor $(a/\bar{\delta})^2$ times a Boltzmann factor. $\bar{\delta}$ is the average distance between the kinks.

In Section 6.5.4 we treated the atomic process in the interphase as a relaxation process. The relationship, which describes the relaxation process on page 286, can be used to define the probability which can replace the factor $(a/\bar{\delta})^2$. In this case the fraction of atoms which have a potential energy E_{pot} which characterizes the solid, can be described by the relationship

$$\left(\frac{a}{\bar{\delta}}\right)^2 = P = f^s = \frac{E_{pot}}{E_{pot}^0} = \sqrt{\frac{V_{growth}\delta}{\pi D_{relax}}} \tag{6.25}$$

where

a = side of a cubic atom
$\bar{\delta}$ = average distance between the kinks
E_{pot} = the strain energy of a 'liquid' atom
E_{pot}^0 = the total heat of solidification, i.e. the difference between a 'liquid' and a 'solid' atom
δ = thickness of the interphase.

The Boltzmann factor is the fraction of atoms, which have energy enough to overcome the activation energy barrier $U_{a\ act}$. The activation energies of the incoming and leaving atoms are different as is shown in Figure 6.24 on page 290. Hence, the fluxes can be written as

$$j_+ = const \times P e^{-\frac{U_{a\ act}}{k_B T}} \tag{6.26}$$

and

$$j_- = Const \times P e^{-\frac{-\Delta G_a + U_{a\ act}}{k_B T}} \tag{6.27}$$

where

P = probability factor
$U_{a\ act}$ = activation energy per atom
$-\Delta G_a$ = difference in Gibbs' free energy between an atom in the melt and an atom in the crystal.

In the case of normal growth in liquids, the *temperature* of the liquid plays an important role. At equilibrium, the growth rate and the net flux are both zero. This occurs at a temperature equal to the melting point T_M and we obtain

$$j_+^{eq} = const \times P e^{-\frac{U_{a\ act}}{k_B T_M}} = j_-^{eq} = Const \times P e^{-\frac{-\Delta G_a + U_{a\ act}}{k_B T_M}} \tag{6.28}$$

At equilibrium $-\Delta G_a^{eq} = 0$ which gives a relationship between the constants in Equations (6.26) and (6.27) are equal

$$const = Const$$

Rate of Normal Growth

The net flux of atoms towards the crystal causes the growth. We combine Equations (6.24), (6.25), (6.26), (6.27) and (6.28) and obtain the growth rate as a function of temperature

$$V_{growth} = a^3 \left(const \times P e^{-\frac{U_{a\ act}}{k_B T}} - const \times P e^{-\frac{-\Delta G_a + U_{a\ act}}{k_B T}} \right)$$

or after rearrangement

$$V_{\text{growth}} = a^3 const \times P\, e^{-\frac{U_{\text{a act}}}{k_{\text{B}}T}} \left(1 - e^{-\frac{-\Delta G_{\text{a}}}{k_{\text{B}}T}}\right) \tag{6.29}$$

If the undercooling $(T_{\text{M}} - T)$ is small, $-\Delta G_{\text{a}}$ is small compared to $k_{\text{B}}T$ and the exponential function in Equation (6.29) can be developed in series. In this case we obtain

$$V_{\text{growth}} \approx a^3 \times const \times P e^{-\frac{U_{\text{a act}}}{k_{\text{B}}T}} \left(-\frac{\Delta G_{\text{a}}}{k_{\text{B}}T}\right) \tag{6.30}$$

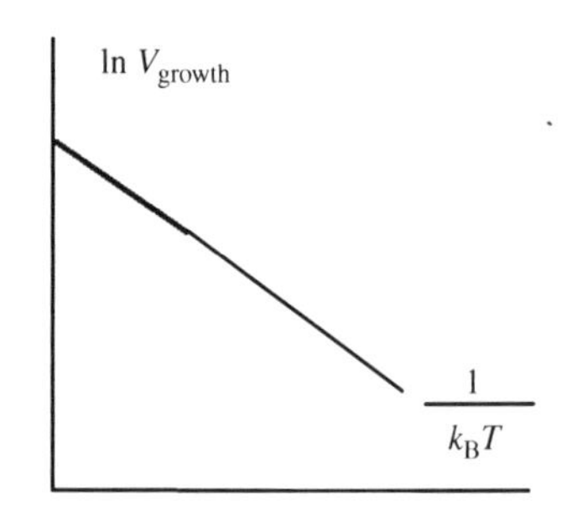

Figure 6.25 Graphical determination of the activation energy $U_{\text{a act}}$.

In this case the driving force is *not* the total driving force.

Instead it is only the difference in Gibbs' free energy between the liquid and the solid in contact with the interphase. To obtain the total Gibbs' free energy you must add the relaxation energy.

The Gibbs' free energy of the solid phase is related to the potential energy of the atoms. This was discussed on pages 285–288 and the total Gibbs' free energy will be derived on pages 293–294.

The activation energy can be determined by measuring the growth rate at several temperatures and plotting the logarithm of the growth rate as a function of $1/k_{\text{B}}T$. The activation energy is derived from the slope of the straight line (Figure 6.25).

Alternative Derivation of Normal Growth Rate

Alternatively Equation (6.30) can be derived in a more direct way without regarding the flux at equilibrium. Instead we use the absolute reaction theory, applied by Turnbull [5] on crystal growth, and consider the atomic motion at the interphase and the probability of atom jumps to and from the melt and the crystal surface, respectively, at the actual temperature T. The method is essentially the same, but it gives a better understanding of the signification of the constant in Equations (6.29) and (6.30).

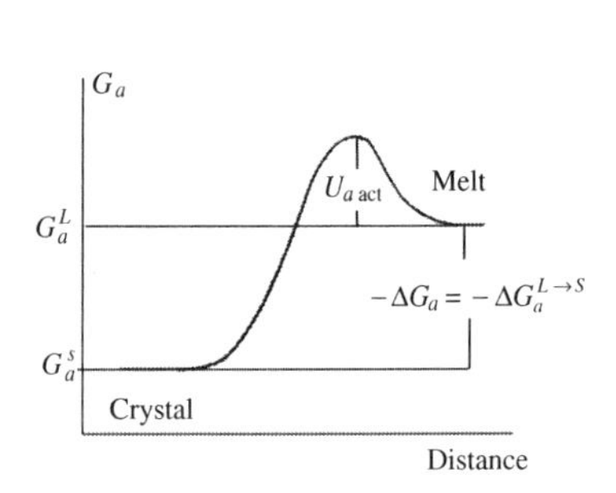

Figure 6.24 Gibbs' free energy of the interphase between a crystal with a rough interphase and its liquid phase as a function of position.

The atoms are in incessant motion relative to each other. The higher the temperature is the more violent will be the motion of the atoms and the higher will be their vibration frequency ν. The kinetic energies of the atoms are not the same but obey the Boltzmann distribution law. In order to find the growth rate we consider the flux of atoms to and from the interphase both towards the liquid side and the crystal side. At a given temperature a steady state is developed.

There is a continuous flux of atoms from the melt via the interphase to the crystal surface. Every second each atom makes ν attempts to escape from the melt to the interphase but only the atoms, that have enough kinetic energy to overcome the energy barrier $U_{\text{a act}}$, are successful and reach the crystal surface (Figure 6.24). Hence, the number of atoms per unit area and unit time, which pass the barrier will be

$$j^{\text{L}\to\text{s}} = \frac{\nu}{a^2} e^{-\frac{U_{\text{a act}}}{k_{\text{B}}T}} \tag{6.31}$$

The remaining atoms are rejected back into the melt. All the incoming atoms with energies > the energy barrier $U_{\text{a act}}$ become incorporated into kinks in the crystal lattice. The upper part of the Boltzmann distribution (above $U_{\text{a act}}$) describes the energy distribution of the adsorbed atoms (ad-atoms).

There is also a flux of atoms in the opposite direction. Each of the incorporated atoms performs ν attempts per unit time to leave the crystal surface. Only the fraction of these incorporated atoms which have kinetic energies > the sum of $U_{\text{a act}}$ and the energy difference between the melt and the crystal, are able to return to the melt. The flux of leaving atoms will be

$$j^{\text{s}\to\text{L}} = \frac{\nu}{a^2} e^{-\frac{-\Delta G_{\text{a}}^{\text{L}\to\text{s}} + U_{\text{a act}}}{k_{\text{B}}T}} \tag{6.32}$$

The net flux of atoms towards the crystal surface will be

$$j^{\mathrm{L}\to\mathrm{s}} - j^{\mathrm{s}\to\mathrm{L}} = \frac{\nu}{a^2}\left(\mathrm{e}^{-\frac{U_{\mathrm{a\ act}}}{k_{\mathrm{B}}T}} - \mathrm{e}^{-\frac{-\Delta G_{\mathrm{a}}^{\mathrm{L}\to\mathrm{s}} + U_{\mathrm{a\ act}}}{k_{\mathrm{B}}T}} \right) \tag{6.33}$$

This flux causes the growth of the solidification front. If we combine the Equations (6.24) on page 290 and (6.33) on page 293 and introduce the probability factor $a^2/\bar{\delta}^2$ (page 290) we obtain

$$V_{\mathrm{growth}} = \frac{a^3\nu}{a^2} P\,\mathrm{e}^{-\frac{U_{\mathrm{a\ act}}}{k_{\mathrm{B}}T}} \left(1 - \mathrm{e}^{-\frac{-\Delta G_{\mathrm{a\ interphase}}^{\mathrm{L}\to\mathrm{s}}}{k_{\mathrm{B}}T}} \right) \tag{6.34}$$

If $-\Delta G_{\mathrm{a}}^{\mathrm{L}\to\mathrm{s}} \ll k_{\mathrm{B}}T$ series expansion can be used

$$V_{\mathrm{growth}} = \frac{a^3\nu}{a^2} P\mathrm{e}^{-\frac{U_{\mathrm{a\ act}}}{k_{\mathrm{B}}T}} \frac{-\Delta G_{\mathrm{a}}^{\mathrm{L}\to\mathrm{s}}}{k_{\mathrm{B}}T} \tag{6.35}$$

in agreement with Equation (6.30) on page 292. The Equations (6.29) and (6.35) become identical when the constant in Equation (6.29) is replaced by the atomic vibration frequency ν.

The right-hand side of relationship (6.35) can be described as a constant multiplied by the gradient of the total driving force $-\Delta G_{\mathrm{interphase}}^{\mathrm{L}\to\mathrm{s}}$ over the interphase with thickness δ. We can consider the transfer of liquid atoms to the solid as a flux of atoms. Shewmon showed (page 140 in [6]) that the growth rate can be written as

$$V_{\mathrm{growth}} = M^{\mathrm{diff}} \frac{-\Delta G_{\mathrm{interphase}}^{\mathrm{L}\to\mathrm{s}}}{\delta}$$

where M^{diff} is the mobility of the atoms. According to Shewmon $M^{\mathrm{diff}} = D_{\mathrm{interphase}}/RT$ where $D_{\mathrm{interphase}}$ is the diffusion coefficient of the atoms over the interphase. With the aid of this relationship Equation (6.35) can alternatively be written as

$$V_{\mathrm{growth}} = \frac{D_{\mathrm{interphase}}}{\delta} \frac{-\Delta G_{\mathrm{interphase}}^{\mathrm{L}\to\mathrm{s}}}{RT} P \tag{6.36}$$

where
$D_{\mathrm{interphase}}$ = diffusion coefficient of the interphase
δ = thickness of the interphase
$-\Delta G_{\mathrm{interphase}}^{\mathrm{L}\to\mathrm{s}}$ = total driving force of solidification over the interphase (J/kmol).

This relationship will be used in the following discussion and more in detail in connection with alloys on page 313.

6.6.3 *Kinetic Coefficient of Growth Rate and Heat of Solidification with Consideration to Relaxation*

The relaxation process of atoms during a solidification process is shown in Figure 6.21b on page 288 as the potential energy versus time or position. The potential energy of the atoms is a part of their internal energy, i.e. the enthalpy. Consequently a change of the internal energy is accompanied by a change of the Gibbs' free energy.

Gibbs' free energy is used to analyze the solidification process as described in Section 6.6.1 and Figure 6.24 on page 290. Figure 6.26 on page 294 shows the solidification process as a combination of Figures 6.21b on page 288 and Figure 6.24. In Figure 6.26 the Gibbs' free energy is plotted versus time.

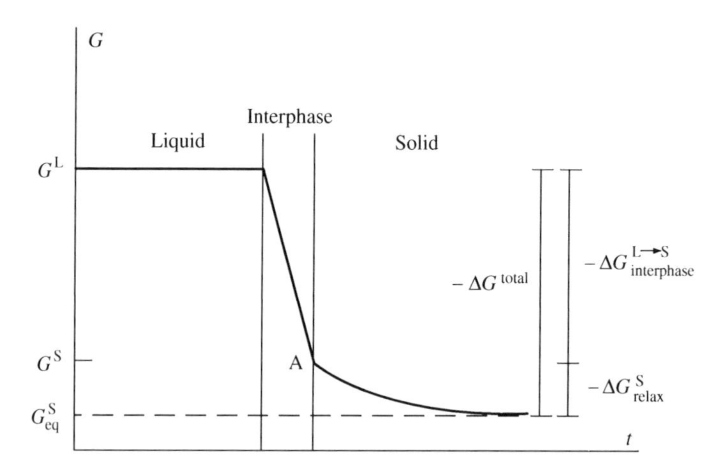

Figure 6.26 Gibbs' free energy of the atoms in the vicinity of the liquid/solid interphase.

Figure 6.26 shows that the total driving force for solidification consists of two parts, one for rearranging the 'liquid' atoms into an ordered solid lattice, and the other for the relaxation process in the solid. The total driving force can be written (1 kmol instead of 'per atom')

$$-\Delta G^{total} = (-\Delta G^{L\to s}_{interphase}) + (-\Delta G^{s}_{relax}) \tag{6.37}$$

The first term on the right-hand side of Equation (6.37) can be derived from Equation (6.36) on page 293.

The second term on the right-hand side of Equation (6.37) is obtained from Equation (6.21) on page 288 where the enthalpy ($-H_0$) is replaced by the Gibbs' total free energy as is shown in Figure 6.26.

Inserting these expressions into Equation (6.37) gives

$$-\Delta G^{total} = \frac{\delta RTV_{growth}}{D_{interphase}P} + (-\Delta G^{total})\sqrt{\frac{V_{growth}\delta}{\pi D_{relax}}} \tag{6.38a}$$

where P is defined by Equation (6.25) on page 291

$$P = \sqrt{\frac{V_{growth}\delta}{\pi D_{relax}}} \tag{6.25}$$

After reduction Equation (6.38a) can be written as

$$-\Delta G^{total} = \left[\frac{RT}{D_{interphase}}\pi D_{relax} + (-\Delta G^{total})\right]\sqrt{\frac{V_{growth}\delta}{\pi D_{relax}}} \tag{6.38b}$$

Kinetic Coefficient of Growth Rate

In order to find a relationship between the growth rate and the undercooling we want to solve V_{growth} from Equation (6.38b). If we square the equation we obtain

$$\frac{V_{growth}\delta}{\pi D_{relax}} = \frac{(-\Delta G^{total})^2}{\left(\frac{RT}{D_{interphase}}\pi D_{relax} + (-\Delta G^{total})\right)^2}$$

which gives

$$V_{\text{growth}} = \frac{(-\Delta G^{\text{total}})^2 \pi D_{\text{relax}}}{\delta\left(\frac{RT}{D_{\text{interphase}}}\pi D_{\text{relax}} + (-\Delta G^{\text{total}})\right)^2} \tag{6.39a}$$

The 'liquid' atoms diffuse across the interphase and the rearrangement of the atoms is described by the diffusion coefficient $D_{\text{interphase}}$. However, the rearrangement is coupled to the secondary relaxation process. The Gibbs' total free energy is described by Equation (6.23) on page 290 (multiplied by N_A)

$$-\Delta G^{\text{total}} = (-\Delta H_0)\frac{T_M - T}{T_M} \tag{6.23}$$

By experience and a rule of thumb it is known that the heat of fusion for metals is roughly equal to $2RT_M$.

With these approximations the sum in the denominator of Equation (6.39) can be written as

$$\frac{D_{\text{relax}}}{\delta D_{\text{interphase}}}\pi RT + 2RT_M\frac{T_M - T}{T_M}$$

and it seems reasonable to assume that the second term can be neglected in comparison with the first one. Hence, Equation (6.39a) can approximately be written as

$$V_{\text{growth}} = \frac{D_{\text{relax}}}{\pi\delta(RT)^2\left(\frac{D_{\text{relax}}}{D_{\text{interphase}}}\right)^2}(-\Delta G^{\text{total}})^2$$

which can be written [with the aid of Equation (6.23) above] as

$$V_{\text{growth}} = \frac{\frac{D^2_{\text{interphase}}}{D_{\text{relax}}}}{\pi\delta(RT)^2}\left[\frac{(-\Delta H_0)(T_M - T)}{T_M}\right]^2 = const \times (\Delta T)^2 \tag{6.39b}$$

or

$$V_{\text{growth}} = \beta_{\text{kin}} \times (\Delta T)^2 \tag{6.40}$$

where β_{kin} is a constant and is called the kinetic coefficient of the growth rate. Identification gives

$$\beta_{\text{kin}} = \frac{(-\Delta H_0)^2 D^2_{\text{interphase}}}{\pi\delta(RTT_M)^2 D_{\text{relax}}} \tag{6.41}$$

Heat of Solidification

Equation (6.21) on page 288 shows the remaining part of the total heat of solidification left in the solid (Figure 6.26 on page 294). The remaining heat $(-H_{\text{strain}})$, that is initially bound as strain energy and successively is released at relaxation, will be

$$-\Delta H_{\text{strain}} = -\Delta H_{\text{released}} = -\Delta H_0\sqrt{\frac{V_{\text{growth}}\delta}{\pi D_{\text{relax}}}} \tag{6.21}$$

The total heat of solidification equals $(-H_0)$ but only the fraction $[(-\Delta H_0) - (-\Delta H_{\text{strain}})]$ is released promptly and measured as heat of solidification

$$-\Delta H_{prompt} = -\Delta H_0 - (-\Delta H_{strain}) = -\Delta H_0 - \left(-\Delta H_0 \sqrt{\frac{V_{growth}\delta}{\pi D_{relax}}}\right)$$

or

$$-\Delta H_{prompt} = -\Delta H_0 \left(1 - \sqrt{\frac{V_{growth}\delta}{\pi D_{relax}}}\right) \tag{6.42}$$

This quantity is the measured heat of solidification.

It is urgent to discuss the relative sizes of $(-\Delta H_0)$ and $(-\Delta H_{strain})$. If the relaxation process is very fast (D_{relax} large), the measured heat of solidification is close to the total heat of solidification: $(-\Delta H_0)$.

If the relaxation process is slow (D_{relax} small), $(-\Delta H_{prompt})$ varies considerably with the growth rate.

Example 6.2

The constant D_{relax} is a measure of the rapidity of the relaxation process in metals after solidification. Plot $(-\Delta H_{prompt})/(-\Delta H_0)$ as a function of $\sqrt{V_{growth}}$ for the two cases $D_{relax} = 10^{-6}\,m^2/s$, and $D_{relax} = 10^{-13}\,m^2/s$.

$(-\Delta H_{prompt})$ is the value of the heat of fusion measured at experiments. $(-\Delta H_0)$ is the total heat of fusion, which includes $(-\Delta H_{prompt})$ and the strain energy that is released during the relaxation process.

$D_{relax} = 10^{-6}\,m^2/s$ corresponds to the magnitude of the thermal diffusivity for most metals. $D_{relax} = 10^{-13}\,m^2/s$ corresponds to the magnitude of the self-diffusion coefficient in cubic FCC metals. It is reasonable to assume that these values are the limits for D_{relax}.

The thickness of the interphase is 0.50 nm. Draw the two curves for the velocity interval 0–4 × 10^{-4} m/s.

Solution and Answer:

Equation (6.42) can be written as

$$\frac{-\Delta H_{prompt}}{-\Delta H_0} = 1 - \sqrt{\frac{\delta}{\pi D_{relax}}} \times \sqrt{V_{growth}}$$

Inserting the numerical values (large D_{relax}) into Equation (6.41) gives

$$\frac{-\Delta H_{prompt}}{-\Delta H_0} = 1 - \sqrt{\frac{0.50 \times 10^{-9}}{\pi \times 10^{-6}}} \times \sqrt{V_{growth}} = 1 - 1.26 \times 10^{-2}\sqrt{V_{growth}} \tag{1'}$$

Analogous calculations for the small value of D_{relax} gives

$$\frac{-\Delta H_{prompt}}{-\Delta H_0} = 1 - 39.9\sqrt{V_{growth}} \tag{2'}$$

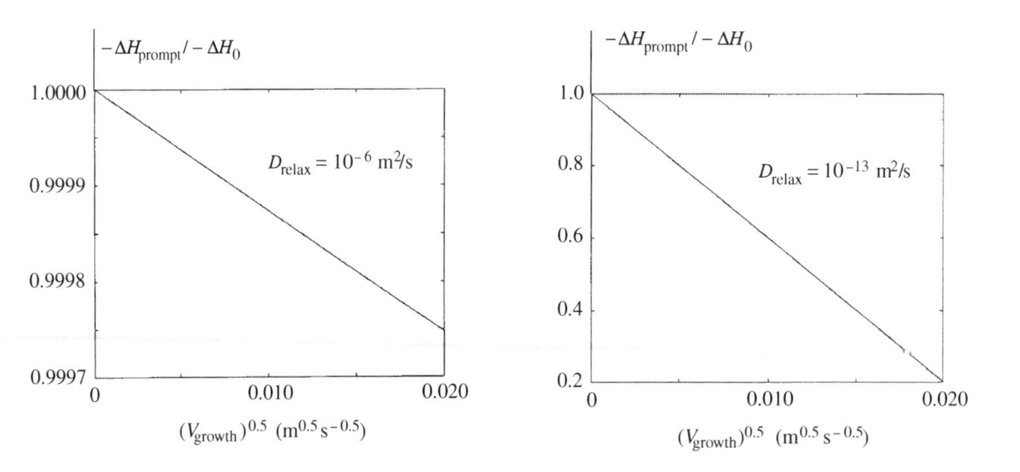

A comparison between the two curves shows that

- The larger D_{relax} is, the larger will be the ratio $-\Delta H_{prompt}/-\Delta H_0$ and the larger will be $-\Delta H_{prompt}$.
- At small values of $D_{relax}(-\Delta H_{prompt})$ varies considerably with $\sqrt{V_{growth}}$.

6.7 Layer Crystal Growth of Smooth Surfaces in Liquids

Comparison between Layer Growth in Vapours and Liquids

In Section 6.6 we have discussed normal growth on rough surfaces in melts and noticed the similarities and differences between normal growth in vapours and in melts (Section 6.4.1). It is not surprising that there also are many similarities between the atomic processes at the interfaces during crystal growth on smooth surfaces and a vapour phase and a melt, respectively. The main similarities are

1. Crystal growth on smooth surfaces in melts and liquids and in vapours is much more difficult than that on rough surfaces.
2. The mechanism of crystal growth on smooth surfaces in melts and liquids is the same as the one in vapours. The growth occurs by layer growth. The layers are formed by 'two-dimensional' nucleation or by spiral growth owing to screw dislocations.
3. The crystal surface is terraced and formed by steps of constant length and height. The height of the steps can be one or several atomic layers (Figure 6.27).
4. The terraces are smooth while the step edges have rough interphases. The layers grow laterally at the step edges and the growth on the smooth surfaces can be completely neglected.

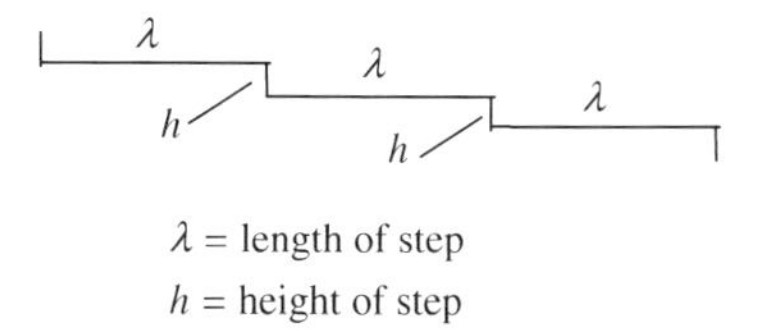

Figure 6.27 Layer growth on smooth crystal surfaces.

By decanting the melt or removing it by other methods from the growing crystals, it has been possible to study the crystal surfaces. Microphotographs of such surfaces prove that such surfaces are terraced. They will be further discussed in connection with cellular growth in Chapter 8. An example is given in Figure 6.16.

As we found on page 279 there are also differences between normal crystal growth in vapours and in liquids and melts. The same is also true for crystal growth on smooth surfaces.

The two main differences are:

1. The growth rate is a function of temperature or supersaturation instead of pressure.
2. The strain, owing to surface tension in the liquid and differences in distances between solid atoms and liquid atoms, respectively, has to be taken into consideration.

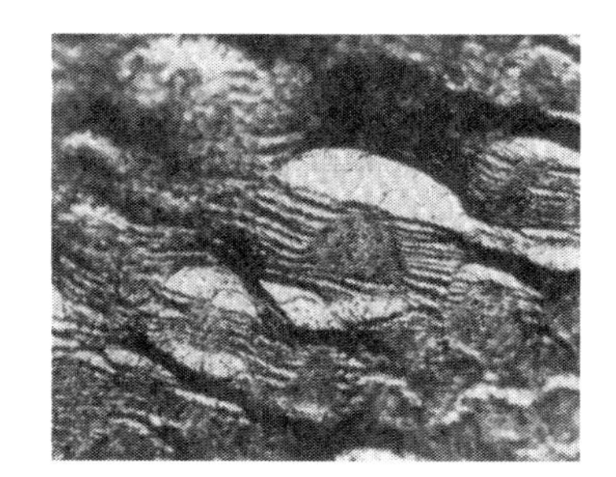

Figure 6.16 Microphotograph of a decanted cellular solidification front.

6.7.1 Chadwick's Hard Sphere Model

Since a long time it is well known that crystal growth occurs preferably in certain crystal orientations under different freezing conditions. This is for instance the case in all faceted crystals. As was mentioned on page 203 in Chapter 5

- Faceted crystals are bounded by the crystallographic planes which have the lowest growth rate.

For metals with face-centred cubic crystal structures it is in most cases the (111) plane and/or the (100) plane which have the lowest growth rate. All other planes grow faster than these two.

The first attempt to explain why, was given by Chalmers and his co-workers [7] in the 1950s but the result was doubtful. At the beginning of the 1960s it was suggested that the phenomenon should be attributed to changes in interphase kinetics with orientation, anisotropy of solid-liquid interfacial energy and anisotropy of heat flow.

At the end of the 1960s Chadwick [8] developed a simple model based on hard spheres. In order to take the different conditions in solids and liquids into consideration, Chadwick assumed that the solid and liquid atoms differ in size, i.e. radius. The model was used to simulate and make quantitative measurements of the rates of crystal growth perpendicular to the (111) and (100) planes in an FCC crystal structure. The hard sphere model is discussed more extensively in Section 1.3 in Chapter 1 in [1].

The results of the experiments were interpreted in terms of atomic packing in the solid and liquid phases at their common interface. Chadwick's theory, based on his hard-sphere simulation experiments, could be extended to other types of crystal structures and types of growth. The outlines of his unifying qualitative theory of crystal growth and the experimental evidence will partly be reported and discussed below.

Chadwick's experiments and his analysis constitute the basis for our discussions on orientation dependence of crystal growth. Chadwick used balls of equal size in his experiments, which means that effects of volume changes of atoms on crystal growth were neglected. These effects have been discussed in the preceding section and will partly be neglected below.

6.7.2 *Surface Packing Parameter. Volume Package Parameter*

In his experiments Chadwick used a heap of equal balls, packed at random and consequently with irregular density. These balls simulated a liquid phase. The balls were brought into contact with a simulated solid phase, a layer of balls, regularly packed and glued in a pattern corresponding to some crystallographic plane. The filling of the 'liquid balls' was either *cautious*, resulting in a close-packed liquid with maximum density, or *incautious* which corresponds to a loose-packed liquid structure with lower density than the close-packed liquid. Chadwick let the equipment vibrate and studied and analyzed the simulated crystal growth.

During his analysis Chadwick introduced a new concept, the dimensionless *surface packing parameter* p_S which is defined and applied on pages 270–272. In terms of crystal planes the definition can be written as

- Surface-packing parameter P_S = the fraction of the area of an (hkl) plane which is occupied by atoms in the plane.

Example 6.1 on pages 270–272 shows how the surface packing parameter can be calculated for some common types of crystals.

Several parameters can be used to describe the tightness of crystalline structure. One of them is the *volume-packing parameter* p_V which is defined as

- Volume-packing parameter = the fraction of a crystal volume, which is occupied by atoms.

The packing fraction can be calculated from a unit cell.

$$p_V = \frac{\frac{4}{3}\pi r^3 Z}{V_c} \tag{6.43}$$

where

p_V = volume packing parameter
Z = number of atoms with radius r per unit cell
V_c = volume of unit cell.

The volume packing parameter, which is dimensionless, is related to the surface packing parameter. The relationship depends on the type of crystal structure. For the most closely packed planes of FCC, BCC, diamond cubic and ideal HCP ($c/a = \sqrt{8/3}$) crystals the relationship between p_S and p_V is given by

$$p_S = \sqrt{3/2}p_V \tag{6.44}$$

The p_S values are characteristics of the crystal structures and do *not* depend on material and material constants.

Values of p_V and p_S for the most common crystal structures are given in Table 6.4. The volume-packing parameter of the interior of a melt is $p_v = 0.637$. The liquid surface is normally more close-packed than the interior and the accepted value of the surface-packing parameter of a close-packed melt is $p_S = 0.80$.

Table 6.4 Values of volume-packing parameters and surface-packing parameters of some common crystal structures. From G. A. Chadwick, W. A. Miller in Metal Science Journal, Vol. 1, The Metals Society, 1967 © Maney www.maney.co.uk.

Type of structure	p_V	p_S (hkl)		
		($h_1\ k_1\ l_1$)	($h_2\ k_2\ l_2$)	($h_3\ k_3\ l_3$)
FCC	0.741	0.907 (111)	0.785 (100)	0.556 (110)
BCC	0.680	0.833 (110)	0.589 (100)	0.481 (211)
HCP	0.741	0.907 (0001)	0.801* $(10\bar{1}0)$	0.566 $(11\bar{2}0)$
Diamond cubic	0.340	0.417 (110)	0.340 (111)	0.295 (100)
Tetragonal	0.535	0.772 (100)	0.772 (100)	0.383 (101)
Liquid (Close-packed random heap)	0.637	0.80	0.80	0.80

*The $(10\bar{1}0)$ planes are corrugated. Corrugated planes are more closely packed than the coplanar planes.

6.7.3 *Reticular Density*

Owing to differences in coordination numbers in a solid metal and the corresponding melt, it is necessary to account for the different radii of the metal atoms in the two phases. For this reason Chadwick introduced the concept *reticular density* in real systems instead of surface packing parameter.

- Reticular density = the number of atoms per unit area of a crystal plane.

The SI unit of reticular density is m^{-2}. It is related to the surface-packing parameter p_S by the relationship

$$D_S(\text{hkl}) = \frac{p_S(\text{hkl})}{\pi r^2} \tag{6.45}$$

where r is the radius of the atoms. Hence, the reticular density depends on material constants of the element in question.

The difference in size between the atoms in the solid and liquid phases is taken into account when reticular densities are used. Equation (6.45) is applied on the metal atoms in the solid and the melt, respectively, and the two equations are divided

$$\frac{D_S^L}{D_S^s(\text{hkl})} = \frac{p_S^L}{p_S^s(\text{hkl})}\left(\frac{r_s}{r_L}\right)^2 \tag{6.46}$$

where r_s and r_L are the atomic radii of the solid and liquid, respectively.

Ratios of the liquid reticular density/solid reticular density of the most common crystal structures are shown in Table 6.5.

Table 6.5 Ratio of liquid surface-packing parameter/solid surface-packing parameter and ratio of reticular liquid density/reticular solid density) for some crystal planes of some common metals. From G. A. Chadwick, W. A. Miller in Metal Science Journal, Vol. 1, The Metals Society, 1967 © Maney www.maney.co.uk.

Structure Type of Metal	$\frac{p_S^L}{p_S^s(h_1k_1l_1)}$	$\frac{D_S^L}{D_S^s(h_1k_1l_1)}$	$\frac{p_S^L}{p_S^s(h_2k_2l_2)}$	$\frac{p_S^L}{p_S^s(h_2k_2l_2)}$
FCC	(111)		(100)	
Pb	0.88	0.95	1.02	1.10
Ag	0.88	0.94	1.02	1.09
Au	0.88	0.93	1.02	1.08
Al	0.88	0.86	1.02	1.00
BCC	(110)		(100)	
Li	0.96	0.91	1.36	1.29
Na	0.96	0.92	1.36	1.30
K	0.96	0.96	1.36	1.36
Rb	0.96	0.98	1.36	1.39
Cs	0.96	0.94	1.36	1.33
Tl	0.96	0.99	1.36	1.40
HCP	(0001)		$(10\bar{1}0)$	
Mg	0.89	0.84	0.99	0.93
Zn	0.88	0.75	1.13	0.97
Cd	0.88	0.86	1.15	1.10
Tetragonal	(100)		(110)	
Sn	1.04	0.90	1.47	1.27

6.7.4 *Influence of Orientation on Layer Growth*

The surface-packing parameter and the reticular density of a crystal plane influence the growth rate of the crystal in the corresponding direction and hence its shape. Faceted crystals show the fact that

- The rate of crystal growth depends on orientation.

If this were not true, all crystals would be spherical.

In Sections 6.5 and 6.6 above we have seen that strain influences the rate of crystal growth in melts. Strain owing to difference in atomic size causes tension forces or pressure forces, which act on the interphase. This phenomenon can be discussed in terms of Chadwick's hard sphere model and his concepts surface density parameter and reticular density of atoms in crystal planes.

If the reticular density of the liquid is *smaller* than that of the solid, it means that the atoms in the solid are more closely packed than the liquid atoms. Pressure forces act within the interphase and compress the liquid or strain the solid during the solidification process.

This can be expressed very simply by means of the ratio of the reticular densities.

- If $\frac{D_S^L}{D_S^s(h_1k_1l_1)} < 1$ the strain of the interphase *contract* the melt and *expand* the solid.

  If the reticular density of the liquid is *larger* than that of the solid it means that the atoms in the solid are less closely packed than the liquid atoms. Tension forces act within the interphase and extend the liquid before solidification can occur.

- If $\frac{D_S^L}{D_S^s(h_1k_1l_1)} > 1$ the strain of the interphase *expand* the melt and *contract* the solid.

Chadwick's Experimental Results

The hard sphere simulation experiments of crystal growth in FCC structures gave the following results:

- Growth normal to the (111) plane from a *close-packed* liquid is *slower* than that normal to the (100) plane and all other planes.

This result can be explained if we consider the surface packing parameter in the crystal planes and in the liquid.

- *Cautious* filling gives a close-packed liquid and corresponds to a slow growth rate in real systems. The relaxation in the liquid makes it possible to maintain the maximum density of the liquid.

In cautious filling corresponds to a loose-packed liquid and is equivalent to fast crystal growth when the relaxation in the liquid is too small to keep the density of the liquid constant.

Density Barrier

The surface packing parameters for the (111) and (100) planes of FCC crystal structures are given in Table 6.4 on page 299. They are calculated to 0.907 respectively 0.785. The surface packing of a close-packed random hard-sphere liquid (cautious filling) with an atomically smooth surface is ~0.80. These differences in surface packing parameters between the two planes and the liquid (the first part of Table 6.4, columns 1 and 3 confirm these results) explain why the (100) plane propagates faster as the (111) plane in the close-packed liquid.

When the liquid close to the (111) plane solidifies, it has to *contract* in the plane of the interphase in order to *increase* its surface packing parameter from ~0.80 to 0.907. The strain within the interface generates pressure forces, which compress the liquid and expand the solid. The process in the solid is very hard to perform. The expansion occurs by formation of vacancies, which is difficult. Hence, this process is fairly slow.

When the liquid close to the (100) plane solidifies, it has to *expand* in the plane of the interface in order to *decrease* its surface packing parameter from ~0.80 to 0.785. The strain within the interphase causes tension forces, which try to expand the liquid and contract the solid. This process is easier than that described above.

Hence, there is a 'density barrier' against growth in the $<111>$ direction in a crystal of FCC structure owing to the denser surface packing of the liquid compared to that of the (100) plane. No corresponding 'density barrier' exists in the $<100>$ direction. Consequently, crystal growth in the $<100>$ direction is faster than that in the $<111>$ direction.

The topic can alternatively be expressed in terms of energy. This discussion is given below.

Influence of Relaxation on Growth Direction

The stronger the forces between the atoms are the more *energy* will be required to pull them apart and the more energy will be released when a liquid atom is incorporated into a stable site in a crystal.

The capture of liquid atoms is therefore easier the more shallow the energy well is. This property is described by the relaxation process during the growth.

The (111) plane in FCC metals has the highest reticular density of all FCC planes. The (100) plane is the next most densely packed plane and it has a high reticular density. The relaxation process becomes more difficult the denser the atoms are packed. The vibrations of the atoms become more restricted in all directions when the atoms are close-packed. In the loosely packed planes the vibrations occur easier in all directions and the relaxation process becomes faster.

Hence, the relaxation of the liquid atoms, incorporated in the crystal surface, is easier in the more shallow energy well of the (100) plane than in the deeper energy well of the (111) plane. The growth rate is therefore faster in the $<100>$ direction than in the $<111>$ direction because the incorporation of atoms is easier in the (100) plane than in the (111) plane. This explanation confirms the result given above.

Equation (6.21) on page 288 shows that the relaxation depends on the diffusion constant D_{relax}. It is well known that the diffusion constant depends on orientation. Consequently the driving force over the interphase $(-\Delta G^{L \rightarrow s}_{interphase})$ is also an anisotrop quantity, i.e. depends on direction. Example 6.2 on pages 296–297 shows clearly that

- A high value of D_{relax} corresponds to a low relaxation energy and consequentely to a high driving force over the interphase and vice versa.

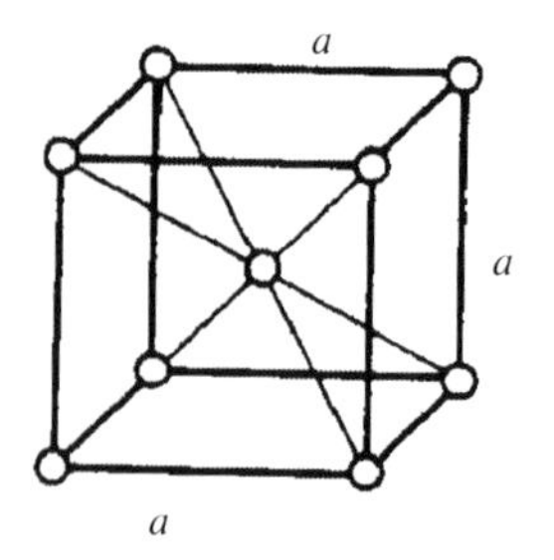

Figure 6.28 Unit cell of a BCC crystal.

As a first attempt to describe the diffusion coefficient as a function of direction, we assume that it is proportional to the squared distance between the atomic planes of the crystal in the direction in question (pages 260–262 in Chapter 5 in [1]).

Consider for example a crystalline material with BCC structure. According to Figure 6.28 the distance between consecutive planes parallel to the (100) plane is equal to the lattice constant a. The distance between planes parallel to the (111) plane is equal to half the principal diagonal, $a\sqrt{3}/2$. Hence, the diffusion coefficient D_{hkl} for growth in the [100] and [111] directions can be written as

$$D_{100} = D_0 \times a^2 \quad \text{and} \quad D_{111} = D_0 \times \left(\frac{a\sqrt{3}}{2}\right)^2$$

which gives $D_{111} = \frac{3}{4} \times D_{100}$

i.e. the diffusion coefficient is smaller in the [111] direction than in the [100] direction.

The same is true for the diffusion coefficient D_{relax}.

A general observation is that

- The smaller the diffusion coefficient is, the larger will be the remaining Gibbs' free energy in the solid.

This statement is also valid for relaxation, i.e. for D_{relax}.

Example 6.3

Pötschke and Klein [9] have analyzed the crystallization rate of Cu crystals in some different crystallographic directions as a function of the undercooling required for solidification. Some of their results are presented in the figures below.

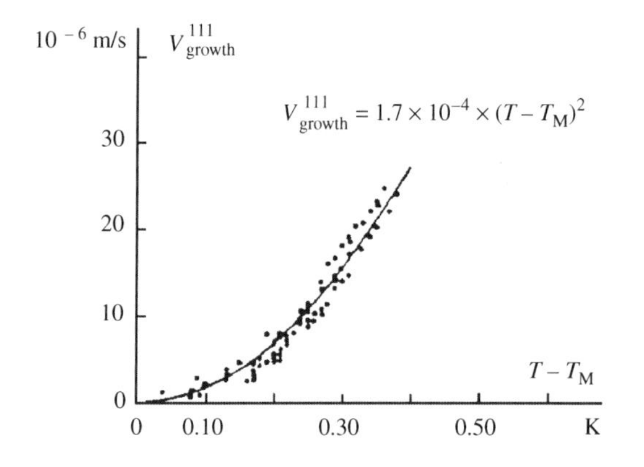

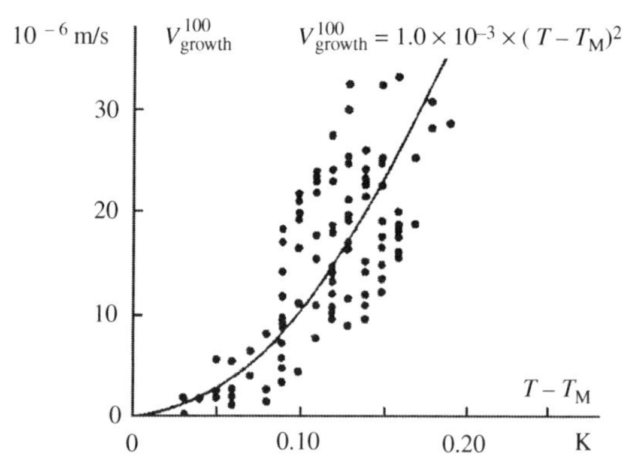

a) Derive the experimental values of the kinetic coefficient β_{kin} of the growth rate in the crystallographic directions [111] and [100] from the figures and calculate the values of D_{111} and D_{100} from Equation (6.41) on page 295 and the material constants of Cu listed below.
   The thickness of the interphase is assumed to be 0.51 nm.
   Material constants of Cu:
   $T_M = 1083\,°C = 1356\,K$
   $-\Delta H_0 = 206\,kJ/kg = 1.31 \times 10^4\,kJ/kmol$.
b) Calculate the ratio D_{111}/D_{100} of the diffusion constants obtained in a) and compare it with the corresponding ratio, derived on pages 301–302.
c) Compare the calculated D values with the diffusion constant in metallic melts, which is of the magnitude $10^{-9}\,m^2/s$. Comment the comparisons in b) and c).

Solution:

a)
The left figure in the text gives the relationship

$$V_{growth} = \beta_{kin} \times (\Delta T)^2 = 1.7 \times 10^{-4} \times (\Delta T)^2$$

We solve the diffusion coefficient D from Equation (6.41)

$$\beta_{kin} = \frac{D(-\Delta H_0)^2}{\pi\delta(RTT_M)^2} \qquad (1')$$

and obtain for the [111] direction

$$D_{111} \approx \beta_{kin}^{111} \frac{\pi\delta R^2 T_M^4}{(-\Delta H_0)^2}$$

Inserting the numerical values gives

$$D_{111} \approx 1.7 \times 10^{-4} \times \frac{\pi \times 0.51 \times 10^{-9} \times (8.31 \times 10^3)^2 \times 1356^4}{(1.31 \times 10^7)^2} = 1.7 \times 10^{-4} \times 2.18 \times 10^{-3}$$

or

$$D_{111} = 3.7 \times 10^{-7}\ \mathrm{m^2/s}$$

In the same way we obtain for the [100] direction with the aid of the right figure in the text

$$D_{100} \approx \beta_{kin}^{100} \frac{\pi\delta R^2 T_M^4}{(-\Delta H_0)^2}$$

Inserting the numerical values gives $D_{100} = 2.2 \times 10^{-6}\ \mathrm{m^2/s}$

Answer:

a) $D_{111} = 3.7 \times 10^{-7}\ \mathrm{m^2/s}$ $D_{100} = 2.2 \times 10^{-6}\ \mathrm{m^2/s}$.
b) D_{111}/D_{100} is roughly between 0.1 and 0.2. This value is about five times lower than the theoretical value 0.75 derived on page 302.
c) The calculated D values are about 100 times larger than the magnitude of the diffusion constants in metal melts.

It is worth while to remember that Equation (6.41) is based on the assumption that $D_{interphase} = D_{relax}$. This assumption may be doubtful. The mechanisms of diffusion is extensively treated in Chapter 5 in [1]. It is based on jumps of atoms from one site to another, in most cases by exchange of tvo atoms or jump to an empty site. The relaxation process is no diffusion process in common sense but rather rearrangement of atoms. This could be one reason for the observed discrepancies.

Another reason could be impurities in the melt, which may have a considerable influence on the kinetics.

When the orientation gives a low growth rate, a large amount of energy is stored in the solid and is successively emitted to the surroundings. The relaxation in the solid occurs either by an anelastic relaxation process or by a creep process when the strain is too large for anelasticity. In the latter case the temperature is high and the creep process in the solid occurs by formation and condensation of vacancies.

In the former case the relaxation energy is identical with the elastic energy or the work derived [Equations (6.6) and (6.7) on page 281]

$$W = Y\varepsilon^2 \frac{M}{\rho} \tag{6.7}$$

where
W = elastic strain energy per kmol of the interface
Y = modulus of elasticity of the solid
ε = relative linear stretch length of the interphase caused by the strain of the interphase
M = atomic weight of the metal
ρ = density of the interphase.

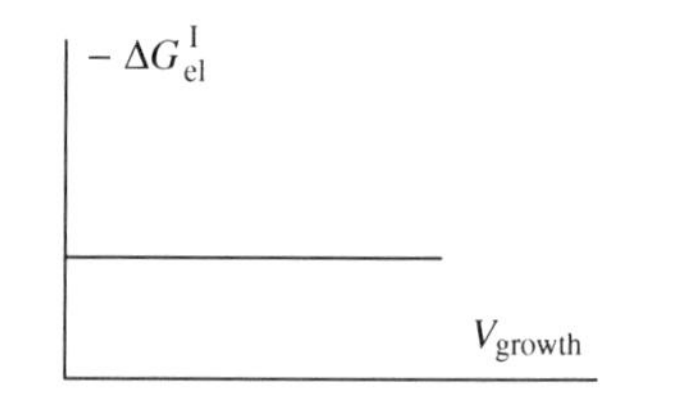

Figure 6.18 Elastic strain energy per kmol of the metal interphase as a function of the growth rate.

The relaxation energy is independent of the growth rate of the crystal (Figure 6.18). Hence, $-\Delta G^{\mathrm{I}}_{\mathrm{el}}$ appears as a horizontal line if it is plotted as a function of V_{growth}. The modulus of elasticity Y is an anisotropic quantity. Its value in the [111] direction is much higher than in all other directions. Y decreases with temperature.

The elastic strain energy can be compared with the remaining energy after the relaxation process over the interphase. Equation (6.21) on page 288.

$$-\Delta H_{\mathrm{strain}} = -\Delta H_0 \sqrt{\frac{V_{\mathrm{growth}}\delta}{\pi D_{\mathrm{relax}}}} \tag{6.21}$$

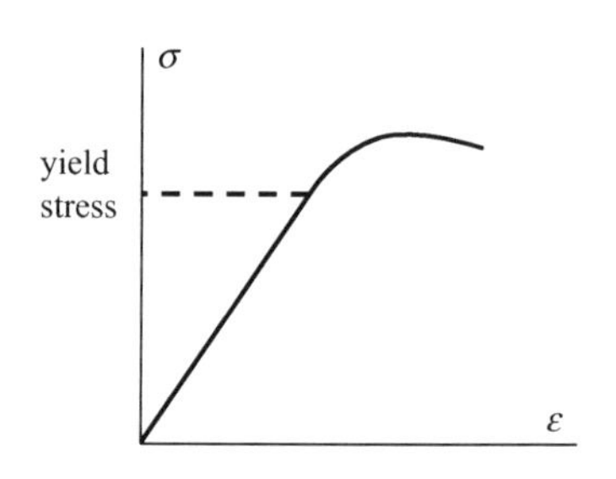

Figure 6.29 Yield stress is the limiting stress for elastic deformation. Above this value the deformation becomes plastic.

As the enthalpy is equal to the strain energy, the value of the strain energy can be inserted into Equation (6.7) and the value of ε can be calculated. By use of Hooke's law the stress can be calculated. If the stress is less than the yield stress (Figure 6.29) the relaxation energy equals the value given in Equation (6.7). If ε exceeds the yield stress vacancies are formed during a creep process.

6.7.5 *Formation of Vacancies*

In Chapter 5 we found that strain, owing to difference in size between lattice atoms and absorbed atoms entering the lattice from outside, causes screw dislocations and spiral steps. We also found that formation and diffusion of vacancies result in relaxation of the strain. Hence, the vacancy concentration in the solid phase is very important for crystal growth in vapours.

The influence of vacancies in the solid phase on the solidification of a liquid or melt is also considerable. There are two reasons for this, the density difference between the liquid and the solid in combination with difference in atomic size of the two phases. Below we will discuss the mechanism of deformation by formation of vacancies in the material.

Equilibrium Concentration of Vacancies as a Function of Temperature

The formation of vacancies has been discussed Section 2.7.1 in Chapter 2. Vacancies are always present in crystals and their concentration is strongly dependent on the temperature. The equilibrium concentration of vacancies at temperature T can be written (Equation (2.92) on page 85 in Chapter 2)

$$x^{\mathrm{eq}}_{\mathrm{vac}} = \mathrm{e}^{-\frac{\Delta H^{\mathrm{a}}_{\mathrm{vac}} - T\Delta S^{\mathrm{a}}_{\mathrm{vac}}}{k_{\mathrm{B}}T}} = \mathrm{e}^{-\frac{\Delta G^{\mathrm{a}}_{\mathrm{vac}}}{k_{\mathrm{B}}T}} \tag{6.47}$$

where

x_{vac}^{eq} = equilibrium mole fraction of vacancies at temperature T
$-\Delta G_{vac}^{a}$ = energy required to create a vacancy and the energy necessary for moving an atom from a site within the crystal to a site in the liquid phase far from the interface
k_B = Boltzmann's constant.

As Equation (6.47) shows, $-\Delta G_{vac}^{a}$ can be written as an enthalpy term and an entropy term. The enthalpy term is of the magnitude 0.7 eV for most metals at their melting point and the entropy term is about a tenth of this value. From Equation (6.47) it follows that the mole fraction of vacancies in metals at their melting points is of the order 10^{-4}. The vacancy concentration drops rapidly when the temperature decreases.

At most solidification processes in metals vacancies are formed in the solid. The number of vacancies will be larger than the equilibrium value given above, owing to kinetic effects.

Mechanism of Vacancy Formation at Solidification of Metal Melts

Chalmers [10] has discussed the formation of vacancies during the growth process and the vacancy concentration in the solid phase compared to the equilibrium concentration given by Equation (6.47). He described the vacancy formation in the following way.

Consider a rough crystal surface growing in a liquid or melt either by normal growth or by step growth in case of a smooth surface. The rough interphase contains many vacancies. A vacant site within the interphase will rapidly be filled with an atom, which normally comes from the liquid. However, one of the atoms in the crystal lattice may fill the kink by a diffusive jump and create a vacancy in the crystal lattice. The empty site could be filled by a neighbour atom (A or B in Figure 6.30). The result is that a vacancy in the crystal has been formed.

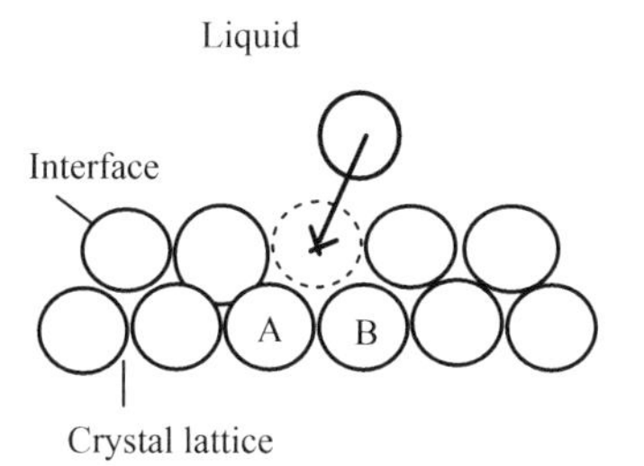

Figure 6.30 Formation of a vacancy during solidification. If atom A or B fills the vacant site in the interphase, a vacancy will be formed inside the crystal lattice.

Chalmers compared the activation energy $U_{a\,act}^{L}$ for an atom jump from the liquid to the interface with the activation energy $U_{a\,act}^{diff}$ for a jump from the crystal lattice. The relative probability P of the two competing processes can be written as

$$P = \frac{e^{-\frac{U_{a\,act}^{diff}}{k_B T}}}{e^{-\frac{U_{a\,act}^{L}}{k_B T}}} \tag{6.48}$$

Chalmers found that the probability for the former process was 10^3:1 compared to the latter. This means that a supersaturation of vacancies in the crystal lattice equal to $10^{-3}/10^{-4} = 10$ times the equilibrium value is to be expected. However, owing to the 'random walk' diffusion of the vacancies, they become rapidly trapped by

- grain boundaries
- edge dislocations
- annihilation at the crystal surface
- aggregation to discs or clusters

unless a driving force in the shape of strain exists in the solid.

Molar Gibbs' Free Energy of Vacancy Formation

Vacancies can be formed in many other ways than that suggested by Chalmers. It shows anyway how easily vacancies become trapped in the crystal lattice during a solidification process. In Chapter 5 on pages 242–244 we found that vacancies also become trapped in the crystal lattice during crystal growth in vapours.

Over a liquid/solid interface there is a relaxation process and part of the potential energy is left in the solid. This energy has been calculated as a Gibbs' free energy increase of the solid [Equation (6.7) on page 281] in case of elastic energy. In case of a creep process, vacancies are formed.

The Gibbs' free energy of the solid will increase if the vacancy concentration is larger than the equilibrium value. The increase of the Gibbs' free energy per kmol of the material owing to vacancy formation, can be calculated by the relationship Equation 2.95 on page 86 in Chapter 2, where y_{vac} has been replaced by the mole fraction x_{vac}, which is reasonable for low concentrations of vacancies):

$$-\Delta G_{\mathrm{vac}} \approx x_{\mathrm{vac}} RT \ln \frac{x_{\mathrm{vac}}}{x_{\mathrm{vac}}^{\mathrm{eq}}} \tag{6.49}$$

where

x_{vac} = mole fraction of vacancies in the crystal
$x_{\mathrm{vac}}^{\mathrm{eq}}$ = mole fraction of vacancies in the crystal at equilibrium, i.e. in the absence of strain and density differences at temperature T
$-\Delta G_{\mathrm{vac}}$ = Gibbs' free energy of formation of vacancies of concentration x_{vac}.

Comparison between the Mechanisms of Deformation

On pages 281–284 we have studied three mechanisms of deformation as a possible way to overcome the volume change at solidification. The mechanism, which requires lowest energy supply, is most likely to occur in reality. To find the proper alternative we have to consider the energy conditions in the three cases and compare them.

Most common metals have high melting points. At these temperatures no values of the modulus of elasticity Y are available and it is doubtful whether elastic deformation is a realistic alternative. For this reason it has been excluded here.

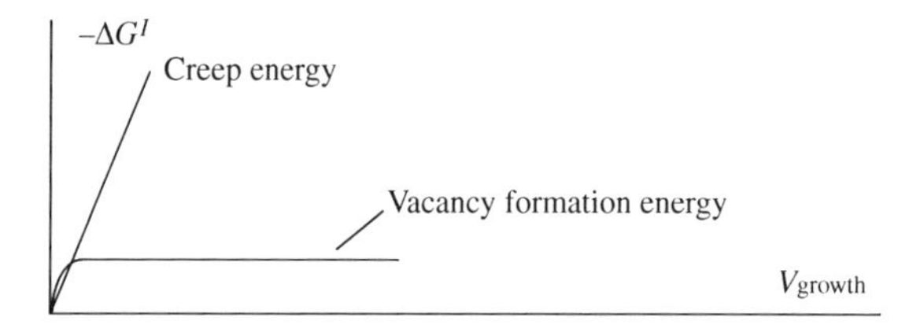

Figure 6.31 Gibbs' free energy change per kmol of the interface as a function of growth rate of a pure metal.

Values of the viscosity constant η of metal melts are available in tables and creep seems to be a possible alternative to vacancy formation to overcome strain in metals For this reason the curves for creep and vacancy formation for a metal have been plotted in the same diagram in Figure 6.31. It shows that

- The most likely process to overcome the volume change during a solidification process is *deformation by vacancy formation* unless the growth rate of solidification is very low.
- At very low growth rates creep is the most likely mechanism. The transition growth rate is found at the intersection point between the creep and vacancy curves.

Application on Relaxation Energy

The condition for vacancy formation is discussed above. By introduction of the expression for the relaxation energy, Equation (6.21) on page 288 into Equation (6.49) above the fraction of vacancies can be calculated as a function of the growth rate

$$-\Delta H_0 \sqrt{\frac{V_{\mathrm{growth}} \delta}{\pi D_{\mathrm{relax}}}} = x_{\mathrm{vac}} RT \ln \frac{x_{\mathrm{vac}}}{x_{\mathrm{vac}}^{\mathrm{eq}}} \tag{6.50}$$

6.7.6 *Rate of Layer Growth in Case of Strain. Length and Height of Steps at Layer Growth*

Origin of Layer Growth

The analysis of Chadwick's results and the statements of the relationship between reticular densities and interphase strain on page 300. offer a possibility to explain the origin of layer growth.

It is known that the steps on the surface of an FCC crystal are most frequently bounded by (111) and (100) planes and sometimes by (110) planes. The strain patterns of the planes depend on the values of the reticular densities for the different planes according to the discussion on page 300. They are listed in Table 6.6.

Table 6.6 Strain patterns of some crystal planes

	Crystal plane	Pressure forces contract the	Tension forces expand the
Type I	$\frac{D_S^L}{D_S^s(h_1k_1l_1)} < 1$		
	(111) FCC	melt	crystal
	(110) BCC	melt	crystal
Type II	$\frac{D_S^L}{D_S^s(h_1k_1l_1)} > 1$		
	(100) BCC	crystal	melt
	(001) FCC	crysta	melt
	all other planes	crystal	melt

The energy law is the principle, which determines the structure of the crystal surface

- The configuration which gives to lowest possible total energy of the interface is the most stable state of the surface and the one, which is formed in reality.

An energy analysis confirms that the most likely state in the case of layer growth is a terraced surface of the crystal. It is the strain pattern of the crystal planes, which gives the terraced structure. The terrace surfaces of the steps have a strain pattern of type I (smooth surface) and the step edge surfaces belong to type II (rough surface) (Figure 6.32).

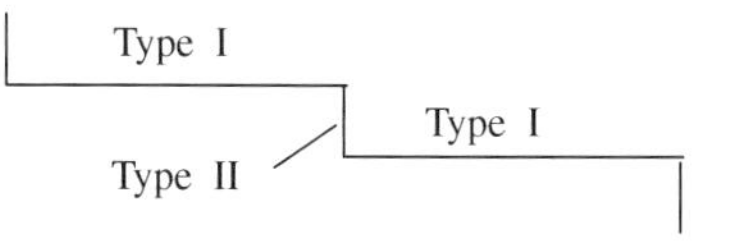

Figure 6.32 Strain patterns at steps of layer growth.

These circumstances make it possible to determine the length and height of steps at layer growth, which will be discussed below.

According to Equation (6.34) on page 293 the growth rate can be written as a function of the total driving force of solidification $-\Delta G^{L\rightarrow s}$:

$$V_{growth} = \frac{a^3}{a^2}\nu P e^{-\frac{U_{a\ act}}{k_B T}} \frac{-\Delta G^{L\rightarrow s}}{RT} \tag{6.51}$$

Equation (6.51) is completely analogous with Equation (6.28) on page 291. Both are valid for rough surfaces.

Mechanism of Layer Growth in Liquids

The mechanism of layer growth of crystals in vapours has been penetrated thoroughly in Chapter 5 Section 5.6.8 on pages 235–236. The theory of crystal growth in vapours is based on the assumption that atoms from the vapour phase are adsorbed along the terraces, diffuse towards the rough step edge and become incorporated into kinks there.

However, it is highly unlikely that liquid atoms become adsorbed on the terraces and diffuse to the steps. It is much more likely that they become incorporated directly at the terraces and steps and that the growth rates differs in the two cases.

Crystals in vapours and liquids are strained because the vapour atoms, the liquid atoms and the adsorbed atoms (ad-atoms) differ in size compared to the lattice atoms. This fact was used in Chapter 5 to calculate the length and height of steps in the case of crystal growth in vapours.

As we will se later in this section it is also possible to calculate length and height of steps at layer growth of crystals in liquids and melts (Figure 6.27). The method is based on application of the strain conditions and values of the reticular densities at the interface crystal/liquid discussed on page 300.

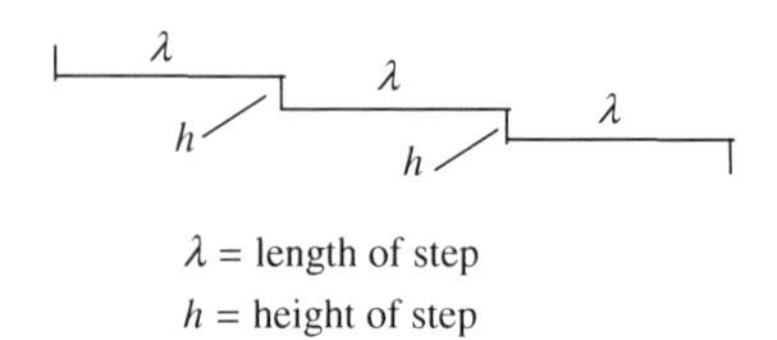

Figure 6.27 Layer growth on smooth crystal surfaces.

Length and Height of Single Steps

In an FCC metal the terraces consist of (111) planes and (100) planes form the steps. In the previous section (Table 6.4 on page 299 and pages 297–300) we found that the (111) interphase was strained, owing to the surface packing parameter and reticular density differences between the solid and the liquid. The growth is obstructed owing to the straining of the solid. The solid at the step in contact with the (100) interface is not strained or much less strained than the solid close to the (111) interphase. Consequently the growth rate will be faster along the $<100>$ direction than along the $<111>$ direction.

As long as the smooth (111) interphase is strained, no or very few atoms can be adsorbed along the terraces. The growth will only occur by adsorption of liquid atoms into kinks at the rough step edge. During the growth, a *relaxation of the lattice* will occur simultaneously. The strain is reduced by the relaxation process at the step edge, which is the most likely process.

On pages 301–302 we discussed the dependence of the diffusion constant with direction. We found that D_{111} is only 75% of D_{100}. During layer growth of crystals the temperatures inside the crystal and also along all interphases are constant. The total driving force is the same in all directions but the growth rate in the $<111>$ direction is smaller than that in the 100 direction owing to the difference in diffusion constants.

For the same reason we have $D_{\text{relax}}^{111} < D_{\text{relax}}^{100}$. Equation (6.21) on page 288 shows that a small value of D_{relax} means a large value of $(-\Delta H_{\text{relax}})$ and vice versa.

As $D_{\text{relax}}^{111} < D_{\text{relax}}^{100}$ the strain energy or relaxation energy, stored in the growing (111) plane will be larger than that, stored in the growing (100) plane. Figure 6.26 and Equation (6.37) on page 294 shows that a large relaxation energy means a small driving force over the interphase and vice versa. Hence, the growth of the vertical rough (100) surface generally ought to grow faster than the smooth (111) surface. However, one necessary condition must be fulfilled.

- The relaxation energy, stored in the (111) plane, must come down to the same level of relaxation energy, stored in the growing in (100) plane, before the step can grow along the (111) plane.

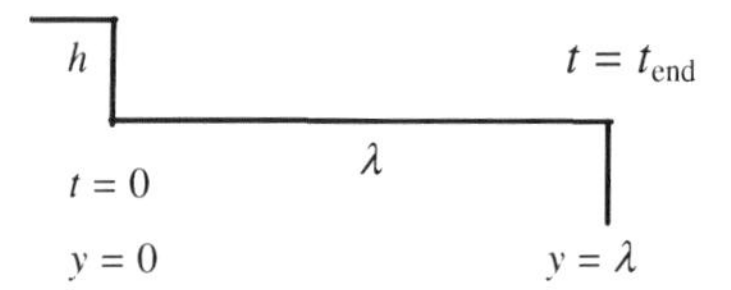

Figure 6.33 Step growth of the rough vertical surface.
λ = length of step
h = height of step.

Consider the growth of the step in Figure 6.33. The relaxation energy of the rough vertical surface is $(-\Delta H_{\text{relax}}^{100})$ at $y = 0$ and $t = 0$. With the aid of Equation (6.21) the relaxation energy can be written as

$$-\Delta H_{\text{relax}}^{100} = -\Delta H_0 \sqrt{\frac{V_{\text{growth}}^{100}\delta}{\pi D_{\text{relax}}^{100}}} \qquad (6.52)$$

At this occasion and the surface grows and the solidification front grows to the right and $-(\Delta H_{\text{relax}}^{111})_{t=0} = -\Delta H_{\text{relax}}^{100}$ or

$$-\Delta H_0 \sqrt{\frac{V_{\text{growth}}^{111}\delta}{\pi D_{\text{relax}}^{111}}} = -\Delta H_0 \sqrt{\frac{V_{\text{growth}}^{100}\delta}{\pi D_{\text{relax}}^{100}}} \qquad (6.53)$$

At time $t = t_{end} = \lambda/V_{growth}^{100}$ the initially high relaxation energy $-\Delta H_{relax}^{111}$ has declined to $-\Delta H_{relax}^{100}$. With the aid of Equation (6.20) on page 286 this condition can be written as

$$\left(-\Delta H_{relax}^{111}\right)_{t=t_{end}} = \left(\frac{(-\Delta H_0) \times \delta}{\sqrt{\pi D_{relax}^{111}}} \frac{\exp(-\lambda^2/4Dt)}{\sqrt{t}}\right)_{t=t_{end}} \tag{6.54}$$

or

$$\left(-\Delta H_{relax}^{111}\right)_{t=t_{end}} = \frac{(-\Delta H_0) \times \delta}{\sqrt{\pi D_{relax}^{111}}} \exp \frac{\dfrac{-\lambda^2}{4D_{relax}^{111} \frac{\lambda}{V_{growth}^{100}}}}{\sqrt{\dfrac{\lambda}{V_{growth}^{100}}}} \tag{6.55}$$

The growth condition can be written as

$$-\Delta H_{relax}^{100} = \left(-\Delta H_{relax}^{111}\right)_{t=t_{end}}$$

or

$$-\Delta H_0 \sqrt{\frac{V_{growth}^{100} \delta}{\pi D_{relax}^{100}}} = -\Delta H_0 \frac{\delta \sqrt{V_{growth}^{100}}}{\sqrt{\pi \lambda D_{relax}^{111}}} \exp \frac{-\lambda V_{growth}^{100}}{4D_{relax}^{111}} \tag{6.56}$$

Equation (6.56) is valid for growth of the rough step edge. By replacing subscript 100 by 111 and subscript 111 by 100 and the length λ by h we obtain a new equation which is valid for growth of the smooth step edge. It can be written as

$$-\Delta H_0 \sqrt{\frac{V_{growth}^{111} \delta}{\pi D_{relax}^{111}}} = (-\Delta H_0) \frac{\delta \sqrt{V_{growth}^{111}}}{\sqrt{\pi h D_{relax}^{100}}} \exp \frac{-h V_{growth}^{111}}{4D_{relax}^{100}} \tag{6.57}$$

Equations (6.56) and (6.57) can be simplified to

$$\sqrt{\frac{\lambda D_{relax}^{111}}{\delta D_{relax}^{100}}} = \exp \frac{-\lambda V_{growth}^{100}}{4D_{relax}^{111}} \tag{6.58}$$

and

$$\sqrt{\frac{h D_{relax}^{100}}{\delta D_{relax}^{111}}} = \exp \frac{-h V_{growth}^{111}}{4D_{relax}^{100}} \tag{6.59}$$

The step length λ and the step height h can be determined from Equations (6.58) and (6.59). After relaxation the smooth interface is free from strain and a new growth process starts and new steps are successively formed.

6.8 Normal Crystal Growth in Binary Alloys

In Sections 6.6 and 6.7 we have studied crystal growth on rough and smooth surfaces, respectively, in liquids and melts. In this section we will discuss the solidification process in alloys, which is more complex than that in pure metals. The complication is that the solid phase has a different composition than the liquid.

The driving force of crystal growth in binary alloys and the growth rate will be derived below by consideration of the mass transport across the interphase and the motion of the solidification front.

6.8.1 *Driving Force of Solidification*

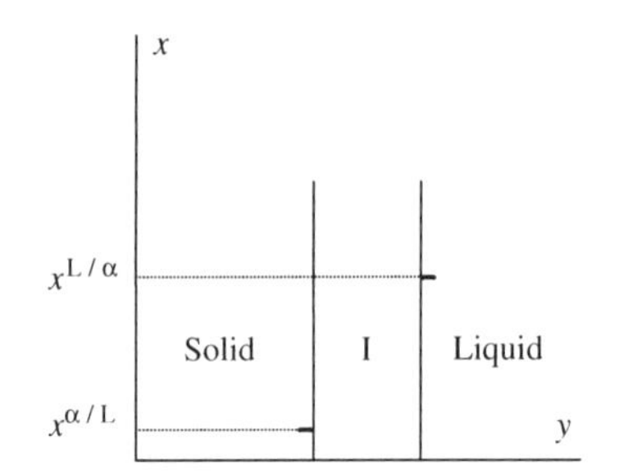

Figure 6.34 Concentrations of solute atoms at the interphase of a solidifying binary alloy ($k < 1$).

During the solidification process both solvent atoms A and solute atoms B are transported across the interphase and the concentration x of the solute is different in the solid α phase and the liquid phase L. This is shown in Figure 6.34 for a binary alloy with a partition coefficient $k < 1$.

Figure 6.35 shows the phase diagram of a binary A–B alloy and the corresponding Gibbs' free energy diagram.

Consider a binary alloy melt of composition x_0^L which solidifies at temperature T. The local compositions on the two sides of the interphase are $x^{\alpha/L}$ and $x^{L/\alpha}$. The driving forces of a solvent atom A and a solute atom B to be transferred across the interphase from the liquid to the solid are $-\Delta G_A^{L\rightarrow\alpha}$ and $-\Delta G_B^{L\rightarrow\alpha}$, respectively.

The mathematical derivation of the driving forces of each element and of the total driving force will be given on pages 310–313.

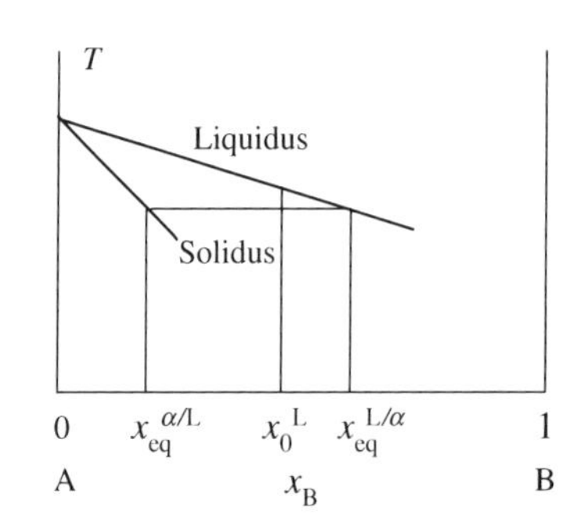

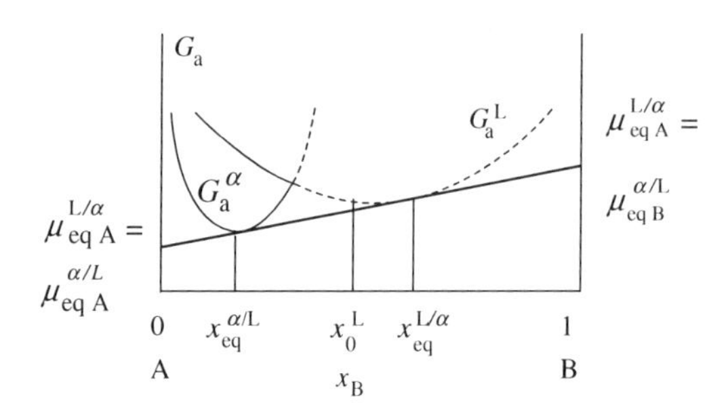

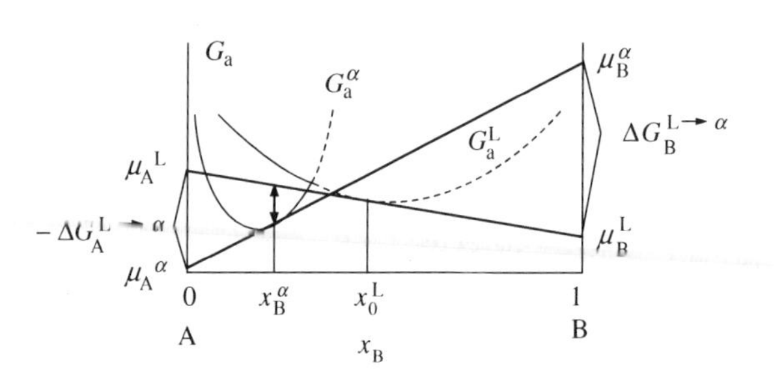

Figure 6.35a Upper figure: Schematic phase diagram of a binary alloy AB.

Figure 6.35b Middle figure: Gibbs' free energy curves of the melt and the solid phase. Construction of the chemical potentials, i.e. the Gibbs' free energies, and the solute concentrations of the solid and the melt during solidification at equilibrium.

Figure 6.35c Lower figure: Analogous constructions at non-equilibrium. The total driving force $-\Delta G^{L\rightarrow\alpha}$ corresponding to an arbitrary solute concentration x_B^α is marked in the figure with a thick line with two arrows.

Equilibrium corresponds to the common tangent of the two curves in Figure 6.35b on page 310 (Chapter 2 pages 54–58). The composition of the solid phase does not necessarily correspond to the equilibrium composition. Such a case is shown in Figure 6.35c together with the equilibrium case.

In Chapter 2 on pages 76–78 the driving force of solidification of a binary alloy was derived as a function of the chemical potentials and mole fractions of the two elements. The driving force of solidification for each element was found to be! (Chapter 2, Equation (2.76) on page 78)

$$-\Delta G^{L\rightarrow\alpha} = \left(1 - x^{\alpha/L}\right)\left(\mu_A^L - \mu_A^\alpha\right) + x^{\alpha/L}\left(\mu_B^L - \mu_B^\alpha\right) \tag{6.60}$$

The driving forces of the solvent atoms A and solute atoms B

$$-\Delta G_A^{L\rightarrow\alpha} = \mu_A^L - \mu_A^\alpha \tag{6.61}$$

$$-\Delta G_B^{L\rightarrow\alpha} = \mu_B^L - \mu_B^\alpha \tag{6.62}$$

can be introduced into Equation (6.60) instead of the chemical potentials which gives

$$-\Delta G^{L\rightarrow\alpha} = \left(1 - x_B^\alpha\right)\left(-\Delta G_A^{L\rightarrow\alpha}\right) + x_B^\alpha\left(-\Delta G_B^{L\rightarrow\alpha}\right) \tag{6.63}$$

where

$-\Delta G^{L\rightarrow\alpha}$ = total driving force of solidification

$-\Delta G_i^{L\rightarrow\alpha}$ = driving force of the process of transferring an A and B atoms, respectively, across the interphase from the liquid to the solid (i = A, B)

x_B^α = mole fraction of B in the solid phase

μ_i^L, μ_i^α = corresponding chemical potentials (i = A, B).

Figure 6.35 and Equations (6.61) show that the driving force for the A atoms is *positive*, which means that they are transferred from the liquid to the solid. According to Figure 6.35 and Equation (6.62) the driving force of the B atoms is *negative*, which means that they move from the solid to the liquid phase.

In order to derive a mathematical expression for the total driving force we use the expressions given in Chapter 2 for the Gibbs' free energy curves in Figure 6.35.

Derivation of the Driving Force of Solidification

Consider an alloy with the concentration x_0^L of the solute B in a diluted solution of atoms A and B. We assume that 'Raoult's and 'Henry's laws (pages 27–28 in Chapter 1) are valid for atoms A and B, respectively. This will simplify the derivation of the partial and total driving forces considerably. Equation (6.63) gives the total driving force. According to Equation (6.61) and Figure 6.35 the driving force of the A atoms can be written as

> Consider a *dilute* solution of atoms B in atoms A. The relationships between the activities a and mole fractions are:
>
> Henry's law: $a_B = \gamma x_B$ is valid for B atoms.
> γ is a material constant.
> Raoult's law: $a_A = 1 \times x_A$ is valid for A atoms

$$-\Delta G_A^{L\rightarrow\alpha} = \mu_A^{0\,L} + RT\ln\left(1 - x_0^L\right) - \left[\mu_A^{0\,\alpha} + RT\ln\left(1 - x_B^\alpha\right)\right] \tag{6.64}$$

where

R = gas constant

T = temperature

x_0^L = mole fraction of B in the liquid

x_B^α = mole fraction of B in the solid phase

$\mu_{A,B}^{0\,L,\alpha}$ = chemical potential of A and B, respectively, in the liquid phase and the α phase, respectively.

Below we will need series development (dilute solution) of the ln function. In addition we will drop the subscript B. For small values of the concentration of B atoms we obtain for instance

$$\ln x_{\mathrm{A}}^{\mathrm{L}} = \ln\left(1-x_{\mathrm{B}}^{\mathrm{L}}\right) \approx -x_{\mathrm{B}}^{\mathrm{L}} = -x^{\mathrm{L}} \quad \text{and} \quad \ln x_{\mathrm{A}}^{\alpha} = \ln\left(1-x_{\mathrm{B}}^{\alpha}\right) \approx -x_{\mathrm{B}}^{\alpha} = -x^{\alpha}$$

Corresponding expressions are valid for any superscript.

At equilibrium the chemical potentials are equal at the interface. As 'Raoult's law is valid for A atoms we obtain (a corresponding expression is valid for B atoms).

$$\mu_{\mathrm{A}}^{0\,\mathrm{L}} + RT\,\ln x_{\mathrm{A\,eq}}^{\mathrm{L}/\alpha} = \mu_{\mathrm{A}}^{0\,\alpha} + RT\,\ln x_{\mathrm{A\,eq}}^{\alpha/\mathrm{L}} \tag{6.65}$$

where

$x_{\mathrm{A\,eq}}^{\mathrm{L}/\alpha}$ = mole fraction of A in the liquid at equilibrium close to the interphase

$x_{\mathrm{A\,eq}}^{\alpha/\mathrm{L}}$ = mole fraction of A in the solid at equilibrium close to the interphase.

Equation (6.65) can be written:

$$\mu_{\mathrm{A}}^{0\,\mathrm{L}} - \mu_{\mathrm{A}}^{0\,\alpha} = RT\,\ln x_{\mathrm{A\,eq}}^{\alpha/\mathrm{L}} - RT\,\ln x_{\mathrm{A\,eq}}^{\mathrm{L}/\alpha}$$

or after simplification

$$\mu_{\mathrm{A}}^{0\mathrm{L}} - \mu_{\mathrm{A}}^{0\,\alpha} \approx RT\left[-x_{\mathrm{eq}}^{\alpha/\mathrm{L}} - \left(-x_{\mathrm{eq}}^{\mathrm{L}/\alpha}\right)\right] \tag{6.66}$$

The expression in Equation (6.66) is introduced into Equation (6.64), which is simplified by series development

$$-\Delta G_{\mathrm{A}}^{\mathrm{L}\to\alpha} \approx RT\left[\left(x_{\mathrm{eq}}^{\mathrm{L}/\alpha} - x_{\mathrm{eq}}^{\alpha/\mathrm{L}}\right) - x_{0}^{\mathrm{L}} + x^{\alpha}\right]$$

or

$$-\Delta G_{\mathrm{A}}^{\mathrm{L}\to\alpha} \approx RT\left[\left(x_{\mathrm{eq}}^{\mathrm{L}/\alpha} - x_{0}^{\mathrm{L}}\right) - \left(x_{\mathrm{eq}}^{\alpha/\mathrm{L}} - x^{\alpha}\right)\right] \tag{6.67}$$

Analogously as for A atoms [compare with Equation (6.64)] we have for B atoms (Henry's law is valid)

$$-\Delta G_{\mathrm{B}}^{\mathrm{L}\to\alpha} = \mu_{\mathrm{B}}^{0\mathrm{L}} + RT\,\ln\gamma_{0}^{\mathrm{L}} x_{0}^{\mathrm{L}} - \left(\mu_{\mathrm{B}}^{0\,\alpha} + RT\,\ln\gamma_{0}^{\alpha} x_{\mathrm{B}}^{\alpha}\right) \tag{6.68}$$

The chemical potentials are equal at the interphase at equilibrium:

$$\mu_{\mathrm{B}}^{0\mathrm{L}} + RT\,\ln\gamma_{0}^{\mathrm{L}} x_{\mathrm{B\,eq}}^{\mathrm{L}/\alpha} = \mu_{\mathrm{B}}^{0\,\alpha} + RT\,\ln\gamma_{0}^{\alpha} x_{\mathrm{B\,eq}}^{\alpha/\mathrm{L}}$$

or

$$\mu_{\mathrm{B}}^{0\mathrm{L}} - \mu_{\mathrm{B}}^{0\,\alpha} = RT\,\ln\gamma_{0}^{\alpha} x_{\mathrm{B\,eq}}^{\alpha/\mathrm{L}} - RT\,\ln\gamma_{0}^{\mathrm{L}} x_{\mathrm{B\,eq}}^{\mathrm{L}/\alpha} \tag{6.69}$$

If we insert this expression into Equation (6.68) and drop subscript B we obtain

$$-\Delta G_{\mathrm{B}}^{\mathrm{L}\rightarrow\alpha} = RT\ \ln\gamma_0^{\alpha} x_{\mathrm{eq}}^{\alpha/\mathrm{L}} - RT\ \ln\gamma_0^{\mathrm{L}} x_{\mathrm{eq}}^{\mathrm{L}/\alpha} + RT\ \ln\gamma_0^{\mathrm{L}} x_0^{\mathrm{L}} - RT\ \ln\gamma_0^{\alpha} x^{\alpha}$$

which can be transformed into

$$-\Delta G_{\mathrm{B}}^{\mathrm{L}\rightarrow} = RT\ \ln\frac{x_{\mathrm{eq}}^{\alpha/\mathrm{L}} x_0^{\mathrm{L}}}{x_{\mathrm{eq}}^{\mathrm{L}/\alpha} x^{\alpha}} = RT\ \ln\left(1 + \frac{x_{\mathrm{eq}}^{\alpha/\mathrm{L}} x_0^{\mathrm{L}}}{x_{\mathrm{eq}}^{\mathrm{L}/\alpha} x^{\alpha}} - 1\right)$$

Series development of the ln function gives

$$-\Delta G_{\mathrm{B}}^{\mathrm{L}\rightarrow\alpha} \approx RT\left(\frac{x_{\mathrm{eq}}^{\alpha/\mathrm{L}} x_0^{\mathrm{L}}}{x_{\mathrm{eq}}^{\mathrm{L}/\alpha} x^{\alpha}} - 1\right) \tag{6.70}$$

The expressions in Equations (6.67) and (6.71) are inserted into Equation (6.63) (page 311), which gives the final expression for the total driving force of solidification where subscript B is dropped

$$-\Delta G_{\mathrm{total}}^{\mathrm{L}\rightarrow\alpha} \approx (1 - x^{\alpha/\mathrm{L}})RT\left[\left(x_{\mathrm{eq}}^{\mathrm{L}/\alpha} - x_0^{\mathrm{L}}\right) - \left(x_{\mathrm{eq}}^{\alpha/\mathrm{L}} - x^{\alpha}\right)\right] + x^{\alpha/\mathrm{L}} RT\left(\frac{x_{\mathrm{eq}}^{\alpha/\mathrm{L}} x_0^{\mathrm{L}}}{x_{\mathrm{eq}}^{\mathrm{L}/\alpha} x^{\alpha}} - 1\right) \tag{6.71}$$

and at low concentrations of B one obtains

$$-\Delta G_{\mathrm{total}}^{\mathrm{L}\rightarrow\alpha} = RT\left(x_{\mathrm{eq}}^{\mathrm{L}/\alpha} - x_0^{\mathrm{L}}\right) \tag{6.72}$$

6.8.2 *Rate of Normal Growth in Binary Alloys*

When a pure metal melt solidifies the process occurs in two steps, one for the incorporation of liquid atoms into the crystal lattice via the rough interphase and one for the reorganization of the solidified atoms from a disordered state to the ordered crystal lattice structure. These processes are controlled by the temperature as a function of position. Normal growth in pure metals has been discussed in Sections 6.4 and 6.6.

During the solidification of a binary alloy there are also a kinetic process at the interphase and a relaxation process in the solid, just as in a pure metal melt (Figure 6.26 on page 294). In addition it is necessary to consider other facts, which influence the processes. Separate equations are required for A and B atoms. The reasons for this are

1. There are two kinds of atoms A and B with different material constants.
2. Concentration differences cause diffusion of the atoms A and B at the interphase during the kinetic process and in the crystal during the rearrangement of atoms in connection with the relaxation process (Figure 6.36).

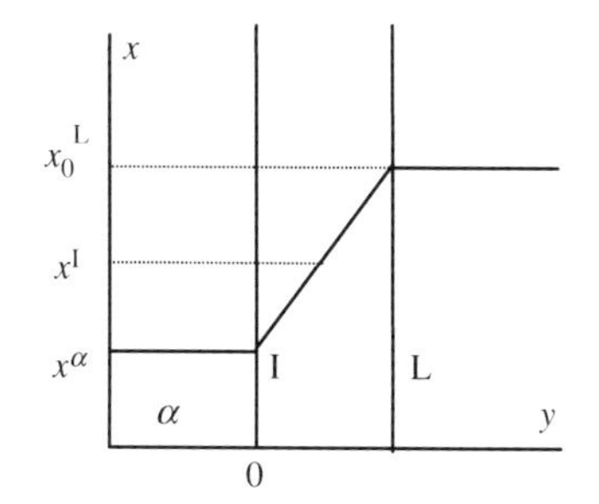

Figure 6.36 Liquid/solid interphase.

The Kinetic Process. Mass Transport across the Interphase

The mass transport is a result of the incorporation of A and B atoms via the rough interphase into the crystal and the simultaneous diffusion of the A and B atoms in the interphase.

Diffusion of A and B Atoms due to Concentration Differences in the Interphase

The mass transport at solidification can be expressed as the flux of atoms, i.e. the number of atoms which pass the interphase per unit area and unit time. Shewmon [7] use this concept in the so-called phenomenological mass transport equation. According to this equation the growth rate of the two components in a binary alloy across the interphase can be written as

$$V_{\text{growth}} = M_{\text{A}}^{\text{diff}} \frac{-\Delta G_{\text{A diff int}}^{\text{L}\rightarrow\alpha}}{\delta(x_0^{\text{L}} - x^{\alpha})} \tag{6.73}$$

$$V_{\text{growth}} = M_{\text{B}}^{\text{diff}} \frac{-\Delta G_{\text{B diff int}}^{\text{L}\rightarrow\alpha}}{\delta(x^{\alpha} - x_0^{\text{L}})} \tag{6.74}$$

where

$-\Delta G_{\text{A, B diff int}}^{\text{L}\rightarrow\alpha}$ = driving force across the interphase for diffusion of A and B atoms, respectively

$M_{\text{A,B}}^{\text{diff}}$ = mobility of the A and B atoms, respectively

δ = thickness of the interphase.

Equations (6.73) and (6.74) give an equality which can be reduced to

$$M_{\text{A}}^{\text{diff}}\left(-\Delta G_{\text{A diff int}}^{\text{L}\rightarrow\alpha}\right) = -M_{\text{B}}^{\text{diff}}\left(-\Delta G_{\text{B diff int}}^{\text{L}\rightarrow\alpha}\right) \tag{6.75}$$

As the mobilities $M_{\text{A}}^{\text{diff}}$ and $M_{\text{B}}^{\text{diff}}$ are positive quantities Equation (6.75) shows that the driving forces have opposite signs, which is confirmed by Figure 6.35 on page 310. The A and B atoms diffuse in opposite directions.

The overall driving force of the concentration diffusion of the A and B atoms is

$$-\Delta G_{\text{diff int}}^{\text{L}\rightarrow\alpha} = (1 - x^{\alpha})\left(-\Delta G_{\text{A diff int}}^{\text{L}\rightarrow\alpha}\right) + x^{\alpha}\left(-\Delta G_{\text{B diff int}}^{\text{L}\rightarrow\alpha}\right) \tag{6.76}$$

where x^{α} is the fraction of B atoms in the solid phase.

Incorporation of A and B Atoms into the Rough Crystal Surface

With the aid of the model presented on pages 290–293 and used for a pure metal it is possible to find an expression for the growth rate of the solidification front at normal growth in an alloy. For the alloying element B we obtain

$$x^{\alpha} V_{\text{growth}} = \frac{a^3 \nu}{a^2} P_{\text{B}} \exp\left(-\frac{U_{\text{a act}}}{k_{\text{B}} T}\right)\left[x^{\text{L}} - x^{\alpha} \exp\left(-\frac{-\Delta G_{\text{B kin int}}^{\text{L}\rightarrow\alpha}}{RT}\right)\right] \tag{6.77}$$

An analogous equation is obtained for element A if we replace subscript B by subscript A, x^{L} by $(1 - x^{\text{L}})$, x^{α} by $(1 - x^{\alpha})$ and P_{B} by P_{A}.

$$(1 - x^{\alpha}) V_{\text{growth}} = \frac{a^3 \nu}{a^2} P_{\text{A}} \exp\left(-\frac{U_{\text{a act}}}{k_{\text{B}} T}\right)\left[(1 - x^{\text{L}}) - (1 - x^{\alpha}) \exp\left(-\frac{-\Delta G_{\text{A kin int}}^{\text{L}\rightarrow\alpha}}{RT}\right)\right] \tag{6.78}$$

The potential energies of the two elements A and B are different in the liquid and in the solid. This is also true for the enthalpies. Therefore, the probabilities also differ, owing to Equation (6.25) on page 291. For this reason the probabilities P in Equations (6.77) and (6.78) have subscripts.

If we add Equations (6.77) and (6.78) we obtain

$$\begin{aligned} V_{\text{growth}} &= \frac{a^3\nu}{a^2}P_{\text{B}}\exp\left(-\frac{U_{\text{a act}}}{k_{\text{B}}T}\right)\left[x^{\text{L}} - x^{\alpha}\exp\left(-\frac{-\Delta G_{\text{B kin int}}^{\text{L}\to\alpha}}{RT}\right)\right] \\ &+ \frac{a^3\nu}{a^2}P_{\text{A}}\exp\left(-\frac{U_{\text{a act}}}{k_{\text{B}}T}\right)\left[(1-x^{\text{L}}) - (1-x^{\alpha})\exp\left(-\frac{-\Delta G_{\text{A kin int}}^{\text{L}\to\alpha}}{RT}\right)\right] \end{aligned} \tag{6.79}$$

If the driving forces are small compared to $k_{\text{B}}T$, Equation (6.79) can be simplified by series expansion, which gives after reduction

$$\begin{aligned} V_{\text{growth}} &= \frac{a^3\nu}{a^2}\exp\left(-\frac{U_{\text{a act}}}{k_{\text{B}}T}\right) \\ &\times\left((x^{\text{L}} - x^{\alpha})(P_{\text{B}} - P_{\text{A}}) + (1-x^{\alpha})P_{\text{A}}\frac{-\Delta G_{\text{A kin int}}^{\text{L}\to\alpha}}{RT} + x^{\alpha}P_{\text{B}}\frac{-\Delta G_{\text{B kin int}}^{\text{L}\to\alpha}}{RT}\right) \end{aligned} \tag{6.80}$$

If $P_{\text{A}} = P_{\text{B}}$ Equation (6.79) can be written as

$$V_{\text{growth}} = \frac{a^3\nu}{a^2}P\exp\left(-\frac{U_{\text{a act}}}{k_{\text{B}}T}\right)\left((1-x^{\alpha})\frac{-\Delta G_{\text{A kin int}}^{\text{L}\to\alpha}}{RT} + x^{\alpha}\frac{-\Delta G_{\text{B kin int}}^{\text{L}\to\alpha}}{RT}\right) \tag{6.81}$$

With the aid of Equations (6.63) on page 311 and Equation (6.81) we can express the growth rate in a compact form as a function of the overall driving force of solidification $-\Delta G_{\text{overall kin int}}^{\text{L}\to\alpha}$

$$V_{\text{growth}} = \frac{a^3\nu}{a^2}P\exp\left(-\frac{U_{\text{a act}}}{k_{\text{B}}T}\right) \times \frac{-\Delta G_{\text{kin int}}^{\text{L}\to\alpha}}{RT} \tag{6.82}$$

Equation (6.78) is completely analogous with Equation (6.35) on page 293 for pure elements.
In analogy with Equation (6.36) Equation (6.82) can be written as

$$V_{\text{growth}} = \frac{D_{\text{interphase}}}{\delta}\frac{-\Delta G_{\text{kin int}}^{\text{L}\to\alpha}}{RT}P \tag{6.83}$$

where

$$-\Delta G_{\text{kin int}}^{\text{L}\to\alpha} = \left[(1-x^{\alpha})\left(-\Delta G_{\text{A kin int}}^{\text{L}\to\alpha}\right) + x^{\alpha}\left(-\Delta G_{\text{B kin int}}^{\text{L}\to\alpha}\right)\right]$$

6.8.3 *Kinetic Coefficient of Growth Rate for Binary Alloys with Consideration to Relaxation*

Relaxation and Rearrangement of the Incorporated Atoms in the Crystal

At solidification the incorporated atoms become transformed into disordered solid atoms at the surface of the crystal. Two parallel processes will occur:

1. Rearrangement of position of the incorporated A and B atoms from the disordered state to the ordered state of the crystal lattice. This is the relaxation process, when the atoms loose potential energy which is emitted to the surroundings.
2. Diffusion of the A and B atoms in order to achieve equilibrium with the lattice atoms, i.e. to reach lowest possible potential energy.

The driving force for the relaxation process for a binary alloy is obtained analogously with the derivation for a pure metal, described on page 293. For a binary alloy the relaxation term given by Equation (6.21) on page 288 where the enthalpy $(-\Delta H_0)$ has been replaced by the total Gibbs' free energy $-\Delta G^{total}$ for the solidification process of the alloy.

$$-\Delta G^{\alpha}_{relax} = -\Delta G^{total}\sqrt{\frac{V_{growth}\delta}{\pi D_{relax}}} \tag{6.84}$$

where $-\Delta G^{total}$ in the case of the alloy equals the sum of the expressions given in Equation (6.76) on page 314, Equation (6.83) and Equation (6.84).

$$-\Delta G^{total} = -\Delta G^{L\to\alpha}_{diff\ int} + -\Delta G^{L\to\alpha}_{kin\ int} + (-\Delta G^{total})\sqrt{\frac{V_{growth}\delta}{\pi D_{relax}}} \tag{6.85}$$

or

$$-\Delta G^{total} = -\Delta G^{L\to\alpha}_{overall\ int} + (-\Delta G^{total})\sqrt{\frac{V_{growth}\delta}{\pi D_{relax}}} \tag{6.86}$$

where

$$-\Delta G^{L\to\alpha}_{overall\ int} = -\Delta G^{L\to\alpha}_{diff\ int} + \left(-\Delta G^{L\to\alpha}_{kin\ int}\right) \tag{6.87}$$

The first term on the right-hand side of Equation (6.87) is much smaller than the second one, especially for small fractions of the alloying element.

Provided that we neglect the first term in Equation (6.85) and assume that $P_A = P_B$ we can solve the second term from Equation (6.83) and insert the expression into Equation (6.85). The result is

$$-\Delta G^{total} = \frac{\delta RT\, V_{growth}}{D_{interphase} P} + (-\Delta G^{total})\sqrt{\frac{V_{growth}\delta}{\pi D_{relax}}} \tag{6.88}$$

Equation (6.88) looks formally identical with Equation (6.38a) on page 294 but the meaning of the quantities differ in the two cases.

Kinetic Coefficient of Growth Rate

Provided that the driving force for the concentration diffusion in the interphase is neglected in Equation (6.86) the derivation of β_{kin} will be completely analogous with the one performed for pure metals on pages 294–295.

For derivation of β_{kin} we use Equation (6.88)

$$-\Delta G^{total} = \frac{\delta RT\, V_{growth}}{D_{interphase} P} + (-\Delta G^{total})\sqrt{\frac{V_{growth}\delta}{\pi D_{relax}}} \tag{6.88}$$

where P is defined by Equation (6.25) on page 294

$$P = \sqrt{\frac{V_{growth}\delta}{\pi D_{relax}}} \tag{6.25}$$

If this expression is inserted into Equation (6.88) we obtain after reduction

$$-\Delta G^{\text{total}} = \left[\frac{RT}{D_{\text{interphase}}}\pi D_{\text{relax}} + (-\Delta G^{\text{total}})\right]\sqrt{\frac{V_{\text{growth}}\delta}{\pi D_{\text{relax}}}} \tag{6.89}$$

Equation (6.89) looks identical with Equation (6.38b) on page 294.

If we solve V_{growth} from Equation (6.89) we obtain

$$V_{\text{growth}} = \frac{(-\Delta G^{\text{total}})^2 \pi D_{\text{relax}}}{\delta\left(\frac{RT}{D_{\text{interphase}}}\pi D_{\text{relax}} + (-\Delta G^{\text{total}})\right)^2} \tag{6.90}$$

For analogous reasons as those given on pages 294–295 the first term in the denominator of Equation (6.90) can be neglected in comparison with the second one.

$$V_{\text{growth}} = \frac{D_{\text{relax}}}{\pi\delta(RT)^2\left(\frac{D_{\text{relax}}}{D_{\text{interphase}}}\right)^2}(-\Delta G^{\text{total}})^2 \tag{6.91}$$

The total Gibbs' free energy is described by Equation (6.72) on page 313 (multiplied by N_A) gives an expression of $(-\Delta G^{\text{total}})$:

$$-\Delta G^{\text{total}} = RT\left(x_{\text{eq}}^{L/\alpha} - x_0^L\right) \tag{6.72}$$

Inserting this expression for the total Gibbs' free energy into Equation (6.91) gives

$$V_{\text{growth}} = \frac{\frac{D^2_{\text{interphase}}}{D_{\text{relax}}}}{\pi\delta}\left(x_{\text{eq}}^{L/\alpha} - x_0^L\right)^2 = const \times \left(x_{\text{eq}}^{L/\alpha} - x_0^L\right)^2 \tag{6.92}$$

or

$$V_{\text{growth}} = \beta_{\text{kin}} \times \left(x_{\text{eq}}^{L/\alpha} - x_0^L\right)^2 \tag{6.93}$$

where β_{kin} is a constant and is called the kinetic coefficient of the growth rate. Identification gives

$$\beta_{\text{kin}} = \frac{D^2_{\text{interphase}}}{\pi\delta D_{\text{relax}}} \tag{6.94}$$

Equation (6.93) gives the growth rate as a function of supersaturation. With the aid of the slope $m = dT/dx$ of the liquidus line we can alternatively obtain the growth rate as a function of undercooling instead of supersaturation.

$$V_{\text{growth}} = \beta_{\text{kin}} \times (\Delta x)^2 = const \times \Delta T^2 \tag{6.95}$$

where the constant equals

$$const = \frac{\beta_{\text{kin}}}{m^2} \tag{6.96}$$

where m is the slope of the liquidus line.

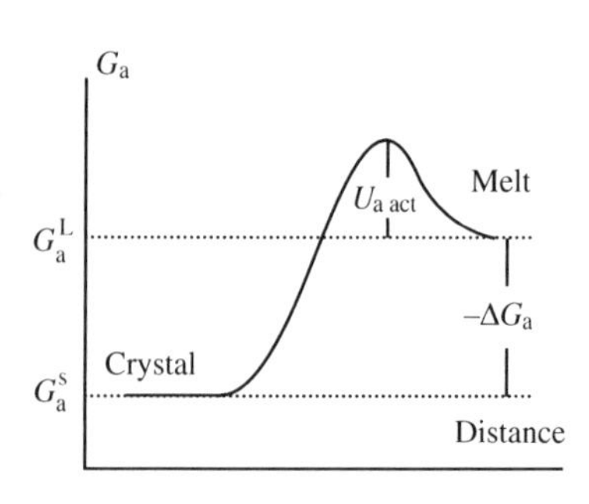

Figure 6.37 Gibbs' free energy levels of the system crystal–melt.

6.9 Diffusion-Controlled Growth of Planar Crystals in Binary Alloys

Consider a growing crystal in a supersaturated alloy melt (Figure 6.37). The growth process depends on a number of items such as interface kinetics, mass transport and heat transport.

If the activation energy of incorporating new atoms from the melt into the crystal lattice is *high*, the *kinetic process* at the interface controls the growth. This case has been discussed in Section 6.7.

If the activation energy is *low*, the overall rate of crystal growth is determined by *heat transport* and/or *mass transport* instead.

If the heat transport is fast the *mass transport*, i.e. *diffusion*, will dominate the overall rate of the crystal growth. This case will be discussed in Sections 6.9 and 6.10 for planar and spherical growths, respectively. For spherical growth we will analyze the effect of interphase kinetics in addition to the diffusion process.

6.9.1 *General Case of Diffusion-Controlled Planar Growth in Alloys*

Exact Solution

Fick's second law:

$$\frac{dc}{dt} = D\left(\frac{\partial^2 c}{\partial x^2} + \frac{\partial^2 c}{\partial y^2} + \frac{\partial^2 c}{\partial z^2}\right)$$

The rate of the solidification process and the composition of the melt as a function of position can be calculated by solving Fick's second law of solidification and applying the boundary conditions which are valid in the present case.

In the case of planar growth one coordinate y is enough. If we use mole fraction as a measure of the composition of the alloy Fick's law of diffusion can be written as

$$\frac{\partial x^L}{\partial t} = D\frac{\partial^2 x^L}{\partial y^2} \tag{6.97}$$

where

y = coordinate
x^L = mole fraction of solute in the melt at distance y
D = diffusion constant of the solute in the melt
t = time.

Figure 6.38 Concentration of solute in the melt in front of the interface if $k > 1$.

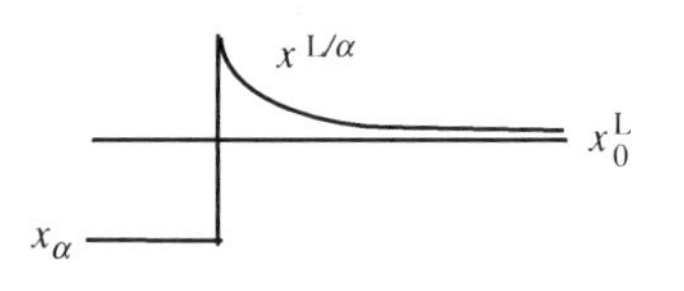

Figure 6.39 Concentration of solute in the melt in front of the interface if $k < 1$.

Two cases have to be considered corresponding to the partition coefficients $k > 1$ and $k < 1$. The following designations, introduced in the Figures 6.38 and 6.39, will be used below:

$x^{L/\alpha}$ = mole fraction of solute in the melt at the interphase
x^L = mole fraction of solute in the melt at coordinate y
x_0^L = mole fraction of solute in the melt far from the solidification front.

A mass balance can be performed at the interphase. The amount of alloying atoms which diffuse away from ($k < 1$) and towards ($k > 1$), respectively, the interphase per unit time equals the rate, adjusted for concentration unit, by which the alloying atoms are transferred from the melt to the solid α phase. According to Fick's first law we obtain for a binary system

Fick's first law:

$$\frac{dm}{dt} = -DA\frac{dc}{dy}$$

$$\frac{dm}{dt} = -\frac{DA}{V_m^L}\left(\frac{dx^L}{dy}\right)_{y=r_c} = \frac{V_{growth}A}{V_m^\alpha}(x^{L/\alpha} - x^{\alpha/L}) \tag{6.98}$$

where

m = mass of diffusing atoms
y = coordinate along axis normal to the solidification front
r_c = coordinate of solidification front (interphase)
A = cross-section area
V_{growth} = growth rate of planar interphase
V_m^L = molar volume of the melt
V_m^α = molar volume of the solid α phase.
$x^{L/\alpha}$ = mole fraction of solute in the melt at the interphase
$x^{\alpha/L}$ = mole fraction of solute in the solid at the interphase.

To find the concentration gradient at the solidification front we have to use the solution of Fick's second law. Zener [11] and Frank [12] showed independently of each other that in the case of a growing planar crystal the solution of Fick's second law

$$\frac{\partial x}{\partial t} = D\frac{\partial^2 x}{\partial y^2} \tag{6.99}$$

can be written as

$$x^{L} - x_0^{L} = \left(x^{L/\alpha} - x_0^{L}\right)\frac{\phi\left(\frac{y}{\sqrt{Dt}}\right)}{\phi\left(\frac{r_c}{\sqrt{Dt}}\right)} \tag{6.100}$$

The function ϕ is defined with the aid of

$$\phi(z) = \int_z^{\infty} e^{-\frac{u^2}{4}} du = \sqrt{\pi}\left(1 - \operatorname{erf}\frac{z}{2}\right) \tag{6.101}$$

where z is an arbitrary variable and not one of the Cartesian coordinates (xyz).

Equation (6.100) represents the exact solution of Equation (6.99). The growth rate at the interphase can be expressed as a function of the dimensionless constant s, coupled to the solution of the differential equation and defined as

$$s = \frac{y}{\sqrt{Dt}} \tag{6.102}$$

As s is constant, y must be proportional to $\sqrt{t}$. We square Equation (6.102):

$$s^2 Dt = y^2 \tag{6.103}$$

Taking the derivative of Equation (6.103) with respect to time gives

$$s^2 D = 2y\frac{dy}{dt} \tag{6.104}$$

The growth rate is

$$V_{growth} = \left(\frac{dy}{dt}\right)_{y = r_c} \tag{6.105}$$

Equations (6.104) and (6.105) give the relationship

$$\frac{s^2}{2} = \frac{V_{growth}\, r_c}{D} \tag{6.106}$$

The concentration gradient, evaluated from Equation (6.100) by taking the derivative of x^L with respect to y, is inserted into Equation (6.98) which gives in combination with Equations (6.101), (6.102) and (6.106)

$$\frac{V_m^{\alpha}}{V_m^{L}}\frac{x^{L/\alpha} - x_0^{L}}{x^{L/\alpha} - x^{\alpha/L}} = \frac{\sqrt{\pi}}{2} s\, e^{\frac{s^2}{4}}\left(1 - \operatorname{erf}\frac{s}{2}\right) \tag{6.107}$$

The left-hand side of Equation (6.107) represents *the supersaturation* of the melt. It is denoted Ω.

$$\Omega = \frac{V_m^{\alpha}}{V_m^{L}}\frac{x^{L/\alpha} - x_0^{L}}{x^{L/\alpha} - x^{\alpha/L}} \tag{6.108}$$

Equation (6.107) gives an implicite relationship between supersaturation and growth rate. Equations (6.107) and (6.108) show that every value of Ω corresponds to a specific value of s and hence indirectly of V_{growth}.

The exact solution (6.100) is shown graphically in the diagram below and will be discussed briefly together with an alternative solution.

Approximate Solution

Engberg et al [13] have introduced a quasi-stationary model which approximates the true solution in a satisfactory way. They applied Fick's first law.

$$\frac{dm}{dt} = -\frac{DA}{V_m^L}\frac{dx^L}{dy} \tag{6.109}$$

The quasi-stationary approximation implies that dm/dt is constant. It can be expressed as a function of the quantities we have introduced above.

The approximate solution can be written as

$$\frac{s^2}{2} = \frac{V_{growth}\, r_c}{D} = \frac{1}{2}\frac{\Omega^2}{1-\Omega} \tag{6.110}$$

According to Equation (6.106) the left-hand side of Equation (6.110) is a measure of the growth rate V_{growth}. Ω and consequently also the right-hand side of Equation (6.110), is an implicite function of the supersaturation.

The exact solution Equation (6.100) and the approximate solution Equation (6.110) are plotted in Figure 6.40. In addition, the two asymptotes of the approximate solution have also been plotted in the figure.

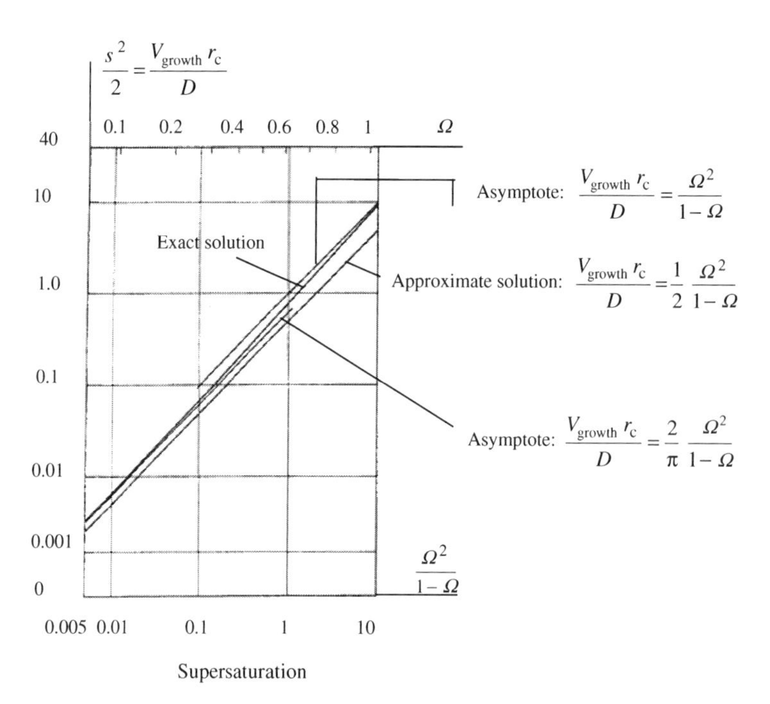

Figure 6.40 The rate of planar crystal growth as an implicite function of supersaturation.

For *small* values of Ω, ($\Omega \approx 0$) the approximate solution has the asymptote

$$\frac{s^2}{2} = \frac{V_{growth}\, r_c}{D} \approx \frac{2}{\pi}\frac{\Omega^2}{1-\Omega} \tag{6.111}$$

For *large* values of Ω ($\Omega \approx 1$) the approximate solution has the asymptote

$$\frac{s^2}{2} = \frac{V_{\text{growth}}\, r_{\text{c}}}{D} = \frac{1}{1-\Omega} \approx \frac{\Omega^2}{1-\Omega} \tag{6.112}$$

The exact solution which shows the planar crystal growth is obviously rather complicated. In next section we will analyze the simplest and most concrete special case, planar crystal growth at stationary conditions and constant growth rate.

6.9.2 *Planar Growth at Stationary Conditions and Constant Growth Rate in Alloys*

Consider a planar crystal interface surrounded by a supersaturated melt. We want to calculate the concentration profile of the alloying elements ahead of the solidification front in the melt as a function of the distance from the interface and the constant growth rate at stationary conditions.

Steady State Conditions

Stationary conditions are characterized by the following conditions:

1. The diffusion of alloying elements in the solid is neglected in comparison with the diffusion in the melt.
2. The equilibrium partition coefficient k, the ratio of the concentrations of the solid and the melt in equilibrium with each other, is constant (Figure 6.41).
3. The composition of the melt in bulk and in the solid are equal and constant

   and in addition

4. The planar solidification front moves forward with a constant rate.

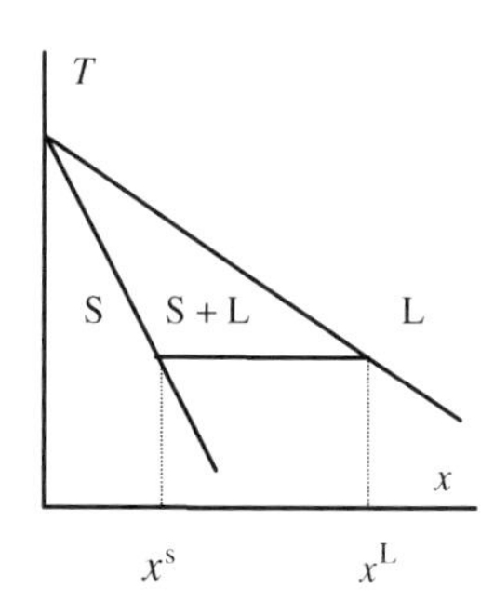

Figure 6.41 Schematic phase diagram of a binary alloy. The figure shows that $k_0 < 1$.

Solid diffusion is usually much lower than liquid diffusion, which makes assumption 1 very reasonable. Assumption 2 can be written as

$$\frac{x^{\text{s/L}}}{x^{\text{L/s}}} = k_0 = const \tag{6.113}$$

where
$x^{\text{s/L}}$ = mole fraction of alloying elements in the solid close to the interphase
$x^{\text{L/s}}$ = mole fraction of alloying elements in the melt close to the interphase
k_0 = the equilibrium partition coefficient.

Assumptions 1–3 result in a concentration profile such as that in Figure 6.42. The concentration of alloying element is constant in the solid and equal to the concentration in the melt.

$$x^{\text{s/L}} = x_0^{\text{L}} \tag{6.114}$$

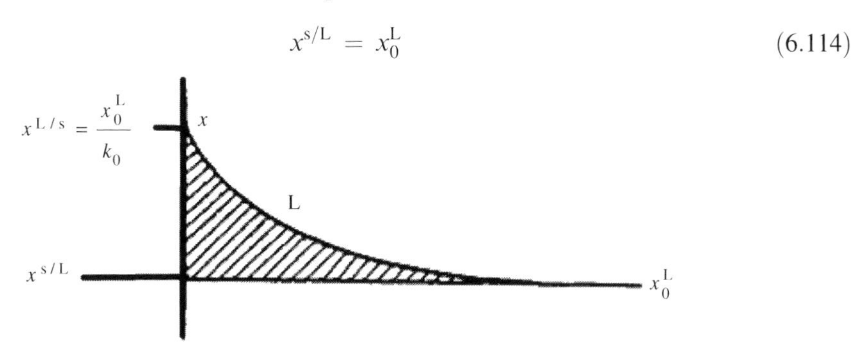

Figure 6.42 Concentration profile of alloying element if $k_0 < 1$.

The concentration of the alloying elements in the melt close to the interphase can be expressed as a function of the concentration in the melt in bulk and the partition coefficient. Equations (6.113) and (6.114) give

$$x^{\text{L/s}} = \frac{x_0^{\text{L}}}{k_0} \tag{6.115}$$

Concentration Distribution of the Alloying Element in the Melt

If $k_0 < 1$ the solute concentration in the melt decreases (otherwise it increases as is shown in Figure 6.38a) on page 318 from x_0^L/k_0 close to the interphase x_0^L far from the solidification front. The excess of solute atoms, left over when the melt solidifies, diffuse back into the melt. The driving force of the diffusion is the concentration difference of the solute in the melt.

Simultaneously the solidification front advances. The concentration distribution ahead of the front is constant and moves forward with the same velocity as the front. At stationary conditions the amount of solute released at the solidification is equal to the amount that disappears by diffusion.

Consider a box with two parallel surfaces perpendicular to the diffusion flow of solute atoms at a distance $\mathrm{d}y$ (Figure 6.43). If we use Fick's first law we obtain the *net flow of diffusing solute atoms* into the box (outgoing flows give negative contributions)

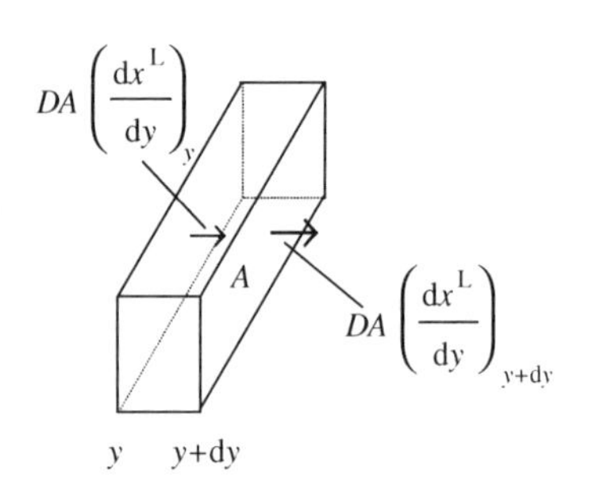

Figure 6.43 Flow of solute through volume element with cross-sectional area A and length $\mathrm{d}y$.

$$-\left(\frac{\mathrm{d}m}{\mathrm{d}t}\right)_{y+\mathrm{d}y}+\left(\frac{\mathrm{d}m}{\mathrm{d}t}\right)_{y}=-\left(-DA\left(\frac{\mathrm{d}x^{L}}{\mathrm{d}y}\right)_{y+\mathrm{d}y}\right)+\left(-DA\left(\frac{\mathrm{d}x^{L}}{\mathrm{d}y}\right)_{y}\right) \tag{6.116}$$

or after expansion in series

$$-\left(\frac{\mathrm{d}m}{\mathrm{d}t}\right)_{y+\mathrm{d}y}+\left(\frac{\mathrm{d}m}{\mathrm{d}t}\right)_{y}=DA\left[\left(\frac{\mathrm{d}x^{L}}{\mathrm{d}y}\right)_{y}+\mathrm{d}y\frac{\mathrm{d}^2x^{L}}{\mathrm{d}y^2}\right]-DA\left(\frac{\mathrm{d}x^{L}}{\mathrm{d}y}\right)_{y}$$

After reduction we obtain the net amount of solute into the volume element $A\mathrm{d}y$ per unit time owing to diffusion $= A\mathrm{d}y\, D\dfrac{\mathrm{d}^2x^{L}}{\mathrm{d}y^2}$

At the time t the amount m of solute in the box is

$$m = A\mathrm{d}y(x^{L})_{y}$$

The whole distribution profile moves with the velocity

$$V_{\mathrm{growth}} = \mathrm{d}y/\mathrm{d}t.$$

At the time $t+\mathrm{d}t$ the amount of solute in the box is $A\mathrm{d}y$ and you get

$$m+\mathrm{d}m = A\mathrm{d}y(x^{L})_{y+\mathrm{d}y}$$

The net increase of solute in time $\mathrm{d}t$ is

$$\mathrm{d}m = A\mathrm{d}y\left[(x^{L})_{y+\mathrm{d}y}-(x^{L})_{y}\right]=A\mathrm{d}y\left[\left((x^{L})_{y}+\mathrm{d}y\left(\frac{\mathrm{d}x^{L}}{\mathrm{d}y}\right)\right)-(x^{L})_{y}\right]=A\mathrm{d}y\frac{\mathrm{d}x^{L}}{\mathrm{d}y}\mathrm{d}y$$

The net increase of solute per unit time owing to motion of the concentration profile is then

$$\frac{\mathrm{d}m}{\mathrm{d}t}=A\mathrm{d}y\frac{\mathrm{d}y}{\mathrm{d}t}\frac{\mathrm{d}x^{L}}{\mathrm{d}y}=A\mathrm{d}yV_{\mathrm{growth}}\frac{\mathrm{d}x^{L}}{\mathrm{d}y} \tag{6.117}$$

Under steady state conditions the total change per unit time must be zero. We add the net flow of diffusing atoms and the net increase owing to the motion of the concentration profile and obtain after division with $A\mathrm{d}y$

$$D\frac{\mathrm{d}^2x^{L}}{\mathrm{d}y^2}+V_{\mathrm{growth}}\frac{\mathrm{d}x^{L}}{\mathrm{d}y}=0 \tag{6.118}$$

where V_{growth} is the velocity of the solidification front. The solution of this differential equation of second order is

$$x^L = A + Be^{-\frac{V_{growth}y}{D}} \qquad (6.119)$$

The constants A and B are evaluated from the boundary conditions:

$$y = \infty \quad \text{and} \quad x^L = x_0^L \quad \Rightarrow \quad A = x_0^L$$

$$y = 0 \quad \text{and} \quad x^L = \frac{x^s}{k_0} = \frac{x_0^L}{k_0} \quad \Rightarrow \quad B = \frac{x_0^L}{k_0} - x_0^L$$

where
x_0^L = mole fraction of the alloying element in the melt far from the solidification front
k_0 = the partition coefficient at equilibrium.

Hence, the concentration of alloying elements in the melt ahead of the solidification front can be written as

$$x^L = x_0^L\left(1 + \frac{1-k_0}{k_0}e^{-\frac{V_{growth}y}{D_L}}\right) \qquad (6.120)$$

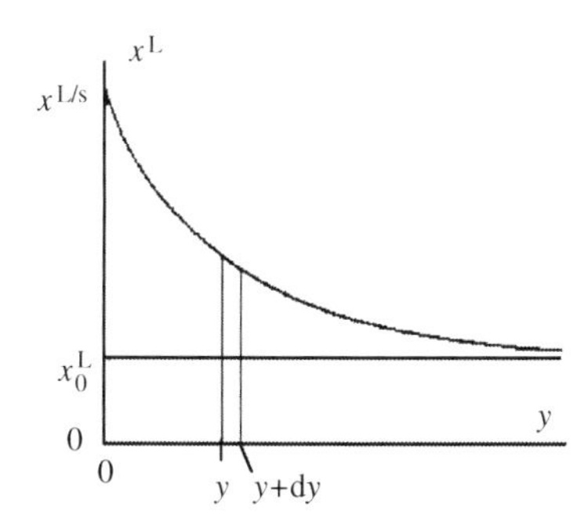

Figure 6.44 Mole fraction of alloying element in the melt as a function of position ahead of the solidification front under stationary conditions.

The function is shown in Figure 6.44 for $k_0 < 1$.

Example 6.4

An alloy was allowed to solidify with a planar solidification front. During the experiment the solidification rate was suddenly increased by stronger cooling for some time. Afterwards the sample was etched metallographically and examined. A band of segregation was found at the level where the solidification rate had been increased.

a) Explain the band of segregation in the sample.
b) Is the amount of alloying element larger or smaller in the band of segregation than in the rest of the sample? The partition coefficient < 1.

Solution and Answer

a)
The concentration of the alloying element ahead of the solidification front varies with the solidification rate V_{growth} according to Equation (6.115). When V_{growth} is increased, the concentration distribution in the melt ahead of the solidification front changes, not abruptly but over a certain distance. The corresponding concentration of alloying element in the solid phase is proportional to the concentration in the melt and will therefore also change.

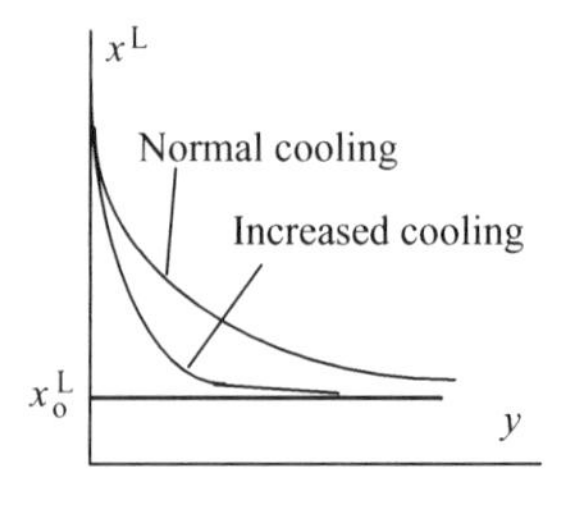

b)
By means of Equation (6.108) we can conclude that if $k_0 < 1$ and V_{growth} increases then x^{L} will decrease more rapidly than before. The temporary concentration curve is more narrow and lower than the normal curve. A smaller amount of the alloying element is ejected back into the melt by diffusion at higher growth rate (stronger cooling) than before changing, because the temporary 'shark fin' has a smaller area than the normal one. Consequently an increased amount of the alloying element is left in the solid.

Hence, the band of segregation contains a higher concentration of the alloying element than the rest of the sample.

6.10 Diffusion-Controlled Growth of Spherical Crystals in Alloys

The growth process of spherical crystals will be studied in order to calculate the concentration profile of the alloying element and the growth rate of the spherical crystals. Initially the general solution will be presented. It is followed by an analysis of the special case when the flow of melt atoms towards the spherical interphase is stationary.

Consider a spherical crystal precipitating in a supersaturated binary melt. The process is shown in Figure 6.45. The upper figure represents the phase diagram of the binary system. Below the equilibrium solidus and liquidus lines the ones corresponding to a supersaturated or undercooled melt are drawn.

The lower figure is constructed from the upper one and represents the concentration profile of the spherical crystal.

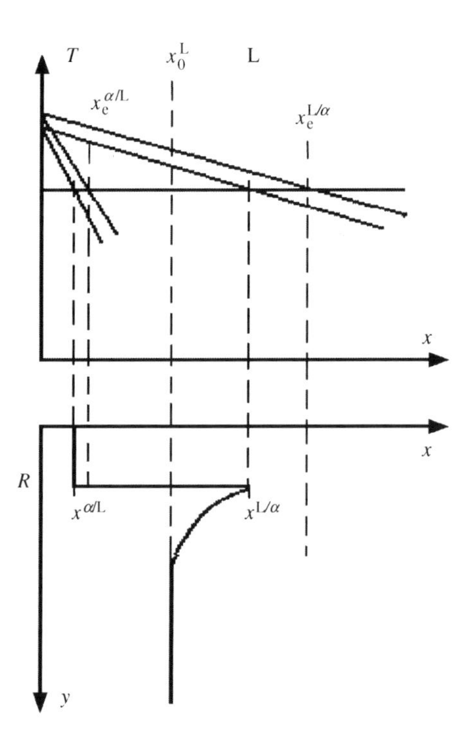

Figure 6.45
Upper figure:
Schematic phase diagram of binary system. The upper lines represent the equilibrium.

Lower figure:
Concentration of the spherical particle versus the distance from the centre of the particle with radius R.

6.10.1 *General Case of Diffusion Controlled Growth of Spherical Crystals in Alloys*

Exact Solution

The rate of the solidification process and the composition of the melt as a function of position can be calculated by solving Fick's second law of solidification.

Fick's second law:

$$\frac{\mathrm{d}c}{\mathrm{d}t} = D\left(\frac{\partial^2 c}{\partial x^2} + \frac{\partial^2 c}{\partial y^2} + \frac{\partial^2 c}{\partial z^2}\right)$$

The solution for spherical particles is found in exactly the same way as that for planar growth. The equations and the solution are the same or small modifications of the ones which are valid for the planar case and given in Section 6.9.1.

The material balance at the interphase is the same in both cases

$$\frac{\mathrm{d}m}{\mathrm{d}t} = -\frac{DA}{V_\mathrm{m}^\mathrm{L}}\left(\frac{\mathrm{d}x^\mathrm{L}}{\mathrm{d}r}\right) = \frac{V_\mathrm{growth}A}{V_\mathrm{m}^\alpha}\left(x^{\mathrm{L}/\alpha} - x^{\alpha/\mathrm{L}}\right) \tag{6.121}$$

where

m = mass of the diffusion atoms
D = diffusion constant of the melt
A = cross-section area
V_growth = rate of crystal growth ($\mathrm{d}r_\mathrm{c}/\mathrm{d}t$)
r = radial coordinate = coordinate of solidification front (interphase)
V_m^L = molar volume of the melt
V_m^α = molar volume of the solid α phase
x^L = mole fraction of the melt at coordinate y
$x^{\mathrm{L}/\alpha}$ = mole fraction of alloying element in the melt close to the interphase
$x^{\alpha/\mathrm{L}}$ = mole fraction of alloying element in the solid close to the interphase.

The only difference is that the coordinate y is replaced by the radial coordinate r.

Zener's solution of Fick's law $\frac{\mathrm{d}c}{\mathrm{d}t} = D\left(\frac{\partial^2 c}{\partial x^2} + \frac{\partial^2 c}{\partial y^2} + \frac{\partial^2 c}{\partial z^2}\right)$ for the spherical case is

$$x^\mathrm{L} - x_0^\mathrm{L} = \left(x^{\mathrm{L}/\alpha} - x_0^\mathrm{L}\right)\frac{\phi\left(\frac{r}{\sqrt{Dt}}\right)}{\phi\left(\frac{r_\mathrm{c}}{\sqrt{Dt}}\right)} \tag{6.122}$$

where

r_c = radius of the spherical crystal
$x^{\mathrm{L}/\alpha}$ = mole fraction of solute in the melt at the interphase.

The function ϕ is defined by the relationship

$$\phi(z) = \int_z^\infty \frac{1}{u^2}\mathrm{e}^{-\frac{u^2}{4}}\,\mathrm{d}u = \frac{\mathrm{e}^{-\frac{z^2}{4}}}{z} - \frac{\sqrt{\pi}}{2}\left(1 - \mathrm{erf}\,\frac{z}{2}\right) \tag{6.123}$$

where z is an arbitrary variable and *not* one of the Cartesian coordinates (xyz).

In analogy with the planar case on pages 318–321 we define the constant s as

$$s = \frac{r}{\sqrt{Dt}} \tag{6.124}$$

We proceed in the same way as on page 320 but with the coordinate r instead of y, and obtain

$$V_\mathrm{growth} = \left(\frac{\mathrm{d}r}{\mathrm{d}t}\right)_{r\,=\,r_\mathrm{c}} \tag{6.125}$$

and also

$$\frac{s^2}{2} = \frac{V_\mathrm{growth}\,r_\mathrm{c}}{D} \tag{6.126}$$

Equation (6.126) will formally be the same as Equation (6.106) on page 319 but with the difference that the definition of the constant s is different in the planar and the spherical case.

The definition of the *supersaturation* of the melt is the same in planar and spherical growth:

$$\Omega = \frac{V_m^\alpha}{V_m^L} \frac{x^{L/\alpha} - x_0^L}{x^{L/\alpha} - x^{\alpha/L}} \tag{6.108}$$

The effect of the curvature on the values x^α and $x^{L/\alpha}$ can be neglected.

The relationship (6.107) on page 319 between Ω and s in the planar case is also formally the same as that in the spherical case below

$$\frac{V_m^\alpha}{V_m^L} \frac{x^{L/\alpha} - x_0^L}{x^{L/\alpha} - x^{\alpha/L}} = \frac{\sqrt{\pi}}{2} s\, e^{\frac{s^2}{4}} \left(1 - \text{erf}\frac{s}{2}\right) \tag{6.127}$$

but it is important to keep in mind that s in this case is defined by Equation (6.124) on page 315.

Equation (6.122) on page 325 represents the exact solution in the spherical case.

Approximate Solution

In the same way as in the planar case it is possible to derive an approximate solution, i.e. a relationship between the super-saturation Ω and the growth rate. The result is

$$\frac{s^2}{2} = \frac{V_{\text{growth}}\, r_c}{D} = \frac{\Omega}{1 - \frac{1}{4}\left(\Omega + \sqrt{\Omega^2 + 8\Omega}\right)} \tag{6.128}$$

For *small* values of Ω ($\Omega \approx 0$) the exact solution can be replaced by

$$\frac{s^2}{2} = \frac{V_{\text{growth}}\, r_c}{D} \approx \Omega \tag{6.129}$$

It is used in many applications.

For *large* values of Ω ($\Omega \approx 1$) the exact solution can be replaced by

$$\frac{s^2}{2} = \frac{V_{\text{growth}}\, r_c}{D} \approx \frac{3}{2} \frac{\Omega}{1 - \Omega} \tag{6.130}$$

A diagram, similar to that given in Figure 6.40 on page 320 can be drawn.

6.10.2 *Diffusion-Controlled Growth Rate as a Function of Crystal Size*

Fick's second law:

$$\frac{dc}{dt} = D\left(\frac{\partial^2 c}{\partial x^2} + \frac{\partial^2 c}{\partial y^2} + \frac{\partial^2 c}{\partial z^2}\right)$$

It is reasonable and usual to make the approximation that the diffusion field around the crystal or particle is given by Fick's second law under the assumption that the flow of melt atom towards the spherical interphase is stationary, i.e. constant in time. By solving Fick's second law we can find an approximate expression for the growth rate of the crystal as a function of its size.

For small crystals with radii close to the critical radius of nucleation it is necessary to take two complications into consideration:

- The growth rate is strongly dependent on the surface tension of the liquid or the melt when the radius of the crystal is small.
- At small sizes of the radius the growth rate of the crystal is limited by the rate of incorporation of solidifying atoms into the crystal lattice.

Starting with the approximate solution we will build up the complete solution of the problem stage by stage.

Stage 1. Approximate Solution

To find the concentration x^{L} of the alloying element in the melt outside the crystal, expressed in mole fraction, we apply Fick's second law.

$$\frac{\partial x^{\mathrm{L}}}{\partial t} = D\left(\frac{\partial^2 x^{\mathrm{L}}}{\partial x^2} + \frac{\partial^2 x^{\mathrm{L}}}{\partial y^2} + \frac{\partial^2 x^{\mathrm{L}}}{\partial z^2}\right) = const \tag{6.131}$$

Owing to spherical symmetry the mole fraction must be a function of the distance r from the crystal centre only. Hence, it is advisable to use spherical coordinates for the special case that x^{L} is independent of the coordinates θ and φ. If we transform Equation (6.131) into spherical coordinates we obtain

$$\frac{1}{r^2}\frac{\mathrm{d}}{\mathrm{d}r}\left(r^2\frac{\mathrm{d}x^{\mathrm{L}}}{\mathrm{d}r}\right) = const \tag{6.132}$$

The solution of this differential equation of the second order is

$$x^{\mathrm{L}} = A + \frac{B}{r} \tag{6.133}$$

where
x^{L} = mole fraction of the alloying element in the melt
r = distance from the centre of the growing crystal.

Two boundary conditions determine the constants A and B.

$$x^{\mathrm{L}} = x_0^{\mathrm{L}} \quad \text{for} \quad r = \infty \quad \Rightarrow \quad A = x_0^{\mathrm{L}}$$
$$x^{\mathrm{L}} = x_{\mathrm{eq}}^{\mathrm{L}/\alpha} \quad \text{for} \quad r = r_{\mathrm{c}} \quad \Rightarrow \quad B = r_{\mathrm{c}}\left(x_{\mathrm{eq}}^{\mathrm{L}/\alpha} - x_0^{\mathrm{L}}\right)$$

where
r_{c} = radius of the growing crystal
x_0^{L} = initial concentration of the alloying element in the melt and
= mole fraction of alloying element in the melt far from the crystal
$x_{\mathrm{eq}}^{\mathrm{L}/\alpha}$ = mole fraction of alloying element in the melt close to the interphase at equilibrium between the solid crystal and the melt.

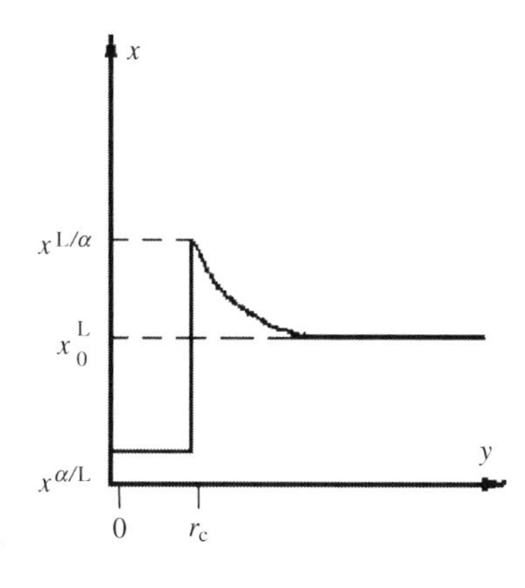

Figure 6.46 Mole fraction of alloying element as a function of distance from the centre of the particle.

If we insert the values of A and B into Equation (6.133) we obtain

$$x^{\mathrm{L}} = x_0^{\mathrm{L}} + \frac{r_{\mathrm{c}}\left(x_{\mathrm{eq}}^{\mathrm{L}/\alpha} - x_0^{\mathrm{L}}\right)}{r} \tag{6.134}$$

The mole fraction of the alloying element in the melt decreases with the distance from the centre of the crystal (Figure 6.46). The rate is

$$\frac{\mathrm{d}x^{\mathrm{L}}}{\mathrm{d}r} = -\frac{1}{r^2}r_{\mathrm{c}}\left(x_{\mathrm{eq}}^{\mathrm{L}/\alpha} - x_0^{\mathrm{L}}\right) \tag{6.135}$$

A mass balance is performed at the interphase. The mass of the alloying atoms which diffuse away from the shell with radius r_{c} per unit time is, according to Fick's first law, equal to.

$$\frac{\mathrm{d}m}{\mathrm{d}t} = -D \times 4\pi r_{\mathrm{c}}^2\left(\frac{\mathrm{d}c^{\mathrm{L}}}{\mathrm{d}r}\right)_{r=r_{\mathrm{c}}} = -D \times 4\pi r_{\mathrm{c}}^2\frac{-\left(x_{\mathrm{eq}}^{\mathrm{L}/\alpha} - x_0^{\mathrm{L}}\right)}{r_{\mathrm{c}}}\frac{1}{V_{\mathrm{m}}^{\mathrm{L}}} \tag{6.136}$$

Fick's first law:

$$\frac{\mathrm{d}m}{\mathrm{d}t} = -DA\frac{\mathrm{d}c}{\mathrm{d}y}$$

The factor $1/V_m^L$ is introduced to express the concentration in mole fraction instead of wt%. At steady state it equals the amount of alloying atoms released per unit time when the shell solidifies with the rate dr_c/dt

$$\frac{x_{eq}^{L/\alpha} - x_{eq}^{\alpha/L}}{V_m^{\alpha}} \frac{4\pi r_c^2 \, dr_c}{dt} = -D \times 4\pi r_c^2 \frac{-\left(x_{eq}^{L/\alpha} - x_0^L\right)}{r_c} \frac{1}{V_m^L} \tag{6.137}$$

where

m = mass of diffusing atoms.
$x_{eq}^{L/\alpha}$ = mole fraction of B in the melt at the interphase at equilibrium
$x_{eq}^{\alpha/L}$ = mole fraction of B in the solid at the interphase at equilibrium
x_0^L = initial mole fraction of B in the melt far from the crystal
r_c = radius of the growing crystal
r = radial coordinate or distance from the centre of the growing crystal
D = diffusion constant of the alloying element B in the melt
V_m^L = molar volume of the melt
V_m^{α} = molar volume of the solid phase.

Equation (6.137) describes the growth of a spherical crystal in a binary system. The equation can be rewritten as

$$V_{growth} = \frac{dr_c}{dt} = D \frac{x_{eq}^{L/\alpha} - x_0^L}{x_{eq}^{L/\alpha} - x_{eq}^{\alpha/L}} \frac{V_m^{\alpha}}{V_m^L} \frac{1}{r_c} \tag{6.138}$$

Equation (6.138) gives the growth rate of the crystal as a function of r_c.

Stage 2. Attention to Surface Tension

The growth rate is strongly influenced by the surface tension. It influences the equilibrium concentration $x_{eq}^{L/\alpha}$ which is highly dependent of the surface tension. The smaller the radius is the stronger will be the dependence. The minimum value of the crystal radius is the critical radius of nucleation r^*.

In Section 6.8 we found that for dilute solutions the driving force of solidification is proportional to the deviation of the concentration from equilibrium. The total driving force for solidification can be written as

$$-\Delta G^{L \to \alpha} = RT\left(x_{eq}^{L/\alpha} - x_0^L\right) \tag{6.139}$$

The right-hand side of Equation (6.139) can formally be split up into two terms.

$$x_{eq}^{L/\alpha} - x_0^L = \left(x_{eq}^{L/\alpha} - x_{r_c}^{L/\alpha}\right) + \left(x_{r_c}^{L/\alpha} - x_0^L\right) \tag{6.140}$$

The *second* term $\left(x_{r_c}^{L/\alpha} - x_0^L\right)$ represents the driving force of diffusion of atoms from the interphase into the liquid.

The *first* term $\left(x_{eq}^{L/\alpha} - x_{r_c}^{L/\alpha}\right)$ in Equation (6.140) represents the driving force of creation of new surfaces. The size of this driving force can be calculated from the equality of the chemical potentials, which is the fundamental condition of equilibrium (Equation (2.45) on page 72 in Chapter 2)

$$\mu_A^L = \mu_A^\alpha \tag{6.141}$$

where
A = solvent atoms
B = alloying element
μ_B^L = chemical potential of B in the melt
μ_B^α = chemical potential of B in the solid phase.

With the aid of the relationships (2.77) in Chapter 2 on page 78 (activity $a = \gamma x$) and including the influence of the surface tension we can transform Equation (6.141) into

$$\mu_A^{0L} + RT \ln\gamma^{L/\alpha}\left(1 - x_B^{L/\alpha}\right) = \mu_A^{0\,\alpha} + RT \ln\gamma^{\alpha/L}\left(1 - x_B^{\alpha/L}\right) + \frac{2\sigma V_A^\alpha}{r_c} \tag{6.142}$$

where
R, T = gas constant, temperature
r_c = radius of the crystal
μ_A^{0L} = chemical potential of A in the melt
$\mu_A^{0\,\alpha}$ = chemical potential of A in the solid phase
$\gamma_B^{L/\alpha}$ = activity coefficient of B in the melt at the interphase
$\gamma_B^{\alpha/L}$ = activity coefficient of B in the solid phase at the interphase
V_A^α = partial molar volume of component A in the α phase
σ = surface tension between melt and solid.

We use Equation (6.142) three times, once for any radius r_c and once for the critical radius r^* of nucleation and finally once for a large crystal ($r_c \approx \infty$) at equilibrium, in combination with Raoult's law. ($\gamma_A^{L/\alpha} \approx 1$ and $\gamma_A^{\alpha/L} \approx 1$) are material constants, see pages 28–29), and obtain

$$\mu_A^{0L} + RT \ln\left(1 - x_{r_c}^{L/\alpha}\right) = \mu_A^{0\,\alpha} + RT \ln\left(1 - x_{eq}^{\alpha/L}\right) + \frac{2\sigma V_A^\alpha}{r_c} \tag{6.143}$$

$$\mu_A^{0L} + RT \ln\left(1 - x_0^L\right) = \mu_A^{0\,\alpha} + RT \ln\left(1 - x_0^{\alpha/L}\right) + \frac{2\sigma V_A^\alpha}{r^*} \tag{6.144}$$

$$\mu_A^{0L} + RT \ln\left(1 - x_{eq}^{L/\alpha}\right) = \mu_A^{0\,\alpha} + RT \ln\left(1 - x_{eq}^{\alpha/L}\right) + 0 \tag{6.145}$$

where
r^* = critical radius of nucleation
$x_{eq}^{L/\alpha}$ = equlibrium mole fraction of B in the melt at the interphase.

Equation (6.145) is subtracted from each of Equations (6.143) and (6.144) which gives

$$RT\left[\ln\left(1 - x_{r_c}^{L/\alpha}\right) - \ln\left(1 - x_{eq}^{L/\alpha}\right)\right] = \frac{2\sigma V_A^\alpha}{r_c} \tag{6.146}$$

$$RT\left[\ln\left(1 - x_0^L\right) - \ln\left(1 - x_{eq}^{L/\alpha}\right)\right] = \frac{2\sigma V_A^\alpha}{r^*} \tag{6.147}$$

If the solution of B is dilute the left-hand side of Equations (6.146) and (6.147) can be expanded in series. After this operation the new equations are divided with each other and we obtain

$$\frac{-x_{r_c}^{L/\alpha} + x_{eq}^{L/\alpha}}{-x_0^L + x_{eq}^{L/\alpha}} = \frac{r^*}{r_c} \tag{6.148}$$

which can easily be transformed into

$$x_{\text{eq}}^{\text{L}/\alpha} - x_{r\text{c}}^{\text{L}/\alpha} = \frac{r^*}{r_\text{c}}\left(x_{\text{eq}}^{\text{L}/\alpha} - x_0^{\text{L}}\right) \tag{6.149}$$

The value of $\left(x_{\text{eq}}^{\text{L}/\alpha} - x_{r\text{c}}^{\text{L}/\alpha}\right)$ in Equation (6.149) is introduced into Equation (6.140) which gives

$$x_{\text{eq}}^{\text{L}/\alpha} - x_0^{\text{L}} = \frac{r^*}{r_\text{c}}\left(x_{\text{eq}}^{\text{L}/\alpha} - x_0^{\text{L}}\right) + \left(x_{r\text{c}}^{\text{L}/\alpha} - x_0^{\text{L}}\right) \tag{6.150}$$

Equation (6.150) is solved for $\left(x_{r\text{c}}^{\text{L}/\alpha} - x_0^{\text{L}}\right)$ which gives

$$\left(x_{r\text{c}}^{\text{L}/\alpha} - x_0^{\text{L}}\right) = \left(x_{\text{eq}}^{\text{L}/\alpha} - x_0^{\text{L}}\right)\left(1 - \frac{r^*}{r_\text{c}}\right) \tag{6.151}$$

The value of $\left(x_{r\text{c}}^{\text{L}/\alpha} - x_0^{\text{L}}\right)$, is proportional to the driving force of the diffusion process. It corresponds to $\left(x_{\text{eq}}^{\text{L}/\alpha} - x_0^{\text{L}}\right)$ in Equation (6.138) on page 328. If the latter expression is replaced by the former one in Equation (6.138) we obtain the growth rate as a function of the crystal size, corrected for influence of surface tension

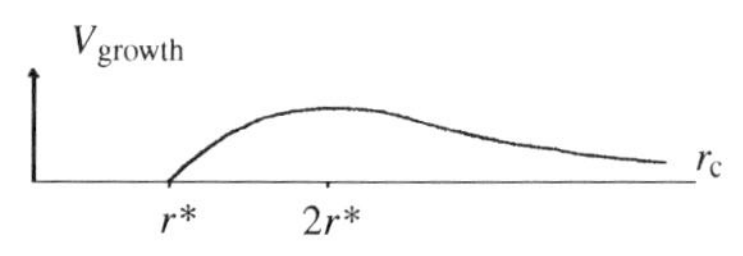

Figure 6.47 The growth rate of a spherical crystal as a function of crystal size corrected for influence of surface tension.

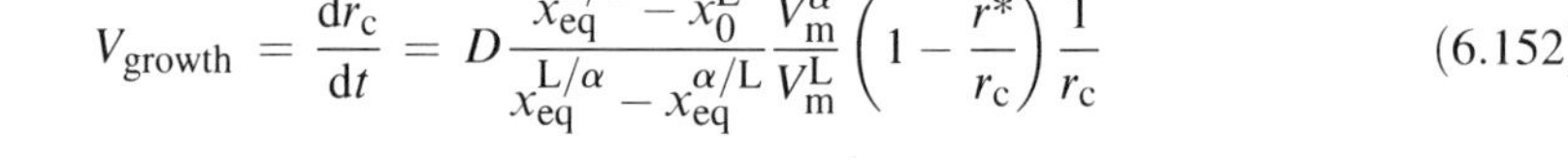

$$V_{\text{growth}} = \frac{\text{d}r_\text{c}}{\text{d}t} = D\frac{x_{\text{eq}}^{\text{L}/\alpha} - x_0^{\text{L}}}{x_{\text{eq}}^{\text{L}/\alpha} - x_{\text{eq}}^{\alpha/\text{L}}}\frac{V_\text{m}^{\alpha}}{V_\text{m}^{\text{L}}}\left(1 - \frac{r^*}{r_\text{c}}\right)\frac{1}{r_\text{c}} \tag{6.152}$$

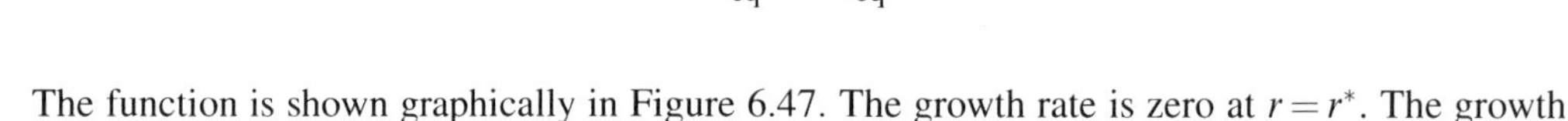

The function is shown graphically in Figure 6.47. The growth rate is zero at $r = r^*$. The growth starts by random statistical fluctuations. There is a maximum in growth rate at $r = 2r^*$.

Equations (6.152) above and (6.138) on page 328 differ by a factor $(1 - r^*/r_\text{c})$, which implicitely represents the influence of surface tension [Equations (4.4) on page 169 and (4.12) on page 172 in Chapter 4]. Equation (6.152) becomes identical with Equation (6.138) if $r_\text{c} \gg r^*$.

Stage 3. Attention to Kinetics of Crystal Growth

Equation (6.138) on page 328 shows that the growth rate increases with decreasing values of the crystal radius. Equation (6.138) is not valid for very small values of the radius. In these cases the rate of incorporation of atoms into the crystal lattice may be the slowest process and may control the growth rate of the crystal. The kinetic process has to be taken into consideration.

The kinetic process during crystal growth of alloys has been discussed on pages 316–317 and we showed that the growth rate is proportional to the square of the undercooling, i.e. the square of the driving force. For an alloy the concentration difference in the melt replaces the undercooling, which gives the relationship

$$\frac{\text{d}r_\text{c}}{\text{d}t} = \beta_{\text{kin}}\left(x_{\text{eq}}^{\text{L}/\alpha} - x_{r\text{c}}^{\text{L}/\alpha}\right)^2 \tag{6.153}$$

where

$x_{r\text{c}}^{\text{L}/\alpha}$ = mole fraction of B atoms in the liquid at the interphase.

β_{kin} is a growth constant, called *kinetic coefficient*, which describes the kinetic process at the interphase. An expression of β_{kin} when relaxation is considered is discussed in the box below.

$$\beta_{\text{kin}} = \frac{D_{\text{interphase}}^2}{\pi\delta\, D_{\text{relax}}} \tag{6.94}$$

Derivation of β_{kin} for Pure Metals and Binary Alloys

Pure Metal

The driving force of solidification is proportional to the undercooling which can be written as

$$-\Delta G_{a}^{total} = (-\Delta H_{a})\frac{T_{M} - T}{T_{M}}$$

Eq. (6.23) page 290

By experience and a 'rule of thumbs' the molar heat of fusion ($-\Delta H$) for metals is roughly equal to $2RT$.

The growth rate of the solidification front depends on the fluxes of 'liquid' atoms entering and leaving the solid/liquid interphase, i.e. diffusion of A and B atoms, and relaxation of the incorporated A and B atoms down to the same energy as the lattice atoms.

The total Gibbs' free energy is the sum of two terms, one for the diffusion into and out of the interphase and one for the relaxation process in the solid.

$$-\Delta G^{total} = \frac{\delta RTV_{growth}}{D_{interphase}P} + (-\Delta G^{total})\sqrt{\frac{V_{growth}\delta}{\pi D_{relax}}}$$

Eq. (6.38a) page 294

Binary Alloys

For binary alloys the driving force of solidification is a function of *concentration differences* instead of *temperature differences*. The total driving force valid for a binary alloy is written as

$$-\Delta G_{total}^{L \to \alpha} \approx RT\left(x_{eq}^{L/\alpha} - x_{0}^{L}\right)$$

Eq. (6.72) p.313

In binary alloys both A and B atoms become incorporated into the ordered solid state, e.g. diffuse through the interphase, rearrange and become lattice atoms. Their excess energies are gradually transferred to the surrounding in a relaxation process.

The total driving force is the sum of two terms which correspond to the processes described above.

$$-\Delta G^{total} = \frac{\delta RTV_{growth}}{D_{interphase}P} + (-\Delta G^{total})\sqrt{\frac{V_{growth}\delta}{\pi D_{relax}}}$$

Eq. (6.38a) page 294

where δ is the thickness of the interphase and P is the probability factor

$$P = \sqrt{\frac{V_{growth}\delta}{\pi D_{relax}}}$$

Eq. (6.25) page 294

Inserting the expression for P into Equation (6.38a) gives

$$-\Delta G^{total} = \left[\frac{RT}{D_{interphase}}\pi D_{relax} + (-\Delta G^{total})\right]\sqrt{\frac{V_{growth}\delta}{\pi D_{relax}}}$$

Eq. (6.38b) page 294

Equation (6.38b) is squared as it is and combined with Equation (6.23) on page 290. The detailed calculations are given, step by step, on page 295. The resulting formula is

$$V_{growth} = \frac{\frac{D_{interphase}^{2}}{D_{relax}}}{\pi\delta(RT)^{2}}\left[\frac{(-\Delta H_{0})(T_{M} - T)}{T_{M}}\right]^{2}$$

Eq. (6.39b) page 295

or

$$V_{\text{growth}} = \beta_{\text{kin}} \times (\Delta T)^2$$

Identification of the last two equations gives finally the expression for the kinetic coefficient

$$\beta_{\text{kin}} = \frac{(-\Delta H_0)^2 D^2_{\text{interphase}}}{\pi\delta(RTT_{\text{M}})^2 D_{\text{relax}}}$$

Eq. (6.41) page 295

where δ is the thickness of the interphase and P is the probability factor. We assume that $P_0 = P_{\text{A}} = P_{\text{B}}$ and insert the expression for P

$$P = \sqrt{\frac{V_{\text{growth}}\delta}{\pi D_{\text{relax}}}}$$

Eq. (6.25) page 291

The result is

$$-\Delta G^{\text{total}} = \left[\frac{RT}{D_{\text{interphase}}}\pi D_{\text{relax}} + \left(-\Delta G^{\text{total}}\right)\right]\sqrt{\frac{V_{\text{growth}}\delta}{\pi D_{\text{relax}}}}$$

Eq. (6.89) page 317

Inserting numerical values into Equation (6.89) it is obvious that the first term inside the brackets is much larger than the second one because $D_{\text{interphase}} \ll D_{\text{relax}}$. The second term can be neglected and it is found

$$V_{\text{growth}} = \frac{D^2_{\text{interphase}}}{\pi\delta \cdot D_{\text{relax}}}\left(x^{\text{L}/\alpha}_{\text{eq}} - x^{\text{L}}_0\right)^2$$

Eq. (6.92) page 317

on page 317 that Equation (6.89) can be written as

or

$$V_{\text{growth}} = \beta_{\text{kin}}\left(x^{\text{L}/\alpha}_{\text{eq}} - x^{\text{L}}_0\right)^2$$

Eq. (6.93) page 317

Identification between gives

$$\beta_{\text{kin}} = \frac{D^2_{\text{interphase}}}{\pi \cdot \delta D_{\text{relax}}}$$

Eq. (6.94) page 317

In analogy with the preceding section we split up the driving force/*RT* into two parts:

$$x^{\text{L}/\alpha}_{\text{eq}} - x^{\text{L}}_0 = \left(x^{\text{L}/\alpha}_{\text{eq}} - x^{\text{L}/\alpha}_{r\text{c}}\right) + \left(x^{\text{L}/\alpha}_{r\text{c}} - x^{\text{L}}_0\right) \tag{6.155}$$

The *first* term on the right-hand side of Equation (6.155) is the driving force of the kinetic process at the crystal surface, proportional to the deviation from the equilibrium. The *second* term represents the driving force of diffusion of the alloying element in the liquid or melt.

If the expressions in Equations (6.153) and (6.148), respectively, are inserted into the left-hand side of Equation (6.155) we obtain

$$x^{\text{L}/\alpha}_{\text{eq}} - x^{\text{L}}_0 = \sqrt{\frac{\text{d}r_{\text{c}}}{\text{d}t}}\sqrt{\frac{1}{\beta_{\text{kin}}}} + \frac{\text{d}r_{\text{c}}}{\text{d}t}\frac{r_{\text{c}}}{D}\frac{V^{\text{L}}_{\text{m}}}{V^{\alpha}_{\text{m}}}\left(x^{\text{L}/\alpha}_{\text{eq}} - x^{\alpha/\text{L}}_{\text{eq}}\right)$$

or

$$\frac{\mathrm{d}r_c}{\mathrm{d}t} + \sqrt{\frac{1}{\beta_{kin}} \frac{D}{r_c} \frac{V_m^\alpha}{V_m^L} \frac{1}{x_{eq}^{L/\alpha} - x_{eq}^{\alpha/L}}} \sqrt{\frac{\mathrm{d}r_c}{\mathrm{d}t}} - \frac{D}{r_c} \frac{V_m^\alpha}{V_m^L} \frac{x_{eq}^{L/\alpha} - x_0^L}{x_{eq}^{L/\alpha} - x_{eq}^{\alpha/L}} = 0 \tag{6.156}$$

This second degree equation has the positive solution

$$\sqrt{\frac{\mathrm{d}r_c}{\mathrm{d}t}} = -A + \sqrt{A^2 + C_3} \tag{6.157}$$

where

$$A = \frac{1}{2}\sqrt{\frac{1}{\beta_{kin}} \frac{D}{r_c} \frac{V_m^\alpha}{V_m^L} \frac{1}{x_{eq}^{L/\alpha} - x_{eq}^{\alpha/L}}}$$

and

$$C_3 = \frac{D}{r_c} \frac{V_m^\alpha}{V_m^L} \frac{x_{eq}^{L/\alpha} - x_0^L}{x_{eq}^{L/\alpha} - x_{eq}^{\alpha/L}}$$

Hence, the growth rate of the crystal will be

$$V_{growth} = \frac{\mathrm{d}r_c}{\mathrm{d}t} = \left(-A + \sqrt{A^2 + C_3}\right)^2 \tag{6.158}$$

The expression (6.158) is the growth rate as a function of the crystal size, due to the kinetic process, i.e. the rate of incorporation of atoms at the crystal surface.

A large value of β_{kin} (small value of A) reduces the influence of the kinetic process. At $\beta_{kin} = \infty$ ($A = 0$), the process is infinitely fast and does not influence the total growth rate and Equation (6.158) becomes identical with Equation (6.138) on page 328.

Stage 4. Attention to both Surface Tension and Kinetics of Crystal Surface

In a case where the surface tension is of a magnitude large enough to reduce the growth rate and the kinetic process is slow we can proceed as follows.

In analogy with the procedure in the preceding sections the driving force/RT can formally be parted into three driving forces:

$$x_{eq}^{L/\alpha} - x_0^L = \left(x_{eq}^{L/\alpha} - x_{r_c}^{L/\alpha}\right) + \left(x_{r_c}^{L/\alpha} - x_{kin}^{L/\alpha}\right) + \left(x_{kin}^{L/\alpha} - x_0^L\right) \tag{6.159}$$

The *first* term on the right-hand side represents the driving force to create new surfaces.

The *second* term corresponds to the kinetic process at the crystal surface according to stage 3 and includes in this case also the influence of the surface tension according to stage 2.

The *third* term represents the diffusion process in the melt.

Substitutions on the right-hand side of Equation (6.159) by expressions from Equations (6.149), (6.153) and Equation (6.148) in the mentioned order, gives

$$x_{eq}^{L/\alpha} - x_0^L = \frac{r^*}{r_c}\left(x_{eq}^{L/\alpha} - x_0^L\right) + \sqrt{\frac{1}{\beta_{kin}}}\sqrt{\frac{\mathrm{d}r_c}{\mathrm{d}t}} + \frac{\mathrm{d}r_c}{\mathrm{d}t} \frac{r_c}{D} \frac{V_m^L}{V_m^\alpha}\left(x_{eq}^{L/\alpha} - x_{eq}^{\alpha/L}\right) \tag{6.160}$$

After reduction Equation (6.160) gives the positive solution of the second degree equation

$$\sqrt{\frac{\mathrm{d}r_c}{\mathrm{d}t}} = -A + \sqrt{A^2 + C_4} \tag{6.161}$$

where

$$A = \frac{D}{2}\sqrt{\frac{1}{\beta_{\text{kin}}}\frac{V_{\text{m}}^{\alpha}}{V_{\text{m}}^{\text{L}}}\frac{1}{x_{\text{eq}}^{\text{L}/\alpha} - x_{\text{eq}}^{\alpha/\text{L}}}} \times \frac{1}{r_{\text{c}}}$$

and

$$C_4 = D\frac{V_{\text{m}}^{\alpha}}{V_{\text{m}}^{\text{L}}}\frac{x_{\text{eq}}^{\text{L}/\alpha} - x_0^{\text{L}}}{x_{\text{eq}}^{\text{L}/\alpha} - x_{\text{eq}}^{\alpha/\text{L}}} \times \frac{1}{r_{\text{c}}}\left(1 - \frac{r^*}{r_{\text{c}}}\right) = C_3 \times \left(1 - \frac{r^*}{r_{\text{c}}}\right)$$

where r^* is the critical radius of nucleation.

Hence, the growth rate of the crystal as a function of the crystal radius will be

$$V_{\text{growth}} = \frac{\text{d}r_{\text{c}}}{\text{d}t} = \left(-A + \sqrt{A^2 + C_4}\right)^2 \tag{6.162}$$

If $\beta_{\text{kin}} \to \infty$ (A = 0) and $r_{\text{c}} \gg r^*$ Equation (6.162) will be identical with Equation (6.148). The only difference between the growth rates (6.158) and (6.162) is the factor $(1 - r^*/r_{\text{c}})$ in the C_4 term.

The function given in Equation (6.162) is shown in Figure 6.48. It shows the effect of different values of the constant β_{kin} on the growth rate as a function of crystal size. The influence of the parameter β_{kin} is obviously considerable.

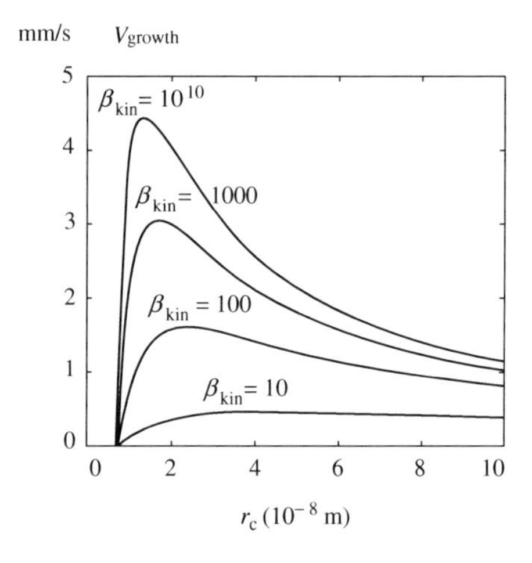

Figure 6.48 The growth rate of a spherical crystal as a function of crystal size with attention to the kinetic process at the crystal surface.

The curves are based on the following values inserted into Equation (6.162).
$D = 10 \times 10^{-9}$ m²/s
$\sigma = 0.1$ J/m²
$x_{\text{eq}}^{\text{L}/\alpha} - x_{\text{eq}}^{\alpha/\text{L}} = 0.800$
$x_{\text{eq}}^{\text{L}/\alpha} - x_0^{\text{L}} = 0.010$

The figure also shows that the maximum growth rate occurs at larger values of r_{c} when β_{kin} decreases. To determine the maximum value of the growth rate we take the derivative of Equation (6.161) with respect to r_{c} and put the derivative equal to zero. The maximum value found in this way is

$$(r_{\text{c}})_{\text{max}} = 2r^* + \frac{2A_0}{\sqrt{C_4^0}}\sqrt{r^*} \tag{6.163}$$

where

$$A_0 = \frac{D}{2}\sqrt{\frac{1}{\beta_{\text{kin}}}\frac{V_{\text{m}}^{\alpha}}{V_{\text{m}}^{\text{L}}}\frac{1}{x_{\text{eq}}^{\text{L}/\alpha} - x_{\text{eq}}^{\alpha/\text{L}}}}$$

and

$$C_4^0 = D\frac{V_{\text{m}}^{\alpha}}{V_{\text{m}}^{\text{L}}}\frac{x_{\text{eq}}^{\text{L}/\alpha} - x_0^{\text{L}}}{x_{\text{eq}}^{\text{L}/\alpha} - x_{\text{eq}}^{\alpha/\text{L}}}$$

Equation (6.163) shows that if $\beta_{\text{kin}} \to \infty$, the limit of r_{c} is $2r^*$, in agreement with Figure 6.48.

Calculation of the Maximum Growth Rate

We start with Equation (6.161) and the expressions of A_0 and C_4^0 given on page 333.
In order to simplify the calculations we will calculate the maximum of $\sqrt{V_{growth}}$ instead of V_{growth} as the two functions have the same maximum.

$$y = \sqrt{\frac{dr_c}{dt}} = -A + \sqrt{A^2 + C_4} \tag{1'}$$

which can be written as

$$y = -\frac{A_0}{r} + \sqrt{\frac{A_0^2}{r^2} + C_4^0 \frac{1}{r}\left(1 - \frac{r^*}{r}\right)} \tag{2'}$$

where

$$A_0 = \frac{D}{2}\sqrt{\frac{1}{\beta_{kin}}\frac{V_m^\alpha}{V_m^L}\frac{1}{x_{eq}^{L/\alpha} - x_{eq}^{\alpha/L}}} \tag{3'}$$

and

$$C_4^0 = D\frac{V_m^\alpha}{V_m^L}\frac{x_{eq}^{L/\alpha} - x_0^L}{x_{eq}^{L/\alpha} - x_{eq}^{\alpha/L}} \tag{4'}$$

We take the derivative of y with respect to r and put it equal to zero.

$$\frac{dy}{dr} = \frac{A_0}{r^2} + \frac{-\frac{2A_0^2}{r^3} + C_4^0\left[-\frac{1}{r^2}\left(1 - \frac{r^*}{r}\right) + \frac{1}{r}\frac{r^*}{r^2}\right]}{2\sqrt{\frac{A_0^2}{r^2} + C_4^0\frac{1}{r}\left(1 - \frac{r^*}{r}\right)}} = 0$$

or, after rearrangement and multiplication by the denominator,

$$-\frac{A_0}{r^2} \times 2\sqrt{\frac{A_0^2}{r^2} + C_4^0\frac{1}{r}\left(1 - \frac{r^*}{r}\right)} = -\frac{2A_0^2}{r^3} + C_4^0\left(-\frac{1}{r^2} + \frac{2r^*}{r^3}\right)$$

The equation is multiplied by r^2 and then squared.

$$(2A_0)^2\left(\frac{A_0^2}{r^2} + C_4^0\frac{1}{r} - C_4^0\frac{r^*}{r^2}\right) = \frac{4A_0^4}{r^2} + (C_4^0)^2\left(-1 + \frac{2r^*}{r}\right)^2 - 4\frac{A_0^2}{r}C_4^0\left(-1 + \frac{2r^*}{r}\right)$$

After reduction we obtain

$$4A_0^2\frac{r^*}{r} = C_4^0\left(-1 + \frac{2r^*}{r}\right)^2 \tag{5'}$$

Equation (5′) is multiplied by r^2 and the square root is taken of both sides.

$$\frac{2A_0\sqrt{r^*}}{\sqrt{C_4^0}} = \pm(2r^* - r) \tag{6'}$$

The two roots of Equation (6′) are

$$r_1 = 2r^* + \frac{2A_0\sqrt{r^*}}{\sqrt{C_4^0}} \quad \text{and} \quad r_2 = 2r^* - \frac{2A_0\sqrt{r^*}}{\sqrt{C_4^0}}$$

The relationship (3′) and Figure 6.48 on page 334 shows that when β_{kin} decreases A_0 and r_{max} increases. For this reason we can conclude that r_{max} must be equal to r_1. Hence, we obtain

$$r_{max} = 2r^* + \frac{2A_0\sqrt{r^*}}{\sqrt{C_4^0}}$$

6.10.3 *Crystal Size and Growth Rate at Diffusion-Controlled Growth as Functions of Time*

The crystal size and the growth rate as functions of time can be calculated by integration of Equations (6.138) and (6.162), respectively, i.e. in the simplest and the most complete case above.

Simplest Case. No Corrections

Integration of Equation (6.138) at page 328

$$\int_{r_0}^{r_c} r_c \, \mathrm{d}r_c = D \frac{x_{eq}^{L/\alpha} - x_0^L}{x_{eq}^{L/\alpha} - x_{eq}^{\alpha/L}} \frac{V_m^\alpha}{V_m^L} \int_0^t \mathrm{d}t \tag{6.164}$$

gives

$$r_c^2 - r_0^2 = 2D \frac{x_{eq}^{L/\alpha} - x_0^L}{x_{eq}^{L/\alpha} - x_{eq}^{\alpha/L}} \frac{V_m^\alpha}{V_m^L} t \tag{6.165}$$

where r_0 is the initial size of the crystal radius. r_0 is small and can often be neglected. The relationship between the radius and the time is therefore parabolic.

If we neglect r_0 and solve r_c in Equation (6.165) we obtain

$$r_c = \sqrt{2D \frac{x_{eq}^{L/\alpha} - x_0^L}{x_{eq}^{L/\alpha} - x_{eq}^{\alpha/L}} \frac{V_m^\alpha}{V_m^L}} \sqrt{t} \tag{6.166}$$

Taking the time derivative of Equation (6.166) gives the growth rate of the crystal as a function of time

$$V_{growth} = \frac{\mathrm{d}r_c}{\mathrm{d}t} = \frac{1}{2} \sqrt{2D \frac{x_{eq}^{L/\alpha} - x_0^L}{x_{eq}^{L/\alpha} - x_{eq}^{\alpha/L}} \frac{V_m^\alpha}{V_m^L}} \sqrt{\frac{1}{t}} \tag{6.167}$$

Example 6.5

Cast iron is sometimes inoculated with graphite. Consider a cast iron melt with a carbon content of 3.2 wt%. The temperature of the melt is 1200 °C. The melt contains graphite nodules with an average radius of 1.0 mm. How long time does it take before all the graphite nodules have disappeared completely?

Material constants can be found in standard tables. The phase diagram of the system Fe–C is also available in standard publications.

Solution

The graphite nodules are supposed to float freely in the melt. Hence, it is reasonable to assume that the effective diffusion distance is at least equal to the size of the graphite nodules. The radius r_c of the is known and Equation (6.165) can be used to find the time when they disappear completely.

The concentrations are given in weight per cent and not in mole fraction, which is a difficulty. However, the two units will be approximately proportional because graphite is a light element compared to iron. Hence, we use the approximate formula (6.165) solved for t

$$t \approx \frac{1}{2D} \frac{c_{eq}^{L/\alpha} - c_{eq}^{\alpha/L}}{c_{eq}^{L/\alpha} - c_0^L} \frac{\rho^\alpha}{\rho^L} \left(0 - r_0^2\right) \tag{1'}$$

Material constants:
$D_{carbon} = 2.5 \times 10^{-8}\ m^2/s$
$r_0 = 3.0 \times 10^{-3}\ m$
$\rho_L \approx \rho_{Fe} = 7.0 \times 10^3\ kg/m^3$
$\rho_{graphite} = \rho_\alpha = 2.2 \times 10^3\ kg/m^3$

The molar volumes are equal to the inverted values of the densities, which have been inserted into Equation (1′).

The material constants, found in tables, are listed in the margin.
The solid α phase consists of pure graphite. Therefore, $c_{eq}^{\alpha/L} = c^{graphite} = 100\ wt\%$.
The initial concentration of carbon in the melt is given in the text: $c_0^L = 3.2\ wt\%$.
To find $c_{eq}^{L/\alpha}$ we use the phase diagram for the system Fe–C.

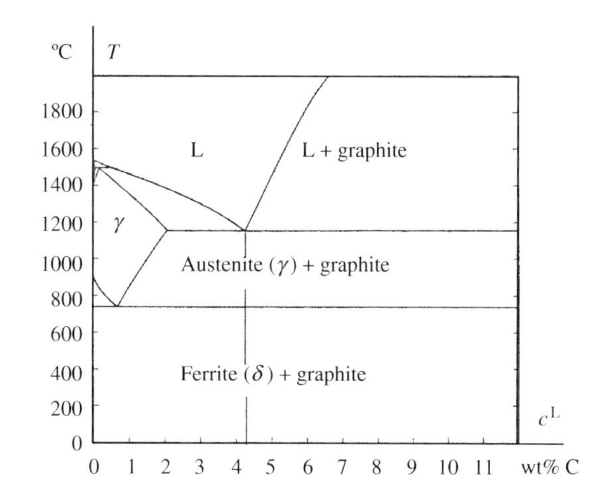

Figure 1 Phase diagram of the system Fe–C. From Binary Alloy Phase Diagrams Vol. 1, Ed. Thaddeus B. Massalski, American Society for Metals 1986. Reprinted with permission of ASM International. All rights reserved. www.asminternational.org

In the phase diagram Fe–C we draw the line $T = 1200\ °C$. The intersection between this line and the equilibrium curve between the melt and the solid graphite phase corresponds to

$$c_{eq}^{L/\alpha} = 4.3\ wt\%.$$

Inserting all these values into Equation (1′) we obtain

$$t = \frac{1}{2 \times 2.5 \times 10^{-8}} \frac{4.3 - 100}{4.3 - 3.2} \frac{2.2 \times 10^3}{7.0 \times 10^3} (-0.0010^2) \frac{1}{60} = 9.0\ min$$

Answer: All graphite nodules will disappear in about 9 min.

Case with Combined Corrections for Surface Tension and Kinetics at the Crystal Surface

To find a relationship between the crystal radius and time in the most complete case we have to integrate Equation (6.162)

$$V_{growth} = \frac{dr_c}{dt} = 2A^2 + C_4 - 2A\sqrt{A^2 + C_4} \tag{6.168}$$

Inserting the expressions A_0 and C_4^0 on page 334 into Equation (6.168) gives

$$\frac{dr_c}{dt} = 2\left(\frac{D}{2}\sqrt{\frac{1}{\beta_{kin}}\frac{V_m^\alpha}{V_m^L}\frac{1}{x_{eq}^{L/\alpha} - x_{eq}^{\alpha/L}}} \times \frac{1}{r_c}\right)^2 + D\frac{V_m^\alpha}{V_m^L}\frac{x_{eq}^{L/\alpha} - x_0^L}{x_{eq}^{L/\alpha} - x_{eq}^{\alpha/L}} \times \frac{1}{r_c}\left(1 - \frac{r^*}{r_c}\right)$$
$$- D\sqrt{\frac{1}{\beta_{kin}}\frac{V_m^\alpha}{V_m^L}\frac{1}{x_{eq}^{L/\alpha} - x_{eq}^{\alpha/L}}}$$
$$\times \frac{1}{r_c}\sqrt{\left(\frac{D}{2}\sqrt{\frac{1}{\beta_{kin}}\frac{V_m^\alpha}{V_m^L}\frac{1}{x_{eq}^{L/\alpha} - x_{eq}^{\alpha/L}}} \times \frac{1}{r_c}\right)^2 + D\frac{V_m^\alpha}{V_m^L}\frac{x_{eq}^{L/\alpha} - x_0^L}{x_{eq}^{L/\alpha} - x_{eq}^{\alpha/L}} \times \frac{1}{r_c}\left(1 - \frac{r^*}{r_c}\right)} \tag{6.169}$$

or after introduction of the summarizing function $f(r_c)$ and integration

$$t = \int_0^t dt = \int_{r_0}^{r_c} \frac{dr_c}{f(r_c)} \tag{6.170}$$

Equation (6.170) represents the time t as a function of the crystal radius r_c. It is practically impossible to solve r_c as an explicite function of t but the relationship between r_c and t can be obtained by numerical integration. The result is shown graphically in Figure 6.49. It shows the radius of a crystal as a function of time for different values of the kinetic coefficient β_{kin}.

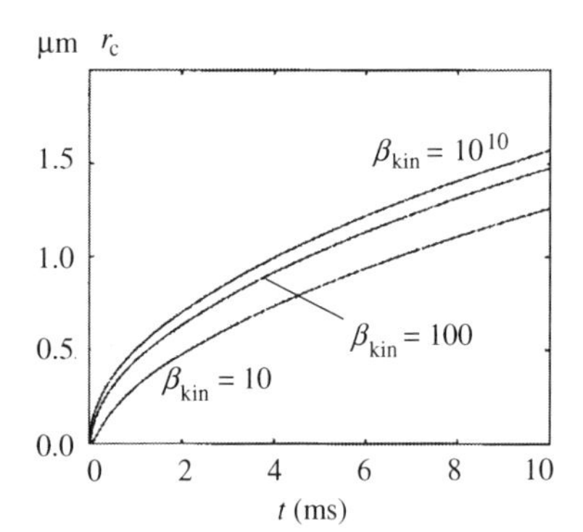

Figure 6.49 The radius of a crystal as a function of time. The kinetic coefficient β_{kin} is a parameter. All values except β_{kin} are the same as in Figure 6.48.

The figure shows that r_c is approximately a parabolic function of t for large values of β_{kin}.

6.11 Impingement

In most precipitation processes, such as crystallization and formation of droplets in vapours and liquids, numerous particles are formed simultaneously instead of a single separate particle. Their compositions and also their growth rates are influenced by nearby crystals or droplets. This influence is called *impingement*.

In this section we will discuss impingement of alloy particles and study the distribution of the alloying element in particles which may be solid spherical crystals or droplets of another liquid than the parent liquid. Many models have been used to describe impingement. One of the simpler models is that used below.

6.11.1 Theory of Impingement

The discussion will be restricted to the two simplest cases, diffusion-controlled particle growth.

The discussion will be performed in terms of solid crystals but it should be kept in mind that the theory also is valid for precipitated liquid droplets.

Diffusion-Controlled Particle Growth without Nucleation in a Slightly Supersaturated Liquid

Consider a slightly supersaturated liquid binary alloy with the components A and B. The partition constant is assumed to be larger than 1 ($k > 1$), which means that the mole fraction of the alloying element B is larger in the growing crystals than in the liquid. After formation in the liquid, the nuclei will probably obtain a spherical shape and their growth will be diffusion-controlled. We also assume that the liquid alloy contains N spherical particles per unit volume, formed at the time $t = 0$. No further crystals are formed.

At a minor supersaturation and in the absence of impingement the growth rate can be described by Equation (6.126) on page 325. It can be written in the following way

$$\frac{\mathrm{d}r_c}{\mathrm{d}t} = \frac{D}{r_c}\,\frac{x_0^{L} - x_{eq}^{L/\alpha}}{x_{eq}^{\alpha/L} - x_{eq}^{L/\alpha}}\,\frac{V_m^{\alpha}}{V_m^{L}} \qquad (6.171)$$

where

r_c = radius of B-rich spherical particles
V_m^{α} = molar volume of the spherical particles (α phase)
V_m^{L} = molar volume of the liquid (L phase)
x_0^{L} = initial mole fraction of B in the liquid
$x_{eq}^{L/\alpha}$ = mole fraction of B in the liquid in equilibrium with the solid α phase
$x_{eq}^{\alpha/L}$ = mole fraction of B in the solid α phase in equilibrium with the liquid.

If there are so many particles per unit volume that there is an impingement between the particles, the distribution profile of solute B will change compared to normal diffusion-controlled growth.

A comparison between Figures 6.50a and b shows that the concentration of solute x_0^{L} in the homogenous liquid will decrease during the precipitation of B-rich crystals. This means that the supersaturation decreases and the growth rate also decreases with time. Therefore, the mole fraction in the liquid x_0^{L} will not keep constant and we change its designation to $x^{L}(t)$ in our further calculations.

$x^{L}(t)$ can be derived from the following material balance

$$\underbrace{N\frac{4\pi r_c^3}{3}\,\frac{x_{eq}^{\alpha/L} - x_{eq}^{L/\alpha}}{V_m^{\alpha}}}_{\text{amount of solute in the particle}} = \underbrace{\left(1 - N\frac{4\pi r_c^3}{3}\right)\frac{x_0^{L} - x^{L}(t)}{V_m^{L}}}_{\text{amount of solute from the melt}} \qquad (6.172)$$

where

N = the number of crystals per unit volume
$x^{L}(t)$ = average mole fraction of B in the liquid at time t.

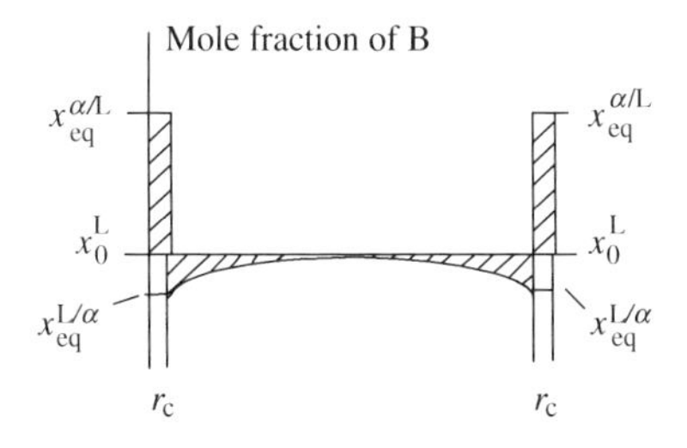

Figure 6.50a Concentration of solute B during growth of (half) a spherical particle in the absence of impingement as a function of distance from the centre of the particle.

The material balance requires that the shaded areas above and under the x_0^{L} line are equal.

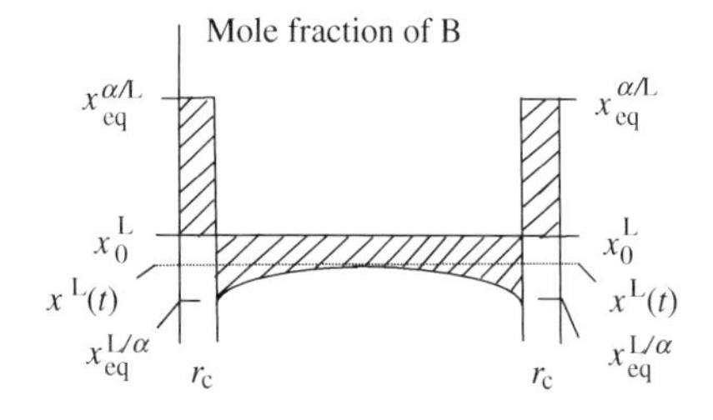

Figure 6.50b Concentration of solute B at time t during growth of a spherical particle in the presence of impingement, as a function of distance from the centre of the particle. The liquid will be depleted of the alloying element and its mole fraction of B will decrease. The material balance requires that the shaded areas above and under the x_0^{L} line are equal.

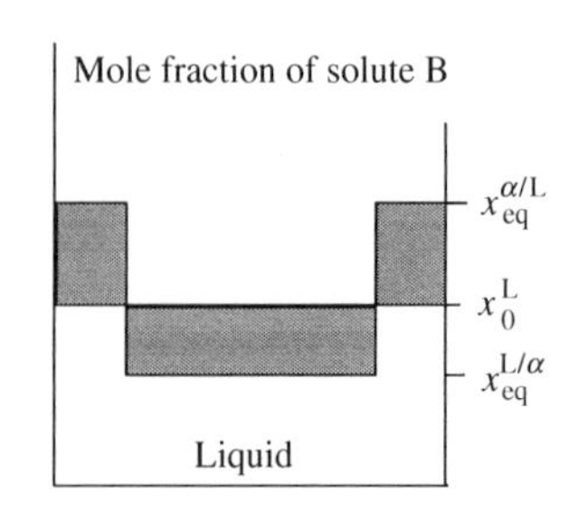

Figure 6.51 Concentration profile in a precipitated crystal in the presence of many large crystals per unit volume.

The material balance requires that the shaded areas above and under the x_0^L line are equal.

If x_0^L in Equation (6.171) is replaced by $x^L(t)$, solved from Equation (6.172), we obtain

$$\frac{dr_c}{dt} = \frac{D}{r_c}\frac{V_m^\alpha}{V_m^L}\left(\frac{x_0^L - x_{eq}^{L/\alpha}}{x_{eq}^{\alpha/L} - x_{eq}^{L/\alpha}} - \frac{V_m^L}{V_m^\alpha}\frac{N\dfrac{4\pi r_c^3}{3}}{1 - N\dfrac{4\pi r_c^3}{3}}\right) \tag{6.173}$$

A comparison between Equations (6.173) and (6.171) shows, that if the second term in the brackets in Equation (6.173) is neglected, the equations will be identical. It is relevant to do this simplification if there are *very few* and/or *very small* particles.

If there are *many large* particles the growth rate of each particle will be small and the left-hand side of Equation (6.173) becomes approximately zero. This implies that the crystal growth ceases and the particle size will be determined by the number of particles, i.e. by the lever rule, which is equal to Equation (6.172) if $x^L(t)$ is substituted by $x_{eq}^{L/\alpha}$. This approximation is reasonable when the difference $x_{eq}^{L/\alpha} - x_0^L$ is small. This case corresponds to the concentration profile given in Figure 6.51.

As an example on impingement the precipitation process of droplets in a Zn-Bi alloy is given below. Figure 6.52 shows a photo of the microstructure of a Zn alloy with 0.03 at% Bi. The figure shows Bi-rich areas in a Zn-rich matrix.

The phase diagram of the Zn-Bi system is given in Figure 6.53. At higher temperatures than 770 K the liquid phase is homogenous. When the temperature decreases, the homogenous liquid phase splits up into two separate liquid phases, i.e. Bi-rich droplets are precipitated in a Zn-rich matrix.

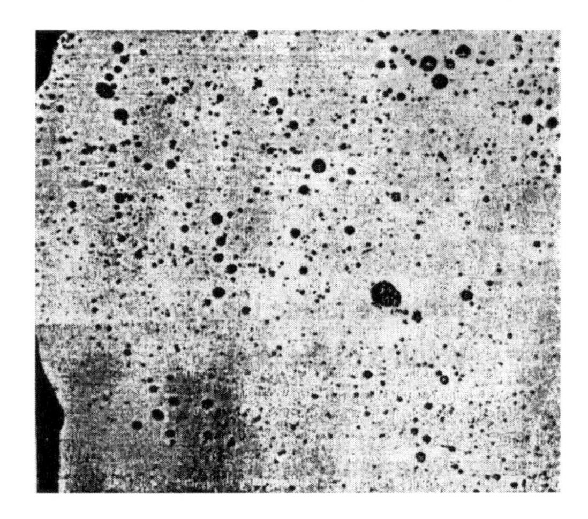

Figure 6.52 Precipitated Bi-rich droplets in a Zn-rich liquid.

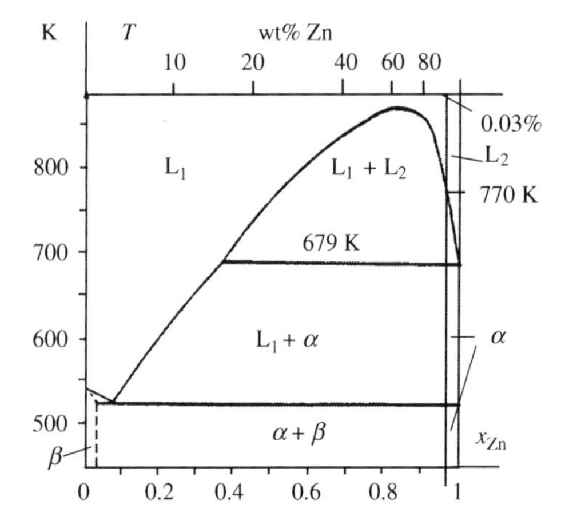

Figure 6.53 The phase diagram of the system Zn–Bi. From selected values of the Thermodynamic Properties of Binary Alloys, R. Hultgren et al, American Society for Metals 1973. Reprinted with permission of ASM International. All rights reserved. www.asminternational.org.

6.12 Precipitation of Pores

Sometimes the formation of crystals in melts are accompanied by a simultaneous precipitation of gas. This phenomenon is sometimes a severe complication in casting industry (Chapter 9 in [14].

Pores are often observed in cast metals. They may form during cooling of the melt and/or during the solidification process. In general, various types of pores can be related to the gas content solved in the melt before casting.

The pores can be classified into three groups:

- rounded pores
- elongated pores
- shrinkage pores.

Rounded pores form in the melt during cooling when the gas concentration in the melt is larger than the maximum concentration of the gas which can be solved in the melt at the present temperature. The pores are normally spherical and usually located in the upper sections of the castings, owing to the buoyancy forces.

Elongated pores form when the initial gas concentration in the melt is less than the maximum concentration of the gas, which can be solved in the melt. This type of pores may nucleate at the solidification front, owing to enrichment of the gas and will grow by a eutectic-like reaction.

Shrinkage pores form in the last parts of the melt during the solidification process. Their shapes are influenced by the dendrite structure. Shrinkage pores form owing to the pressure drop caused by solidification shrinkage.

The discussion below will be restricted to the growth of rounded pores, owing to dissolved gas in the melt.

6.12.1 Growth of Rounded Pores in Melts

Rounded pores form in the melt when the gas concentration in the melt is larger than the solubility of the gas at the actual temperature. The size of the pores is determined by the concentration of the dissolved gas in the melt at various temperatures and the difference in solubility of the gas in the melt and the solid phase, respectively.

Consider a pore in a melt at distance h from its upper surface (Figure 6.54). The pressure of the pore is determined by the surface tension forces and the compressive forces in the melt according to the equation

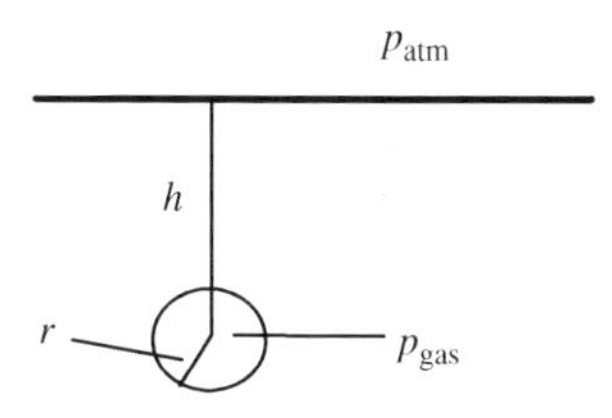

Figure 6.54 Gas pore in a liquid, close to the surface.

$$p_{gas} = p_{atm} + \rho\, gh + \frac{2\sigma}{r} \tag{6.174}$$

where

p_{gas} = total gas pressure inside the pore
p_{atm} = atmospheric pressure (**not** partial pressure of the gas)
h = distance from the pore to the free surface of the melt
σ = surface tension between melt and gas
r = radius of the spherical pore.

In order to find the growth rate of the pore we assume that the pore is situated close to the surface of the melt and neglect the second term in Equation (6.174).

$$p_{gas} = p_{atm} + \frac{2\sigma}{r} \tag{6.175}$$

We also assume that the gas is diatomic and practically totally dissociated at the surface of the melt, owing to the high temperature.

$$G_2(g) \leftrightarrow 2G$$

In the pore the gas has primarily precipitated as diatomic molecules. We use the general gas law to calculate the number of kmole of the gas

$$n_{G_2} = \frac{p_{gas} V_{pore}}{RT} \tag{6.176}$$

where

p_{gas} = total pressure inside the pore
V_{pore} = volume of the pore
R = gas constant
T = absolute temperature.

Combining Equations (6.175) and (6.176) we can calculate the number of kmole of gas *atoms* G in the pore as a function of the pore radius:

$$n_G = 2n_{G_2} = \frac{2\left(p_{atm} + \frac{2\sigma}{r}\right)\frac{4\pi r^3}{3}}{RT} \tag{6.177}$$

If we take the derivative of Equation (6.177) with respect to time we obtain a relationship between the increase of the number of kmoles of atoms per unit time in the pore and the increase of the radius of the pore per unit time

$$\frac{dn_G}{dt} = \frac{8\pi}{3R}\frac{3p_{atm}r^2 + 4\sigma r}{T}\frac{dr}{dt} \tag{6.178}$$

We can obtain an alternative expression of the increase of the number of kmoles of gas atoms per unit time by considering the diffusion of gas atoms from the melt into the pore. We use Fick's first law

$$\frac{dm}{dt} = -DA\frac{dc}{dy} \tag{6.179}$$

The increase of mass per unit time is proportional to the cross-section area and the concentration gradient dc/dy. We apply the law ($y = r$) and notice that the pore grows with a rate proportional to the concentration gradient. Gas atoms diffuse into the pore.

$$\frac{dn_G}{dt} = -D_G \times 4\pi r^2 \frac{x_G^L - x_G^I}{V_m r} \tag{6.180}$$

where

n_G = number of kmoles of gas atoms in the pore
V_m = molar volume
D_G = diffusion constant of the gas atoms in the melt
x_G^L = the initial concentration of gas atoms in the melt, expressed in mole fractions
x_G^I = concentration of gas atoms at the interface melt/pore.

We combine Equations (6.178) and (6.180) and solve dr/dt.

$$\frac{dr}{dt} = \frac{3D_G}{2V_m}\frac{x_G^L - x_G^I}{3p_{atm}r + 4\sigma}RT \tag{6.181}$$

Equation (6.181) is the desired expression for the growth rate of rounded pores. Equation (6.181) is useful only if the gas concentrations in the melt and in the pore are known.

x_G^L is the initial concentration of the gas atoms in the melt. x_G^I is strongly temperature dependent and can be calculated from the relationship

$$x_{gas}^I = A\sqrt{\frac{p_{atm} + \frac{2\sigma}{r}}{p_0}}\,e^{-\frac{B}{T}} \tag{6.182}$$

where

A, B = constants
p_0 = standard pressure of the gas, usually 1 atm.

The theory given above is important to casting industry, where the precipitation of gas pores in some cases causes severe problems, for example

- hydrogen in steel and iron alloys
- hydrogen in aluminium alloys
- nitrogen in steel and iron alloys
- oxygen in steel and iron alloys
- carbon monoxide in steel and iron alloys.

Several methods have been developed to prevent formation of pores in metal melts ([14] in Chapter 9).

Summary

■ Crystal Growth in Liquids and Melts

Crystals are produced by cooling a liquid or melt below its melting point, which gives an undercooled or supersaturated liquid. The type of solidification mode is determined by the cooling conditions and by the properties of the element or alloy.

Modern methods of crystal growth in a crucible are the Bridgman method (crucible moved through a temperature gradient region) and the dragging method to produce single crystals.

Zone melting is a very common method to refine crystals, for example in the metal and semiconductor industries.

■ Mechanisms of Crystal Growth in Liquids and Melts

The mechanisms of crystal growth in liquids depend strongly on the structure of the solid/liquid interface. Depending on the surface structure normal growth on rough surfaces and layer growth on smooth surfaces occur.

The contact between the solid and the liquid causes strain across and along the interface during the crystal growth. The stress forces, caused by the strain, vary with the crystallographic direction as they depend on the packing of the different crystal planes.

In terms of the Chadwick model, strain arises owing to the difference in atomic diameter of the metal atoms in the solid and the melt, respectively. In alloys the solid and the melt differ in composition, which causes strain. Surface tension between the melt and the crystal has also to be considered.

■ Deformation Properties of Solid Materials

Stress: $\sigma = F/A$ Strain: $\varepsilon = \Delta l/l$

Hooke's law:

$$\sigma_{\text{tension}} = Y\varepsilon$$

Deformation energy per unit volume:

$$w_{\text{el}} = \frac{Y\varepsilon^2}{2}$$

The Gibbs' free energy change per kmol owing to the elastic strain:

$$-\Delta G_{\text{el}}^{\text{I}} = Y\varepsilon^2 \frac{M}{\rho}$$

Elastic and Plastic Materials
Ideal elastic materials
are characterized by

- the strain is proportional to the applied stress
- the applied stress which instantly gives the corresponding strain
- when the stress is removed the strain disappears instantly and goes back to zero.

Anelastic materials
So-called anelastic materials show the same properties as ideal elastic materials but with the difference that part of the strain is instantaneous and the rest is delayed, both at application and removal of the stress.

Plastic and viscoelastic materials
Plastic and linear viscoplastic materials do not return to their original shape when the stress is removed. There will be a permanent deformation.

A material is often elastic for small values of the stress and becomes plastic beyond a characteristic value of the stress, the yield stress, which is a material constant.

■ Volume Changes during Crystal Growth

The volume differences between the melt and the crystal causes primarily instability over the interface. Adaption to a stable state occurs by *deformation.* There are three possible mechanisms of deformation:

- elastic deformation
- creep
- formation of vacancies.

Elastic Deformation
Strain energy per unit volume: $w_{\mathrm{el}} = \dfrac{Y\varepsilon^2}{2}$

Strain energy per kmol: $-\Delta G_{\mathrm{el}}^{\mathrm{I}} = Y\varepsilon^2 \dfrac{M}{\rho}$

Creep
Creep is a plastic deformation, which moves through a material.
Creep can occur in two ways:

- by generation of vacancies followed by condensation of vacancies
- by rearrangement of atoms, where several atoms change their positions simultaneously.

The Gibbs' free visco-plastic creep energy per kmol:

$$-\Delta G_{\mathrm{creep}}^{\mathrm{I}} = \eta\varepsilon^2 \frac{V_{\mathrm{growth}}}{d} \frac{M}{\rho}$$

Formation of Vacancies
A rough interphase contains many vacancies. If a vacant site in the interphase is filled with an atom from the crystal lattice a vacancy will be formed within the crystal.

Equilibrium concentration of vacancies as a function of temperature:

$$x_{\mathrm{vac}}^{\mathrm{eq}} = \mathrm{e}^{-\frac{\Delta H_{\mathrm{vac}}^{\mathrm{a}} - T\,\Delta S_{\mathrm{vac}}^{\mathrm{a}}}{k_B T}} = \mathrm{e}^{-\frac{\Delta G_{\mathrm{vac}}^{\mathrm{a}}}{k_{\mathrm{B}} T}}$$

Two processes occur to overcome strain caused by the volume differences between the solid and the liquid

- vacancy formation
- visco-elastic relaxation.

Visco-elastic relaxation is the most probable mechanism at very low growth rates, otherwise vacancy formation with vacancy creep. The transition growth rate is found at the intersection point between the visco-elastic relaxation and vacancy creep in the $-\Delta G^{I} - V_{growth}$ curves.

■ Relaxation

Definition of relaxation time

$$\delta = \delta_0 \, e^{-\frac{t}{\tau}}$$

Relaxation Processes during Anelastic Crystal Growth in Metal Melts

Liquid atoms have higher potential energy than solid atoms. The process of transferring liquid atoms into solid atoms in their lowest potential energy state can be described as a two-step relaxation process, one at the interphase liquid-solid and one in the solid. The former is much faster than the latter owing to the presence of 'liquid' atoms.

The rearrangement of the atoms in the solid can be described as a relaxation process in terms of energy deviation from the equilibrium state.

$$\Delta E_{pot} = \Delta E_{pot}^{0} \times C e^{-\frac{t}{\tau}}$$

One way to describe the potential energy deviation is to regard the change of the heat of solidification.

Relaxation of the Interphase

The solution of the differential equation

$$\frac{\partial(-\Delta H)}{\partial t} = D_{relax} \frac{\partial^2(-\Delta H)}{\partial y^2}$$

can be written as

$$-\Delta H = \frac{(-\Delta H_0) \times (1 \times \delta)}{\sqrt{\pi D_{relax}}} \frac{\exp(-y^2/4Dt)}{\sqrt{t}}$$

which results in the strain energy per unit volume

$$-\Delta H_{strain} = -\Delta H_0 \sqrt{\frac{V_{growth}\delta}{\pi D_{relax}}}$$

■ Normal Crystal Growth at Rough Surfaces in Pure Metals

Driving Force of Solidification

$$-\Delta G_a^{L \to s} = -(G_a^s - G_a^L) = G_a^L - G_a^s = -\Delta G_a$$

The driving force of solidification is proportional to the undercooling of the melt.

$$-\Delta G_{\mathrm{a}}^{\mathrm{L}\rightarrow\mathrm{s}} = (-\Delta H_{\mathrm{a}}^{\mathrm{fusion}})\frac{T_{\mathrm{M}} - T}{T_{\mathrm{M}}}$$

Rate of Normal Growth

Solidification of a 'liquid' atom means that it becomes adsorbed at the surface and incorporated into the crystal lattice. The probability of adsorption of a 'vapour' atom with the area a^2 into kinks with the average size $\overline{\delta}^2$ has been described as the ratio $(a/\overline{\delta})^2$ times a Boltzmann factor.

In the case of solidification of a 'liquid' atom the atomic process in the interphase has been treated as a relaxation process. The relationship, which describes the relaxation process can be used to define the probability which can replace the factor $(a/\overline{\delta})^2$. In this case the fraction of atoms which have the potential energy E_{pot} can be described by the relationship

$$\left(\frac{a}{\overline{\delta}}\right)^2 = P = f^{\mathrm{s}} = \frac{E_{\mathrm{pot}}}{E_{\mathrm{pot}}^0} = \sqrt{\frac{V_{\mathrm{growth}}\delta}{\pi D_{\mathrm{relax}}}}$$

Rate of normal growth ($-\Delta G_{\mathrm{a\ interphase}}^{\mathrm{L}\rightarrow\mathrm{s}} \ll k_B T$):

$$V_{\mathrm{growth}} = \frac{a^3}{a^2}\nu P \mathrm{e}^{-\frac{U_{\mathrm{a\ act}}}{k_{\mathrm{B}}T}} \frac{-\Delta G_{\mathrm{a\ interphase}}^{\mathrm{L}\rightarrow\mathrm{s}}}{k_{\mathrm{B}}T}$$

The activation energy U_{a} can be determined by plotting ln V_{growth} versus $1/T$.

The growth rate is often is presented as

$$V_{\mathrm{growth}} = \frac{D_{\mathrm{interphase}}}{\delta}\frac{-\Delta G_{\mathrm{interphase}}^{\mathrm{L}\rightarrow\mathrm{s}}}{RT}P$$

Total Driving Force of Solidification with Consideration to Relaxation

The total driving force consists of two parts, one for rearranging the 'liquid' atoms into an ordered solid lattice, and the other for the relaxation process in the solid to remove strain. The total driving force can be written as

$$-\Delta G^{\mathrm{total}} = (-\Delta G_{\mathrm{interphase}}^{\mathrm{L}\rightarrow\mathrm{s}}) + (-\Delta G_{\mathrm{relax}}^{\mathrm{s}})$$

or with the aid of earlier expressions and reduction

$$-\Delta G^{\mathrm{total}} = \left[\frac{RT}{D_{\mathrm{interphase}}}\pi D_{\mathrm{relax}} + \left(-\Delta G^{\mathrm{total}}\right)\right]\sqrt{\frac{V_{\mathrm{growth}}\delta}{\pi D_{\mathrm{relax}}}}$$

where

$$P = \sqrt{\frac{V_{\mathrm{growth}}\delta}{\pi D_{\mathrm{relax}}}}$$

Kinetic Coefficient of the Growth Rate in Pure Metals

Relationship between V_{growth} and undercooling:

$$V_{\mathrm{growth}} = \beta_{\mathrm{kin}} \times (\Delta T)^2$$

Kinetic coefficient of the growth rate:

$$\beta_{kin} = \frac{(-\Delta H_0)^2 D_{interphase}^2}{\pi\delta(RTT_M)^2 D_{relax}}$$

Heat of Solidification

The total heat of solidification equals $(-H_0)$ but only the fraction $[(-\Delta H_0) - (-\Delta H_{strain})]$ is released promptly and measured as heat of solidification.

$$(-\Delta H_0) = (-\Delta H_{prompt}) + (-\Delta H_{strain})$$

The remaining heat $(-H_{strain})$, that is initially bound as strain energy and successively is released at relaxation, is

$$-\Delta H_{strain} = -\Delta H_{released} = -\Delta H_0\sqrt{\frac{V_{growth}\delta}{\pi D_{relax}}}$$

Measured heat of solidification:

$$-\Delta H_{prompt} = -\Delta H_0\left(1 - \sqrt{\frac{V_{growth}\delta}{\pi D_{relax}}}\right)$$

■ Layer Crystal Grooth at Smooth Surfaces

- Crystal growth on smooth surfaces is more complex than that on rough surfaces both in vapours and liquids.
- The crystal surface is terraced and consists of steps of equal length and height. The height of the steps can be one or several atomic distances. Layers are formed by 'two-dimensional' nucleation or most frequently by spiral growth owing to screw dislocations.
- The terraces are smooth while the step edges have rough interphases. The layers grow laterally at the step edges.

The main differences between layer growth in vapours and liquids are:

- The growth rate is a function of temperature or supersaturation instead of pressure.
- The strain, owing to surface tension in the liquid and difference in atom size between solid atoms and liquid atoms, has to be taken into consideration.

■ Chadwick's Hard Sphere Model

Chadwick's hard sphere model illustrates crystal growth in a very concrete way. He introduced new concepts of atomic density and studied among other things the growth rate of FCC crystals as a function of orientation.

Surface-Packing Parameter

ρ_S (hkl) = the fraction of the area of an (hkl) plane which is occupied by atoms in the plane.

Volume-Packing Parameter

ρ_V = the packing density of atoms within the unit cell.

$$\rho_V = \frac{\frac{4}{3}\pi r^3 Z}{V_c}$$

Reticular Density

$D_S(\mathrm{hkl})$ = the number of atoms per unit area of a crystal plane.

Reticular density is related to the surface-packing parameter ρ_S by the relationship

$$D_S(\mathrm{hkl}) = \frac{\rho_S(\mathrm{hkl})}{\pi r^2}$$

The difference in size of the atoms in the solid and liquid phases are taken into account when reticular densities are used.

$$\frac{D_S^L}{D_S^s(\mathrm{hkl})} = \frac{\rho_S^L}{\rho_S^s(\mathrm{hkl})}\left(\frac{r_s}{r_L}\right)^2$$

Influence of Orientation on Layer Growth

The rate of crystal growth depends on orientation.
If this were not true, all crystals would be spherical.

$$\text{If} \quad \frac{D_S^L}{D_S^s(\mathrm{h_1k_1l_1})} < 1$$

the strain of the interface causes *pressure* forces, which *contract* the melt and *expand* the solid.

$$\text{If} \quad \frac{D_S^L}{D_S^s(\mathrm{h_1k_1l_1})} > 1$$

the strain of the interface causes *tension* forces, which *expand* the melt and *contract* the solid.

Owing to these conditions a density barrier to crystal growth exists in some directions, which makes crystal growth faster in some crystallographic directions than in others.

Influence of Relaxation on Growth Rate in Different Growth Directions

The driving force of solidification consists of two parts, one for the interphase solidification and one for the rearrangement of the atoms in the solid.

$$-\Delta G^{\mathrm{total}} = \left(-\Delta G_{\mathrm{interphase}}^{L\rightarrow s}\right) + \left(-\Delta G_{\mathrm{relax}}^{s}\right)$$

where

$$-\Delta G_{\mathrm{relax}} = \left(-\Delta G^{\mathrm{total}}\right)\sqrt{\frac{V_{\mathrm{growth}}\delta}{\pi D_{\mathrm{relax}}}}$$

Obviously

- A high value of D_{relax} corresponds to a low relaxation energy and consequently to a high driving force over the interphase and vice versa.

As the diffusion coefficient depends on orientation this is also true for the relaxation energy, the driving force over the interphase and the growth rate.

■ Origin of Layer Growth. Rate of Layer Growth in Case of Strain. Length and Height of Steps at Layer Growth

The energy law is the basic law that determines the structure of the crystal.

- The configuration which gives the lowest possible total energy of the interface is the most stable state of the surface and the one, which is formed in reality.

An energy analysis confirms that the most likely state in the case of layer growth is a terraced surface of the crystal. It is the strain pattern of the crystal planes, which gives the terraced structure. The terrace surfaces of the steps have a strain pattern of type I (smooth surface) and the step edge surfaces belong to type II (rough surface).

Rate of Step-Edge Growth

The step edge is a rough surface and the growth rate corresponds to normal growth:

$$V_{\text{growth}} = \frac{a^3 \nu}{a^2} P \, e^{-\frac{U_{\text{a act}}}{RT}} \frac{-\Delta G^{\text{L} \to \text{s}}}{RT}$$

Length and Height of Single Steps

Crystals in vapours and liquids are strained because the vapour atoms, the liquid atoms and the adsorbed atoms (adatoms) differ in size compared to the lattice atoms.

It is possible to calculate the length and height of the steps at layer growth of crystals in liquids. The method is based on application of the relaxation theory, the strain conditions and the values of the reticular densities at the interface crystal/liquid.

A crystal with FCC structure is used as an example.

Condition for Growth of the Rough Step Edge

– The relaxation energy, stored in the (111) plane, must come down to the same level of relaxation energy, stored in the growing in (100) plane, before the step can grow along the (111) plane.

Application of this condition gives an equation that can be simplified to

$$\sqrt{\frac{\lambda D_{\text{relax}}^{111}}{\delta D_{\text{relax}}^{100}}} = \exp \frac{-\lambda V_{\text{growth}}^{100}}{4 D_{\text{relax}}^{111}}$$

Condition for Growth of the Smooth Step Edge

– The relaxation energy, stored in the (100) plane, must come down to the same level of relaxation energy, stored in the growing in (111) plane, before the step can grow along the (100) plane.

Application of this condition on the smooth edge gives the equation that can be simplified to

$$\sqrt{\frac{h D_{\text{relax}}^{100}}{\delta D_{\text{relax}}^{111}}} = \exp \frac{-h V_{\text{growth}}^{111}}{4 D_{\text{relax}}^{100}}$$

The step length λ and the step height h can be determined from the above equations.

After relaxation the smooth interface is free from strain and a new growth process starts and new steps are successively formed.

■ Crystal Growth in Binary Alloys

The solidification process in alloys is more complex than that in pure metals. The complication is that the solid phase has a different composition than the liquid.

Driving Force of Solidification

$$-\Delta G^{L\rightarrow\alpha} = (1-x^{\alpha})\left(-\Delta G_{A}^{L\rightarrow\alpha}\right) + x^{\alpha}\left(-\Delta G_{B}^{L\rightarrow\alpha}\right)$$

where

$$-\Delta G_{A}^{L\rightarrow\alpha} \approx RT\left[\left(x_{eq}^{L/\alpha} - x_{0}^{L}\right) - \left(x_{eq}^{\alpha/L} - x^{\alpha}\right)\right]$$

$$-\Delta G_{B}^{L\rightarrow\alpha} \approx RT\left(\frac{x_{eq}^{\alpha/L}x_{0}^{L}}{x_{eq}^{L/\alpha}x^{\alpha}} - 1\right)$$

■ Normal Crystal Growth at Rough Surfaces in Binary Alloys

Total Driving Force of Solidification

The driving force of solidification controls the growth rate of the solidification process and has a strong influence on the structure of the solidified material. The larger the driving force is, the smaller will be the crystals and the better will be the properties of the material.

The solidification process consists of two main processes, one at the interphase and one (relaxation) in the solid. Each of them consist of two subprocesses, owing to diffusion of the A and B atoms of the alloy. The total driving force can be written as

$$-\Delta G^{total} = -\Delta G_{overall\ int}^{L\rightarrow\alpha} + \left(-\Delta G_{relax}^{\alpha}\right)$$

where

$$-\Delta G_{overall\ int}^{L\rightarrow\alpha} = -\Delta G_{diff\ int}^{L\rightarrow\alpha} + \left(-\Delta G_{kin\ int}^{L\rightarrow\alpha}\right)$$

where

$$-\Delta G_{diff\ int}^{L\rightarrow\alpha} = (1-x^{\alpha})\left(-\Delta G_{A\ diff\ int}^{L\rightarrow\alpha}\right) + x^{\alpha}\left(-\Delta G_{B\ diff\ int}^{L\rightarrow\alpha}\right)$$

and

$$-\Delta G_{kin\ int}^{L\rightarrow\alpha} = \frac{V_{growth}\delta\, RT}{D_{interphase}P}$$

Rate of Normal Growth

The Kinetic Process. Mass Transport over the Interphase

Diffusion of A and B atoms due to Concentration Differences in the Interphase:

$$V_{growth} = M_{A}^{diff}\frac{-\Delta G_{A\ diff\ int}^{L\rightarrow\alpha}}{\delta\left(x_{0}^{L} - x^{\alpha}\right)}$$

$$V_{growth} = M_{B}^{diff}\frac{-\Delta G_{B\ idiff\ int}^{L\rightarrow\alpha}}{\delta\left(x^{\alpha} - x_{0}^{L}\right)}$$

The overall driving force of the concentration diffusion of the A and B atoms is

$$-\Delta G_{diff\ int}^{L\rightarrow\alpha} = (1-x^{\alpha})\left(-\Delta G_{A\ diff\ int}^{L\rightarrow\alpha}\right) + x^{\alpha}\left(-\Delta G_{B\ diff\ int}^{L\rightarrow\alpha}\right)$$

Incorporation of A and B Atoms into the Rough Crystal Surface

$$V_{\text{growth}} = \frac{a^3}{a^2}\nu P_{\text{B}}\,\mathrm{e}^{-\frac{U_{\text{a act}}}{RT}}\left[x^{\text{L}} - x^{\alpha}\exp\left(-\frac{-\Delta G^{\text{L}\to\alpha}_{\text{B kin int}}}{RT}\right)\right]$$
$$+\frac{a^3}{a^2}\nu P_{\text{A}}\mathrm{e}^{-\frac{U_{\text{a act}}}{RT}}\left[\left(1-x^{\text{L}}\right) - \left(1-x^{\alpha}\right)\exp\left(-\frac{-\Delta G^{\text{L}\to\alpha}_{\text{A kin int}}}{RT}\right)\right]$$

Provided that the driving forces are small compared to RT and $P_{\text{A}} = P_{\text{B}}$ the expression can be simplified to

$$V_{\text{growth}} = \frac{a^3}{a^2}\nu P\,\mathrm{e}^{-\frac{U_{\text{a act}}}{RT}}\left(\left(1-x^{\alpha}\right)\frac{-\Delta G^{\text{L}\to\alpha}_{\text{A kin int}}}{RT} + x^{\alpha}\frac{-\Delta G^{\text{L}\to\alpha}_{\text{B kim int}}}{RT}\right)$$

or as a function of the overall driving force of solidification

$$V_{\text{growth}} = \frac{a^3}{a^2}\nu P\,\mathrm{e}^{-\frac{U_{\text{a act}}}{RT}} \times \frac{-\Delta G^{\text{L}\to\alpha}_{\text{kin int}}}{RT}$$

Analogous to pure metals the relationship is valid:

$$V_{\text{growth}} = \frac{D_{\text{interphase}}}{\delta}\frac{-\Delta G^{\text{L}\to\alpha}_{\text{kin int}}}{RT}P$$

where

$$-\Delta G^{\text{L}\to\alpha}_{\text{kin int}} = \left[\left(1-x^{\alpha}\right)\left(-\Delta G^{\text{L}\to\alpha}_{\text{A kin int}}\right) + x^{\alpha}\left(-\Delta G^{\text{L}\to\alpha}_{\text{B kin int}}\right)\right]$$

Kinetic Coefficient of Growth Rate for Binary Alloys with Consideration to Relaxation

Relaxation and Rearrangement of the Incorporated Atoms in the Crystal

At solidification the incorporated atoms become transformed into disordered solid atoms at the surface of the crystal. Two parallel processes will occur.

1. Rearrangement of position of the incorporated A and B atoms from the disordered state to the ordered state of the crystal lattice. This is the relaxation process, when the atoms loose potential energy, which is emitted to the surroundings.
2. Diffusion of the A and B atoms in order to achieve equilibrium with the lattice atoms, i.e. to obtain lowest possible potential energy.

Driving force of the relaxation process:

$$-\Delta G^{\alpha}_{\text{relax}} = -\Delta G^{\text{total}}\sqrt{\frac{V_{\text{growth}}\delta}{\pi D_{\text{relax}}}}$$

where

$$-\Delta G^{\text{total}} = -\Delta G^{\text{L}\to\alpha}_{\text{diff int}} + \left(-\Delta G^{\text{L}\to\alpha}_{\text{kin int}}\right) + \left(-\Delta G^{\text{total}}\right)\sqrt{\frac{V_{\text{growth}}\delta}{\pi D_{\text{relax}}}}$$

or, if $(-\Delta G^{L\rightarrow\alpha}_{\text{diff int}})$ is neglected and $P_A = P_B = P$,

$$-\Delta G^{\text{total}} = \frac{\delta\, RT\, V_{\text{growth}}}{D_{\text{interphase}} P} + \left(-\Delta G^{\text{total}}\right)\sqrt{\frac{V_{\text{growth}}\delta}{\pi D_{\text{relax}}}}$$

where

$$P = \sqrt{\frac{V_{\text{growth}}\delta}{\pi D_{\text{relax}}}}$$

Kinetic Coefficient of Growth Rate in Alloys

$$V_{\text{growth}} = \beta_{\text{kin}} \times (\Delta x)^2$$

where

$$\beta_{\text{kin}} = \frac{D^2_{\text{interphase}}}{\pi\, \delta\, D_{\text{irelax}}}$$

or in terms of concentration and slope of the liquidus line

$$V_{\text{growth}} = \beta_{\text{kin}} \times (\Delta x)^2 = const \times (\Delta T)^2$$

$$const = \frac{\beta_{\text{kin}}}{m^2}$$

m = slope of the liquidus line.

■ Diffusion-Controlled Growth in Binary Alloys

The growth of crystals in a supersaturated alloy melt depends on a number of processes such as the interface kinetics, the mass transport and the heat transport.

- If the activation energy of incorporating new atoms from the melt into the crystal lattice is *high*, the *kinetic process* at the interface controls the growth.
- If the activation energy is *low*, the overall rate of crystal growth is determined by the *heat* and/or *mass transport* instead.
- If the heat transport is fast the *mass transport*, i.e. *diffusion*, will dominate the overall rate of the crystal growth.

Diffusion Controlled Growth of Planar Crystals in Alloys
The exact solution of the general case is presented in the text.

Special case:

Planar Growth at Stationary Conditions and Constant Growth Rate in Alloys
Concentration distribution of alloying element in the melt:

$$x^L = x_0^L \left(1 + \frac{1-k_0}{k_0} e^{-\frac{V_{\text{growth}} y}{D}}\right)$$

■ Diffusion-Controlled Growth of Spherical Crystals in Binary Alloys

The exact solution of the general case is presented in the text

Diffusion Controlled Growth Rate as a Function of Crystal Size

Stage 1: No corrections:

$$V_{\text{growth}} = \frac{dr_c}{dt} = D\frac{x_{eq}^{L/\alpha} - x_0^L}{x_{eq}^{L/\alpha} - x_{eq}^{\alpha/L}}\frac{V_m^\alpha}{V_m^L}\frac{1}{r_c}$$

Stage 2: Attention to surface tension:

$$V_{\text{growth}} = \frac{dr_c}{dt} = D\frac{x_{eq}^{L/\alpha} - x_0^L}{x_{eq}^{L/\alpha} - x_{eq}^{\alpha/L}}\frac{V_m^\alpha}{V_m^L}\left(1 - \frac{r^*}{r_c}\right)\frac{1}{r_c}$$

Stage 3: Attention to kinetic process at the interface:

$$V_{\text{growth}} = \frac{dr_c}{dt} = \left(-A + \sqrt{A^2 + C_3}\right)^2$$

where

$$A = \frac{1}{2}\sqrt{\frac{1}{\beta_{\text{kin}}}\frac{V_m^\alpha}{V_m^L}\frac{D}{x_{eq}^{L/\alpha} - x_{eq}^{\alpha/L}}} \times \frac{1}{r_c}$$

$$C_3 = D\frac{V_m^\alpha}{V_m^L}\frac{x_{eq}^{L/\alpha} - x_0^L}{x_{eq}^{L/\alpha} - x_{eq}^{\alpha/L}} \times \frac{1}{r_c} \quad \text{and} \quad \beta_{\text{kin}} = \frac{D_{\text{interphase}}^2}{\pi\,\delta\,D_{\text{irelax}}}$$

Stage 4: Attention to surface tension and kinetic process at the interface:

$$V_{\text{growth}} = \frac{dr_c}{dt} = \left(-A + \sqrt{A^2 + C_4}\right)^2$$

A and β_{kin}: the same as above

$$C_4 = D\frac{V_m^\alpha}{V_m^L}\frac{x_{eq}^{L/\alpha} - x_0^L}{x_{eq}^{L/\alpha} - x_{eq}^{\alpha/L}} \times \frac{1}{r_c}\left(1 - \frac{r^*}{r_c}\right) = C_3 \times \left(1 - \frac{r^*}{r_c}\right)$$

Crystal Size and Growth Rate of Diffusion Controlled Growth as a Function of Time

No corrections:

$$r_c^2 - r_0^2 = 2D\frac{x_{eq}^{L/\alpha} - x_0^L}{x_{eq}^{L/\alpha} - x_{eq}^{\alpha/L}}\frac{V_m^\alpha}{V_m^L} \times t$$

or, if r_0 is neglected:

$$r_c = \sqrt{2D\frac{x_{eq}^{L/\alpha} - x_0^L}{x_{eq}^{L/\alpha} - x_{eq}^{\alpha/L}}\frac{V_m^\alpha}{V_m^L}} \times \sqrt{t}$$

Growth rate of the spherical crystal as a function of time:

$$V_{\text{growth}} = \frac{dr_c}{dt} = \frac{1}{2}\sqrt{2D\frac{x_{\text{eq}}^{L/\alpha} - x_0^L}{x_{\text{eq}}^{L/\alpha} - x_{\text{eq}}^{\alpha/L}}\frac{V_m^\alpha}{V_m^L}} \times \sqrt{\frac{1}{t}}$$

Attention to the surface tension and kinetic process at the interface:
Numerical solution given as a figure in the text.

■ Impingement

In most precipitation processes, such as crystallization and formation of droplets in vapours and liquids, a great number of particles are formed simultaneously instead of a single free particle. Nearby crystals or droplets influence their compositions and also their growth rates. This influence is called *impingement.*

Impingement changes the distribution of the alloying element and the growth rate of the particles.

Growth Rate without Impingement

$$V_{\text{growth}} = \frac{dr_c}{dt} = \frac{D}{r_c}\frac{x_0^L - x_{\text{eq}}^{L/\alpha}}{x_{\text{eq}}^{\alpha/L} - x_{\text{eq}}^{L/\alpha}}\frac{V_m^\alpha}{V_m^L}$$

If there are *many particles* the crystal growth decreases, i.e. x_0^L will not keep constant but varies with time and has to be replaced by $x^L(t)$. It can be calculated from the material balance

$$N\frac{4\pi r_c{}^3}{3}\frac{x_{\text{eq}}^{\alpha/L} - x_{\text{eq}}^{L/\alpha}}{V_m^\alpha} = \left(1 - N\frac{4\pi r_c{}^3}{3}\right)\frac{x_0^L - x^L(t)}{V_m^L}$$

Influence of Impingement on Growth Rate

$$V_{\text{growth}} = \frac{dr_c}{dt} = \frac{D}{r_c}\frac{V_m^\alpha}{V_m^L}\left(\frac{x_0^L - x_{\text{eq}}^{L/\alpha}}{x_{\text{eq}}^{\alpha/L} - x_{\text{eq}}^{L/\alpha}} - \frac{V_m^L}{V_m^\alpha}\frac{N\frac{4\pi r_c{}^3}{3}}{1 - N\frac{4\pi r_c{}^3}{3}}\right)$$

If there are *very few* and/or *very small* particles the impingement can be neglected.

■ Precipitation of Pores

Sometimes the formation of crystals in melts are accompanied by a simultaneous precipitation of gas. In general, various types of pores can be related to the gas content solved in the melt before casting.

The pores can be classified into three groups:

- rounded pores
- elongated pores
- shrinkage pores.

Growth Rate of Rounded Pores in Metal Melts

$$\frac{dr}{dt} = \frac{3D_G}{2V_m}\frac{x_G^L - x_G^I}{3p_{atm}r + 4\sigma}RT$$

The mole fraction of the gas at the interface melt/pore is strongly temperature dependent.

$$x_G^I = A\sqrt{\frac{p_{atm} + \frac{2\sigma}{r}}{p_0}}\,e^{-\frac{B}{T}}$$

Exercises

6.1 At solidification of metal melts the surface becomes strained because the solidified 'liquid' atoms become 'solid' atoms but keep initially their 'liquid atomic' distances, which exceed the interatomic distances in the crystal lattice. The 'liquid' atoms have a higher energy than the lattice atoms. The excess energy is stored in the crystal lattice and is called relaxation energy. The relaxation energy is successively emitted as heat to the surroundings (the crystal lattice). The distance between the atoms change with time and can be described by the following equation

$$a - a_{lattice} = (a_0 - a_{lattice})e^{-\frac{t}{\tau}} \quad (1)$$

where

$a_{lattice}$ = interatomic distance in the crystal lattice
a_0 = the initial distance between the solidified 'liquid' atoms at $t = 0$
a = distance between the former 'liquid' atoms at time t
τ = relaxation time, i.e. the time after which the interatomic distance has been reduced by a factor 1/e.

a) At what time does the relaxation finish, i.e. at what time do the liquid atoms obtain the same interatomic distances as the lattice atoms?
b) How long time, expressed as number of relaxation times, is required to reduce the interatomic distance to half of $(a_0 - a_{lattice})$?
c) How long time, expressed as number of relaxation times, is required to reduce $(a - a_{lattice})$ to 1 per cent of its initial value?
d) By what factor has $(a_0 - a_{lattice})$ been reduced after 5τ?

6.2 The growth rate at normal growth in pure metals is derived in Section 6.6.2 in Chapter 6. According to Equation (6.29) on page 292 it can be written as

$$V_{growth} = a^3 const \times P\, e^{-\frac{U_{a\,act}}{k_B T}}\left(1 - e^{-\frac{-\Delta G_a}{k_B T}}\right) \quad (1)$$

The constant ahead of the exponential factor is often presented in the literature as

$$a^3 const \times P = \frac{D_0}{\delta} \quad (2)$$

where
D_0 = the diffusion constant
δ = the jump distance of the diffusing 'solute' atom
$U_{a\,act}$ = activation energy for diffusion.

[Diffusion has been treated extensively in Chapter 5 in the book 'Physics of Functional Materials'.]

a) Inserting Equation (2) into Equation (1) gives

$$V_{\text{growth}} = \frac{D_0}{\delta} \exp \frac{-U_{\text{a act}}}{k_B T}\left(1 - \exp \frac{-\Delta G_a}{k_B T}\right) \tag{3}$$

Use the relationship (3) and plot the growth rate as a function of the undercooling required for solidification of a pure Cu melt with no consideration of the strain energy.
Material constants of Cu:

$M = 63.54$ kg/kmol
$\rho = 8.93 \times 10^3$ kg/m^3
$-\Delta H = 13.0 \times 10^6$ J/kmol
$T_M = 1356$ K
$V_m = 7.11 \times 10^{-3}$ m^3/kmol
$D_0 = 1.46 \times 10^{-7}$ m^2/s
$\delta = 3.0 \times 10^{-9}$ m
$U_{act} = 4.066 \times 10^7$ J/kmol
$Y = 6.5 \times 10^{10}$ N/m^2
$\varepsilon = 0.060$

b) During the growth the atoms become strained over the interface due to the density difference between the solid and the melt. The strain can be considered as an additive to the activation energy for diffusion.
Analyze the effect of the strain on the growth rate by plotting V_{growth} as a function of the undercooling with consideration of the elastic strain.

6.3 A series of solidification experiments have been performed on single crystals of pure copper. The samples were decanted during the solidification process and their solid/liquid interphase structures were studied.
The structure of the interphase consisted of steps. The horizontal steps consisted of {111} facets and the vertical steps were {100} facets. The ratio of the length to the height λ/h of the steps was 100:1.
Calculate the ratio of the relaxation (diffusion) constants D_{relax} in the $<100>$ and $<111>$ directions for copper.

6.4 When a metal melt is cooled unidirectionally it solidifies along a planar solidification front. It moves forward with the growth rate V_{growth}. If the heat of solidification is removed at a constant cooling rate then V_{growth} is also constant.

a) The molar heat of fusion can be measured quantitatively. Observations show that the measured $(-\Delta H)$ depends on the cooling rate and consequently also on the growth rate of solidification.
Start with the molar heat of fusion and set up an expression for the total driving force of solidification $(-\Delta G^{L\rightarrow s})$. It is the sum of two terms, which ones? Show them in a figure. Set up the relationship between them and explain the significance of the two terms.

b) Set up the relationship between the strain energy and the growth rate. When the melt solidifies solidification shrinkage appears and the interphase becomes strained. What happens to the strain energy during the cooling and how is it related to the heat of fusion?

c) Set up the relationship between the total molar heat of fusion and the molar strain energy. Discuss how the tabulated total heat of fusion values can be determined.

d) Copper is an FCC metal. Use the information in Example 6.2 on pages 296–297 in Chapter 6 and the material constants for copper given in the text of Exercise 6.2 above to determine the expected value of the ratio $(-\Delta H_{\text{strain}})/(-\Delta H_0)$ for the growth rate of solidification $V_{\text{growth}} = 1 \times 10^{-3}$ m/s provided that $D_{\text{relax}} = D_0$.

6.5 When diffusion-controlled solidification processes are studied, it is common to assume that the crystals formed immediately obtain equilibrium composition. This is not always the case when the growth rate is high.

Consider a binary alloy as an example. The diffusion constants of the two components A and B in the melt are equal. This means that the driving forces for diffusion of A and B must be equal. The phase diagram of the alloy is shown in the figure. The initial composition of the solidifying melt is x_0^{L}. The melt starts to solidify at temperature T_0. At equilibrium the melt has the composition $x_{\mathrm{eq}}^{\mathrm{L}/\alpha}$ and the solid phase has the composition $x_{\mathrm{eq}}^{\alpha/\mathrm{L}}$. During the solidification process equilibrium is not achieved and the solid phase has the composition x^{α}, that deviates from $x_{\mathrm{eq}}^{\alpha/\mathrm{L}}$.

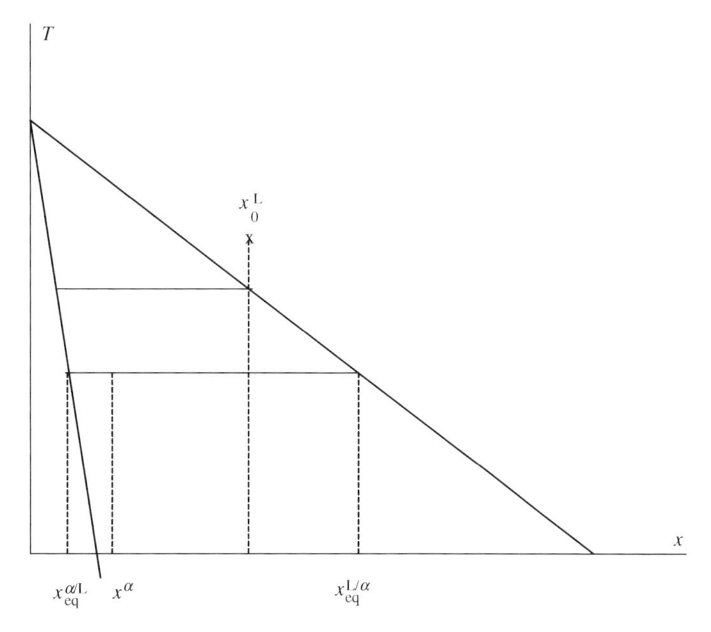

Figure 6.5–1

How far from equilibrium composition the solid is, compared to the equilibrium composition, can be studied by plotting $(x_{\mathrm{eq}}^{\mathrm{L}/\alpha} - x^{\alpha})$ and $(x_{\mathrm{eq}}^{\mathrm{L}/\alpha} - x_{\mathrm{eq}}^{\alpha/\mathrm{L}})$ as functions of $x_{\mathrm{eq}}^{\mathrm{L}/\alpha}$. At your help you have the following information:

- Equation (6.75) on page 314 in Chapter 6 is valid here.
- $M_{\mathrm{A}}^{\mathrm{diff}}\left(-\Delta G_{\mathrm{A\ diff\ int}}^{\mathrm{L}\to\alpha}\right) = -M_{\mathrm{B}}^{\mathrm{diff}}\left(-\Delta G_{\mathrm{B\ diff\ int}}^{\mathrm{L}\to\alpha}\right)$ where the subscript 'int' stands for interphase.
- The driving force of solidification of a binary alloy melt consists of two terms, one for each component

$$-\Delta G_{\mathrm{total}}^{\mathrm{L}\to\alpha} = -\Delta G_{\mathrm{A}}^{\mathrm{L}\to\alpha} + \left(-\Delta G_{\mathrm{B}}^{\mathrm{L}\to\alpha}\right)$$

where

$$-\Delta G_{\mathrm{A}}^{\mathrm{L}\to\alpha} \approx RT\left[\left(x_{\mathrm{eq}}^{\mathrm{L}/\alpha} - x_0^{\mathrm{L}}\right) - \left(x_{\mathrm{eq}}^{\alpha/\mathrm{L}} - x^{\alpha}\right)\right] \quad \text{and} \quad -\Delta G_{\mathrm{B}}^{\mathrm{L}\to\alpha} \approx RT\left(\frac{x_{\mathrm{eq}}^{\alpha/\mathrm{L}} x_0^{\mathrm{L}}}{x_{\mathrm{eq}}^{\mathrm{L}/\alpha} x^{\alpha}} - 1\right)$$

Plot the two functions in the same diagram. The k-value of the alloy is $k = x_{\mathrm{eq}}^{\alpha/\mathrm{L}} / x_{\mathrm{eq}}^{\mathrm{L}/\alpha} \approx 0.10$. Comment your result.

6.6 Commercially produced diamonds are often made from a Ni-C alloy under high pressure and high temperature. The basic material is a mixture of graphite, nickel and diamond powder. The figures show the phase diagram of Ni-C at two different pressures.

a) Diamonds are often produced in a closed chamber at a pressure of 65 kbar. Perform a preparatory theoretical planning of the diamond production. Use the figures for an optimal choice of temperature. Check your suggested temperature with the temperature given in the answer. The height of the closed chamber is 100 cm.

b) Calculate the growth rate of crystalline diamond, produced at the pressure 65 kbar and the temperature given in the answer in a). How long time is required to produce a diamond layer of 1 mm?

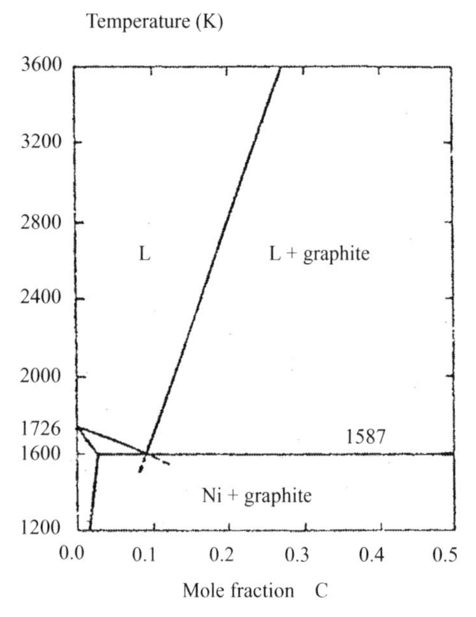

Figure 6.6–1 Phase diagram of the system Ni–C at standard pressure.

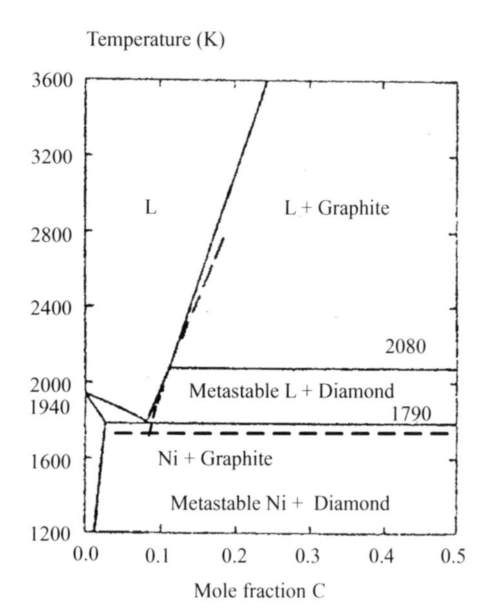

Figure 6.6–2 Phase diagram of the system Ni–C at the pressure 65 kbar.

Material constants:

$D_{\underline{C}}^{L} = 1.0 \times 10^{-9}\ m^2/s$

$V_m^{diamond} = 4.0 \times 10^{-3}\ m^3/kmol$

$V_m^{L} = 6.6 \times 10^{-3}\ m^3/kmol$

6.7 a) The system Bi-Zn contains a miscibility gap (see the phase diagram below). An alloy, which contains 30 wt% Zn is homogenized at 850 K and is subsequently cooled with a rate of 100 K/min. When the miscibility gap is reached, Zn precipitates droplets. These droplets continue to grow during the further cooling.

The diffusion constant of carbon in the melt is $1 \times 10^{-9}\ m^2/s$.

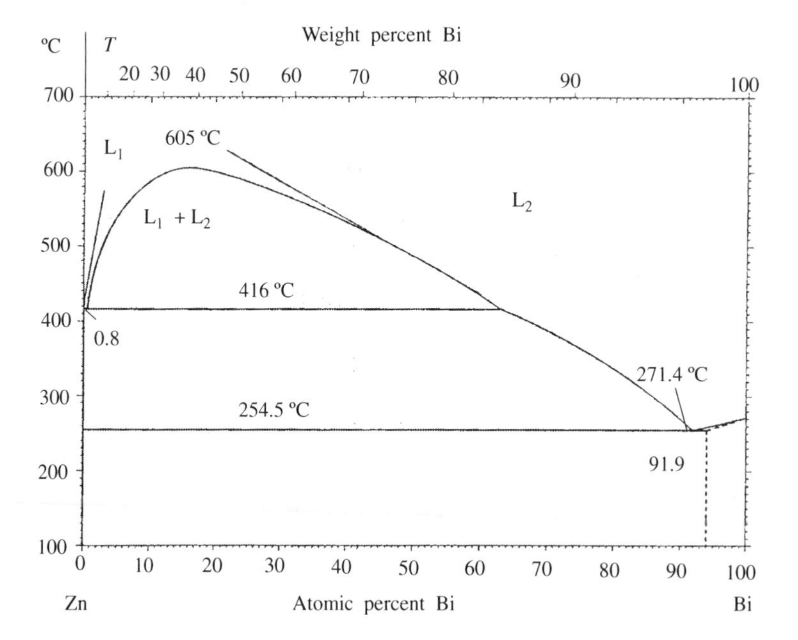

Figure 6.7–1

Calculate the maximum droplet size when the monotectic temperature is reached. The Bi content in the Zn droplets can be neglected. You may approximate the liquidus lines with straight lines.

Material constants:

D^{L}_{Zn} $= 1.0 \times 10^{-9}\,m^2/s$

M_{Zn} $= 65.37\,kg/kmol$

M_{Bi} $= 209.0\,kg/kmol$

Molar volume of Zn $= 9.22 \times 10^{-3}\,m^3/kmol$

Molar volume of Bi $= 20.9 \times 10^{-3}\,m^3/kmol.$

b) Consider the same alloy as in a). Several droplets normally form at the temperature when the melt reaches the immiscibility gap. The droplets interact and the growth is influenced by the impingement in the melt.

Calculate the size of the droplets as a function of the number of droplets per unit volume at the monotectic temperature 416 °C provided that the number is large enough to achieve equilibrium in the melt.

6.8 The figure shows an isothermal cut of the phase diagram of the ternary system A–BC.

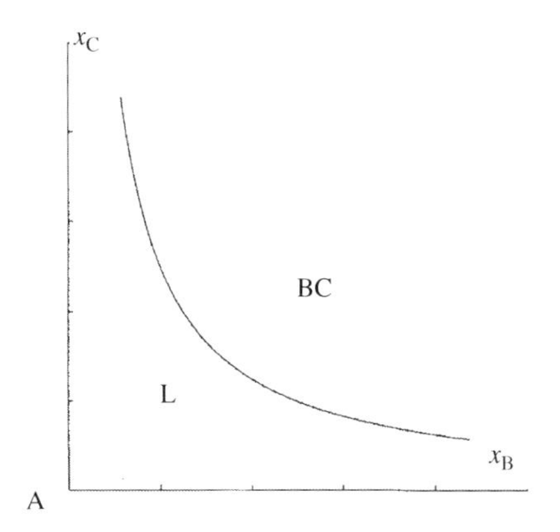

Figure 6.8–1

The diagram shows the equibility line between the melt and the intermediate phase BC. The solubility product can be described by the relationship

$$x_B^{L/BC} x_C^{L/BC} = 1.0 \times 10^{-3}$$

When crystals of the phase BC are precipitated from the melt they show a spherical morphology because the growth rate of B and C are equal.

Derive an analytical expression for the growth rate of the spherical crystals as a function of their size (radius) and the initial composition of the melt $x^{L}_{0\,B}$. and $x^{L}_{0\,C}$. Neglect the influences of the surface tension and the kinetics.

The diffusion constants of B and C atoms in the melt are

$$D^{L}_{B} = D^{L}_{C} = D^{L} = 1.0 \times 10^{-9} m^2/s.$$

References

1. H. Fredriksson, U. Åkerlind, *'Physics of Functional Materials'*, Chichester, England, John Wiley & Sons, 2008.
2. K. A. Jackson, *Liquid Metals and Solidification*, A.S.M, Cleveland, Ohio, US, 1958, 174.
3. D. E. Temkin, *Crystallization Processes*, English Translation by Consultants Bureau, New York, 1966.
4. C. Elbaum, B. Chalmers, *Canadian Journal of Physics*, **33**, 1955, 196.
5. D. Turnbull, *Thermodynamics in Physical Metallurgy*, A.S.M, Metals Park, Ohio, 1949, page 282.
6. P. Shewmon, *Diffusion in Solids, The Minerals*, Metals & Materials Society, Pennsylvania, US, 1989, page 140.
7. B. Chalmers, *Trans AIME*, **200**, 1956, 519.
8. G. A, Chadwick, A hard-sphere model of crystal growth, *Metal Science Journal*, **1**, 1967, 132–139.
9. F. J. Klein, J. Pötschke, The influence of impurities on the crystallization of supercooled copper melts, *Journal of Crystal Growth*, **46**, 1979, 112–118.
10. B. Chalmers, *Principles of Solidification*, New-York, John Wiley & Sons, 1964, page 52.
11. C. Zener, Theory of growth of spherical precipitates from solid solutions, *Journal of Applied Physics*, **20**, 1949, 950–953.
12. F. C. Frank, Radially-symmetric phase growth controlled by diffusion, *Proc. Royal Society*, **201A**, 1950, 586–599.
13. G. Engberg, M. Hillert, A. Odén, Estimation or the rate of diffusion-controlled growth by means of quasi-stationary model, *Scandinavian Journal of Metallurgy*, **4**, 1975, 93–96.
14. H. Fredriksson, U. Åkerlind, *'Materials Processing during Casting'*, Chichester, England, John Wiley & Sons, 2005.

7

Heat Transport during Solidification Processes. Thermal Analysis

Solidification and Crystallization Processing in Metals and Alloys, First Edition. Hasse Fredriksson and Ulla Åkerlind.

7.1 Introduction

In the preceding chapters the mechanisms of crystallization, the structures and compositions of crystals were discussed. We found that mass transport has a great influence on the composition of crystals.

When crystals are formed energy of formation is released. The heat of formation must be removed, otherwise the temperature will increase above the melting point and the crystallization process will stop. The rate of heat removal controls the solidification time and the temperature distribution in the material and its surroundings. Hence, heat transport is a *very* important part of the crystallization process.

Most material constants depend on temperature, which directly or indirectly controls the rate of crystallization processes and the quality and properties of the solid products.

The application of heat equations on solidification of metals is in most cases very complicated. The number of variables is large and the knowledge of the solidification conditions, especially in two-phase region may be poor.

Today commercial computer programs are available for calculation of the temperature profile and the rate of solidification which may be physically and mathematically very complicated. It is in most cases difficult to find an analytical solution of the heat equation. Computer calculations may give reasonable approximate solutions.

Fortunately it is possible in some cases to make reasonable simplifying assumptions and find good approximate solutions. These simplifications are often based on the assumption that the slowest one in a series of subprocesses controls the total process. Hence, the slowest subprocess has to be identified in each case.

In this chapter we will introduce the basic concepts and laws of heat transport together with applications in some special cases, which frequently appear in connection with heat transport in metals. Suitable simplifications will be introduced as indicated above and the corresponding approximate solutions will be derived for future use in Chapter 8.

The two most important experimental methods to study solidification processes, controlled unidirectional solidification and thermal analysis, will be discussed in this chapter. The heat transport controls the processes in both methods. Solidification in one direction with constant growth rate and a constant temperature gradient offers a useful method for single crystal production.

Thermal analysis is very useful for accurate measurements of chemical and physical properties such as the heat of fusion and the heat capacity of a material. The heat of fusion is not constant but varies with the cooling rate. This effect is discussed at the end of the chapter.

7.2 Basic Concepts and Laws of Heat Transport

There are three ways of heat transport

- conduction
- radiation
- convection.

In the case of solidifying metal melts, heat conduction will be the most important way of heat transport.

7.2.1 *Heat Conduction*

Stationary Conditions

Basic Law of Heat Conduction

The basic law of heat conduction during stationary conditions can be written as

$$\frac{dQ}{dt} = -\lambda A \frac{dT}{dy} \tag{7.1}$$

where
Q = amount of heat
t = time
y = coordinate
λ = thermal conductivity
A = cross-section area
T = temperature.

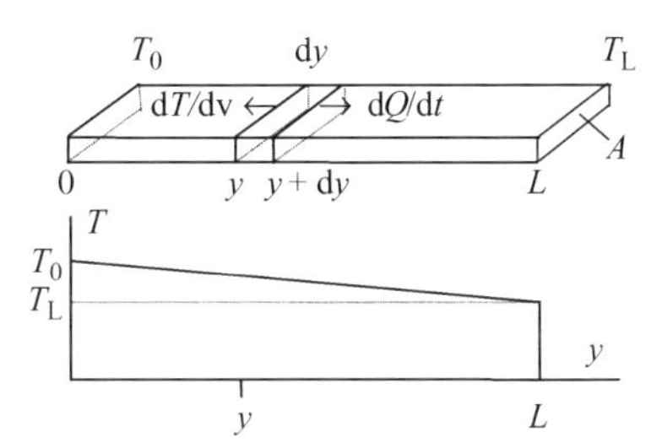

Figure 7.1 Temperature as a function of position but not of time (stationary conditions).

The heat flow is proportional to the temperature gradient and flows in the opposite direction compared to the gradient.

Thermal energy always flows from higher to lower temperature. The temperature difference serves as the driving force of the heat flow. If we introduce $q = Q/A$ we obtain the heat flux

$$\frac{dq}{dt} = -\lambda \frac{dT}{dy} \quad \text{Fourier's first law} \tag{7.2}$$

Thermal Conduction through Several Layers Coupled in Series

The thermal flow in a heat-conducting layer can be compared to an electrical current through a resistor, where the heat flow corresponds to the current I and the temperature difference to the potential difference U.

At stationary conditions we obtain (Figure 7.1)

$$\frac{dQ}{dt} = -\frac{\lambda A}{L}(T_L - T_0) \quad \text{to be compared to} \quad I = \frac{U}{R}$$

The 'thermal resistance' of the layer, corresponding to the electrical resistance, can be written $L/\lambda A$.

If several layers are in contact with each other they can be considered as coupled in series (Figure 7.2) and the total 'thermal resistance' will be

$$\frac{L}{\lambda A} = \frac{L_1}{\lambda_1 A_1} + \frac{L_2}{\lambda_2 A_2} + \frac{L_3}{\lambda_3 A_3} + \ldots\ldots \tag{7.3}$$

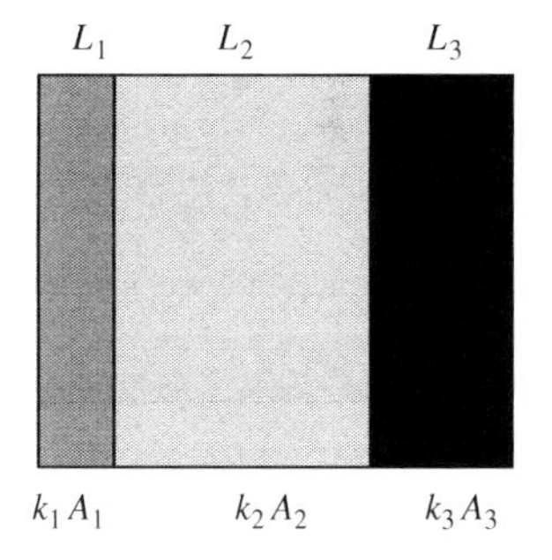

Figure 7.2 Heat-conducting layers, coupled in series. For simplicity the cross-sectional areas are drawn equal in the figure.

Heat Transfer across the Interface between Two Materials

At stationary conditions, there is a constant temperature difference across the interface between the two materials (Figure 7.3). The heat transferred per unit time from the warmer to the colder side is proportional to the temperature difference.

$$\frac{dQ}{dt} = -hA(T_2 - T_1) \tag{7.4}$$

where
h = heat transfer constant or heat transfer coefficient.

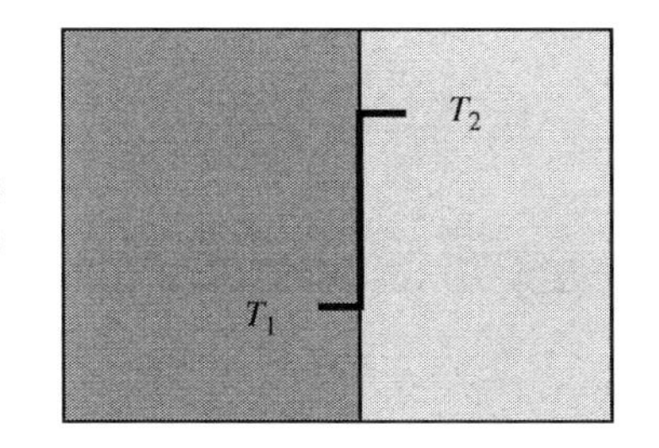

Figure 7.3 Thermal conduction across an interface between two materials.

Poor contact between the materials will create a thin layer of air between them. Air is a poor heat conductor and the heat transfer will decrease considerably when an air layer appears.

We apply Fourier's law (Equation (7.2) on page 363) and the heat transfer Equation (7.4) divided by A.

$$\frac{dq}{dt} = -h(T_2 - T_1) = -\lambda\frac{T_2 - T_1}{\delta} \tag{7.5}$$

Equation (7.5) contains two expressions for the heat flux. Division with the factor $(T_2 - T_1)$ gives

$$h = \frac{\lambda}{\delta} \tag{7.6}$$

where
h = heat transfer coefficient of air
λ = thermal conductivity of air
δ = thickness of air layer.

Non-Stationary Conditions

General Law of Heat Conduction

If the thermal conduction is *transient,* i.e. the temperature is a function of both position and time, the basic Equation (7.2) of thermal conduction is no longer valid. It has to be replaced by a more general equation.

Consider the volume element in Figure 7.4. The heat amount dQ_y flows into the element through the surface A_y during the time dt. Simultaneously the heat amount dQ_{y+dy} leaves the element through the surface A_{y+dy}. The difference stays within the volume element and causes its temperature to increase by the amount dT.

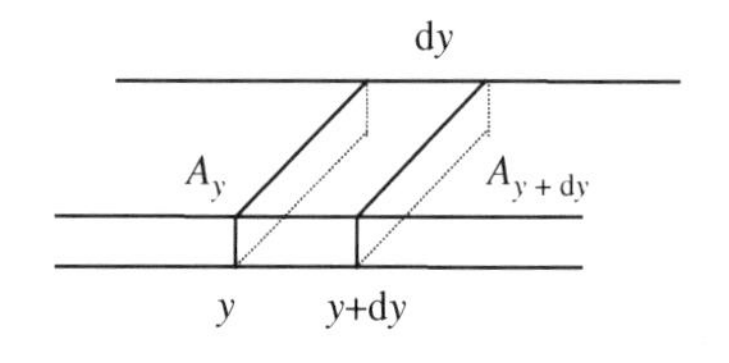

Figure 7.4 Volume element.

The law of energy conservation gives

$$dQ_y - dQ_{y+dy} = -\left[\lambda\rho A_y\left(\frac{\partial T}{\partial y}\right)_y - \lambda\rho A_{y+dy}\left(\frac{\partial T}{\partial y}\right)_{y+dy}\right]dt = c_p\rho\, Ady dT \tag{7.7}$$

We also have $A = A_y = A_{y+dy}$
After simplification and reorganization, Equation (7.7) can be written as

$$\frac{\left(\frac{\partial T}{\partial y}\right)_{y+dy} - \left(\frac{\partial T}{\partial y}\right)_y}{dy} = \frac{\rho c_p}{\lambda}\frac{\partial T}{\partial t} \tag{7.8}$$

After a limes transition we obtain

$$\frac{\partial T}{\partial t} = \alpha\frac{\partial^2 T}{\partial y^2} \quad \text{Fourier's second law} \tag{7.9}$$

where α is the *thermal diffusivity*.

$$\alpha = \frac{\lambda}{\rho c_p} \tag{7.10}$$

where
α = thermal diffusivity or coefficient of thermal diffusion
λ = thermal conductivity
ρ = density
c_p = heat capacitivity at constant pressure.

Equation (7.9) is the *general equation of thermal conduction* in one dimension.

Temperature Gradient

It is well known that heat moves spontaneously from higher to lower temperature, never in the opposite direction. The *temperature gradient* is a vector in the direction of increasing temperature. This direction is normal to each of the isothermal surfaces, which join all points of the same temperature (Figure 7.5a).

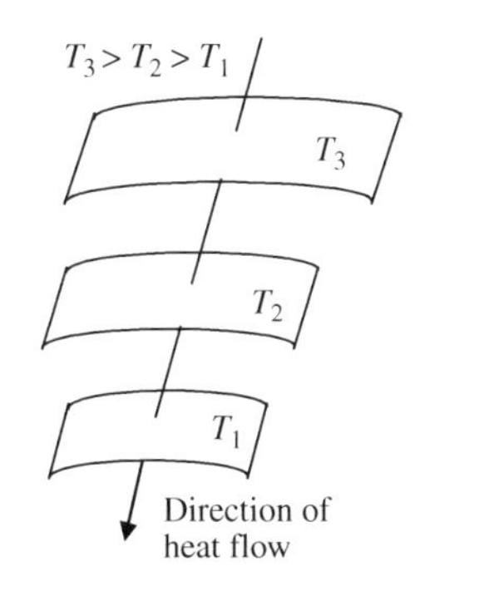

Figure 7.5a The direction of heat flow in a temperature field and the temperature gradient always have opposite directions.

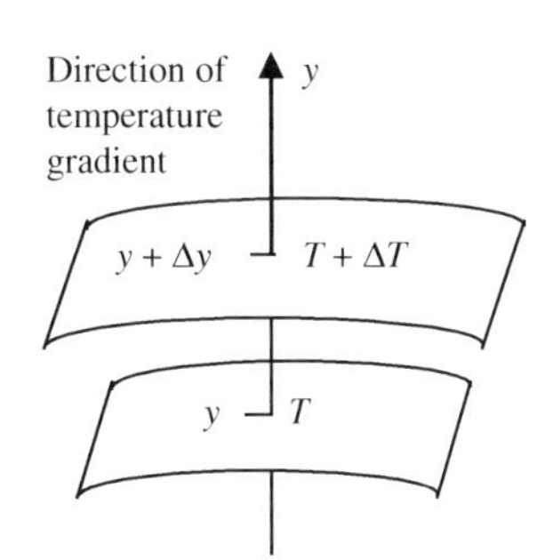

Figure 7.5b Direction of the temperature gradient in one dimension grad $T = \mathrm{d}T/\mathrm{d}y$.

Consider two isothermal surfaces with temperatures T respectively $T + \Delta T$ (Figure 7.5b). We introduce the normal to the two surfaces as coordinate axis and define the temperature gradient as

$$\text{grad}\, T = \lim \frac{\Delta T}{\Delta y}\hat{y} = \frac{\mathrm{d}T}{\mathrm{d}y}\hat{y} \tag{7.11}$$

The temperature is a *scalar quantity* and grad T is a vector. As heat flows from higher to lower temperature, the temperature gradient and the derivative of temperature have opposite directions compared to the heat flow

7.2.2 *Thermal Radiation*

All bodies emit electromagnetic radiation or *thermal radiation* to their surroundings. At the same time they absorb such radiation from the surroundings.

Figure 7.6 shows the energy radiation from a perfectly black body as a function of the wavelength (Planck's radiation law). The higher the temperature is, the more energy will be emitted and the more the radiation maximum will be moved towards shorter wavelengths.

The function in Figure 7.6 can be integrated graphically, i.e. the area under the curve is calculated. The result is an expression for the total radiation of energy per unit time and unit area from the body.

Boltzmann calculated the total thermal flux theoretically. The result is Stefan-Boltzmann's law. During the time interval $\mathrm{d}t$ a perfectly black body with temperature T and area A emits the energy

$$\mathrm{d}W = \varepsilon\sigma_{\mathrm{B}}AT^4\mathrm{d}t \tag{7.12}$$

where σ_{B} is a constant.

All real surfaces radiate less than a perfectly black body. This is taken into account by introduction of a dimensionless factor ε (< 1) called *emissivity* into Equation (7.12). The shinier the surface is and the more it reflects radiation, the lower will be the value of ε.

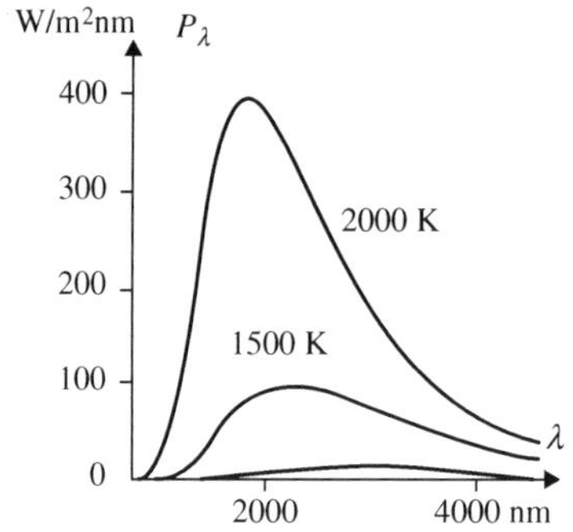

Figure 7.6 The energy emission per unit time and unit area within the wavelength interval λ and $\lambda + \mathrm{d}\lambda$ at some different temperatures.

If the absorption of radiation from the surroundings is taken into account the net radiation is

$$\mathrm{d}W_{\mathrm{total}} = \varepsilon\sigma_{\mathrm{B}}A\left(T^4 - T_0^4\right)\mathrm{d}t \tag{7.13}$$

For metal melts ($T \approx 1800$ K) the absorption from the surroundings ($T_0 \approx 300$ K) can be neglected.

The third possibility of heat transport is convection, which involves simultaneous mass and heat transport. It will be treated in next section.

7.3 Convection

Thermal conduction and thermal radiation results in transport of energy but not of material. Convection involves transport of material together with its heat content. There is a distinction between *free* or *natural convection* and *forced convection.*

Forced convection means that the convection is controlled by action from outside. Examples of external action, that affects the convection, are water-cooling and mechanical stirring of a melt. In the case of a steel melt it can be alternatively be stirred by a rotating magnetic field. *Free* or *natural convection* in a fluid is owing to existing density differences within the fluid without any action from outside.

The energy $\mathrm{d}q$ per unit area transferred from the surface during the time interval $\mathrm{d}t$ to a surrounding flowing cooling medium can be written as

$$\frac{\mathrm{d}q}{\mathrm{d}t} = h_{\mathrm{con}}(T - T_0) \tag{7.14}$$

where

q = energy per unit area
h_{con} = heat transfer number of convection
T = temperature of the surface
T_0 = temperature of cooling medium.

The heat transfer number of convection depends on the speed of the flowing cooling medium, the geometry of its channel and the shape and size of the surface.

After an initial introduction to boundary layer theory, the laws of forced and free convection, respectively, will be discussed briefly below.

7.3.1 *Boundary Layer Theory*

The convection phenomenon is closely related to *hydrodynamics* and the so-called *boundary layer theory*, which will be shortly discussed below.

Hydrodynamics

The motion of a fluid is often very complex. The flow pattern can be described graphically by stream lines, which indicate the direction of the particle flow in the fluid. They are drawn in such a way that the number of stream lines per unit area perpendicular to the flow is a measure of the magnitude of the particle velocity.

In the simplest case, the flow is *laminar,* otherwise *turbulent*. A few dimensionless numbers are defined to describe the behaviour of fluids. These numbers are used to give significant information on the flow pattern. They are listed and defined in Table 7.1 for future use. For further information textbooks in hydrodynamics should be consulted.

Table 7.1 Definitions and significance of some dimensionless numbers of interest in fluid dynamics. The quantities, appearing in the definition equations, are explained in Table 7.2.

Number	Designation	Definition equation	Physical significance	Dimension
Nussel's number (forced convection)	Nu	$\mathrm{Nu} = \frac{hL}{\lambda}$	Ratio of heat transfer across an interface to conductive heat transfer	zero
Grashof's number (free convection)	$\mathrm{Gr_{heat}}$	$\mathrm{Gr_{heat}} = \frac{g\,L^3\,\beta\Delta T}{\nu_{\mathrm{kin}}^2}$	Ratio of buoyancy force to viscous force acting on the fluid, in terms of thermal expansion	zero
Grashof's number (free convection)	$\mathrm{Gr_{mass}}$	$\mathrm{Gr_{mass}} = \frac{g\,L^3\,\beta'\Delta C}{\nu_{\mathrm{kin}}^2}$	Ratio of buoyancy force to viscous force acting on the fluid, in terms of concentration expansion	zero
Prandl's number (forced and free convection)	Pr	$\mathrm{Pr} = \frac{c_{\mathrm{p}}\,\eta}{\lambda} = \frac{\nu_{\mathrm{kin}}}{\alpha}$	Ratio of kinematic viscosity coefficient to thermal diffusivity	zero
Raleigh's number (free convection)	$\mathrm{Ra_{heat}}$	$\mathrm{Ra_{heat}} = \frac{g\,L^3\beta\Delta T}{\nu_{\mathrm{kin}}\alpha}$ $\mathrm{Ra_{heat}} = \mathrm{Gr_{heat}\,Pr}$	Ratio of heat transfer by convection to heat transfer by conduction, in terms of thermal expansion	zero
Raleigh's number (free convection)	$\mathrm{Ra_{mass}}$	$\mathrm{Ra_{mass}} = \frac{g\,L^3\beta''\Delta C}{\nu_{\mathrm{kin}}\alpha}$ $\mathrm{Ra_{mass}} = \mathrm{Gr_{mass}Pr}$	Ratio of heat transfer by convection to heat transfer by conduction, in terms of concentration expansion	zero
Reynold's number (fluid flow)	Re	$\mathrm{Re} = \frac{u_0 L}{\nu_{\mathrm{kin}}}$	Ratio of inertial force to viscous force acting on the fluid	zero
Schmidt's number (forced and free convection)	Sc	$\mathrm{Sc} = \frac{\nu_{\mathrm{kin}}}{D}$	Ratio of kinematic viscosity coefficient and the diffusion coefficient of the fluid	zero

Table 7.2 Definitions of quantities appearing in Table 7.1.

Quantity	Designation	Definition equation	Unit
Acceleration owing to gravity	g		m/s^2
Concentration: wt%	c		zero
Mole fraction	x		zero
Density	ρ		kg/m^3
Temperature	T		K
Thermal conductivity	λ	$\frac{dQ}{dt} = -\lambda A \frac{dT}{dy}$	$kg\ m/s^3K$
Heat transfer coefficient	h	$\frac{dQ}{dt} = -hA\Delta T$	kg/s^3K
Thermal expansion coefficient	β	$\beta = -\frac{1}{\rho}\frac{\partial \rho}{\partial T}$	K^{-1}
Concentration expansion coefficient	β'	$\beta' = \frac{1}{\rho}\frac{\partial \rho}{\partial C}$	zero
	β''	$\beta'' = \frac{1}{\rho}\frac{\partial \rho}{\partial x}$	zero
Thermal diffusivity	α	$\frac{\partial T}{\partial t} = \alpha \frac{\partial^2 T}{\partial y^2}$	m^2/s
Thermal capacitivity	c	$\alpha = \frac{\lambda}{\rho c}$	$m^2/s^2\ K$
Velocity	u	$\Delta Q = cm\Delta T$	m/s
Dynamic viscosity coefficient	η	$F = \eta A \frac{du}{dy}$	kg/ms
Kinematic viscosity coefficient	ν_{kin}	$\nu_{kin} = \frac{\eta}{\rho}$	m/s
Diffusion coefficient	D	$\frac{\partial x}{\partial t} = D\frac{\partial^2 x}{\partial y^2}$	m^2/s

Boundary Layer Theory

When a fluid flows along and parallel to a plate at rest a stationary flow pattern develops. The simplest way to treat such a system theoretically is to use the so-called boundary layer theory. The theory leads to general differential equations, which can be used to derive expressions of the velocity, the temperature and the composition of the fluid as functions of position under different given circumstances.

The procedures are indicated below for the case of fluid flow along a horizontal flat plate and the result will be applied on forced convection.

Velocity Boundary Layer

Consider a large horizontal stationary planar solid plate in a binary liquid, which moves parallel to the planar surface with the constant velocity u_∞ (Figure 7.7). We introduce a three-dimensional coordinate system according to the figure. Close to the plate, the particle velocity is zero and increases as a function of the distance y from the plate until it approaches the velocity u_∞ of the undisturbed fluid.

The undisturbed fluid is supposed to be ideal, i.e. free from viscosity. However, there must be friction forces in the fluid as the flow rate close to the plate is zero and equal to u_∞ far from the plate.

All the viscous forces are supposed to be concentrated to a thin layer close to the plate and the rest of the fluid is assumed to be an ideal fluid. Outside this *boundary layer* the velocity of the fluid is constant and equal to u_∞. The change of the flow rate occurs entirely within the boundary layer.

The flow rate increases with the distance y from the plate up to the value u_∞ but it is also a function of the distance z along the plate. Owing to viscosity in the fluid within the boundary layer, the velocity at a given value of y will decrease when z increases.

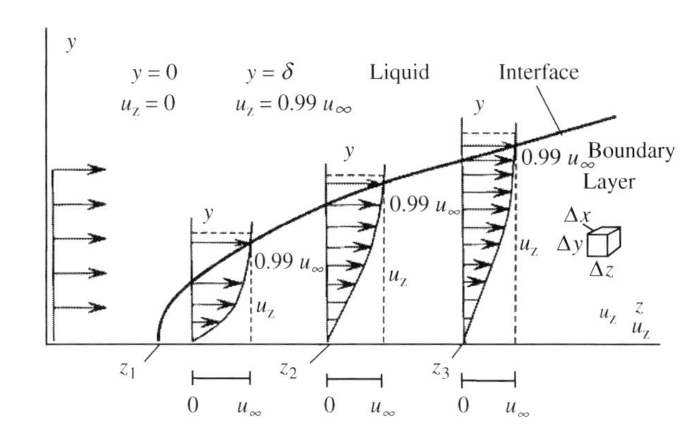

Figure 7.7 Velocity boundary layer in the vicinity of a solid plate.

In addition to the zy coordinate system with the boundary layer $z\delta$ curve, three overlapping u_z-y systems have been added, one for each of three different z values (z_1, z_2 and z_3).

Figure 7.7 can be used to define the value of y when the velocity u_z equals 99% of the velocity u_∞ of the undisturbed liquid at a given value of z.

δ = the distance from the solid plate when

$$\delta u_z = 0.99\, u_\infty \tag{7.15}$$

Figure 7.7 shows that the thickness δ of the boundary layer varies with coordinate z. Hence, a *velocity boundary layer* of variable thickness is developed below the curve. Outside the boundary layer, the liquid moves with the constant velocity u_∞ free from viscous forces. All the viscous effects, velocity gradients and shear stresses, are concentrated to the thin boundary layer.

In order to find the general differential equation for calculation of the velocity field $u(y, z)$, i.e. the flow rate as a function of position, we consider a small volume element $\Delta x \Delta y \Delta z$ within the boundary layer.

There is no velocity component in the x direction. Hence, we only have to consider the flow through the xy planes and the xz planes of the volume element. The net flow into the volume element must be zero for continuity reasons.

$$\frac{\partial u_z}{\partial z}\Delta z \Delta x \Delta y + \frac{\partial u_y}{\partial y}\Delta y \Delta x \Delta z = 0$$

or

$$\frac{\partial u_z}{\partial z} = -\frac{\partial u_y}{\partial y} \tag{7.16}$$

As no net force acts on the volume element its momentum must be conserved. Its x component is zero. The z component of the momentum is given by

$$\rho\left(u_z\frac{\partial u_z}{\partial z} + u_y\frac{\partial u_z}{\partial y}\right) = -\frac{\partial p}{\partial z} + \eta\left(\frac{\partial^2 u_z}{\partial z^2} + \frac{\partial^2 u_z}{\partial y^2}\right) \tag{7.17}$$

where

p = pressure of the fluid
ρ = density of the fluid
η = viscosity coefficient of the fluid.

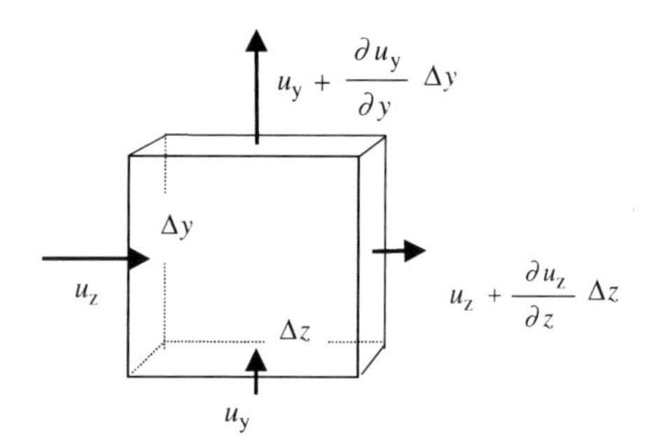

Figure 7.8 Volume element.

Similarly the y component can be written as

$$\rho\left(u_z\frac{\partial u_y}{\partial z}+u_y\frac{\partial u_y}{\partial y}\right)=-\frac{\partial p}{\partial y}+\eta\left(\frac{\partial^2 u_y}{\partial z^2}+\frac{\partial^2 u_y}{\partial y^2}\right) \tag{7.18}$$

Equations (7.16), (7.17) and (7.18) are the general equations which can be used for calculation of the properties of the velocity boundary layer. We will return to them in connection with forced convection in Section 7.3.2.

Thermal Boundary Layer

If the liquid and solid surface temperatures differ, a temperature gradient is developed which results in a *thermal boundary layer*. Outside of the thermal boundary layer the temperature of the liquid is constant and an analogous curve to that in Figure 7.7 can be drawn.

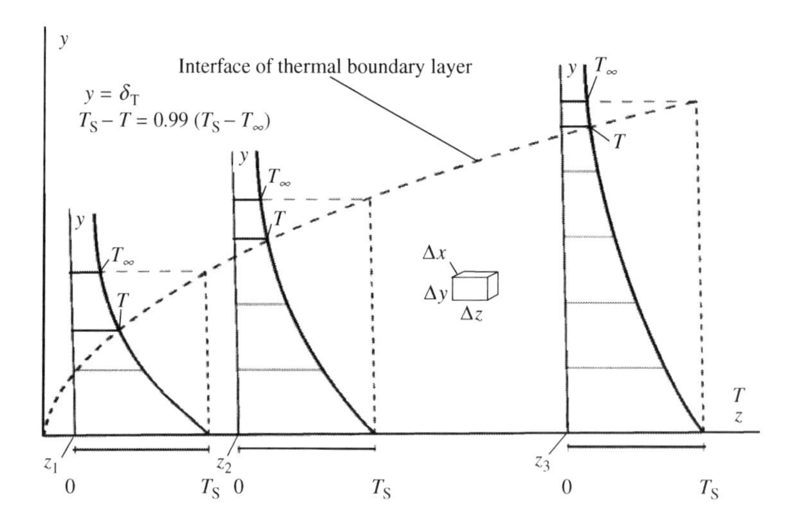

Figure 7.9 Thermal boundary layer at the vicinity of a solid plate for the case that the plate has a higher temperature than the fluid.

In addition to the zy coordinate system with the dotted boundary layer $z\delta_T$ curve, three overlapping Ty systems have been added, one for each of three different z values (z_1, z_2 and z_3). $T_S > T_\infty$.

We will discuss the case when the temperature of the plate is constant (T_S) and the temperature of the fluid far from the plate is T_∞. The temperature of the plate is supposed to be higher than that of the fluid.

The temperature at a certain value of distance y from the plate increases with distance z. The thickness of the boundary layer obviously increases with distance z. The thickness δ_T of the temperature boundary layer at a given value of z is defined analogously to that of the velocity boundary layer:

$\delta_T =$ the distance from the solid plate when

$$T_S - T = 0.99(T_S - T_\infty) \tag{7.19}$$

where
T = temperature at $y = \delta_T$
T_S = temperature of the solid surface
T_∞ = temperature of the undisturbed liquid flow.

δ_T varies with the coordinate z. The thermal boundary layer is analogous to but not identical with the velocity boundary layer, i.e. $\delta(z) \neq \delta_T(z)$.

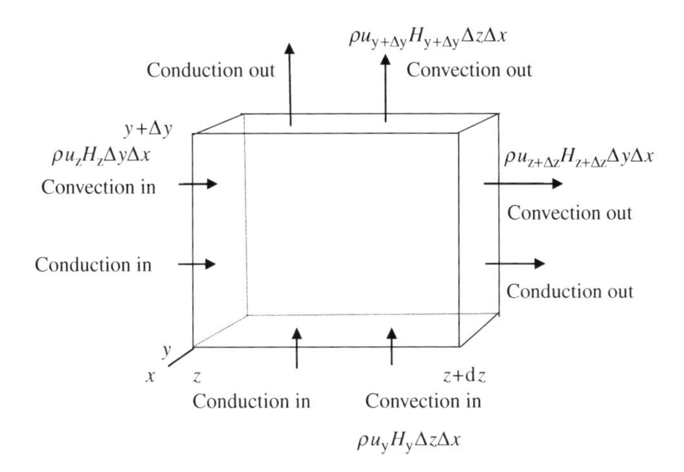

Figure 7.10 Volume element.

Consider the volume element within the boundary layer given in Figure 7.10. If the plate has a higher temperature than the fluid heat is transferred from the plate to the fluid by convection and conduction. In addition to Equations (7.16), (7.17) and (7.18) a heat balance for the volume element must be valid.

No heat is transferred through the yz planes of the volume element. The net heat flux into the volume element must be zero at stationary conditions:

$$\underbrace{\left[\left(\frac{\partial q_y}{\partial t}\right)_y - \left(\frac{\partial q_y}{\partial t}\right)_{y+\Delta y}\right]\Delta z\Delta x}_{\text{Vertical net conduction heat flow}} + \underbrace{\left[\rho u_y H_y - \rho u_{y+\Delta y} H_{y+\Delta y}\right]\Delta z\Delta x}_{\text{Vertical net convection heat flow}} +$$

$$\underbrace{\left[\left(\frac{\partial q_z}{\partial t}\right)_z - \left(\frac{\partial q_z}{\partial t}\right)_{z+\Delta z}\right]\Delta y\Delta x}_{\text{Horizontal net conduction heat flow}} + \underbrace{\left[\rho u_z H_z - \rho u_{z+\Delta z} H_{z+\Delta z}\right]\Delta y\Delta x}_{\text{Horizontal net convection heat flow}} = 0 \tag{7.20}$$

where H is the enthalpy per unit mass of the fluid. Equation (7.20) can be written with the aid of Fourier's first law (Equation (7.2) on page 363), after series expansion and division by the factor $\Delta x\ \Delta y\Delta z$

$$-\lambda\left[\frac{\partial^2 T}{\partial y^2} + \frac{\partial^2 T}{\partial z^2}\right] + \frac{\partial(\rho u_z H_z)}{\partial z} + \frac{\partial(\rho u_y H_y)}{\partial y} = 0 \tag{7.21}$$

Provided that the thermal conductivity and the density of the fluid are constants we obtain

$$-\lambda\left(\frac{\partial^2 T}{\partial z^2} + \frac{\partial^2 T}{\partial y^2}\right) + \rho\left[u_z\frac{\partial H}{\partial z} + u_y\frac{\partial H}{\partial y} + H\left(\frac{\partial u_z}{\partial z} + \frac{\partial u_y}{\partial y}\right)\right] = 0 \tag{7.22}$$

The last term in Equation (7.22) is equal to zero for fluids with constant density and $u_x = 0$ because of the continuity Equation (7.16) on page 369. With the aid of the definition of the heat capacitivity at constant pressure

$$c_p = \left(\frac{\partial H}{\partial T}\right)_p \tag{7.23}$$

we can transform Equation (7.22) into

$$\lambda\left(\frac{\partial^2 T}{\partial z^2} + \frac{\partial^2 T}{\partial y^2}\right) = \rho c_p \left[u_z \frac{\partial T}{\partial z} + u_y \frac{\partial T}{\partial y}\right] \tag{7.24}$$

If we introduce the thermal diffusivity $\alpha = \lambda/\rho c_p$ into Equation (7.24) it can be written as

$$u_z \frac{\partial T}{\partial z} + u_y \frac{\partial T}{\partial y} = \alpha\left(\frac{\partial^2 T}{\partial z^2} + \frac{\partial^2 T}{\partial y^2}\right) \tag{7.25}$$

Equation (7.25) is the general differential equation that controls the heat flow at convection together with Equations (7.15), (7.16) and (7.18). It will be discussed further in connection with forced convection in Section 7.3.2.

Concentration Boundary Layer

If the concentration of the solute at the solid surface differs from that in the liquid, a solute concentration boundary layer is developed. It is defined in complete analogy with the temperature boundary layer. The concept will not be used in this chapter. It will be introduced into Chapter 8 in connection with growth of single crystals.

Comparison between the Convection Boundary Layers

The velocity boundary layer is related to surface friction and friction in the liquid which is always present. Thermal boundary layers and concentration boundary layers are only present if temperature and concentration differences exist between the solid plate and the liquid, respectively. They are related to convection heat transfer and convection mass transfer, respectively.

The three quantities δ, δ_T and δ_c are not equal. These concepts of the boundary layers will be applied below in connection with forced and free convection, respectively.

7.3.2 *Forced Convection*

Forced convection is controlled by external circumstances. It is used for several purposes, for example mixing components in a liquid or cooling with water or air as a cooling agent.

In nearly all casting processes, liquids are poured into a mould. The solidification process starts during this procedure and is influenced by the convection in the melt caused by filling the mould. In many crystallization processes, for example the Czochralski process (Chapter 8), the crystals are rotated during the growth process. The rotation itself causes forced convection in the melt.

Thickness of Velocity Boundary Layer

The concept of forced convection means, that the velocity of the fluid can be varied, which affects the shape of the boundary layer. It is hard to find exact solutions of the differential equations and some reasonable assumptions have to be accepted. As the boundary layer is assumed to be very thin and in contact with the solid planar surface, Prandl found it reasonable to suggest that $u_y \ll u_z$ and that $\partial u_z/\partial y$ is large in comparison with $\partial u_z/\partial z$. If so, all terms that contain u_y and its derivatives are small and can be neglected. This simplifies the problem considerably. If, in addition, there is no pressure gradient within the boundary layer, the solution involves the following expression of the thickness of the boundary layer as a function of z:

$$\delta = 5.0 \times \sqrt{\frac{\nu_{kin}}{u_{\infty}}}\sqrt{z} \tag{7.26}$$

where
ν_{kin} = the kinematic viscosity coefficient
u_{∞} = velocity of the undisturbed fluid
z = distance along the plate in the flow direction.

δ is a parabolic function of z. For a given value of z the thickness increases with increasing kinematic viscosity and decreases with increasing flow rate of the fluid. Alternatively, it can be expressed as a function of Reynold's number if we assume that the characteristic length is equal to z (Table 7.1 on page 367)

$$\delta = \frac{5.0}{\sqrt{\text{Re}_z}} z \tag{7.27}$$

Equation (7.27) is often used for calculation of the thickness of the velocity boundary layer. Other approximations lead to similar results with somewhat lower value of the constant than 5.0 in Equation (7.27).

Heat Transfer from a Horizontal Planar Plate at a Uniform Constant Temperature. Heat Transfer Coefficient

The basic equations to find the properties of the thermal layer are Equations (7.16), (7.17) and (7.25). The last one controls the heat transport.

$$u_z \frac{\partial T}{\partial z} + u_y \frac{\partial T}{\partial y} = \alpha \left(\frac{\partial^2 T}{\partial z^2} + \frac{\partial^2 T}{\partial y^2} \right) \tag{7.25}$$

The corresponding momentum boundary equation in the absence of a pressure gradient in the y direction is

$$u_z \frac{\partial u_z}{\partial z} + u_y \frac{\partial u_z}{\partial y} = \nu_{kin} \left(\frac{\partial^2 u_z}{\partial z^2} + \frac{\partial^2 u_z}{\partial y^2} \right) \tag{7.28}$$

where α has been replaced by the kinematic viscosity coefficient ν_{kin} (Table 7.2, page 368). If $\nu_{kin} = \alpha$, Equations (7.25) and (7.28) become identical and the temperature and velocity boundary layers coincide ($\delta = \delta_T$). The ratio of ν_{kin} and α is called *Prandl's number*. It can be used to characterize the heat transport from the planar plate to the fluid.

$$\text{Pr} = \frac{\nu_{kin}}{\alpha} = \frac{\eta}{\rho}\frac{\rho c_p}{\lambda} = \frac{\eta c_p}{\lambda} \tag{7.29}$$

Molten metals, that often have low heat capacitivities and high thermal conductivities, have small values of Prandl's number. In this case the dominating mechanism of heat transport is conduction.

In fluids with low thermal conductivites, the Prandl numbers are high and the dominating heat transport occurs by convection.

Heat Transport Coefficient

The solution of the thermal and momentum boundary equations for laminar fluid flow within the interval $0.6 \leq \mathrm{Pr} \leq 50$ gives the local heat transfer coefficient h_z

$$h_z = 0.332\,\lambda \times \mathrm{Pr}^{0.343} \times \sqrt{\frac{u_\infty}{\nu_{\mathrm{kin}}}} z^{-0.5} \tag{7.30}$$

The average heat transfer coefficient is obtained as

$$\overline{h_{\mathrm{L}}} = \frac{1}{L}\int_0^L h_z\,\mathrm{d}z \tag{7.31}$$

where L is the extension of the plate in the z direction. Inserting the expression of h_z into Equation (7.31) we obtain

$$\overline{h_{\mathrm{L}}} = 0.664\,\lambda \times \mathrm{Pr}^{0.343} \times \sqrt{\frac{u_\infty}{\nu_{\mathrm{kin}}}} L^{-0.5} \tag{7.32}$$

which is twice the local heat transfer coefficient at $z = L$.

Another comparison between thermal conduction and heat transport by convection can be made as follows. If the heat transfer from the solid surface (temperature T_{S}) occurred only by convection, the heat flux per unit area would be

$$\left(\frac{\mathrm{d}q}{\mathrm{d}t}\right)_{\mathrm{convection}} = -h(T_\infty - T_{\mathrm{S}}) \tag{7.33}$$

If the heat transfer would occur by heat conduction only across the same length L we would obtain, according to Fourier's first law

$$\left(\frac{\mathrm{d}q}{\mathrm{d}t}\right)_{\mathrm{conduction}} = -\lambda\frac{T_\infty - T_{\mathrm{S}}}{L} = \frac{\lambda(T_{\mathrm{S}} - T_\infty)}{L} \tag{7.34}$$

If we divide Equation (7.33) with Equation (7.34) we obtain

$$\frac{\left(\frac{\mathrm{d}q}{\mathrm{d}t}\right)_{\mathrm{convection}}}{\left(\frac{\mathrm{d}q}{\mathrm{d}t}\right)_{\mathrm{conduction}}} = \frac{hL}{\lambda} = \mathrm{Nu} \tag{7.35}$$

The ratio of the heat fluxes turns out to be equal to Nussel's number. Nussel's number is therefore a measure of the mechanism of heat transport in the fluid. A high value of h and/or a small value of λ, i.e. a *high* Nussel's number, indicate that heat transport by *convection* is the dominating mechanism.

If the characteristic length is set to z, the distance to the leading edge of the plate we can define the local Nussel number as

$$\mathrm{Nu}_z = \frac{hz}{\lambda} \tag{7.36}$$

Combining Equations (7.32) and (7.36) we obtain

$$\mathrm{Nu}_z = 0.664 \times \mathrm{Pr}^{0.343} \times \sqrt{\frac{u_z z}{\nu_{\mathrm{kin}}}} = 0.664 \times \mathrm{Pr}^{0.343} \times \mathrm{Re}_z^{0.5} \tag{7.37}$$

and similarly

$$\mathrm{Nu_L} = 0.664 \times \mathrm{Pr}^{0.343} \times \sqrt{\frac{u_z L}{\nu_{\mathrm{kin}}}} = 0.664 \times \mathrm{Pr}^{0.343} \times \mathrm{Re}_{\mathrm{L}}^{0.5} \tag{7.38}$$

where Re_z is the local Reinhold's number. Hence, for heat transfer by forced convection from or to a horizontal planar plate Nussel's number is a function of Prandl's number and Reinhold's number only.

By combining Equation (7.38) with the definition of Nussel's number, Equation (7.35), we can conclude that the heat transfer by convection increases with the velocity of the liquid.

7.3.3 Natural Convection

Free or natural convection is less powerful than forced convection but still of great importance in metal melts, for example in many solidification processes such as ingot solidification. As we will see in Chapter 8 it is essential in zone refining.

The driving force of natural convection is a *density difference*, normally caused by a *temperature gradient* in the liquid. The density of a metal melt decreases with increasing temperature, owing to the volume expansion of the melt. As an example, we will consider the motion in a solidifying ingot. Heat is removed through the mould and hence the solidification front is cooled. Its temperature decreases and the density of the melt close to the front increases. The melt sinks downwards and initiates of the natural convection. The cooled melt is replaced by warmer melt with lower density, which moves upwards. A ring of cyclic motions appears in the melt, which keeps it in continuous motion and transports heat from the interior of the melt to the solidification front. The process is shown in Figure 7.11.

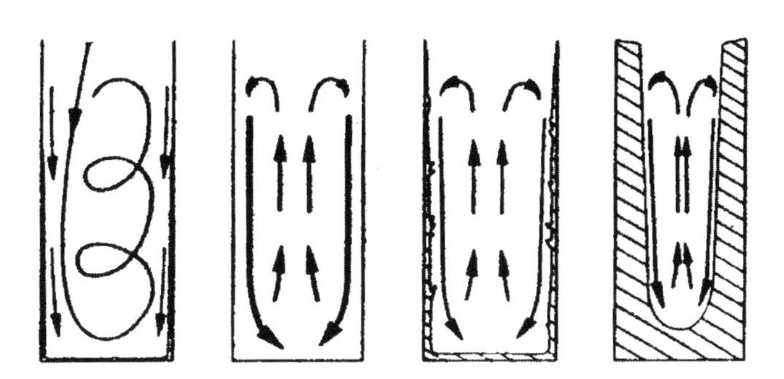

Figure 7.11 Natural convection in an ingot during the solidification process.

The temperature distribution in the liquid controls the velocity of the flowing melt and consequently the heat transport. To obtain an estimation of the magnitude of the heat transport within the melt we have to study the temperature distribution in the melt and the flow rate caused by temperature differences.

Natural Convection at a Planar Vertical Surface

Eckert and Drake [1], who developed the theory of boundary layers, have analyzed several typical cases of natural convection, for example at a planar vertical surface, between two parallel walls and two concentric cylinders, respectively, of different temperatures. Here we will discuss the simplest model, the planar vertical wall, and apply Eckert and Drake's mathematical model on metal melts.

We will calculate the thickness δ of the velocity boundary layer and determine the temperature and the flow rate as a function of position within the boundary layer.

If the vertical wall is colder than the metal melt heat is transferred to the wall and the melt flows downwards. Eckert and Drake assumed that the flow is located to the thin boundary layer of varying thickness. The thickness of the boundary layer is zero at the top and increases downwards along the vertical wall.

The driving force, which keeps the natural convection in action, is the temperature differences in the melt. Eckert and Drake assumed that the temperature within the boundary layer increases from the solidus temperature T_S at the vertical wall to T_m the temperature of the melt inside the boundary layer. The temperature T_S of the interface is assumed to be constant along the whole vertical surface.

The approximate temperature distribution can be described mathematically by the function

$$T = T_m - (T_m - T_S)\left(1 - \frac{y}{\delta}\right)^2 \tag{7.39}$$

where

δ = the thickness of the velocity boundary layer

T_m = temperature of the melt inside the boundary layer.

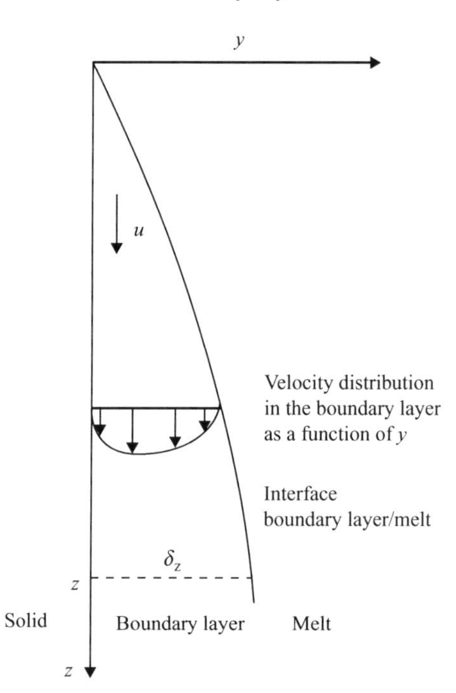

Figure 7.12 Velocity boundary layer at natural convection.

Obviously the function fulfills the boundary conditions:

$$\begin{array}{ccc} T = T_S & \text{and} & T = T_m \\ y = 0 & & y = \delta \end{array}$$

Eckert and Drake derived an expression the flow rate within the boundary layer on the basis of the temperature distribution given above within the boundary layer

$$u = u_0(z)\frac{y}{\delta}\left(1 - \frac{y}{\delta}\right)^2 \tag{7.40}$$

where $u_0\,(z)$ is a function of z, the depth under the free surface of the melt. The dimension of $u_0\,(z)$ is velocity.

The functions in Equations (7.39) and (7.40) are shown graphically in Figure 7.13.

Figure 7.13 shows that the temperature increases continuously from the vertical wall to the end of the boundary layer while the flow rate has a maximum within this region. By normal

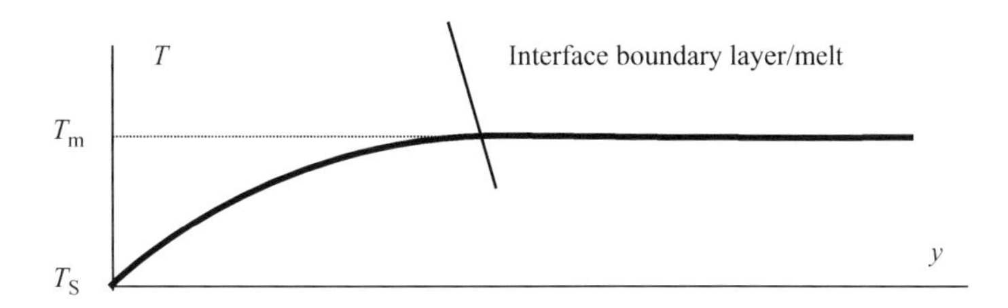

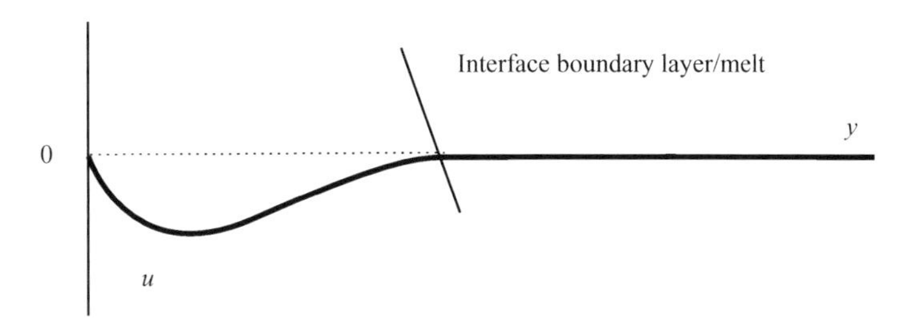

Figure 7.13 Natural convection. Temperature and flow rate in the thin boundary layer as functions of the distance from the vertical wall.
Compare with Figure 7.12.

maximum/minimum examination of the function, the coordinates of the maximum value of the flow rate can be calculated:

$$u_{\max} = \frac{4}{27} \times u_0(z) \quad \text{for} \quad y = \frac{\delta}{3} \tag{7.41}$$

In our calculations above we have made the approximation that $\delta \approx \delta_T$. This is not completely in agreement with reality but it is an approximation, which simplifies our further calculations considerably. In the end, good agreement between theory and experimental results is achieved, which justifies the approximation.

If the Equations (7.39) and (7.40) for temperature and flow rate are completed with a momentum equation (the total momentum is constant) and a heat flow equation (the energy law), it is possible to evaluate $u_{\max}$ and δ as functions of z, T_S and T_L. The derivation will be omitted here. The expression for $u_{\max}$ is of minor interest but the expression for the thickness of the boundary layer will be given here:

$$\delta(z) = 3.93 \times \left(\frac{\nu_{\text{kin}}}{\alpha}\right)^{-\frac{1}{2}} \left(\frac{20}{21} + \frac{\nu_{\text{kin}}}{\alpha}\right)^{\frac{1}{4}} \left[\frac{g\beta(T_m - T_s)}{\nu_{\text{kin}}^2}\right]^{-\frac{1}{4}} z^{\frac{1}{4}} \tag{7.42}$$

where
α = thermal diffusivity of the melt
β = volume dilatation coefficient of the melt
ν_{kin} = kinematic viscosity coefficient of the melt.

α, β and ν_{kin} are material constants (see Table 7.2 on page 368). For each alloy melt the expression for the maximum interface thickness can be written as

$$\delta(z) = B\left[\frac{g}{z}(T_m - T_s)\right]^{-\frac{1}{4}} \tag{7.43}$$

where B is summarized material constant. Examples of B values of some common metal melts are given in Table 7.3.

Table 7.3 Values of B

Metal	$B \times 10^2\ \text{m}^{\frac{3}{4}}\,\text{K}^{\frac{1}{4}}$
Fe	5.2
Al	11.9
Cu	13.8

When the expression of δ is known the temperature and flow rate of the melt within the boundary layer as functions of position can easily be found by use of Equations (7.39) and (7.40). Hence, the desired information has been achieved.

7.4 Theory of Heat Transport at Unidirectional Solidification

7.4.1 Unidirectional Solidification of Alloys at Ideal Heat Transport

The properties of solid alloy materials are strongly influenced by the solidification process. The solidification process of a molten alloy depends mainly on the material properties of the alloy and the mould and on the cooling conditions. The more rapid the crystallization process is, the better will be the mechanical properties of the product in most cases.

Good thermal conductivities of the alloy and the mould material and by strong cooling, i.e. by a high temperature gradient, promote the crystallization process. Another property, which is very important and out of control, is the width of the solidification interval of the alloy.

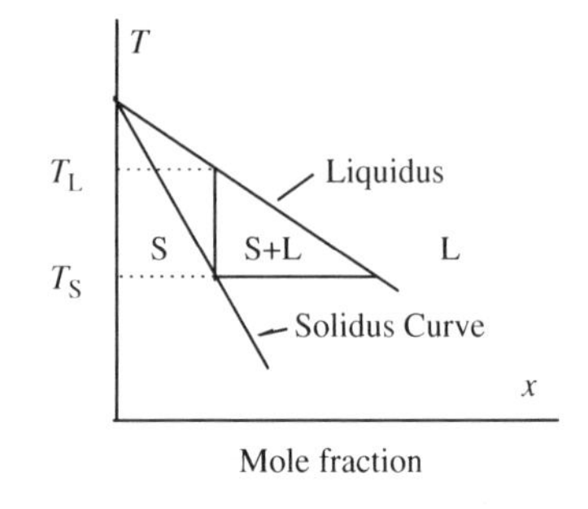

Figure 7.14 Schematic phase diagram of a binary alloy. The solidification interval $= T_L - T_S$. The curves can often be replaced by straight lines.

Modes of Solidification

The width of the solidification interval of an alloy at equilibrium conditions is defined as the difference between its liquidus and solidus temperatures. It is a function of the composition of the alloy as is shown in Figure 7.14.

Pure metals and eutectic mixtures have very narrow solidification intervals, i.e. the liquidus and solidus lines almost coincide.

Most alloys have wide solidification intervals. The shape of the solidification front is very different in the two cases as shown in Figures 7.15 and 7.16.

Alloys with narrow solidification intervals and good thermal conductivites, which solidify in chill moulds with strong cooling, solidify rapidly. Less favourable conditions are wide solidification intervals, poor thermal conductivity of the alloy and the mould, for example a sand mould and/or weak cooling.

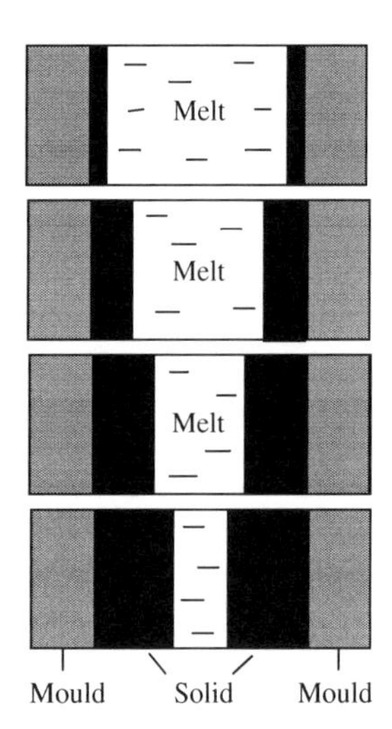

Figure 7.15 Solidification process of a pure metal or a eutectic mixture under the influence of strong cooling. Reproduced from Elsevier © 1970.

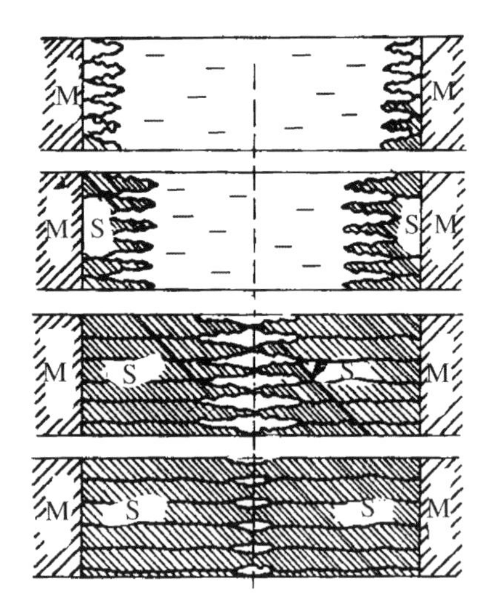

Figure 7.16 Solidification process of an alloy with a wide solidification interval or an alloy under the influence of poor cooling. The two-phase areas are indicated in the figure. A high temperature gradient gives a small distance between the liquidus and solidus solidification fronts. Reproduced from Elsevier © 1970.

Temperature Distribution and Solidification Rate of an Alloy in a Metal Mould at Ideal Cooling[1]

In order to analyze the solidification process, i.e. the influence the various parameters, it is necessary to solve the equation for the heat flow, i.e. Equation 7.9 on page 364.

[1] The content in the following sections (pages 378–383) is treated in detail on pages 64–70 in Chapter 4 in [2].

$$\frac{\partial T}{\partial t} = \alpha \frac{\partial^2 T}{\partial y^2} \qquad \text{Fourier's second law} \tag{7.9}$$

The solution of this partial differential equation of second order is the temperature T as a function of position y and time t. The solution contains two arbitrary constants, which are determined by given boundary conditions.

If a small amount of heat is distributed in an infinitely large body the solution of Equation (7.9) will be

$$T = A + \frac{B}{\sqrt{t}} e^{-\frac{y^2}{4\alpha t}} \tag{7.44}$$

We make the following assumptions

1. The contact between melt and mould is good, which means that there is no resistance against heat transport over the mould/metal interface.
2. The metal is not superheated.
3. The metal melt has a narrow solidification interval.
4. The volume of the mould is very large, approximately unlimited.

The schematic temperature profile in Figure 7.17 has been drawn according to these assumptions. The temperatures of metal and mould will adjust to equilibrium, i.e. they are equal.

The time is zero when the metal melt is poured into the mould. The position of the y axis is shown in Figure 7.17. y is zero at the interface between melt and mould, positive in the melt and negative in the mould.

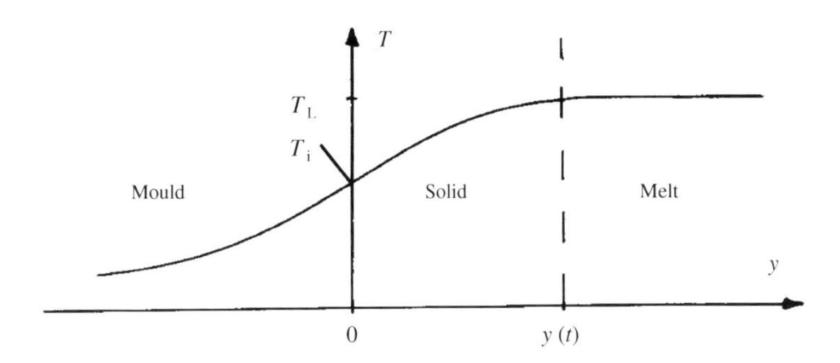

Figure 7.17 Approximate temperature distribution in a solidifying but not superheated metal melt in a metal mould.

Since the metal and the mould are made of different materials with different properties, the differential Equation (7.9) above must be solved separately for each of them.

The solutions can be written as

$$T_{\text{mould}} = A_{\text{mould}} + B_{\text{mould}} \operatorname{erf}\left(\frac{y}{\sqrt{4\alpha_{\text{mould}} t}}\right) \tag{7.45a}$$

$$T_{\text{metal}} = A_{\text{metal}} + B_{\text{metal}} \operatorname{erf}\left(\frac{y}{\sqrt{4\alpha_{\text{metal}} t}}\right) \tag{7.45b}$$

where

y = coordinate of the solidification front
α = thermal diffusivity (page 364)
t = time
T_{mould} = temperature of the mould at position ($y < 0$)
T_{metal} = temperature of the metal at position ($y > 0$)
A, B = integration constants.

There are five boundary conditions for determination of the four integration constants and the parameter λ, defined by the relationship

$$y_L(t) = \lambda\sqrt{4\alpha_{metal}t} \tag{7.46}$$

where $y_L(t)$ is the position of the advancing solidification front and. $\alpha_{metal} = \dfrac{\lambda_{metal}}{\rho_{metal}c_p^{metal}}$ is the diffusivity of the metal.

The erf function (page 67 in Chapter 4 in [2]) is a distribution function, defined by the relationship

$$\operatorname{erf}(\lambda) = \frac{2}{\sqrt{\pi}}\int_0^{\lambda} e^{-y^2}\,dy \tag{7.47}$$

The details of the solution will be omitted here. The whole solution is given on pages 67–70 in Chapter 4 in [2].

The parameter λ is found by solving Equation (7.48)

$$\frac{c_p(T_L - T_0)}{-\Delta H} = \sqrt{\pi}\lambda\, e^{\lambda^2}\left(\sqrt{\frac{\lambda_{metal}\rho_{metal}c_p^{metal}}{\lambda_{mould}\rho_{mould}c_p^{mould}}} + \operatorname{erf}(\lambda)\right) \tag{7.48}$$

with the aid of known material constants, either graphically by use the curves in Figures 7.18a and 7.18b or by trial and error. A solved example is given on page 72 in Chapter 4 in [2].

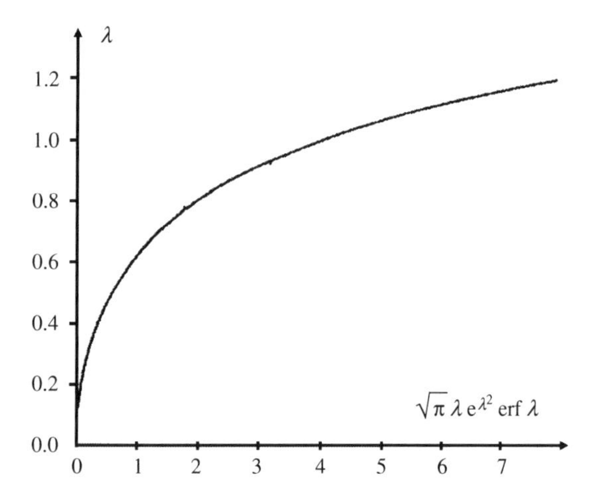

Figure 7.18a The erf function $\sqrt{\pi}\lambda e^{\lambda^2}\operatorname{erf}\lambda$ as a function of λ.

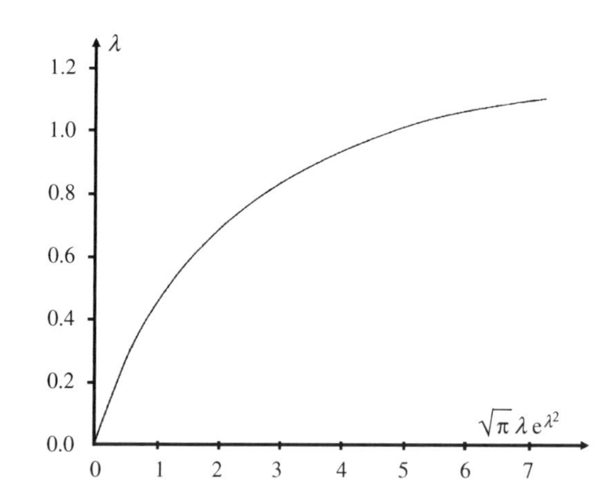

Figure 7.18b The function $\sqrt{\pi}\lambda e^{\lambda^2}$ as a function of λ.

When the value of the parameter λ is known, erf λ are inserted into the expressions of the integration constants

$$B_{metal} = \frac{T_L - T_0}{\dfrac{\lambda_{metal}\sqrt{\alpha_{mould}}}{\lambda_{mould}\sqrt{\alpha_{metal}}} + \operatorname{erf}(\lambda)} \tag{7.49a}$$

$$B_{mould} = \frac{T_L - T_0}{\dfrac{\lambda_{metal}\sqrt{\alpha_{mould}}}{\lambda_{mould}\sqrt{\alpha_{metal}}} + \operatorname{erf}(\lambda)}\,\frac{\lambda_{metal}\sqrt{\alpha_{mould}}}{\lambda_{mould}\sqrt{\alpha_{metal}}} \tag{7.49b}$$

$$A_{\text{mould}} = A_{\text{metal}} = T_0 + \frac{T_L - T_0}{\dfrac{\lambda_{\text{metal}}\sqrt{\alpha_{\text{mould}}}}{\lambda_{\text{mould}}\sqrt{\alpha_{\text{metal}}}} + \operatorname{erf}(\lambda)} \frac{\lambda_{\text{metal}}\sqrt{\alpha_{\text{mould}}}}{\lambda_{\text{mould}}\sqrt{\alpha_{\text{metal}}}} \tag{7.50}$$

where

λ = parameter

λ_{metal} = conductivity of heat of the metal

λ_{mould} = conductivity of heat of the mould

T_L = temperature of the melt

T_0 = temperature of the surroundings.

The temperatures are known. Then the temperature distribution is given by the relationships (7.45) (page 379). When λ and the temperature distribution are known, other quantities can be calculated, for example the rate of solidification, by taking the derivative of Equation (7.46) with respect to t:

$$\frac{\mathrm{d}y_L(t)}{\mathrm{d}t} = \lambda\sqrt{\frac{4\alpha_{\text{metal}}}{t}} \tag{7.51}$$

7.4.2 *Unidirectional Solidification of Alloys at Poor Heat Transport*

The procedure to find the temperature distribution in the alloy and in the mould is the same as that in Section 7.4.1. The heat Equation (7.9) has to be solved. However, the assumptions and boundary conditions differ of course in the two cases.

As a typical application we choose the case of an alloy, which solidifies in a dry sand mould. We make the following assumptions:

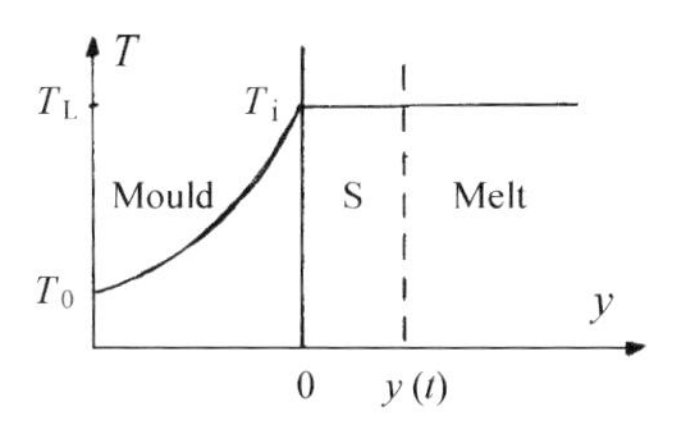

Figure 7.19 Schematic temperature profile at solidification of an alloy in a dry sand mould.

1. The conductivity of the alloy is very large compared to that of the sand mould.
2. The temperature of the sand mould wall is equal to the constant temperature T_L of the melt during the whole solidification process.
3. At a large distance from the interface alloy/sand mould the temperature of the mould equals room temperature T_0.

The temperature of the alloy is simple in this case. It is constant and equal to the temperature of the melt T_L.

$$T_{\text{metal}} = T_L = const \tag{7.52}$$

The temperature of the sand mould as a function of position and time is found by calculations of the heat conduction in the mould. The solution of the heat equation can be written as

$$T_{\text{mould}} = A_{\text{mould}} + B_{\text{mould}} \operatorname{erf}\left(\frac{y}{\sqrt{4\alpha_{\text{mould}}t}}\right) \tag{7.53}$$

Introduction of boundary conditions gives the temperature distribution in the dry sand mould

$$T_{\text{mould}}(y,t) = T_i - (T_i - T_0)\operatorname{erf}\left(\frac{y}{\sqrt{4\alpha_{\text{mould}}t}}\right) \tag{7.54}$$

where T_i is the temperature of the interface mould/solid metal. For given values of y and t the erf function (Equation (7.47) on page 380) can be found in tables.

Solidification Rate and Thickness of Solidifying Layer as a Function of Time in a Dry Sand Mould

The heat flow through the interface consists entirely of heat of fusion because the temperature T_L in the solid phase and the melt is constant. The heat flux can be written in two ways:

$$\frac{dq}{dt} = \rho_{metal}(-\Delta H)\left(\frac{dy_L}{dt}\right)_{y=0} = \lambda_{mould}\left(\frac{dT_{mould}}{dt}\right)_{y=0} \tag{7.55}$$

where
dy_L/dt = solidification rate
$-\Delta H$ = heat of fusion of the alloy (J/kg)
ρ_{metal} = density of the alloy.

If we take the derivative of Equation (7.54) with respect to y and introduce the derivative into Equation (7.55) we obtain an expression of the solidification rate (dy/dt) at the interface. The thickness of the solidifying layer as a function of time can be calculated by integration of the solidification rate with respect to t.

$$y_L(t) = \frac{2}{\sqrt{\pi}}\frac{T_i - T_0}{\rho_{metal}(-\Delta H)}\sqrt{\lambda_{mould}\rho_{mould}c_p^{mould}}\sqrt{t} \tag{7.56}$$

- The thickness of the solidified layer is proportional the square root of time.

If we assume that each area unit of the mould surface absorbs heat independently of its position and shape we obtain

$$t_{total} = C\left(\frac{V_{metal}}{A}\right)^2 \quad \text{Chvorinov's rule} \tag{7.57}$$

V_{metal} is the volume of the casting and A is its area. The constant C is equal to

$$C = \frac{\pi}{4}\left(\frac{\rho_{metal}(-\Delta H)}{(T_i - T_0)}\right)^2 \frac{1}{\lambda_{mould}\rho_{mould}c_p^{mould}}$$

Chvorinov's rule is applied on page 80 in [2].

Solidification Rate and Solidification Time of an Alloy in a Mould with Poor Contact between Mould and Alloy

Owing to solidification shrinkage the solidified alloy shell will loose the contact with the mould and an air gap arises.

The temperature at the interface becomes discontinuous as is seen in Figure 7.20.

Analogous calculations of the temperature distribution give among other things.

The temperature of the solid alloy at the interface:

$$T_{i\,metal} = \frac{T_L - T_0}{1 + \frac{h}{\lambda}y_L(t)} + T_0 \tag{7.58}$$

The velocity of the solidification front:

$$\frac{dy_L}{dt} = \frac{T_L - T_0}{\rho(-\Delta H)}\frac{\lambda}{y_L}\left(1 - \frac{1}{1 + \frac{h}{\lambda}y_L}\right) \tag{7.59}$$

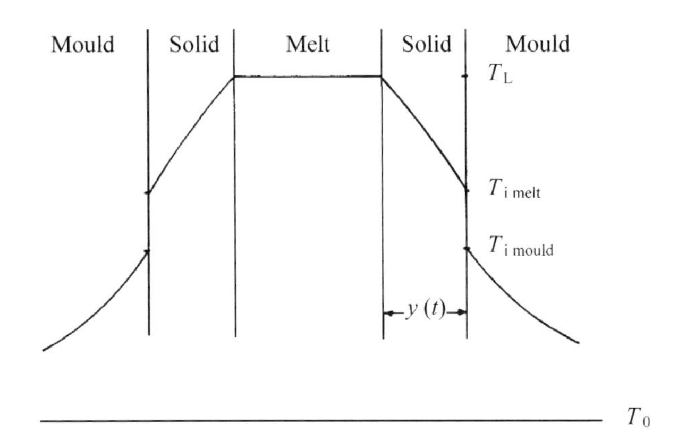

Figure 7.20 Temperature distribution in the case of poor contact between mould and metal at time t. T_L = temperature of melt;

$T_{i\ melt}$ = temperature of the solid metal at the interface;

$T_{i\ mould}$ = temperature of the mould at the interface;

T_0 = temperature of the surroundings.

The solidification time:

$$t = \frac{\rho(-\Delta H)}{T_L - T_0}\frac{y_L}{h}\left(1 + \frac{h}{2\lambda}y_L\right) \tag{7.60}$$

More detailed information and derivation of the formulas are given on page 74 in Chapter 4 in [2].

Example 7.1

In an experimental work by Zekely on the effect of natural convection on the solidification process in an ingot, the shape of the solidification front, shown in the figure in the margin was observed.

The height of the ingot was 1.5 m. Use the data given below and try to explain the shape of the solidification front by calculation of the position of the solidification front y_L (t) as a function of z.

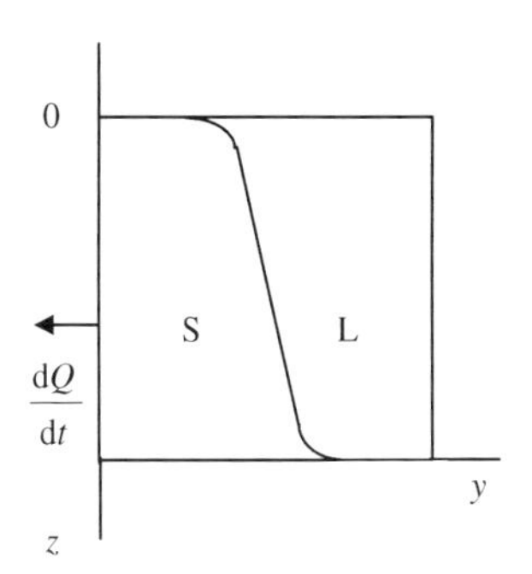

Known data

$$\begin{aligned} \rho_{Fe} &= 7.5 \times 10^3\ \text{kg/m}^3 & T_s &= 1500^\circ\text{C} \\ -\Delta H &= 2.72 \times 10^3\ \text{J/kg} & T_L &= 1550^\circ\text{C} \\ \lambda_s &= 30\ \text{J/msK} & h &= 1.0 \times 10^3\ \text{J/m}^2\ \text{sK} \\ \lambda_L &= 10\ \text{J/msK} & g &= 10\ \text{m/s}^2. \end{aligned}$$

Solution

The temperature profile in an ingot during solidification is shown in the figure below. The temperature distribution in the solid is assumed to be linear. The temperature in the liquid near the solid is assumed to vary linearly within a boundary layer: $\delta = f(z)$ where z is the depth under the free surface of the ingot.

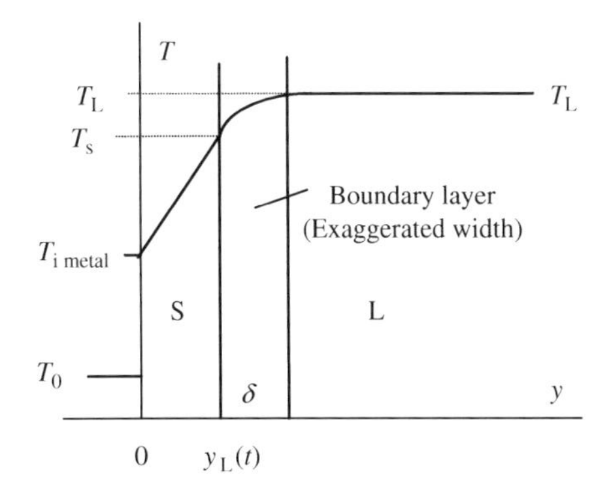

The heat flow across the solidification front can be written as

$$\lambda_s \frac{T_s - T_{i\,metal}}{y_L} = \lambda_L \frac{T_L - T_s}{\delta} + \rho(-\Delta H)\frac{dy_L}{dt} \tag{1'}$$

heat per unit area and unit time through the solid	heat per unit area and unit time through the boundary layer	heat of fusion per unit area and unit time

Owing to large temperature differences within the melt at different positions of the ingot strong natural convection will appear in the melt, which has to be considered. Equation (7.43) on page 377 gives an expression of the thickness of the boundary layer δ as a function of z:

$$\delta(z) = B\left[\frac{g}{z}(T_L - T_s)\right]^{-\frac{1}{4}} \tag{2'}$$

Table 7.3 on page 377 gives $B_{Fe} = 5.2 \times 10^2\ m^{\frac{3}{4}}\ K^{\frac{1}{4}}$.

The temperature of the solidified metal near the air gap close to the mould can be written (according to Figure 7.20 on page 383 and Equation (7.58) on page 382)

$$T_{i\,metal} = \frac{T_L - T_0}{1 + \frac{h}{\lambda} y_L(t)} + T_0 \tag{3'}$$

The expressions for δ and T_{imetal} are inserted into Equation (1′), which gives a differential equation containing y_L and dy_L/dt and known quantities. z is a parameter, which can be varied.

The differential equation can be solved numerically which gives

$$y_L = f(t, z)$$

where y_L is the coordinate for the solidification front. If y_L is plotted versus various values of z within the range 0–1.5 m at a constant value of t we obtain the shape of the solidification front at the given time. If other values of t are chosen we obtain the shape of the solidification front for different times as a set of curves of the type

$$z = F(y_L, t)$$

where t is a parameter.

Answer:

Calculations with the given data result in the plotted curves shown below.

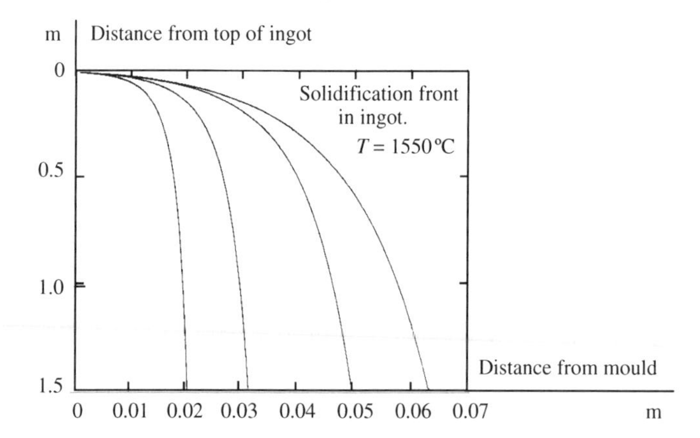

The solidification front is obviously far from planar as a result of the convection in the melt. The agreement between the experimental and calculated shapes of the solidification front is good for small values of z.

However, for large values of z no agreement is obtained. Large values of z corresponds to large values of δ. In this case the first term in Equation (1′) can be neglected compared to the second term and the value of dy_L/dt becomes large, in agreement with the curves. The model obviously is not good enough at the bottom of the ingot. The reason is that Equation (2′) is not valid for large values of z and that the bottom forces the liquid along the bottom in the y direction.

Instead some sort of symmetry is to be expected. If z is replaced by $(1.5 - z)$, Equation (2′) is valid for large values of z. We also have to consider the different temperature conditions. At the top a convection flow of hot melt is directed to the solidification front which forces it away to low values of y_L. At large values of z a convection flow of cooler melt leaves the interface region. The result is that the solidification front is forced to high values of y_L. The sketch in the figure shows the appearance of the solidification front. It is in agreement with the expected solidification front in the text of this exercise.

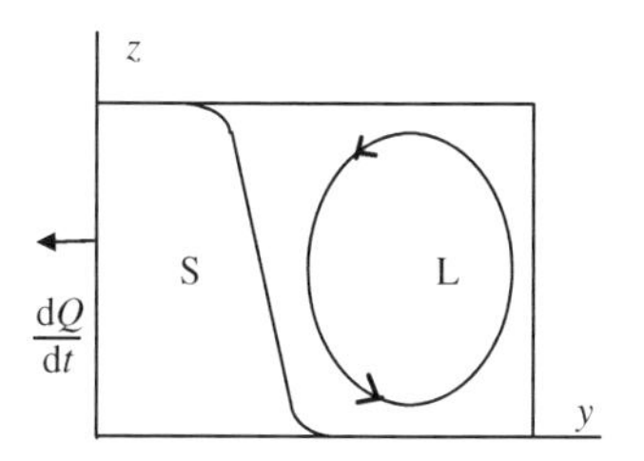

7.5 Production of Single Crystals by Unidirectional Solidification

The most frequent method to produce products with extremely good mechanical and other physical properties is production of single crystals by unidirectional solidification of a melt. This method, originally invented by Bridgman in 1925 [3], has been briefly described in Chapter 6. The method has been developed further and a number of variants have been introduced.

The advantages of the group of Bridgman methods are that they rapidly produce crystals with good dimensional accuracy, the technique is simple and requires comparatively little control and supervision. An important and favourable fact is also that the products require minimal finishing.

Other methods to produce single crystals are zone melting methods and the Czochralski method [4]. Mass transport is more important than heat transport in these methods. These two groups of methods will be introduced and analyzed in Chapter 8.

7.5.1 Single Crystal Production by Bridgman Technique

A crucible is moved downwards through a furnace, normally at a rate between 1–30 mm per hour. The temperature of the furnace is slightly above the melting point of the metal in the crucible. The solidification starts at the bottom of the crucible and the solidification front moves slowly upwards. Alternatively the crucible can be fixed and the furnace moved upwards or its temperature can be changed in a way which is equivalent to this relative motion.

It is very important that neither the melt nor its vapour reacts chemically with the crucible material. Coatings of for example vapour-deposited carbon or oil are used to reduce the risk. If the thermal expansion of the crystal and the crucible are very different, the crystal may be strained which should be avoided.

Both effects can be avoided if a horizontal water-cooled boat of the crystal material is used as a crucible. In this case the equipment is designed to let a vertical solidification front move horizontally (Kapitza [5]).

The crystals grown by the Bridgman method are almost always rather imperfect. The dislocation density normally exceeds $10^4\,cm^{-2}$.

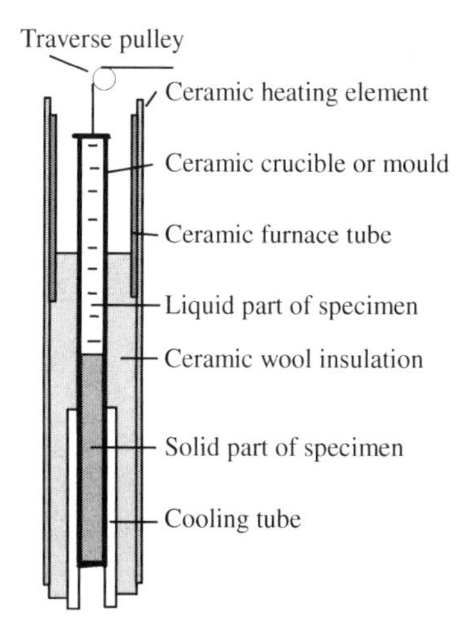

Figure 7.21 Equipment for production of single crystals according to the Bridgman technique.

7.5.2 Solidification of a Metal Sample inside a Long Ceramic Tube

Temperature Distribution in the Melt

A metal sample is included in a long ceramic tube which is continuously moved from an upper hot zone in a furnace into a lower water-cooled zone. We want to calculate the temperature distribution within the sample and start with the liquid part. We assume that the temperature distribution as a function of position is stationary and corresponds to equilibrium everywhere.

Consider a metal sample in the shape of a cylindrical rod which is moved downwards with a constant velocity V_0 through a temperature gradient created by electrical heating of the ceramic tube followed by water-cooling as is shown in Figure 7.21 on page 385. A simple sketch is given in Figure 7.22.

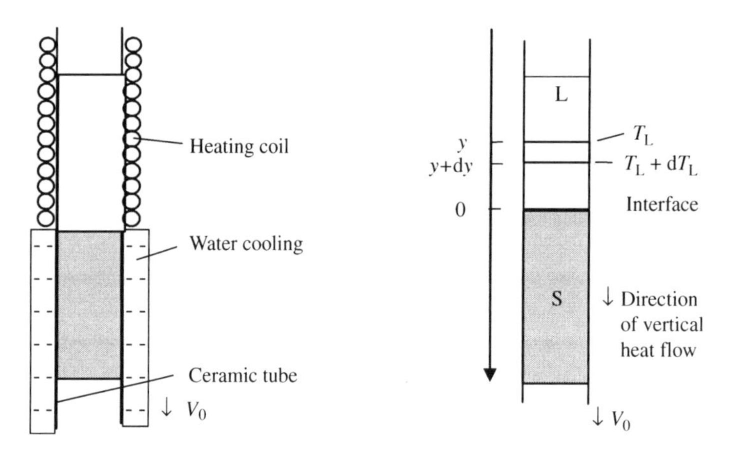

Figure 7.22 Unidirectional solidification of a metal rod in a temperature gradient.

Figure 7.23 Heat balance in the liquid in a unidirectionally solidifying sample.

We introduce a coordinate axis, directed downwards and with $y = 0$ at the interface, and consider a small cylindrical element inside the liquid according to Figure 7.23. The coordinate y is negative everywhere in the liquid. The heat flux is directed downwards. The temperature decreases with increasing y. It is important to observe that the temperature T_L is the *variable* temperature of the volume element when it is *liquid.*

At stationary conditions the sum of net heat exchange inside the element, owing to the motion of the rod, and the heat fluxes, which *enter* the element through the cylindrical surface and the upper circular surface, must be equal to the heat flux, which *leaves* the element through the lower circular surface. The heat balance can therefore be written as

$$-V_0 \pi r^2 \rho_L c_p^L dT_L - h_L 2\pi r\, dy\, (T_L - T_F) - \lambda_L \pi r^2 \left(\frac{dT_L}{dy}\right)_y = -\lambda_L \pi r^2 \left(\frac{dT_L}{dy}\right)_{y+dy} \tag{7.61}$$

where

V_0 = constant velocity of the rod
h_L = heat transfer coefficient
r = radius of the metal rod
y = vertical coordinate
T_F = temperature inside the furnace
T_L = temperature of the metal element at position y
V_0 = pulling rate of the rod
λ_L = thermal conductivity of the metal melt
ρ_L = density of the melt
c_p^L = capacity of heat of the melt at constant pressure
α_L = thermal diffusivity of the melt ($\lambda_L/\rho_L c_p^L$).

Equation (7.61) can be transformed into

$$\frac{2h_\mathrm{L}}{r\rho_\mathrm{L}c_\mathrm{p}^\mathrm{L}}(T_\mathrm{F}-T_\mathrm{L}) = -\alpha_\mathrm{L}\frac{\left(\frac{\mathrm{d}T_\mathrm{L}}{\mathrm{d}y}\right)_{y+\mathrm{d}y}-\left(\frac{\mathrm{d}T_\mathrm{L}}{\mathrm{d}y}\right)_y}{\mathrm{d}y}+V_0\frac{\mathrm{d}T_\mathrm{L}}{\mathrm{d}y} = -\alpha_\mathrm{L}\frac{\mathrm{d}^2T_\mathrm{L}}{\mathrm{d}y^2}+V_0\frac{\mathrm{d}T_\mathrm{L}}{\mathrm{d}y} \tag{7.62}$$

As T_F is constant Equation (7.62) can be transformed into

$$\frac{\mathrm{d}^2(T_\mathrm{F}-T_\mathrm{L})}{\mathrm{d}y^2}-\frac{V_0}{\alpha_\mathrm{L}}\frac{\mathrm{d}(T_\mathrm{F}-T_\mathrm{L})}{\mathrm{d}y} = \frac{2h_\mathrm{L}}{r\rho_\mathrm{L}c_\mathrm{p}^\mathrm{L}\alpha_\mathrm{L}}(T_\mathrm{F}-T_\mathrm{L}) \tag{7.63}$$

Solution approach: $T_\mathrm{F}-T_\mathrm{L} = A\mathrm{e}^{\eta_1 y}+B\mathrm{e}^{\eta_2 y}$ where η_1 and η_2 are roots to the characteristic equation

$$\eta^2-\frac{V_0}{\alpha_L}\eta = \frac{2h_\mathrm{L}}{r\rho_\mathrm{L}c_\mathrm{p}^\mathrm{L}\alpha_\mathrm{L}}$$

The roots are solved from the equation of second degree:

$$\eta_{1,2} = \frac{V_0}{2\alpha_\mathrm{L}}\pm\sqrt{\left(\frac{V_0}{2\alpha_\mathrm{L}}\right)^2+\frac{2h_\mathrm{L}}{r\rho_\mathrm{L}c_\mathrm{p}^\mathrm{L}\alpha_\mathrm{L}}}$$

Hence, the solution is:

$$T_\mathrm{F}-T_\mathrm{L} = A\mathrm{e}^{\left[\frac{V_0}{2\alpha_\mathrm{L}}+\sqrt{\left(\frac{V_0}{2\alpha_\mathrm{L}}\right)^2+\frac{2h_\mathrm{L}}{r\rho_\mathrm{L}c_\mathrm{p}^\mathrm{L}\alpha_\mathrm{L}}}\right]y}+B\mathrm{e}^{\left[\frac{V_0}{2\alpha_\mathrm{L}}-\sqrt{\left(\frac{V_0}{2\alpha_\mathrm{L}}\right)^2+\frac{2h_\mathrm{L}}{r\rho_\mathrm{L}c_\mathrm{p}^\mathrm{L}\alpha_\mathrm{L}}}\right]y}$$

Boundary conditions:

$$\begin{aligned} &T_\mathrm{L}=T_\mathrm{F} \quad \text{for} \quad y=-\infty \quad \text{which gives } B=0 \\ &T_\mathrm{L}=T_\mathrm{M} \quad \text{for} \quad y=0 \quad \text{which gives } A=T_\mathrm{F}-T_\mathrm{M} \end{aligned}$$

Hence, the temperature distribution in the melt will be

$$T_\mathrm{L} = T_\mathrm{F}-(T_\mathrm{F}-T_\mathrm{M})\mathrm{e}^{\left[\frac{V_0}{2\alpha_\mathrm{L}}+\sqrt{\left(\frac{V_0}{2\alpha_\mathrm{L}}\right)^2+\frac{2h_\mathrm{L}}{r\rho_\mathrm{L}c_\mathrm{p}^\mathrm{L}\alpha_\mathrm{L}}}\right]y} \tag{7.64}$$

where $y<0$ and T_M is the melting-point temperature of the alloy. The function (7.64) is shown in Figure 7.24.

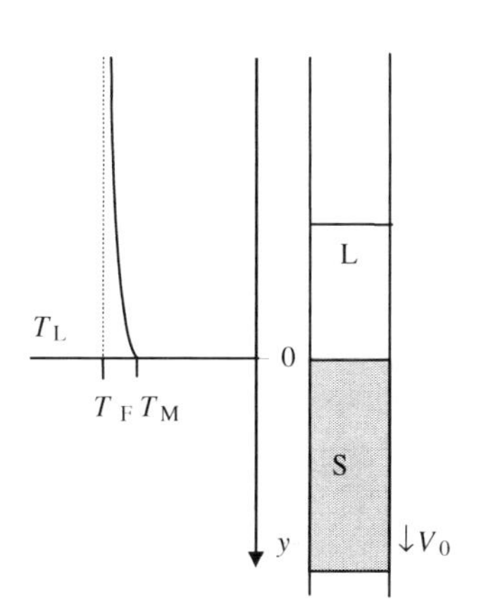

Figure 7.24 Temperature distribution in the liquid part of a rod at unidirectional solidification.

Temperature Distribution in the Solid

The temperature distribution in the solid part of the rod can be found by setting up a similar heat balance as above [Equation (7.61) on page 386]. We use the same coordinate system as above. The coordinate y is positive everywhere in the solid part of the rod. The temperature decreases with increasing y, because the heat flows downwards. It is important to observe that the temperature T_s is the *variable* temperature of the volume element when it is *solid.*

Consider the volume element in the solid in Figure 7.25. At stationary conditions the sum of the net exchange inside the element, owing to the motion of the rod and the heat flux, which *enters* the element through the upper circular surface, must be equal to the sum of the heat flux which *leaves* the cylindrical surface and the lower circular surface. The heat balance can be written as

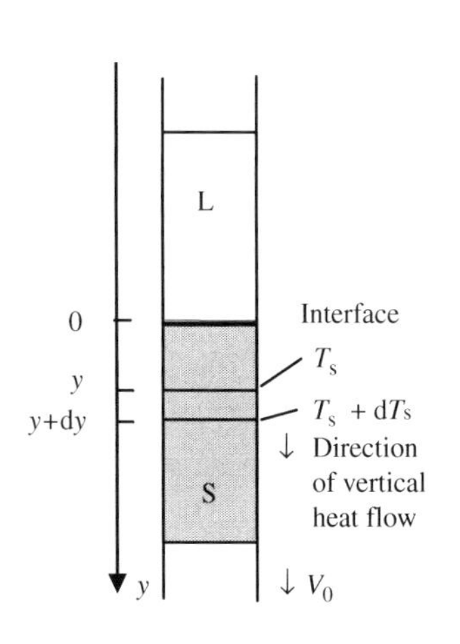

Figure 7.25 Heat balance in the solid in a unidirectionally solidifying sample.

$$-V_0\pi r^2\rho_s c_p^s dT_s - \lambda_s\pi r^2\left(\frac{dT_s}{dy}\right)_y = -h_s 2\pi r\, dy(T_0 - T_s) - \lambda_s\pi r^2\left(\frac{dT_s}{dy}\right)_{y+dy} \tag{7.65}$$

where
h_s = heat transfer coefficient
T_s = temperature on the solid rod at position y
T_0 = temperature of the surroundings
λ_s = thermal conductivity of the solid metal.

As T_0 is constant Equation (7.65) can be transformed into

$$\frac{d^2(T_s - T_0)}{dy^2} - \frac{V_0}{\alpha_s}\frac{d(T_s - T_0)}{dy} = \frac{2h_s}{r\rho_s c_p^s \alpha_s}(T_s - T_0) \tag{7.66}$$

which is analogous to Equation (7.63).

Solution approach: $T_s - T_0 = Ae^{\eta_1 y} + Be^{\eta_2 y}$where η_1 and η_2 are roots to the characteristic equation

$$\eta^2 - \frac{V_0}{\alpha_s}\eta = \frac{2h_s}{r\rho_s c_p^s \alpha_s}$$

The roots are solved from the equation of second degree

$$\eta_{1.2} = \frac{V_0}{2\alpha_s} \pm \sqrt{\left(\frac{V_0}{2\alpha_s}\right)^2 + \frac{2h_s}{r\rho_s c_p^s \alpha_s}}$$

Hence, the solution is

$$T_s - T_0 = Ae^{\left[\frac{V_0}{2\alpha_s} + \sqrt{\left(\frac{V_0}{2\alpha_s}\right)^2 + \frac{2h_s}{r\rho_s c_p^s \alpha_s}}\right]y} + Be^{\left[\frac{V_0}{2\alpha_s} - \sqrt{\left(\frac{V_0}{2\alpha_s}\right)^2 + \frac{2h_s}{r\rho_s c_p^s \alpha_s}}\right]y} \tag{7.67}$$

where A and B are constants which have to be determined by boundary conditions. We choose the boundary conditions:

$T_s = T_0$ for $y = +\infty$ which gives $A = 0$
$T_s = T_M$ for $y = 0$ which gives $B = T_M - T_0$

Hence, the temperature distribution in the solid will be

$$T_s = T_0 + (T_M - T_0)e^{\left[\frac{V_0}{2\alpha_s} - \sqrt{\left(\frac{V_0}{2\alpha_s}\right)^2 + \frac{2h_s}{r\rho_s c_p^s \alpha_s}}\right]y} \quad y > 0 \tag{7.68}$$

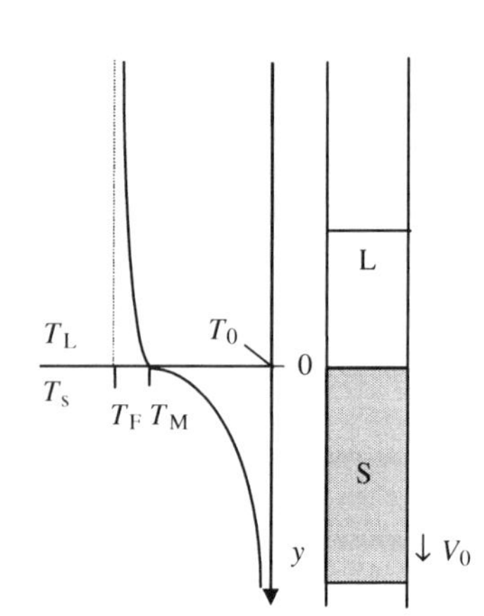

Figure 7.26 Temperature distribution in the liquid and solid parts of a rod on unidirectional solidification.

The function is shown in Figure 7.26.

A third boundary condition must necessarily be fulfilled: the heat balance at the interface.

Heat Balance at the Interface. Equilibrium Condition

Consider a small volume element of the rod at the interface. The solidification process occurs at the interface. As the temperature fields in the liquid and the solid are stationary, heat can not accumulate anywhere. Therefore, the sum of the heat entering the volume element $\pi r^2 dy$ and the heat generated inside the volume element must be equal to the heat leaving the element during the same time dt.

We assume that dt permits complete solidification of the volume element. The energy balance can be written as

$$\pi r^2 \rho_s \mathrm{d}y(-\Delta H) \quad - \pi r^2 \rho_s c_p^s \mathrm{d}y \mathrm{d}T_s \quad - \lambda_L \pi r^2 \left(\frac{\mathrm{d}T_L}{\mathrm{d}y}\right)_{y=0} \mathrm{d}t =$$

amount of heat generated inside the volume element when it solidifies

amount of heat in the volume element which corresponds to the temperature change dT_s

amount of heat entering the volume element from the liquid through the upper circular surface during the time dt

$$= -h_s 2\pi r \mathrm{d}y(T_s - T_0)\mathrm{d}t \quad - \lambda_0 \pi r^2 \left(\frac{\mathrm{d}T_s}{\mathrm{d}y}\right)_{y=0+\mathrm{d}y} \mathrm{d}t$$

amount of heat leaving the volume element through the cylindrical surface to the surroundings during the time dt

amount of heat leaving the volume element to the solid through the lower circular surface during the time dt

The equation is divided by dt which gives

$$\pi r^2 \rho_s(-\Delta H)\frac{\mathrm{d}y}{\mathrm{d}t} - \pi r^2 \rho_s c_p^s \mathrm{d}y \frac{\mathrm{d}T_s}{\mathrm{d}t} - \lambda_L \pi r^2 \left(\frac{\mathrm{d}T_L}{\mathrm{d}y}\right)_{y=0} = -h_s 2\pi r(T_0 - T_s)\mathrm{d}y - \lambda_s \pi r^2 \left(\frac{\mathrm{d}T_s}{\mathrm{d}y}\right)_{y=0+\mathrm{d}y} \tag{7.69}$$

where dy/dt is the pulling rate V_0 and dT_s/dt is the cooling rate.

The second term is small compared with the first term and can be neglected (d$y \approx 0$). If the interface area is insulated (Figure 7.27) no heat passes the cylindrical surface, which means that $h_s \approx 0$ and that the fourth term in Equation (7.69) will vanish. In this case Equation (7.69) can be written (dy/d$t = V_0$)

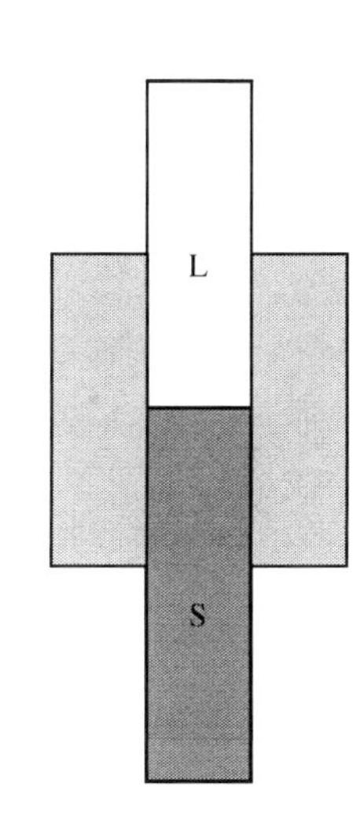

Figure 7.27 Cross-section of an insulated interface zone on unidirectional solidification.

$$\pi r^2 V_0 \rho_L(-\Delta H) - \lambda_L \pi r^2 \left(\frac{\mathrm{d}T_L}{\mathrm{d}y}\right)_{y=0} = -\lambda_s \pi r^2 \left(\frac{\mathrm{d}T_s}{\mathrm{d}y}\right)_{y=0}$$

or

$$V_0 \rho_L(-\Delta H) - \lambda_L \left(\frac{\mathrm{d}T_L}{\mathrm{d}y}\right)_{y=0} = -\lambda_s \left(\frac{\mathrm{d}T_s}{\mathrm{d}y}\right)_{y=0} \tag{7.70}$$

This relationship must be satisfied if the interface is insulated. The heat flux through the solid is higher than that through the melt because of the heat of fusion which has to be transported away.

To find the expressions for the derivatives in Equation (7.70) we take the derivative of Equations (7.64) and (7.68) with respect to y and insert $y = 0$. The result is

$$\left(\frac{\mathrm{d}T_L}{\mathrm{d}y}\right)_{y=0} = -(T_F - T_M)\left(\frac{V_0}{2\alpha_L} + \sqrt{\left(\frac{V_0}{2\alpha_L}\right)^2 + \frac{2h_L}{r\rho_L c_p^L \alpha_L}}\right) \tag{7.71}$$

and

$$\left(\frac{\mathrm{d}T_s}{\mathrm{d}y}\right)_{y=0} = (T_M - T_0)\left(\frac{V_0}{2\alpha_s} - \sqrt{\left(\frac{V_0}{2\alpha_s}\right)^2 + \frac{2h_s}{r\rho_s c_p^s \alpha_s}}\right) \tag{7.72}$$

When these expressions are inserted into Equation (7.70) we obtain

$$\rho_L V_0(-\Delta H) + \lambda_L(T_F - T_M)\left(\frac{V_0}{2\alpha_L} + \sqrt{\left(\frac{V_0}{2\alpha_L}\right)^2 + \frac{2h_L}{r\rho_L c_p^L \alpha_L}}\right)$$
$$= -\lambda_s(T_M - T_0)\left(\frac{V_0}{2\alpha_s} - \sqrt{\left(\frac{V_0}{2\alpha_s}\right)^2 + \frac{2h_s}{r\rho_s c_p^s \alpha_s}}\right) \tag{7.73}$$

Equation (7.73) is the relationship between T_0, T_F and V_0 which has to be satisfied if equilibrium is to be maintained. Two of these quantities can be chosen arbitrarily and determine the third one. Equilibrium means a stationary state when the temperature distribution is constant and the velocity of the rod is equal but anti-parallel to the growth rate at the solidification front.

If this condition is not fulfilled the solidification front will no longer be planar but be either convex or concave. Fluctuations in temperature caused by convection loops in the melt may also affect the orientation of the interface. It is easier to maintain a planar front if the heat transfer to the surroundings is decreased. For this reason an equipment like that in Figure 7.21 is to be preferred to that in Figure 7.22. These topics will be further discussed in Example 7.2 on page 392.

7.5.3 *Experimental Investigation of Heat Flow in Controlled Unidirectional Solidification of Metals*

The theory given above for the unidirectional solidification of a metal rod in a ceramic tube is only a rough approximation. Reality is much more complex and experimental measurements involve many sources of errors.

In 1980 Clyne [5] made a careful experimental study of directional solidification of commercially pure aluminium in ceramic crucibles with the equipment shown in Figure 7.28. His results and conclusions will be discussed shortly below.

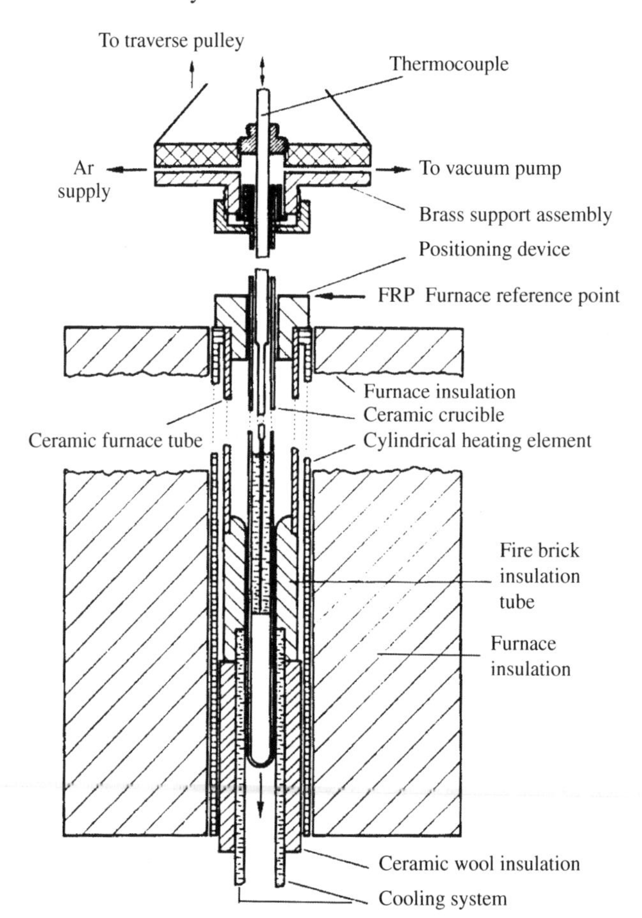

Figure 7.28 Clyne's experimental equipment used to determine interface movements and other thermal characteristics during directional solidification of commercially pure aluminium.

Cylindrical specimens of three various lengths (80, 130 and 160 mm) of commercially pure aluminium in a tube (crucible) of Al_2O_3 were slowly monitored through a furnace at three different traverse velocities (0.001–1.0 mm/s) followed by a cooling device.

During each solidification run the thermal environment was standardized by furnace control and coolant rate settings. Thermocouples served as temperature instruments on line during the solidification runs and special measurements were performed without a specimen to measure the thermal environment separately. The position of the interface was measured during the solidification runs with the aid of a dipstick relative to the furnace reference point (FRP in Figure 7.28).

The ideal steady state would be that when the specimen moves *downwards* with a given traverse velocity, the interface of the sample would move *upwards* with the same velocity, i.e. the position of the interface would be constant relative to the furnace.

Clyne's results showed that this ideal case is difficult to achieve. He found that the velocity of the interface was approximately constant but roughly 30% higher than the traverse velocity, especially at the beginning of an experiment. The consequence was that the interface initially moved considerably, relative to the furnace (FRP point in Figure 7.28), as is shown in Figure 7.29.

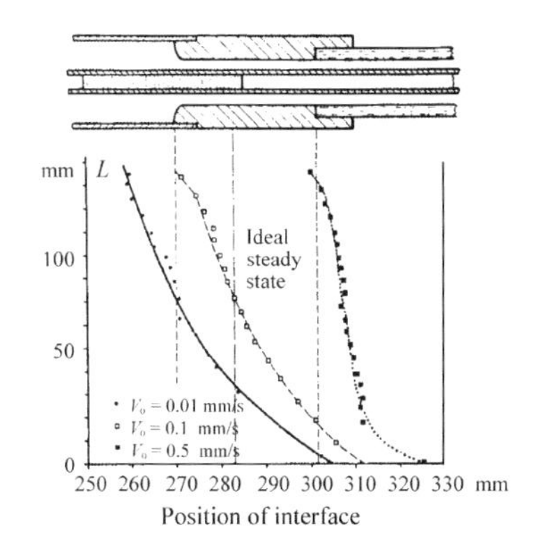

Figure 7.29 Solidified length L as a function of position of the interface (distance from FRP in Figure 7.28).

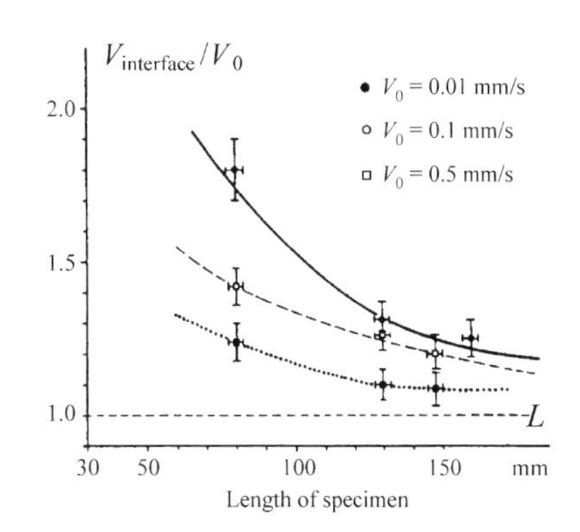

Figure 7.30 Interface velocity/crucible velocity as a function of length L of the specimen at various crucible velocities.

The shape of the specimen also influences the position of the interface. Figure 7.30 shows the ratio interface velocity/crucible velocity as a function of length of the specimen and crucible velocity.

The conclusion is that *increasing length* and *high traverse speed* promotes an approach to the ideal ratio = 1.

Several sources of error are associated with temperature measurements. The dimensions and the positions of the wires to the thermocouples influence the result of the measurements. A symmetric arrangement of the wires can partly eliminate the last effect.

In order to reduce the systematic errors during the experiments, a cylindrical insulator was inserted around the solid/liquid interface as shown in Figure 7.27. In this case, there was no radial heat transport, only along the rod in the zone of insulation.

It is important that the solidification front remains within the insulated area during the experiment. The condition for a stationary solidification front is

$$V_0 \lambda_L (-\Delta H) - \lambda_L \left(\frac{\partial T_L}{\partial y} \right)_{y=0} = -\lambda_s \left(\frac{\partial T_s}{\partial y} \right)_{y=0} \tag{7.70}$$

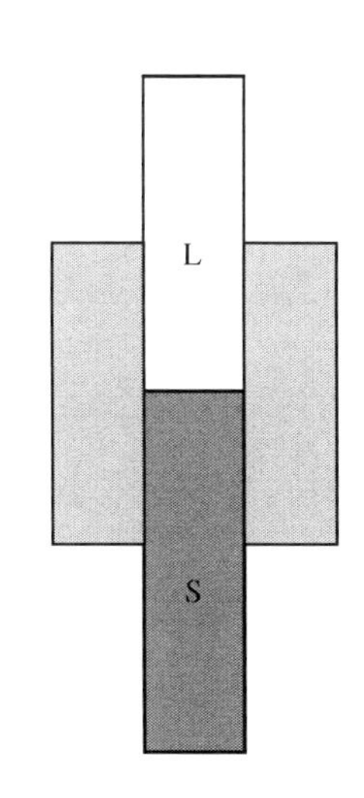

Figure 7.27 Cross-section of an insulated interface zone on unidirectional solidification.

The consequences of this condition was discussed on pages 388–390.

The temperature distribution in case of no insulation has been analyzed on pages 385–389. In Example 7.2 below a rough model of temperature distribution in rod with an insulated solidification region will be discussed.

Example 7.2

One of the most difficult problems when using the Bridgman technique to grow single crystals is to establish a planar solidification front and maintain a steady state. To achieve maximum control of the experimental conditions the solidification region is insulated.

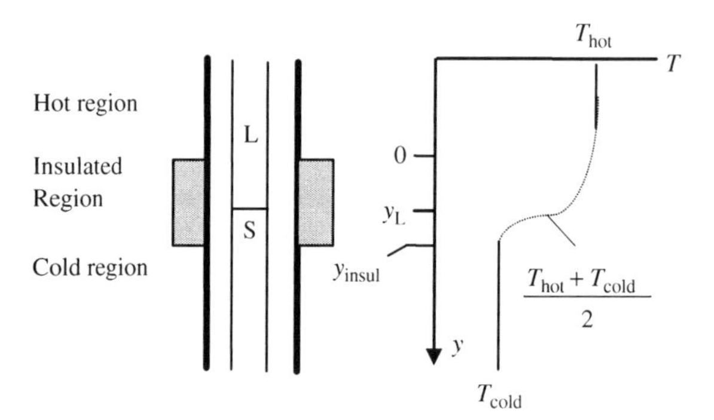

In an equipment such as that in the figure the solidification front is allowed to move inside the insulated region in order to control and initiate adjustments of the temperature gradients. In such a case it is very inconvenient to choose the mobile solidification front as the origin of the coordinate system. Instead the border between the hot zone and the insulated region is chosen as origin, $y = 0$.

Assume that the heat transfer coefficient h is zero within the insulated periphery area and very large elsewhere, both in the hot and cold parts of the experimental equipment. The temperature at the solidification front y_L is assumed to be constant and equal to

$$T_L = \frac{T_{hot} + T_{cold}}{2}$$

The temperatures at the hot and cold zones are chosen in such a way that a planar growth front is established.

a) Find expressions of the temperature gradient in the liquid and in the solid at the solidification front ($y = y_L$) as a function of the crucible velocity V_0.
b) Discuss the effect of the crucible velocity V_0 on the temperature gradient in the liquid and in the solid at the solidification front.
c) Find an equation for calculation of the position of the solidification front inside the insulated region.

Solution:

a) ***Temperature gradient in the liquid at the interface***
To find the temperature gradient in the liquid we apply Equation (7.63) on page 387

$$\frac{d^2(T_F - T_L)}{dy^2} - \frac{V_0}{\alpha_L}\frac{d(T_F - T_L)}{dy} = \frac{2h_L}{r\rho_L c_p^L \alpha_L}(T_F - T_L)$$

and insert $h = 0$ and $T_F = T_{hot}$. The result is

$$\alpha_L \frac{\partial^2(T_{hot} - T_L)}{\partial y^2} - V_0 \frac{\partial(T_{hot} - T_L)}{\partial y} = 0 \qquad (1')$$

The solution of this equation is of the type

$$T_{hot} - T_L = Ae^{\eta_1 y} + Be^{\eta_2 y}$$

where η_1 and η_2 are the roots of the characteristic equation

$$\alpha_L \eta^2 - V_0 \eta = 0$$

The roots are $\eta_1 = \frac{V_0}{\alpha_L}$ and $\eta_2 = 0$

Hence, the solution can be written as

$$T_{hot} - T_L = A e^{\frac{V_0}{\alpha_L} y} + B \tag{2'}$$

The boundary conditions for determination of the constants A and B are:

$$y = 0 \quad \text{and} \quad y = y_L$$

$$T_L = T_{hot} \qquad T_L = \frac{T_{hot} + T_{cold}}{2}$$

These pairs of values are inserted into Equation (2′) and we obtain

$$0 = A + B \quad \text{and} \quad T_{hot} - \frac{T_{hot} + T_{cold}}{2} = A e^{\frac{V_0}{\alpha_L} y_L} + B$$

and the result is

$$A = \frac{T_{hot} - T_{cold}}{2\left(e^{\frac{V_0}{\alpha_L} y_L} - 1\right)} \text{ and } B = -\frac{T_{hot} - T_{cold}}{2\left(e^{\frac{V_0}{\alpha_L} y_L} - 1\right)}$$

The solution can be written as

$$T_{hot} - T_L = \frac{T_{hot} - T_{cold}}{2\left(e^{\frac{V_0}{\alpha_L} y_L} - 1\right)} \left(e^{\frac{V_0}{\alpha_L} y} - 1\right) \quad y \leq y_L \tag{3'}$$

Taking the derivative with respect to y gives the temperature gradient

$$\text{grad}\, T_L = \frac{dT_L}{dy} = \frac{-(T_{hot} - T_{cold})}{2\left(e^{\frac{V_0}{\alpha_L} y_L} - 1\right)} \frac{V_0}{\alpha_L} e^{\frac{V_0}{\alpha_L} y} \tag{4'}$$

Inserting $y = y_L$ into Equation (4′) gives the desired value of $(\text{grad}\, T_L)_{y=y_L}$.

b) ***Temperature gradient in the solid at the interface***
Analogously we apply Equation (7.66) on page 388

$$\frac{d^2(T_s - T_0)}{dy^2} - \frac{V_0}{\alpha_s}\frac{d(T_s - T_0)}{dy} = \frac{2h_s}{r\rho_s c_p^s \alpha_s}(T_s - T_0)$$

and insert $h = 0$ and $T_0 = T_{cold}$. For $y > y_{insul}$ the result is

$$\alpha_s \frac{\partial^2(T_s - T_{cold})}{\partial y^2} - V_0 \frac{\partial(T_s - T_{cold})}{\partial y} = 0 \tag{5'}$$

The solution of this equation is of the type

$$T_s - T_{cold} = A e^{\eta_1 y} + B e^{\eta_2 y}$$

where η_1 and η_2 are the roots of the characteristic equation

$$\alpha_s \eta^2 - V_0 \eta = 0$$

The roots are $\eta_1 = \dfrac{V_0}{\alpha_s}$ and $\eta_2 = 0$

The solution can be written as

$$T_s - T_{cold} = A e^{\frac{V_0}{\alpha_s} y} + B \qquad (6')$$

The boundary conditions for determination of the constants A and B are

$$y = y_{insul} \quad \text{and} \quad y = y_L$$

$$T_s = T_{cold} \qquad T_s = \frac{T_{hot} + T_{cold}}{2}$$

These pairs of values are inserted into Equation (6′) and we obtain the equation system

$$0 = A e^{\frac{V_0}{\alpha_s} y_{insul}} + B$$

$$\frac{T_{hot} + T_{cold}}{2} - T_{cold} = A e^{\frac{V_0}{\alpha_L} y_L} + B$$

and the result is

$$A = \frac{T_{hot} - T_{cold}}{2\left(e^{\frac{V_0}{\alpha_s} y_L} - e^{\frac{V_0}{\alpha_s} y_{insul}}\right)} \quad \text{and} \quad B = -\frac{T_{hot} - T_{cold}}{2\left(e^{\frac{V_0}{\alpha_s} y_L} - e^{\frac{V_0}{\alpha_s} y_{insul}}\right)} e^{\frac{V_0}{\alpha_s} y_{insul}}$$

The solution can be written $(y > y_L)$

$$T_s - T_{cold} = \frac{T_{hot} - T_{cold}}{2\left(e^{\frac{V_0}{\alpha_s} y_L} - e^{\frac{V_0}{\alpha_s} y_{insul}}\right)} \left(e^{\frac{V_0}{\alpha_s} y} - e^{\frac{V_0}{\alpha_s} y_{insul}}\right) \qquad (7')$$

Taking the derivative of Equation (7′) with respect to y gives the temperature gradient

$$\text{grad } T_s = \frac{dT_s}{dy} = \frac{T_{hot} - T_{cold}}{2\left(e^{\frac{V_0}{\alpha_s} y_L} - e^{\frac{V_0}{\alpha_s} y_{insul}}\right)} \frac{V_0}{\alpha_s} e^{\frac{V_0}{\alpha_s} y} \qquad (8')$$

Inserting $y = y_L$ into Equation (4′) gives the desired value of $(\text{grad } T_L)_{y=y_L}$

c) We apply an equation analogous to Equation (7.70) on page 389

$$\rho_L V_0(-\Delta H) - \lambda_L \left(\frac{dT_L}{dy}\right)_{y=y_L} = -\lambda_s \left(\frac{dT_s}{dy}\right)_{y=y_L} \qquad (9')$$

or

$$\rho_L V_0(-\Delta H) - \lambda_L (\text{grad } T_L)_{y=y_L} = -\lambda_s (\text{grad } T_s)_{y=y_L} \qquad (10')$$

Answer:

a)

$$(\text{grad } T_L)_{y=y_L} = \frac{dT_L}{dy} = \frac{-(T_{hot} - T_{cold})}{2\left(e^{\frac{V_0}{\alpha_L} y_L} - 1\right)} \frac{V_0}{\alpha_L} e^{\frac{V_0}{\alpha_L} y_L}$$

$$(\text{grad } T_s)_{y=y_L} = \frac{dT_s}{dy} = \frac{T_{hot} - T_{cold}}{2\left(e^{\frac{V_0}{\alpha_s} y_L} - e^{\frac{V_0}{\alpha_s} y_{insul}}\right)} \frac{V_0}{\alpha_s} e^{\frac{V_0}{\alpha_s} y_L}$$

Both gradients are negative, in agreement with the fact that the temperature decreases when y_L and y_s increase (downwards).

b) An increase of V_0 results in steeper T versus y curves, i.e. a strong increase of the magnitude of the temperature gradient at the interface, both in the liquid and the solid.

c) The position y_L of the interface is found by solving the equation

$$\rho_L V_0(-\Delta H) - \lambda_L(\text{grad } T_L)_{y=y_L} = -\lambda_s(\text{grad } T_s)_{y=y_L} \quad (6')$$

numerically after introduction of the expressions for the derivatives, given in a).

To achieve stationary conditions it is very important to keep the solidification front within the insulated region. It must be chosen long enough to avoid the situation when the interface is forced down into the cold region.

7.6 Thermal Analysis

Thermal analysis is a generally accepted concept which can be defined as follows:

- Thermal analysis is the measurement of some physical property of a material as a function of its temperature.

The measurements constitute the basis for calculation of desired quantities and relationships. It is a very old and at the same time very new and modern analytical method. Its roots go back to Ancient Greece centuries before Christ. It was frequently used ever since and long before the nature of heat was known. It contributed to the development of measurement of temperature and heat in the 19th century and to the construction of devices for such measurements.

Thermal analysis is very modern in the sense that there has been an enormous development of the instrumentation especially during the 1990s which strongly has improved the sensitivity and accuracy of the measurements. Keywords are small samples and high accuracy. Thermal analysis has many applications, primarily in the field of chemistry, but also in many other fields.

The definition of thermal analysis given above is very wide as it includes all sorts of measurements connected with temperature and heat. It includes a number of different techniques for measurement of various properties. In this chapter we will present two thermal analysis techniques

- DTA = differential thermal analysis
- DSC = differential scanning calorimetry.

7.6.1 Differential Thermal Analysis

Clyne's experiments (page 390) illustrate the fact that thermal measurements are difficult and afflicted with errors. The main sources of errors in temperature measurements are heat losses to the surroundings.

These difficulties are almost eliminated by the *differential technique*. The instrument measures the *difference* in temperature between the sample and a reference, which are exposed to the same heating process in a furnace. If, for example, a transformation in the sample is to be studied, a substance with about the same 'thermal mass' as the sample but with no transformation in the relevant temperature range is chosen as reference material. The heat losses to the surroundings and the influence of the thermocouple wires will be about the same in both cases and do not influence the difference in temperature very much.

Principle of DTA

DTA is an elegant way of solving the experimental difficulties. It was introduced by the Brittish scientist Roberts-Austen. The first DTA curve was published in 1899. The idea of differential temperature measurements is the basis of the development of commercial devices, which started in the 1940s. Figure 7.31 shows the principle of an instrument for differential thermal analysis.

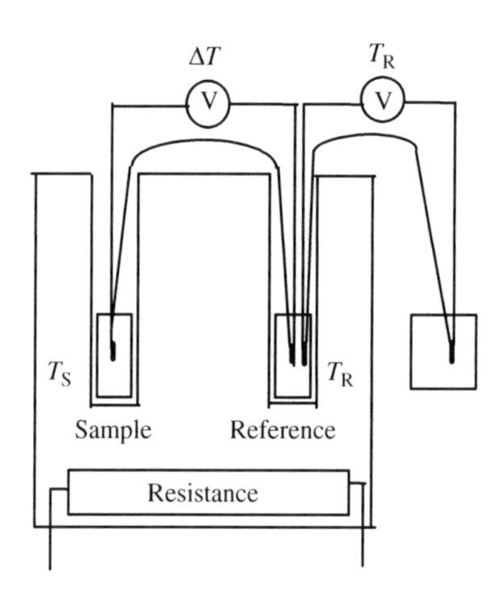

Figure 7.31 Principle of DTA.

The sample S, the substance to be examined, and the reference R, an inert material, are symmetrically included in a block with high thermal conductivity that can be heated or cooled at a constant rate. The temperature difference between R and S is measured as a function of time by means of a thermocouple. The temperatures of S and R are also measured separately. The instrument has to be calibrated by means of a known temperature difference.

Figure 7.32 shows a typical result of a DTA measurement. The transformation of sample S from solid to liquid phase is studied. Initially R and S are heated and their temperatures increase. When the temperatures are plotted as a function of time, two straight lines are obtained. The slopes depend on their heat capacities, which are nearly but not completely equal. Therefore, the two straight lines diverge slightly.

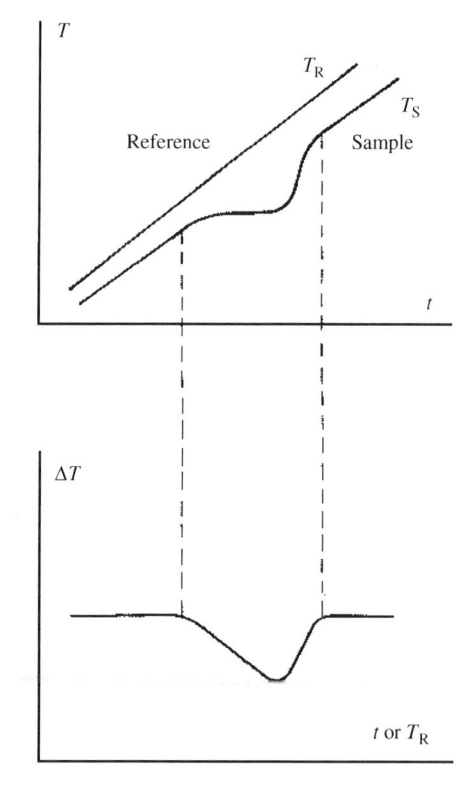

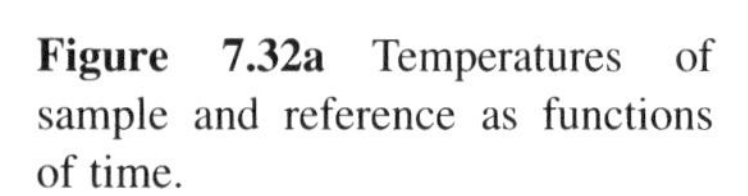

Figure 7.32a Temperatures of sample and reference as functions of time.

Figure 7.32b Temperature difference $|T_S - T_R|$ as a function of time t or temperature T_R.

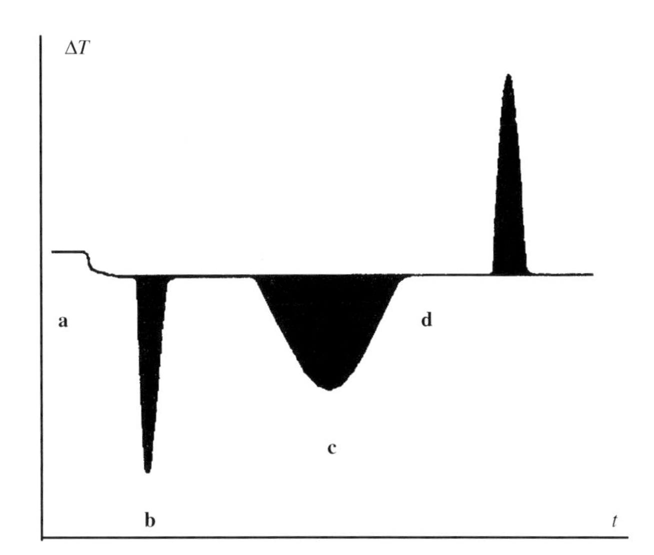

Figure 7.33 Typical DTA curves: (a) A change of the horizontal baseline caused by a second-order transition.

(b) An endothermic peak indicating a fusion or melting transition.

(c) An endothermic peak caused by a dissociation or decomposition reaction.

(d) An exothermic peak owing to a crystalline phase change.

When the melting point of the sample S is reached, it absorbs heat and its temperature remains constant until the whole sample is molten. The temperature of the sample increases again and the heating curve of the sample becomes nearly parallel to the reference curve.

Figure 7.33 presents some other typical DTA curves, which show four different types of transitions. Endothermic transitions give curves with minima, exothermic reactions are manifested by maxima.

7.6.2 *Theory of Differential Thermal Analysis*

Differential Thermal Analysis with a Reference Sample

The theory, given below, deals with the solidification process. It is in principle valid for any transformation or reaction in the sample. For future use we will introduce and define the following concepts given in Table 7.4 below.

Table 7.4 Nomenclature

Definition	Corresponding term/definition applied on solidification processes
Transformation or reaction	Solidification
Fraction transformed $= f$	Fraction solidified $= ft$
Reaction rate $= df/dt$	Fraction of solid phase formed per unit time $= df/dt$
Heat of reaction	Heat of fusion

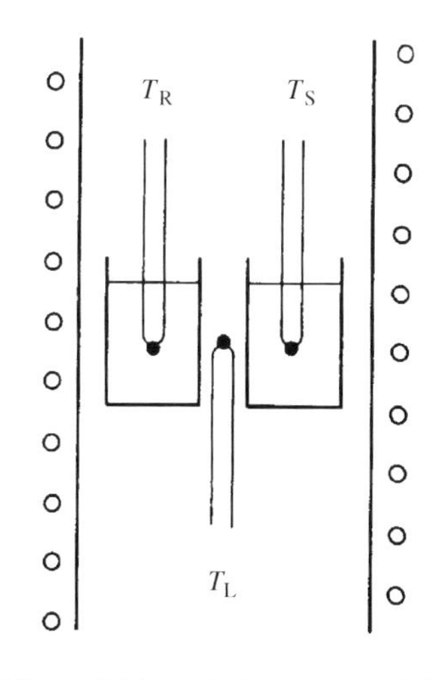

Figure 7.34 Principal sketch of differential thermal analysis (DTA) equipment.

Figure 7.34 shows schematically the basic equipment for differential thermal analysis. Two samples, one for analysis and one as a reference, each connected to a thermocouple, are heated in a furnace and allowed to cool and the cooling curves are registered.

As a basis for the theoretical discussion, we will use a concrete example and study the cooling curve of a high speed tool steel sample of type M2. As Figure 7.35 shows the high speed steel sample pass through three reactions, indicated by three peaks in the DTA curve during the cooling cycle. In order they are

1. Primary precipitation of ferrite: $L \rightarrow \delta$ (1440 °C)
2. A peritectic reaction: $L + \delta \rightarrow \gamma$ (1268 °C)
3. A eutectic reaction $L \rightarrow \gamma +$ carbide (1242 °C)

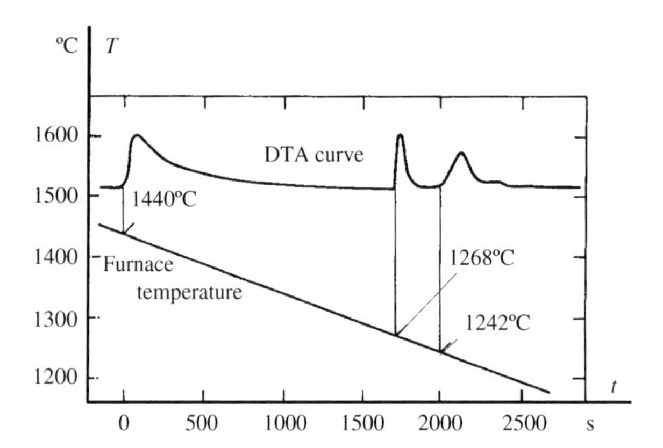

Figure 7.35 DTA curve and furnace temperature as functions of time for a high-speed tool steel of type M2.

Before setting up the heat balance for the sample to be analyzed we will discuss the heat losses to the surroundings. The sample looses heat by radiation and conduction.

$$\text{Heat loss} = Ah(T_S - T_F) + A\sigma_B\varepsilon\left({T_S}^4 - {T_F}^4\right) \tag{7.74}$$

where
A = area of the sample
h = heat transfer coefficient between the crucible and the surroundings
σ_B = Stefan-Boltzmann's constant
ε = emissivity of the surface of the sample
T_S = temperature of the sample
T_F = temperature of the furnace.

If the temperature difference between the sample and the furnace is comparably small, T_F is approximately equal to T_S and we can simplify Equation (7.74):

$$T_S^4 - T_F^4 = \left({T_S}^2 + {T_F}^2\right)(T_S + T_F)(T_S - T_F)$$

or

$$T_S^4 - T_F^4 \approx 4T_F^3(T_S - T_F) \tag{7.75}$$

Hence, the heat loss can be written as

$$\text{Heat loss} \approx \left(Ah + 4A\sigma_B\varepsilon T_F^3\right)(T_S - T_F) \approx F_{\text{loss}} \times (T_S - T_F)$$

where the factor

$$F_{\text{loss}} \approx \left(Ah + 4A\sigma_B\varepsilon T_F^3\right) \tag{7.76}$$

Temperature Difference between Sample and Furnace as a Function of Time in Regions Free from Phase Transformations

If we neglect the heat absorbed by the crucible the heat balance of the sample can, in its most general form, be written as

$$\frac{dQ}{dt} - F_{\text{loss}}(T_S - T_F) = \rho_S V_S c_p^S \frac{dT_S}{dt} - \frac{df}{dt}\rho_S V_S(-\Delta H_S) \tag{7.77}$$

Supplied heat	Losses to the surroundings	Increase of the temperature of the sample	Melting, i.e. decrease of fraction solid ($df/dt < 0$)

where

$\frac{dQ}{dt}$ = amount of heat absorbed by the sample per unit time
F_{loss} = quantity defined by Equation (7.76)
ρ_S = density of the sample
V_S = volume of the sample
c_p^S = heat capacitivity of the sample material
f = fraction solid
$-\Delta H_S$ = heat of fusion of the sample material (J/kg).

We consider the special case of cooling with no supply of heat and no solidification or melting of the sample, which we call the normal part of the cooling curve.

$$0 - F_{loss}(T_S - T_F)_{normal} = \rho_S V_S c_p^S \left(\frac{dT_S}{dt}\right)_{normal} - 0$$

or

$$\left(\frac{dT_S}{dt}\right)_{normal} = \frac{-F_{loss}}{\rho_S V_S c_p^S}(T_S - T_F)_{normal} \tag{7.78}$$

where both T_S and T_F are functions of time. As Figure 7.35 on page 398 shows the furnace temperature is lowered linearly with time which can be expressed as

$$T_F = T_\beta - \beta t \tag{7.79}$$

where β is the cooling rate of the furnace and T_β the initial furnace temperature. If we introduce this expression for T_F into Equation (7.78) we obtain

$$\frac{dT_S}{dt} + \frac{F_{loss}}{\rho_S V_S c_p^S}(T_S - T_\beta + \beta t) = 0 \tag{7.80}$$

The general solution and a particular solution to this differential equation are found and the total solution is

$$T_S = \underbrace{T_\beta e^{-\frac{F_{loss}}{\rho_S V_S c_p^S}t}}_{\text{General solution}} + \underbrace{\left(\frac{\rho_S V_S c_p^S}{F_{loss}}\beta - \beta t + T_\beta\right)}_{\text{Particular solution}} \tag{7.81}$$

or

$$(T_S - T_F)_{normal} = T_\beta e^{-\frac{F_{loss}}{\rho_S V_S c_p^S}t} + \frac{\rho_S V_S c_p^S}{F_{loss}}\beta \tag{7.82}$$

The temperature difference increases with increasing volume and density of the sample and increasing heat capacitivity, decreasing value of F_{loss} and increasing cooling rate. The corresponding equation for the reference is obtained if subscript 'S' is replaced by 'R'. The two cooling curves both approach asymptotically a constant value when time increases.

It is desirable that the temperatures of the sample to be analyzed and the reference sample are equal in the absence of phase transformations. This is the case if the products $\rho\, Vc_p/F_{loss}$ are made equal for the two samples.

Calculation of Reaction Rate, Heat of Fusion and Fraction of Solid Phase

In order to find the reaction rate (see Table 7.4 on page 397) we have to consider the heat balance of the sample in the solidification region. No heat is supplied and we apply Equation (7.77). The heat balance of the sample to be analyzed, can be written as

$$0 - F_{\text{loss}}(T_{\text{S}} - T_{\text{F}}) = \rho_{\text{S}} V_{\text{S}} c_{\text{p}}^{\text{S}} \frac{\mathrm{d}T_{\text{S}}}{\mathrm{d}t} - \frac{\mathrm{d}f}{\mathrm{d}t} \rho_{\text{S}} V_{\text{S}}(-\Delta H_{\text{S}})$$

or

$$F_{\text{loss}}(T_{\text{S}} - T_{\text{F}}) = \rho_{\text{S}} V_{\text{S}} c_{\text{p}}^{\text{S}} \left(-\frac{\mathrm{d}T_{\text{S}}}{\mathrm{d}t}\right) + \frac{\mathrm{d}f}{\mathrm{d}t} \rho_{\text{S}} V_{\text{S}}(-\Delta H_{\text{S}}) \quad (7.83)$$

The heat losses consist of cooling heat and heat of fusion.

During the experiment the temperature difference between the sample and the furnace is not measured, only the temperature difference between the two samples. This difficulty is eliminated by substituting $(T_{\text{S}} - T_{\text{F}})$ by $(T_{\text{S}} - T_{\text{R}}) + (T_{\text{R}} - T_{\text{F}})$ in Equation (7.83) where T_{R} is the temperature of the reference sample.

$$F_{\text{loss}}(T_{\text{S}} - T_{\text{R}}) + F_{\text{loss}}(T_{\text{R}} - T_{\text{F}}) = \rho_{\text{S}} V_{\text{S}} c_{\text{p}}^{\text{S}} \left(-\frac{\mathrm{d}T_{\text{S}}}{\mathrm{d}t}\right) + \frac{\mathrm{d}f}{\mathrm{d}t} \rho_{\text{S}} V_{\text{S}}(-\Delta H_{\text{S}}) \quad (7.84)$$

To derive the heat balance for the reference sample we apply Equation (7.83) and replace subscript 'S' by subscript 'R' and keep in mind that no phase transformation occurs in the reference sample.

$$F_{\text{loss}}(T_{\text{R}} - T_{\text{F}}) = \rho_{\text{R}} V_{\text{R}} c_{\text{p}}^{\text{R}} \left(-\frac{\mathrm{d}T_{\text{R}}}{\mathrm{d}t}\right) \quad (7.85)$$

The second term in Equation (7.84) is substituted by the expression in Equation (7.85) which gives

$$F_{\text{loss}}\,(T_{\text{S}} - T_{\text{R}}) + \rho_{\text{R}} V_{\text{R}} c_{\text{p}}^{\text{R}} \left(-\frac{\mathrm{d}T_{\text{R}}}{\mathrm{d}t}\right) = \rho_{\text{S}} V_{\text{S}} c_{\text{p}}^{\text{S}} \left(-\frac{\mathrm{d}T_{\text{S}}}{\mathrm{d}t}\right) + \frac{\mathrm{d}f}{\mathrm{d}t} \rho_{\text{S}} V_{\text{S}}(-\Delta H_{\text{S}}) \quad (7.86)$$

or

$$\frac{\mathrm{d}f}{\mathrm{d}t} = \frac{F_{\text{loss}}(T_{\text{S}} - T_{\text{R}}) + \rho_{\text{S}} V_{\text{S}} c_{\text{p}}^{\text{S}} \dfrac{\mathrm{d}T_{\text{S}}}{\mathrm{d}t} - \rho_{\text{R}} V_{\text{R}} c_{\text{p}}^{\text{R}} \dfrac{\mathrm{d}T_{\text{R}}}{\mathrm{d}t}}{\rho_{\text{S}} V_{\text{S}}(-\Delta H_{\text{S}})} \quad (7.87)$$

where F_{loss} is determined by heating a sample with known properties and allowing it to cool in the furnace.

Some approximations of Equation (7.87) are often introduced in the scientific literature. One of them is to assume that F_{loss}, defined by Equation (7.76) on page 398, is a constant. In fact it is no constant because the constants involved depend on temperature and especially because the temperature T_{F} decreases with time. Another approximation is to assume that the two last terms in the numerator in Equation (7.87) cancel. In this case Equation (7.87) is simplified to

$$\frac{\mathrm{d}f}{\mathrm{d}t} = \frac{F_{\text{loss}}}{\rho_{\text{S}} V_{\text{S}}(-\Delta H_{\text{S}})} (T_{\text{S}} - T_{\text{R}}) \quad (7.88)$$

where $F_{\text{loss}} \approx \left(Ah + 4A\sigma_{\text{B}}\varepsilon T_{\text{F}}^3\right)$

Figure 7.36 shows the reaction rate as a function of temperature evaluated from the DTA curve of the high speed tool steel given in Figure 7.35.

The simplifications, which result in Equation (7.88), are not generally valid and are doubtful in many cases. For this reason it is better to use Equation (7.87) to estimate the reaction rate d*f*/d*t* from the cooling curves.

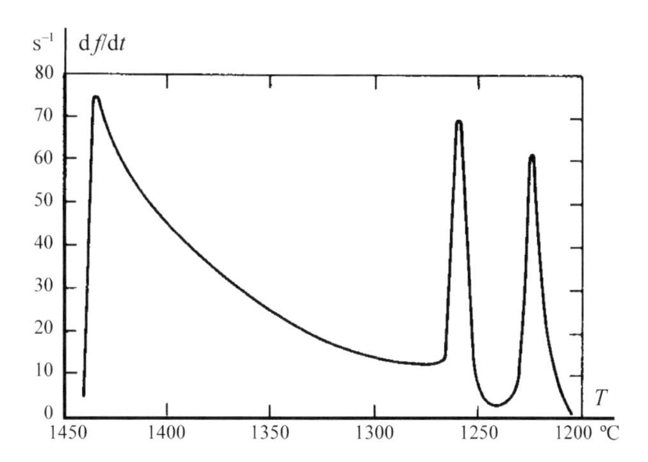

Figure 7.36 Reaction rate as a function of temperature calculated from Equation (7.88) on the basis of the DTA curve in Figure 7.35.

By graphical integration of the curve corresponding to Equation (7.87) or (7.88) from t_{start} to t, the *fraction solidified phase f* can be evaluated (see also page 403).

Equation (7.88) is often used for determining the *heat of fusion* $(-\Delta H_s)$. It is obtained by graphical integration of the curve corresponding to Equation (7.88) from t_{start} $(f=0)$ to t_{end} $(f=1)$ (see also page 403).

Differential Thermal Analysis without a Reference Sample

Differential thermal analysis is also possible in the absence of a reference sample. As a basis for the theoretical discussion below we will use the DTA curve for the Al-5%Cu alloy in Figure 7.37a.

If the sample had gone through no phase transformations, the temperature of the sample would have followed the dashed line in Figure 7.37a. This line is referred to as T_{ref}. It corresponds to the cooling curve of the reference sample during a differential thermal analysis. T_{ref} can be calculated as a function of time by solving the differential Equation (7.78) on page 399 with the subscript 'S' replaced by 'ref'. We also drop the sample indices and the subscript 'normal', which means no phase transformations.

$$\frac{dT_{ref}}{dt} = \frac{-F_{loss}}{\rho V c_p}(T_{ref} - T_F) \tag{7.89}$$

The designations are analogous to those in the preceding section. The solution of the differential Equation (7.89) can be written as

$$T_{ref} = T_F + Ce^{-\frac{F_{loss}}{\rho V c_p}t} \tag{7.90}$$

if we assume for simplicity that T_F and the expression F_{loss}, defined by Equation (7.76) on page 398, are approximately constant during the solidification.

The constant C is determined by a boundary condition. The solidification starts at the time t_{start} and is finished at the time t_{sol}. For simplicity we choose a coordinate system with $t=0$ at the end of the solidification process and obtain the following boundary condition:

At $t = t_{sol} = 0$ the reference temperature is $T_{ref} = T_{ref}(t_{sol})$.

These values, inserted into Equation (7.90), give

$$C = T_{ref}(t_{sol}) - T_F$$

and we obtain

$$T_{ref} = T_F + [T_{ref}(t_{sol}) - T_F]e^{-\frac{F_{loss}}{\rho V c_p}t} \tag{7.91}$$

The extrapolation of the dashed T_{ref} line coincides with the temperature line T_S of the sample at a certain time t_{end} after the completion of the solidification process. As we will see below, this assumption allows calculation of the heat of fusion ($-\Delta H_S$), the solidification rate and the fraction solid as functions of time.

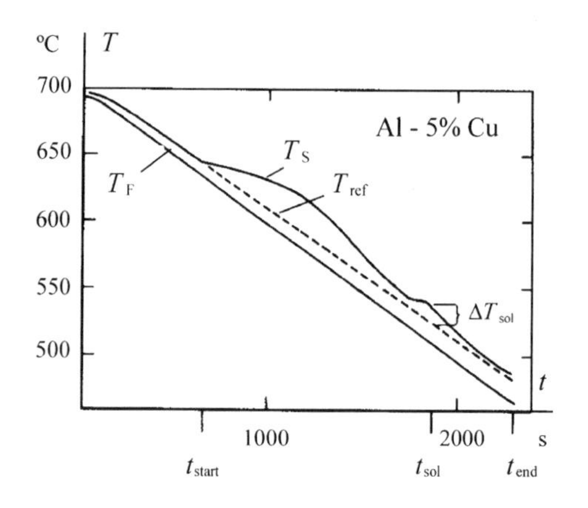

Figure 7.37a Thermal analysis of an Al–Cu alloy containing 5% Cu, when the furnace is cooled at a rate of 0.1 K/s.

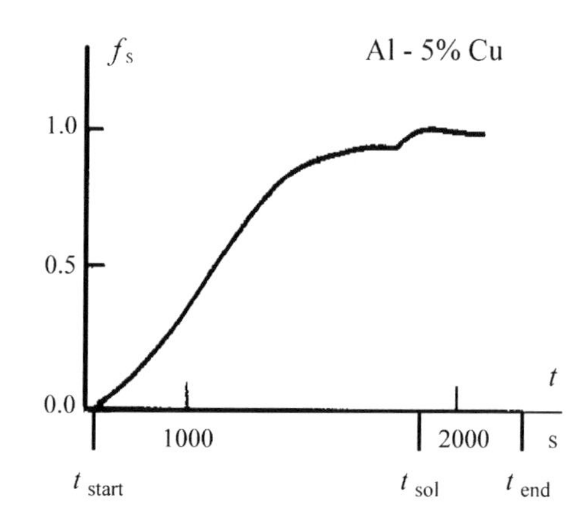

Figure 7.37b The fraction solidified f_s calculated from the DTA curve in Figure 7.37a as a function of time.

Analogously we can solve the differential Equation (7.78) on page 399 which is valid for the sample to be analyzed.

$$\left(\frac{dT_S}{dt}\right)_{normal} = \frac{-F_{loss}}{\rho_S V_S c_p^S}(T_S - T_F)_{normal} \tag{7.78}$$

If T_F and F_{loss} are assumed to be constant, integration of Equation (7.78) gives

$$T_S = T_F + Ce^{-\frac{F_{loss}}{\rho V c_p}t} \tag{7.92}$$

We obtain the boundary condition from Figure 7.37a
At $t = t_{sol} = 0$ the temperature of the sample is

$$T_S = T_{ref}(t_{sol}) + \Delta T_{sol}$$

These values, inserted into Equation (7.92), give

$$C = [T_{ref}(t_{sol}) + \Delta T_{sol}] - T_F$$

and we obtain

$$T_S = T_F + [T_{ref}(t_{sol}) + \Delta T_{sol} - T_F]e^{-\frac{F_{loss}}{\rho V c_p}t} \tag{7.93}$$

If we subtract Equation (7.91) from Equation (7.93) we obtain

$$T_S - T_{ref} = \Delta T_{sol}e^{-\frac{F_{loss}}{\rho V c_p}t} \tag{7.94}$$

Equation (7.94) shows that the T_S and T_{ref} curves approach each other asymptotically. The difference converges more rapidly towards zero when the time coefficient is large, i.e. for small values of ρ, V and c_p and large values of F_{loss}.

Calculation of Heat of Fusion, Reaction Rate and Fraction of Solid Phase
If we replace T_R by T_{ref} Equation (7.88) on page 400 can be written as

$$\mathrm{d}f = \frac{F_{loss}}{\rho_S V_S(-\Delta H_S)}(T_S - T_{ref})\mathrm{d}t \tag{7.95}$$

By integrating this equation with respect to the time between t_{start} and t_{end} we obtain

$$\int_{t_{start}}^{t_{end}} \mathrm{d}f = 1 = \frac{\int_{t_{start}}^{t_{end}} F_{loss}(T_S - T_{ref})\mathrm{d}t}{\rho_S V_S(-\Delta H_S)} \tag{7.96}$$

This equation can be used to determine *the heat of fusion* of the sample material:

$$-\Delta H_S = \frac{\int_{t_{start}}^{t_{end}} F_{loss}(T_S - T_{ref})\mathrm{d}t}{\rho_S V_S} \tag{7.97}$$

The same equation is obtained if we assume that F_{loss} is constant and integrate Equation (7.87) on page 399 within the same time limits, at first for $T = T_S$ and then for $T = T_{ref}$ and finally subtract the new equations.

In order to derive *the reaction rate* (see Table 7.4 on page 397) we introduce the expression of $(-\Delta H_S)$ into Equation (7.87):

$$\frac{\mathrm{d}f}{\mathrm{d}t} = \frac{F_{loss}(T_S - T_{ref}) + \dfrac{\mathrm{d}T_S}{\mathrm{d}t} - \dfrac{\mathrm{d}T_{ref}}{\mathrm{d}t}}{\int_{t_{start}}^{t_{end}} F_{loss}(T_S - T_{ref})\mathrm{d}t} \tag{7.98}$$

which is well adapted for practical use.

The *fraction of solid phase* can be derived by integrating Equation (7.98) graphically.

$$f^* = \frac{\int_{t_{start}}^{t^*} F_{loss}(T_S - T_{ref})\mathrm{d}t + T_S{}^* - T_{ref}{}^*}{\int_{t_{start}}^{t_{end}} F_{loss}(T_S - T_{ref})\mathrm{d}t} \tag{7.99}$$

where
f^* = fraction solid at time $t = t^*$
T_S = temperature of the sample at time t
T_{ref} = reference temperature at time t.

7.6.3 *Differential Scanning Calorimetry*

Differential temperature measurement is the oldest DTA technique. In the middle of the 1950s Boersma introduced a new technique, the so-called *heat flux differential scanning calorimetry*. It allows *quantitative measurements of heat*. In 1964 an alternate method, *power compensation differential scanning calorimetry* was introduced. Both these methods of measuring heat differences have been the basis for development of commercial instruments for DSC, *differential scanning calorimetry*.

There is some confusion concerning the classification of the methods and instruments. The main difference between DTA and DSC is that the DTA instrument signal is proportional to the *temperature difference* between the sample and the reference, while the DSC instrument signal is proportional to the differential *thermal power*. However, a DTA equipment can, after calibration, give the same information as a heat flux DSC instrument. For this reason the method is sometimes denoted heat flux DTA.

The principles of heat flux DSC and power compensation DSC will be described below. The distinction between them is shown in Figure 7.38a,b,c and d.

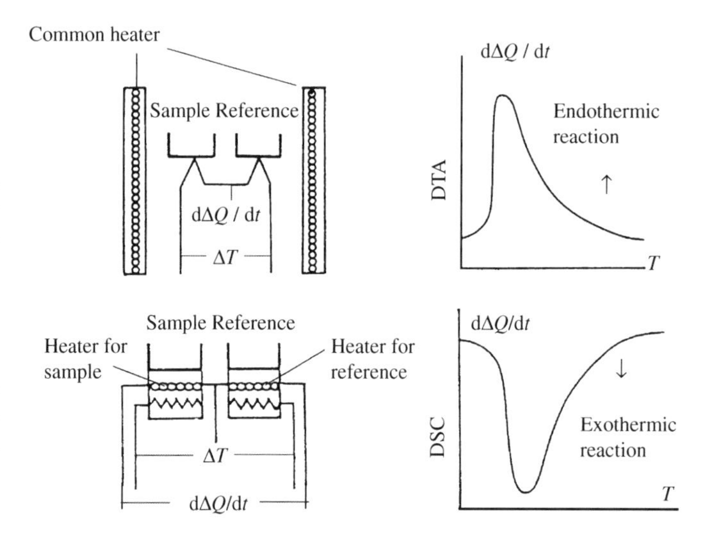

Figure 7.38a + b The principles and typical recordings of heat-flux DSC: Common heater. Temperature difference and power are measured.

Figure 7.38c + d Power-compensation DSC: Separate heaters. No temperature difference. The power difference is measured.

Principle of Heat Flux DSC

The output traces of a differential scanning calorimeter are similar to the ones of a differential thermal analyzer, but the operating principle is different. The great difference is that no direct quantitative heat measurements can be performed with DTA technique while DSC is especially designed for this kind of measurements.

In the sense that the instrument signal is a temperature, difference the equipment should be characterized as a DTA instrument. On the other hand, in the sense that the instrument allows an output of accurate differential thermal power it could as well be designated as a DSC instrument. We have used the last terminology here.

Figure 7.39 shows the principle of a heat flux DSC instrument. The sample and the reference material are located symmetrically in two small equal crucibles on a plate. A precision resistance resistor generates a very well controlled heat flux from the furnace wall to the sample and the reference crucibles. The temperature difference between the sample and the reference material is measured as a function of time or the reference temperature. The temperatures of the sample and the reference material are also measured. The temperatures are measured under the crucibles directly at the plate, which eliminates the influence of the thermal resistance changes in the sample.

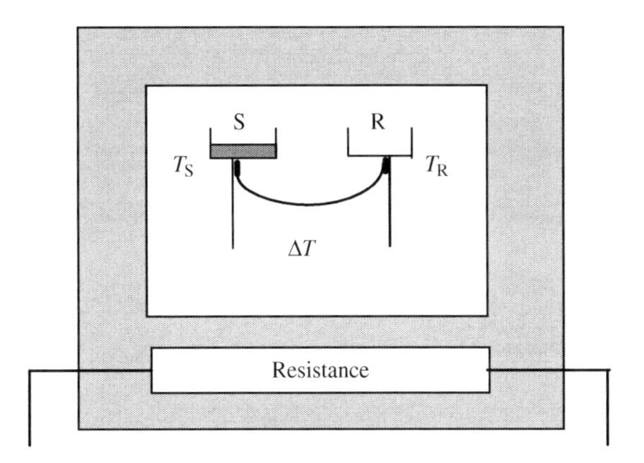

Figure 7.39 Principle of heat-flux DSC, formally called quantitative DTA.

As the electrical heat flux is very well known enthalpy changes in the sample can be measured accurately as a function of time or reference temperature in the following way.

The temperature difference $|T_S - T_R|$ is determined by the voltage U, which is proportional to the difference ΔP in heat flux delivered to the sample and the reference.

$$U = C\Delta P \tag{7.100}$$

The instrument has to be calibrated by means of a sample with some known transformation, for example melting of indium. Figure 7.40 shows a typical DSC curve of a 10 mg sample of In. The calibration process is performed as follows.

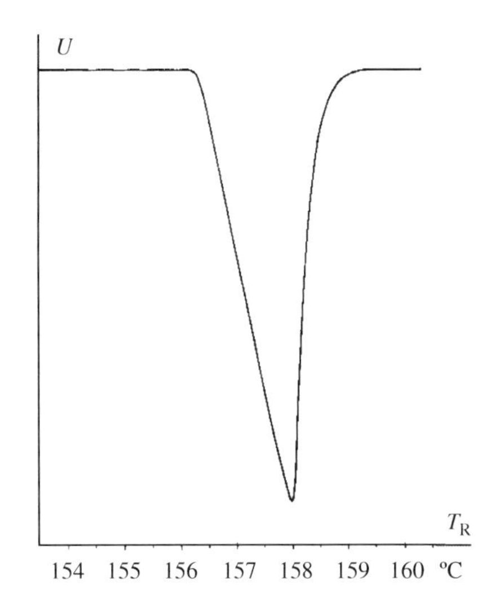

Figure 7.40 Melting curve of indium, registered by means of DSC.

The total heat flux comes from the electrically heated precision resistance resistor.

$$P_{\text{total}} = RI^2 \tag{7.101}$$

A certain constant fraction f of this power goes to the reference crucible, which receives the amount q of heat during the time t and increases its temperature by the amount $(T_R - T_0)$:

$$q = fP_{\text{total}}t = c_p^R m_R (T_R - T_0) \tag{7.102}$$

Hence, the time t and the reference temperature difference $(T_R - T_0)$ are proportional and either t or T_R can be used as a coordinate axis.

In Figure 7.40 the voltage U is measured as a function of T_R which can be transferred to ΔP as a function of time. The area of the peak under the equilibrium line is evaluated graphically. Using Equations (7.100), (7.101) and the differentiated Equation (7.102) gives

$$\text{Area} = \int U \, dT_R = \frac{f P_{total}}{c_p^R m_R} C \int \Delta P dt = \frac{f P_{total}}{c_p^R m_R} C \, m_S(-\Delta H_S) \tag{7.103}$$

Calibration means that the U-T_R curve is measured for a known sample. The area under the equilibrium line is evaluated graphically or numerically and divided by the known value of the product $m_S(-\Delta H_S)$, which gives the value of the product of the last two constant factors on the right-hand side of Equation (7.103). Known values of the four quantities in the first factor give the value of proportionality factor C, which can be used for other measurements with the instrument.

Principle of Power Compensation DSC

Unlike heat flux DSC, sample and reference crucibles are completely separated from each other in power compensation DSC. Each of them has its own separate heating element and temperature sensor. Both sample and reference are heated and a temperature regulation device makes sure that they always have the same temperature. In case of an endothermic reaction in the sample, extra heat is needed to the sample crucible.

If the reaction is exothermic, less heat is required to maintain a constant temperature. By means of electronic circuits additional or less power is sent to the sample holder to keep the sample and the reference at the same temperature. The power change is recorded and in this way the heat flow changes and consequently enthalpy changes are measured.

Modern heat flux and power compensation DSC devices are very sensitive and accurate instruments, which are capable to measure heat fluxes of the magnitude μW. Even minor transformations can easily be registered. As an example we choose the determination of the Curie point of nickel.

The Curie point is the critical temperature when a ferromagnetic substance suddenly becomes paramagnetic (Chapter 6 in [6]). The change is accompanied by a sudden change of the heat capacitivity of the material. A DSC measurement is a very convenient way to find the Curie temperature (Figure 7.41).

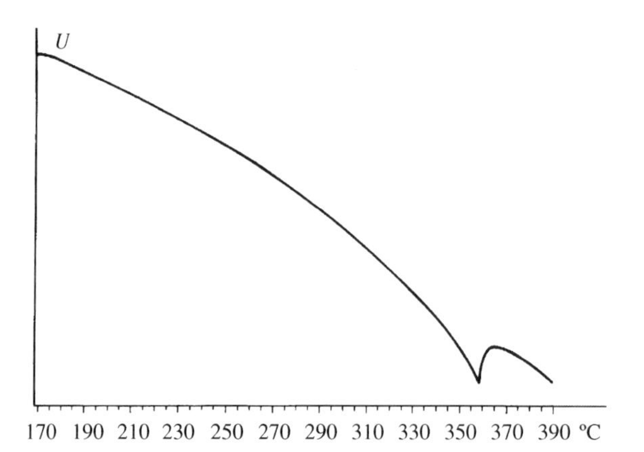

Figure 7.41 Measurement of the Curie point temperature of nickel by means of a DSC measurement.

7.6.4 *Applications of Thermal Analysis*

Figure 7.42 shows a review of the DTA/DSA methods. The development of modern, very sensitive and accurate methods of thermal analysis have greatly increased the possibilities to make reliable measurements of temperature and heat flow. The instruments can be used for most varying applications.

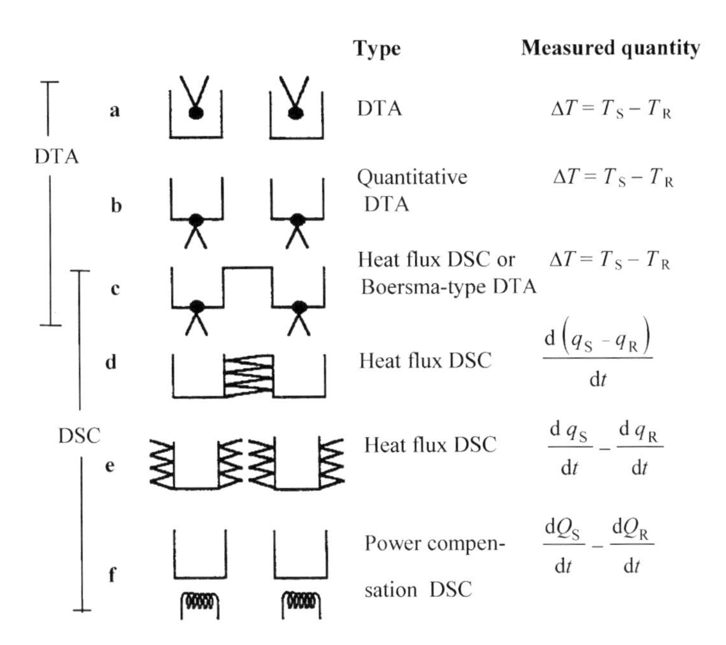

Figure 7.42 Review of the DTA and DSC methods.

All sorts of enthalpy changes associated with *structural changes* such as melting and crystallization processes, vapourization, sublimation and other phase transformations in liquids and solids can be measured.

Thermal capacitivities at constant volume and constant pressure, respectively, melting points, temperatures for phase changes, heat expansion coefficients and enthalpies of phase changes are examples of thermal properties, which can be measured. As a last example it is shown in Figure 7.43 how a fraction curve of a transformation process can be derived from a DTA/DSC plot.

The DTA curves offer possibilities to obtain information of the kinetics of the transition or reaction that is studied. If we assume that the rate of a reaction is proportional to the heat released per unit time, then the partial area (black in Figure 7.43) under the horizontal baseline of the reaction curve divided by the total area equals the fraction of material formed so far.

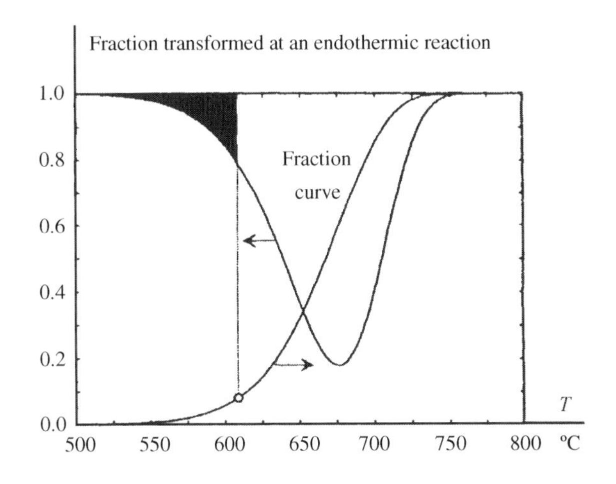

Figure 7.43 Derivation of a fraction curve of a transformation or reaction from DTA/DSC measurements. Reproduced from CRC/Taylor & Francis © 1994.

The curve can be used to derive the fraction curve, which gives information about the reaction kinetics. An example is given in Figure 7.43.

$$f = \frac{\text{partial area}}{\text{total area}} \tag{7.104}$$

By stepwise area integration and calculation of corresponding f values, which are gradually plotted in the diagram the fraction curve is gradually derived as indicated in Figure 7.43.

7.6.5 *Determination of Liquidus Temperatures of Metals and Alloys*

Thermal analysis is a number of methods used for accurate measurements of many various quantities associated with heat. Here we will describe how precision measurements of the equilibrium liquidus temperature can be performed.

For this purpose the horizontal part of the cooling curves is used such as the one in Figure 7.48 on page 411. In Chapter 6 on pages 293–295 we discussed the relationship between the undercooling and the growth rate in terms of the relaxation process, i.e. the quantity D_{relax}, The higher the solidification rate or the cooling rate is, the larger will be the undercooling.

In order to make full use of the accuracy it is customary to take the derivative of the cooling curve (Figure 7.49 on page 411). The temperature at which the derivative changes sign is used as the liquidus temperature.

A series of measurements at different cooling rates are performed in order to determine the equilibrium liquidus temperature. The measured plateau temperatures are plotted as a function of the square root of the cooling rate. The desired liquidus temperature is obtained by extrapolation to zero cooling rate. An example is shown in Figure 7.44.

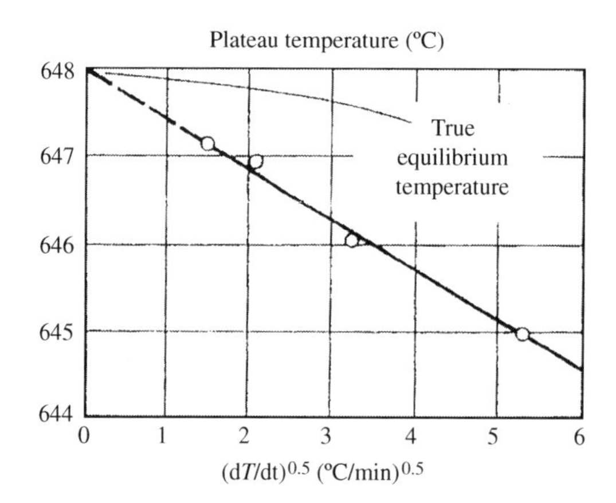

Figure 7.44 Determination of the equilibrium liquidus temperature for an Al–Cu alloy.

7.7 Variable Heat of Fusion of Metals and Alloys

7.7.1 *Experimental Determination of Heat of Fusion*

Modern thermal analysis and numerical methods make it possible to obtain much more accurate cooling curves than earlier methods together with simultaneous temperature registrations. By analysis of a cooling curve, the heat of fusion of a substance can be evaluated. The new experimental possibilities raised a fresh interest for determination of the heat of fusion of metals and alloys. Considerable attention was paid to evaluate the heat of fusion of solidified samples as a function of the cooling rate at the beginning of the 1990s. The results of such research work will be described below.

As a typical example, we will study the pure and a Sr modified Al-Si eutectic alloy. Cooling curves were recorded and their heats of fusion were calculated on basis of the two curves. Three experimental methods were used: DTA, mirror furnace and levitation casting techniques. In addition, the heat of fusion was also determined by DSC for the pure eutectic alloy. DTA has been introduced in Sections 7.6.1 and 7.6.2. The mirror furnace and levitation casting techniques will be introduced below.

DTA

The experimental method has been described in Section 7.6.1. The sample is placed in an Al crucible during the experimental run. The cooling curve and the temperature difference between the sample and the surroundings are recorded simultaneously. Examples of the cooling curves are given on page 410.

Calculation of Heat of Fusion from a Cooling Curve
In Section 7.6.2, the theory of DTA is treated and methods to calculate the heat of fusion, the reaction rate and the fraction of solid phase are indicated. As a complement, we will derive the heat of fusion of the sample from a DTA-recorded cooling curve in terms of cooling rate of the sample, solidification time and temperature of the surroundings. The influence of the crucible is included in the heat balance.

Cooling of the Melt. No Phase Transformation
The calculation of the heat of fusion of the sample is based on the following considerations. The law of energy conservation gives for the case when no phase transformation is going on, i.e. before and after the solidification process:

$$\frac{dQ}{dt} = \rho_{sam} V_{sam} c_p^{sam} \frac{dT_{sam}}{dt} + \rho_{cru} V_{cru} c_p^{cru} \frac{dT_{cru}}{dt} \tag{7.105}$$

where
dQ/dt = heat flow per unit time from the surroundings to the sample and crucible during cooling
$\rho_{sam,cru}$ = density of the sample and the crucible, respectively
$V_{sam,cru}$ = volume of the sample and the crucible, respectively
$c_p^{L\ sam}$ = thermal capacitivity of the liquid sample.
c_p^{cru} = thermal capacitivity of the crucible
$-dT_{sam,cru}/dt$ = cooling rate of the sample and crucible, respectively.

The crucibles and the samples are always very small at these experiments. It is reasonable to assume that temperatures and the cooling rates of the sample and the crucible are *equal*. They are kept *constant* during each experiment. The constant cooling rate is denoted dT_{cool}/dt and is negative.

$$-\frac{dT_{cru}}{dt} = -\frac{dT_{sam}}{dt} = -\frac{dT_{cool}}{dt} \tag{7.106}$$

Provided that convection and radiation of heat can be neglected the heat flow per unit time dQ/dt to the sample and crucible from the surroundings can be written as

$$\left(\frac{dQ}{dt}\right)_{cool} = \rho_{sam} V_{sam} c_p^{L\ sam} \left(-\frac{dT_{cool}}{dt}\right) + \rho_{cru} V_{cru} c_p^{cru} \left(-\frac{dT_{cool}}{dt}\right) \tag{7.107}$$

As the cooling rate is constant dQ/dt is also constant. It is negative because heat is *removed* from the sample. dQ/dt is roughly proportional to the temperature difference $(T - T_0)$. and is determined by the experimental conditions.

Phase Transformation
During the solidification process the conditions change compared to the cooling period. The temperature of the solidifying sample becomes almost constant, which means that very little cooling heat is developed during the solidification time. Instead heat of solidification is currently emitted.

At time t the heat flow per unit time from the surroundings to the sample can be written as

$$\left(\frac{dQ}{dt}\right)_{sol} = \rho_{sam} V_{sam} c_p^{L,\ S} \frac{dT_{sol}}{dt} + \rho_{cru} V_{cru} c_p^{cru} \frac{dT_{sol}}{dt} - \rho_S V_S \frac{df}{dt}(-\Delta H_S) \tag{7.108}$$

where
f = fraction solidified material in the sample at time t
$-\Delta H_S$ = heat of fusion per unit mass of the sample
$c_p^{L,S}$ = heat capacitivity of the melt and solid, respectively.

A more accurate relationship is obtained if we consider the difference in thermal capacitivity in the solid and liquid phases:

$$\left(\frac{dQ}{dt}\right)_{sol} = (1-f)\rho_{sam}V_{sam}c_p^{L\ sam}\frac{dT_{sol}}{dt} + f\rho_{sam}V_{sam}c_p^{s\ sam}\frac{dT_{sol}}{dt} + \rho_{cru}V_{cru}c_p^{cru}\frac{dT_{sol}}{dt} - \rho_{sam}V_{sam}\frac{df}{dt}(-\Delta H_{sam}) \tag{7.109}$$

Often the difference in heat capacitivities in the liquid and the solid of the sample can be neglected.

The total heat flow to the surroundings is determined by the rate of heat removal, which roughly is a function of the temperature difference ($T_M - T_0$) between the melting-point temperature and the temperature of the surroundings.

The temperature of the surroundings during the solidification is often equal to the room temperature and much lower than the sample temperature as the cooling rate is constant.

Because the experimental conditions determine the rate of heat removal and they are the same during the periods of cooling and solidification it is reasonable to assume that heat flow per unit time dQ/dt is equal during the cooling and the solidification processes.

$$\left(\frac{dQ}{dt}\right)_{cool} = \left(\frac{dQ}{dt}\right)_{sol} \tag{7.110}$$

Calculation of the Heat of Fusion

The expressions in Equations (7.107) and (7.108) are inserted into Equation (7.110). The new equation is multiplied by dt and integrated over the solidification time t_{sol}. Using the relationship $\int df = 1$ we obtain

$$-\Delta H_{sam} = \frac{(\rho_{sam}V_{sam}c_p^{L\ sam} + \rho_{cru}V_{cru}c_p^{cru})\left[-\left(\frac{dT_{cool}}{dt} - \frac{dT_{sol}}{dt}\right)\right]}{-\rho_{sam}V_{sam}} \times t_{sol} \tag{7.111}$$

The cooling rate during the solidification time is small and can often be neglected.

Equation (7.111) can be used for calculation of the heat of fusion. T_0 is known. The two cooling rates and t_{sol} are derived from measurements on the recorded curve. Figures 7.45 and 7.46 show some curves recorded by DTA technique. The cooling rate is very low.

Mirror Furnace and Levitation Casting Techniques

As Figure 7.45 shows, the cooling rate at DTA is very low. For rapid cooling processes, other techniques have to be used.

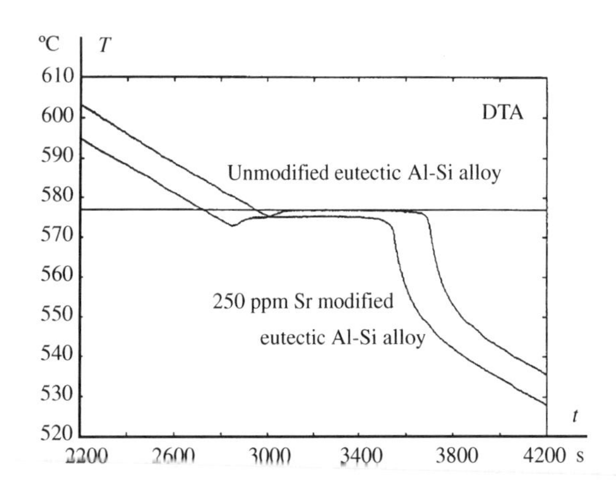

Figure 7.45 Typical cooling curves for pure and a 250 ppm Sr modified eutectic Al–Si alloy during solidification, recorded with DTA technique. Cooling rate 0.035 K/s.

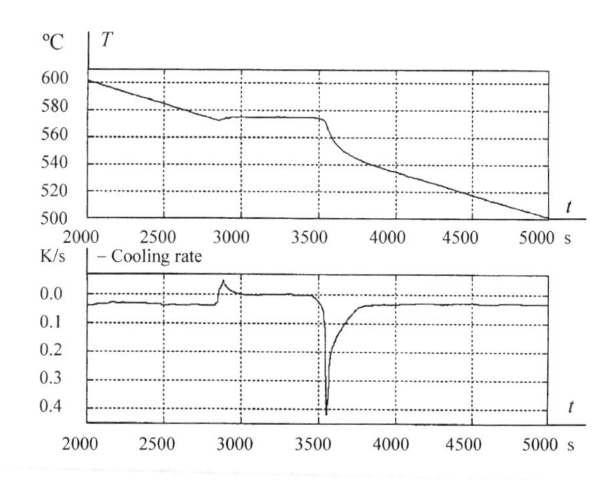

Figure 7.46 Typical cooling curve and its differentiated curve used for calculation of the cooling rate of samples during solidification, recorded with DTA technique.

Mirror Furnace Technique

The mirror furnace technique was developed for cooling times of the magnitude 1–10 seconds. The Microgravity Applications Furnace Facility (MAFF) was constructed for gravitation free experiments in an aircraft, which gives microgravity-free conditions during approximately 20 seconds. The compact equipment consists of a rack with drawers. It contains six mirror furnaces, a PC for control and data storage, a gas cooling system, accelerometers, electronics and a power supply.

Each mirror furnace consists of two high power halogen lamps placed inside two ellipsoidal mirrors with a common focus where the sample and crucible are located (Figure 7.47). The furnace is capable of producing a maximum sample temperature of 1600 °C. A computer controls the power of the lamps via a switch cassette. The cooling rate is much higher than in the DTA case. Some typical cooling curves are shown in Figures 7.48 and 7.49.

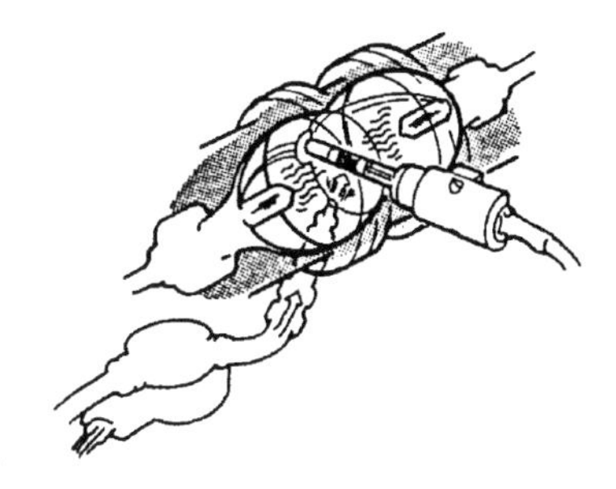

Figure 7.47 Mirror furnace with sample in focus.

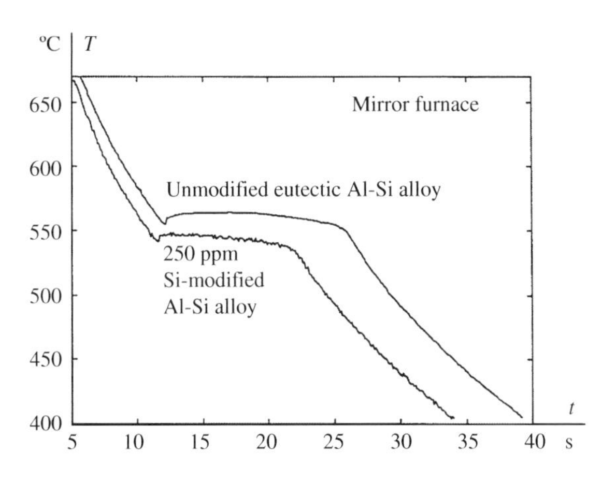

Figure 7.48 Typical cooling curves recorded for a pure eutectic Al–Si alloy and a 250 ppm Sr-modified eutectic Al–Si alloy recorded with mirror furnace technique.

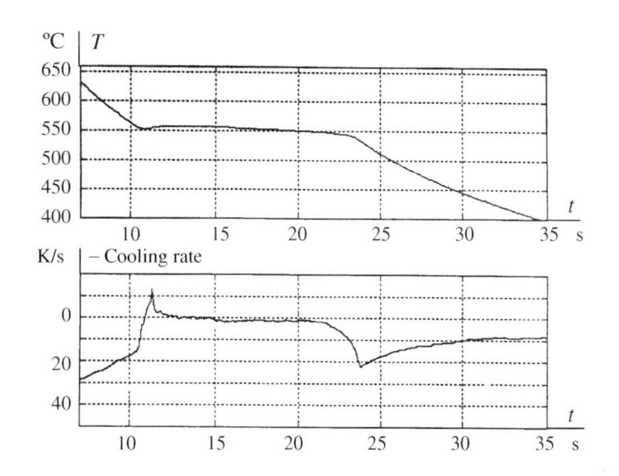

Figure 7.49 Typical cooling curve and its differentiated curve used for calculation of the cooling rate of samples during solidification recorded with mirror furnace technique.

Equation (7.111) can be used to derive the heat of fusion from curves recorded with mirror furnace technique as the sample is kept in a crucible both at DTA and mirror furnace experiments.

Levitation Casting Technique

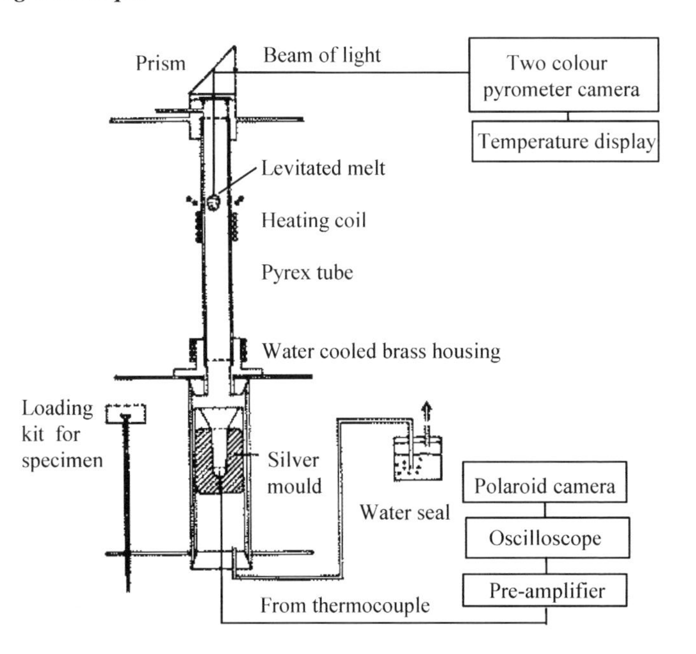

Figure 7.50 Levitation casting equipment. The sample is melted by induction in a He atmosphere and quenched at an Ag mould. An open thermocouple at the bottom of the mould monitors the temperature during solidification. The output is recorded on a memory oscilloscope.

The levitation casting technique is designed for extremely short solidification times of the sample with a lower limit of 15 milliseconds. Consider a sample of about 1.5 grams that is levitation melted by induction in an He atmosphere.

A 0.5 mm thick rectangular Ag mould or a cylindrical copper mould, 5 mm in diameter, is used for quenching (Figure 7.50). The upper parts of the mould have a conical shape. An open thermocouple is placed at the bottom of the mould to monitor the temperature during solidification. The output is recorded on a memory oscilloscope. The cooling rate is of the magnitude $10^5 - 10^6$ K/s.

Some typical cooling curves are shown in Figures 7.51 and 7.52.

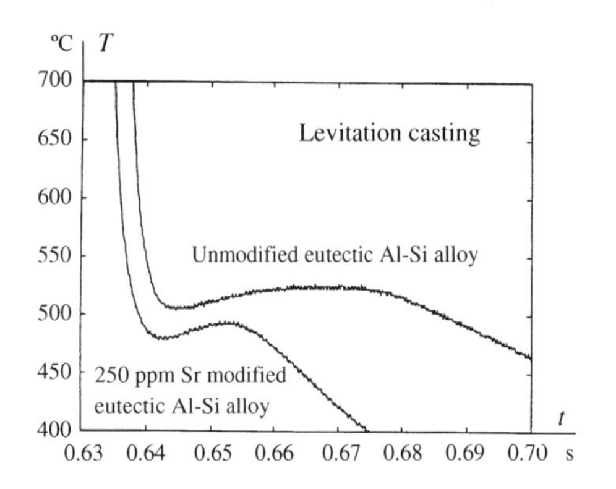

Figure 7.51 Typical cooling curves recorded for a pure eutectic Al–Si alloy and a 250 ppm Sr-modified eutectic Al–Si alloy recorded with levitation casting technique.

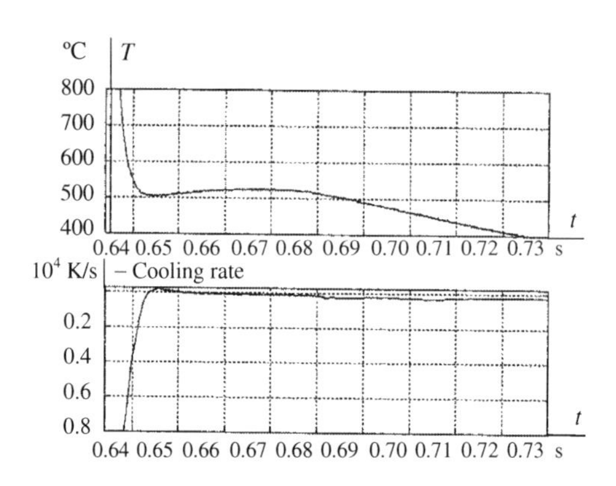

Figure 7.52 Typical cooling curve and its differentiated curve used for calculation of the cooling rate of samples during solidification recorded by use of the levitation casting technique.

Calculation of Cooling Rates from a Mirror Furnace or Levitation Casting Cooling Curve

Figures 7.45+7.46, 7.48+7.49 and 7.51+7.52 show that the cooling rate is remarkably increased in the levitation casting case compared to the mirror furnace technique and to DTA. Both the mirror furnace and levitation casting methods operate with short solidification times. In those experiments, it is therefore reasonable to assume that the heat transfer coefficient controls the heat flux. There are several methods proposed to calculate the heat of fusion. We will use the calculation method, described below, in the case of the mirror furnace and the levitation casting techniques.

No Phase Transformation

The temperatures of the sample and the mould are not the same. The energy law gives for the case of no phase transformation i.e. before and after the solidification of the sample.

$$Ah\left(T_{\text{mould}} - T_{\text{sam}}\right) = \rho V c_{\text{p}} \left(-\frac{\mathrm{d}T}{\mathrm{d}t}\right)_{\text{no ph-tr}} \tag{7.112}$$

where

A = contact area between mould and sample
h = heat transfer coefficient between mould and sample
ρ = density of the sample
V = volume of the sample
c_{p} = heat capacitivity of the sample
$-(\mathrm{d}T/\mathrm{d}t)_{\text{no ph-tr}}$ = cooling rate at the cooling curve in case of no phase transformation.

Phase Transformation

During the phase transformation, i.e. during solidification, the energy law can be written as

$$Ah(T_{\text{mould}} - T_{\text{sam}}) = \rho V c_{\text{p}} \left(-\frac{\text{d}T}{\text{d}t} \right)_{\text{ph-tr}} + \rho V \frac{\text{d}f}{\text{d}t} (-\Delta H_{\text{sam}}) \tag{7.113}$$

The subscript 'ph-tr' refers to the cooling curve with phase transition. By combining Equations (7.112) and (7.113) and integration the following equation can be deduced

$$-\Delta H_{\text{sam}} = -\Delta H_{\text{S}} \int_0^1 \text{d}f = c_{\text{p}} \int_{t_{\text{start}}}^{t_{\text{end}}} - \left[\left(\frac{\text{d}T}{\text{d}t} \right)_{\text{no ph-tr}} - \left(\frac{\text{d}T}{\text{d}t} \right)_{\text{ph-tr}} \right] \text{d}t \tag{7.114}$$

Comparison between Equation (7.114) and Figure 7.53 shows that the bounded area between the two curves in Figure 7.53, i.e. between the cooling curve *with* a phase transformation and the cooling curve *with no* phase transformation represents the heat of fusion.

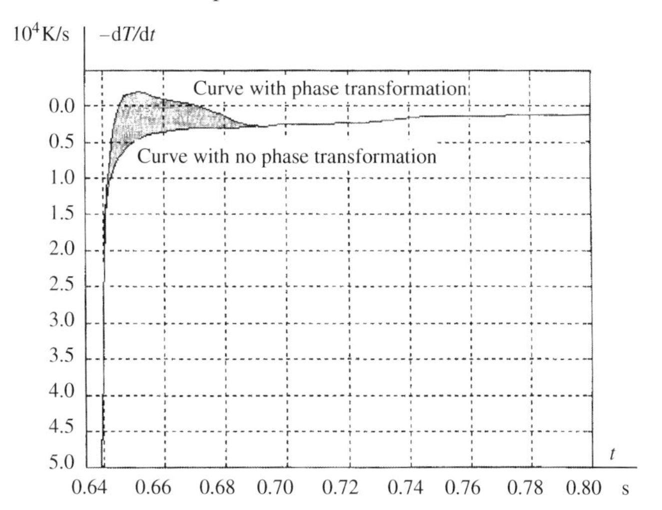

Figure 7.53 Cooling rate as a function of time for an eutectic Al–Si alloy. The curve with no phase transformation is calculated under the assumption that no heat of fusion is released.

Heat of Fusion as a Function of Cooling Rate

Figure 7.54 shows a graphical comprehension of measurements of the heats of fusion of the eutectic Al-Si alloys over a wide range of cooling rates, performed by Fredriksson and Talaat [7],

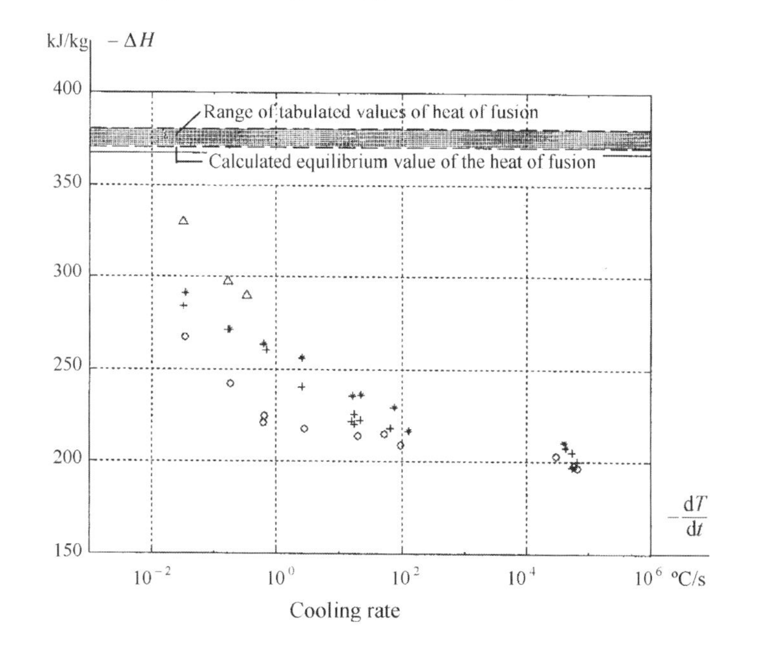

Figure 7.54 Heat of fusion of Al–Si alloys as a function of cooling rate.
* = pure eutectic Al–Si alloy
+ = 150 ppm Sr modified eutectic alloy
o = 250 ppm Sr modified eutectic alloy
Δ = measured heat of fusion by use of DSC for a pure eutectic Al–Si alloy.

Figure 7.54 and a great number of cooling curves, registered with the aid of thermal analysis technique, shows very clearly that

- the heat of fusion decreases with increasing cooling rate
- the measured values of heat of fusion are generally lower than those, tabulated in literature.

7.7.2 *Derivation of the Heat of Fusion as a Function of the Cooling Rate with Consideration to Relaxation*

The generally accepted explanation of the results above is that vacancies are trapped at the solidification front during the solidification process. The more rapidly the solidification front moves the larger will be the concentration of vacancies. The heat of fusion represents the energy required to break the bonds between the lattice atoms. If vacancies are present some bonds are missing and therefore the total energy for melting is reduced.

An alternative hypothesis, which also can explain both the results above satisfactory, is the relaxation theory which has been introduced into Chapter 6. The explanation in terms of relaxation will be discussed below.

Influence of Relaxation on the Heat of Fusion

The liquid atoms do not reach their lowest energy during the solidification process. The larger the solidification rate is, the larger will be the deviation from the equilibrium value of the potential energy. Therefore, the larger the solidification rate is, the higher will be the relaxation energy and the lower will be the heat of fusion.

After every instantaneous solidification process in the interphase (pages 293–294 in Chapter 6) a relaxation process always follows. The liquid atoms reach their lowest potential energy when the relaxation process is finished. The sum of the prompt heat of fusion and the relaxation energy is constant and equal to the total heat of fusion (Figure 6.26 on page 294 in Chapter 6).

In Chapter 6 on pages 293–295 we discussed the determination of the heat of fusion of a metal melt and found that the measured value of the heat of fusion depends on the growth rate V_{growth} of the solidification front [Equation (6.42) on page 296] or the cooling rate, which is proportional to the growth rate.

$$-\Delta H_{prompt} = -\Delta H_0\left(1 - \sqrt{\frac{\delta}{\pi D_{relax}}}\sqrt{V_{growth}}\right) \tag{7.115}$$

- The more rapidly the solidification front moves, i.e. the higher the cooling rate is, the lower will be the measured heat of fusion.

In alloys the heat of fusion, i.e. the change of the potential energy is also influenced by other factors. Figure 7.54 on page 413 and Example 7.3 below show that a small amount of impurities influence the heat of fusion of an alloy. Small amounts of Sr has been added to an eutectic AlSi alloy, which reduces its heat of fusion compared to an unmodified eutectic AlSi alloy.

Example 7.3

The figure shows the cooling curves during solidification for two slightly different Al-13.5%Si alloys. The curve to the right is valid for a sodium modified alloy (cooling rate before solidification = 50 °C/min) and the curve to the left refers to an unmodified alloy (cooling rate before solidification = 35 °C/min). $c_p^L \approx c_p^s$ for both the sample and the crucible.

The experiments were performed in a mirror furnace. As the sample has the shape of a cylinder with a cross-section diameter $D = 7.0$ mm. As the length L of the cylinder $\gg$ the diameter the heat transport in the axial direction can be neglected. The crucible is made of boron nitride, which has a very high thermal conductivity. Therefore, the crucible and the sample have the same temperature. The thickness of the crucible wall is 1.0 mm.

Material constants:

$c_p^{AlSi} = 1.2 \times 10^3$ J/kg $\qquad c_p^{AlSi} = 2.22 \times 10^3$ kg/m^3

$c_p^{cru} = 2.22 \times 10^3$ J/kg $\qquad c_p^{cru} = 2.11 \times 10^3$ kg/m^3

a) Evaluate the heats of fusion evolved during the solidification process of the two alloys.
b) Discuss the evaluation and give an explanation of the results.
c) The heat of fusion of the unmodified Al-13.5%Si alloy, given in literature, is 3.70×10^5 J/kgK. Compare your result with this estimation and give an explanation of the discrepancy.

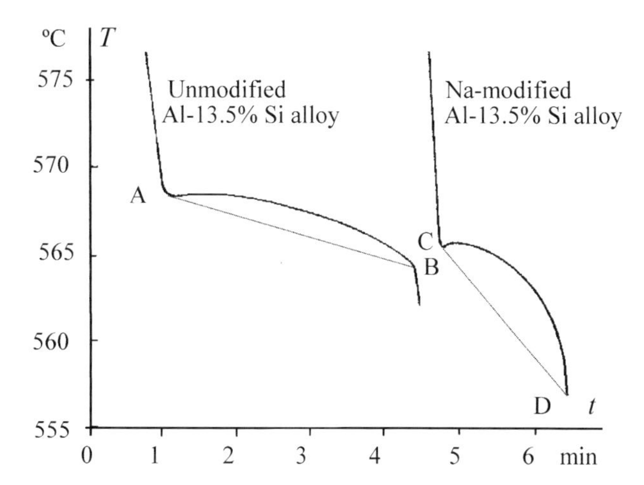

Solution:

We obtain the rate of heat transfer from the crucible and its sample to the surroundings during cooling and during solidification by application of Equations (7.107) and (7.108) on page 409, respectively

$$\left(\frac{dQ}{dt}\right)_{cool} = \rho_{sam} V_{sam} c_p^{L\ sam}\left(-\frac{dT_{cool}}{dt}\right) + \rho_{cru} V_{cru} c_p^{cru}\left(-\frac{dT_{cool}}{dt}\right) \tag{1'}$$

$$\left(\frac{dQ}{dt}\right)_{sol} = \rho_{sam} V_{sam} c_p^{L\ sam}\left(-\frac{dT_{sol}}{dt}\right) + \rho_{cru} V_{cru} c_p^{cru}\left(-\frac{dT_{sol}}{dt}\right) - \rho_{sam} V_{sam}\frac{df}{dt}(-\Delta H_{sam}) \tag{2'}$$

The heat flows per unit time during the cooling period and the solidification period are equal, which gives

$$\rho_{sam} V_{sam} c_p^{L\ sam}\left(-\frac{dT_{cool}}{dt}\right) + \rho_{cru} V_{cru} c_p^{cru}\left(-\frac{dT_{cool}}{dt}\right)$$
$$= \rho_{sam} V_S c_p^{L\ sam}\left(-\frac{dT_{sol}}{dt}\right) + \rho_{cru} V_{cru} c_p^{cru}\left(-\frac{dT_{sol}}{dt}\right) - \rho_{sam} V_{sam}\frac{df}{dt}(-\Delta H_{sam}) \tag{3'}$$

Multiplication of Equation (3′) with dt and integration gives ($f = 1$ for $t = t_{sol}$)

$$(-\Delta H_{sam}) = \frac{\left(\rho_{sam} V_{sam} c_p^{L\ sam} + \rho_{cru} V_{cru} c_p^{cru}\right)\left[-\left(\frac{dT_{cool}}{dt} - \frac{dT_{sol}}{dt}\right)\right]}{-\rho_S V_S} t_{sol} \tag{4'}$$

Numerical values:

$$\frac{V_{\text{cru}}}{V_{\text{sam}}} = \frac{L\left(\frac{\pi d_{\text{cru}}^{2}}{4} - \frac{\pi d_{\text{sam}}^{2}}{4}\right)}{L\frac{\pi d_{\text{sam}}^{2}}{4}} = \frac{d_{\text{cru}}^{2} - d_{\text{sam}}^{2}}{d_{\text{sam}}^{2}} = \frac{9^2 - 7^2}{7^2} = 0.653$$

Readings in the figure in the text give

The average cooling rate during solidification of the unmodified Al-Si alloy: (line AB) is

$$\frac{dT_{\text{sol}}}{dt} = \frac{568 - 564}{4.3 - 1.0} = \frac{4}{3.3} = 1.2\ \text{K/min}$$

The average cooling rate during solidification of the Na-modified Al-Si alloy (line CD) is

$$\frac{dT_{\text{sol}}}{dt} = \frac{565 - 557}{6.4 - 4.8} = \frac{8}{1.6} = 5.0\ \text{K/min}$$

Inserting the numerical values into Equation (4′) after division with V_S in the nominator and denominator gives

Unmodified alloy

$$(-\Delta H_{\text{sam}}) = \frac{-(2.22 \times 10^3 \times 1.2 \times 10^3 + 2.11 \times 10^3 \times 0.653 \times 0.90 \times 10^3)(35 - 1)}{-2.22 \times 10^3} \times 3.3 = 146\,\text{kJ/kg}$$

Modified alloy

$$(-\Delta H_{\text{sam}}) = \frac{-(2.22 \times 10^3 \times 1.2 \times 10^3 + 2.11 \times 10^3 \times 0.653 \times 0.90 \times 10^3)(50 - 5)}{-2.22 \times 10^3} \times 1.6 = 0.94\,\text{kJ/kg}$$

Answer:

a) The heat of fusion for the unmodified alloy is 1.5×10^{2} kJ/kg. The heat of fusion for the modified alloy is 94 kJ/kg.

b) Addition of sodium reduces the heat of fusion considerably, which means that the melting-point temperature also is decreased. van't Hoff's rule says that the melting point decrease of a solution (metal melt) is proportional to the concentration of the solute (alloying element). In this case the decrease is much larger than that calculated by van't Hoff's rule.

c) The heat of fusion of the unmodified alloy is very low, compared to the literature value. The reason is lack of equilibrium during the cooling process. The rapid cooling results in an increase of the potential energy. Sodium increases the potential energy in the solid after solidification. Both effects result in a decrease of the heat of fusion.

7.8 Variable Heat Capacitivity of Metals and Alloys

In Section 7.7 we found that the measured heat of fusion decreases with increasing cooling rate. This phenomenon could be interpreted as an effect of relaxation of the solidified liquid atoms. During the relaxation process the atoms successively loose strain energy until equilibrium is achieved when they finally obtain the same potential energy as the lattice atoms in the solid crystal.

Measured values of the heat capacitivities of solid metals confirm that they vary with the cooling rate. This phenomenon will be analyzed and discussed below.

7.8.1 *Experimental Evidence*

During rapid cooling of a solidified metal melt the internal or potential energy shows a large deviation from the equilibrium value. At very high cooling rates, such as in levitation experiments, an unordered solid phase forms with strained atoms and high potential energy, which show a large deviation from equilibrium. The reason is that there is not enough time for the atoms to adjust their positions.

The following relaxation starts immediately after the beginning of the solidification and goes on after complete solidification when the last solidified atoms relax.

These circumstances also affect the heat capacitivity of the solid at the beginning of the cooling process. If the heat capacitivity were not influenced by the relaxation in the solid, the slope of the 'solid' cooling curve would be steeper than that of the liquid curve because the forces between the atoms are stronger in solids than in liquids.

Figure 7.48 on page 411 recorded with the mirror furnace technique, and especially Figures 7.51 and upper Figure 7.52 on page 412 recorded with the aid of levitation casting technique, show clearly that the slope of the solid curve initially is *less steep* than that of the liquid curve.

The following experiments, based on thermal analysis technique, confirm these observations. In a series of solidification experiments with Cu-base alloys the heat capacitivity was studied as a function of the cooling rate.

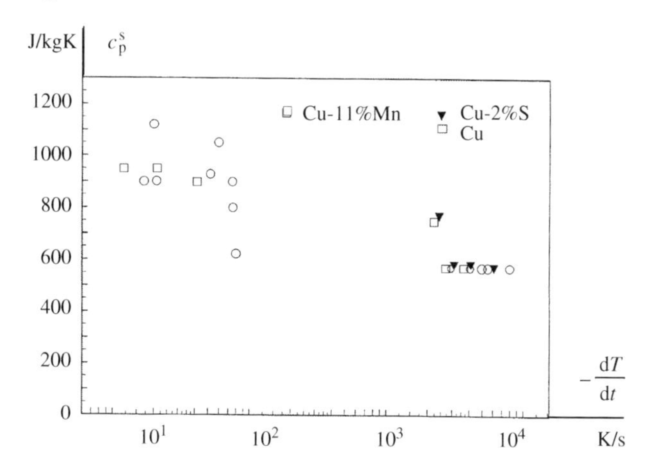

Figure 7.55 Heat capacitivities of some solid Cu-based alloys, cooled with some different cooling rates.

The result is shown in Figure 7.55. The figure shows that

- The higher the cooling rate is, the smaller will be the heat capacitivities evaluated from the cooling curve. The equilibrium value of the heat capacitivity c_p^0 for copper is 500 J/kgK.

This topic will be discussed below.

Basic Information from Cooling Curves

A cooling curve such as the one in Figure 7.56 can roughly be divided into three regions. Region I extends from the initial temperature of the superheated melt to the temperature at which the

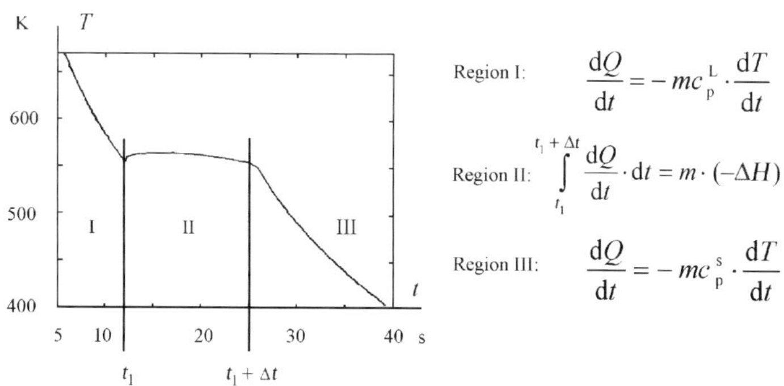

Figure 7.56 By simultaneous registration of temperature and emitted heat per unit time as functions of time c_p^L, c_p^s and the heat of fusion $(-\Delta H)$ (J/kg) can be derived with the aid of the equations given in the figure.

solidification starts. Region II corresponds to the solidification time of the specimen. It is characterized by an almost constant temperature. When the solidification process is finished, the temperature starts to decrease again and the solid specimen cools.

The heat capacitivities of the melt c_p^L and the solid c_p^s and the heat of fusion $-\Delta H$ can be derived as indicated in Figure 7.56 with the aid of the following information (one-dimensional heat flow)

- mass m of the specimen
- heat flow $-\,\mathrm{d}Q/\mathrm{d}t$ from the specimen to the surroundings ($\mathrm{d}Q/\mathrm{d}t$ is negative)
- temperature T of the specimen as a function of time t ($\mathrm{d}T/\mathrm{d}t$ is negative)
- solidification time.

Influence of Relaxation on Heat Capacitivity c_p^s

Experimental conditions during cooling experiments determine the heat flow per unit time to the surroundings, which can be kept constant.

Consider a very thin sample of a liquid metal, which is cooled by a constant heat flow per unit time before, during and after solidification. After complete solidification the temperature of the sample decreases. For simplicity we also assume that the experiment is arranged in such a way that the heat withdrawal is unidirectional (Figure 7.57).

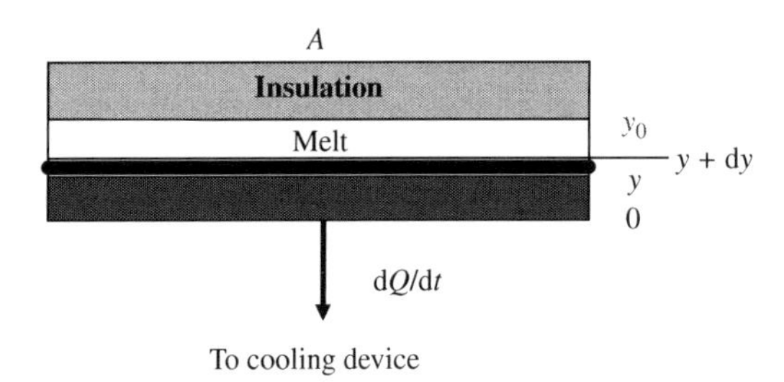

Figure 7.57 Solidifying thin sample with thickness y_s and area A. The small sample is insulated on the top and heat is removed downwards. The growth rate $V_{growth} = \mathrm{d}y/\mathrm{d}t$ is directed upwards.

The sample is kept in a mould when the cooling curve is recorded. Both the sample and the mould must be considered when mass balance equations are set up below. For the sake of simplicity the terms, which belong to the mould, are *not* written in the formulas here, but it is necessary to include them in all practical cases.

Region I. Cooling of the Superheated Melt

As the heat flow per unit time is constant, the cooling rate will also be constant, provided that the heat capacitivity is constant and we obtain

$$-\frac{\mathrm{d}Q}{\mathrm{d}t} = mc_p^L\left(-\frac{\mathrm{d}T}{\mathrm{d}t}\right) \tag{7.116}$$

or

$$CHR = \left(-\frac{\mathrm{d}Q}{\mathrm{d}t}\right)_{\text{const}} = \rho A y_0 c_p^L\left(-\frac{\mathrm{d}T}{\mathrm{d}t}\right) \tag{7.117}$$

where

$CHR = -\frac{dQ}{dt}$ = constant heat rate (J/s) = amount of heat per unit time emitted from the sample to the surroundings (dQ/dt is negative)

m = mass of the sample

ρ = density of the sample

A = cross-section area of the sample, perpendicularly to the direction of the heat flow

y_0 = thickness of the sample

c_p^L = heat capacitivity of the solidified sample (J/kgK)

T = temperature of the sample at time t

$-\frac{dT}{dt}$ = cooling rate (dT/dt is negative).

Region II. Solidification with Consideration to Relaxation

If $-\Delta H_{strain}$ were zero and $-\Delta H_0 = -\Delta H_{prompt}$ there would be no relaxation, as is shown in Figure 7.58.

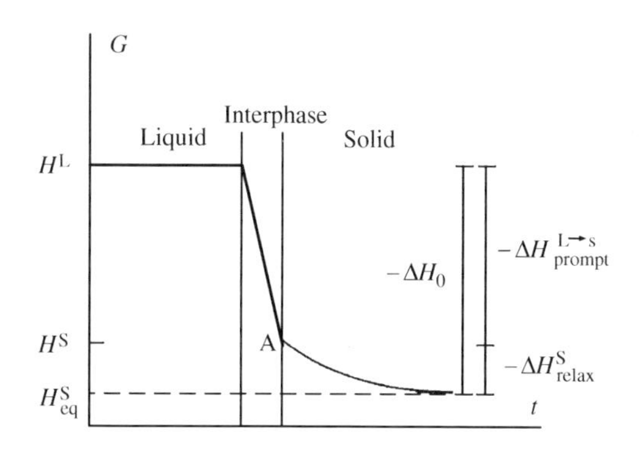

Figure 7.58 Solidification process in terms of potential energy of the atoms. Compare Figure 6.26 on page 294 in Chapter 6.

In case of relaxation, the solidified 'liquid' atoms are strained and the strain energy is successively emitted to the surroundings, The strain energy is released in the solid both during the solidification and during the following cooling.

We have derived one expressions for H_{strain} [Equation 6.21) on page 288] and one for H_{prompt} [Equation (6.42) on page 296 in Chapter 6. They are given in this chapter as Equation (7.118) on page 419 and Equation (7.115)] on page 414, respectively.

$$-\Delta H_{strain} = H_{strain}^{max} = -\Delta H_0 \sqrt{\frac{V_{growth}\delta}{\pi D_{relax}}} \tag{7.118}$$

and

$$-\Delta H_{prompt} = -\Delta H_0 \left(1 - \sqrt{\frac{V_{growth}\delta}{\pi D_{relax}}}\right) \tag{7.115}$$

where

$-\Delta H_{prompt}$ = instant heat of fusion per unit mass

H_{strain} = strain energy per unit mass at time t

H_{strain}^{max} = total strain energy per unit mass before the relaxation process starts

$-\Delta H_0$ = total heat of fusion per unit mass of the metal or alloy

V_{growth} = growth rate of the solidification front at solidification

δ = thickness of the liquid/solid interphase

D_{relax} = material constant coupled to heat transport (m^2/s).

An energy balance per unit time, valid at time t during the solidification, can be written as

$$CHR = V_{\text{growth}}(-\Delta H_{\text{prompt}})\rho A + V_{\text{growth}}(-\Delta H_{\text{strain}})\rho A +$$

emitted heat of fusion per unit time at time t — emitted strain energy per unit time at time t?

$$\rho A y f_s c_p^s\left(-\frac{dT}{dt}\right) + \rho A y(1-f_s)c_p^L\left(-\frac{dT}{dt}\right) \tag{7.119}$$

emitted cooling heat from the solid part of the sample — emitted cooling hear from the liquid part of the sample

where

CHR = constant heat rate to the surroundings
c_p^s = heat capacitivity of the solid part of the sample (J/kgK)
V_{growth} = growth rate of solidification, equal to dy/dt
f_s = solidified fraction of the sample at time t
T = temperature of the sample at time t.

As the temperature is practically constant it is reasonable to neglect the last two terms in the basic Equation (7.119)

Calculation of the Solidification Time

Equations (7.119) and (7.120) are hard to solve, i.e. it is hard to find an exact solution of the solidification time. Instead we will perform a step by step analysis and introduce some approximations.

We assume that

1. the amount of heat removed from the sample and the mould per unit time is constant
2. the sample is thin but has a finite thickness y_s
3. the total solidification time of the thin sample is short compared to the relaxation time of the metal or alloy
4. the temperature is constant during the solidification time.

The third assumption is valid because the sample is very thin. It is reasonable to assume that the relaxation energy released during the total solidification time is practically zero. Hence, no relaxation energy at all is released during the solidification time. Therefore, the second term on the right-hand side of Equation (7.119) can be disregarded in comparison with the first one.

The fourth assumptions means that the last two terms on the right-hand side of Equation (7.119) also can be neglected in comparison with the first term. Hence, there is only one term left on the right-hand side of Equation (7.119).

Inserting the expressions in Equation (7.115) on page 414 for $-\Delta H_{\text{prompt}}$ into the shortened Equation (7.119) gives

$$CHR = V_{\text{growth}}\rho A(-\Delta H_0)\left(1-\sqrt{\frac{V_{\text{growth}}\delta}{\pi D_{\text{relax}}}}\right) \tag{7.120}$$

The rate of heat flow CHR to the surroundings from the sample is constant and known. Equation (7.120) can be used to solve the solidification time t_{sol} because the growth rate is constant as all the other quantities are constant.

$$V_{\text{growth}} = \frac{dy}{dt} = \frac{y_0}{t_{\text{sol}}} \tag{7.121}$$

V_{growth} is replaced by y_0/t_{sol} in Equation (7.120)

$$CHR = \frac{y_0}{t_{sol}}\rho A(-\Delta H_0)\left(1 - \sqrt{\frac{\delta}{\pi D_{relax}}}\sqrt{\frac{y_0}{t_{sol}}}\right) \qquad (7.122)$$

If we multiply Equation (7.122) by $(t_{sol})^{3/2}$ we obtain after some reduction ($y_0\rho A = m$)

$$CHR \times t_{sol}{}^{3/2} = m(-\Delta H_0)\left(t_{sol}{}^{1/2} - \sqrt{\frac{y_0\delta}{\pi D_{relax}}}\right) \qquad (7.123)$$

We introduce $z = \sqrt{t_{sol}}$ into Equation (7.123) and obtain after reduction

$$z^3 - \frac{m(-\Delta H_0)}{CHR}z + \frac{m(-\Delta H_0)}{CHR}\sqrt{\frac{y_0\delta}{\pi D_{relax}}} = 0 \qquad (7.124)$$

Equation (7.125) has three roots and can be solved numerically. The root of interest in this case corresponds to $z = \sqrt{t_{sol}}$ and can be found by comparison with the value of t_{sol} measured from the cooling curve. Hence

$$t_{sol} = z^2 \qquad (7.125)$$

Region III. Cooling after Solidification with Consideration to Relaxation

It is important to remember the approximation we introduced on page X95 that the relaxation time in connection with the solidification of the thin sample is long, compared to the short solidification time of the thin sample. We assumed that no relaxation energy is released during the solidification time. The accumulated relaxation energy is stored as internal energy in the sample and is successively released during the cooling of the sample.

After complete solidification the fraction $f_s = 1$ and the first and last terms in Equation (7.119) will vanish. In this case the energy balance per unit time can be written as

$$CHR = y_0\rho A\frac{d(-\Delta H_{strain})}{dt} + \rho A y_0 \times 1 \times c_p^s\left(-\frac{dT}{dt}\right) \qquad (7.126)$$

Emitted strain energy per unit time at time $t_{sol} + t$ — emitted cooling energy from the solid sample per unit time

If we introduce the dimensions of the sample in analogy with Equation (7.122).

Influence of Relaxation on Heat Capacitivity c_p^s

When the sample has solidified completely the generation of prompt solidification heat ceases abruptly. We have assumed that no relaxation energy is emitted during the solidification time. Therefore, the accumulated strain energy is emitted at the same time as the sample cools.

In thermal analysis experiments, the cooling rate is often very fast and the lattice atoms do not have time to change their potential energy instantly. This results in relaxation.

Provided that there is no relaxation, the heat capacitivity c_p^s is constant and approximately equal to c_p^L at temperatures close to the melting point temperature T_M. We call the constant heat capacitivity c_p^0. At constant rate of heat removal c_p^s is found to be larger than c_p^0 in case of relaxation.

In case of relaxation Equation (7.126) can be written as

$$CHR = m\frac{d(-\Delta H_{strain})}{dt} + mc_p^0\left(-\frac{dT}{dt}\right) = mc_p^s\left(-\frac{dT}{dt}\right) \qquad (7.127)$$

When the relaxation is zero, c_p^s is equal to c_p^0. When relaxation is present, c_p^s must necessarily be smaller than c_p^0 because the sum of the two terms on the right-hand side of Equation (7.127) is constant.

The strain energy obeys the relaxation law, which can be written as

$$-\Delta H_{\text{strain}} = -\Delta H_{\text{strain}}^{\max} e^{-\frac{t}{\tau}} \tag{7.128}$$

where $t = 0$ is the time when the cooling starts.

We take the derivative of Equation (7.128) and insert the derivative into Equation (7.127):

$$CHR = m(-\Delta H_0)\sqrt{\frac{y_0\delta}{\pi D_{\text{relax}}}} e^{-\frac{t}{\tau}} \frac{-1}{\tau} + mc_{\text{p}}^{\text{s}}\left(-\frac{dT}{dt}\right) = mc_{\text{p}}^{\text{s}}\left(-\frac{dT}{dt}\right) \tag{7.129}$$

Integration of Equation (7.129) with respect to time from 0 to t gives relationship between the temperature T and the time t.

$$CHR \times t = m(-\Delta H_0)\sqrt{\frac{y_0\delta}{\pi D_{\text{relax}}}}\left(1 - e^{-\frac{t}{\tau}}\right) + mc_{\text{p}}^{\text{s}}(T_{\text{M}} - T) \tag{7.130}$$

The relaxation time τ is an unknown material constant. It is reasonable to assume that τ is short, compared to the total cooling time. A simple calculation shows that after the time 4τ less than 2% of its strain energy remains in the sample. Hence, more than 98% of the strain energy has been emitted after the time 4τ. During this time c_{p}^{s} is larger than c_{p}^{0} but increases gradually. When the strain energy is gone c_{p}^{s} becomes equal to c_{p}^{0}. The difference of the slope of the cooling curve of the solid sample at $t = 0$ and at a much later time shows the effect of the relaxation.

The high rates of heat removal in thermal analysis greatly facilitates observation of the influence of relaxation on the heat of fusion and the heat capacitivity of metals and alloys.

Due to differences in material constants, cooling curves show a wide variety of appearances. Some examples are shown in Figures 7.45 on page 410 and Figure 7.51 on page 412.

Summary

Basic Concepts and Laws of Heat Transport

Heat is transported with the aid of conduction, radiation and convection.

■ Thermal Conduction

Stationary Conditions

$$\frac{dQ}{dt} = -\lambda A \frac{dT}{dy} \quad \text{or} \quad \frac{dq}{dt} = -\lambda \frac{dT}{dy} \quad \text{Fourier's first law}$$

Heat transfer at stationary conditions:

$$\frac{dQ}{dt} = -hA(T_2 - T_1) \quad \text{or} \quad \frac{dq}{dt} = -h(T_2 - T_1)$$

Non-Stationary Conditions

General law of heat conduction:

$$\frac{\partial T}{\partial t} = \alpha \frac{\partial^2 T}{\partial y^2} \quad \text{Fourier's second law}$$

Temperature gradient in one dimension:

$$\text{grad}\, T = \frac{dT}{dy}\hat{y}$$

The temperature gradient is a vector in the direction of a increasing heat flow.

■ Radiation of Heat

$$dW_{\text{total}} = \varepsilon \sigma_B A(T^4 - T_0^4)dt$$

■ Convection

Heat transfer by convection:

$$\frac{dq}{dt} = h_{\text{con}}(T - T_o)$$

Dimensionless numbers of hydrodynamics together with definitions of the quantities appearing in those numbers are listed in the text.

Boundary Layer Theory

A stationary flow pattern will develop around a planar plate and a fluid in relative motion to the plate.

Velocity Boundary Layer:

All viscous forces are supposed to be concentrated to a thin boundary layer. Outside the boundary layer the fluid flow is undisturbed.

The thickness δ of the velocity boundary layer is a function of the y and z coordinates. At a given value of z

$\delta =$ the distance from the solid plate when $u = 0.99\,u_\infty$.

General differential equations for calculation of the velocity field $u\,(y, z)$ inside the layer are given in the text.

Thermal Boundary Layer:

A thermal boundary layer only develops if there is a temperature difference between the fluid and the plate.

The thickness δ of the thermal boundary layer is a function of the y and z coordinates. At a given value of z

$\delta_T =$ the distance from the solid plate when

$$T_s - T = 0.99(T_s - T_\infty)$$

General differential equations for calculation of the temperature field $T(y, z)$ inside the thermal boundary layer are given in the text.

A solute concentration boundary layer in a concentration field can be defined in analogy with the temperature boundary layer.

Forced Convection

External circumstances control forced convection.

Thickness of the velocity boundary layer:

$$\delta = 5.0 \times \sqrt{\frac{\nu_{\text{kin}}}{u_\infty}} \times \sqrt{z} \quad \text{or} \quad \delta = \frac{5.0}{\sqrt{\text{Re}_z}}\sqrt{z}$$

Heat transfer coefficient:

The solution of the thermal boundary equations for laminar fluid flow within the interval $0.6 \leq \text{Pr} \leq 50$ gives the local heat transfer coefficient h_z

$$h_z = 0.332\,\lambda \times \text{Pr}^{0.343}\sqrt{\frac{u_\infty}{\nu_{\text{kin}}}} \times z^{-0.5}$$

Average heat transfer coefficient:

$$\overline{h_L} = 0.664\,\lambda \times \Pr^{0.343}\sqrt{\frac{u_\infty}{\nu_{kin}}} \times L^{-0.5}$$

Natural Convection
Free or natural convection is less powerful than forced convection but of great importance in metal melts. The driving force of natural convection is a *density difference*, normally caused by a *temperature gradient* in the liquid. The density of a metal melt decreases with increasing temperature, owing to the volume expansion of the melt.

Natural Convection at a Planar Vertical Surface

Approximate temperature distribution: $T = T_m - (T_m - T_s)\left(1 - \frac{y}{\delta}\right)^2$

Flow rate inside the boundary layer: $u = u_0(z)\frac{y}{\delta}\left(1 - \frac{y}{\delta}\right)^2$

Thickness of boundary layer $\delta = \delta_T$

$$\delta(z) = 3.93 \times \left(\frac{\nu_{kin}}{\alpha}\right)^{\frac{-1}{2}}\left(\frac{20}{21} + \frac{\nu_{kin}}{\alpha}\right)^{\frac{1}{4}}\left[\frac{g\beta(T_m - T_s)}{\nu_{kin}^2}\right]^{-\frac{1}{4}} \times z^{\frac{1}{4}}$$

■ Theory of Heat Transport at Unidirectional Solidification

Modes of Solidification
Pure metals or eutectic mixtures under strong cooling solidify with a well defined liquid/solid interface.

Alloys with wide solidification interval or alloys under influence of poor cooling solidify with an interface, that consists of a two-phase zone of liquid + solid, i.e. there are two solidification fronts, one at the beginning and one at the end of the two-phase region.

Unidirectional Solidification at Ideal Cooling
Solution of Fourier's second law results in formulas for temperature distribution given in the text.

Position of the solidification front: $y_L(t) = \lambda\sqrt{4\alpha_{metal}t}$
where λ is a constant.
Velocity of the solidification front: $\frac{dy_L(t)}{dt} = \lambda\sqrt{\frac{4\alpha_{metal}}{t}}$

Unidirectional Solidification at Poor Cooling
Formulas for temperature distribution are given in the text.

Position of solidification front = thickness of solidified shell:

$$y_L(t) = \frac{2}{\sqrt{\pi}}\frac{T_i - T_0}{\rho_{metal}(-\Delta H)}\sqrt{\lambda_{mould}\rho_{mould}c_p^{mould}} \times \sqrt{t}$$

where

Chvorinov's rule: $t_{total} = C\left(\frac{V_{metal}}{A}\right)^2$

Solidification Rate and Solidification Time of an Alloy in a Mould with Poor Contact between Mould and Alloy
Temperature of the solid alloy at the interface:

$$T_{i\,metal} = \frac{T_L - T_0}{1 + \frac{h}{\lambda}y_L(t)} + T_0$$

Velocity of the solidification front: $\frac{dy_L}{dt} = \frac{T_L - T_0}{\rho(-\Delta H)} \frac{\lambda}{y_L} \left(1 - \frac{1}{1 + \frac{h}{\lambda} y_L} \right)$

Solidification time: $t = \frac{\rho(-\Delta H)\, y_L}{T_L - T_0} \frac{y_L}{h} \left(1 + \frac{h}{2\lambda} y_L \right)$

■ Production of Single Crystals by Unidirectional Solidification. Bridgman Method

A crucible is moved downwards through a furnace, normally at a rate between 1-30 mm per hour. The temperature of the furnace is slightly above the melting point of the metal in the crucible. The solidification starts at the bottom of the crucible and the solidification front moves slowly upwards.

The advantages of the group of Bridgman methods are that they rapidly produce crystals with good mechanical strength and good dimensional accuracy, the technique is simple and requires comparatively little control and supervision. An important and favourable fact is also that the products require minimal finishing.

A disadvantage is that the crystals grown by the Bridgman always contain crystal defects, for example dislocations.

Solidification of a Metal Sample inside a Long Ceramic Tube

Temperature distribution in the liquid:

$$T_L = T_F - (T_F - T_M) e^{\left[\frac{V_0}{2\alpha_L} + \sqrt{\left(\frac{V_0}{2\alpha_L} \right)^2 + \frac{2h_L}{r \rho_L c_p^L \alpha_L}} \right] y} \qquad y < 0$$

Temperature distribution in the solid:

$$T_s = T_0 + (T_M - T_0) e^{\left[\frac{V_0}{2\alpha_s} - \sqrt{\left(\frac{V_0}{2\alpha_s} \right)^2 + \frac{2h_s}{r \rho_s c_p^s \alpha_s}} \right] y} \qquad y > 0$$

Heat Balance at the Insulated Interface:

A necessary condition for equilibrium and a planar stationary solidification front is

$$\rho_L V_0(-\Delta H) - \lambda_L \left(\frac{dT_L}{dy} \right)_{y=0} = -\lambda_s \left(\frac{dT_s}{dy} \right)_{y=0}$$

Careful experiments have been performed, which are briefly discussed in the text.

■ Thermal Analysis

Thermal analysis is the measurement of some physical property of a material as a function of its temperature.

Thermal analysis is very modern in the sense that there has been an enormous development of commercial instrumentation during the 1990s, which has strongly improved the sensitivity and accuracy of the measurements. Very small amounts of the specimen are required.

Two types of thermal analysis techniques are used:

- DTA = differential thermal analysis.
- DSC = differential scanning calorimetry.

The main difference between DTA and DSC is that the DTA instrument signal is proportional to the temperature difference between the sample and the reference while the DSC instrument signal is proportional to the differential thermal power.

Two DSC techniques, heat flux DSC (sometimes called heat flux DTA) and power-compensation DSC are used.

A survey of the methods and the measured quantities is given on page 407.

Theory of Differential Thermal Analysis

Differential Thermal Analysis with a Reference Sample

Two samples, one for analysis and one as a reference are connected to a thermocouple, heated in a furnace and allowed to cool while their cooling curves are registered.

Temperature difference between sample and reference:

$$(T_S - T_F)_{normal} = T_\beta \, e^{-\frac{F_{loss}}{\rho_S V_S c_p^S} t} + \frac{\rho_S V_S c_p^S}{F_{loss}} \beta$$

where $F_{loss} \approx \left(Ah + 4A\,\sigma_B \varepsilon T_F^3\right)$

Reaction rate:

$$\frac{df}{dt} = \frac{F_{loss}(T_S - T_R) + \rho_S V_S c_p^S \left(\frac{dT_S}{dt}\right) - \rho_R V_R c_p^R \left(\frac{dT_R}{dt}\right)}{\rho_S V_S(-\Delta H_S)}$$

or with some simplifications

$$\frac{df}{dt} = \frac{F_{loss}}{\rho_S V_S(-\Delta H_S)}(T_S - T_R)$$

Fraction f:

By graphical integration of one of the curves, corresponding to the equations above from t_{start} to t the *fraction solidified f* can be evaluated.

Heat of solidification:

The simplified equation above is often used for determining the *heat of solidification* $(-\Delta H_S)$. It is obtained by graphical integration of the curve corresponding to the simplified equation from t_{start} $(f=0)$ to t_{end} $(f=1)$.

Differential Thermal Analysis with no Reference Sample

Differential thermal analysis is possible also in the absence of a reference sample. If the sample had gone through no phase transformations, the temperature of the sample would have followed a straight line. This line is referred to as T_{ref}. It corresponds to the cooling curve of the reference sample during a differential thermal analysis.

Equation of reference line:

$$T_{ref} = T_F + [T_{ref}(t_{sol}) - T_F] e^{-\frac{F_{loss}}{\rho V c_p} t}$$

Sample temperature:

$$T_S = T_F + (T_{ref}(t_{sol}) + \Delta T_{sol} - T_F) \, e^{-\frac{F_{loss}}{\rho V c_p} t}$$

Heat of fusion:

$$-\Delta H_S = \frac{\int_{t_{start}}^{t_{end}} F_{loss}(T_S - T_{ref}) dt}{\rho_S V_S}$$

Reaction rate:

$$\frac{\mathrm{d}f}{\mathrm{d}t} = \frac{F_{\text{loss}}(T_S - T_{\text{ref}}) + \dfrac{\mathrm{d}T_S}{\mathrm{d}t} - \dfrac{\mathrm{d}T_{\text{ref}}}{\mathrm{d}t}}{\int\limits_{t_{\text{start}}}^{t_{\text{end}}} F_{\text{loss}}(T_S - T_{\text{ref}})\mathrm{d}t}$$

Fraction solidified (obtained by graphical integration):

$$f^* = \frac{\int\limits_{t_{\text{start}}}^{t^*} F_{\text{loss}}(T_S - T_{\text{ref}})\mathrm{d}t + T_S^* - T_{\text{ref}}^*}{\int\limits_{t_{\text{start}}}^{t_{\text{end}}} F_{\text{loss}}(T_S - T_{\text{ref}})\mathrm{d}t}$$

■ Heat of Fusion of Metals and Alloys

Experimental Determination of Heat of Fusion

The heat of fusion is derived from cooling curves of metal samples. Three methods are discussed:

- DTA (described above) (low cooling rates)
- Mirror furnace technique (medium cooling rates)
- Levitation casting technique (high cooling rates).

The cooling rate at DTA is slow. For rapid cooling processes, the other methods can be used.

The sample is exposed to six mirror furnaces, each consisting of two high power halogen lamps placed inside two ellipsoidal mirrors with a common focus where the sample is located. Cooling time 1–10 s.

The levitation casting technique is designed for extremely short solidification times of the sample with a lower limit of 15 milliseconds. A sample of about 1.5 grams was melted by induction in an He atmosphere.

Determination of Heat of Fusion

Case I: Cooling Curves Recorded by Use of DTA Technique:

Calculation:

$$(-\Delta H_S) = c_p^S \dot{T}\, t_{\text{sol}} + \left(c_p^S + \frac{\rho_C V_C}{\rho_S V_S} \cdot c_p^c\right) \frac{\dot{T}^2}{T_0 - T_S} \frac{t_{\text{sol}}^2}{2}$$

$\dot{T}$ and $(T_0 - T_S)$ are derived from the DTA measurements and $(-\Delta H_S)$ is calculated.

- DTA cooling is slow.
- Mirror furnace cooling is intermediate.
- Levitation casting cooling is rapid.

The heat of fusion in the last two cases is calculated by another method.

Case II: Cooling Curves Recorded by Use of Mirror Furnace Technique and Levitation Casting Technique:

Calculation:

$$-\Delta H_S = -\Delta H_S \int\limits_0^1 \mathrm{d}f = c_p \int\limits_{t_{\text{start}}}^{t_{\text{end}}} \left[\left(\frac{\mathrm{d}T}{\mathrm{d}t}\right)_{\text{no ph-tr}} - \left(\frac{\mathrm{d}T}{\mathrm{d}t}\right)_{\text{ph-tr}}\right] \mathrm{d}t$$

The two time derivatives inside the integral sign are recorded as functions of time during the experiment. The heat of fusion is derived graphically as the area between the two curves.

Variable Heat of Fusion of Metals and Alloys

Experimental evidence shows that the heat of fusion decreases with increasing cooling rate.

A very natural and reasonable explanation is that the phenomenon is caused by so-called relaxation in the solidified material. Relaxation is closely related to strain energy.

Relaxation

In a liquid metal or alloy the atoms have a random disrtribution and higher potential energy than lattice atoms in the solid metal or alloy. At solidification, the liquid atoms instantly loose potential energy and become incorporated into the solid but are strained and still distributed at random. They possess strain energy and have a higher potential energy than the lattice atoms in their fixed average positions.

Hence, the solidified liquid atoms are not at equilibrium but loose energy gradually and become stable lattice atoms. This process is called relaxation.

$$-\Delta H_{\text{relax}} = -\Delta H_{\text{strain}} = -\Delta H_0\sqrt{\frac{V_{\text{growth}}\delta}{\pi D_{\text{relax}}}}$$

$$-\Delta H_{\text{total}} = -\Delta H_0$$

Heat of Fusion and Relaxation

Relationship between total heat of fusion, prompt heat of fusion and relaxation energy:

$$-\Delta H_{\text{total}} = -\Delta H_{\text{prompt}} + (-\Delta H_{\text{relax}})$$

Relaxation energy or strain energy as a function of time:

$$-\Delta H_{\text{strain}} = -\Delta H_{\text{strain}}^{\text{max}}\,\mathrm{e}^{-\frac{t}{\tau}}$$

Prompt heat of fusion is the measured heat of fusion at a given growth rate

$$-\Delta H_{\text{prompt}} = -\Delta H_0\left(1 - \sqrt{\frac{V_{\text{growth}}\delta}{\pi D_{\text{relax}}}}\right)$$

The higher the growth rate is, the smaller will be the prompt heat of fusion.

Variable Heat Capacitivityof Metals and Alloys

Cooling Curves in Three Regions

Region I:

$$CHR = -\left(\frac{\mathrm{d}Q}{\mathrm{d}t}\right)_{\text{const}} = mc_{\text{p}}^{\text{L}}\left(-\frac{\mathrm{d}T}{\mathrm{d}t}\right)$$

Region II:

$$CHR = V_{\text{growth}}(-\Delta H_{\text{prompt}})\rho A + V_{\text{growth}}(-\Delta H_{\text{strain}})\rho A + \rho Ayf_{\text{s}}c_{\text{p}}^{\text{s}}\left(-\frac{\mathrm{d}T}{\mathrm{d}t}\right) + \rho Ay(1-f_{\text{s}})c_{\text{p}}^{\text{L}}\left(-\frac{\mathrm{d}T}{\mathrm{d}t}\right)$$

As the temperature is approximately constant, the last two terms can be neglected in most cases.

Region III:

$$CHR = m\frac{\mathrm{d}(-\Delta H_{\text{strain}})}{\mathrm{d}t} + mc_{\text{p}}^{0}\left(-\frac{\mathrm{d}T}{\mathrm{d}t}\right) = mc_{\text{p}}^{\text{s}}\left(-\frac{\mathrm{d}T}{\mathrm{d}t}\right)$$

If the relaxation time τ is much smaller than the solidification time t_{sol} of the thin sample, no relaxation energy is emitted during the solidification time. All the relaxation energy is emitted during the cooling time.

The cooling starts at time $t = 0$.

Material balance at time t:

$$CHR \times t = m(-\Delta H_0)\sqrt{\frac{y_0\delta}{\pi D_{relax}}}\left(1 - e^{-\frac{t}{\tau}}\right) + mc_p^s(T_M - T)$$

At $t = 4\tau$ practically all the relaxation energy is emitted to the surroundings. When the strain energy is gone c_p^s becomes equal to a constant value c_p^0.

During the relaxation time c_p^s is larger than c_p^0 but decreases gradually. At temperatures close to $T_M c_p^0 \approx c_p^L$.

The difference of the slope of the cooling curve of the solid sample at $t = 0$ and at a much later time, shows the effect of the relaxation.

Exercises

7.1 Turbine blades for jet engines are made by precision casting of highly resistant Fe- and Ni-base alloys. The performance and life time of the blades can be increased by aligning the solidification at casting and the following solidification process in such a way that the crystals are directed along the lengthwise direction of the blades. One way to do this is shown in the figure.

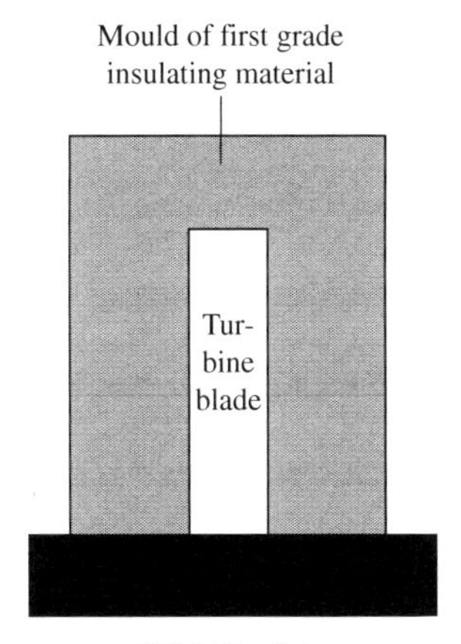

Figure 7.1–1

Calculate the solidification time of a blade with the length of 10 cm. For the calculation the material constants of pure nickel can be used. The melt is not superheated. The contact between the Cu-plate and the casting is assumed to be ideal.

Material constants and other data are found in the table below.

Table 7.1–1

Quantity	Ni (metal)	Cu (mould)
λ	90 W/mK (700 °C)	350 W/mK (200 °C)
ρ^s	8.90×10^3 kg/m^3 (25 °C)	8.96×10^3 kg/m^3 (25 °C)
c_p^s	444 J/kgK (~1100 °C)	397 J/kgK (~400 °C)
$-\Delta H$	310 kJ/kg	
T	$T_L = T_M = 1453$ °C (no excess temperature)	$T_0 = 100$ °C

You will need Figure 7.18b on page 380 in Chapter 7 and the erf z table below.

z	erf(z)	z	erf(z)	z	erf(z)	z	erf(z)
0.00	0.0000	0.40	0.4284	0.80	0.7421	1.40	0.9523
0.05	0.0564	0.45	0.4755	0.85	0.7707	1.50	0.9661
0.10	0.1125	0.50	0.5205	0.90	0.7969	1.60	0.9763
0.15	0.1680	0.55	0.5633	0.95	0.8209	1.70	0.9838
0.20	0.2227	0.60	0.6039	1.00	0.8427	1.80	0.9891
0.25	0.2763	0.65	0.6420	1.10	0.8802	1.90	0.9928
0.30	0.3286	0.70	0.6778	1.20	0.9103	2.00	0.9953
0.35	0.3794	0.75	0.7112	1.30	0.9340	∞	1.0000

7.2 a) Calculate the solidification time when you cast pure copper in a sand mould. The size of the casting is 900 × 100 × 900 mm. The melt is not superheated. Material constants for the sand mould are given in the table below. The temperature of the surroundings is 25 °C.

Material constants
Copper:
$T_L = T_M = 1083\,°C$
$\rho_{Cu} = 8.96 \times 10^3\ kg/m^3$
$-\Delta H = 280\ kJ/kg$
Sand mould:
$\lambda_{mould} = 0.63\ W/mK$
$\rho_{mould} = 1.61 \times 10^3\ kg/m^3$
$c_p^{mould} = 1.05 \times 10^3\ J/kgK$
Steel:
$T_L = T_M = 1535\,°C$
$\rho_{Cu} = 7.88 \times 10^3\ kg/m^3$
$-\Delta H = 272\ kJ/kg$

b) Calculate the solidification time for a steel casting with the same dimensions as in a) that is cast in a sand mould. Discuss and compare the results in a) and b) in terms of driving force of heat transport in the two cases.

7.3 In the search for new amorphous alloys wedge casting is often used to test the ability of the alloys to form amorphous phases. In such an experiment with a wedge-shaped copper mould crystals were formed on half the height of the mould and upwards and an amorphous phase in the bottom up to half the height.

Calculate the critical cooling rate necessary for formation of an amorphous structure of the alloy in question. The casting temperature is 1150 °C. The heat capacity is 420 J/kg K and the heat transfer coefficient is 1200 J/m^{2}sK. The dimensions of the wedge are shown in the figure.

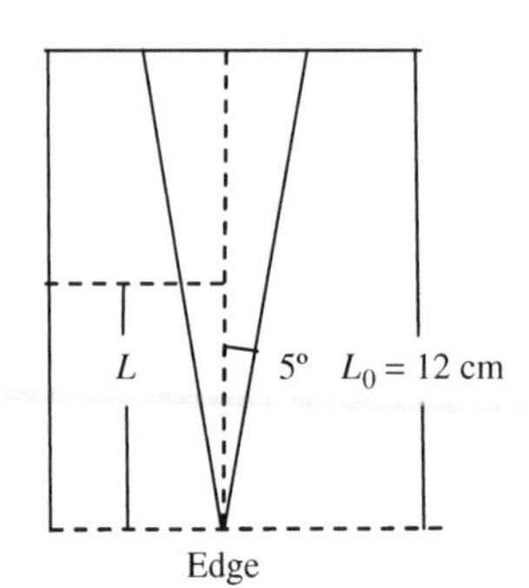

Figure 7.3–1

7.4 The figure shows a cooling curve of a metal melt, derived experimentally with the aid of thermal analysis. The heat capacitivity c_p^s of the solid phase is known and equal to 2.10 J/kg K.

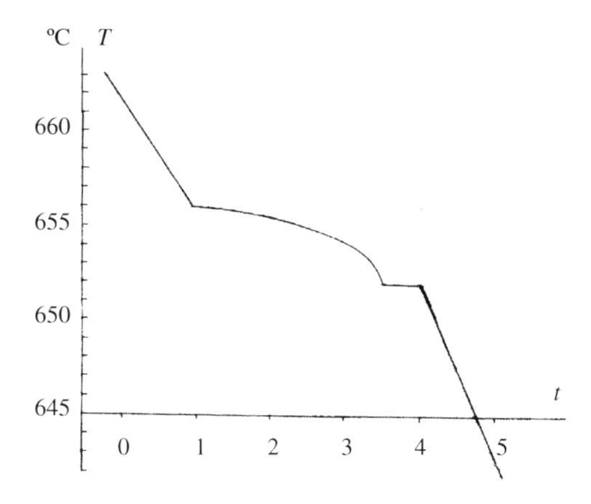

Figure 7.4–1

Use the curve for estimation of c_p^L of the melt. Assume that the evaluated heat emitted from the specimen per unit time is constant. The weight of the sample is 1.0 kg.

7.5 A series of experiments on eutectic Al-Si alloys have been performed in space and on earth with the aid of thermal analysis. The results of two of the experiments are shown graphically below.

a) If you compare the two temperature curves you notice that, although the cooling rates immediately before the solidification process starts, are nearly equal in the two experiments or somewhat higher in the space sample (215 °C/min compared to 200 °C/min on earth), the solidification time of the space sample is longer than that of the reference sample on earth (42 s compared to 34 s).

Evaluate the heat of fusion per kg on earth and in space from the curves.

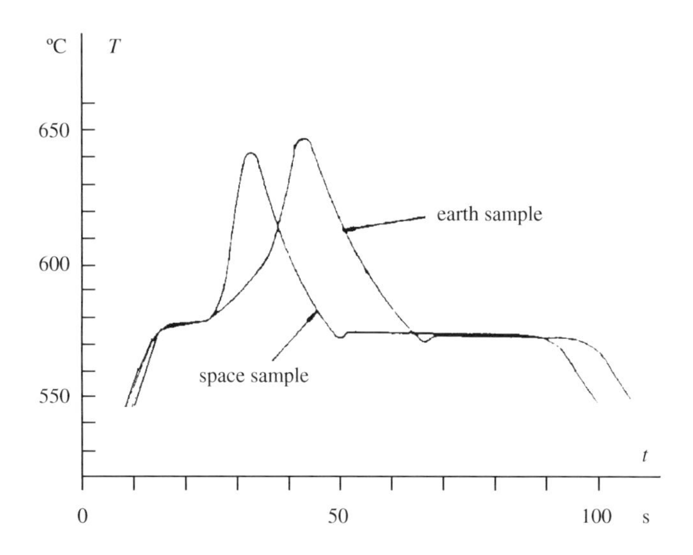

Figure 7.5–1

b) A comparrison between the experimental values in a) and the value given in the literature $-\Delta H = 371$ kJ/kg shows that they differ from the latter. Explain why the higher cooling rate in space can correspond to a longer solidification time in space than on earth.

The heat capacitivity c_p^L of the alloy melt in space and on earth is 888 J/kg K.

7.6 The modern method to find the fraction solidified phase as a function of time and to determine the temperatures when phase reactions start and end is thermal analysis.

A into of crucible with the alloy to be measured, a thermoelement is introduced. The crucible is introduced into a furnace and heated until the alloy becomes molten. During the cooling process the temperature of the alloy and the time are recorded. A temperature-time curve is obtained.

a) Predict and sketch such a cooling curve for an Fe–C alloy with 0.20 wt% C. Assume that the emitted amount of heat per unit time from the sample is constant. The cooling rate of the melt, immediately before the start of the solidification, is 1.0 °C/min. The schematic phase diagram is shown in the figure.

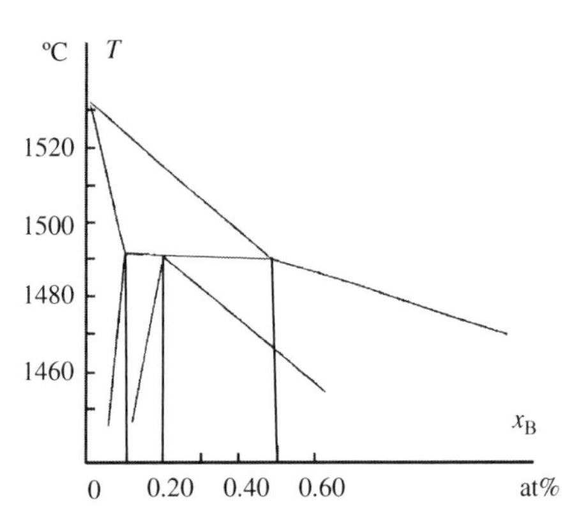

Figure 7.6–1

Material constants:

$$c_p^L \approx c_p^s = c_p = 700\,\text{J/kg K}$$
$$-\Delta H = 272\,\text{kJ/kg}$$

b) Comment the shape of the curve.

References

1. E. R. G. Eckert, R. M. Drake, *Heat and mass transfer*, New York, McGraw-Hill. 1959.
2. H. Fredriksson, U. Åkerlind, *Materials Processing during Casting*, Chichester, England, John Wiley & Sons, 2005.
3. P. W. Bridgman, *Proceedings from the American Academy for Art and Science*, **60**,1925, 303.
4. J. Czochralski, *Zeitschrift der Physik und Chemie*, **92**,1917, 219.
5. T. W. Clyne, Heat flow in controlled directional solidification of metals, *Journal of Crystal Growth*, **50**, 1980, 684–690.
6. H. Fredriksson, U. Åkerlind, *Physics for Functional Materials*, Chichester, England, John Wiley & Sons, 2008.
7. E. Talaat, H. Fredriksson, Solidification mechanism of unmodified and strontium modified Al-Si alloys. *Materials Transaction,* JIM, **41**, 2000, 507–515.

8

Crystal Growth Controlled by Heat and Mass Transport

8.1 Introduction

The mechanisms of crystallization based on mass transport have been discussed in Chapters 5 and 6. However, no attention was paid to heat transfer and temperature distribution during the solidification processes in either case.

Heat of fusion is released when crystals are formed. This heat must be removed otherwise the temperature will increase above the melting point and the crystallization process will stop. The rate of heat removal controls the solidification time and the temperature distribution in the material. Therefore, heat transport is a very important part of the crystallization process.

Solidification and Crystallization Processing in Metals and Alloys, First Edition. Hasse Fredriksson and Ulla Åkerlind.

However, it is not enough to consider the heat transport during crystal formation. Heat and mass transport occur simultaneously and cannot be separated from each other. Some processes, for example the Bridgman method of single crystal production, are mainly controlled by heat transport. This process has been discussed in Chapter 7 for this reason. In other single crystal production methods both the mass and heat transport controls crystallization. It is necessary to analyze the solidification and crystallization processes involved, i.e. to penetrate the influence of heat and mass transport and temperature conditions on crystal growth to make it possible to find the optimal production process in order to produce crystals of good quality.

For this reason we will discuss and analyze zone refining and the Czochralski technique in this chapter by combining the heat and mass transport laws. We will also introduce the concept of cellular growth and study the conditions for this phenomenon in order to facilitate avoidance of this defect in single crystal production.

In addition to optimizing the quality at single crystal production this chapter has a second aim. It is the basis for next chapter, which deals with structures of the crystals formed at crystallization processes. The heat and mass transport laws are essential for the growth conditions that lead to various types of crystal structures.

8.2 Heat and Mass Transports in Alloys during Unidirectional Solidification

In Chapter 7 heat transport in metals was discussed and the temperature distribution connected to unidirectional solidification in presence of temperature differences was calculated. The influence of mass transport was not included, however. In this section a complete treatment is given.

Crystals with a homogeneous composition across and along the crystal are desirable. To achieve this quality, it is of interest to examine the distribution of alloying element during the unidirectional solidification and especially the end effects, which are of importance in zone refining processes.

8.2.1 Distribution of Solute in a Molten Rod during and after Solidification. End Effects

Consider a cylindrical rod with a molten alloy during unidirectional solidification at stationary conditions. The solidification front is planar and its velocity, the growth rate, is constant.

This case has been treated in Chapter 6, Section 6.8.2 on page 313. We assume that the same conditions are valid in the present case:

- The diffusion of alloying elements in the solid is neglected in comparison with the diffusion in the melt. There is no convection in the melt, only diffusion.
- The equilibrium partition coefficient k_0, i.e. the ratio of the concentrations of the solid and the melt in equilibrium with each other, is constant.
- The compositions of the melt in bulk and in the solid are equal and constant.
- The growth rate of the solidification front is constant.

Steady State

When the alloy melt solidifies the concentration of solute in the solid will differ from that of the melt.

If $k_0 < 1$, an excess of solute atoms arises in the solid, close to the interface. Solute atoms become rejected from the interface into the melt. The higher concentration of solute at the solidification front, compared to the rest of the melt, leads to diffusion of solute atoms from the interface into the melt and an equilibrium distribution of solute atoms in the melt is established (Figure 8.1a).

If $k_0 > 1$, a lack of solute atoms arises in the melt, close to the interface. The lower concentration of solute at the solidification front compared to the rest of the melt leads to diffusion of solute atoms from the melt towards the interface. An equilibrium distribution of solute atoms in the melt is established (Figure 8.1b).

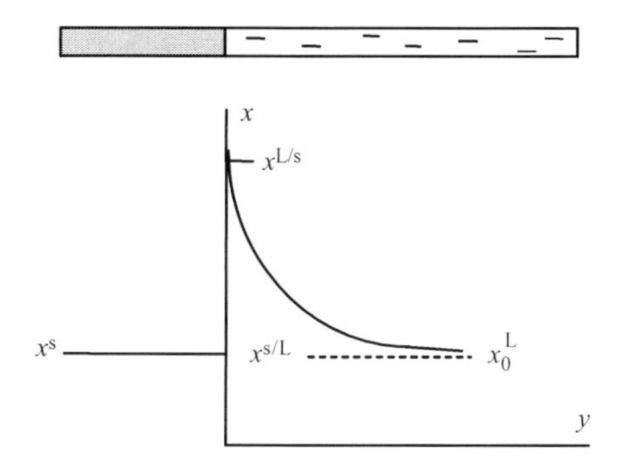

Figure 8.1a Equilibrium distribution of solute in the melt as a function of distance from the interface for $k_0 < 1$.

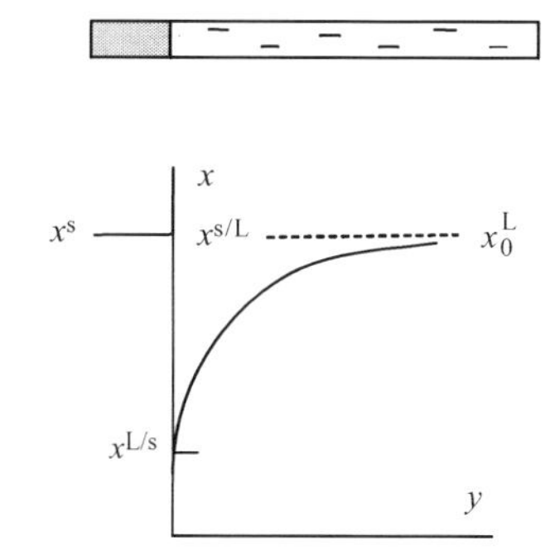

Figure 8.1b Equilibrium distribution of solute in the melt as a function of distance from the interface for $k_0 > 1$.

In Chapter 6 we showed that the equilibrium distribution of solute in the melt as a function of the distance from the interface can be written [Equation 6.120 on page 323 in Chapter 6. It is valid provided that the convection in the melt can be neglected.]

$$x^L = x_0^L \left(1 + \frac{1 - k_0}{k_0} e^{-\frac{V_{growth}}{D_L} y} \right) \tag{8.1}$$

where

y = distance from the interface
x^L = concentration of solute in the melt at distance y
x_0^L = initial concentration of solute in the melt
V_{growth} = growth rate of the solidification front
D_L = diffusion constant of the solute in the melt
$x^{L/s}$ = concentration of solute in the melt close to the interface
$x^{s/L}$ = concentration of solute in the solid close to the interface.
k_0 = partition coefficient, i.e. $x^{s/L}/x^{L/s}$ at the interface.

Initial Transient

In the steady state, the concentration profile of solute in the melt remains unchanged relative to the interface when it advances during the solidification process.

The steady state is based on a mass balance at the interface. For $k_0 < 1$ the number of solute atoms rejected from the solid per unit time is equal to the number of solute atoms which leave the interface per unit time by diffusion into the melt. This equilibrium state is not achieved until the stationary solute distribution in the melt close to the interface has been established. It takes some time to reach this equilibrium distribution. Equation (8.1) is not valid during this initial time interval.

Tiller et al [1] suggested that the solute distribution in the solid during the initial period can be represented by the function

$$x^s = x_0^L \left[(1 - k_0) \left(1 - e^{-\frac{k_0 V_{growth}}{D_L} y} \right) + k_0 \right] \tag{8.2}$$

It has been shown that the distribution given in Equation (8.2) is approximate and the exact distribution is given by a more complicated expression including the erf function (page 380 in

Chapter 7 and page 365 in Chapter 4 in [2]). Equation (8.2) is a satisfactory approximation for most purposes.

Figure 8.2 shows the initial transient at some different times, i.e. the distributions of solute atoms in the solid and the melt that formed during the initial period before equilibrium was established.

The shaded areas represent the concentrations of redistributed solute atoms relative to the initial solute concentration. The lack of solute in the solid equals the excess of solute in the melt. At $t = t_3$ the steady state has been reached and the concentration profile of solute in the melt does not change during the stationary period. The solute concentration in the solid close to the interface equals the initial solute concentration.

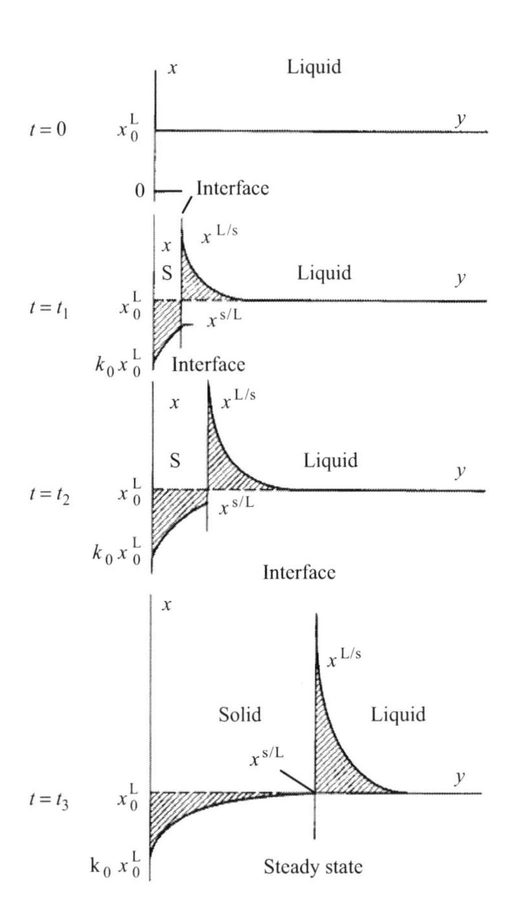

Figure 8.2 Growth of initial transient for $k_0 < 1$.
Alloying element as a function of position at various times when a molten alloy in a cylindrical tube starts to solidify. Bruce Chalmers, courtesy of S. Chalmers.

The shape of the initial transient depends on the value of $k_0 V_{growth}/D_L$. The higher this value is the steeper will be the curve. The smaller the value is the flatter and more extended will be the initial transient.

Final Transient

The steady state in the solidifying rod will be maintained as long as there is sufficient liquid ahead of the interface for the diffusion of solute into the melt and the growth rate is constant. The former condition is not fulfilled at the end of the solidification. The rejected solute atoms must be included in a reduced volume, which corresponds to an increased concentration in the remaining melt. The more the interface approaches the end of the rod the higher will be the concentration of solute if $k_0 < 1$ (Figure 8.3a). The opposite is true if $k_0 > 1$ (Figure 8.3b).

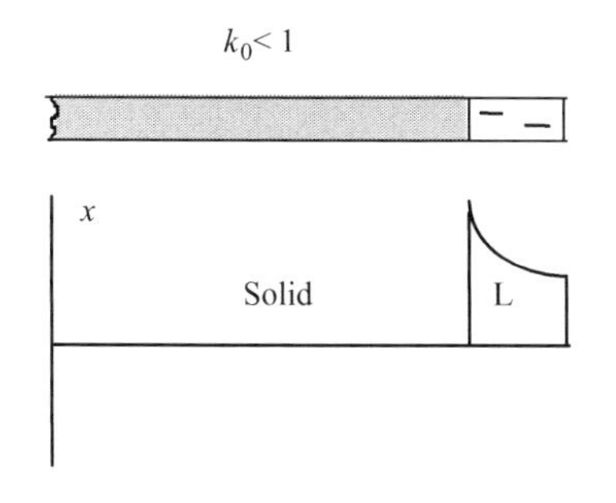

Figure 8.3a Solute distribution in the liquid at the end of the solidification for $k_0 < 1$.

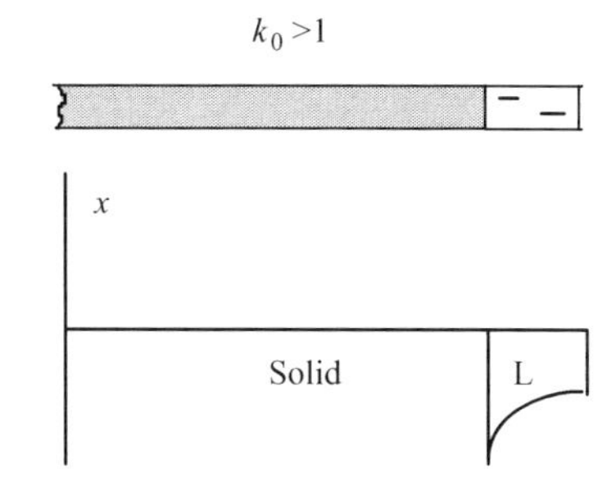

Figure 8.3b Solute distribution in the liquid at the end of the solidification for $k_0 > 1$.

The increased solute concentration in the melt at the end of the solidification corresponds to an increase of solute in the solid at the end of the rod if $k_0 < 1$. This is shown in Figure 8.4a, that shows the solute distribution in the rod after complete solidification. For the case $k_0 > 1$ both transients change sign (Figure 8.4b).

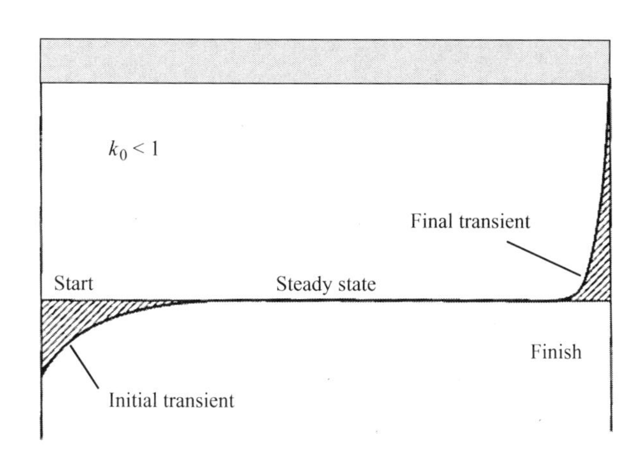

Figure 8.4a Solute distribution in a solidified rod for $k_0 < 1$. Bruce Chalmers, courtesy of S. Chalmers.

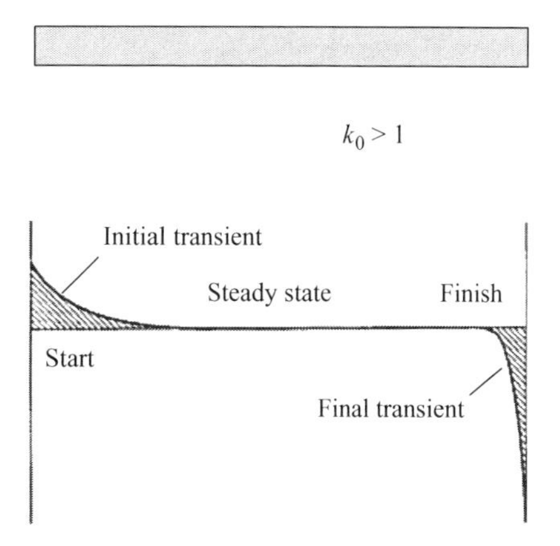

Figure 8.4b Solute distribution in a solidified rod for $k_0 > 1$.

A comparison between the transients in Figures 8.2 and 8.4 shows that the extension of the final transient is more narrow than the initial transient. Besides, impingement effects (Chapter 6 page 338) influence the shape of the final transient.

8.2.2 *Influence of Natural Convection on Crystal Growth*

Natural convection (Chapter 7, Section 7.3.3) always exists in a liquid or melt when temperature differences are present and it can not be avoided.

In Section 8.2.1 we found that the distribution of solute in the melt in the steady state during solidification can be described by Equation (8.1) together with Figure 8.1a if $k_0 < 1$, and Figure 8.1b if $k_0 > 1$. This model is valid only if convection in the melt can be neglected.

It is important to generalize the model and include the influence of convection. This will give a better description of the real solidification process.

Model for Simultaneous Diffusion and Convection in the Melt at Unidirectional Solidification

Several scientists have treated the problem of finding a useful model of simultaneous diffusion and convection in the melt, among them Burton, Prim and Slichter [3]. It is hard to treat convection and diffusion simultaneously, which made it necessary to introduce a simplification. The simplification consists of an *artificial separation of the diffusion process and the convection process.* This mathematical abstraction has proved to be useful.

Wagner introduced the assumption that there is a layer of thickness δ^* beyond which there is sufficient convection to ensure total mixing in the melt, i.e. its solute concentration is the same everywhere, independent of position. Inside the layer δ^* there is no convection at all.

Wagner called the new quantity δ^* the *hypothetical stagnant layer.* His pioneering ideas and general derivation of δ^* were applied on the Bridgman experiment by Burton, Prim and Slichter.

The theory by Burton, Prim and Slichter [3], published in 1953, deals with a *boundary layer* (Figure 8.5). Their model is valid for the steady state and describes the impurity (solute) concentration along the growth direction. It is based on the following assumptions, described by our designations:

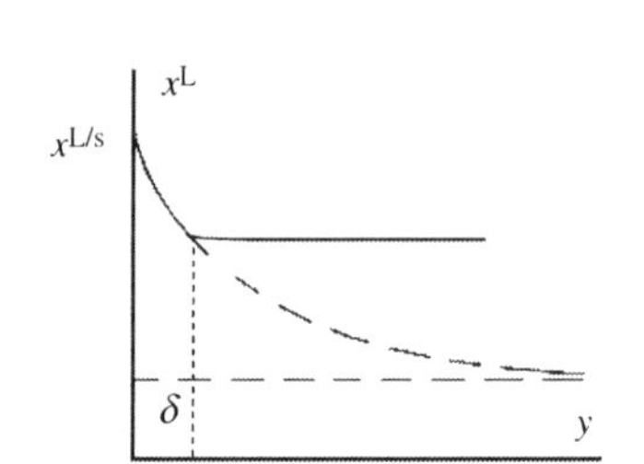

Figure 8.5 Mole fraction of the alloying element in the melt ahead of the solidification front under stationary conditions. The influence of convection outside δ changes the solute concentration in the melt radically.

1. Inside the boundary layer δ the solute in the melt is transported by diffusion only. Outside the layer natural convection causes complete mixing of the melt.
2. The diffusion coefficient D_L in the melt is constant. There is no diffusion of solute in the solid.
3. The solidification front moves with a constant velocity V_{growth}.
4. The partition coefficient k_0 at the solid/liquid interface is constant.
5. The difference in density between the solid and the melt is taken into account.

The influence of convection is described by the width δ of the boundary layer. The stronger the convection is the more will the convection dominate over the undisturbed diffusion in the boundary layer and the thiner will be the boundary layer δ. It is of the magnitude 1 mm or less.

Solute Distribution in the Solid and the Liquid with and without the Influence of Convection

At the calculation of the concentration profile of solute as a function of position, we will as a first approximation neglect the difference in density between the solid and the melt.

In case of no convection Equation (8.1) and Figure 8.6 are valid.

$$x^L = x_0^L \left(1 + \frac{1 - k_0}{k_0} e^{-\frac{V_{growth}}{D_L} y}\right) \tag{8.1}$$

In case of convection in the melt Figure 8.6 and Equation (8.1) are no longer valid, neither *inside* nor *outside* the boundary layer δ.

The corresponding concentration profile in case of convection is shown in Figure 8.7. A comparison between the Figures 8.6 and 8.7 shows that *in presence of convection*

- the solute concentration in the melt outside the boundary layer is constant and *higher* than x_0^L
- the solute concentration in the solid x^s is no longer equal to the initial solute concentration in the melt x_0^L but has a *lower* value
- the solute concentration in the melt inside the boundary layer is *lower* than the corresponding concentration (the same y value) in case of no convection.

For $y > \delta$ the concentration of solute in the melt has a *higher* value than x_0^L as a result of the convection in the melt. Consequently the solute concentration in the solid must be *lower* than x_0^L, even if there is

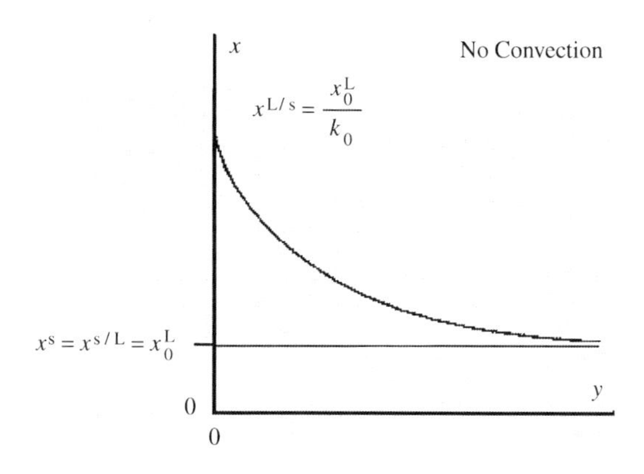

Figure 8.6 Solute distribution in the solid and in the melt when convection is excluded.

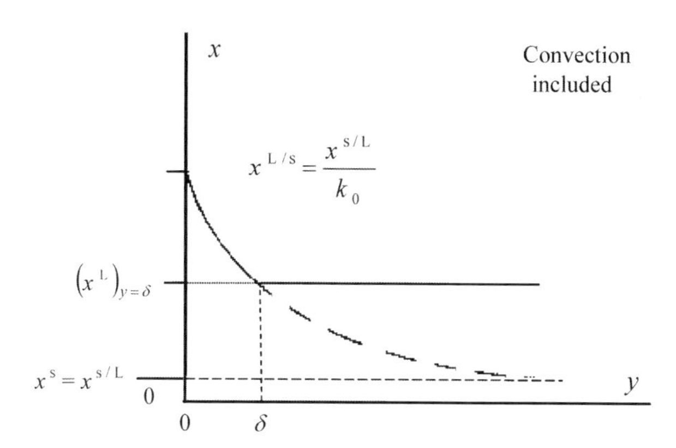

Figure 8.7 Solute distribution in the solid and in the melt when convection is included.

only diffusion and no convection inside the boundary layer, because the boundary layer is very thin and the total amount of solute is constant.

In order to find an expression of the solute concentration in the melt inside the boundary layer in case of convection we have to perform a derivation completely analogous to that in Chapter 6 on pages 317–319 with the only difference that x_0^L has to be replaced by the more general value x^s. Hence, Equation (8.1) can be replaced by

$$x^L = x^s\left(1 + \frac{1-k_0}{k_0}e^{-\frac{V_{growth}}{D_L}y}\right) \qquad y \leq \delta \tag{8.3}$$

where x^s is constant and smaller than x_0^L. Equation (8.3) represents the solute concentration in the melt inside the boundary layer δ as a function of the constant solute concentration in the solid.

Outside the boundary layer δ the solute concentration is constant and equal to the value of x^L at $y = \delta$

$$\left(x^L\right)_{y=\delta} = x^s\left(1 + \frac{1-k_0}{k_0}e^{-\frac{V_{growth}}{D_L}\delta}\right) \qquad y \geq \delta \tag{8.4}$$

The two functions (8.3) and (8.4) are shown in Figure 8.7.

The stronger the convection is the smaller will be δ and the higher will be the solute concentration in the melt outside the boundary layer and the lower will be the solute concentration in the solid.

The solute concentration in the solid phase of the rod, when the solidification process is finished, is of particular interest. We have assumed on page 338 – which is a very reasonable assumption – that there is no solute diffusion in the solid and consequently the constant value

$$x^s = x^{s/L} \tag{8.5}$$

is maintained after cooling.

It is not possible to calculate x^s and x^L separately and express them as functions of the initial solute concentration in the melt x_0^L. The theory given above is valid only for the steady state. As no information is available about the initial and final transients it is impossible to set up a complete material balance.

Effective Partition Coefficient. Relationship between k_{eff} and δ

In order to express the solute concentration in the solid in a simple way we will introduce the *effective partition coefficient* k_{eff}, which is a very useful concept. The effective partition coefficient is defined as *the ratio of the final solute concentration in the solid to the average solute concentration in the melt.*

$$k_{eff} = \frac{x^s}{\overline{x^L}} \tag{8.6}$$

The boundary layer is always thin or very thin and the volume of the boundary layer is mostly small compared to that of the melt. In this case, the average solute concentration in the melt is approximately equal to the solute concentration outside the boundary layer.

$$\overline{x^L} \approx \left(x^L\right)_{y=\delta} = x^s\left(1 + \frac{1-k_0}{k_0} e^{-\frac{V_{growth}}{D_L}\delta}\right) \tag{8.7}$$

which directly gives

$$k_{eff} = \frac{x^s}{\overline{x^L}} = \frac{1}{\left(1 + \frac{1-k_0}{k_0} e^{-\frac{V_{growth}}{D_L}\delta}\right)} \tag{8.8}$$

or

$$k_{eff} = \frac{x^s}{\overline{x^L}} = \frac{k_0}{\left(k_0 + (1-k_0) e^{-\frac{V_{growth}}{D_L}\delta}\right)} \tag{8.9}$$

Equation (8.9) is the desired relationship between k_{eff} and δ. The relationship between the solute concentrations in the solid and the melt and k_{eff} can be written as

$$x^s = k_{eff}\overline{x^L} \tag{8.10}$$

where k_{eff} is equal the expression in Equation (8.9).

Influence of Density Differences on Solute Concentrations

So far we have neglected the fifth condition on page 438. A more careful and complex derivation than that given above shows that Equations (8.3), (8.7) and (8.9) should be replaced by the more exact Equations (8.11), (8.12) and (8.13). The result is

$$x^L = x^s\left(\frac{\rho_L}{\rho_s} + \frac{1 - k_0\frac{\rho_L}{\rho_s}}{k_0} e^{-\frac{V_{growth}}{D_L}y}\right) \qquad y \leq \delta \tag{8.11}$$

$$\overline{x^L} \approx \left(x^L\right)_{y=\delta} = x^s\left(\frac{\rho_L}{\rho_s} + \frac{1 - k_0\frac{\rho_L}{\rho_s}}{k_0} e^{-\frac{V_{growth}}{D_L}\delta}\right) \tag{8.12}$$

$$k_{eff} = \frac{k_0}{k_0\frac{\rho_l}{\rho_s} + \left(1 - k_0\frac{\rho_L}{\rho_s}\right) c^{-\frac{V_{growth}}{D_L}\delta}} \tag{8.13}$$

$$k_0 \leq k_{eff} \leq 1$$

When V_{growth} is small and D_L is large: $k_{eff} \approx k_0$
When V_{growth} is large and D_L is small: $k_{eff} \approx 1$.

Microsegregation

Convection also affects the microsegregation in a solid.

If $k_0 < 1$ and the convection is neglected the solute concentration increases with increasing degree of solidification in the crystal. This phenomenon is described by *Scheil's segregation equation,* which can be derived from a material balance of a solidifying volume element in a rod of given limited length (see the box on page 442). In this case the temperature has to decrease continuously and isothermally during the solidification process.

The solute concentration in the remaining melt is

$$x^L = \frac{x^s}{k_0} = x_0^L (1 - f)^{-(1-k_0)} \quad \text{Scheil's equation} \tag{8.14}$$

where f = solidified fraction of the rod.

Equation (8.14) is only valid when convection can be neglected. If convection is considered k_0 has to be replaced by k_{eff}.

$$x^L = \frac{x^s}{k_{eff}} = x_0^L (1 - f)^{-(1-k_{eff})} \tag{8.15}$$

When the growth rate is small and/or the diffusion coefficient large the exponent in Equation (8.13) becomes small. The exponential term approaches the value 1 which corresponds to $k_{eff} = k_0$. This is the minimum value of k_{eff}.

When $k_{eff} = k_0$ the solute concentration in the solid reaches the eutectic value before complete solidification. This corresponds to a constant solute concentration at the end of the solidification.

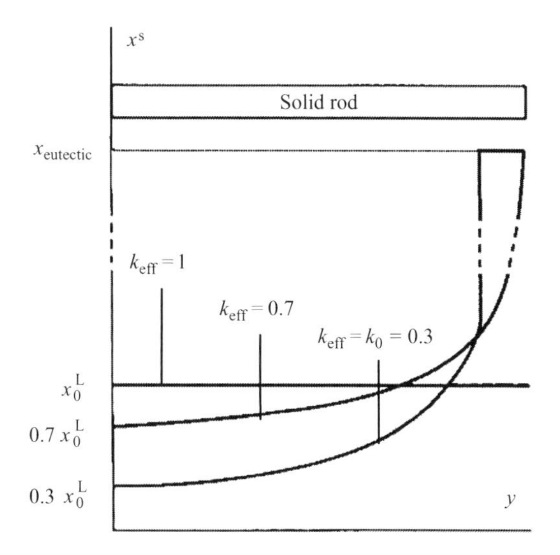

Figure 8.8 Solute concentration in the solidifying rod as a function of position for some different values of k_{eff}. Merton C. Flemings © McGraw Hill 1974.

In the opposite case, slow diffusion and rapid crystal growth in combination with strong convection, the exponential term in Equation (8.13) may be neglected which corresponds to the value $k_{eff} = 1$ if $\rho_L \approx \rho_s$. This is the maximum value of k_{eff}.

Figure 8.8 shows the solute distribution in the solid for these two extreme values and an intermediate value. $k_{eff} = 1$ corresponds to a uniform solute distribution both in the melt and the solid.

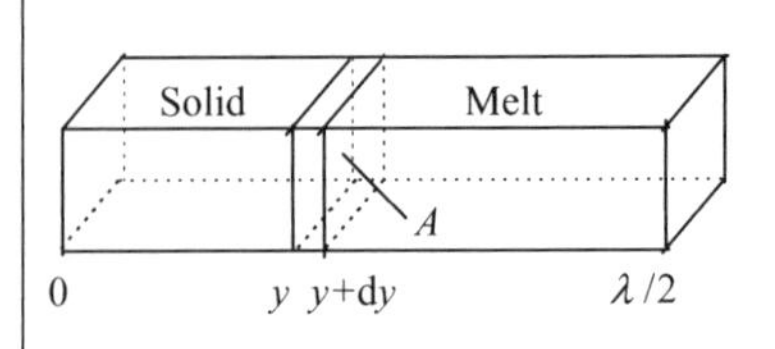

Figure 8.1 V_m = the molar volume.

Derivation of Scheil's Equation

Consider the volume element in the figure. Solidification of the slice Ady requires a decrease of its concentration of alloying element from x^L to x^s.

The amount $(x^L - x^s)\,A\,dy/V_m$ of the alloying element is moved into the melt and increases its concentration by dx^L. This amount of alloying element is brought from the solidified slice Ady to the volume element $A\,(\lambda/2 - y)$ and increases its concentration from x^L to $x^L + dx^L$. The material balance can be written (V_m = the molar volume) as

$$\frac{(x^L - x^s)A\,dy}{V_m} = \frac{A\left(\frac{\lambda}{2} - y\right)dx^L}{V_m} \tag{1'}$$

By division by A/V_mEquation (1′) can be simplified to

$$\left(\frac{\lambda}{2} - y\right)dx^L = \left(x^L - x^s\right)dy \tag{2'}$$

By introduction of k_0 we can eliminate x^s. Integration of Equation (2′) from $y = 0$ to y and x^L from x_0^L to x^L gives

$$\int_{x_0^L}^{x^L} \frac{dx^L}{x^L - k_0 x^L} = \int_0^y \frac{dy}{\frac{\lambda}{2} - y} \tag{3'}$$

where x_0^L = the initial concentration of the alloying element in the melt. After integration we obtain

$$\frac{1}{1-k_0}\ln\frac{x^L}{x_0^L} = -\ln\frac{\lambda/2 - y}{\lambda/2} \tag{4'}$$

Equation (4′) is solved for x^L:

$$x^L = \frac{x^s}{k_0} = x_0^L\left(1 - \frac{2y}{\lambda}\right)^{-(1-k_0)}$$

or if we introduce the solidified fraction of the rod $f = 2y/\lambda$ we obtain

$$x^L = \frac{x^s}{k_0} = x_0^L(1 - f)^{-(1-k_0)} \quad \text{Scheil's equation}$$

8.3 Zone Refining

The aim of zone refining is to reduce the impurities in a solid crystal as much as possible by monitoring a molten zone along the crystal. The impurities (solute) accumulate in the molten zone ahead of the solidification front. When the zone advances the impurities remain in the liquid phase and are carried to the end of the crystal. End effects arise, similar to those discussed in Section 8.2.1.

A comparison between Figures 8.6 and 8.7 on page 439 shows very clearly that convection is a necessary condition for a solute concentration, lower than x_0^L in the solid. Convection is important for zone refining.

Zone melting methods have been introduced briefly in Section 6.2.3 in Chapter 6. In this section we will discuss the theoretical basis of zone refining more closely and examine the conditions for an optimal effect of zone refining to be used in the production of crystals of high quality.

8.3.1 Theoretical Basis of Zone Refining

Concentration of Solute in a Rod after Zone Refining

The appearance of a zone refining equipment is shown in Figure 8.9. The molten zone is moved slowly along the specimen, in this case a rod, with constant velocity.

We want to find the solute concentration in the solid after passage of the molten zone, expressed in terms of the initial solute concentration in the rod and other known quantities. This is done with the aid of a material balance.

The change of the solute content in the molten zone during the time interval dt is equal to the addition of solute when the melt incorporates the volume Ady from the solid side minus the amount of solute, which disappears when the volume Ady solidifies on the other side.

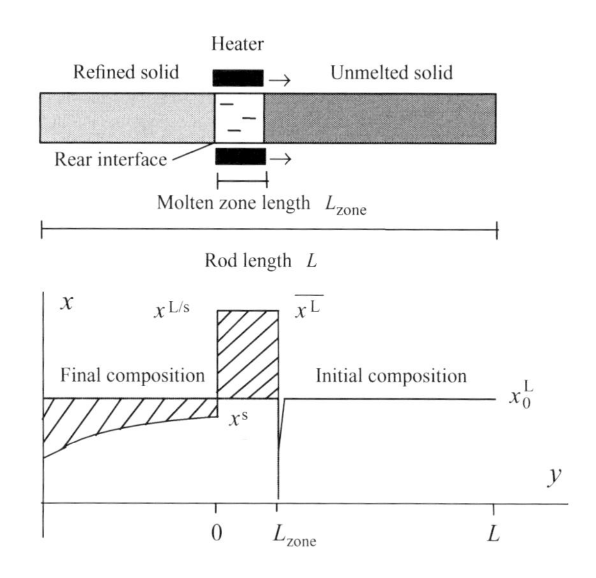

Figure 8.9 Sketch of a zone-refining equipment and of the solute distribution in the rod in case of complete mixing in the molten zone owing to convection.
The marked areas in the figure are suppose to be equal.

$$d\left(AL_{zone}\overline{x^L}\right) = Ady\, x_0^L - Ady\, k_{eff} x^{L/s} \tag{8.16}$$

where

A = cross-section of the rod
$\overline{x^L}$ = mean value of the solute concentration in the molten zone
x_0^L = initial concentration of solute in the rod.
k_{eff} = effective partition coefficient at the interface of the molten rod (see definition on page 440)
$x^{L/s}$ = concentration of solute in the melt close to the interface of the molten rod.

If we assume that the boundary layer is very thin and that the convection is strong enough to result in complete mixing of the melt in the molten zone – which is a very reasonable assumption – the solute concentration in the molten zone will be

$$\overline{x^{L}} = x^{L/s} \tag{8.17}$$

If this value is introduced into Equation (8.16) we obtain after division by A

$$L_{zone} dx^{L/s} = x_0^L dy - k_{eff} x^{L/s} dy$$

After separation of variables the equation can be integrated:

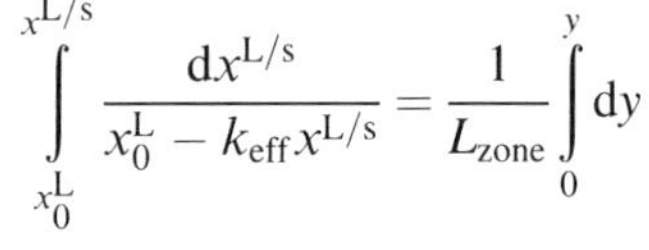

$$\int_{x_0^L}^{x^{L/s}} \frac{dx^{L/s}}{x_0^L - k_{eff} x^{L/s}} = \frac{1}{L_{zone}} \int_0^y dy$$

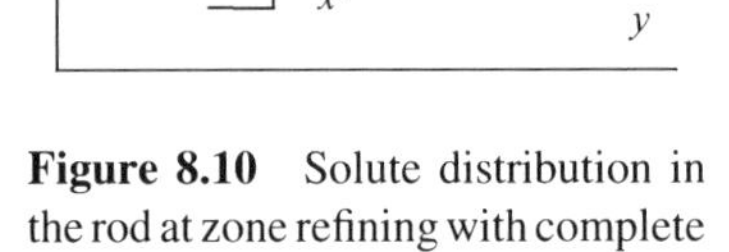

Figure 8.10 Solute distribution in the rod at zone refining with complete mixing.

and we obtain

$$\frac{-1}{k_{eff}} \left[\ln \frac{x_0^L - k_{eff} x^{L/s}}{x_0^L - k_{eff} x_0^L} \right] = \frac{1}{L_{zone}} y$$

or

$$x^{L/s} = \overline{x^{L}} = \frac{x_0^L}{k_{eff}} - \frac{1 - k_{eff}}{k_{eff}} x_0^L e^{-\frac{k_{eff}}{L_{zone}} y} \tag{8.18}$$

Using Equation (8.6) on page 440 we obtain the solute concentration in the solid after zone refining:

$$x^s = x_0^L \left[1 - (1 - k_{eff}) e^{-\frac{k_{eff}}{L_{zone}} y} \right] \tag{8.19}$$

Equation (8.19) is the desired expression of the solute concentration in the rod after zone refining. The lower x^s is the more effective will be the zone refining. The function is shown graphically in Figures 8.11a and 8.11b during and after zone refining.

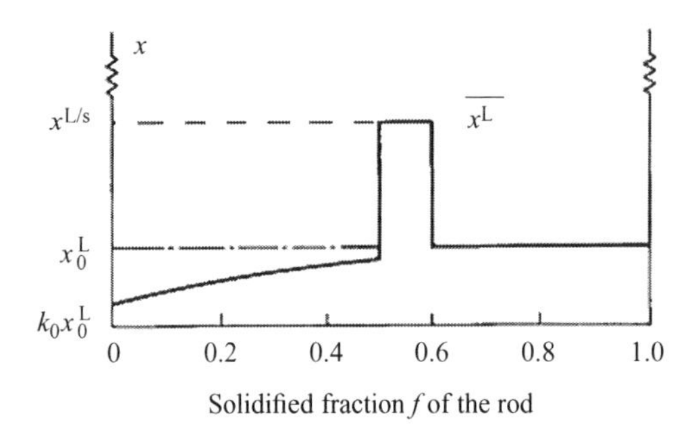

Figure 8.11a Solute distribution in the solid after zone melting as a function of position during zone refining and partial solidification. Merton C. Flemings © McGraw Hill 1974.

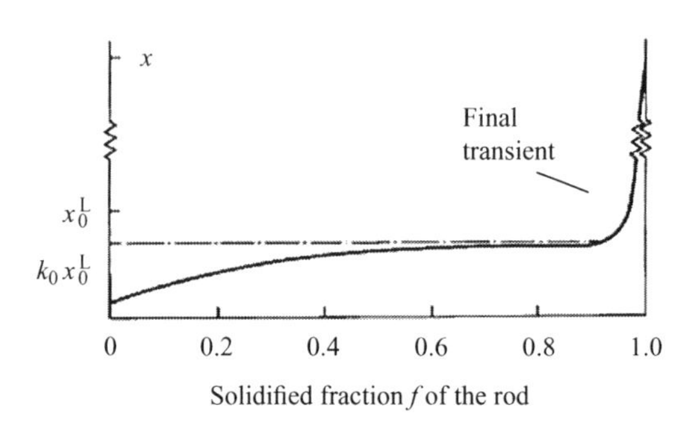

Figure 8.11b Solute distribution in the solid after zone melting as a function of position after end of zone refining and complete solidification. Merton C. Flemings © McGraw Hill 1974.

8.3.2 *Optimal Conditions of Zone Refining*

In order to transfer impurities from the first part to the end of a rod in the best possible way it is of interest to analyze the influence of experimental parameters which can be controlled and choose their values in an optimal way.

The solute concentration distribution in the rod after zone refining depends primarily on

- the convection of the melt in the molten zone
- the zone length
- the pulling rate of the molten zone
- the diffusion rate of the solute in the melt.

Equation (8.20) gives a measure of the efficiency of the zone refining process. The equation can be written as

$$\frac{x^{s}}{x_0^{L}} = \left[1 - (1 - k_{\text{eff}})e^{-\frac{k_{\text{eff}}}{L_{\text{zone}}}y}\right] \tag{8.20}$$

It is desirable to reduce the remaining solute concentration in the solid after the zone refining process as much as possible. From theoretical point of view we can conclude that a minimum value of x^{s}/x_0^{L} is obtained if the last term on the right-hand side of Equation (8.20) is as large as possible. This is the case if the exponential factor is as small as possible, i.e. when the exponent ratio $k_{\text{eff}}/L_{\text{zone}}$ is as small as possible.

The conclusion (page 441) is that

- k_{eff} shall be chosen as close to k_0 as possible.

The statement above is confirmed by a comparison between Figures 8.15 a and 8.15 b on page 448. If we compare the curves for $n = 1$ in the two figures we will find that the solute concentration is considerably lower for $k_{\text{eff}} = 0.10$ than for $k_{\text{eff}} = 0.25$.

The length of the zone can not be chosen especially long for practical reasons. Firstly, a long zone requires very strong convection. Secondly, if L_{zone} is long the final transient zone with high concentration of impurities at the end of the rod will also be long, which is a disadvantage. Besides, Equation (8.20) is not valid for the final transient zone.

Next we will find out how the experimental parameters should be chosen in order to give a small value of k_{eff}. We use the approximate Equation (8.9) on page 440

$$k_{\text{eff}} = \frac{k_0}{k_0 + (1 - k_0)e^{-\frac{V_{\text{growth}}}{D_{\text{L}}}\delta}} \tag{8.9}$$

A small value of k_{eff} implies a large value of the denominator in Equation (8.9), which means that the exponential factor must be large and that the exponent $V_{\text{growth}}\delta/D_{\text{L}}$ must be as small as possible. Hence, we can conclude that k_{eff} will be small for

- a small value of V_{growth}
- a small value of δ
- a large value of D_{L}.

V_{growth} is the solidification rate. It is not an experimental parameter, but it is limited by the rate with which the molten zone is moved and can be assumed to be equal to this quantity. The latter rate is an experimental parameter that easily can be varied. Hence, small values of the pulling rate of the molten zone promotes an effective zone refining.

In Section 8.3.1 we have noticed that convection is an essential mechanism in zone refining. The stronger the convection is the smaller will be δ, i.e. the more narrow will be the boundary zone, free from convection, and the lower will be the remaining impurities in the solid.

A strong diffusion of solute in the melt and consequently a large value of the diffusion coefficient D_L also promotes the efficiency of zone refining. However, D_L is a material constant and not an experimental parameter that can be varied arbitrarily. D_L varies strongly with temperature and can be changed by variation of the temperature of the molten zone.

Hence, the optimal conditions for zone refining are:

- strong convection in the melt
- relatively short zone length, related to the strength of convection in the molten zone
- low pulling rate of the molten zone
- high diffusion rate of the solute in the melt.

8.3.3 *Stability of Molten Zone. Maximum Zone Length*

The length of the molten zone at zone refining can not be chosen arbitrarily or entirely with respect to the refining efficiency, i.e. the impurity distribution in the solid. In addition, we have to consider the mechanical stability of the molten zone, the thermal balance of the process and the effect of surface convection.

Mechanical Stability

Consider a zone melting process in a vertical rod. Convection plays an important role in zone refining. The rod has to be vertical for this reason.

The molten zone hangs like a drop between the two solid parts of the rod during the zone melting process. Its shape is a consequence of the motion of the molten zone and the forces, acting on the drop, are gravity and surface tension. The net forces of gravity and the surface tension must balance each other, otherwise the drop will be unstable and split up into droplets. A liquid cylinder will be unstable if its length exceeds its circumference.

$$\text{Condition for stability}: \quad L < 2\pi r \tag{8.21}$$

where
L = length of the molten cylinder
r = radius of the molten cylinder.

This condition gives an upper limit for the zone length.

The shape of the molten zone depends on the direction of its motion. If it moves upwards, the equilibrium shape corresponds to Figure 8.12a. The shape of the molten zone, that moves downwards, is shown in Figure 8.12b. The value of the meniscus angle φ has to be kept constant in order to maintain a constant diameter of the crystallizing rod.

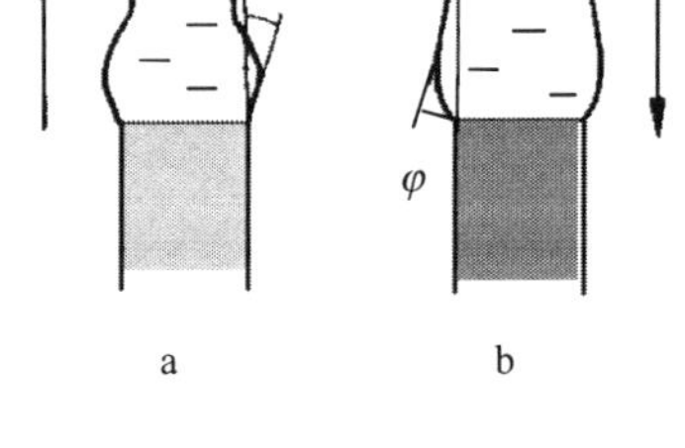

Figure 8.12 Equilibrium shape of molten zone if it moves. a) upwards; b) downwards.

At stationary flow conditions and some special flow patterns, for example a uniformly rotating rod, in combination with a surface tension, independent of position, the following relationship between the hydrostatic pressure and the capillary pressure, caused by the surface tension, will be valid:

$$\sigma_L\left(\frac{1}{R_1}+\frac{1}{R_2}\right) = -\rho_L g\, h + const \tag{8.22}$$

where
σ_L = surface tension of the liquid
$R_{1,2}$ = perpendicular radii of curvature of the drop
h = height above the lower interface ($h = 0$)
ρ_L = density of the liquid
const = constant, which depends of the volume of the molten zone.

Thermal Stability. Shape of Phase Boundary Surface

The temperature distribution of the floating zone and its thermal balance are mainly functions of the thermal conductivities of the melt and the solid, the ratio of the energy flux through the rod ends and the radiation from the rod circumference.

The shape of the phase boundary surface depends on the effect P of the heater that is used for melting.

Figure 8.13 shows the outer and central lengths of the molten zone as functions of the heating power P. Within region A the power is not sufficient to melt the rod at all. At power P_1 the rod starts to melt from outside along a ring (region B). At power P_2 the centre of the rod starts to melt. In region C the inner length increases rapidly but it is shorter than the outer length and the molten zone becomes concave as Figure 8.13 shows.

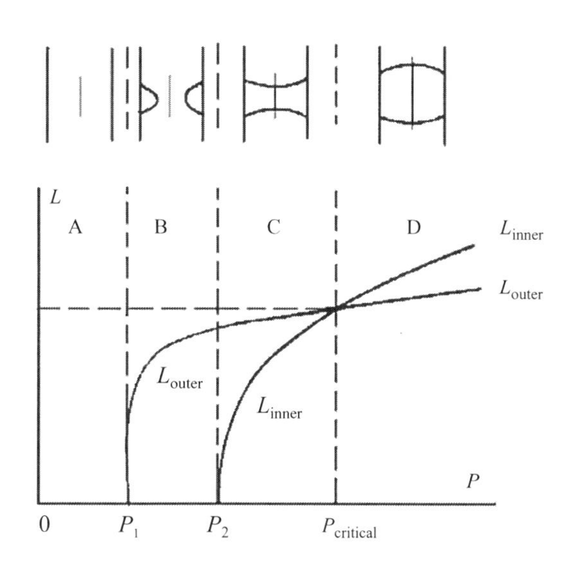

Figure 8.13 Zone shape as a function of heating power. L_{inner} = inner length of drop; L_{outer} = outer length of drop.

If the power is increased further, the inner length will exceed the outer length and the shape of the molten zone becomes convex in region D. The critical power $P_{critical}$ corresponds to the case when the phase boundary surface is planar.

By a convenient design and optimal power of the heater the shape of the boundary surface can be controlled and kept planar.

Surface Convection

In Section 8.3.2 we found that natural convection is a necessary condition for zone melting. In addition to the normal natural convection, caused by temperature and density differences in the melt, there is another type of convection, which is caused by temperature differences at the surface of the melt.

Surface tension is strongly temperature dependent. A hot melt has normally lower surface tension than a cold one. Surface tension can be considered as surface energy per unit area. The most stable state of the melt is the one with the lowest possible energy. Hence, there are surface forces which cause the free surface to move from hot to cold areas. The surface works like a two-dimensional rubber band, which tries to reduce its area as much as possible. The result is a surface flow of melt.

The phenomenon is called *Marangoni convection* (Figure 8.14). It is important for small zones up to 10 mm in diameter. However, under most practical conditions the normal natural convection dominates, especially at large floating zones.

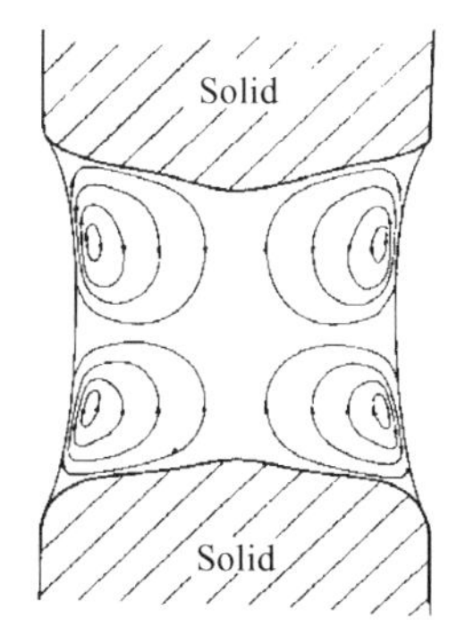

Figure 8.14 Sketch of streamlines of Marangoni convection.

8.3.4 Multiple Zone Refining. Continuous Zone Refining

As mentioned in Section 6.2.3 in Chapter 6 it is normally necessary to repeat the zone melting process several times before the required purity becomes reached. The practical methods of zone refining normally include several or many successive processes.

Repeated fractional crystallization with close zones saves time as the processes at all the zones occur nearly simultaneously with no interruptions for separation of pure and impure fractions. The extreme process is continuous zone refining, which corresponds to an infinite number of zone refining steps.

Multiple Zone Refining

Multiple zone refining equipments consist of a number of narrow molten zones of equal length, each of the appearance shown in Figure 8.9 on page 443. The molten zones are coupled in series and simultaneously monitored slowly over the rod range. The length of each molten zone is short and the number of molten zones is large. Arrangements are made to let all the molten zones, successively pass the material, which will be refined.

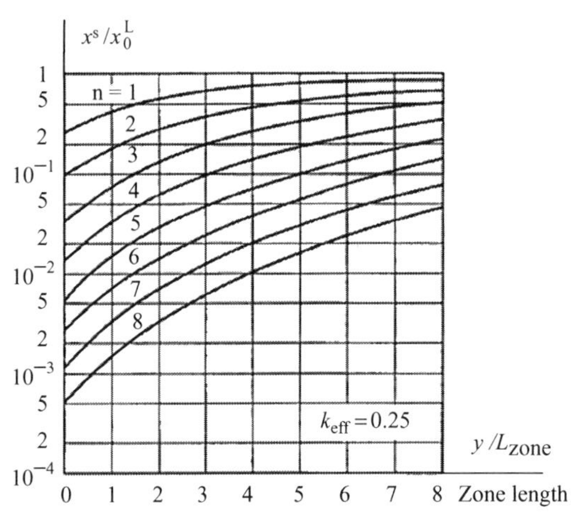

Figure 8.15a Solute concentration after n zone-refining processes as a function of distance, expressed in molten zone lengths L_{zone} for $k_{eff} = 0.25$. Bruce Chalmers, courtesy of S. Chalmers.

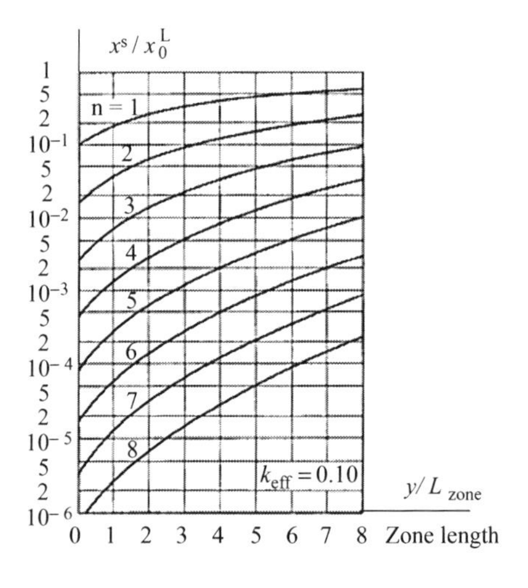

Figure 8.15b Solute concentration after n zone-refining processes as a function of distance, expressed in molten zone lengths L_{zone} for $k_{eff} = 0.10$. Bruce Chalmers, courtesy of S. Chalmers.

The principle of zone refining is to carry impurities from the first part to the end of the rod. For small k values a single process may be sufficient. However, if the k value is high and close to 1 the effect of a single zone refining process is poor. Figures 8.15a and 8.15b show that the refining is greatly improved by multiple zone refining. A comparison of the Figures 8.15a and 8.15b shows that

- the result of zone refining is improved by increasing the number of zone refining processes
- the smaller the effective partition coefficient is the better will be the result.

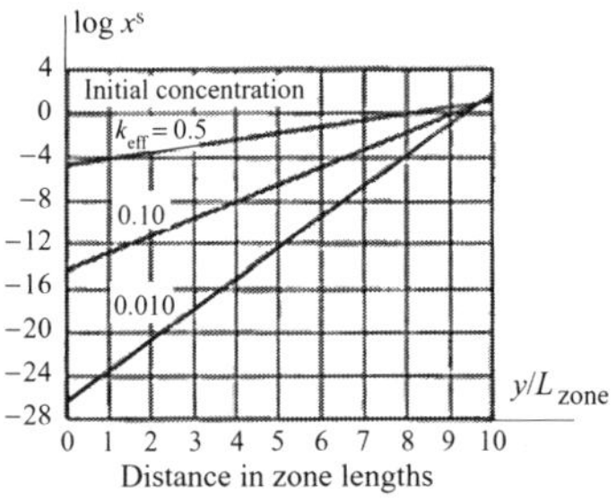

Figure 8.16 Solute distribution as a function of distance, expressed in molten zone lengths, after passage of many zones with the effective partition coefficient as parameter. Bruce Chalmers, courtesy of S. Chalmers.

The last statement is in agreement with the discussion in Section 8.4.2. Figure 8.16 also confirms the same result.

Figure 8.17 shows the fact that hundreds or thousands of zone refining processes are required if the effective partition coefficient is close to 1.

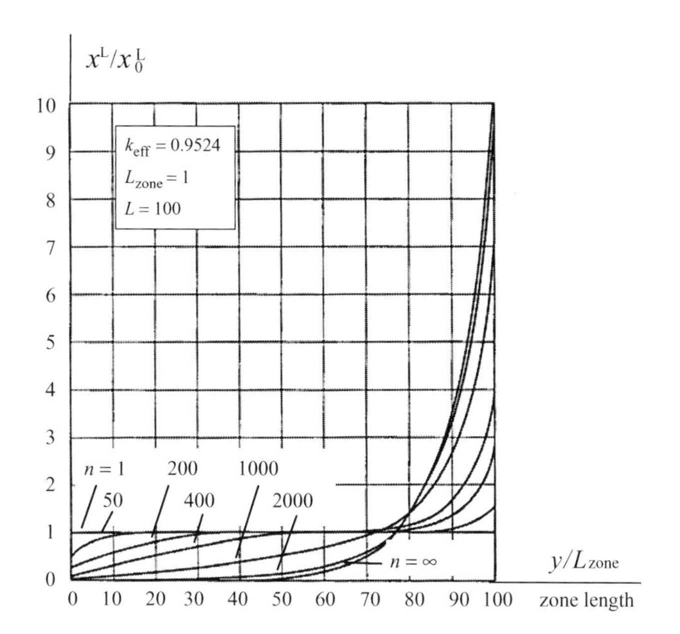

Figure 8.17 Solute concentration in specimen after multiple zone refining for $k_{eff} = 0.95$. Bruce Chalmers, courtesy of S. Chalmers.

Continuous Zone Melting

In the single or multiple zone melting processes described so far the zone melting zone starts in one end and moves slowly to the other end of the specimen. Pfann suggested a continuous zone refining apparatus, which is most useful when a great number of zone melting cycles are required. His apparatus is described in Figure 8.18.

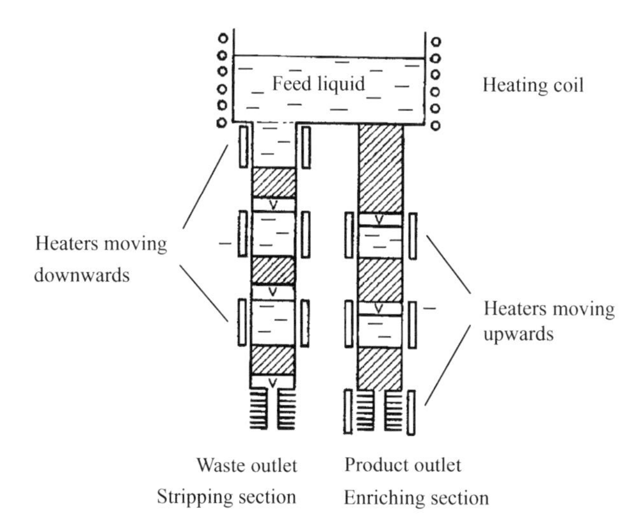

Figure 8.18 Continuous zone refiner.

The heaters consist of induction coils, which move slowly upwards and downwards, respectively, during the refining process and are then quickly carried back to their initial positions for a new cycle.

The material to be refined is fed into the equipment at the middle of the sequence of molten zones. The purified product leaves the equipment at the bottom of the right 'leg', called the *enriching section*. The exit for the material, in which the impurities are concentrated, is the bottom of the left 'leg' or the *stripping section*.

It may be surprising that a mass flow can occur when there are plugs of solid in both legs all the time. The explanation is the following. Waste and product can leave the equipment only when a heater covers the outlet and melts the materials. Well defined amounts of material leaves the legs and are replaced in each case by a void of well defined volume. The void moves upwards when in contact with next molten zone, from a position below it to a position on top of the melt. The solid parts fall through the void when molten and are therefore gradually moved thanks to the voids. The voids and the molten zones move upwards in both legs and the solid parts are displaced downwards.

The process is continuous in the sense that the refined product can be continuously added to the feed liquid, i.e. 100% reflux, until the refined product corresponds to the requirements of purification. Then the outlet material is removed as the final product and another batch of unrefined material is added to the container in the furnace and a new refining circle starts.

8.4 Single Crystal Production by Czochralski Technique

A single crystal production method, which is widely used for growing single crystals, is *crystal pulling* or *Czochralski technique*. High quality crystals of the diameter 50–200 mm grown by this method are commercially available for a number of materials of the magnitude a hundred.

The method is especially useful and widely used for growth of silicon crystals and doped semiconductor materials such as for example gallium arsenide and indium phosphide. The composition of the melt can be varied during growth. The Czochralski technique is a very important method to produce wafers with a great number of integrated circuits. Thanks to the simultaneous mass production of integrated circuits on wafers, the prices of such components are low.

8.4.1 Growth Process

The growth process can be described as follows:

1. The growth material is molten in a crucible and kept at a constant temperature, some degrees above the melting point, about 25°C for materials with high thermal conductivity and less for low-conducting materials. The temperature is controlled by the supplied input of heat to the melt.
2. A seed crystal is attached to a vertical pull rod, rotated slowly and lowered until it touches the melt of the proper temperature. If the seed crystal melts and looses the contact with the melt, the temperature is too high. If the crystal starts to grow the temperature is too low and has to be raised until thermal equilibrium is achieved.
3. The seed crystal is then slowly pulled upwards under simultaneous slow rotation at half the final rate. The crucible also rotates slowly. The pulling rate controls the diameter of the crystal. The rate is adjusted to give new crystals with a diameter slightly smaller than that of the seed crystal.
4. The pulling rate and rotation rates are increased to their final values and a long narrow neck is grown to avoid dislocations in the future crystal.
5. The temperature is then lowered slowly, which increases the diameter of the crystal. The temperature decrease is stopped when the desired diameter is achieved and the growth continues until the desired crystal length is obtained. The crucible is continuously raised during the growth in order to keep the free surface at a constant level.
6. The melt temperature is raised which causes the crystal diameter to decrease slowly to zero. The equipment is left to cool slowly.

An example of the design of a growth equipment according to the Czochralski technique is shown in Figure 8.19 in Section 8.4.2.

8.4.2 *Maximum Pulling Rate of Crystals*

The solidification of melt releases heat of fusion, which has to be transported away from the solid/melt interface. As the melt has a higher temperature than the interface, the heat flux must be conducted away through the crystal. The crystallization process is mainly controlled by two quantities

- the rate of heat removal from the solid/liquid interface
- the pulling rate of the seed when it is withdrawn from the melt.

The equilibrium can be described by the simple heat balance

$$V_{\text{growth}} A \rho_{\text{c}} \, (-\Delta H) = -\lambda_{\text{c}} A \frac{\text{d}T}{\text{d}y} \tag{8.23}$$

or

$$V_{\text{growth}} \rho_{\text{c}} \, (-\Delta H) = -\lambda_{\text{c}} \frac{\text{d}T}{\text{d}y} \tag{8.24}$$

where

V_{growth} = growth rate of the crystal at the solidification front
A = area of solid/liquid interface
ρ_{c} = density of the solid
$-\Delta H$ = heat of fusion per mass unit
λ_{c} = thermal conductivity of the crystal
T = temperature
y = distance from interface, perpendicular to the interface.

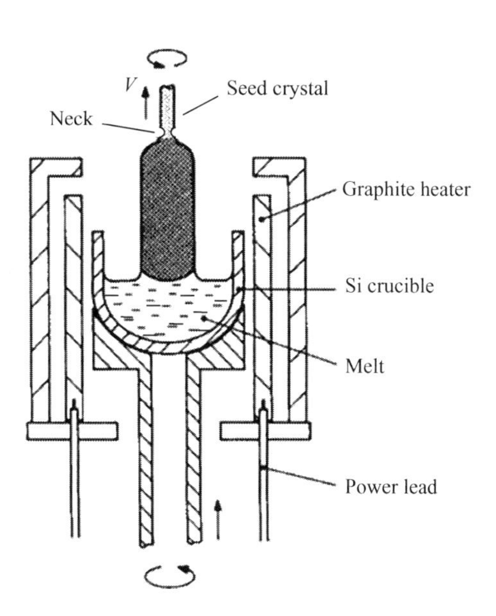

Figure 8.19 Growth equipment for growth of silicon crystals. Best results are obtained if the crystal and the crucible are rotated in the same direction. Reproduced from John Wiley & Sons © 1986.

The pulling rate V (Figure 8.19) has the same size as V_{growth} but opposite direction. The value calculated from Equation (8.24) is the highest possible growth rate. Any higher growth rate leads to an increase of the released amount of heat. A tentative excess heat can not be conducted away from the interface. Its temperature would increase, the solidification process would stop and the crystal would start to remelt.

The temperature gradient at the interface $(\text{d}T/\text{d}y)_{y=0}$ is determined by the heat transport from the growing crystal to the surroundings. The convection around the growing crystal and the radiation to the surroundings mainly control the heat transport.

Calculations show that the upper limit for the growth rates of semiconductor materials cannot exceed a few millimeters per hour.

The slower the growth rate is, the larger crystal will be grown. After a relatively rapid initial pulling rate, which gives a 'neck' (Figure 8.19) immediately below the seed crystal, a slow pulling rate permits growth of crystals with diameters of the magnitude 200–300 mm. Crystals with diameters up to 300 mm have been grown but it is difficult to avoid dislocations if the diameter exceeds 200 mm.

8.4.3 *Fluid Motion in the Crucible*

In the Czochralski process the fluid motion of the melt in the crucible has to be considered.

If a disc of infinite extension rotates on the surface of an infinite fluid solute and thermal boundary layers are developed, which are functions of the dimensionless quantities Schmidt's number Sc and Prandtl's number Pr, respectively, (Chapter 7 page 367) of the fluid. If $\mathrm{Sc} > 100$ and $\mathrm{Pr} < 0.1$ the following relationships are valid for the Czochralski process

$$\delta_D = 1.61 \times \sqrt{\frac{\nu_{\mathrm{kin}}}{\omega}}\sqrt[3]{\mathrm{Sc}} \tag{8.25}$$

$$\delta_T = 1.13 \times \sqrt{\frac{\nu_{\mathrm{kin}}}{\omega}}\sqrt[3]{\mathrm{Pr}} \tag{8.26}$$

where

δ_D = thickness of solute boundary layer (solute flow)
δ_T = thickness of thermal boundary layer (thermal flow)
ν_{kin} = kinematic viscosity coefficient of the fluid
ω = angular velocity of the rotating disc
D_{L} = diffusion constant of the fluid
Sc = Schmidt's number, the ratio of $\nu_{\mathrm{kin}}/D_{\mathrm{L}}$
Pr = Prandl's number, the ratio η/α.
η = dynamic viscosity coefficient of the fluid
α = thermal diffusivity of the fluid.

In practice *the disc,* i.e. the growth area, *has a limited size,* which causes minor errors if Reinhold's number (Chapter 7, page 367) is less than 10. Hence, it is desirable that

$$\mathrm{Re} = \frac{r^2\omega}{\nu_{\mathrm{kin}}} > 10 \tag{8.27}$$

where r is the radius of the crystal and ω is its rate of rotation.

The *limited dimensions of the fluid* cause more serious errors than the limited growth area. If the height of the fluid $h < (\eta/\omega)^{1/2}$, there will be no boundary layer. The boundary layer model of the solute flow and the thermal flow, respectively, is only valid if

$$h \gg \delta_D \quad \text{and} \quad h \gg \delta_T.$$

If the ratio $R/r > 2.1$ the experimental values of the boundary layers agree with the theoretical values. R is the radius of the fluid. This condition is normally satisfied for the Czochralski process.

The convective flow pattern in the fluid depends strongly on the motion of the crystal and the crucible. *A stable flow pattern is obtained only if the crystal and the crucible rotate in the same direction* as is shown in Figure 8.20.

(a)

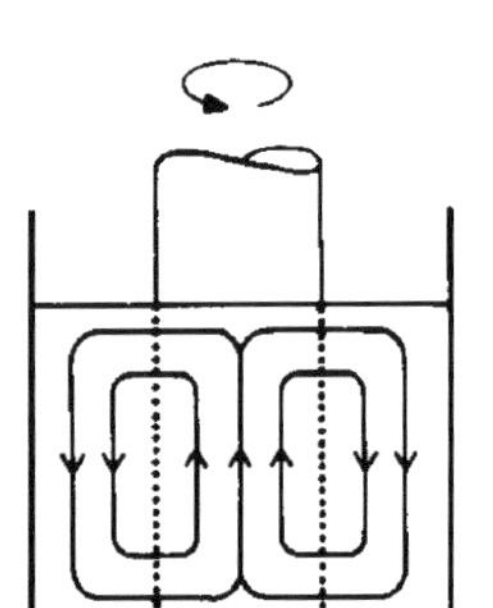

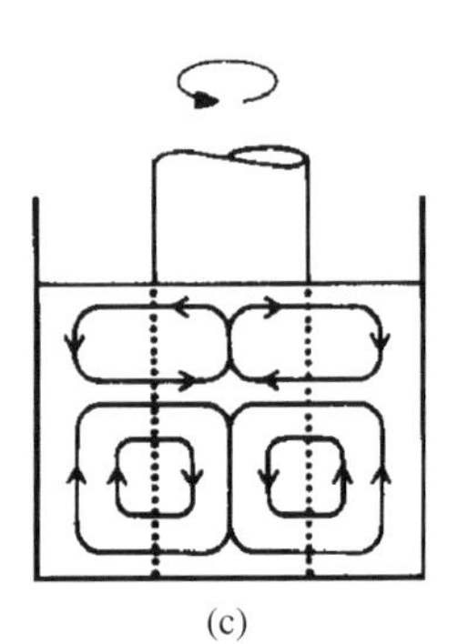

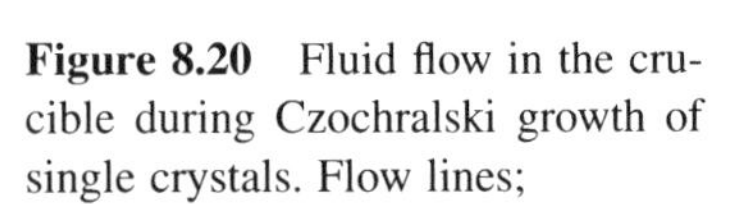

Figure 8.20 Fluid flow in the crucible during Czochralski growth of single crystals. Flow lines;
a) in case of convective flow and no rotation of the crucible;
b) when the crystal and the crucible rotate in the same direction;
c) when the crystal and the crucible rotate in opposite directions.

8.4.4 *Striation*

The Czochralski growth process is an example of crystal growth from a completely mixed melt. However, the thermal convection is hardly perfectly homogeneous which generates defects in the growing crystals. When the rotating crystal passes regions with varying solute concentrations, generated by different local temperatures, a characteristic type of *striation* pattern appears.

Striation is a striped pattern in crystals, which consists of repeated impurity depositions at the interface (Figure 8.21). The concept was originally introduced to describe variations in doping in semiconductor materials but the concept has been generalized to enclose other types of regular fluctuations, such as fluctuations in the structure of eutectics, distribution of oxide clusters in metals and variations in composition during growth and irregularities caused by temperature fluctuations.

In most cases striation is a disadvantage, for example in crystal production by the Czochralski technique, but it can also be used for studies of growth mechanisms. The striation pattern can be visualized e.g. by etching or interference contrast microscopy. By introducing regular pulses of some impurity, the interface position can be registered at given time intervals. The information can be used to find the growth rate and to study stacking faults and dislocation loops.

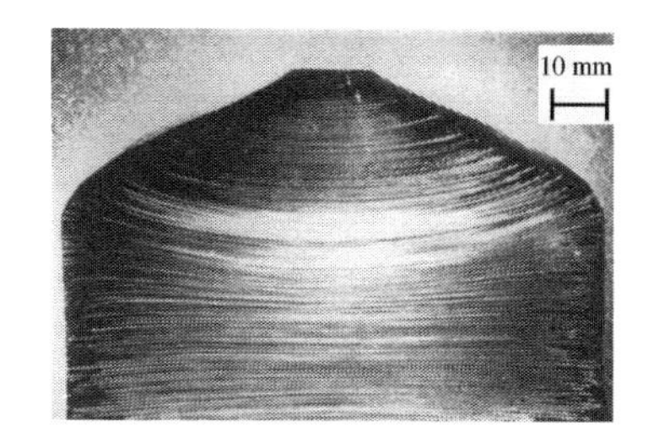

Figure 8.21 Example of striation. First grown part of a Si crystal, grown by the Czochralski technique. The crystal has a striation pattern of oxygen precipitated after annealing at 1000 °C for approximately 8 h.

Methods of Striation Reduction

The striations are formed during changes in the convection mode. There are a number of methods to affect the convection mode and reduce the formation of striations. Reduction of the temperature gradient during growth, application of static magnetic fields in various ways and reduction of gravity are some methods, which can be used.

Another method is to change the geometry of the melt during growth with the aid of a second inner crucible, from which the crystal is pulled (Figure 8.22). The method is used for production of extremely pure crystals by the Czochralski technique.

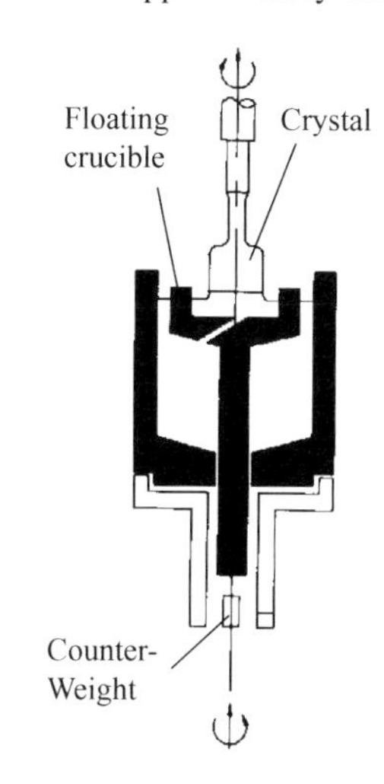

Figure 8.22 Sketch of double crucible used for production of extremely pure crystals by the Czochralski method. The crystal is grown in the inner crucible.

The inner and outer crucibles are connected by a channel, which is small enough to prevent extensive mixing of melt between the two crucibles but large enough to allow flow of melt to replace the melt which disappears during the crystallization process.

Assume that the initial solute concentration of the melt in both crucibles is x_0^L. Then the initial composition of the crystal will be $k_0 x_0^L$ and solute atoms will be rejected back into the melt if $k_0 < 1$. Its concentration close to the interface increases until a steady state is reached when the melt at the interface has the composition x_0^L/k_0 and the solute concentration in the crystal is x_0^L. The counter weight keeps the floating inner crucible in the melt.

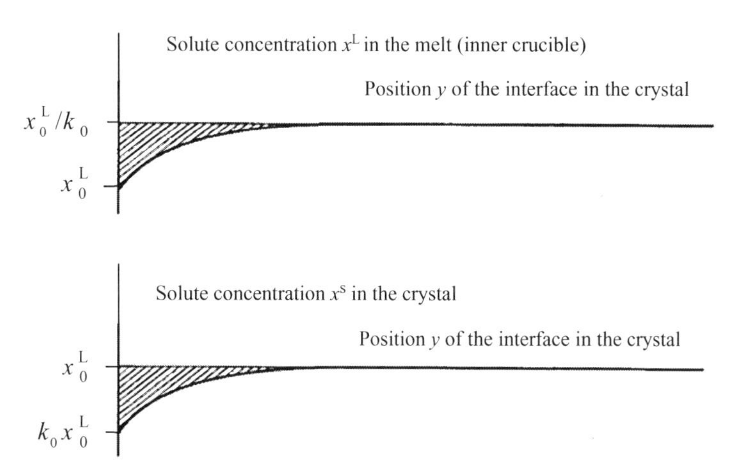

Figure 8.23 Solute profile close to the interface in the melt and in the crystal during growth.

Melt of the composition x_0^L is continuously soaked into the inner crucible. The crystal, which is pulled out of the melt, also has the composition x_0^L as soon as the initial period is over. This is shown in Figure 8.23. If the inner crucible is fed with melt of composition x_0^L/k_0 from the beginning the initial transient can be avoided completely.

The method has the advantage that microsegregation is avoided, which normally causes a gradual increase of the solute concentration in the melt and in the crystal when the amount of melt decreases in the crucible.

8.5 Cellular Growth. Constitutional Undercooling. Interface Stability

8.5.1 Cellular Growth

Transverse sections of solids, unidirectionally grown under special experimental conditions, show a so-called *cellular structure*. Figures 8.24 and 8.25 give two examples of the appearance of different cellular structures.

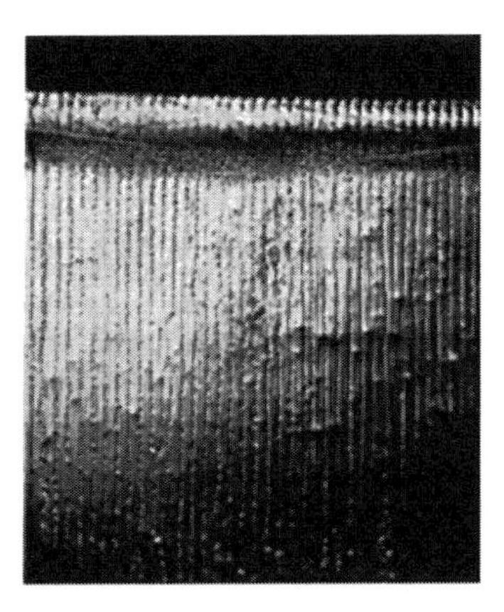

Figure 8.24 Cellular structure of Sn. View of the free top surface of a tin crystal. Bruce Chalmers, courtesy of S. Chalmers.

Figure 8.25 Transverse section of a growing Al crystal perpendicular to the growth direction after rapid decantation of the melt.

Figure 8.24 shows the pattern of a free top surface of a tin crystal with grooves, parallel to the growth direction. Figure 8.25 shows the transverse section of the tin crystal, perpendicular to the growth direction, after rapid decantation of the melt. The appearance reminds of a honeycomb.

The puzzling phenomenon was primarily explained in the middle of the twentieth century when Rutter and Chalmers [4] showed that a planar solidification front under certain circumstances is instable and that cellular structure can be satisfactory explained in terms of so-called *constitutional supercooling* or, more adequately, *undercooling*. Their theory, which is generally accepted today, is shortly described below.

8.5.2 Instability of a Planar Solidification Front

Consider a planar interface in a solidifying alloy melt. The undercooling of the melt ahead of the interface is the driving force of solidification. The temperature distribution in the melt, which controls the solidification process and the stability of the interface, is a function of position, i.e. the distance from the interface.

If a system, which for some reason is disturbed, returns spontaneously to its initial state, the system is said to be *stable*. If the system does not return to its initial state the system is *instable*.

This general statement can be applied to the solidification process at a planar interface. As long as the undercooling increases with the distance from the interface (Figure 8.26) a temporary outgrowth will grow faster than the rest of the interface (Figure 8.27) and form a protuberance. The planar interface is instable and the protuberance grows until it reaches a steady state.

The convexity of such a protuberance leads to

- a higher growth temperature than that of a planar interface
- a lower concentration of solute in the solid than that of a planar interface.

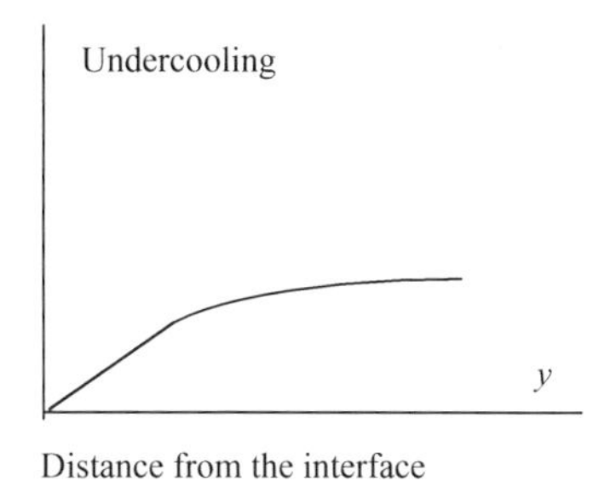

Figure 8.26 Undercooling of the melt as a function of distance from the solidification front.

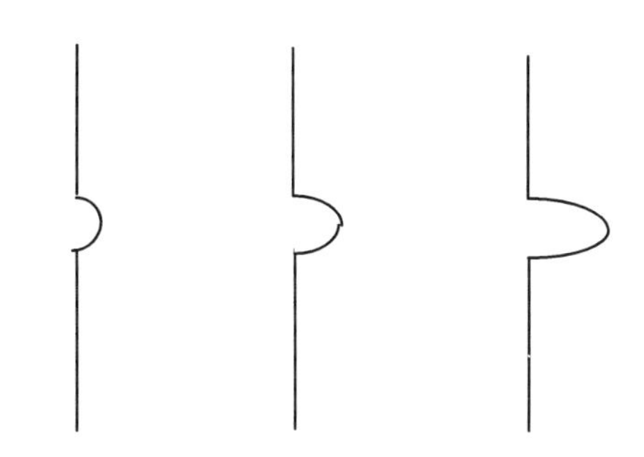

Figure 8.27 If the undercooling increases with distance from the interface, a local disturbance grows faster than the rest of the solidification front and does not disappear. The interface is unstable.

The concentration of solute around the base of the protuberance is therefore higher than in the protuberance itself. This retards solidification in the latter region and the protuberance cannot expand radially. As a consequence of the convex areas outside regions such as A and B in Figure 8.28 new protuberances around the original ones are formed which explains the groove structure in Figure 8.24.

Figures 8.24 and 8.25 and later Figure 8.39 on page 464 show clearly that the structure of cellular growth can have the most different shapes. The varying structures are a consequence of the *magnitude* of the so-called *constitutional undercooling*. Strong undercooling corresponds to hexagonal cells and results in a structure such as that in Figure 8.25, which reminds of a honeycomb. A more detailed discussion is given in Section 8.5.5 on page 463 and also on pages 565–568 in Chapter 9.

There are two alternative methods to analyze the conditions of cellular growth. Either the growth in characteristic directions of maximum growth is studied or perturbation theory is applied. The former method has been used qualitatively above and will be applied in Section 9.7 in Chapter 9. The latter method will be used in Section 8.5.4.

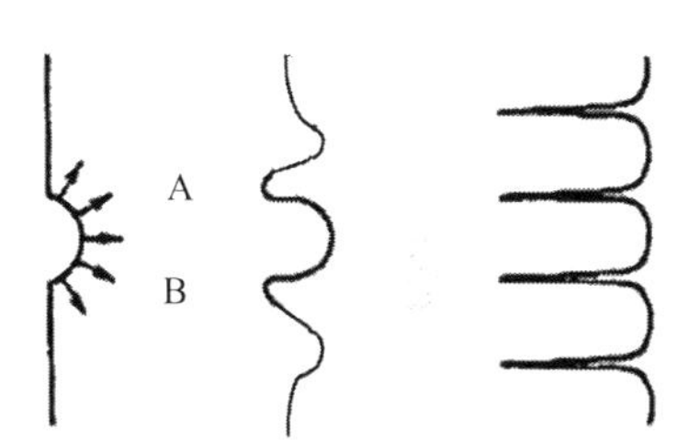

Figure 8.28 Formation of cellular structure. Several secondary maxima and minima are formed at each side of the initial perturbation. The final macroscopic appearance of the perturbations is shown on the right-hand side. Bruce Chalmers, courtesy of S. Chalmers.

8.5.3 *Constitutional Undercooling in Binary Alloys*

In order to derive the size and position of the undercooling in the melt of a solidifying alloy it is necessary to find expressions for the concentration and temperature distributions in the alloy.

Equilibrium Temperature T_L in the Melt as a Function of Position

The phase diagram of the alloy represents a relationship between temperature and solute concentration in the melt. Figure 8.29 shows that the equilibrium temperature of the interface, i.e. the temperature of the melt in equilibrium with the solid can be written as

$$T_L = T_M + m_L x^L \tag{8.28}$$

where

x^L = concentration of solute in the melt
T_L = equilibrium temperature at concentration x^L
T_M = melting point of the pure metal
m_L = slope of liquidus line.

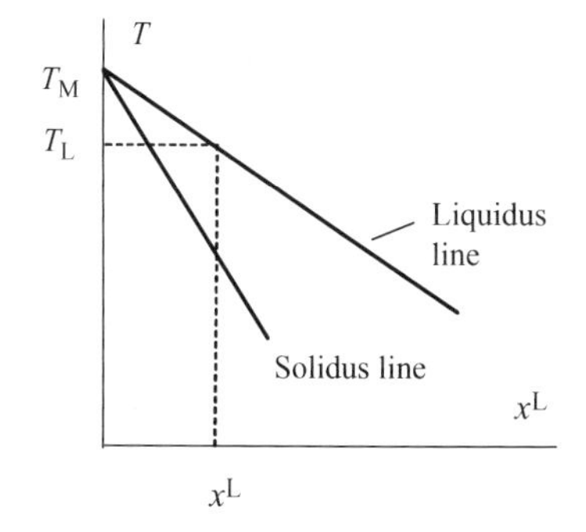

Figure 8.29 Schematic phase diagram of a binary alloy.

To derive the equilibrium temperature as a function of the distance from the interface we have to find the solute concentration as a function of position.

When an alloy melt solidifies the concentration of solute is normally lower in the solid than in the melt. This can be expressed by the usual relationship

$$k_0 = \frac{x^s}{x^L} < 1 \tag{8.29}$$

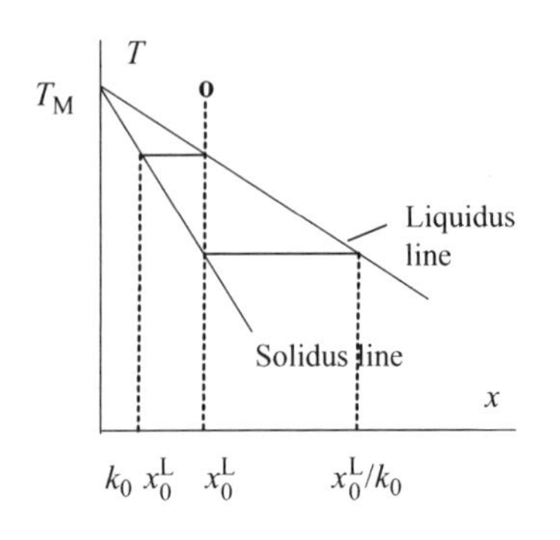

Figure 8.30 Schematic phase diagram of an binary alloy.

Consider a cooling melt with the solute concentration x_0^L. The first precipitated solid has the solute concentration $k_0 x_0^L$ (Figure 8.30). The temperature continues to decrease and the last precipitated solid has the concentration x_0^L. The last liquid has the solute concentration x_0^L/k_0 (Figure 8.30).

When the melt solidifies at the interface an excess of solute atoms obviously arises in the melt close to the interface and a diffusion of solute atoms from the interface into the melt occurs. Equilibrium is achieved when the number of ejected solute atoms from the solid equals the number of atoms diffusing from the melt into the interface per unit time.

According to Equation (8.1) on page 435 the solute concentration in the melt ahead of a planar interface, which moves forward with the constant velocity V_{growth}, as a function of the distance form the interface, can be written as

$$x^L = x_0^L \left(1 + \frac{1 - k_0}{k_0} e^{-\frac{V_{growth}}{D_L} y}\right) \tag{8.30}$$

where

y = distance from the interface in the melt
x^L = solute concentration in the melt at distance y
x_0^L = average solute concentration in the melt before solidification
k_0 = partition coefficient of solute
V_{growth} = growth rate = velocity of solidification front
D_L = diffusion constant of solute atoms in the melt.

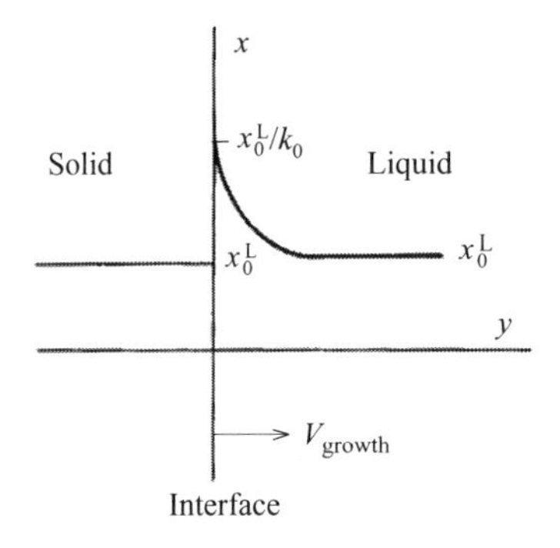

Figure 8.31 Solute concentration as a function of distance from the interface in a solidifying alloy. The solidification front moves with the constant velocity V_{growth}.

The function in Equation (8.30) is plotted in Figure 8.31, which is in agreement with discussion and phase diagram on page 455.

Equation (8.30) is the desired relationship between solute concentration and position, which we need to find the equilibrium temperature in the melt as a function of position. If we combine Equations (8.28) and (8.30) we obtain

$$T_L = T_M + m_L x_0^L \left(1 + \frac{1 - k_0}{k_0} e^{-\frac{V_{growth}}{D_L} y}\right) \tag{8.31}$$

Actual Temperature *T* in the Melt Close to the Interface as a Function of Position

It is reasonable to assume that the solid and the melt are in equilibrium with each other at the interface. The temperature of the melt at the interface can be obtained either from the phase diagram by inserting $x_L = x_0^L/k_0$ into Equation (8.28) or by inserting $y = 0$ into Equation (8.31). The result is

$$T(0) = T_L(0) = T^I = T_M + m_L \frac{x_0^L}{k_0} \tag{8.32}$$

Elsewhere in the melt the actual temperature differs from the equilibrium temperature. To find the actual temperature distribution in the melt we have to solve the general heat equation in the solid and in the melt in accordance with given boundary conditions. However, the most interesting part of the temperature distribution is the region close to the interface where the temperature changes most rapidly. Within this narrow region it is reasonable to use a linear representation of the temperature, i.e. use the approximation

$$T = T(0) + G_L y$$

or, by use of Equation (8.32),

$$T = T_M + m_L \frac{x_0^L}{k_0} + G_L y \tag{8.33}$$

where G_L is the temperature gradient in the melt close to the interface. The temperature gradient is determined by the experimental conditions and material constants. It has been derived in Chapter 7. With the aid of Equation (7.71) on page 389 in Chapter 7 we obtain the following expression of G_L

$$G_L = -\left(\frac{dT_L}{dy}\right)_{y=0} = (T_F - T_M)\left(\frac{V_0}{2\alpha_L} + \sqrt{\left(\frac{V_0}{2\alpha_L}\right)^2 + \frac{2h_L}{r\rho_L c_p^L \alpha_L}}\right) \tag{8.34}$$

where the direction of the y axis has been changed compared to the coordinate system in Chapter 7.

The actual temperature T for different values of y [Equation (8.33)] is shown as the straight line in Figure 8.32.

Constitutional Undercooling

The bent curve in Figure 8.32 shows the equilibrium temperature in the melt as a function of the distance from the interface. It approaches asymptotically the value $T_M + m_L x_0^L$ which is identical with the liquidus temperature T_L at $x_L = x_0^L$.

As Figure 8.32 shows the actual temperature, given by Equation (8.33), is lower than the equilibrium temperature, given by Equation (8.31), within the enclosed temperature interval in Figure 8.32. The melt is undercooled close to the interface. Chalmers introduced the concept *constitutional undercooling* for this phenomenon.

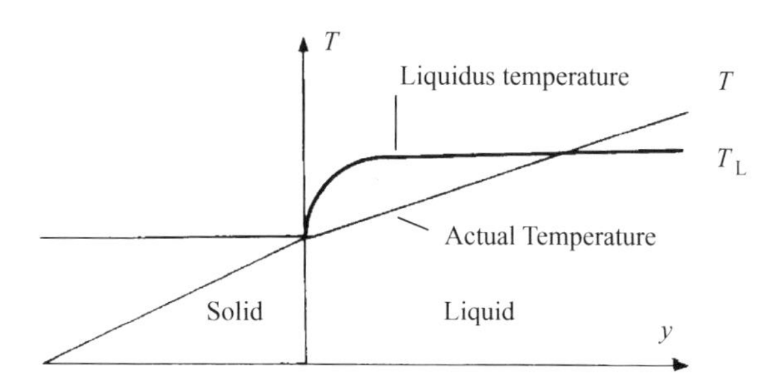

Figure 8.32 Equilibrium temperature and actual temperature in the melt as functions of distance from the interface.

Figure 8.33 shows the equilibrium temperatures and the actual temperatures for different values of the growth rate V_{growth} and the temperature gradient G_L in the melt.

From Figure 8.34 we can conclude that the condition for undercooling is

$$G_L \leq \left(\frac{dT_L}{dy}\right)_{y=0} \tag{8.35}$$

Equation (8.35) can be transformed into a more useful form if we take the derivative of Equation (8.31) with respect to y. The result is

$$\frac{dT_L}{dy} = m_L x_0^L \frac{1-k_0}{k_0}\left(-\frac{V_{growth}}{D_L}\right) e^{-\frac{V_{growth}}{D_L} y} \tag{8.36}$$

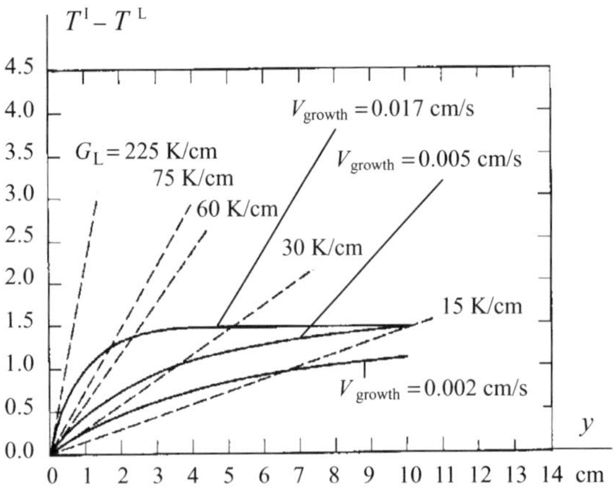

Figure 8.33 Equilibrium temperature and actual temperature in the melt as functions of distance from interface for various values of V_{growth} and G_L. Bruce Chalmers, courtesy of S. Chalmers.

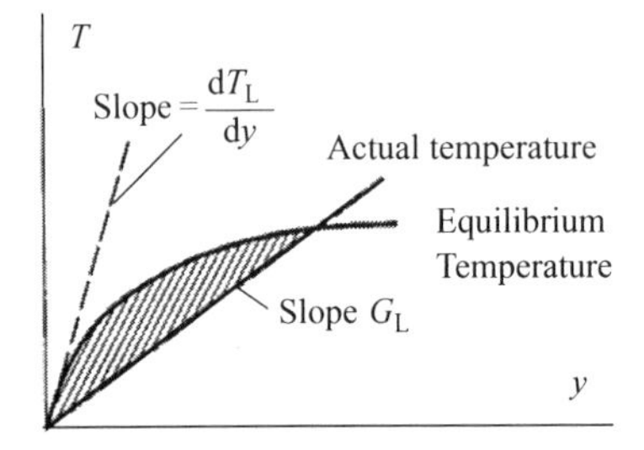

Figure 8.34 Undercooling in front of a planar interface. Bruce Chalmers, courtesy of S. Chalmers.

Hence, the *condition for constitutional undercooling* can be written as

$$\frac{G_L}{V_{growth}} \leq -\frac{m_L}{D_L}\frac{1-k_0}{k_0}x_0^L \tag{8.37}$$

where

G_L = temperature gradient in the melt close to the interface
m_L = slope of liquidus line in the phase diagram
k_0 = partition coefficient of solute
V_{growth} = velocity of solidification front
D_L = diffusion constant of solute atoms in the melt
x_0^L = average concentration in the melt before solidification.

Example 8.1

The condition for constitutional undercooling is given in Equation (8.37). The temperature gradient G_L, given in Equation (8.37), is primarily controlled by the concentration conditions in the liquid, which are related to temperature conditions via the phase diagram of the alloy.

Another and more direct way to find the temperature gradient is to derive it directly from the temperature distribution in the liquid during unidirectional solidification. This temperature distribution has been derived in Section 7.5.2 in Chapter 7 for the case when the interface is not insulated.

a) Derive the expression of the temperature gradient from the known temperature distribution in the liquid in Chapter 7 for the case when the interface is insulated.
b) Introduce this expression into Equation (8.37) and calculate the minimum furnace temperature required to avoid constitutional undercooling.

Solution:

a)

The temperature distribution in the liquid at unidirectional solidification is given by Equation (7.64) on page 387 in Chapter 7

$$T_L = T_F - (T_F - T_M)e^{\left[\frac{V_0}{2\alpha_L} + \sqrt{\left(\frac{V_0}{2\alpha_L}\right)^2 + \frac{2h_{2L}}{r\rho_L c_p^L \alpha_L}}\right]y} \tag{1'}$$

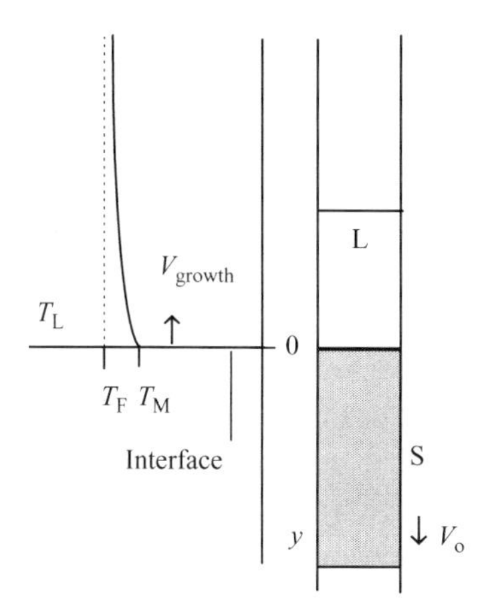

The temperature distribution is shown in the margin figure. Before any operations can be done, we have to change the coordinate system and Equation (1′) in agreement with the new coordinate system, which is used in this section in Chapter 8.

The conditions will be fully compatible if we change direction of the y axis, which means that y is replaced by $(-y)$ in Equation (1′). The growth rate of the solidification front then advances in the positive y direction in both cases since V_{growth} is equal but antiparallel to the pulling rate V_0 (equilibrium condition). V_{growth} is directed upwards in the new coordinate system. In this system the temperature distribution is

$$T_L = T_F - (T_F - T_M)e^{\left[\frac{V_0}{2\alpha_L} + \sqrt{\left(\frac{V_0}{2\alpha_L}\right)^2 + \frac{2h_L}{r\rho_L c_p^L \alpha_L}}\right](-y)} \tag{2'}$$

The interface is insulated and the heat flow through the cylindrical surface can be neglected. Hence, we have the heat transfer coefficient $h_L \approx 0$ and we obtain

$$T_L = T_F - (T_F - T_M)e^{-\frac{V_0}{\alpha_L}y} \tag{3'}$$

We take the derivative of the simplified equation with respect to y:

$$\frac{dT_L}{dy} = -(T_F - T_M)\left(-\frac{V_0}{\alpha_L}\right)e^{-\frac{V_0}{\alpha_L}y} \tag{4'}$$

As $V_{growth} = -V_0$ the expression for the temperature gradient close to the interface $(y = 0)$ will be

$$G_L = \left(\frac{dT_L}{dy}\right)_{y=0} = -\frac{T_F - T_M}{\alpha_L}V_{growth} \tag{5'}$$

b)

The expression in a) is introduced into Equation (8.37) on page 458 and the condition for constitutional undercooling is found to be

$$-\frac{T_F - T_M}{\alpha_L}V_{growth} \leq \frac{m_L x_0^L}{D_L}\frac{1 - k_0}{k_0}V_{growth} \tag{6'}$$

Answer:

a) For the case when the interface is insulated the temperature gradient close to the interface is

$$G_L = -\frac{T_F - T_M}{\alpha_L}V_{growth}$$

b) Constitutional undercooling is avoided in this case if the furnace temperature is higher than the minimum value given by

$$T_F \geq T_M - \alpha_L \frac{m_L x_0^L}{D_L}\frac{1 - k_0}{k_0}$$

8.5.4 *Interface Stability Analysis. Perturbation Theory*

The condition for constitutional undercooling (Equation (8.37) on page 458) provides a useful tool to predict the morphology of an advancing solid/liquid interface. However, the criterion is rather rough and a further development of the theory is desirable for three reasons

1. The effect of surface tension is not taken into consideration.
2. Only the temperature gradient in the liquid, not in the solid, is involved.
3. The constitutional undercooling theory gives no indication of the magnitudes of the perturbations, which will develop if an interface becomes instable.

The theory of a liquid/solid interface instability has been further developed by Mullins and Sereka [5]. The main results of their interface perturbation theory are reported briefly below.

Perturbation Theory

Perturbation analysis permits calculation of the wavelength of the instabilities, which arise at cellular growth. The result is useful for prediction of whether an interface is stable or not. It is also of great value in the theory of dendritic growth.

Perturbation Model

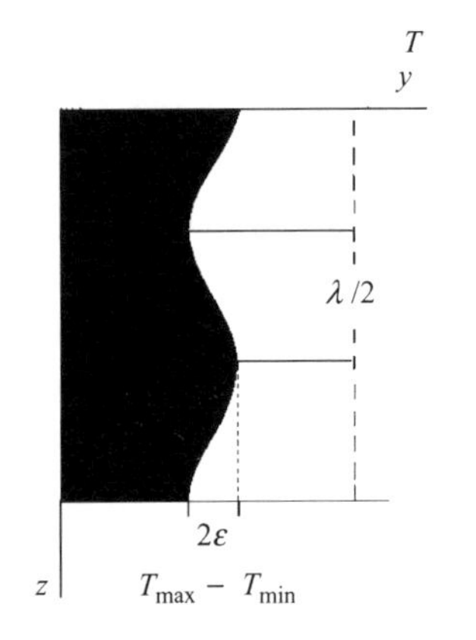

Figure 8.35 Accumulated solute on a perturbed planar interface.

When an advancing planar solidification front is exposed to a shape perturbation, the perturbation theory gives information of the response of the interface. Solute accumulates at the interface and the partition coefficient k_0 is supposed to be < 1 (Figure 8.35). The theory results in differential equations for heat transport and mass transport through the interface. From the solutions of these equations, it is possible to conclude whether the shape perturbation decays or grows with time.

Consider a solid/liquid interface that initially was planar. The planar solidification front is overlapped by a small perturbation of arbitrary profile. Here we will use the shape of a sinus curve, which is the simplest possible and at the same time the most general shape, as any shape can be described by a Fourier series of a great number of overlapping single sinus curves of various frequencies and amplitudes. The sinus wave can be written as

$$y = \varepsilon \sin \omega z \qquad (8.38)$$

where
y = coordinate in the growth direction of the interface
ε = amplitude of the sinus wave
λ = wave length of the sinus wave
ω = wave number of the sinus wave $= 2\pi/\lambda$.
z = coordinate perpendicular to y.

The amplitude ε of the sinus wave is supposed to be very small as its designation indicates. The perturbation is assumed to be so small that it does *not* affect the temperature and solute distributions at all.

The temperature of the liquid or melt close to the interface is given by Equation (8.33) on page 457. An analogous but different equation is assumed to be valid for the interface. We assume that a local equilibrium exists at the interface. The temperature of the perturbed interface is a function of the melting point temperature, the solute concentration at the interface and the curvature of the surface

$$T^{\mathrm{I}} = T_{\mathrm{M}} - m_{\mathrm{L}} x^{\mathrm{I/L}} + const \times K^{\mathrm{I}} \qquad (8.39)$$

where
T^{I} = temperature of the interface
T_{M} = melting point temperature of the alloy
m_{L} = slope of the liquidus line
$x^{\mathrm{I/L}}$ = solute concentration at the interface close to the liquid
K^{I} = curvature of the interface.

The constant in Equation (8.39) is proportional to the surface tension. The curvature K of the interface can be written as

$$K^{\mathrm{I}} = \frac{y''}{\sqrt{\left(1 + (y')^2\right)}}$$

where
$y' = \frac{\mathrm{d}y}{\mathrm{d}z} = \varepsilon\omega \cos \omega z$ and $y'' = \frac{\mathrm{d}^2 y}{\mathrm{d}z^2} = -\varepsilon\omega^2 \sin \omega z$
The curvature K is positive when the interface is concave towards the liquid.

Critical Wavelength

Equation (8.39) is applied on two points of the perturbation curve, a maximum point and a minimum point of the sinus curve. The difference between the two equations is

$$T_{\max} - T_{\min} = m_{\mathrm{L}}\left(x_{\max}^{\mathrm{I/L}} - x_{\min}^{\mathrm{I/L}}\right) - const \times (K_{\max} - K_{\min}) \tag{8.40}$$

The curvatures at the maxima and minima of the interface curve are

$$K_{\max} = -K_{\min} = \left(\frac{-\varepsilon\omega^2 \sin \omega z}{\sqrt{1 + (\varepsilon\omega \cos \omega z)^2}}\right)_{z=\frac{\lambda}{4}} \approx -\varepsilon\omega^2 = -\frac{4\pi^2\varepsilon}{\lambda^2} \tag{8.41}$$

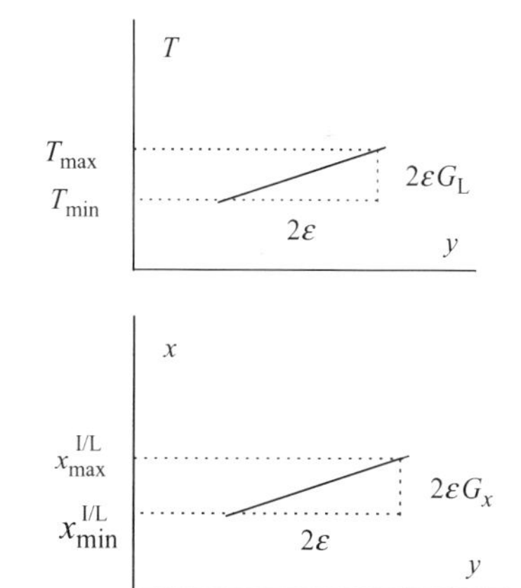

Figure 8.36 Temperature and concentration gradients.

As the temperature and solute distributions are unaffected by the perturbations and the perturbation is very small ($\varepsilon \ll 1$) the temperature and solute differences can be obtained very simply if we introduce the temperature and concentration gradients G_{L} and G_x (Figure 8.36)

$$T_{\max} - T_{\min} = 2\varepsilon G_{\mathrm{L}} \tag{8.42}$$

and

$$x_{\max}^{\mathrm{I/L}} - x_{\min}^{\mathrm{I/L}} = 2\varepsilon G_x \tag{8.43}$$

The values in Equations (8.41), (8.42) and (8.43) are inserted into Equation (8.39) and we obtain

$$2\varepsilon G_{\mathrm{L}} = m_{\mathrm{L}} \times 2\varepsilon G_x - 2 \times const \times \frac{4\,\pi^2\varepsilon}{\lambda^2}$$

or, after reduction

$$\lambda = \lambda_{\mathrm{i}} = 2\pi\sqrt{\frac{const}{m_{\mathrm{L}}G_x - G_{\mathrm{L}}}} \tag{8.44}$$

The expression $(m_{\mathrm{L}}G_x - G_{\mathrm{L}})$ is called the *degree of constitutional undercooling.*
The wavelength λ_{i} is called the *critical wavelength* for reasons which will be evident below.

Rate of Development of a Perturbation at a Constitutionally Undercooled Interface

Detailed perturbation calculations give an expression of the relative rate of development of a perturbation. This expression will only be given here in a simplified form. If we assume that

- the concentration of the solute in the solid is zero
- the temperature gradients in the liquid and the solid are equal

the relative rate of development of a perturbation becomes

$$\frac{\frac{\mathrm{d}\varepsilon}{\mathrm{d}t}}{\varepsilon} = \frac{\dot{\varepsilon}}{\varepsilon} = \frac{V_{\mathrm{growth}}\left(V_{\mathrm{growth}} - b\right)}{m_{\mathrm{L}}G_x}\left(m_{\mathrm{L}}G_x - G_{\mathrm{L}} - const \times \omega^2\right) \tag{8.45}$$

where $\dot{\varepsilon}$ is the time derivative of ε.

We substitute ω^2 by $4\pi^2/\lambda^2$ in Equation (8.45) and obtain

$$\frac{\dot{\varepsilon}}{\varepsilon} = \frac{V_{\text{growth}}\left(V_{\text{growth}} - b\right)}{m_{\text{L}} G_x}\left(m_{\text{L}} G_x - G_{\text{L}}\right)\left(1 - \frac{const}{m_{\text{L}} G_x - G_{\text{L}}}\frac{4\,\pi^2}{\lambda^2}\right) \tag{8.46}$$

where

V_{growth} = growth rate of interface

b = a constant.

If we combine Equations (8.46) and (8.44) we obtain

$$\frac{\dot{\varepsilon}}{\varepsilon} = \frac{V_{\text{growth}}\left(V_{\text{growth}} - b\right)}{m_{\text{L}} G_x}\left(m_{\text{L}} G_x - G_{\text{L}}\right)\left(1 - \frac{\lambda_{\text{i}}^2}{\lambda^2}\right) \tag{8.47}$$

where λ_{i} is the critical wavelength. Equation (8.47) shows that

- the function is positive if $\lambda > \lambda_{\text{i}}$ which means that the perturbation grows with time. The interface is instable.
- the function is negative if $\lambda < \lambda_{\text{i}}$ which means that the perturbation decreases with time. The interface is stable.
- the function is zero for $\lambda = \lambda_{\text{i}}$. The wave profile will be stationary with respect to the unperturbed interface. The perturbation does not grow.

The function in Equation (8.47) has been plotted versus λ in Figure 8.37.

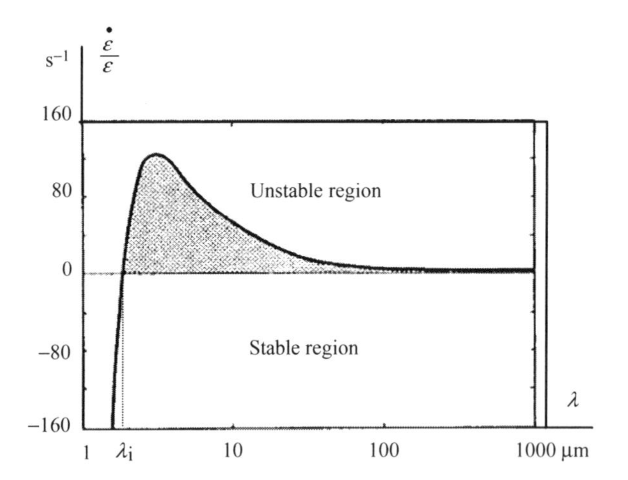

Figure 8.37 Relative rate of development of the amplitude of a perturbation in an AlCu 2wt% alloy.

At $\lambda = \lambda_{\text{i}}$ the relative growth rate of the perturbation is zero. It neither grows nor disappears. The wavelength that corresponds to the fastest relative growth rate, is the most probable wavelength.

By use of Equations (8.42) and (8.43) in combination with the phase diagram for a binary alloy it can be shown that, if the melt and the solid are in equilibrium with each other, the relationship $(G_{\text{L}} = m_{\text{L}} G_{\text{x}})_{\text{critical}}$ is valid. By use of the condition (8.37) on page 458 we obtain

$$\left(G_{\text{L}}\right)_{\text{critical}} = \left(m_{\text{L}} G_x\right)_{\text{critical}} = \frac{m_{\text{L}} x_0^{\text{L}}}{D_{\text{L}}}\frac{1 - k_0}{k_0} V_{\text{growth}} \tag{8.48}$$

Equation (8.48) is valid *only* at equilibrium conditions. The value of $m_{\text{L}} x_0^{\text{L}}/k_0$ can be expressed in terms of temperatures by use of Equation (8.32) on page 456:

$$\frac{m_{\text{L}} x_0^{\text{L}}}{k_0} = T_{\text{M}} - T^{\text{I}} \tag{8.49}$$

If we combine Equations (8.48) and (8.49) we obtain

$$(m_L G_x)_{critical} = \frac{T_M - T^l}{D_L}(1 - k_0)V_{growth} \tag{8.50}$$

If $G_x \ll m_L G_x$ and we assume that the value of $m_L G_x$ is proportional to $(m_L G_x)_{critical}$ we can write Equation (8.44) on page 461:

$$\lambda_i \approx 2\pi\sqrt{\frac{const}{m_L G_x}} = 2\pi\sqrt{\frac{D_L Const}{V_{growth}(T_M - T^l)}} \tag{8.51}$$

This expression of the critical wavelength is very concrete. The critical value increases with the size of the diffusion constant, decreasing growth rate and decreasing temperature difference. To avoid the risk of cellular growth it is desirable to increase λ_i as much as possible.

8.5.5 Constitutional Undercooling and Surface Morphology

As Rutter and Chalmers [4] showed in 1954, constitutional cooling offers an explanation of the transformation of a planar interface into a cellular morphology. They proposed the criterion (8.37) on page 458 and showed that their qualitative experiments agreed well with the theory. Later Tiller, Jackson, Rutter and Chalmers [1], Tiller and Rutter [6] performed quantitative studies, which supported the theory [7, 8].

Rapid heat transfer facilitates and controls the growth in solidifying pure metals. In alloys the slower diffusion of solute is the rate-controlling process. Hence, the study of cellular growth is easier to perform in alloy melts than in pure metal melts.

The necessary condition for instability of a planar interface (Figures 8.26 and 8.27 on page 455) shows that the undercooling increases with the distance from the interface, is fulfilled in the case of constitutional undercooling. This is clearly shown in Figure 8.34 on page 458 where the distance between the (upper) equilibrium temperature and the (lower) actual temperature increases with the distance from the interface when starting at $y = 0$. The undercooling leads to a breakdown of the planar interface in favour of the so-called cellular structure.

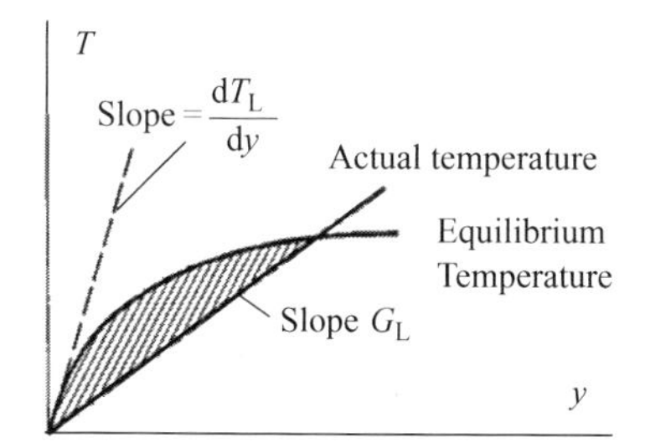

Figure 8.34 Undercooling in front of a planar interface. Bruce Chalmers, courtesy of S. Chalmers.

Experimental studies on alloys confirm that

- cellular structure appears below the critical value of the ratio G_L/V_{growth} (Figure 8.38), i.e. if

$$\frac{G_L}{V_{growth}} \leq \frac{m_L}{D_L}\frac{1 - k_0}{k_0}x_0^L \tag{8.52}$$

- the critical value of the ratio G_L/V_{growth} is proportional to the concentration of the solute in the melt – in agreement with the criterion in Equation (8.37) on page 458 and Equation (8.52).
- the quantitative theory of constitutional undercooling is based on the assumption that if the interface is planar there is no transverse diffusion of solute.

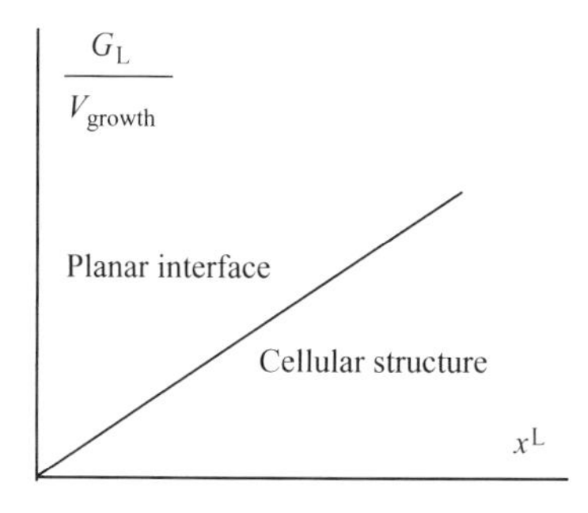

Figure 8.38 If the slope of the line exceeds the critical value $\frac{m_L}{D_L}\frac{1 - k_0}{k_0}$ no cells are formed and the solidification front remains planar.

The Figures 8.39 show that the condition (8.52) is not fulfilled at high degrees of undercooling. Hence, the theory is not valid in those cases and the structures of cellular growth can have various appearances.

The appearance of cellular growth is a matter of magnitude of the constitutional undercooling, which is shown in the Figures 8.39. At small constitutional undercoolings the cellular growth may look like pox. Larger undercoolings result in grooves or elongated cells. Strong undercoolings corresponds to hexagonal cells and results in structures such as the one in Figure 8.25 on page 454, which reminds of a honeycomb.

However, cellular growth is more complicated than the simple theory given above indicates. The conditions, given above, are fulfilled as long as no cells are formed but are no longer valid when new cells are formed. Much research work has been devoted to examination of the geometry of the cells for better understanding of the mechanisms of cellular growth.

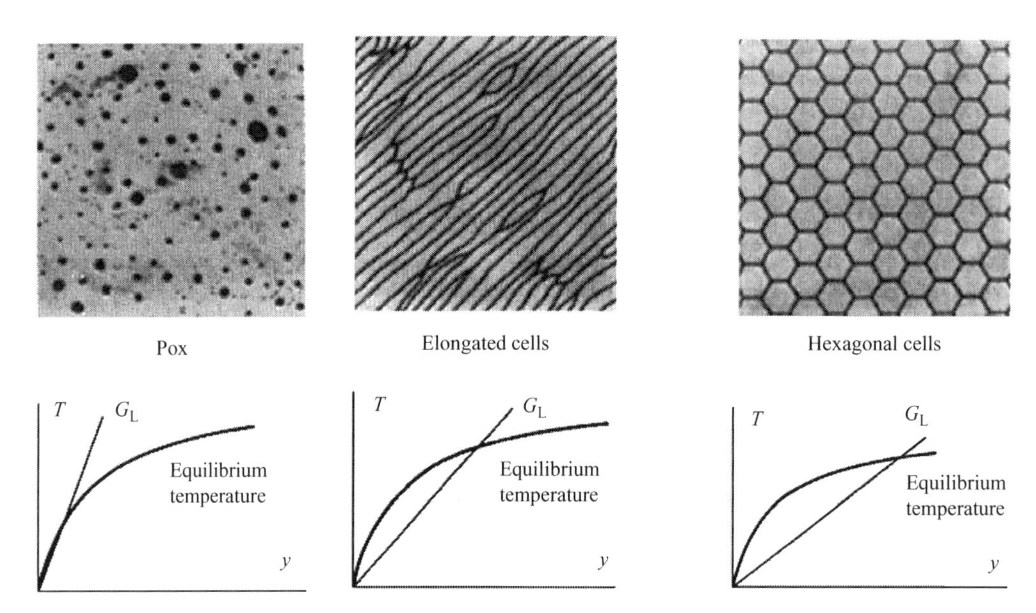

Figure 8.39a Pox structure of cellular growth at small undercooling. Reproduced from Cambridge University Press © 1991.

Figure 8.39b Example of cellular growth at intermediate undercooling. Reproduced from Cambridge University Press © 1991.

Figure 8.39c Hexagonal structure of cellular growth at large undercooling when the planar interface has collapsed. See also Figure 8.40. Reproduced from Cambridge University Press © 1991.

Figure 8.40 Cell structure in a decanted interface. Bruce Chalmers, courtesy of S. Chalmers.

Structure of Cells Formed at Cellular Growth

The structure of the cells has been studied from three points of view

- cell size
- direction of cellular growth
- shape of cell surface.

Cell Size

Quantitative studies of the cell size indicate that the size *decreases* with *increasing temperature gradient* of the melt and *increasing growth rate* (Figure 8.41).

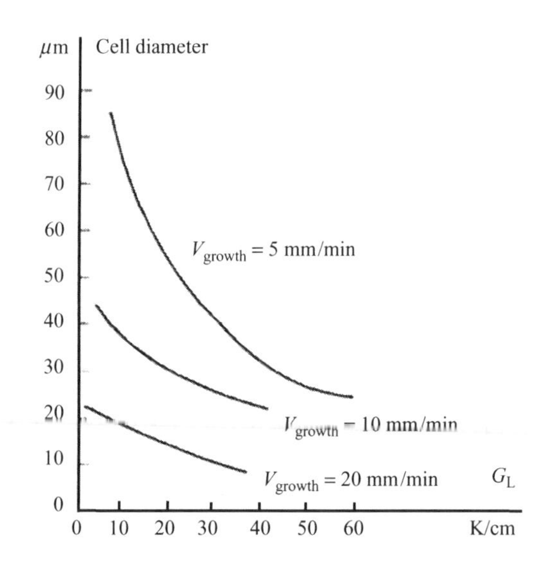

Figure 8.41 Experimental determination of cell diameter as a function of the temperature gradient for two-dimensional zinc cells (after Bocek et al. 1958]). The growth rate of the cells is a parameter. Reproduced from Springer © 1958.

Bolling and Tiller [9] have suggested that the cell size may be determined by the distance the solute atoms can diffuse laterally before they become trapped by the solidification front. Very rough agreement with experimental measurements by Bocek, Kratochvil and Valoual [10] support their theory.

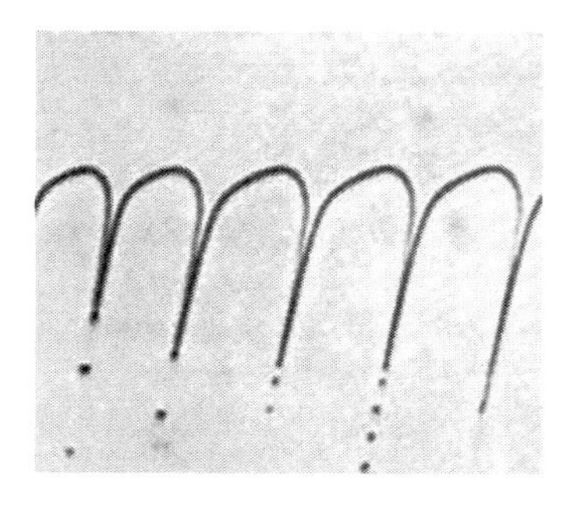

Figure 8.42 Cells with oblique growth direction.

Direction of Cellular Growth

Experimental studies, mainly on tin and lead, show that the direction of cell growth is not always perpendicular to the interface (Figure 8.42). Cell walls have been found, which deviate from the growth direction towards the nearest 'dendrite direction'. The deviation angle depends on the growth rate, the solute concentration and the inclination between the growth direction and the 'dendrite direction'. The deviation is always less than half of this angle.

The deviation angle decreases with decreasing growth rate and decreasing solute concentration.

So far there is no satisfactory quantitative explanation of the direction of cellular growth. It is likely to be associated with anisotropy of the growth rate in analogy with the phenomenon of dendritic growth, which will be discussed in Chapter 9.

Figure 8.43 Typical microphotograph of a decanted cellular interface after solidification. Bruce Chalmers, courtesy of S. Chalmers.

Shape of Cell Surface

Figure 8.43 shows the structure of a cellular interface after rapid decantation of the melt. A thin film of melt is left on the solid, which solidifies and reveals a terraced structure, which was built up during the crystal growth.

Figure 8.44a Observed cell structure in decanted interface. Bruce Chalmers, courtesy of S. Chalmers.

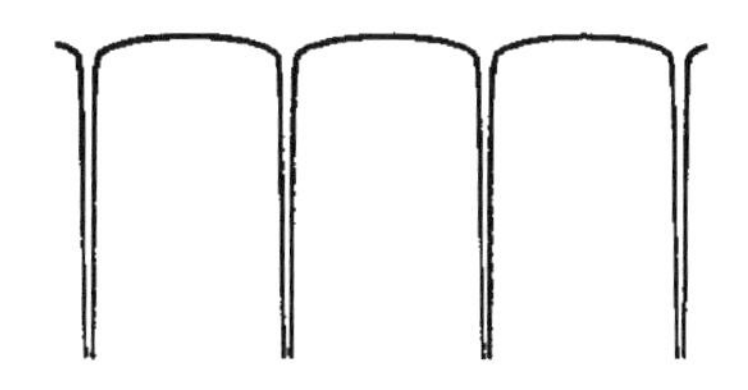

Figure 8.44b Presumed shape of cells during growth. Bruce Chalmers, courtesy of S. Chalmers.

Microphotographs of decanted interfaces indicate a surface profile such as that in Figure 8.44a. The interface between the cells is shallow and comparatively wide. Under the surface the cells may be separated by relatively deep grooves such as those sketched in Figure 8.44b.

The evidence in favour of the model in Figure 8.44b is

- Irregular pores (not shown in the figure) often appear in the cell walls. Such pores are shrinkage pores. They are formed when residual melt in the grooves solidifies.

Summary

■ Heat and Mass Transport during Unidirectional Solidification

Heat and mass transports during solidification of a melt or liquid control the crystallization process and are vital for the quality of the crystals formed.

Distribution of Solute during Solidification

Stationary state: $x^{\mathrm{L}} = \dfrac{x^{\mathrm{s}}}{k_0} = x_0^{\mathrm{L}}\left(1 + \dfrac{1-k_0}{k_0}\mathrm{e}^{-\frac{V_{\mathrm{growth}}}{D_{\mathrm{L}}}y}\right)$

End effects

The initial and final transients are treated in the text.

Influence of Natural Convection on Crystal Growth in Alloys

To simplify the mathematical treatment an artificial separation of the diffusion process and the convection process is made.

- Inside the boundary layer δ the solute flow depends on diffusion only
- Outside the boundary layer δ convection gives complete mixing in the melt.

Inside the boundary layer: $\rho_L \neq \rho_s$

$$x^L = x^s\left(\frac{\rho_L}{\rho_s} + \frac{1 - k_0\frac{\rho_L}{\rho_s}}{k_0}e^{-\frac{V_{growth}}{D_L}y}\right) \quad y \leq \delta$$

Inside the boundary layer: $\rho_L \approx \rho_s$

$$x^L = x^s\left(1 + \frac{1 - k_0}{k_0}e^{-\frac{V_{growth}}{D_L}y}\right) \quad y \leq \delta$$

Outside the boundary layer: $\rho_L \neq \rho_s$

$$\overline{x^L} \approx \left(x^L\right)_{y=\delta} = x^s\left(\frac{\rho_L}{\rho_s} + \frac{1 - k_0\frac{\rho_L}{\rho_s}}{k_0}e^{-\frac{V_{growth}}{D_L}\delta}\right) \quad y \geq \delta$$

Outside the boundary layer: $\rho_L \approx \rho_s$

$$\left(x^L\right)_{y=\delta} = x^s\left(1 + \frac{1 - k_0}{k_0}e^{-\frac{V_{growth}}{D_L}\delta}\right) \quad y \geq \delta$$

Effective Partition Coefficient

Definition: $k_{eff} = \dfrac{x^s}{\overline{x^L}}$

$$k_{eff} = \frac{k_0}{k_0\frac{\rho_L}{\rho_s} + \left(1 - k_0\frac{\rho_L}{\rho_s}\right)e^{-\frac{V_{growth}}{D_L}\delta}}$$

If $\rho_L \approx \rho_s$:

$$k_{eff} = \frac{x^s}{\overline{x^L}} = \frac{k_0}{\left(k_0 + (1 - k_0)e^{-\frac{V_{growth}}{D_L}\delta}\right)}$$

$$k_0 \leq k_{eff} \leq 1$$

When V_{growth} is small and D_L is large: $k_{eff} \approx k_0$
When V_{growth} is large and D_L is small: $k_{eff} \approx 1$

■ Zone Refining

The aim of zone refining is to reduce impurities from a solid crystal as much as possible by monitoring a molten zone along the crystal. The impurities (solute) accumulate in the molten zone ahead of the solidification front. When the zone advances, the impurities remain in the liquid phase and are carried to the end of the crystal.

Convection is a necessary condition for zone refining.

Solute concentration in the solid rod after zone refining:

$$x^s = x_0^L \left[1 - (1 - k_{eff}) e^{-\frac{k_{eff}}{L_{zone}} y} \right]$$

Influence of Microsegregation on Zone Refining
Scheil's equation when convection can be neglected:

$$x^L = \frac{x^s}{k_0} = x_0^L (1 - f)^{-(1-k_0)} \quad 0 \leq f \leq 1$$

Scheil's equation when convection is considered:

$$x^L = \frac{x^s}{k_{eff}} = x_0^L (1 - f)^{-(1-k_{eff})} \quad 0 \leq f \leq 1$$

If k_{eff} is close to 1 the microsegregation will be small in the rod after zone refining ($x^s \approx x_0^L$ independent of y).

Optimal Conditions for Zone Refining

- strong convection in the melt
- relatively short zone length, related to the strength of convection in the molten zone
- low pulling rate of the molten zone
- high diffusion rate of the solute in the melt.

Mechanical and Thermal Stability of the Molten Zone
Length of molten zone:

$$L < 2\pi r$$

It is also important to choose the effect of the heater in such a way that the phase boundary surface becomes planar.

Multiple Zone Refining. Continuous Zone Refining
The result of zone refining is improved by increasing the number of zone refining steps.
The smaller the effective partition coefficient is the better will be the result.
Continuous zone refining can be used when a great number of zone refining steps is required (large value of k_{eff}). The zone refining goes on a certain time and is interrupted when the desired purity of the specimen has been achieved.

■ Single Crystal Production by Czochralski Technique

Crystals of good quality can be produced by the Czochralski process.
A rotating seed crystal in contact with a melt in a rotating crucible at thermal equilibrium is raised slowly under controlled conditions and a crystal of desired length is grown. A stable flow pattern can be obtained if the crystal and the crucible rotate in the same direction.
The method is very important in production of semiconductor circuits and components.

Maximum Growth Rate

$$V_{growth} \, \rho_s(-\Delta H) = \lambda_s \frac{dT}{dy}$$

The maximum growth rate is of the magnitude mm/h.

The slower the growth rate is the larger crystals can be grown.

Striation

Striation is a striped pattern in crystals, which consists of repeated impurity depositions at the interface. It is caused by local variations in composition during growth and irregularities caused by temperature fluctuations.

Several methods are available to reduce the formation of striation. One method is to use an inner floating crucible connected with an outer crucible by a narrow channel.

Melt of the composition x_0^L is continuously soaked into the inner crucible. The crystal, which is pulled out of the melt, also has the composition x_0^L.

The method has the advantage that microsegregation is avoided.

■ Cellular Growth

Cellular growth is a crystal defect, which appears owing to instability of a planar solidification front. The instability is caused by constitutional undercooling.

The cellular structure depends on the magnitude of the constitutional undercooling. Small undercooling gives a pox structure. Medium undercooling results in a stripe pattern. Strong undercooling gives a cell structure, which reminds of a honeycomb.

Constitutional Undercooling in Binary Alloys

See Figure 8.34 on page 458.

The equilibrium temperature of the melt at equilibrium with the solid can be written as

$$T_L = T_M + m_L x^L = T_M + m_L x_0^L \left(1 + \frac{1 - k_0}{k_0} e^{-\frac{V_{growth}}{D_L} y}\right)$$

x^L is the solute concentration in the melt as a function of position ahead of the solidification front which moves with a constant growth rate.

T_L as a function of y is a curve that starts at the origin, increases and approaches a constant asymptotic value.

Condition for Constitutional Undercooling:

The true temperature close to the interface does *not* correspond to equilibrium.

Actual temperature close to the interface solid/liquid can approximately be written with the aid of the temperature gradient G_L:

$$T = T_M + m_L \frac{x_0^L}{k_0} + G_L y$$

Condition for constitutional undercooling:

$$G_L \leq \left(\frac{dT_L}{dy}\right)_{y=0} \quad \text{or} \quad \frac{G_L}{V_{growth}} \leq \frac{m_L}{D_L} \frac{1 - k_0}{k_0} x_0^L$$

Interface Stability Analysis. Perturbation Theory

By studying the rate of development of a sinusoidal perturbation of wavelength λ at a constitutionally undercooled interface information of the stability of the interface can be obtained.

If $\lambda > \lambda_i$ the perturbation increases with time and the interface is instable.

If $\lambda < \lambda_i$ the perturbation decreases with time and the interface is stable.

Formulas of the relative perturbation growth rate and the critical wavelength λ_i are given in the text.

Constitutional Undercooling and Surface Morphology
Cellular structure instead of planar growth occur in case of constitutional undercooling, i.e. if the ratio fulfills the condition

$$\frac{G_L}{V_{growth}} \leq \frac{m_L}{D_L} \frac{1-k_0}{k_0} x_0^L$$

The size of the cells *decreases* with *increasing temperature gradient* in the melt and *increasing growth rate.*
The cell walls form deep grooves during growth.
The direction of cell growth is not always perpendicular to the interface.
The reason for this is assumed to be associated with anisotropy of the growth rate.

Exercises

8.1 In semiconductor industry materials free from impurities are of vital importance. Suppose that you have a bar of germanium of 10 cm length and have to reduce its concentration of an impurity by a factor 2. You need a bar of 5 cm length of the purified material.

Examine with the aid of the figure below if this is possible by use of zone refining in an apparatus where the width of the molten zone is 1.0 cm. The partition coefficient of the impurity is $k = 0.70$. How many cycles are required and what will be the impurity concentration at the beginning of the bar after the zone refining?

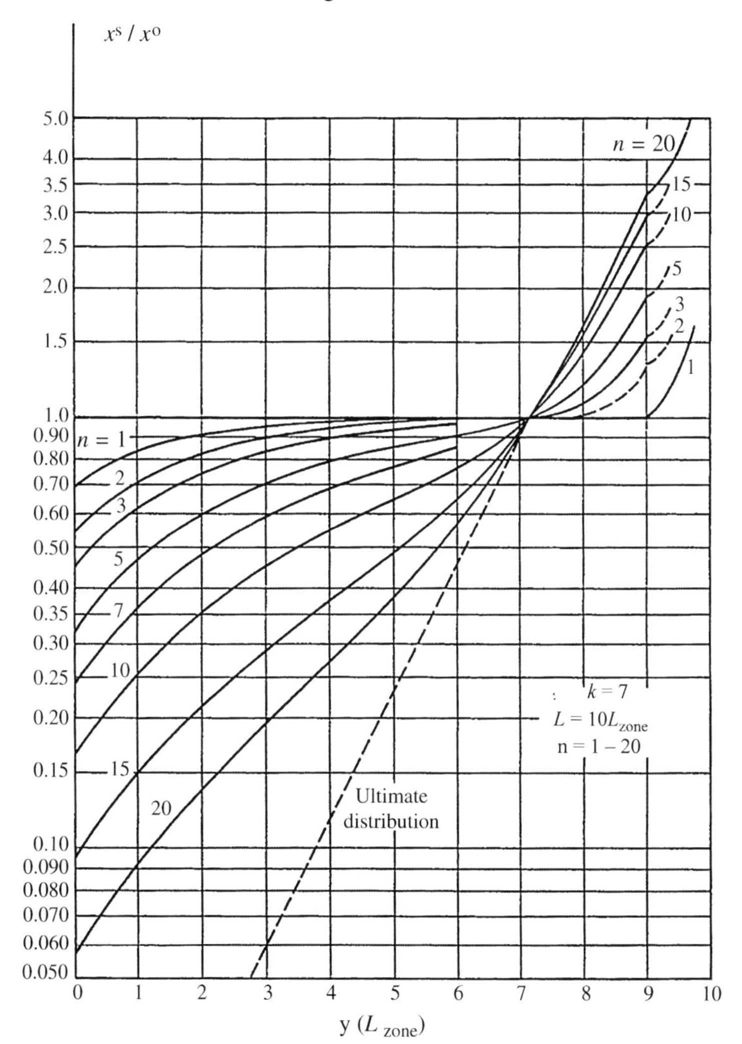

Figure 8.1–1

8.2 Zone refining is described by the figure below.

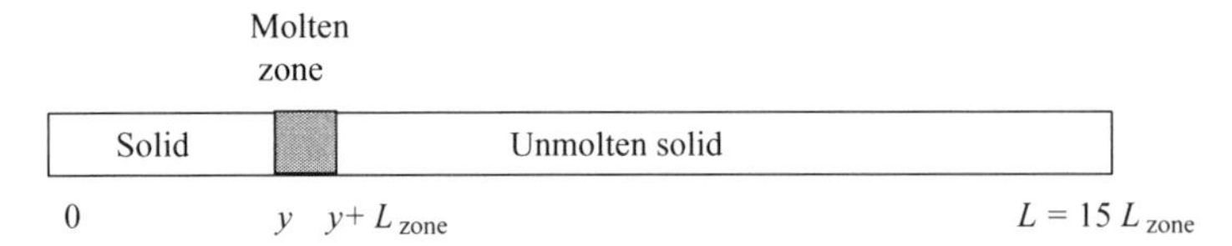

Figure 8.2–1

a) Calculate the composition distribution in a bar made of an Al 0.5at%Cu alloy after passage of one zone and show the function graphically. The length L of the bar is 15 cm and $L/L_{zone} = 15$ where L_{zone} is the width of the molten zone. The convection is strong during the zone refining.

b) Discuss the influence of convection on the Cu concentration distribution. The phase diagram of the system Al-Cu is given below.

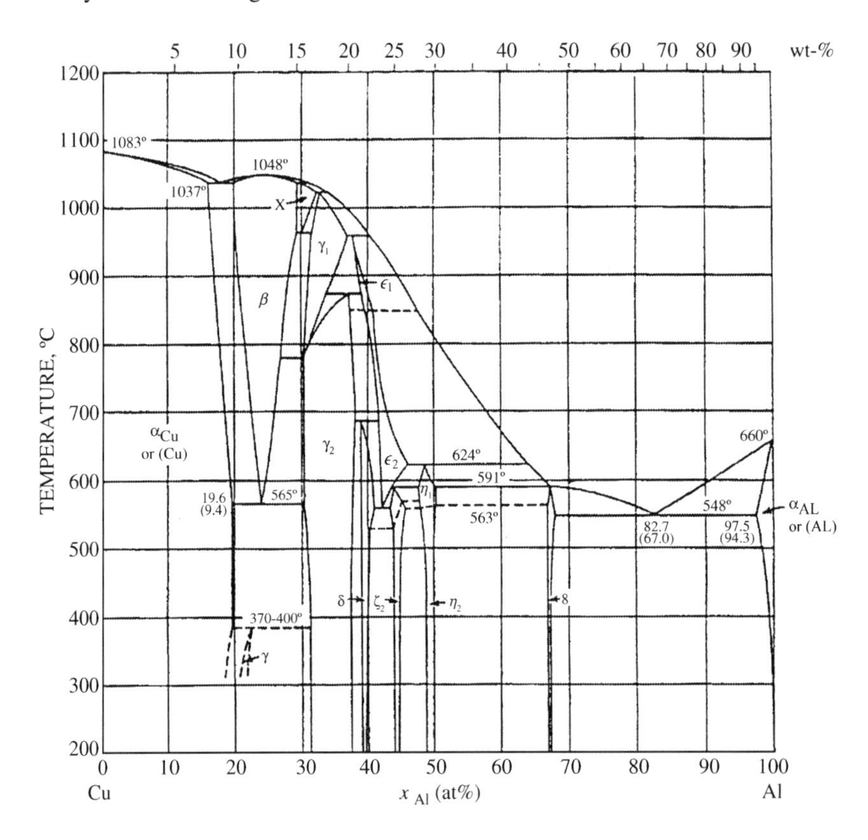

Figure 8.2–2

8.3 Zone melting of a rod of an alloy with the partition coefficient $k = 0.10$ is to be performed. The initial impurity concentration before zone melting is 0.50 at%. The initial width of the molten zone is 0.50 cm. After zone melting of a distance of 10 cm the width of the molten zone is suddenly increased to twice its original value.

Calculate the concentration distribution of the impurity in the rod after the zone melting.

8.4 Liquid metal films or zones can be carried along a solid material in a temperature gradient. This motion causes dissolution of solid metal and reprecipitation of solid metal in the temperature gradient. The effect is shown by the following experiment.

A slice of eutectic Al-Cu composition is placed and connected to two pure Al rods by welding. The specimen is placed in a furnace with a temperature gradient of 20 °C/mm. The experimental conditions are sketched in Figure 8.4–1 below, which shows the temperature of the Al rods as functions of the distance from the left end, $y = 0$.

$D_L = 1.0 \times 10^{-9}$ m^2/s. Part of the phase diagram of Al-Cu is given in Figure 2.

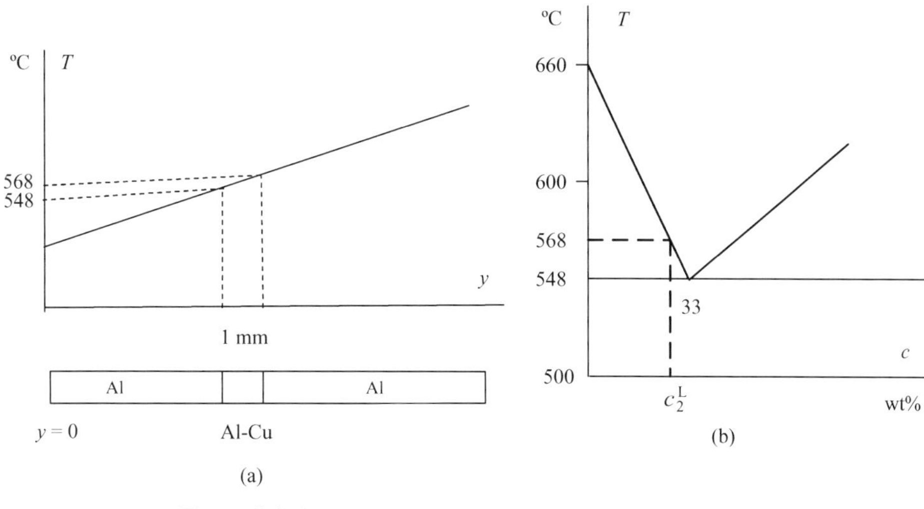

Figure 8.4–1 **Figure 8.4–2**

a) Show schematically the composition of the specimen at the start of the experiment.
b) Estimate the velocities of the two L/s boundaries at the start of the experiment.
c) Explain why the hot L/s boundary moves faster than the cold one.

8.5 The figure below shows the phase diagram of the system A–B. You have a melt of an alloy AB with the composition x_B and let it cool with a constant cooling rate of 10 °C/min. When the liquidus temperature has been reached α crystals are nucleated and grow during the continued cooling.

Material constants:
diffusion constant of B in the melt $D_L = 1.0 \times 10^{-9}$ m²/s
partition coefficient $k = 0.10$. (The solidus line is not drawn in the figure.)

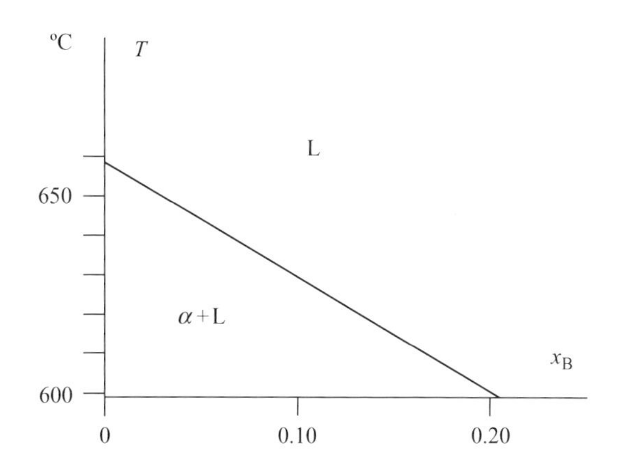

Figure 8.5–1

Calculate the size (radius) of the spherical crystal as a function of time and show the function graphically. You may assume that the growth is spherically symmetric and that the removal of solidification heat is so rapid that it does not influence the crystal growth. Neglect also the influence of the surface tension.

8.6 A Ge-Ga ingot, which contains 10 ppm Ga, solidifies at a growth rate of 8.0×10^{-3} cm/s with negligible convection. The diffusion coefficient in the melt is $D_L = 5.0 \times 10^{-5}$ cm²/s and the partition coefficient $k = 0.10$.

a) Calculate the distribution of gallium in the melt ahead of the solidification front.
b) The rate of solidification is then raised abruptly by a factor of 5 and kept constant at this new growth rate. Sketch what will happen with the composition of the solid phase.

8.7 When an Sn-Pb alloy solidifies there may be a transition from a planar solidification front to an unstable front, depending on the experimental conditions. Figure 1 below shows that the transition depends both on the composition of the alloy and on the ratio of the temperature gradient G and the growth rate.

Calculate the diffusion coefficient with the aid of Figure 1 and the phase diagram of the Sn-Pb system. The molar heat of fusion of lead = 7.00×10^6 J/kmol.

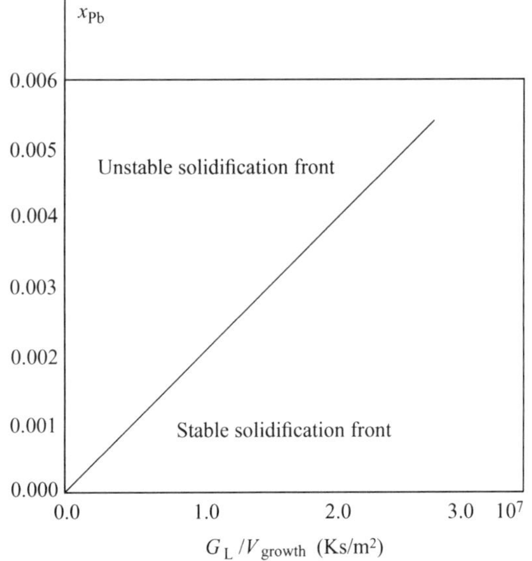

Figure 8.7–1 Mapping of a stable and an unstable solidification front, respectively, of a number of Sn–Pb alloys. The line in the figure represents the border line between the unstable and stable solidification front regions.

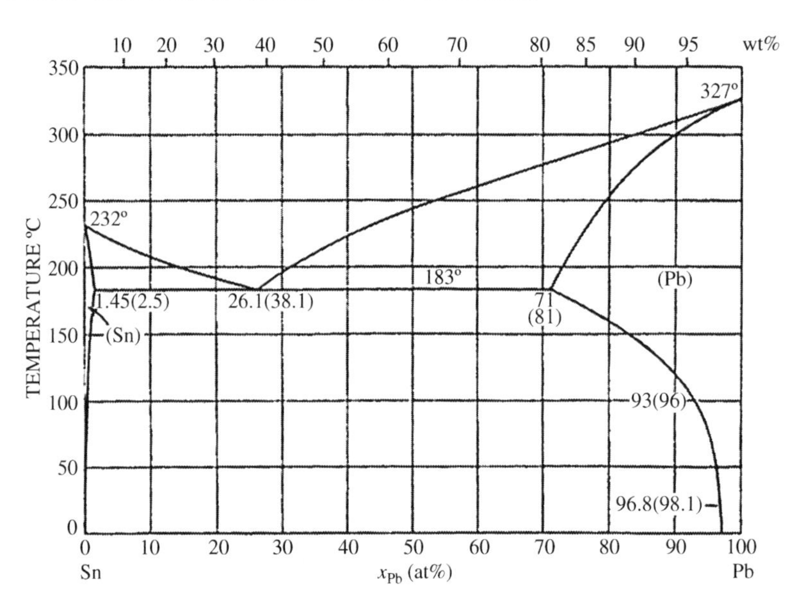

Figure 8.7–2 Phase diagram of the Sn–Pb system.

8.8 Consider a solidifying alloy melt. The solidification front is planar and moves forward with the constant growth rate $V_{growth.}$ With the aid of the phase diagram for the alloy components (Figure 8.29 on page 455 in Chapter 8) the equilibrium temperature in the melt close to the solid/liquid interface can be written as [Equation (8.28) on page 455 in Chapter 8]

$$T_L = T_M + m_L x^L \tag{1}$$

where x^L is the equilibrium concentration of the solute in the melt close to the solid/liquid interface. m_L is the slope of the liquidus line. The function is shown in Figure 8.8–1.

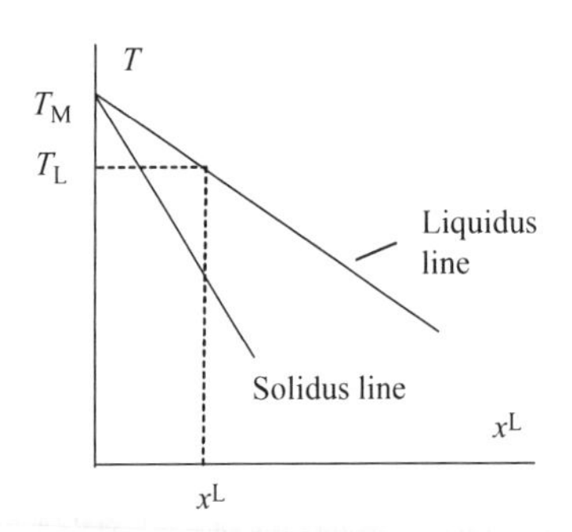

Figure 8.8–1 Schematic phase diagram of a binary alloy.

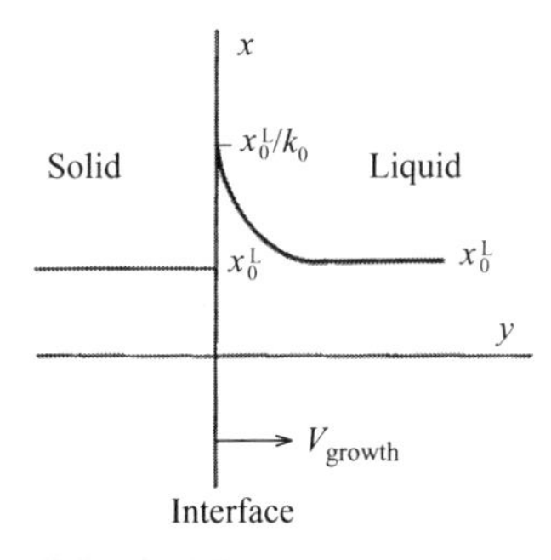

Figure 8.8–2 Solute concentration as a function of distance from the interface in a solidifying alloy.

According to Equation (8.1) on page 435 in Chapter 8 the solute concentration in the melt ahead of the moving planar interface can be written as

$$x^{\mathrm{L}} = x_0^{\mathrm{L}}\left(1 + \frac{1-k_0}{k_0}\mathrm{e}^{-\frac{V_{\mathrm{growth}}}{D_{\mathrm{L}}}y}\right) \tag{2}$$

where k_0 is the partition constant in the melt close to the interface and y is the distance to the interface. The function is shown in Figure 8.8–2 (Figure 8.31 on page 456 in Chapter 8).

The corresponding temperature distribution in the melt ahead of the interface is obtained by replacing x^{L} in Equation (1) by the expression in Equation (2).

$$T_{\mathrm{L}} = T_{\mathrm{M}} + m_{\mathrm{L}}x_0^{\mathrm{L}}\left(1 + \frac{1-k_0}{k_0}\mathrm{e}^{-\frac{V_{\mathrm{growth}}}{D_{\mathrm{L}}}y}\right) \tag{3}$$

Equation (3) represents the equilibrium temperature T_{L} as a function of the distance y from the interface in the melt. It is shown graphically in Figure 3 (Figure 8.32 on page 457).

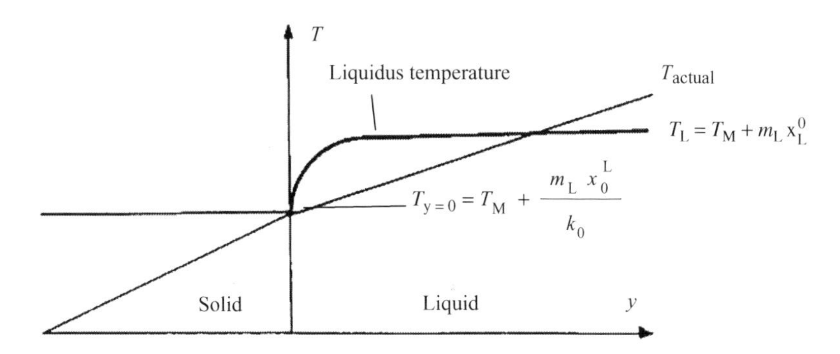

Figure 8.8–3 Equilibrium temperature and actual temperature in the melt as functions of distance from the interface.

Equation (3) and Figure 3 show that the equilibrium curve approaches asymptotically the line

$$T_{\mathrm{L}} = T_{\mathrm{M}} + m_{\mathrm{L}}x_0^{\mathrm{L}} \tag{4}$$

for large values of y.

Inserting $y = 0$ into Equation (3) gives

$$T = T_{\mathrm{M}} + \frac{m_{\mathrm{L}}x_0^{\mathrm{L}}}{k_0} \tag{5}$$

However, the actual temperature of the melt is determined by the experimental conditions and can be solved from the general heat equation and a number of boundary conditions. In the vicinity of the interface ($y = 0$) the actual temperature can be approximated by the line in Figure (3). The slope of the line is the temperature gradient at $y = 0$

$$T = G_{\mathrm{L}} = \left(\frac{\mathrm{d}T}{\mathrm{d}y}\right)_{y=0} \tag{6}$$

According to Figure 3 and Equations (4′) and (5′) the equation for the actual temperature line can be written as

$$T = T_{\mathrm{M}} + m_{\mathrm{L}}\frac{x_0^{\mathrm{L}}}{k_0} + G_{\mathrm{L}}y \tag{7}$$

a) In the region where the actual temperature is smaller than the equilibrium temperature T_L you have so-called constitutional cooling. One condition for the stability of the planar solidification front is to avoid constitutional cooling.

Derive the condition for avoiding constitutional cooling in terms of G_L as a function of the slope m_L, the partition coefficient k_0, the diffusion constant D_L, the solute concentration far from the interface and the growth rate V_{growth}.

b) Calculate G_L for the alloy Al 0.5 at%Cu at the growth rate 1.0×10^{-2} mm/s.
Material constants for the alloy are:

$$D_L = 2 \times 10^{-9}\ \mathrm{m^2/s}$$

$$k_0 = 0.145$$

$$m_L = -636\ \mathrm{K/mol\ fraction.}$$

References

1. W. A. Tiller, K. A. Jackson, J. W. Rutter, B. Chalmers, *Acta Metallurgica*, **1**, 1953, 428.
2. H. Fredriksson, U. Åkerlind, *Materials Processing during Casting*, Chichester, England, John Wiley & Sons, 2005.
3. J. A. Burton, R. C. Prim, W. P. Slichter, *Journal of Chemical Physics*, **21**, 1953, 1987.
4. J. W. Rutter, B. Chalmers, *Canadian Journal of Physics*, **31**, 1951, 15.
5. W. W. Mullins, R. F. Sekerka, *Journal of Applied Physics*, **34**, 1963, 323.
6. W. A. Tiller, J. W. Rutter, *Canadian Journal of Physics*, **34**, 1956, 96.
7. J. W. Rutter, in: *Liquid metals and solidification*, American Society for Metals, 1958, 250.
8. B. Chalmers, *Principles of Solidification*, John Wiley & Sons, 1964.
9. G. F. Bolling, W. A. Tiller, *Journal of Applied Physics*, **31**, 1960, 2040.
10. M. Bocek, P. Kratochvil, M. Valoual. Czech. *Journal of Physics*, **8**, 1958, 557.

9

Faceted and Dendritic Solidification Structures

Solidification and Crystallization Processing in Metals and Alloys, First Edition. Hasse Fredriksson and Ulla Åkerlind.

9.1 Introduction

Crystal structures of solids are usual in nature. Common examples are the regular and beautiful patterns of snow flakes and minerals of the earth crust.

There are several types of crystal structures. Two of them will be treated in this chapter. The snow flakes represent *dendrite structures* with their branched structure.

The minerals appear in various kinds of *faceted crystal structures*, i.e. structures with planar surfaces, which form various angles with each other. Flashing examples are big, cut and polished diamonds and other precious stones in jewels. The subgroups of faceted crystal structures have been discussed in Chapter 1 of the book 'Physics of Functional Materials' [1]. There the nomenclature of planes and directions in crystals are defined (pages 15–16 in Section 1.4.3 in [1]). This basic knowledge will frequently be applied in the present chapter. A repetition is highly recommended in [1] or other book on crystal structure.

Metals and alloys also have crystalline structures. Many times the crystals are small and below the visibility limit of the human eye. By use of scanning electron microscopes, it is nowadays possible to catch a true glimpse of the hidden and complex structures of metals and alloys.

Faceted crystals and faceted growth in metals and alloys are discussed in Sections 9.2–9.4 in this chapter. Dendrite crystals and dendritic growth are treated in Sections 9.5 and 9.6. Faceted and dendritic growth in metals and alloys are extensively discussed in this chapter together with the resulting structures of such growth.

Many different morphologies of crystals, grown in melts and vapours, have been observed. The most common types of crystals formed in metal melts are *dendrites*, composed of branches, and *faceted crystals*, which can be cubic, octahedral, columnar or plate-shaped. Transitions between dendrite and faceted structures have also been observed and are discussed in Section 9.7.

9.2 Formation of Faceted Crystals

Faceted crystals form in saturated solutions or, in the case of metals and alloys, in melts or from vapour phases. The shape of the crystal depends on the lowest driving force to form the crystal.

- A faceted crystal always forms in such a shape that is given by the lowest driving force of formation.

The driving force of a faceted crystal consists of several parts caused by

- the surface tension between its facets and the melt
- the kinetics, i.e. the growth rate at the solidification front as a function of the undercooling of the melt
- the mass transport, i.e. the diffusion of atoms in the melt.

These parts will be discussed below for pure metals and alloys separately. The last part is important in alloys and refers to diffusion of the alloying element in the melt. In Section 9.4.4 simple theory of faceted growth in binary alloys with cubic structure will be discussed.

9.3 Growth of Faceted Crystals in Pure Metal Melts

As an introduction, we will summarize the results from Chapter 3, which are valid for faceted crystals and apply them on pure metals.

During the whole chapter crystal directions and crystal planes are presented. In these cases the international nomenclature is used, which is as follow: General direction <xxx>, general plane {xxx}, special direction [xxx] and special plane (xxx).

9.3.1 Growth Directions of Faceted Crystals in Pure Metal Melts

The shapes of faceted crystals have been analyzed in Section 3.5 in Chapter 3 and several examples of faceted crystals have been given above. Their structures can all be explained by a few criterions, which are valid for faceted growth.

- The equilibrium shape of a crystal of constant volume is that at which Gibbs' free energy of the crystal has a minimum.

  This stability condition is called the 'Gibbs–Curie–Wulff's theorem. It has been derived in Section 3.5.1 on pages 145–149 in Chapter 3.

$$\frac{\sigma_1}{h_1} = \frac{\sigma_2}{h_2} = \frac{\sigma_3}{h_3} = \cdots = \frac{\sigma_o}{h_o} \tag{9.1}$$

where
σ_i = surface energy per unit area of crystal surface i
h_i = distance from the centre of the crystal to the crystal surface i.

- The shape of a faceted crystal is determined by the growth of the surfaces, which have the lowest growth rate.

This last rule represents a very useful criterion for qualitative explanations of crystal growth.

The growth rate of the crystal is strongly anisotropic, i.e. it varies with the crystallographic direction. This means that the crystallographic planes with the lowest growth rate form the crystal surfaces.

If the ratios of the growth rates in different crystallographic directions remain constant the shape of the crystal will be unchanged (Example 9.1). Otherwise, the crystal will change its shape when it grows. Such a case is shown below on page 479.

The rule above has been applied in Figure 5.4 on page 203 in Chapter 5, reproduced here as Figure 9.1. The growth rate is faster in the [110] direction than in the [010] and [100] directions. The surface with the fastest growth disappears and is no stable surface of the crystal.

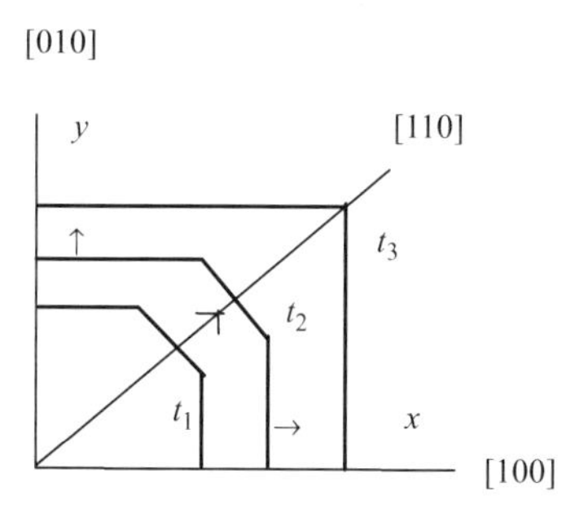

Figure 9.1 Crystal surfaces of part of a crystal at three different times. $t_1 < t_2 < t_3$. The surface with the fastest growth disappears.

Example 9.1

Calculate a relationship between the growth rates V_{100} and V_{111} in an a) FCC crystal and b) BCC crystal. In both cases the crystal keeps its shape when it grows.

Solution:

a) FCC crystal.

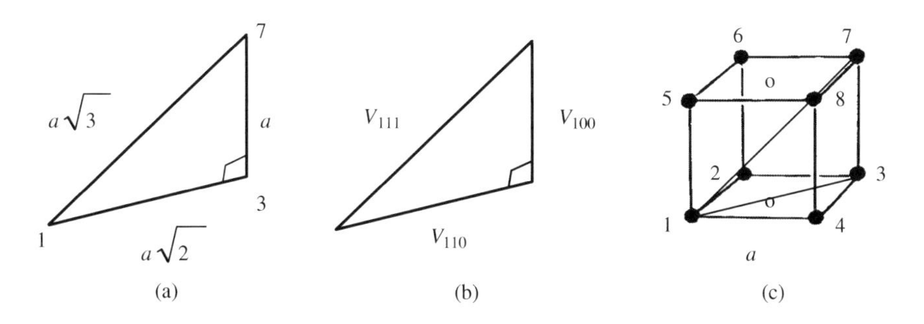

For simplicity, the atoms at the centres of the four vertical surfaces of the FCC cube have been omitted in this figure.

The sides of the right-angle triangle in figure (a) are the side of the cube, the bottom diagonal and the principal diagonal of the cube.

The middle figure (b) shows the growth rates of the planes (100), (110) and (111) i.e. V_{100}, V_{110} and V_{111}. They form a triangle, that has the same angles as the triangle in figure (a). Hence, we obtain

$$\frac{a\sqrt{3}}{a} = \frac{V_{111}}{V_{100}}$$

which gives

$$V_{111} = V_{100}\sqrt{3}$$

b) BCC crystal.

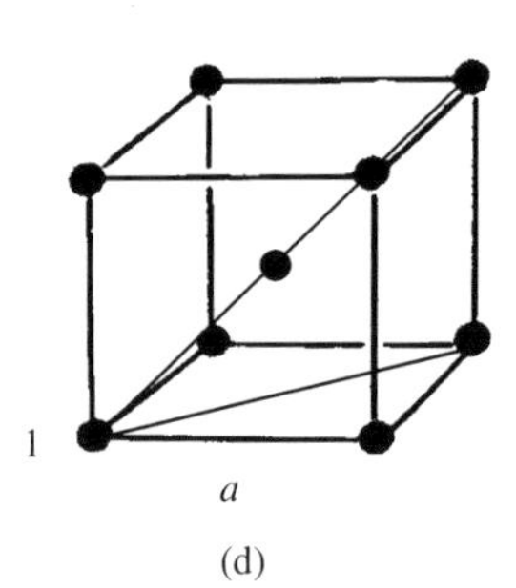

(d)

The atom in the centre of the cube makes no difference. The solution will be exactly the same as that in a).

Answer:
In both cases the relationship is $V_{111} = V_{100}\sqrt{3}$.

9.3.2 *Influence of Interface Kinetics on Faceted Crystals in Pure Metal Melts*

The concept 'interface kinetics' refers to the kinetics of a reaction, that occurs at the interface between two phases. In the present case the reaction is the solidification process at the faces of faceted crystals.

In Chapters 5 and 6 we have penetrated the interface kinetics of the solidification process. We found that the interface structure and the type of crystal surface influence the reaction (growth) rate. A consequence of this is that the interface reaction is anisotropic and that the growth rate varies in different crystallographic directions (Figure 9.1 on page 477). Each metal has its own particular structure and crystallographic properties.

A case of anisotropy is shown in Figure 9.2. The figure shows a cubic crystal where one corner is cut. Obviously the roughness of the interface depends on the crystallographic orientation.

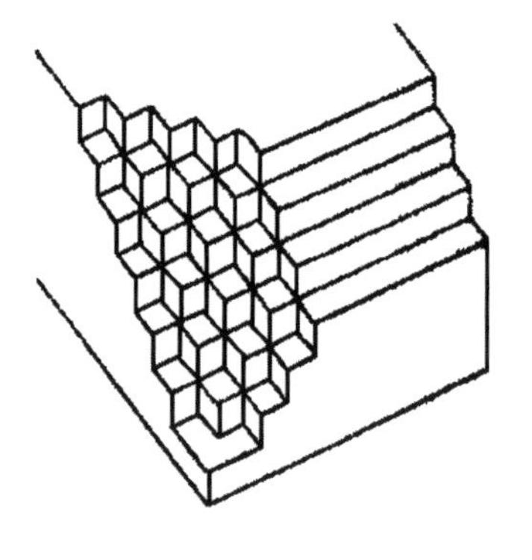

Figure 9.2 Interface roughness as a function of orientation.

The growth rate of a crystal surface, determined by an interface reaction, can be described by the following basic relationship Equation (6.40) on page 295 in Chapter 6).

$$V_{\text{growth}} = \frac{dr_{\text{hkl}}}{dt} = \beta_{\text{kin}}\left(T_{\text{M}} - T_{\text{growth}}\right)^2 \tag{9.2}$$

where
V_{growth} = growth rate of the surface
r_{hkl} = distance from the midpoint of a faceted surface [crystal plane (hkl)] to the centre of the crystal
β_{kin} = growth constant of the facet
T_{M} = melting point of the pure metal melt
T_{growth} = temperature of the melt at the solidification front
$T_{\text{M}} - T_{\text{growth}}$ = undercooling.

The growth constant β_{kin} has different values in different crystallographic directions. The faces with the smallest growth constant determine the growth morphology (page 477).

Growth of Faceted Cubic Crystals. Cases 1, 2 and 3

The growth of cubic crystals is anisotropic, which means that the growth rate varies with the crystallographic direction. Only the faceted surfaces or simply facets will survive which corresponds to the lowest growth rate.

In cubic crystals all other growth directions have normally much higher growth rates than those in the directions $\langle 100 \rangle$ and $\langle 111 \rangle$. For this reason the only facets, which normally occur in such crystals, are $\{100\}$ and $\{111\}$ planes.

The ratio of the growth rates in the $\langle 100 \rangle$ and $\langle 111 \rangle$ directions determines the shape of the crystal when it grows. Three cases are possible.

Case 1: $V_{111} = V_{100}\sqrt{3}$

Analogously a comparison between V_{111} and the resultant of three V_{100} vectors in perpendicular directions gives

$$V_{111} = V_{100}\sqrt{3} \tag{9.3}$$

Example 9.1 on page 477 shows that Equation (9.3) is the condition for constant shape. If this condition is fulfilled for a cubic crystal, it will keep its shape during growth (Figure 9.3). This case occurs seldom or never in practice.

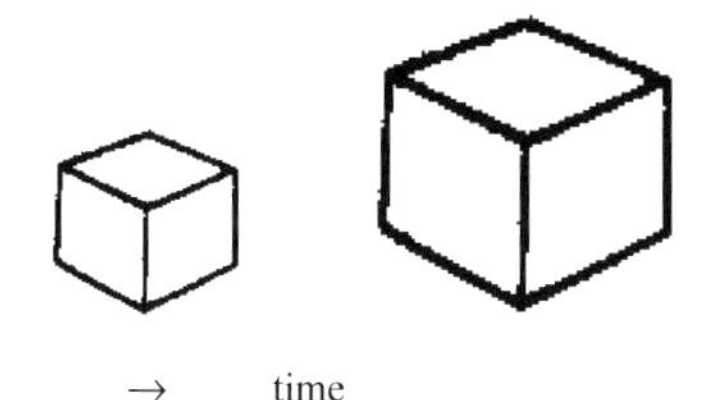

Figure 9.3 Growth of cubic crystals when $V_{111} = V_{100}\sqrt{3}$.

Case 2: $V_{111} < V_{100}\sqrt{3}$

$$V_{111} < V_{100}\sqrt{3} \tag{9.4}$$

If the condition in Equation (9.4) is fulfilled for a cubic crystal, it will change its shape during growth. V_{111} is not large enough to maintain the cubic shape. Figure 9.4 shows the result of a series of calculations, which give the shape of the crystal as a function of time or increasing size. At the calculations, we assume that $\beta_{111} < \beta_{100}$ where β is the growth constant of a facet.

In Figure 9.4 the $\{111\}$ surfaces gradually grow at the expense of the $\{100\}$ facets which finally disappear completely. Then the crystal will be bounded by merely $\{111\}$ facets and form an octahedron if the solidification process is not interrupted in advance.

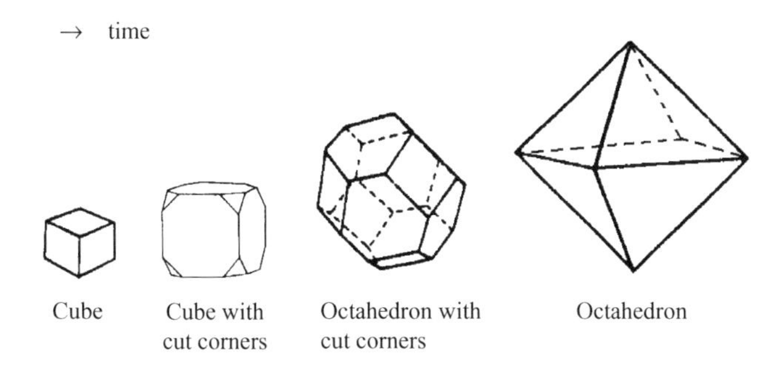

Figure 9.4 Change of crystal shape with time and increasing size for a cubic crystal. when $V_{111} < V_{100}\sqrt{3}$. Reproduced from Elsevier © 1974.

This result can also be understood in terms of the first criterion on page 477. Consider a small cubic crystal with the property $V_{111} < V_{100}\sqrt{3}$. At each size, the crystal shape of lowest possible total energy is most probable. This is the reason for the successive change of shape with increasing size. The shape of the final crystals is a regular octahedron that is bounded by $\{111\}$ facets only.

Case 3: $V_{111} > V_{100}\sqrt{3}$

$$V_{111} > V_{100}\sqrt{3} \tag{9.5}$$

Consider a small cubic crystal of a pure metal, which grows in a melt of the same metal. The relationship $V_{111} > V_{100}\sqrt{3}$ is valid for the crystals of the metal. The {100} facets grow gradually and the {111} facets never appear. The shape for all the intermediate crystals is a cube, bounded by {100} facets only. This optimal shape is not always achieved in reality, though.

→ time

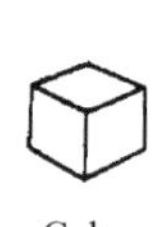

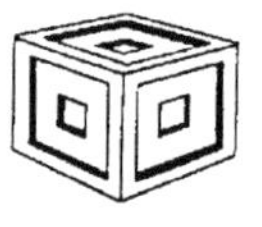

Cube | Cube | Hopper crystal

Figure 9.5 Change of crystal shape with time and increasing size for a cubic crystal when $V_{111} > V_{100}\sqrt{3}$.

For cubic crystals and other types of crystals, for example octahedrons larger than a certain size, a new phenomenon appears which is indicated in Figure 9.5. The surface of such crystals is no longer planar but contains steps at the corners and along the edges. Such crystals are called *hopper crystals* and will be discussed more closely on pages 484–485.

An example of a case 3 metal is bismuth. Hopper crystals have been observed for Bi, which has diamond structure (see pages 484–485).

The three cases are summarized in Table 9.1.

Table 9.1 Influence of interface kinetics on growth of cubic crystals

Condition	$V_{111} = V_{100}\sqrt{3}$	$V_{111} < V_{100}\sqrt{3}$	$V_{111} > V_{100}\sqrt{3}$
Final shape	Cube	Octahedron	Cubic hopper crystal

It should be noticed that we have only considered the kinetics above. When diffusion also is taken into account, the intermediate shape in Figure 9.4 on page 479, i.e. an octahedron with the corners cut, may be the final and most stable shape. This topic will be discussed in Section 9.4.4.

Growth of Faceted Octahedron Crystals. Cases I, II and III

Case I: $V_{100} = V_{111}\sqrt{3}$

$$V_{111} = V_{100}\sqrt{3} \tag{9.3}$$

Equation (9.3) represents the condition for unchanged growth of a cube. We want to find the corresponding relationship between V_{111} and V_{100} which has to be fulfilled for growth of the octahedron with unchanged shape.

The growth rates V_{111} and V_{100} of the crystal are proportional to the distances OM and OA, respectively, in Figure 9.6. The point M is the centre of mass of the equilateral octahedron facet (Figure 9.6). Simple geometrical calculations confirm that the condition for growth of the octahedron crystal with unchanged shape is

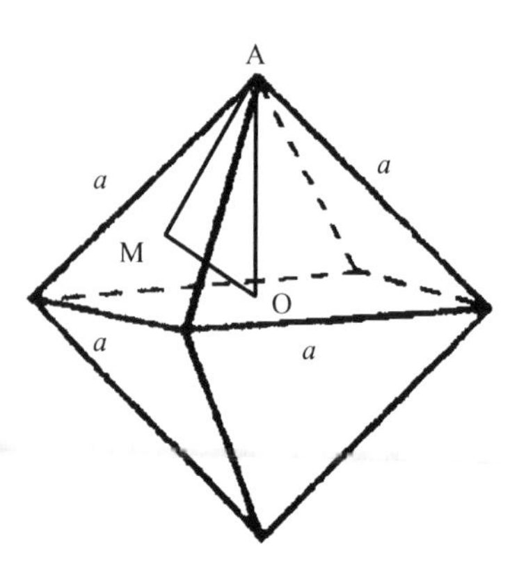

Figure 9.6 Octahedron.

$$V_{100} = V_{111}\sqrt{3} \tag{9.6}$$

The Equations (9.3) and (9.6) are reversed, i.e. the subscripts 111 and 100 exchange positions.

This relationship is equal to the reversed condition, valid for growth of a cubic crystal with unchanged shape.

Case II: $V_{100} < V_{111}\sqrt{3}$

For symmetry reasons there probably exists a simple relationship between V_{100} and V_{111} that is valid for the reversed order of shape change, shown in Figure 9.4, when a small octahedron crystal grows. We guess that the relationship can be written as

$$V_{100} < V_{111}\sqrt{3} \tag{9.7}$$

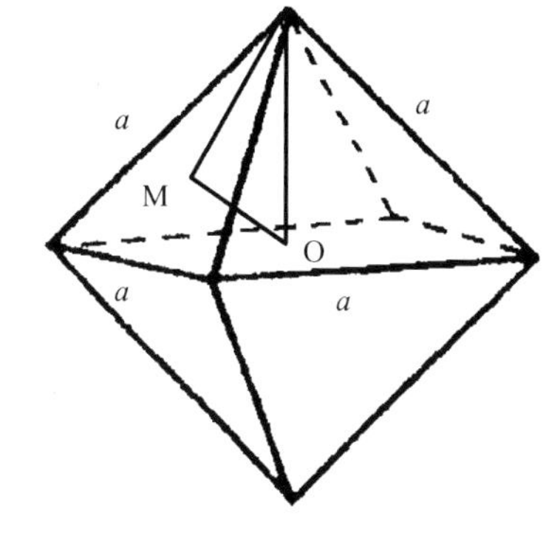

Figure 9.6 Octahedron.

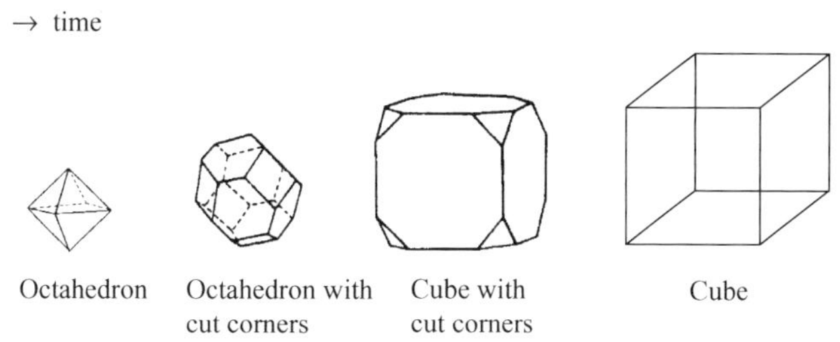

→ time

Octahedron | Octahedron with cut corners | Cube with cut corners | Cube

Figure 9.7 Change of crystal shape with time and increasing size for a cubic crystal. when $V_{100} < V_{111}\sqrt{3}$.

Consider an octahedron crystal where V_{111} is proportional to the distance OM (Figure 9.6). If V_{100} of the crystal is smaller than the growth rate which is proportional to the distance OA, V_{100} is not large enough to maintain the octahedron shape when the crystal grows. {100} facets form at the corners of the octahedron crystal and grow at the expense of the {111} facets which finally disappear completely as is shown in Figure 9.7.

Case III: $V_{100} > V_{111}\sqrt{3}$

There are metals and alloys where the crystals, which nucleate in a melt at solidification, have octahedron structure and the property that

$$V_{100} > V_{111}\sqrt{3} \tag{9.8}$$

when the crystals grow. In analogy with growing cubic structures, which have the property reported in case 3 for cubic crystals, the octahedron crystals do not change their shape but form octahedron hopper crystals.

The corners grow with a growth rate larger than V_{100} in case 1. The growth rate in the <111> direction in the middle of the facets is not enough to maintain the shape of the crystals. The condition (9.8) does not permit formation of {100} facets.

Instead additional layers are built up first of all on the octahedron corners but also along the edges and gradually also on the {111} facets. The reason is that the difference in growth rate is largest at the corners and smallest close to the midpoints of the facets. Figure 9.8 tries to illustrate the bottom of the well, formed at the facets with only two instead of numerous steps. The walls of the well result from layer growth, which has been discussed in Chapter 6.

The three cases are summarized in Table 9.2.

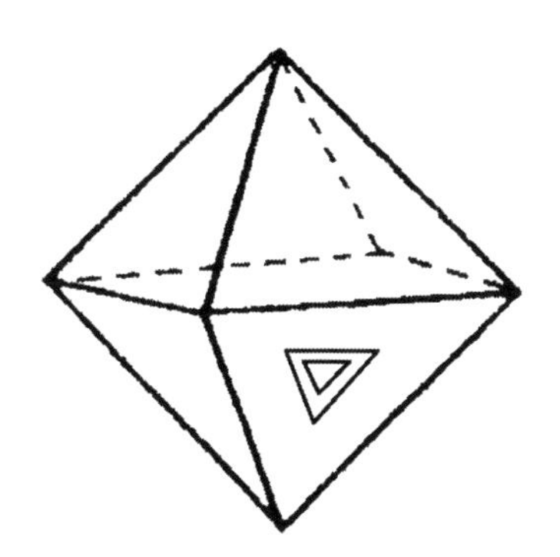

Figure 9.8 Octahedron with a minor well. All the facets have similar wells but they are not drawn in the figure.

Table 9.2 Influence of interface kinetics on growth of octahedron crystals

Condition	$V_{100} = V_{111}\sqrt{3}$	$V_{100} < V_{111}\sqrt{3}$	$V_{100} > V_{111}\sqrt{3}$
Final shape	Octahedron	Cube	Octahedron hopper crystal

The step structure is discussed below in connection with an experimental case of faceted crystal growth in a pure metal melt.

9.3.3 *Step-like Crystal Surfaces. Formation of Hopper Crystals*

Hopper crystals are closely related to step formation and step growth, which have been treated in Chapter 5 on pages 235–237 and in Chapter 6, Section 6.7.4 on pages 299–309. As an introduction and a basis for the discussion of hopper crystals below we will report two microscopic studies of the solid/liquid interface of single crystals of lead.

Step Formation and Step Growth

Elbaum and Chalmers [2] examined the structure of the solid/liquid interface of single crystals and bicrystals (crystals with different crystallographic orientation relative to the growth direction) of Pb in 1956 by rapid decanting of the remaining liquid during growth and use of a light microscope.

They observed a step-like structure in all crystals when either the {111} plane or the {100} plane formed a small angle with the interface, less than a critical angle of 15–20°. When the angle was increased the step pattern became closer and disappeared for angles > the critical angle. Elbaum and Chalmers verified that the steps consisted of small parts of close-packed {111} or {100} planes. The unidirectional solidification was performed at temperature gradients between 1000–4000 K/m and a growth rate between 0.2–10 mm/min.

The measured lengths λ of the steps varied between 0.001–0.01 mm and their heights h were usually less than 0.001 mm.

Rosenberg and Tiller [3] found a similar step-like structure on the growing surfaces of single crystals of ultra-pure zone-refined lead, when the {111} and {100} planes, respectively, formed small angles with the liquid/solid interface. They confirmed Elbaum and Chalmers's angle observations and showed that the steps had a fine structure (Figure 9.9).

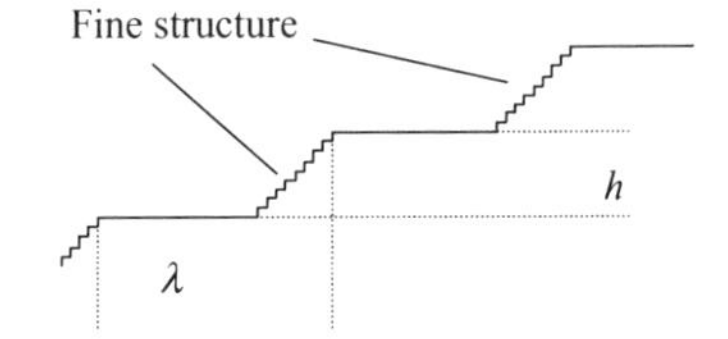

Figure 9.9 Fine structure of a step-like crystal structure.

Example 9.2

At an experiment with unidirectional solidification of an FCC metal, steps were observed on the (111) plane, slightly tilted relative to the growth direction. With the aid of the growth law (Equation (9.2) on page 478), the known undercooling and measurements of the height and length of the steps, the growth constants β_{111} and β_{100} can be calculated.

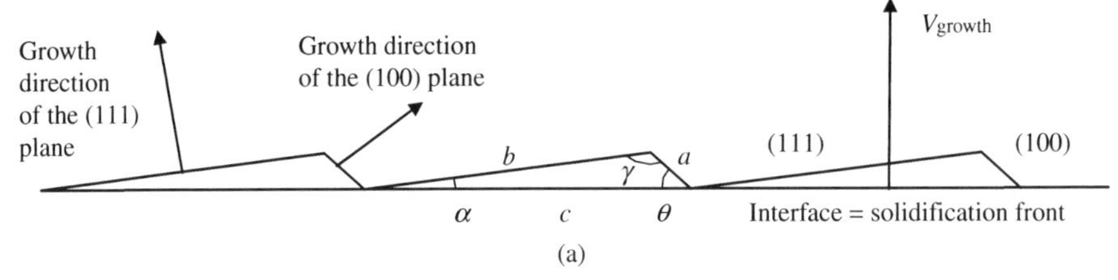

(a)

The angle α between the (111) plane and the interface, i.e. the solidification front, was measured to 10°.

Calculate the growth rate constant $\beta_{\perp}$, perpendicular to the solidification front, in terms of the growth constants β_{111} and β_{100} and other known quantities.

Solution

The growth rate V_{growth} can be calculated from two expressions of the volume of melt which solidifies per unit time.

$$cV_{growth} = a\frac{dz_a}{dt} + b\frac{dz_b}{dt} \tag{1'}$$

where a, b and c are the sides in the triangle in the figure in the text. z is a coordinate in the growth direction. The time derivatives can be calculated with the aid of Equation (9.2) on page 478.

$$\frac{dz_a}{dt} = \beta_{100}(T_M - T_{melt})^2 \tag{2'}$$

$$\frac{dz_b}{dt} = \beta_{111}(T_M - T_{melt})^2 \tag{3'}$$

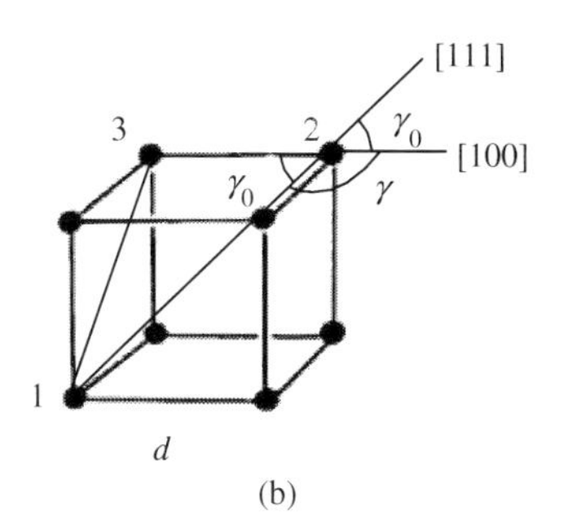

(b)

The triangle sides are not known but it is possible to determine the angles in the triangle. The angle α is given in the text. The angle γ is the obtuse angle between the {111} and {100} planes or between the directions {111} and {100}. The latter angle can easily be found geometrically.

With the aid of the 90° triangle 123 we obtain

$$\tan \gamma_0 = \sqrt{2}$$

which gives $\gamma_0 = 54.7°$ and $\gamma \approx 180° - 55° = 125°$

The angle θ in the figure in the text will be $= 180° - 125° - 10° = 45°$

With the aid of the sinus theorem we obtain the ratios *a/c* and *b/c*:

$$\frac{\sin \alpha}{a} = \frac{\sin \theta}{b} = \frac{\sin \gamma}{c} \tag{4'}$$

or

$$\frac{b}{c} = \frac{\sin \theta}{\sin \gamma} \quad \text{and} \quad \frac{a}{c} = \frac{\sin \alpha}{\sin \gamma} \tag{5'}$$

Substitution of the expressions in Equations (2′), (3′) and (5′) into Equation (1′) gives

$$V_{growth} = \left(\frac{\sin \alpha}{\sin \gamma}\beta_{100} + \frac{\sin \theta}{\sin \gamma}\beta_{111}\right)(T_M - T_{melt})^2 \tag{6'}$$

Inserting the numerical values gives

$$V_{growth} = \left(\frac{\sin 10^o}{\sin 125^o}\beta_{100} + \frac{\sin 45^o}{\sin 125^o}\beta_{111}\right)(T_M - T_{melt})^2$$

Equation (6′) is the growth law for growth perpendicular to the solidification front.

$$V_{growth} = \beta_{\perp}(T_M - T_{melt})^2$$

Answer:
The growth constant is

$$\beta_{\perp} = \frac{\sin \alpha}{\sin \gamma}\beta_{100} + \frac{\sin \theta}{\sin \gamma}\beta_{111} = \frac{\sin 10^o}{\sin 55^o}\beta_{100} + \frac{\sin 45^o}{\sin 55^o}\beta_{111}$$

Step growth is closely related to interface kinetics. The latter is extensively discussed in Chapter 6. We found there that relaxation over the interface is part of the kinetic process. The 'liquid' atoms release energy when their potential energy changes from the value in the liquid down to the lower value in the crystal lattice after solidification.

The relaxation process is slowest in the direction of the normal to the most densely packed crystal planes. In FCC metals, such as lead, the most densely packed planes are in order the {111} and {100} planes. Hence, the steps of such faceted crystals are likely to be built up by {111} and {100} planes.

The height and length relaxation of the steps depend on the relaxation process. The theory of this topic is given in Section 5.6.8 on pages 235–236 in Chapter 5 and in Section 6.7.6 on pages 307–309 in Chapter 6. The steps grow mainly in the ⟨100⟩ directions the growth rate is therefore controlled by the kinetics in these directions. The steps form by heterogeneous nucleation on irregularities at the interface or at the corners of the crystals. Hopper crystals form in this way, which will be discussed below.

Formation of Hopper Crystals

Elbaum and Chalmers' experiments [2] mentioned above show that steps are formed when the {111} and {100} planes, respectively, form small angles with the solid/liquid interface in lead. Only elements, which belong to case 3 (cubic structure)) and case III (octahedron structure) or obey a corresponding condition for other structure types are able to form hopper crystals.

The formation of a hopper crystal can be described as follows.

As case 3 is valid the growth rate of the <111> directions exceeds the growth rate of the <100> directions with more than a factor $\sqrt{3}$. The growth at the corners of the cube will therefore be more rapid than that at the middle of each surface. Steps are formed at the corners and extend along the edges of the cube, which is shown in Figure 9.6. Wells are formed at the centres of the faceted surfaces.

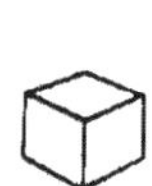
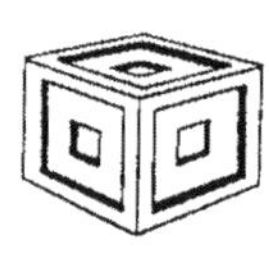
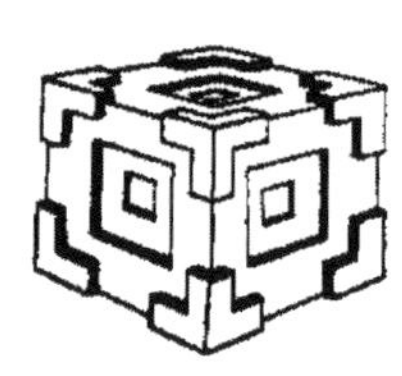

Figure 9.10 Formation of a hopper crystal. $V_{111} > V_{100}\sqrt{3}$.

The number of steps is much larger in reality than Figure 9.10 shows. This is also shown in Figures 9.11 and 9.12.

The Figures 9.11 and 9.12 show two hopper crystals with very deep wells. They have been picked up from a pure bismuth melt during the growth process.

Figure 9.11 Longish bismuth hopper crystal seen from above. The height of the crystal is 3 cm.

Figure 9.12 Well-developed big bismuth hopper crystal.

Figure 9.11 shows mainly one of the facets of the hopper crystal, where a great number of steps have been formed and cover the whole facet except its centre. The step growth has been favoured on and concentrated to this particular facet owing to strong cooling in a direction perpendicular to the facet, which results in a highly elongated crystal instead of a cube. This shape gives the observer the clear impression of a deep well and the crystal is a true but extreme hopper crystal.

Figure 9.12 shows another Bi hopper crystal with a strongly developed step structure and big coarse subcrystals, which cover the edges and corners.

The steps of hopper crystals are primarily formed at the corners because the atoms are more easily incorporated there, owing to a faster relaxation of the excess of potential energy there than elsewhere. The reason is the crystal geometry at the corner and also the crystallographic direction [according to page 480 <111> have the corners the fastest growth rate in the crystals].

The steps grow along primarily at the corners and secondly along the edges and spred finally over the planar facets. The centres have the minimum growth rate and a hole is formed, a so called hopper.

The growth rate is controlled by the kinetics at the rough edges of the steps. The smooth steps grow by layer growth and their lengths are related to the relaxation process of the atoms. The growth of the steps depend on the supersaturation. It is largest at the corners and has minima at the midpoints of the facets at the corners new hoppers starts to form similar to the illustrations shown on the right hand Figure 9.10. This topic will be discussed on pages 501–502.

9.4 Growth of Faceted Crystals in Alloy Melts

9.4.1 Formation of Hopper Crystals in Alloy Melts

The growth of crystals in pure metal melts has been treated in Section 9.3.2 on pages 478–481. The same laws are valid for growth of crystals in alloy melts but with the difference that concentration differences instead of temperature differences control the growth rates.

Hopper crystals in alloys form with the aid of the same mechanism as in pure metals. We discussed the formation of simple hopper crystals in pure metals on pages 480–481 and 484. The hopper effect is more pronounced in alloys than in pure metals because the difference between the growth rates at the corners and at the midpoints of the facets are larger in alloys than in pure metals.

Hopper crystals also appear in other types of crystals than those with cubic and octahedron structures.

As a typical example of growth of faceted crystals in alloys we will discuss the morphology of primary silicon crystals formed in a hyper-eutectic Al–Si alloy melt. Hyper-eutectic means that the Si-concentration exceeds the eutectic Si concentration.

9.4.2 Growth Morphologies in Al-Si Alloys

The study of the growth morphologies of Al-Si alloys, referred below, has been performed by Fredriksson, Hillert and Lange [4]. Primarily formed Si-crystals in hyper-eutectic Al-Si alloys, floating freely in the melt, can be visualized by deep-etching and studied and photographed with the aid of a scanning electron microscope (SEM).

The phase diagram of the Al-Si system is shown in Figure 9.13. The morphologies of the precipitated Si crystals varies with

- the cooling rate
- the nucleation frequency
- small additions of alkali metals such as sodium or strontium.

Silicon crystals have a cubic lattice and the same morphologies as many metallic systems with cubic lattices. Figures 9.14 and 9.15 show SEM pictures of Si crystals formed in an ingot solidified at a cooling rate of roughly 10 K/minute.

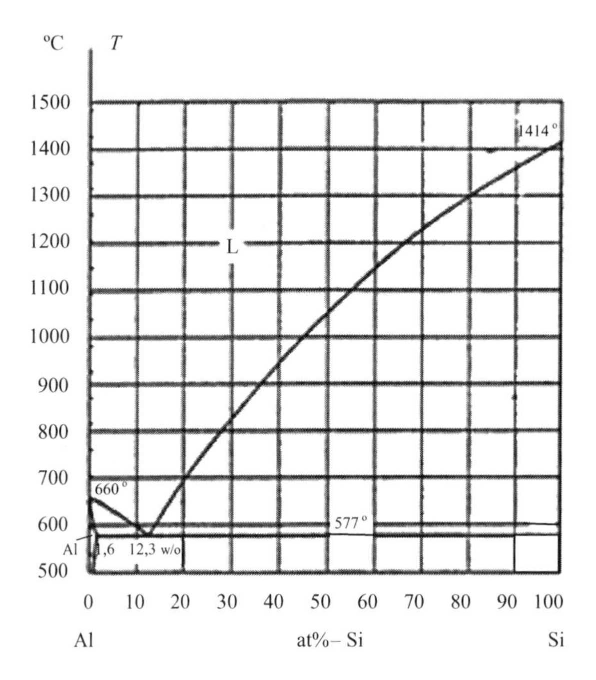

Figure 9.13 Phase diagram of the system Al–Si. From Binary Alloy Phase Diagrams Vol. 1, Ed. Thaddeus B. Massalski, American Society for Metals 1986. Reprinted with permission of ASM International. All rights reserved. www.asminternational.org.

Figure 9.14 Si crystal with nearly ideal octahedral shape formed in an ingot solidified at a cooling rate of roughly 10 K per minute. Reproduced from CRC/Taylor & Francis © 1985.

Figure 9.15 Non-ideal Si crystal formed in an ingot solidified at a cooling rate of roughly 10 K per minute. An embryo of a hopper crystal (case III page X10) is shown in the figure. The corners are elongated. Reproduced from CRC/Taylor & Francis © 1985.

All the crystals in the ingot were formed with {111} facets. The ideal shape is an octahedron. Figure 9.14 shows an example of a nearly perfect crystal. Figure 9.15 shows a case when the crystal deviates from the ideal morphology. Its {111} facets contain holes and the corners are elongated.

A similar sample, modified with a *high* sodium content, gave primary Si-crystals of the type shown in Figure 9.16. The crystals contain both {111} and {100} facets. The {100} facets are dominant.

A sample with a *low* content of sodium gave the morphology shown in Figures 9.17a and 9.17b. The crystals are bounded by {111} and {100} facets. The {111} facets are dominant.

Figure 9.1 and the nearby growth rate law on page 477 can be applied to interpret the pictures.

For Si crystals with no sodium content (Figures 9.14 and 9.15) the growth rate must have a minimum for the {111} facets, which form the octahedron facets.

For the alloy with high sodium content in Figure 9.16 the growth rate is probably lowest for the cube planes or the {100} facets (compare Section 6.7.4 in Chapter 6).

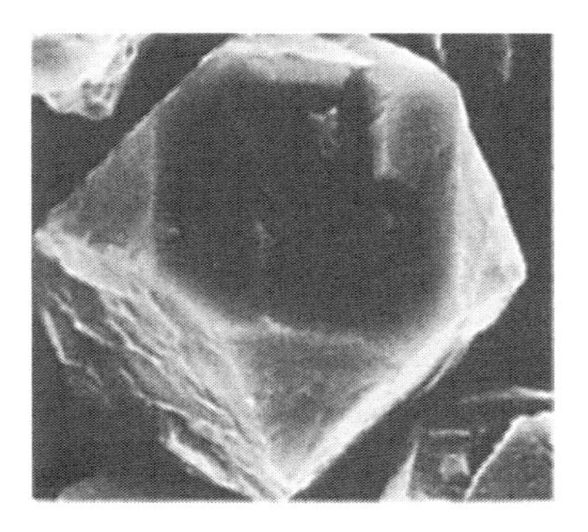

Figure 9.16 Faceted Si crystal with {100} and {111} facets. High sodium content. Reproduced from CRC/Taylor & Francis © 1985.

Figure 9.17a Faceted Si crystal with {100} and {111} facets. Low sodium content. Reproduced from CRC/Taylor & Francis © 1985.

Figure 9.17b Si crystal from the same sample as in Figure 9.17 a. The facets can vary in size (case III). Reproduced from CRC/Taylor & Francis © 1985.

When the crystal in Figure 9.17a grows, the {100} facets will disappear and the {111} facets will dominate. This process is nearly fulfilled for the crystal in Figure 9.17b.

9.4.3 *Model Experiments of Faceted Growth of Transparent Flat Crystals*

The concentration field of the solute around faceted crystals in an alloy melt controls their growth. There are no experimental methods to study the concentration field around a faceted alloy crystal. For this reason it is necessary to perform model experiments with the aid of studies of transparent crystals growing in a saturated solution. Berg [5] and Humphreys-Owen [6] have performed such experiments in order to obtain better understanding of the growth of faceted crystals in solutions. Both of them used optical interference methods to study the variation of the solute concentration around the growing crystals.

We do not refer the details of the experiments but the principle was to study the concentration field around a small, very thin crystal of sodium chlorate surrounded by a drop of a supersaturated solution of a solute. The flat crystal and the solution were surrounded by two parallel semitransparent planar mirrors and exposed to monochromatic green light (the Hg line $\lambda = 546.1$ nm). In a microscope a fringe system was observed in the solution outside the crystal.

The light was reflected towards the mirrors and the optical path was $n \times 2d$ where n is the refractive index of the solution and d is the distance between the parallel semitransparent mirrors. The refractive index depends on the concentration of the solute. The circles around the crystal represent iso-concentration lines. The average solute concentration is known. Knowledge of the refractory index as a function of the solute concentration and the iso-concentration pattern permit calculation of the solute concentration and the concentration gradient as a function of position, i.e. the concentration field.

An example is given in Figure 9.18, which originates from an experiment of the type mentioned above The figure shows that the circles are practically equidistant, which means that the concentration gradient is approximately constant.

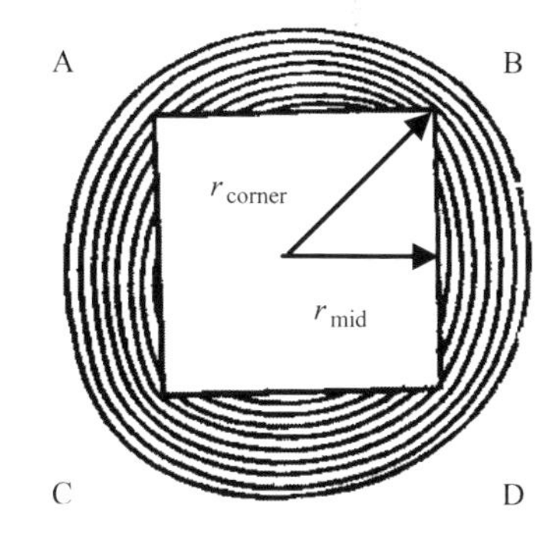

Figure 9.18 Isoconcentration lines in the solution around a flat square crystal. Reproduced with permission from Royal Society of London © 1999.

9.4.4 *Simple Theory of Faceted Growth in Binary Alloys with Cubic Structure*

Influence of Interface Kinetics, Diffusion and Surface Tension on Faceted Growth

In Section 9.3 we pointed out that the formation of faceted crystals is determined by the interface reaction. This is true only if the growth is not limited by the rate of transport of matter (alloys) or heat (pure metals) in the surrounding matrix. In order to analyze the general growth process in alloys one has to consider

- the interface reaction (kinetics)
- the diffusion process of the alloying element in the melt along and perpendicular to the facets of growing crystals
- the effect of surface tension on the growth rate.

Influence of Interface Kinetics on Faceted Growth in Alloy Melts

In Section 9.3.2 we discussed the influence of interface kinetics on faceted growth in pure metals. In case of alloys the influence is analogous but there is a difference. In pure metal melts the growth is controlled by a *temperature difference*. In the case of alloys the growth is controlled by a *concentration difference* of the alloying element (Equations (6.95) and (6.96) on page 317 in Chapter 6).

$$V_{\text{growth}} = \left(\frac{\mathrm{d}r_{\text{mid}}}{\mathrm{d}t}\right)_{\text{kin}} = \beta_{\text{kin}}\left(x_{\text{eq}}^{\text{L}/\alpha} - x_{\text{mid}}^{\text{L}/\alpha}\right)^2 \tag{9.9}$$

where

V_{growth} = growth rate of the surface
r_{mid} = distance from the midpoint of a faceted surface [crystal plane (hkl)] to the centre of the crystal
β_{kin} = growth constant of the facet
$x_{\text{eq}}^{\text{L}/\alpha}$ = equilibrium concentration of solute in the liquid at the interface at a given temperature (see Figures 9.19a and b)
$x_{\text{mid}}^{\text{L}/\alpha}$ = solute concentration of liquid at the midpoint of a facet (hkl) at distance r_{mid} from the centre.

Influence of Diffusion on Faceted Growth in Alloy Melts

The concentration of the alloying element has minima at the midpoints and maxima at the corners if $k > 1$. The driving force of the interface kinetics is proportional to the supersaturation at the surface, which controls the growth process. The supersaturation is smallest at the midpoints and largest at the corners. The opposite is true for the driving force of the diffusion process.

The influence of the diffusion process will be extensively discussed below.

Influence of Surface Tension on Faceted Growth in Alloy Melts

Nucleation of crystals depends on the surface tension between the crystal and the melt. This topic has been treated in Chapter 4. The influence of surface tension in combination with diffusion of the solute towards the facets of the growing crystals has been discussed in Chapter 6.

Anisotropy of the surface tension influences the shape of the crystal morphology during solidification. However, for small faceted crystals one has found that the diffusion field is nearly spherical (Section 9.4.3). For large crystals the effect of the surface tension can approximately be neglected ($r^* \ll r_c$, see page 331 in Chapter 6).

Driving Force of Solidification at Faceted Growth

Crystals grow when melt close to the facets solidifies. The driving force of the kinetic process at the interface is closely related to the concentration distribution of the alloying element in the melt close to the facet.

We will start with the concentration field around a flat faceted metal crystal with two square facets. The results of the model experiments in Section 9.4.3 can be applied.

It is hard to make an exact calculation of the concentration field of the solute around a faceted crystal in a melt. For this reason we have to introduce several approximations which will be listed on page 492.

Driving Force of Solidification

Below we will disregard the influence of surface tension on the driving force and only consider the effect of interface kinetics and diffusion.

From Equation (6.72) on page 313 in Chapter 6 we can conclude that, for dilute solutions with a small solubility in the solid phase, the total driving force of solidification is proportional to the deviation of the given concentration of the solute in the melt from equilibrium value of the solute at the interface, i.e. the concentration difference $|x_0^{\mathrm{L}} - x_{\mathrm{eq}}^{\mathrm{L}/\alpha}|$. The equilibrium value can be found with the aid of the phase diagram of the alloy system when x_0^{L} is known. As is shown in Figures 9.19a and b we have to distinguish between two cases, depending on the value of the partition coefficient k.

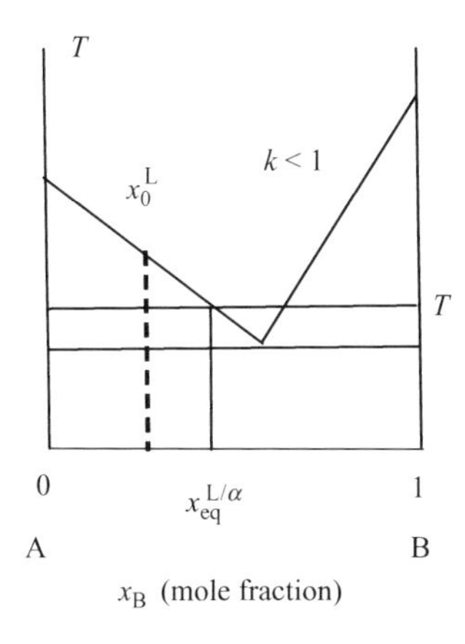

Figure 9.19a Phase diagram of the system AB.$x_{\mathrm{eq}}^{\mathrm{L}/\alpha}$ = equilibrium concentration of solute B close to the facet at temperature T when the partition coefficient $k > 1$.

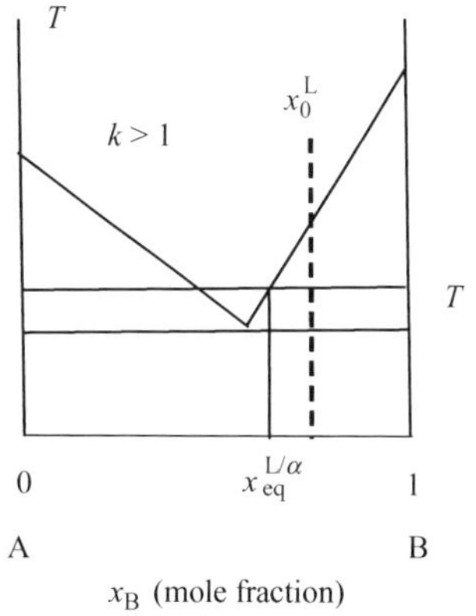

Figure 9.19b Phase diagram of the system AB $x_{\mathrm{eq}}^{\mathrm{L}/\alpha}$ = equilibrium concentration of solute B close to the facet at temperature T when the partition coefficient $k < 1$.

The driving force is always a positive quantity, which is a necessary condition for the occurrence of a process. For this reason we have to separate the two cases.

For alloys with $k > 1$ the average solute concentration x_0^{L} is larger than $x_{\mathrm{eq}}^{\mathrm{L}}$ and the driving force can be written as

$$\text{Driving force} = const\left(x_0^{\mathrm{L}} - x_{\mathrm{eq}}^{\mathrm{L}/\alpha}\right) = const\left[\underbrace{\left(x_0^{\mathrm{L}} - x_{\mathrm{kin}}^{\mathrm{L}/\alpha}\right)}_{\text{diffusion process}} + \underbrace{\left(x_{\mathrm{kin}}^{\mathrm{L}/\alpha} - x_{\mathrm{eq}}^{\mathrm{L}/\alpha}\right)}_{\text{kinetic process}}\right] \tag{9.10a}$$

where

$const$ = constant, equal RT at small concentrations

x_0^{L} = solute concentration in the melt far from the facet

$x_{\mathrm{eq}}^{\mathrm{L}/\alpha}$ = solute concentration in the melt close to the interface at equilibrium (see Figures 9.19a and b).

$x_{\mathrm{kin}}^{\mathrm{L}/\alpha}$ = concentration of solute in the melt close to the facet, required for the kinetic process.

For alloys with $k < 1$ the average solute concentration $x_{\mathrm{eq}}^{\mathrm{L}/\alpha}$ is larger than x_0^{L} and the driving force can be written as

$$\text{Driving force} = const\left(x_{\mathrm{eq}}^{\mathrm{L}/\alpha} - x_0^{\mathrm{L}}\right) = const\left[\underbrace{\left(x_{\mathrm{eq}}^{\mathrm{L}/\alpha} - x_{\mathrm{kin}}^{\mathrm{L}/\alpha}\right)}_{\text{kinetic process}} + \underbrace{\left(x_{\mathrm{kin}}^{\mathrm{L}/\alpha} - x_0^{\mathrm{L}}\right)}_{\text{diffusion process}}\right] \tag{9.10b}$$

Solute Distribution in the Melt Close to the Interface

The concentration profile of the solute in the melt in front of the facet is shown in Figures 9.20a and b.

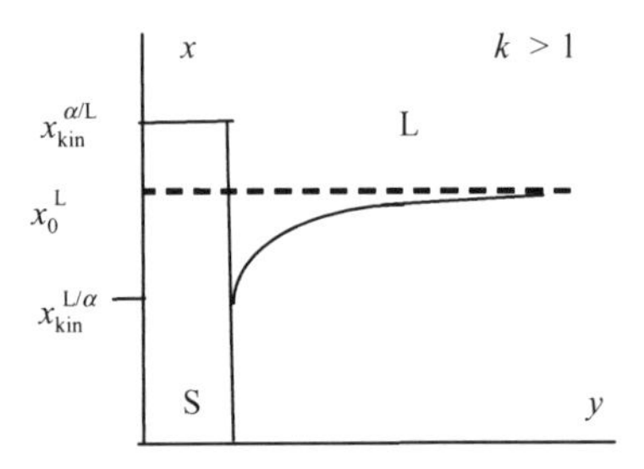

Figure 9.20a Concentration of solute in the melt as a function of distance y ahead of the interface if $k > 1$.

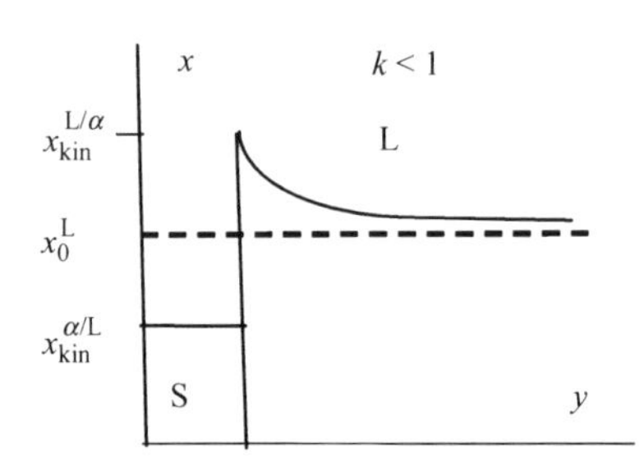

Figure 9.20b Concentration of solute in the melt as a function of distance y ahead of the interface if $k < 1$.

Consider a case when the partition coefficient > 1 and apply Equation (9.10a) and the curves in Figures 9.18a and 9.19a on a vertical facet of the thin square crystal (left figure). From the curve in the right figure you can read the concentration differences, which are proportional to the driving forces of the kinetic reaction and the horizontal diffusion at the middle of the facet and the upper and lower corners, respectively.

With the aid of Figures 9.21a and b we can find the concentrations required for the kinetic reaction at the middle of the facet and at the upper corner.

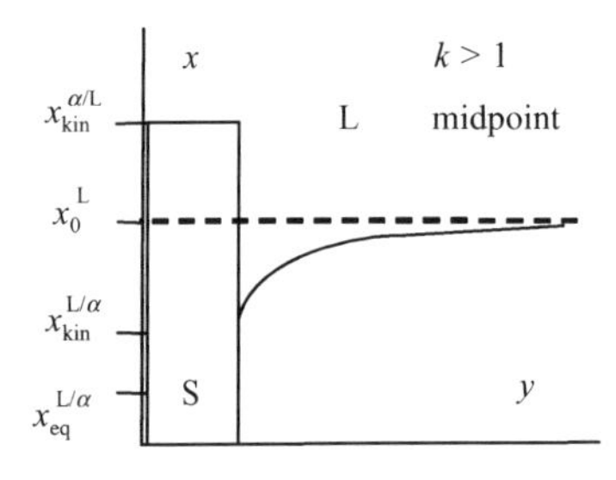

Figure 9.21a Application of Figure 9.20a on the middle of the facet when k $>$ 1. The concentration difference $\left(x_{kin}^{L/\alpha} - x_{eq}^{L/\alpha}\right)$ shows that the driving force for the kinetic reaction is comparatively small.

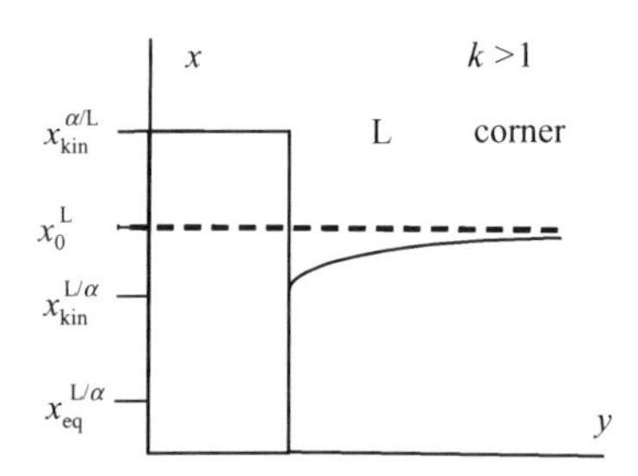

Figure 9.21b Application of Figure 9.20a on the upper corner of the facet when k $>$ 1. The concentration difference $\left(x_{kin}^{L/\alpha} - x_{eq}^{L/\alpha}\right)$ shows that the driving force for the kinetic reaction at the corner is much larger than that at the midpoint of the facet.

The concentration difference, required for the kinetic reaction, are derived in Figures 9.21a and b. A comparison between them shows that the driving force for the kinetic reaction, $\left(x_{kin}^{L/\alpha} - x_{eq}^{L/\alpha}\right)$, is much larger at the corner than at the midpoint of the facet. The result is a much faster growth at the corner than at the midpoint of the facet. when $k > 1$.

According to Equation (9.10a) the sum of the concentration differences for the kinetic reaction and the horizontal diffusion equals $\left(x_0^L - x_{eq}^{L/\alpha}\right)$. Hence, the driving force for diffusion is much larger at the midpoint of the facet than that at the corner when $k > 1$.

The opposite is true when $k < 1$.

Solute Distribution and Solute Gradient of a Facet

All the concentrations mentioned above are plotted in Figure 9.22a and the concentration differences can easily be derived. when $k > 1$.

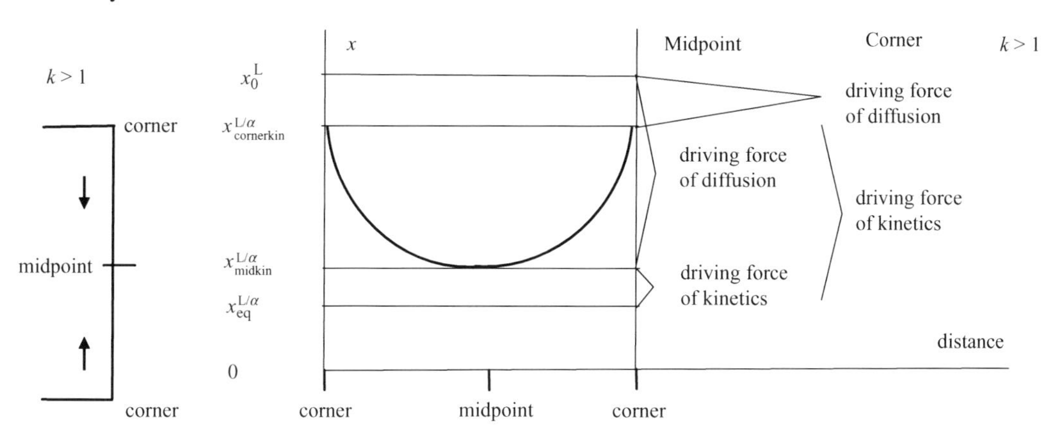

Figure 9.22a Review of concentrations at the middle of a facet and at the corners, valid for $k > 1$. The parabolic curve represents the solute concentration along the crystal interface. The concentration differences, required for the kinetic reaction and for horizontal diffusion towards the vertical surface can be derived from the figure.

Figure 9.22a shows that the driving force for diffusion of the alloying element towards the facet surface is *small* at the corner and *large* at the midpoint of the facet.

Observations of growing crystals show that the facets keep their planar shape. This can be explained by the existence of vertical diffusion along the solid facet surface. As the concentration $x^{\alpha/L}_{kin}$ is much larger at the corner than at the midpoint for the case when $k > 1$, the horizontal mass transport in the melt is smaller at the midpoint than at the corners and the vertical mass transport of solute atoms is directed from the corners towards the midpoint of the facet. The direction of the vertical diffusion has been marked by arrows to the left in Figure 9.22a.

If $k < 1$ the opposite is true and the vertical mass transport is directed from the midpoint towards the corners, Figure 9.22b is analogous to Figure 9.22a and it is valid for $k < 1$.

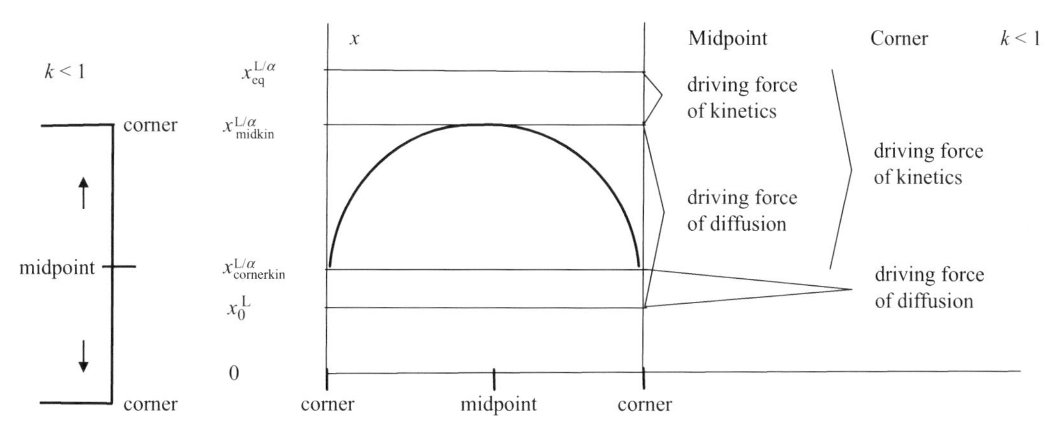

Figure 9.22b Review of concentrations at the middle of a facet and at the corners, valid for $k < 1$. The parabolic curve represents the solute concentration along the crystal interface. The concentration differences, required for the kinetic reaction and for horizontal diffusion towards the vertical surface can be derived from the figure.

If the growth rates perpendicular to the facet at the middle of a facet and at the corners are equal (i.e. $V_{corner}/V_{facet} = \sqrt{3}$), the planar growth of the facets is maintained. If this condition is not valid hopper crystals may develop. Hence, it is urgent to find expressions for the growth rates and compare them in order to understand the formation of hopper crystals better.

Concentration Field around a Faceted Cubic Crystal. Growth Rate of Cubic Crystals

In order to examine the net effect of mass transport and interface kinetics on crystal growth, a treatment based on the theory of diffusion growth of a spherical particle, given in Chapter 6, will be used.

It has been pointed out on page 479 in this chapter and on page 203 in Chapter 5 that a faceted crystal is bounded by its slowest growing facets. If the growth process were completely interface-controlled, Equation (9.9) on page 488 would describe the size and shape of the crystal. Otherwise the diffusion process, i.e. mass transport and the effect of surface tension also influence the growth process and makes it much more complicated than a growth process entirely controlled by the kinetics.

It is hard to make an exact calculation of the concentration field of the solute around a faceted crystal in a melt. For this reason we have to introduce several approximations.

1. The iso-concentration surfaces around the crystals are assumed to be spherical, which gives a spherically symmetrical concentration field.
2. The facets of the crystals are assumed to be perfectly planar.
3. The interface kinetic theory of solidification with respect to relaxation given in Chapter 6, can be applied.
4. The effect of surface tension on the growth, treated in Chapter 6, can be applied. However, the calculations will be simpler if the surface tension can be neglected. The influence of surface tension is larger for small crystals than for large ones (> 1 μm).

In Chapter 6 we analyzed the influences on the solidification process in four stages. The alternatives were: diffusion only, diffusion with influence of the surface tension, diffusion with influence of the

Table 9.3 Growth rate of midpoint of a facet as a function of r_{mid}

	Equation in Chapter 6	Equation
Stage 1. Diffusion only (from Fick's second law)		
$V_{\text{growth}} = \dfrac{dr_{\text{mid}}}{dt} = D\dfrac{x_{\text{eq}}^{L/\alpha} - x_0^{L}}{x_{\text{eq}}^{L/\alpha} - x_{\text{eq}}^{\alpha/L}}\dfrac{V_{\text{m}}^{\alpha}}{V_{\text{m}}^{L}}\dfrac{1}{r_{\text{mid}}}$	(6.138) on page 328	(9.11)
Stage 2. Diffusion with attention to surface tension		
$V_{\text{growth}} = \dfrac{dr_{\text{mid}}}{dt} = D\dfrac{x_{\text{eq}}^{L/\alpha} - x_0^{L}}{x_{\text{eq}}^{L/\alpha} - x_{\text{eq}}^{\alpha/L}}\dfrac{V_{\text{m}}^{\alpha}}{V_{\text{m}}^{L}}\left(1 - \dfrac{r^*}{r_{\text{mid}}}\right)\dfrac{1}{r_{\text{mid}}}$	(6.152) on page 330	(9.12)
r^* = the critical radius of nucleation.		
Stage 3. Diffusion with attention to kinetics of crystal growth		
$V_{\text{growth}} = \dfrac{dr_{\text{mid}}}{dt} = \left(-A + \sqrt{A^2 + C_3}\right)^2$	(6.158) on page 333	(9.13)
where $A = \dfrac{D}{2}\sqrt{\dfrac{1}{\beta_{\text{kin}}}\dfrac{V_{\text{m}}^{\alpha}}{V_{\text{m}}^{L}}\dfrac{1}{x_{\text{eq}}^{L/\alpha} - x_{\text{eq}}^{\alpha/L}} \times \dfrac{1}{r_{\text{mid}}}}$ and $C_3 = \dfrac{D}{r_{\text{mid}}}\dfrac{V_{\text{m}}^{\alpha}}{V_{\text{m}}^{L}}\dfrac{x_{\text{eq}}^{L/\alpha} - x_0^{L}}{x_{\text{eq}}^{L/\alpha} - x_{\text{eq}}^{\alpha/L}}$		
r^* = the critical radius of nucleation		
Stage 4. Diffusion with attention to both surface tension and kinetics of crystal surface		
$V_{\text{growth}} = \dfrac{dr_{\text{mid}}}{dt} = \left(-A + \sqrt{A^2 + C_4}\right)^2$	(6.162) on page 334	(9.14)
where $A = \dfrac{D}{2}\sqrt{\dfrac{1}{\beta_{\text{kin}}}\dfrac{V_{\text{m}}^{\alpha}}{V_{\text{m}}^{L}}\dfrac{1}{x_{\text{eq}}^{L/\alpha} - x_{\text{eq}}^{\alpha/L}} \times \dfrac{1}{r_{\text{mid}}}}$		
$C_4 = D\dfrac{V_{\text{m}}^{\alpha}}{V_{\text{m}}^{L}}\dfrac{x_{\text{eq}}^{L/\alpha} - x_0^{L}}{x_{\text{eq}}^{L/\alpha} - x_{\text{eq}}^{\alpha/L}} \times \dfrac{1}{r_{\text{mid}}}\left(1 - \dfrac{r_{\text{mid}}^*}{r_{\text{mid}}}\right) = C_3 \times \left(1 - \dfrac{r_{\text{mid}}^*}{r_{\text{mid}}}\right)$		
r^* = the critical radius of nucleation.		

kinetic reaction, and finally diffusion with simultaneous influence of the surface tension and the kinetic reaction.

These alternatives offer a wide choice of possible alternative calculations on faceted crystal growth. It is beyond the scope of this book to penetrate this topic in detail. We restrict the discussion below to a review of the possibilities and some examples. The equations for faceted crystal growth, which correspond to the four alternatives in Chapter 6, are listed in Table 9.3. It gives a review of this and all the other alternatives which are analogous to those, studied in Chapter 6,

Tables 9.1 and 9.2 can be used to predict simple rules for changes of shape when two types of crystals grow, It should be pointed out that these rules may be violated when the influence of diffusion and surface tension is taken into account, This topic will be discussed briefly in Section 9.4.5.

Solute Distribution Field

The choice of solute concentration field is very important for predicting the growth and size of a faceted crystal. The model experiments described on pages 485–487 showed that a constant solute gradient is approximately relevant close to the crystals. The experiments also show that the concentration field.can be regarded to be approximately spherically symmetrical, even if the crystals shows no such symmetry.

As a basis for the derivation of the solute field around faceted crystals we will use the same solute distribution as that given in Chapter 6. As was shown on pages 489–490 it is necessary to distinguish between the two cases $k > 1$ and $k < 1$. The constant solute gradient in Figure 9.23 is negative for $k < 1$ and corresponds to the tangent of the curve at $r = r_{mid}$.

If we apply Equation (6.134) on page 327 in Chapter 6 we obtain the concentration of solute around the faceted crystal (Figure 9.23):

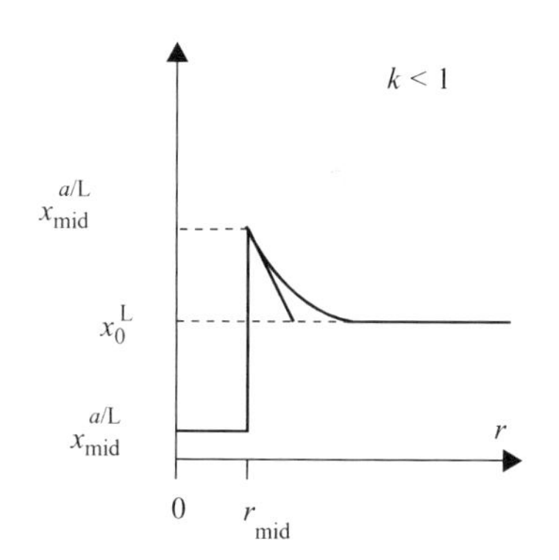

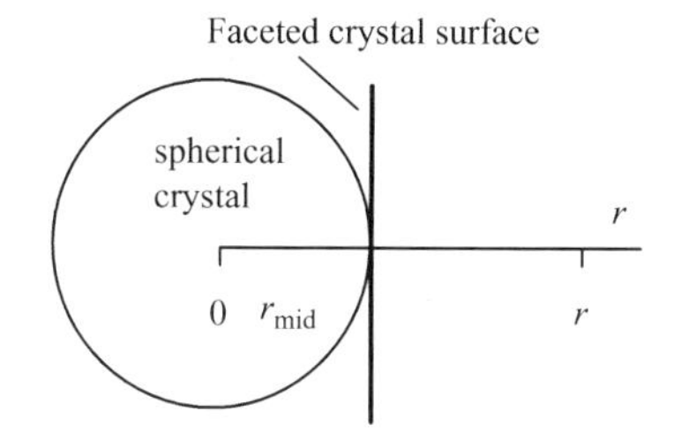

Figure 9.23 Concentration of solute around a spherical faceted crystal. The concentration gradient at the interface is shown in the figure.

$$x^L = x_0^L + \frac{r_{mid}\left(x_{eq}^{L/\alpha} - x_0^L\right)}{r} \tag{9.15}$$

where

r_{mid} = distance between the centre of the crystal and the planar facet, i.e. the midpoint of the facet
$x_{eq}^{L/\alpha}$ = equilibrium concentration of solute in the melt outside the midpoint of the facet at the given temperature (See Figures 9.19a and b)
x_0^L = concentration of solute in the melt far from the crystal
r = distance from the centre of the crystal to a point in the liquid along the normal to the facet
x^L = solute concentration at a point in the melt at distance r from the centre of the crystal.

The concentration gradient within the boundary layer dx^L/dr at the midpoint of a facet can be obtained by taking the derivative of Equation (9.15):

$$\frac{dx^L}{dr} = -\frac{r_{mid}\left(x_{eq}^{L/\alpha} - x_0^L\right)}{r^2} \tag{9.16}$$

Equations (9.15) and (9.16) are valid when the diffusion process is important. The constant gradient close to the crystals is valid only when the *kinetic term of the driving force dominates the solidification process*.

In the box below an example is given, which shows how approximate calculations of crystal growth can be performed. The calculations are based on a linearly decreasing solute concentration field in the melt in agreement with Figure 9.18 on page 487.

Basic Equations for Calculation of Growth and Shape of Faceted Three-Dimensional Crystals when only Interface Kinetics and Mass Transport are Considered

The aim of this box is to give a survey of the basic equations which represent the principles used for calculation of the shape of a faceted crystal as a function of time or size of the crystal.

The equations contain many material constants. Information from the phase diagram of the alloy in question is also required. It is beyond the scope of this book to carry out real calculations for any special crystal. We want merely to supply a general idea of how to perform such calculations.

Growth of the Midpoint of a Facet ($k < 1$)

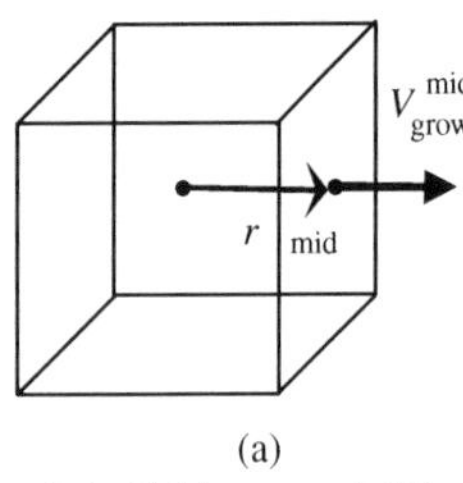

(a)

Figure 1a′. Cubic crystal. The midpoint of a facet and the growth rate in the midpoint are marked in the figure.

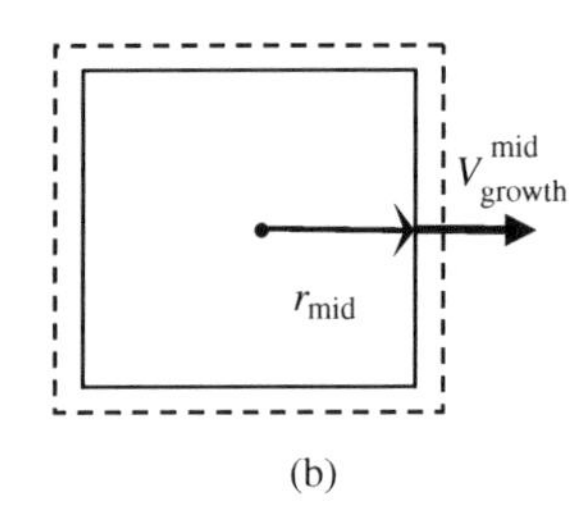

(b)

Figure 1b′. The cubic crystal and the boundary layer in the melt, seen from above.

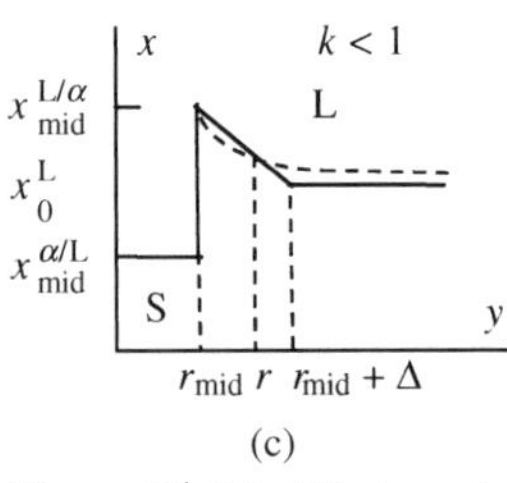

(c)

Figure 2′. Modified version of lower Figure 9.23. Solute concentration along the axis through the centre and the midpoint of a facet.

Consider a facet of a cubic crystal (Figures 1a and 1b). The vertical facet grows with the growth rate V_{growth}. We assume that the concentration field of the solute in the melt close to the crystal consists of a boundary layer of thickness Δ. Inside the boundary layer the concentration decreases approximately linearly (Figure 2′). Outside the boundary layer the solute concentration is constant and equal to x_0^L.

Figure 2′ shows that the solute concentration within the boundary layer can be written as

$$x^L = x_{mid}^{L/\alpha} + \frac{x_{mid}^{L/\alpha} - x_0^L}{\Delta}(r - r_{mid}) \tag{1'}$$

where

Δ = thickness of boundary layer
r = distance along the axis from the midpoint to an arbitrary point within the boundary layer
x^L = solute concentration at point r
$x_{mid}^{L/\alpha}$ = concentration of solute in the melt at the midpoint close to the facet, required for the kinetic process.

Material Balance. Calculation of the Thickness of the Boundary Layer

We set up a material balance at the solid/liquid interface for the solidification process and obtain at equilibrium

$$A = B + C \tag{2'}$$

where

A = excess amount of solute (above x_0^L) emitted from the solid crystal into the melt at solidification
B = excess amount of solute in the white shell which surrounds the crystal
C = excess amount of solute within the volume V (the dark melt between the outer shell and the cube. Figure 3′).

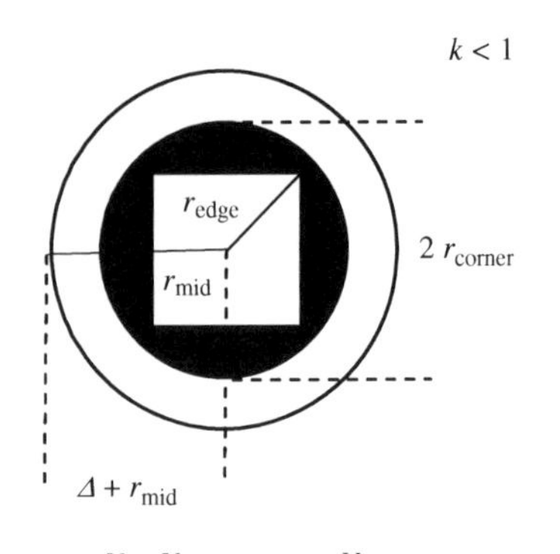

(d)

Figure 3′. Cubic crystal surrounded by a spherical shell which corresponds to the boundary layer.

$r_{edge} = r_{mid}\sqrt{2}$

$r_{corner} = r_{mid}\sqrt{3}$

A can be written as

$$A = (2r_{mid})^3 \times \frac{x_0^L - x_{eq}^{\alpha//L}}{V_m^s} \tag{3'}$$

The temperature T and x_0^{L} are known. With the aid of the phase diagram $x_{\mathrm{eq}}^{\mathrm{L}/\alpha}$ can be derived [see Figure (9.19b) on page 489].

The partition coefficient can also be derived from the phase diagram and hence $x_{\mathrm{eq}}^{\alpha/\mathrm{L}}$ can be calculated. The molar volumes are known material constants. Consequently Δ is the only unknown quantity and can be solved from Equation (1′) as function of r_{mid}, which changes with time.

Calculation of B and C is more complicated as the concentration in the melt varies with the distance from the midpoint of the facet. The amount of excess solute in the white shell in Figure 3′ can be written as

$$B = \int \left(x_{\mathrm{mid}}^{\mathrm{L}/\alpha} - x_0^{\mathrm{L}}\right)\mathrm{d}V = \int_{r_{\mathrm{corner}}}^{r_{\mathrm{mid}}+\Delta} \left[x_{\mathrm{mid}}^{\mathrm{L}/\alpha} - \left(x_{\mathrm{mid}}^{\mathrm{L}/\alpha} - x_0^{\mathrm{L}}\right)\frac{r - r_{\mathrm{mid}}}{\Delta} - x_0^{\mathrm{L}}\right] 4\pi r^2 \mathrm{d}r$$

or

$$B = 4\pi \frac{x_{\mathrm{mid}}^{\mathrm{L}/\alpha} - x_0^{\mathrm{L}}}{\Delta} \int_{r_{\mathrm{corner}}}^{r_{\mathrm{mid}}+\Delta} \left[(r_{\mathrm{mid}} + \Delta)r^2 - r^3\right]\mathrm{d}r$$

or

$$B = 4\pi \frac{x_{\mathrm{mid}}^{\mathrm{L}/\alpha} - x_0^{\mathrm{L}}}{\Delta} \left|(r_{\mathrm{mid}} + \Delta)\frac{r^3}{3} - \frac{r^4}{4}\right|_{r_{\mathrm{corner}}}^{r_{\mathrm{mid}}+\Delta}$$

or

$$B = 4\pi \frac{x_{\mathrm{mid}}^{\mathrm{L}/\alpha} - x_0^{\mathrm{L}}}{\Delta} \left|(r_{\mathrm{mid}} + \Delta)\frac{r^3}{3} - \frac{r^4}{4}\right|_{r_{\mathrm{corner}}}^{r_{\mathrm{mid}}+\Delta}$$

or

$$B = 4\pi \frac{x_{\mathrm{mid}}^{\mathrm{L}/\alpha} - x_0^{\mathrm{L}}}{\Delta} \left[(r_{\mathrm{mid}} + \Delta)\left(\frac{(r_{\mathrm{mid}} + \Delta)^3}{3} - \frac{r_{\mathrm{corner}}^3}{3}\right) - \frac{(r_{\mathrm{mid}} + \Delta)^4}{4} + \frac{r_{\mathrm{corner}}^4}{4}\right]$$

which can be reduced to

$$B = \frac{\pi}{3}\frac{x_{\mathrm{mid}}^{\mathrm{L}/\alpha} - x_0^{\mathrm{L}}}{\Delta}\left[(r_{\mathrm{mid}} + \Delta)^4 - 4r_{\mathrm{corner}}^3(r_{\mathrm{mid}} + \Delta) + 3r_{\mathrm{corner}}^4\right] \tag{4′}$$

The amount C of solute in the volume V can be derived by similar calculations as above. Integration of the solute contents in each volume element gives the total amount of excess solute. The concentration of the solute in the volume element $x^{\mathrm{L}}(r)\,\mathrm{d}V$ can be found with the aid of Equation (1′). Figure 3′ shows that the integration limits are complicated and it is favourable to use numerical calculations.

$$C = \int_V \left(x_{\mathrm{mid}}^{\mathrm{L}/\alpha} - x_0^{\mathrm{L}}\right)\mathrm{d}V = \int_V \left(x_{\mathrm{mid}}^{\mathrm{L}/\alpha} - \left(x_{\mathrm{mid}}^{\mathrm{L}/\alpha} - x_0^{\mathrm{L}}\right)\frac{r - r_{\mathrm{mid}}}{\Delta}\right) - x_0^{\mathrm{L}}\Big)\mathrm{d}V \tag{5′}$$

where $r = \sqrt{x^2 + y^2 + z^2}$ and $\mathrm{d}V = \mathrm{d}x\ \mathrm{d}y\ \mathrm{d}z$

When the expressions of A, B and C are inserted into Equation (2′) a value of the thickness Δ of the boundary layer can be calculated.

Growth Equation of the Crystal

We apply Equation (9.10b) on page 489, which can be written as

$$x_{eq}^{L/\alpha} - x_0^L = \left[\underbrace{\left(x_{eq}^{L/\alpha} - x_{kin}^{L/\alpha}\right)}_{\text{kinetic process}} + \underbrace{\left(x_{kin}^{L/\alpha} - x_0^L\right)}_{\text{diffusion process}} \right]. \tag{6'}$$

With the aid of Equations (9.9) on page 488 and Fick's first law) Equation (6′) can be written as

$$x_{eq}^{L/\alpha} - x_0^L = \left[\sqrt{\frac{V_{growth}^{mid}}{\beta_{kin}}} + \frac{V_{growth}^{mid}\left(x_{eq}^{L/\alpha} - x_{eq}^{\alpha/L}\right)\Delta}{D} \frac{V_m^L}{V_m^\alpha} \right] \tag{7'}$$

where $V_{growth}^{mid} = \frac{dr_{mid}}{dt}$.

x_0^L, $x_{eq}^{L/\alpha}$ and $x_{eq}^{\alpha/L}$ are known. The molar volumes, β_{kin} and the diffusion constant D are known material constants. Δ can be calculated numerically from Equation (2′) in combination with Equations (3′), (4′) and (5′). Equation (7′) is a second order equation of $\sqrt{V_{growth}^{mid}}$ which can be solved as a function of r_{mid}. As $V_{growth}^{mid} = \frac{dr_{mid}}{dt}$ we have a differential equation from which r_{mid} can be solved as a function of time. The equation has to be solved numerically because of Δ. Hence, V_{growth}^{mid}, r_{mid} and Δ can be plotted as functions of time and also V_{growth}^{mid} as a function of r_{mid}.

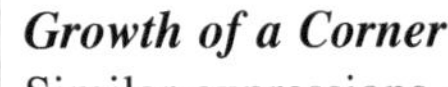

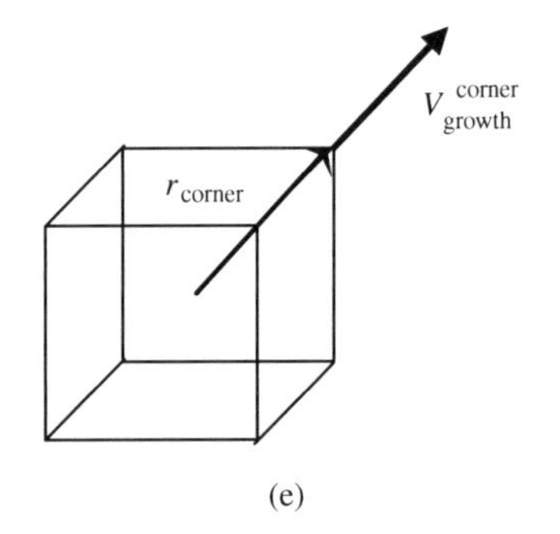

Figure 4′. Direction of corner growth.

Growth of a Corner

Similar expressions can be derived for facet growth at the corners of cubic crystals. Equation (1′) is the same as above. Equation (7′) can be replaced by

$$x_0^L - x_{eq}^{L/\alpha} = \left[\sqrt{\frac{V_{growth}^{corner}}{\beta_{kin\ corner}}} + \frac{V_{growth}^{corner}\left(x_{eq}^{L/\alpha} - x_{eq}^{\alpha/L}\right)\left(\Delta - (r_{mid} - r_{corner})\right)}{D} \frac{V_m^L}{V_m^\alpha} \right]. \tag{8'}$$

where $r_{corner} = r_{mid}\sqrt{3}$ (Figure 4′).

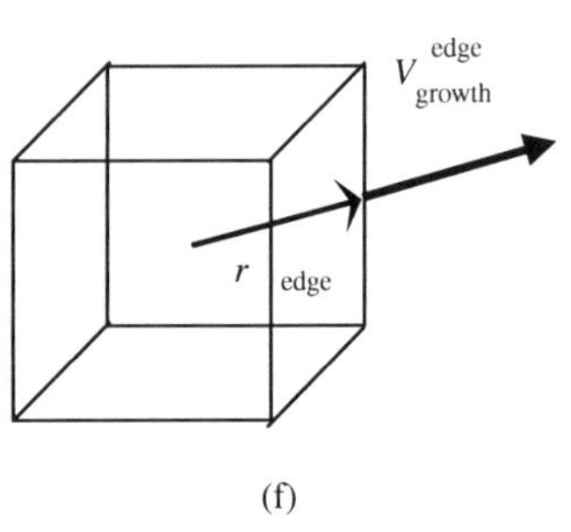

Figure 5′. Direction of edge growth.

Growth of the Midpoint of an Edge

The corresponding expressions for the midpoint of an edge are;
Equation (1′) is the same as above. Equation (7′) can be replaced by

$$x_0^L - x_{eq}^{L/\alpha} = \left[\sqrt{\frac{V_{growth}^{edge}}{\beta_{kin\ edge}}} + \frac{V_{growth}^{edge}\left(x_{eq}^{L/\alpha} - x_{eq}^{\alpha/L}\right)\left(\Delta - (r_{mid} - r_{edge}\right)}{D} \frac{V_m^L}{V_m^\alpha} \right] \tag{9'}$$

where $r_{edge} = r_{mid}\sqrt{2}$ (Figure 5′).

Growth Rate at the Midpoint of a Facet when the Mass Transport Dominates the Solidification Process

The growth rate of the facets can be obtained by applying the simplest case without corrections, Equation (6.138) on page 328 in Chapter 6. It is adapted to faceted growth [Equation (9.11) in Table 9.3 on page 492] and equal to the advancement of the midpoint per unit time. It can be written as

$$V_{\text{growth}} = \left(\frac{\mathrm{d}r}{\mathrm{d}t}\right)_{r=r_{\text{mid}}} = D\frac{x_{\text{kin mid}}^{\text{L}/\alpha} - x_0^{\text{L}}}{x_{\text{eq}}^{\text{L}/\alpha} - x_{\text{eq}}^{\alpha/\text{L}}}\frac{V_{\text{m}}^{\alpha}}{V_{\text{m}}^{\text{L}}}\frac{1}{r_{\text{mid}}} \tag{9.11}$$

where

V_{growth} = growth rate of the facet in a direction perpendicular to the facet
D = diffusion constant of the solute in the melt
$V_{\text{m}}^{\text{L},\alpha}$ = molar volume of liquid and crystal, respectively
$x_{\text{kin mid}}^{\text{L}/\alpha}$ = solute concentration at the midpoint of the facet
$x_{\text{eq}}^{\text{L}/\alpha}$ = equilibrium value of the solute concentration in the melt at the interface outside the midpoint of the facet at the given temperature(see Figures 9.19a and b) on page 489
$x_{\text{eq}}^{\alpha/\text{L}}$ = equilibrium concentration of solute at the given temperature in the crystal at the midpoint at distance r_{mid} from the centre of the crystal
r_{mid} = distance from the centre of the crystal to the planar facet.

If the kinetic driving force is neglected Equation (9.11) can be rewritten and simplified to

$$V_{\text{growth}} = \frac{D}{x_{\text{eq}}^{\text{L}/\alpha} - x_{\text{eq}}^{\alpha/\text{L}}}\frac{V_{\text{m}}^{\alpha}}{V_{\text{m}}^{\text{L}}}\frac{x_{\text{mid}}^{\text{L}/\alpha} - x_0^{\text{L}}}{\Delta} \tag{9.17}$$

where the last factor is the negative concentration gradient within the boundary layer according to Figure 3′ in the box on page 494.

Growth Rate at the Corner of a Facet when the Mass Transport Dominates the Solidification Process

In analogy with Equation (9.17) the growth rate at the corner of the crystal can be written as

$$V_{\text{growth}} = \frac{D}{x_{\text{eq}}^{\text{L}/\alpha} - x_{\text{eq}}^{\alpha/\text{L}}}\frac{V_{\text{m}}^{\alpha}}{V_{\text{m}}^{\text{L}}}\frac{x_{\text{corner}}^{\text{L}/\alpha} - x_0^{\text{L}}}{\delta} \tag{9.18}$$

where the last factor is the concentration gradient in the boundary layer outside the corner (Figure 9.24).

It should be noticed that the facet is planar and the growth of the facet is also planar. Otherwise its shape would be changed during growth, which is not the case at normal faceted growth of crystals. Therefore, the growth rate at *every* point of the facet is *perpendicular to the facet* and *constant*.

Hence, the component of the growth rate at the corner, perpendicular to the facet, must be equal to the velocity at the midpoint of the facet. The condition for this can be written as

$$\frac{x_{\text{corner}}^{\text{L}/\alpha} - x_0^{\text{L}}}{\sqrt{3}\delta} = \frac{x_{\text{mid}}^{\text{L}/\alpha} - x_0^{\text{L}}}{\Delta} \tag{9.19a}$$

Division with the factor $1/\sqrt{3}$ in Equation (9.19) times the total growth rate of the corner gives the component which is perpendicular to the growing facet.

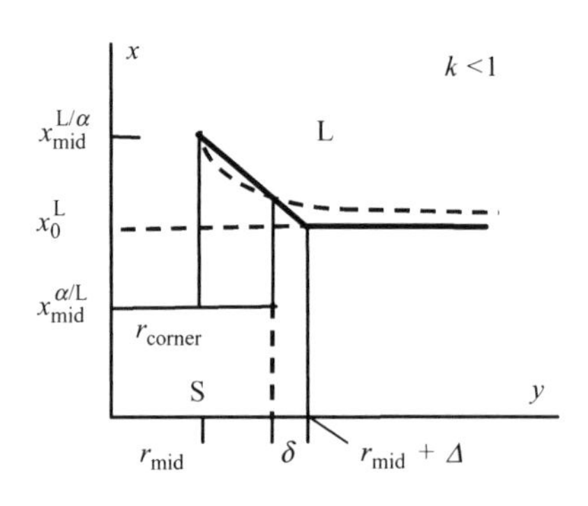

Figure 9.24 The concentration gradient outside the corner (see Figure 3′ in the box on page 494) is

$$\frac{x_{corner}^{L/\alpha} - x_0^L}{\delta}$$

where $\delta = r_{mid} + \Delta - r_{corner}$.

The condition for formation of hopper crystals is

$$\frac{x_{corner}^{L/\alpha} - x_0^L}{\sqrt{3}\delta} > \frac{x_{mid}^{L/\alpha} - x_0^L}{\Delta} \tag{9.19b}$$

This condition can be achieved because δ can be considerably smaller than Δ.

Once more we want to emphasize that the constant gradient close to the crystals (Figure 9.18 on page 487) is valid only when the kinetic term of the driving force dominates the solidification process. When diffusion is most important, Equations (9.15) and (9.16) on page 493 are valid.

Example 9.3

Figures 9.4 on page 479 shows the gradual changes of an alloy crystal shape when the solidification process changes from kinetic (interface) control to diffusion control. When size of the crystal increases, the driving force for diffusion in the <100> and <111> directions change with the crystal size. At a special size of the crystal the crystal the growth rates in the two directions [100] and [111] become equal.

Derive expressions for the ratio of the extension of the crystal in the two directions [100] and [111], i.e. r_{100}/r_{111} and the ratio β_{100}/β_{111} for the case when the growth rates in the two directions becomes equal at the transition from interface control to diffusion control.

The diffusion coefficient D_L of the solute in the melt and the growth constants for the interface kinetics β_{100} and β_{111} in the <100> and <111> directions are supposed to be known. The partition coefficient $k < 1$. The difference in molar volumes between the melt and the crystal can be neglected.

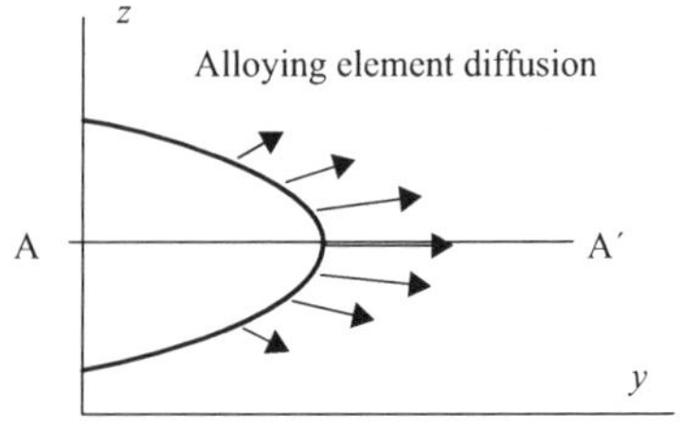

Phase diagram Al-Si.

Solution:

In the present case it is necessary to consider both the kinetic and diffusion processes at solidification. As $k < 1$ the total driving force can be written as (Equation 9.10b on page 489)

$$const\left(x_{eq}^{L/\alpha} - x_0^L\right) = const\left[\underbrace{\left(x_{eq}^{L/\alpha} - x_{kin}^{L/\alpha}\right)}_{\text{kinetic process}} + \underbrace{\left(x_{kin}^{L/\alpha} - x_0^L\right)}_{\text{diffusion process}}\right] \tag{1'}$$

Kinetic Process

The kinetic process is controlled by the relationship [in analogy with Equation (9.9) on page 488]

$$V_{growth} = \left(\frac{dr}{dt}\right)_{kin} = \beta_{kin}\left(x_{eq}^{L/\alpha} - x_{kin}^{L/\alpha}\right)^2 \tag{2'}$$

which can be transformed into

$$x_{eq}^{L/\alpha} - x_{kin}^{L/\alpha} = \sqrt{\frac{V_{growth}}{\beta_{kin}}} \tag{3'}$$

Diffusion Process

In analogy with Equation (9.15) on page 493 the concentration field around the crystal can be described by the general relationship

$$x^L = A + \frac{B}{r} \tag{4'}$$

The constants A and B can be determined by the boundary conditions $x^L = x_0^L$ for $r = \infty$ which gives $A = x_0^L$.

As a second boundary condition we chose

$$x^L = x_{111}^{L/\alpha} \quad \text{for} \quad r = r_{111}$$

which gives

$$x^L = x_{111}^{L/\alpha} = x_0^L + \frac{B}{r_{111}}$$

or

$$B = r_{111}\left(x_{111}^{L/\alpha} - x_0^L\right).$$

We insert the values of A and B into Equation (2′) and obtain

$$x^L = x_0^L + \frac{r_{111}\left(x_{111}^{L/\alpha} - x_0^L\right)}{r} \tag{5'}$$

According to Fick's first law

$$V_{growth} = -\frac{D_L}{V_m^L}\frac{dx^L}{dr} \tag{6'}$$

We take the derivative of Equation (5′):

$$\frac{dx^L}{dr} = -\frac{r_{111}\left(x_{111}^{L/\alpha} - x_0^L\right)}{r^2} \tag{7'}$$

which is inserted into Equation (6′). Hence, we obtain

$$V_{growth} = \frac{D_L}{V_m^L}\frac{r_{111}\left(x_{111}^{L/\alpha} - x_0^L\right)}{r^2} \tag{8'}$$

Calculations

We apply Equation (1′) in combination with Equations (3′) and (8′) on the midpoint of a facet and on a corner, respectively.

$$x_{\text{eq}}^{\text{L}/\alpha} - x_0^{\text{L}} = \sqrt{\frac{V_{\text{growth}}^{100}}{\beta_{\text{kin}}^{100}}} + \frac{V_{\text{growth}}^{100}}{D_{\text{L}}} \frac{r_{111}\left(x_{111}^{\text{L}/\alpha} - x_0^{\text{L}}\right)}{r_{100}^2} V_{\text{m}}^{\text{L}} \tag{9′}$$

and

$$x_{\text{eq}}^{\text{L}/\alpha} - x_0^{\text{L}} = \sqrt{\frac{V_{\text{growth}}^{111}}{\beta_{\text{kin}}^{111}}} + \frac{V_{\text{growth}}^{111}}{D_{\text{L}}} \frac{r_{111}\left(x_{111}^{\text{L}/\alpha} - x_0^{\text{L}}\right)}{r_{111}^2} V_{\text{m}}^{\text{L}} \tag{10′}$$

Elimination of $\left(x_{\text{eq}}^{\text{L}/\alpha} - x_0^{\text{L}}\right)$ gives

$$\sqrt{\frac{V_{\text{growth}}^{100}}{\beta_{\text{kin}}^{100}}} + \frac{V_{\text{growth}}^{100}}{D_{\text{L}}} \frac{r_{111}\left(x_{111}^{\text{L}/\alpha} - x_0^{\text{L}}\right)}{r_{100}^2} V_{\text{m}}^{\text{L}} = \sqrt{\frac{V_{\text{growth}}^{111}}{\beta_{\text{kin}}^{111}}} + \frac{V_{\text{growth}}^{111}}{D_{\text{L}}} \frac{r_{111}\left(x_{111}^{\text{L}/\alpha} - x_0^{\text{L}}\right)}{r_{111}^2} V_{\text{m}}^{\text{L}} \tag{11′}$$

According to the text we know that $V_{\text{growth}}^{100} = V_{\text{growth}}^{111}$. In addition, the velocities are equal *at the transition boundary between kinetic-controlled and diffusion-controlled growth*. This means we obtain two more equalities:

$$\sqrt{\frac{V_{\text{growth}}^{100}}{\beta_{\text{kin}}^{100}}} = \sqrt{\frac{V_{\text{growth}}^{111}}{\beta_{\text{kin}}^{111}}} \tag{12′}$$

$$\frac{V_{\text{growth}}^{100}}{D_{\text{L}}} \frac{r_{111}\left(x_{111}^{\text{L}/\alpha} - x_0^{\text{L}}\right)}{r_{100}^2} = \frac{V_{\text{growth}}^{111}}{D_{\text{L}}} \frac{r_{111}\left(x_{111}^{\text{L}/\alpha} - x_0^{\text{L}}\right)}{r_{111}^2} \tag{13′}$$

Equation (12′) can be written as

$$\frac{V_{\text{growth}}^{100}}{\beta_{\text{kin}}^{100}} = \frac{V_{\text{growth}}^{111}}{\beta_{\text{kin}}^{111}} \tag{14′}$$

As the growth rates are equal we can conclude that the growth constants also must be equal.

$$\beta_{\text{kin}}^{100} = \beta_{\text{kin}}^{111} \tag{15′}$$

Equation (13′) can be written as

$$\frac{1}{r_{100}^2} = \frac{1}{r_{111}^2} \tag{16′}$$

which means that

$$r_{100} = r_{111} \tag{17′}$$

Answer:

The condition $V_{\text{growth}}^{100} = V_{\text{growth}}^{111}$ gives

$$\beta_{\text{kin}}^{100} = \beta_{\text{kin}}^{111} \quad \text{and} \quad r_{100} = r_{111}.$$

The latter relationship corresponds to a shape halfway between a cube and an octahedron. The corners in the <111> directions a cut and have turned into facets.

9.4.5 *Influence of Simultaneous Interface Kinetics, Mass Transport and Surface Tension on Faceted Growth in Alloys*

Above we have used the simplest equation for diffusion of the solute and neglected the influence of the kinetics on the growth of faceted crystals. Below we will briefly discuss the case when the interface kinetics, the mass transport and the surface tension are considered simultaneously.

Approximate Expression for the Growth Rate of a Facet

In all crystal processes, and also in any other heat or mass transport processes, the slowest process determines will control the growth rate of the overall process. In the present case three processes influence the growth of faceted crystals. The relative influence of the processes on the crystal growth rates in different directions varies during growth. This means that the shape of the crystals changes with the size of the crystals.

The growth of faceted crystals is complicated and depends on many variables and involves also material constants, which varies widely for alloys. No general mathematical model has been found so far. For his reason this topic will only be briefly discussed below with reference to the basic theory for spherical crystal growth in Chapter 6 as a rough approximations.

The influences of the interface kinetics, the mass transport, i.e. the diffusion, and the surface tension on the growth of faceted crystals, occur simultaneously. This case has been treated in Chapter 6 for spherical crystals. We set up the same equation here [Equation (6.162) on page 334 in Chapter 6] but adapt and it for non-spherical crystal [Equation (9.14) in Table 9.3 on page 492].

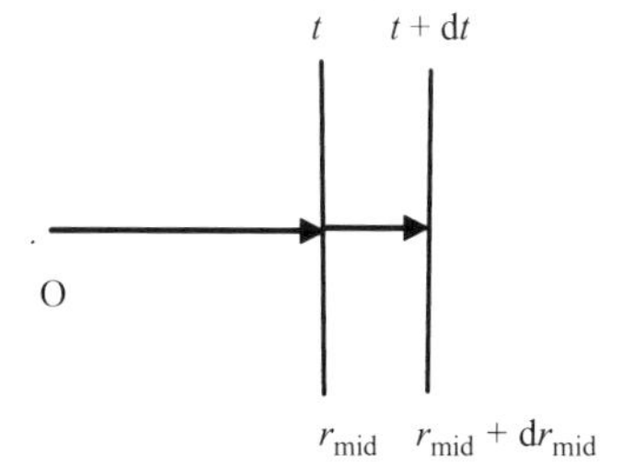

Figure 9.25 Facet of a growing crystal.

$$V_{growth} = \frac{dr_{mid}}{dt} = \left(-A + \sqrt{A^2 + C_4}\right)^2 \tag{9.14}$$

where

$$A = \frac{D}{2}\sqrt{\frac{1}{\beta_{kin}}\frac{V_m^\alpha}{V_m^L}\frac{1}{x_{eq}^{L/\alpha} - x_{eq}^{\alpha/L}}} \times \frac{1}{r_{mid}}$$

$$C_4 = D\frac{V_m^\alpha}{V_m^L}\frac{x_{eq}^{L/\alpha} - x_0^L}{x_{eq}^{L/\alpha} - x_{eq}^{\alpha/L}} \times \frac{1}{r_{mid}}\left(1 - \frac{r_{mid}^*}{r_{mid}}\right)$$

β_{kin} = growth constant or kinetic coefficient (page 332 in Chapter 6)
D = diffusion coefficient of the alloying element in the melt
V_m^L = molar volume of the melt
V_m^α = molar volume of the solid phase α
$x_{eq}^{L/\alpha}$ = equlibrium concentration of solute in the melt close to the midpoint of the facet at the given temperature (see Figures 19a and b)
$x_{eq}^{\alpha/L}$ = equlibrium concentration of solute in the solid crystal close to the midpoint of the facet at the given temperature
x_0^L = initial mole fraction of B in the melt far from the interface
r_{mid} = distance from the centre of the growing crystal to the midpoint of the facet
r_{mid}^* = critical distance of nucleation of a crystal measured from the centre of the growing crystal to the middle of the facet.

Numerical calculations are necessary when Equation (9.14) is used. This was also the case in Chapter 6 on pages 337–338.

Change of Shape at Growth of Faceted Cubic Crystals

Figure 9.4 on page 479 shows the fact that a growing faceted crystal may change its shape radically during growth. For each size the growing crystal obtains the shape which corresponds to the lowest possible total energy.

Fredriksson and West [7] have performed some calculations based on this principle for a growing crystal, which follows case 2 (page 479). Their result is shown in Figure 9.26.

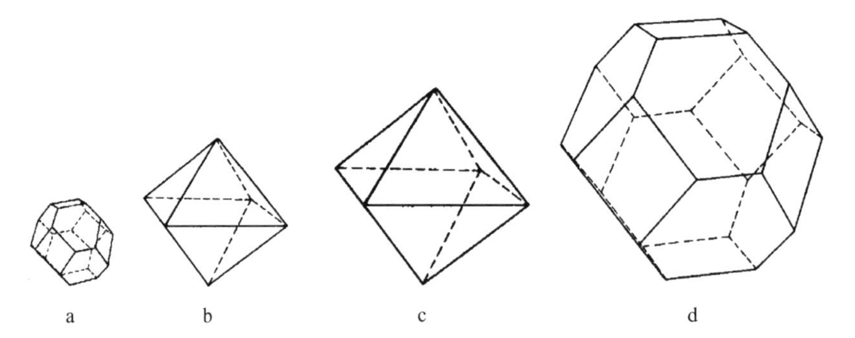

Figure 9.26 Calculated shapes of a growing crystal as a function of size.

Figure 9.26 shows four different morphologies of the same crystal at different sizes.

- Figure 9.26a shows shape and the relative size of the crystal when the anisotropy in the surface tension dominates over the other effects.
- The Figures 9.26b and c illustrate the shape and relative size when the interface kinetics dominates the growth process.
- Figure 9.26d corresponds to the shape and relative size when the mass transport, i.e. the diffusion starts to control the growth process.

The concentration field around crystals is also affected by their size Figure 9.18 on page 487 shows the concentration field around a small crystal. When the crystal grows the spherical field becomes deformed. The field will be weaker at the corners and slightly flatter near the midpoints of the facets. This means that the shape transition may be displaced and occur at other values than the calculated sizes.

Step Growth

During part of the growth period when the interface kinetics controls the crystal growth, the interface obtains a step structure of the same kind as the structure, which appears when hopper crystals are formed in pure metals (page 484). The steps are created at the corners and grow along the edges and surfaces. The lengths and heights of the steps are determined by the relaxation of the potential energy during the growth. The mechanism for this has been discussed in Chapter 6 on pages 307–309.

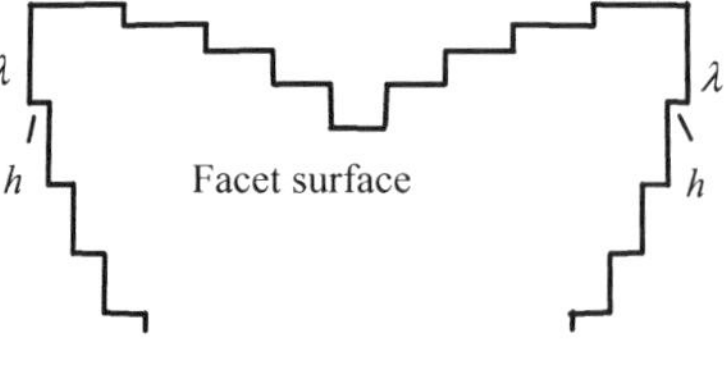

Figure 9.27 Cross-section of a growing cubic crystal. O is the centre of the crystal.

The effect is strongly exaggerated in the figure and the proportions between the corners and the crystal are out of scale.

The lengths and heights of the steps vary from the corners to the midpoints. Close to the corners the length of the steps are longer and their heights are smaller than those at the midpoints. According to the discussions about step growth during crystal growth in a vapour phase in Chapter 5 (pages 235–236), we can conclude that the relaxation process determines the length and height of the steps. The relaxation process of a step has to be finished before a new step can be formed at the corner. Since no new steps are formed at the midpoints or elsewhere, we realize that the relaxation there can not be finished unless the steps are long enough to permit complete relaxation before a new step is formed at the corner. New steps can only form at the corners. The result is that the growth rate is lower at the midpoints than at the corners, which is shown in Figure 9.27. The ratio λ/h decreases with the distance from the corner.

The interface of a growing faceted alloy crystal is shown in Figure 9.28a. When the crystal grows, the sizes of the {111} steps increase. The macroscopic direction of the surface will no longer be perpendicular to the {100} plane but will be tilted an angle α (Figure 9.28a).

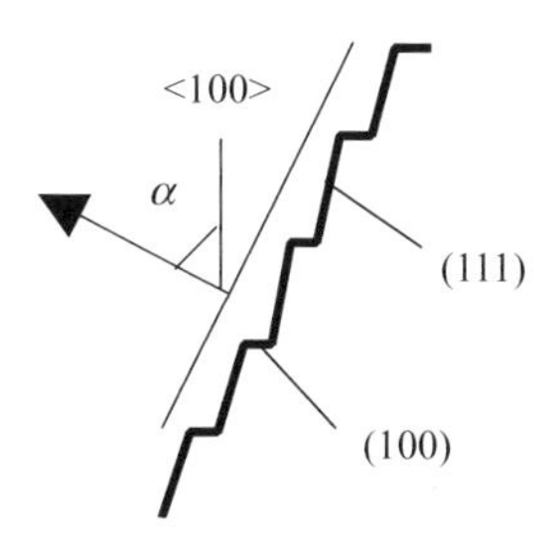

Figure 9.28a Interface of a growing faceted crystal.

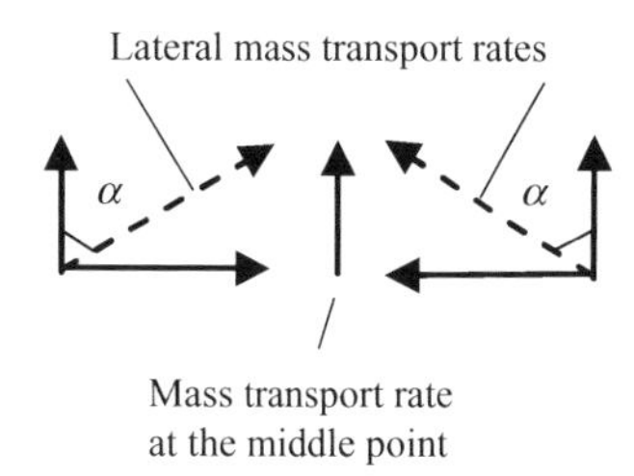

Figure 9.28b At the midpoint of the faceted surface the mass transport rate is equal to the vertical component of the lateral mass transport rate.

The mass transport is normally perpendicular to the macroscopic plane of the crystal surface but this is not the case at step growth. The mass transport vectors from the sides meet at the midpoint of the faceted surface. The net effect is a slower mass transport at the midpoint than at the sides. Figure 9.28b shows that the mass transport is decreased by a factor cos α and is perpendicular to the {100} plane. If the angle α is large, the mass transport rate at the middle point is considerably *lower* than the lateral mass transport rate.

9.5 Growth of Dendrite Crystals

9.5.1 Formation of Dendrite Crystals

Most metals solidify by a primary precipitation of dendrites. This type of crystals was early observed in cast metals, among others by Grignon and Tschernoff. Both of them studied dendrites in shrinkage cavities of steel ingots. Well developed dendrites are often found near the top of large ingots. A photograph of such crystals, found by Tschernoff, is shown in Figure 9.29.

Figure 9.29 Dendrite found by Tschernoff in 1870. Height 35 cm. Reproduced from CRC/Taylor & Francis © 1959.

Both Grignon and Tschernoff drew pictures of their observations. These are shown in Figures 9.30 and 9.31. These figures were intended to be simplifications and we may guess that the aim was to show the main principles, which control the growth of dendrites.

The pictures in Figures 9.30 and 9.31 show interesting similarities and differences. They both show that dendrites develop into thin branches growing in three perpendicular directions. The main difference between the pictures is that Grignon arranged the branches in rows while Tschernoff placed them evenly spaced in three dimensions.

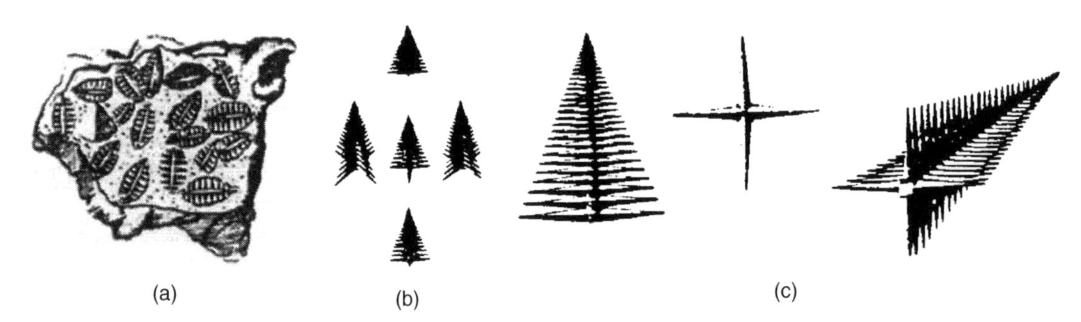

Figure 9.30 Drawings of dendrites according to Grignon 1775. Reproduced from Grignon, M.. 1775. Reproduced from Grignon, M.. 1776. Reproduced from Grignon, M.. 1777.

A careful analysis of the Tschernoff crystal drawing in Figure 9.29 and of similar crystals shows that a dendrite is much more complex than the drawings by Grignon and Tschernoff show. A dendrite crystal may instead be drawn like that in Figure 9.32. The figure shows that a dendrite has a very complex structure. The formation of dendrites has been extensively discussed in the scientific literature through the years. There are many separate as well as integrated phenomena associated with the understanding of the formation process.

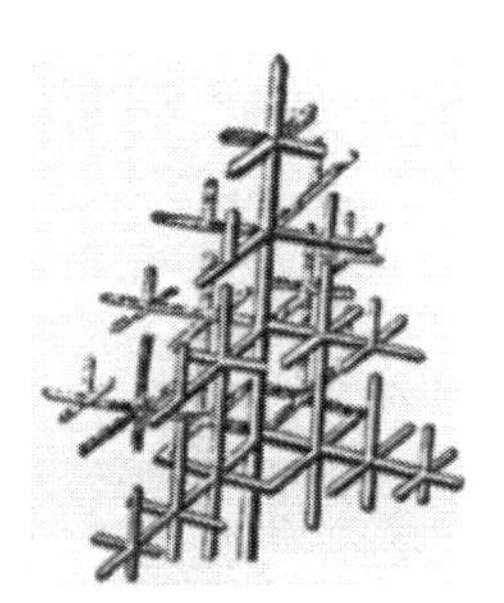

Figure 9.31 Structure of a dendrite crystal of austenite according to Tschernoff. Reproduced from CRC/Taylor & Francis © 1959.

Figure 9.32 Principle drawing of a dendrite according to modern ideas.

The present idea is that a dendritic network of the type in Figure 9.32 is formed in a solidifying alloy melt. When the concentration of the alloying element in the melt has reached the eutectic composition, the dendritic growth is interrupted. The remaining melt solidifies with a eutectic structure in the interdendritic regions.

Numerous photographs of metals and alloys, in which the growth process has been interrupted, confirm this conception. Figure 9.33 gives an example. It shows a single dendrite formed in a quenched, unidirectionally solidified sample of a steel alloy, containing 0.4 wt% C.

The coarse dendrite structure to the left in the micrograph was formed before quenching. The dendrite arms are dark and the space between them is bright. The reason for this is that different parts of the sample etch differently depending on concentration variations of the alloying element in the structure. The micrograph shows that the growth occurs at a leading tip. Immediately behind the tip new arms form, which usually are classified as secondary dendrite arms.

On the next pages we will introduce the modern view on the growth process of dendrites and analyze it theoretically.

Figure 9.33 Dendritic growth in steel, which contains 0.4 wt% C.

9.5.2 *Undercooling and Solidification Rate at Dendritic Growth in Pure Metal Melts. Ivantsov's Model*

The driving force of dendritic growth in a pure metal is the undercooling of the melt. It consists of three parts:

1. Driving force owing to the curvature of the tip and the surface tension:
   The difference between the melting-point temperature at the tip T_M^{tip} and the melting-point temperature T_M of a planar interface.
2. Kinetic driving force owing to the kinetic reaction at the interface:
   The temperature difference between the temperature close to the interface without (T_M) and with ($T_M^{L/\alpha}$) the kinetic reaction.
3. Thermal driving force:
   The temperature difference between the temperature of the melt close to the interface $T^{L/\alpha}$ and the temperature T_0 of the melt far from the interface.

The undercooling can be expressed as the sum of the three terms:

$$\Delta T_{total} = \Delta T_{surface} + \Delta T_{kin} + \Delta T_{thermal} \tag{9.20}$$

or

$$\Delta T_{total} = (T_M^{tip} - T_M) + (T_M - T^{L/\alpha}) + (T^{L/\alpha} - T_0) \tag{9.21}$$

where

T_M^{tip} = melting-point temperature of the curved tip
T_M = melting-point temperature of the solid at the planar interface
$T^{L/\alpha}$ = temperature of the melt close to the interface
T_0 = temperature of the melt far from the interface
$\Delta T_{surface}$ = change of temperature of the melt close to the tip owing to the curvature of the dendrite tip and the surface tension
ΔT_{kin} = change of temperature at the interface owing to the kinetic reaction
$\Delta T_{thermal}$ = temperature difference between the temperatures $T^{L/\alpha}$ and T_0, which causes heat transfer through the melt
ΔT_{total} = difference between the melting-point temperature T_M, and the temperature T_0 far from the interface.

The First Term $\Delta T_{surface}$

The first term $\Delta T_{surface} = T_M^{tip} - T_M$ depends on the radius of the tip and the surface tension between the melt and the solid phase.

In most theoretical investigations this effect is considered and it will be discussed on pages 507–508.

The Second Term ΔT_{kin}

The second term in Equation (9.21) ΔT_{kin} is obtained from the expression

$$V_{growth} = \beta_{kin} \Delta T_{kin}^2 \tag{9.22}$$

where β_{kin} is the growth constant. Hence, we obtain

$$\Delta T_{kin} = \sqrt{\frac{V_{growth}}{\beta_{kin}}} \tag{9.23}$$

The Third Term $\Delta T_{\text{thermal}}$

Growth Rate as a Function of Undercooling in Pure Metal Melts

Below we will discuss the third term $\Delta T_{\text{thermal}}$ in Equation (9.21), i.e. the relationship between the undercooling and the growth rate of the solidification front, i.e. the growth rate of the dendrite tips. A very simple model will be used here. In Section 9.5.4 we will come back with a more correct model, valid for both pure metals and alloys. The topic will be further discussed on page 521.

In the 1950s Fisher [8] suggested a simple model for dendritic growth. He made the following assumptions

- The tip is a hemisphere and isotropic
- the kinetic driving force is zero
- the tip looses heat by radial conduction and moves forward with its shape unchanged.

Consider the heat transport through the hemispheric interface (Figures 9.34a and b). Fourier's law, Equation (7.1) on page 362 in Chapter 7 gives

$$\frac{dQ}{dt} = -\lambda \times 2\pi\rho_{\text{tip}}^2 \frac{T^{L/\alpha} - T_0}{\rho_{\text{tip}}} \tag{9.24}$$

where

$\frac{dQ}{dt}$ = heat flow per unit time through the tip

λ = thermal conductivity of the melt

ρ_{tip} = radius of the hemispherical tip

$T^{L/\alpha}$ = temperature of the melt close to the tip

T_0 = temperature of the melt far from the interface.

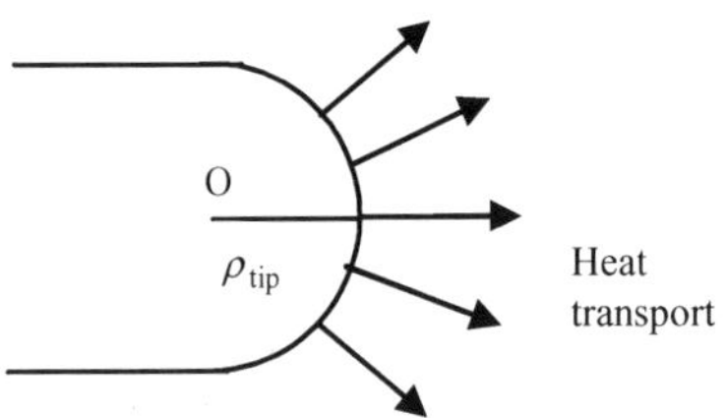

Figure 9.34a Emission of heat from a dendrite tip into a pure metal melt.

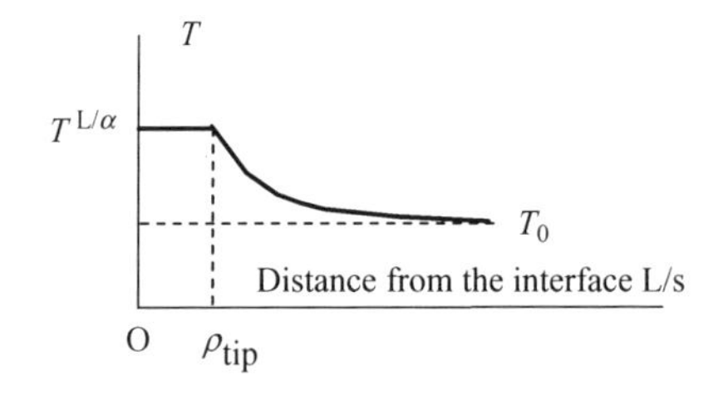

Figure 9.34b Temperature distribution in the melt around a growing dendrite tip. The temperature of the melt at the interface melt/tip is $T^{L/\alpha}$.

dQ/dt is also equal to loss of solidification heat of the solidified volume per unit time:

$$\frac{dQ}{dt} = -V_{\text{growth}}\pi\rho_{\text{tip}}^2 \frac{-\Delta H_m}{V_m^\alpha} \tag{9.25}$$

where

V_{growth} = growth rate of the tip and the solidification front

$1/V_m^\alpha$ = density of the melt

V_m^α = molar volume of the solid alloy

$-\Delta H_m$ = molar heat of fusion of the metal (J/kmol).

Comparison of the Equations (9.24) and (9.25) gives

$$V_{\text{growth}} = \frac{2\lambda V_{\text{m}}^{\alpha}}{\rho_{\text{tip}}(-\Delta H_{\text{m}})}(T^{\text{L}/\alpha} - T_0) \tag{9.26}$$

The radius of the tip is unknown. We need another relationship to find an expression for ρ_{tip}. The undercooling owing to the surface tension σ and the curvature K of the tip, can be calculated with the aid of Equation (1.46) on page 12 in Chapter 1. The calculations are given in the box

Calculation of the Undercooling as a Function of the Surface Tension σ and Curvature of the two Dendrite Tips

Equation (1.46) describes the temperature difference between the two solid α phases i.e. two dendrite tips with different curvatures 0 and K.

$$T_{\text{M}}(p_{\text{K}}^{\alpha}) - T_{\text{M}}(p_0^{\alpha}) = \frac{T_{\text{M}}\left(V_{\text{M}}^{\alpha} - V_{\text{M}}^{\text{L}}\right)}{-\Delta H_{\text{m}}} 2\sigma K \tag{1'}$$

where

$T_{\text{M}}(p_{\text{K}}^{\alpha})$ = melting point at curvature K of the tip and constant liquid pressure p^{L}
$T_{\text{M}}(p_0^{\alpha})$ = melting point at curvature 0 of the tip and constant liquid pressure p^{L}.

We choose the temperature $T^{\text{L}/\alpha}$ of the melt close to the interface as reference temperature and obtain

$$\left[T_{\text{M}}(p_{\text{K}}^{\alpha}) - T^{\text{L}/\alpha}\right] - \left[T_{\text{M}}(p_0^{\alpha}) - T^{\text{L}/\alpha}\right] = \frac{T_{\text{M}}\left(V_{\text{M}}^{\alpha} - V_{\text{M}}^{\text{L}}\right)}{-\Delta H_{\text{m}}} 2\sigma K \tag{2'}$$

If the interface is planar K will be zero. Inserting $K = 0$ into Equation (2′) we obtain the trivial solution

$$T_{\text{M}}(p_{\text{K}}^{\alpha}) - T^{\text{L}/\alpha} = 0 \tag{3'}$$

If $K \neq 0$ it can be replaced by $1/\rho_{\text{tip}}$ in Equation (2′) which gives

$$T_{\text{M}}(p_{\text{K}}^{\alpha}) - T^{\text{L}/\alpha} = \frac{T_{\text{M}} V_{\text{M}}^{\alpha}}{-\Delta H_{\text{m}}} 2\sigma K = \frac{2\sigma T_{\text{M}} V_{\text{M}}^{\alpha}}{(-\Delta H_{\text{m}})\rho_{\text{tip}}} \tag{4'}$$

We introduce the shorter name T_{M} instead of $T_{\text{M}}(p_{\text{K}}^{\alpha})K = 1/\rho_{\text{tip}}$ and solve ρ_{tip} from Equation (2′) in the box we obtain:

$$\rho_{\text{tip}} = \frac{2\sigma T_{\text{M}} V_{\text{m}}^{\alpha}}{(-\Delta H_{\text{m}})(T_{\text{M}} - T^{\text{L}/\alpha})} \tag{9.27}$$

where

σ = surface tension of the melt
T_{M} = melting point at curvature K of the tip.

If we insert the value of ρ_{tip} into Equation (9.26) above, we obtain after reduction

$$V_{\text{growth}} = \frac{\lambda}{\sigma T_{\text{M}}}(T_{\text{M}} - T^{\text{L}/\alpha})(T^{\text{L}/\alpha} - T_0) \tag{9.28}$$

The interface temperature $T^{L/\alpha}$ is unknown. Fisher assumed that the correct value of V_{growth} is the maximum value of the function in Equation (9.28). If we take the derivative of V_{growth} with respect to $T^{L/\alpha}$ and the derivative is set equal to zero, we obtain

$$T^{L/\alpha} = \frac{T_M + T_0}{2} \tag{9.29}$$

which corresponds to a maximum of V_{growth}:

$$V_{growth} = \frac{\lambda}{\sigma T_M} \frac{(T_M - T_0)^2}{4}$$

or

$$V_{growth} = const(\Delta T)^2 \tag{9.30}$$

where
$\Delta T = T_M - T_0 =$ the total undercooling of the melt.

The simple theory indicates that the growth rate increases with the square of the undercooling.

Experimental Examination

Several metallurgists tested Fisher's theory, among them Walker [9], who investigated Ni and Co melts. He cooled a given amount of the metal melt to the homogeneous nucleation temperature. Solidification was initiated on the surface of the melt and the time interval between the solidification at two points in the melt at a given distance and the growth rate could be calculated. The radiation from the hot melt was observed with the aid of quartz rods, placed at the two observation points.

The rods transmitted the light to photocells connected to an oscilloscope. It showed the radiation spectrum as a function of the wavelength for each rod. When the solidification front passed a quartz rod the curve suddenly changed, because the temperature of the solid phase is lower than that of the melt, and the time for each passage could be registered.

Experiments were performed for many values of the undercooling and the corresponding growth rates were plotted versus the square of the undercooling. Figure 9.35 shows Walker's result. Later experiments have confirmed his results. Measurements on other metals have also been performed.

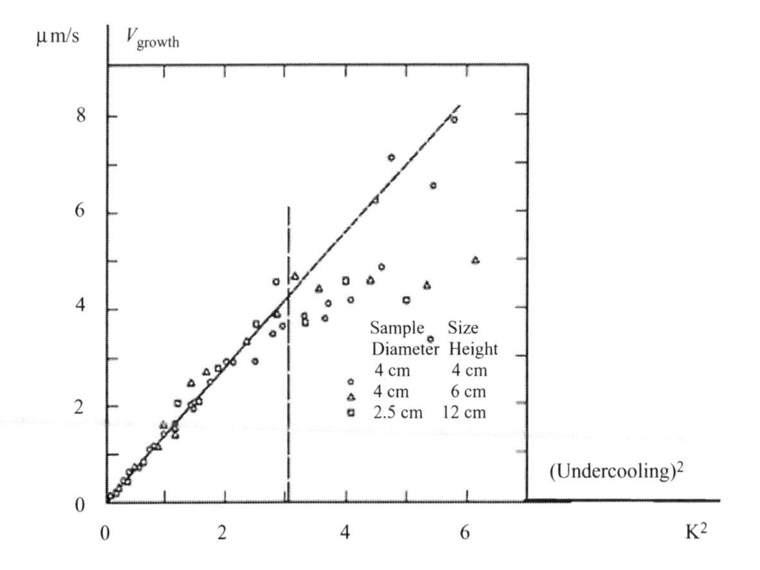

Figure 9.35 Growth rate as a function of undercooling in a pure Ni melt. (Walker, 1956.) Bruce Chalmers, courtesy of S. Chalmers.

Figure 9.35 shows that the agreement with the theory is acceptable for *low* undercoolings, i.e. the experiments verify the square law:

$$V_{\text{growth}} = const(\Delta T)^2 \tag{9.30}$$

At *higher* undercoolings the experiments show that a linear growth law describes the experimental results better. The change of growth law is indicated in the figure. However, the constant differs with a factor more than 10 between theory and experiments,

The radius of the tip has also been investigated experimentally. It was found that the true radius is much larger than the radius according to Fisher's theory (pages 507–508). The conclusion is that the kinetics of the interface is of importance. If the interface kinetics is taken into consideration, the derived radius becomes larger and better agreement between theory and experiments is achieved.

Below we will discuss different ways to obtain a better agreement between theory and experiments.

Temperature Field around a Growing Dendrite in Pure Metal Melts. Ivantsov's Model

Fick's second law:

$$\frac{1}{D_{\text{L}}}\frac{\partial x^{\text{L}}}{\partial t} = \frac{\partial^2 x^{\text{L}}}{\partial x^2} + \frac{\partial^2 x^{\text{L}}}{\partial y^2} + \frac{\partial^2 x^{\text{L}}}{\partial z^2}$$

The mathematics of dendritic growth has been discussed during a long time in the scientific literature. Papapetrou [10] suggested in the middle of the 1930s that the shape of a dendrite tip is parabolic. Later Ivantsov [11] presented a theoretical treatment of this proposal by solving Fick's second law for a growing needle with a parabolic shape. Ivantsov's model will be discussed on pages 516–518 (mass transport) and on page 522 (heat transport) in connection with the theory of dendritic growth in alloys. Here we just give the expression Ivantsov derived. It is called the *Invantsov's function.*

$$I(Pe) = Pe\, e^{Pe} E_I(Pe) \tag{9.31}$$

where Pe is Peclet's number. It is defined by the relationship

$$Pe = \frac{V_{\text{growth}}\, \rho_{\text{tip}}}{2D_{\text{L}}} \qquad \text{Peclet's number} \tag{9.32}$$

D_{L} is the diffusion coefficient of the alloying element in the melt. Ivantsov's function for a temperature field [the left-hand side of Equation (9.33)] is defined as

$$I(Pe) = \frac{C_{\text{p}}(T_{\text{interface}} - T_{\text{melt}})}{-\Delta H_{\text{m}}} \tag{9.33}$$

where C_{p} is the molar heat capacity and $-\Delta H_{\text{m}}$ is the molar heat of fusion of the alloy.

E_{I} in Equation (9.31) is the so-called *exponential integral function.* It is defined by Equation (9.43) on page 516. ρ_{tip} is the tip radius of the dendrite.

The relationship (9.31) gives only the product $V_{\text{growth}}\, \rho_{\text{tip}}$ at a given undercooling and not separate values of the two quantities. The product $I(Pe) = Pe\, e^{Pe} E_{\text{I}}(Pe)$ can be series expanded (see pages 521–522) and the first approximate value is $I(Pe) = Pe$.

In order to find a relationship between the growth rate and the tip radius, different types of stability criteria have been used. One example of such a stability criterion is the assumption that the tip adopts its most stable shape related to influences from the surroundings. Examples of such influences are temperature gradients or surface perturbations.

The most common relationship between V_{growth} and ρ_{tip} is the expression given by Kurz and Fisher [12]

$$\rho_{tip} = Const \frac{1}{\sqrt{V_{growth}}} \tag{9.34}$$

where the constant in most cases is given by

$$Const = \sqrt{\frac{2\pi\lambda\sigma T_0}{(-\Delta H)(T^{L/\alpha} - T_{melt})}} \tag{9.35}$$

where

λ = thermal conductivity of the melt
σ = surface tension between the melt and the solid
T_0 = temperature of the melt far from the interface
$-\Delta H_m$ = molar heat of fusion
$T^{L/\alpha}$ = temperature of the melt close the interface
T_{melt} = temperature of the melt far from the interface.

Introduction of the relationship (9.34) into Equation (9.26) on page 507 gives a simple relationship between the growth rate and the undercooling. This relationship gives a reasonably good agreement between the theory and the experiments. However, the assumptions can be questioned and the theory does not give the answer on the question why dendritic growth occurs in particular directions.

In addition to the temperature field around the tip, we have to consider the concentration field around the tip in alloy melts. An improvement of the theory of dendritic growth is obtained if the influences on the undercooling of the surface tension and the kinetics are introduced. This is analogous with the treatment of the theory of faceted growth and will be discussed on the next pages.

9.5.3 Temperature Field, Undercooling and Solidification Rate at Dendritic Growth in Alloy Melts

Undercooling and Solidification Rate at Dendritic Growth in Alloy Melts

The driving force of dendritic growth in an *alloy* melt is the undercooling of the melt ahead of the solidification front or, more strictly, the interface undercooling relative to the temperature of the bulk liquid.

The new contribution to the undercooling in an alloy, compared to that of a pure metal, is the important effect of mass transport, i.e. the diffusion of the alloying element.

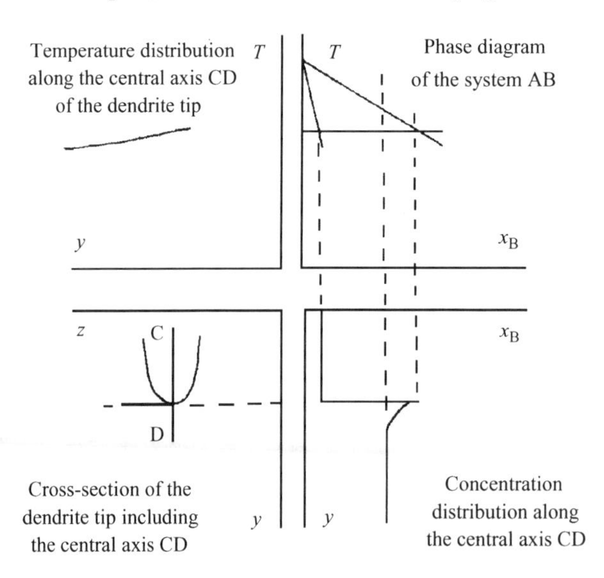

Figure 9.36 Concentration and temperature distributions in front of a growing dendrite.

Figure 9.36 shows the concentration and temperature distributions in front of a growing dendrite tip. The growth rate of a dendrite with such distributions has been analyzed theoretically by many authors.

Total Undercooling in Binary Alloys

The undercooling consists of four parts for alloys. In addition to the three parts discussed above for pure metals a fourth part, related to the concentration of the alloying element, has to be added.

$$\Delta T_{total} = \Delta T_{surface} + \Delta T_{kin} + \Delta T_{mass} + \Delta T_{thermal} \tag{9.36}$$

where

ΔT_{total} = difference between the melting-point temperature $T_{M'}$ and the temperature T_0 in the melt far from the interface

$\Delta T_{surface}$ = change of temperature of the melt close to the tip owing to the curvature of the dendrite tip and the surface tension (see pages 506–508)

ΔT_{kin} = change of temperature owing to the kinetic reaction (see page 505)

ΔT_{mass} = temperature difference owing to mass transport of the alloying element

$\Delta T_{thermal}$ = temperature difference between the temperatures $T^{L/\alpha}$ and T_0, which causes heat transfer through the melt (see page 505 and Figure 9.34b).

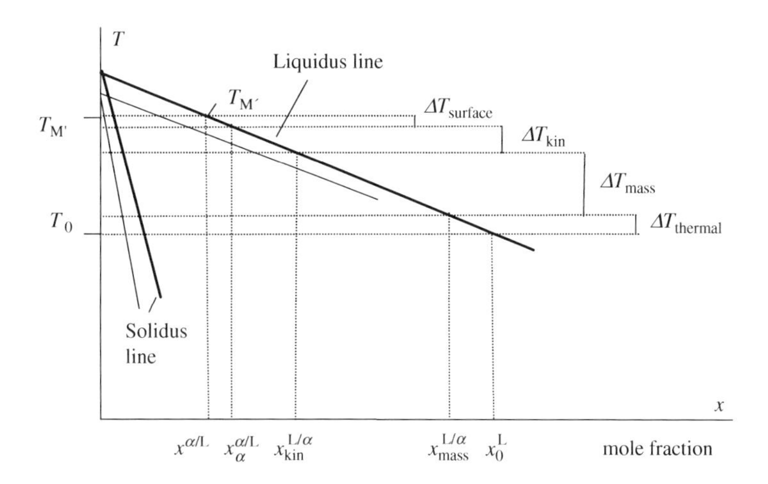

Figure 9.37 Part of the phase diagram of a binary alloy.
$T_{M'}$ = melting point temperature of the tip
$T^{L/\alpha}$ = temperature of the melt at the interface
T_0 = temperature far from the interface.

The quantities in Equation (9.36) are shown in the phase diagram in Figure 9.37.

Influence of Various Parameters on Undercooling at Dendritic Growth in Binary Alloy Melts

At dendritic growth of alloys the undercooling depends on

- the cooling rate and therefore the growth rate
- the concentration of the alloying element in the melt
- the type of alloying element.

The influences of these parameters have been examined experimentally.

Influence of Cooling Rate and Concentration of the Alloying Element on Undercooling

The cooling rate influences the heat removal from the solidification front and hence the solidification process, i.e. the growth rate and the undercooling.

Bäckerud [13] has studied the influence of cooling rate and solute concentration on the undercooling of a number of Al-Cu alloys. The Figures 9.38 and 9.39 show some of his results.

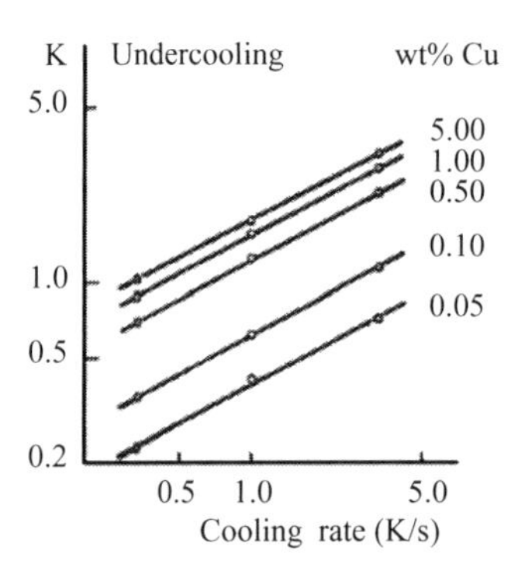

Figure 9.38 Undercoolings of a series of solidifying Al–Cu alloys of different compositions as a function of the cooling rate. © Skanaluminium 1986.

Figure 9.39 Undercoolings of a series of solidifying Al–Cu alloys at various cooling rates as a function of composition. © Skanaluminium 1986.

Figure 9.38 shows the undercooling as a function of the cooling rate of a series of Al-Cu alloys. Figure 9.39 shows the result of the same investigation plotted in another way to show the undercooling as a function of the concentration of the alloying element.

Figures 9.38 and 9.39 show that

- the undercooling increases with increasing cooling rate and increasing concentration of the alloying element.

Influence of Type of Alloying Element on Undercooling in Binary Alloys

Figure 9.33 on page 504 shows the undercooling as a function of the square root of the cooling rate for pure Al and a series of Al alloys with different alloying elements with the same concentration (2 at%). Obviously the type of alloying element influences undercooling strongly. All the alloying elements increase the undercooling.

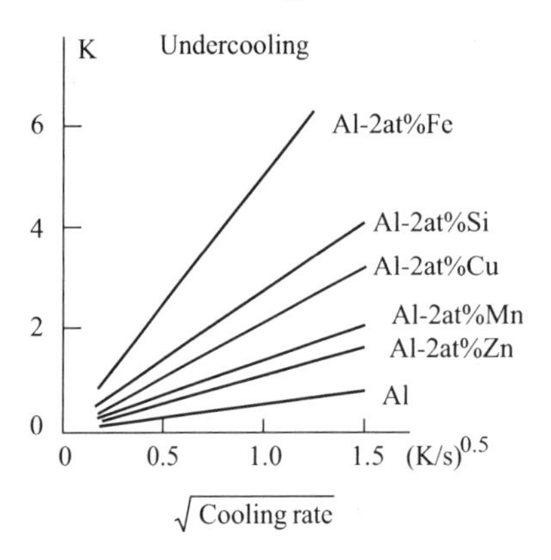

Figure 9.40 Undercooling as a function of cooling rate in pure Al and in five Al alloys. © Skanaluminium 1986.

Table 9.4 Total undercooling as a function of k

Sample	k	ΔT_{total} (°C)
Al-2at%Mn	0.8	0.5
Al-2at%Zn	0.5	0.8
Al-2at%Cu	0.17	1.5
Al-2at%Si	0.13	2.0
Al-2at%Fe	0.022	4.0

Table 9.4 shows that the smaller the partition coefficient k of the alloy is, the larger will be the undercooling.

Different alloying elements means that both the slope of the liquidus line and the partition coefficient change. Both these quantities influence the total undercooling.

- A low slope of the liquidus line and a decrease of the partition coefficient k increase the undercooling owing to mass transport of the alloying element and hence result in an increase of the total undercooling.

These topics will be analyzed more closely in Section 9.5.4 after Example 9.4 below.

Example 9.4

Burden and Hunt [14] have performed numerous experiments on dendritic growth in Al-Cu alloys. Some results of their experiments with the alloy Al-2.0wt%Cu are shown in the Figures 1′ and 2′.

The Figures 1′ and 2′ show the undercooling and the concentration of Cu in the central part of the primary stems, respectively, as functions of the growth rate. Figure 3′ shows part of the phase diagram of the system Al-Cu.

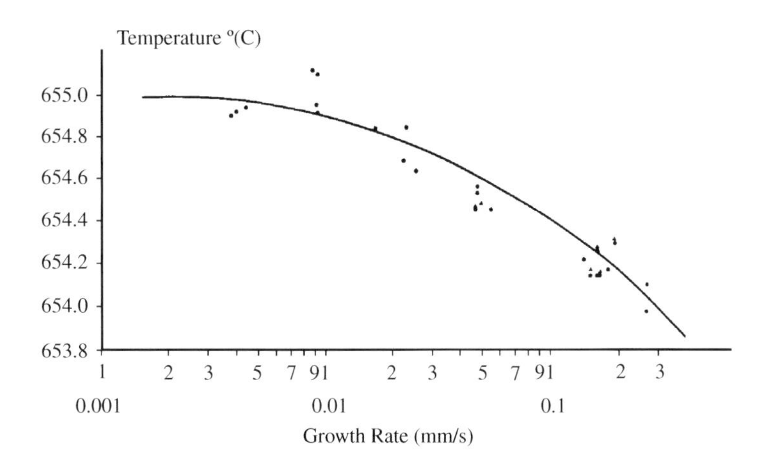

Figure 1′ Relationship between the temperature of the melt and the growth rate when the alloy Al-2.0 wt%Cu alloy solidifies and becomes quenched.

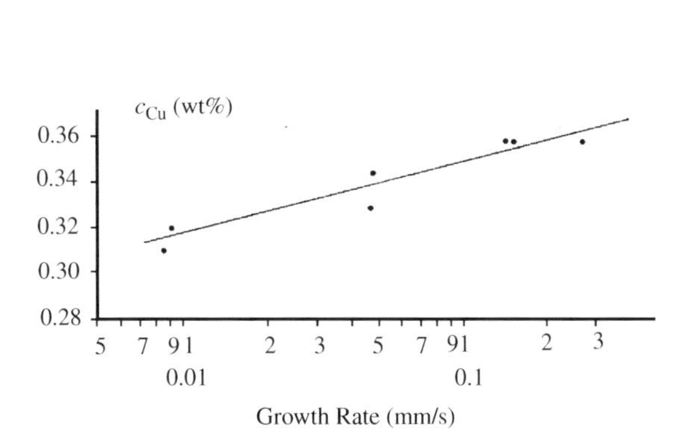

Figure 2′ Relationship between the growth rate and the solute concentration in the central part of the primary stems of the dendrites formed on solidification and quenching of Al–Cu alloys. Reproduced from Elsevier © 1974.

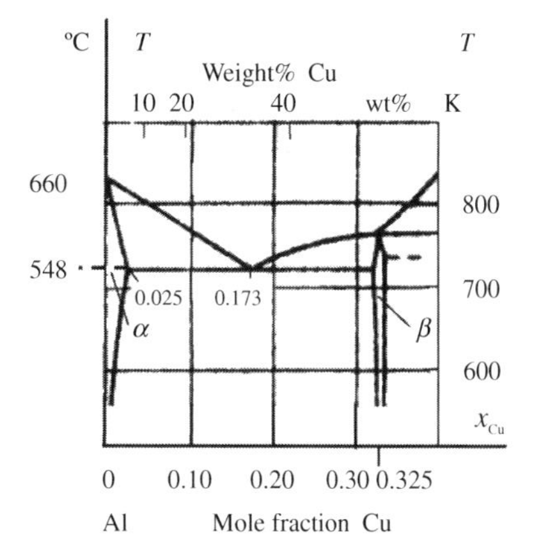

Figure 3′ Part of the phase diagram of the system Al–Cu.

Do the measured Cu concentrations in the primary stems follow the equilibrium conditions at the solid/liquid interface during the solidification process? Motivate your answer and discuss the reason for the experimental behaviour.

Solution

The phase diagram corresponds to the equilibrium conditions during solidification. From the phase diagram you can derive the inverted slope of the solidus line $\approx \dfrac{0-5}{660-548} \approx 0.05\,\text{wt\%/°C}$.

The corresponding calculation on the experimental values can be obtained from the Figures 1 and 2. Figure 1′ shows that the undercooling increases with approximately 654.9 − 654.4 = 0.5 °C when the growth rate increases from 0.01 to 0.1 mm/s.
Figure 2′ shows that the same change of the growth rate corresponds to an increase of the Cu concentration from 0.31 to 0.34 wt%, i.e. by 0.03 wt%. Hence, the measured increase of the concentration per centigrade will be

$$\frac{\Delta c}{\Delta T} \approx \frac{0.03}{0.5} \approx 0.06\ \text{wt\%/°C}$$

The experimental value of the increase of the concentration per centigrade is larger than the theoretical value. Therefore, the solidification process does not occur at equilibrium conditions. This conclusion is supported by calculation of the partition coefficient of the alloy, From the phase diagram we obtain at equibrilium (eutectic solidification)

$$k_{\text{exp}} \approx \frac{0.025}{0.173} \approx 0.14$$

An experimental value of the partition coefficient is obtained if we divide the measured final solid Cu concentration with the Cu concentration of the initial melt.

$$k_{\text{exp}} \approx \frac{c^{\text{s}}_{\text{final}}}{c^{\text{L}}_{\text{ave}}} = \frac{0.34}{2.0} \approx 0.17$$

The Cu concentration in the dendrite stem is larger than expected from the equilibrium value.

Answer:
The measured Cu concentrations indicate that the solidification process does not occur at equilibrium conditions,
The reason is that the kinetic process is rapid and the relaxation process is slower than the kinetic process. There is not enough time for the following relaxation process. The result is that the solidification process does not occur at equilibrium conditions but not very far from equilibrium conditions.

9.5.4 Theory of Dendrite Growth in Alloy Melts. Ivantsov's Model for the Concentration Field around a Paraboloid Tip

Figure 9.33 Dendritic growth in steel, which contains 0.4 wt% C.

The solute in the dendrites in Figure 9.33 diffuses into the melt and reduces the etching effect on the melt near the dendrites, which results in the white areas. The etching pattern of the dendrite in Figure 9.33 reveals that there is a *concentration field* in the melt in front of the primary tip. The thickness of the boundary layer ahead of the tip is of the same order as the radius of the dendrite tip. Since heat is evolved at the tip, there will also be a *temperature field* around the tip.

Undercooling as a Function of Concentrations, Surface Tension and Interface Kinetics in Binary Alloys

With the aid of common thermodynamic relationships and the phase diagram in Figure 9.37 on page 511, the temperature differences on page 511 can be expressed as functions of concentrations and growth rates.

The following equations are obtained for each of the terms in Equation (9.36)

$$\Delta T_{\text{total}} = \Delta T_{\text{surface}} + \Delta T_{\text{kin}} + \Delta T_{\text{mass}} + \Delta T_{\text{thermal}} \tag{9.36}$$

namely

$$\Delta T_{\text{total}} = \frac{x^{L/\alpha} - x_0^L}{m_L} \quad (9.37)$$

$$\Delta T_{\text{surface}} = \frac{x^{L/\alpha} - x_\sigma^{L/\alpha}}{m_L} = \frac{2\sigma_{\text{hkl}} T^{L/\alpha} V_m^\alpha}{\rho_{\text{tip}}(-\Delta H_m)} \quad (9.38)$$

$$\Delta T_{\text{kin}} = \frac{x_\sigma^{L/\alpha} - x_{\text{kin}}^{L/\alpha}}{m_L} = \left(\frac{V_{\text{growth}}}{\beta_{\text{hkl}}}\right)^{\frac{1}{n}} \quad (9.39)$$

$$\Delta T_{\text{mass}} = \frac{x_{\text{kin}}^{L/\alpha} - x_{\text{mass}}^{L/\alpha}}{m_L} \quad (9.40a)$$

$$\Delta T_{\text{thermal}} = \frac{x_{\text{mass}}^{L/\alpha} - x_0^L}{m_L} \quad (9.41)$$

where

m_L = slope of the liquidus line in the phase diagram
x_0^L = average concentration of the alloying element in the alloy
$x^{L/\alpha}$ = concentration of the alloying element in the melt close to a planar interface
$x_\sigma^{L/\alpha}$ = concentration of the alloying element in the melt close to the curved interface
σ_{hkl} = surface tension in the crystallographic direction hkl
ρ_{tip} = radius of the hemispherical tip
V_m^α = molar volume of the solid phase
β_{hkl} = growth constant in the crystallographic direction hkl
$x_{\text{kin}}^{L/\alpha}$ = concentration of the alloying element in the melt related to the interface kinetics
$x_{\text{tmass}}^{L/\alpha}$ = concentration of the alloying element in the melt related to the mass transport of the alloying element
V_{growth} = growth rate of the dendritic tip
x_0^L = concentration of the alloying element in the melt far from the interface.

Normally ΔT_{mass} is much larger than $\Delta T_{\text{thermal}}$ and the latter can often be neglected. This is the case in our further analysis in this chapter. When $\Delta T_{\text{thermal}}$ is neglected

$$\Delta T_{\text{mass}} = \frac{x_{\text{kin}}^{L/\alpha} - x_0^L}{m_L} \quad (9.40b)$$

Both the surface tension and also the growth constant are anisotropic and follow the crystallographic directions in analogy with the conditions for faceted growth.

Concentration Field around a Dendrite Tip in Binary Alloys

Ivantsov's theoretical analysis predicts that the dendrite tip will have the shape of a *paraboloid of revolution* (Figure 9.41a).

If the partition coefficient $k < 1$, Figure 9.41b shows the solubility of the alloying element is lower in the tip than in the melt ahead of the solidification front. Therefore, the solute atoms close to the interface diffuse into the melt. The Figure 9.41b shows the concentration of the solute in the melt along the line AA' in Figure 9.41a as a function of position.

The dendritic growth is influenced by the *anisotropy of the surface energy* and of the *interface kinetics*. This must be considered in a more accurate theoretical analysis.

Figure 9.41a Emission of alloying element from a growing dendrite tip.

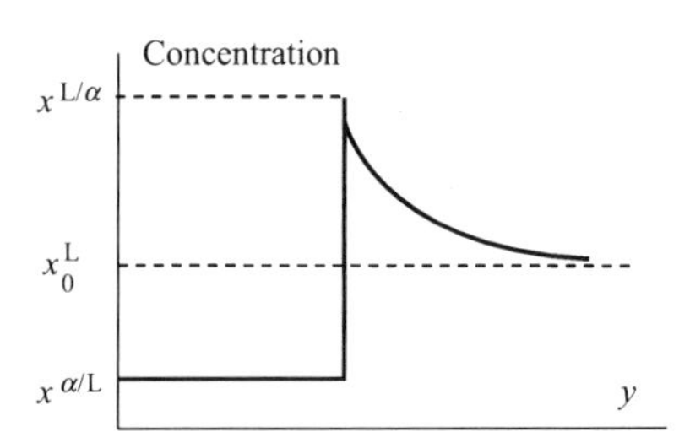

Figure 9.41b Concentration of the alloying element inside and around a growing dendrite tip.

Ivantsov's Model for the Concentration Field around a Paraboloid Tip

It has been known for a long time that a paraboloid of revolution is a better approximation of the shape of a dendrite tip than a hemisphere. Papapetrou [10) showed in 1936 that a rod with a growing tip has a parabolic profile at steady state conditions. Ten years later Ivantsov [11] developed a model for the concentration field around a growing tip, based on Papapetrou's observation.

Ivantsov 's model is based on two assumptions:

1. The concentration of the alloying element is constant along the surface of the tip.
2. The conditions are stationary. This means that the concentration field and the growth rate are independent of time.

His model will be described but not derived below in cartesian coordinates.

Cartesian Coordinates

The diffusion field is derived as a solution of Fick's second law in three dimensions

$$\frac{1}{D_L}\frac{\partial x^L}{\partial t} = \frac{\partial^2 x^L}{\partial x^2} + \frac{\partial^2 x^L}{\partial y^2} + \frac{\partial^2 x^L}{\partial z^2} \tag{9.42}$$

where

x, y, z = coordinates
x^L = concentration of the alloying element in the melt as a function of the coordinates
D_L = diffusion coefficient of the alloying element in the melt.

The concentration can be expressed in mole fraction units or weight per cent (wt%). Here we will use the first alternative.

According to Ivantsov's model the concentration field around a growing dendrite tip can be written as

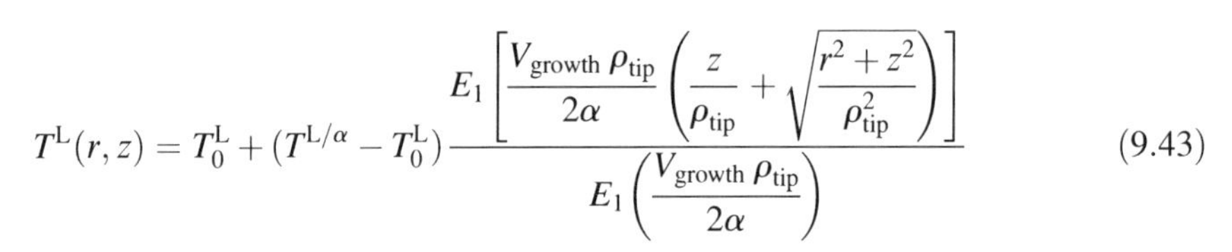

$$T^L(r,z) = T_0^L + (T^{L/\alpha} - T_0^L)\frac{E_1\left[\frac{V_{growth}\rho_{tip}}{2\alpha}\left(\frac{z}{\rho_{tip}} + \sqrt{\frac{r^2+z^2}{\rho_{tip}^2}}\right)\right]}{E_1\left(\frac{V_{growth}\rho_{tip}}{2\alpha}\right)} \tag{9.43}$$

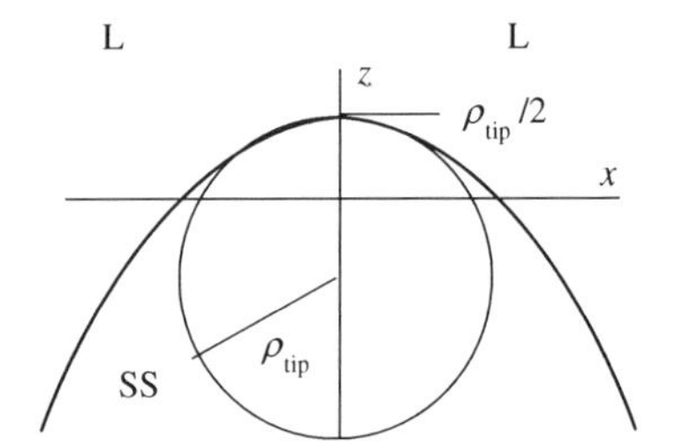

Figure 9.42 Parabolic tip with the radius of curvature $= \rho_{tip}$ at the tip. $z_{tip} = \rho_{tip}/2$.

where the E_1 function (*exponential integral function*) is defined by the relationship

$$E_1(u) = \int_u^\infty \frac{e^{-u}}{u} du \tag{9.44}$$

and

$x^{L/\alpha}$ = concentration of the alloying element in the melt close to the top of the dendrite tip
x_0^L = concentration of the alloying element in the melt far from the dendrite tip
V_{growth} = growth rate in the z direction
ρ_{tip} = radius of curvature at the tip.

Parabolic Coordinates

The concentration distribution in the melt around the dendrite tip can alternatively be described with the aid of parabolic coordinates, α, β and φ (Figure 9.43). They are defined by the relationships:

$$x = \rho_{tip}\alpha\beta\cos\varphi \tag{9.45}$$

$$y = \rho_{tip}\alpha\beta\sin\varphi \tag{9.46}$$

$$z = \rho_{tip}\frac{\alpha^2 - \beta^2}{2} \tag{9.47}$$

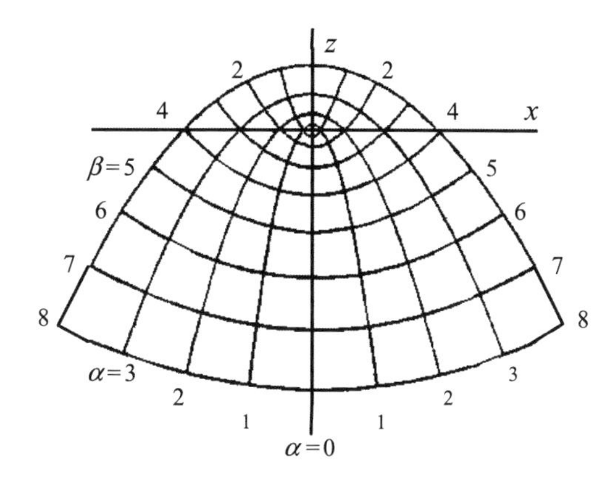

Figure 9.43 Definition of circular-parabolic coordinates.

The origin has the coordinates (0, 0), i.e. $\alpha = 0$ and $\beta = 0$.

$\alpha = 1$ and $\beta = 0$ corresponds to the surface of the growing dendrite tip. Hence $\frac{z}{\rho_{tip}} = \frac{\alpha^2 - \beta^2}{2} = \frac{1}{2}$. Reproduced from Elsevier © 1968.

In order to express the concentration distribution with the aid of parabolic coordinates we calculate part of the numerator of Equation (9.43):

$$\left(\frac{z}{\rho_{tip}} + \sqrt{\frac{x^2 + y^2 + z^2}{\rho_{tip}^2}}\right) = \frac{\alpha^2 - \beta^2}{2} + \sqrt{\alpha^2\beta^2\cos^2\varphi + \alpha^2\beta^2\sin^2\varphi + \frac{(\alpha^2 - \beta^2)^2}{4}}$$

$$= \frac{\alpha^2 - \beta^2}{2} + \sqrt{\alpha^2\beta^2 + \frac{(\alpha^2 - \beta^2)^2}{4}} = \frac{\alpha^2 - \beta^2}{2} + \frac{\alpha^2 + \beta^2}{2} = \alpha^2$$

This expression is inserted into Equation (9.43), which gives

$$x^{L}(\alpha) = x_0^{L} + (x^{L/\alpha} - x_0^{L})\frac{E_1\left[\frac{V_{growth}\,\rho_{tip}}{2D_L}\alpha^2\right]}{E_1\left(\frac{V_{growth}\,\rho_{tip}}{2D_L}\right)} \tag{9.48}$$

The ratio $V_{growth}\,\rho_{tip}/2D_L$ is equal to Péclet's number [Equation (9.32) on page 509]. It is equal to the distance $\rho_{tip}/2$ of the tip from the origin divided by the so-called *diffusion distance* D/V_{growth}.

$$V_{growth}\,\rho_{tip} = 2D_L Pe \tag{9.49}$$

The advantage with this representation in parabolic coordinates is that the concentration field around the tip can be expressed by a single coordinate.

$\alpha = 1$ corresponds to the surface of the tip. If we introduce $\alpha = 1$ into Equations (9.45), (9.46), and (9.47) we obtain the equation of the surface in parameter representation.

$$x = \rho_{\text{tip}}\,\beta\cos\varphi \tag{9.50}$$

$$y = \rho_{\text{tip}}\,\beta\sin\varphi \tag{9.51}$$

$$z = \rho_{\text{tip}}\frac{1-\beta^2}{2} \tag{9.52}$$

If the parameter β is eliminated we obtain the equation of the tip surface. The Equations (9.50) and (9.51) are squared and added, which gives

$$x^2 + y^2 = \rho_{\text{tip}}^2\beta^2 \tag{9.53}$$

Elimination of β^2 between Equations (9.52) and (9.53) gives

$$z = \frac{\rho_{\text{tip}}}{2} - \frac{x^2+y^2}{2\rho_{\text{tip}}} \tag{9.54}$$

Equation (9.54) is the equation of the surface of the tip.
At the tip $x^2 + y^2 = \rho_{\text{tip}}^2\beta^2 = 0$, which means that $z = \rho_{\text{tip}}/2$.

Cylindrical Coordinates
Another representation of the concentration field is obtained if we introduce cylindrical coordinates r, φ and z (Figure 9.44) into Equation (9.43).

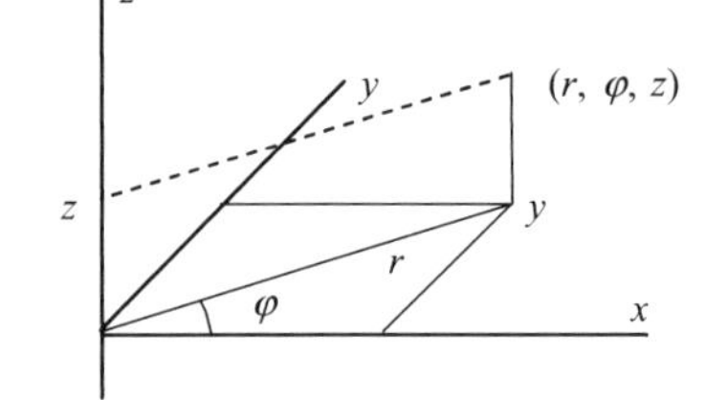

Figure 9.44 Cylindrical and Cartesian coordinate systems.

$$x = r\cos\varphi \tag{9.55}$$

$$y = r\sin\varphi \tag{9.56}$$

$$z = z \tag{9.57}$$

In this representation the concentration distribution can be written as

$$x^{\text{L}}(r,z) = x_0^{\text{L}} + \left(x^{\text{L}/\alpha} - x_0^{\text{L}}\right)\frac{E_1\left[\dfrac{V_{\text{growth}}\,\rho_{\text{tip}}}{2D_{\text{L}}}\left(\dfrac{z}{\rho_{\text{tip}}} + \sqrt{\dfrac{r^2+z^2}{\rho_{\text{tip}}^2}}\right)\right]}{E_1\left(\dfrac{V_{\text{growth}}\,\rho_{\text{tip}}}{2D_{\text{L}}}\right)} \tag{9.58}$$

The validity of this equation can be checked by introduction of $r = 0$ and $z = \rho_{\text{tip}}/2$ into Equation (9.58), which gives $x^{\text{L}} = x^{\text{L}/\alpha}$.
An application of Equation (9.58) is given in Example 9.5 below.

Example 9.5

A dendrite with a paraboloid tip grows in an alloy melt with the concentration x_0^{L}. The radius of curvature ρ_{tip} of the tip is constant but not known. The concentration of the melt close to the tip is known and equal to $x^{\text{L}/\alpha}$.

a) Use Equation (9.58) to derive an expression for the concentration gradient of the alloying element along the central axis ($r = 0$) in the melt as a function of z, known quantities and Péclet's number Pe. In addition, set up an expression for the concentration gradient at the tip.

b) Determine the ratio $\dfrac{x^{\text{L}/\alpha} - x_0^{\text{L}}}{x^{\text{L}/\alpha} - x^{\alpha/\text{L}}}$ as a function of Pe at the tip.

Solution

a)

Along the central axis $r = 0$. The desired quantity is the value of $\frac{\mathrm{d}x^{\mathrm{L}}(0,z)}{\mathrm{d}z}$ at the tip. Hence, we need the derivative of the function $x^{\mathrm{L}}(0, z)$ with respect to z. With the concentrations expressed in mole fractions Equation (9.58) can be written as

$$x^{\mathrm{L}}(0,z) = x_0^{\mathrm{L}} + (x^{\mathrm{L}/\alpha} - x_0^{\mathrm{L}})\frac{E_1[Pe\,g(0,z)]}{E_1(Pe)} \tag{1'}$$

where

$$g(0,z) = \left(\frac{z}{\rho_{\mathrm{tip}}} + \sqrt{\frac{r^2+z^2}{\rho_{\mathrm{tip}}^2}}\right)_{r=0} = \frac{2z}{\rho_{\mathrm{tip}}} \tag{2'}$$

and Equation (9.32) on page 509

$$Pe = \frac{V_{\mathrm{growth}}\,\rho_{\mathrm{tip}}}{2D_{\mathrm{L}}} \tag{3'}$$

Taking the derivative of Equation (1′) with respect to z gives

$$\frac{\mathrm{d}x^{\mathrm{L}}(0,z)}{\mathrm{d}z} = \frac{\partial x^{\mathrm{L}}(0,z)}{\partial E_1[Pe\,g(0,z)]}\,\frac{\partial E_1[Pe\,g(0,z)]}{\partial g(0,z)}\,\frac{\mathrm{d}g(0,z)}{\mathrm{d}z} \tag{4'}$$

If the variable u is replaced by the function $g(0, z)$ in Equation (9.44) on page 516 we obtain

$$\frac{\partial E_1[Pe\,g(0,z)]}{\partial g(0,z)} = -\frac{e^{-Pe\,g(0,z)}}{Pe\,g(0,z)}\,\frac{\partial[Pe\,g(0,z)]}{\partial g(0,z)}$$

or, because Pe is a constant,

$$\frac{\partial E_1[Pe\,g(0,z)]}{\partial g(0,z)} = -\frac{e^{-Pe\,g(0,z)}}{Pe\,g(0,z)}Pe = -\frac{e^{-Pe\,g(0,z)}}{g(0,z)} \tag{5'}$$

The first and the last derivatives in Equation (4′) are obtained by taking the derivative of Equations (1′) and (2′) with respect to z. These derivatives and the derivative in Equation (5′) are inserted into Equation (4′), which gives the concentration gradient as a function of z:

$$\frac{\mathrm{d}x^{\mathrm{L}}(0,z)}{\mathrm{d}z} = \frac{x^{\mathrm{L}/\alpha} - x_0^{\mathrm{L}}}{E_1(Pe)}\,-\frac{\mathrm{e}^{-Pe\frac{2z}{\rho_{\mathrm{tip}}}}}{\frac{2z}{\rho_{\mathrm{tip}}}}\,\frac{2}{\rho_{\mathrm{tip}}} \tag{6'}$$

Equation (6′) is valid for $z \geq \frac{\rho_{\mathrm{tip}}}{2}$.

The value of the concentration gradient at the tip is obtained if we insert $z = \rho_{\mathrm{tip}}/2$ into Equation (9.54) on page 518 into Equation (6′).

$$\left(\frac{\mathrm{d}x^{\mathrm{L}}(0,z)}{\mathrm{d}z}\right)_{z=\frac{\rho_{\mathrm{tip}}}{2}} = \frac{x^{\mathrm{L}/\alpha} - x_0^{\mathrm{L}}}{E_1(Pe)}\left(-\frac{e^{-Pe}}{1}\right)\frac{2}{\rho_{\mathrm{tip}}} \tag{7'}$$

b)

A mass balance at the tip gives

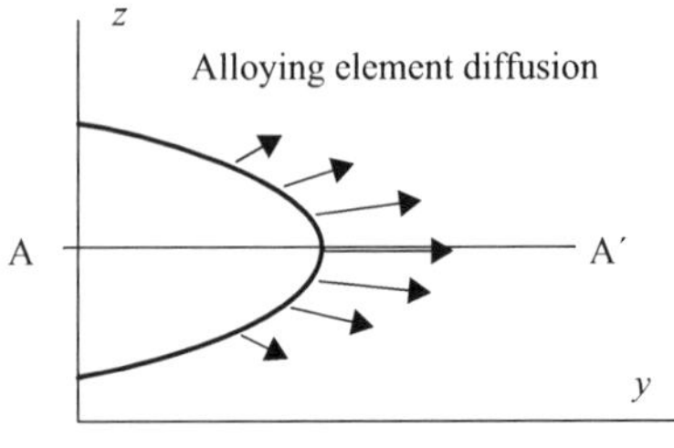

Figure 9.5 Mass balance at diffusion through a volume at the tip.

$$(x^{L/\alpha} - x^{\alpha/L})V_{growth} = -D_L\left(\frac{dx^L(0,y)}{dy}\right)_{z=\frac{\rho_{tip}}{2}} \qquad y \geq \frac{\rho_{tip}}{2} \qquad (8')$$

This is Fick's first law applied on a volume element along the line AA' in the figure at steady state growth.

If we insert the value of the concentration gradient [Equation (7′)] into Equation (8′) we obtain

$$(x^{L/\alpha} - x^{\alpha/L})V_{growth} = -D_L\frac{x^{L/\alpha} - x_0^L}{E_1(Pe)}\left(-e^{-Pe}\right)\frac{2}{\rho_{tip}} \qquad (9')$$

Reduction and combination with Equation (3′) gives

$$\frac{x^{L/\alpha} - x_0^L}{x^{L/\alpha} - x^{\alpha/L}} = E_1(Pe)e^{Pe}\frac{V_{growth}}{D_L}\frac{\rho_{tip}}{2} = Pe\, e^{Pe}\, E_1(Pe) \qquad (10')$$

where $E_1(Pe) = \int\limits_{Pe}^{\infty}\frac{e^{-u}}{u}du$

Answer:

a) The concentration along the central axis in the melt is

$$\frac{dx^L(0,z)}{dz} = \frac{x^{L/\alpha} - x_0^L}{E_1(Pe)} - \frac{e^{-Pe\frac{2z}{\rho_{tip}}}}{\frac{2z}{\rho_{tip}}}\frac{2}{\rho_{tip}} \qquad \text{for} \quad z \geq \frac{\rho_{tip}}{2}$$

The concentration gradient at the tip is

$$\left(\frac{dx^L(0,z)}{dz}\right)_{z=\frac{\rho_{tip}}{2}} = \frac{x^{L/\alpha} - x_0^L}{E_1(Pe)}\left(-e^{-Pe}\right)\frac{2}{\rho_{tip}}$$

b) $\frac{x^{L/\alpha} - x_0^L}{x^{L/\alpha} - x^{\alpha/L}}$ is equal to $Pe\, e^{Pe}E_1(Pe)$

where $E_1(Pe) = \int\limits_{Pe}^{\infty}\frac{e^{-u}}{u}du$ [Equation (9.44) on page 516]

and

$I(Pe) = Pe\, e^{Pe}E_1(Pe)$ [Equation (9.31) on page 509]

Various Approximations of Pe and Ivantsov's Function I(Pe)

Peclet's number Pe is defined as $V_{growth}\,\rho_{tip}/2D_L$ (page 509).

Example 9.5 above shows that if the partition k and the concentrations x_0^L and $x^{L/\alpha}$ are known, it is possible to derive Pe numerically from the equation in the answer of Example 9.5b.

$$\frac{x^{L/\alpha} - x_0^L}{x^{L/\alpha} - x^{\alpha/L}} = Pe\, e^{Pe}E_1(Pe) \qquad (9.59)$$

When Pe has been found with the aid of Equation (9.59) or an approximate value is used, the product $V_{growth}.\rho_{tip}$ can be solved from Equation (9.32) on page 509 if the diffusion constant D_L is known.

$$V_{growth}\,\rho_{tip} = 2D_L Pe \tag{9.49}$$

Unfortunately the growth rate V_{growth} and radius of curvature ρ_{tip} cannot be calculated separately. In order to find another relationship between the growth rate and the tip radius we have to consider the interface kinetics and the interface surface tension. This topic will be discussed on pages 525–529 and in Example 9.6 on page 526.

When Pe is known, the exponential integral function E_1 (Pe), defined on page X64, can be calculated. Values of E_1 (Pe) as a function of Pe are listed and available in tables.

The product $Pe\,e^{Pe}E_1(Pe)$ is Ivantsov's function and denoted by $I\,(Pe)$. It can be written as an infinite fraction: (see Kurz and Fisher [12])

$$I(Pe) = \frac{x^{L/\alpha} - x_0^L}{x^{L/\alpha} - x^{\alpha/L}} = Pe\,e^{Pe}E_1(Pe) = \cfrac{Pe}{Pe + \cfrac{1}{1 + 1Pe + \cfrac{2}{1 + \cfrac{2}{Pe + \cdots}}}} \tag{9.60}$$

Instead of using the tables of $E_1(Pe)$ various approximations of Ivantsov's function can be used. Figure 9.45 shows a number of approximations, which correspond to successive abbreviations, and in the end to the true value of the function.

$$I_0 = Pe \qquad \text{(line 1)} \tag{9.61}$$

$$I_1 = \frac{Pe}{Pe + 1} \qquad \text{(line 1 + 2)} \tag{9.62}$$

$$I_2 = \frac{Pe}{Pe + \frac{1}{1+1}} = \frac{2Pe}{2Pe + 1} \qquad \text{(lines 1 + 2 + 3)} \tag{9.63}$$

and so on. All I values are < 1. Equation (9.61) is only valid for small supersaturations.

$$I_\infty = Pe\,e^{Pe}E_1(Pe) \tag{9.64}$$

Some of these approximations are shown in Figure 9.45.

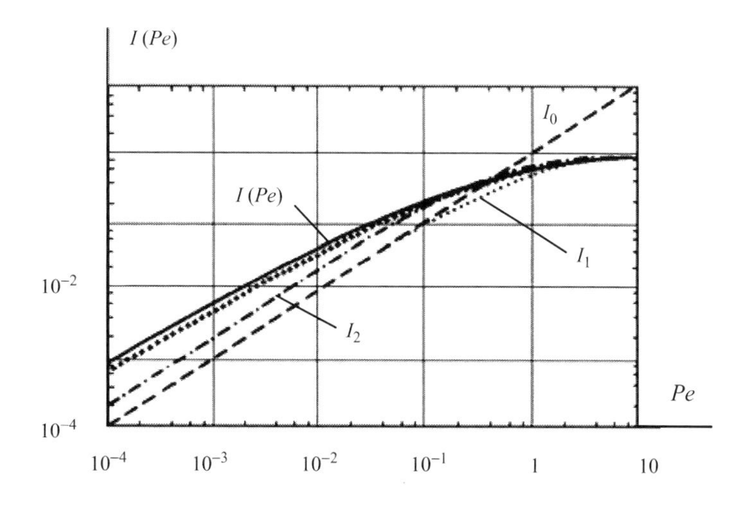

Figure 9.45 $I\,(Pe)$ as a function of Pe (full line curve) and some approximations of this curve. Reproduced with permission from Trans Tech © 1992.

The full line curve represents the function $I\,(Pe). = Pe\,e^{Pe}E_1(Pe)$ and the dashed curves show the approximations $I_0\,(Pe)$, $I_1\,(Pe)$ and $I_2\,(Pe)$. The dotted curve close to $I\,(Pe)$ is the function

$$I = \frac{Pe\,e^{Pe}E_1(Pe)}{\ln Pe\,e^{Pe}E_1(Pe)} \tag{9.65}$$

which is a very good approximation for $I(Pe)$.

Temperature Field around a Dendrite Tip

This topic has been discussed briefly for pure metals on pages 505–508. The present model is valid for both alloys and pure metals.

The growth of dendrite tips in an alloy is influenced by the diffusion of the alloying element. The concentration distribution controls the distribution of the solute.

In a pure metal there is no solute. The growth of the dendrite arms is instead controlled by the temperature field around the growing tip.

The theory of the temperature field around a growing tip in a pure metal melt is essentially the same as the theory for the concentration field around a growing tip in an alloy melt. The equations, which correspond to Equations (9.58) and (9.59), are

$$T^{\mathrm{L}}(r,z) = T_0^{\mathrm{L}} + (T^{\mathrm{L}/\alpha} - T_0^{\mathrm{L}})\frac{E_1\left[\dfrac{V_{\mathrm{growth}}\,\rho_{\mathrm{tip}}}{2\alpha}\left(\dfrac{z}{\rho_{\mathrm{tip}}} + \sqrt{\dfrac{r^2+z^2}{\rho_{\mathrm{tip}}^2}}\right)\right]}{E_1\left(\dfrac{V_{\mathrm{growth}}\,\rho_{\mathrm{tip}}}{2\alpha}\right)} \tag{9.66}$$

and

$$I(Pe) = \frac{C_{\mathrm{p}}(T^{\mathrm{L}/\alpha} - T_0^{\mathrm{L}})}{-\Delta H_{\mathrm{m}}} = Pe\,e^{Pe}\,E_1(Pe) \tag{9.67}$$

where

$\alpha = \dfrac{\lambda}{\rho C_{\mathrm{p}}}$ = thermal diffusivity or coefficient of thermal diffusion of the melt (m^2/s)

Superscript α = solid phase

λ = thermal conductivity of the melt

ρ = density of the melt

C_{p} = molar heat capacitivity of the melt. Capacitivity C_{p} means capacity per unit mass.

The properties of the function $Pe\,e^{Pe}\,E_1(Pe)$ have been discussed on pages 521.

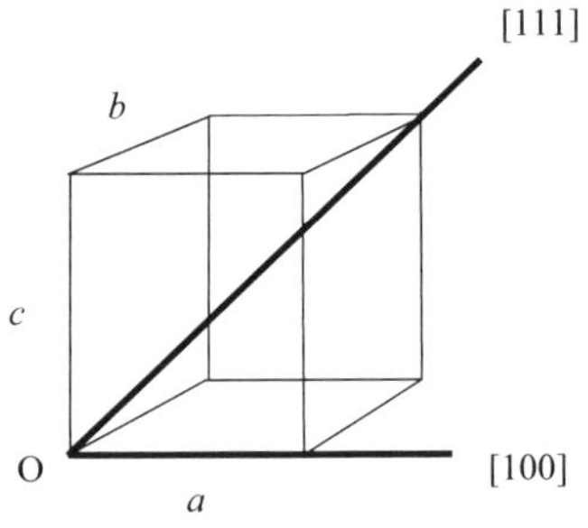

Anisotropy of the Surface Energy

From the theory of surface structure in Chapter 3 we know that the surface tension is smallest/next smallest in the <111> and <100> directions and larger in all other directions. For this reason these directions are of particular interest and will be the subject to special attention here and in the rest of this chapter.

The variation of σ is analogous in both the <100> and <111> directions and it is therefore enough to discuss one of these cases, for example the <100> direction. Close to this direction (small values of θ) the surface tension can be described by the relationship

$$\sigma = \sigma_0(1 + A\sin\theta) \tag{9.68}$$

Figure 9.46 <100> and <111> directions in a crystal. Compare Section 1.4.3 in Chapter 1 in [1]. [100] [111] are the particular directions of <100> and <111> shown in the figure.

where
σ = surface tension in direction θ.
A = positive factor, varying from zero and upwards, which describes the degree of a nisotropy
σ_0 = the surface tension in the $\langle 100 \rangle$ directions or other arbitrary crystallographic directions
θ = small angle describing the deviation from the <100> or other direction.

Anisotropy of the Interface Kinetics

At low undercooling the growth rate of the solidification front can be described by a constant β times the undercooling raised to a power of n.

$$V_{\text{growth}} = \beta(\Delta T)^n \tag{9.69}$$

where
V_{growth} = growth rate
β = growth constant
ΔT = undercooling
n = dimensionless number.

The constant β, and also the number n, varies with orientation.

Close to the <100> directions or other directions, the variation of the constant β with the crystallographic direction can be described by the relationship

$$\beta = \beta_0(1 + B \sin \theta) \tag{9.70}$$

where
β = growth constant in the direction θ
β_0 = value of growth constant in the <100> directions or other crystallographic directions
B = positive factor, varying from zero and upwards, which describes the degree of anisotropy.

From the known variation of β with the surface structure roughness it can be concluded that β is smallest in the <111> directions and second smallest in the <100> directions. See pages 15–16 in [1].

Driving Force of Dendritic Growth in Binary Alloy Melts

The driving force of the dendrite tip growth is proportional to the concentration difference between the melt close to the interface and the melt far from the interface. The concentration difference can be written as the sum of four terms, which represent the added influence of several causes on the driving force. The effect of the thermal transport is neglected in this case as was discussed on page 515. According to Equation (9.36) and Equations (9.37)–(9.40) on page 515 we obtain

$$x^{L/\alpha} - x_0^L = (x^{L/\alpha} - x_\sigma^{L/\alpha}) + (x_\sigma^{L/\alpha} - x_{\text{kin}}^{L/\alpha}) + (x_{\text{kin}}^{L/\alpha} - x_0^L) \tag{9.71}$$

where
$x^{L/\alpha}$ = concentration of solute in the liquid close to the interface in equilibrium with the solid
x_0^L = concentration of solute in the liquid far from the interface.
$x_\sigma^{L/\alpha}$ = concentration of solute in the liquid close to the interface when the surface tension is considered
$x_{\text{kin}}^{L/\alpha}$ = concentration of solute in the liquid close to the interface when the interface kinetic reaction is considered.

Each of the three terms in Equation (9.71) will be discussed below.

The First Term $x^{L/\alpha} - x_{\sigma}^{L/\alpha}$

The first term on the right-hand side of Equation (9.71), is the change in composition of the liquid at the interface, owing to surface tension. The so-called Gibbs-Thompson equation relates the difference $\left(x^{L/\alpha} - x_{\sigma}^{L/\alpha}\right)$ to the curvature of the dendrite tip, i.e.

$$x^{L/\alpha} - x_{\sigma}^{L/\alpha} = m_L \frac{\sigma T_M}{-\Delta H_m} K \tag{9.72}$$

where

m_L = slope of the liquidus line in the phase diagram
σ = surface tension
$-\Delta H_m$ = molar heat of fusion
T_M = melting-point temperature
K = curvature of the interface.

The average curvature of a parabolic tip has been derived by Kotler and Tiller [15]:

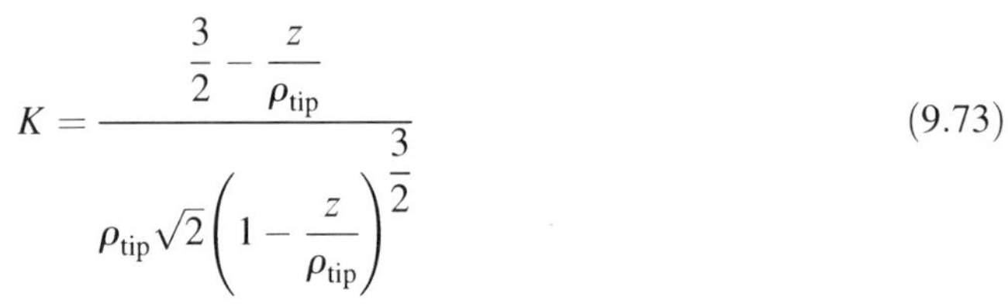
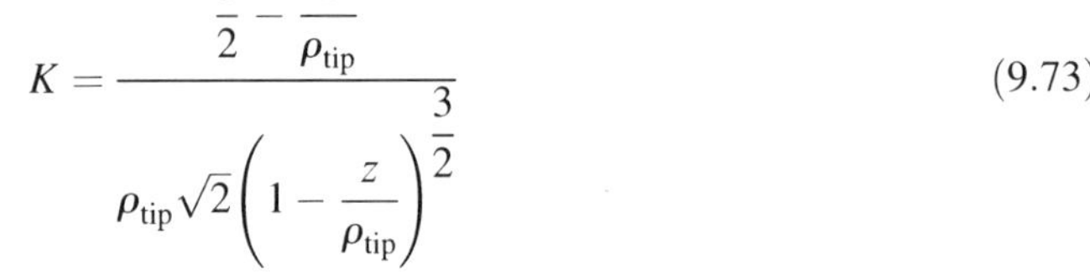

$$K = \frac{\dfrac{3}{2} - \dfrac{z}{\rho_{tip}}}{\rho_{tip}\sqrt{2}\left(1 - \dfrac{z}{\rho_{tip}}\right)^{\frac{3}{2}}} \tag{9.73}$$

where z is defined in Figure 9.43. ρ_{tip} is the radius of curvature of the tip.

If $z/\rho_{tip} \ll 1$ (z small) the average curvature of the interface will be

$$K = \frac{3}{2\sqrt{2}} \frac{1}{\rho_{tip}} \tag{9.74}$$

For the <100> and <111> directions we obtain

$$K_{111} = \frac{3}{2\sqrt{2}} \frac{1}{\rho_{111}} \quad \text{and} \quad K_{100} = \frac{3}{2\sqrt{2}} \frac{1}{\rho_{100}} \tag{9.75}$$

Figure 9.43 Definition of circular-parabolic coordinates.
The origin has the coordinates (0, 0), i.e. $\alpha = 0$ and $\beta = 0$.
$\alpha = 1$ and $\beta = 0$ corresponds to the surface of the growing dendrite tip.
Hence $\dfrac{z}{\rho_{tip}} = \dfrac{\alpha^2 - \beta^2}{2} = \dfrac{1}{2}$

The Second Term $x_{\sigma}^{L/\alpha} - x_{kin}^{L/\alpha}$

The second term of Equation (9.71) is proportional to the driving force owing to the reaction at the interface. It can be described by the relationship

$$x_{\sigma}^{L/\alpha} - x_{kin}^{L/\alpha} = m_L \left(\frac{V_{\perp}}{\beta}\right)^{\frac{1}{n}} \tag{9.76}$$

where

$V_{\perp}$ = growth rate perpendicular to the surface
β = growth constant
n = constant number.

$V_{\perp}$ corresponds to the growth rate in various crystallographic directions in the further discussion.

The Third Term $x_{kin}^{L/\alpha} - x_0^L$

The third term $x_{kin}^{L/\alpha} - x_0^L$ in Equation (9.71) corresponds to the concentration difference in the liquid. It represents the driving force of diffusion ahead of the growing dendrite tip and corresponds to the driving force of diffusion in the melt. If we neglect the first and second terms in comparison with the third term, which is tempting and not unreasonable, we obtain $x_{kin}^{L/\alpha} - x_0^L \approx x^{L/\alpha} - x_0^L$. This relationship in combination with Equation (9.59) on page 520 gives

$$x_{\text{kin}}^{\text{L}/\alpha} - x_0^{\text{L}} \approx x^{\text{L}/\alpha} - x_0^{\text{L}} = (x^{\text{L}/\alpha} - x^{\alpha/\text{L}}) Pe\, \text{e}^{\text{Pe}} E_1(Pe) \tag{9.77}$$

where

$x^{\alpha/\text{L}}$ = equilibrium concentration of the solute in the solid phase close to the interface at equilibrium with the liquid

Pe = $V_{\text{growth}}\, \rho_{\text{tip}}/2D_{\text{L}}$ = Peclet's number

$E_1\ (Pe)$ = exponential integral of Peclet's number.

Please observe that the *Equation* (9.77) *is only valid at the tip*. Elsewhere Equation (9.58) on page 518 should be used.

Total Driving Force of Solidification in Binary Alloys

The total concentration difference, which is proportional to the driving force of the solidification process, is obtained by inserting the expressions in Equations (9.72), (9.76) and (9.77) into Equation (9.71)

$$x^{\text{L}/\alpha} - x_0^{\text{L}} = m_{\text{L}} \frac{\sigma T_{\text{M}}}{-\Delta H_{\text{m}}} K + m_{\text{L}} \left(\frac{V_{\perp}}{\beta} \right)^{\frac{1}{n}} + (x^{\text{L}/\alpha} - x^{\alpha/\text{L}}) Pe\, \text{e}^{Pe} E_1(Pe) \tag{9.78}$$

Application of Ivantsov's Model on an FCC Structure

The surface of the tip is built up by a number of small {100} and {111} facets which grow rapidly and form the tip. Therefore, it is reasonable that we can apply Ivantsov's theory on an FCC crystal growing in the <100> directions and in the <111> directions, respectively. In addition this will show that the abstract theory can generate concrete results.

We apply Equation (9.78) on the growth of {100} and {111} facets. The result is

$$x^{\text{L}/\alpha} - x_0^{\text{L}} \approx m_{\text{L}} \frac{T_{\text{M}}}{-\Delta H_{\text{m}}} \frac{3}{2\sqrt{2}} \frac{\sigma_{100}}{\rho_{100}^{\text{tip}}} + m_{\text{L}} \left(\frac{V_{100}}{\beta_{100}} \right)^{\frac{1}{n}} + (x^{\text{L}/\alpha} - x^{\alpha/\text{L}}) [Pe\, \text{e}^{Pe} E_1(Pe)]_{100} \tag{9.79}$$

where $Pe_{111} = \dfrac{V_{111}\rho_{111}^{\text{tip}}}{2D_{\text{L}}}$

and

$$x^{\text{L}/\alpha} - x_0^{\text{L}} \approx m_{\text{L}} \frac{T_{\text{M}}}{-\Delta H_{\text{m}}} \frac{3}{2\sqrt{2}} \frac{\sigma_{111}}{\rho_{111}^{\text{tip}}} + m_{\text{L}} \left(\frac{V_{111}}{\beta_{111}} \right)^{\frac{1}{n}} + (x^{\text{L}/\alpha} - x^{\alpha/\text{L}}) [Pe\, \text{e}^{Pe} E_1(Pe)]_{111} \tag{9.80}$$

where $Pe_{111} = \dfrac{V_{111}\rho_{111}^{\text{tip}}}{2D_{\text{L}}}$.

The left-hand side of Equations (9.79) and (9.80) are equal. This must also be valid for the right-hand side expressions. This condition can be written as

$$m_{\text{L}} \frac{T_{\text{M}}}{-\Delta H_{\text{m}}} \frac{3}{2\sqrt{2}} \frac{\sigma_{100}}{\rho_{100}^{\text{tip}}} + m_{\text{L}} \left(\frac{V_{100}}{\beta_{100}} \right)^{\frac{1}{n}} + (x^{\text{L}/\alpha} - x^{\alpha/\text{L}}) \left[Pe\, \text{e}^{Pe} E_1(Pe)\right]_{100}$$
$$= m_{\text{L}} \frac{T_{\text{M}}}{-\Delta H_{\text{m}}} \frac{3}{2\sqrt{2}} \frac{\sigma_{111}}{\rho_{111}^{\text{tip}}} + m_{\text{L}} \left(\frac{V_{111}}{\beta_{111}} \right)^{\frac{1}{n}} + (x^{\text{L}/\alpha} - x^{\alpha/\text{L}}) \left[Pe\, \text{e}^{Pe} E_1(Pe)\right]_{111} \tag{9.81}$$

The product of the growth rate and the tip radius for the two directions, i.e. $V_{100}\rho_{100}^{\text{tip}}$ and $V_{111}\rho_{111}^{\text{tip}}$, *must be equal*. The reason for this condition is given.

In Example 9.5 on pages 518–520 we found that the left-hand side of the ratio

$$\frac{x^{L/\alpha} - x_0^L}{x^{L/\alpha} - x^{\alpha/L}} = Pe\, e^{Pe} E_1(Pe) \qquad (9.59 \text{ on page } 520)$$

obviously is independent of direction. This must also be true for the right-hand side of the relationship above. It is very well known that the growth rate depends strongly on the direction of the growth. This is valid for faceted growth as well as dendritic growth. Sections 9.3 and 9.4 show clearly that V_{growth} is not constant. If the growth rate were independent of direction no faceted crystals could exist, only spherical crystals.

1. The explanation came when Ivantsov developed a theoretical model. He studied the mass and heat transport by solving Fick's second law for dendritic growth. It is described on page 509. He introduced the so-called Ivantsov's function.

$$I(Pe) = Pe\, e^{Pe} E_I(Pe) \qquad (9.31 \text{ on page } 509)$$

where Pe is *Peclet's number Pe*. It is defined as

$$Pe = \frac{V_{growth}\rho_{tip}}{2D_L} \qquad (9.32 \text{ on page } 509)$$

D_L is the diffusion coefficient of the alloying element in the melt. It is a material constant. Instead of V_{growth} it is the product $V_{growth} \times \rho_{tip}$ that is constant. You never can solve the growth rate and the tip radius separately.

2. We know that V_{growth} depends on direction (see for example the figure in the box above). D_L is a constant. The only possibility for Pe to be independent of direction is that the product $V_{growth}\, \rho_{tip}$ is *constant* and hence *independent of direction.*
3. The only way to fulfil this condition is that *the tip radius* ρ_{tip} *depends on direction* in such a way that the product of the two factors is constant and independent of direction.

The condition gives the condition

$$V_{growth}^{100}\, \rho_{100}^{tip} = V_{growth}^{111}\, \rho_{111}^{tip} \qquad (9.82)$$

The condition for constant shape of an octahedron crystal during growth is [Equation (9.6) on page 480]

$$V_{100} = \sqrt{3}V_{111} \qquad (9.6)$$

Example 9.6

Material constants for Al
$T_M = 931$ K
$-\Delta H_m = 11.1 \times 10^6$ J/kmol
$\beta_{100} = 1.0$ m/s
$\frac{\beta_{100}}{\beta_{111}} = 2$ or 10
$\sigma_{100} = 0.10$ J/m^2
$\frac{\sigma_{100}}{\sigma_{111}} = 1.1$ or 1.5 or 2.0

a) Figure 9.4 on page 479 shows a sequence of a growing cubic crystal for the case when $V_{111} < \sqrt{3}\, V_{100}$. Explain why the last shape is an octahedron.
b) We want to study how the growth rate, for example V_{100}, depends on the growth rate constant β [see Equation (9.76) on page 524] and the surface tension σ for an Al crystal in the shape of an octahedron. Assume that Equation (9.6) is valid and plot V_{100} as a function of ρ_{100}^{tip} for the two values $\beta_{100}/\beta_{111} = 2$ and $\beta_{100}/\beta_{111} = 10$.

   For each of these ratios you draw three curves, which correspond to the parameter values $\sigma_{100}/\sigma_{111} = 2$, 1.5 and 1.1.

   Perform the calculations for Al, that has FCC structure. Material constants are given in the margin. Assume that $n = 2$.
c) What conclusions can you draw from the two curves in b)?

Solution

a)

The condition (page 479) $V_{100} = \sqrt{3}\, V_{111}$ implies that the shape of the FCC crystal does not change. The balance is disturbed if $V_{100} < \sqrt{3}\, V_{111}$ and the new growth rate of the {111} facets grow in a *slower* way than the balance value.

The general rule is that the most stable facet is that with the slowest growth rate (page 477). Consequently the {111} facets grow at the expense of the {100} facets. In the end the {100} facets will disappear and the shape of the crystal will be an octahedron with only {111} facets.

b)

Equation (9.82) on page 526 implies that $\mathrm{Pe}_{100} = \mathrm{Pe}_{111}$ and consequently

$$\left(Pe\, \mathrm{e}^{Pe} E_1(Pe)\right)_{100} = \left(Pe\, \mathrm{e}^{Pe} E_1(Pe)\right)_{111} \tag{1'}$$

For this reason the third terms on the left-hand side and the right-hand side in Equation (9.81) cancel and it can be written as

$$\frac{T_{\mathrm{M}}}{-\Delta H_{\mathrm{m}}} \frac{3}{2\sqrt{2}} \frac{\sigma_{100}}{\rho_{100}^{\mathrm{tip}}} + \left(\frac{V_{100}}{\beta_{100}}\right)^{\frac{1}{n}} = \frac{T_{\mathrm{M}}}{-\Delta H} \frac{3}{2\sqrt{2}} \frac{\sigma_{111}}{\rho_{111}^{\mathrm{tip}}} + \left(\frac{V_{111}}{\beta_{111}}\right)^{\frac{1}{n}} \tag{2'}$$

Equation (2′) can be transformed into

$$\frac{T_{\mathrm{M}}}{-\Delta H_{\mathrm{m}}} \frac{3}{2\sqrt{2}} \frac{\sigma_{100}}{\rho_{100}^{\mathrm{tip}}} \left(1 - \frac{\sigma_{111}\, \rho_{100}^{\mathrm{tip}}}{\sigma_{100}\, \rho_{111}^{\mathrm{tip}}}\right) = \left(\frac{V_{100}}{\beta_{100}}\right)^{\frac{1}{n}} \left[\left(\frac{V_{111}\, \beta_{100}}{V_{100}\, \beta_{111}}\right)^{\frac{1}{n}} - 1\right]$$

or

$$\left(\frac{V_{100}}{\beta_{100}}\right)^{\frac{1}{n}} = \frac{\dfrac{T_{\mathrm{M}}}{-\Delta H_{\mathrm{m}}} \dfrac{3}{2\sqrt{2}} \dfrac{\sigma_{100}}{\rho_{100}^{\mathrm{tip}}} \left(1 - \dfrac{\sigma_{111}\, \rho_{100}^{\mathrm{tip}}}{\sigma_{100}\, \rho_{111}^{\mathrm{tip}}}\right)}{\left(\dfrac{V_{111}\, \beta_{100}}{V_{100}\, \beta_{111}}\right)^{\frac{1}{n}} - 1}$$

or

$$V_{100} = \beta_{100} \left[\frac{\dfrac{T_{\mathrm{M}}}{-\Delta H_{\mathrm{m}}} \dfrac{3}{2\sqrt{2}} \sigma_{100} \left(1 - \dfrac{\sigma_{111}\, \rho_{100}^{\mathrm{tip}}}{\sigma_{100}\, \rho_{111}^{\mathrm{tip}}}\right)}{\left(\dfrac{V_{111}\, \beta_{100}}{V_{100}\, \beta_{111}}\right)^{\frac{1}{n}} - 1}\right]^{n} \frac{1}{\rho_{100}^{\mathrm{tip}\, n}} \tag{3'}$$

The ratio of the tip radii can be calculated from Equations (9.82) and (9.6):

$$\frac{\rho_{100}^{\mathrm{tip}}}{\rho_{111}^{\mathrm{tip}}} = \frac{V_{111}}{V_{100}} = \frac{1}{\sqrt{3}} \tag{4'}$$

Inserting the relationships in Equations (9.6) and (4′) into Equation (3′) gives

$$V_{100} = \beta_{100}\left[\frac{\dfrac{T_M}{-\Delta H}\dfrac{3}{2\sqrt{2}}\sigma_{100}\left(1-\dfrac{\sigma_{111}}{\sigma_{100}}\dfrac{1}{\sqrt{3}}\right)}{\left(\dfrac{1}{\sqrt{3}}\dfrac{\beta_{100}}{\beta_{111}}\right)^{\frac{1}{n}}-1}\right]^n \frac{1}{\rho_{100}^{\text{tip}^n}} \tag{5'}$$

Equation (5′) represents the growth rate of a dendrite tip in the ⟨100⟩ directions as a function of the radius of curvature of its paraboloid tip in this direction. This function is plotted for the alternative values of the ratios of the β and σ values. The plots are given in the answer.

Answer:

a) See the solution above.
b) See Equation (5′) above.

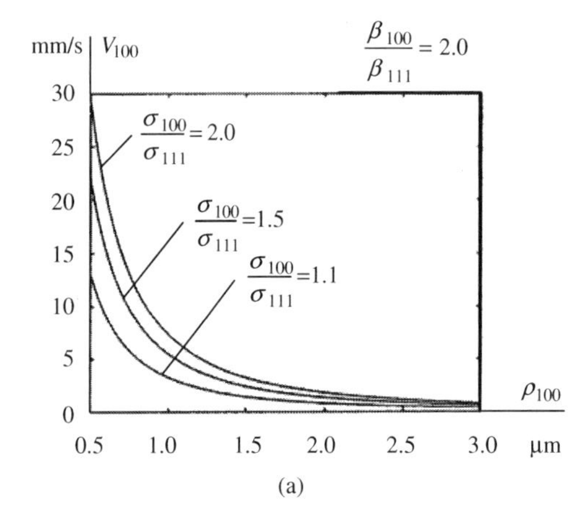

Growth rate of a dendrite in the <100> directions as a function of the radius of curvature of the top of the parabolic tip. Both vary with direction but the product is constant. The growth constant $\beta_{100} = 1.0$ m/s. The growth constant $\beta_{111} = 0.5$ m/s.

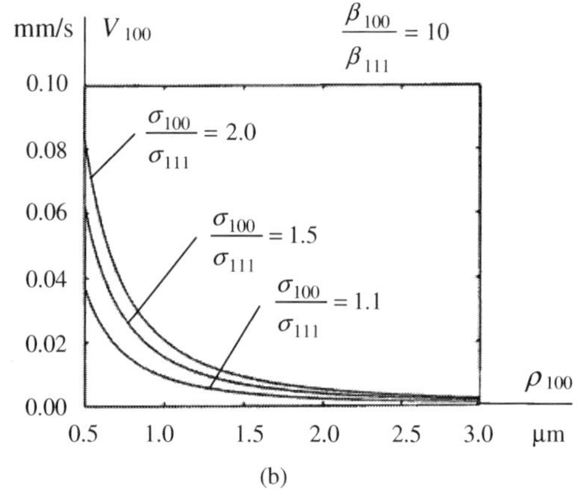

Growth rate of a dendrite in the <100> directions as a function of the radius of curvature of the parabolic tip. Both vary with direction but the product is constant. The growth constant $\beta_{100} = 1.0$ m/s. The growth constant $\beta_{111} = 0.5$ m/s.

c)

- The growth rate V_{100} depends strongly on the ratio the β values. The larger β_{100}/β_{111} is, the smaller will be V_{100}.
- The larger the ratio $\sigma_{100}/\sigma_{111}$ is, the larger will be V_{100}.
- At increasing growth rate the tip radius approaches a small finite value, never to zero.

Equation (5′) represents one relationship between the growth rate and the radius of the dendrite tip. The value of n is often close to 2. In this case the relationship can be written as

$$V_{\text{growth}}^{\text{tip}} = const \times \frac{1}{\rho_{\text{tip}}^2} \tag{9.83}$$

which is similar to Kurz and Fisher's relationship (9.34) on page 510. However, Equation (9.83) depends only on the interphase kinetics and the anisotropy of the surface tension. If we insert the relationship (9.83) into Equation (9.79) the supersaturation as a function of the growth rate can be found.

As the designation indicates, $V^{\text{tip}}_{\text{growth}}$ is the growth rate of the growing dendrite tip, which is equal to the growth rate of the solidification front. Normally it is called V_{growth} with no indices at all. It should not be mixed up with the direction dependent growth rate described below.

The growth rate, which occurs in for example Equation (9.82) i.e. $V^{100}_{\text{growth}}\rho^{\text{tip}}_{100} = V^{111}_{\text{growth}}\rho^{\text{tip}}_{111}$, represents the growth rate at the crystal surface. It is perpendicular to the surface and varies with direction and position on the surface. As stated on page 526, the size of the tip radius also varies with direction because the product of growth rate and tip radius is constant.

9.6 Development of Dendrites

9.6.1 *Predendritic Solidification*

Figure 9.47 shows a puzzling phenomenon which occasionally can be observed at the surface of chill-cast alloy melts. Biloni and Chalmers [16] have described and studied the appearance and composition of the observed surface patterns in Al-Cu and Pb-Sn alloys.

Discs surrounded by two concentric rings were observed. Radial grooves with periodically varying composition start at the rings and form later regions of parallel grooves in various crystallographic directions. Dendrites emanate from these regions outside the second rings. For this reason the phenomenon is called *predendritic solidification*.

Figure 9.47 shows that each ring is the origin of an embryo. This can be concluded from the directions of the dendrites and the grain boundaries, that are the dark lines where the dendrites change orientation in an Al-Cu alloy.

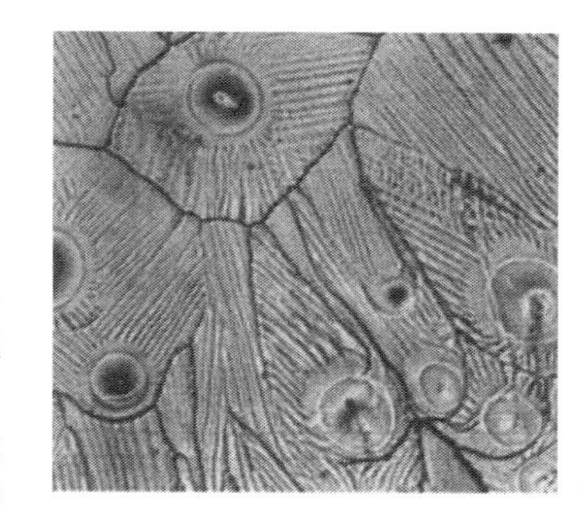

Figure 9.47 Discs and rings formed during pre-dendritic solidification in an Al–Cu alloy. From the Solidification of Metals, Iron and Steel Institute, 1967 © Maney. www.maney.co.uk.

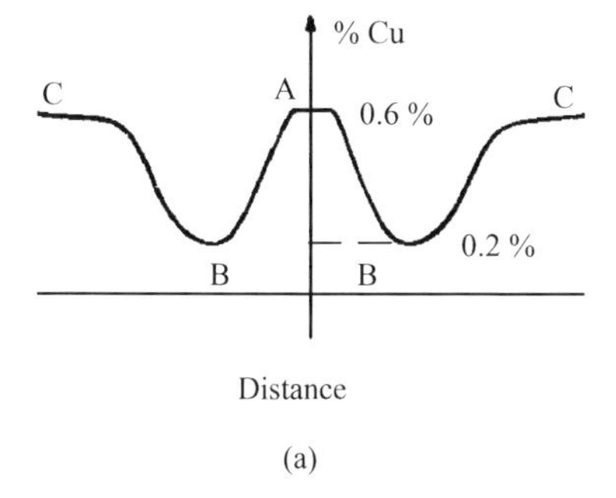

Figure 9.48a Concentration profile of the inner ring around a disc in an Al–Cu alloy. Reproduced from Springer © 1965.

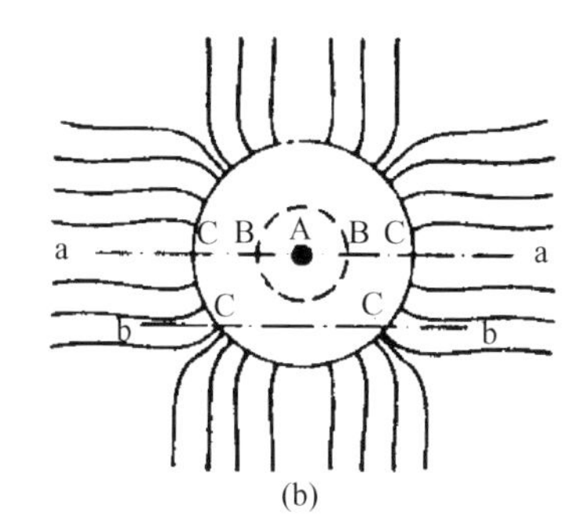

Figure 9.48b Sketch of a disc (black central spot A) surrounded by two rings B and C. Grooves emanate from the outer ring. Reproduced from Springer © 1965.

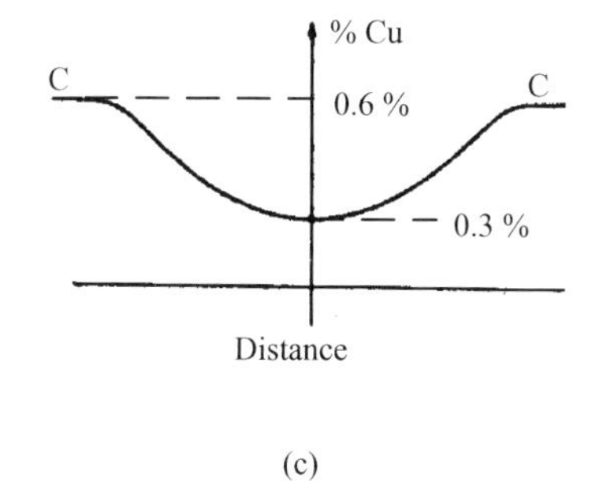

Figure 9.48c Concentration profile of the outer ring around a disc in an Al–Cu alloy. Reproduced from Springer © 1965.

At high undercooling (laboratory experiments) the formation process is so rapid that the composition of the discs is the same as that of the melt, i.e. there is no diffusion of the alloying element and consequently no microsegregation (Section 9.6.5).

At intermediate cooling, e.g. in ingots, only the central parts of the discs have the same composition as the melt. The composition of the rings varies as Figures 9.48 a, b and c show. The inner ring shows initially a concentration minimum and increases then to the same concentration of the solute as the melt.

9.6.2 *Development of Dendrites*

Figure 9.33 on page 504 shows that dendritic growth occurs with a primary tip and secondary arms formed immediately behind the tip. The mechanism of formation of these secondary dendrite arms has been discussed many times in the scientific literature.

Figure 9.49 shows a dendrite tip of Pb in a shrinkage cavity of a small Pb-Sb ingot. The figure shows that the tip is slightly faceted and that side branches are formed behind the tip. According to the model

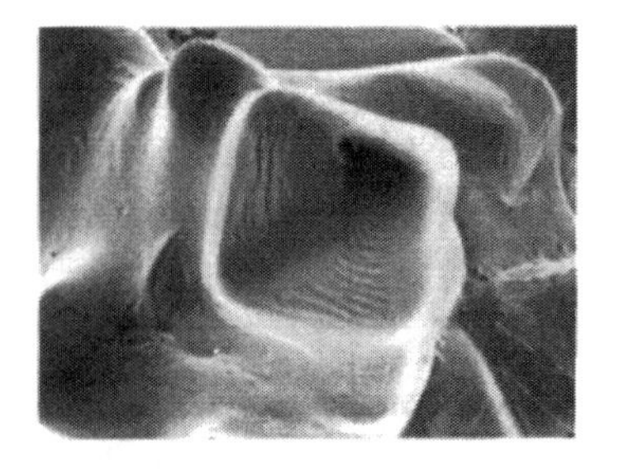

Figure 9.49 Dendrite tip of Pb grown in a Pb–Sn alloy.

proposed in Section 9.5.4 on page 514 the tip and its nearby region has grown and is formed as an isothermal and iso-concentrated parabolic surface.

- The isothermal growth of a dendrite is stabilized by the anisotropy of the surface energies and also of the interface reaction.

The faceted shape of the tip and the formation of side branches is a result of the anisotropy of the surface tension and of the reaction at the interface. This anisotropy implies that at some distances behind the tip the parabolic shape of the dendrite stem cannot be achieved. This is the reason for the formation of the faceted shape of the stem close to the tip, and also for the formation of side branches (Figure 9.50).

The diffusion field around a side arm may also influence the dendrite formation. The segregation of solute atoms in the liquid around a side branch will overlap the one around the primary branch and therefore slow down its sideways growth close to the side branch. The result is that a convex region close to the tip is formed.

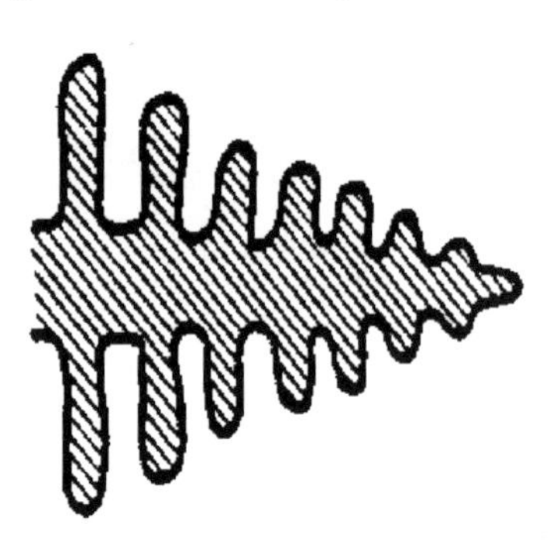

Figure 9.50 Ideal development of the side branches, which grow side by side.

The convex region will automatically develop into a new side branch. Figure 9.33 on page 504 shows this process. Figure 9.33 indicates that the growth of the main stem is quite different from that of the side branches near the tip. This is shown more in detail in Figure 9.29 on page 503. From Figures 9.33 and 9.49 the following conclusion can be drawn:

- The side branches grow much more slowly than the main tip.

One possible explanation to this difference in growth rate is that the side branches are so close together that the concentration fields around neighbouring tips overlap and give lower resulting concentration gradients than that around the main tip.

As long as the side branches grow side by side, the growth process will be more or less like the edgewise growth of a plate. While these conditions are valid, each dendrite arm cannot make use of the strong tip effect of an individual tip, but only a weaker tip effect of roughly the same magnitude as the one at the edge of a plate. Accordingly, the growth rate will be lower for the side branches.

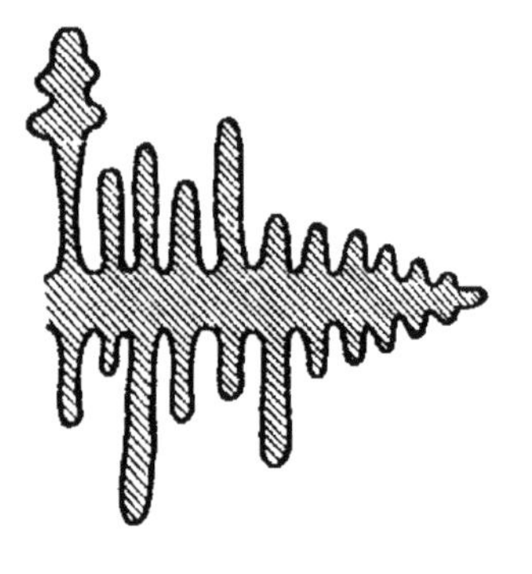

Figure 9.51 Development of a leading tip from a side branch.

When the secondary dendrite arms are growing side by side, there is a strong competition between them. This competition will sooner or later result in a situation when one of the side branches can make use of the three-dimensional tip effect and its growth rate will increase. In addition, the growth of the neighbours will decrease, depending on the influence of the heat and concentration fields around the 'winning' dendrite branch. This is shown in Figure 9.33 on page 504 and more detailed in Figure 9.51. The 'winning' tip will gradually achieve the same growth rate as the main stem.

The formation of side branches just behind the dendrite tip is caused by an automatic growth process. On the other hand, the chance of a side branch to grow ahead of the others and formnew main stems may depend on fluctuations. It is most probably a result of a variation in length of the individual side branches. We have earlier pointed out that as soon as one of the branches has grown ahead of the others, it will grow faster than the rest. The difference in length may be initiated by many factors, which cause different sorts of instabilities in the system, such as for example temperature fluctuations owing to natural convection in the liquid.

9.6.3 *Formation of Primary Stems*

The growth process of secondary arms, which have grown ahead of the others, will be more and more similar to the growth process of the primary stems. This is shown in Figure 9.52, which shows how the spacings between the sidway-growing tips increase with time. It is obvious that there is an incessant competition between the growing tips.

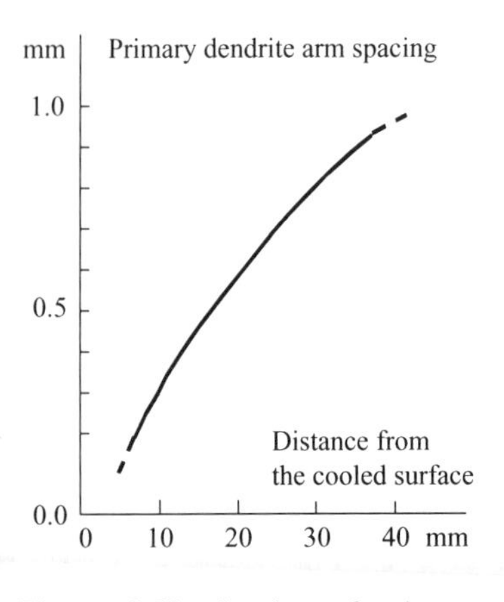

Figure 9.52 Spacing of primary dendrite arms as a function of distance from a cooled surface.

It is a well confirmed fact that the primary dendrite arm spacing increases with the distance from a cooled surface during a unidirectional solidification experiment, for instance during the growth towards the centre of a casting or ingot. This is clearly shown in Figure 9.52 where a continuous increase of the distance between the primary dendrite arms can be observed.

Hultgren [17] suggested that the distance between the *secondary* dendrite arms might be determined by overlapping *concentration* fields from two perpendicular tips. The large distance between two parallel dendrite arms, compared with the small distance between the secondary dendrite arms, indicates that the distance between *parallel* arms is determined by overlapping *temperature* fields.

According to the mathematical model of growing rods, the ratio of the distance between the primary and secondary dendrite arms should be proportional to $\sqrt{\alpha/D}$[1]. This factor varies between 10 and 100 for different alloys.

The temperature distribution around the growing tips determines the most stable distance between the tips in a growing bunch of dendrite stems. The competition between the growing dendrite tips, related to the temperature distribution around the tips, causes the continuous increase of the distance between the primary dendrite arms, which is shown in Figure 9.52. If the distance is smaller than the most stable one, it will be adjusted through a mechanism in which one of the dendrites stops to grow and all others adjust (increase) their distances.

The competition between growing parallel stems is shown in Figure 9.53. It is obvious that the average distance increases when some tips are stopped.

The stop process is caused by fluctuations in the system, for instance by local temperature variations, owing to convection. The temperature field around a growing tip is shown in Figure 9.34b on page 506.

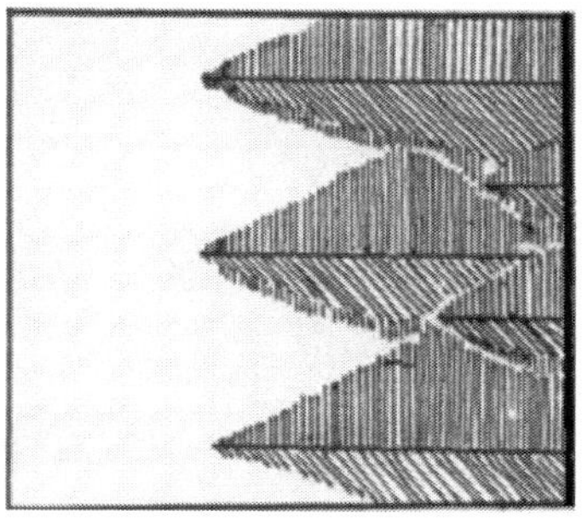

Figure 9.53 The secondary arms are found in two perpendicular planes. Both planes are perpendicular to the growth direction of the tip.

Superdendrites

The distances between primary dendrites and ditto secondary dendrites in large ingots of steel have been studied carefully by Larén and Fredriksson [18]. The results of their measurements are shown in Figures 9.54a and b.

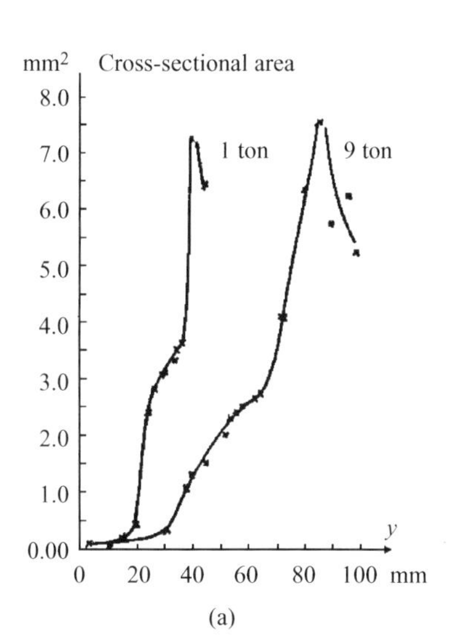

Figure 9.54a Transverse section area of columnar crystals in two ingots as functions of the distance from the ingot surface.

Reproduced with permission from the Scandinavian Journal of Metallurgy, Blackwell.

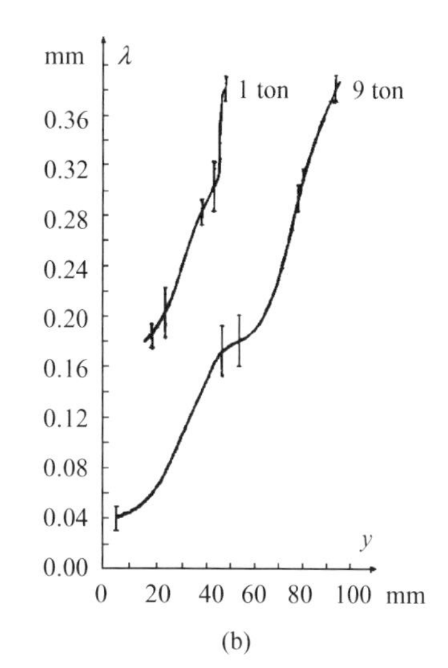

Figure 9.54b The spacing of secondary plates in the columnar zone of the ingots in Figure 9.54a as functions of the distance from the ingot surface.

Reproduced with permission from the Scandinavian the Journal of Metallurgy, Blackwell.

The curve in Figure 9.54b does not correspond to the simple parabolic relationship between the dendrite arm distance and the distance from the cooling surface. The reason is that the growth conditions and consequently also the structure morphologies are not constant. The dendrite arms seem to grow in two or three different steps. For each step the relationship $v_{growth}\lambda_{den}^2 = const$ is valid, but with different constants for each step.

[1] The diffusivity $\alpha = \lambda/\rho c_p$ where λ = thermal conductivity, ρ = density and c_p = heat capacitivity at constant pressure of the alloy. D = diffusion constant.

(c)

Figure 9.54c Superdendrites in a solidifying Al–Cu ingot after Bolling. From Solidification, American Society for Metals 1971. Reprinted with permission of ASM International.

Close to the interface there is a subzone with cellular-like dendrite structure with weakly developed secondary arms cellular crystals (with no secondary arms), that are formed at high growth rates and large temperature gradients. A second subzone of normal dendrites succeeds this zone. At larger distances from the ingot a third subzone of *superdendrites* can be identified. Superdendrites are characterized by very large distances between their primary arms. They were first discussed by Cole and Bolling [19] in 1968.

Bolling showed the formation of superdendrites by disturbing the growth of dendrites by electrical pulses in the melt. By doing so he could mark the solidification front during solidification of Al-Cu ingots. Figure 9.54c shows three different vertical positions of the front. The left solidification front shows a number of normal dendrite tips, growing side by side at the front. At the right front there are only a few of them left leaving the superdendrites alone.

The formation of superdendrites seems to be related to the temperature of the melt in which the dendrites grow. The formation of superdendrites, i.e. the sudden increase of the dendrite arm distance at $y \sim 50$ mm in Figure 9.54b was found to correspond to the time when all superheat (excess temperature above the melting point) had disappeared.

9.6.4 *Growth Directions at Dendritic Growth*

In Section 9.2.1 we have summarized two simple rules for faceted growth. A similar discussion for dendritic growth will be given below.

Unidirectional Growth

At unidirectional solidification primary dendritic stems grow in the direction parallel to the temperature gradient. Primary stems are also formed spontaneously in other directions but loose the competition and only the parallel dendrites survive. An example of this is the following.

Two dendrites grow in a melt with a constant temperature gradient in the y direction. One of them grows in the direction of the temperature gradient. The other one grows in an inclined direction, which forms an angle of 45° with the gradient.

The growth rate of a dendrite can be written as a function of the undercooling of the melt

$$V_{\text{growth}} = \mu(T_{\text{L}} - T)^n \tag{9.84}$$

where

n = a constant

T_{L} = liquidus temperature

T = temperature of the melt at the dendrite tip

μ = growth constant which includes influence of surface tension, interface kinetics and mass transport.

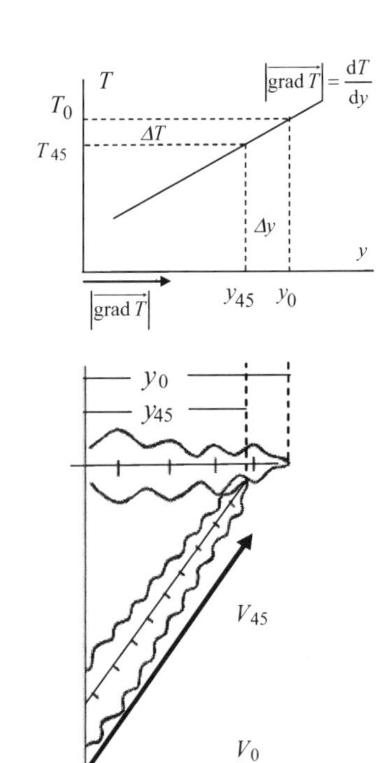

Figure 9.55 Two competing dendrites.

Figure 9.55 shows that the growth rate $V_{45} \gg V_0$. Obviously $y_0 > y_{45}$. The parallel dendrite is ahead of the inclined dendrite. It means that the parallel dendrite will stop the inclined dendrite as the space is already occupied by the parallel dendrite when the inclined dendrite tip arrives.

Anisotropic Dendritic Growth

Under other circumstances than those above dendritic growth shows similarities with faceted growth. Depending on the cooling conditions faceted growth can change into dendritic growth and vice versa. There are cases when the dendritic growth rate depends on the growth direction in analogy with the faceted growth rate. We will discuss this type of anisotropic dendritic growth below.

Experience shows that most alloys, which have a cubic structure, solidify with dendritic structure in <100> directions. However, deviations from the normal structure have been found among others in Al-alloys and in Cu–Pb alloys. An example of each type of alloy will be discussed below.

Tip Effect at Dendritic Crystal Growth

Primary Arms

It is well known that the preferred directions of dendritic growth in cubic metals are the cube directions <100> but there are other observations, which show that alternative directions also occur in some cases.

The dendritic growth of copper dendrites in Cu–Pb alloy melts during natural cooling has been studied and the crystallographic directions of the individual branches and side-branches have been determined.

Spontaneous developments of facets on the dendrite tips occur at the end of the precipitation. As early as in 1888 Lehman observed this phenomenon, which is called the *tip effect*.

With the aid of scanning electron microscopes and deep-etching technique is possible to study a combination of dendritic and faceted growth. The facets in the specimen can be visualized by removing the eutectic phase, which consists of almost pure Pb, by deep-etching in hot hydrocloric acid, and then be photo-graphed in a scanning microscope.

Figure 9.56 gives a typical example of a dendrite arm, where facets have developed on the most protruding tips at the top of the microphotograph. These facets form almost perfect octahedrons. It is quite evident that that these facets represent the octahedral planes {111}. The vertices represent the cubic directions <100>. It can be concluded from Figure 9.56 that the growth direction of the dendrite arm upwards in the microphotograph is a cubic direction.

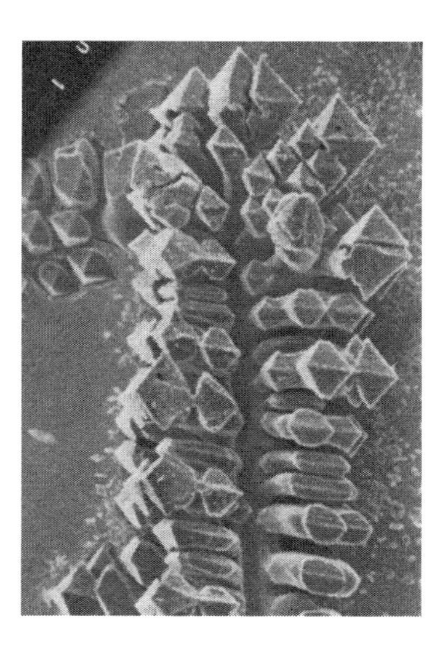

Figure 9.56 Left figure. Dendrite arm of Cu growing in a <100> direction (upwards) in a Cu–Pb melt. Scanning electron micrograph.

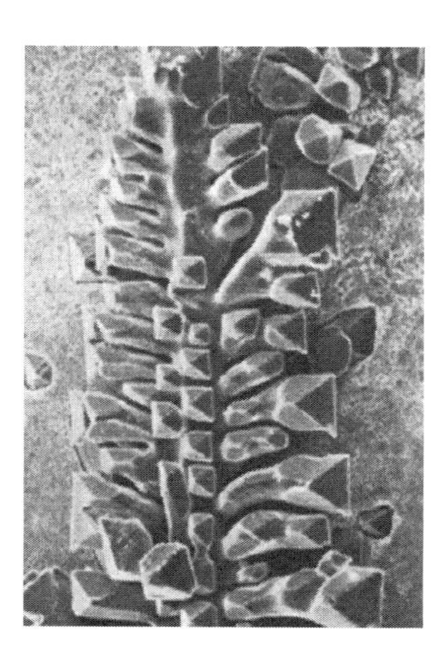

Figure 9.57 Right figure. Dendrite arm of Cu growing in a ⟨110⟩ direction (upwards) in a Cu–Pb melt. Scanning electron micrograph.

Figure 9.57 shows another example taken from the same specimen, a dendrite arm which grows in a <110> direction. This case was found relatively often. In fact the two types of growth directions <100> and <110> seem to be of equal importance in this particular alloy.

This is confirmed in Figure 9.58 which shows a star-like crystal in the same specimen, formed by the development of dendrite arms of both the <100> and <110> types. The crystallographic identity of each arm, revealed by the facets, is in reasonable agreement with its direction in space

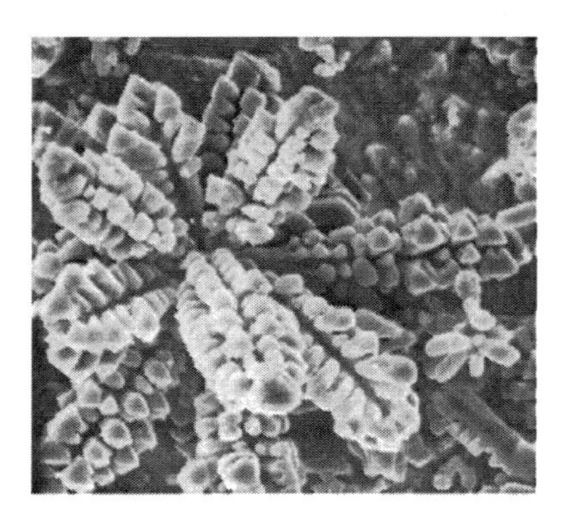

Figure 9.58 Main dendrite arms of Cu developed in <100> and <110> directions. Scanning electron micrograph.

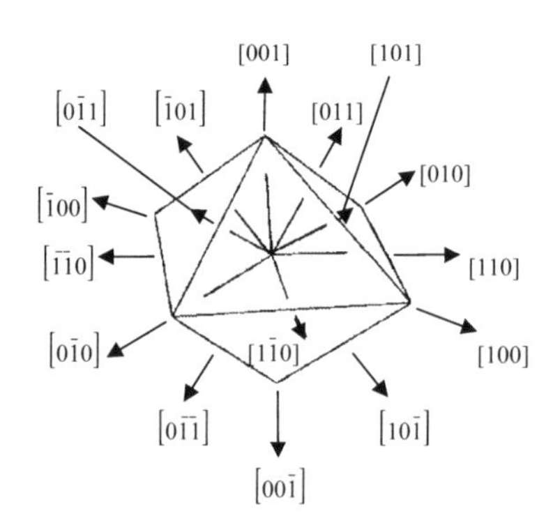

Figure 9.59 Possible main branches of the <100> and <110> types.

(Figure 9.59). In the fully developed ideal case a star-like crystal of this type can have 18 arms, 6 of the <100> type and 12 of the <110> type.

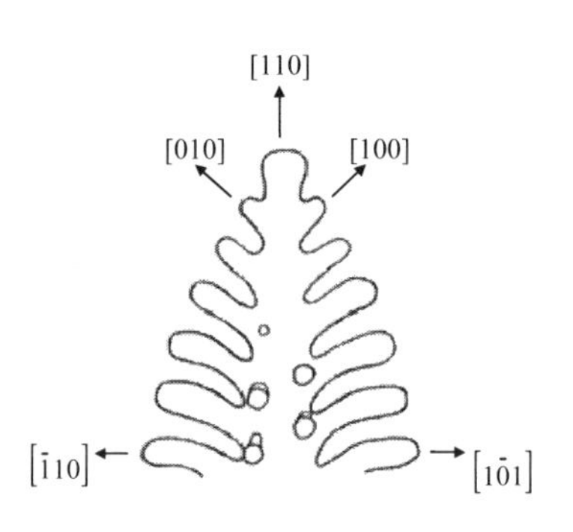

Figure 9.60 Schematic shape of dendrite arm, which grows in the <110> direction, before change to facet conditions.

Side Branches

The dendrite arms growing in the cubic directions develop side branches in four cubic directions, perpendicular to the main arm. Figure 9.60 shows schematically the development of side branches on the <110> dendrite arm in Figure 9.57 before the facets appeared.

Figure 9.60 shows that the side branches initially form at 45° angle to the <110> arm. They later turn away perpendicularly from the arm and approach the <$\bar{1}$10> and <1$\bar{1}$0> directions, probably as a result of mutual competition. All these growth directions fall in a common plane, the cubic plane (001), which gives the dendrite a rather plate-like shape.

The development of side branches, perpendicular to the cubic plane {001}, starts much later, in spite of the fact that this also is a cubic direction and should normally be a favourable growth direction. For the star-like crystals in Figures 9.58 and 9.59 the kinetics is such that the growth conditions are fulfilled in both directions. This matter will be further discussed below.

Tip Formation at Dendritic Growth

In analogy with Cu–Pb alloys, which have been treated in the preceding section, a close connection between faceted and dendritic growth is found in Al alloys. A comparison between the Figures 9.61 and 9.62 shows that the strength of the connection varies.

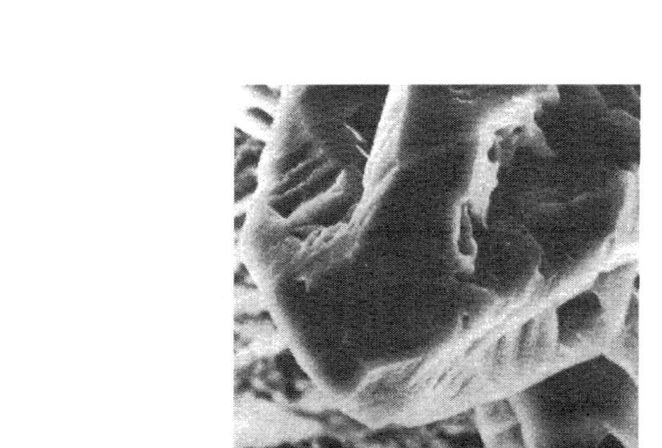

Figure 9.61 The shape of the tip on one of the dendrite arms, which belong to a star-like Si-crystal, precipitated in an unmodified Al–Si alloy, solidified at a high cooling rate.

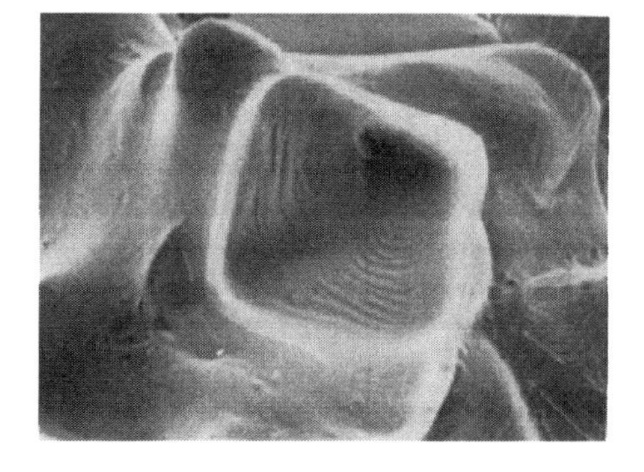

Figure 9.62 A tip of a Pb dendrite, found in a shrinkage cavity of a Pb–Sb ingot.

The dendrite tip in Figure 9.61 is not as faceted as the one in Figure 9.62. The tip in Figure 9.61 is more rounded than the tip in Figure 9.62, which has a clearly faceted tip. If its growth direction is supposed to be in a ⟨100⟩ direction the tip is bounded by {111} and {110} planes.

If growth occurs in an ⟨110⟩ direction then the tip will be bounded by {100} and {111} planes (Figure 9.63). With the aid of the maximum growth theory we can find the criteria, which have to be fulfilled for dendritic growth with a faceted tip in a ⟨110⟩ direction.

The condition for stationary growth and shape of a faceted octahedron crystal is a constant ratio between the growth rates in the three crystal directions ⟨100⟩, ⟨110⟩ and ⟨111⟩. In analogy with case I on page 479 we obtain the relationships

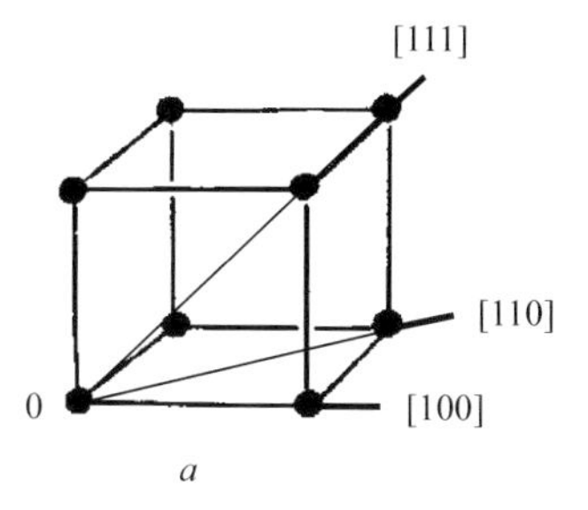

Figure 9.63 The simple cubic structure is representative for both the FCC and BCC structures.

$$V_{100} = \sqrt{2}\, V_{110} = \sqrt{3}\, V_{111} \tag{9.76}$$

The relationship between V_{110} and V_{100} is obtained by comparison of the growth vector V_{110} and the resultant of the two perpendicular V_{100} vectors in the direction $< V_{110} >$. The relationship between V_{111} and V_{100} is obtained in an analogous way.

Dendritic Growth in Aluminum Alloys

The structure shown in Figure 9.64 occurs in many aluminium alloys. The figure shows an Al dendrite from a hypereutectic Al-Si alloy with FCC structure.

In the central part of Figure 9.64 there is a region, where two perpendicular primary dendrite arm cross. The secondary arms have initially grown perpendicularly to the primary arms. In the cross-section region the secondary arms grow in a direction, which form an angle of 45° with both primary stems.

It is reasonable to assume that the primary dendrite grow in <100> directions. Then the secondary dendrite arms in the cross-section region grow in a <110> direction. This type of change in growth direction has been reported for Al-Sn alloys and occurs also in Al-Cu alloys. This shows that the growth directions in aluminium base alloys are either <100> or ⟨110⟩ or both simultaneously. Similar observations have been made for Cu dendrites.

Figure 9.64 Primary Al dendrite precipitated in a rapidly solidified hypereutectic Al-Si alloy.

Feathery Grain Growth of Aluminum Alloys

A better word for the phenomenon called feathery grain growth is *twinned crystal growth* (Chapter 1, pages 29–30 in [1]). The feathery grains consist of large colonies of twinned dendritic crystals. This type of growth is unique for aluminium. It is desirable to avoid it at both mould casting and continuous or semi-continuous casting of aluminium as it deteriorates the quality of the solidified metal. Once the twinned crystal growth has started it has a tendency to dominate over the normal dendritic growth (Figure 9.65).

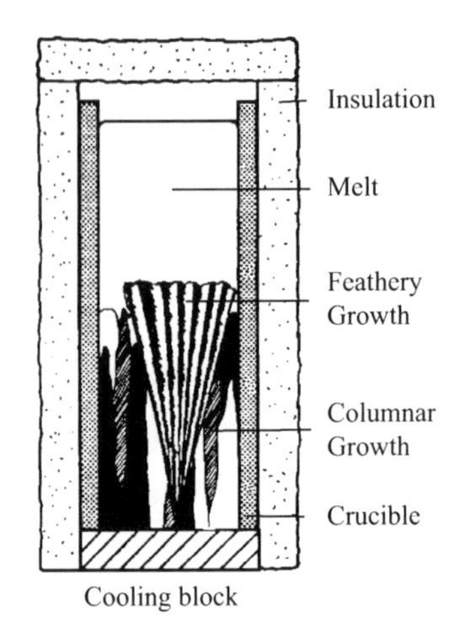

Figure 9.65 Directional solidification of aluminium with formation of twinned crystals. © Skanaluminium 1986.

The grains consist of thin parallel lamellae with a thickness of ≈ 0.1 mm. They develop side by side and form aggregates of lamellae over large areas, up to 100 cm^2 if the space is avai-lable. The feathery grains often develop twin planes parallel to the casting direction and not necessarily in the direction of the heat flow as ordinary dendrites.

Figure 9.66a shows a typical example of a twinned crystal structure. The dendrites have grown in a <112> direction with branches, which form an angle of 54° with the primary stems. The latter ones grow in two <110> directions (Figure 9.66b).

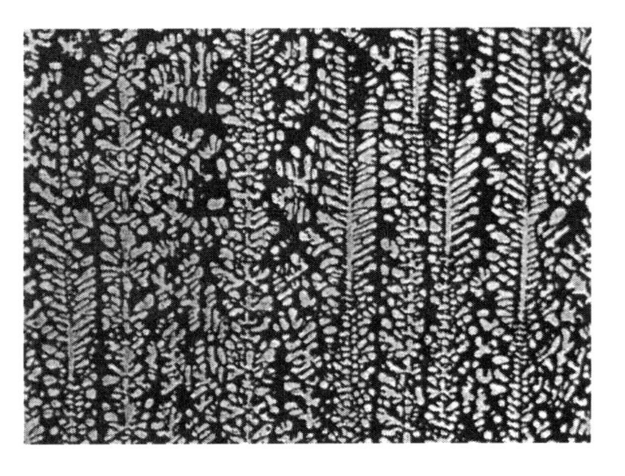

Figure 9.66a Structure of a solidified Al–8.6%Si alloy with dominating twinned dendrite arms. © Skanaluminium 1986.

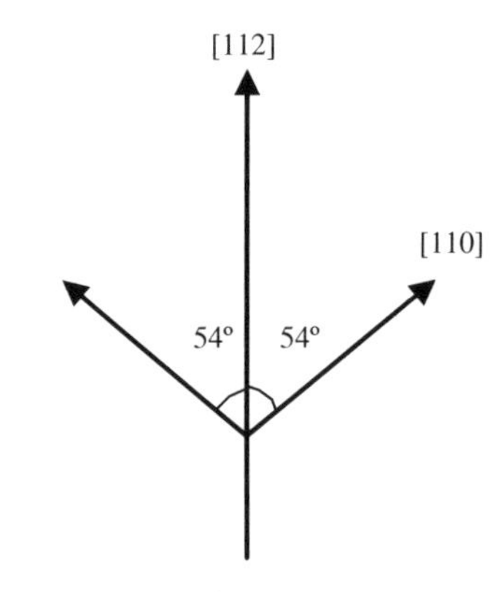

Figure 9.66b Growth directions of the dendrite arms in Figure 9.66a. © Skanaluminium 1986.

Feathery grain growth is enhanced by the following factors:

- high temperature gradients in the melt at the solidification front
- high cooling rates
- presence of some alloying elements or impurities for example Ti (worst), Mg, Cu, Zn and Si.
- poor grain refinement.

Measures to minimize the influence of these factors should be taken.

Theory of Twinned Crystal Growth

The feathery grains consist of a large number of parallel thin lamellae. Each lamella consists of two crystals in twin orientation. Examination by means of the etch-pit method and ordinary X-ray methods has shown that the orientation of the twin boundary plane is {111} and that the growth direction occurs preferably in one of the six <112> directions. The coupling of the twinned crystals to large aggregates works only if the crystals have alternating growth directions in the boundary plane as is shown in Figure 9.67 and 9.68.

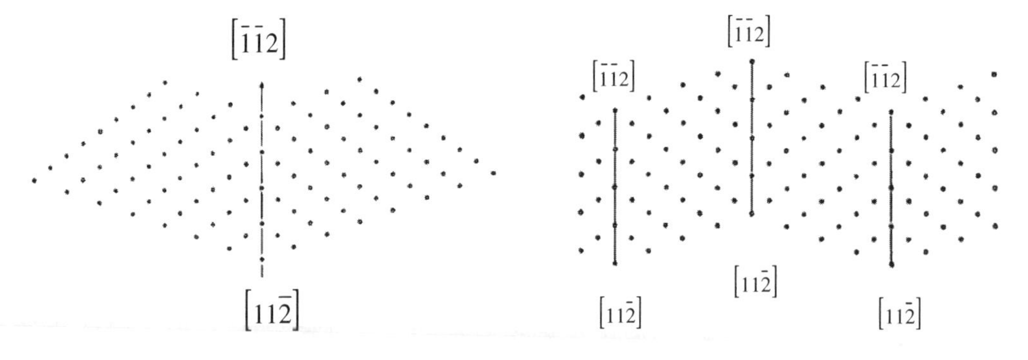

Figure 9.67 The atomic arrangement of a pair of twinned crystals.

Figure 9.68 The atomic arrangement of two coupled twin crystals.

Growth Direction of Twinned Aluminum Crystals

Earlier in this chapter we have seen that the growth direction of dendrites is related to the crystallographic directions of faceted crystals of the metal. This is also true for aluminium.

The <100> directions are usually the optimal growth directions for dendritic growth of individual aluminium crystals. The close-packed {111} faces have the lowest surface energy. Figure 9.69 shows that the crystal has a tip in one of the <100> directions surrounded by four {111} faces. The higher the anisotropy of the surface energy is, the higher will be the growth rate.

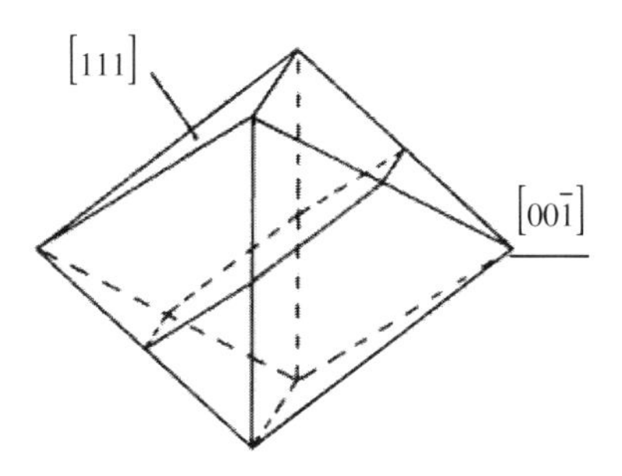

Figure 9.69 Equilibrium shape of an isolated Al crystal.

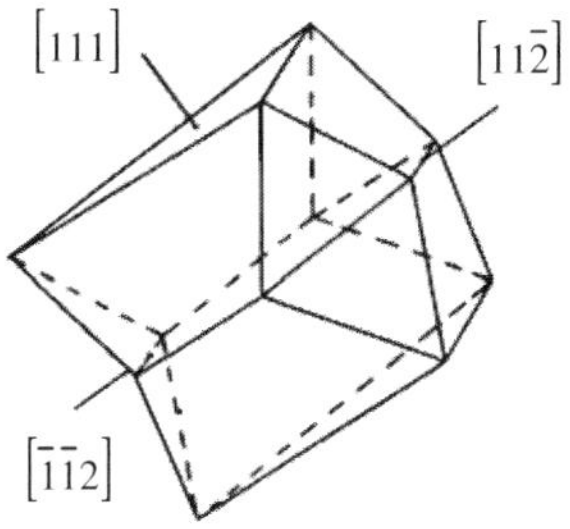

Figure 9.70 Equilibrium shape of a twinned crystal in case of negligible twin boundary energy.

Figure 9.70 shows the corresponding equilibrium shape of a twinned Al crystal. It has been obtained from Figure 9.69 by rotating the lower half of the crystal 180° around the [111] axis.

The growth of the crystal occurs in the <112> direction. There are two possible directions, one with a tip and the other with a reentrant edge. Examination of the feathery grains of aluminium with polarized light shows clearly that the growth occurs in the tip direction.

9.6.5 *Microsegregation in Dendritic Structures*

The phenomenon of microsegregation during formation of dendrites in an alloy has been discussed on pages 441–442 in Chapter 8, where the simplest and most common model of microsegregation is described.

After the formation of a dendrite crystal the further solidification will continue by coarsening and growth of the already formed dendrite arms as is shown in Figure 9.71. During the solidification process the concentration of the alloying element in the melt either or decreases, depending on the difference in solubility in the solid and liquid phases. If the partition coefficient increases $k < 1$ the alloying element will be enriched in the melt during the solidification process.

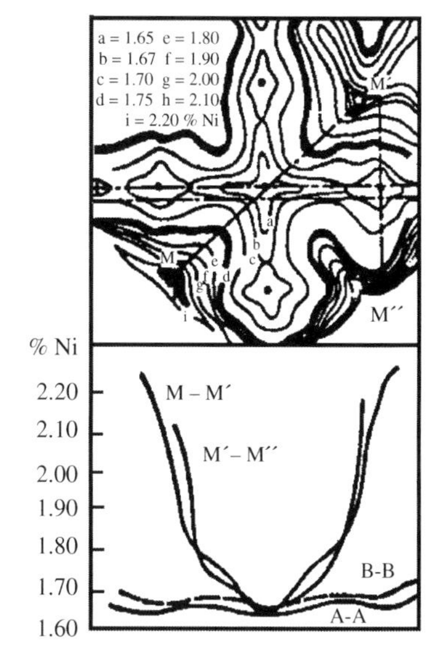

Figure 9.71 Microsegregation in a dendrite cross.

Upper figure: Isoconcentration curves.

Lower figure: Measured concentrations along some different lines through the dendrite cross.

Models of Microsegregation

A number of simplified models have been presented in the scientific literature. Scheils'model is the oldest and simplest one. It has been derived in Chapter 8 on page 442 and is described shortly below.

Scheil's equation gives the concentration of the alloying element as a function of the average concentration of the original melt and a function of f, the fraction of solidification. Scheil assumed that the diffusion of the alloying element in the solid phase is zero. He also assumed that the partition coefficient k, equal to the ratio x^s/x^L, is constant, i.e. the solidus and liquidus lines in the phase diagram of the alloy are straight.

$$x^s = kx_0^L(1-f)^{-(1-k)} \tag{9.85}$$

It is well known that this relationship is a reasonable way to describe the microsegregation when the diffusion coefficient is very low, less than 10^{-15} m^2/s at the solidification temperature.

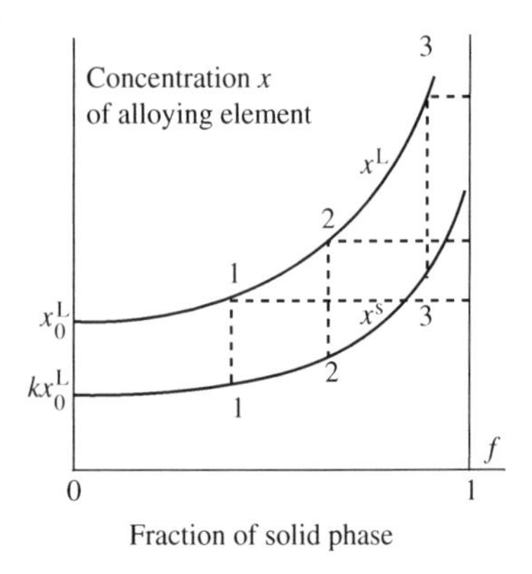

Figure 9.72 Concentration of the alloying element as a function of the fraction of solidification according to Scheil's equation.

A comparison between Scheil's model and experimental results differ from each other in two ways.

1. Figure 9.72 and Equation (9.85) on page 537 show that the concentration of the alloying element approaches infinity at the end of the solidification process. When the concentration in the melt becomes very high the alloying element starts to diffuse back into the solid phase in reality. It is also common that the solidification ends with a eutectic reaction. The back diffusion influences the volume fraction of the eutectic structure.
2. The concentration of the alloying element at the centres of the dendrites increases with time, due to the back diffusion.

If the back diffusion, i.e. the diffusion of the alloying element into the solid phase, is taken into account *Scheil's modified equation* is obtained.

$$x^s = kx_0^L\left(1 - \frac{f}{1 + D_s\frac{4\theta}{\lambda_{den}^2}k}\right)^{-(1-k)} \tag{9.86}$$

where the correction term $B = D_s\frac{4\theta}{\lambda_{den}^2}k$ is small when the diffusion constant in the solid phase D_s is small. If D_s is equal to zero Equations (9.85) and (9.86) become identical.

The dendrite morphology and the partition of the alloying element inside the dendrites are very complex as is shown in Figure 9.71 on page 537. Analytical models are often very useful to show physical phenomena. However, it has been impossible so far to find an analytical model, which shows the distribution of the alloying element in complex geometries such as that shown in Figure 9.71. It is also hard to perform numerical calculations of the microsegregation in dendrites.

The models of microsegregation which include back diffusion have to describe the subsequent diffusion in the solid phase and the resulting distribution of the alloying element within the solid. Consideration of the complete effect of back diffusion requires an overall mass balance for the solid and liquid phases. The calculations can only be performed with the aid of numerical methods. A number of such models have been suggested in the scientific literature.

However, even the most advanced numerical models do not describe the complex geometry of a dendrite structure properly enough. A fairly good description can often be found by use of the simple numerical model presented below.

Simple Numerical Method for Calculation of the Micro-segregation in Primary Dendrite Arms

The distances between adjacent primary dendrite arms are about 100 times larger than the secondary arm spacings. The back diffusion becomes much faster in the secondary arms than in the primary arms owing to the shorter distances in the thinner secondary dendrite arms than in the thicker primary arms. Below we will discuss microsegregation with attention to back diffusion in primary stems.

Simple Model of Microsegregation in Primary Dendrite Arms

Consider an array of primary dendrite arms in the shape of infinitely thin parallel plates at the distance λ_{den} from each other in a melt with the initial concentration x_0^L of the alloying element (Figure 9.73).

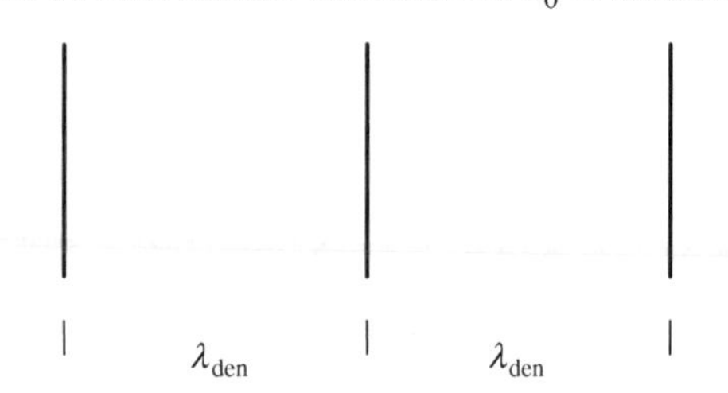

Figure 9.73 One-dimensional model of thin primary dendrite arms.

The dendrites grow and their widths increase. The distribution of the alloying element in the solid phase is uneven and is smoothed out owing to back diffusion of the alloying element in the solid phase.

The concentration of the alloying element is a function of both position and time. It can be found by solving Fick's second law for the diffusion process. The solution can in this case be expressed as a Fourier series with a fundamental tone and a number of overtones. The amplitudes of the overtones rapidly decay with time. In the end only the fundamental tone with wavelength λ is left.

Fick's second law:

$$\frac{\partial x}{\partial t} = D\frac{\partial^2 x}{\partial y^2}$$

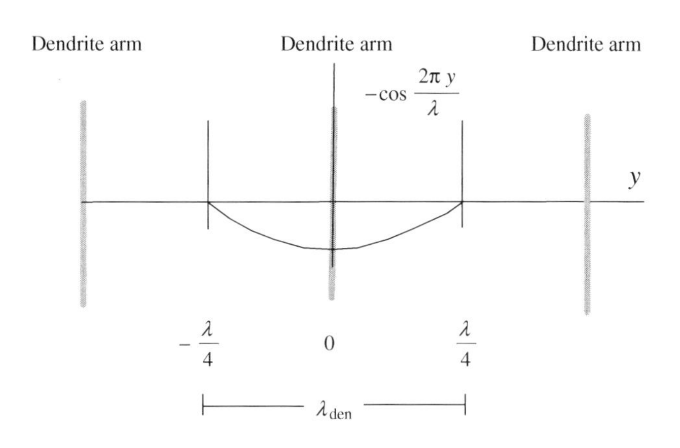

Figure 9.74 $-\cos\frac{2\pi y}{\lambda}$ is the position-dependent part of Equation (9.77) which describes the concentration of the alloying element in a dendrite as a function of position. $-\cos\frac{2\pi y}{\lambda}$ is zero for $y = \frac{\lambda}{4}$.

If the origin ($y = 0$) is located at the centre of a dendrite arm the solution (Figure 9.74), i.e. the concentration of the alloying element in the solid phase can be written as

$$x(y, t) = x^{s/L} - (x^{s/L} - x_{min})e^{-\frac{4\pi^2 D_s}{\lambda^2}t}\cos\frac{2\pi y}{\lambda} \tag{9.87}$$

where

$x(y, t)$ = concentration of the alloying element (mole fraction)
$x^{s/L}$ = concentration of the alloying element in the solid at the interface at time t
x_{min} = concentration of the alloying element at the centre of the dendrite at time t
D_s = diffusion constant of the alloying element in the solid
λ = wavelength of the fundamental tone
y = distance from the centre of the dendrite, which corresponds to minimum concentration of the solute
f = fraction of solid phase at time t.

Figure 9.74 shows that a dendrite distance λ_{den} can only include *half* a wavelength λ of the fundamental tone.

$$\frac{\lambda}{2} = \lambda_{den} \tag{9.88a}$$

or

$$\lambda = 2\lambda_{den} \tag{9.88b}$$

To avoid confusion we will keep λ_{den}, which has a physical signification, and eliminate λ in Equation (9.87). Hence, the concentration distribution in the solidified dendrite can be written as

$$x(y,t) = x^{s/L} - (x^{s/L} - x_{min})e^{-\frac{\pi^2 D_s}{\lambda_{den}^2}t} \cos\frac{\pi y}{\lambda_{den}^2} \quad (9.89)$$

Equations (9.88) and (9.89) are valid *if and only if* the dendrite arms are completely solidified. During the solidification process at time t the dendrite is *partly* solidified with the fraction f_0 of λ_{den}. Then the width of the solid dendrite is only $f_0\ \lambda_{den}$ and λ_{den} in Equation (9.89) has to be replaced by $f_0\ \lambda_{den}$:

$$x(y,t) = x^{s/L} - (x^{s/L} - x_{min})e^{-\frac{\pi^2 D_s}{f_0^2 \lambda_{den}^2}t} \cos\frac{\pi y}{f_0 \lambda_{den}} \quad (9.90)$$

where f_0 represents the solidification fraction of the dendrite at time t.

Equation (9.90) is the general basic relationship which describes the microsegregation in a solidifying dendrite arm with the width $f_0\ \lambda_{den}$ including the back diffusion as a function of time and position. It will be used to describe the concentration of the solute in the dendrite arm during a solidification process.

The solidification process includes simultaneous coarsening of the dendrite and diffusion of the alloying element in the solid phase. At $t = 0$ the melt with the composition x_0^L starts to solidify. The Figures 9.75 below show the concentration profile at two intermediate solidification fractions and the time t_3 at the end of the solidification process. The maximum y values at these three occasions are called y_1, y_2 and y_3.

A comparison between the Figures 9.75 shows that a considerable redistribution of the alloying element occurs when the solidification fraction gradually increases as a consequence of the back diffusion. The figures show that $x^{L/s}$ and $x^{s/L}$ successively change. The concentration x_{min} at the centre of the dendrite is *not* a constant but increases gradually until the solidification process is ended.

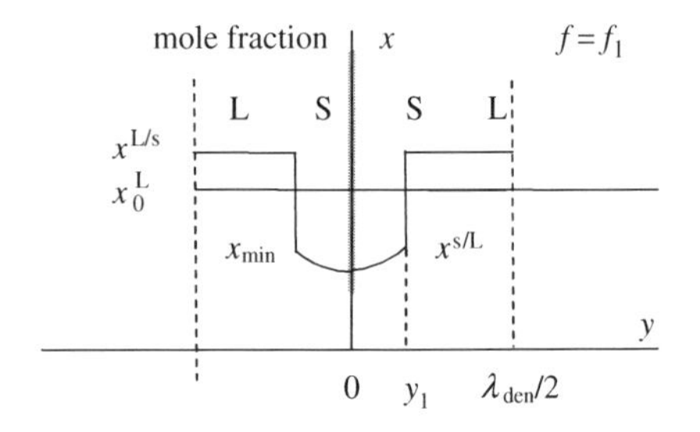

Figure 9.75a Concentration profile of the alloying element in the partly solidified dendrite at solidification fraction f_1.

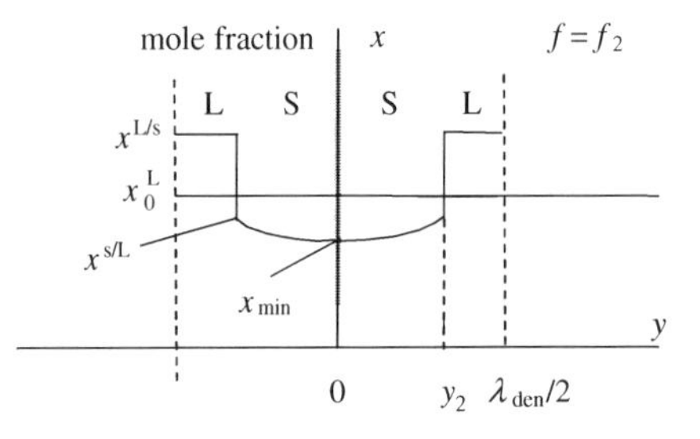

Figure 9.75b Concentration profile of the alloying element in the partly solidified dendrite at solidification fraction f_2.

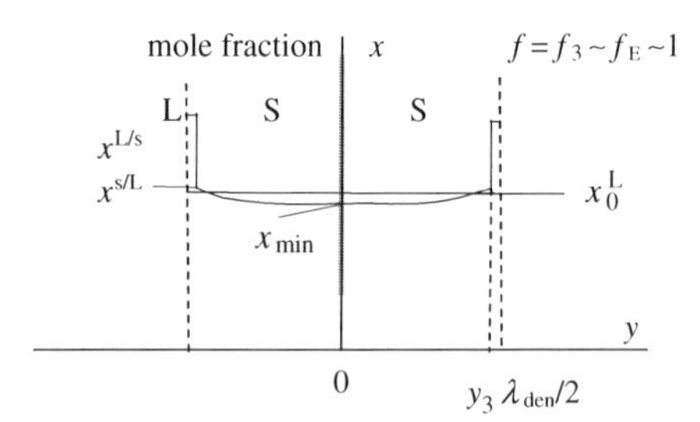

Figure 9.75c Concentration profile of the alloying element in the nearly solidified dendrite. f_E = the eutectic composition.

The microsegregation as a function of time can be calculated quantitatively if the solidification conditions are known. It is also necessary to set up a quantitative condition for the redistribution of the alloying element. This will be done below.

Material Balance of the Alloying Element during Diffusion in the Solid

The diffusion of the solute in the melt is very rapid and it is most reasonable to assume that the concentration of the alloying element is the same everywhere in the melt, independent of time.

As can be concluded from Equation (9.90) the concentration distribution of the alloying element depends of time, temperature, diffusion constant and concentration of the alloying element at the solid/liquid interface.

Time is the independent variable and several of the quantities above depend on the solidification conditions. Therefore, it is not possible to find a general solution but some conditions must always be fulfilled.

We assume that the phase diagram of the alloy is known and that the solidification process at the interface occurs at equilibrium. In this case it is reasonable to assume that the partition coefficient is constant.

$$k = \frac{x^{s}}{x^{L}} \tag{9.91}$$

The condition (9.91) is valid independently of the solidification conditions. Another condition, which has to be fulfilled, is the material balance. At the back diffusion the total amount of the alloying element in the melt and the solid must be constant. The material balance can be expressed quantitatively with the aid of Equations (9.90), (9.91) and Figure 9.76. We apply Equation (9.90) on a dendrite with the solidification fraction f_0.

The excess of alloying element in the melt, which surrounds the dendrite must exactly balance the loss of alloying element in the solid dendrite. Hence, the marked areas in Figure 9.76 must be equal, which gives the relationship

$$\left(x^{L/s} - x_0^L\right)\left(\frac{\lambda_{den}}{2} - y_{max}\right) = \int_0^{y_{max}} \left[x_0^L - x(y,t)\right] dy$$
$$= \int_0^{y_{max}} \left[x_0^L - x^{s/L} + \left(x^{s/L} - x_{min}\right) e^{-\frac{\pi^2 D_s}{f_0^2 \lambda_{den}^2} t} \cos\frac{\pi y}{f_0 \lambda_{den}} \right] dy \tag{9.92}$$

where $y_{max} = f_0 \dfrac{\lambda_{den}}{2}$.

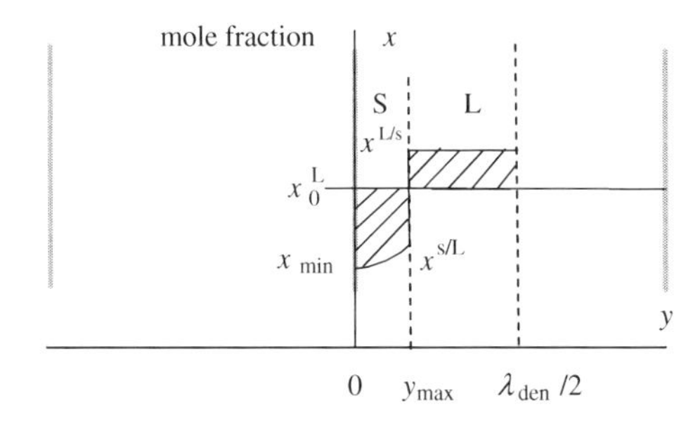

Figure 9.76 One-dimensional model of a solidifying primary dendrite arm with its centre at the origin. Concentration of the alloying element as a function of position, i.e. the distance from the centre of the dendrite. The width of half the solid plate is $y_{max} = f\,\lambda_{den}/2$.

We obtain a more general expression if we replace y by $f\,\lambda_{den}/2$ in Equation (9.92).

$$f = \frac{y}{\frac{\lambda_{den}}{2}} \tag{9.93a}$$

or

$$y = f\frac{\lambda_{den}}{2} \qquad f \leq f_0 \tag{9.93b}$$

where $y_{max} = f_0\,\lambda_{den}/2$ which corresponds to $f = f_0$.

Equation (9.92) can then be written as

$$\left(x^{L/s} - x_0^L\right)\left(\frac{\lambda_{den}}{2} - f_0\frac{\lambda_{den}}{2}\right) =$$

$$= \int_0^{f_0} \left[x_0^L - x^{s/L} + \left(x^{s/L} - x_{min}\right) e^{-\frac{\pi^2 D_s}{f_0^2 \lambda_{den}^2} t} \cos\frac{\pi f \frac{\lambda_{den}}{2}}{f_0 \lambda_{den}} \right] \frac{\lambda_{den}}{2} df$$

Division with $\frac{\lambda_{den}}{2}$ gives

$$(x^{L/s} - x_0^L)(1 - f_0) = \int_0^{f_0} \left[x_0^L - x^{s/L} + (x^{s/L} - x_{min}) e^{-\frac{\pi^2 D_s}{f_0^2 \lambda_{den}^2} t} \cos\frac{f}{f_0}\frac{\pi}{2} \right] df \qquad (9.94)$$

Equation (9.94) is the desired material balance, which always has to be fulfilled during solidification.

Numerical Calculations of the Microsegregation with the Aid of the Simple Model

We want to find the microsegregation as a function of time and position. With the aid of Equation (9.93b) Equation (9.90) can be transformed into

$$x(f, t) = x^{s/L} - (x^{s/L} - x_{min}) e^{-\frac{\pi^2 D_s}{f_0^2 \lambda_{den}^2} t} \cos\frac{f}{f_0}\frac{\pi}{2} \qquad (9.95)$$

The material balance, Equation (9.94), must be fulfilled during the whole solidification time.

Solidification processes depend primarily on the temperature distribution in the solid. In this case solidification and diffusion occur simultaneously and results in a process, which is not stationary.

Known quantities in Equations (9.94) and (9.95) are λ_{den} and D_s. There are five variable quantities: time t, $x^{L/s} x^{s/L}$, f_0 and x_{min} and in addition the temperature T, which we choose as the independent variable instead of time t.

To find the solution it is necessary to find t, $x^{L/s}$, $x^{s/L}$, f_0 and x_{min} as functions of time and present the result graphically with the time as a parameter. For this purpose we need and have at our disposal:

- the phase diagram
- a known relationship between the temperature and time
- the material balance for the alloying element.

Below we will describe the calculation of related values of the unknown quantities.

Step 1

- Start at $t = 0$, the concentration x_0^L and temperature T_L in the phase diagram (Figure 9.77). The temperature is lowered an arbitrary amount ΔT_1. This process requires the time Δt_1, which can not be found in the phase diagram.

  The values of $x^{s/L}$ and $x^{L/s}$, that correspond to the temperature difference ΔT_1, can be read directly or calculated from the phase diagram.
- The time difference Δt_1 is derived with the aid of the known temperature-time relationship. As $t = 0$ at the beginning we call this time t_1.
- The solidification fraction f_1 and the corresponding x_{min} are the two remaining unknown quantities to be determined for the temperature difference ΔT_1. This is done with the aid of Equations (9.90), (9.94), (9.95) and iteration. The detailed process is described in the box below.
- The calculated values of $x^{s/L}$, t_1, f_1 and $(x_{min})_1$ are inserted into Equation (9.95) together with the given values of D_s and λ_{den}. This equation is the desired function $x(y, t_1)$ for the temperature difference ΔT_1. It is plotted as a function of y with t_1 as a parameter in a diagram.

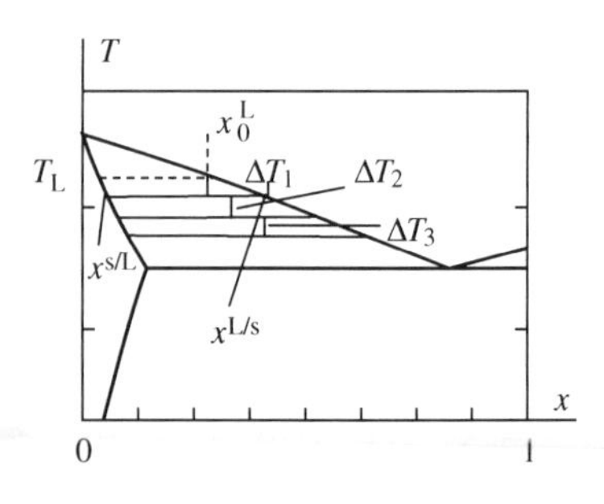

Figure 9.77 Phase diagram of a binary system.

Step 2

A new value ΔT_2 is chosen (Figure 9.77) and the same cycle as in step 1 is repeated. The new function $x(y, t_2)$ is plotted into the same diagram. The corresponding time represents a parameter and should be marked close to the curve in the figure.

Further steps

More steps are calculated until the solidification process is ended or the eutectic temperature is reached. The multitude of curves is the complete answer of the problem. The calculation process is described in the box below.

In Example 9.7 simple concrete application is given to demonstrate the calculation method.

Calculation of f_1 and $(x_{min})_1$ for Temperature interval ΔT_1

Loop 1

1. We start at $t = 0$. Then $x_{min} = kx_0^L$ and $f = f_0$ can be calculated with the aid of the material balance (9.94) on page 542

$$\left(x^{L/s} - x_0^L\right)(1 - f_0) = \int_0^{f_0} \left[x_0^L - x^{s/L} + \left(x^{s/L} - kx_0^L\right) e^{-\frac{\pi^2 D_s}{f_0^2 \lambda_{den}^2} \times 0} \cos\frac{f}{f_0}\frac{\pi}{2} \right] df \qquad (1')$$

This gives a primary value f_0. (The true value of f_0 at $t = 0$ is zero. The preliminary f_0 is only a start value for calculation of f_1.)

2. f_0, $t = t_1$ and $x_{min} = kx_0^L$ are inserted into Equation (9.87) on page 539. t_1 is given by the cooling rate when ΔT_1, is selected.

$$x(f, t_1) = x^{s/L} - \left(x^{s/L} - kx_0^L\right) e^{-\frac{\pi^2 D_s}{f_0^2 \lambda_{den}^2} t_1} \cos\frac{f}{f_0}\frac{\pi}{2} \qquad (2')$$

Equation (2′) is used for calculation of a new and better value of x_{min} than kx_0^L when $f = 0$. The new value is

$$x_{min}' = x(0, t_1) = x^{s/L} - (x^{s/L} - kx_0^L) e^{-\frac{\pi^2 D_s}{f_0^2 \lambda_{den}^2} t_1} \cos\frac{0}{f_0} \times \frac{\pi}{2} \qquad (3')$$

Loop 2

New start. Insert $t = t_1$ and x_{min}' into the material balance (9.94) and calculate a new and better value of f, equal to f_0'.

$$\left(x^{L/s} - x_0^L\right)(1 - f_0') = \int_0^{f_0'} \left[x_0^L - x^{s/L} + (x^{s/L} - x_{min}') e^{-\frac{\pi^2 D_s}{f_0^2 \lambda_{den}^2} t_1} \cos\frac{f}{f_0}\frac{\pi}{2} \right] df \qquad (4')$$

f_0' is compared to f_0. If the difference is small we accept the values f_0' and x_{min}' as the correct values of f_1 and $(x_{min})_1$ at step ΔT_1.

3. If f_0' and f_0 differ too much we have to repeat the loop once more and use f_0' and x_{min}' to calculate a better value x_{min}'' and use t_1 and x_{min}'' for calculation of a new and better value f_0''. If $f_0'' \approx f_0'$ we accept the values f_0'' and x_{min}'' otherwise we have to run another loop and calculate x_{min}'' and f_0''. We go on until we have found acceptable values of f_1 and $(x_{min})_1$.

Calculation of f and x_{min} for the intervals ΔT_2, ΔT_3 etc.

New concentration values $x^{s/L}$ and $x^{L/s}$ are read from the phase diagram and used in the equations for the temperature interval ΔT_2.

The same procedure as that in the interval ΔT_1 is repeated for the interval ΔT_2. The loops start at $t = 0$ and the time Δt_2 is replaced by t_2 as the aim is to find the function $x(f, t_2)$, which is independent of $x(f, t_1)$ and all the other functions.

The process is repeated for all the temperature intervals until the eutectic temperature is reached.

Example 9.7

A specimen of the alloy Al-4.5wt%Cu has solidified by unidirectional dendritic solidification. The temperature gradient was 1.7×10^3 K/m and the growth rate of the dendrites was 1.0×10^{-4} m/s. The constant growth rate is valid for the dendrite tips and not for the widening of the primary stems. After the solidification the dendrite distance was measured to 76 μm.

The diffusion constant D_s of the alloying element in the solid phase varies strongly with the temperature. For the present Al-Cu alloy the temperature dependence can be described by the relationship

$$D_s = D_s^0 e^{-\frac{Q}{RT}} = 8.4 \times 10^{-6} e^{-\frac{16430}{T}} \text{ m}^2/\text{s}$$

Calculate the microsegregation in the dendrite arms, i.e. the concentration of the alloying element as a function of the distance from the centre of the dendrite at various times during the solidification. Plot the functions graphically with the time as parameter.

The phase diagram of the binary system Al-Cu is given below. The liquidus temperature of the alloy is 646 °C.

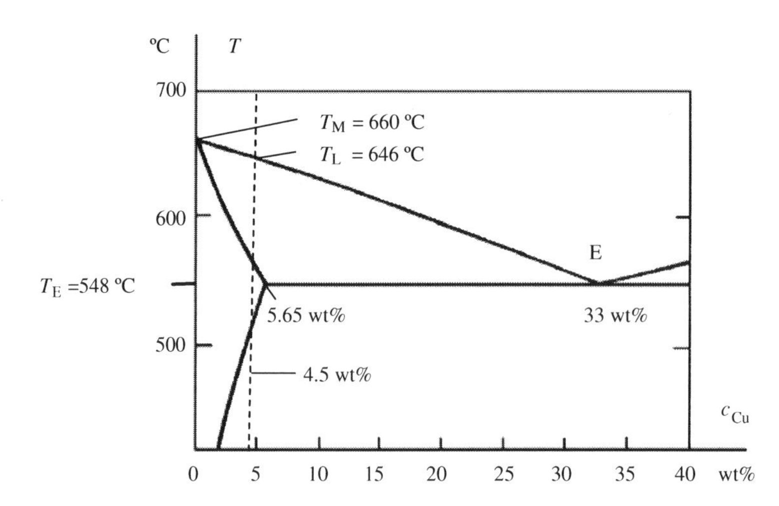

Solution

The large figure on next page is a sketch of the experimental equipment at two different times and a volume element during the solidification.

The upper figure in the margin shows the position of the two phase region of the volume element as a function of time.

The lower figure in the margin shows the concentration of the alloying element in a dendrite arm as a function of position with the time as a parameter.

The general solution of this problem has been discussed on pages 538–544. Below we will list some of the equations, which are valid in this particular case.

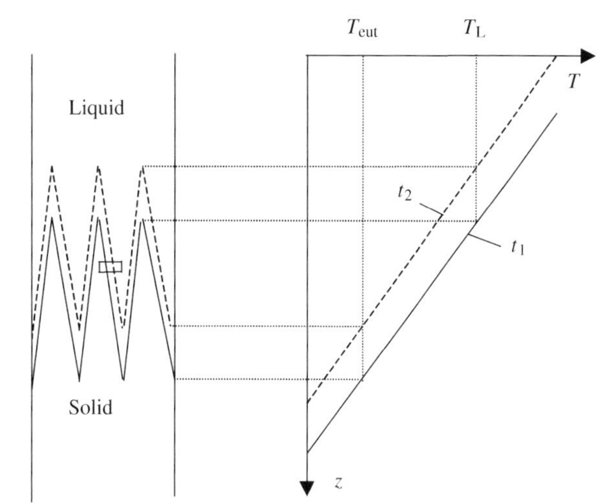

Solidification front (left figure) and temperature distribution in a unidirectional solidfication experiment.

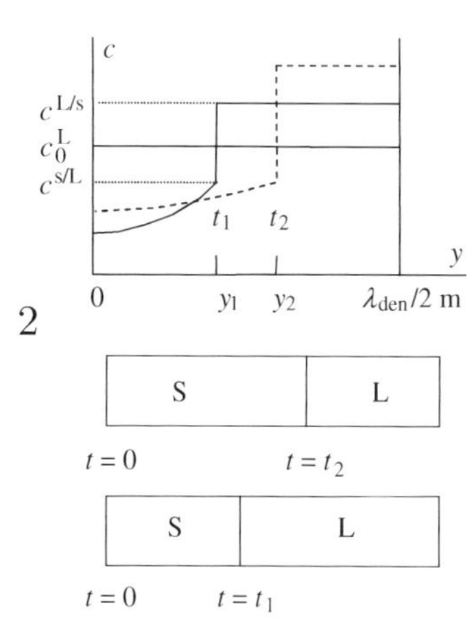

The upper figure shows the concentration of the alloying element in half a dendrite as a function of distance from the centre.

The lower figure is an enlargement of the small rectangle in the right figure at two different times.

1. The solidification interval $T_L - T_E = 646 - 548\ °C$ is divided into five ΔT intervals and the corresponding $c^{s/L}$ and $c^{L/s}$ values are calculated by linear interpolation from the phase diagram.
2. The relationship between ΔT and Δt can be calculated with the aid of the known temperature gradient and the constant growth rate V_{growth}, given in the text.

$$\frac{dT}{dt} = \frac{dT}{dz}\frac{dz}{dt} = 1.7 \times 10^3 \text{K/m} \times 1.0 \times 10^{-4} \text{m/s} \qquad (1')$$

which gives

$$dt = 5.9\, dT \qquad (2')$$

The solidification time is obtained with the aid of Equation (1′) with $dt = t_{sol}$ and $dT = T_L - T_E$:

$$t_{sol} = 5.9(T_L - T_E) \qquad (3')$$

which gives

$$t_{sol} = 5.9(T_L - T_E) = 5.9(646 - 548) = 578 \text{ s} \qquad (4')$$

3. The partition constant can be calculated with the aid of the phase diagram. We assume that the solidus and liquidus lines are straight and use the concentration values at the eutectic temperature.

$$k = \frac{5.65}{33.0} = 0.17 \qquad (5')$$

Calculations

We use wt% instead of mole fractions because of the units in the phase diagram.

$$t_{sol} = 578\text{ s} \qquad \frac{\lambda_{den}}{2} = \frac{76}{2} = 38\,\mu\text{m} \quad \text{and} \quad c_0^L = 4.5\text{ wt\%}$$

$$D_s = D_s^0 e^{-\frac{Q}{RT}} = 8.4 \times 10^{-6} e^{-\frac{16430}{T+273}}\ \text{m}^2/\text{s}\ (T \text{ measured in } ^\circ\text{C})$$

The concentrations are calculated by linear interpolation with the aid of the solidus and liquidus lines.

$$c^{L/s} = c_0^L + (c_E - c_0^L)\frac{T_L - T}{T_L - T_E} = 4.5 + 28.5 \times \frac{(646 - T)}{646 - 548} \quad (T \text{ measured in } ^\circ\text{C})$$

$$c^{s/L} = c_0^s \frac{T_M - T}{T_M - T_E} = 5.65\frac{(660 - T)}{660 - 548} \quad (T \text{ measured in } ^\circ\text{C})$$

$$c_0^L = 4.5\text{ wt-\%} \quad k = 0.17$$

$$y_{max} = f_0 \frac{\lambda_{den}}{2} \qquad t = f_0 t_{sol}$$

Table 1′ Calculation of the concentrations of the alloying element in the solid respectively liquid and the average diffusion constant.

T °C	ΔT $T_L - T$ °C	$c^{L/s}$ wt%	$c^{s/L}$ wt%	Temperature interval °C	$T_{average}$ °C	D_s^{ave} 10^{-14} m²/s
646.0	0	4.5	0.71			
				ΔT_1	636.2	11.9
626.4	19.6	10.2	1.70			
				ΔT_2	616.6	8.0
606	39.2	15.9	2.68			
				ΔT_3	597.0	5.25
586	58.8	21.6	3.67			
				ΔT_4	577.4	3.40
566	78.4	27.3	4.66			
				ΔT_5	557.8	2.16
548	98	33.0	5.65			

After calculation of the basic quantities we follow the recommended solution procedure in the box on page 109. Integration of Equation (1′) in the box for calculation of f_0 gives

$$\left(c^{L/s} - c_0^L\right)(1 - f_0) = \left(c_0^L - x^{s/L}\right) f_0 + \left(c^{s/L} - kc_0^L\right) \times 1 \times \sin\frac{\pi}{2} \times \frac{2}{\pi} f_0 \qquad (6')$$

with the aid of the values in Table 2′.

Table 2′ Calculation of f_0 at $t=0$

ΔT °C	c_0^{L} wt%	$c^{\mathrm{s/L}}$ wt%	$c^{\mathrm{L/s}}$ wt%	$c^{\mathrm{L/s}}-c_0^{\mathrm{L}}$ wt%	$c_0^{\mathrm{L}}-c^{\mathrm{s/L}}$ wt%	$c^{\mathrm{s/L}}-kc_0^{\mathrm{L}}$ wt%	f_0
ΔT_1	4.5	1.70	10.2	5.7	2.80	0.94	0.626
ΔT_2	4.5	2.68	15.9	11.4	1.82	1.92	0.789
ΔT_3	4.5	3.67	21.6	17.1	1.59	2.91	0.832
ΔT_4	4.5	4.66	27.3	22.8	−0.16	3.90	0.908
ΔT_5	4.5	5.65	33.0	28.5	−1.15	4.89	0.936

Next we calculate $c_{\mathrm{min}}{}'$ with the aid of Equation (3′) in the box on page 543.

$$c_{\mathrm{min}}{}' = c(0,t) = c^{\mathrm{s/L}} - (c^{\mathrm{s/L}} - kc_0^{\mathrm{L}})\mathrm{e}^{-\frac{\pi^2 D_{\mathrm{s}}^{\mathrm{ave}}}{f_0^2 \lambda_{\mathrm{den}}^2} t_1} \times 1 \qquad (7')$$

The $D_{\mathrm{s}}^{\mathrm{ave}}$ values are calculated in Table 1′ on page 546. The resulting c'_{min}’ values are given in Table 3′.

Table 3′ Calculation of $c_{\mathrm{min}}{}'$

ΔT °C	f_0	$c^{\mathrm{s/L}}$ wt%	$c^{\mathrm{s/L}} - kc_0^{\mathrm{L}}$ wt%	ΔT_1 s	$-\frac{\pi^2 D_{\mathrm{s}}^{\mathrm{ave}}}{f_0^2 \lambda_{\mathrm{den}}^2} t_1$	$\mathrm{e}^{-\frac{\pi^2 D_{\mathrm{s}}^{\mathrm{ave}}}{\lambda_{\mathrm{den}}^2} t_1}$	c'_{min} wt%
ΔT_1	0.626	1.70	0.94	116	−0.0602	0.942	0.814
ΔT_2	0.789	2.68	1.92	231	−0.0507	0.951	0.854
ΔT_3	0.832	3.67	2.91	347	−0.0450	0.956	0.888
ΔT_4	0.908	4.66	3.90	462	−0.0326	0.968	0.885
ΔT_5	0.936	5.65	4.89	578	−0.0244	0.976	0.877

Loop 1 is now completed. A better value $f_0{}'$ can be calculated from Equation (4′) in the box on page 543. After integration we obtain the equation

$$(c^{\mathrm{L/s}} - c_0^{\mathrm{L}})(1 - f_0{}') = (c_0^{\mathrm{L}} - c^{\mathrm{s/L}})f_0{}' + (c^{\mathrm{s/L}} - c_{\mathrm{min}}{}') \times 1 \times \frac{2}{\pi} f_0{}' \qquad (8)$$

The only difference between Equations (8′) and (6′) is that kc_0^{L} has been replaced by c'_{min} in Equation (8′). The calculations are listed in Table 4′.

Table 4′ Calculation of f'_0

ΔT °C	c_0^{L} wt%	$c^{\mathrm{s/L}}$ wt%	$c^{\mathrm{L/s}}$ wt%	$c^{\mathrm{L/s}}-c_0^{\mathrm{L}}$ wt%	$c_0^{\mathrm{L}} - c^{\mathrm{s/L}}$ wt%	$c^{\mathrm{s/L}}-c'_{\mathrm{min}}$ wt%	f'_0	y_{max} µm
ΔT_1	4.5	1.70	10.2	5.7	2.80	0.89	0.629	23.9
ΔT_2	4.5	2.68	15.9	11.4	1.82	1.83	0.792	30.1
ΔT_3	4.5	3.67	21.6	17.1	1.59	2.78	0.836	31.8
ΔT_4	4.5	4.66	27.3	22.8	−0.16	3.78	0.910	34.6
ΔT_5	4.5	5.65	33.0	28.5	−1.15	4.77	0.938	35.6

A comparison between f_0 in Table 2′ and $f_0{}'$ in Table 4′ shows that they differ less than 1%, which is less than the accuracy of many of the measurements. Therefore, further loops and calculations are not required. $c_{\mathrm{min}}{}'$ and $f_0{}'$ are accepted as the final values. We can now list the functions $c(y, t)$ and plot them in a diagram.

Answer:

The desired functions for the five temperature intervals, defined in Table 1′ on page 546 are

$$c(y,t) = c^{s/L} - (c^{s/L} - c_{min})e^{-\frac{\pi^2 D_s^{ave}}{f_0^2 \lambda_{den}^2}t} \cos\frac{\pi y}{f'_0 \lambda_{den}}$$

$$c(y,t) = 1.70 - 0.89\,e^{-5.14\times 10^{-4}t}\cos 0.0657\,y \qquad 0 < t < 116\,s \qquad 0 < y < 24\,\mu m$$

$$c(y,t) = 2.68 - 1.83\,e^{-2.18\times 10^{-4}t}\cos 0.0522\,y \qquad 116\,s < t < 231\,s \qquad 0 < y < 30\,\mu m$$

$$c(y,t) = 3.67 - 2.78\,e^{-1.28\times 10^{-4}t}\cos 0.0494\,y \qquad 231\,s < t < 347\,s \qquad 0 < y < 32\,\mu m$$

$$c(y,t) = 4.66 - 3.78\,e^{-7.02\times 10^{-5}t}\cos 0.0454\,y \qquad 347\,s < t < 462\,s \qquad 0 < y < 34.6\,\mu m$$

$$c(y,t) = 5.65 - 4.77\,e^{-4.19\times 10^{-5}t}\cos 0.0441\,y \qquad 462\,s < t < 578\,s \qquad 0 < y < 35.6\,\mu m$$

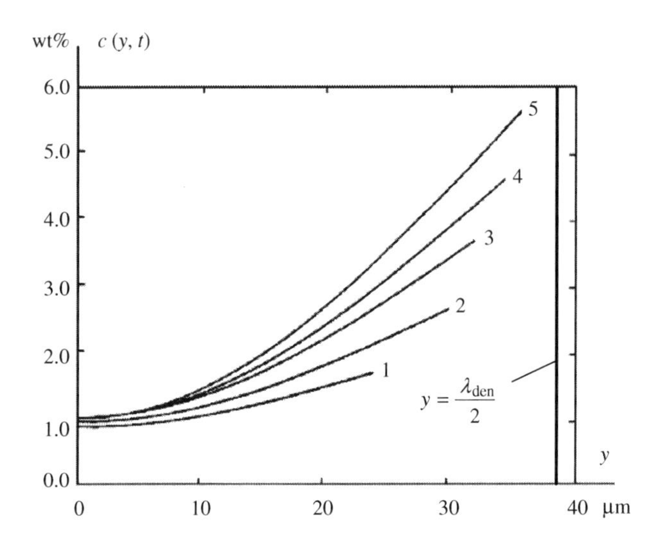

The functions in the figure are:

$$y = 38\,\mu m$$

$$c(y,t) = 1.70 - 0.84\cos 0.0657y \qquad 0 < y < 24\,\mu m$$

$$c(y,t) = 2.68 - 1.74\cos 0.0522y \qquad 0 < y < 30\,\mu m$$

$$c(y,t) = 3.67 - 2.66\cos 0.0494y \qquad 0 < y < 32\,\mu m$$

$$c(y,t) = 4.66 - 3.66\cos 0.0454y \qquad 0 < y < 34.6\,\mu m$$

$$c(y,t) = 5.65 - 4.66\cos 0.0441y \qquad 0 < y < 35.6\,\mu m$$

9.6.6 Dendrite Arm Spacings in Alloy Crystals

The structure of a dendrite is shown in Figure 9.32 on page 504. A very important feature of the quality of a material is the coarseness of its microstructure. The quality of cast materials is controlled in many respects by the distance between its adjacent primary respectively secondary dendrite arms.

Formation of Dendrites

The formation of primary and secondary dendrite arms has been discussed in Section 9.6.3 on pages 530–531. The concentration field of the alloying element around the growing tip controls the secondary dendrite arm distance. The extension of the concentration gradient around the tip is of the same magnitude as the radius of the dendrite tip. The distance between the adjacent secondary dendrite arms is proportional to the tip radius.

The temperature field around the tip determines the distance between two adjacent parallel primary dendrite stems. On page 531 we mentioned that this distance is proportional to the tip radius times the factor $\sqrt{\alpha/D_L}$, where α is the thermal diffusivity (page 531) and D_L the diffusion constant of the alloying element in the melt. For both primary and secondary arms there is an incessant competition between the dendrite arms during the growth.

Branching and Engulfing of Primary Dendrite Arms

The primary dendrite arms grow in parallel stems. The growth direction agrees with the direction of the temperature gradient. These columnar dendrites adjust their primary spacings during growth.

If the distances between the primary stems is too large, they may decrease by *branching*, which means that the distance between the stems is large and offers the possibility that some secondary dendrite arms grow exceptionally rapidly and become transformed into primary arms (Figure 9.78). In this way the average distance between the primary arms decreases. This process is very slow, though.

If the dendrite arm distances are too small they may be adjusted by *engulfing*, which means that the temperature fields overlap too much and one arm looses the competition between the stems and disappears as in Figure 9.55 on page 532. The reason is often a slightly wrong growth direction. This case is shown in Figure 9.79. The topic is also discussed in Section 9.6.4 on page 532).

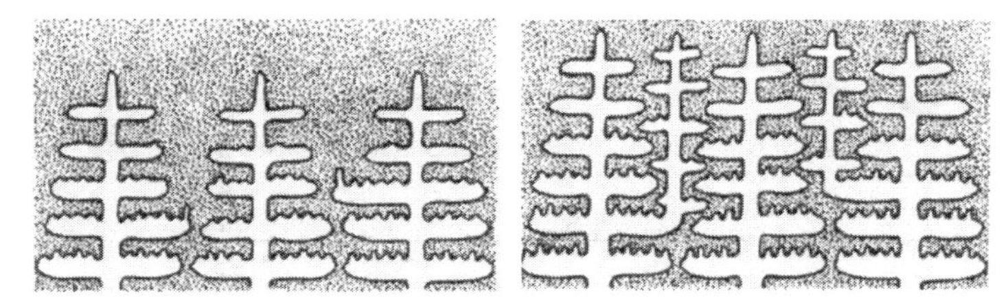

Figure 9.78 Branching, i.e. formation of new primary dendrite arms, provided that the growth rate increases gradually. Merton C. Flemings © McGraw Hill 1974.

Figure 9.79 Engulfing of two primary dendrite arms. See also Figure 9.55 on page 532. Reproduced from Springer © 2002.

Spacings of Primary Dendrite Arms

The spacings of the primary dendrite arms have been discussed in many publications. The most common method of examination has been to plot the dendrite arm distance as a function of the alloy composition at a constant cooling rate as a parameter.

In this way Modolfo et al [20] have examined the influence of an alloying element on the distance between parallel primary dendrite arms. Their results are shown in Figure 9.80. It is obvious from the figure that

- The dendrite arm spacing depends on the cooling rate and the concentration of the alloying element.

The higher the cooling rate is the smaller will be the dendrite distance. Figure 9.80 shows that λ_{den} has a minimum for a specific Cu concentration in Al-Cu alloys.

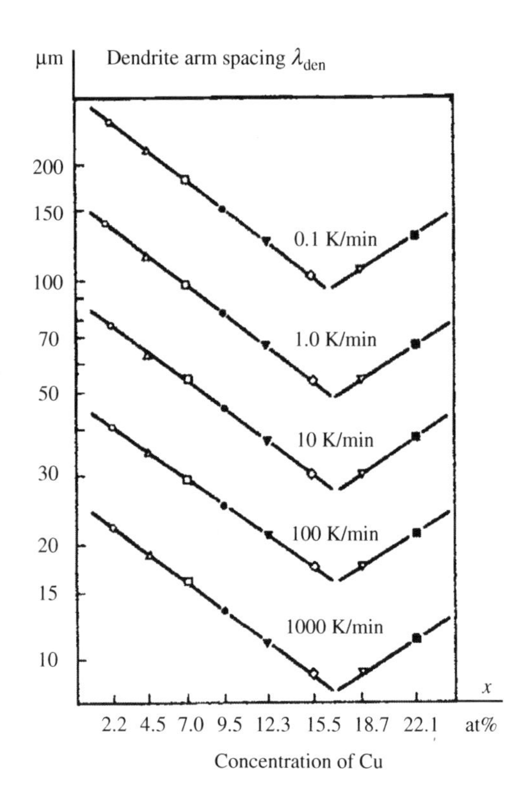

Figure 9.80 Dendrite arm spacing in the system Al–Cu as a function of the Cu concentration for different cooling rates of the alloy. Reproduced from John Wiley & Sons © 1962.

Another way to present the dependence of the dendrite arm spacing on the cooling rate is to plot the spacing as a function of the solidification time t_{sol}. There exists a simple relationship between the cooling rate and the solidification time:

$$t_{sol} = \Delta T_s / -\frac{dT}{dt} \tag{9.96}$$

where

ΔT_s = solidification interval, i.e. the time required for solidification of a distance equal to the dendrite spacing λ_{den}

$-dT/dt$ = cooling rate.

Spacings of Secondary Dendrite Arms

Figure 9.33 on page 504 and Figures 9.50 and 9.51 on page 530 show the formation of secondary dendrite arms and the competition between them during growth. Some of the arms win and grow ahead of the other ones, some of them remelt and others merge. The process is complex and the result is that

- The secondary dendrite arms have no equal and constant distances between adjacent arms.

However, it makes sense to form an average secondary arm distance. It is important to find the secondary dendrite arm spacings. Several methods are used to find the average secondary dendrite arm distance and relate it to experimentally measurable quantities. Some of them will be reported below.

Relationship between the Average Dendrite Arm Spacing and Solidification Time

A wide experience shows that the following empirical relationship often is a good approximation for description of the dendrite arm distance λ_{den} as a function of solidification time t_{sol}

$$\lambda_{\text{den}} = const\ t_{\text{sol}}^{n} \tag{9.97}$$

where n is a number, characteristic for the alloy. For steel we have $1/3 < n < 1/2$. The relationship is valid for both primary and average secondary dendrite arm spacings.

Relationship between the Average Dendrite Arm Spacing and Growth Rate

Another empirical relationship, which is often used, is

$$V_{\text{growth}}\,\lambda_{\text{den}}^{2} = const \tag{9.98}$$

The value of the constant depends on the specific alloy. The relationship is valid for both primary and average secondary dendrite arms of an alloy but the value of the constant is different in the two cases. It has been verified for a large number of alloys.

Figure 9.81 and extensive experimental evidence show that a relationship similar to Equation (9.99) relates the dendrite arm spacing and the cooling rate at solidification.

$$\text{Cooling rate} = GV_{\text{growth}} \tag{9.99}$$

If the temperature gradient G is constant the cooling rate is proportional to the growth rate. The higher the cooling rate is the denser will be the dendrite structure.

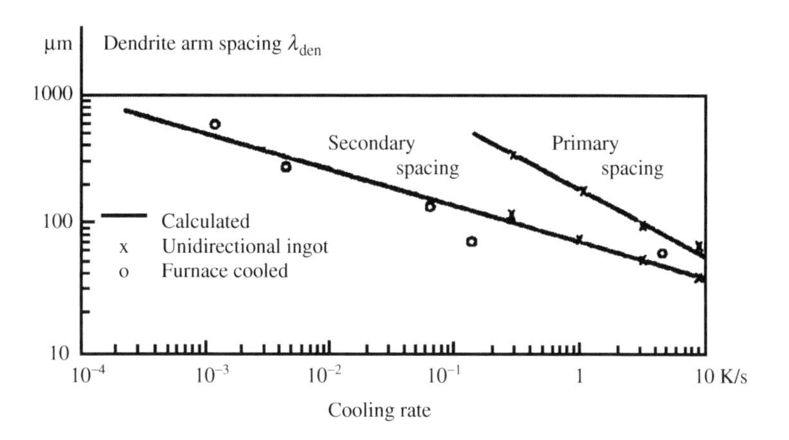

Figure 9.81 Dendrite arm spacing as a function of cooling rate of an Fe–25%Ni alloy. Merton C. Flemings © McGraw Hill 1974.

A high cooling rate results in a finer microstructure and a better quality of the material than a low cooling rate.

9.6.7 Coarsening of Secondary Dendrite Arms in Alloy Crystals

The final primary dendrite arm spacings, which can be measured in completely solidified castings, are normally the same as those that form initially. In addition the stems thicken and grow with time during the solidification process.

The thickening process of primary dendrite arms has been studied in connection with the microsegregation in primary stems in Section 9.6.5 on pages 537–544.

Below we will discuss coarsening of secondary dendrite arms. Figure 9.33 on page 504 shows that their thickness varies. The secondary dendrite arms have no specific and constant distance between adjacent arms and they are not of the same thickness.

The secondary arm spacings increase with time. It is well verified that a coarsening process occurs during the further cooling of the sample when the melt between the dendrites solidifies. The conditions for coarsening of primary and secondary dendrite arms differ in many respects.

- The secondary dendrite arm spacings are *much smaller* than those of the primary dendrite arms.
- The secondary dendrite arms are *much thinner* than the primary stems.
- The coarsening rate of the secondary dendrite arms is *very rapid.*

These differences result in very different growth conditions for primary and secondary coarsening. As will be shown below each set of secondary arms merge into one and form a continuous plate. This never happens to the primary stems as the distances between them are too large and the process too slow.

The coarsening process of secondary dendrite arms occurs with the aid of two simultaneous mechanisms, *remelting* and *back diffusion.*

Remelting of Secondary Dendrite Arms

The *remelting process* is a matter of mass transport. Diffusion of the alloying element from a thinner dendrite arm into neighbouring thicker dendrite arms leads to thickening of the thicker dendrite arms at the expense of the thinner dendrite arm, which disappears. There are three mechanisms of remelting of thin dendrite arms. This one is called *ripening*.

The initial secondary dendrite arms change during growth. Some of them become unstable and remelt, while others continue to grow. The surface energy and the curvature of the dendrite tips plays an important role at the remelting process. Three different models have been suggested to explain the remelting mechanism.

Model 1
Flemings et al [21] considered a row of parallel equal cylindrical rods. If one of them differs from the others and has a smaller radius than the others the smaller radius of curvature leads to a lower melting point of this dendrite arm than that of the others. It disappears by melting and gradual transfer of material from the thin arm to the thicker ones (Figure 9.82a).

Model 2
Flemings et al [21] discussed the case when one of the dendrite arms has a root, which is smaller than normal. Owing to a smaller radius of curvature at the root the melting point is lower there, which results in remelting and separation of the dendrite arm from the primary stem (Figure 9.82b).

Figure 9.82a Model 1. Merton C. Flemings © McGraw Hill 1974.

Figure 9.82b Model 2. Merton C. Flemings © McGraw Hill 1974.

Figure 9.82c Model 3. Merton C. Flemings © McGraw Hill 1974.

Model 3
Flemings [21] studied a single dendrite arm with slightly thinner thickness than the others. The double radius of curvature of the tip surface leads to a lower melting point of the tip than that along the cylindrical surface of the dendrite arm, which has only one radius of curvature (the other one is infinite). Figure 9.82c shows how the dendrite arm melts off starting from the tip to the root.

Back Diffusion

When a melt solidifies alloying atoms are thrown out in the melt because their equilibrium solubility is lower in the solid than in the melt. The ratio $x^{s/L}/x^{L/s}$ equals the partition constant k and the concentration of the alloying element in the melt increases. If the alloying atoms diffuse into the interior of the solid phase their concentration at the interface will decrease while their concentration in the melt is very high.

This situation leads to back diffusion, i.e. more alloying atoms diffuse from the melt into the solid phase than in the opposite direction owing to solidification. This enhances the solidification process at the interface as the system tries to achieve equilibrium. Back diffusion is said to 'consume' the melt.

Back diffusion into the secondary dendrite arms causes consumption of the melt between the dendrites for reasons described above. The result is that the dendrite arms merge and form a continuous so-called plate. This mechanism is shown in Figures 9.83.

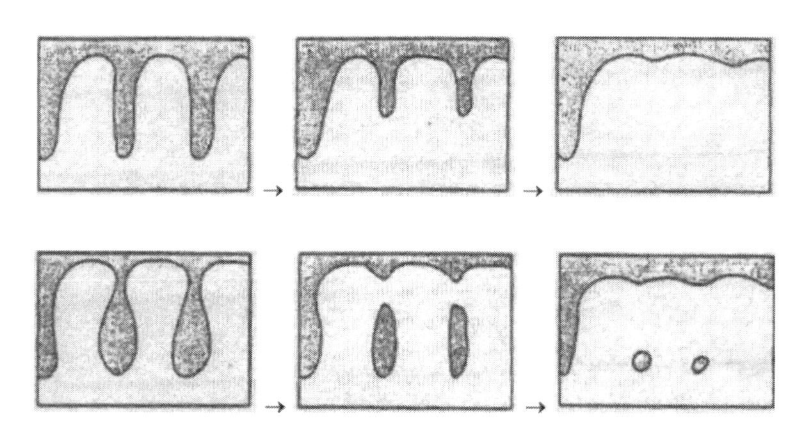

Figure 9.83 Two coalescence mechanisms, which cause the dendrite arms to merge into a secondary plate.

The rate of the coarsening process increases with increasing growth rate. It is most rapid at the beginning of the process.

Back Diffusion Coarsening

Figure 9.29 on page 503 shows a photo of two primary stems. Their secondary dendrite arms are perpendicular to the primary arms and are located in two perpendicular planes, both perpendicular to the growth direction of the primary stems.

The primary arms consist of growing parallel tips and the secondary arms are formed just behind the tip. The fact that they lie in the same plane indicates that there is a competition between the secondary arms. Some of them grow ahead of the other ones. The result is secondary arms of different length and thickness. Between the arms there is a small amount of melt left. Much more melt is located in the space between the primary arms.

During the solidification process back diffusion will appear. The back diffusion occurs both in the primary and the secondary dendrite arms. It is much faster in the secondary arms than in the primary ones because of the shorter distances between the adjacent arms corresponds to shorter diffusion distances.

The consequence of the back diffusion in the secondary arms is that the small amounts of melt between the arms are consumed. This process occurs much faster for the smaller and thinner secondary arms than for the thicker and larger ones. This results in a process where the smaller dendrite arms merge with the thicker ones into a plate.

Assume that the fraction of melt between the secondary dendrite arms along a path parallel with the primary stem is g. The amount of the alloying element along this path is constant during the solidification process, i.e. no melt enters from the surroundings during the continued solidification process. The melt between the arms solidifies during the cooling. The cooling process is controlled by the ordinary relationships, given by the phase diagram of the alloy and the back diffusion into the solid.

The fraction of melt can be calculated in the same manner as the one we used for back diffusion into the primary arms in Section 9.6.5. However, in this case the dendrite arm distances are much shorter than in the case of the primary arms.

Consider an array of secondary dendrite arms and in particular the three arms in Figure 9.84. The diffusion distance from the surface to the centre of the dendrite is considerably shorter in the thin dendrite than in the thicker ones. The thinner dendrites remelt and the thicker dendrites grow. The melt between the dendrites is consumed. In the end the thin dendrites disappear completely and the two thickened dendrites merge.

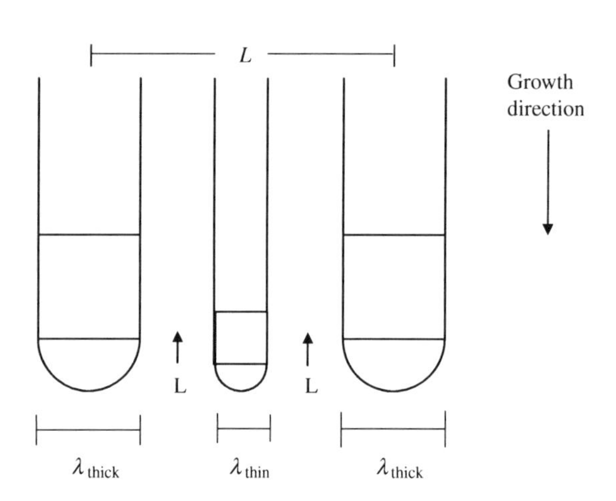

Figure 9.84 Dendrite arms in a melt.

When the interdendritic regions have solidified completely the concentration profile of the alloying element in the secondary plate has the shape, which corresponds to the dashed curve in Figure 9.85b.

The two maxima correspond to the last solidified melt.

A homogenization process will eliminate the maximum in the former liquid regions (Figure 9.85a). $x^{s/L}$ will increase with decreasing temperature The final concentration profile is of the type shown by the continuous curve in Figure 9.85b.

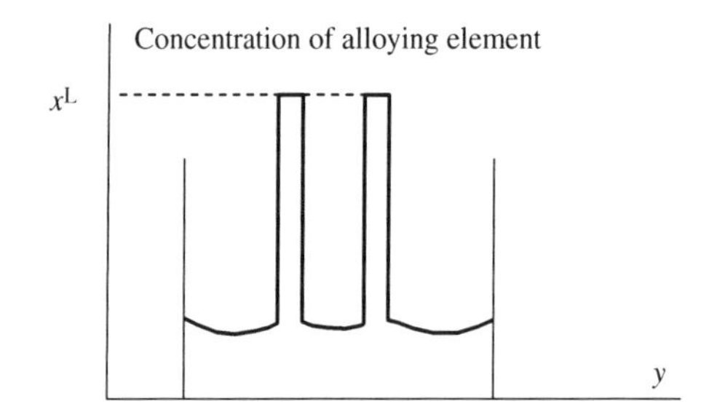

Figure 9.85a Concentration distribution of the alloying element before completed solidification of the melt between the secondary dendrites.

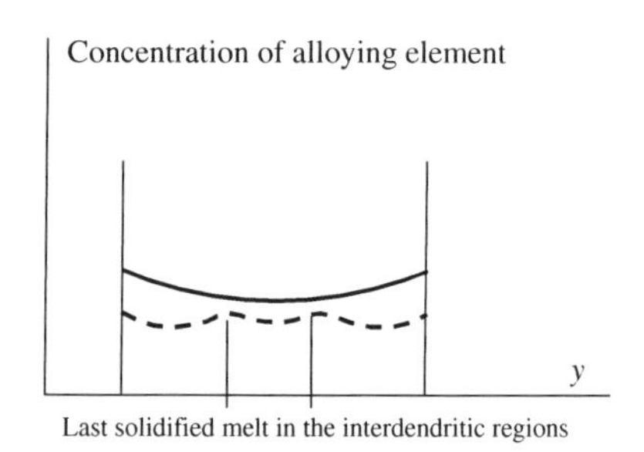

Figure 9.85b Concentration distribution of the alloying element immediately after solidification of the melt between the secondary dendrite arms (dashed curve) and after the following simultaneous solidification process (full line).

In this way the secondary dendrite arms merge and a so-called secondary plate is formed. Hultgren [17] was the first one, who was aware of such secondary plates and introduced the concept. Figures 9.86 and 9.87 show primary and secondary plates in a steel alloy. The primary plates and the secondary arms merge and the secondary plates merge with tertiary arms, formed at the secondary arms.

Figure 9.86a shows a vertical section, which we call the yz plane, through columnar grains, parallel to the growth direction. The vertical line in the middle of Figure 9.86a is a primary side plate. This direction has been chosen as z axis. The y axis cuts a secondary side plate built up of tertiary arms formed on the secondary arms.

Figure 9.87a shows a cut, perpendicular to the growth direction, through a columnar grain. Two types of secondary plates have been formed, perpendicular to each other and to the z direction. They form characteristic crosses, which lie in the same plane. This is in agreement with the normal dendritic structure (Figure 9.32 on page 504).

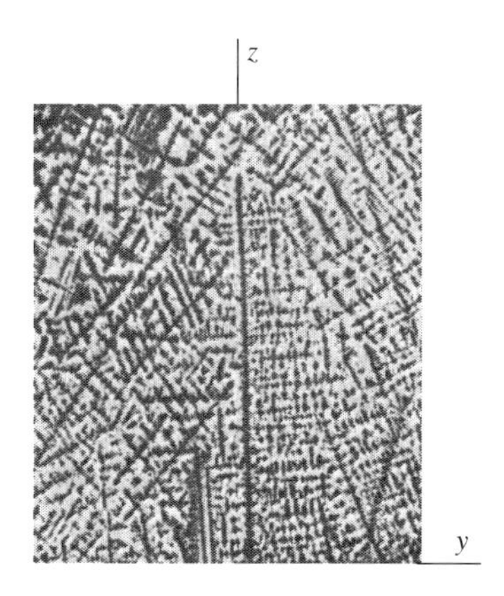

Figure 9.86a Cut through the yz-plane of a steel ingot. The cross-section happens to contain a primary side plate, which is the vertical line in the middle of the figure.

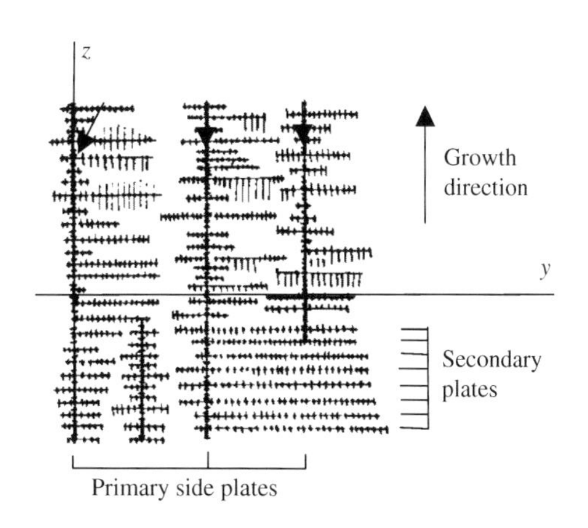

Figure 9.86b Primary side plates in the yz-plane parallel to the z-axis and secondary side plates in planes parallel to the xy-plane and parallel to the y-axis.

Figure 9.87a Cut through the xy-plane of the steel ingot in Figure 9.86. The characteristic crosses, which are shown in the structure are secondary side plates in two directions (x and y) perpendicular to the growth direction z.

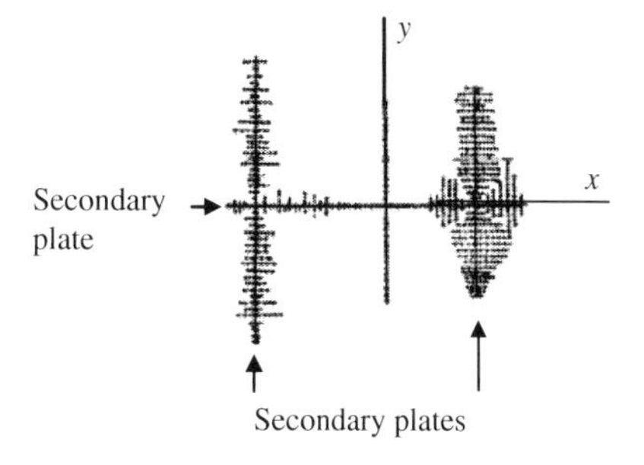

Figure 9.87b Secondary side planes in the xy-direction, one in the x-axis and two parallel to the y-axis. The coordinate system is the same in Figures 9.86 b and 9.87 b. The z-axis points in the direction out of the paper. The primary stems are directed in the z-direction and cut the secondary plates in two points along the x-axis.

Simplified Theory of Coarsening of Secondary Dendrite Arms

For careful calculations on dendrite coarsening it is important to take the back diffusion into account. Simplified calculations, where the back diffusion model is replaced with the lever rule, are sketched below. Back diffusion has been treated on page 553.

Consider a binary alloy melt with a given undercooling ΔT.

The dendrite grows with a tip, which forms the primary stem. Behind the tip the secondary arms are formed. The arms are not exactly equidistant owing to remelting and ripening effects. The arm spacings are distributed along a Gauss function around an average value.

The average value of the distance λ_{den} between the secondary arms is determined experimentally. Between the secondary arms a small fraction of liquid is left.

Secondary arms are developed on both sides of each primary stem. Their concentration profile can be approximated by a square wave. In Figure 9.88 a symmetry line has been drawn through the secondary arms and through the corresponding square wave.

The lower square wave in Figure 9.88 shows the concentration distribution of the alloying element in the secondary arms along the central line as a function of distance.

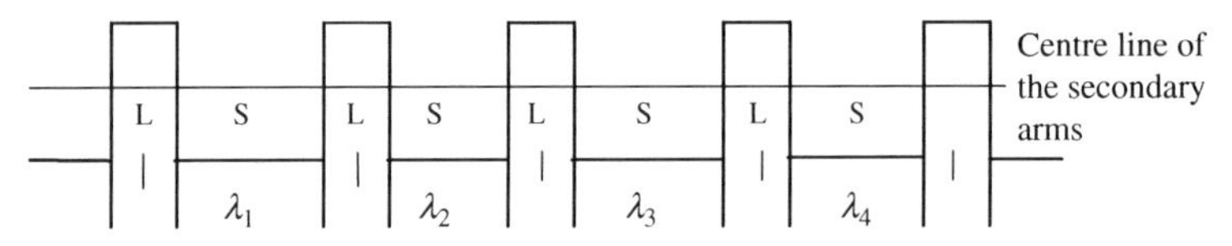

Figure 9.88. Top figure: Approximate model of the secondary arms.

Bottom figure: Concentration of the alloying element in the wide secondary arms and the intermediate melt as a function of position.

The average concentration can be calculated from the relationship

$$\overline{x} = g\, x^{L/\alpha} + (1 - g)x^{\alpha/L} \tag{9.100}$$

where

$\overline{x}$ = average concentration of the alloying element
g = fraction of melt between the secondary arms
$x^{L/\alpha}$ = concentration of the alloying element in the melt
$x^{\alpha/L}$ = concentration of the alloying element in the solid phase α.

Figure 9.89 shows how the concentrations of the alloying element at the interface, i.e. in the melt and in the solid phase, can be read from the phase diagram.

If we consider a single specific secondary arm and its characteristic distance λ_i the average concentration over the distance λ_i will vary. The wider the arm is the higher will be the average concentration over the interval λ_i. It can be described with the aid of the function

$$\overline{x_i} = \left(g x^{L/\alpha} + (1 - g)x^{\alpha/L}\right)\frac{\lambda_i}{\overline{\lambda}} \tag{9.101}$$

Next we want to find out how the concentration of the alloying element increases when the solidification process continuous.

We have to take the diffusion of the alloying element in the solid phase into consideration because the thickness of the arms, i.e. the diffusion distances, is small. For this reason Scheil's equation (Section 9.6.5 page 537) can not be used because it neglects the diffusion in the solid phase completely. The best model is Sheil's modified equation, which includes the back diffusion (Section 9.6.5 page 537). For simplicity we will use the other extreme, the lever rule, which assumes rapid instead of zero diffusion in the solid phase.

Lever rule:

$$x^{\mathrm{L}} = \frac{x_{\mathrm{initial}}}{1 - f(1-k)}$$

where
k = partition constant
f = fraction of solid phase.

In the present case the initial concentration is the concentration given in Equation (9.101). Hence, we obtain the average concentration in the secondary arm distance λ_{i} at the solidification fraction f

$$\overline{x_{\mathrm{i}}^{f}} = \frac{\left(g\,x^{\mathrm{L}/\alpha} + (1-g)x^{\alpha/\mathrm{L}}\right)\frac{\lambda_{\mathrm{i}}}{\overline{\lambda}}}{1 - f(1-k)} \tag{9.102}$$

where
g = initial fraction of the melt
f = fraction of solid
k = partition constant
$\overline{x_{\mathrm{i}}^{\mathrm{f}}}$ = the average concentration of the alloying element within the distance λ_{i} (melt + a secondary arm).

With the aid of Equation (9.102) we can derive the average concentration for different λ_{i} values when the solidification is complete, i.e. when $f = 1$. At this concentration the arms of size λ_{i} will disappear. From the phase diagram (Figure 9.89 we can obtain the temperature when this happens. Figure 9.90 shows how the temperature T_{S} can be constructed when $\overline{x_{\mathrm{i}}^{f=1}}$ is known.

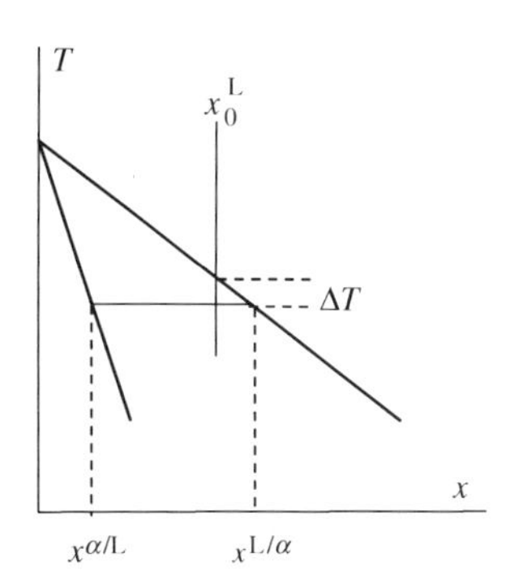

Figure 9.89 Phase diagram of a binary alloy.

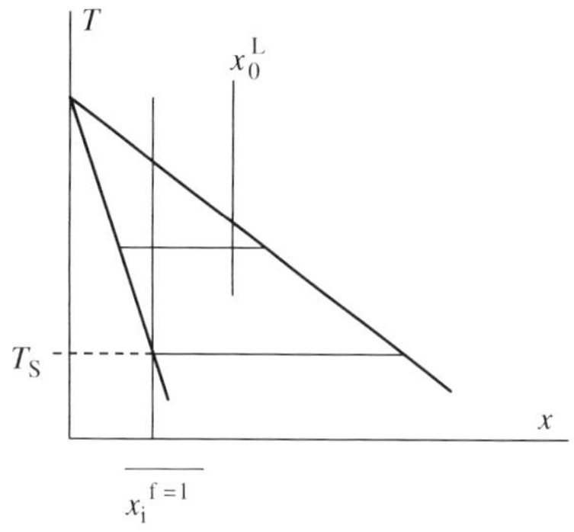

Figure 9.90 Phase diagram of a binary alloy.

$\overline{x_{\mathrm{i}}^{f=1}}$ varies with the value of λ_{i}. Hence, T_{S} is a function of λ_{i}.

We want to find the time required for total solidification, i.e. for change from $f = 0.9$ to $f = 1$ for different λ_{i} values. For this purpose we derive Equation (9.102) with respect to time.

$$\frac{d\overline{x_{\mathrm{i}}^{f}}}{\mathrm{d}t} = -\frac{\left(gx^{\mathrm{L}/\alpha} + (1-g)x^{\alpha/\mathrm{L}}\right)\frac{\lambda_{\mathrm{i}}}{\overline{\lambda}}}{[1 - f(1-k)]^2}[-(1-k)]\frac{\mathrm{d}f}{\mathrm{d}t} \tag{9.103}$$

The expression on the left-hand side can be written

$$\frac{\mathrm{d}\overline{x_{\mathrm{i}}^{f}}}{\mathrm{d}t} = \frac{\mathrm{d}T}{\mathrm{d}t}\frac{\mathrm{d}\overline{x_{\mathrm{i}}^{f}}}{\mathrm{d}T} \tag{9.104}$$

$-\mathrm{d}T/\mathrm{d}t$ represents the cooling rate C, which is determined by the experimental conditions. $\mathrm{d}\overline{x_i^f}/\mathrm{d}T$ is the inverted slope m_L of the liquidus line in the phase diagram of the alloy. These quantities are introduced into Equation (9.103). In can be transformed to

$$\mathrm{d}t = \frac{1}{C\frac{1}{m_L}} \frac{\left(g\, x^{L/\alpha} + (1-g)x^{\alpha/L}\right)\frac{\lambda_i}{\bar{\bar{\lambda}}}}{[1-f(1-k)]^2}(1-k)\mathrm{d}f \tag{9.105}$$

Equation (9.105) can be integrated, which gives

$$\int_0^t \mathrm{d}t = \int_{0.9}^{1.0} \frac{m_L}{C} \frac{\left(g\, x^{L/\alpha} + (1-g)x^{\alpha/L}\right)\frac{\lambda_i}{\bar{\bar{\lambda}}}}{[1-f(1-k)]^2}(1-k)\mathrm{d}f \tag{9.106}$$

Equation (9.106) can be used to calculate the time required to dissolve dendrite arms with the characteristic length λ_i or to merge two dendrite arms of the same length

Theory of Dendrite Coarsening

Empirical Relationships

In addition to the relationships mentioned on page 551 there is another empirical relation, which is valid for many alloys. The dendrite arm spacing is related to the solidification growth rate with the aid of the relationship (Equation (9.98) on page 551)

$$V_{\text{growth}}\, \lambda_{\text{den}}^2 = const \tag{9.98}$$

where
V_{growth} = growth rate of solidification
λ_{den} = dendrite arm distance.

If we replace V_{growth} by $\mathrm{d}\lambda_{\text{den}}/\mathrm{d}t$ in Equation (9.98) and integrate the new equation we obtain

$$\frac{\mathrm{d}\lambda_{\text{den}}}{\mathrm{d}t}\lambda_{\text{den}}^2 = const$$

or

$$\int_{\lambda_0}^{\lambda_{\text{den}}} \lambda_{\text{den}}^2 \mathrm{d}\lambda_{\text{den}} = const \int_0^t \mathrm{d}t \tag{9.107}$$

If $\lambda_{\text{den}} \gg \lambda_0$ Equation (9.107) can be written after integration

$$\lambda_{\text{den}} = \lambda_0 + Const \times t \tag{9.108}$$

where
λ_{den} = dendrite arm distance at time t
$Const$ = coarsening constant
t = time.

Equation (9.108) is entirely empirical. Flemings [21] has derived a similar relationship from known and basic theoretical relationships. His resulting relationship depends strongly on the geometry and other assumptions, which have to be made.

Theoretical Calculation of the Coarsening Constant

In 1990 Mortensen [22] published some predictions concerning the rate of dendrite arm coarsening for some alloys and a non-metallic material. His calculations and necessary approximations concerning Al-Cu alloys will be briefly described below.

Mortensen performed his calculations in two steps.

- Primarily he derived an expression for the coarsening constant in case of isothermal coarsening instead of solidification.
- Secondly he assumed that the instantaneous coarsening rate is identical to the isothermal rate of coarsening and calculated the cumulative coarsening by integration of the isothermal coarsening after dendritic solidification of the sample.

Mortensen's calculations resulted in the relationship

$$\lambda_{\text{den}}^3 - \lambda_0^3 = Ct \tag{9.109}$$

where

λ_{den} = average dendrite arm spacing (Figure 9.91)
λ_0 = dendrite arm spacing at the beginning
C = coarsening constant.

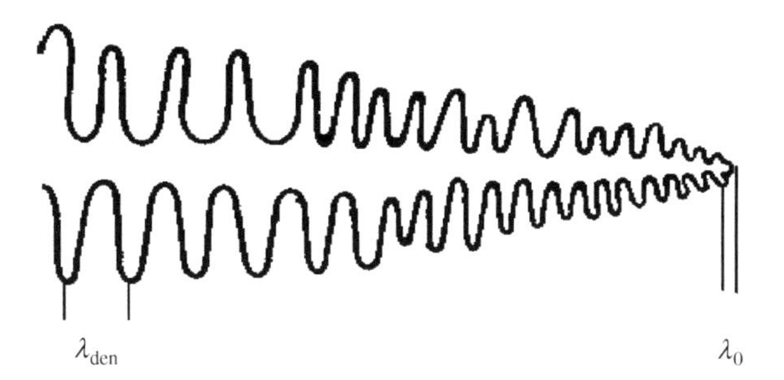

Figure 9.91 Dendrite arm. Merton C. Flemings © McGraw Hill 1974.

The coarsening constant was found to be

$$C = \frac{27BD_{\text{L}}}{4m_{\text{L}}c^{\text{L}}(1-k)g_{\text{solid}}(1-\sqrt{g_{\text{solid}}})} \tag{9.110}$$

where

B = a constant, the so-called Gibbs-Thomson coefficient$_t$
D_{L} = diffusion constant of the solute in the melt
g_{solid} = volume fraction of solid in the volume analyzed
m_{L} = slope of the liquidus line in the phase diagram of the alloy
c = concentration of the alloying element
k = partition coefficient of the alloy.

Coarsening During Solidification

In order to calculate the total coarsening during the whole solidification time it is necessary to introduce a number of additional approximations.

- The total coarsening is obtained by integration of the instantaneous coarsening at time t multiplied by the time interval dt. The integration should be extended over the whole solidification time.
- The instantaneous coarsening during solidification is assumed to be identical with the isothermal coarsening, derived above. Hence, Equation (9.110) can be used for the integration.

$$\lambda_{\text{den}}^2 \mathrm{d}\lambda_{\text{den}} = \frac{27BD_{\text{L}}}{4m_{\text{L}}c^{\text{L}}(1-k)g_{\text{solid}}(1-\sqrt{g_{\text{solid}}})} \cdot \mathrm{d}t \tag{9.111}$$

Two quantities on the right-hand side of Equation (9.111), c^{L} and g_{solid} vary with time during the solidification process.

- In most casting processes it reasonable to assume that the rate of heat removal is constant

$$\frac{\mathrm{d}g_{\text{solid}}}{\mathrm{d}t} = \frac{g_{\text{solid}}}{t} = \frac{1}{t_{\text{final}}} \tag{9.112}$$

which can be written as

$$\mathrm{d}t = t_{\text{final}} \mathrm{d}g_{\text{solid}} \tag{9.113}$$

- It is also reasonable to assume that the diffusion in the solid phase is small and can be neglected. In this case Scheil's equation is valid.

$$c^{\text{L}} = c_0(1-g_{\text{solid}})^{-(1-k)} \tag{9.114}$$

Introduction of the expressions in Equations (9.113) and (9.114) into Equation (9.111) gives

$$\lambda_{\text{den}}^2 \mathrm{d}\lambda_{\text{den}} = \frac{27BD_{\text{L}}t_{\text{final}}}{4m_{\text{L}}c_0(1-k)} \frac{\mathrm{d}g_{\text{solid}}}{(1-g_{\text{solid}})g_{\text{solid}}(1-\sqrt{g_{\text{solid}}})} \tag{9.115}$$

- Special care has to be used to find proper integration limits of the integral. The lower limit is chosen to 0.1, which means that the value of the integral within the interval $0 < g_{\text{solid}} \leq 0.1$ is neglected in comparison with the total value of the integral. This is a reasonable assumption as the dendrites are very thin at the beginning.

  The upper integration limit of the integral depends on the type of binary alloy. In the case of the alloy Al-4.5 wt% Cu the upper solidification fraction equals 0.93 when the eutectic point is reached and the dendrite solidification is finished.

Integration of Equation (9.115) gives for the alloy Al-4.5 wt% Cu

$$\int_{\lambda_0}^{\lambda_{\text{den}}} \lambda_{\text{den}}^2 \mathrm{d}\lambda_{\text{den}} = \int_{0.1}^{0.93} \frac{27BD_{\text{L}}t_{\text{final}}}{4m_{\text{L}}c_0 \cdot (1-k)} \frac{\mathrm{d}g_{\text{solid}}}{(1-g_{\text{solid}})g_{\text{solid}}(1-\sqrt{g_{\text{solid}}})} \tag{9.116}$$

If λ_0 is assumed to be ~ 0 and the other numerical values of the material and other constants are inserted the result can be written

$$\lambda_{\text{den}} = 12 t_{\text{final}}^{\frac{1}{3}} \quad (\mu\text{m, seconds}) \tag{9.117}$$

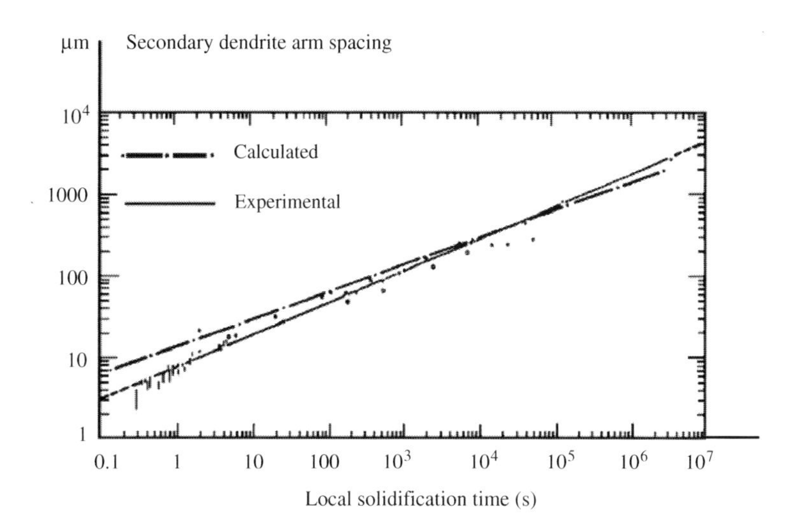

Figure 9.92. Secondary dendrite spacing of the alloy Al–4.5wt-% Cu as a function of the local solidification time.

Local solidification time is defined as the time interval between beginning (at the liquidus temperature) and end (at the solidus temperature) of solidification. Reproduced with permission from Springer © 1991.

This function is shown in Figure 9.92 and compared with experimental values. The agreement is satisfactory as Mortensen mention an uncertainty of a factor 2 in the theoretical calculations.

It should be pointed out that Equations (1′) and (2′) are special for the particular Al-Cu alloy and *not* generally valid. For this reason the have been given particular numbers.

However, if Scheil's equation is valid the shape of the relationship between the dendrite arm spacing and the solidification time

$$\lambda_{\text{den}}^3 - \lambda_0^3 = Ct \tag{9.118}$$

will be generally valid even if the properties of the alloys vary.

Influence of Coarsening on Dendritic Structure

The coarsening process during solidification leads to other effects than simple increase of the dendrite arm spacing. It also promotes

- reduction of the microsegregation in the dendritic arms
- change of the dendrite morphology.

The first effect occurs because some 'early' dendrites shrink and dissolve later. New arms reprecipitate later with a higher concentration of the alloying element than before and deviations from the normal microsegregation pattern arise.

The second effect is caused by the formation of plates in the space between the cylindrical dendrite arms. Coarsening is supposed to be the most important reason for plate formation.

9.6.8 *Influence of Convection on Dendritic Growth Rates*

The basic theory of free or natural convection has been treated in Chapter 7 on pages 366–377. It will be used in this section where the influence of free convection on growth rate is studied. The influence of forced convection on growth rate is also briefly discussed.

Free Convection

It is well known that free or natural convection is very important in solidifying ingots. For this reason we will discuss the influence of the free convection on the concentration (alloys) and temperature

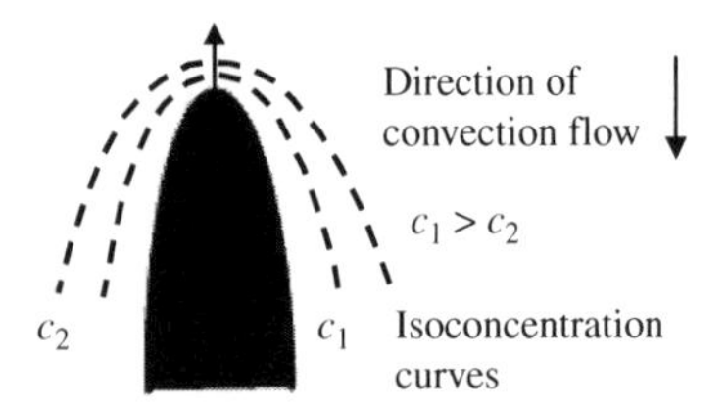

Figure 9.93a Growing tip surrounded by a streaming alloy melt. Some isoconcentration curves are drawn in the vicinity of the tip.

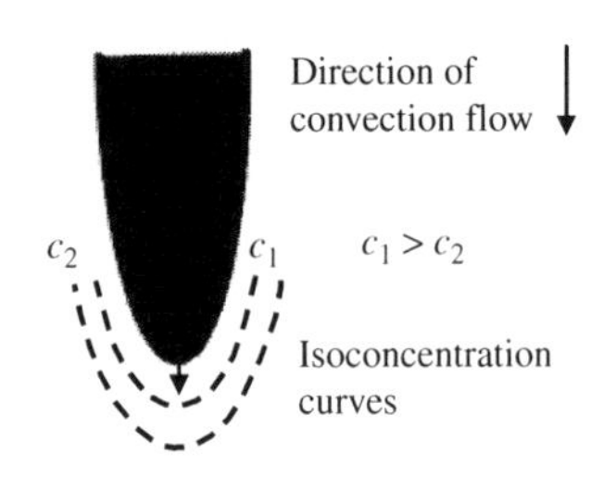

Figure 9.93b Growing tip surrounded by a streaming alloy melt. Some isoconcentration curves are drawn in the vicinity of the tip.

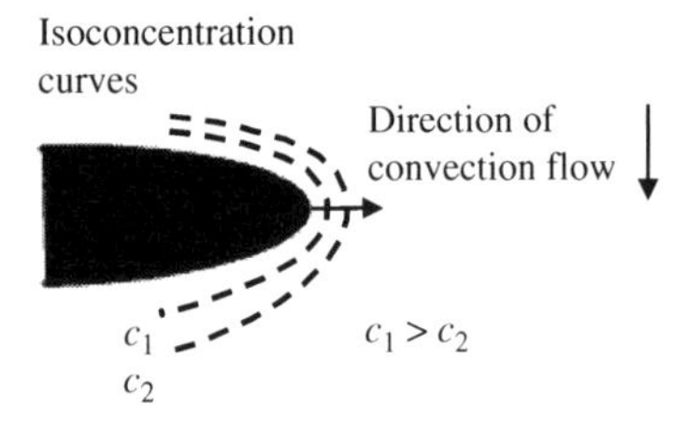

Figure 9.93c Growing tip surrounded by a streaming alloy melt. Some isoconcentration curves are drawn in the vicinity of the tip.

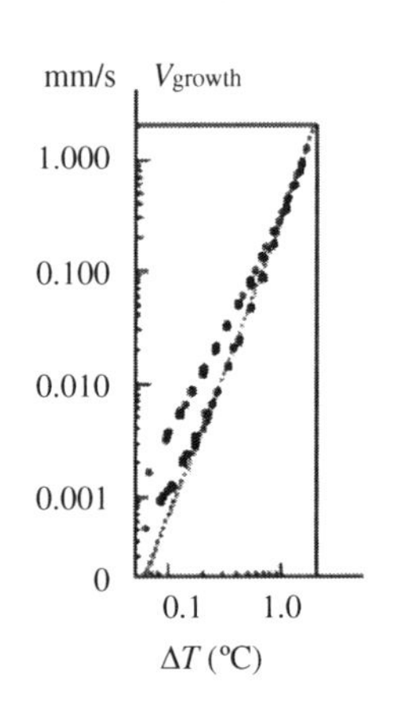

Figure 9.94 Dendritic tip velocity as a function of undercooling with and without influence of convection. © ISIJ (Iron and Steel Institute of Japan) 1995.

fields (pure metal melts) around a growing dendritic tip with special attention to application on solidifying ingots.

Theoretical calculations predict that convection should only have a minor influence on the very tip of a growing dendrite. However, it may be worth while to discuss qualitatively the effect of the *orientation* of the tip relative to gravity. Three different orientations will be discussed.

a) The tip is growing in a direction opposite to gravity.
b) The tip is growing in the direction of gravity.
c) The tip is growing in a direction perpendicular to the direction of the gravity field.

The temperature field (pure metal) or the concentration field (alloy) around a growing tip in a metal melt causes a lower temperature/concentration of the melt ahead of the tip. The density field around the tip, i.e. the shapes of the isodensity curves near the tip, depends on the growth direction of the tip and the flow direction of the melt. Some examples will be discussed below.

When a dendrite tip grows vertically and upwards (Figure 9.93a) the temperature or concentration gradient around the tip will be larger, owing to the downwards flow in the melt, than in the absence of convection. This means that

- The tip will grow more rapidly and at a lower undercooling or concentration difference the larger the temperature/concentration gradient close to the tip is.

When a dendrite tip grows vertically and downwards (Figure 9.93b) the temperature or concentration gradient around the tip will be smaller, owing to the downwards flow in the melt, than in the absence of convection. This means that

- The tip will grow more slowly and at a higher undercooling or concentration difference the smaller the temperature/concentration gradient close to the tip is.

When the tip grows perpendicularly to gravity (Figure 9.93c) the temperature or concentration gradient will be larger at the upper side of the tip and smaller at the lower side than in the absence of convection in the melt. The result is that the dendrite becomes tilted upwards.

These predictions have been verified by experimental evidence and model experiments.

Influence of Convection on Undercooling

The driving force of convection is the density gradient in the melt caused by a temperature gradient. Glickman et al [23] have investigated the influence of convection on the relationship between undercooling and growth rate at dendritic growth.

Figure 9.94 shows the result of two experiments on identical specimens, one performed on earth under normal conditions and the other performed in a space shuttle in the absence of gravity. The growth rate as a function of undercooling is only influenced by convection at low undercoolings, less than 1 K.

Influence of Free Convection on Growth Direction

Figure 9.95 shows part of the columnar zone of an ingot. It can be clearly seen that the columnar zone has not grown perpendicularly to the vertical cooling surface (to the left in the figure) but is directed slightly upwards.

The explanation is probably that convection makes the boundary layer thinner on the side of the tip, which is turned towards the convection flow (Figure 9.96). This promotes the growth in the vertical and upward direction and the result is that the dendritic growth will tilt upwards. This is in agreement with the observations in Figure 9.95 and the discussion above (Figure 9.93c).

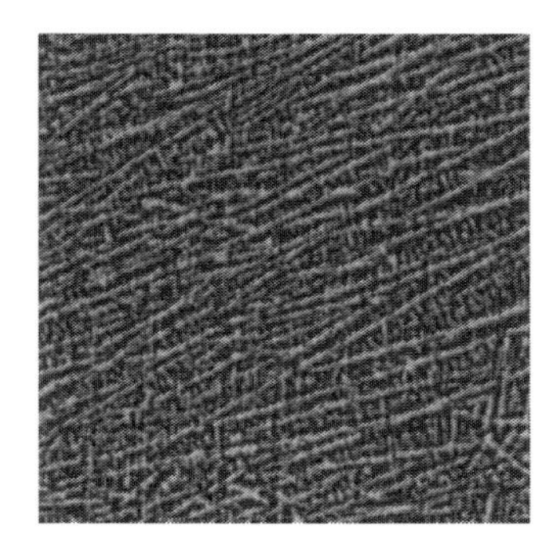

Figure 9.95 Columnar zone of an ingot under the influence of free convection. Vertical cooling surface to the left.

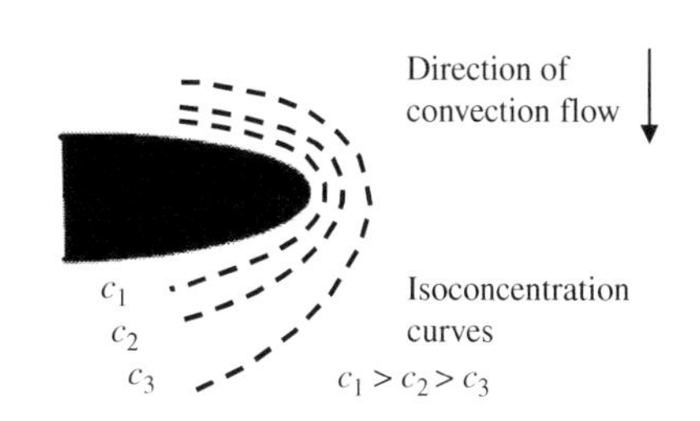

Figure 9.96 Isoconcentration curves around a growing dendrite tip in the columnar zone in an ingot.

Forced Convection

As an example of the influence of forced convection on the growth rate we will discuss the solidification of an Al-0.5wt%Cu ingot. The experiments discussed below have been performed by Fredriksson et al [24].

Influence of Forced Convection on an Al-0.5wt%Cu Ingot
Alloy melts of the composition given above were poured into the type of mould in Figure 9.97 and allowed to solidify under various conditions. After solidification half a slice was cut from the medium region of each ingot and examined after etching.

The melt was stirred by a stirrer with small wings. The stirring speeds could be varied from zero up to 970 rounds per minute.

The results of the experiments are reported below.

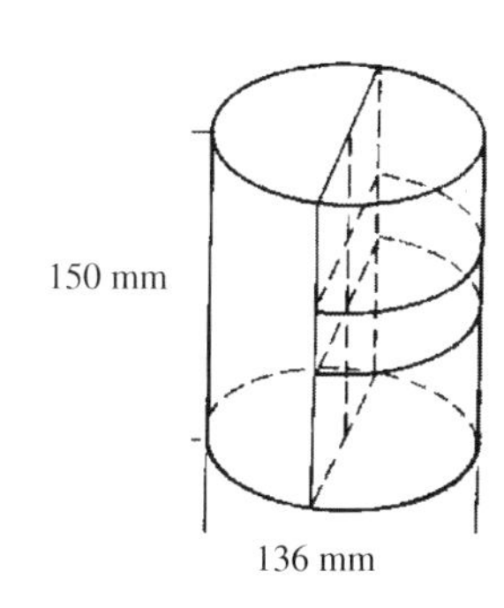

Figure 9.97 Mould.

Macrostructure
Figures 9.98a-d show the influence of convection on the macrostructure of the ingot.

Figure 9.98a Left figure. Macrostructure of an unstirred pure Al ingot.

Figure 9.98b Right figure. Macrostructure of an Al–0.5 wt%Cu ingot, stirred 200 r.p.m.

Figure 9.98c Left figure. Macrostructure of an Al-0.5 wt-%Cu ingot, stirred 970 r.p.m.

Figure 9.98d Right figure. Macrostructure of an Al–0.5 wt%Cu ingot. Delayed stirring (40 s after pouring) 970 r.p.m.

The pure Al ingot shows a coarse columnar zone and there is no equaxed zone at all. Stirring changes the pattern. Stirring shortens the columnar zone considerably and extended equiaxed zones appear. The columnar crystals are no longer perpendicular to the periphery but grow at a certain angle relative to the radius of the ingot. Stirring obviously makes the structure of the ingot more fine-grained.

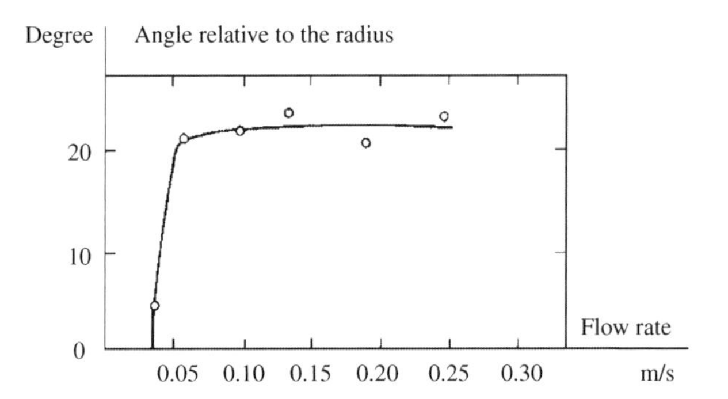

Figure 9.99 Tilting angle of the columnar crystals as a function of the flow rate of the periphery flow rate at the solidification front.

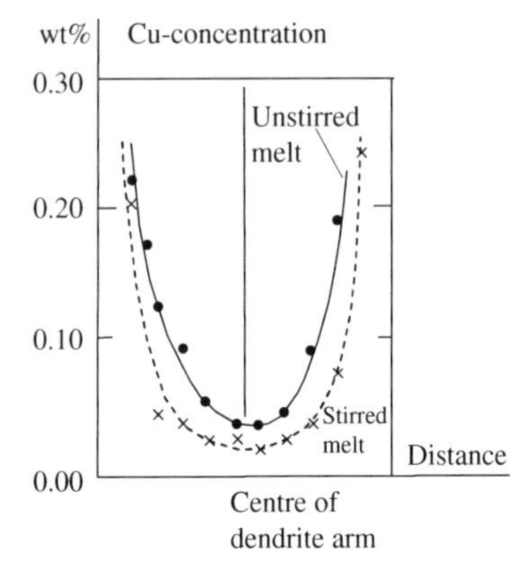

Figure 9.100 Cu concentration in the central part of a columnar dendrite stem, in an unstirred and in a stirred Al–0.5wt% Cu melt (970 r.p.m), respectively, as a function of position at stirring.

Figure 9.99 shows that the flow rate of the melt has very little influence on the angle beyond a certain value of the flow rate.

Microsegregation

The distribution of the alloying element Cu within the dendrite arms, i.e. the microsegregation, is changed by the forced convection.

The effect is shown in Figure 9.100 The minimum of the Cu-concentration in the centre becomes lower in a stirred ingot and more eutectic structure is formed between the arms.

Influence of Convection on the Equiaxed Crystal Zone

At reocasting processes (thixomoulding (see pages 11–13 in [25]) and at vigorous stirring of the melt a large number of equiaxed crystals are formed. The morphology of the crystals deviates from the normal dendritic pattern. They are more rounded and rosette-like than ordinary equiaxed crystals.

The large number of dendrites is formed when the dendrites split up by melting off secondary dendrite arms from the primary stems.

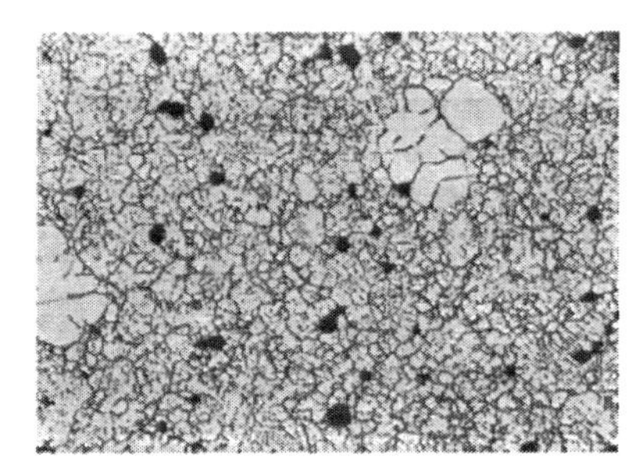

Figure 9.101a Microstructure of the Al–Cu melt after 100 s. The black lumps in the figure are pores.

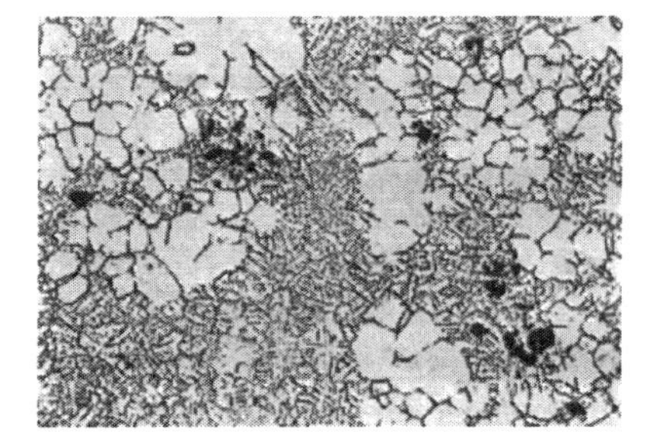

Figure 9.101b Microstructure of the Al–Cu melt after 202 s. The white areas in Figures 9.101 a and b are rosette-like crystals.

Figure 9.101a shows two bright rosette-like crystals floating freely in the melt. The fine dendritic structure in the figure represents the quenched melt. The experiment shows that the number of rosette-like crystals increases with time and that they form clusters (Figure 9.101b). After a long time of stirring the number of free rosette-like crystals decreases and the number of clustered spherical crystals increases.

The concentration of the alloying element inside the rosette-like crystals has been measured as a function of the distance from the dendrite centre. One exemple of the result is shown in Figure 9.102.

Just as in the columnar zone, the Cu distribution inside the dendritic stems, i.e. the microsegregation, differs as a result of stirring (lower curve in Figure 9.102). Figure 9.102 shows that, in the Al-Cu system, the Cu concentration is lower in the crystals grown in the stirred melt than those grown in the unstirred melt.

However, in other alloys the opposite effect has been observed. In the Al-Si system, a higher Si concentration has been observed in crystals grown in stirred melt than in crystals grown under unstirred conditions.

The convection influences the concentration distribution of the alloying elements in the crystals. Figures 9.93a–c on page 562 show that the mass transport to the crystals changes during the growth process. This may partly explain the effect on the concentration distribution shown in the Figures 9.100 and 9.102.

Another reason may be the fact that the atomic structure in the melt changes owing to convection. The atomic clusters in the melt may break and the energy be released as heat. Hence, the Gibbs' free energy of the melt will increase. The result is that the (liquidus) temperature of the melt increases. The increase of the melting point may (will) initiate crystal multiplication by melting off secondary arms.

Change of the partition coefficient k with the increasing temperature also contributes to the explanation of the results shown in Figure 9.102.

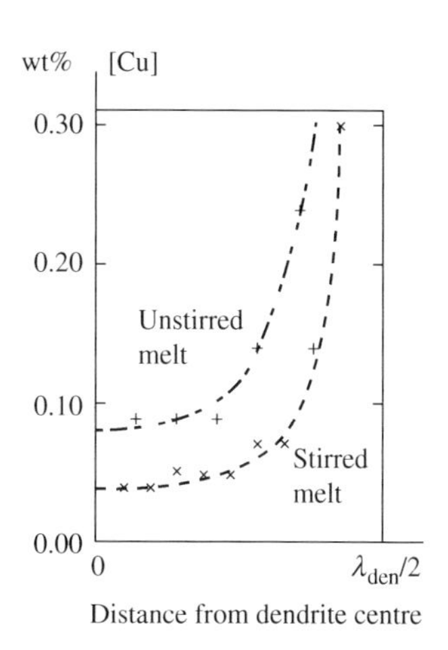

Figure 9.102 Cu-concentration in the central part of a dendrite stem in the equiaxed zone of an unstirred and a stirred Al–0.5wt%Cu melt (970 r.p.m.), respectively, as a function of position.

9.7 Transitions between Structure Types in Alloys

When an alloy melt solidifies the structure of the solid phase can vary very much. Earlier in this chapter we have mainly discussed faceted and dendritic growth and faceted and dendritic morphologies. In earlier chapters crystal growth with a planar solidification front and also cellular crystal growth and structure has been treated extensively.

The structure of the solid phase is entirely a result of the solidification conditions. Many parameters influence the structure. The most important of them are

- composition of the alloy
- temperature
- temperature gradient
- growth rate.

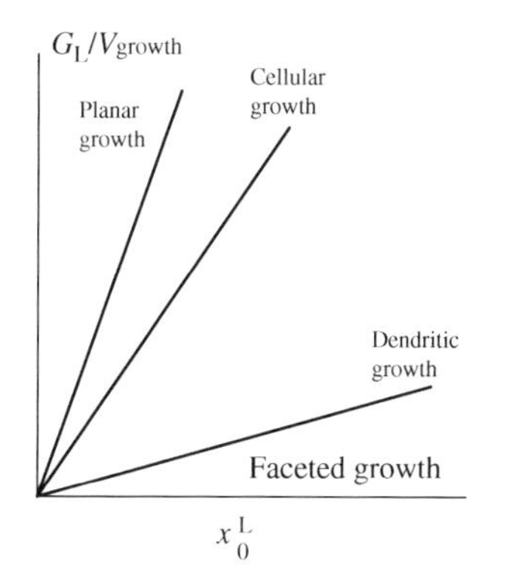

Figure 9.103 The ratio G_L/V_{growth} as a function of the solute concentration of a solidifying binary alloy melt where

G_L = temperature gradient in the melt close to the solidification front

V_{growth} = growth rate of the solid phase.

If these parameters are changed during the solidification the structure of the material will also change. We will devote the last section of this chapter to a brief review of the possible transitions from one structure to another. Figure 9.103 shows schematically the various regions of planar, cellular, dendritic and faceted growth. Below we will try to find the boundary conditions for transition from one type of structure to another and give some concrete examples.

9.7.1 Transitions between Planar and Cellular Growths

At high temperature gradient and low growth rate the solidification front will be planar.

If the growth rate is increased and the temperature gradient is decreased the planar solidification front breaks down, owing to some temporary fluctuation, which causes a perturbation in the planar solidification front (Figure 9.104). The planar solidification front becomes cellular, which means that it consists of elongated cells in a dense-packed pattern. One example is given in Figure 8.25 on page 454 in Chapter 8.

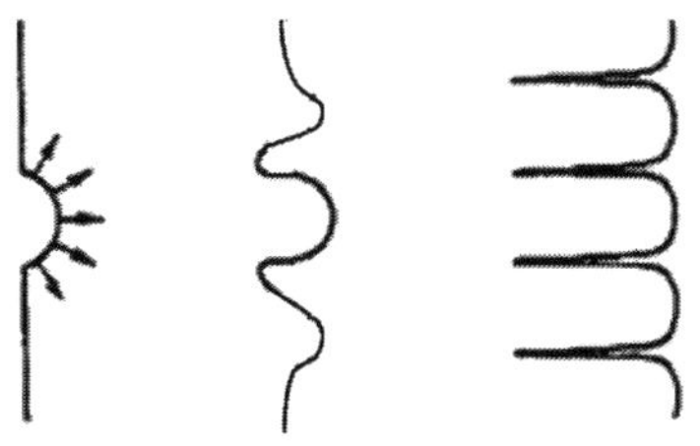

Figure 9.104 Formation of cellular structure. (Identical with Figure 8.28 on page 455 in Chapter 8.) Bruce Chalmers, courtesy of S. Chalmers.

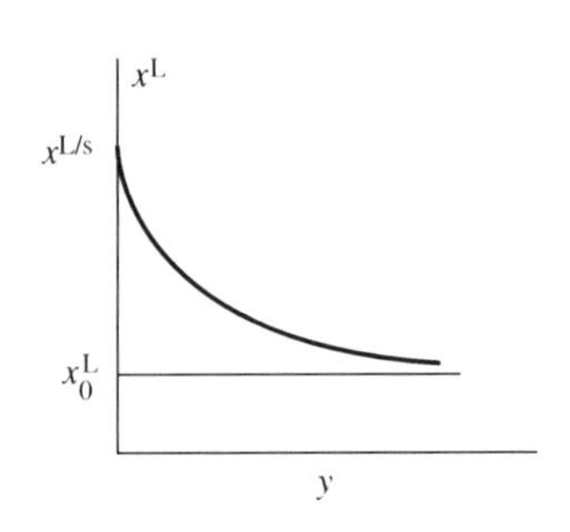

Figure 9.105 Solute concentration in the melt as a function of position.

Planar Growth at Stationary Conditions

Planar growth at stationary conditions has been extensively treated in Chapter 8. On page 455–459 in Chapter 6 we find Equation (6.120), which describes the concentration of the alloying element in the melt as a function of position (Figure 9.105).

$$x^L = x_0^L\left(1 + \frac{1-k_0}{k_0}\mathrm{e}^{-\frac{V_{growth}y}{D_L}}\right) \tag{9.119}$$

The temperature of the melt ahead of the solidification front is a function of the concentration of the alloying element and the position. With the aid of the phase diagram we obtain for small y values

$$T = T_M + m_L x^L \tag{9.120}$$

where m_L is the slope of the liquidus line in Figure 9.106.

x^L in Equation (9.117) is replaced by the expression in Equation (9.116) which gives

$$T = T_M + m_L x_0^L\left(1 + \frac{1-k_0}{k_0}\mathrm{e}^{-\frac{V_{growth}y}{D_L}}\right) \tag{9.121}$$

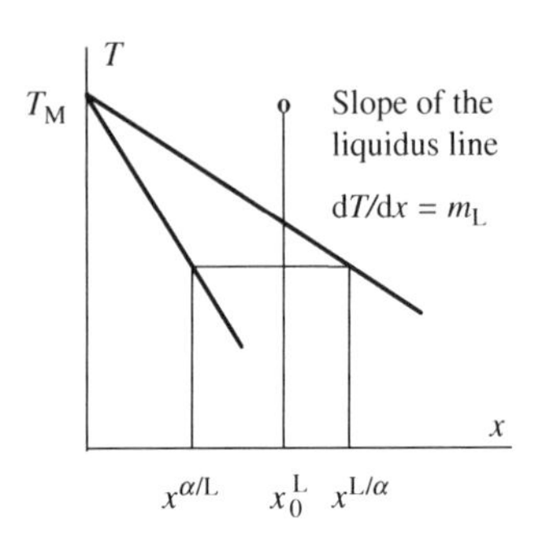

Figure 9.106 Phase diagram of a binary alloy.

The temperature gradient can be calculated by taking the derivative of T with respect to y.

$$G = \frac{\mathrm{d}T}{\mathrm{d}y} = m_L x_0^L \frac{1-k_0}{k_0}\mathrm{e}^{-\frac{V_{growth}y}{D_L}}\left(-\frac{V_{growth}}{D_L}\right) \tag{9.122}$$

Close to the interface $y = 0$, which gives the temperature gradient at the interface:

$$G_L = -\frac{m_L x_0^L}{D_L}\frac{1-k_0}{k_0}V_{growth} \tag{9.123}$$

If $k_0 < 1 m_L$ is negative. If $k_0 > 1 m_L$ is positive. Hence, the temperature gradient is always a positive quantity.

Cellular Growth

Cellular growth and interface instability has been treated in Section 8.5 in Chapter 8. It will be further discussed in Section 9.7.2.

Determination of the Boundary Condition for Transition from Planar Growth to Cellular Growth and Vice Versa

A necessary condition for solidification is undercooling of the melt close to the interface. Cellular growth is related to so-called constitutional undercooling (Figure 9.107), which has been treated on pages 456–458 in Chapter 8. It is shown there that the condition for cellular growth [Equation (8.35) on page 457 in Chapter 8] is

$$G_L \leq \left(\frac{\mathrm{d}T}{\mathrm{d}y}\right)_{y=0} \tag{9.124}$$

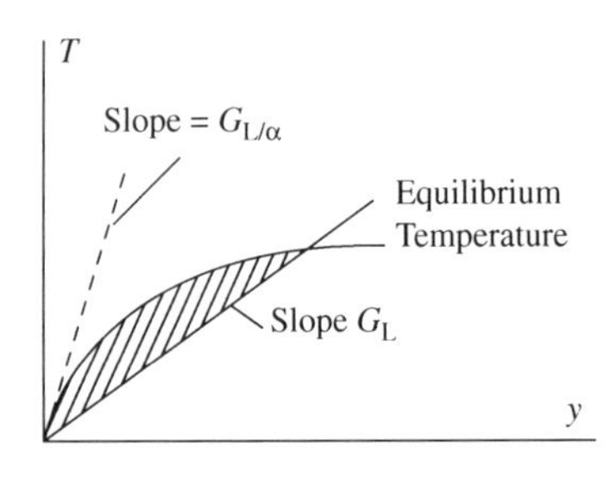

Figure 9.107 Undercooling in front of a planar interface. (Identical with Figure 8.34 on page 458 in Chapter 8.) Bruce Chalmers, courtesy of S. Chalmers.

where the right-hand side of Equation (9.121) is identical with the expression in Equation (8.37) in Chapter 8. Hence, the condition for cellular growth instead of planar growth is

$$\frac{G_L}{V_{growth}} \leq -\frac{m_L}{D_L}\frac{1-k_0}{k_0}x_0^L \tag{9.125}$$

The size and appearance of the elongated cells is a function of the growth rate and the temperature gradient. The influence of the undercooling on the cellular structure is shown in the Figures 8.39 on page 464 in Chapter 8.

9.7.2 *Transitions between Cellular and Dendritic Growths*

Some Experimental Evidence

Figure 9.108 shows a schematic transition from planar growth to cellular growth. Obviously the ratio G_L/V_{growth} is a vital parameter when the transition from one mode to another is concerned.

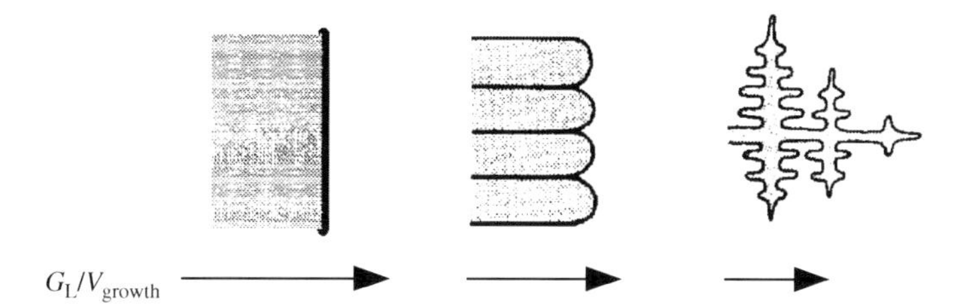

Figure 9.108 Schematic transition from a planar solidification front via cellular growth to dendritic growth.
The decreasing lengths of the arrows symbolize the decrease of G_L/V_{growth}. The directions of the arrows show the growth direction in the three cases. © Skanaluminium 1986.

If the growth rate is increased and/or if the temperature gradient is decreased even more than the middle arrow indicate, the cellular growth will be changed into dendritic growth. Figure 9.108 also shows a schematic transition from a cellular structure to a typical dendritic structure.

Campell [26] has studied transitions from cellular to dendritic structure in low-carbon steel. Figure 9.109 shows a real transition from a coarse cellular structure to a fine dendritic structure as a function of the growth rate or rather the cooling rate.

Elongated cells in a length intersection of the specimen are shown in the lower part of the figure. The transition from cellular to dendritic growth is shown in the middle of the figure. The figure shows that the transition to the dendritic structure in the upper part of the figure occurs rather abruptly.

Two research groups, Burden and Hunt [27] and in particular Miyata et al [28] have investigated the transition from cellular to dendritic growth in Al-Cu alloys very carefully.

Figures 9.110a, b and c show a transition from cells to dendrites in an unidirectionally solidified sample. The Figures 9.110 show the cell structure (a) and the dendrite structure (c) with well

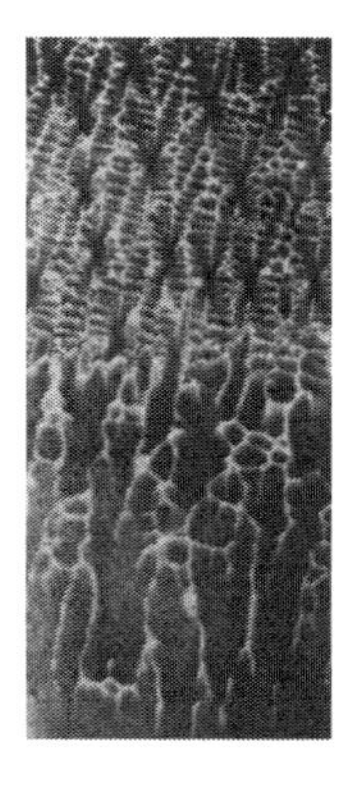

Figure 9.109 The structure of a low-alloy steel, which was exposed to a cooling rate, that was increased downwards during the solidification.
Increase of the cooling rate gives a higher growth rate.
The structure changes suddenly from cellular (left) to dendritic (right). Reproduced from Elsevier © 2003.

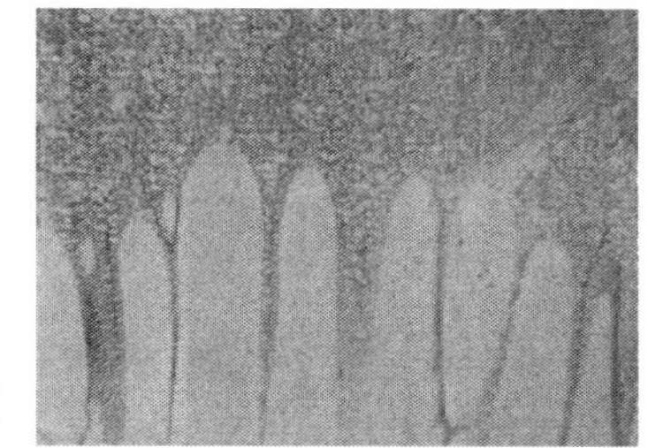

Figure 9.110a Cell structure: $G_L = 4.1$ K/m $V_{growth} = 3.00$ μm/s. From Metallurgical Transactions A, 16A, AIME © Kluwer 1985 with kind permission from Springer Science+Business Media B.V.

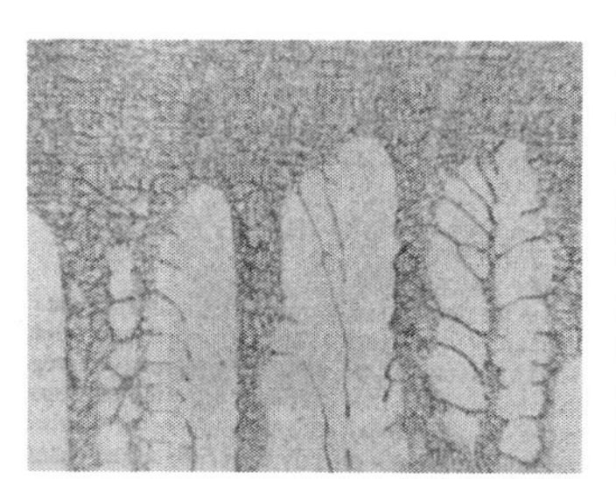

Figure 9.110b Intermediate structure. $G_L = 4.5$ K/m $V_{growth} = 4.95$ μm/s. From Metallurgical Transactions A, 16A, AIME © Kluwer 1985 with kind permission from Springer Science +Business Media B.V.

Figure 9.110c Dendritic structure. $G_L = 4.5$ K/m $V_{growth} = 10.5$ μm/s. From Metallurgical Transactions A, 16A, AIME © Kluwer 1985 with kind permission from Springer Science +Business Media B.V.

developed secondary arms. Figure 9.110b shows an intermediate structure with more irregular secondary dendrite arms.

Figure 9.111a shows how the tip radius and the distances between the secondary arms were measured. Miyata et al measured the radii of the tips both for the cells and the dendrites at the solidification front. The result is shown in Figure 9.111b.

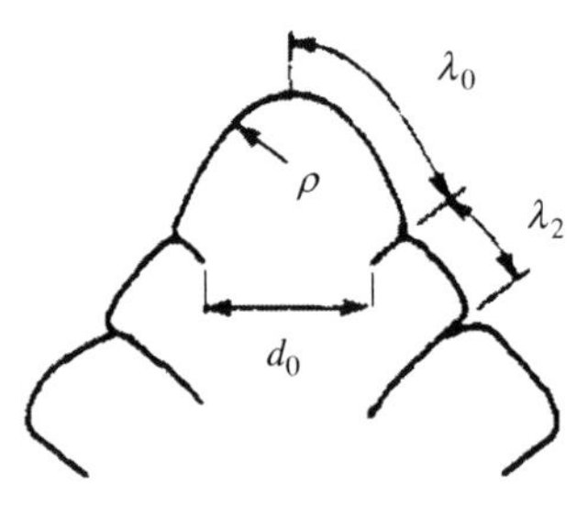

Figure 9.111a Definition of measured characteristics of a dendrite tip.
ρ = tip radius.
d_0 = core diameter of dendrite stem.
λ_0 = half length of the tip arc.
λ_2 = the first secondary arm spacing.
From Metallurgical Transactions A, 16A, AIME © Kluwer 1985 with kind permission from Springer Science+Business Media B.V.

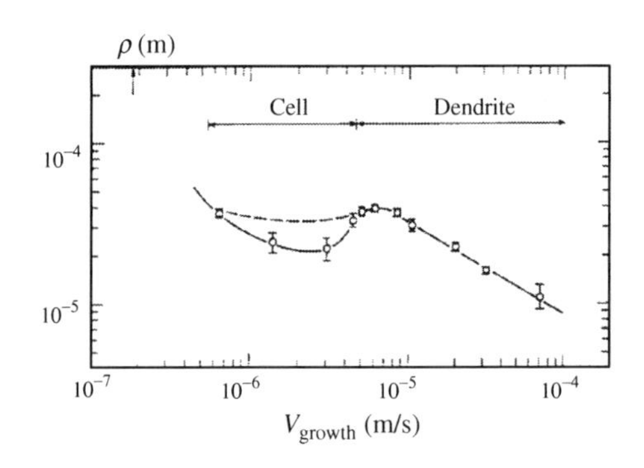

Figure 9.111b Tip radius as a function of the growth rate. The regions for cells and dendrites are marked in the figure. From Metallurgical Transactions A, 16A, AIME © Kluwer 1985 with kind permission from Springer Science+Business Media B.V.

Miyata et al [28] also measured the primary arm spacings and the distances between the cell stems and also the core diameters of the primary dendrites and the cell stems as functions of the growth rate. The results are shown if Figures 9.112 and 9.113, respectively.

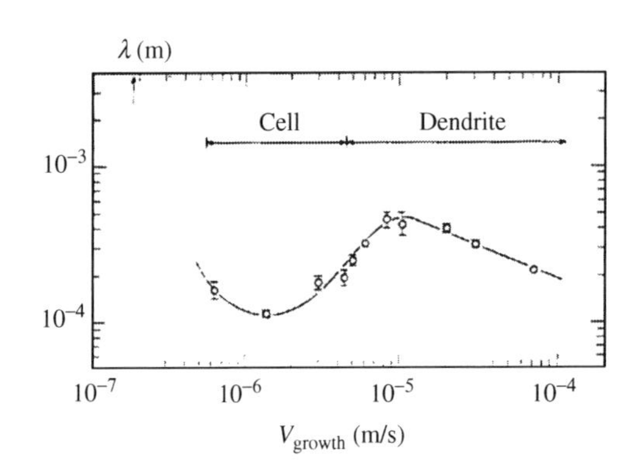

Figure 9.112 Primary arms spacing and distance between the cells as a function of the growth rate. From Metallurgical Transactions A, 16A, AIME © Kluwer 1985 with kind permission from Springer Science+Business Media B.V.

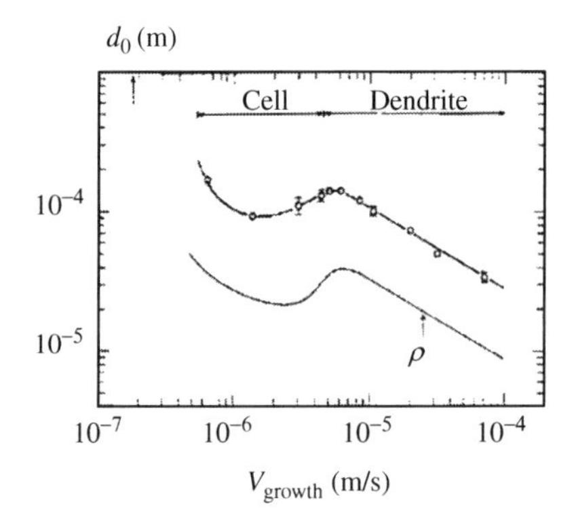

Figure 9.113 Core diameters of the primary dendritic stems and the cells as a function of the growth rate. From Metallurgical Transactions A, 16A, AIME © Kluwer 1985 with kind permission from Springer Science+Business Media B.V.

The figures show that both the cell radii and the distances between the cell stems decrease with increasing growth rate. At the transition to dendritic growth the cell radius and the distance between the stems suddenly increase and then decreases again with increasing growth rate. The secondary arms follow the same behaviour as the tip radius and the primary arms.

Dendritic Growth in a Temperature Gradient

The essential outlines of the theory of dendritic growth have been described in Section 9.5.4 on pages 514–529. Ivantsov's model for the concentration field around a paraboloid tip was introduced. A theoretical expressions for total driving force of solidification was set up [Equation (9.78) on page 525]. This equation involves terms which describe the influence of surface tension and interface kinetics. The radius of the growing dendrite tip and the growth rate of the solidification front are the most important quantities among other quantities such as concentrations of alloying elements in the melt and material constants. The basic equation is the expression for the driving force of solidification, which is proportional to

$$x^{\mathrm{L}/\alpha} - x_0^{\mathrm{L}} = \underbrace{m_{\mathrm{L}} \frac{\sigma T_{\mathrm{M}}}{-\Delta H_{\mathrm{m}}} K}_{\text{surface tension}} + \underbrace{m_{\mathrm{L}} \left(\frac{V_{\perp}}{\beta} \right)^{\frac{1}{n}}}_{\text{interface kinetics}} + \underbrace{(x^{\mathrm{L}/\alpha} - x^{\alpha/\mathrm{L}}) Pe\, \mathrm{e}^{Pe} E_1(Pe)}_{\text{supersaturation according to Ivantsov's model}} \tag{9.78}$$

We also applied Ivantsov's theory on an FCC crystal growing in the <100> directions and in the <111> directions, respectively. In practice this meant that we introduced the direction-dependent material constants valid for growth in the <100> and <111> directions into Equation (9.78). The result was

$$x^{\mathrm{L}/\alpha} - x_0^{\mathrm{L}} \approx m_{\mathrm{L}} \frac{T_{\mathrm{M}}}{-\Delta H_{\mathrm{m}}} \frac{3}{2\sqrt{2}} \frac{\sigma_{100}}{\rho_{100}^{\mathrm{tip}}} + m_{\mathrm{L}} \left(\frac{V_{100}}{\beta_{100}} \right)^{\frac{1}{n}} + (x^{\mathrm{L}/\alpha} - x^{\alpha/\mathrm{L}}) [Pe\, \mathrm{e}^{Pe} E_1(Pe)]_{100} \tag{9.79}$$

where $Pe_{100} = \frac{V_{100}\,\rho_{100}^{\mathrm{tip}}}{2D_{\mathrm{L}}}$ [application of Equation (9.32) on page 509] and

$$x^{\mathrm{L}/\alpha} - x_0^{L} \approx m_{\mathrm{L}} \frac{T_{\mathrm{M}}}{-\Delta H_{\mathrm{m}}} \frac{3}{2\sqrt{2}} \frac{\sigma_{111}}{\rho_{111}^{\mathrm{tip}}} + m_{\mathrm{L}} \left(\frac{V_{111}}{\beta_{111}} \right)^{\frac{1}{n}} + (x^{\mathrm{L}/\alpha} - x^{\alpha/\mathrm{L}}) [Pe\, \mathrm{e}^{Pe} E_1(Pe)]_{111} \tag{9.80}$$

where $Pe_{111} = \frac{V_{111}\,\rho_{111}^{\mathrm{tip}}}{2D_{\mathrm{L}}}$ [application of Equation (9.32) on page 509].

Further calculations proved that the two products $V_{100}\rho_{100}^{\mathrm{tip}}$ and $V_{111}\rho_{111}^{\mathrm{tip}}$ are equal and constant (pages 524–526).

Application of Ivantsov's model on an FCC crystal, i.e. Equations (9.79) and (9.80), gave another relationship between the tip radius ρ and the growth rate V_{growth}, independent of Equation (9.32). The relationship derived in Example 9.6 on page 526 was written as

$$V_{100} = \beta_{100} \left[\frac{\frac{T_{\mathrm{M}}}{-\Delta H} \frac{3}{2\sqrt{2}} \sigma_{100} \left(1 - \frac{\sigma_{111}}{\sigma_{100}} \frac{1}{\sqrt{3}} \right)}{\left(\frac{1}{\sqrt{3}} \frac{\beta_{100}}{\beta_{111}} \right)^{\frac{1}{n}} - 1} \right]^{n} \frac{1}{\rho_{100}^{\mathrm{tip}\,n}} \tag{9.123}$$

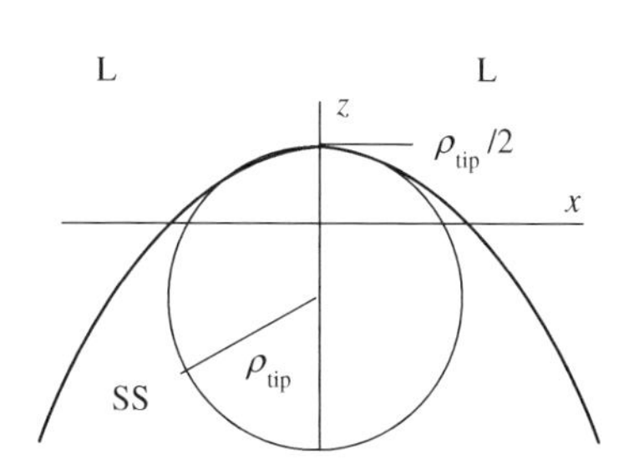

Figure 9.42 Parabolic tip with the radius of curvature $= \rho_{\mathrm{tip}}$ at the tip. $z_{\mathrm{tip}} = \rho_{\mathrm{tip}}/2$.

Application of Ivantsov's Model on Dendritic Growth in a Melt with a Temperature Gradient
When an alloy melt is exposed to a temperature gradient the concentration field in the vicinity of the growing tip is not the same as that valid for a melt at a constant temperature. This means that the Equations (9.79) and (9.80) can not be applied as they are but have to be modified.

In an experiment with a temperature gradient G_{L} in a <100> direction, the solute concentration in the melt at the solidification front in equilibrium, with the solid phase can be written as

$$x_{\text{eq}\,z}^{\text{L}/\alpha} = x_{\text{eq}\,0}^{\text{L}/\alpha} + m_{\text{L}} G_{\text{L}} z \tag{9.124}$$

where

$x_{\text{eq}\,z}^{\text{L}/\alpha}$ = concentration of solute at distance z from the tip along the paraboloid axis at temperature T
$x_{\text{eq}\,0}^{\text{L}/\alpha}$ = concentration of solute at the tip at temperature T_0
m_{L} = slope of the liquidus line in the phase diagram
G_{L} = temperature gradient in the melt along the paraboloid axis
z = distance from the tip to an outer point along the paraboloid axis.

If we replace $x^{\text{L}/\alpha}$ in Equation (9.79) by the expression in Equation (9.124), i.e. by $x_{\text{eq}\,z}^{\text{L}/\alpha} = x_{\text{eq}\,0}^{\text{L}/\alpha} + m_{\text{L}} G_{\text{L}} z$ we obtain

$$x_{\text{eq}\,0}^{\text{L}/\alpha} + m_{\text{L}} G_{\text{L}} z_{100} - x_0^{\text{L}} = m_{\text{L}} \frac{T_{\text{M}}}{-\Delta H} \frac{3}{2\sqrt{2}} \frac{\sigma_{100}}{\rho_{100}^{\text{tip}}} + m_{\text{L}} \left(\frac{V_{100}}{\beta_{100}}\right)^{\frac{1}{n}} + (x^{\text{L}/\alpha} - x^{\alpha/\text{L}})[Pe\, \text{e}^{Pe} E_1(Pe)]_{100} \tag{9.125}$$

The corresponding replacement in Equation (9.80) gives analogously

$$x_{\text{eq}\,0}^{\text{L}/\alpha} + m_{\text{L}} G_{\text{L}} z_{111} - x_0^{L} = m_{\text{L}} \frac{T_{\text{M}}}{-\Delta H} \frac{3}{2\sqrt{2}} \frac{\sigma_{111}}{\rho_{111}^{\text{tip}}} + m_{\text{L}} \left(\frac{V_{111}}{\beta_{111}}\right)^{\frac{1}{n}} + (x^{\text{L}/\alpha} - x^{\alpha/\text{L}})[Pe\, \text{e}^{Pe} E_1(Pe)]_{111} \tag{9.126}$$

Calculations, analogous to those in Example 9.6 on pages 526–528, give the modified version of Equation (9.123). It is

$$V_{100} = \beta_{100} \left[\frac{\frac{T_{\text{M}}}{-\Delta H} \frac{3}{2\sqrt{2}} \frac{\sigma_{100}}{\rho_{100}^{\text{tip}}} \left(1 - \frac{\sigma_{111}}{\sigma_{100}} \frac{1}{\sqrt{3}}\right) + m_{\text{L}} G_{\text{L}} (z_{111} - z_{100})}{\left(\frac{1}{\sqrt{3}} \frac{\beta_{100}}{\beta_{111}}\right)^{\frac{1}{n}} - 1} \right]^{n} \tag{9.127}$$

Cellular Growth

As is shown in Figure 9.103 cellular growth occurs at intermediate conditions between the two boundaries planar growth/cellular growth and cellular growth/dendritic growth. The first boundary is emanates from the well-defined condition of constitutional undercooling and is given by Equation (9.125) on page 566. The boundary between cellular growth and dendritic growth is more difficult to identify and express analytically.

In Section 9.7.1 we pointed out that the ratio of a high temperature gradient and a low growth rate results in planar growth. When the ratio $G_{\text{L}}/V_{\text{growth}}$ is decreased the growth will be cellular owing to formation of perturbations. Within the cellular region in Figure 9.103 on page 565 we know that

$$\frac{G_{\text{L}/\alpha}}{V_{\text{growth}}} \leq -\frac{m_{\text{L}}}{D_{\text{L}}} \frac{1 - k_0}{k_0} x_0^{\text{L}} \tag{9.122}$$

Equation (9.122) shows that cellular growth is diffusion-controlled at the upper boundary.

At further decrease of $G_{\text{L}}/V_{\text{growth}}$ the conditions approach those which are valid for dendritic growth. Equation (9.122) gives the upper boundary for cellular growth but gives no information about the lower boundary. The latter is identical with the upper boundary for dendritic growth. It may be worth while to compare the growth conditions near the boundary between dendritic and cellular growths.

At dendritic growth the total driving force depends on surface tension, interface kinetics and mass transport (diffusion) [Equation (9.78) on page 525]. It is reasonable to assume that this is the case also for cellular growth near the boundary cellular/dendritic growth.

According to experimental observations the cells have a much larger radius than the dendrite tips. The curvature K in Equation (9.78) is therefore smaller for cells than for dendrites. Hence, the influence of surface tension is small at cellular growth and can be neglected.

At the lower boundary for cellular growth V_{growth} is the same for cellular and dendritic growths. Hence, the influence of interface kinetics will be the same for the two structures.

Transition from Cellular Growth to Dendritic Growth and Vice Versa

Consider Equation (9.127). The nominator is finite. The growth rate will approach infinity when the denominator decreases closer and closer to zero. Hence, the condition

$$\left(\frac{1}{\sqrt{3}}\frac{\beta_{100}}{\beta_{111}}\right)^{\frac{1}{n}} - 1 = 0 \qquad (9.128)$$

corresponds to a breakdown of the dendritic growth. Hence, Equation (9.128) has to be fulfilled at the boundary between cellular and dendritic growths, n is often a value close to 2. In this case the boundary condition can be written as

$$\frac{\beta_{100}}{\beta_{111}} = \sqrt{3} \qquad (9.129)$$

9.7.3 *Transitions between Dendritic and Faceted Growths*

So far no generally accepted theory is known for transitions between dendritic and faceted growth but the phenomenon is well known from experimental point of view. Transitions have been observed in both directions. Some examples will be given below and the factors, which influence the type of growth will be discussed.

Transition from Dendritic to Faceted Growth

In Section 9.6.4 on pages 532–537 the so-called tip effect has been discussed. During natural cooling a metal melt normally solidifies by dendritic growth. It has been observed at the end of the solidification process that faceted crystals are formed on the secondary dendrite tips. Figure 9.56 shown again below shows a typical example.

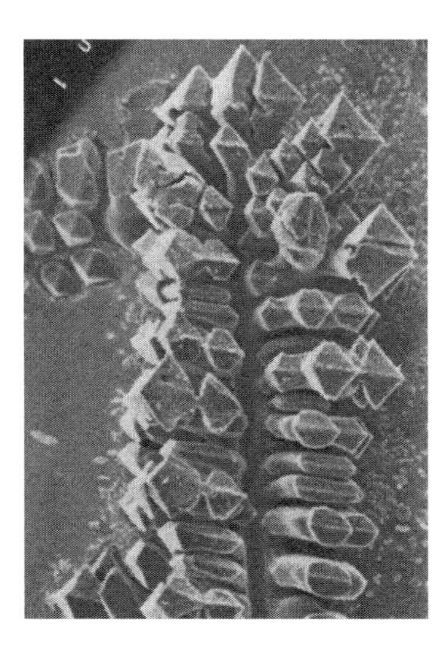

Figure 9.56 Left figure. Dendrite arm of Cu growing in a <100> direction (upwards) in a Cu–Pb melt. Scanning electron micrograph.

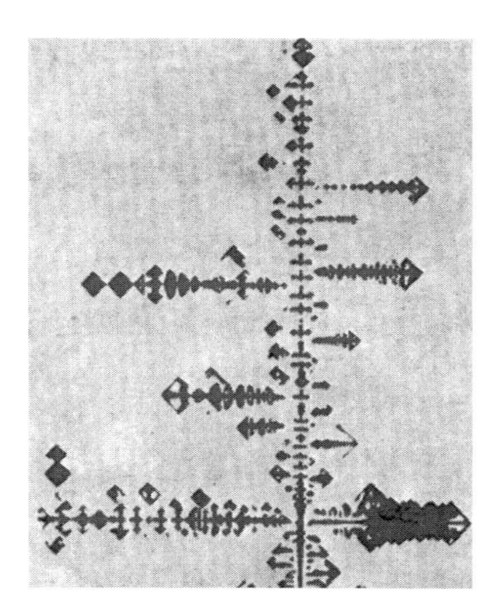

Figure 9.114 Dendrites of a Cr_2O_3 crystal, precipitated from a steel melt. Facets have been formed at a late stage of solidification.

The change of growth mode results in formation of arrow-heads around the dendrite tips. This phenomenon is very common.

From a comparison between Figures 9.56 and 9.114 we can conclude that the initial growth is dendritic in both cases, i.e. when the primary stem is formed, and the tendency of faceted growth comes at the end of the solidification process.

A comparison between the two figures on page at the bottom of the last page shows that, for a given value of the tip radius,

- the growth rate decreases strongly with increasing values of the ratio of the kinetic growth constants β in the directions <100> and <111>.

According to Equation (9.83) on page 528

$$V_{\text{growth}}^{\text{tip}} = const \times \frac{1}{\rho_{\text{tip}}^2} \tag{9.83}$$

It is obvious that

- the tip radius decreases with increasing growth rate.

If we use the condition that dendrites growing with a tip radius $< 1\ \mu\text{m}$ (the distance which can be resolved in a normal light microscope) we can read from the left figure on page 528 that a transition will occur at $V_{\text{growth}} = 3\text{–}7$ mm/s if $\beta_{100}/\beta_{111} = 2$. This is a reasonable magnitude of V_{growth} according to experimental experience.

The transition growth rate varies of course from alloy to alloy depending on their material constants.

Transition from Faceted to Dendritic Growth

Above several examples have been given where dendritic growth changes into faceted growth. The faceted growth implies also a change of growth direction. The faceted crystals grow in other directions than the dendrite tip.

The transition from faceted to dendritic growth of Al_2Cu crystals in an Al-Cu melt can simply be induced by change of the growth rate caused by increase of the cooling rate. Such a case is shown in Figure 9.115. Al_2Cu has an ortorombic faceted structure and the columnar crystals grow in the <001> direction. The dendrites mainly grow in the <111> direction.

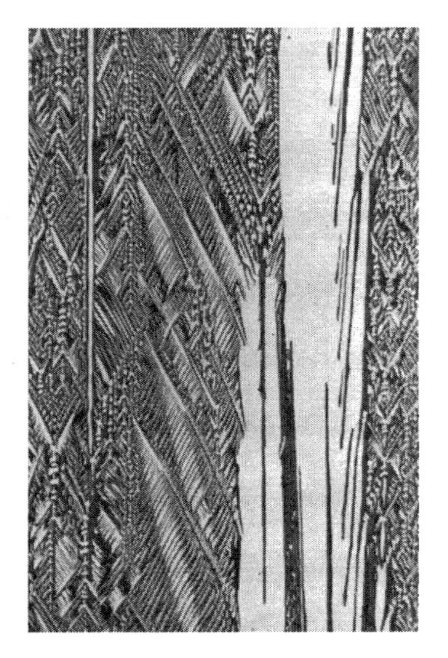

Figure 9.115 Primary precipitation of Al_2Cu from an Al–Cu melt at a controlled solidification experiment. A sudden increase of the cooling rate caused a change of the growth rate and simultaneously a change from faceted growth to dendritic growth.

The white strings to the left in Figure 9.115 are faceted columnar crystals of Al_2Cu. They are of the same type as the Bi crystals shown in Figures 9.11 and 9.12 on pages 484–485. When the cooling rate suddenly is increased a dendritic structure of Al_2Cu appears. The primary arms are shown as white strings, which form an angle of 30° with the white columnar crystals in the central part of the figure.

The primary dendrite stems show the growth direction of the dendrites. The secondary dendrite arms emanate from the primary arms (thick white strings) and form an angle with them. The dendrite looks like a fish skeleton in the figure.

The controlled solidification experiment in Figure 9.115 indicates that

- A transition from faceted growth to dendritic growth is promoted by increase of the growth rate.

Transition from faceted to dendritic growth has also been found in hypoeutectic Al-Si alloys. A comparison between the Figures 9.14–9.17 on pages 486–488 and Figure 9.61 on page 534 show additional examples of this transition. The last figure shows a faceted tip of Si formed in a melt of same alloy as the Si crystals in the Figures 9.14–9.17 but formed at stronger cooling.

The transition from faceted to dendritic growth does not always occur. The transition from faceted to dendritic growth indicated in Figure 9.103 on page 565 seldom occurs. The Equation (5′) on page 528 in Example 9.6 shows that the transition from faceted growth to dendritic growth occurs if

$$\frac{\beta_{100}}{\beta_{111}} = \sqrt{3} \tag{9.130}$$

which indicates that the kinetics at the interfaces must be growth related and that the anisotropy in the kinetic reaction must increase with increasing growth rate.

Summary

■ Faceted Solidification Structures

Faceted crystals form in saturated solutions or, in the case of metals and alloys, in melts. The shape of the crystal depends on the total energy of the crystal.

- A faceted crystal always has such a shape that its total energy is as low as possible.

The total energy of a faceted crystal depends on

- the surface tension between its facets and the melt
- the kinetics, i.e. the growth rate at the solidification front as a function of the undercooling of the melt
- the mass transport, i.e. the diffusion in the melt.

Faceted Growth in Pure Metal Melts

Growth Directions of Faceted Growth in Pure Metal Melts
Gibbs'–Curie–Wulff's theorem:

$$\frac{\sigma_1}{h_1} = \frac{\sigma_2}{h_2} = \frac{\sigma_3}{h_3} = \cdots = \frac{\sigma_o}{h_o}$$

The most stable surface of a faceted crystal is the one, that has the lowest growth rate.

Influence of Interface Kinetics on Faceted Growth in Pure Metal Melts

$$V_{growth} = \frac{dr_{hkl}}{dt} = \beta_{kin}\left(T_M - T_{growth}\right)^2$$

Growth of Faceted Cubic Crystals in Pure Metal Melts. Hopper Crystals
In cubic crystals all other growth directions have normally much higher growth rates than those in the directions <100> and <111>. For this reason the only facets, which normally occur in such crystals, are {100} and {111} planes.

Growth of cubic crystals:
If $V_{111} = V_{100}\sqrt{3}$ the crystal grows and keeps its shape.
If $V_{111} < V_{100}\sqrt{3}$ the crystal changes shape during growth and ends up as an octahedron with only {111} facets.
If $V_{111} > V_{100}\sqrt{3}$ the crystal keeps its shape during growth but hopper crystals are formed.
Cubic and other crystals with terraces at the corners and along the edges are called hopper crystals.

Growth of octahedron crystals:
If $V_{100} = V_{111}\sqrt{3}$ the crystal grows and keeps its shape.
If $V_{100} < V_{111}\sqrt{3}$ the crystal changes shape during growth and ends up as an octahedron with only {111} facets.
If $V_{100} > V_{111}\sqrt{3}$ the crystal keeps its shape during growth but hopper crystals are formed.

■ Faceted Growth in Alloys

The growth depends on

- the interface reaction (kinetics)
- the diffusion process of the alloying element in the melt along and perpendicular to the facets of growing crystals
- the effect of surface tension on the growth rate. It can be neglected for large crystals.

Driving Force of Solidification

For alloys with $k > 1$ the average solute concentration x_0^L is larger than x_{eq}^L and the driving force can be written as

$$\text{Driving force} = const\left(x_0^{L} - x_{eq}^{L/\alpha}\right) = const\left[\underset{\text{diffusion process}}{\left(x_0^{L} - x_{kin}^{L/\alpha}\right)} + \underset{\text{kinetic process}}{\left(x_{kin}^{L/\alpha} - x_{eq}^{L/\alpha}\right)}\right]$$

For alloys with $k < 1$ the average solute concentration $x_{eq}^{L/\alpha}$ is larger than x_0^L and the driving force can be written as

$$\text{Driving force} = const\left(x_{eq}^{L/\alpha} - x_0^{L}\right) = const\left[\underset{\text{kinetic process}}{\left(x_{eq}^{L/\alpha} - x_{kin}^{L/\alpha}\right)} + \underset{\text{diffusion process}}{\left(x_{kin}^{L/\alpha} - x_0^{L}\right)}\right]$$

See Figures 9.22a and b on page 491.

Influence of Interface Kinetics on Faceted Growth in Binary Alloys

In pure metals the growth is controlled by a *temperature difference*. In case of alloys the growth is controlled by a *concentration difference* of the alloying element.

$$V_{growth} = \left(\frac{dr_{mid}}{dt}\right)_{kin} = \beta_{kin}\left(x_{eq}^{L/\alpha} - x_{mid}^{L/\alpha}\right)^2$$

When the kinetic process dominates, the solute distribution around the crystal decreases linearly with increasing distance from the centre as experiments with transparent crystals show,

Influence of Mass Transport on Faceted Growth in Binary Alloys

$$V_{growth} = \frac{dr_{mid}}{dt} = D\frac{x_{eq}^{L/\alpha} - x_0^{L}}{x_{eq}^{L/\alpha} - x_{eq}^{\alpha/L}}\frac{V_m^{\alpha}}{V_m^{L}}\frac{1}{r_{mid}}$$

A survey of mixed influence of kinetics, diffusion with relaxation and surface tension is given in Table 3 on page 492.

When the diffusion process dominates the solute distribution around the crystal is

$$x^{L} = x_0^{L} + \frac{r_{mid}\left(x_{eq}^{L/\alpha} - x_0^{L}\right)}{r}$$

The concentration gradient within the boundary layer dx^L/dr at the midpoint of a facet can be written as

$$\frac{dx^{L}}{dr} = -\frac{r_{mid}\left(x_{eq}^{L/\alpha} - x_0^{L}\right)}{r^2}$$

Growth Rates at the Corner of a Crystal and at the Midpoint of a Facet of a Three-dimensional Crystal

As is shown in the Figures 9.23a and b on page 493 the kinetic process dominates at the corner and the diffusion process dominates at the midpoint.

Condition for planar growth of a facet:

$$\frac{x_{\text{corner}}^{L/\alpha} - x_0^{L}}{\sqrt{3}\,\delta} = \frac{x_{\text{mid}}^{L/\alpha} - x_0^{L}}{\Delta}$$

Condition for formation of hopper crystals:

$$\frac{x_{\text{corner}}^{L/\alpha} - x_0^{L}}{\sqrt{3}\,\delta} > \frac{x_{\text{mid}}^{L/\alpha} - x_0^{L}}{\Delta}$$

Δ = width of a boundary layer with linearly decreasing solute concentration
$\delta = r_{\text{mid}} + \Delta - r_{\text{corner}}$

■ Dendritic Solidification Structures

Primary parallel stems are formed at distances λ_{den}.

Behind the tip secondary dendrite arms are formed in two directions perpendicular to the primary stems. Behind their tips tertiary dendrite tips are formed. All the tips grow with time.

Undercooling and Solidification Rate at Dendritic Growth in Pure Metal Melts

The driving force is controlled by the undercooling, i.e. the temperature difference between the tip of the dendrite and the melt ahead of the tip.

$$\Delta T_{\text{total}} = \Delta T_{\text{surface}} + \Delta T_{\text{kin}} + \Delta T_{\text{thermal}}$$

or

$$\Delta T_{\text{total}} = (T_{\text{M}}^{\text{tip}} - T_{\text{M}}) + (T_{\text{M}} - T^{L/\alpha}) + (T^{L/\alpha} - T_0)$$

The first term depends on the surface tension and the radius of the tip. It can often be neglected.

The second term comes from the growth law of dendritic growth in pure metal melts.

$$\Delta T_{\text{kin}} = T_{\text{M}} - T^{L/\alpha} = \sqrt{\frac{V_{\text{growth}}}{\beta_{\text{kin}}}}$$

If the tip is supposed to be hemispherical and the surface and the kinetic driving forces are neglected, the growth rate will be

$$V_{\text{growth}} = const(T_{\text{M}} - T_0)^2$$

Undercooling and Solidification Rate at Dendritic Growth in Binary Alloys

$$\Delta T_{\text{total}} = \Delta T_{\text{surface}} + \Delta T_{\text{kin}} + \Delta T_{\text{mass}} + \Delta T_{\text{thermal}}$$

$$\Delta T_{\text{total}} = \frac{x^{L/\alpha} - x_0^{L}}{m_{\text{L}}}$$

$$\Delta T_{\text{surface}} = \frac{x^{L/\alpha} - x_{\sigma}^{L/\alpha}}{m_{\text{L}}} = \frac{2\sigma_{\text{hkl}} T^{L/\alpha} V_{\text{m}}^{\alpha}}{\rho_{\text{tip}}(-\Delta H_{\text{m}})}$$

$$\Delta T_{\text{kin}} = \frac{x_{\sigma}^{L/\alpha} - x_{\text{kin}}^{L/\alpha}}{m_L} = \left(\frac{V_{\text{growth}}}{\mu_{\text{hkl}}}\right)^{\frac{1}{n}}$$

$$\Delta T_{\text{mass}} = \frac{x_{\text{kin}}^{L/\alpha} - x_{\text{mass}}^{L/\alpha}}{m_L}$$

$$\Delta T_{\text{thermal}} = \frac{x_{\text{mass}}^{L/\alpha} - x_0^L}{m_L}$$

Ivantsov's Model for the Concentration Field around a Paraboloid Tip

Ivantsov's model is based on two assumptions:

- The concentration of the alloying element is constant along the surface of the tip.
- The conditions are stationary, i.e. the concentration field and the growth rate are independent of time.

The diffusion field is derived as a solution of Fick's second law.

$$x^L(x,y,z) = x_0^L + (x^{L/\alpha} - x_0^L)\frac{E_1\left[\frac{V_{\text{growth}}\,\rho_{\text{tip}}}{2D_L}\left(\frac{z}{\rho_{\text{tip}}} + \sqrt{\frac{x^2+y^2+z^2}{\rho_{\text{tip}}^2}}\right)\right]}{E_1\left(\frac{V_{\text{growth}}\,\rho_{\text{tip}}}{2D_L}\right)}$$

where the E_1 function (exponential integral function) is

$$E_1(u) = \int_u^{\infty} \frac{e^{-u}}{u}\,du$$

Ivantsov's function is defined as

$$I(Pe) = \frac{C_p(T_{\text{interface}} - T_{\text{melt}})}{-\Delta H_m}$$

Ivantsov derived the relationship

$$I(Pe) = Pe\, e^{Pe} E_1(Pe)$$

where Peclet's number

$$Pe = \frac{V_{\text{growth}}\,\rho_{\text{tip}}}{2D_L}$$

The solute concentration as a function of the parabolic ρ_{tip} and z can be written (cylindrical coordinates) as

$$x^L(r,z) = x_0^L + (x^{L/\alpha} - x_0^L)\frac{E_1\left[\frac{V_{\text{growth}}\,\rho_{\text{tip}}}{2D_L}\left(\frac{z}{\rho_{\text{tip}}} + \sqrt{\frac{r^2+z^2}{\rho_{\text{tip}}^2}}\right)\right]}{E_1\left(\frac{V_{\text{growth}}\,\rho_{\text{tip}}}{2D_L}\right)}$$

Temperature Field around a Dendrite Tip in Pure Metal Melts

The corresponding expressions for pure metals are:

$$T^L(r,z) = T_0^L + (T^{L/\alpha} - T_0^L)\frac{E_1\left[\frac{V_{\text{growth}}\,\rho_{\text{tip}}}{2\alpha}\left(\frac{z}{\rho_{\text{tip}}} + \sqrt{\frac{r^2+z^2}{\rho_{\text{tip}}^2}}\right)\right]}{E_1\left(\frac{V_{\text{growth}}\,\rho_{\text{tip}}}{2\alpha}\right)}$$

and

$$I(Pe) = \frac{C_p(T^{L/\alpha} - T_0^L)}{-\Delta H_m} = E_1(Pe)e^{Pe}Pe$$

Anisotropy of the Surface Energy in Pure Metal Melts

$$\sigma = \sigma_0(1 + A\sin\theta)$$

Anisotropy of the Interface Kinetics in Pure Metal Melts

The growth constant in the growth law

$$V_{growth} = \beta(\Delta T)^n = \beta\left(\frac{\Delta x^L}{m_L}\right)^n$$

shows anisotropy

$$\beta = \beta_0(1 + B\sin\theta)$$

Driving Force of Dendritic Growth in Binary Alloys

$$x^{L/\alpha} - x_0^L = (x^{L/\alpha} - x_\sigma^{L/\alpha}) + (x_\sigma^{L/\alpha} - x_{kin}^{L/\alpha}) + (x_{kin}^{L/\alpha} - x_{/mass}^{L/\alpha}) + (x_{mass}^{L/\alpha} - x_{/0}^L)$$

where

The first term is $x^{L/\alpha} - x_\sigma^{L/\alpha} = m_L \dfrac{\sigma T_M}{-\Delta H_m} K$

The second term is $x_\sigma^{L/\alpha} - x_{kin}^{L/\alpha} = m_L\left(\dfrac{V_\perp}{\beta}\right)^{\frac{1}{n}}$

The third and fourth terms can be written with the aid of Ivantsov's expression:

$$x_{kin}^{L/\alpha} - x_0^{L/\alpha} \approx x^{L/\alpha} - x_0^{L/\alpha} = (x^{L/\alpha} - x^{\alpha/L})Pe\,e^{Pe}E_1(Pe)$$

The total driving force of solidification can be written as

$$x^{L/\alpha} - x_0^L = m_L \frac{\sigma T_M}{-\Delta H_m} K + m_L\left(\frac{V_\perp}{\beta}\right)^{\frac{1}{n}} + (x^{L/\alpha} - x^{\alpha/L})\left(Pe\,e^{Pe}E_1(Pe)\right)$$

Ivantsov's model is applied on an FCC crystal structure in the text on page 525.

■ Development of Dendrites

The dendrites emanate from circular predendritic regions.

The isothermal growth of a dendrite is stabilized by anisotropy of the surface energy and also of the interface reaction.

The side branches grow much more slowly than the main tip.

Growth Directions of Dendrites in a Temperature Gradient

$$V_{growth} = \mu(T_L - T)^n$$

There is always a competition between various parallel stems. Dendrites growing in the direction of the temperature gradient will win.

Tip Effect at Dendritic Crystal Growth

Spontaneous developments of facets on the dendrite tips occur at the end of the precipitation.

The most common growth directions for the facets in FCC and BCC metals are <100> but <110> are also found.

Twinned crystal growth is treated briefly in the text.

■ Microsegregation in Dendritic Structures in Alloys

A model of iterative calculation of the solute concentration in primary stems as a function of time and position. It is described and applied in the text.

$$x(y,t) = x^{s/L} - (x^{s/L} - x_{min})e^{-\frac{4\pi^2 D_s}{\lambda^2}t} \cos\frac{2\pi y}{\lambda}$$

where $0 < t <$ solidification time.

■ Dendrite Arm Spacings in Alloys

Branching and Engulfing Primary Stems

The primary dendrite arms grow in parallel stems.

Distance between primary parallel stems:

$$\lambda_{den} = Const\sqrt{\frac{\alpha}{D}}\rho_{tip}$$

The stems grow in the direction of the temperature gradient and adjust their spacings during growth.

If the dendrite arm distances are too small they may be adjusted by *engulfing,* which means that the temperature fields overlap too much and one arm looses the competition between the stems and disappears.

If the dendrite arm distances are too large a secondary dendrite arm is turned into a primary stem and neighbouring stems adjust slowly their distances.

Spacings of Primary Dendrite Arms

The dendrite arm spacing depends on the cooling rate and the concentration of the alloying element.

The higher the cooling rate is, the smaller will be the dendrite distance. λ_{den} has a minimum value for a specific solute concentration in the alloy.

Relationship between cooling rate and solidification time:

$$t_{sol} = \Delta T_s / -\frac{dT}{dt}$$

Spacings of Secondary Dendrite Arms

There is always a competition between the secondary arms. Some win the competition and grow ahead of their neighbours, others remelt and disappear.

The secondary dendrite arms have no equal and constant distances between adjacent arms.

It makes sense to calculate and use an average value of λ_{den}.

Relationships between λ_{den} and Measured Quantities

– Relationship between the average dendrite arm spacing and the solidification time:

$$\lambda_{den} = const\, t_{sol}^{n}$$

– Relationship between the average dendrite arm spacing and the growth rate:

$$V_{growth}\lambda_{den}^{2} = const$$

This relationship is valid for both primary and secondary dendrite spacings.

– General relationship between cooling rate, temperature gradient and growth rate:

$$\text{Cooling rate} - \frac{dT}{dt} = GV_{\text{growth}} = \frac{dT}{dy}\frac{dy}{dt}$$

If the temperature gradient G is constant the cooling rate is proportional to the growth rate.
The higher the cooling rate is, the denser will be the dendrite structure.
A high cooling rate results in a finer microstructure and a better quality of the material than a low cooling rate.

■ Coarsening of Secondary Dendrite Arms in Alloys

The conditions for coarsening of primary and secondary dendrite arms differ in many respects.

- The secondary dendrite arm spacings are *much smaller* than those of the primary dendrite arms.
- The secondary dendrite arms are *much thinner* than the primary stems.
- The coarsening rate of the secondary dendrite arms is *very* rapid.

The coarsening process of secondary dendrite arms occurs with the aid of two simultaneous mechanisms, *remelting* and *back diffusion.*

Remelting of Secondary Dendrite Arms

The *remelting process* is a matter of mass transport. Diffusion of the alloying element from a thinner dendrite arm into neighbouring thicker dendrite arms leads to thickening of the thicker dendrite arms at the expense of the thinner dendrite arms, which disappear. The three mechanisms of remelting of thin dendrite arms are called *ripening* and shown graphically in the text.

Back Diffusion

When a melt solidifies, and $k < 1$, the alloying atoms are thrown out in the melt if their equilibrium solubility is lower in the solid than in the melt. The ratio $x^{s/L}/x^{L/s}$ is the partition constant k and the concentration of the alloying element in the melt increases. If the alloying atoms at the solid surface diffuse into the interior of the solid phase their concentration at the interface will decrease while their concentration in the melt is very high.

This situation leads to back diffusion, i.e. more alloying atoms diffuse from the melt into the solid phase than in the opposite direction owing to solidification. This enhances the solidification process at the interface as the system tries to achieve equilibrium. Back diffusion is said to 'consume' the melt.

Back diffusion into the secondary dendrite arms causes consumption of the melt between the dendrites for reasons described above. The result is that the dendrite arms merge and form a continuous so-called plate.

In this way the secondary dendrite arms merge and a so-called secondary plate is formed. The primary plates and the secondary arms merge and the secondary plates merge with tertiary arms, formed at the secondary arms. Two types of secondary plates are formed, perpendicular to each other. They form characteristic 'crosses' in the structure.

Simplified Theory of Coarsening of Secondary Dendrite Arms

The secondary dendrites thicken during solidification.

$$\int_0^t dt = \int_{0.9}^{1.0} \frac{m_L}{C} \frac{\left(gx^{L/\alpha} + (1-g)x^{\alpha/L}\right)\dfrac{\lambda_i}{\lambda}}{[1 - f(1-k)]^2}(1-k)df$$

Relationship between Coarsening and Solidification Time

Empirical relationship:

$$\lambda_{den} = const\, t^{\frac{1}{3}}$$

Theoretical calculations give

$$\lambda_{den}^3 - \lambda_0^3 = Ct$$

■ Influence of Convection on Dendritic Growth

The influence of free and forced convection on dendritic growth is discussed in the text.

The influence of free convection depends on the growth direction relative to gravity.

Forced convection changes the macrostructure of the solid alloy. Stirring during solidification influences the solute distribution of the solute results in a more fine-grained microstructure.

Transitions between Structure Types in Alloys

The structure of the solid phase is entirely a result of the solidification conditions. Many parameters influence the structure. The most important of them are

- composition of the alloy
- temperature
- temperature gradient in the melt
- growth rate of the solidification front.

If these parameters are changed during the solidification the structure of the material will also change.

Planar growth	$\leftrightarrow$	Cellular growth	$\leftrightarrow$	Dendritic growth	$\leftrightarrow$	Faceted growth

At high temperature gradient G_L and low growth rate V_{growth} the solidification front will be planar.

When the ratio G_L/V_{growth} successively is changed the solidification mode changes as shown below.

When G_L/V_{growth} decreases the $\rightarrow$ sign is valid.

When G_L/V_{growth} increases the $\leftarrow$ sign is valid.

The transition boundaries are discussed in the text. In most cases no quantitative analytic conditions are known.

Exercises

9.1 Consider the octahedron crystal in Figure 9.9 on page 482 in the text book. The side OM of the inner triangle is perpendicular to one of the {111} facets of the crystal and proportional to the growth rate V_{111}. Analogously the side OA represents one of the <100> directions and is proportional to the growth rate V_{100}.

Prove with the aid of geometrical calculations that the condition for growth of the octahedron crystal with unchanged shape is

$$V_{100} = V_{111}\sqrt{3}$$

The side of the octahedron crystal is equal to a.

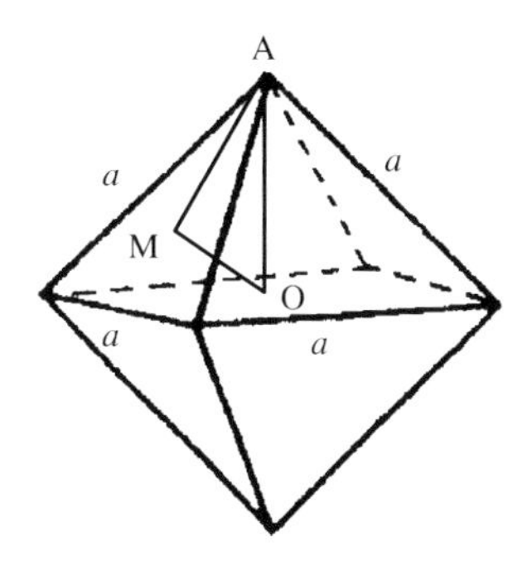

Figure 9.1–1

9.2 The β phase of an AB alloy has a hexagonal structure.

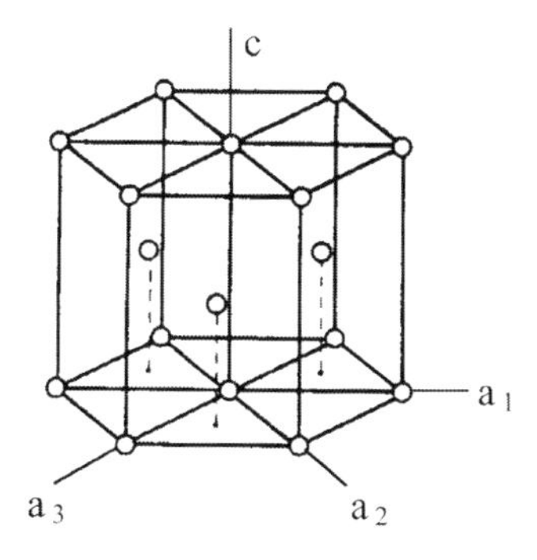

Figure 9.2–1

Experiments show that during the primary precipitation of β in an AB alloy the growth rates in the base plane (a and b directions) are linearly proportional to the undercooling.

$$V_{\text{growth}}^{\text{a}} = 1.0 \times 10^{-6} \times \qquad \Delta T \text{m/s}$$

The growth rate in the perpendicular c direction follows a parabolic growth law:

$$V_{\text{growth}}^{\text{c}} = 0.10 \times 10^{-6} \times \Delta T^2 \qquad \text{m/s}$$

a) Plot the growth rates as functions of undercooling for the a and c directions of the hexagonal crystals in a diagram.
b) Make schematic drawings which show how the morphologies of the β crystals change as a function of the undercooling for the three values $\Delta T = 5$ K, 10 K and 15 K.

9.3 A large number of morphologies of primarily precipitated Si crystals in eutectic AlSi alloys have been observed. The observed faceted crystals are often octahedrons, i.e. bound by {111} planes. The growth rate of the {111} facets is controlled by the kinetics. In all other directions the growth rate is very large and diffusion controlled.

a) Compare the growth rates in the <111> directions with the growth rates in the <100> directions. The kinetics in the <111> directions are given by the relationship

$$V_{\text{growth}} = 1.0 \times 10^{-3} \times (x^{L/\alpha} - x^{\alpha/L}) \qquad \text{m/s}$$

where the concentrations are the solute concentrations in the melt and crystal, respectively, at the interface.

The growth rates in the <100> directions are diffusion controlled. The diffusion constant D_L is 2.0×10^{-9} m²/s.

b) What is the size of a crystal when the growth rates in the <111> and <100> directions are equal at the eutectic temperature?

The initial Si concentration $x_0^L = 0.15$ (mole fraction).
The eutectic Si concentration $x_E^L = 0.123$.
The phase diagram of the Al-Si system is given below.

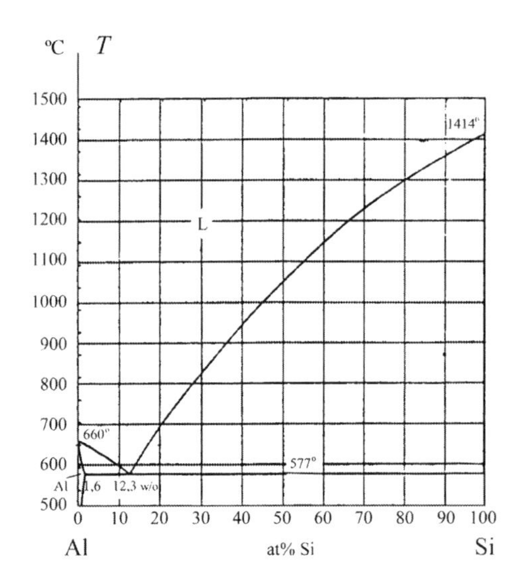

Figure 9.3–1

9.4 In a series of unidirectional solidification experiments with an Al-Cu alloy with the Cu concentration $x_0^L = 0.021$ one observed that the tip radius was 9 µm at a growth rate of the solidification front equal to 1×10^{-4} m/s and an undercooling of 5 K.

The kinetic growth rate is described by a parabolic growth law. Derive the kinetic controlled and the diffusion controlled undercoolings, respectively, and the growth constant β_{kin} for the kinetic process.

The diffusion constant D_L is 2.0×10^{-9} m²/s. The phase diagram of the system Al-Cu is given below.

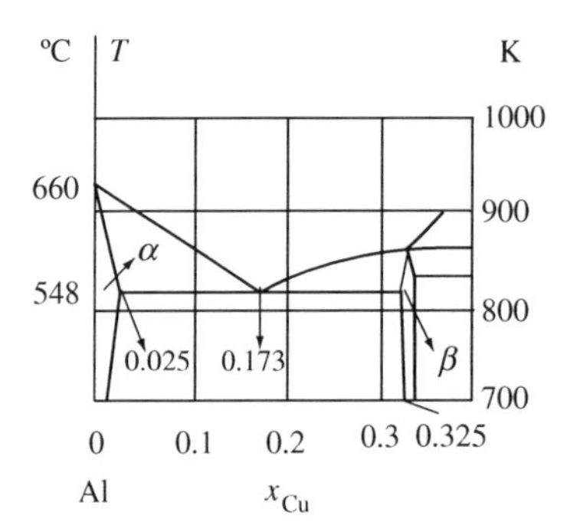

Figure 9.4–1

9.5 You want to estimate the segregation ratio S of Cr, Ni and Mo in an austenitic stainless steel. The ingot that you want to cast is large and you are interested in predicting the segregation of the elements at the centre of the ingot. You know that the cooling rate at the centre will be 5.0 °C/min and that the solidification interval is 40 °C for the alloy. In the literature you have found the dendrite arm spacing 150 µm at the specific cooling rate.

Perform the calculation of the segregation ratios for an alloy with 17 at % Cr, 13 at% Ni and 2 at% Mo. The partition coefficients of Cr, Ni and Mo for austenite and their melts are:

$$x_{Cr}^{\gamma/L} = 0.85 \quad x_{Cr}^{\gamma} = 7.5 \times 10^{-13} \text{m}^2/\text{s}$$

$$x_{Ni}^{\gamma/L} = 0.90 \quad x_{Ni}^{\gamma} = 2.0 \times 10^{-13} \text{m}^2/\text{s}$$

$$x_{Mo}^{\gamma/L} = 0.65 \quad x_{Mo}^{\gamma} = 7.5 \times 10^{-13} \text{m}^2/\text{s}$$

9.6 Solidification experiments on dendritic growth and microsegregation were performed with an AlCu alloy with a Cu concentration of $x_0^L = 0.025$. The composition in the centres of the primary dendrite arms were measured and also the solidification times and the distance between the dendrite arms.

In two of the experiments the following values for the Cu concentration at the centres of the dendrite arms in the table were obtained:

x_{Cu}^s	t_{sol} (s)	distance between the primary arms (μm)
0.013	850	105
0.015	560	75

The Cu concentrations in the dendrite arms deviate from the equilibrium values. They can not be calculated from expected equilibrium values (Scheil's modified equation).

Calculate the Cu concentration values for back diffusion in the centres of the dendrite arms and find out whether the results can be explained by back diffusion or not. If the theory and the experimental results are not compatible, discuss the result.

The diffusion coefficient of Cu in the solid alloy is 1.5×10^{-13} m²/s. The phase diagram of the system Al-Cu is given below.

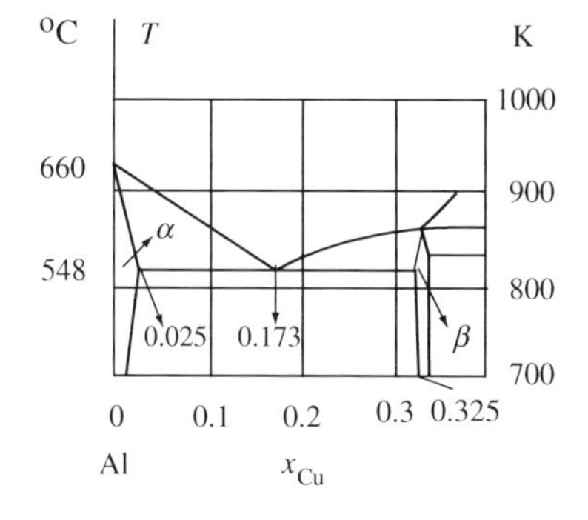

Figure 9.6–1

9.7 The secondary dendrite arm coarsening is discussed in Chapter 9 on pages 556–561. Three different mechanisms are discussed. All of them result in a solution, that can be expressed as [Equation (9.109) on page 559 in Chapter 9]

$$\lambda_{den}^3 - \lambda_0^3 = Ct$$

where

λ_{den} = secondary dendrite arm spacing at time t

λ_0 = dendrite arm distance at time $t = 0$

C = a constant.

Hogan et al has analyzed the coarsening process in AlCu alloys. Their experimental results are shown in the figure below. The results follow the law given above reasonably well.

Derive a value of the constant C. Compare the result with the value calculated from an experiment with a surface tension driven coarsening process, which gave the value $C \approx 1$ provided that the λ values are expressed in the unit 1 μm.

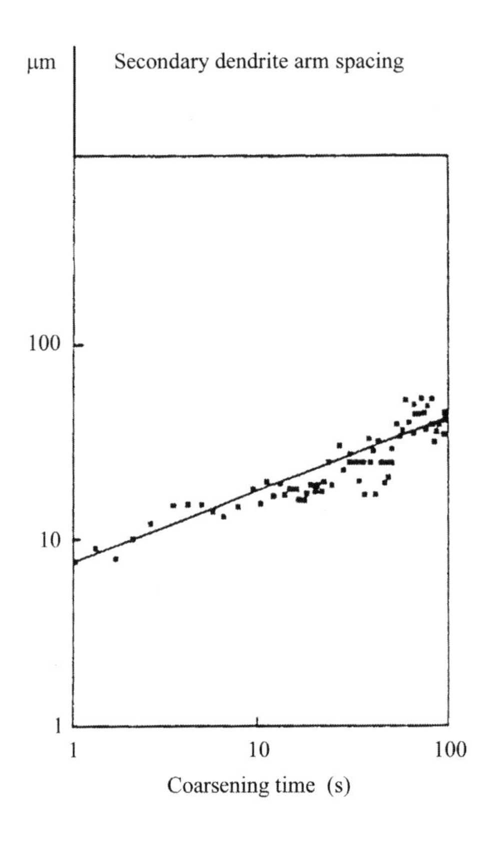

Figure 9.7–1

9.8 Two series of experiments with strong stirring during solidification of an Al alloy with 5.6wt%Cu were performed. During the experiments the samples were quenched at two specified temperatures. Their concentrations of solute Cu in the solid phase and the fraction of the solid phase at quenching were measured. The results are given in the table.

T_{quench} (°C)	c_{Cu} (wt%) at quenching	Volume fraction f of solid
630	0.52	0.70
620	0.92	0.82

On page 191 in [1] it was shown that Scheil's equation alternatively can be written in terms of temperature as

$$1 - f = \left(\frac{T_{\text{M}} - T}{T_{\text{M}} - T_0} \right)^{\frac{-1}{1-k}}$$

where

T_{M} = melting-point temperature of pure aluminium
T_{quench} = quench temperature
T_0 = temperature of the alloy when it starts to solidify
k = partition coefficient derived from the phase diagram below, based on wt%.

Compare the experimental values of the fraction solid phase and the Cu concentration in this phase at quenching with the values calculated from Scheil's equation [the equation above and Equation (9.85) on page 537 in Chapter 9, respectively]. Discuss your results.

The phase diagram of CuAl is given below.

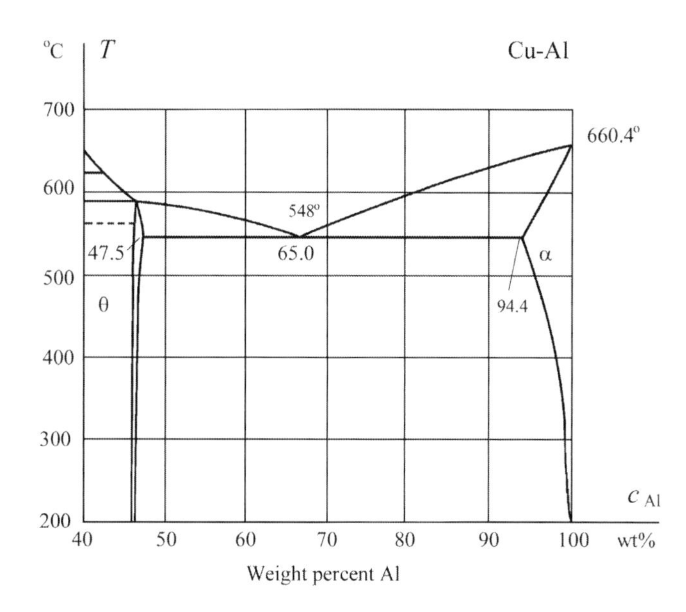

Figure 9.8–1

References

1. H. Fredriksson, U. Åkerlind, *'Physics of Functional Materials'* Chichester, England, John Wiley & Sons, 2008. X74, X96.
2. C. Elbaum, B. Chalmers, *Canadian Journal of Physics*, **33**, 1955, 196.
3. Rosenberg, W. A. Tiller, *Acta Metallurgica*, **5**, 1957, 565.
4. H Fredriksson, M. Hillert, N. Lange, *Journal of the Institute of Metals*, **101**, 1973, 285.
5. W. F. Berg, *Proc. Royal Society*, **164A**, 1938, 79.
6. S. P. F. Humphreys-Owen, *Proc. Royal Society*, **197A**, 1938, 218.
7. R. West, H. Fredriksson, On the mechanism of faceted growth, *Journal of Material Science*, **20**, 1985, 1061–1068.
8. J. C. Fisher, in: *Principles of Solidification*, J. Wiley & Sons, 1964, page 105.
9. J. L. Walker, in: *Principles of Solidification*, J. Wiley & Sons, 1964, page 114.
10. A. Papapetrou, Untersuchungen ű ber Dendritisches Wachstum von X63 Kristallen, *Zeitschrift Die Kristallographie*, **92**, 1935, 89–130.
11. G. P. Ivantsov, *Growth of Crystals*, Consultants Bureau, New York, **1**, 1958, 76.
12. W. Kurz, D. J. Fisher, *Fundamentals of Solidification*. Aedermansdorf, Trans Tech Publications, Switzerland, 1992, 74.
13. L. Bäckerud, *Solidification Characteristics of Aluminium Alloys*, Skanaluminium, Oslo, 1986.
14. M. H. Burden, J. Hunt, Cellular and dendritic growth, *Journal of Crystal Growth*, **22**, 1974, 99–108.
15. G. R. Kotler, W. A. Tiller, Stability of the needle crystals, *Journal of Crystal Growth*, **2**, 1968, 287–307.
16. H. Biloni, B. Chalmers, Predendritic solidification, *Transaction of the Metallurgical Society of AIME*, **233**, 1965, 373–378.
17. A. Hultgren, Crystallization and segregation phenomena in 1.10 percent carbon steel ingots, *Journal of the Iron and Steel Institute*, **120**, 1929, 69–111.
18. I. Larén, H. Fredriksson, Relations between ingot size and Microsgregations, *Scandinavian Journal of Metallurgy*. **1**, 1972, 59–68.
19. G. S. Cole, G. F. Bolling, Transaction of Metallurgy Society. *AIME*, **242**, 1968, 153–154.
20. J. A. Horwath, L. F. Modolfo, Dendritic Growth, *Acta Metallurgica*, **10**, 1962, 1037–1042.
21. T. Z. Kattamis, J. C. Coughlin, M. C. Flemings, Influence of coarsening on dendrite arm spacing of Al-Cu alloys, *Transactions of the Metallurgical Society of AIME*, **239**, 1967.

22. A. Mortensen, On the rate of dendrite arm coarsening, *Metallurgical Transaction*, **22A**, 1991, 569–573.
23. S. C. Huang, M. E. Glickman, *Acta Metallurgica.* **29**, 1981, 717.
24. H. Fredriksson, N. Mahallawy, M Taha, X. Liu, G. Wänglöv, The effect of stirring on the solidification process in metals, *Scandinavian Journal of Metallurgy* **15**, 1986, 127–137.
25. H. Fredriksson, U. Åkerlind, *'Materials Processing during Casting'*, Chichester, England, John Wiley & Sons, 2005.
26. J. Campell, *Castings*, Oxford, Butterworth, 2003, 134.
27. M. H. Burden, J. Hunt, Cellular and dendritic growth, *Journal of Crystal Growth*, **22**, 1974, 99–108.
28. Y. Miyata, T. Suzuki, J. Uno, Cellular and Dendric Growth, *Metallurgical Transaction* **16A**, 1985, 1799–1814.

10

Eutectic Solidification Structures

Solidification and Crystallization Processing in Metals and Alloys, First Edition. Hasse Fredriksson and Ulla Åkerlind.

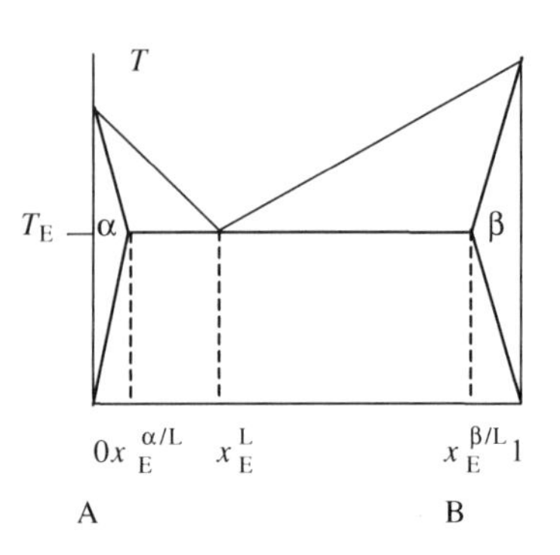

Figure 10.1 Schematic phase diagram of a binary alloy.

$$\text{Fraction of } \alpha = \frac{x_E^{\beta/L} - x_E^L}{x_E^{\beta/L} - x_E^{\alpha/L}}$$

$$\text{Fraction of } \beta = \frac{x_E^L - x_E^{\alpha/L}}{x_E^{\beta/L} - x_E^{\alpha/L}}$$

10.1 Introduction

Eutectic solidification of a liquid is defined as a simultaneous precipitation of two or more phases via a eutectic reaction

$$L \leftrightarrow \alpha + \beta$$

at constant temperature. The solid phases α and β have approximately constant compositions and occur at a constant ratio (Figure 10.1).

Several technically important alloys solidify in this way. During the last decades of the 20th century, eutectic alloys have been used as composite materials for many technical purposes. In all these cases it is important to understand the solidification process and its influence on the alloy structure in order to obtain optimal material properties. In this section the basic concept of eutectic reactions will be discussed. This knowledge will be used to analyze the structures of cast iron and silumin.

10.2 Classification of Eutectic Structures

Eutectic reactions in melts often result in solids with widely different morphologies. Owing to their appearances, they have been given descriptive names such as lamellar, rod, Chinese script, spiral and nodular eutectic structures. Some examples are given in Figures 10.2a-f.

Figure 10.2a Rod-like and plate-like eutectic structures in Cu-Ag.

Figure 10.2b Irregular plate-like eutectic structure in Al-Si.

Figure 10.2c Spiral eutectic structure in Zn-Mg-Zn_2.

Figure 10.2d Irregular plate-like eutectic structure in Cd-Bi.

Figure 10.2e Irregular eutectic structure in Co-TaC.

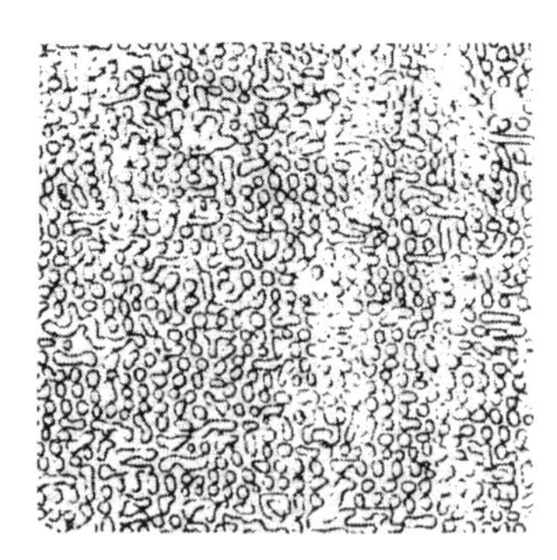

Figure 10.2f Rod-like eutectic structure in $FeFe_2B$.

As two or three of these morphologies can appear in the same system, there are other ways to classify eutectic structures. One way is to classify the eutectic as *normal* or *degenerate* based on its ability to establish cooperation between the two phases during the precipitation.

At a *normal eutectic reaction*, there is a *close cooperation* between the two phases. They grow with a common interface in the melt. The two phases form lamellar- or rod-shaped aggregates, which grow perpendicularly to the interface.

Normal eutectic growth implies that *the two phases grow at the same rate*. Normal eutectic growth will be discussed in Section 10.3.

Characteristic of *degenerate eutectic growth* is that *no cooperation between the phases* is established. A lamellar or rod eutectic cannot be formed if one phase always grows ahead of the other. Degenerate eutectic growth will be discussed briefly in Section 10.4.

In a *degenerate* eutectic reaction, *one of the two phases grows faster than the other one.* Consequently the *growth mechanism* of the single phases determines the structure of the eutectic. The crucial factor is to a large extent the conditions at the interface, i.e. the interface kinetics [a relationship between the growth rate and the concentration difference, for instance $V_{\text{growth}} = \mu(x^{\text{L}} - x_0^{\text{L}})$] of the two phases.

10.3 Normal Eutectic Growth

10.3.1 Normal Lamellar Eutectic Growth

Some binary eutectic structures show a high degree of order between the two precipitated phases, especially lamellar and rod eutectics. Rapid-cooling experiments on such eutectics show that they are formed in a solidification process where the two phases precipitate side by side with a common, almost planar solidification front. The two phases cooperate with each other during the growth. Such a coupled crystallization of the two phases requires less energy and is therefore more favourable than non-coupled crystallization. It will occur faster than precipitation of each phase separately.

Figure 10.3 shows a sketch of a lamellar eutectic structure in a binary alloy containing A and B atoms. A simultaneous precipitation of two phases occurs, an A rich α phase and a B rich β phase. The compositions of the two phases are roughly given by the indicated concentrations in the phase diagram in Figure 10.4.

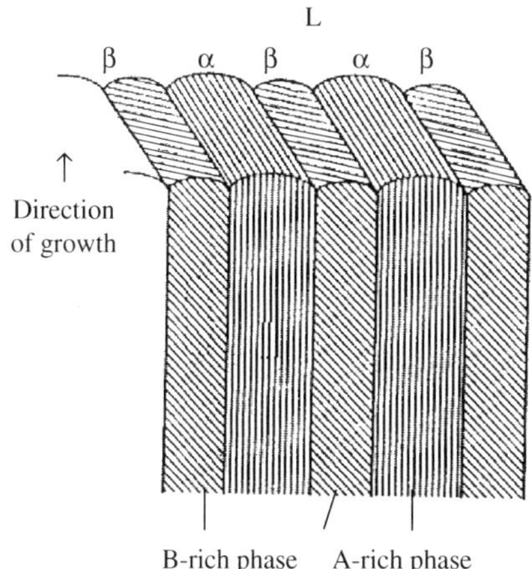

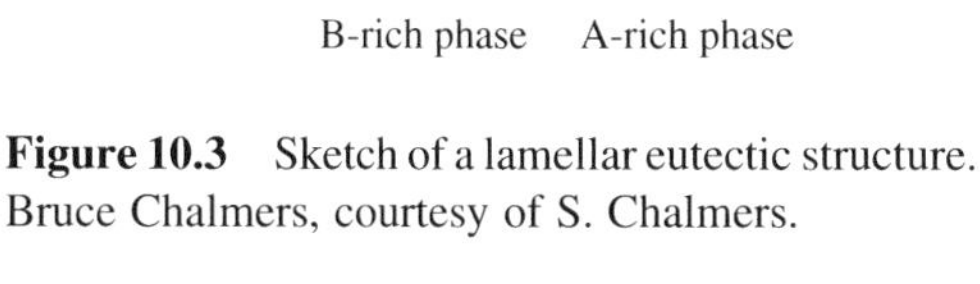

Figure 10.3 Sketch of a lamellar eutectic structure. Bruce Chalmers, courtesy of S. Chalmers.

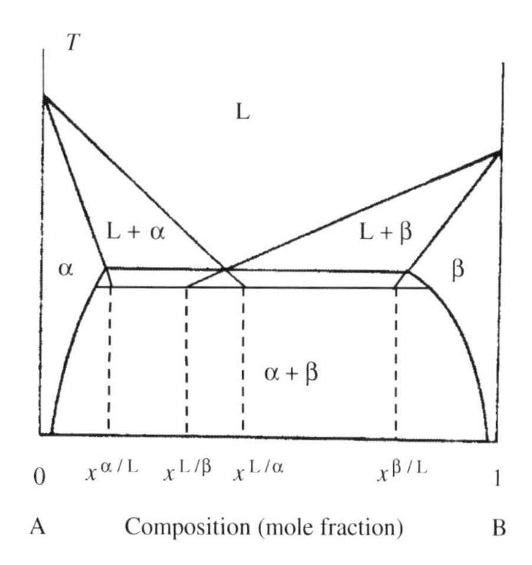

Figure 10.4 Binary eutectic phase diagram.

The solidification front is approximately planar but Figure 10.3 shows that it contains parallel lowered groves perpendicular to the growth direction. They are the meeting lines, intersections between the meeting planes of the two phases and the solidification front, where the liquid and the α and β phases join. The three phases are in *thermodynamic equilibrium* with each other.

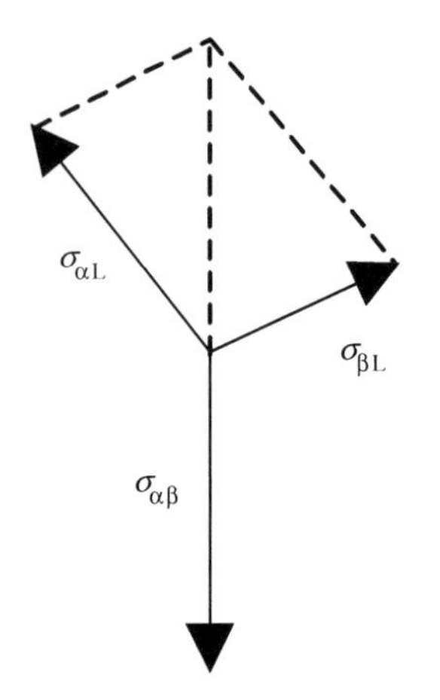

Figure 10.5 Surface tension forces in equilibrium with each other.

The condition for this is that the resultant surface tension force is zero (Figure 10.5). One way to achieve this is *curved* α and β surfaces, which explains the appearance of the lamellar structure.

Another possibility is formation of lattice defects. They change the free energies of the two phases. This results in change of the positions of the liquidus and solidus lines in the phase diagram. During the solidification of an α lamella, B atoms will continuously be rejected back into the melt ahead of the α lamellae front (compare Figure 6.33 on page 308 in Chapter 6). The melt ahead of the β lamellae will be enriched of A atoms for the same reason. Owing to the inhomogeneous atom distribution in the melt close to the interface, B atoms will diffuse from the α lamellae to the β lamellae and A atoms in the opposite direction, which is shown in Figure 10.6. A stationary state is developed.

In Chapter 2 on page 76 we found that the phase boundary curvature influences the phase diagram of the binary alloy (Figure 2.30). The liquidus line is lowered and its slope becomes steeper. It is also possible that lattice defects are formed during the solidification process, which changes the positions of the liquidus and solidus lines.

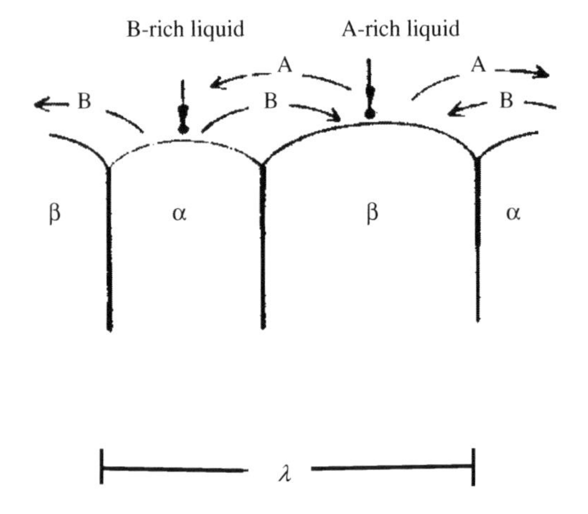

Figure 10.6 Lateral diffusion is a vital part of the growth mechanism of lamellar eutectics. Bruce Chalmers, courtesy of S. Chalmers.

In the present case, it means that the temperature of the solidification front is *lower* than the eutectic temperature. Owing to the curvature of the lamellae the compositions of the two phases are also changed from the values given in Figure 10.4.

These topics will be discussed in connection with the theory of normal lamellar growth in Section 10.3.2. Normal rod eutectic growth, which is the other type of normal eutectic growth, will be discussed in Section 10.3.3. In both cases, the concentration and temperature distributions ahead of the solidification front control the eutectic growth. The diffusion of atoms in the melt and the balance between the surface tension forces at the common intersection line between the three phases lead to a steady state concentration profile in the melt ahead of the solidification front and a temperature distribution, closely related to the concentration distribution.

In Chapter 6 (Section 6.8.2 on page 313) we have discussed planar growth at stationary conditions and constant growth rate and derived the concentration distribution in the melt ahead of the solidification front. The same basic differential equation (Equation (6.119) on page 323 in Chapter 6) in its general form is valid in the case of eutectic growth.

$$D\left(\frac{\partial^2 x^L}{\partial x^2} + \frac{\partial^2 x^L}{\partial y^2} + \frac{\partial^2 x^L}{\partial z^2}\right) + V_{growth}\frac{\partial x^L}{\partial z} = 0 \tag{10.1}$$

where

D = diffusion constant of the A and B atoms in the melt

V_{growth} = the solidification rate in the z direction

x^L = mole fraction of solute in the melt.

In the case of lamellar eutectic growth, Equation (10.1) in its two-dimensional form will be used. For rod eutectics the three-dimensional version has to be applied.

10.3.2 *Simplified Theory of Normal Lamellar Eutectic Growth*

In the analysis below we will consider an alloy with a phase diagram like that in Figure 10.4. We assume that the eutectic reaction in this system occurs in the way described in Section 10.3.1 and shown in Figure 10.6.

Lamellae of the α and β phases grow side by side in the melt. During the growth B atoms will diffuse from the melt, close to the α phase, to the melt ahead of the β phase. A atoms diffuse in the opposite direction according to Figure 10.6. Primarily Hillert [1] and later Jackson and Hunt [2] have carefully investigated this double diffusion pattern. A modification of their theory is given below in a simplified form. In particular we want to show

- that the volume fractions of the two phases α and β are not necessarily equal
- the influence of different slopes of the liquidus lines in the phase diagram on the concentration distribution of the solute in the two phases α and β
- the effect of the interface kinetics on the concentration distribution
- the effect of the surface tension on the concentration distribution
- that the last two effects can be different for the two phases.

Concentration Profile of the Melt

Equation (10.1), which describes the solute distribution in the melt in front of a moving planar interface, can be written as

$$\frac{\partial^2 x^{\mathrm{L}}}{\partial y^2} + \frac{\partial^2 x^{\mathrm{L}}}{\partial z^2} + \frac{V_{\mathrm{growth}}}{D}\frac{\partial x^{\mathrm{L}}}{\partial z} = 0 \tag{10.2}$$

We have to solve this equation in order to find the concentration profile as a function of y and z in the melt. The general solution can be written as

$$x^{\mathrm{L}} = x_0^{\mathrm{L}} + A\,\mathrm{e}^{-\frac{V_{\mathrm{growth}} z}{D}} + B\,\mathrm{e}^{-C z}\sin\frac{2\pi y}{\lambda_{\mathrm{wave}}} \tag{10.3}$$

where

x_0^{L} = the initial composition of the melt
λ_{wave} = wavelength of the sinus function
y = coordinate (see Figure 10.7)
z = coordinate and growth direction.

The first two terms are associated with the planar interface as a whole and the third term concerns the periodic structure of the lamellae. A and B are constants, which have to be determined with the aid of boundary conditions. The solution contains only two arbitrary constants and the positive growth constant C has to be determined.

If we take the derivative of Equation (10.3) twice with respect to y and once and twice with respect to z and insert the derivatives into Equation (10.2) we obtain a second order differential equation of C. The solution of this equation is in this case

$$C = \frac{V_{\mathrm{growth}}}{2D} + \sqrt{\left(\frac{V_{\mathrm{growth}}}{2D}\right)^2 + \left(\frac{2\pi}{\lambda_{\mathrm{wave}}}\right)^2} \tag{10.4}$$

At low growth rates ($V_{\mathrm{growth}}/2D \ll 2\pi/\lambda_{\mathrm{wave}}$) C will be $\approx 2\pi/\lambda_{\mathrm{wave}}$ and we obtain

$$x^{\mathrm{L}} = x_0^{\mathrm{L}} + A\,\mathrm{e}^{-\frac{V_{\mathrm{growth}}z}{D}} + B\,\mathrm{e}^{-\frac{2\pi z}{\lambda_{\mathrm{wave}}}} \sin\frac{2\pi y}{\lambda_{\mathrm{wave}}} \tag{10.5}$$

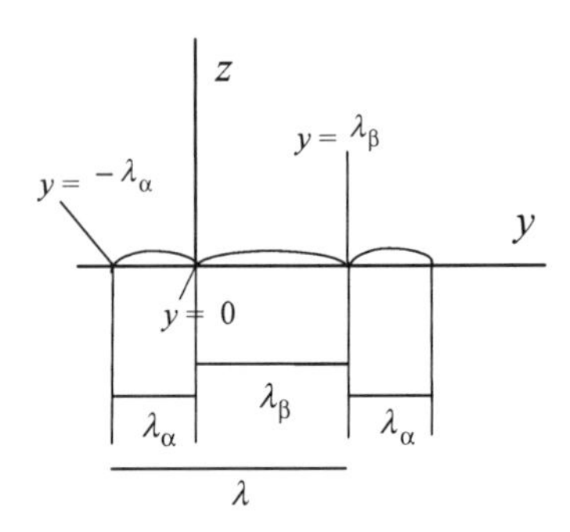

Figure 10.7 $\lambda = \lambda_\alpha + \lambda_\beta$.

We must distinguish between the mathematical wavelength λ_{wave} and the main lamellar distance λ (Figure 10.7). λ_{wave} is the length of a whole wavelength. Hence, $\lambda_{\mathrm{wave}} = 2\lambda$ which will be used later.

The constants A and B are determined by considering the flux conditions for the two eutectic phases. The value of V_{growth}/D normally gives a rapid decrease of the solute concentration along the z axis.

The first two terms in Equation (10.5) describe the concentration profile in the xz plane in analogy with Equation (6.121) on page 325 in Chapter 6. The third term describes the periodic change of solute concentration in the y direction.

In front of the α lamellae and the β lamellae a steady state is developed in terms of solute concentration in the melt ahead of the liquid/solid interface. The concentration profile is shown in Figure 10.8.

If there were no lateral diffusion of A and B atoms, the concentration profile of the melt ahead of the α and β interfaces would be the dotted square function in Figure 10.8 with a discontinuous change at each α/β interface.

However, the lateral diffusion in the melt results in a smooth concentration profile of the B atoms and no discontinuities at the interfaces as Figure 10.8 shows.

Figure 10.8 also indicates that the simple relationships in Equations (10.3) or (10.5) can not describe the concentration profile properly. The reason is that there is one thickness λ_α of the α phase and another thickness λ_β of the β phase. To describe the concentration profile in Figure 10.8 adequately, Fourier series have to be used.

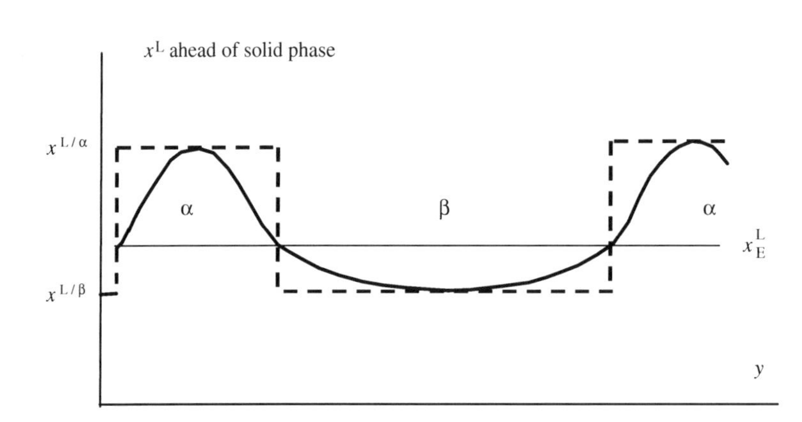

Figure 10.8 Solute concentration (B atoms) in the melt ahead of the solid/liquid interface as a function of y. The dotted square function represents the solute concentration (B atoms) ahead of the α and β lamellae as a function of y with *no* attention to the lateral diffusion. The continuous curve represents the solute concentration (B atoms) as a function of y *with* attention to the lateral diffusion and concentration equilibrium at the α/β interfaces.

In the further analysis we will simplify this demand by choosing *two* different equations, one for the α phase and another one for the β phase. This approach is shown in Figure 10.9.

We apply Equation (10.5) on each phase. In these cases $\lambda_{\mathrm{wave}} = 2\lambda_\alpha$ and $\lambda_{\mathrm{wave}} = 2\lambda_\beta$, respectively, (Margin Figure 10.7 above).

$$x^{\mathrm{L}} = x_0^{\mathrm{L}} + A_\alpha \mathrm{e}^{-\frac{V_{\mathrm{growth}}^{\alpha}z}{D}} + B_\alpha \mathrm{e}^{-\frac{2\pi z}{2\lambda_\alpha}} \sin\frac{2\pi y}{2\lambda_\alpha} \tag{10.6}$$

According to Figure 10.9 we have $-\lambda_\alpha \leq y \leq 0$.

$$x^{\mathrm{L}} = x_0^{\mathrm{L}} + A_\beta \mathrm{e}^{-\frac{V_{\mathrm{growth}}^{\beta}z}{D}} + B_\beta \mathrm{e}^{-\frac{2\pi z}{2\lambda_\beta}} \sin\frac{2\pi y}{2\lambda_\beta} \tag{10.7}$$

According to Figure 10.9 we have $0 \leq y \leq \lambda_\beta$.

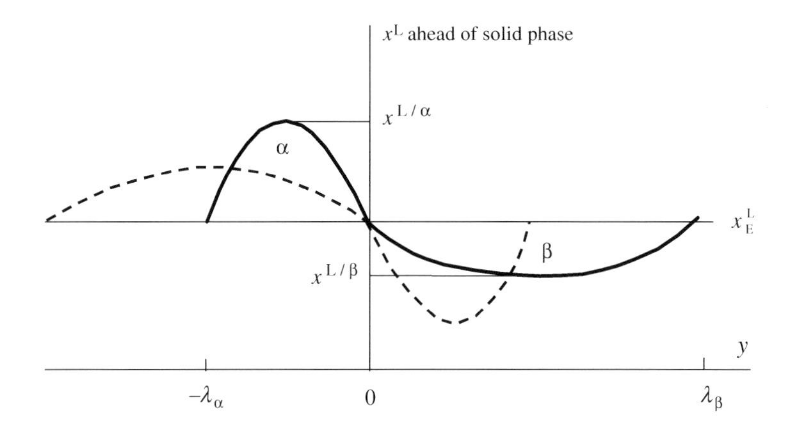

Figure 10.9 Solute concentration (B atoms) in the melt ahead of the solid/liquid interface as a function of y with attention to the lateral diffusion. $z = 0$.

The solute concentrations ahead of each phase (α and β) are represented by different equations.

The following boundary conditions are valid.

For the α phase:	For the β phase:
$z = 0,\ y = 0 \Rightarrow x^L = x_0^L$	$z = 0,\ y = 0 \Rightarrow x^L = x_0^L$
$z = 0,\ y = \dfrac{-\lambda_\alpha}{2} \Rightarrow x^L = x^{L/\alpha}$	$z = 0,\ y = \dfrac{\lambda_\beta}{2} \Rightarrow x^L = x^{L/\beta}$

The expressions for the α phase are inserted into Equation (10.6)

$$x_0^L = x_0^L + A_\alpha \times 1 + B_\alpha \times 1 \times \sin 0 \qquad \Rightarrow$$

and

$$A_\alpha = 0$$

$$x^{L/\alpha} = x_0^L + A_\alpha \times 1 + B_\alpha \times 1 \times \sin\left(-\frac{\pi}{2}\right) \qquad \Rightarrow$$

$$B_\alpha = x_0^L - x^{L/\alpha} \tag{10.8}$$

The expressions for the β phase are introduced into Equation (10.7)

$$x_0^L = x_0^L + A_\beta \times 1 + B_\beta \times 1 \times \sin 0 \qquad \Rightarrow$$

and

$$A_\beta = 0$$

$$x^{L/\beta} = x_0^L + A_\beta \times 1 + B_\beta \times 1 \times \sin\frac{\pi}{2} \qquad \Rightarrow$$

$$B_\beta = x^{L/\beta} - x_0^L \tag{10.9}$$

We insert the calculated A and B values into Equations (10.6) and (10.7), respectively, and obtain

$$x^L = x_0^L - \left(x^{L/\alpha} - x_0^L\right) e^{-\frac{\pi z}{\lambda_\alpha}} \sin\frac{\pi y}{\lambda_\alpha} \qquad (-\lambda_\alpha \le y \le 0) \tag{10.10}$$

$$x^L = x_0^L + \left(x^{L/\beta} - x_0^L\right) e^{-\frac{\pi z}{\lambda_\beta}} \sin\frac{\pi y}{\lambda_\beta} \qquad (0 \le y \le \lambda_\beta) \tag{10.11}$$

In Equations (10.10) and (10.11) the initial composition of the liquid is often set equal to the eutectic composition. However, it is not necessary that x_0^L is exactly equal to x_E^L. This more general case will be discussed below.

Volume Fractions of the α and β Phases

In order to find the mole fractions of the α and β phases we go back to the phase diagram in Figure 10.4 on page 589. The compositions of the two phases $x^{\alpha/L}$ and $x^{\beta/L}$ are close to $x_E^{\alpha/L}$ and $x_E^{\beta/L}$, respectively, (see Figure 10.1) on page 588. The lever rule is applied to find the volume fractions of the two phases.

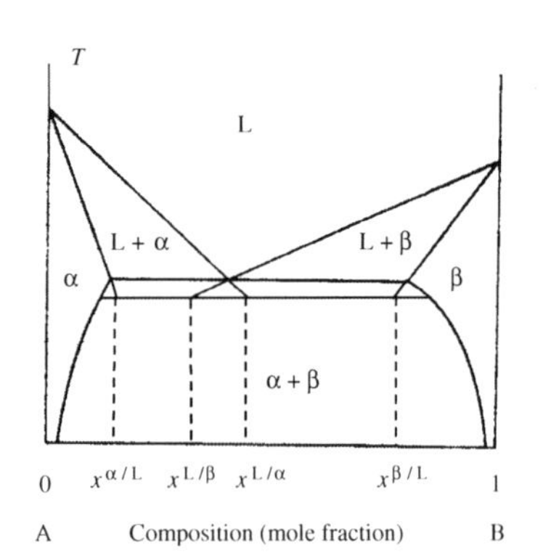

Figure 10.4 Binary eutectic phase diagram.

$$f_\alpha = \frac{x^{\beta/L} - x_0^L}{x^{\beta/L} - x^{\alpha/L}} \approx \frac{x_E^{\beta/L} - x_0^L}{x_E^{\beta/L} - x_E^{\alpha/L}} \tag{10.12}$$

$$f_\beta = \frac{x_0^L - x^{\alpha/L}}{x^{\beta/L} - x^{\alpha/L}} \approx \frac{x_0^L - x_E^{\alpha/L}}{x_E^{\beta/L} - x_E^{\alpha/L}} \tag{10.13}$$

As x_0^L is known and often equal to x_E^L the volume fractions of the two phases can be calculated with the aid of Equations (10.12) and (10.13).

In addition we have

$$f_\alpha = \frac{V_\alpha}{V_\alpha + V_\beta} = \frac{\lambda_\alpha bL}{\lambda_\alpha bL + \lambda_\beta bL} = \frac{\lambda_\alpha}{\lambda}$$

or

$$\lambda_\alpha = f_\alpha \lambda \tag{10.14}$$

In the same way we obtain

$$\lambda_\beta = f_\beta \lambda \tag{10.15}$$

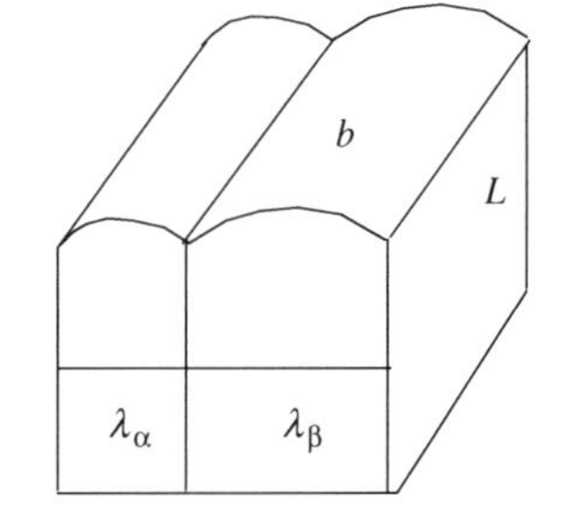

Figure 10.10 α and β lamellae.

By multiplying the volume fractions of the α and β phases with the main lamellar thickness λ we obtain the widths of the α and β phases.

Growth Rates of the α and β Phases

The α and β lamellae grow in the z direction. The solidification process is closely related to the diffusion of the solute (the alloying element) at the solidification front.

The relationship between the growth rate and the composition of the melt at the interfaces α/L and β/L, respectively, controls the mass transport. Fick's first law describes the mass transport, i.e. the diffusion of the solute, at the interface.

We apply Fick's first law (see Equation (6.92) on page 317 in Chapter 6) on the α and β phases

> Fick's First Law
>
> $$\frac{dm}{dt} = -DA\frac{dc}{dz}$$
>
> which can be transformed into
>
> $$\frac{dm}{dt} = -\frac{DA}{V_m^L}\frac{dx^L}{dz}$$

$$\frac{V_{\text{growth}}^\alpha}{V_m^\alpha}\left(x^{L/\alpha} - x^{\alpha/L}\right) = -\frac{D}{V_m^L}\left(\frac{dx^L}{dz}\right)_{z=0}^\alpha \tag{10.16}$$

$$\frac{V_{\text{growth}}^\beta}{V_m^\beta}\left(x^{L/\beta} - x^{\beta/L}\right) = -\frac{D}{V_m^L}\left(\frac{dx^L}{dz}\right)_{z=0}^\beta \tag{10.17}$$

The concentration gradients are found by taking the derivatives of Equations (10.10) and (10.11) on page 593 with respect to z. The concentration gradients vary with the y coordinate along the α and β interfaces. The mass flux over each lamella is found by integration of dx^L/dz with respect to y. The sinus functions in Equations (10.10) and (10.11) have to be replaced by their *average values across the cross-section areas* of the α and β lamellae, respectively, perpendicular to the growth direction. The areas are $b\lambda_\alpha$ and $b\lambda_\beta$, respectively, (see Figure 10.10). The calculations are performed in the boxes on the pages 595 and 596.

Derivation of $\left(\dfrac{\mathrm{d}x^{\mathrm{L}}}{\mathrm{d}z}\right)^{\alpha}_{z=0}$

The derivative is obtained by taking the derivative of Equation (10.10) with respect to z. The y region, which is valid in this case, is shown in the upper figure.

$$\frac{\mathrm{d}x^{\mathrm{L}}}{\mathrm{d}z} = -\left(x^{\mathrm{L}/\alpha} - x_0^{\mathrm{L}}\right)\mathrm{e}^{-\frac{\pi z}{\lambda_\alpha}}\,\frac{-\pi}{\lambda_\alpha}\,\sin\frac{\pi y}{\lambda_\alpha} \qquad (1')$$

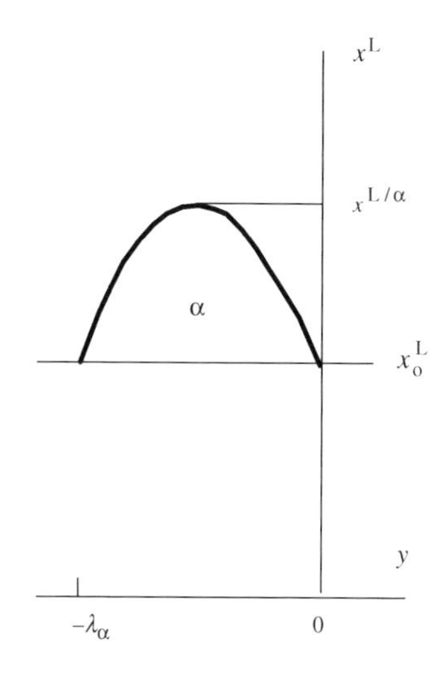

where the average value of the sinus function over the area $b\lambda_\beta$ is calculated from the expression

$$\overline{\sin\frac{\pi y}{\lambda_\alpha}} = \frac{2\displaystyle\int_{-\frac{\lambda_\alpha}{2}}^{0}\sin\frac{\pi y}{\lambda_\alpha}\,b\,\mathrm{d}y}{2\displaystyle\int_{-\frac{\lambda_\alpha}{2}}^{0} b\,\mathrm{d}y} = \frac{2\left[-\cos\dfrac{\pi y}{\lambda_\alpha}\,\dfrac{\lambda_\alpha}{\pi}\right]_{-\frac{\lambda_\alpha}{2}}^{0}}{\lambda_\alpha} = \frac{2[-1-(-0)]}{\lambda_\alpha}\frac{\lambda_\alpha}{\pi} = \frac{-2}{\pi}$$

Inserting $z = 0$ and the value of the average sinus function over the area $b\lambda_\alpha$ into Equation (1′) gives

$$\left(\frac{\mathrm{d}x^{\mathrm{L}}}{\mathrm{d}z}\right)^{\alpha}_{z=0} = -\left(x^{\mathrm{L}/\alpha} - x_0^{\mathrm{L}}\right)\frac{-\pi}{\lambda_\alpha}\,\frac{-2}{\pi}$$

or

$$\left(\frac{\mathrm{d}x^{\mathrm{L}}}{\mathrm{d}z}\right)^{\alpha}_{z=0} = -\left(x^{\mathrm{L}/\alpha} - x_0^{\mathrm{L}}\right)\frac{2}{\lambda_\alpha} \qquad (2')$$

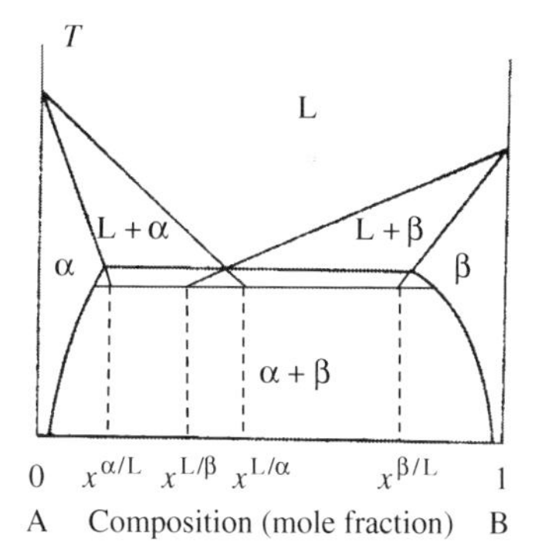

If we combine Equations (10.16) + (2′) and Equations (10.17) + (4′), respectively, we obtain

$$\frac{V^{\alpha}_{\text{growth}}}{V^{\alpha}_{\mathrm{m}}}\left(x^{\mathrm{L}/\alpha} - x^{\alpha/\mathrm{L}}\right) = -\frac{D}{V^{\mathrm{L}}_{\mathrm{m}}}\left[-\left(x^{\mathrm{L}/\alpha} - x_0^{\mathrm{L}}\right)\right]\frac{2}{\lambda_\alpha}$$

$$\frac{V^{\beta}_{\text{growth}}}{V^{\beta}_{\mathrm{m}}}\left(x^{\mathrm{L}/\beta} - x^{\beta/\mathrm{L}}\right) = -\frac{D}{V^{\mathrm{L}}_{\mathrm{m}}}\left(x^{\mathrm{L}/\beta} - x_0^{\mathrm{L}}\right)\frac{-2}{\lambda_\beta}$$

or, after reduction

$$V^{\alpha}_{\text{growth}} = \frac{2D}{\lambda_\alpha}\,\frac{x^{\mathrm{L}/\alpha} - x_0^{\mathrm{L}}}{x^{\mathrm{L}/\alpha} - x^{\alpha/\mathrm{L}}}\,\frac{V^{\alpha}_{\mathrm{m}}}{V^{\mathrm{L}}_{\mathrm{m}}} \qquad (10.18)$$

$$V^{\beta}_{\text{growth}} = \frac{2D}{\lambda_\beta}\,\frac{x^{\mathrm{L}/\beta} - x_0^{\mathrm{L}}}{x^{\mathrm{L}/\beta} - x^{\beta/\mathrm{L}}}\,\frac{V^{\beta}_{\mathrm{m}}}{V^{\mathrm{L}}_{\mathrm{m}}} \qquad (10.19)$$

(a)

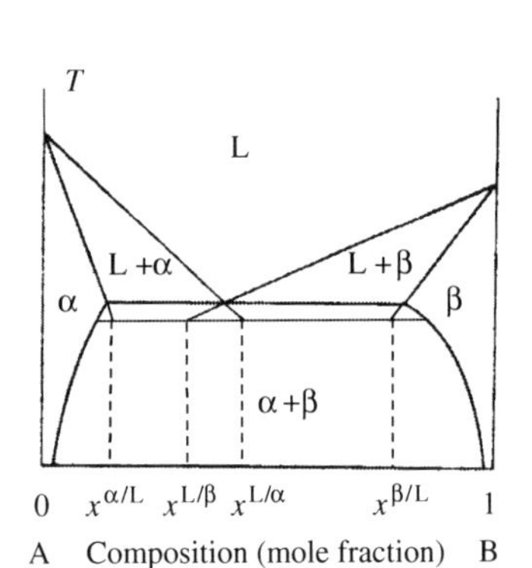

(b)

Derivation of $\left(\frac{\mathrm{d}x^{\mathrm{L}}}{\mathrm{d}z}\right)^{\beta}_{z=0}$

The derivative is obtained by taking the derivative of Equation (10.11) with respect to z. The y region, which is valid in this case, is shown in the upper figure.

$$\frac{\mathrm{d}x^{\mathrm{L}}}{\mathrm{d}z} = \left(x^{\mathrm{L}/\beta} - x_0^{\mathrm{L}}\right) e^{-\frac{\pi z}{\lambda_\beta}} \frac{-\pi}{\lambda_\beta} \sin\frac{\pi y}{\lambda_\beta} \qquad (3')$$

The average value of the sinus function over the area $b\lambda_\beta$ is calculated from the expression

$$\overline{\sin\frac{\pi y}{\lambda_\beta}} = \frac{2\int_0^{\frac{\lambda_\beta}{2}} \sin\frac{\pi y}{\lambda_\beta} b\,\mathrm{d}y}{2\int_0^{\frac{\lambda_\beta}{2}} b\,\mathrm{d}y} = \frac{2\left[-\cos\frac{\pi y}{\lambda_\beta}\frac{\lambda_\beta}{\pi}\right]_0^{\frac{\lambda_\beta}{2}}}{\lambda_\beta} = \frac{2\left[0-(-1)\frac{\lambda_\beta}{\pi}\right]}{\lambda_\beta} = \frac{2}{\pi}$$

Inserting $z = 0$ and the value of the average sinus function into Equation (3′) gives

$$\left(\frac{\mathrm{d}x^{\mathrm{L}}}{\mathrm{d}z}\right)^{\beta}_{z=0} = \left(x^{\mathrm{L}/\beta} - x_0^{\mathrm{L}}\right) \frac{-\pi}{\lambda_\beta}\frac{2}{\pi}$$

or

$$\left(\frac{\mathrm{d}x^{\mathrm{L}}}{\mathrm{d}z}\right)^{\beta}_{z=0} = \left(x^{\mathrm{L}/\beta} - x_0^{\mathrm{L}}\right) \frac{-2}{\lambda_\beta} \qquad (4')$$

where
$V^{\mathrm{i}}_{\mathrm{growth}}$ = growth rate in the z direction of the two lamellae (i = α and β, respectively)
D = diffusion constant of the B atoms in the melt
λ_{i} = width of the lamella i (i = α and β, respectively)
$x^{\mathrm{L/i}}$ = concentration of B atoms in the melt close to the interface L/i (i = α and β, respectively)
$x^{\mathrm{i/L}}$ = concentration of B atoms in the solid phase i close to the interface L/i (i = α and β, respectively)
x_0^{L} = initial concentration of B atoms in the melt.

The main growth rate V_{growth} is controlled by the experimental conditions, i.e. the heat flow.

Driving Force and Supersaturation

The primary cause of solidification is undercooling. In Chapter 2 on pages 43–46 it is stated that the driving force of solidification is proportional to the undercooling. The temperature is coupled to concentration via the phase diagram. Hence, the driving force can be expressed as a function of concentration differences, i.e. in terms of supersaturation.

The mechanism of lamellar eutectic growth, shown in Figure 10.6 on page 590, is closely related to diffusion. The driving force is proportional to the difference between the concentration close to the surface and the concentration far from the interface [Equation (6.137) on page 328 in Chapter 6] The larger the concentration difference is the more rapid will be the solidification and growth rate. Equation (10.18) and (10.19) confirm this statement for lamellar eutectic growth.

Below we will discuss three factors, which influence the driving force of lamellar eutectic growth

- the surface tension
- the speed of the kinetic reaction at the interface
- the diffusion in the melt.

Such influences have been extensively discussed on pages 313–320 in Chapter 6 for a spherical crystal. An equation similar to Equation (6.138) in Chapter 6 on page 333 will be used here for the α phase

$$x^{L/\alpha} - x_0^L = \left(x^{L/\alpha} - x_\sigma^{L/\alpha}\right) + \left(x_\sigma^{L/\alpha} - x_{kin}^{L/\alpha}\right) + \left(x_{kin}^{L/\alpha} - x_0^L\right) \quad (10.20)$$

The corresponding equation for the β phase can be written with the aid of reflection of the phase diagram (every concentration x is replaced by $(1-x)$ and indices α and β are exchanged. This procedure will be frequently used below)

$$\left[(1 - x^{L/\beta}) - (1 - x_0^L)\right] = \left[(1 - x^{L/\beta}) - (1 - x_\sigma^{L/\beta})\right] + \\ + \left[(1 - x_\sigma^{L/\beta}) - (1 - x_{kin}^{L/\beta})\right] + \left[(1 - x_{kin}^{L/\beta}) - (1 - x_0^L)\right]$$

or

$$x^{L/\beta} - x_0^L = \left(x^{L/\beta} - x_\sigma^{L/\beta}\right) + \left(x_\sigma^{L/\beta} - x_{kin}^{L/\beta}\right) + \left(x_{kin}^{L/\beta} - x_0^L\right) \quad (10.21)$$

A comparison between Equation (6.154) on page 332 and Equation (6.136) on page 327, both in Chapter 6, and the margin equation, shows the influence of the kinetic reaction at the interface. If the interface reaction is slow (small value of the growth constant μ), its influence must be considered. This general case will be treated on page 604.

On the other hand, if the interface reaction is very fast (large value of μ) its contribution to the total supersaturation can be neglected. In this particular case we obtain

$$x^{L/\alpha} - x_0^L = \left(x^{L/\alpha} - x_\sigma^{L/\alpha}\right) + \left(x_\sigma^{L/\alpha} - x_0^L\right) \quad (10.22)$$

$$(1 - x^{L/\beta}) - (1 - x_0^L) = \left[(1 - x^{L/\beta}) - (1 - x_\sigma^{L/\beta})\right] + \left[(1 - x_\sigma^{L/\beta}) - (1 - x_0^L)\right] \quad (10.23)$$

Growth of Spherical Crystals

Diffusion only:

$$\frac{dr_c}{dt} = D\frac{x_e^{L/\alpha} - x_0^L}{x_e^{L/\alpha} - x_e^{\alpha/L}}\frac{V_m^\alpha}{V_m^L}\frac{1}{r_c}$$

Diffusion + Kinetic reaction:

$$\frac{dr_c}{dt} = \frac{x_e^{L/\alpha} - x_0^L}{\frac{1}{\mu} + \frac{r_c}{D}\left(x_e^{L/\alpha} - x_e^{\alpha/L}\right)\frac{V_m^L}{V_m^\alpha}}$$

where μ is the growth constant and

$$V_{growth} = \mu\left(x_e^{L/\alpha} - x_0^L\right)$$

Supersaturation Required for Surface Tension

The *first* terms in Equations (10.22) and (10.23) refer to the creation of new curved surfaces. This effect is related to the surface tension balance between the three phases α, β and liquid L (Figure 10.5 on page 590).

The upper surfaces of the solid α and β lamellae in Figure 10.11 do not change during the lamellar growth. The contact area between the α and β lamellae increases by the amount bdz when the solidification front moves by the length dz.

Consider an α lamella. The surface energy required to increase the contact area between the lamellae on *each* side of the α lamella by the amount bdz is $\sigma_{\alpha\beta}bdz$ where $\sigma_{\alpha\beta}$ is the surface energy per unit area.

The required energy (*two* surfaces) is equal to the volume fraction of the driving force of surface formation $(-\Delta G_m^{L\rightarrow\alpha})$. The driving force can be written as $(-\Delta G_m^{L\rightarrow\alpha}) = RT(x^{L/\alpha} - x_\sigma^{L/\alpha})$. Hence, we obtain for the α lamella

$$2\sigma_{\alpha\beta}bdz = \frac{f_\alpha\lambda bdz}{V_m^\alpha}\left(-\Delta G_m^{L\rightarrow\alpha}\right) = \frac{f_\alpha\lambda bdz}{V_m^\alpha}RT\left(x^{L/\alpha} - x_\sigma^{L/\alpha}\right) \quad (10.24a)$$

where the molar volume V_m^α is introduced to find the number of kmoles of the volume element $f_\alpha\lambda bdz$.

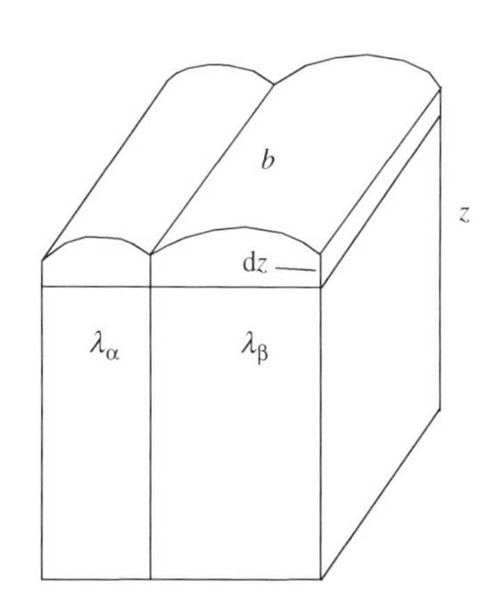

Figure 10.11 α and β lamellae.

In the same way we obtain for the β lamella

$$2\sigma_{\alpha\beta} b\mathrm{d}z = \frac{f_\beta \lambda b\mathrm{d}z}{V_\mathrm{m}^\beta}\left(-\Delta G_\mathrm{m}^{\mathrm{L}\to\beta}\right) = \frac{f_\beta \lambda b\mathrm{d}z}{V_\mathrm{m}^\beta} RT\left[\left(1 - x^{\mathrm{L}/\beta}\right) - \left(1 - x_\sigma^{\mathrm{L}/\beta}\right)\right] \qquad (10.24\mathrm{b})$$

Equations (10.24a and b) can be divided by the factor $b\mathrm{d}z$ and we obtain

$$\frac{RT\left(x^{\mathrm{L}/\alpha} - x_\sigma^{\mathrm{L}/\alpha}\right)}{V_\mathrm{m}^\alpha} = \frac{2\sigma_{\alpha\beta}}{f_\alpha \lambda} \qquad (10.25)$$

$$\frac{RT\left[\left(1 - x^{\mathrm{L}/\beta}\right) - \left(1 - x_\sigma^{\mathrm{L}/\beta}\right)\right]}{V_\mathrm{m}^\beta} = \frac{2\sigma_{\alpha\beta}}{f_\beta \lambda} \qquad (10.26)$$

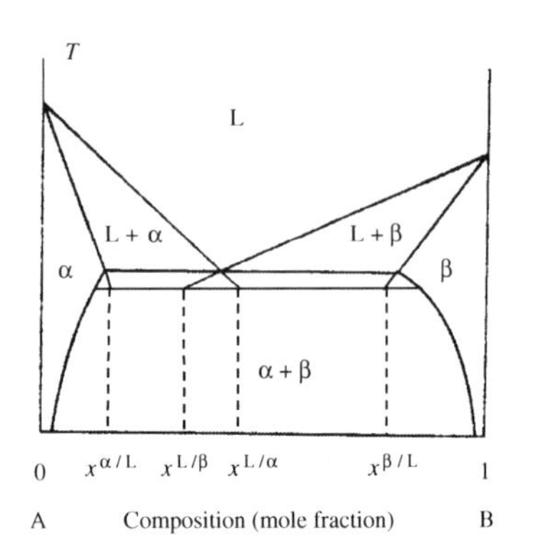

Figure 10.4 Binary eutectic phase diagram.

Supersaturation Required for Diffusion

The *second* terms in Equations (10.22) and (10.23) on page 597 describe the supersaturation necessary for diffusion of the alloy components in the melt. These contributions can be derived by application of Equations (10.18) and (10.19) on page 595 in combination with the relationships $\lambda_\alpha = f_\alpha \lambda$ [Equation (10.14) and Figure 10.4] and $\lambda_\beta = f_\beta \lambda$ [Equation (10.15)] on page 594.

$$V_\mathrm{growth}^\alpha = \frac{2D}{f_\alpha \lambda} \frac{x_\sigma^{\mathrm{L}/\alpha} - x_0^\mathrm{L}}{x_\sigma^{\mathrm{L}/\alpha} - x^{\alpha/\mathrm{L}}} \frac{V_\mathrm{m}^\alpha}{V_\mathrm{m}^\mathrm{L}} \qquad (10.27\mathrm{a})$$

$$V_\mathrm{growth}^\beta = \frac{2D}{f_\beta \lambda} \frac{\left(1 - x_\sigma^{\mathrm{L}/\beta}\right) - \left(1 - x_0^\mathrm{L}\right)}{\left(1 - x_\sigma^{\mathrm{L}/\beta}\right) - \left(1 - x^{\beta/\mathrm{L}}\right)} \frac{V_\mathrm{m}^\beta}{V_\mathrm{m}^\mathrm{L}} \qquad (10.27\mathrm{b})$$

The denominators in the two equations can be simplified to

$$x_\sigma^{\mathrm{L}/\alpha} - x^{\alpha/\mathrm{L}} \approx x_\mathrm{E}^\mathrm{L} - x_\mathrm{E}^{\alpha/\mathrm{L}} = f_\beta\left(x_\mathrm{E}^{\beta/\mathrm{L}} - x_\mathrm{E}^{\alpha/\mathrm{L}}\right)$$

$$\left(1 - x_\sigma^{\mathrm{L}/\beta}\right) - \left(1 - x^{\beta/\mathrm{L}}\right) \approx \left(1 - x_\mathrm{E}^\mathrm{L}\right) - \left(1 - x_\mathrm{E}^{\beta/\mathrm{L}}\right) \approx x_\mathrm{E}^{\beta/\mathrm{L}} - x_\mathrm{E}^\mathrm{L} = f_\alpha\left(x_\mathrm{E}^{\beta/\mathrm{L}} - x_\mathrm{E}^{\alpha/\mathrm{L}}\right)$$

and Equations (10.27a) and (10.27b) can be written as

$$V_\mathrm{growth}^\alpha = \frac{2D}{f_\alpha f_\beta \lambda} \frac{x_\sigma^{\mathrm{L}/\alpha} - x_0^\mathrm{L}}{x_\mathrm{E}^{\beta/\mathrm{L}} - x_\mathrm{E}^{\alpha/\mathrm{L}}} \frac{V_\mathrm{m}^\alpha}{V_\mathrm{m}^\mathrm{L}} \qquad (10.28)$$

$$V_\mathrm{growth}^\beta = \frac{2D}{f_\alpha f_\beta \lambda} \frac{\left(1 - x_\sigma^{\mathrm{L}/\beta}\right) - \left(1 - x_0^\mathrm{L}\right)}{x_\mathrm{E}^{\beta/\mathrm{L}} - x_\mathrm{E}^{\alpha/\mathrm{L}}} \frac{V_\mathrm{m}^\beta}{V_\mathrm{m}^\mathrm{L}} \qquad (10.29)$$

Total Supersaturation

The total supersaturation is the sum of the contributions owing to surface tension and diffusion.

$$x^{\mathrm{L}/\alpha} - x_0^\mathrm{L} = \left(x^{\mathrm{L}/\alpha} - x_\sigma^{\mathrm{L}/\alpha}\right) + \left(x_\sigma^{\mathrm{L}/\alpha} - x_0^\mathrm{L}\right) \qquad (10.22)$$

$$\left(1 - x^{\mathrm{L}/\beta}\right) - \left(1 - x_0^\mathrm{L}\right) = \left[\left(1 - x^{\mathrm{L}/\beta}\right) - \left(1 - x_\sigma^{\mathrm{L}/\beta}\right)\right] + \left[\left(1 - x_\sigma^{\mathrm{L}/\beta}\right) - \left(1 - x_0^\mathrm{L}\right)\right] \qquad (10.23)$$

The expressions for the concentration differences in Equations (10.25)+(10.28) and (10.26)+(10.29), respectively, are introduced into the equations above.

$$x^{\mathrm{L}/\alpha} - x_0^\mathrm{L} = \frac{1}{\lambda}\frac{2\sigma_{\alpha\beta}}{RT}\frac{V_\mathrm{m}^\alpha}{f_\alpha} + \frac{V_\mathrm{growth}^\alpha f_\alpha f_\beta \lambda}{2D}\left(x_\mathrm{E}^{\beta/\mathrm{L}} - x_\mathrm{E}^{\alpha/\mathrm{L}}\right)\frac{V_\mathrm{m}^\mathrm{L}}{V_\mathrm{m}^\alpha} \qquad (10.30)$$

$$x_0^\mathrm{L} - x^{\mathrm{L}/\beta} = \frac{1}{\lambda}\frac{2\sigma_{\alpha\beta}}{RT}\frac{V_\mathrm{m}^\beta}{f_\beta} + \frac{V_\mathrm{growth}^\beta f_\alpha f_\beta \lambda}{2D}\left(x_\mathrm{E}^{\beta/\mathrm{L}} - x_\mathrm{E}^{\alpha/\mathrm{L}}\right)\frac{V_\mathrm{m}^\mathrm{L}}{V_\mathrm{m}^\beta} \qquad (10.31)$$

At normal eutectic growth the two phases grow with the same growth rate:

$$V^{\alpha}_{\text{growth}} = V^{\beta}_{\text{growth}} = V_{\text{growth}} \tag{10.32}$$

This relationship will be used henceforth.

Equations (10.30) and (10.31) contain two unknown quantities: λ and $x^{L/\alpha}$ or $x^{L/\beta}$. The last two concentrations can be replaced by the temperature T as independent variable. Hence, there are two independent variables, λ and T.

Undercooling

Consider a solidifying alloy melt with the concentration x_0^L. As Figure 10.12 shows the undercoolings ΔT_α^L and ΔT_β^L of the melt ahead of the solidification fronts at the α and β phases are different. As a measure of the undercooling we will introduce

$$\Delta T_E = T_E - T \tag{10.33}$$

where T is the temperature of the melt at the solidification front. Equations (10.30) and (10.31) can be written as

$$x^{L/\alpha} - x_0^L = \frac{A_\alpha}{\lambda} + B_\alpha V_{\text{growth}}\,\lambda \tag{10.34a}$$

$$x_0^L - x^{L/\beta} = \frac{A_\beta}{\lambda} + B_\beta V_{\text{growth}}\,\lambda \tag{10.34b}$$

where

$$A_\alpha = \frac{2\sigma_{\alpha\beta}}{RT}\frac{V_m^\alpha}{f_\alpha} \quad \text{and} \quad B_\alpha = \frac{f_\alpha f_\beta V_m^L\left(x_E^{\beta/L} - x_E^{\alpha/L}\right)}{2D}\frac{1}{V_m^\alpha}$$

$$A_\beta = \frac{2\sigma_{\alpha\beta}}{RT}\frac{V_m^\beta}{f_\beta} \quad \text{and} \quad B_\beta = \frac{f_\alpha f_\beta V_m^L\left(x_E^{\beta/L} - x_E^{\alpha/L}\right)}{2D}\frac{1}{V_m^\beta}$$

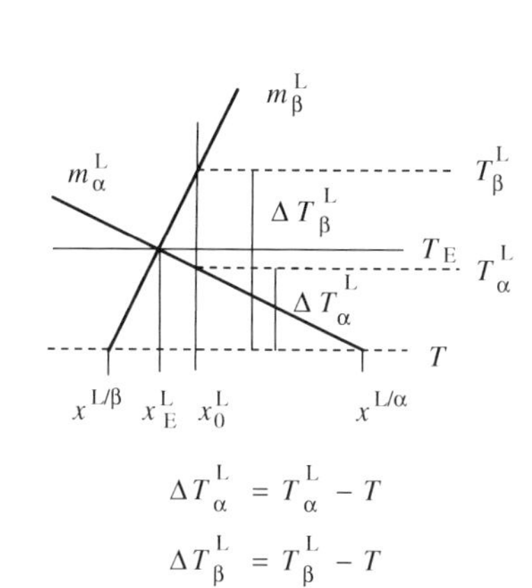

Figure 10.12 Part of the phase diagram close to the eutectic point (temperature T versus concentration x).

The left-hand side of Equations (10.34a) and (10.34b) can be split up into two terms:

$$\left(x^{L/\alpha} - x_E^L\right) + \left(x_E^L - x_0^L\right) = \frac{A_\alpha}{\lambda} + B_\alpha V_{\text{growth}}\lambda \tag{10.35}$$

$$\left(x_E^L - x^{L/\beta}\right) - \left(x_E^L - x_0^L\right) = \frac{A_\beta}{\lambda} + B_\beta V_{\text{growth}}\lambda \tag{10.36}$$

With the aid of the slopes m_α^L and m_β^L of the liquidus lines in the phase diagram and the undercooling $\Delta T_E = (T_E - T)$ we obtain

$$\frac{-\Delta T_E}{m_\alpha^L} + \left(x_E^L - x_0^L\right) = \frac{A_\alpha}{\lambda} + B_\alpha V_{\text{growth}}\lambda \tag{10.37}$$

$$\frac{\Delta T_E}{m_\beta^L} - \left(x_E^L - x_0^L\right) = \frac{A_\beta}{\lambda} + B_\beta V_{\text{growth}}\lambda \tag{10.38}$$

Equation (10.37) is multiplied by $(-m_\alpha^L)$ and Equation (10.38) is multiplied by m_β^L. The resulting equations are

$$\Delta T_E - m_\alpha^L\left(x_E^L - x_0^L\right) = -m_\alpha^L\frac{A_\alpha}{\lambda} - m_\alpha^L B_\alpha V_{\text{growth}}\lambda \tag{10.39}$$

$$\Delta T_E - m_\beta^L\left(x_E^L - x_0^L\right) = m_\beta^L\frac{A_\beta}{\lambda} + m_\beta^L B_\beta V_{\text{growth}}\lambda \tag{10.40}$$

The Equations (10.39) and (10.40) are added and the undercooling ΔT_E is derived. The result is

$$\Delta T_E = \frac{1}{2}\left(m_\alpha^L + m_\beta^L\right)\left(x_E^L - x_0^L\right) + \frac{1}{2\lambda}\left(m_\beta^L A_\beta - m_\alpha^L A_\alpha\right) + \frac{V_{growth}\,\lambda}{2}\left(m_\beta^L B_\beta - m_\alpha^L B_\alpha\right) \qquad (10.41)$$

The expressions for A_α, A_β, B_α and B_β on page 599 between Equations (10.34) and (10.35) are inserted into Equation (10.41). After reduction we obtain

$$\Delta T_E = \frac{1}{2}\left(m_\alpha^L + m_\beta^L\right)\left(x_E^L - x_0^L\right) + \frac{1}{2\lambda}\frac{2\sigma_{\alpha\beta}}{RT}\left(\frac{m_\beta^L V_m^\beta}{f_\beta} - \frac{m_\alpha^L V_m^\alpha}{f_\alpha}\right) + \frac{V_{growth}\,\lambda}{2}\frac{f_\alpha f_\beta V_m^L\left(\left(x_E^{\beta/L} - x_E^{\alpha/L}\right)\right.}{2D}\left(\frac{m_\beta^L}{V_m^\beta} - \frac{m_\alpha^L}{V_m^\alpha}\right) \qquad (10.42)$$

Equation (10.42) represents the undercooling ($T_E - T$) as a function of known quantities and the growth rate of the solidification front.

Relationship between Growth Rate and Main Lamellar Distance

The growth rate $V_{growtth}$ is in most cases determined by external factors mainly the heat flow, which often follows a parabolic growth law

$$V_{growth} = const/\sqrt{t} = Const/l \qquad (10.43)$$

where

l = thickness of the solidified shell
t = time.

The undercooling is adapted to the growth rate with the aid of Equation (10.42). Below we will derive a relationship between the main lamellar distance λ and the growth rate.

The energy law, applied on a solidification process, says that *the process will win, which requires a minimum of energy*. This means minimum undercooling. The growth rate is determined by external factors and is not influenced by the main lamellar distance. Instead the lamellar distance has to adapt the value, which agrees with the minimum condition of the undercooling.

If we take the derivative of Equation (10.41) with respect to λ we obtain

$$\frac{d\Delta T_E}{d\lambda} = 0 - \frac{1}{2\lambda^2}\left(m_\beta^L A_\beta - m_\alpha^L A_\alpha\right) + \frac{V_{growth}}{2}\left(m_\beta^L B_\beta - m_\alpha^L B_\alpha\right)$$

The condition for a minimum is that the derivative is zero and that the second derivative is positive. The latter condition is fulfilled in this case. The former condition can be written as

$$-\frac{1}{2\lambda^2}\left(m_\beta^L A_\beta - m_\alpha^L A_\alpha\right) + \frac{V_{growth}}{2}\left(m_\beta^L B_\beta - m_\alpha^L B_\alpha\right) = 0$$

which gives

$$V_{growth}\,\lambda^2 = \frac{m_\beta^L A_\beta - m_\alpha^L A_\alpha}{m_\beta^L B_\beta - m_\alpha^L B_\alpha} \qquad (10.44)$$

After introduction of the expressions A_α, A_β, B_α and B_β on page 600 into Equation (10.44) and reduction we obtain

$$V_{growth}\,\lambda^2 = \frac{\dfrac{2\sigma_{\alpha\beta}}{RT}}{\dfrac{f_\alpha f_\beta V_m^L\left(x_E^{\beta/L} - x_E^{\alpha/L}\right)}{2D}} \cdot \frac{\dfrac{m_\beta^L V_m^\beta}{f_\beta} - \dfrac{m_\alpha^L V_m^\alpha}{f_\alpha}}{\dfrac{m_\beta^L}{V_m^\beta} - \dfrac{m_\alpha^L}{V_m^\alpha}} \qquad (10.45)$$

Apart from the λ^2 factor the right-hand side of Equation (10.45) consists of known quantities, which are constant. Table 10.1 and the equation can be written as

$$V_{\text{growth}}\,\lambda^2 = C_{\text{lamella}} \tag{10.46}$$

where

$$C_{\text{lamella}} = \frac{2\sigma_{\alpha\beta}2D}{RT}\,\frac{A_{\text{lamella}}}{C_{0\,\text{lamella}}B_{\text{lamella}}} \tag{10.47}$$

The main lamellar distance is adapted to the growth rate with the aid of Equation (10.46). The growth rate is determined by experimental conditions i.e. the heat flow and the cooling conditions.

Equation (10.46) has the same shape as the corresponding relationship for dendritic growth. The only difference is the interpretation of λ, which is the dendrite arm distance and the main lamellar distance, respectively.

Table 10.1 Constants of normal lamellar eutectic growth

$A_{\text{lamella}} = \dfrac{m_\beta^{\text{L}} V_{\text{m}}^\beta}{f_\beta} - \dfrac{m_\alpha^{\text{L}} V_{\text{m}}^\alpha}{f_\alpha}$
$B_{\text{lamella}} = \dfrac{m_\beta^{\text{L}}}{V_{\text{m}}^\beta} - \dfrac{m_\alpha^{\text{L}}}{V_{\text{m}}^\alpha}$
$C_{0\,\text{lamella}} = f_\alpha f_\beta V_{\text{m}}^{\text{L}}\left(x_{\text{E}}^{\beta/\text{L}} - x_{\text{E}}^{\alpha/\text{L}}\right)$

Example 10.1

At an experiment with unidirectional solidification of a Sn-Pb alloy with eutectic composition the phases show normal lamellar eutectic growth at the eutectic temperature.

a) Calculate the value of the constant in Equation (10.46) with the aid of the phase diagram and the given material and other constants.
b) Compare your result with the values for some other alloys.

Experience shows that the surface tension between two solid metal surfaces varies from 0.05 J/m^2 to 0.25 J/m^2. In this case, it is reasonable to choose the value 0.1 J/m^2 for the surface tension between the solid α and β lamellae.

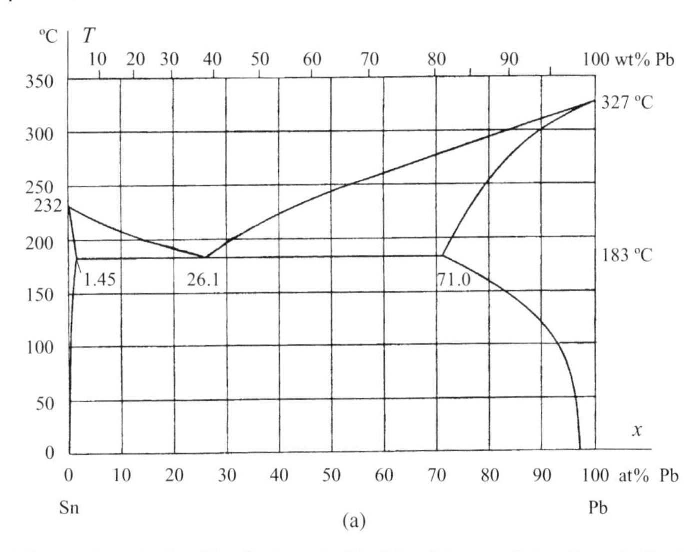

The α phase is Sn. The β phase is Pb. Max Hansen © McGraw Hill 1958.

Material constants

$\sigma_{\alpha\beta} = 0.1$ J/m^2
$D = 3 \times 10^{-9}$ m^2/s
Experimental molar volumes at temperature 183 °C:
$V_{\text{m}}^{\text{L}} = 17.4 \times 10^{-3}$ m^3/kmol
$V_{\text{m}}^{\alpha} = 16.9 \times 10^{-3}$ m^3/kmol
$V_{\text{m}}^{\beta} = 19.2 \times 10^{-3}$ m^3/kmol

Solution:

a)

The eutectic solidification starts at the temperature $T_{\text{E}} = 183\,°\text{C}$ and the alloy has eutectic composition.

The constant C_{lamella} in Equation (10.46) can be calculated with the aid of Equation (10.47):

$$C_{\text{lamella}} = \frac{2\sigma_{\alpha\beta} \times 2D}{RT f_\alpha f_\beta V_{\text{m}}^{\text{L}}\left(x_{\text{E}}^{\beta/\text{L}} - x_{\text{E}}^{\alpha/\text{L}}\right)}\;\frac{\dfrac{m_\beta^{\text{L}} V_{\text{m}}^\beta}{f_\beta} - \dfrac{m_\alpha^{\text{L}} V_{\text{m}}^\alpha}{f_\alpha}}{\dfrac{m_\beta^{\text{L}}}{V_{\text{m}}^\beta} - \dfrac{m_\alpha^{\text{L}}}{V_{\text{m}}^\alpha}} \tag{1'}$$

If numerical values are inserted into Equation (1') the right-hand side of the equation gives the desired value of the constant. Hence, we must find numerical values of all the quantities. Many of them are obtained from the phase diagram.

The slopes of the liquidus lines can be derived with the aid of the phase diagram. The construction is shown in the figure below.

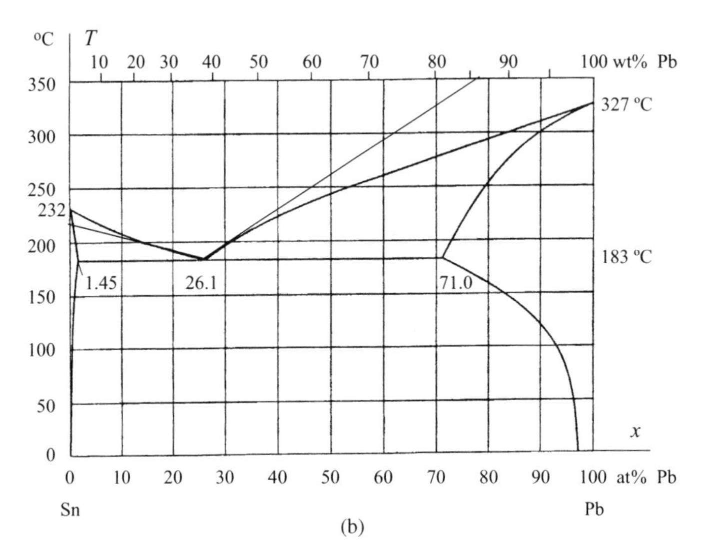

(b)

(Max Hansen © McGraw Hill 1958.)

The result is

$$m_\alpha^L = \frac{183 - 217}{0.261} = -130\text{ K} \quad \text{and} \quad m_\beta^L = \frac{350 - 183}{0.785 - 0.261} = 319\text{ K}$$

The values below are read from the phase diagram.

$$T_E = 183 + 273 = 456\text{ K}$$

$$x_0^L = x_E^L = 0.261 \text{ (mole fraction)}$$

$$x_E^{\alpha/L} = 0.015 \text{ (mole fraction)}$$

$$x_E^{\beta/L} = 0.710 \text{ (mole fraction)}$$

In addition we have

$$x^{L/\alpha} \approx x^{L/\beta} \approx x_E^L = 0.261 \text{ (mole fraction)}$$

f_α and f_β are calculated from the phase diagram with the aid of the lever rule. The α phase consists of Sn and the β phase of Pb.

$$f_\alpha = \frac{71.0 - 26.1}{71.0 - 1.5} = 0.646 \quad \text{and} \quad f_\beta = \frac{26.1 - 1.5}{71.0 - 1.5} = 0.354$$

All the values are inserted into Equation (1'):

$$C_{\text{lamella}} = \frac{2\sigma_{\alpha\beta}2D}{RT\, f_\alpha f_\beta V_m^L \left(x_E^{\beta/L} - x_E^{\alpha/L}\right)} \frac{\dfrac{m_\beta^L V_m^\beta}{f_\beta} - \dfrac{m_\alpha^L V_m^\alpha}{f_\alpha}}{\dfrac{m_\beta^L}{V_m^\beta} - \dfrac{m_\alpha^L}{V_m^\alpha}}$$

or

$$C_{\text{lamella}} = \frac{2\sigma_{\alpha\beta}2D}{RT}\frac{A}{C_{\text{o}}B} = \frac{2 \times 0.1 \times 2 \times 3 \times 10^{-9}}{8.31 \times 10^3 \times 456} \times \frac{20.7}{2.76 \times 10^{-3} \times 2.43 \times 10^4}$$
$$= 0.976 \times 10^{-16}\ \text{m}^3/\text{s}$$

Experimental values of $C_{\text{lamella}} = V_{\text{growth}}\lambda^2$ for some eutectic alloys, read from a diagram with logarithim scales on both axes (P Haasen) [3]

Alloy	$V_{\text{growth}}\,\lambda^2$ (m³/s)
Al-Zn	$\approx 0.8 \times 10^{-16}$
Sn-Zn	$\approx (0.5 - 0.6) \times 10^{-16}$
Sn-Pb	$\approx 0.4 \times 10^{-16}$

Answer:

a) The value of the constant C_{lamella} for the eutectic SnPb alloy is 1×10^{-16} m³/s.
b) See the margin table. The calculated value seems to be fairly reasonable.

Relationship between Growth Rate and Undercooling

With the aid of Equation (10.42) on page 600 and Equation (10.46) on page 601 we can derive a simple relationship between the growth rate V_{growth} and the undercooling ΔT_{E}. Equation (10.42) can be written as

$$\Delta T_{\text{E}} = \frac{1}{2}\left(m_{\alpha}^{\text{L}} + m_{\beta}^{\text{L}}\right)\left(x_{\text{E}}^{\text{L}} - x_0^{\text{L}}\right) + \frac{1}{2\lambda}\frac{2\sigma_{\alpha\beta}}{RT}\left(\frac{m_{\beta}^{\text{L}}V_{\text{m}}^{\beta}}{f_{\beta}} - \frac{m_{\alpha}^{\text{L}}V_{\text{m}}^{\alpha}}{f_{\alpha}}\right)$$
$$+ \frac{V_{\text{growth}}\lambda}{2}\frac{f_{\alpha}f_{\beta}V_{\text{m}}^{\text{L}}\left(x_{\text{E}}^{\beta/\text{L}} - x_{\text{E}}^{\alpha/\text{L}}\right)}{2D}\left(\frac{m_{\beta}^{\text{L}}}{V_{\text{m}}^{\beta}} - \frac{m_{\alpha}^{\text{L}}}{V_{\text{m}}^{\alpha}}\right) \tag{10.42}$$

If we replace $V_{\text{growth}}\,\lambda^2$ by the constant C_{lamella} and replace the main lamellar distance λ by $\sqrt{C_{\text{lamella}}/V_{\text{growth}}}$ in Equation (10.42) we obtain

$$\Delta T_{\text{E}} = \frac{1}{2}\left(m_{\alpha}^{\text{L}} + m_{\beta}^{\text{L}}\right)\left(x_{\text{E}}^{\text{L}} - x_0^{\text{L}}\right) + \frac{\sqrt{V_{\text{growth}}}}{2\sqrt{C_{\text{lamella}}}}\frac{2\sigma_{\alpha\beta}}{RT}\left(\frac{m_{\beta}^{\text{L}}V_{\text{m}}^{\beta}}{f_{\beta}} - \frac{m_{\alpha}^{\text{L}}V_{m}^{\alpha}}{f_{\alpha}}\right)$$
$$+ \frac{\sqrt{V_{\text{growth}}}\sqrt{C_{\text{lamella}}}}{2}\frac{f_{\alpha}f_{\beta}V_{\text{m}}^{\text{L}}\left(x_{\text{E}}^{\beta/\text{L}} - x_{\text{E}}^{\alpha/\text{L}}\right)}{2D}\left(\frac{m_{\beta}^{\text{L}}}{V_{\text{m}}^{\beta}} - \frac{m_{\alpha}^{\text{L}}}{V_{\text{m}}^{\alpha}}\right) \tag{10.48}$$

The first term in Equation (10.48) is small and can normally be neglected. In this case, Equation (10.48) can be written as

$$\Delta T_{\text{E}} = const_{\text{lamella}} \times \sqrt{V_{\text{growth}}} \tag{10.49}$$

or

$$V_{\text{growth}} = Const_{\text{lamella}} \times (\Delta T_{\text{E}})^2 \tag{10.50}$$

The predictions of the theory can be checked by comparison with experiments. As an example, we choose the eutectic alloy SnPb and use the values and results in Example 10.1.

Chadwick [4] has studied the undercooling as a function of the growth rate experimentally for a eutectic SnPb alloy and plotted the undercooling as a function of $\sqrt{V_{\text{growth}}}$. The plot is shown in Figure 10.13. The value of the constant can alternatively be calculated from Equation (10.48) when the first term is zero.

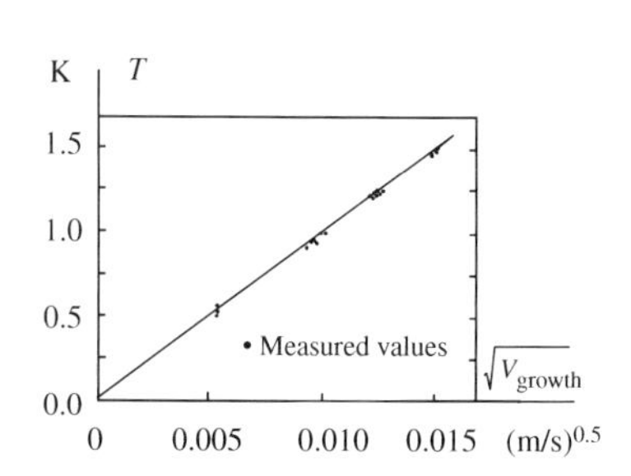

Figure 10.13 Undercooling ΔT_{E} as a function of $\sqrt{V_{\text{growth}}}$ for the eutectic alloy SnPb.

The slope of the line in the figure is $100\ \text{K}\,\text{m}^{-0.5}\,\text{s}^{0.5}$.

$$const_{\text{lamella}} = \frac{\Delta T_{\text{E}}}{\sqrt{V_{\text{growth}}}} \approx \frac{\frac{\sigma_{\alpha\beta}}{RT}\left(\frac{m_{\beta}^{\text{L}}V_{\text{m}}^{\beta}}{f_{\beta}} - \frac{m_{\alpha}^{\text{L}}V_{\text{m}}^{\alpha}}{f_{\alpha}}\right)}{\sqrt{C_{\text{lamella}}}} + \frac{\sqrt{C_{\text{lamella}}}f_{\alpha}f_{\beta}V_{\text{m}}^{\text{L}}\left(x_{\text{E}}^{\beta/\text{L}} - x_{\text{E}}^{\alpha/\text{L}}\right)}{4D}\left(\frac{m_{\beta}^{\text{L}}}{V_{\text{m}}^{\beta}} - \frac{m_{\alpha}^{\text{L}}}{V_{\text{m}}^{\alpha}}\right) \tag{10.51}$$

$const_{lamella}$ for eutectic SnPb is

$$\sqrt{\frac{0.1 \times 20.7 \times 2.43 \times 10^4 \times 2.76 \times 10^{-2}}{8.31 \times 10^3 \times 456 \times 3 \times 10^{-9}}}$$

or

$$const_{SnPb} = 1.1 \times 10^2 \text{ Km}^{-0.5} \text{ s}^{0.5}$$

The expression for the constant in Equation (10.51) can be transformed into a more compact shape with the aid of the shortenings $A_{lamella}$, $B_{lamella}$ and $C_{0\ lamella}$, introduced on page 601 and used in Example 10.1 on page 601. The calculations give

$$const_{lamella} = \sqrt{\frac{\sigma_{\alpha\beta} A_{lamella}\ B_{lamella}\ C_{0\ lamella}}{RTD}} \tag{10.52}$$

The value of the constant for eutectic SnPb has been calculated in the margin box.

The agreement between the experimental value and the theoretical calculation is satisfactory. We must keep in mind that the numerical values of the surface tension and the diffusion constant are not very accurate.

With the aid of Equations (10.49) and (10.52) Equation (10.48) can be written as

Table 10.1 Constants of normal lamellar eutectic growth

$$A_{lamella} = \frac{m_\beta^L V_m^\beta}{f_\beta} - \frac{m_\alpha^L V_m^\alpha}{f_\alpha}$$

$$B_{lamella} = \frac{m_\beta^L}{V_m^\beta} - \frac{m_\alpha^L}{V_m^\alpha}$$

$$C_{0\ lamella} = f_\alpha f_\beta V_m^L \left(x_E^{\beta/L} - x_E^{\alpha/L}\right)$$

$$\Delta T_E = \frac{1}{2}\left(m_\alpha^L + m_\beta^L\right)\left(x_E^L - x_0^L\right) + const_{lamella}\sqrt{V_{growth}} \tag{10.53}$$

or

$$\Delta T_E = \frac{1}{2}\left(m_\alpha^L + m_\beta^L\right)\left(x_E^L - x_0^L\right) + \sqrt{\frac{\sigma_{\alpha\beta} A_{lamella}\ B_{lamella}\ C_{0\ lamella}}{RTD}}\sqrt{V_{growth}} \tag{10.54}$$

Influence of Interface Kinetics on Growth Rate

The expression in Equation (10.45) on page 600 represents the growth rate of the solidification process under the influence of surface tension and diffusion but with no influence of interface kinetics. In many eutectic reactions one of the solid phases (let us call it α) obtains a faceted structure, which indicates a strong interface kinetic influence. In such cases, it is necessary to pay attention to the interface kinetics.

Relationship between Growth Rate and Undercooling when Interface Kinetics is Considered

In order to analyze the effect of interface kinetics on the growth rate theoretically we have to go back to Equations (10.20) and (10.21) on page 597. No change is necessary for the β phase. The second term in Equation (10.20) can be written as

Basic formula

$$V_{growth} = \mu_{kin}^\alpha \left(x_\sigma^{L/\alpha} - x_{kin}^{L/\alpha}\right)^{n_{kin}^\alpha}$$

For $n_{kin}^\alpha = 1$ we obtain

$$x_\sigma^{L/\alpha} - x_{kin}^{L/\alpha} = \frac{V_{growth}}{\mu_{kin}^\alpha}$$

For $n_{kin}^\alpha = 2$ we obtain

$$x_\sigma^{L/\alpha} - x_{kin}^{L/\alpha} = \sqrt{\frac{V_{growth}}{\mu_{kin}^\alpha}}$$

$$x_\sigma^{L/\alpha} - x_{kin}^{L/\alpha} = \left(\frac{V_{growth}}{\mu_{kin}^\alpha}\right)^{\frac{1}{n_{kin}^\alpha}} \tag{10.55}$$

where μ_{kin}^α is the kinetic growth constant. n_{kin}^α is a constant number between 1 and 2. Theory predicts that $n_{kin}^\alpha = 1$ but experiments indicate that n_{kin}^α is close to 2 for most alloys. Hence, it is reasonable to use the lower expression in the margin box. If this expression is introduced into Equation (10.20), Equation (10.30) on page 598 will be replaced by

$$x^{L/\alpha} - x_0^L = \frac{2\sigma_{\alpha\beta}}{f_\alpha \lambda}\frac{V_m^\alpha}{RT} + \sqrt{\frac{V_{growth}}{\mu_{kin}^\alpha}} + \frac{V_{growth} f_\alpha f_b \lambda}{2D}\left(x_E^{\beta/L} - x_E^{\alpha/L}\right)\frac{V_m^L}{V_m^\alpha} \tag{10.56}$$

We use the same procedure as on pages 599–600 to derive a new equation for the undercooling, that will replace Equation (10.42). The result is

$$\Delta T_E = \frac{1}{2}\left(m_\alpha^L + m_\beta^L\right)\left(x_E^L - x_0^L\right) + \frac{1}{2\lambda}\frac{2\sigma_{\alpha\beta}}{RT}\left(\frac{m_\beta^L V_m^\beta}{f_\beta} - \frac{m_\alpha^L V_m^\alpha}{f_\alpha}\right)$$
$$-\frac{m_\alpha^L}{2}\sqrt{\frac{V_{growth}}{\mu_{kin}^\alpha}} + \frac{V_{growth}\lambda}{2}\frac{f_\alpha f_\beta V_m^L\left(x_E^{\beta/L} - x_E^{\alpha/L}\right)}{2D}\left(\frac{m_\beta^L}{V_m^\beta} - \frac{m_\alpha^L}{V_m^\alpha}\right) \quad (10.57)$$

Equation (10.57) is the modified relationship between the undercooling and the growth rate when the kinetic reaction at the interface is considered. In analogy with Equation (10.53) Equation (10.57) can alternatively be written as

$$\Delta T_E = \frac{1}{2}\left(m_\alpha^L + m_\beta^L\right)\left(x_E^L - x_0^L\right) + \left(const_{lamella} - \frac{m_\alpha^L}{2\sqrt{\mu_{kin}^\alpha}}\right)\sqrt{V_{growth}} \quad (10.58)$$

Relationship between Growth Rate and Main Lamellar Distance when Interface Kinetics is Considered

To find the relationship between the growth rate and the main lamellar distance we proceed in the same way as on page 600. The minimum of ΔT_E is obtained if we take the derivative of Equation (10.54) with respect to λ and set the derivative equal to zero while V_{growth} is assumed to be constant.

$$\frac{d\Delta T_E}{d\lambda} = 0 + \frac{-1}{\lambda^2}\frac{\sigma_{\alpha\beta}}{RT}\left(\frac{m_\beta^L V_m^\beta}{f_\beta} - \frac{m_\alpha^L V_m^\alpha}{f_\alpha}\right) - 0 + \frac{V_{growth}}{2}\frac{f_\alpha f_\beta V_m^L\left(x_E^{\beta/L} - x_E^{\alpha/L}\right)}{2D}\left(\frac{m_\beta^L}{V_m^\beta} - \frac{m_\alpha^L}{V_m^\alpha}\right) = 0$$

which gives

$$V_{growth}\,\lambda^2 = \frac{2\sigma_{\alpha\beta}2D}{RT f_\alpha f_\beta V_m^L\left(x_E^{\beta/L} - x_E^{\alpha/L}\right)}\,\frac{\dfrac{m_\beta^L V_m^\beta}{f_\beta} - \dfrac{m_\alpha^L V_m^\alpha}{f_\alpha}}{\dfrac{m_\beta^L}{V_m^\beta} - \dfrac{m_\alpha^L}{V_m^\alpha}} \quad (10.59)$$

This is the same relationship as Equation (10.45) on page 600 which is valid when the interface kinetics is neglected.

The relationship $V_{growth}\lambda^2 = C_{lamella}$ is valid as long as the diffusion in the melt dominates over the kinetic reaction at the interface. If the diffusion term is much smaller than the kinetic term, the former can be neglected and the relationship is no longer valid.

The Equations (10.49) and (10.50) on page 603 are no longer valid as the undercooling ΔT_E contains a kinetic term. The interface kinetics has a large effect on the undercooling.

10.3.3 Normal Rod Eutectic Growth

Experiments have shown that eutectics, in which one of the phases is a minority phase, often grow as rod eutectics. In addition, when lamellar eutectics are forced to change growth direction (change of direction of heat flow) the lamellar aggregates break down and rods are formed.

Figure 10.14 shows a photograph of a typical rod eutectic structure. Other examples are shown in Figure 10.2a and 2f on page 588.

Figure 10.15 shows a sketch of a rod eutectic structure. It can shortly be described as a number of thin cylindrical parallel rods of an α phase, regularly distributed in a matrix of a β phase. The growth occurs out of the plane of the paper towards the reader.

A simplified theory of normal rod eutectic growth is given below. It is analogous to the theory given for lamellar growth in Section 10.3.2.

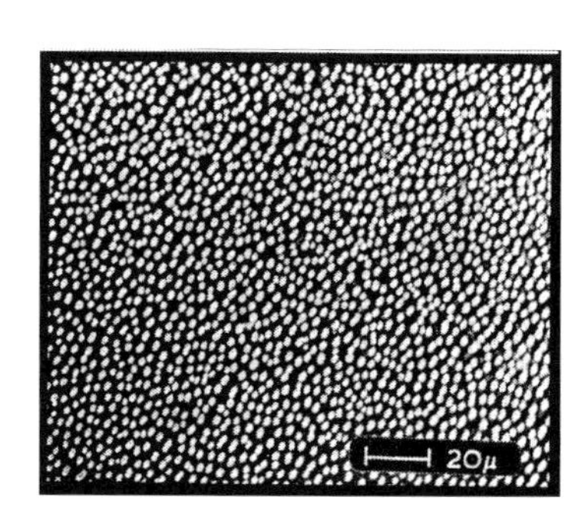

Figure 10.14 Rod eutectic structure.

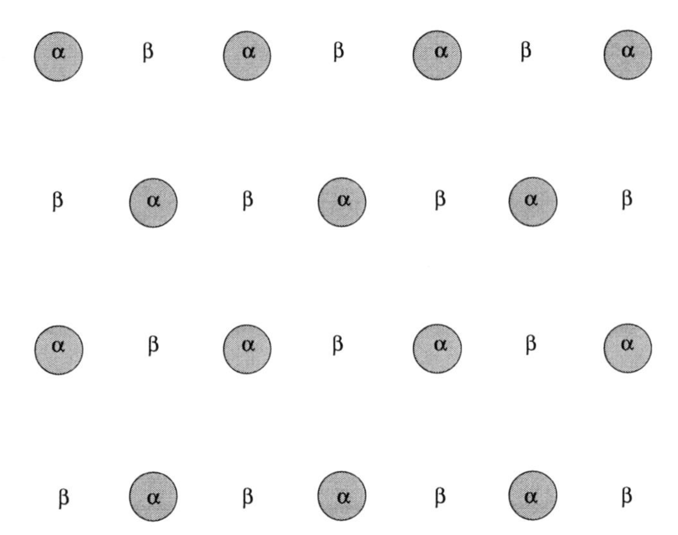

Figure 10.15 Rod eutectic structure shown from above. The cylinders with the α phase and the growth rate are perpendicular to the plane of the paper. The growth rate is directed upwards.

10.3.4 Simplified Theory of Normal Rod Eutectic Growth

Concentration Profile of the Melt

Jackson and Hunt [2] have performed an analysis for eutectics when the minor phase adopts a rod structure rather than a lamellar structure. The two-dimensional Equation (10.2) on page 591 does not work for three-dimensional rods. It has to be replaced by Equation (10.1) on page 590. The solution of this differential equation in three dimensions

$$D\left(\frac{\partial^2 x^{\mathrm{L}}}{\partial x^2}+\frac{\partial^2 x^{\mathrm{L}}}{\partial y^2}+\frac{\partial^2 x^{\mathrm{L}}}{\partial z^2}\right)+\frac{V_{\mathrm{growth}}}{D}\frac{\partial x^{\mathrm{L}}}{\partial z}=0 \tag{10.1}$$

is more complicated than Equation (10.3) on page 591, the solution at lamellar growth. Bessel functions instead of cosine or sinus functions are involved in the general solution.

We will introduce the same approximation as in the case of normal lamellar eutectic growth.

- Two solutions of the diffusion equation instead of one will be derived, one for each phase.

In the present case we will in addition assume that

- The cross-sections of the cylinders of α phase are circles. Hence, we have rotational symmetry, which makes it convenient and much simpler to use cylindrical coordinates instead of Cartesian coordinates.

Figure 10.16 shows the rod eutectic structure, which grows perpendicularly to the plane of the paper. A atoms diffuse from the interface L/β to the interface L/α and B atoms diffuse in the opposite direction.

The β phase or the matrix covers the whole cross-section except the circular cross-section areas of the α phase. Each cylinder is surrounded by a regular hexagon. As is shown in Figure 10.16 the hexagons cover together the whole area. Each hexagon is approximately assumed to be a circle with the same area as the hexagon. Hence, the β phase is represented by all the haxagons minus all the cross-section areas of the α phase.

The concentration of the alloying element varies in the α and the β cylinders. Figure 10.17 gives an idea of the concentration profile along the line AA′ in Figure 10.16.

Below we will calculate the concentration of the alloying element as a function of position.

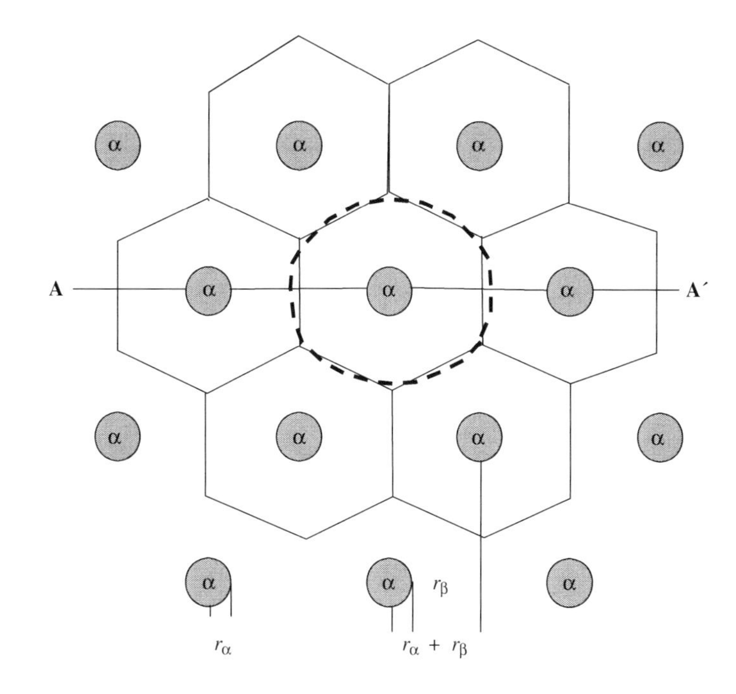

Figure 10.16 r_α = radius of the α cylinders. $r_\alpha + r_\beta$ = outer radius of the β cylinders.

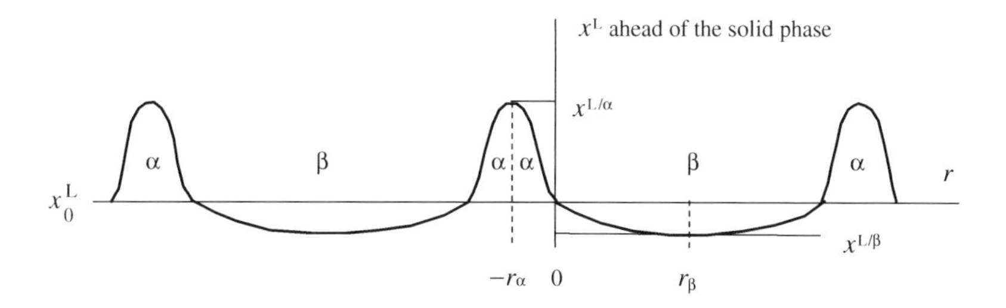

Figure 10.17 Sketch of the concentration distribution of the alloying element at normal rod eutectic growth. The figure shows the concentration along the line AA′ in Figure 10.16.

Calculation of the Concentration Distribution

If the diffusion equation or Fick's second law is solved for each of the two phases we obtain the distribution field of the alloying element.

If we introduce cylindrical coordinates (r, φ, z) [Equations (9.55), (9.56) and (9.57) on page 518 in Chapter 9] instead of Cartesian coordinates (x, y, z coordinates), Equation (10.1) on page 590 can be written as

$$\frac{\partial^2 x^L}{\partial r^2} + \frac{1}{r^2}\frac{\partial^2 x^L}{\partial \varphi^2} + \frac{\partial^2 x^L}{\partial z^2} + \frac{V_{\text{growth}}}{D}\frac{\partial x^L}{\partial z} = 0 \tag{10.60}$$

Owing to the rotation symmetry of the α and β phases x^L is independent of φ. Therefore, the second partial derivative with respect to φ is zero and Equation (10.60) becomes

$$\frac{\partial^2 x^L}{\partial r^2} + \frac{\partial^2 x^L}{\partial z^2} + \frac{V_{\text{growth}}}{D}\frac{\partial x^L}{\partial z} = 0 \tag{10.61}$$

The only difference between Equation (10.61) and Equation (10.2) on page 591 is that y has been replaced by r in the former equation.

We choose the same coordinate system as for lamellar eutectic growth to show the concentration profile along an axis through the centres of the rods. It is shown in Figure 10.17. If we replace y by r in Equation (10.3) on page 591, the new equation must be the solution of Equation (10.61).

$$x^{\mathrm{L}} = x_0^{\mathrm{L}} + A\mathrm{e}^{-\frac{V_{\mathrm{growth}} z}{D}} + B\mathrm{e}^{-Cz} \sin \frac{2\pi r}{\lambda_{\mathrm{wave}}} \tag{10.62}$$

where
x_0^{L} = the initial composition of the melt
λ_{wave} = wavelength of the sinus function
r = coordinate (see Figure 10.18)
z = coordinate and growth direction.

The growth constant C is determined in the same way as for normal lamellar growth (page 591). The value of the constant C is the same in this case.

$$C = \frac{V_{\mathrm{growth}}}{2D} + \sqrt{\left(\frac{V_{\mathrm{growth}}}{2D}\right)^2 + \left(\frac{2\pi}{\lambda_{\mathrm{wave}}}\right)^2} \tag{10.4}$$

At low growth rates ($V_{\mathrm{growth}}/2D \ll 2\pi/\lambda_{\mathrm{wave}}$) $C \approx 2\pi/\lambda_{\mathrm{wave}}$ and we obtain

$$x^{\mathrm{L}} = x_0^{\mathrm{L}} + A\mathrm{e}^{-\frac{V_{\mathrm{growth}} z}{D}} + B\mathrm{e}^{-\frac{2\pi z}{\lambda_{\mathrm{wave}}}} \sin \frac{2\pi r}{\lambda_{\mathrm{wave}}} \tag{10.63}$$

We must relate the mathematical wavelength λ_{wave} and the radii of the α and β phases. λ_{wave} is the length of one whole wavelength. According to Figure 10.18 the relationships will be $\lambda_{\mathrm{wave}} = 2r_\alpha$ for the α phase and $\lambda_{\mathrm{wave}} = 4r_\beta$ for the β phase.

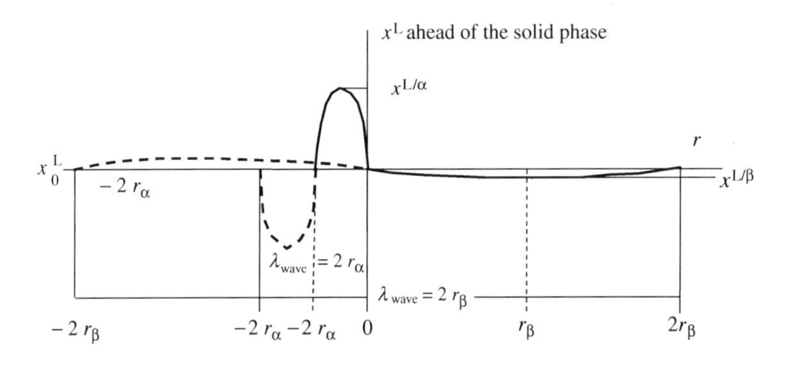

Figure 10.18 Concentration distribution of the alloying element at normal rod eutectic growth.

Hence, the solutions for the α and β phases can be written as

$$x^{\mathrm{L}} = x_0^{\mathrm{L}} + A_\alpha\mathrm{e}^{-\frac{V_{\mathrm{growth}} z}{D}} + B_\alpha\mathrm{e}^{-\frac{2\pi z}{4r_\alpha}} \sin \frac{2\pi r}{2r_\alpha} \tag{10.64}$$

valid within the interval $-2r_\alpha \leq r \leq 0$ and

$$x^{\mathrm{L}} = x_0^{\mathrm{L}} + A_\beta\mathrm{e}^{-\frac{V_{\mathrm{growth}} z}{D}} + B_\beta\mathrm{e}^{-\frac{2\pi z}{4r_\beta}} \sin \frac{2\pi r}{2r_\beta} \tag{10.65}$$

valid within the interval $0 \leq r \leq 2r_\beta$.

The values of the A and B constants can be calculated with the aid of the boundary conditions.

For the α phase:	For the β phase:
$z = 0,\ r = 0 \Rightarrow x^{\mathrm{L}} = x_0^{\mathrm{L}}$	$z = 0,\ r = 0 \Rightarrow x^{\mathrm{L}} = x_0^{\mathrm{L}}$
$z = 0,\ r = -r_\alpha \Rightarrow x^{\mathrm{L}} = x^{\mathrm{L}/\alpha}$	$z = 0,\ r = r_\beta \Rightarrow x^{\mathrm{L}} = x^{\mathrm{L}/\beta}$

These values are introduced into Equations (10.64) and (10.65), respectively, and the constants are calculated in the same way as for lamellar eutectic growth (page 593). The result is

$$A_\alpha = 0 \quad A_\beta = 0$$

$$B_\alpha = -\left(x^{L/\alpha} - x_0^L\right) \quad B_\beta = x^{L/\beta} - x_0^L$$

Inserting these values into Equations (10.64) and (10.65), respectively, gives

$$x^L = x_0^L - \left(x^{L/\alpha} - x_0^L\right) e^{-\frac{2\pi z}{4 r_\alpha}} \sin \frac{2\pi r}{4 r_\alpha} \tag{10.66}$$

valid within the interval $-2r_\alpha \leq r \leq 0$ and

$$x^L = x_0^L + \left(x^{L/\beta} - x_0^L\right) e^{-\frac{2\pi z}{4 r_\beta}} \sin \frac{2\pi r}{4 r_\beta} \tag{10.67}$$

valid within the interval $0 \leq r \leq 2r_\beta$.

Change of Coordinate System

The coordinate system used above is not very convenient, as it does not enable us to make use of the symmetry of the unit cell shown in Figure 10.16 on page 607. To achieve this advantage we have to use a coordinate system with the centre of a rod as origin (Figure 10.19).

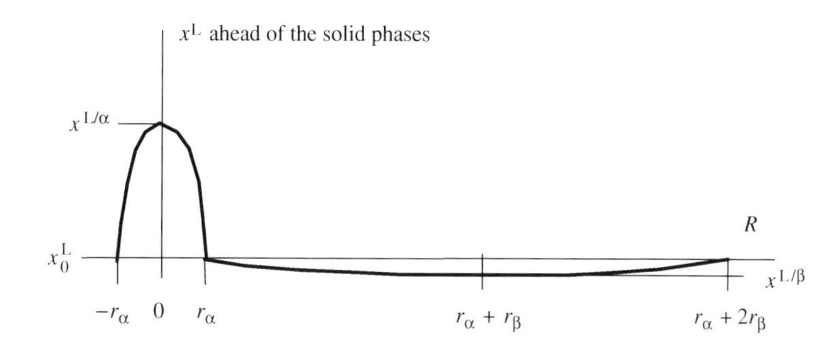

Figure 10.19 Concentration distribution of the alloying element at rod eutectic growth.

$$r = R - r_\alpha \tag{10.68}$$

$(R - r_\alpha)$ (see Figure 10.20) we obtain the concentration profiles in the α and β phases.

$$x^L = x_0^L - \left(x^{L/\alpha} - x_0^L\right) e^{-\frac{2\pi z}{4 r_\alpha}} \sin \frac{2\pi (R - r_\alpha)}{4 r_\alpha} \tag{10.69}$$

$$0 \leq r \leq r_\alpha$$

and

$$x^L = x_0^L + \left(x^{L/\beta} - x_0^L\right) e^{-\frac{2\pi z}{4 r_\beta}} \sin \frac{2\pi (R - r_\beta)}{4 r_\beta} \tag{10.70}$$

$$r_\alpha \leq R \leq r_\alpha + r_\beta$$

These relationships will be used in the calculations below

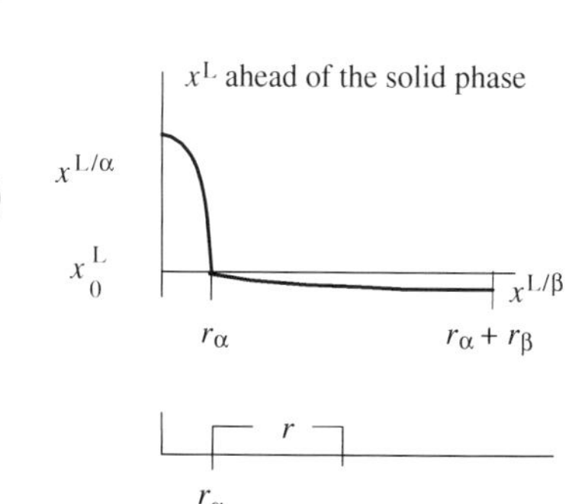

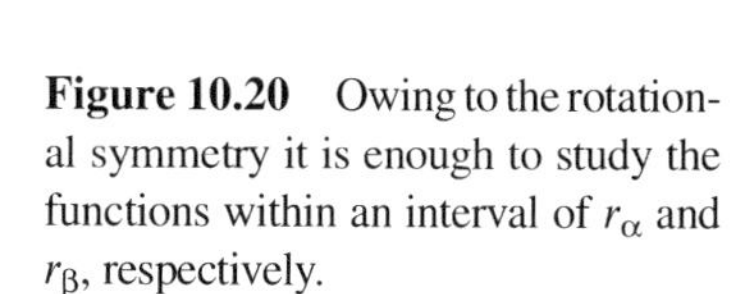

Figure 10.20 Owing to the rotational symmetry it is enough to study the functions within an interval of r_α and r_β, respectively.

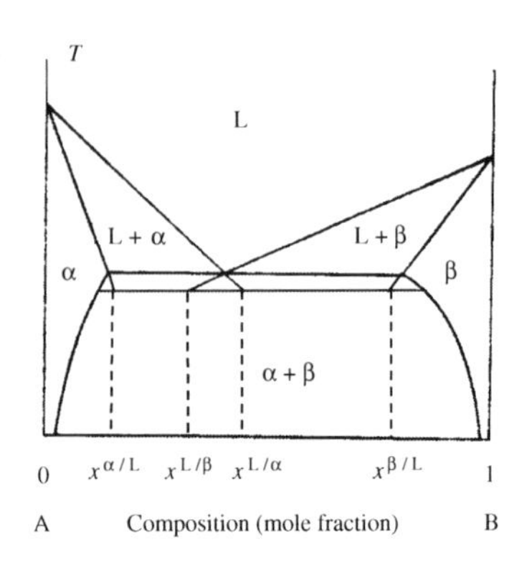

Figure 10.4 $x_E^{\alpha/L}$ and $x_E^{\beta/L}$ represent the end points of the eutectic lines. These points are not marked in the figure.

Volume Fractions of the α and β phases

The mole fractions of the α and β phases will be the same as those derived from the phase diagram of the alloy on page 594 or from Figure 10.1 on page 588.

$$f_\alpha = \frac{x^{\beta/L} - x_0^L}{x^{\beta/L} - x^{\alpha/L}} \approx \frac{x_E^{\beta/L} - x_0^L}{x_E^{\beta/L} - x_E^{\alpha/L}} \tag{10.12}$$

$$f_\beta = \frac{x_0^L - x^{\alpha/L}}{x^{\beta/L} - x^{\alpha/L}} \approx \frac{x_0^L - x_E^{\alpha/L}}{x_E^{\beta/L} - x_E^{\alpha/L}} \tag{10.13}$$

As x_0^L is known and equal to x_E^L (see Figure 10.4) on page 589 the volume fractions of the two phases can be calculated with the aid of Equations (10.12) and (10.13).

The volume fractions can also be expressed in terms of r_α and r_β. According to Figure 10.21 we obtain

$$f_\alpha = \frac{V_\alpha}{V_\alpha + V_\beta} = \frac{\pi r_\alpha^2 L}{\pi (r_\alpha + r_\beta)^2 L} = \left(\frac{r_\alpha}{r_\alpha + r_\beta}\right)^2 \tag{10.71}$$

and

$$f_\beta = \frac{V_\beta}{V_\alpha + V_\beta} = \frac{\pi L\left[(r_\alpha + r_\beta)^2 - r_\alpha^2\right]}{\pi L (r_\alpha + r_\beta)^2} = 1 - \left(\frac{r_\alpha}{r_\alpha + r_\beta}\right)^2 \tag{10.72}$$

where L is the height of the cylinder.

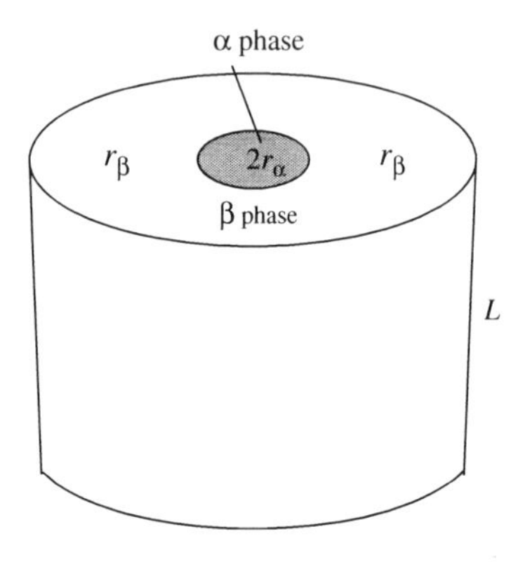

Figure 10.21 α phase rod surrounded by β phase.

Growth Rates of the α and β Phases

The growth rates of the α and β phases at normal rod eutectic growth can be calculated in the same way as for normal lamellar eutectic growth (page 594).

Fick's first law describes the mass transport, i.e. the diffusion of the solute, at the interface (page 594).

$$\frac{V_{\text{growth}}^\alpha}{V_m^\alpha}\left(x^{L/\alpha} - x^{\alpha/L}\right) = -\frac{D}{V_m^L}\left(\frac{dx^L}{dz}\right)_{z=0}^\alpha \tag{10.16}$$

$$\frac{V_{\text{growth}}^\beta}{V_m^\beta}\left(x^{L/\beta} - x^{\beta/L}\right) = -\frac{D}{V_m^L}\left(\frac{dx^L}{dz}\right)_{z=0}^\beta \tag{10.17}$$

The concentration gradients are found by taking the derivatives of Equations (10.69) and (10.70) with respect to z. The concentration gradients vary with the r coordinate along the α and β interfaces. The mass flux across each phase is found by integration of dx^L/dz with respect to r.

The sinus functions in Equations (10.69) and (10.70) have to be replaced by their average values across the *cross-section area* of the α and β cylinders, respectively. The calculations are performed in the boxes on pages 611 and 612.

If we combine Equations (10.16) + (2′) and Equations (10.17) + (4′), respectively, we obtain

$$\frac{V_{\text{growth}}^\alpha}{V_m^\alpha}\left(x^{L/\alpha} - x^{\alpha/L}\right) = -\frac{D}{V_m^L}\left(x^{L/\alpha} - x_0^L\right)\frac{-\left(2 - \frac{4}{\pi}\right)}{r_\alpha} \tag{10.73}$$

$$\frac{V_{\text{growth}}^\beta}{V_m^\beta}\left(x^{L/\beta} - x^{\beta/L}\right) = -\frac{D}{V_m^L}\left(x^{L/\beta} - x_0^L\right)\frac{-\left(2r_\alpha + \frac{4r_\beta}{\pi}\right)}{r_\beta(2r_\alpha + r_\beta)} \tag{10.74}$$

or after rearrangement

$$V_{\text{growth}}^{\alpha} = 2D\,\frac{2-\dfrac{4}{\pi}}{r_\alpha}\,\frac{x^{L/\alpha}-x_0^{L}}{x^{L/\alpha}-x^{\alpha/L}}\,\frac{V_m^{\alpha}}{V_m^{L}} \tag{10.75}$$

$$V_{\text{growth}}^{\beta} = 2D\,\frac{2r_\alpha+\dfrac{4r_\beta}{\pi}}{r_\beta(2r_\alpha+r_\beta)}\,\frac{x^{L/\beta}-x_0^{L}}{x^{L/\beta}-x^{\beta/L}}\,\frac{V_m^{\beta}}{V_m^{L}} \tag{10.76}$$

Derivation of $\left(\dfrac{\mathrm{d}x^{L}}{\mathrm{d}z}\right)_{z=0}^{\alpha}$

The derivative is obtained by derivation of Equation (10.66) with respect to z.

$$\frac{\mathrm{d}x^{L}}{\mathrm{d}z} = -\left(x^{L/\alpha}-x_0^{L}\right)e^{-\frac{\pi z}{2r_\alpha}}\,\frac{-\pi}{2r_\alpha}\,\overline{\sin\frac{\pi(R-r_\alpha)}{2r_\alpha}} \tag{1'}$$

To find the average value of the sinus function we choose the rotation-symmetrical area element $2\pi R\,\mathrm{d}R$.

$$\overline{\sin\frac{\pi(R-r_\alpha)}{2r_\alpha}} = \frac{\int_0^{r_\alpha}\sin\frac{\pi(R-r_\alpha)}{2r_\alpha}\times 2\pi R\mathrm{d}R}{\int_0^{r_\alpha}2\pi R\mathrm{d}R} = \frac{\int_0^{r_\alpha}\sin\frac{\pi(R-r_\alpha)}{2r_\alpha}\times 2\pi R\mathrm{d}R}{[\pi R^2]_0^{r_\alpha}}$$

Partial integration over the cross-section area of the rod gives

$$\overline{\sin\frac{\pi(R-r_\alpha)}{2r_\alpha}} = \frac{\left[\left(-\cos\frac{\pi(R-r_\alpha)}{2r_\alpha}\,\frac{2r_\alpha}{\pi}\right)2\pi R\right]_0^{r_\alpha} - \int_0^{r_\alpha}\left(-\cos\frac{\pi(R-r_\alpha)}{2r_\alpha}\,\frac{2r_\alpha}{\pi}\right)2\pi\mathrm{d}R}{\pi r_\alpha^2}$$

which gives

$$\overline{\sin\frac{\pi(R-r_\alpha)}{2r_\alpha}} = \frac{(-1)\times 4r_\alpha^2 - \left[-\sin\frac{\pi(R-r_\alpha)}{2r_\alpha}\,\frac{2r_\alpha}{\pi}\,\frac{2r_\alpha}{\pi}\,2\pi\right]_0^{r_\alpha}}{\pi {r_\alpha}^2} = \frac{-4r_\alpha^2+\dfrac{8r_\alpha^2}{\pi}}{\pi r_\alpha^2}$$

Inserting $z=0$ and the value of the average sinus function into Equation (1′) gives

$$\left(\frac{\mathrm{d}x^{L}}{\mathrm{d}z}\right)_{z=0}^{\alpha} = -\left(x^{L/\alpha}-x_0^{L}\right)\frac{-\pi}{2r_\alpha}\,\frac{-4+\dfrac{8}{\pi}}{\pi}$$

or

$$\left(\frac{\mathrm{d}x^{L}}{\mathrm{d}z}\right)_{z=0}^{\alpha} = \left(x^{L/\alpha}-x_0^{L}\right)\frac{-\left(2-\dfrac{4}{\pi}\right)}{r_\alpha} \tag{2'}$$

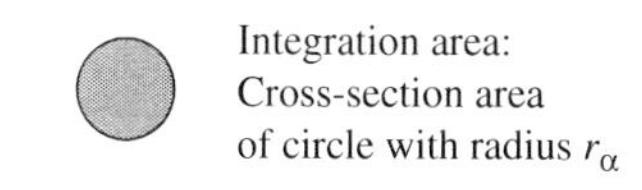

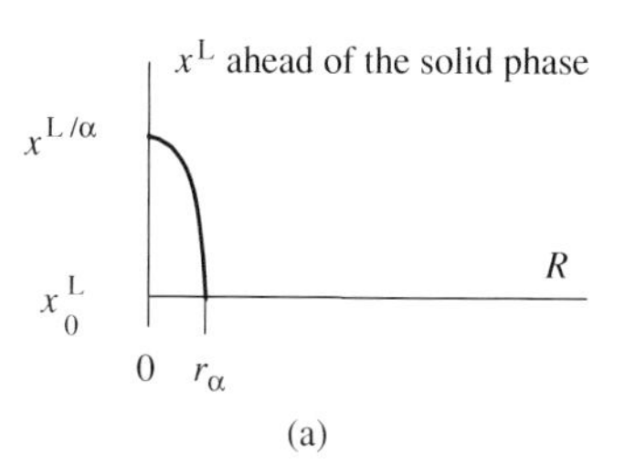

(a)

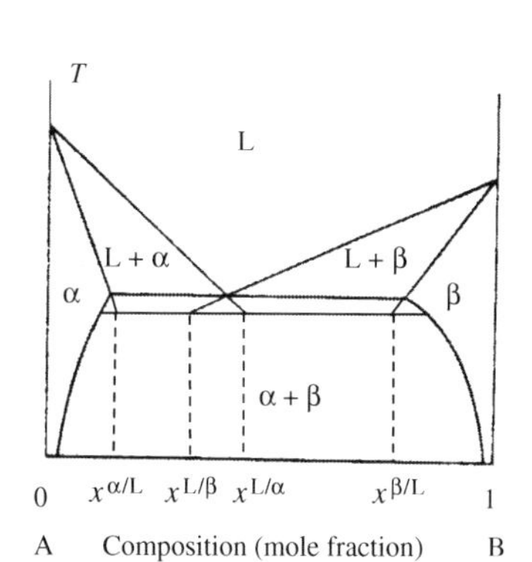

(b)

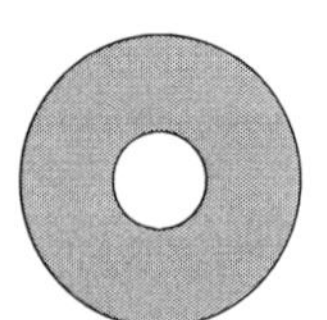

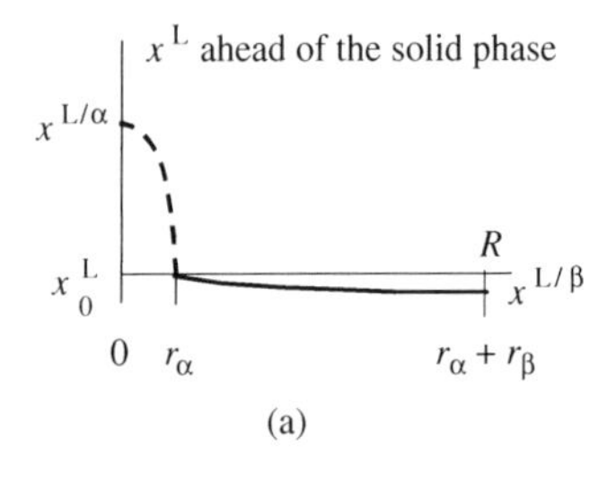

(a)

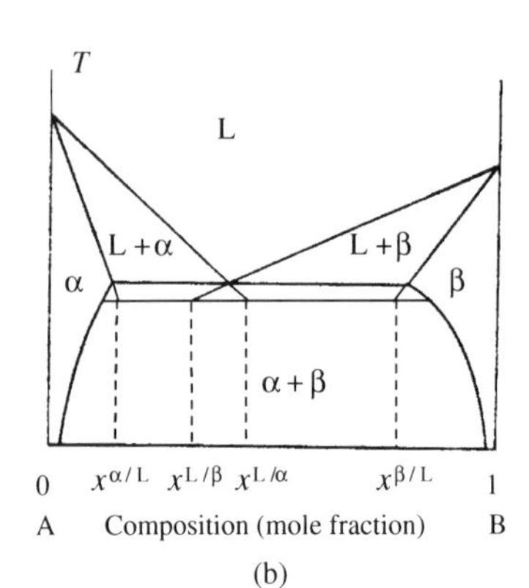

(b)

Derivation of $\left(\frac{dx^L}{dz}\right)^\beta_{z=0}$

The derivative is obtained by derivation of Equation (10.67) with respect to z.

$$\frac{dx^L}{dz} = \left(x^{L/\beta} - x_0^L\right)\left(e^{-\frac{\pi z}{2r_\beta}} \frac{-\pi}{2r_\beta} \overline{\sin\frac{\pi(R-r_\alpha)}{2r_\beta}}\right) \tag{3'}$$

To find the average value of the sinus function we choose the rotation-symmetrical area element $2\pi R dR$.

$$\overline{\sin\frac{\pi(R-r_\alpha)}{2r_\beta}} = \frac{\int_{r_\alpha}^{r_\alpha+r_\beta} \sin\frac{\pi(R-r_\alpha)}{2r_\beta} \times 2\pi R dR}{\int_{r_\alpha}^{r_\alpha+r_\beta} 2\pi R dR} = \frac{\int_{r_\alpha}^{r_\alpha+r_\beta} \sin\frac{\pi(R-r_\alpha)}{2r_\beta} \times 2\pi R dR}{\left[\pi R^2\right]_{r_\alpha}^{r_\alpha+r_\beta}}$$

Partial integration over the cross-section area of the hollow cylinder gives

$$\overline{\sin\frac{\pi(R-r_\alpha)}{2r_\beta}} = \frac{\left[-\cos\frac{\pi(R-r_\alpha)}{2r_\beta}\frac{2r_\beta}{\pi} \times 2\pi R\right]_{r_\alpha}^{r_\alpha+r_\beta} - \int_{r_\alpha}^{r_\alpha+r_\beta} -\cos\frac{\pi(R-r_\alpha)}{2r_\beta}\frac{2r_\beta}{\pi} \times 2\pi dR}{\pi\left[(r_\alpha+r_\beta)^2 - r_\alpha^2\right]}$$

which gives

$$\overline{\sin\frac{\pi(R-r_\alpha)}{2r_\beta}} = \frac{\left[-0 + 1 \times \frac{2r_\beta}{\pi} \times 2\pi r_\alpha\right] - \left[-\sin\frac{\pi(R-r_\alpha)}{2r_\beta}\frac{2r_\beta}{\pi}\frac{2r_\beta}{\pi} \times 2\pi\right]_{r_\alpha}^{r_\alpha+r_\beta}}{\pi r_\beta(2r_\alpha + r_\beta)}$$

$$\overline{\sin\frac{\pi(R-r_\alpha)}{2r_\beta}} = \frac{4r_\alpha r_\beta + 1 \times \left(\frac{2r_\beta}{\pi}\right)^2 \times 2\pi}{\pi r_\beta(2r_\alpha + r_\beta)} = \frac{4r_\alpha r_\beta + \frac{8r_\beta^2}{\pi}}{\pi r_\beta(2r_\alpha + r_\beta)}$$

Inserting $z = 0$ and the value of the average sinus function into Equation (3′) gives

$$\left(\frac{dx^L}{dz}\right)^\beta_{z=0} = \left(x^{L/\beta} - x_0^L\right)\frac{-\pi}{2r_\beta}\frac{4r_\alpha + \frac{8r_\beta}{\pi}}{\pi(2r_\alpha + r_\beta)}$$

or

$$\left(\frac{dx^L}{dz}\right)^\beta_{z=0} = \left(x^{L/\beta} - x_0^L\right)\frac{-\left(2r_\alpha + \frac{4r_\beta}{\pi}\right)}{r_\beta(2r_\alpha + r_\beta)} \tag{4'}$$

Driving Force and Supersaturation

The theory, given on pages 596–597, and the Equations (10.20)–(10.23) are general and therefore valid for both lamellar eutectic and rod eutectic growth.

Supersaturation Required for Surface Tension

The supersaturation required for surface tension of the α and β phases for normal rod eutectic growth can be derived in the same way as the one for normal lamellar eutectic growth.

The upper curved surfaces of the solid α and β phases in Figure 10.22 do not change during the growth. The contact area between the solid the α and β phases increases by the amount $2\pi r_\alpha dz$ when the solidification front moves by the length dz.

Consider an α rod (Figure 10.22). The surface energy required to increase the contact area between the α and β phases is $\sigma_{\alpha\beta} \times 2\pi r_\alpha dz$ where $\sigma_{\alpha\beta}$ is the surface energy per unit area.

The required energy is equal to the volume fraction times the driving force of surface formation $(-\Delta G_m^{L\rightarrow\alpha}) = RT(x^{L/\alpha} - x_\sigma^{L/\alpha})$. Hence, we obtain for the α lamella

$$\sigma_{\alpha\beta} 2\pi r_\alpha dz = \frac{\pi r_\alpha^2 dz}{V_m^\alpha}\left(-\Delta G_m^{L\rightarrow\alpha}\right) = \frac{\pi r_\alpha^2 dz}{V_m^\alpha} RT\left(x^{L/\alpha} - x_\sigma^{L/\alpha}\right) \tag{10.77a}$$

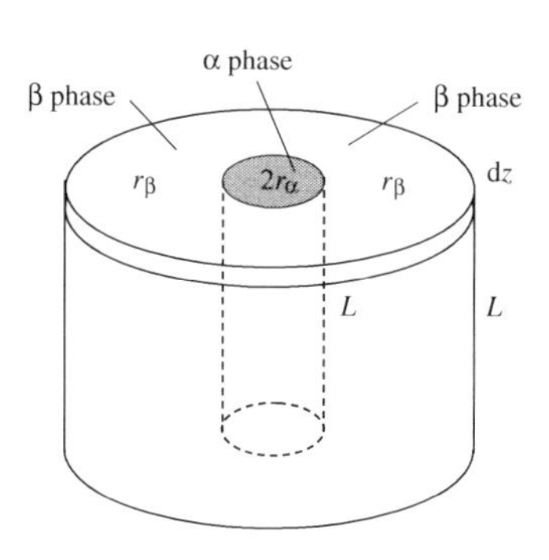

Figure 10.22 α phase rod surrounded by β phase.

where the molar volume V_m^α is introduced to find the number of kmoles of the volume element $\pi r_\alpha^2 dz$.

In the same way we obtain for the hollow cylinder (β phase)

$$\sigma_{\alpha\beta} \times 2\pi r_\alpha dz = \frac{\pi\left[\left(r_\alpha^2 + r_\beta^2\right) - r_\alpha^2\right] dz}{V_m^\beta} RT\left[(1 - x^{L/\beta}) - (1 - x_\sigma^{L/\beta})\right] \tag{10.77b}$$

Equations (10.77a) and (10.77b) can be divided by the factor πdz and we obtain

$$\frac{RT\left(x^{L/\alpha} - x_\sigma^{L/\alpha}\right)}{V_m^\alpha} = \frac{2\sigma_{\alpha\beta}}{r_\alpha} \tag{10.78}$$

$$\frac{RT\left[(1 - x^{L/\beta}) - (1 - x_\sigma^{L/\beta})\right]}{V_m^\beta} r_\beta = \frac{2\sigma_{\alpha\beta} r_\alpha}{r_\beta(2r_\alpha + r_\beta)} \tag{10.79}$$

Supersaturation Required for Diffusion

The supersaturation required for diffusion of the alloying element in the melt for normal lamellar eutectic growth has been derived in Section 10.3.2 on page 591–596. The resulting relationships between the growth rates and the supersaturation $(x^{L/\alpha} - x_0^L)$ are the Equations (10.18) and (10.19) on page 595.

Analogous calculations have been performed above for normal rod eutectic growth. Application of the Equations (10.75) and (10.76) on page 611 gives

$$V_{growth}^\alpha = 2D \frac{2 - \dfrac{4}{\pi}}{r_\alpha} \frac{x_\sigma^{L/\alpha} - x_0^L}{x_\sigma^{L/\alpha} - x^{\alpha/L}} \frac{V_m^\alpha}{V_m^L} \tag{10.80a}$$

$$V_{growth}^\beta = 2D \frac{2r_\alpha + \dfrac{4r_\beta}{\pi}}{r_\beta(2r_\alpha + r_\beta)} \frac{\left(1 - x_\sigma^{L/\beta}\right) - \left(1 - x_0^L\right)}{\left(1 - x_\sigma^{L/\beta}\right) - \left(1 - x^{\beta/L}\right)} \frac{V_m^\beta}{V_m^L} \tag{10.80b}$$

With the aid of the relationships on page 598

$$x_\sigma^{L/\alpha} - x^{\alpha/L} \approx x_E^L - x_E^{\alpha/L}$$

$$\left(1 - x_\sigma^{L/\beta}\right) - \left(1 - x^{\beta/L}\right) \approx x_E^{\beta/L} - x_E^L$$

the Equations (10.80a) and (10.80b) can be written as

$$V_{\text{growth}}^{\alpha} = 2D \frac{2 - \frac{4}{\pi}}{r_{\alpha}} \frac{x_{\sigma}^{\text{L}/\alpha} - x_0^{\text{L}}}{x_{\text{E}}^{\text{L}} - x_{\text{E}}^{\alpha/\text{L}}} \frac{V_{\text{m}}^{\alpha}}{V_{\text{m}}^{\text{L}}} \tag{10.81}$$

$$V_{\text{growth}}^{\beta} = 2D \frac{2r_{\alpha} + \frac{4r_{\beta}}{\pi}}{r_{\beta}(2r_{\alpha} + r_{\beta})} \frac{\left(1 - x_{\sigma}^{\text{L}/\beta}\right) - \left(1 - x_0^{\text{L}}\right)}{\left(1 - x_{\text{E}}^{\text{L}}\right) - \left(1 - x_{\text{E}}^{\beta/\text{L}}\right)} \frac{V_{\text{m}}^{\beta}}{V_{\text{m}}^{\text{L}}} \tag{10.82}$$

Total Supersaturation

The total supersaturation is the sum of the contributions owing to surface tension and diffusion.

$$x^{\text{L}/\alpha} - x_0^{\text{L}} = \left(x^{\text{L}/\alpha} - x_{\sigma}^{\text{L}/\alpha}\right) + \left(x_{\sigma}^{\text{L}/\alpha} - x_0^{\text{L}}\right) \tag{10.22}$$

$$\left(1 - x^{\text{L}/\beta}\right) - \left(1 - x_0^{\text{L}}\right) = \left[\left(1 - x^{\text{L}/\beta}\right) - \left(1 - x_{\sigma}^{\text{L}/\beta}\right)\right] + \left[\left(1 - x_{\sigma}^{\text{L}/\beta}\right) - \left(1 - x_0^{\text{L}}\right)\right] \tag{10.23}$$

The expressions for the concentration differences in Equations (10.78) + (10.81) and (10.79) + (10.82), respectively, are introduced into Equations (10.22) and (10.23).

$$x^{\text{L}/\alpha} - x_0^{\text{L}} = \frac{2\sigma_{\alpha\beta}}{r_{\alpha}} \frac{V_{\text{m}}^{\alpha}}{RT} + \frac{V_{\text{growth}}^{\alpha}\left(x_{\text{E}}^{\text{L}} - x_{\text{E}}^{\alpha/\text{L}}\right)}{2D} \frac{r_{\alpha}}{2 - \frac{4}{\pi}} \frac{V_{\text{m}}^{\text{L}}}{V_{\text{m}}^{\alpha}} \tag{10.83}$$

$$x_0^{\text{L}} - x^{\text{L}/\beta} = \frac{2\sigma_{\alpha\beta} r_{\alpha}}{r_{\beta}(2r_{\alpha} + r_{\beta})} \frac{V_{\text{m}}^{\beta}}{RT} + \frac{V_{\text{growth}}^{\beta}\left(x_{\text{E}}^{\beta/\text{L}} - x_{\text{E}}^{\text{L}}\right)}{2D} \frac{r_{\beta}(2r_{\alpha} + r_{\beta})}{2r_{\alpha} + \frac{4r_{\beta}}{\pi}} \frac{V_{\text{m}}^{\text{L}}}{V_{\text{m}}^{\beta}} \tag{10.84}$$

At normal eutectic growth the two phases grow with the same growth rate. According to Equation (10.32) on page 599 we have

$$V_{\text{growth}}^{\alpha} = V_{\text{growth}}^{\beta} = V_{\text{growth}} \tag{10.32}$$

This relationship will be used henceforth.

The number of unknown quantities seems to be four: r_{α}, r_{β}, $x^{\text{L}/\alpha}$ and $x^{\text{L}/\beta}$. r_{β} can be replaced by the product of r_{α} and r_{β}/r_{α}. The ratio r_{β}/r_{α} is a function of the known fraction f_{α} (Equation (10.71) on page 610). The temperature T can replace the mole fractions $x^{\text{L}/\alpha}$ and $x^{\text{L}/\beta}$ as independent variable. Hence, we have two independent variables, r_{α} and T.

Undercooling

In order to find the undercooling we proceed in the same way as for lamellar growth on pages 599–600. As a measure of the undercooling we will use $\Delta T_{\text{E}} = T_{\text{E}} - T$.

Equations (10.83) and (10.84) can be written as

$$x^{\text{L}/\alpha} - x_0^{\text{L}} = \frac{A_{\alpha}}{r_{\alpha}} + B_{\alpha} V_{\text{growth}} r_{\alpha} \tag{10.85}$$

$$x_0^{\text{L}} - x^{\text{L}/\beta} = \frac{A_{\beta}}{r_{\beta}} + B_{\beta} V_{\text{growth}} r_{\beta} \tag{10.86}$$

where

$$A_{\alpha} = 2\sigma_{\alpha\beta} \frac{V_{\text{m}}^{\alpha}}{RT} \quad \text{and} \quad B_{\alpha} = \frac{x_{\text{E}}^{\text{L}} - x_{\text{E}}^{\alpha/\text{L}}}{2D} \frac{1}{2 - \frac{4}{\pi}} \frac{V_{\text{m}}^{\text{L}}}{V_{\text{m}}^{\alpha}}$$

$$A_\beta = 2\sigma_{\alpha\beta}\frac{r_\alpha}{2r_\alpha + r_\beta}\frac{V_m^\beta}{RT} \quad \text{and} \quad B_\beta = \frac{x_E^{\beta/L} - x_E^L}{2D}\frac{(2r_\alpha + r_\beta)}{2r_\alpha + \frac{4r_\beta}{\pi}}\frac{V_m^L}{V_m^\beta}$$

The left-hand side of Equations (10.85) and (10.86) can be split up into two terms:

$$\left(x^{L/\alpha} - x_E^L\right) + \left(x_E^L - x_0^L\right) = \frac{A_\alpha}{r_\alpha} + B_\alpha V_{growth} r_\alpha \tag{10.87}$$

$$\left(x_E^L - x^{L/\beta}\right) - \left(x_E^L - x_0^L\right) = \frac{A_\beta}{r_\beta} + B_\beta V_{growth} r_\beta \tag{10.88}$$

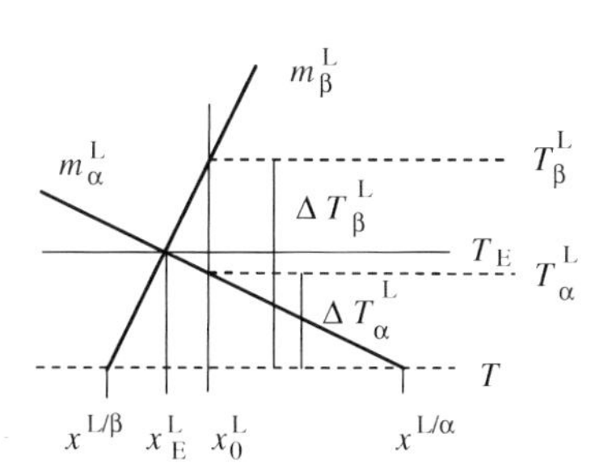

Figure 10.12 Part of the phase diagram close to the eutectic point (temperature T versus concentration x).

With the aid of the slopes m_α^L and m_β^L of the liquidus lines in the phase diagram and the undercooling $\Delta T_E = T_E - T$ we obtain

$$\frac{-\Delta T_E}{m_\alpha^L} + \left(x_E^L - x_0^L\right) = \frac{A_\alpha}{r_\alpha} + B_\alpha V_{growth} r_\alpha \tag{10.89}$$

$$\frac{\Delta T_E}{m_\beta^L} - \left(x_E^L - x_0^L\right) = \frac{A_\beta}{r_\beta} + B_\beta V_{growth} r_\beta \tag{10.90}$$

Equation (10.89) is multiplied by $(-m_\alpha^L)$ and Equation (10.90) is multiplied by m_β^L. The resulting equations are

$$\Delta T_E - m_\alpha^L\left(x_E^L - x_0^L\right) = -m_\alpha^L\frac{A_\alpha}{r_\alpha} - m_\alpha^L B_\alpha V_{growth} r_\alpha \tag{10.91}$$

$$\Delta T_E - m_\beta^L\left(x_E^L - x_0^L\right) = m_\beta^L\frac{A_\beta}{r_\beta} + m_\beta^L B_\beta V_{growth} r_\beta \tag{10.92}$$

The Equations (10.91) and (10.92) are added and the undercooling ΔT_E is calculated. The result is

$$\Delta T_E = \frac{1}{2}\left(m_\alpha^L + m_\beta^L\right)\left(x_E^L - x_0^L\right) + \frac{1}{2}\left(\frac{m_\beta^L}{r_\beta}A_\beta - \frac{m_\alpha^L}{r_\alpha}A_\alpha\right) + \frac{V_{growth}}{2}\left(m_\beta^L r_\beta B_\beta - m_\alpha^L r_\alpha B_\alpha\right) \tag{10.93}$$

The expressions for A_α, A_β, B_α and B_β are inserted into Equation (10.93)

$$\Delta T_E = \frac{1}{2}\left(m_\alpha^L + m_\beta^L\right)\left(x_E^L - x_0^L\right) + \frac{1}{2}\left(\frac{m_\beta^L}{r_\beta}2\sigma_{\alpha\beta}\frac{r_\alpha}{2r_\alpha + r_\beta}\frac{V_m^\beta}{RT} - \frac{m_\alpha^L}{r_\alpha}2\sigma_{\alpha\beta}\frac{V_m^\alpha}{RT}\right)$$
$$+ \frac{V_{growth}}{2}\left(m_\beta^L r_\beta\frac{x_E^{\beta/L} - x_E^L}{2D}\frac{2r_\alpha + r_\beta}{2r_\alpha + \frac{4r_\beta}{\pi}}\frac{V_m^L}{V_m^\beta} - m_\alpha^L r_\alpha\frac{x_E^L - x_E^{\alpha/L}}{2D}\frac{1}{2 - \frac{4}{\pi}}\frac{V_m^L}{V_m^\alpha}\right)$$

which can be written as

$$\Delta T_E = \frac{1}{2}\left(m_\alpha^L + m_\beta^L\right)\left(x_E^L - x_0^L\right) + \frac{\sigma_{\alpha\beta}}{RT}\frac{1}{r_\alpha}\left(m_\beta^L V_m^\beta\frac{r_\alpha^2}{r_\beta(2r_\alpha + r_\beta)} - m_\alpha^L V_m^\alpha\right)$$
$$+ \frac{V_{growth} r_\alpha V_m^L}{4D}\left(\frac{m_\beta^L}{V_m^\beta}\frac{r_\beta}{r_\alpha}\frac{2r_\alpha + r_\beta}{2r_\alpha + \frac{4r_\beta}{\pi}}\left(x_E^{\beta/L} - x_E^L\right) - \frac{m_\alpha^L}{V_m^\alpha}\frac{x_E^L - x_E^{\alpha/L}}{2 - \frac{4}{\pi}}\right) \tag{10.94}$$

Equation (10.94) gives the undercooling as a function of the growth rate at the solidification process and known quantities, among them r_β/r_α. The equation is analogous to Equation (10.42) for lamellar eutectic growth.

Relationship between Growth Rate and Rod Radius

Equation (10.94) can be written as

$$\Delta T_E = \frac{1}{2}\left(m_\alpha^L + m_\beta^L\right)\left(x_E^L - x_0^L\right) + \frac{A}{r_\alpha} + BV_{growth}r_\alpha \tag{10.95}$$

where

$$A = \frac{\sigma_{\alpha\beta}}{RT}\left(m_\beta^L V_m^\beta \frac{1}{\frac{r_\beta}{r_\alpha}(2r_\alpha + \frac{r_\beta}{r_\alpha})} - m_\alpha^L V_m^\alpha\right)$$

and

$$B = \frac{V_m^L}{4D}\left(\frac{m_\beta^L}{V_m^\beta}\frac{r_\beta}{r_\alpha}\frac{2+\frac{r_\beta}{r_\alpha}}{2+\frac{4r_\beta}{\pi r_\alpha}}\left(x_E^{\beta/L} - x_E^L\right) - \frac{m_\alpha^L}{V_m^\alpha}\frac{x_E^L - x_E^{\alpha/L}}{2-\frac{4}{\pi}}\right)$$

The expressions A and B contain known quantities and the ratio r_β/r_α but not the radii r_β and r_α separately. The ratio r_β/r_α is related to the fraction f_α of the α phase by the Equation (10.71) on page 610. The relationship can be written as

$$\frac{r_\beta}{r_\alpha} = \sqrt{\frac{1}{f_\alpha} - 1} \tag{10.96}$$

As in the case of normal eutectic lamellar growth the minimum condition for the undercooling ΔT_E is valid for normal rod eutectic growth. *The solidification process, which requires a minimum of undercooling, will always win.* This is a consequence of the energy law (page 600). The growth rate is determined by external factors (page 600) and is not influenced by the main lamellar distance. Instead the rod radius has to adapt the value, which agrees with the minimum condition of the undercooling.

We take the derivative of Equation (10.95) with respect to r_α and set the derivative equal to zero. A and B are constants because the ratio r_β/r_α is known [Equation (10.96)].

$$\frac{\Delta T_E}{dr_\alpha} = -\frac{A}{{r_\alpha}^2} + BV_{growth} = 0$$

which gives

$$V_{growth}{r_\alpha}^2 = \frac{A}{B} \tag{10.97}$$

Introduction of the expressions for A and B and some reduction results in

$$V_{growth}r_\alpha^2 = \frac{\sigma_{\alpha\beta} \times 4D\left(\frac{m_\beta^L V_m^\beta}{\frac{r_\beta}{r_\alpha}(2+\frac{r_\beta}{r_\alpha})} - m_\alpha^L V_m^\alpha\right)}{RT\, V_m^L\left(\frac{m_\beta^L}{V_m^\beta}\frac{r_\beta}{r_\alpha}\frac{2+\frac{r_\beta}{r_\alpha}}{2+\frac{4r_\beta}{\pi r_\alpha}}\left(x_E^{\beta/L} - x_E^L\right) - \frac{m_\alpha^L}{V_m^\alpha}\frac{x_E^L - x_F^{\alpha/L}}{2-\frac{4}{\pi}}\right)} \tag{10.98}$$

where

$$\frac{r_\beta}{r_\alpha} = \sqrt{\frac{1}{f_\alpha} - 1} \tag{10.96}$$

Equation (10.98) can be written as

$$V_{\text{growth}} {r_\alpha}^2 = C_{\text{rod}} \tag{10.99}$$

Equation (10.99) is analogous to the corresponding relationship (10.46) on page 601 for lamellar eutectic growth. C_{rod} can be written as

$$C_{\text{rod}} = \frac{4\sigma_{\alpha\beta} D}{RTV_m^{\text{L}}} \frac{A_{\text{rod}}}{B_{\text{rod}}} \tag{10.100}$$

where the expressions A_{rod} and B_{rod} are

$$A_{\text{rod}} = \frac{r_\beta}{r_\alpha} \frac{m_\beta^{\text{L}} V_{\text{m}}^\beta}{2 + \dfrac{r_\beta}{r_\alpha}} \qquad A_{\text{rod}} = \frac{r_\beta}{r_\alpha} \frac{m_\beta^{\text{L}} V_{\text{m}}^\beta}{\dfrac{r_\beta}{r_\alpha}\left(2 + \dfrac{r_\beta}{r_\alpha}\right)} - m_\alpha^{\text{L}} V_{\text{m}}^\alpha$$

and

$$B_{\text{rod}} = \frac{m_\beta^{\text{L}}}{V_{\text{m}}^\beta} \frac{r_\beta}{r_\alpha} \frac{2 + \dfrac{r_\beta}{r_\alpha}}{2 + \dfrac{4 r_\beta}{\pi r_\alpha}} \left(x_{\text{E}}^{\beta/\text{L}} - x_{\text{E}}^{\text{L}}\right) - \frac{m_\alpha^{\text{L}}}{V_{\text{m}}^\alpha} \frac{x_{\text{E}}^{\text{L}} - x_{\text{E}}^{\alpha/\text{L}}}{2 - \dfrac{4}{\pi}}$$

Relationship between Growth Rate and Undercooling

With the aid of Equation (10.94) on page 615 and Equation (10.99) above we can derive a simple relationship between the growth rate V_{growth} and the undercooling ΔT_{E}.

$$\begin{aligned}\Delta T_{\text{E}} = {} & \frac{1}{2}\left(m_\alpha^{\text{L}} + m_\beta^{\text{L}}\right)\left(x_{\text{E}}^{\text{L}} - x_0^{\text{L}}\right) + \frac{\sigma_{\alpha\beta}}{RT}\frac{1}{r_\alpha}\left(m_\beta^{\text{L}} V_{\text{m}}^\beta \frac{1}{\dfrac{r_\beta}{r_\alpha}\left(2 + \dfrac{r_\beta}{r_\alpha}\right)} - m_\alpha^{\text{L}} V_{\text{m}}^\alpha\right) \\ & + \frac{V_{\text{growth}} r_\alpha V_{\text{m}}^{\text{L}}}{4D}\left(\frac{m_\beta^{\text{L}}}{V_{\text{m}}^\beta}\frac{r_\beta}{r_\alpha}\frac{2 + \dfrac{r_\beta}{r_\alpha}}{2 + \dfrac{4 r_\beta}{\pi r_\alpha}}\left(x_{\text{E}}^{\beta/\text{L}} - x_{\text{E}}^{\text{L}}\right) - \frac{m_\alpha^{\text{L}}}{V_{\text{m}}^\alpha}\frac{x_{\text{E}}^{\text{L}} - x_{\text{E}}^{\alpha/\text{L}}}{2 - \dfrac{4}{\pi}}\right)\end{aligned} \tag{10.94}$$

If we replace $V_{\text{growth}} {r_\alpha}^2$ by the constant C_{rod} and replace the radius r_α by $\sqrt{C_{\text{rod}}/V_{\text{growth}}}$ in Equation (10.94) we obtain

$$\begin{aligned}\Delta T_{\text{E}} = {} & \frac{1}{2}\left(m_\alpha^{\text{L}} + m_\beta^{\text{L}}\right)\left(x_{\text{E}}^{\text{L}} - x_0^{\text{L}}\right) + \frac{\sqrt{V_{\text{growth}}}}{\sqrt{C_{\text{rod}}}}\frac{\sigma_{\alpha\beta}}{RT}\left(m_\beta^{\text{L}} V_{\text{m}}^\beta \frac{1}{\dfrac{r_\beta}{r_\alpha}\left(2 + \dfrac{r_\beta}{r_\alpha}\right)} - m_\alpha^{\text{L}} V_{\text{m}}^\alpha\right) \\ & + \frac{\sqrt{V_{\text{growth}} C_{\text{rod}}}\, V_{\text{m}}^{\text{L}}}{4D}\left(\frac{m_\beta^{\text{L}}}{V_{\text{m}}^\beta}\frac{r_\beta}{r_\alpha}\frac{2 + \dfrac{r_\beta}{r_\alpha}}{2 + \dfrac{4 r_\beta}{\pi r_\alpha}}\left(x_{\text{E}}^{\beta/\text{L}} - x_{\text{E}}^{\text{L}}\right) - \frac{m_\alpha^{\text{L}}}{V_{\text{m}}^\alpha}\frac{x_{\text{E}}^{\text{L}} - x_{\text{E}}^{\alpha/\text{L}}}{2 - \dfrac{4}{\pi}}\right)\end{aligned} \tag{10.101}$$

where

$$\frac{r_\beta}{r_\alpha} = \sqrt{\frac{1}{f_\alpha} - 1} \tag{10.96}$$

As the first term in Equation (10.101) is small it can normally be neglected. In this case, Equation (10.101) can be written as

$$\Delta T_E = const_{rod}\sqrt{V_{growth}} \tag{10.102}$$

or

$$V_{growth} = Const_{rod} \times (\Delta T_E)^2 \tag{10.103}$$

where

$$const_{rod} = \frac{\sigma_{\alpha\beta}}{RT\sqrt{C_{rod}}}\left(\frac{m_\beta^L V_m^\beta}{\frac{r_\beta}{r_\alpha}\left(2+\frac{r_\beta}{r_\alpha}\right)} - m_\alpha^L V_m^\alpha\right) + \frac{\sqrt{C_{rod}}V_m^L}{4D}\left(\frac{m_\beta^L}{V_m^\beta}\frac{r_\beta}{r_\alpha}\frac{2+\frac{r_\beta}{r_\alpha}}{2+\frac{4r_\beta}{\pi r_\alpha}}\left(x_E^{\beta/L} - x_E^L\right) - \frac{m_\alpha^L}{V_m^\alpha}\frac{x_E^L - x_E^{\alpha/L}}{2-\frac{4}{\pi}}\right) \tag{10.104}$$

Equations (10.102) and (10.103) are analogous to Equations (10.49) and (10.50) on page 603 for lamellar eutectic growth. If we introduce the expression in Equation (10.100) on page 617 into Equation (10.104) it can be transformed into

$$const_{rod} = \sqrt{\frac{\sigma_{\alpha\beta}A_{rod}B_{rod}V_m^L}{RTD}} \tag{10.105}$$

where

$$A_{rod} = \frac{m_\beta^L V_m^\beta}{\frac{r_\beta}{r_\alpha}\left(2+\frac{r_\beta}{r_\alpha}\right)} - m_\alpha^L V_m^\alpha$$

$$B_{rod} = \frac{m_\beta^L}{V_m^\beta}\frac{r_\beta}{r_\alpha}\frac{2+\frac{r_\beta}{r_\alpha}}{2+\frac{4r_\beta}{\pi r_\alpha}}\left(x_E^{\beta/L} - x_E^L\right) - \frac{m_\alpha^L}{V_m^\alpha}\frac{x_E^L - x_E^{\alpha/L}}{2-\frac{4}{\pi}}$$

In analogy with lamellar eutectic growth Equation (10.101) can be written as

$$\Delta T_E = \frac{1}{2}\left(m_\alpha^L + m_\beta^L\right)\left(x_E^L - x_0^L\right) + const_{lamella}\sqrt{V_{growth}} \tag{10.106}$$

Influence of Interface Kinetics on Growth Rate

V_{growth} in Equation (10.98) on page 616 represents the growth rate of the phases when the influence of surface tension and diffusion but the effect of interface kinetics is not considered. In many eutectic reactions one of the solid phases (let us call it α) obtains a faceted structure, which indicates a strong interface kinetic influence. In such cases, it is necessary to pay attention to the interface kinetics.

Relationship between Growth Rate and Undercooling when Interface Kinetics is Considered
In order to analyze the effect of interface kinetics on the growth rate theoretically we have to go back to Equations (10.20) and (10.21) on page 597. For the β phase no change is necessary. The second term in Equation (10.20) can be written as

$$x_{\sigma}^{L/\alpha} - x_{kin}^{L/\alpha} = \left(\frac{V_{growth}}{\mu_{kin}^{\alpha}}\right)^{\frac{1}{n_{kin}^{\alpha}}} \tag{10.55}$$

where μ_{kin}^{α} is the kinetic growth constant. n_{kin}^{α} is a constant number between 1 and 2. Theory predicts (see the margin box) that $n_{kin}^{\alpha} = 1$ but experiments indicate that n_{kin}^{α} is close to 2 for most alloys. Hence, it is reasonable to use the lower expression in the margin box. If this expression is introduced into Equation (10.20) on page 597. Equation (10.83) on page 614 will be replaced by

$$x^{L/\alpha} - x_0^{L} = \frac{2\sigma_{\alpha\beta}}{r_{\alpha}}\frac{V_m^{\alpha}}{RT} + \sqrt{\frac{V_{growth}}{\mu_{\alpha}}} + \frac{V_{growth}\left(x_E^{L} - x_E^{\alpha/L}\right)}{2D}\,\frac{r_{\alpha}}{2-\dfrac{4}{\pi}}\,\frac{V_m^{L}}{V_m^{\alpha}} \tag{10.107}$$

Basic formula:

$$V_{growth} = \mu_{kin}^{\alpha}\left(x_{\sigma}^{L/\alpha} - x_{kin}^{L/\alpha}\right)^{n_{kin}^{\alpha}}$$

For $n_{kin}^{\alpha} = 1$ we obtain

$$x_{\sigma}^{L/\alpha} - x_{kin}^{L/\alpha} = \frac{V_{growth}}{\mu_{kin}^{\alpha}}$$

For $n_{kin}^{\alpha} = 2$ we obtain

$$x_{\sigma}^{L/\alpha} - x_{kin}^{L/\alpha} = \sqrt{\frac{V_{growth}}{\mu_{kin}^{\alpha}}}$$

We use the same procedure as on pages 614–616 to derive a new equation for the undercooling, which will replace Equation (10.94) on page 615. The result is

$$\Delta T_E = \frac{1}{2}\left(m_{\alpha}^{L} + m_{\beta}^{L}\right)\left(x_E^{L} - x_0^{L}\right) + \frac{1}{r_{\alpha}}\frac{\sigma_{\alpha\beta}}{RT}\left(m_{\beta}^{L}V_m^{\beta}\frac{1}{\dfrac{r_{\beta}}{r_{\alpha}}\left(2+\dfrac{r_{\beta}}{r_{\alpha}}\right)} - m_{\alpha}^{L}V_m^{\alpha}\right)$$
$$-\frac{m_{\alpha}^{L}}{2}\sqrt{\frac{V_{growth}}{\mu_{kin}^{\alpha}}} + \frac{V_{growth}r_{\alpha}V_m^{L}}{4D}\left(\frac{m_{\beta}^{L}}{V_m^{\beta}}\frac{r_{\beta}}{r_{\alpha}}\frac{2+\dfrac{r_{\beta}}{r_{\alpha}}}{2+\dfrac{4r_{\beta}}{\pi r_{\alpha}}}\left(x_E^{\beta/L} - x_E^{L}\right) - \frac{m_{\alpha}^{L}}{V_m^{\alpha}}\frac{x_E^{L} - x_E^{\alpha/L}}{2-\dfrac{4}{\pi}}\right) \tag{10.108}$$

where

$$\frac{r_{\beta}}{r_{\alpha}} = \sqrt{\frac{1}{f_{\alpha}}} - 1 \tag{10.96}$$

Equation (10.108) is the modified relationship between undercooling and growth rate when the kinetic reaction at the interface is considered. In analogy with Equation (10.106), Equation (10.108) can alternatively be written as

$$\Delta T_E = \frac{1}{2}\left(m_{\alpha}^{L} + m_{\beta}^{L}\right)\left(x_E^{L} - x_0^{L}\right) + \left(const_{rod} - \frac{m_{\alpha}^{L}}{2\sqrt{\mu_{kin}^{\alpha}}}\right)\sqrt{V_{growth}} \tag{10.109}$$

Equation (10.109) is analogous to Equation (10.58) on page 605 for lamellar eutectic growth.

Relationship between Growth Rate and Rod Radius when Interface Kinetics is Considered

To find the relationship between the growth rate and the main rod distance we proceed in the same way as on pages 616–617.

The minimum of ΔT_E is obtained if we take the derivative Equation (10.108) with respect to r_{α} and set the derivative to zero while V_{growth} is regarded as a constant.

The minimum condition for the undercooling will be

$$\frac{\Delta T_E}{dr_{\alpha}} = 0 - \frac{1}{r_{\alpha}^2}\frac{\sigma_{\alpha\beta}}{RT}\left(m_{\beta}^{L}V_m^{\beta}\frac{1}{\dfrac{r_{\beta}}{r_{\alpha}}\left(2+\dfrac{r_{\beta}}{r_{\alpha}}\right)} - m_{\alpha}^{L}V_m^{\alpha}\right) - 0$$
$$+\frac{V_{growth}V_m^{L}}{4D}\left(\frac{m_{\beta}^{L}}{V_m^{\beta}}\frac{r_{\beta}}{r_{\alpha}}\frac{2+\dfrac{r_{\beta}}{r_{\alpha}}}{2+\dfrac{4r_{\beta}}{\pi r_{\alpha}}}\left(x_E^{\beta/L} - x_E^{L}\right) - \frac{m_{\alpha}^{L}}{V_m^{\alpha}}\frac{x_E^{L} - x_E^{\alpha/L}}{2-\dfrac{4}{\pi}}\right) = 0$$

which gives

$$V_{\text{growth}} r_{\alpha}{}^{2} = \frac{\sigma_{\alpha\beta} \times 4D \left(m_{\beta}^{\text{L}} V_{\text{m}}^{\beta} \dfrac{1}{\dfrac{r_{\beta}}{r_{\alpha}}\left(2 + \dfrac{r_{\beta}}{r_{\alpha}}\right)} - m_{\alpha}^{\text{L}} V_{\text{m}}^{\alpha} \right)}{RT\, V_{\text{m}}^{\text{L}} \left(\dfrac{m_{\beta}^{\text{L}}}{V_{m}^{\beta}} \dfrac{r_{\beta}}{r_{\alpha}} \dfrac{2 + \dfrac{r_{\beta}}{r_{\alpha}}}{2 + \dfrac{4 r_{\beta}}{\pi r_{\alpha}}} \left(x_{\text{E}}^{\beta/\text{L}} - x_{\text{E}}^{\text{L}}\right) - \dfrac{m_{\alpha}^{\text{L}}}{V_{\text{m}}^{\alpha}} \dfrac{x_{\text{E}}^{\text{L}} - x_{\text{E}}^{\alpha/\text{L}}}{2 - \dfrac{4}{\pi}} \right)} \qquad (10.110)$$

Equation (10.110) is identical with Equation (10.98) on page 616, which is valid when the interface kinetics is neglected.

The relationship $V_{\text{growth}} \lambda^{2} = C_{\text{rod}}$ is valid as long as the diffusion in the melt dominates over the kinetic reaction at the interface. If the diffusion term is much smaller than the kinetic term, the former can be neglected and Equation (10.110) is no longer valid.

The simplified Equations (10.102) and (10.103) on page 618 are not valid as the undercooling ΔT_{E} contains a kinetic term. The interface kinetics has a strong effect on the undercooling.

Comparison between Normal Lamellar and Rod Eutectic Growths

A comparison between normal lamellar eutectic growth and normal rod eutectic growth shows striking similarities. Many of the formulas are analogous but more complex in the case of rod eutectic growth than at lamellar growth. The reason is that the geometry of the structure is more complicated for rod eutectic growth than for lamellar growth.

In Section 10.8 we will shortly discuss the transition from lamellar eutectic growth to rod eutectic growth and vice versa.

10.4 Degenerate and Coupled Eutectic Growth

In Sections 10.3 we have only considered the possibility that the two phases grow with the same rate. Here we will discuss the alternative when one of the phases grows more rapidly than the other. Such growth is called *degenerate eutectic growth* or *irregular eutectic growth*. Irregular eutectic growth implies that the growth of the two phases occur simultaneously but without coupling to each other.

Degenerate eutectic growth also includes the case when there is a coupling between the phases even if they grow with different growth rates. i.e. follow different growth laws. Such so-called *coupled growth* will be discussed in Section 10.4.3 on page 624.

10.4.1 Types of Instable Eutectic Interfaces

At degenerate eutectic growth in binary alloys the cooperation between the two eutectic phases no longer exists. The almost planar eutectic solid/liquid interface becomes instable, which results in other types of microstructures of the solidified alloy than normal eutectic structures.

There are two types of eutectic solid/liquid interface instability, so-called *single-phase instability* and *two-phase instability*.

Single-phase instabilities develop when only one of the phases becomes instable. The growths of the phases result in two different structures.

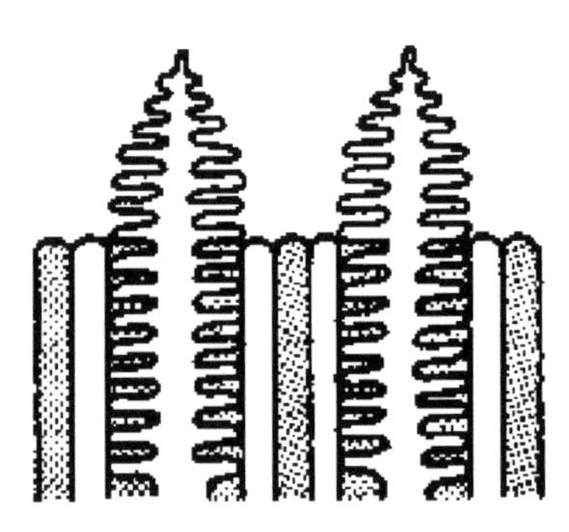

Figure 10.23 Single-phase instability of a planar eutectic solid/liquid interface. Reproduced with permission from Trans Tech © 1992.

Figure 10.23 shows the structure that arises from growth of an instable primary phase, for example dendrites or faceted crystals. The instable phase grows ahead of the eutectic solidification front.

Two-phase instabilities of planar eutectic interfaces either depend on a variation in the lamellar distance (Figure 10.24a and Section 10.4.2 below) or on the influence of a third element, which destabilizes the whole structure because of a long-range diffusion boundary layer, formed ahead of the solid/liquid interface.

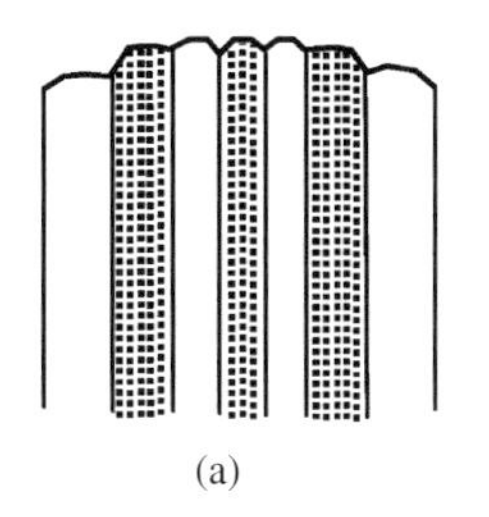
(a)

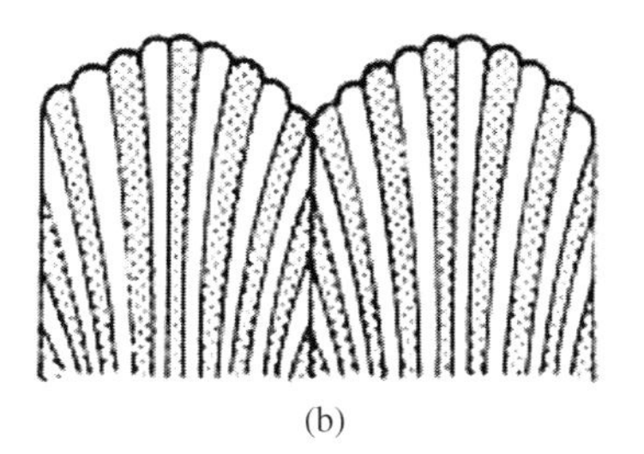
(b)

Figure 10.24 Two-phase instability of a planar eutectic solid/liquid interface. Reproduced with permission from Trans Tech © 1992.

It may result in formation of two-phase eutectic cells or dendrites (Figure 10.24b). This case will be discussed more extensively in Section 10.5 on page 635. If there is no cooperation between the phases an irregular structure is formed (Figure 10.36 on page 635).

10.4.2 *Shape of Solidification Front and Lamellar Spacing Adjustment at Coupled Growth*

Relationship between Undercooling and Lamellar Spacing

The necessary undercooling for eutectic growth is composed of one component owing to diffusion and another component associated with surface tension and the curvature of the lamellae. According to Equation (10.42) on page 600 the eutectic undercooling can be written as the sum of two terms

$$\Delta T = K_1/\lambda + K_2 V_{\text{growth}}\lambda \tag{10.111}$$

where
ΔT = undercooling
$K_1\, K_2$ = constants, specific for each alloy
V_{growth} = growth rate of the solidification front
λ = lamellar spacing.

The function (10.111) is shown in Figure 10.25. The figure shows that the undercooling ΔT does not uniquely define the spacing λ as several λ values are possible. The curve shows a minimum close to twice the smallest existing value λ_c. The true point $2\lambda_c$ is close to and slightly to the right of the minimum point. However, it is very difficult to adjust the lamellar distances experimentally in most cases.

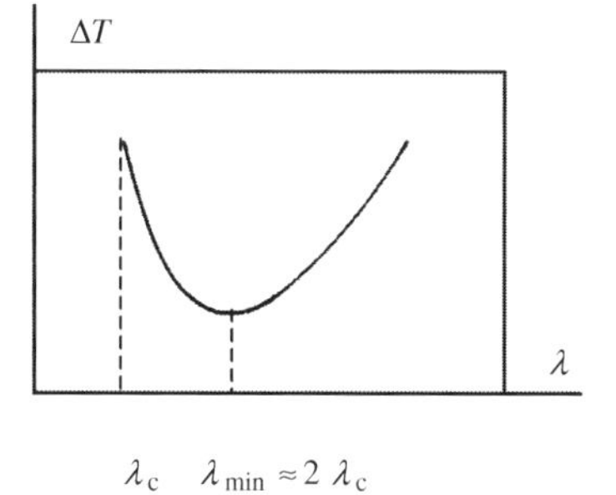

Figure 10.25 Undercooling as a function of lamellar spacing for a eutectic alloy at a certain growth rate.

Puls and Kiraldy [5] proposed that the eutectic may contain a spectrum of lamellar spacings. Carlberg and Fredriksson [6] analyzed this assumption theoretically and assumed that the lamellar spacing spectrum can be described by the following relationship

$$\lambda = \lambda_0 \pm \frac{\Delta\lambda}{2}\sin\frac{2\pi}{n\lambda_0}y \tag{10.112}$$

where
λ_0 = selected lamellar spacing
$\Delta\lambda$ = spread of lamellar spacing
$n\lambda_0$ = magnitude of the interval between two consequtive lamellae with wavelength λ_0
y = coordinate.

The expression in Equation (10.112) is introduced into Equation (10.111), which gives

$$\Delta T = \frac{K_1}{\lambda_0 \pm \frac{\Delta\lambda}{2}\sin\frac{2\pi}{n\lambda_0}y} + K_2 V_{\text{growth}}\left(\lambda_0 \pm \frac{\Delta\lambda}{2}\sin\frac{2\pi}{n\lambda_0}y\right) \tag{10.113}$$

Shape of the Solidification Front

With the aid of Equation (10.113) we can describe the character of the solidification front in a unidirectional solidification experiment with a constant temperature gradient.

Consider as an example the case of an Al-$CuAl_2$ alloy, which is allowed to solidify at a rate of 10^{-5} m/s in a temperature gradient of 10^4 K/m. For this alloy the constants in Equation (10.113) have the values $K_1 = 5.1 \times 10^{-7}$ Km and $K_2 = 1.3 \times 10^{10}$ Ks/m^2. The wavelength is assumed to be $10\lambda_0$ where λ_0 is the optimum value predicted by the theory.

The results of the calculations are given in Figure 10.26, which shows the calculated shape of the solidification front for the Al-$CuAl_2$ alloy at a growth rate of 1×10^{-5} m/s at different values of $\Delta\lambda$.

Figure 10.26 Calculated shape of solidification front when (a) $\Delta\lambda = 2 \times 10^{-7}$ m (left); (b) $\Delta\lambda = 4 \times 10^{-7}$ m (right). Reproduced from Elsevier © 1977.

The conclusion is that

- The solidification front becomes smoother when $\Delta\lambda$ decreases.

Comparison with other alloys shows in addition that

- The solidification front becomes smoother when the ratio K_1/K_2 decreases.

It has also been observed that the change in lamellar spacing is associated with the fault frequency of the eutectic structure. The lamellae adjust there distances in the optimal way, which corresponds to the fastest possible growth rate at the given undercooling.

During the solidification process there is a competition between different parts of the solidification front. The regions which lead have the opportunity to spread out at the expense of the back regions.

Influence of the Growth Rate on the Shape of the Solidification Front

If the growth rate is gradually increased from a given initial rate under constant conditions, the lamellar spacing will deviate from the optimum value $2\lambda_c$. Calculations can be performed, which show the changes of the shape of the solidification front when the growth rate varies. An example is shown in Figure 10.27. The figure shows how the shape of the solidification front in the Al-$CuAl_2$ alloy in Figure 10.26b changes when the growth rate increases.

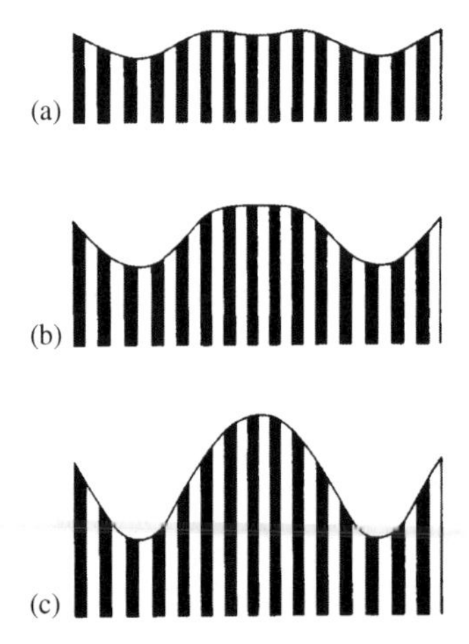

Figure 10.27 Calculated shape of the solidification front at increasing growth rate for the case in Figure 10.26b when $\Delta\lambda = 4 \times 10^{-7}$ m and (a) $V_{growth} = 1.10 \times 10^{-5}$ m/s; (b) $V_{growth} = 1.25 \times 10^{-5}$ m/s; (c) $V_{growth} = 1.50 \times 10^{-5}$ m/s. © Elsevier 1977.

The calculations show that the solidification front becomes unstable. Both the frequency and the amplitude of the roughness increases with increasing growth rate. The figure and comparison with alloys with other ratios of K_1/K_2 shows that

- The roughness of the solidification front increases with the lamellar variation $\Delta\lambda$ and the value of the ratio K_1/K_2.

  and

- The region with the optimal lamellar spacing grows faster than other regions.

Owing to the instability of the solidification front areas of faster growth spread out ahead of other areas, which results in a modification of the lamellar spacing. The rate of the variations depends on the experimental conditions, including the growth rate, temperature gradient and the convection pattern in the molten alloy.

Discontinuous Changes of Lamellar Spacing in Alloys owing to Change of Solidification Rate

Sudden discontinuous changes of the lamellar spacing have been observed in alloys. Figure 10.28 shows an example.

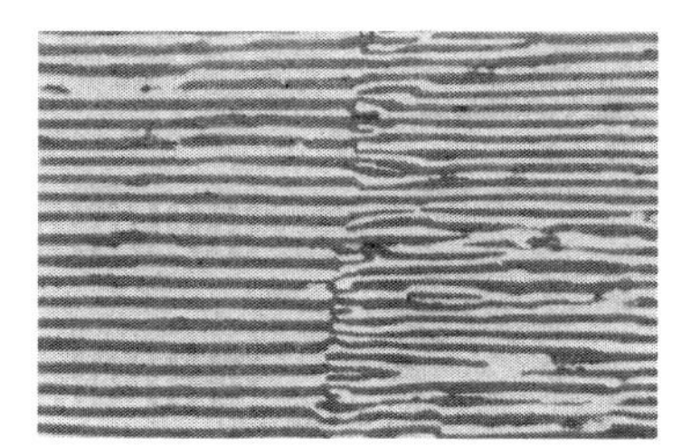

Figure 10.28 Longitudinal cross-section of an Al–$CuAl_2$ eutectic with a sudden and drastic decrease of the lamellar spacing. © Elsevier 1977.

This phenomenon can be explained in the following way. The undercooling as a function of the lamellar spacing for some different growth rates are shown in Figures 10.29a and b. The undercooling required for the solidification process obviously increases with increasing growth rate. Two mechanisms for the sudden reduction of the lamellar spacing are possible.

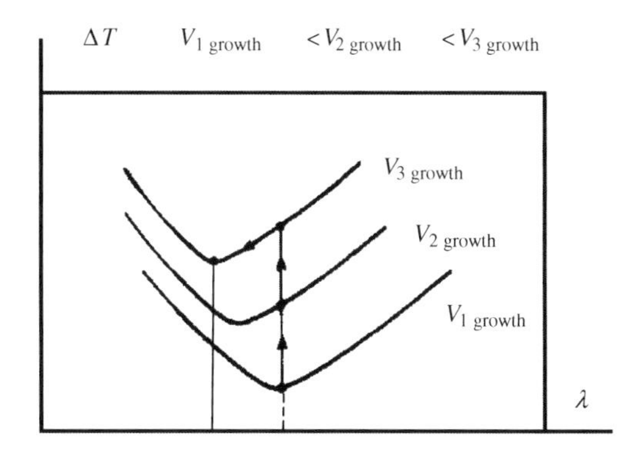

Figure 10.29a Undercooling as a function of lamellar spacing for some different growth rates. © Elsevier 1977.

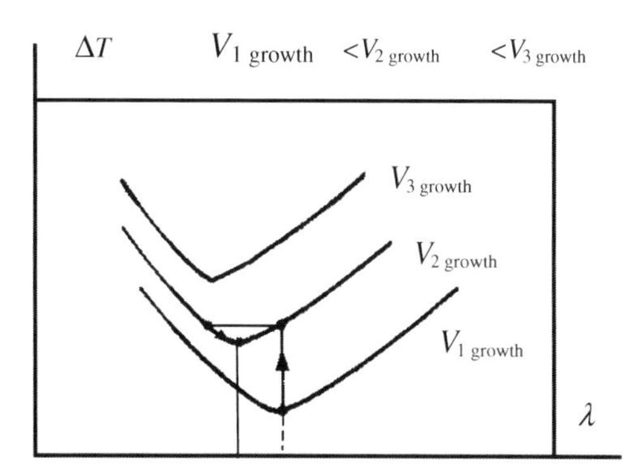

Figure 10.29b Undercooling as a function of lamellar spacing for some different growth rates. © Elsevier 1977.

The curves show the undercooling as a function of the lamellar spacing at various growth rates. At the growth rate V_1 the equilibrium lamellar spacing corresponds to the minimum value of the undercooling (Figure 10.29a). When the growth rate is increased the equilibrium becomes disturbed and the undercooling increases, primarily with no change of the lamellar spacing. At a certain value V_3 the system diverges so much from a new equilibrium that it is achieved by a discontinuous 'jump' to the lamellar spacing, which corresponds the minimum undercooling at growth rate V_3.

Figure 10.29b shows the alternative process. The equilibrium at growth rate V_1, originally corresponding to the optimal conditions, is disturbed when the growth rate is increased. At a higher growth rate V_2 a discontinuous 'jump' to a much smaller lamellar distance occurs, immediately followed by another 'jump' down to a larger lamellar distance and a new optimal undercooling. A new equilibrium and a new lamellar distance have been achieved.

The energy law, the material constants of the alloy and the external conditions determine the mechanism, that is valid in the individual case.

10.4.3 Competitive Eutectic Growth and Growth of Primary Phases. Coupled Growth

Competitive Eutectic Growth. Growth of Primary Phases

In Sections 10.3.2 and 10.3.3 normal lamellar and rod eutectic structures, respectively, have been penetrated. When an alloy melt of arbitrary composition gradually cools, phases of different structures will precipitate. The phases may be a primary α or β phase, normally with dendritic structure or a plate-like structure of faceted crystals, and/or a normal eutectic phase, in most cases with a lamellar structure.

The three possible phases are always competing and the resulting structure depends on the cooling conditions and the type and composition of the alloy. In Section 10.4.1 on page 620 it was mentioned that primary dendritic structure and normal eutectic structure can occur together. These topics will be discussed below.

The precipitated structure depends on the temperature and growth rates of the solid phases. Each structure has its own characteristic relationship between growth rate and temperature, which can be shown as a curve in a diagram. With the aid of these so-called *growth laws*, regions of different structures can be identified and plotted into the phase diagram of the alloy.

Growth Laws of the α and β Phases

The structure of each phase grows according to its own growth law. Growth laws are normally expressed as relationships between the growth rate and the undercooling.

Relationships between growth rate and composition have been derived in Chapter 9. Simple examples are Equations (9.9) on page 488 for faceted growth and Equation (9.84) on page 532 for dendritic growth, both in Chapter 9. These relationships can be used to find the growth laws. With the aid of the slopes of the liquidus line in the phase diagram composition differences can be replaced by temperature differences.

For the α and β phases the following laws are generally valid.

$$V_{\text{growth}}^{\alpha} = \mu_{\alpha}\left(x^{\text{L}/\alpha} - x_0^{\text{L}}\right)^{n_{\alpha}} \tag{10.114}$$

$$V_{\text{growth}}^{\beta} = \mu_{\beta}\left(x_0^{\text{L}} - x^{\text{L}/\beta}\right)^{n_{\beta}} \tag{10.115}$$

where

$V_{\text{growth}}^{\alpha\beta}$ = growth rate of the α and β phases, respectively, at temperature T

T = actual temperature of the melt

x_0^{L} = concentration of the alloying element B in the melt

$x^{\text{L}/\alpha,\beta}$ = concentration of alloying element B in the melt ahead of the α and β phases, respectively, when they solidify at temperature T

$n_{\alpha,\beta}$ = a constant, characteristic for the α and β phases, respectively.

The values of n_{α} and n_{β} are normally numbers between 1 and 2 for most alloys. Equations (10.114) and (10.115) are generally valid. The diffusion of the alloying element in the melt or the kinetic reaction at the interface between the liquid and solid phases controls the growth. The shape of the equations will be the same in both cases.

With the aid of the slopes of the liquidus lines in the phase diagram, the concentration differences can be replaced by temperature differences. From Figure 10.12 we obtain the relationships

$$x^{L/\alpha} - x_0^L = \frac{T_\alpha^L - T}{-m_\alpha^L} \tag{10.116}$$

$$x_0^L - x^{L/\beta} = \frac{T_\beta^L - T}{m_\beta^L} \tag{10.117}$$

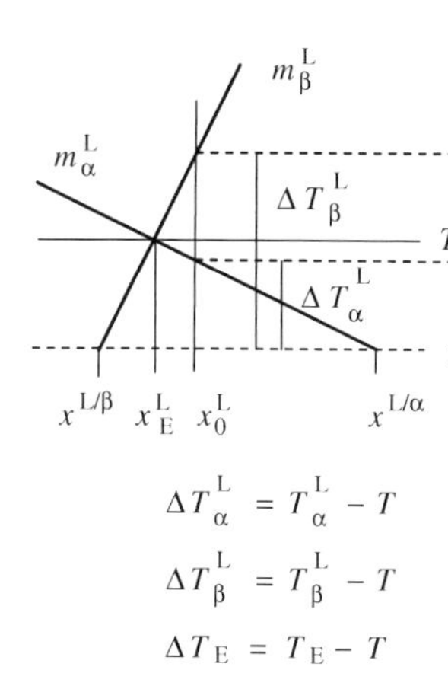

Figure 10.12 Part of the phase diagram close to the eutectic point (temperature T versus concentration x).

where

$T_{\alpha,\beta}^L$ = the temperature at which the solid phase α and β, respectively, started to precipitate in the melt with concentration x_0^L

$m_{\alpha,\beta}^L$ = slope of the liquidus line of the α and β phases, respectively.

The growth laws of the α and β phases are obtained by introduction of the concentration differences in Equations (10.116) and (10.117) into Equations (10.114) and (10.115), respectively

$$V_{\text{growth}}^\alpha = \frac{\mu_\alpha}{\left(-m_\alpha^L\right)^{n_\alpha}}\left(T_\alpha^L - T\right)^{n_\alpha} \tag{10.118}$$

$$V_{\text{growth}}^\beta = \frac{\mu_\beta}{\left(m_\beta^L\right)^{n_\beta}}\left(T_\beta^L - T\right)^{n_\beta} \tag{10.119}$$

where

$\mu_{\alpha,\beta}$ = growth constant of the α and β phases, respectively

$\Delta T_{\alpha,\beta}^L = T_{\alpha,\beta}^L - T$ = undercooling ahead of phases α and β, respectively.

It should be noted that *undercoolings* and *growth rates* are always positive quantities. The slopes of the liquidus lines are frequently included in the growths constants. Here we will call them μ_{eff}^α and μ_{eff}^β, respectively.

Growth Law of the Eutectic Phase

The growth law of the eutectic phase at normal eutectic lamellar growth, i.e. a relationship between the undercooling and the growth rate, is derived in Section 10.3.2. It is given in two versions.

The simplest version, Equation (10.53) on page 604, is valid when the kinetic reaction at the interface is rapid and its influence on the growth rate can be neglected.

If the kinetic reaction at the solidification front influences the growth a kinetic term has to be added. This case has been treated in Section 10.3.2. on pages 591–605 for normal eutectic growth.

Equation (10.58) on page 605, must be used when the kinetic reaction of one phase is slow and has to be considered together with the surface tension and the diffusion of the alloying element in the melt. It is more general and is given here.

$$\Delta T_E = \frac{1}{2}\left(m_\alpha^L + m_\beta^L\right)\left(x_E^L - x_0^L\right) + \left(const_{\text{lamella}} - \frac{m_\alpha^L}{2\sqrt{\mu_{\text{kin}}^\alpha}}\right)\sqrt{V_{\text{growth}}^E} \tag{10.58}$$

where

$$const_{\text{lamella}} = \sqrt{\frac{\sigma_{\alpha\beta}A_{\text{lamella}}B_{\text{lamella}}C_{0\ \text{lamella}}}{RTD}}$$

$const_{\text{lamella}}$ is always positive. The second term in Equation (10.58) always dominates over the first one.

Table 10.1 Constants of normal lamellar eutectic growth

$$A_{\text{lamella}} = \frac{m_\beta^L V_m^\beta}{f_\beta} - \frac{m_\alpha^L V_m^\alpha}{f_\alpha}$$

$$B_{\text{lamella}} = \frac{m_\beta^L}{V_m^\beta} - \frac{m_\alpha^L}{V_m^\alpha}$$

$$C_{0\ \text{lamella}} = f_\alpha f_\beta V_m^L\left(x_E^{\beta/L} - x_E^{\alpha/L}\right)$$

$\Delta T_{\alpha}^{L} = T_{\alpha}^{L} - T$

$\Delta T_{\beta}^{L} = T_{\beta}^{L} - T$

$\Delta T_{E} = T_{E} - T$

Figure 10.12 Part of the phase diagram close to the eutectic point (temperature T versus concentration x).

Undercooling as a Function of Growth Rate and Composition of the Alloy

The undercooling of the α and β phases and the eutectic phase depends on two variables, the growth rate and the composition of the alloy.

The expressions for the undercooling ΔT_{α}^{L} and ΔT_{β}^{L} are solved from Equations (10.118) and (10.119), respectively

$$\Delta T_{\alpha}^{L} = T_{\alpha}^{L} - T = -m_{\alpha}^{L}\left(\frac{V_{\text{growth}}^{\alpha}}{\mu_{\alpha}}\right)^{\frac{1}{n_{\alpha}}} \tag{10.120}$$

$$\Delta T_{\beta}^{L} = T_{\beta}^{L} - T = m_{\beta}^{L}\left(\frac{V_{\text{growth}}^{\beta}}{\mu_{\beta}}\right)^{\frac{1}{n_{\beta}}} \tag{10.121}$$

The temperatures T_{α}^{L} and T_{β}^{L} are found in the phase diagram at the intersections of the line $x = x_{0}^{L}$ and the liquidus lines of the α and β phases, respectively. Figure 10.12 shows that the following relationships are valid

$$T_{\alpha}^{L} = T_{E} + m_{\alpha}^{L}(x_{0}^{L} - x_{E}^{L}) \tag{10.122}$$

$$T_{\beta}^{L} = T_{E} + m_{\beta}^{L}(x_{0}^{L} - x_{E}^{L}) \tag{10.123}$$

These expressions are introduced into Equations (10.120) and (10.121), respectively, which gives

$$\Delta T_{\alpha}^{L} = T_{E} + m_{\alpha}^{L}(x_{0}^{L} - x_{E}^{L}) - T = -m_{\alpha}^{L}\left(\frac{V_{\text{growth}}^{\alpha}}{\mu_{\alpha}}\right)^{\frac{1}{n_{\alpha}}} \tag{10.124}$$

$$\Delta T_{\beta}^{L} = T_{E} + m_{\beta}^{L}(x_{0}^{L} - x_{E}^{L}) - T = m_{\beta}^{L}\left(\frac{V_{\text{growth}}^{\beta}}{\mu_{\beta}}\right)^{\frac{1}{n_{\beta}}} \tag{10.125}$$

For the eutectic phase Equation (10.58) is valid, which gives the undercooling of the eutectic phase:

$$\Delta T_{E} = T_{E} - T = \frac{1}{2}\left(m_{\alpha}^{L} + m_{\beta}^{L}\right)\left(x_{E}^{L} - x_{0}^{L}\right) + \left(const_{\text{lamella}} - \frac{m_{\alpha}^{L}}{2\sqrt{\mu_{\text{kin}}^{\alpha}}}\right)\sqrt{V_{\text{growth}}^{E}} \tag{10.126}$$

The Equations (10.124), (10.125) and (10.126) can be used to study the influence on undercooling of growth rate and composition.

Influence on Undercooling of Growth Rate

The growth rate is normally determined by the experimental conditions, i.e. by the heat flow, and can be regarded as constant at stationary conditions or as a function of the heat flow if it changes during the solidification process.

Hence, the growth rates is known and the undercoolings of the three phases are determined by the Equations (10.124), (10.125) and (10.126). They show that

- When the growth rate is increased the undercooling increases.

This is true for all the structures, i.e. α, β and eutectic structure.

Influence on Undercooling of Composition

It can be concluded from Figure 10.12 above and Equations (10.124) and (10.125) that if x_{0}^{L} is *increased* the undercooling will *decrease* for the α phase and *increase* for the β phase.

Competitive Growth. Criterion of Winning Phase

There is always a competition between the three possible structures α, β and eutectic structure. The eutectic structure is a mixture of the α and β phases growing in eutectic proportions as normal lamellar or rod eutectics.

- The phase, that requires the *lowest* undercooling, i.e. corresponds to the *highest* temperature T will win. It corresponds to the *lowest* growth rate at the given temperature.

The temperature T in Equations (10.124), (10.125) and (10.126) is solved, which gives for the α phase:

$$T = T_{\mathrm{E}} + m_{\alpha}^{\mathrm{L}}\left(x_0^{\mathrm{L}} - x_{\mathrm{E}}^{\mathrm{L}}\right) + m_{\alpha}^{\mathrm{L}}\left(\frac{V_{\mathrm{growth}}^{\alpha}}{\mu_{\alpha}}\right)^{\frac{1}{n_{\alpha}}} \tag{10.127}$$

for the β phase:

$$T = T_{\mathrm{E}} + m_{\beta}^{\mathrm{L}}\left(x_0^{\mathrm{L}} - x_{\mathrm{E}}^{\mathrm{L}}\right) - m_{\beta}^{\mathrm{L}}\left(\frac{V_{\mathrm{growth}}^{\beta}}{\mu_{\beta}}\right)^{\frac{1}{n_{\beta}}} \tag{10.128}$$

for the eutectic phase:

$$T = T_{\mathrm{E}} + m\left(x_0^{\mathrm{L}} - x_{\mathrm{E}}^{\mathrm{L}}\right) - \left(const_{\mathrm{lamella}} - \frac{m_{\alpha}^{\mathrm{L}}}{2\sqrt{\mu_{\mathrm{kin}}^{\alpha}}}\right)\sqrt{V_{\mathrm{growth}}^{\mathrm{E}}} \tag{10.129}$$

where

$$m = \frac{1}{2}\left(m_{\alpha}^{\mathrm{L}} + m_{\beta}^{\mathrm{L}}\right)$$

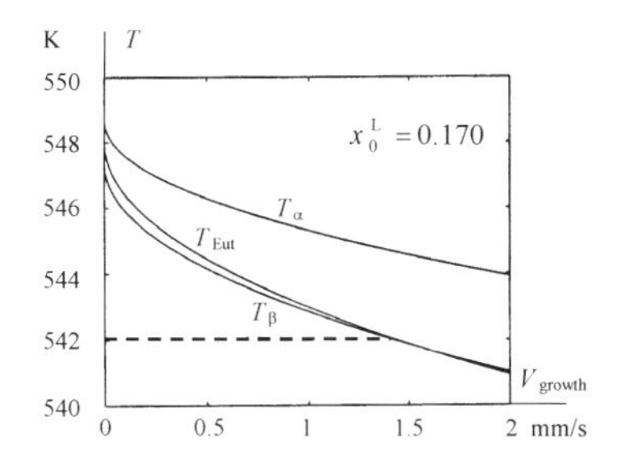

Figure 10.30a Temperature as a function of growth rate for the α and β curves and the eutectic curve for a hypoeutectic Al–$CuAl_2$ alloy with composition $x_0^{\mathrm{L}} = 0.170$. The β and the eutectic curves intersect at $T = 542$ K.

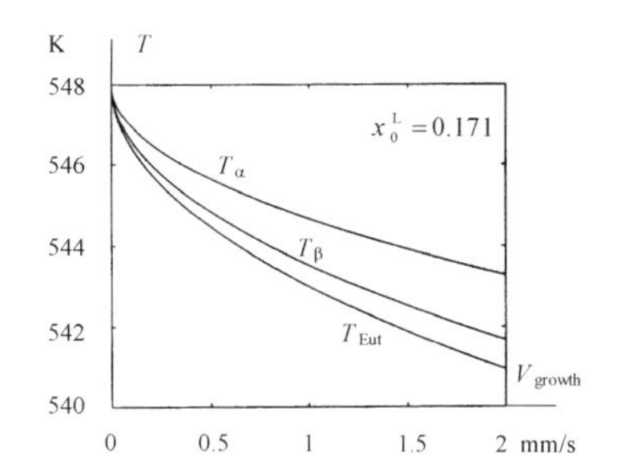

Figure 10.30b Temperature as a function of growth rate for the α and β curves and the eutectic curve for a eutectic Al–$CuAl_2$ alloy with composition $x_0^{\mathrm{L}} = 0.171$. No intersections at all between the three curves.

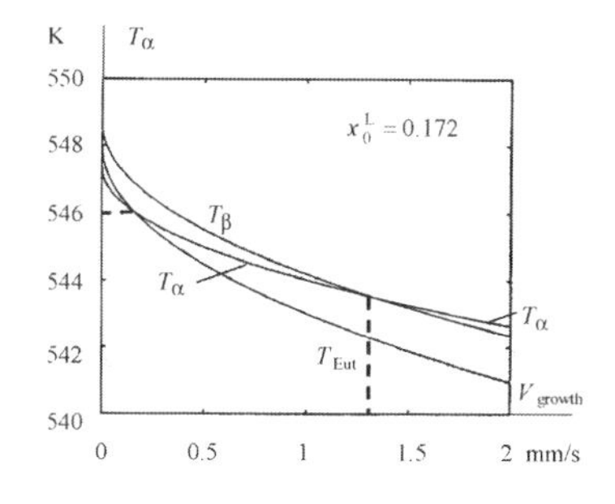

Figure 10.30c Temperature as a function of growth rate for the α and β curves and the eutectic curve for a hypereutectic Al–$CuAl_2$ alloy with composition $x_0^{\mathrm{L}} = 0.172$. The α and the eutectic curves intersect at $T = 546$ K.

If we plot T for a given composition of the alloy as a function of V_{growth} for the three phases in the same diagram and apply the principle given above, we can use the plot to predict which one of the phases will be the winning phase. To demonstrate the method we will use Al-$CuAl_2$ alloys as an example.

Figures 10.30a, b and c show the temperatures of the phases as functions of the growth rate for three different initial compositions of the molten alloy, a hypoeutectic melt ($x_0^{\mathrm{L}} < x_{\mathrm{E}}^{\mathrm{L}}$), the eutectic melt ($x_0^{\mathrm{L}} = x_{\mathrm{E}}^{\mathrm{L}}$) and a hypereutectic melt ($x_0^{\mathrm{L}} > x_{\mathrm{E}}^{\mathrm{L}}$). The curves are based on the material constants given in Example 10.2 on pages 628–632.

Figure 10.30a show that the hypoeutectic alloy solidifies with α structure, i.e. as α dendrites, at all growth rates within the interval given in the figure. The α curve diverges even more from the other curves at all reasonable growth rates and no intersect will ever occur.

Figure 10.30b shows that the eutectic alloy precipitates with α structure at all growth rates because it corresponds to the highest temperature and lowest undercooling.

Figure 10.30c shows that the hypereutectic alloy precipitates with β structure, i.e. as β dendrites at low growth rates. At growth rates $> 1.3 \times 10^{-3}$ m/s α dendrites starts to precipitate.

Coupled Growth. Coupled Zone

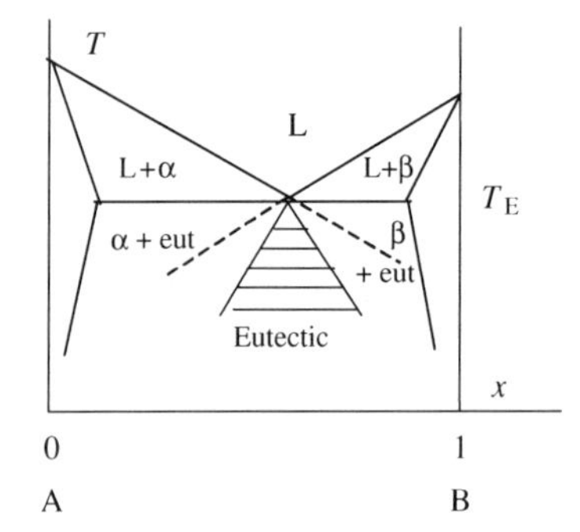

Figure 10.31 Symmetrical coupled zone. A symmetrical coupled zone is obtained when the growth rates of the primary α and β phases are the same at the same undercooling.

Above we have found that there is always a competition between the primary phases α and β and the eutectic phase. We want to find the region in the phase diagram within which the *eutectic structure* wins. This region is called the *coupled zone* or *the region of coupled growth*. In most cases, it arises from normal lamellar eutectic growth.

The *coupled zone* is defined as the region with a eutectic structure, where the α and β phases cooperate during growth. Figure 10.31 shows that the composition within the coupled zone varies. It is always located below the eutectic line.

If the growth rates of primary α phase and primary β phase are approximately the same at the same undercooling, the coupled zone will be symmetrically located below the eutectic temperature. This case is shown in Figure 10.31. The coupled zone can never exceed the region, which is limited by the prolonged liquidus lines.

The coupled zone and the primary phases are separated by two *boundary curves* that separate the α phase from the eutectic phase and the β phase from the eutectic phase, respectively. Along the boundary curves the two phases cooperate (Figure 10.23 on page 620).

- At each point on the boundary curves the growth rates of the two adjacent structures must be equal.

With the aid of this condition, the known growth laws of the three phases, the phase diagram and the material constants of the alloy it is possible to find the boundary curves. It can be done either by calculation of the equations of the boundary curves or graphically.

To demonstrate the procedures we will apply both methods on a simple example, the eutectic Al-$CuAl_2$ alloy below.

Al is the α phase
$CuAl_2$ is the β phase.
$n_\alpha = 2$ (α phase)
$n_\beta = 2$ (β phase)
$\mu_{eff}^{\alpha} = 9 \times 10^{-5}$ m/s
$\mu_{eff}^{\beta} = 5 \times 10^{-5}$ m/s
$m_\alpha^L = -648$ K*
$m_\beta^L = 670$ K*
$T_E = 548\,°C = 821$ K
$x_E^{\alpha/L} = 0.025$
$x_E^L = 0.171$
$x_E^{\beta/L} = 0.319$
$V_m^\alpha = 10.6 \times 10^{-3}$ m³/kmol
$V_m^\beta = 11.0 \times 10^{-3}$ m³/kmol
$V_m^L = 9.76 \times 10^{-3}$ m³/kmol
$\sigma_{\alpha,\beta} = 0.2$ J/m²
$D = 3 \times 10^{-9}$ m²/s

*Calculated from an enlarged phase diagram of the Al-Cu system.

Calculation of the Boundary Curves of the Coupled Zone

Example 10.2

Find the region of coupled growth for the eutectic Al-$CuAl_2$ alloy and plot the coupled zone in the phase diagram. The growth laws of the α and β phases, Equations (10.118) and (10.119), can be written as

$$V_{growth}^{\alpha} = \frac{\mu_\alpha}{\left(-m_\alpha^L\right)^{n_\alpha}} \left(T_\alpha^L - T\right)^{n_\alpha} = \mu_{eff}^{\alpha} \left(T_\alpha^L - T\right)^{n_\alpha}$$

$$V_{growth}^{\beta} = \frac{\mu_\beta}{\left(m_\beta^L\right)^{n_\beta}} \left(T_\beta^L - T\right)^{n_\beta} = \mu_{eff}^{\beta} \left(T_\beta^L - T\right)^{n_\beta}$$

The growth Equation (10.126) on page 626 is valid for the eutectic phase of the alloy Al-$CuAl_2$. The kinetic reaction at the interface is supposed to be rapid and can be neglected. The solid alloy has a normal lamellar structure.

Required material constants are given in the margin. The growth constants and the $n_{\alpha,\beta}$ values below have been determined by experiments. Part of the phase diagram of the system Al-Cu is given on page 629.

Solution

Growth Law of the Eutectic Phase
In this case the kinetic term in Equation (10.126) is missing. Hence, the growth law for the eutectic phase can be written as

$$T_E - T = \frac{m_\alpha^L + m_\beta^L}{2}\left(x_E^L - x_0^L\right) - const_{lamella}\sqrt{V_{growth}^E}$$

which can be transformed into

$$V_{growth}^E = \frac{\left(T_E + m\right)\left(x_0^L - x_E^L\right) - T\right)^2}{const_{lamella}^2} = \mu_{eff}^E\left[T_E + m\left(x_0^L - x_E^L\right) - T\right]^2 \qquad (1')$$

where m is the average of the slopes of the liquidus lines of the α and β phases.

Calculation of μ_{eff}^E
For future use we start to calculate the value of μ_{eff}^E. The relationship between the constant in Equation (1′) and μ_{eff}^E is

$$\mu_{eff}^E = \frac{1}{const_{lamella}^2}$$

As the alloy Al-CuAl$_2$ has a normal lamellar eutectic structure the constant in question is Equation (10.52) given on page 604

$$const_{lamella} = \sqrt{\frac{\sigma_{\alpha\beta}A_{lamella}B_{lamella}C_{0\ lamella}}{RTD}} \qquad (2')$$

where $\sigma_{\alpha,\beta}$ is the surface tension between the α and β phases and D is the diffusion constant of the alloying element in the melt. The remaining quantities in Equation (2′) are defined in the margin. μ_{eff}^E can be calculated from the relationship

$$\mu_{eff}^E = \frac{1}{const_{lamella}^2} = \frac{RTD}{\sigma_{\alpha\beta}A_{lamella}B_{lamella}C_{0\ lamella}} \qquad (3')$$

We will need the fractions of the α and β phases in the eutectic structure. They are obtained from the enlarged phase diagram and application of the lever rule (see Figure 10.1 on page 588). With the aid of the given numerical values we obtain [Equations (10.12) and (10.13) on page 594]

$$f_\alpha = \frac{x_E^{\beta/L} - x_E^L}{x_E^{\beta/L} - x_E^{\alpha/L}} = \frac{0.319 - 0.171}{0.319 - 0.025} = 0.503$$

$$f_\beta = \frac{x_E^L - x_E^{\alpha/L}}{x_E^{\beta/L} - x_E^{\alpha/L}} = \frac{0.171 - 0.025}{0.319 - 0.025} = 0.497$$

Now the quantities in Equation (3′) can be calculated:

$$A_{lamella} = \frac{m_\beta^L V_m^\beta}{f_\beta} - \frac{m_\alpha^L V_m^\alpha}{f_\alpha} = \frac{670 \times 11.0 \times 10^{-3}}{0.497} - \frac{(-648) \times 10.6 \times 10^{-3}}{0.503} = 28.5\,\frac{\text{Km}^3}{\text{kmol}}$$

$$B_{lamella} = \frac{m_\beta^L}{V_m^\beta} - \frac{m_\alpha^L}{V_m^\alpha} = \frac{670}{11.0 \times 10^{-3}} - \frac{-648}{10.6 \times 10^{-3}} = 1.22 \times 10^5\ \text{K kmol/m}^3$$

$$C_{0\ lamella} = f_\alpha f_\beta V_m^L\left(x_E^{\beta/L} - x_E^{\alpha/L}\right)$$

Table 10.1 Constants of normal lamellar eutectic growth

$$A_{lamella} = \frac{m_\beta^L V_m^\beta}{f_\beta} - \frac{m_\alpha^L V_m^\alpha}{f_\alpha}$$

$$B_{lamella} = \frac{m_\beta^L}{V_m^\beta} - \frac{m_\alpha^L}{V_m^\alpha}$$

$$C_{0\ lamella} = f_\alpha f_\beta V_m^L\left(x_E^{\beta/L} - x_E^{\alpha/L}\right)$$

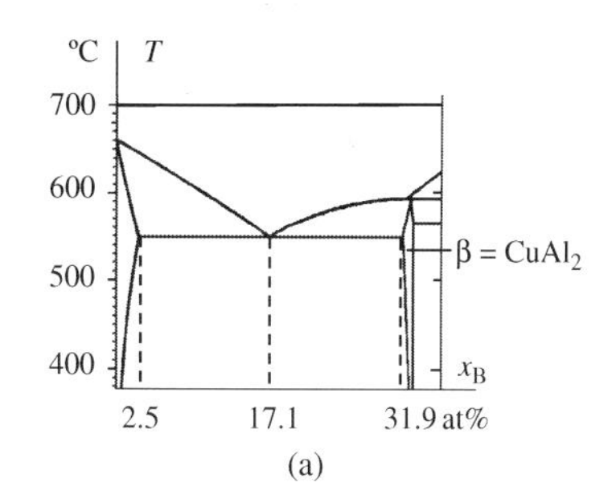

Part of the phase diagram of the system Al–Cu. From Binary Alloy Phase Diagrams Vol. 1, Ed. Thaddeus B. Massalski, American Society for Metals 1986. Reprinted with permission of ASM International. www.asminternational.org.

or

$$C_{0\,\text{lamella}} = 0.503 \times 0.497 \times 9.76 \times 10^{-3} \times (0.319 - 0.025) = 7.17 \times 10^{-4}\ \text{m}^3/\text{kmol}$$

Inserting the numerical values into Equation (3′) gives

$$\mu_{\text{eff}}^{\text{E}} = \frac{8.31 \times 10^3 \times 821 \times 3 \times 10^{-9}}{0.2 \times 28.5 \times 1.22 \times 10^5 \times 7.17 \times 10^{-4}} = 4 \times 10^{-5}\ \text{m/K}^2\text{s}$$

Growth Laws of the α and β Phases
The α phase
As $n_\alpha = 2$ the growth law of the α phase, Equation (10.118), can be written as

$$V_{\text{growth}}^{\alpha} = \frac{\mu_\alpha}{\left(-m_\alpha^{\text{L}}\right)^2}\left(T_\alpha^{\text{L}} - T\right)^2 = \mu_{\text{eff}}^{\alpha}\left(T_\alpha^{\text{L}} - T\right)^2 \tag{4′}$$

The right-hand side of Equation (4′) is positive even if the slope of the liquidus line of the α phase is negative. We introduce the expression into Equation (10.122) on page 626 for T_α^{L} into Equation (4′) and obtain

$$V_{\text{growth}}^{\alpha} = \mu_{\text{eff}}^{\alpha}\left(T_{\text{E}} + m_\alpha^{\text{L}}\left(x_0^{\text{L}} - x_{\text{E}}^{\text{L}}\right) - T\right)^2 \tag{5′}$$

The β phase
As $n_\beta = 2$ the growth law of the β phase can be written as

$$V_{\text{growth}}^{\beta} = \frac{\mu_\beta}{\left(m_\beta^{\text{L}}\right)^2}\left(T_\beta^{\text{L}} - T\right)^2 = \mu_{\text{eff}}^{\beta}\left(T_\beta^{\text{L}} - T\right)^2 \tag{6′}$$

We introduce the expression into Equation (10.123) on page 626 for T_β^{L} into (6′) and obtain

$$V_{\text{growth}}^{\beta} = \mu_{\text{eff}}^{\beta}\left[T_{\text{E}} + m_\beta^{\text{L}}\left(x_0^{\text{L}} - x_{\text{E}}^{\text{L}}\right) - T\right]^2 \tag{7′}$$

Boundary Curve between the α Phase and the Eutectic Phase
The basic condition $V_{\text{growth}}^{\alpha} = V_{\text{growth}}^{\text{E}}$ is applied and we obtain with the aid of Equations (5′) and (1′):

$$\mu_{\text{eff}}^{\alpha}\left[T_{\text{E}} + m_\alpha^{\text{L}}\left(x_o^{\text{L}} - x_{\text{E}}^{\text{L}}\right) - T_{\text{intersect}}\right]^2 = \mu_{\text{eff}}^{\text{E}}\left[T_{\text{E}} + m\left(x_o^{\text{L}} - x_{\text{E}}^{\text{L}}\right) - T_{\text{intersect}}\right]^2 \tag{8′}$$

Equation (8′) is the desired boundary curve. When x_0^{L} varies (different composition of the alloy) the point $(x_0^{\text{L}}, T_{\text{intersect}})$ moves along the curve in the phase diagram. At $x_0^{\text{L}} = x_{\text{E}}^{\text{L}}$ the temperature $T_{\text{intersect}}$ is equal to T_{E}.

We solve $T_{\text{intersect}}$ as a function of x_0^{L}. There are two alternatives.

$$T_{\text{E}} + m_\alpha^{\text{L}}\left(x_0^{\text{L}} - x_{\text{E}}^{\text{L}}\right) - T_{\text{intersect}} = \pm\sqrt{\frac{\mu_{\text{eff}}^{\text{E}}}{\mu_{\text{eff}}^{\alpha}}}\left[T_{\text{E}} + m\left(x_0^{\text{L}} - x_{\text{E}}^{\text{L}}\right) - T_{\text{intersect}}\right] \quad \text{page 627} \tag{9′}$$

If we introduce $a = \sqrt{\frac{\mu_{\text{eff}}^{\text{E}}}{\mu_{\text{eff}}^{\alpha}}}$ the solutions can be written as

$$T_{\text{intersect}} = T_{\text{E}} + \frac{m_\alpha^{\text{L}} \mp a\,m}{1 \mp a}\left(x_0^{\text{L}} - x_{\text{E}}^{\text{L}}\right) \tag{10′}$$

The two solutions to Equation (8′) represent two straight lines. Only one of them is the correct boundary line. To find the correct alternative we examine the slopes of the two lines. The slope of the boundary line must be steeper than the liquidus line as the coupled zone is always *smaller* than the region between the prolonged liquidus lines.

The slopes are

$$\text{Slope 1} = \frac{m_\alpha^L - a\,m}{1 - a} \quad \text{and} \quad \text{Slope 2} = \frac{m_\alpha^L + a\,m}{1 + a}$$

We insert the numerical values

$$m = 11 \quad \text{and} \quad a = \sqrt{\frac{\mu_{\text{eff}}^E}{\mu_{\text{eff}}^\alpha}} = \sqrt{\frac{4 \times 10^{-5}}{9 \times 10^{-5}}} = \frac{2}{3} = 0.67$$

and obtain, for comparison with the slope of the liquidus line $m_\alpha^L = -648$ K, the slope of the boundary line:

$$\text{Slope 1} = \frac{-648 - 0.67 \times 11}{1 - 0.67} = -1945 \text{ K}$$

or

$$\text{Slope 2} = \frac{-648 + 0.67 \times 11}{1 + 0.67} = -384 \text{ K}$$

Obviously solution 1 [upper sign in Equation (10′)] is the boundary line. It is plotted in the phase diagram in the answer.

Boundary Curve between the β Phase and the Eutectic Phase

Completely analogous calculations can be performed for the β phase. Even if m_β^L is positive in this case we do not take one of the alternative solutions for granted from the beginning.

The basic condition $V_{\text{growth}}^\beta = V_{\text{growth}}^E$ is applied on the β phase.

With the aid of Equations (7′) and (1′) we obtain

$$T_E + m_\beta^L(x_0^L - x_E^L) - T_{\text{intersect}} = \pm\sqrt{\frac{\mu_{\text{eff}}^E}{\mu_{\text{eff}}^\beta}}\left[T_E + m(x_0^L - x_E^L) - T_{\text{intersect}}\right]$$

or

$$T_{\text{intersect}} = T_E + \frac{m_\beta^L \mp b\,m}{1 \mp b}(x_0^L - x_E^L) \tag{11′}$$

where

$$b = \sqrt{\frac{\mu_{\text{eff}}^E}{\mu_{\text{eff}}^\beta}} = \sqrt{\frac{4}{5}} = 0.89$$

The slope of the boundary line (11′) is

$$\text{Slope 1} = \frac{670 - 0.89 \times 11}{1 - 0.89} = 6000 \text{ K}$$

or

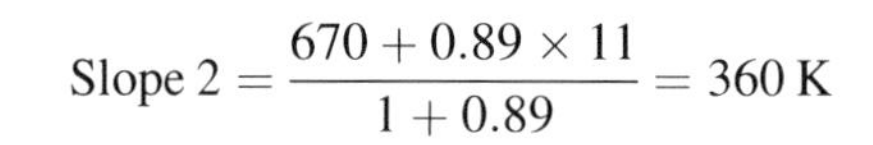

$$\text{Slope } 2 = \frac{670 + 0.89 \times 11}{1 + 0.89} = 360\ \text{K}$$

The slope of the liquidus line is $m_{\beta}^{L} = 670$ K.

Obviously the upper sign in Equation (10′) corresponds to the boundary line.

Answer:

The equations of the two boundary lines are

$$T_{\text{intersect}} = T_{\text{E}} + \frac{m_{\alpha}^{L} - a\,m}{1 - a}(x_0^L - x_E^L) \quad \text{where} \quad a = \sqrt{\frac{\mu_{\text{eff}}^{\text{E}}}{\mu_{\text{eff}}^{\alpha}}}$$

$$T_{\text{intersect}} = T_{\text{E}} + \frac{m_{\beta}^{L} - b\,m}{1 - b}(x_0^L - x_E^L) \quad \text{where} \quad b = \sqrt{\frac{\mu_{\text{eff}}^{\text{E}}}{\mu_{\text{eff}}^{\beta}}}$$

The lines have been plotted in the phase diagram in the figure for the given numerical values. The coupled zone is shown in the margin figure.

(b)

From Binary Alloy Phase Diagrams Vol. 1, Ed. Thaddeus B. Massalski, American Society for Metals 1986. Reprinted with permission of ASM International. All rights reserved. www.asminternational.org.

If one of the phases grows at lower undercooling than the other one, there will be an asymmetrical coupled zone below the horizontal eutectic point in the phase diagram. The asymmetry is also by influenced by the slopes of the liquidus lines.

- The coupled zone is found on the same side of the eutectic point as the phase with the steepest slope of the liquidus line.

In the case of Al-$CuAl_2$ the region of coupled growth is symmetrical. If there are large differences between the growth constants of the two phases the coupled zone will be highly asymmetrical. An example of an asymmetrical coupled zone is shown in Figure 10.34b on page 634 and in the Figures 10.35a and b on page 634.

Growth Constants and Growth Equations

$\mu_{\text{eff}}^{\alpha}$ is related to an equation of the type (10.118) on page 625

$$V_{\text{growth}}^{\alpha} = \mu_{\text{eff}}^{\alpha}(T_{\alpha}^{L} - T)^{n_{\alpha}} = \frac{\mu_{\alpha}}{(-m_{\alpha}^{L})^{n_{\alpha}}}(T_{\alpha}^{L} - T)^{n_{\alpha}}$$

The relationship between μ_{α} and $\mu_{\text{eff}}^{\alpha}$, for the special case $n_{\alpha} = 2$, is

$$\mu_{\text{eff}}^{\alpha} = \frac{\mu_{\alpha}}{(-m_{\alpha}^{L})^{2}}$$

For the β phase an analogous relationship is valid when $n_{\beta} = 2$

$$\mu_{\text{eff}}^{\beta} = \frac{\mu_{\beta}}{(-m_{\beta}^{L})^{2}}.$$

__Graphical Method to Find the Coupled Zone__

The graphical way to find the boundary curves of the coupled zone is based on the close relationship between the boundary curves of the coupled zone and the temperature versus growth rate curves of the three phases. The method can be described with the aid of a concrete example. We will use the alloy Al-$CuAl_2$ for an easy comparison with the method in Example 10.2.

1. The temperature as a function of the growth rate is derived for the three phases. In the case of Al-$CuAl_2$ Equation (10.127) on page 627 can be applied with $n_{\alpha} = 2$ and Equation (10.128) with $n_{\beta} = 2$. By use of $\mu_{\text{eff}}^{\alpha}$ instead of μ_{α} they can be written as

$$T = T_{\text{E}} + m_{\alpha}^{L}(x_0^L - x_E^L) + \frac{\sqrt{V_{\text{growth}}^{\alpha}}}{\sqrt{\mu_{\text{eff}}^{\alpha}}} \tag{10.127}$$

$$T = T_{\text{E}} + m_{\beta}^{L}(x_0^L - x_E^L) - \frac{\sqrt{V_{\text{growth}}^{\beta}}}{\sqrt{\mu_{\text{eff}}^{\beta}}} \tag{10.128}$$

These curves are drawn in Figures 10.30 on page 627.

Equation (10.129) on page 627 is valid for the eutectic phase. As the kinetic reaction can be neglected in this case, Equation (10.129) will be

$$T = T_E + m(x_0^L - x_E^L) - const \times V_{growth}^{\beta} \tag{10.129}$$

2. The eutectic T-V_{growth} curve and the corresponding β curve are plotted in a diagram for a given value of the concentration $x_0^L = 0.166$. The eutectic point corresponds to $x_0^L = 0.171$. In another diagram the eutectic curve and the α curve are plotted. T is plotted as a function of V_{growth} for the composition $x_0^L = 0.176$. The diagrams are shown in Figures 10.32a and 10.32c.
3. Figure 10.32a shows that the eutectic curve and the β curve intersect at the temperature 517 K. $V_{growth}^{\alpha} = V_{growth}^{E}$ at this temperature and the point (0.166, 517) must be a point on the left boundary curve.

   Analogously the intersection point $T = 538$ K is read from the diagram in Figure 10.32c when $V_{growth}^{\beta} = V_{growth}^{E}$. Hence, the intersection point (0.176, 538) must be a point on the right boundary curve.
4. The procedure described above is repeated for a considerable number of other concentrations and the intersection temperatures are plotted into Figure 10.32b. This gives the whole region of coupled growth. From the Figures 10.30a and c on page 627 we obtain the two intersection points (0.170, 542) and (0.172, 546). The intersection points (Table 10.2) are plotted in Figure 10.32b. The lines in Figure 10.32b are the calculated boundary lines in Example 10.2.

Table 10.2 Intersection points:

x_0^L	$T_{intersect}$
0.166	517 K
0.170	542 K
0.171	548 K
0.172	546 K
0.176	538 K

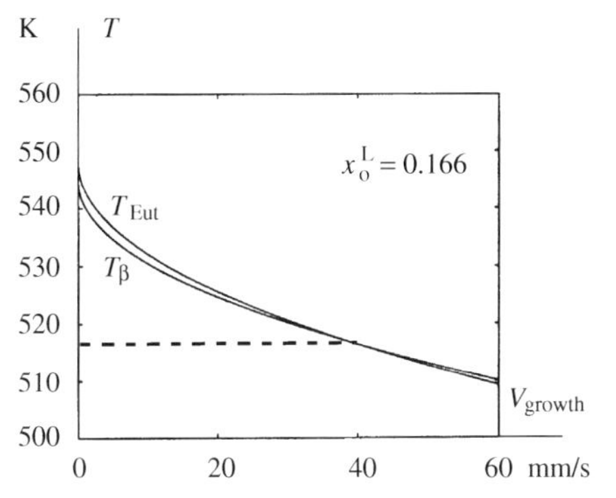

Figure 10.32a $T_{intersect} = 517$ K.

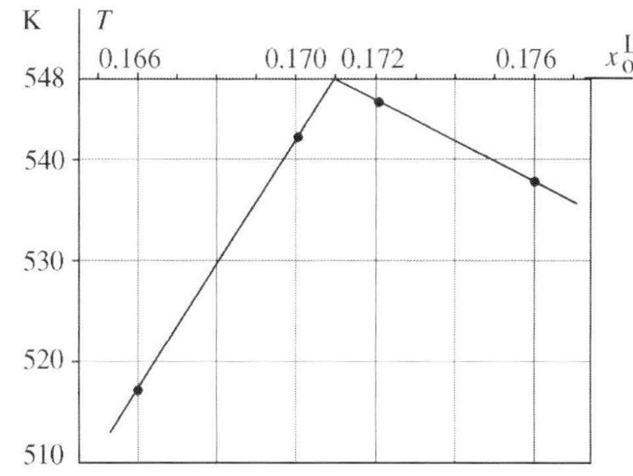

Figure 10.32b Coupled zone in the phase diagram.

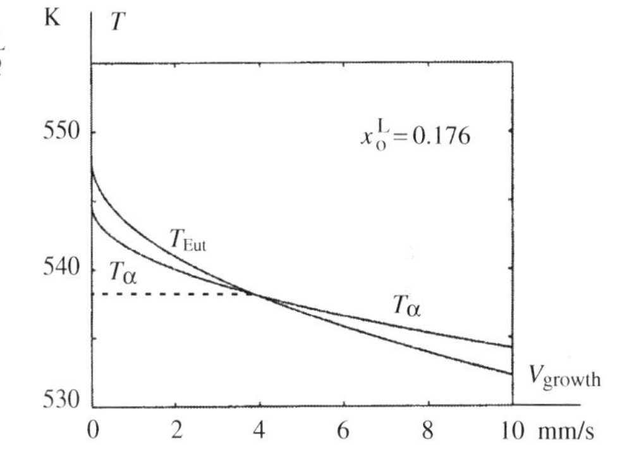

Figure 10.32c $T_{intersect} = 525$ K.

The two methods are actually the same. The graphic way is more visual and clear but much slower than the mathematical method, demonstrated in Example 10.2.

Effect of a Temperature Gradient on the Coupled Zone

The discussion given above is valid for solidification at low temperature gradients in the melt. This is the case in most casting processes.

At unidirectional solidification processes, such as in a Bridgman furnace, the extension of the coupled zone is also influenced by the temperature gradient.

Figure 10.33 gives an example of an extended region of the coupled zone.

The widening of the upper part of the coupled zone in Figure 10.33 is an effect of the temperature gradient, that appears during unidirectional solidification experiments. In this case the diffusion field changes in the same way as in the case of cellular growth in binary systems and suppresses the formation of a dendritic structure. The result is that the volume fractions of the α and β phases in the eutectic structure correspond to the original composition of the alloy.

Two concrete and technically important examples of coupled growth will be discussed more extensively later, Fe-C eutectics in Sections 10.6 and Al-Si eutectics in Section 10.7.

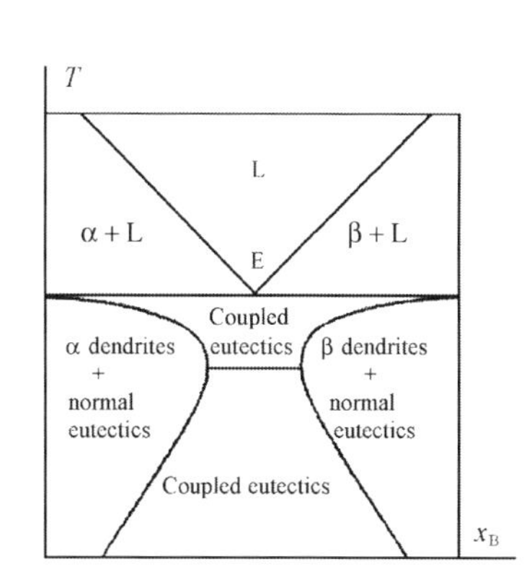

Figure 10.33 Phase diagram of binary alloy with zones of coupled eutectic and dendritic structure. Reproduced with permission from Trans Tech © 1992.

Solidification Processes in the Coupled Zone

The boundary lines of the coupled zone represent the temperature-composition region where a primary phase and the eutectic phase cooperate and grow simultaneously *at the same rate*. Inside the coupled zone the growth results in a normal eutectic structure of the lamellar or rod type. This region always lies below the eutectic equilibrium temperature as is shown in the Figures 10.34a and b.

If the melting points of the eutectic components differ very much or if one of the phases grows anisotropically, the coupled zone will be asymmetrical relative to the eutectic point as is shown in Figure 10.34b. In this case the undercooling of a melt of eutectic composition results in solidification outside the coupled zone, i.e. the two phases cannot grow simultaneously at a low temperature even if the composition is eutectic. The nucleation of the two phases controls the further course of the crystallization.

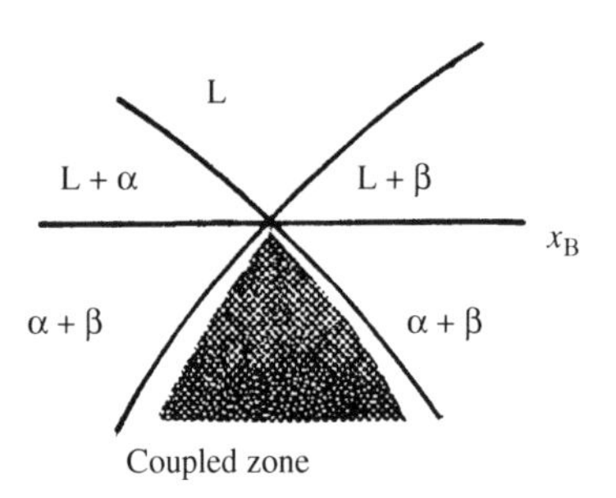

Figure 10.34a Coupled zone in a symmetrical system. From The Solidification of Metals, Iron and Steel Institute, 1967 © Maney www.maney.co.uk.

Figure 10.34b Coupled zone in an asymmetrical system. The coupled zone is always located on the side of the steepest liquidus line. From The Solidification of Metals, Iron and Steel Institute, 1967 © Maney www.maney.co.uk.

Observations show that the coupled zone in such cases does not start at the eutectic point, i.e. eutectic composition and temperature. It starts at a lower temperature and at a composition, which deviates from the eutectic one as is shown in the Figures 10.35. The solidification can be described in the following way.

In a *hypoeutectic* alloy ($x_0^L < x_E^L$) the α-dendrites nucleate first, at point 1 in Figure 10.35a, above the eutectic temperature. The temperature drops and the composition of the melt changes along the liquidus line until point 2 is reached. At point 2 β-dendrites nucleate, owing to high undercooling. The precipitation of β-dendrites changes the composition of the melt to the left, towards lower B-concentration. Then coupled growth of β- and α-eutectics starts and occurs inside the coupled zone.

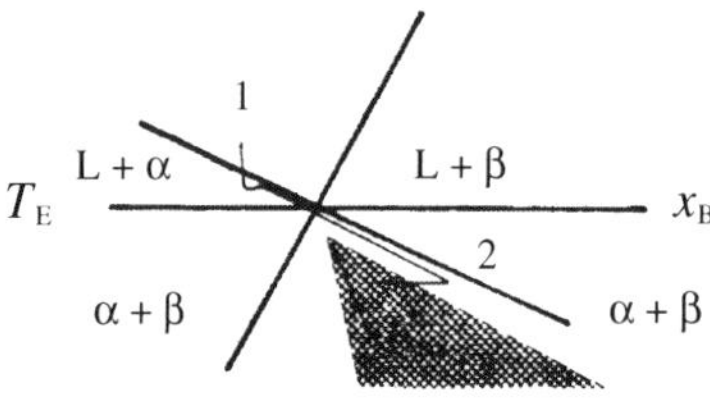

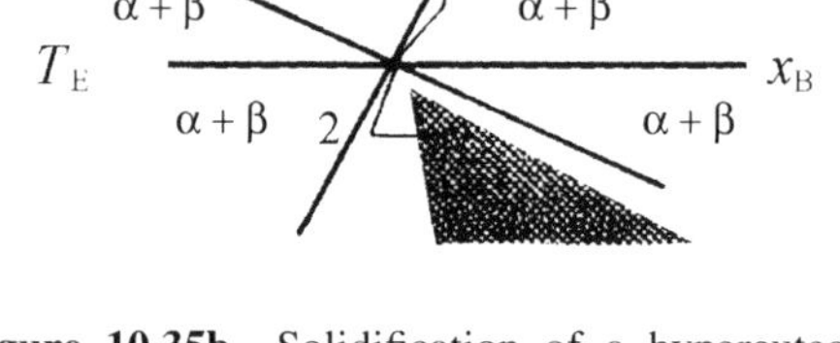

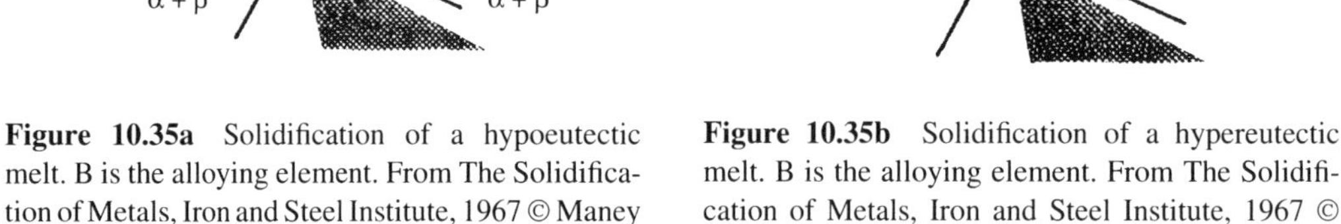

Figure 10.35a Solidification of a hypoeutectic melt. B is the alloying element. From The Solidification of Metals, Iron and Steel Institute, 1967 © Maney www.maney.co.uk.

Figure 10.35b Solidification of a hypereutectic melt. B is the alloying element. From The Solidification of Metals, Iron and Steel Institute, 1967 © Maney www.maney.co.uk.

On the other hand, in a *hypereutectic* melt ($x_0^L > x_E^L$) the β-dendrites nucleate first, for example at point 1 in Figure 10.35b. The B-concentration of the melt becomes displaced towards the left, away from the coupled zone. α-dendrite nucleation starts at point 2, when the supersaturation of the α-phase and the undercooling are high enough. The composition of the melt becomes displaced to the right and coupled growth of β- and α-eutectics starts when the composition is brought into the coupled zone.

10.4.4 *Irregular Eutectic Growth*

In Section 10.3 normal lamellar and rod eutectic growth have been treated extensively and a simplified theory is given in the Sections 10.3.2 for lamellar eutectic growth and in Section 10.3.3 for rod eutectic growth.

The definition of normal eutectic growth is that the two phases cooperate and grow with the same growth rate. The condition for this is that the kinetic reaction at the interface can be disregarded for both phases. If this is the case we obtain comparatively simple relationships between the growth rate and the lamellar distance in the case of lamellar growth (Equation (10.46) on page 601) and between the growth rate and the rod radius at rod eutectic growth (Equation (10.99) on page 617):

$$V_{\text{growth}}\,\lambda^2 = C_{\text{lamella}} \tag{10.46}$$

and

$$V_{\text{growth}}\,r_\alpha^2 = C_{\text{rod}} \tag{10.99}$$

In both cases a simple relationship is valid between the undercooling and the growth rate. They are Equation (10.50) on page 603 and Equation (10.103) on page 618.

$$V_{\text{growth}} = Const_{\text{lamella}} \times (\Delta T_{\text{E}})^2 \tag{10.103}$$

$$V_{\text{growth}} = Const_{\text{rod}} \times (\Delta T)_{\text{E}}^2 \tag{10.50}$$

When the kinetic reaction is taken into consideration Equation (10.53) on page 604 is no longer valid. Instead a kinetic term, based on the growth Equation (10.55) on pages 604, is introduced into Equation (10.53), which gives Equation (10.58) on page 605:

$$\Delta T_{\text{E}} = \frac{1}{2}\left(m_\alpha^{\text{L}} + m_\beta^{\text{L}}\right)\left(x_{\text{E}}^{\text{L}} - x_0^{\text{L}}\right) + \left(const_{\text{lamella}} - \frac{m_\alpha^{\text{L}}}{2\sqrt{\mu_{\text{kim}}^{\alpha}}}\right)\sqrt{V_{\text{growth}}} \tag{10.58}$$

where $const_{\text{lamella}}$ is given in Equation (10.52) on page 604.

Analogously Equation (10.109) on page 619 will replace Equation (10.106) on page 618:

$$\Delta T_{\text{E}} = \frac{1}{2}\left(m_\alpha^{\text{L}} + m_\beta^{\text{L}}\right)\left(x_{\text{E}}^{\text{L}} - x_0^{\text{L}}\right) + \left(const_{\text{rod}} - \frac{m_\alpha^{\text{L}}}{2\sqrt{\mu_{\text{kim}}^{\alpha}}}\right)\sqrt{V_{\text{growth}}} \tag{10.109}$$

where $const_{\text{rod}}$ is given in Equation (10.105) on page 618.

In both cases a kinetic term is added in the second term. The more these terms dominate over the other ones the more will deviate the growth from normal eutectic growth and change into *non-coupled growth* or *irregular eutectic growth*.

Figure 10.2b on page 588 gives an example of an irregular eutectic structure. Such structures are normally very complex, which is shown in Figure 10.36.

Two most important cast alloys are the Fe–C and Al-Si alloys, which both have irregular structures. We will come back to Fe–C and Al-Si eutectics in Sections 10.6 and 10.7.

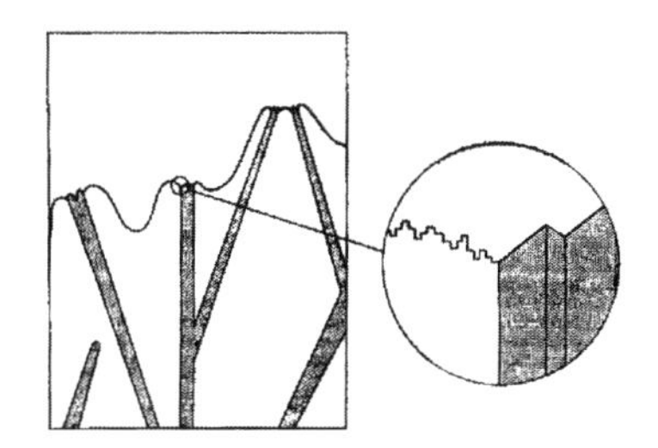

Figure 10.36 Solidification front at formation of an irregular eutectic structure. Reproduced with permission from Trans Tech © 1992.

10.5 Structures of Ternary Alloys

The basis for understanding of the structure of a ternary alloy is the ternary phase diagram. The material constants of the alloy determine the shape of the phase diagram.

The phase diagram of a ternary alloy with the components A, B and C is three-dimensional. The temperature axis is vertical. The phase diagram contains the three binary phase diagrams of the systems A–B, A-C and B-C (Figure 10.37a) in the vertical planes, which contain the lines of the 'bottom' triangle, shown in Figure 10.37b.

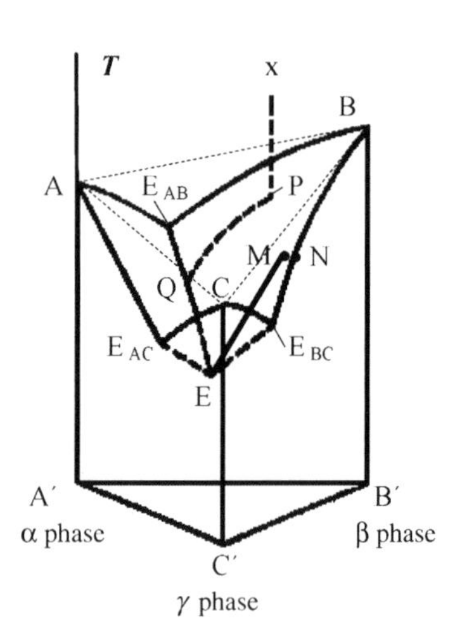

Figure 10.37a When a melt with the composition, which corresponds to the point x, cools it starts to solidify at point P. Its composition changes, owing to microsegregation, along the line PQ. The composition point then slides along the eutectic line down the ternary eutectic point E. This motion describes the solidification process.

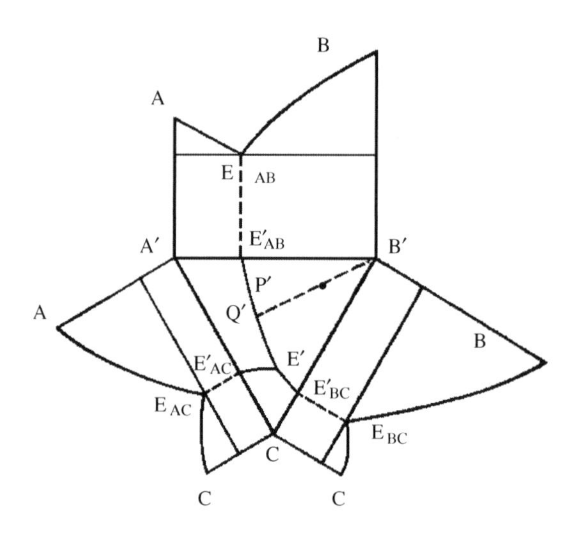

Figure 10.37b The binary phase diagrams and the bottom plane A′B′C′ are situated in the same plane, when the three planes A′ABB′, C′CBB′ and A′ACC′ are turned down into the paper plane. The eutectic lines $E'_{AB} - E'$, $E'_{AC} - E'$ and $E'_{BC} - E'$ meet at E′, the projection of the ternary eutectic point E in Figure 10.37a. At point E the melt and the phases α, β and γ are in equilibrium with each other. The eutectic temperature T_E is the lowest possible temperature of the melt.

The basic properties of ternary alloys can be found in Chapter 7 of [7] or in any other comparable literature.

10.5.1 *Structure Formation in a Simple Ternary Alloy during the Solidification Process*

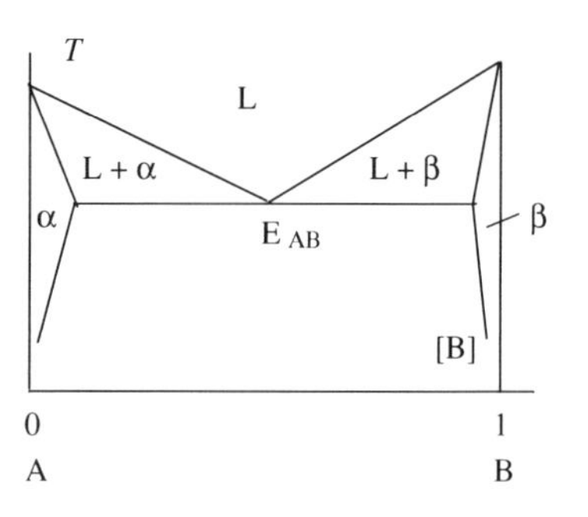

Figure 10.38 Phase diagram of the binary alloy AB.

Most alloys of technical interest do not consist of binary or eutectic alloys. They often contain a third and even a fourth element. In addition to primary and secondary phases, a three-component alloy also forms a three-phase eutectic structure.

A simple binary alloy with the components A and B (Figure 10.38) can only contain *two* phases according to the lever rule. The phases are a *pure primary phase*, either α or β depending on the composition of the alloy, and a *eutectic phase*. At solidification primary precipitation of a solid α or β phase occurs. Its composition changes gradually during the solidification and follows the liquidus line until the eutectic point E_{AB} is reached. At this point, the eutectic mixture of phases α and β precipitates. Obviously two structures occur in the solid alloy, a primary single-phase structure and the binary eutectic structure.

A simple ternary alloy with the components A, B and C normally contains *three* phases, a *pure primary phase*, either α, β or γ, followed by a *binary eutectic phase*, both depending on the composition of the alloy, and a *ternary eutectic phase*.

Below we will discuss the solidification process of a simple ternary alloy more in detail and indicate how the fractions of the three structures in the solid alloy can be derived with the aid of a study of the microsegregation during cooling.

Fractions of the Structures in a Simple Ternary Alloy

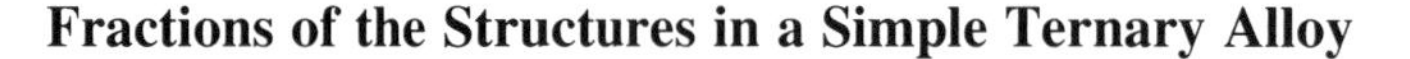

For an alloy with the initial composition, which corresponds to an arbitrary point P in the two-phase region in Figure 10.37a, the solidification starts with a primary precipitation of pure component A, normally as α-dendrites.

- Structure Type I: $L \rightarrow \alpha$

During the continued solidification the compositions of the precipitated α phase and the liquid gradually change until the point Q on the eutectic valley line E_{AB}-E in Figure 10.37a is reached. Then the eutectic reaction

- Structure Type II: $L \rightarrow \alpha + \beta$

follows. The process is analogous to the one for binary alloys. The eutectic reaction goes on until the composition of the melt corresponds to point E. The composition change of the melt during this process can be described as a motion along the eutectic line E_{AB}-E from point Q to the ternary eutectic point E.

At the ternary eutectic point E the eutectic reaction changes to

- Structure Type III: $L \rightarrow \alpha + \beta + \gamma$

The three new precipitated solid phases form together the ternary eutectic composition, which corresponds to point E.

Analogous processes in alloys with other initial compositions, where the precipitation starts with either β or γ precipitation, occur in the same way.

Composition of the Melt during the Primary Dendrite Solidification

The pathway in a ternary system can be calculated by the expressions, which are valid for binary alloys. In order to simplify the calculations and make them more illustrative we will assume that there is no back diffusion in the primarily precipitated phases (α, β or γ dendrites) during the solidification.

To describe the microsegregation during the primary precipitation of a solid phase in a ternary system, for example for components A and B along the path PQ in Figure 10.37a, we use Scheil's equation (Chapter 8 on page 442). It can be applied simultaneously on the two components B and C:

On the way from P to Q only α phase is precipitated: $L \rightarrow \alpha$. For this reason the partition coefficients $k^{\alpha/L}$ for the elements B and C occur in Equations (10.130) and (10.131).

$$x_B^L = x_B^{0\,L}(1 - f_\alpha)^{-\left(1 - k_B^{\alpha/L}\right)} \tag{10.130}$$

$$x_C^L = x_C^{0\,L}(1 - f_\alpha)^{-\left(1 - k_C^{\alpha/L}\right)} \tag{10.131}$$

where

$x_{B,C}^L$ = concentration of alloying elements B and C, respectively, in the melt at a arbitrary point along the line PQ
$x_{B,C}^{0\,L}$ = initial concentration of elements B and C, respectively, in the melt
f_α = volume fraction of solid α phase
$k_{B,C}^{\alpha/L}$ = partition coefficient for elements B and C, respectively, between the solid α phase and the melt.

As the fraction f_α is the same for both components at a given temperature, f_α is the same in both equations. This gives the following relationship between f_α and the concentrations of the two components

$$1 - f_\alpha = \left(\frac{x_B^L}{x_B^{0\,L}}\right)^{-\frac{1}{1-k_B^{\alpha/L}}} = \left(\frac{x_C^L}{x_C^{0\,L}}\right)^{-\frac{1}{1-k_C^{\alpha/L}}} \tag{10.132}$$

The Equations (10.132) are derived by solving $(1-f_\alpha)$ from the Equations (10.130) and (10.131). When the temperature decreases f_α increases and controls the concentrations of the three components, via Equations (10.128) and (10.129) and the relationship

$$x_A^L = 1 - x_B^L - x_C^L \tag{10.133}$$

The relationship (10.132), which is valid between the fraction f_α and the concentrations of the alloying elements during the reaction, means that the solidification process follows the line PQ in Figure 10.37a. The α precipitation lasts until point Q is reached.

If the solidification occurs at equilibrium conditions and the back diffusion is neglected the relationship (10.132) proves to be a good description of the concentration of the components B and C in the melt during the solidification of the melt.

The relationships give no information about the reaction temperatures. The temperatures are read for each set of concentrations of the alloying elements from the phase diagram or can be derived with the aid of the thermodynamic relationships, which are used for calculation of the phase diagram.

Composition of the Melt during the Binary Eutectic Solidification

At point Q the precipitation of α phase stops and is replaced by precipitation of two phases $L \rightarrow \alpha + \beta$ when the temperature decreases. The composition of the melt changes during the eutectic reaction owing to segregation, and follows the eutectic line from point Q down to the ternary eutectic point E in Figure 10.37a. The α and β phases are precipitated simultaneously along the line QE. If the volume fraction of each phase and the corresponding partition coefficients are known, the following relationships describe the segregation path during the eutectic reaction

$$x_B^{L\,bin} = x_B^{0\,L\,bin}(1-f_{\alpha+\beta})^{-\left[\left(1-k_B^{\alpha/L}\right)f_\alpha + \left(1-k_B^{\beta/L}\right)f_\beta\right]} \tag{10.134}$$

$$x_C^{L\,bin} = x_C^{0\,L\,bin}(1-f_{\alpha+\beta})^{-\left[\left(1-k_C^{\alpha/L}\right)f_\alpha + \left(1-k_C^{\beta/L}\right)f_\beta\right]} \tag{10.135}$$

where

$x_{B,C}^{L\,bin}$ = concentration of elements B and C, respectively, in the melt at an arbitrary point on the line QE during the binary eutectic reaction

$x_{B,C}^{0\,L\,bin}$ = concentration of elements B and C, respectively, in the melt at the start of the binary eutectic reaction (point Q)

$f_{\alpha+\beta}$ = fraction solidified binary eutectic, i.e. the $(\alpha+\beta)$ structure, in the melt

$f_{\alpha,\,\beta}$ = volume fraction of the α and β phases, respectively, in the binary eutectic structure.

It is reasonable to assume that the partition coefficients and also the volume fractions in the eutectic structure are almost constant. If we assume that they are constant, the k and f values in Equations (10.134) and (10.135) can be evaluated from the binary phase diagrams. The k values are derived with the aid of the solidus and liquidus lines. The volume fractions are found by use of the lever rule at the eutectic temperature in the binary phase diagrams. The values are assumed to be the same in the ternary phase diagram.

Ternary Eutectic Solidification

In a three-component alloy at the ternary eutectic point E the composition of the melt remains constant during the rest of the solidification process. At point E the remaining melt solidifies at the constant temperature T_E by the ternary eutectic reaction:

$$L \rightarrow \alpha + \beta + \gamma.$$

During the path PQ, the fraction f_α of the initial volume solidifies. During the path QE the fraction $f_{\alpha+\beta}$ of the remaining melt, i.e. the volume fraction $(1 - f_\alpha)$, solidifies. Hence, the total fraction of solid f_E of the *initial* volume at point E will be

$$f_E = f_\alpha + f_{\alpha+\beta}(1 - f_\alpha) \tag{10.136}$$

The remaining fraction $(1 - f_E)$ of the melt solidifies with the constant ternary eutectic structure $(\alpha + \beta + \gamma)$. It can be calculated with the aid of Equations (10.130), (10.131), (10.134), (10.135) and the binary phase diagrams of the component systems.

$$f_{\alpha+\beta+\gamma} = 1 - f_\alpha - f_{\alpha+\beta}(1 - f_\alpha) \tag{10.137}$$

10.5.2 Types of Structures in Ternary Alloys

Above the compositions of the regions of types I, II and III, which can be identified in a simple solid ternary alloy, have been discussed. Below the structures of the regions I, II and III will be classified and discussed.

As will be evident below, the structures of ternary alloys are composed of different phases, resulting from various combinations of regular, degenerate and irregular eutectic growth, in combination with dendrites or other primary phases, and also two-phase cells. The structures can be described with the aid of these components. The most common structure regions will be described and characterized on pages 643–646, where we use the concept:

- Single-phase dendrites = primary α, β or γ dendrites.
- Two-phase cells/dendrites = cells/dendrites of binary eutectic structure.

Influence of a Third Element on Binary Eutectic Growth

It is well known that addition of a second element to a solidifying pure metal melt results in a dramatic change of the solidification pattern. The planar solidification front of the pure element breaks down and is replaced by colonies of the type seen in Figures 10.39a and b.

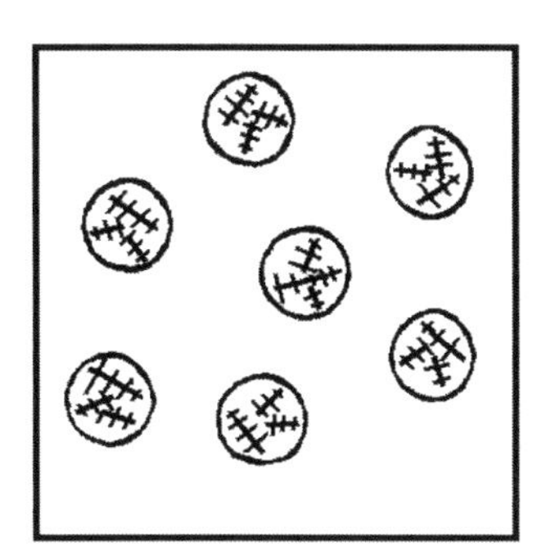

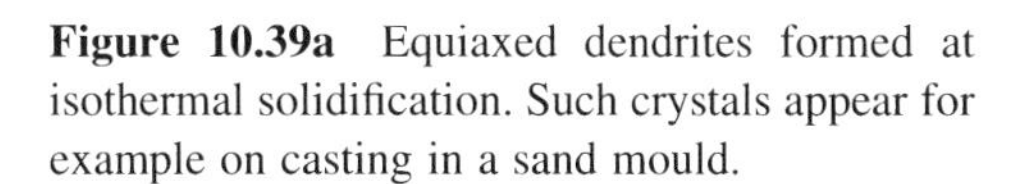

Figure 10.39a Equiaxed dendrites formed at isothermal solidification. Such crystals appear for example on casting in a sand mould.

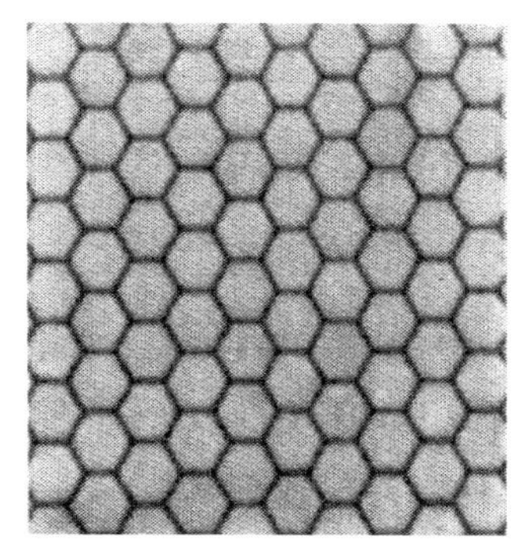

Figure 10.39b Hexagonal cells, which appear at large undercooling when the planar interface has collapsed.

Figure 10.39b shows a cross-section of a cellular structure formed in a unidirectionally solidified sample (see Chapter 8).

Experimental observations show that *addition of a third element* to a binary eutectic alloy also changes the solidification pattern radically. The eutectic structure at *isothermal* solidification is similar to that of the equiaxed crystals in Figure 10.39a. In this case spherical cells are formed (Figure 10.40a).

At *unidirectional* solidification of a ternary sample, a structure of two-phase cells is formed, similar to the single phase structure in Figure 10.39b. Figure 10.40b shows a cross-section of such a two-phase cell structure. The cells may branch in the same way as dendrites.

By unidirectional solidification it is possible to stabilize the planar eutectic solidification front for the *binary* system if the criterion for the ratio of the temperature gradient and the growth rate G/V_{growth} [Equation (8.52) on page 463 in Chapter 8) for *single-phase solidification* is fulfilled. A planar solidification front is maintained if

$$\left(\frac{G_L}{V_{growth}}\right) \geq \frac{m_L}{D_L}\frac{1-k_0}{k_0}x_0^L \tag{10.138}$$

This relationship is only valid for binary alloys. An analogous relationship will be derived for ternary eutectic solidification below.

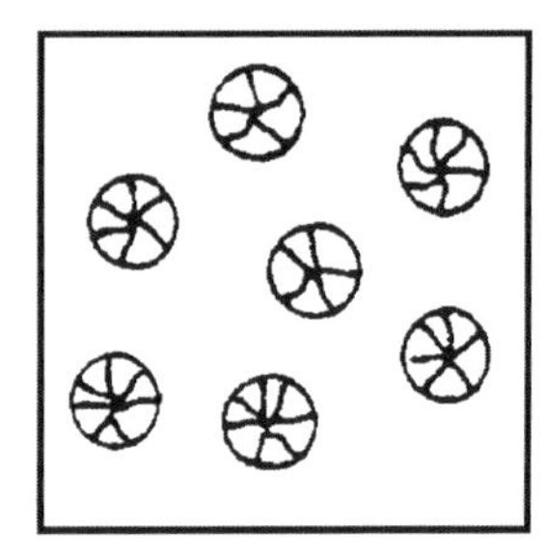

Figure 10.40a Spherical cells with eutectic structure formed at isothermal solidification of a ternary alloy.

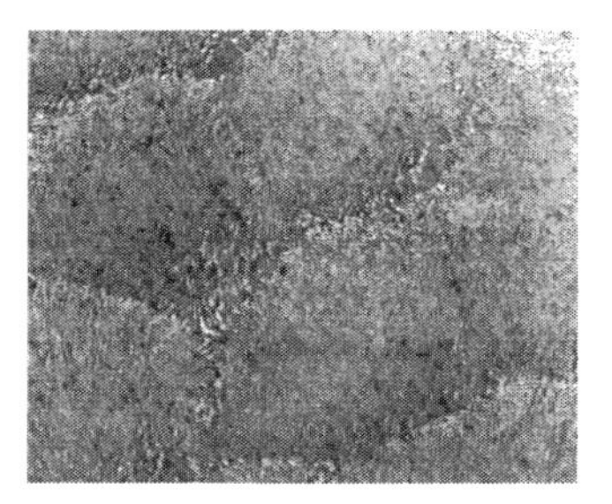

Figure 10.40b Cross-section of a two-phase cell structure formed at unidirectional solidification of a ternary alloy.

The main differences between binary and ternary systems are:

1. If the fraction of the alloying addition is the same, the two-phase cells in a ternary system are much coarser than those in a binary system.
2. The coarseness of the two-phase cells in a ternary alloy decreases with increasing fraction of the alloying element, in contradiction to ordinary cell structure in binary alloys.
3. The growth law, that is valid for binary dendritic growth, controls by no means the growth of the two-phase cells in a ternary alloy.

Simple Theory of Influence of a Third Element on Binary Eutectic Growth

Consider a binary eutectic alloy AB, which gives a eutectic structure of two phases α and β during solidification. A small amount of a third element C is added to the alloy melt. The partition coefficients of C between the α phase/the melt L and the β phase/the melt L are $k_C^{\alpha/L}$ and $k_C^{\beta/L}$, respectively. The volume fractions of the two phases are f_α and f_β.

Provided that the partitions coefficients < 1 the third element C will be enriched in the melt ahead of the solidification fronts of the solid phases. The solidification process will be similar to the planar solidification process, discussed in Chapter 6 on page 307. To simplify the problem we will only consider the case when the average diffusion distance of element C is much larger than the diffusion distances of elements A and B, which only diffuse between the lamellae.

Equation (6.120) on page 323 in Chapter 6 describes the concentration of element C in the melt ahead of the solidification front. It contains two constants A and B.

$$x_C^L = A + B\,e^{-\frac{V_{growth}y}{D}} \tag{10.139}$$

where D is the diffusion constant of element C in the melt.

In this case the boundary conditions will be $y = \infty$, $x_C^L = x_0^L$ and, according to condition 3, Equation (6.116) on page 322 in Chapter 6. We also have $x_C^s = x_0^L$, which gives

$$y = 0 \quad \text{and} \quad x_C^L = \frac{x_C^s}{k} = \frac{x_0^L}{f_\alpha k_C^{\alpha/L} + f_\beta k_C^{\beta/L}}$$

where

$$k = f_\alpha k_C^{\alpha/L} + f_\beta k_C^{\beta/L} \tag{10.140}$$

is the average partition coefficient between the melt and the solid phases. Inserting these values into Equation (10.139) we obtain

$$A = x_0^L \quad \text{and} \quad B = x_0^L \frac{1 - \left(f_\alpha k_C^{\alpha/L} + f_\beta k_C^{\beta/L}\right)}{f_\alpha k_C^{\alpha/L} + f_\beta k_C^{\beta/L}}$$

Hence, the concentration of element C in the melt ahead of the solidification front will be

$$x_C^L = x_0^L \left[1 + \frac{1 - \left(f_\alpha k_C^{\alpha/L} + f_\beta k_C^{\beta/L}\right)}{f_\alpha k_C^{\alpha/L} + f_\beta k_C^{\beta/L}} \mathrm{e}^{-\frac{V_{\text{growth}}\, y}{D}}\right] \tag{10.141}$$

Provided that the average diffusion distance of element C is much larger than the diffusion distances of elements A and B, we obtain the concentrations of element C in the melt, ahead of the two solid phases, with the aid of Equation (10.141)

$$x_C^\alpha = k_C^{\alpha/L} x_C^L = k_C^\alpha x_0^L \left[1 + \frac{1 - \left(f_\alpha k_C^{\alpha/L} + f_\beta k_C^{\beta/L}\right)}{f_\alpha k_C^{\alpha/L} + f_\beta k_C^{\beta/L}} \mathrm{e}^{-\frac{V_{\text{growth}}\, y}{D}}\right] \tag{10.142}$$

$$x_C^\beta = k_C^{\beta/L} x_C^L = k_C^\beta x_0^L \left[1 + \frac{1 - \left(f_\alpha k_C^{\alpha/L} + f_\beta k_C^{\beta/L}\right)}{f_\alpha k_C^{\alpha/L} + f_\beta k_C^{\beta/L}} \mathrm{e}^{-\frac{V_{\text{growth}}\, y}{D}}\right] \tag{10.143}$$

The concentration of element C as a function of position in the melt is similar to the curve in Figure 6.40 on page 320 in Chapter 6.

Instability of the Solidification Front owing to a Third Element

As a measure of the instability of the solidification front, when it is exposed to temperature fluctuations, we will use the temperature gradient in the melt in the direction of the solidification front.

$$\frac{\mathrm{d}T_L}{\mathrm{d}y} = \frac{\mathrm{d}T_L}{\mathrm{d}x_C^L} \frac{\mathrm{d}x_C^L}{\mathrm{d}y} \tag{10.144}$$

where the first factor is the slope of the binary eutectic line, which corresponds to the line $E_{AB}-E$ in Figure 10.37a for instance. Hence, the first factor can be written as

$$\frac{\mathrm{d}T_L}{\mathrm{d}x_C^L} = m_{E_{AB}-E}^L \tag{10.145}$$

The second factor in Equation (10.144) is obtained by taking the derivative of Equation (10.141) with respect to the coordinate y. The resulting temperature gradient will be

$$\frac{\mathrm{d}T_{\mathrm{L}}}{\mathrm{d}y} = m^{\mathrm{L}}_{\mathrm{E_{AB}-E}} x^{\mathrm{L}}_0 \frac{1-\left(f_\alpha k_{\mathrm{C}}^{\alpha/\mathrm{L}} + f_\beta k_{\mathrm{C}}^{\beta/\mathrm{L}}\right)}{f_\alpha k_{\mathrm{C}}^{\alpha/\mathrm{L}} + f_\beta k_{\mathrm{C}}^{\beta/\mathrm{L}}} \left(-\frac{V_{\mathrm{growth}}}{D}\right) \mathrm{e}^{-\frac{V_{\mathrm{growth}} y}{D}} \tag{10.146}$$

At the solidification front ($y = 0$) we obtain

$$\left(\frac{\mathrm{d}T_{\mathrm{L}}}{\mathrm{d}y}\right)_{y=0} = m^{\mathrm{L}}_{\mathrm{E_{AB}-E}} x^{\mathrm{L}}_0 \frac{1-\left(f_\alpha k_{\mathrm{C}}^{\alpha/\mathrm{L}} + f_\beta k_{\mathrm{C}}^{\beta/\mathrm{L}}\right)}{f_\alpha k_{\mathrm{C}}^{\alpha/\mathrm{L}} + f_\beta k_{\mathrm{C}}^{\beta/\mathrm{L}}} \left(-\frac{V_{\mathrm{growth}}}{D}\right) \tag{10.147}$$

At low concentrations x^{L}_0 and/or low growth rate V_{growth} the instability will be low, which means that the solidification front will be and remain planar. If these conditions are not fulfilled the eutectic front becomes unstable and cells will be formed.

Hence, the theory confirms the experimental observations that the planar solidification front may break down and be replaced by a cellular profile (condition on page 324).

Types of Structures in Ternary Alloys

From the discussion above, we can conclude that ternary eutectic structures can have many different appearances depending on

1. the shape of the phase diagram, i.e. the material constants
2. the composition of the alloy
3. the solidification process, i.e. the cooling conditions
4. the kinetics at the interface α/L or β/L.

In *casting processes* the temperature gradients are *small* during the solidification. At *unidirectional solidification processes* the gradients are *high*. At unidirectional solidification, the phases or substructures in the growing solid sample are formed in sequences when the temperature gradually decreases and the growth temperature gradually is reached. At such processes, elongated structures are formed, which mainly consist of cells with other substructures in the cell boundaries than in the cell cores. During casting processes more irregular structures with no texture are formed.

Below we will discuss different types of solidification modes and the corresponding structures. We will use a simple ternary phase diagram of the type shown in Figure 10.37a on page 636. A projection of the three-dimensional phase diagram is shown in Figure 10.41, which is drawn after a model of by McCartney, Hunt and Jordan [8], published in 1979.

The figures and letters represent the various regions of different structures, which appear at cooling. The structures will be described in the text below. It should be emphasized that the regions are three-dimensional (see Figures 10.37a and Figure 10.37b both on page 636), not planar as in Figure 10.41 on page 643. The regions also extend in the perpendicular temperature direction.

The three phases α, β and γ, which represent single phases of the components A, B and C, respectively, are marked by 1 in Figure 10.41 on page 643. The areas, which are marked by 2, represent two-phase regions. Similarly, areas marked by 3 correspond to three-phase regions. The signification and description of regions 4 and 5 will be explained below.

In order to show structures formed at low temperature gradients, examples are taken from casting samples of Al-Si-Cu alloys. The projection of part of the phase diagram of the Al-Si-Cu system is shown in Figure 10.42 on page 643.

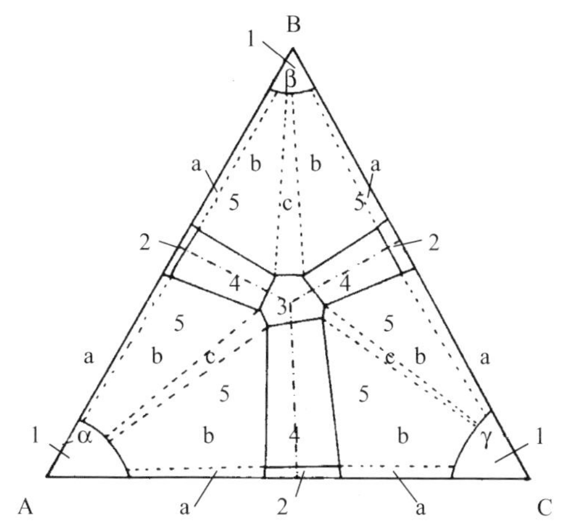

The three single phases α, β and γ are marked by black, white and dashed areas, respectively, in the Figures 10.44a – 10.49a.

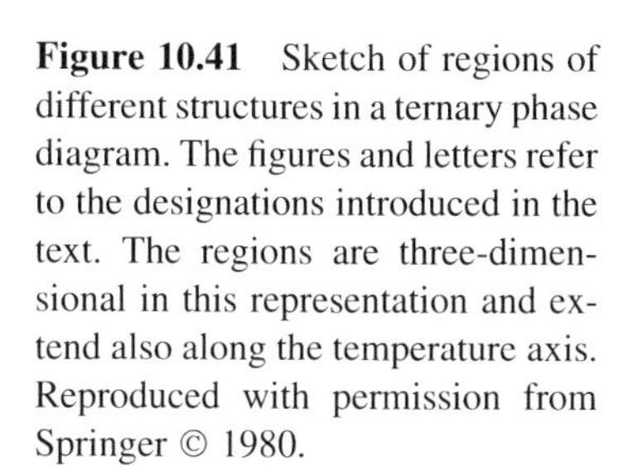

Figure 10.41 Sketch of regions of different structures in a ternary phase diagram. The figures and letters refer to the designations introduced in the text. The regions are three-dimensional in this representation and extend also along the temperature axis. Reproduced with permission from Springer © 1980.

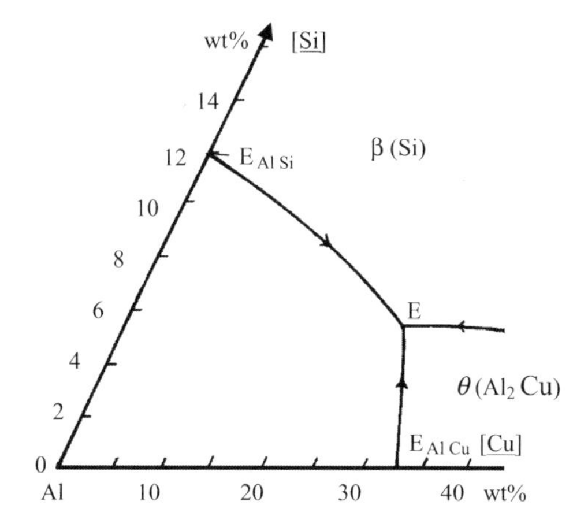

Figure 10.42 The α phase corner of the ternary phase diagram of the system Al–Si–Cu. © Metallurgical Services (Planned Products (Metallurgy) Ltd) 1961.

Figure 10.42 shows the Al corner in the phase diagram of the system Al-Si-Cu. It corresponds to the A-corner in Figure 10.41. The primary precipitated α phase consists of Al and is a single-phase region.

Hexagonal cells

Figure 10.43 Single-phase cells in a cellular single-phase region.

Regions 1

The three regions 1 represent single-phase regions. They consist of either nearly pure component A, B or C, respectively.

The structure is normally dendritic but cellular structure also occurs (Figure 10.43).

Regions 2

Each of the three regions 2 contains two liquidus surfaces, which intersect along a so-called eutectic 'valley', which runs downwards (towards lower temperature). Hence, each region 2 must consist of a two-phase region (Figure 10.44a). The two-phase regions normally have a lamellar or rod-like structure (Figures 10.44b). The region corresponds to the coupled zone and the structure varies with the cooling rate.

With a small amount of a third element a cellular structure is formed (Figure 10.44c). The bent curves in the figure are perpendicular to the lamellae. They represent the solidification front and show the cellular shape of the lamellae.

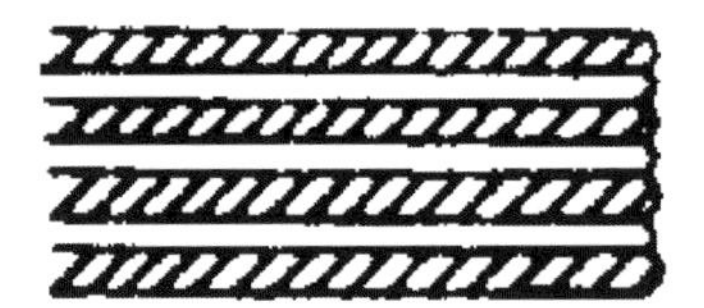

Figure 10.44a Sketch of two-phase normal eutectic structure of lamellar type. Reproduced with permission from Springer © 1980.

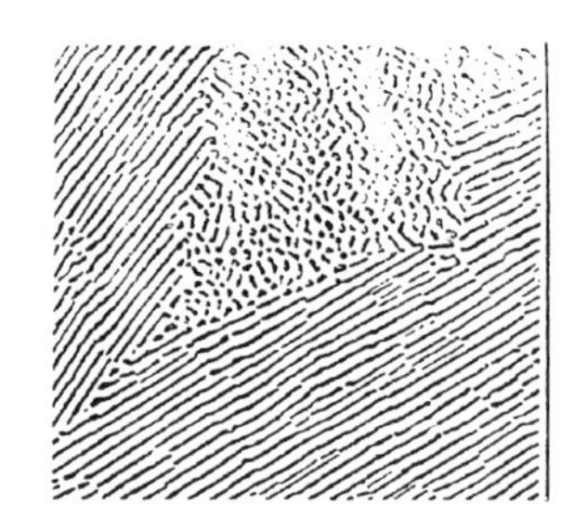

Figure 10.44b (identical with Figure 10.2a). Rod-like and plate-like (lamellar) eutectic structures in Cu–Ag (page 589).

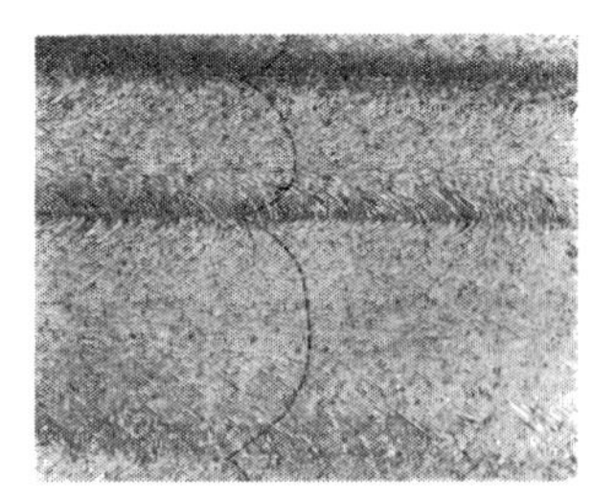

Figure 10.44c Cellular structure in a two-phase eutectic alloy with a small amount of a third element. Bruce Chalmers, courtesy of S. Chalmers.

Region 3

The path from the ternary point E to a point within the bottom triangle A'B'C' in Figure 10.37a, corresponds to precipitation of a pure three-phase eutectic structure, without any primary single phases or two-phase binary eutectic structures. The structure, which forms, occurs in region 3 in Figure 10.41 on page 643.

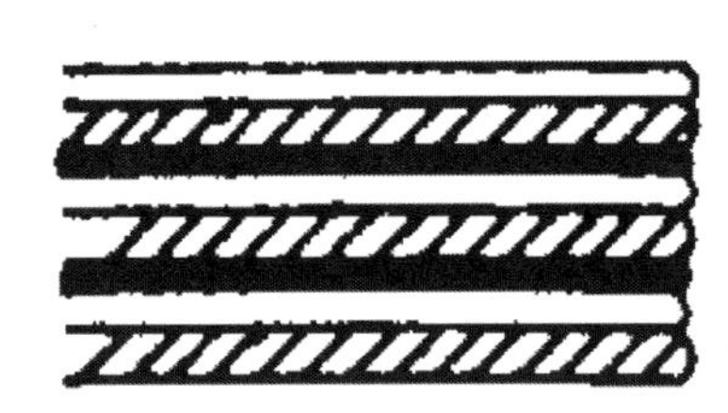

Figure 10.45a Principal sketch of a regular lamellar three-phase eutectic structure, formed on unidirectional solidification. Reproduced with permission from Springer © 1980.

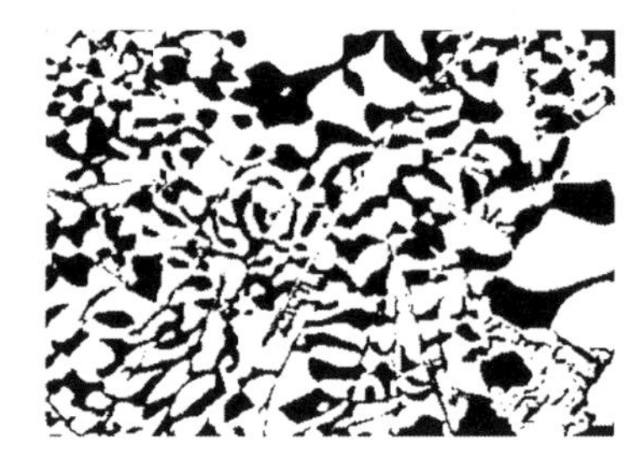

Figure 10.45b Three-phase eutectic structure in an Al 35%–Cu 4% Si, formed at isothermal solidification. © Metallurgical Services (Planned Products (Metallurgy) Ltd) 1961.

Ternary eutectic structure may form by normal eutectic growth, i.e. all three phases grow with the same rate in a coupled mode. The three-phase ternary eutectic structure grows along a planar solidification front in the same way as the coupled two-phase binary lamellar eutectic structures in region 2. Just as in region 2 the growth of the phases is coupled. The growth rate and consequently also the structure vary with the cooling rate. The structure in region 3 corresponds to the coupled zone.

The type of structure in region 3 is shown in Figures 10.45. The structure may be more complex than the regular lamellar or rod-like structure sketched in Figure 10.45a. It may alternatively arise as an irregular eutectic structure of the type shown in Figure 10.45b.

Regions 4

During unidirectional solidification of alloys with compositions equal or close to any of the eutectic valleys, two-phase eutectic cells are formed. They constitute region 4. The third element is enriched at the phase boundaries. When the temperature decreases to T_E and the composition simultaneously reaches the ternary eutectic point E, a three phase eutectic reaction occurs.

The solidification process corresponds to the pathway Q-E in Figure 10.37a on page 636 and results in a structure of type region 4.

The ternary eutectic structure forms a shell around each two-phase eutectic cell. This structure is a case of coupled growth between normal eutectic growth and two-phase cells.

Figure 10.46a Principal sketch, which illustrates coupled growth between two-phase cells and a ternary (three-phase) normal eutectic structure.

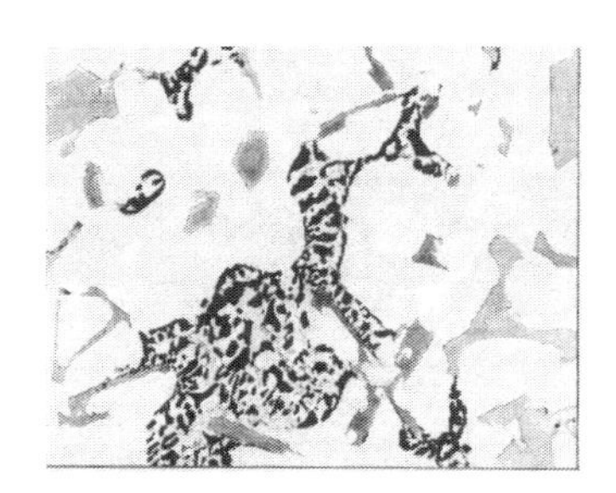

Figure 10.46b Structure of an Al 10%–Cu 12%–Si alloy.
White areas = α phase (Al).
Grey areas = $(\alpha + \beta)$ phase (Al + Si).
Black areas = θ phase ($CuAl_2$).
© Metallurgical Services (Planned Products (Metallurgy) Ltd) 1961.

The resulting structure is schematically shown in Figure 10.46a. Figure 10.46b gives an example of a real structure, that corresponds to Figure 10.46a. It has solidified under isothermal conditions.

Regions 5

Far from the eutectic valleys single phase cells or dendrites are likely to appear, followed by the structures, described as regions 2, 4 and 3. Region 5 can be divided into three subgroups as is shown in Figure 10.41 on page 643. The subgroups can be characterized as follows:

Region 5a – between Regions 1 and 2

Primary single-phase dendrites are followed by a two-phase binary eutectic structure. The structure is schematically shown in Figure 10.47a. The similarity between the Figures 10.23 and 10.47a is striking. They actually show the same thing.

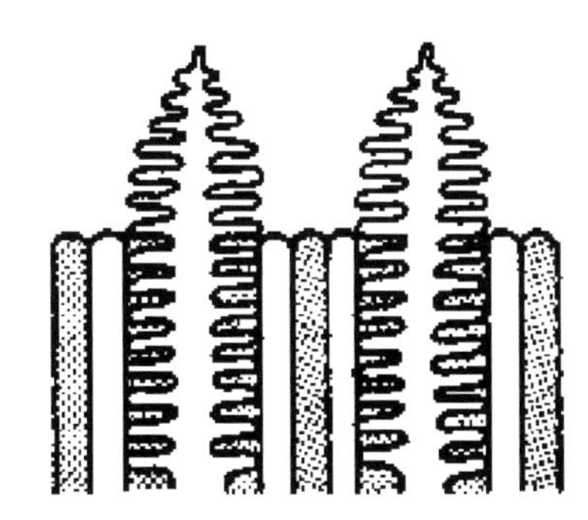

Figure 10.23 Coupled growth between a primary phase of dendrites and two-phase binary eutectic cells. Reproduced with permission from Trans Tech © 1992.

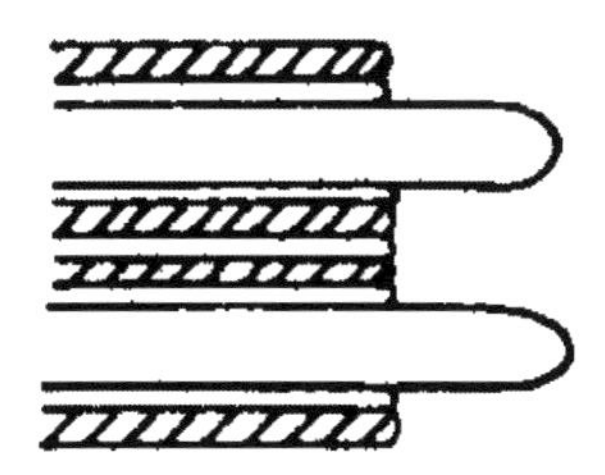

Figure 10.47a Sketch of primary single-phase cells followed by two-phase binary eutectic cells.

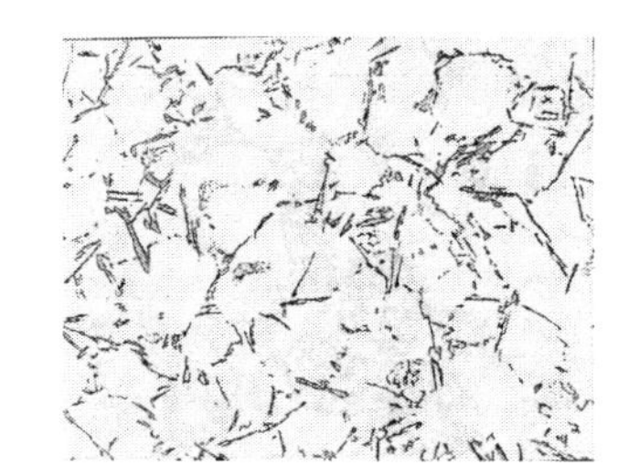

Figure 10.47b Structure of an Al 7%–Si alloy. The ternary structure 5 a looks the same. © Metallurgical Services (Planned Products (Metallurgy) Ltd) 1961.

The region is close to the vertical binary phase diagram and the structure is very similar to the corresponding binary structure. Hence, the solidification process is close to the pathway N–E_{BC} in Figure 10.37a (page 636). N is a point on the liquidus line in the binary phase diagram.

Region 5b – between Regions 2 and 4

Single-phase dendrites can be followed by two-phase binary eutectic structure and finally by three-phase ternary eutectic structure. Figure 10.48a shows schematically this type of structure, which is very common. Figure 10.48b shows a real example.

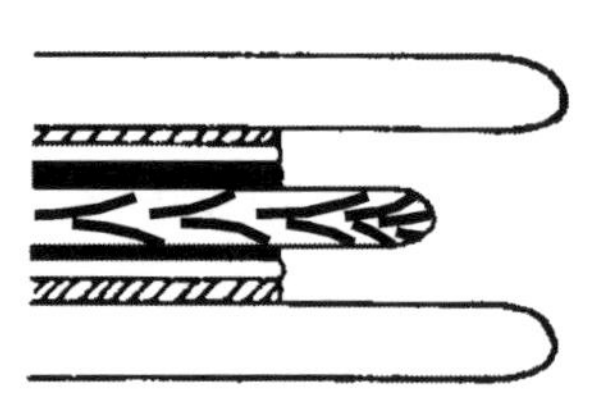

Figure 10.48a Single-phase cells, followed by two-phase binary eutectic cells (in the middle), followed by a normal ternary eutectic structure.

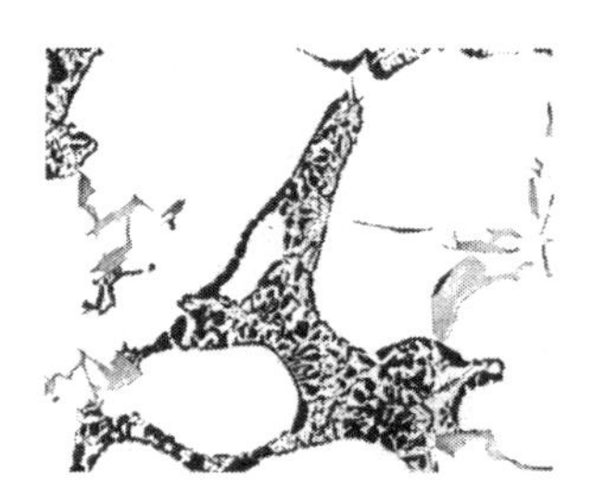

Figure 10.48b Structure of an Al 10%Cu 6% Si alloy
White areas = α phase (Al).
Grey areas = $(\alpha + \beta)$ phase (Al + Si).
Black areas = θ phase (Al_2Cu).

The solidification process corresponds to the pathway P-Q-E in Figure 10.37a (page 636).

Region 5c – between Regions 1 and 3
In some alloys single-phase dendrites can be followed directly by three-phase ternary eutectic structure. Such a structure is shown in Figures 10.49a and 10.49b.

The solidification process corresponds to the pathway M-E in Figure 10.37a.

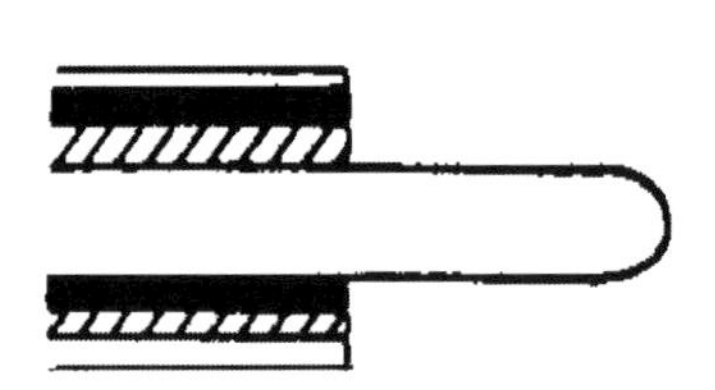

Figure 10.49a Single-phase cells followed by three-phase ternary eutectic structure.

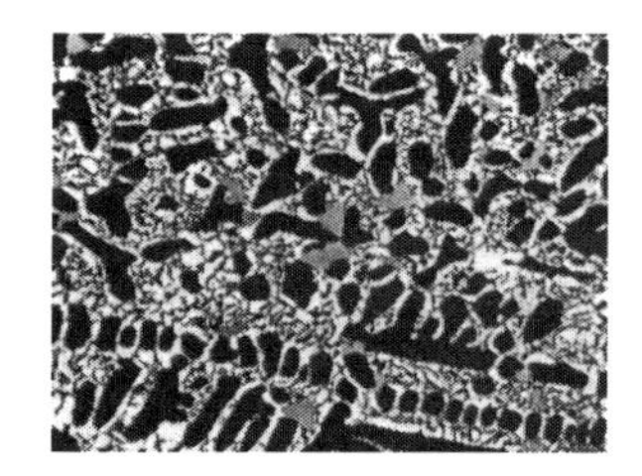

Figure 10.49b Structure of an Al, 30% Cu, 6.5% Si alloy.

10.6 Solidification of Fe-C Eutectics

FeC-alloys with a carbon composition > 2 wt% $\underline{C}$[1] are called cast iron. Cast iron is one of the most common and most frequently used alloys in the world. It is cheap and strong. The art of cast iron production has been known for more than 3000 years. Today cast iron is of utmost technical importance.

The phase diagram of Fe–C contains two eutectic points. One is stable and the eutectic reaction is

$$L \rightarrow \gamma + \text{graphite}$$

Cast iron, which has solidified according to the stable eutectic reaction is often called grey cast iron owing to the colour of its surface.

The other eutectic point is metastable and the eutectic reaction occurs around 6 °C below the stable eutectic reaction. The metastable reaction is

$$L \rightarrow \gamma + Fe_3C$$

Eutectic cast iron starts to solidify when the melt has reached the temperature 1153 °C. The eutectic structure consists of a mixture of austenite and graphite. Further cooling leads to

[1] Underlined chemical symbol means than the atoms (in this case carbon) are dissolved in the solid phase.

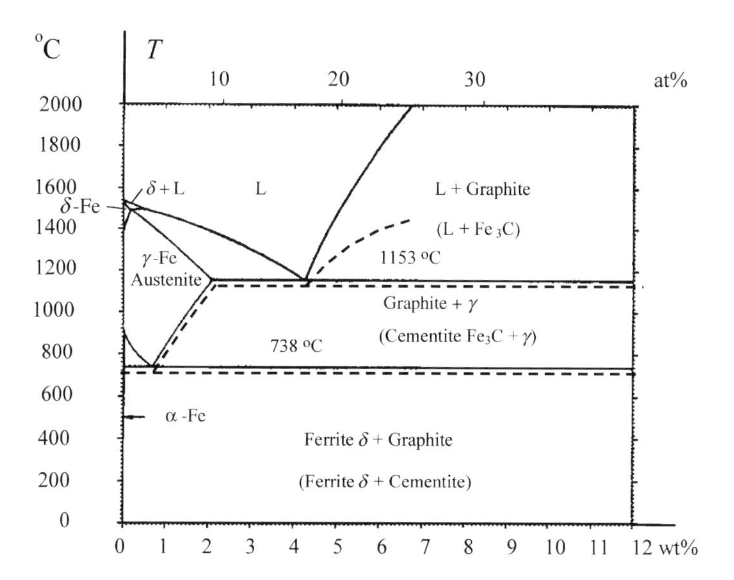

Figure 10.50 Phase diagram of the system Fe–C.

The stable eutectic reaction corresponds to full lines. The lines that belong to the unstable eutectic reaction are dashed. Adapted from Constitution of Binary Alloys, 2nd Edition, Max Hansen, McGraw Hill 1958.

either a graphite or cementite precipitation and a decrease of the carbon concentration in the austenite. At and below 738 °C austenite is transformed either into pearlite or, at low cooling rates, into ferrite and graphite.

From metallurgical point of view, cast iron is a eutectic alloy, which shows a variety of different structures. As is shown from the phase diagram the carbon concentration of eutectic cast iron is 4.3 wt%.

Below we will discuss grey cast iron and particularly nodular cast iron.

10.6.1 Eutectic Growth of Grey Cast Iron

Types of Grey Iron Structures

Grey cast iron usually shows a great number of morphologies. Different types of graphite structure are roughly classified as A, B, C and D graphite.

A *graphite* is a very coarse flake graphite in a matrix of austenite (γ iron) (Figure 10.51a). D graphite is a very fine graphite in an austenite matrix, so-called undercooled graphite. The formation of the different kinds of graphite structure depends on the composition of the melt and the cooling rate.

Figure 10.51a Structure of flake graphite (A graphite).

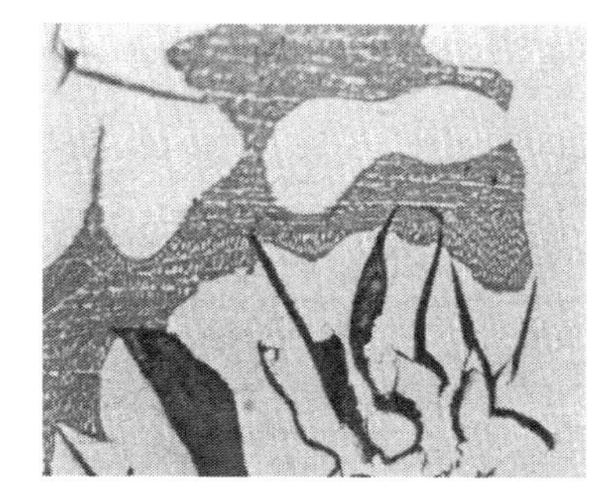

Figure 10.51b Growing eutectic colony of flake graphite. The (black) graphite grows ahead of the (white) austenite into the (grey) quenched structure of the melt.

Figure 10.51c A scanning electron micrograph of graphite flakes in a deeply etched flake graphite sample.

We will not try to identify and list all the different factors, which influence the morphology of the graphite. Instead we will demonstrate some important differences in the growth conditions between the two extreme types A and D.

Metallographic examination of cast iron shows that when it solidifies in the A mode the leading phase is flake graphite (Figure 10.51b and c). The austenite solidifies as dendrites, which is a degenerate eutectic reaction.

Figure 10.52a Transition from central fine-grained D graphite to outer flake graphite.

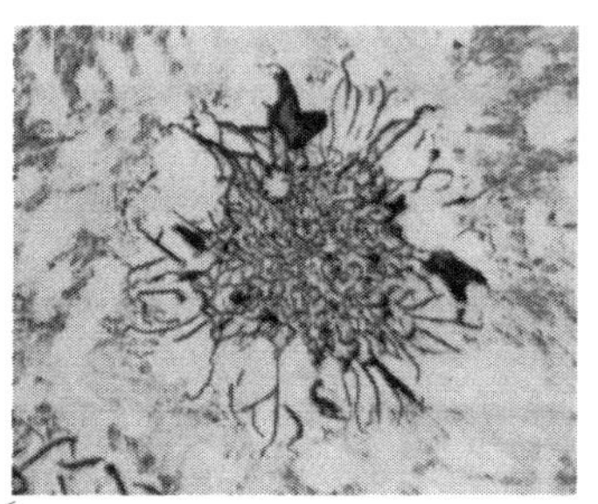

Figure 10.52b Fine-grained D graphite recorded with the aid of an ordinary optical microscope.

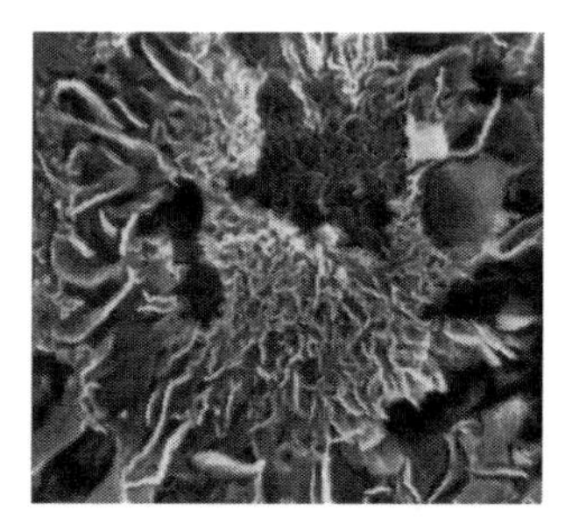

Figure 10.52c A scanning electron micrograph of undercooled graphite in a deeply etched sample.

The D mode grey cast iron has been shown to grow by a normal eutectic reaction. Figure 10.52a shows a growing eutectic colony of fine-grained D graphite. Figure 10.52b shows that change of the growth conditions leads to an abrupt change the solidification mode from D to A mode. The undercooled graphite forms pen-shaped graphite. An example is shown in Figure 10.52c.

Growth of Primary Austenite and Graphite

Figure 10.51b shows the structure of regions, where primary graphite is the fastest growing phase, The figure indicates that the formation of type A graphite occurs at a degenerate eutectic reaction where the graphite is the leading phase. In order to find growth rate and the region in the phase diagram, where this type of structure is formed, we will analyze the growth of primary austenite and primary graphite.

The growth rate for ordinary diffusion-controlled dendritic growth, which is valid for austenite in this case is known to be

Dendritic growth of A:

$$V_{\text{austenite}} = 2.2 \times 10^{-6} \times (\Delta T)^2 \text{m/s} \quad (10.148)$$

Figure 10.53 shows that flake graphite grows as plates. The growth law for such plates is known to be primary graphite growth:

$$V_{\text{graphite}} = 2.2 \times 10^{-11} \times (\Delta T)^4 \text{m/s} \quad (10.149)$$

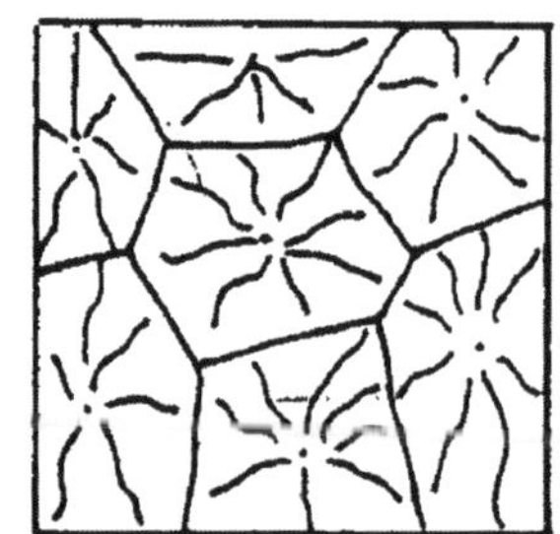

Figure 10.53a, b and c Eutectic growth in grey cast iron.

Comparison between the Growth Rates of Austenite and Graphite
The growth rates of austenite dendrites and graphite flakes vary with the carbon concentration. For a comparison between them we will use the schematic phase diagram of the system FeC in Figure 10.54. The bent line below the eutectic point represents the boundary line along which the growth rates of the primary phases austenite and graphite are equal.

In a *hypoeutectic* ($x_C < x_E$) FeC alloy, austenite is predicted to grow much faster than graphite for all reasonable values of undercooling, according to Equations (10.148) and (10.149).

However, in a *hypereutectic* ($x_C > x_E$) FeC alloy, the phase diagram shows that the undercooling is much larger for graphite than for austenite, $\Delta T^{L \rightarrow gr} >> \Delta T^{L \rightarrow \gamma}$. Hence, graphite will be the fastest growing phase down to a certain critical temperature. T_{cr}.

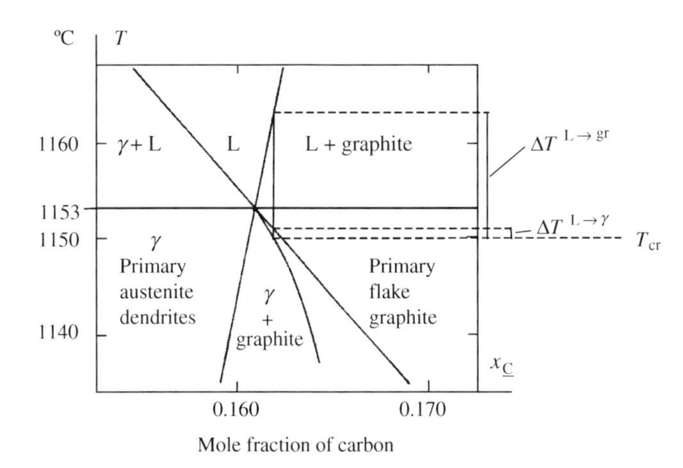

Figure 10.54 Phase diagram of the system Fe–C.

The bent line of equal growth rate of primary austenite and primary graphite has been plotted in the diagram. Reproduced with permission from Springer © 1975.

The line of equal growth in Figure 10.54 separates the regions where austenite and graphite, respectively, has the highest growth rate. Solidification will result in the so-called flake graphite structure to the right of this line. To the left of the line primary austenite dendrites will form. When the carbon concentration in the melt between the dendrite arms reaches the value that corresponds to the bent line, flake graphite will precipitate.

Coupled Zone of Fe–C Eutectics

The transition from type A graphite (flake graphite precipitated by a degenerated eutectic reaction) to type D graphite (undercooled graphite precipitated by a normal eutectic reaction) will be discussed here in the light of the coupled zone concept.

In Section 10.4.3 on page 624 the coupled zone of the two eutectic phases α and β in the phase diagram of a binary alloy AB was introduced.

The aim of this section is to describe how the coupled zone of Fe–C eutectics can be derived and plotted into the phase diagram.

In Section 10.3.2 normal lamellar eutectic growth has been treated. If the influence of the kinetic reaction at the solidification front is neglected, the two phases show normal eutectic growth and a symmetrical coupled zone.

In the present case, the kinetics can not be neglected but has to be included in the calculations. As Fe–C eutectics show degenerate growth we have reason to predict that the coupled zone in the phase diagram of Fe–C will be asymmetrical.

Calculation of the Coupled Zone of Fe–C Eutectics
On pages 628–633 we have described how the coupled zone can be calculated for alloys, that show normal eutectic growth. The calculation procedure was used in Example 10.2 for the eutectic Al-$CuAl_2$ alloy and the result was given on page 632. The procedure was shown graphically with the aid of the Figures 10.32 on pages 632–633.

Analogous calculations have to be performed for the Fe–C system (Fredriksson [9]). The calculations will be based on the theory given on pages 604–605, where the kinetic reaction is considered. The relationship (10.58) on page 605 is used as growth law for the eutectic phase in combination with the growth laws (10.148) and (10.149) for the primary phases.

The calculations have been performed for three different values of the growth rate: $\mu = 10^{-2}$ m/s, $\mu = 10^{-3}$ m/s and $\mu = \infty$.

Figure 10.55 shows the coupled zone in the phase diagram for the case of unidirectional solidification. The shaded area corresponds to $\mu = 0.01$ m/s, the small black area represents $\mu = 0.001$ m/s and the thin outer lines indicate the extension of the coupled zone for $\mu = \infty$.

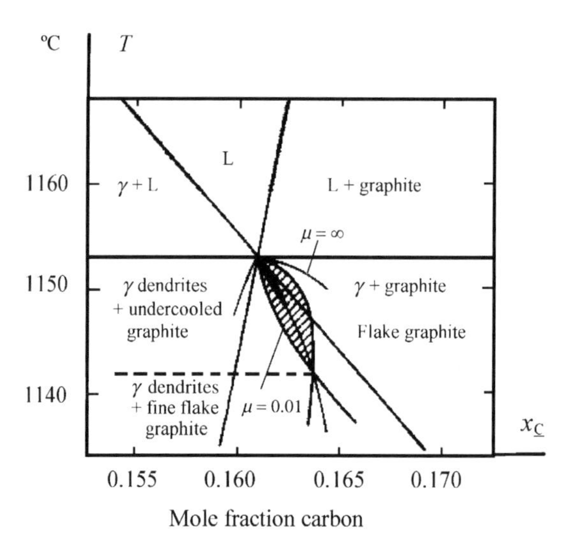

Figure 10.55 The theoretically predicted coupled zone in the stable Fe–C phase diagram. γ = austenite. Reproduced with permission from Springer © 1975.

Undercooled graphite is formed by a cooperative eutectic reaction. Figure 10.55 shows that undercooled graphite has the highest growth rate within the coupled zone (compare Figure 10.54).

The extension of the coupled zone depends on a large number of factors. Undercooled graphite is known to be favoured by low sulfur and oxygen concentrations in the liquid. This fact may be explained as follows.

Oxygen and sulfur are two surface-active elements, which lower the surface tension between graphite and liquid. This affects the kinetic reaction and reduces the kinetic growth constant μ for graphite. Figure 10.55 shows that the region for coupled growth decreases with μ. This means that formation of undercooled graphite is reduced in presence of oxygen and sulfur.

In a hypoeutectic melt γ dendrites will precipitate first according to the discussion on page 634. In a hypereutectic melt γ dendrites with a different composition will precipitate primarily (page 634). In both cases, the composition changes gradually towards the coupled zone, owing to microsegregation, until the composition has reached the coupled region. Coupled growth of austenite and graphite follows and an undercooled graphite structure is formed.

10.6.2 *Eutectic Growth of Nodular Cast Iron*

Figure 10.56 Structure of nodular cast iron.

B graphite is a very common type of grey cast iron. It is called spheroidal graphite or *nodular graphite*. The nodular structure of cast iron results in products with measurable degrees of ductility, which has extended the usefulness of nodular cast iron enormously.

Figure 10.56 shows the eutectic structure of nodular cast iron. It consists of spherical graphite particles embedded in a matrix of ferrite or austenite. Austenite is formed primarily but is later transformed into ferrite and graphite.

The purpose of the section below is to study the solidification process of nodular cast iron.

Formation of Graphite Nodules

Observations

Graphite nodules nucleate in eutectic cast iron when the temperature decreases below the eutectic temperature. They grow gradually by diffusion of carbon atoms from the surroundings. The microstructure of the solidified eutectic nodular cast iron in Figure 10.56 shows a mixture of small and big spherical graphite nodules. New nodules are formed in the melt all the time and grow continuously.

Figures 10.57a and b show cross-sections of solidified nodular cast iron. The solidification process has suddenly been interrupted by quenching the sample into brine. The remaining melt solidifies as cementite and austenite. The figures show that small graphite nodules nucleate in the melt. The big ones consist of a core of graphite surrounded by a spherical shell of solid austenite.

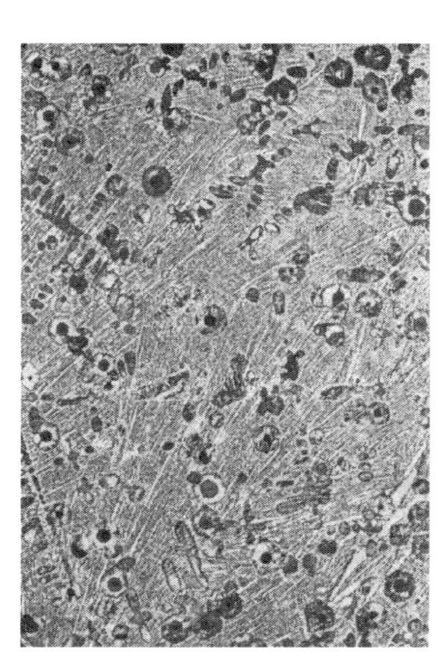

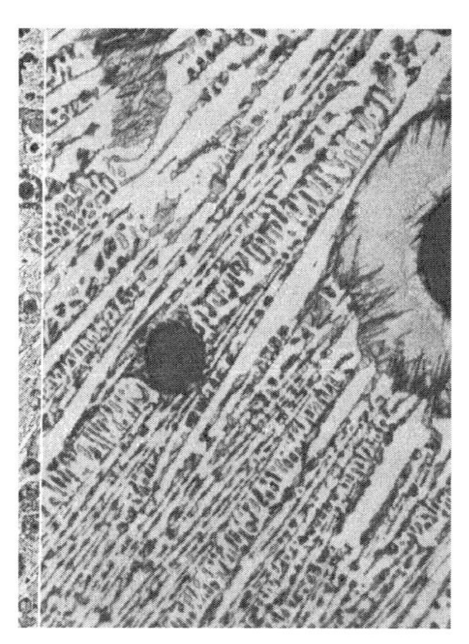

Figure 10.57a (left figure) Microstructure of a eutectic nodular cast iron sample, quenched 6 s after the start of the eutectic reaction. From Solidification Process of Nodular Cast Iron, S.-E. Wetterfall et al, Journal of the Iron and Steel Institute, 1972 © Maney www.maney.co.uk.

Figure 10.57b (right figure) Graphite nodules with and without an austenite shell floating in the melt in Figure 10.57a. From Solidification Process of Nodular Cast Iron, S.-E. Wetterfall et al, Journal of the Iron and Steel Institute, 1972 © Maney www.maney.co.uk.

Figure 10.57a shows many such nodules of approximately the same size in the melt. Figure 10.57b shows an enlarged part of the picture in Figure 10.57a. A thick austenite layer surrounds the graphite core of the big nodule on the right-hand side of the Figure 10.57b. The graphite nodule in the centre of the photograph is growing freely in the melt.

The heat of fusion as a function of the fraction of solid phase has been studied for nodular cast iron and flake graphite.

The measurements showed that the heat of fusion was much lower than the tabulated value at the beginning of the solidification process and increased at the end for nodular cast iron. No corresponding effect could be observed for flake graphite.

This puzzling topic will be discussed in Section 10.6.3 on pages 655–656.

Interpretation of the Observations. Formation of Nodules

The observations on nodular growth above can be interpreted as follows and give a qualitative idea of the formation of nodules.

In the cooling eutectic melt dendrites are formed together with free graphite nodules, which grow by diffusion of carbon atoms from the melt to the interface melt/graphite. New nuclei are nucleated incessantly and grow.

The growth of a nodule in direct contact with the melt is interrupted when the nodule happen to get in touch with an austenite dendrite owing to floatation.

An austenite shell is formed around the graphite core, which continues to grow owing to diffusion of carbon atoms from the melt through the austenite shell. The carbon atoms precipitate as graphite at the interface graphite/austenite. The graphite nodule continues to grow to a much larger size than it had in the melt. However, the growth rate through the austenite shell becomes very much lower than during the earlier direct diffusion of carbon atoms from the melt to the nodule. As the slowest process always determines the overall rate of a number of successive subprocesses, the diffusion through the austenite layer is most important of the two diffusion processes. The thickness of the austenite shell also grows.

Quantitative calculations of the growth process can also be performed. An example is given below.

Example 10.3

In a eutectic cast iron melt nodules are formed when the temperature of the melt is below the eutectic temperature. The figure shows the known concentration distribution of carbon in and around a nodule at a given temperature.

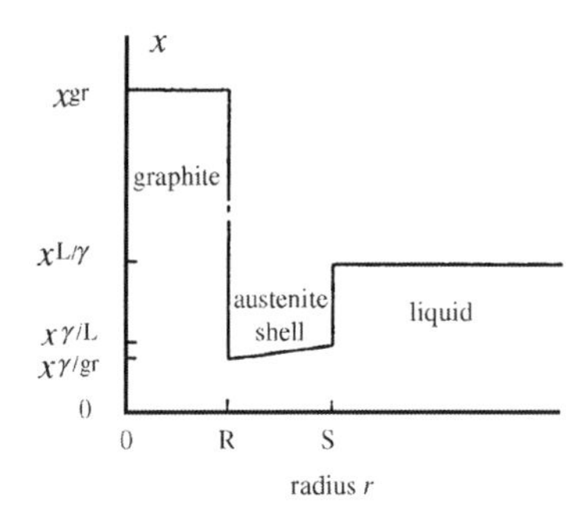

Figure 1′. Inner radius $R = \text{OR}$. Outer radius $S = \text{OS}$. Both depend on time.

a) Find the growth rate dR/dt of the spherical graphite grain as a function of the carbon concentration in the austenite shell, the radius R, the outer radius S and the diffusion constant D_γ of carbon in solid austenite and known quantities.
b) Find the radius R as a function of time t.
c) Find an expression for the ratio S/R and calculate the numerical value with the aid of the values given below. You may assume that the radius R_0 at time t is small and can be neglected.

Material constants:

$$V_m^L = 7.1 \times 10^{-3} \text{m}^3/\text{kmol} \; x^{gr} = 1$$

$$V_m^\gamma = 7.0 \times 10^{-3} \text{m}^3/\text{kmol} \; x^L = 0.174$$

$$V_m^{gr} = 5.5 \times 10^{-3} \text{m}^3/\text{kmol} \; x^\gamma = 0.086$$

Solution:

a)

C atoms diffuse from the melt through the austenite shell to the graphite core. We apply Fick's first law on the austenite shell with radius $r = R$.

Fick's first law:

$$\frac{dm}{dt} = -DA\frac{dc}{dy}$$

or

$$\frac{dm}{dt} = -\frac{DA}{V_m}\frac{dx}{dy}$$

$$\frac{dm}{dt} = -\frac{D_\gamma A}{V_m^\gamma}\left(\frac{dx^\gamma}{dr}\right)_{r=R} \tag{1'}$$

where

V_m^γ = the molar volume of austenite
A = area of the austenite shell surface
x = carbon concentration (mole fraction).

The derivative of x^{γ} with respect to r, i.e. the slope of the austenite line in Figure 1′, can be written as

$$\left(\frac{\mathrm{d}x^{\gamma}}{\mathrm{d}r}\right)_{r=R} = \frac{x^{\gamma/\mathrm{L}} - x^{\gamma/\mathrm{gr}}}{S - R} \tag{2'}$$

The amount of C atoms, that arrive at the graphite core and increase its volume with the amount $A\mathrm{d}R$ per unit time can alternatively be written as

$$\frac{\mathrm{d}m}{\mathrm{d}t} = -A\frac{\mathrm{d}R}{\mathrm{d}t}\frac{x^{\mathrm{gr}} - x^{\gamma/\mathrm{gr}}}{V_{\mathrm{m}}^{\mathrm{gr}}} \tag{3'}$$

Combining Equations (1′), (2′) and (3′) gives

$$-\frac{D_{\gamma}A}{V_{\mathrm{m}}^{\gamma}}\frac{x^{\gamma/\mathrm{L}} - x^{\gamma/\mathrm{gr}}}{S - R} = -A\frac{\mathrm{d}R}{\mathrm{d}t}\frac{x^{\mathrm{gr}} - x^{\gamma/\mathrm{gr}}}{V_{\mathrm{m}}^{\mathrm{gr}}}$$

or

$$\frac{\mathrm{d}R}{\mathrm{d}t} = \frac{D_{\gamma}}{S - R}\frac{x^{\gamma/\mathrm{L}} - x^{\gamma/\mathrm{gr}}}{x^{\mathrm{gr}} - x^{\gamma/\mathrm{gr}}}\frac{V_{\mathrm{m}}^{\mathrm{gr}}}{V_{\mathrm{m}}^{\gamma}} \tag{4'}$$

b)
Next we will eliminate S and find R as a function of t.
Elimination of S requires another equation in combination with Equation (4′). We will use a carbon balance. For this purpose we use Figure 2′.

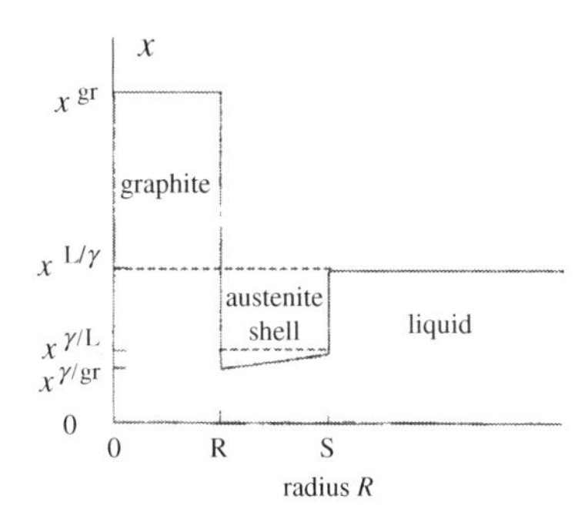

Figure 2′. Inner radius $R = \mathrm{OR}$. Outer radius $S = \mathrm{OS}$. Both depend on time.

C atoms diffuse from the melt, through the austenite shell to the growing graphite sphere. The melt is close to the eutectic temperature and therefore its concentration x^{L} is constant. The C concentration of pure graphite is constant and equal to 1.

The amount of carbon, required to increase the radius of the graphite sphere from R_0 to R, is taken from the melt and transported by diffusion through the austenite layer. The graphite core consists of pure carbon and the austenite shell is depleted on C but the average concentration must be equal to x^{L}.

This material balance is indicated in Figure 2′ by the dashed lines in the graphite and liquid regions, respectively.

$$\frac{4\pi(r^3 - R_0^3)}{3}\frac{x^{\mathrm{gr}} - x^{\mathrm{L}}}{V_{\mathrm{m}}^{\mathrm{gr}}} \approx \frac{4\pi(S^3 - R^3)}{3}\frac{x^{\mathrm{L}} - x^{\gamma/\mathrm{L}}}{V_{\mathrm{m}}^{\mathrm{L}}} \tag{5'}$$

$R_0 \approx 0$ and can be neglected. Hence, we obtain

$$S^3 - R^3 = R^3 \frac{x^{\text{gr}} - x^{\text{L}}}{x^{\text{L}} - x^{\gamma/\text{L}}} \frac{V_{\text{m}}^{\text{L}}}{V_{\text{m}}^{\text{gr}}}$$

or

$$S = R \sqrt[3]{\left(1 + \frac{x^{\text{gr}} - x^{\text{L}}}{x^{\text{L}} - x^{\gamma/\text{L}}} \frac{V_{\text{m}}^{\text{L}}}{V_{\text{m}}^{\text{gr}}}\right)} \tag{6'}$$

We apply Equation (4′) and insert the S value, which gives

$$\frac{\mathrm{d}R}{\mathrm{d}t} = \frac{D_\gamma}{R \sqrt[3]{\left(1 + \frac{x^{\text{gr}} - x^{\text{L}}}{x^{\text{L}} - x^{\gamma/\text{L}}} \frac{V_{\text{m}}^{\text{L}}}{V_{\text{m}}^{\text{gr}}}\right)} - \text{R}} \frac{x^{\gamma/\text{L}} - x^{\gamma/\text{gr}}}{x^{\text{gr}} - x^{\gamma/\text{gr}}} \frac{V_{\text{m}}^{\text{gr}}}{V_{\text{m}}^{\gamma}}$$

or after separation of the variables

$$R\mathrm{d}R = \frac{D_\gamma}{\sqrt[3]{\left(1 + \frac{x^{\text{gr}} - x^{\text{L}}}{x^{\text{L}} - x^{\gamma/\text{L}}} \frac{V_{\text{m}}^{\text{L}}}{V_{\text{m}}^{\text{gr}}}\right)} - 1} \frac{x^{\gamma/\text{L}} - x^{\gamma/\text{gr}}}{x^{\text{gr}} - x^{\gamma/\text{gr}}} \frac{V_{\text{m}}^{\text{gr}}}{V_{\text{m}}^{\gamma}} \mathrm{d}t$$

Integration gives

$$\int_0^{\text{R}} R\mathrm{d}R = \frac{D_\gamma}{\sqrt[3]{\left(1 + \frac{x^{\text{gr}} - x^{\text{L}}}{x^{\text{L}} - x^{\gamma/\text{L}}} \frac{V_{\text{m}}^{\text{L}}}{V_{\text{m}}^{\text{gr}}}\right)} - 1} \frac{x^{\gamma/\text{L}} - x^{\gamma/\text{gr}}}{x^{\text{gr}} - x^{\gamma/\text{gr}}} \frac{V_{\text{m}}^{\text{gr}}}{V_{\text{m}}^{\gamma}} \int_0^t \mathrm{d}t$$

or

$$R = \sqrt{\left[\frac{D_\gamma}{\sqrt[3]{\left(1 + \frac{x^{\text{gr}} - x^{\text{L}}}{x^{\text{L}} - x^{\gamma/\text{L}}} \frac{V_{\text{m}}^{\text{L}}}{V_{\text{m}}^{\text{gr}}}\right)} - 1} \frac{x^{\gamma/\text{L}} - x^{\gamma/\text{gr}}}{x^{\text{gr}} - x^{\gamma/\text{gr}}} \frac{V_{\text{m}}^{\text{gr}}}{V_{\text{m}}^{\gamma}}\right] \times t} \tag{7'}$$

c) S/R can be calculated from Equations (6’).

$$\frac{S}{R} = \sqrt[3]{\left(1 + \frac{x^{\text{gr}} - x^{\text{L}}}{x^{\text{L}} - x^{\gamma/\text{L}}} \frac{V_{\text{m}}^{\text{L}}}{V_{\text{m}}^{\text{gr}}}\right)} = \sqrt[3]{\left(1 + \frac{1 - 0.174}{0.174 - 0.086} \times \frac{7.1}{5.5}\right)} = 2.36$$

Answer:

a) $\dfrac{dR}{dt} = \dfrac{D_\gamma}{S-R}\dfrac{x^{\gamma/L} - x^{\gamma/gr}}{x^{gr} - x^{\gamma/gr}}\dfrac{V_m^{gr}}{V_m^{\gamma}}$

b) $R = \sqrt{\left[\dfrac{D_\gamma}{\sqrt[3]{\left(1 + \dfrac{x^{gr} - x^{L}}{x^{L} - x^{\gamma/L}}\dfrac{V_m^{L}}{V_m^{gr}}\right)} - 1}\dfrac{x^{\gamma/L} - x^{\gamma/gr}}{x^{gr} - x^{\gamma/gr}}\dfrac{V_m^{gr}}{V_m^{\gamma}}\right]} \times \sqrt{t}$

c) The ratio S/R is 2.4. The outer radius of the austenite shell is 2.4 times the graphite radius, independent of time.

10.6.3 Influence of Lattice Defects on Heats of Fusion of Near Eutectic Nodular Cast Iron and Flake Graphite

Experiments and Results

In 2003 Fredriksson, Tinoco et al [10] investigated the heat of fusion of near eutectic nodular cast iron and flake graphite as a function of the cooling rate with the aid of thermal analysis technique.

A number of solidification experiments were performed on cast iron samples at different cooling rates. A DTA furnace (Chapter 7 pages 397–388 and 408–411) was used for low cooling rates (0.08 K/s, 0.15 K/s and 0.35 K/s). For cooling rates between 30–55K/s (Chapter 7 411–412) a mirror furnace was used. Three alloys were examined, one of flake graphite and two of nodular cast iron.

Both the flake graphite and nodular alloys showed very clearly decreasing heat of fusion with increasing cooling rate. At high cooling rates the effect was stronger for nodular cast iron than for flake graphite. Figure 10.58 shows these results.

Additional experiments were performed in order to examine the discrepancy between flake graphite and nodular cast iron at high cooling rates. The heat of fusion was measured at various fractions of solidification by interrupting the solidification process with the aid of quenching the samples.

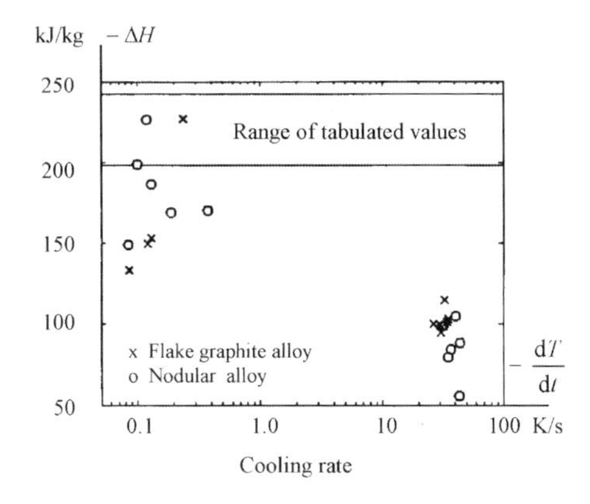

Figure 10.58 Measured heat of fusion as a function of cooling rate for flake graphite and nodular eutectic cast iron.

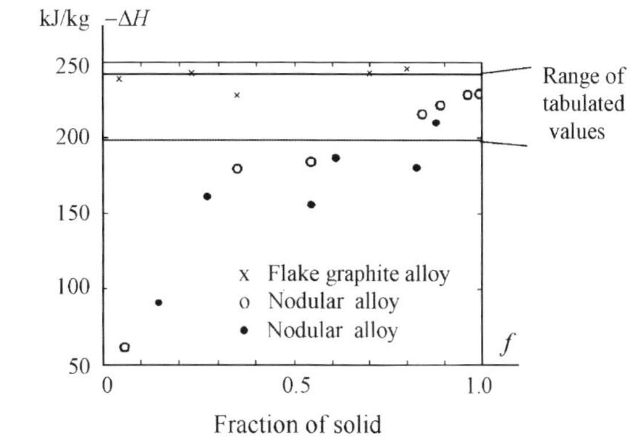

Figure 10.59 Measured heat of fusion as a function of fraction of solid flake graphite and nodular eutectic cast iron.

Figures 10.58 and 10.59 shows the result of these experiments. The heat of fusion of the flake graphite sample was found to be independent of the fraction of solidification, while the heat of fusion varied strongly with the fraction of solid in the two samples of nodular cast iron.

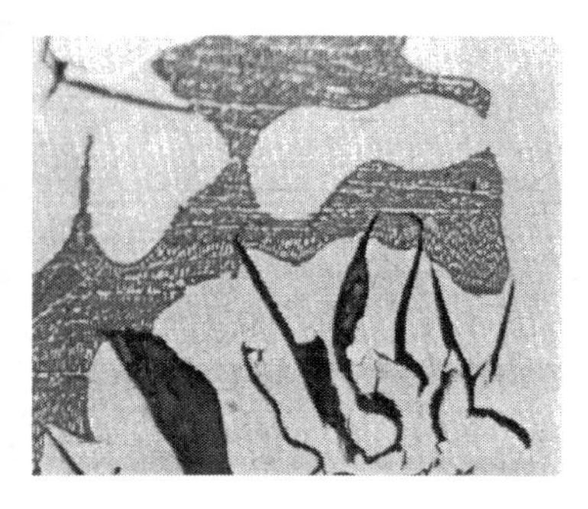

Figure 10.51b Structure of lamellar flake graphite. Grey area = liquid.

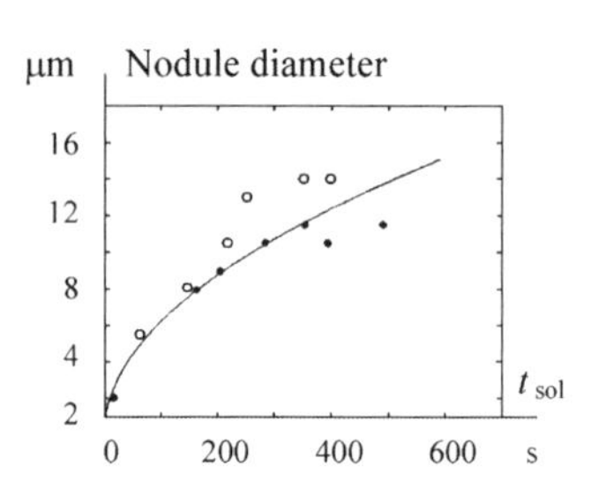

Figure 10.60 Nodular diameter as a function of time for the two nodular alloys at the cooling rate 0.15 K/s.

The difference in behaviour between flake graphite and nodular cast iron can be explained if the presence of vacancies in the materials and the difference in structure of the two types of cast iron are considered.

Formation Mechanisms of Flake Graphite and Nodular Cast Iron

Lamellar flake graphite grows as cells of graphite and austenite. The growth occurs mainly radially with a rather uneven solidification front (Figure 10.51b). The graphite leads the growth and the austenite is mainly formed behind the graphite flakes. The growth is no normal eutectic growth but can be characterized as primary graphite precipitation followed by formation of a secondary austenite phase (page 649).

The nodular eutectic structure of cast iron is formed in a completely different way.

The graphite nodules are formed in the melt and grow freely to a certain size and become later surrounded by a solid austenite shell. The continued growth of the nodules occur by diffusion of carbon atoms from the melt through the austenite shell to the spherical graphite core (Example 10.3 on page 652).

Experiments show that more than 70% of the fraction of graphite in the nodules is formed by diffusion of carbon through the austenite shell. is faster The smaller the nodules are the faster will be the growth rate of the nodules as Figure 10.60 shows.

Interpretation of the Experimental Results

On pages 412–414 in Chapter 7 the reason for decreasing heat of fusion with increasing cooling rate was discussed. Formation of lattice defects, mainly vacancies, requires energy. The higher the concentration of vacancies is, the easier it will be to break the rest of the bonds between the atoms in the crystal lattice and the lower will be the heat of fusion.

This general effect is valid for both flake graphite and eutectic nodular cast iron. The remaining problem is to explain why the heat of fusion at high cooling rates is lower for nodular cast iron than for flake graphite. Figure 10.58 shows clearly this additional effect.

The molar volume of graphite is much larger than the molar volumes of carbon dissolved in austenite and in the melt. Hence, the austenite shell around the nodules must expand by plastic deformation during the growth of the graphite nodules.

At the high temperature in question, the deformation process will occur by creep, i.e. by gradual vacancy formations and vacancy condensations. The plastic deformation rate is faster for small nodules than for larger ones (Figure 10.60). Therefore, more vacancies per unit time are formed at the beginning of the solidification process (low fraction of solid) than at the end, which results in a low heat of fusion for nodular cast iron for low fractions of solid, i.e. at the beginning of the solidification process.

When the fraction of solid increases the vacancy formation will gradually decrease and the heat of fusion increases and approaches the region with tabulated values of heat of fusion. This effect concerns only nodular cast iron and is absent for flake graphite. For this reason the heat of fusion for flake graphite is independent of the fraction of solid as shown in Figure 10.59.

The vacancy formation also affects the heat of fusion as a function of the cooling rate, owing to plastic deformation of the austenite shell.

When the cooling rate is increased both the growth rate and the nucleation rate of new nodules increases. More nodules are formed per unit time and the vacancy formation increases, owing to the plastic deformation process of the austenite shell. Hence, the growth of nodules supplies an additional contribution of vacancies. This effect explains why the heat of fusion is lower for nodular cast iron than for flake graphite at high cooling rates (Figure 10.59). At lower cooling rates the effect is negligible.

Thermodynamic Calculations

Fredriksson, Stjerndahl and Tinoco [11] have performed thermodynamic calculations on the growth of nodules in the cast iron melt.

When the nodules initially are formed by diffusion of carbon through the melt they are small and grow rapidly. When the shell around the nodules has been formed they grow by diffusion of carbon through the austenite shell. This process gives a large number of lattice defects. The diffusion of carbon through the austenite and the deformation process of the expanding austenite shell control the growth of nodules surrounded by an austenite shell.

- The growth rate of the nodule cavity is expressed in terms of its radius with and without the austenite shell, the diffusion constant for carbon in austenite and the concentration profile of carbon.
- A thermodynamic expression for the total driving force is set up and the fraction of vacancies, needed for the plastic deformation of the austenite shell, is derived.

Figure 10.61 shows a schematic representation of the free-energy curves. With the aid of the 'common tangent' method the driving force can be constructed graphically in the same way as on pages 77–78 in Chapter 2.

From the thermodynamic expressions the non-equilibrium phase diagram of the system Fe–C can be derived. Figure 10.62 shows the change of the phase diagram. The calculations explain the growth behaviour of nodules. They grow freely in the melt until they have reached a certain size.

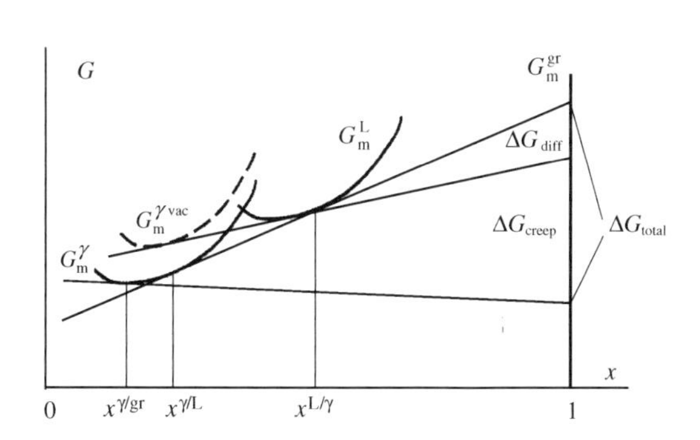

Figure 10.61 Sketch of the free-energy curves at a temperature below the eutectic temperature of Fe–C. Driving force $= -\Delta G_{total}$.

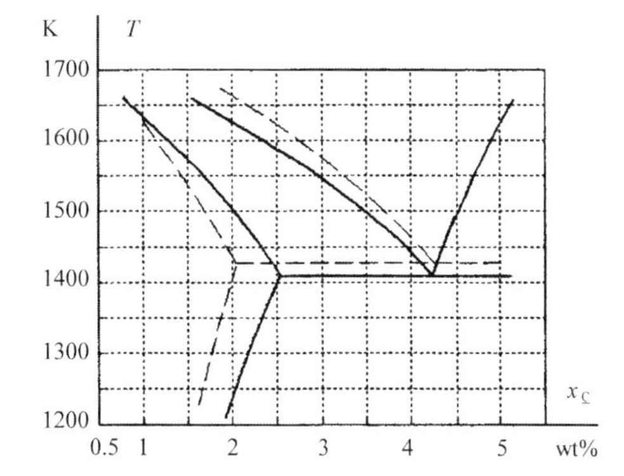

Figure 10.62 Stable (dashed lines) and non-stable (full lines) phase diagram of Fe–C.

Small nodules have high growth rates. A tentative precipitation of an austenite shell is prohibited as there is no driving force for the process. The reason is that the growth rate would be so high that it would cause numerous lattice defects per unit volume in the presumed austenite layer, that eliminate the driving force for precipitation of an austenite shell. Instead free nodules are formed together with austenite dendrites.

When the nodules have achieved a certain size the growth rate has decreased so much that the reduced number of defects per unit volume no longer eliminates the driving force. The precipitation of an austenite layer becomes possible and starts.

The calculation of the non-equilibrium phase diagrams is supported by measurements of the volume fractions of the different phases.

10.6.4 Eutectic Growth of White Cast Iron

When a eutectic iron melt solidifies at *low* growth rates, grey iron is formed and the first precipitated phase is graphite.

White cast iron is formed at *high* growth rates. It consists of a mixture of cementite, Fe_3C, and austenite, which often is called ledeburite. The eutectic structure can appear in several different morphologies, depending on the composition of the melt. Figure 10.63 shows a typical structure of ledeburite.

Figure 10.63 shows a dendrite structure of austenite together with a plate-like structure of cementite. The growth rates of the cementite phase are very fast at the edges of the plate. For this reason the cementite and austenite phases do not cooperate during growth.

The remaining liquid between the cementite plates and the dendrites normally solidifies by growth in a direction, perpendicular to the plates as shown in Figure 10.63.

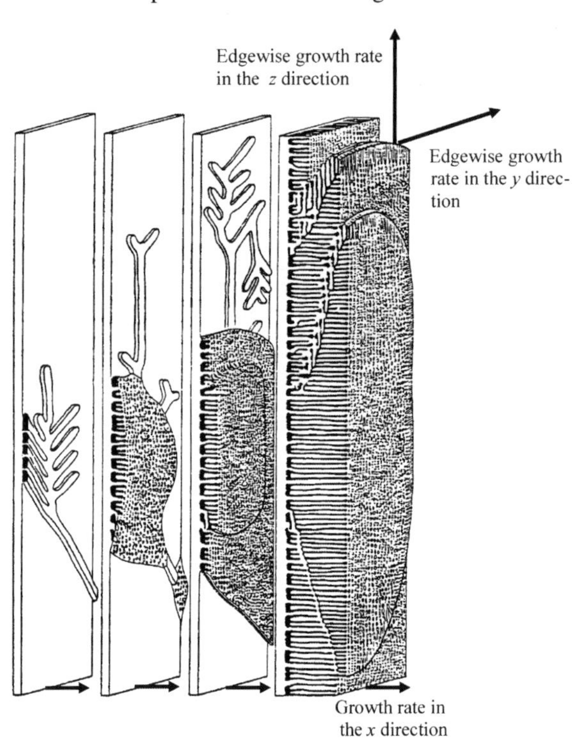

Figure 10.63 Precipitation of cementite
Dendrites and austenite in white iron. Sketch of a typical structure.

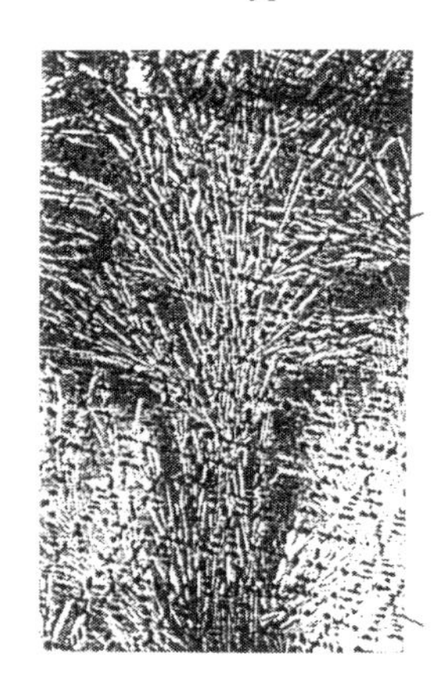

Figure 10.64 Structure of white cast iron. The broom structure is caused by crystal multiplication of fragile cementite flakes.

When a cementite flake grows perpendicularly to its plane, the growth rate is lower than in the former direction and a good cooperation between the cementite and austenite phases is established. A normal eutectic structure with parallel rods in a matrix of austenite is formed. Hence, white iron consists of a mixture of two different eutectic structures, one with no co-operation and one with a normal rod eutectic structure.

The cementite flakes are fragile, which makes white iron brittle. For this reason, high growth rates over a certain critical value are not used in order to avoid ledeburite formation in castings. However, white iron is wear-resistant and for this reason it is often used in applications where this property is important.

The desire to avoid white iron formation is the reason why cast iron is inoculated. Inoculation with Fe-Si favours the nucleation of grey cast iron. Silicon is also added to the melt in order to increase the temperature difference between the stable and metastable eutectic temperatures (Figure 10.50 on page 647).

10.7 Solidification of Al-Si Eutectics

10.7.1 Solidification of Unmodified and Sr-Modified Al-Si Eutectics

Al-Si alloys constitute a large percentage of the shaped castings produced in foundries. The structures of Al-Si alloys depend strongly on the casting process and also on modification.

Modification means that small fractions of Na or Sr are added to the Al-Si alloys. The purpose is to achieve a finer eutectic structure of the alloys. Most Al-Si alloys are modified for this reason. Figures 10.65a and b show the effect.

One of the factors, which affect the structure of the alloy most of all, is the cooling rate. In this section, we will examine the influence of the cooling rate on the properties of unmodified and modified Al-Si alloys.

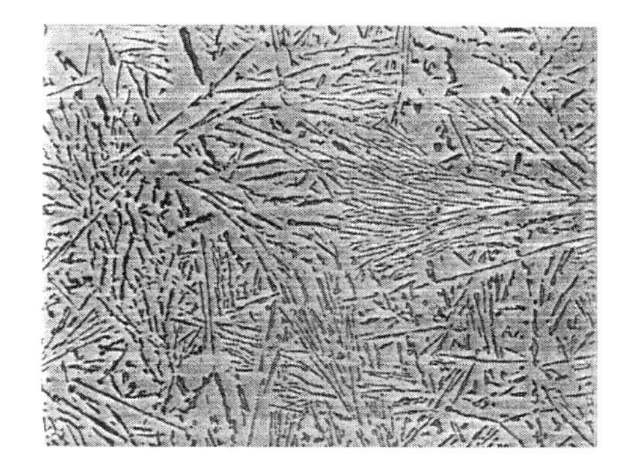

Figure 10.65a Unmodified Al–Si alloy.

Figure 10.65b Al–Si alloy modified with Na.

Thermal analysis methods, particularly DTA (differential thermal analysis) and DSC (differential scanning calorimetry) are often used to study solidification processes. Some results from such studies are given below.

These techniques have been described in Chapter 7. A wide range of cooling rates from 0.03 K/s to 10^5 K/s was used. For cooling rates within the interval 10000 K/s – 200 K/s the levitation technique was used, within the interval 200 K/s – 1 K/s the mirror furnace technique was applied and the DTA technique were used for cooling rates less than 1 K/s.

Eutectic Al-12.6 wt% Si, Al-14 wt% Si and Al-16 wt% Si alloys were used. Two different Sr concentrations, 150 ppm and 250 ppm, were used to modify each alloy.

Microstructure as a Function of Cooling Rate and Modification

Fredriksson and Talaat [12] have studied the influence of cooling rate and modification on the microstructures of Al-Si alloys. Some of their results are reported here. Figures 10.66a–d and 10.67 a–d show the microstructures of the eutectic unmodified Al-Si alloy and the eutectic Al-Si alloy, modified with 250 ppm at various cooling rates.

Each pair of alloys was cooled at cooling rates of the same magnitude but not exactly the same cooling rate. The figures are not reproduced at exactly the same magnification. To obtain a fair comparison and a possibility to evaluate the result quantitatively, the average particle distances were derived and plotted as a function of the cooling rate. The result is given in Figure 10.68 on page 661.

Figure 10.66a shows a faceted plate-like microstructure for the unmodified Al-Si eutectic. At increased cooling rate the micro-structure of this alloy shows precipitation of polyhedron Si-crystals in a matrix of a primary Al-rich phase and a eutectic microstructure. The coarse eutectic structure (Figure 10.66b) becomes finer when the cooling rate is further increased (Figure 10.66c). At an extremely high cooling rate a further refinement of the microstructure is obtained with a remarkable increase of the amount of dendrites of Al-rich phase. This is shown in the enlarged Figure 10.66d.

The Sr-modified eutectic Al-Si alloys in Figures 10.67 show much finer structures, even at low cooling rates, than the corresponding unmodified alloy. As the cooling rate is increased the modified fibrous silicon becomes finer with the increased volume fraction of Al-rich phase. At high cooling rates the modified eutectic Al-Si alloy shows a typical hypoeutectic dendritic microstructure, i.e. a very fine interdendritic structure as the Figures 10.67c and 10.67d show. No major difference in structure exists between the Figures 10.66d and 10.67d. This is confirmed by Figure 10.68.

From the Figures 10.66, 10.67 and 10.68 we can conclude that

1. The microstructure becomes finer when the eutectic alloy is modified.
2. The microstructure becomes finer when the cooling rate is increased.
3. The modified and unmodified structures look the same at high cooling rates.

Figure 10.66a (left figure) Eutectic unmodified Al–Si alloy. Cooling rate: 0.167 K/s.
E.-B. Talaat, H. Fredriksson in Materials Transactions © The Japan Institute of Metals 2000.

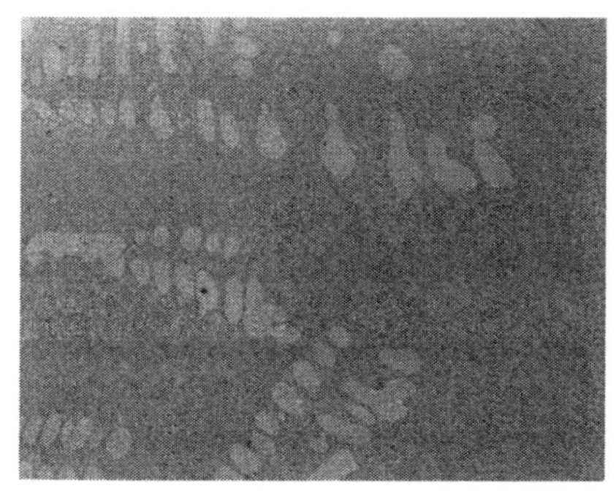

Figure 10.67a (right figure) Eutectic Al–Si alloy modified with 250 ppm Sr. Cooling rate: 0.616 K/s.
E.-B. Talaat, H. Fredriksson in Materials Transactions © The Japan Institute of Metals 2000.

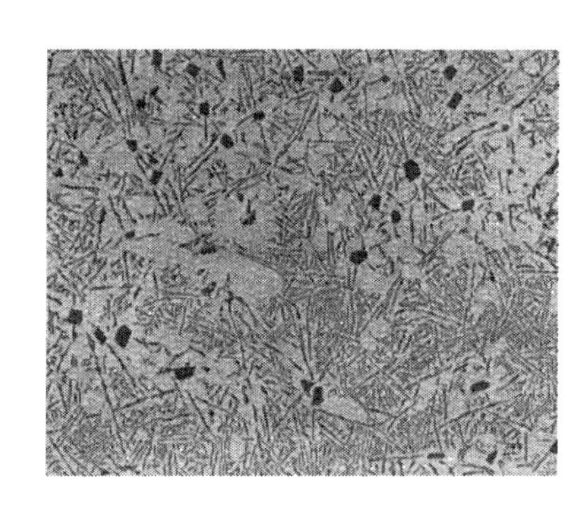

Figure 10.66b (left figure) Left figure. Eutectic unmodified Al–Si alloy. Cooling rate: 2.64 K/s.
E.-B. Talaat, H. Fredriksson in Materials Transactions © The Japan Institute of Metals 2000.

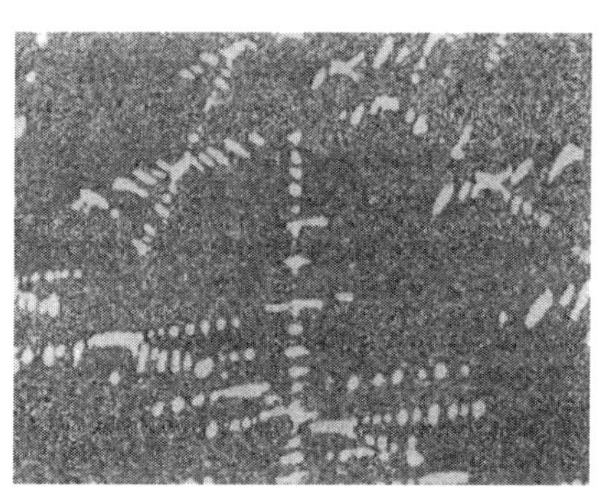

Figure 10.67b (right figure) Eutectic Al–Si alloy modified with 250 ppm Sr. Cooling rate: 2.8 K/s.
E.-B. Talaat, H. Fredriksson in Materials Transactions © The Japan Institute of Metals 2000.

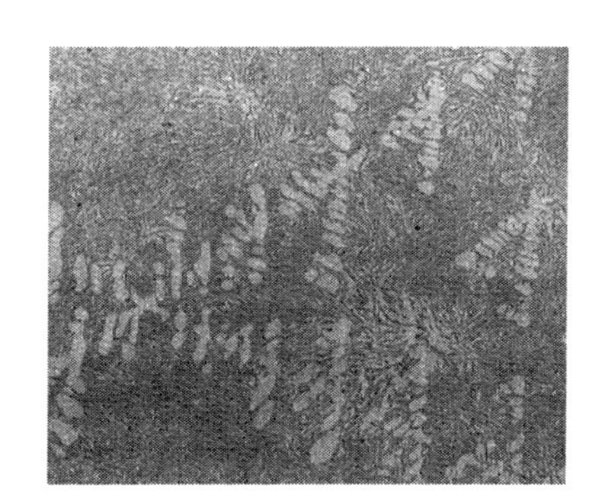

Figure 10.66c (left figure) Eutectic unmodified Al–Si alloy. Cooling rate: 16.8 K/s.
E.-B. Talaat, H. Fredriksson in Materials Transactions © The Japan Institute of Metals 2000.

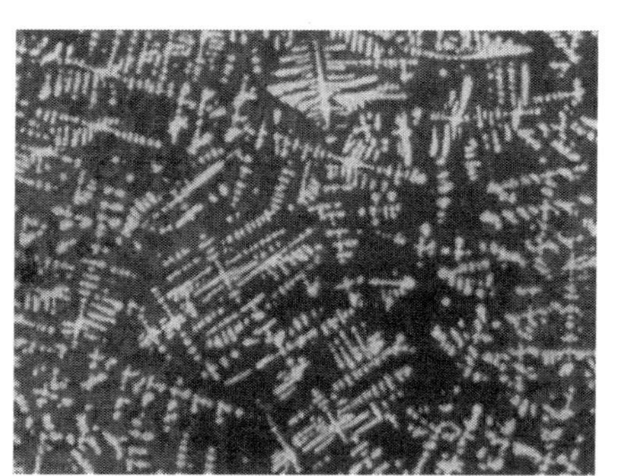

Figure 10.67c (right figure) Eutectic Al–Si alloy modified with 250 ppm Sr. Cooling rate: 52.4 K/s.
E.-B. Talaat, H. Fredriksson in Materials Transactions © The Japan Institute of Metals 2000.

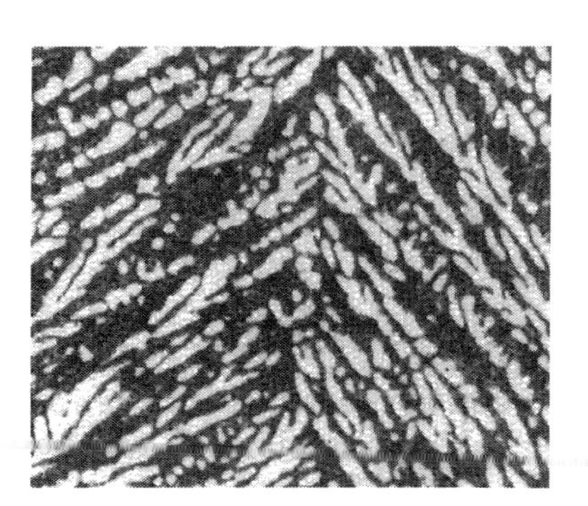

Figure 10.66d (left figure) Left figure. Eutectic unmodified Al–Si alloy. Cooling rate: 5.5×10^4 K/s.
E.-B. Talaat, H. Fredriksson in Materials Transactions © The Japan Institute of Metals 2000.

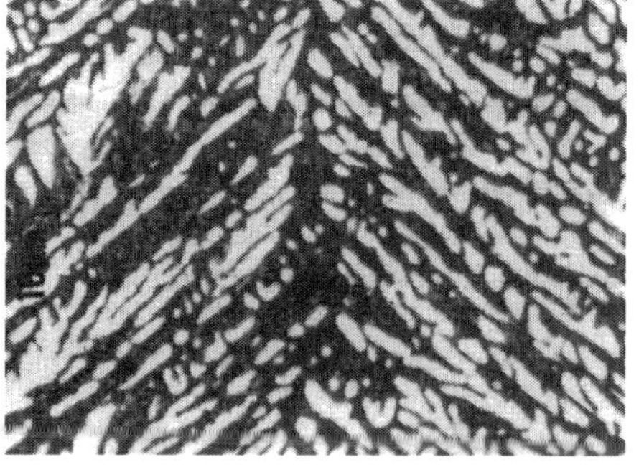

Figure 10.67d (right figure) Eutectic Al–Si alloy modified with 250 ppm Sr. Cooling rate: 6.5×10^4 K/s.
E.-B. Talaat, H. Fredriksson in Materials Transactions © The Japan Institute of Metals 2000.

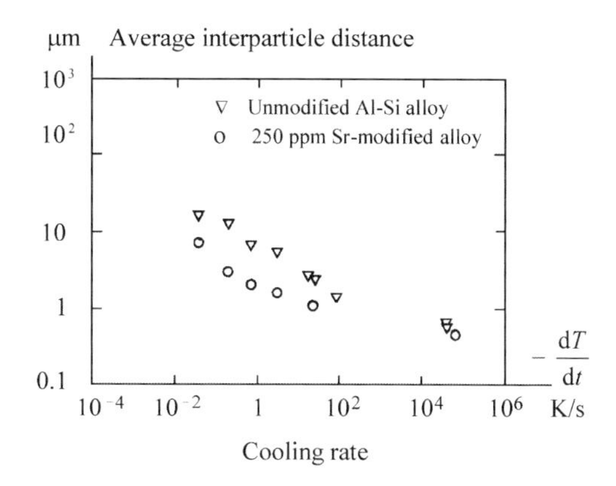

Figure 10.68 The average particle distances in Figures 10.66 and 10.77 as a function of the cooling rate. E.-B. Talaat, H. Fredriksson in Materials Transactions © The Japan Institute of Metals 2000.

Si Fraction in Al-Si Eutectics as a Function of Cooling Rate

The volume fraction of the eutectic structure and the volume fraction of Al dendrites were determined by conventional linear analysis of photographs of the samples. The Si fraction in the Al-rich phase could be calculated from the volume fraction measurements, the composition of the Al phase and the lever rule. The average Si fraction in the eutectic structure was also determined with the aid of microprobe measurements.

The Si fraction was determined for samples of unmodified and modified Al-Si eutectics solidified at various cooling rates. The results are shown in the Figures 10.69a, b and c.

At the highest cooling rates the silicon fraction in the eutectic structure was around 20 at% in all three cases and independent of the degree of modification.

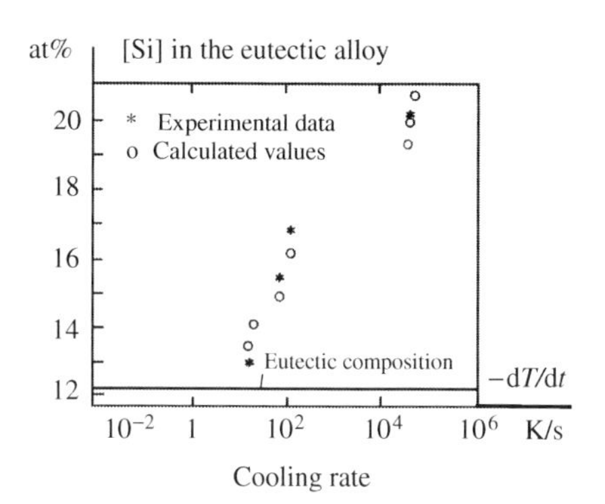

Figure 10.69a Si concentration (at%) as a function of the cooling rate. Unmodified eutectic structure. E.-B. Talaat, H. Fredriksson in Materials Transactions © The Japan Institute of Metals 2000.

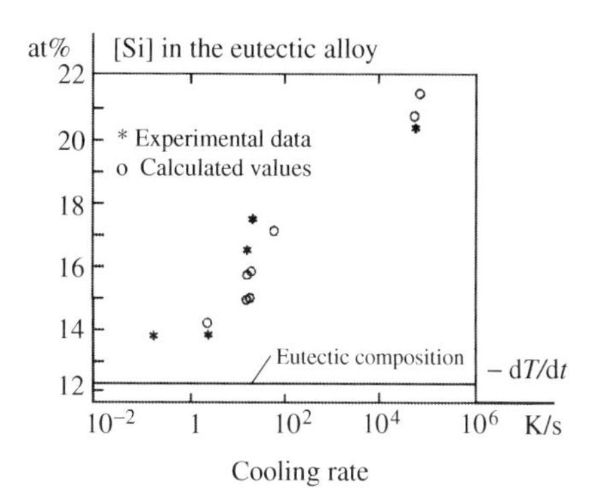

Figure 10.69b Si concentration (at%) as a function of the cooling rate. Modified eutectic structure 150 ppm Sr. E.-B. Talaat, H. Fredriksson in Materials Transactions © The Japan Institute of Metals 2000.

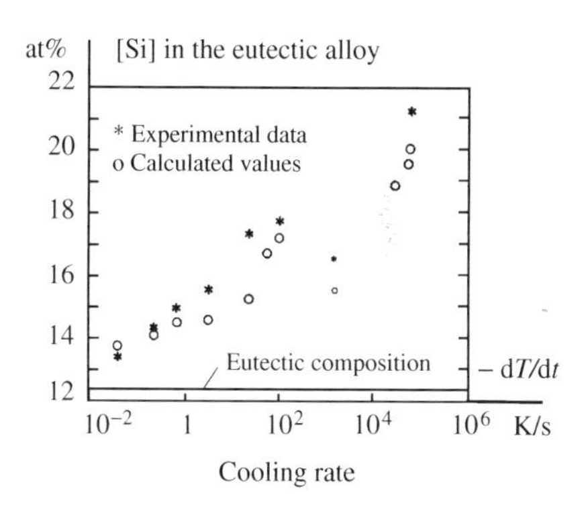

Figure 10.69c Si concentration (at%) as a function of the cooling rate. Modified eutectic structure 250 ppm Sr. E.-B. Talaat, H. Fredriksson in Materials Transactions © The Japan Institute of Metals 2000.

10.7.2 Influence of Lattice Defects on Eutectic Al-Si Alloys

Heat of Fusion

It is well known that alloys have values of the heat of fusion, which depend on the cooling rate. In most cases, the effect is small and can be neglected. In the case of Al-Si alloys the effect cannot be neglected and will be discussed below.

The heat of fusion can be calculated from a cooling curve, i.e. recordings of the temperature as a function of time (Chapter 7 pages 408–414). The cooling curve and the temperature difference between the sample and the surroundings can be measured with the aid of DTA technique for a sample placed in an Al crucible. Examples of such cooling curves are given in Figure 10.70a, b and c. The results of the determinations are summarized in Figure 10.71.

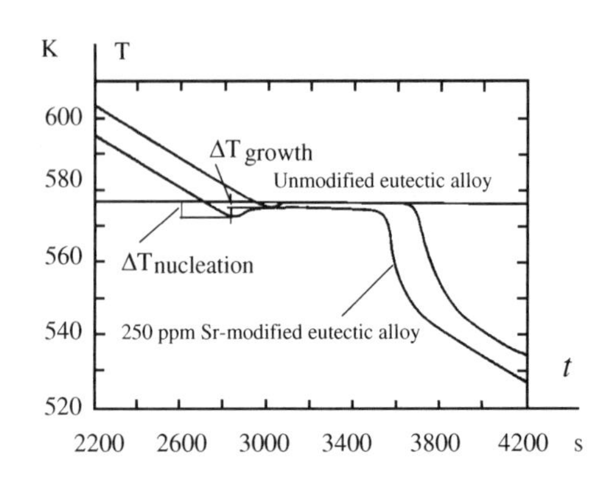

Figure 10.70a Cooling curves for unmodified and 250 ppm Sr-modified eutectic Al–Si alloys, recorded with the aid of DTA-technique. E.-B. Talaat, H. Fredriksson in Materials Transactions © The Japan Institute of Metals 2000.

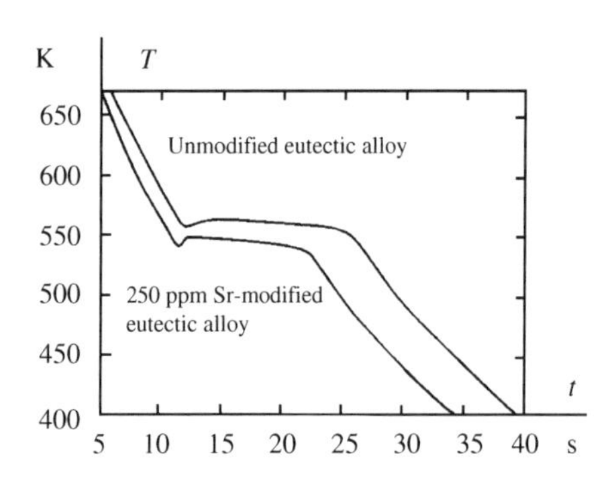

Figure 10.70b Cooling curves for unmodified and 250 ppm Sr-modified eutectic Al–Si alloys, solidified in a mirror furnace. E.-B. Talaat, H. Fredriksson in Materials Transactions © The Japan Institute of Metals 2000.

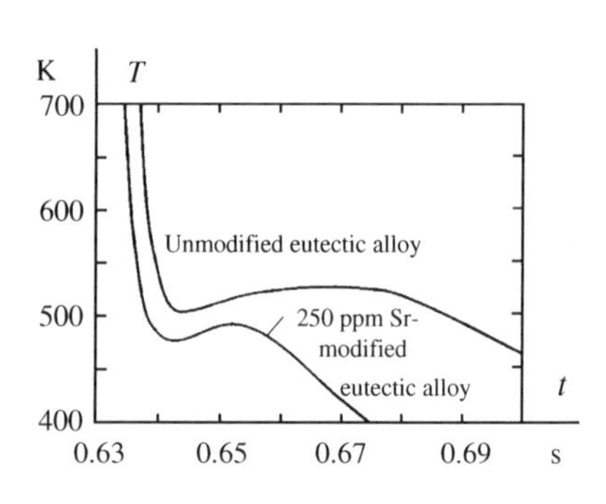

Figure 10.70c Cooling curves for unmodified and 250 ppm Sr-modified eutectic Al–Si alloys, recorded with the aid of levitation casting technique. E.-B. Talaat, H. Fredriksson in Materials Transactions © The Japan Institute of Metals 2000.

Figure 10.71 shows graphically the measurements of the heat of fusion for eutectic Al-Si alloys over a wide range of cooling rates.

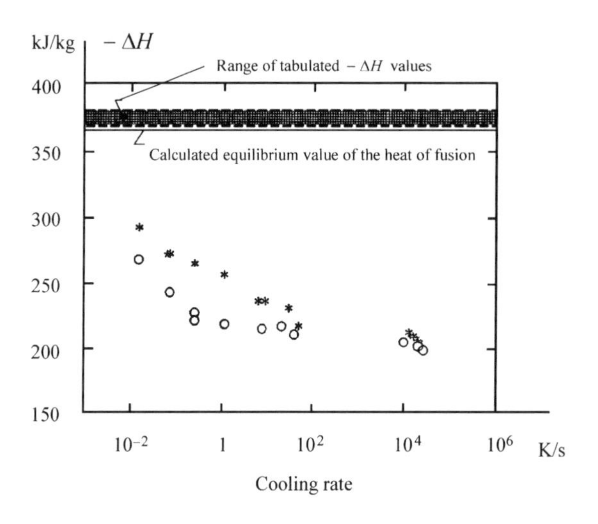

Figure 10.71 Heat of fusion as a function of the cooling rate for eutectic Al–Si alloys. E.-B. Talaat, H. Fredriksson in Materials Transactions © The Japan Institute of Metals 2000

The following conclusions are obvious:

1. Extrapolation shows that at cooling rates $> 10^{-3}$ K/s the heat of fusion starts to decrease with increasing cooling rate.
2. The decrease is faster in the modified alloy than in the unmodified one.
3. The heat of fusion is lower in the modified alloy than in the unmodified alloy at cooling rates between 10^{-2}–10^{2} K/s.
4. At the highest cooling rates, the heat of fusion is the same in the unmodified and the modified alloy. The reduced value of the heat of fusion is in this case about 60% of the tabulated value, which corresponds to a cooling rate equal to zero.

The following explanation to the reduction of the heat of fusion for Al-Si is the same as that for cast iron Fe–C (page 655). The number of lattice defects, mainly vacancies, increases with increasing growth rate. The fraction of defects is exceptionally large in graphite and also in silicon.

The heat of fusion ($-\Delta H$) represents the energy required to melt 1 mass unit (1 kg) or 1 kmol of the substance, i.e. the total energy required to break the bonds between the atoms in the crystal lattice. If vacancies are present, some bonds are missing and less energy is required to break the remaining ones. Hence, the required energy decreases with increasing vacancy concentration; i.e. increased growth rate and cooling rate.

Equilibrium of vacancies can only exist when the growth rate at the solidification front is zero. This is impossible to achieve in practice. Any growth rate results in a higher vacancy concentration than the equilibrium concentration and results in a lower heat of fusion than the tabulated value. The latter can never be achieved in reality, only be calculated.

Coupled Zone of Eutectic Al-Si Alloys

Four different types of microstructures have been observed in the Al-Si alloys when they solidify (Figures 10.72). In *hypoeutectic* Al-Si alloys primary Al rich dendrites are formed (region A in the phase diagram), followed by either a plate-like faceted eutectic structure (region B in the phase diagram) or a non-faceted eutectic structure, formed by coupled growth of aluminium and silicon (region C in the phase diagram). In *hypereutectic* Al-Si alloys primary silicon (region D in the phase diagram) precipitates, followed by either of the two structures corresponding to regions B and C.

The undercooling at the solidification can be derived from the cooling curves in the Figures 10.70. The information of about the undercooling, the composition of the alloy and the measured composition of the normal eutectic structure can be used to map the regions of different morphologies in the phase diagram. They are shown in the phase diagram for the unmodified and the two Sr-modified Al-Si alloys in the Figures 10.72.

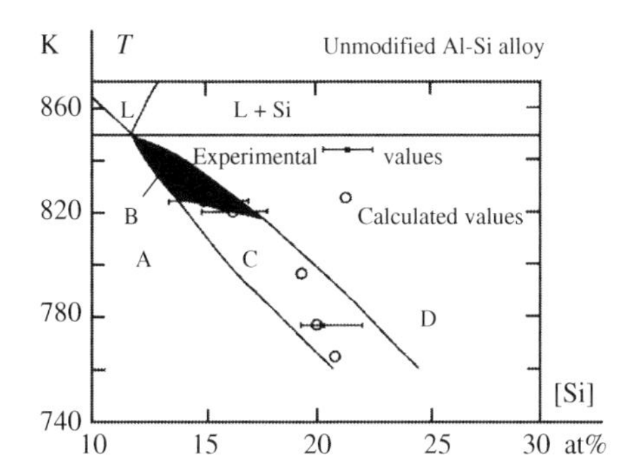

Figure 10.72a Coupled zone of the Al–Si system estimated for unmodified alloys. E.-B. Talaat, H. Fredriksson in Materials Transactions © The Japan Institute of Metals 2000.

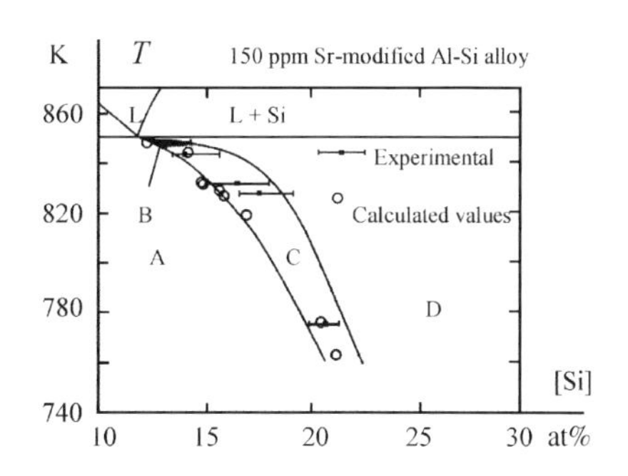

Figure 10.72b Coupled zone of the Al–Si system estimated for alloys modified with 150 ppm Sr. E.-B. Talaat, H. Fredriksson in Materials Transactions © The Japan Institute of Metals 2000.

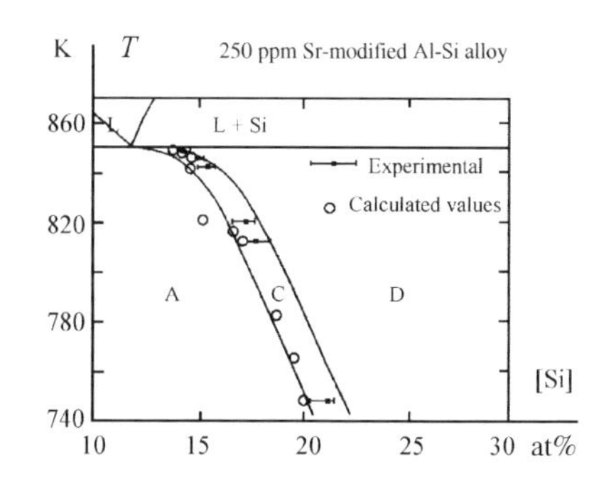

Figure 10.72c Coupled zone of the Al–Si system estimated for alloys modified with 250 ppm Sr. E.-B. Talaat, H. Fredriksson in Materials Transactions © The Japan Institute of Metals 2000.

Region A corresponds to the Al rich primary phase, which is precipitated in hypoeutectic Al-Si alloys. Region B shows the temperature-composition region where the plate-like faceted eutectic structure is formed. The region, where the rod-like non-faceted eutectic structure exists, is marked by C. D represents the region where the primary precipitation of faceted silicon crystals occurs.

A comparison between the Figures 10.72a, b and c shows that region B, which is large in the unmodified alloy, is suppressed in the 150 ppm Sr-modified alloy and absent in the 250 ppm Sr-modified alloy. The effect of Sr-modification on the structure of the alloy is strong.

Influence of Fraction of Vacancies on the Phase Diagram of the Al-Si system

As a rough summary of the results above we can state that

1. Sr modification has a strong influence on the structure of Al-Si alloys.
2. The heat of fusion decreases with increasing cooling rate during the solidification process. The heat of fusion is lower for modified Al-Si alloys than for unmodified ones for small and medium cooling rates.
3. The coupled zone in the Al-Si system is strongly influenced by the degree of Sr-modification.

The decrease of the heat of fusion can be satisfactorily explained as an effect of lattice defects, which mainly consist of vacancies. It is reasonable to assume that there is a coupling between the concentration of vacancies and the Sr-modification. This idea is supported by the discussion on the extension of the coupled zone in the Al-Si phase diagram and the strong influence the Sr-modification has on the region B in Figure 10.72.

- The basic and common reason to all the phenomena discussed in this section is likely to be the fraction of lattice defects, mainly the vacancies, which are present in alloys.

The lattice defects will also influence the very phase diagram of the Al-Si system. As an end of this section we will introduce some theoretical calculations by Fredriksson and Talaat [12]. We omit the calculations and give their results for three different vacancy concentrations in the Figures 10.73b–d. They show the phase diagram as a function of the vacancy concentrations. The vacancy concentrations are chosen is such a way that they are related to the decrease of the heat of fusion in Figure 10.71.

A comparison between the Figures 10.73a–d in this theoretical study shows that the phase diagram changes drastically with the vacancy concentration.

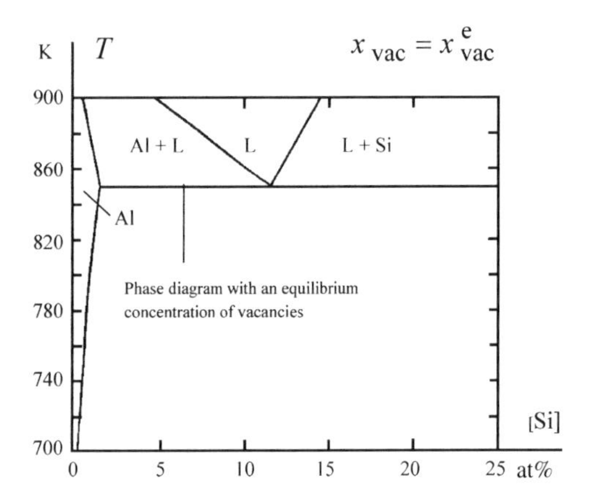

Figure 10.73a Phase diagram in case of equilibrium fraction of vacancies. E.-B. Talaat, H. Fredriksson in Materials Transactions © The Japan Institute of Metals 2000.

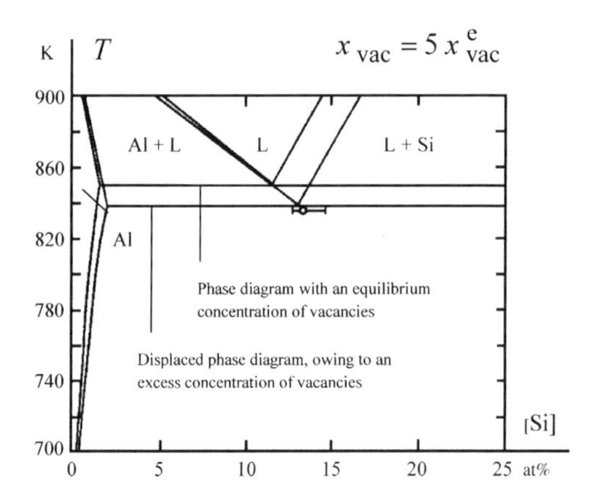

Figure 10.73b Phase diagram at increased fraction of vacancies. Cooling rate = 17 K/s. Undercooling = 13 K. E.-B. Talaat, H. Fredriksson in Materials Transactions © The Japan Institute of Metals 2000.

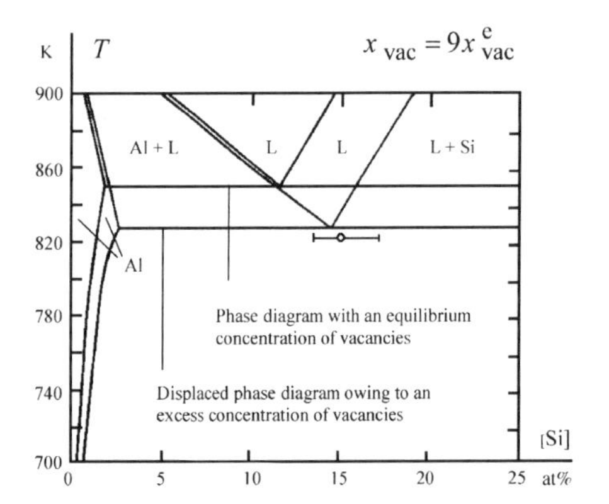

Figure 10.73c Phase diagram at strongly increased fraction of vacancies. Cooling rate = 80 K/s. Undercooling = 25.5 K. E.-B. Talaat, H. Fredriksson in Materials Transactions © The Japan Institute of Metals 2000.

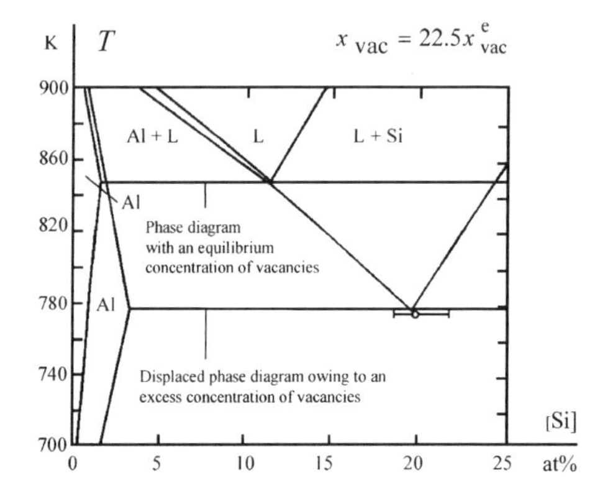

Figure 10.73d Phase diagram at very strongly increased fraction of vacancies. Cooling rate = 130 K/s. Undercooling = 29 K. E.-B. Talaat, H. Fredriksson in Materials Transactions © The Japan Institute of Metals 2000.

10.8 Transition between Normal Lamellar and Rod Eutectic Growth

In Section 10.4.3 on page 624 we discussed competitive eutectic growth.

- When several structures are competing at a given growth rate the structure, that requires the lowest under-cooling $\Delta T_E = T_E - T$ will win.
- At a given undercooling. the structure with the highest growth rate will win.

This fact can be used to analyze the change from normal lamellar to normal rod eutectic growth and vice versa.

Transition from Normal Lamellar to Normal Rod Eutectic Growth and Vice Versa

As a basis for a discussion we have the two equations on pages 603 and 614:

$$\Delta T_{\text{lamella}} = const_{\text{lamella}}\sqrt{V_{\text{growth}}} \quad (10.49)$$

$$\Delta T_{\text{rod}} = const_{\text{rod}}\sqrt{V_{\text{growth}}} \quad (10.102)$$

The structure with the lowest undercooling will win. As the growth rate is the same in both cases we can compare the *constants* instead of the *undercoolings* and use the two equa-tions (pages X26 and X47)

$$const_{\text{lamella}} = \sqrt{\frac{\sigma_{\alpha\,\beta} A_{\text{lamella}} B_{\text{lamella}} C_{0\ \text{lamella}}}{RTD}} \quad (10.52)$$

$$const_{\text{rod}} = \sqrt{\frac{\sigma_{\alpha\beta} A_{\text{rod}} B_{\text{rod}} V_{\text{m}}^{\text{L}}}{RTD}} \quad (10.105)$$

where the constants, introduced on pages X22 and X47, respectively, are listed in the boxes below. We also need Equation (10.96) on page X44 where f_α is the eutectic fraction of the α phase.

Constants of normal lamellar eutectic growth:

$$A_{\text{lamella}} = \frac{m_\beta^{\text{L}} V_{\text{m}}^\beta}{f_\beta} - \frac{m_\alpha^{\text{L}} V_{\text{m}}^\alpha}{f_\alpha}$$

$$B_{\text{lamella}} = \frac{m_\beta^{\text{L}}}{V_{\text{m}}^\beta} - \frac{m_\alpha^{\text{L}}}{V_{\text{m}}^\alpha}$$

$$C_{0\ \text{lamella}} = f_\alpha f_\beta V_{\text{m}}^{\text{L}}\left(x_{\text{E}}^{\beta/\text{L}} - x_{\text{E}}^{\alpha/\text{L}}\right)$$

Constants of normal rod eutectic growth:

$$A_{\text{rod}} = \frac{m_\beta^{\text{L}} V_{\text{m}}^\beta}{\frac{r_\beta}{r_\alpha}\left(2 + \frac{r_\beta}{r_\alpha}\right)} - m_\alpha^{\text{L}} V_{\text{m}}^\alpha \qquad \frac{r_\beta}{r_\alpha} = \sqrt{\frac{1}{f_\alpha}} - 1$$

$$B_{\text{rod}} = \frac{m_\beta^{\text{L}}}{V_{\text{m}}^\beta}\frac{r_\beta}{r_\alpha}\frac{2 + \frac{r_\beta}{r_\alpha}}{2 + \frac{4r_\beta}{\pi r_\alpha}}\left(x_{\text{E}}^{\beta/\text{L}} - x_{\text{E}}^{\text{L}}\right) - \frac{m_\alpha^{\text{L}}}{V_{\text{m}}^\alpha}\frac{x_{\text{E}}^{\text{L}} - x_{\text{E}}^{\alpha/\text{L}}}{2 - \frac{4}{\pi}}$$

If we calculate the constants $const_{\text{lamella}}$ and $const_{\text{rod}}$ for the known value of f_α of the eutectic alloy it is possible to predict the type of structure obtained at normal eutectic growth. The lowest value corresponds to the winning growth mode.

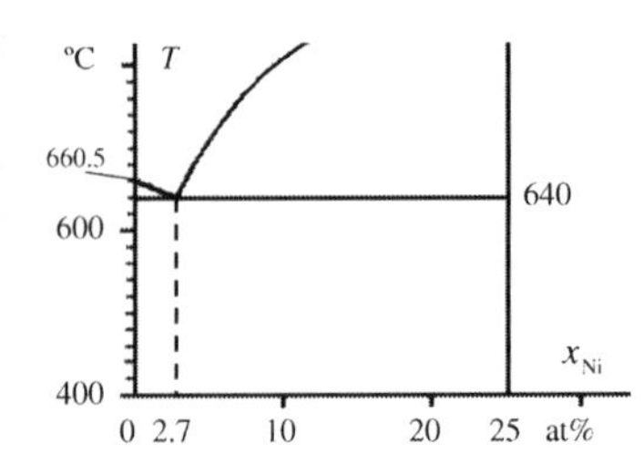

$T_E = 640\,°C = 913\,K$
$x^{\alpha/L} = 0.00$
$x_E^L = 0.027$
$x^{\beta/L} = 0.25$
$V_m^\alpha = 10.6\times10^{-3}\,m^3/kmol$
$V_m^\beta = 8.81\times10^{-3} m^3/kmol$
$V_m^L = 10.2\times10^{-3}\,m^3/kmol$
$\sigma_{\alpha,\beta} = 0.15\,J/m^2$
$D = 3\times10^{-9}\,m^2/s$

Example 10.4

The eutectic alloy Al-$NiAl_3$ is one of the few eutectic alloys, which solidifies with normal rod eutectic structure. Prove that this is in agreement with theoretical calculations.

Required material constants are given in the margin. Part of the phase diagram of the Al-Ni system is shown in the figure.

Solution:

To predict the solidification structure of the alloy we will calculate $const_{lamella}$ and $const_{rod}$ and compare them.

With the aid of the phase diagram we will calculate the slopes of the liquidus lines and the fractions of the α and β phases.

Al is the α phase and $NiAl_3$ is the β phase.

The slopes of the liquidus lines are obtained from readings in the phase diagram:

$$m_\alpha^L = \frac{640 - 660.5}{0.027 - 0} = -759 \quad \text{and} \quad m_\beta^L = \frac{640 - 520}{0.027 - 0} = 4444$$

The eutectic fractions of the α and β phases are derived with the aid of the lever rule.

$$f_\alpha = \frac{x_E^{\beta/L} - x_E^L}{x_E^{\beta/L} - x_E^{\alpha/L}} = \frac{0.25 - 0.027}{0.25 - 0} = 0.892$$

$$f_\beta = \frac{x_E^L - x_E^{\alpha/L}}{x_E^{\beta/L} - x_E^{\alpha/L}} = \frac{0.027 - 0}{0.25 - 0} = 0.108$$

Calculation of $const_{lamella}$

If we insert

$$A_{lamella} = \frac{m_\beta^L V_m^\beta}{f_\beta} - \frac{m_\alpha^L V_m^\alpha}{f_\alpha} = \frac{4444 \times 8.81 \times 10^{-3}}{0.108} - \frac{(-759) \times 10.6 \times 10^{-3}}{0.892} = 371\ \text{Km}^3/\text{kmol}$$

$$B_{lamella} = \frac{m_\beta^L}{V_m^\beta} - \frac{m_\alpha^L}{V_m^\alpha} = \frac{4444}{8.81 \times 10^{-3}} - \frac{-759}{10.6 \times 10^{-3}} = 5.76 \times 10^5\ \text{K kmol/m}^3$$

$$C_{o\ lamella} = f_\alpha f_\beta V_m^L (x_E^{\beta/L} - x_E^{\alpha/L}) = 0.892 \times 0.108 \times 10.2 \times 10^{-3} \times (0.25 - 0) = 2.46 \times 10^{-4}\ \text{m}^3/\text{kmol}$$

and other given values (page X122) into Equation (10.52) we obtain

$$const_{lamella} = \sqrt{\frac{\sigma_{\alpha\beta} A_{lamella} B_{lamella} C_{o\ lamella}}{RTD}} = \sqrt{\frac{0.15 \times 371 \times 5.76 \times 10^5 \times 2.46 \times 10^{-4}}{8.31 \times 10^3 \times 913 \times 3 \times 10^{-9}}}$$

or

$$const_{lamella} = 590\ Km^{-0.5}\ s^{0.5}$$

Calculation of $const_{rod}$

$const_{\text{rod}}$ is calculated analogously.

$$\frac{r_\beta}{r_\alpha} = \sqrt{\frac{1}{f_\alpha}} - 1 = \sqrt{\frac{1}{0.892}} - 1 = 0.0588$$

This value is used to calculate A_{rod} and B_{rod}:

$$A_{\text{rod}} = \frac{m_\beta^L V_m^\beta}{\frac{r_\beta}{r_\alpha}\left(2 + \frac{r_\beta}{r_\alpha}\right)} - m_\alpha^L V_m^\alpha = \frac{4444 \times 8.81 \times 10^{-3}}{0.0588 \times 2.0588} - (-759) \times 10.6 \times 10^{-3}$$

which gives

$$A_{\text{rod}} = 315\ \text{Km}^3/\text{kmol}$$

and

$$B_{\text{rod}} = \frac{m_\beta^L}{V_m^\beta}\frac{r_\beta}{r_\alpha}\frac{2 + \frac{r_\beta}{r_\alpha}}{2 + \frac{4r_\beta}{\pi r_\alpha}}\left(x_E^{\beta/L} - x_E^L\right) - \frac{m_\alpha^L}{V_m^\alpha}\frac{x_E^L - x_E^{\alpha/L}}{2 - \frac{4}{\pi}}$$

$$B_{\text{rod}} = \frac{4444}{8.81 \times 10^{-3}} \times 0.0588 \times \frac{2.0588}{2 + \frac{4}{\pi} \times 0.0588} \times (0.25 - 0.027) - \frac{-759}{10.6 \times 10^{-3}} \times \frac{0.027 - 0}{2 - \frac{4}{\pi}}$$

which gives

$$B_{\text{rod}} = 3.90 \times 10^3\ \text{K kmol/m}^3$$

These values and other given values (page X122) are inserted into Equation (10.105), which gives:

$$const_{\text{rod}} = \sqrt{\frac{\sigma_{\alpha\beta} A_{\text{rod}} B_{\text{rod}} V_m^L}{RTD}} = \sqrt{\frac{0.15 \times 315 \times 3.90 \times 10^3 \times 10.2 \times 10^{-3}}{8.31 \times 10^3 \times 913 \times 3 \times 10^{-9}}}$$

or

$$const_{\text{rod}} = 290\ \text{Km}^{-0.5}\,\text{s}^{0.5}$$

Answer:

$const_{\text{rod}} < const_{\text{lamella}}$. The conclusion is that the alloy Al-$NiAl_3$ solidifies with normal rod eutectic structure because this structure requires lower undercooling than normal lamellar eutectic structure.

The result in Example 10.4 agrees with the earlier statement on page 620 and experimental evidence that alloys with a minority phase normally grow as rod eutectics.

Experimental observations of the structures of alloys show that the lamellar structure is the most frequent one. Calculations according to the theory, which correspond to those in Example 10.4, predict in nearly all cases that the rod eutectic structure should be more favoured that the lamellar structure. There are several explanations to these discrepancies, however:

- The reason why a lamellar structure often is observed instead of a rod structure at experiments is related to an anisotropy of the crystals, which influences the surface tension between the two solid phases. This topic will be discussed briefly on page 668 in terms of surface energy.
- In addition, the cooling conditions also influence the type of structures, as we will see below.
- Small fractions of impurities are also known to be able to change the type of structure.

Influence of Growth Rate on Type of Normal Eutectic Growth

There are several experimental observations of a transition from lamellar to rod eutectic structure when the solidification conditions are changed. Such transitions have been observed for instance for Sn-Zn, Al-$CuAl_2$ and Al-$NiAl_3$ alloys.

Figure 10.74 results from measurements of the undercooling ΔT as a function of the solidification rate of the eutectic reaction in the system Sn-Zn.

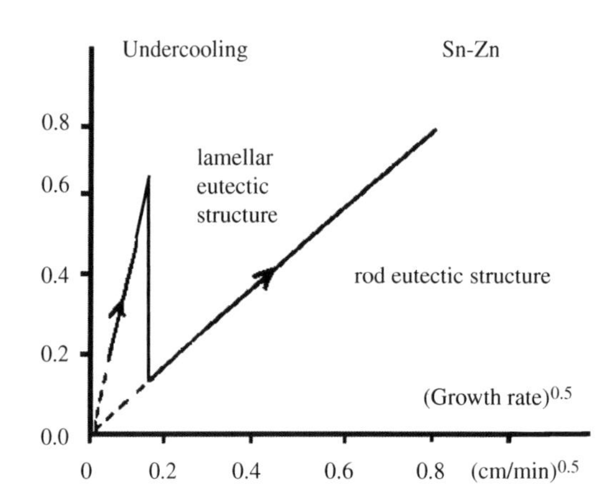

Figure 10.74 Undercooling as a function of growth rate at eutectic growth. Scripta Metallurgica 3, (1969), 249.

Figure 10.74 shows that a low growth rate favours lamellar eutectic structure. An increased solidification rate favours a transition to a rod eutectic structure. At the transition from a lamellar to a rod eutectic structure the undercooling suddenly decreases very much.

Influence of Crystal Orientation on Eutectic Lamellar Surface Energy

In connection with solid/solid interfaces in Chapter 3 on page 141 we discussed interfaces between solid crystalline surfaces. Three different types of interfaces were defined: coherent, semicoherent and incoherent interfaces.

Coherent interfaces show a good fitting between the two different crystal lattices. Incoherent interfaces lack such fitting between the atoms in the two surfaces. Semicoherent interfaces show some match but no complete fitting.

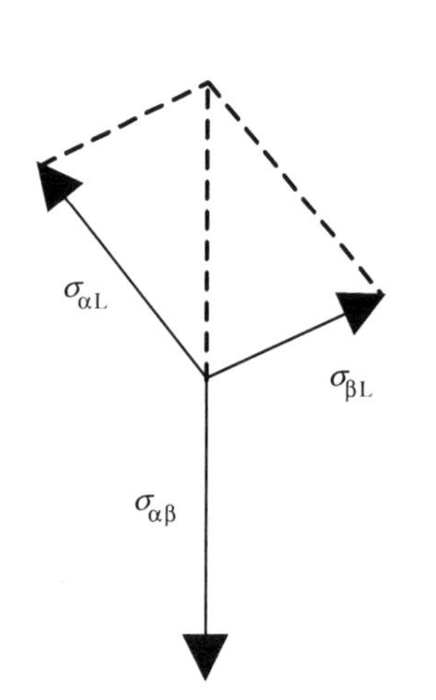

Figure 10.5 Surface tension forces in equilibrium with each other.

Coherent interfaces have much lower surface tensions than those of incoherent surfaces. In the theory presented in Section 10.3.2 the surface tension $\sigma_{\alpha\beta}$ between the two lamella surfaces α and β, which are formed during the eutectic reaction, is involved in the Equations (10.25) and (10.26) on page 598. The equations show that the smaller the surface tension is, the smaller will be the supersaturation (concentration difference). Small supersaturation corresponds to a small undercooling. According to Equations (10.49) and (10.52) on pages 603 and 604, respectively, we can conclude that

At a given growth rate the *undercooling* decreases when the *surface tension* decreases.

The same arguments and conclusions hold for rod eutectic growth as for lamellar eutectic growth.

At normal eutectic growth the cooperation between the phases is good. Consequently the lamellar eutectic structure is likely to be formed with a good atomic match between the two phases during the solidification process.

Fletcher [13] has tried to calculate the surface energy of the interface between two crystals. His results are shown in Figure 10.75a. The crystal orientations in the figure are given as the ratio of the atomic distances for each crystal type at the contact surface.

Figure 10.75a shows that there are minima in the surface energy when the ratios are 0.5, 1.0, 1.5 and 2.0 (Figure 10.75b). Experiments have confirmed that the two phases in a lamellar eutectic alloy usually have misfit ratios, which correspond to the lowest value of the surface energy between the phases. The best fit is obviously 1.0, which corresponds to the lowest possible surface energy.

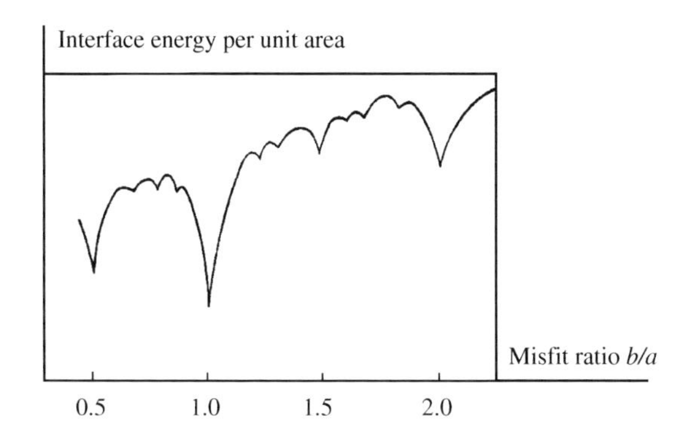

Figure 10.75a Calculated surface energy of a one-dimensional surface as a function of the misfit ratio b/a according to Fletcher. a and b are the interatomic distances in the two types of crystals.

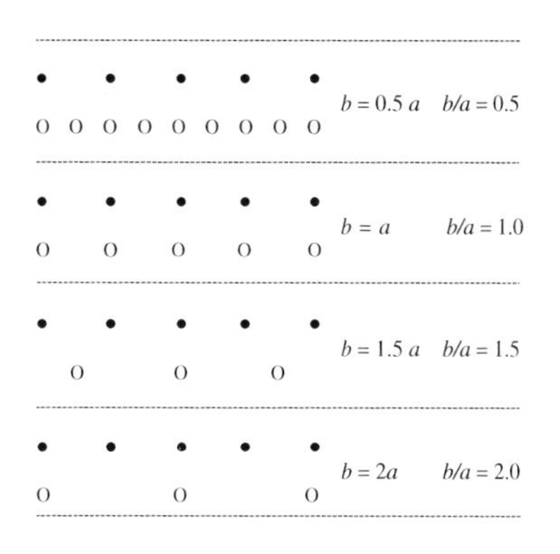

Figure 10.75b Misfit ratio at the interface of crystals A and B. Upper row of atoms corresponds to crystal A. Lower row of atoms corresponds to crystal B.

The ratios $\Delta T_{\text{rod}}/\Delta T_{\text{lamella}}$ and $const_{\text{rod}}/const_{\text{lamella}}$ are independent of the surface tension according to the Equations (10.127) and (10.105) on page 632. However, on page 668 we found that low surface energy leads to low undercooling. Formation of new surfaces at the deep minima results in low surface energy and reduces the undercooling and the growth rate according to the Equations (10.49) and (10.102). This favours lamellar growth and contributes to explain the fact that lamellar growth dominates over rod eutectic growth.

Summary

Classification of Eutectic Structures

Normal Eutectic Structure of a Binary Alloy:

The two phases cooperate during growth and obey the same growth law. Their growth rates are equal.

$$V^{\alpha}_{\text{growth}} = V^{\beta}_{\text{growth}} = V_{\text{growth}}$$

Degenerate Eutectic Structure of a Binary Alloy:

Coupled Growth:

The two phases cooperate but obey different growth laws.

Irregular Growth:

The two phases do not cooperate at all.

Eutectic reactions in melts often result in solids with widely different morphologies

Normal Lamellar Eutectic Growth

The phases α and β grows as parallel lamellae with the same growth rate.

Concentration of the Melt ahead of the Solid Phases

$$x^{\text{L}} = x_0^{\text{L}} - \left(x^{\text{L}/\alpha} - x_0^{\text{L}}\right) e^{-\frac{\pi z}{\lambda_\alpha}} \sin \frac{\pi y}{\lambda_\alpha} (-\lambda_\alpha \leq y \leq 0)$$

$$x^{\text{L}} = x_0^{\text{L}} + \left(x^{\text{L}/\beta} - x_0^{\text{L}}\right) e^{-\frac{\pi z}{\lambda_\beta}} \sin \frac{\pi y}{\lambda_\beta} (0 \leq y \leq \lambda_\beta)$$

Volume Fractions of the α and β Phases

$$f_\alpha = \frac{x^{\beta/L} - x_0^L}{x^{\beta/L} - x^{\alpha/L}} \approx \frac{x_E^{\beta/L} - x_0^L}{x_E^{\beta/L} - x_E^{\alpha/L}} = \frac{\lambda_\alpha}{\lambda}$$

$$f_\beta = \frac{x_0^L - x^{\alpha/L}}{x^{\beta/L} - x^{\alpha/L}} \approx \frac{x_0^L - x_E^{\alpha/L}}{x_E^{\beta/L} - x_E^{\alpha/L}} = \frac{\lambda_\beta}{\lambda}$$

Growth Rates of the α and β Phases

$$V_{\text{growth}}^\alpha = \frac{2D}{\lambda_\alpha} \frac{x^{L/\alpha} - x_0^L}{x^{L/\alpha} - x^{\alpha/L}} \frac{V_m^\alpha}{V_m^L}$$

$$V_{\text{growth}}^\beta = \frac{2D}{\lambda_\beta} \frac{x^{L/\beta} - x_0^L}{x^{L/\beta} - x^{\beta/L}} \frac{V_m^\beta}{V_m^L}$$

At normal lamellar eutectic growth

$$V_{\text{growth}}^\alpha = V_{\text{growth}}^\beta$$

Relationship between Growth Rate and Main Lamellar Distance

$$V_{\text{growth}} \lambda^2 = C_{\text{lamella}}$$

$$C_{\text{lamella}} = \frac{2\sigma_{\alpha\beta} 2D}{RT} \frac{A_{\text{lamella}}}{C_{0\,\text{lamella}} B_{\text{lamella}}}$$

where

$$A_{\text{lamella}} = \frac{m_\beta^L V_m^\beta}{f_\beta} - \frac{m_\alpha^L V_m^\alpha}{f_\alpha}$$

$$B_{\text{lamella}} = \frac{m_\beta^L}{V_m^\beta} - \frac{m_\alpha^L}{V_m^\alpha}$$

$$C_{0\,\text{lamella}} = f_\alpha f_\beta V_m^L \left(x_E^{\beta/L} - x_E^{\alpha/L} \right)$$

Relationship between Growth Rate and Undercooling

$$\Delta T_E = \frac{1}{2}\left(m_\alpha^L + m_\beta^L\right)\left(x_E^L - x_0^L\right) + \left(const_{\text{lamella}} - \frac{m_\alpha^L}{2\sqrt{\mu_{\text{kin}}^\alpha}} \right) \sqrt{V_{\text{growth}}}$$

$$const_{\text{lamella}} = \sqrt{\frac{\sigma_{\alpha\beta} A_{\text{lamella}} B_{\text{lamella}} C_{o\,\text{lamella}}}{RTD}}$$

The undercooling is influenced by the surface tension, the diffusion and the interface kinetics. If the interface kinetics in the α phase can be neglected the term that includes μ_{kin}^{α}, disappears.

If the composition of the alloy is eutectic the first term vanishes and the undercooling becomes proportional to the square root of the growth rate.

Normal Rod Eutectic Growth

A eutectic, in which one of the phases appears to be a minority phase, often grows as a rod eutectic.

Rod eutectic structure can shortly be described as a number of thin cylindrical parallel rods of an α phase, regularly distributed in a matrix of a β phase.

The theory of normal rod eutectic growth is analogous to normal lamellar eutectic growth.

$$x^{L} = x_{0}^{L} - \left(x^{L/\alpha} - x_{0}^{L}\right) e^{-\frac{2\pi z}{4r_{\alpha}}} \sin \frac{2\pi(R - r_{\alpha})}{4r_{\alpha}}$$

$$x^{L} = x_{0}^{L} + \left(x^{L/\beta} - x_{0}^{L}\right) e^{-\frac{2\pi z}{4r_{\beta}}} \sin \frac{2\pi(R - r_{\beta})}{4r_{\beta}}$$

Volume Fractions of the α and β Phases

$$f_{\alpha} = \frac{x^{\beta/L} - x_{0}^{L}}{x^{\beta/L} - x^{\alpha/L}} \approx \frac{x_{E}^{\beta/L} - x_{0}^{L}}{x_{E}^{\beta/L} - x_{E}^{\alpha/L}} = \left(\frac{r_{\alpha}}{r_{\alpha} + r_{\beta}}\right)^{2}$$

$$f_{\beta} = \frac{x_{0}^{L} - x^{\alpha/L}}{x^{\beta/L} - x^{\alpha/L}} \approx \frac{x_{0}^{L} - x_{E}^{\alpha/L}}{x_{E}^{\beta/L} - x_{E}^{\alpha/L}} = 1 - \left(\frac{r_{\alpha}}{r_{\alpha} + r_{\beta}}\right)^{2}$$

Growth Rates of the α and β Phases

$$V_{growth}^{\alpha} = 2D \frac{2 - \frac{4}{\pi}}{r_{\alpha}} \frac{x^{L/\alpha} - x_{0}^{L}}{x^{L/\alpha} - x^{\alpha/L}} \frac{V_{m}^{\alpha}}{V_{m}^{L}}$$

$$V_{growth}^{\beta} = 2D \frac{2r_{\alpha} + \frac{4r_{\beta}}{\pi}}{r_{\beta}(2r_{\alpha} + r_{\beta})} \frac{x^{L/\beta} - x_{0}^{L}}{x^{L/\beta} - x^{\beta/L}} \frac{V_{m}^{\beta}}{V_{m}^{L}}$$

At normal rod eutectic growth

$$V_{growth}^{\alpha} = V_{growth}^{\beta}$$

Relationship between Volume Fraction and Rod Radius

$$\frac{r_{\beta}}{r_{\alpha}} = \sqrt{\frac{1}{f_{\alpha}}} - 1$$

Relationship between Growth Rate and Rod Radius

$$V_{\text{growth}} r_{\alpha}^{2} = C_{\text{rod}}$$

$$C_{\text{rod}} = \frac{2\sigma_{\alpha\beta} 2D}{RT} \frac{A_{\text{rod}}}{V_{\text{m}}^{\text{L}} B_{\text{rod}}}$$

$$A_{\text{rod}} = \frac{m_{\beta}^{\text{L}} V_{\text{m}}^{\beta}}{\frac{r_{\beta}}{r_{\alpha}}\left(2 + \frac{r_{\beta}}{r_{\alpha}}\right)} - m_{\alpha}^{\text{L}} V_{\text{m}}^{\alpha}$$

$$B_{\text{rod}} = \frac{m_{\beta}^{\text{L}}}{V_{\text{m}}^{\beta}} \frac{r_{\beta}}{r_{\alpha}} \frac{2 + \frac{r_{\beta}}{r_{\alpha}}}{2 + \frac{4r_{\beta}}{\pi r_{\alpha}}} \left(x_{\text{E}}^{\beta/\text{L}} - x_{\text{E}}^{\text{L}}\right) - \frac{m_{\alpha}^{\text{L}}}{V_{\text{m}}^{\alpha}} \frac{x_{\text{E}}^{\text{L}} - x_{\text{E}}^{\alpha/\text{L}}}{2 - \frac{4}{\pi}}$$

Relationship between Growth Rate and Undercooling

$$\Delta T_{\text{E}} = \frac{1}{2}\left(m_{\alpha}^{\text{L}} + m_{\beta}^{\text{L}}\right)\left(x_{\text{E}}^{\text{L}} - x_{\text{o}}^{\text{L}}\right) + \left(const_{\text{rod}} - \frac{m_{\alpha}^{\text{L}}}{2\sqrt{\mu_{\text{kin}}^{\alpha}}}\right)\sqrt{V_{\text{growth}}}$$

$$const_{\text{rod}} = \sqrt{\frac{\sigma_{\alpha\beta} A_{\text{rod}} B_{\text{rod}} V_{\text{m}}^{\text{L}}}{RTD}}$$

The undercooling is influenced by the surface tension, the diffusion and the interface kinetics. If the interface kinetics in the α phase can be neglected the term, which includes $\mu_{\text{kin}}^{\alpha}$, disappears.

If the composition of the alloy is eutectic the first term vanishes and the undercooling becomes proportional to the square root of the growth rate.

Degenerate Eutectic Growth

A eutectic contains a spectrum of lamellar spacings.

$$\lambda = \lambda_0 \pm \frac{\Delta\lambda}{2} \sin \frac{2\pi}{n\lambda} y$$

The roughness of the solidification front increases with the lamellar variation $\Delta\lambda$. The region with the optimal lamellar spacing grows faster than other regions.

Sudden changes of the lamellar spacing can be explained with the aid of energy arguments.

Growth of Primary Phases. Competitive Growth

The primary phases α and β of a solidifying binary alloy occur normally as dendrites but other structures may appear.Growth rates of the primary phases::

$$V_{\text{growth}}^{\alpha} = \mu_{\alpha}\left(x^{\text{L}/\alpha} - x_{0}^{\text{L}}\right)^{n\alpha}$$

$$V_{\text{growth}}^{\beta} = \mu_{\beta}\left(x_{0}^{\text{L}} - x^{\text{L}/\beta}\right)^{n\beta}$$

or in terms of undercooling

$$V_{\text{growth}}^{\alpha} = \frac{\mu_{\alpha}}{\left(-m_{\alpha}^{\text{L}}\right)^{n\alpha}}\left(T_{\alpha}^{\text{L}} - T\right)^{n\alpha}$$

$$V_{\text{growth}}^{\beta} = \frac{\mu_{\beta}}{\left(m_{\beta}^{\text{L}}\right)^{n\beta}}\left(T_{\beta}^{\text{L}} - T\right)^{n\beta}$$

There is a competition between the different structures. The structure, which requires the lowest undercooling, i.e. the highest temperature T will win

Coupled Growth

Two primary phases may cooperate during growth in spite of the fact that they obey different growth laws. This phenomenon is called *coupled growth.*

The *coupled zone* is defined as the region in the phase diagram with a eutectic-like structure, i.e. where the α and β phases cooperate during growth.

If the growth rates of primary α phase and primary β phase are approximately the same at the same under-cooling, the coupled zone will be symmetrically located below the eutectic temperature.

If one of the phases grows at lower undercooling than the other, there will be an asymmetric, coupled zone below the horizontal eutectic temperature line in the phase diagram.

A true eutectic structure has the same composition everywhere. The composition of a coupled growth structure varies within the limits of the coupled zone.

Along the line of equal growth rate, the so-called eutectic line in the phase diagram both phases grow with equal growth rate. This is not the case in the rest of the coupled zone.

Calculation of the Coupled Zone

1. The grows laws of the primary phases and the eutectic phase are set up.
2. The left boundary line is obtained from the condition:

   $$V^{\beta}_{\text{growth}} = V^{\text{E}}_{\text{growth}}$$

   You obtain a relationship between T and x^{L}_{0}, which is the equation of the left boundary line.
3. The right boundary line is obtained from the condition:

   $$V^{\alpha}_{\text{growth}} = V^{\text{E}}_{\text{growth}}$$

   You obtain a relationship between T and x^{L}_{0}, which is the equation of the right boundary line.

Irregular Eutectic Growth

No cooperation between the phases occurs at growth in regions outside the coupled zone. Such growth is called degenerate eutectic growth or irregular eutectic growth.

Structures of Ternary Alloys

The structures of ternary alloys depend mainly on the ternary phase diagram (page 636), i.e. the material constants, the composition of the alloy and the cooling conditions.

Compositions and Fractions of Structures in a Simple Ternary Alloy

Structure Type I L → α

Composition of the melt during the primary dendrite solidification

$$x^{\text{L}}_{\text{B}} = x^{0\text{L}}_{\text{B}}(1 - f_{\alpha})^{-\left(1-k^{\alpha/\text{L}}_{\text{B}}\right)}$$

$$x^{\text{L}}_{\text{C}} = x^{0\text{L}}_{\text{C}}(1 - f_{\alpha})^{-\left(1-k^{\alpha/\text{L}}_{\text{C}}\right)}$$

$$x^{\text{L}}_{\text{A}} = 1 - x^{\text{L}}_{\text{B}} - x^{\text{L}}_{\text{C}}$$

Volume fraction of phase α:

$$1 - f_{\alpha} = \left(\frac{x^{\text{L}}_{\text{B}}}{x^{0\text{L}}_{\text{B}}}\right)^{-\frac{1}{1-k^{\alpha/\text{L}}_{\text{B}}}} = \left(\frac{x^{\text{L}}_{\text{C}}}{x^{0\text{L}}_{\text{C}}}\right)^{-\frac{1}{1-k^{\alpha/\text{L}}_{\text{C}}}}$$

Structure Type II L → α + β
Composition of the melt during the binary eutectic solidification:

$$x_{\mathrm{B}}^{\mathrm{L\,bin}} = x_{\mathrm{B}}^{0\,\mathrm{L\,bin}}(1 - f_{\alpha+\beta})^{-\left[\left(1-k_{\mathrm{B}}^{\alpha/\mathrm{L}}\right)f\alpha+\left(1-k_{\mathrm{B}}^{\beta/\mathrm{L}}\right)f_{\beta}\right]}$$

$$x_{\mathrm{C}}^{\mathrm{L\,bin}} = x_{\mathrm{C}}^{0\,\mathrm{L\,bin}}(1 - f_{\alpha+\beta})^{-\left[\left(1-k_{\mathrm{C}}^{\alpha/\mathrm{L}}\right)f\alpha+(1-k_{\mathrm{C}}^{\beta/\mathrm{L}})f_{\beta}\right]}$$

Total solidified fraction of the initial volume at the ternary eutectic point E ⊗ page X76).

$$f_{\mathrm{E}} = f_{\alpha} + f_{\alpha+\beta}(1 - f_{\alpha})$$

Structure Type III L → α + β + γ
Fraction of ternary eutectic structure:

$$f_{\alpha+\beta+\gamma} = 1 - f_{\alpha} - f_{\alpha+\beta}(1 - f_{\alpha})$$

Structures of Ternary Alloys
Structures of ternary alloys are complex and contain a manifold of different structures.
Examples of such structures are given in the text.

Structures of Fe–C Eutectics

FeC alloys with a carbon composition >2 wt% C are called cast iron.
Cast iron is an irregular eutectic alloy, which shows a variety of different structures, among them flake graphite and nodular cast iron. The carbon concentration of cast iron is 4.3 wt%.

Flake Graphite
Flake graphite is a coarse structure of the primary phases austenite and graphite, which grow in colonies.
The graphite phase grows faster than the austenite phase.
Growth law:

$$V_{\mathrm{graphite}} = Const\,(\Delta T)^4 \text{ Eutectic colony growth}$$

Austenite grows as dendrites.
Growth law:

$$V_{\mathrm{austenite}} = const\,(\Delta T)^2$$

As the growth laws differ, flake graphite is a degenerate eutectic structure. It grows with coupled growth between the phases.

Nodular Cast Iron
Eutectic nodular cast iron consists of spherical graphite particles embedded in a matrix of austenite.
Graphite particles nucleate in a eutectic cast iron melt at temperatures below the eutectic temperature. They grow to a certain size and become then enclosed in a shell of austenite.
The particles grow by diffusion of carbon atoms from the melt through the austenite shell.
The heat of fusion of nodular cast iron decreases with increasing growth rate, owing to formation of lattice defects, mainly vacancies, in the austenite shell during the solidification process.

The heat of fusion increases with increasing fraction of solidification in nodular cast iron. No such effect is observed in flake graphite.

Structures of Al-Si Eutectics

Sr-modified Al-Si alloys show a much finer structure than the ones of unmodified Al-Si alloys.

Like nodular cast iron Al-Si alloys have a heat of fusion, which decreases with increasing growth rate.

The reason is formation of lattice defects, mainly vacancies, at the solidification front during growth.

Coupled Zone of Al-Si Alloys

Four types of microstructures are found in eutectic Al-Si alloys. They are shown in the phase diagram on page X118.

Region A: Al-rich primary dendrites.
Region B: Plate-like eutectic structure.
Region C: Non-faceted eutectic structure formed by coupled growth of Al and Si.
Region D: Primary Si, followed by either of the structures B or C, provided that the alloy is hypereutectic.

The region B is large in the unmodified alloy. It is strongly suppressed in a modified Al-Si alloy and disappears completely at a strongly modified alloy.

The vacancy concentration influences the phase diagram of Al-Si very strongly. The higher the vacancy concentration is in the alloy the lower will be the eutectic temperature.

Transitions between Normal Lamellar and Normal Rod Eutectic Growth

The criterion for normal lamellar eutectic growth is that lowest undercooling wins, i.e. $const_{\text{lamella}} < const_{\text{rod}}$

On the other hand, if $const_{\text{rod}} < const_{\text{lamella}}$ normal rod eutectic growth will win.

Lamellar eutectic structure is much more common in alloys than rod eutectic structure in contradiction to the theoretically predicted structures. The reason may be influence of surface tension, growth rate or impurities.

Exercises

10.1 Tassa and Hunt have measured the undercooling ΔT as a function of the solidification rate V_{growth} for a eutectic Al-Cu alloy and some near-eutectic Al-Cu alloys. Figure 1 below shows their results.

Calculate the undercooling as a function of the growth rate of a eutectic Al-Cu alloy with the aid of the theory given in Chapter 10, and compare your result with the experimental data. The alloys show normal lamellar growth.

At your help you have the phase diagram of the system Al-Cu (Figure 2) and some material constants.

Material constants:

surface tension between the two phases $\sigma_{\alpha/\beta} = 0.20\ \text{J/m}^2$
diffusion constant in the melt $D_L = 3 \times 10^{-9}\ \text{m}^2/\text{s}$
molar weight of Al = 27.0 kg/kmol
density $\rho_{Al} = 2.70 \times 10^3\ \text{kg/m}^3$
$V_m^L = 10.6 \times 10^{-3}\ \text{m}^3/\text{kmol}$
$V_m^{Al_2Cu} = 9.35 \times 10^{-3}\ \text{m}^3/\text{kmol}$
eutectic temperature $T_E = 821\ \text{K}$.

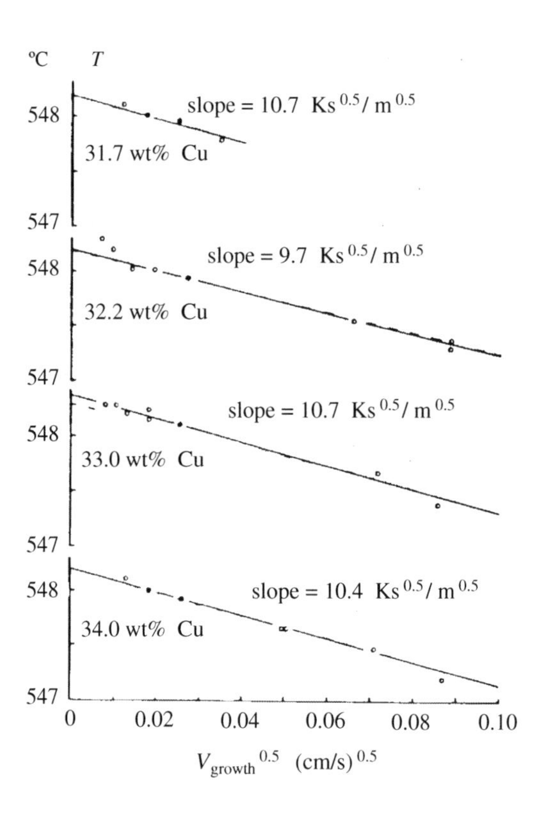

Figure 10.1–1 Temperature in the vicinity of the eutectic temperature-versus $1/\sqrt{V_{growth}}$ for some Al–Cu alloys with Cu concentrations 31.7, 32.2, 33.0 and 34.0 wt %. The lines represent the measurements. The Cu concentration is a parameter.

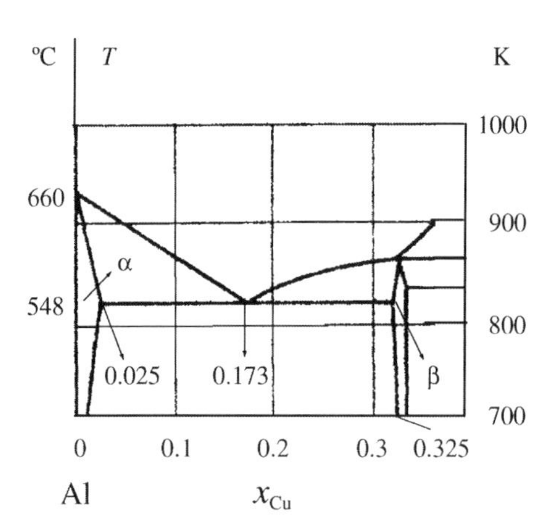

Figure 10.1–2 Principle sketch of the phase diagram of the binary system Al–Cu.

10.2 At a solidification experiment with the eutectic Cu–Pb alloy liquid Pb rods (α phase) form in a liquid Cu matrix (β phase). The spacing r_α, equal to the rod radius, was found to be proportional to the inverse of the square root of the solidification rate V_{growth}. The constant product was equal to

$$V_{growth} r_\alpha^2 = 2 \times 10^{-14} \text{m}^3/\text{s}$$

Compare this value with the theoretical value obtained from the theory of rod eutectic growth. The phase diagram of the system Cu–Pb is given in the figure.

Material constants:

surface tension between the Pb-rich liquid and Cu $\sigma_{\alpha/\beta} = 0.32\ \mathrm{J/m^2}$.
density of the Pb-rich liquid $\rho_\alpha = 9.80 \times 10^3\ \mathrm{kg/m^3}$
molar weight of Pb = 207.19 kg/kmol
density of Cu = $8.93 \times 10^3\ \mathrm{kg/m^3}$
molar weight of Cu = 63.54 kg/kmol
diffusion coefficient $D_L = 2 \times 10^{-9}\ \mathrm{m^2/s}$
$V_m^L = 9.31 \times 10^{-3}\ \mathrm{m^3/kmol}$
$V_m^{L_2} = 9.0 \times 10^{-3}\ \mathrm{m^3/kmol}$.

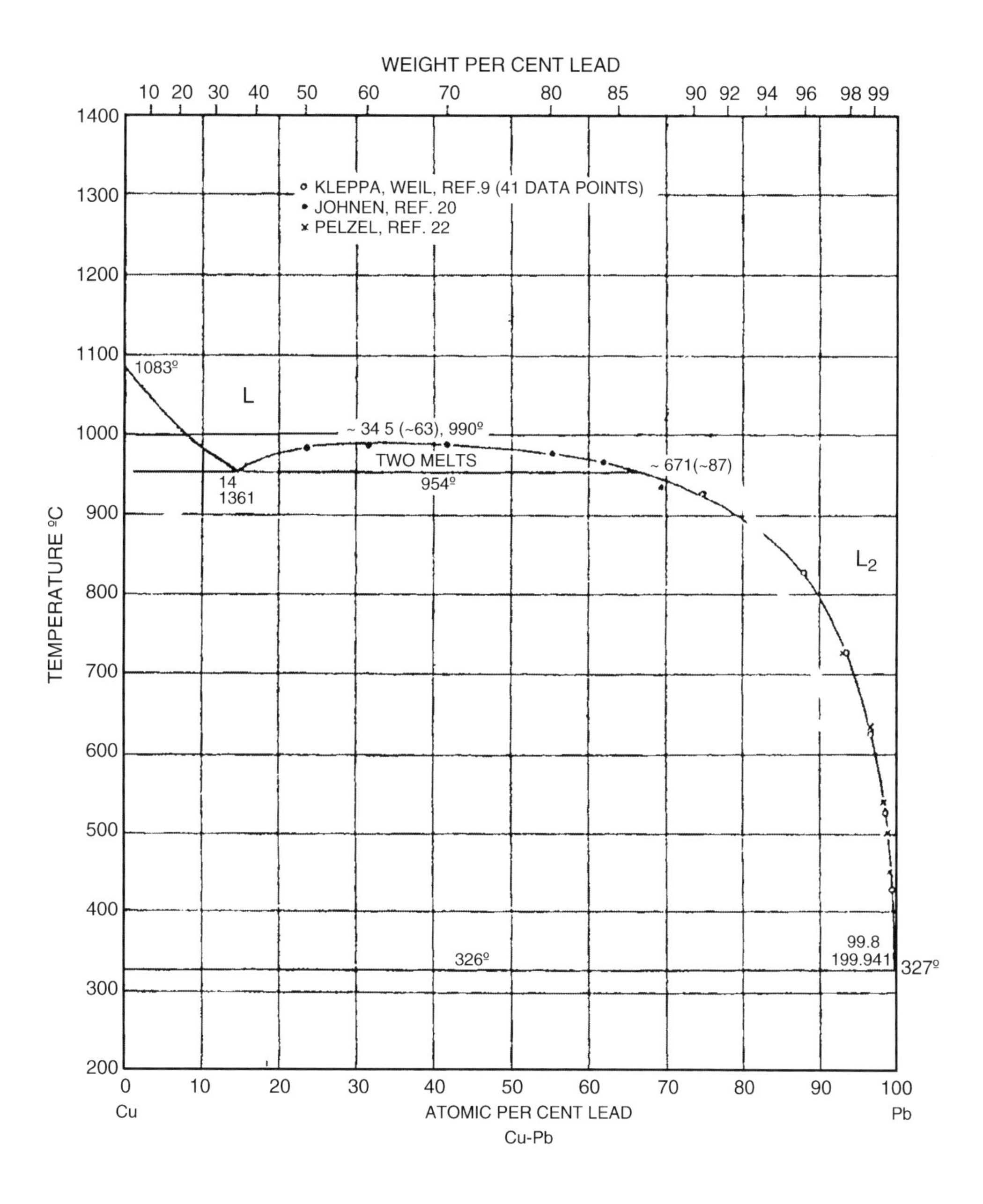

Figure 10.2–1

10.3 Eutectic alloys have specific properties and their structure with a coupled zone depends strongly on the solidification process when they were formed.

Consider a eutectic melt in a bar-shaped mould, which is insulated on five surfaces and cooled unidirectionally on the last one, as is shown in Figure 10.3–1.

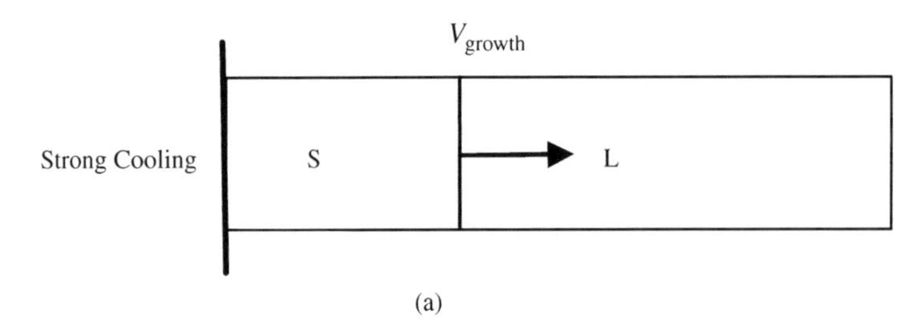

Figure 10.3–1

Both the growth rate and the thickness of the solidified layer vary with time. The thickness s as a function of time is described by the following relationship:

$$s = 1.2 \times 10^{-2}\sqrt{t}\ \text{(SI units)}$$

Experimental evidence shows that the relationship between the solidification rate and the undercooling for eutectic growth can be written as

$$V_{growth} = 1.0 \times 10^{-4} \times (\Delta T)^2$$

In the phase diagram in Figure 10.3–1 the so-called coupled zone for eutectic growth has been introduced. It can be used to predict the structure of the cooling bar in Figure 10.3–2. The phase diagram shows that the coupled zone follows a vertical line until the undercooling is 8 K (dashed temperature line at 542 °C) and then suddenly changes direction.

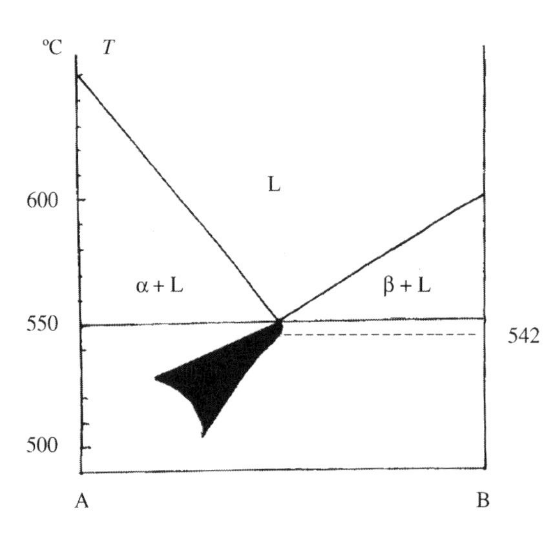

Figure 10.3–2

a) Derive a relationship between the growth rate V_{growth} of the solidification front and the thickness s of the solid phase. Plot V_{growth} as a function of s in a diagram.
b) It is well known that the structure of an alloy varies strongly with the growth rate. It is obvious that the abrupt change of the shape of the coupled zone is connected with the undercooling of 8 K. Calculate the critical growth rate and the critical position of the solidification front s_{cr} at $T = 542°C$. The latter represents the transition temperature from one structure to another.

   Predict the structures of the precipitated phases above and below this temperature. At your help you have the diagram in a).

10.4 Dendritic growth and eutectic growth can be described by the growth law

$$V_{\text{growth}} = \mu(\Delta x)^2$$

where Δx is the supersaturation measured as mole fraction and μ is the growth constant.

Consider an alloy with the two primary phases α and β. The phases follow the growth law above and their μ values are

$$\mu_{\alpha} = 1.0 \times 10^{-4} \text{ m/s}$$

$$\mu_{\beta} = 1.0 \times 10^{-5} \text{ m/s}$$

a) Draw the coupled zone in the phase diagram below for $\mu_{\text{E}} = 1.0 \times 10^{-5}$ m/s.
b) Justify your answer.
c) Calculate the value of μ_{E} at which the coupled zone disappears.

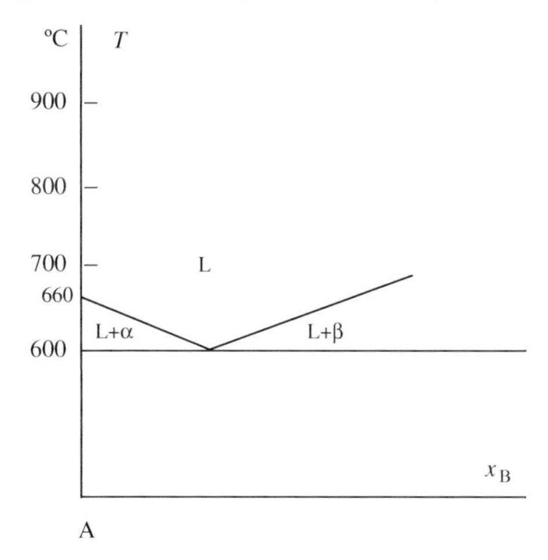

Figure 10.4–1

10.5 The phosphide eutectic called steadite is formed at the boundaries in eutectic cast iron as a consequence of microsegregation of phosphorous during the solidification process. The concentration of the steadite influences strongly the properties of the cast iron. For this reason it is essential to know how the steadite concentration varies with the P concentration.

Derive the concentration of the phosphide eutectic as a function of the P concentration x_{P}^{0} in cast iron eutectic, provided that the partition coefficient between the γ phase and the melt is 0.20 for phosphorous and that the P concentration in steadite is 10.1 at%.

The phase diagram of the system Fe–C is given in the figure.

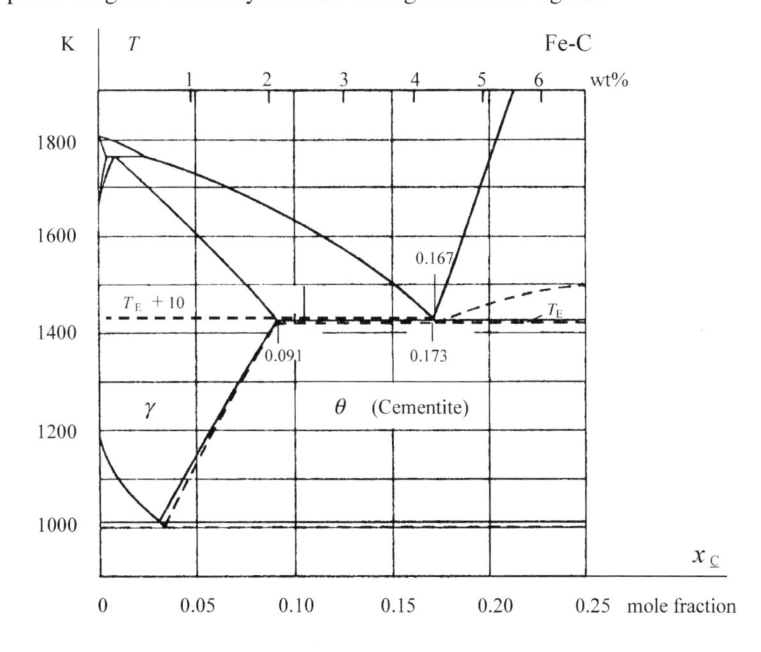

Figure 10.5–1

10.6 The undercooling as a function of the solidification rate of a eutectic Sn-Zn alloy has been measured. The result is shown in the figure. The figure shows that a transition from a lamellar eutectic to a rod eutectic structure is obtained when the solidification rate increases.

Find a reasonable explanation for this transition.

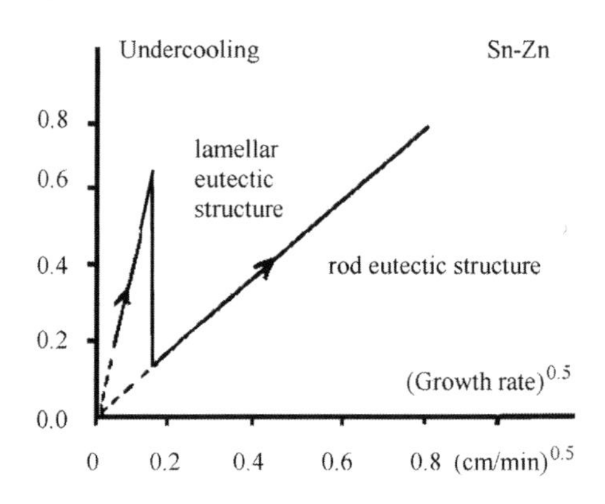

Figure 10.6–1

References

1. M. Hillert, Jernkontorets Annaler 1957.
2. K. A. Jackson, J. D. Hunt. Lamellar and rod eutectic growth, *Transactions of the Metallurgical Society of AIME*, **236**, 1966, 1129–1142.
3. P. Haasen, *'Physical Metallurgy*, Cambridge Press, Oxford, 1996, page 78.
4. G. A. Chadwick, *Journal Institute of Metals*, **91**, 1963, 169.
5. M. P. Puls, J.S. Kiraldy, *Metallurgical Transaction*, **3**, 1972, 2777.
6. T. Carlberg, H. Fredriksson, On the mechanism of lamellar spacing adjustment in eutectic alloys, *Journal of Crystal Growth*, **42**, 1977, 526–535.
7. H. Fredriksson, U. Åkerlind, *'Materials Processing during Casting'* (Wiley 2005).
8. D. G. McCartney, J. D. Hunt, R. M. Jordan, The structures expected in a simple ternary eutectic system, Part 1. Theory. *Metallurgical Transactions*, **11A**, 1980, 1243–1257.
9. H Fredriksson, The coupled zone in grey cast iron, *Metallurgical Transactions*, **6A**, 1975, 1658–1659.
10. J. Tinoco, P. Delvasto, O. Quintero, H. Fredriksson, *International Journal of Cast Metals Research*, **16**, 2003, 53–58.
11. H Fredriksson, J. Stjerndahl, J. Tinoco, On the solidification of nodular cast iron and its relation to the expansion and contraction, *Materials Science and Engineering*, **A413/414**, 2005, 363–372.
12. E. Talaat, H. Fredriksson, Solidification Mechanism of Unmodified and strontium modified Al-Si alloys. *Materials Transactions, JIM*, **41**, 2000, 507–515.

11

Peritectic Solidification Structures

11.1 Introduction

At *peritectic solidification* a liquid reacts with a primarily precipitated phase and a new secondary solid phase is formed.

Peritectic solidifications are very common when metals solidify. They occur in Fe–C and in many other technically and commercially important alloys, for example in Fe-Ni alloys, and in many important steel alloys such as stainless steel and low-carbon alloy steel. Peritectic solidification is also common in Cu-Sn and Cu-Zn alloys.

Peritectic growth occurs with the aid of a *peritectic reaction* at the solidification front. Later a *peritectic transformation* follows. These two mechanisms will be explained and discussed in Sections 11.2 and applied on iron-base alloys in Section 11.3.

Solidification and Crystallization Processing in Metals and Alloys, First Edition. Hasse Fredriksson and Ulla Åkerlind.

Metatectic reaction is the name of a process where a primarily precipitated phase decomposes into a liquid phase and a new solid phase. This type of reaction is discussed in Section 11.5.

11.2 Peritectic Reactions and Transformations

Definition of a Peritectic Reaction

A *peritectic reaction* = a reaction where a liquid phase L reacts with a primarily precipitated phase α and forms a new secondary phase β.

$$L + \alpha \rightarrow \beta$$

A peritectic reaction, which occurs in a binary alloy can be described in its phase diagram. The peritectic reaction has a very characteristic appearance, which is shown in Figure 11.1.

Normally the three phases, α, β and liquid L stay in contact with each other during the peritectic reaction but physical contact is not really necessary as we will see below on pages 683–684.

Definition of a Peritectic Transformation

A *peritectic transformation* = the secondary process, which follows after a peritectic reaction, when the α phase is transformed into β phase.

The β phase insulates the α phase from the liquid phase. The transformation requires B atoms as the β phase has a higher concentration of solute atoms than the α phase. The B atoms diffuse through the β phase to the α phase.

$$\alpha + \text{B atoms} \rightarrow \beta$$

We will use these definitions below. Kerr et al [1] were the first ones, who made a distinction between a peritectic *reaction* and a peritectic *transformation*.

Appearance of a Peritectic Reaction and Transformation in a Binary Phase Diagram

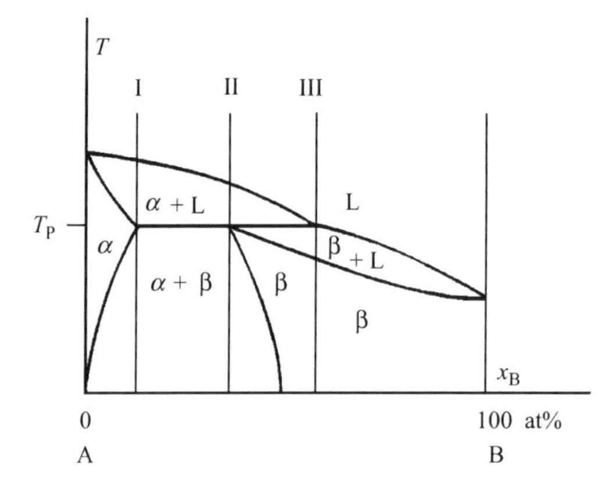

Figure 11.1 Principle sketch of a peritectic reaction

$$L + \alpha \rightarrow \beta$$

in the phase diagram of a binary alloy.

Depending on the shape of the phase diagram the peritectic reaction in the phase diagram can alternatively be mirror-inverted. (See Figure 11.9c on page 687.)

Phase diagrams are very instructive tools to describe phase reactions and transformations. Figure 11.1 shows a schematic phase diagram with a peritectic reaction. The diagram shows that all alloys with compositions to the left of line I will solidify as α crystals under equilibrium conditions. Similarly, all alloys with compositions to the right of line III will solidify as β crystals when the temperature is gradually lowered.

Alloys with compositions between the lines II and III will primarily solidify as α crystals, which transform into stable β crystals at and below the peritectic temperature. Alloys with compositions between lines I and II solidify primarily as α crystals. A fraction of the α crystals transforms into β crystals when the temperature reaches the peritectic temperature and decreases further.

The volume fraction of each phase is given by the lever rule if the alloy solidifies under equilibrium conditions. However, in most cases the lever rule will not give the correct volume fractions of the different phases. The reason is that the kinetics at the solidification front in addition to the diffusion rate in the solid phases determines the time to achieve the state of equilibrium. In many cases, the equilibrium will not be reached in a reasonable time, which explains the deviations from the lever rule.

Peritectic reactions and transformations will be studied below and the influence of different nucleation conditions on the secondary phase will be discussed. Peritectic reactions in multi-component systems will be treated briefly. In addition, we will later consider the possibility of precipitation of metastable β crystals instead of α crystals in alloys with compositions to the left of line III in Figure 11.1.

11.2.1 Peritectic Reactions of Types 1 and 2

Two different types of peritectic reactions are described in the scientific literature.

1. Nucleation and growth of β crystals on the primary α phase, when the three phases α, β and liquid L stay in contact with each other. Simultaneously the growing β phase gradually separates the α phase from the liquid phase L.
2. Nucleation and growth of β crystals when a liquid layer L separates the primary α phase and the β phase.

The latter case can either be an effect of different densities of the two phases or a result of the influence of the surface tension.

Type 1

During type 1 reactions, when the three phases stay in contact with each other, the nucleation of the secondary β phase occurs between the primary α phase and the liquid. In addition, a lateral growth of β phase around the α phase follows when B atoms diffuse from the β phase towards the α phase (Figure 11.2a). Part of the α phase will dissolve and form a β phase.

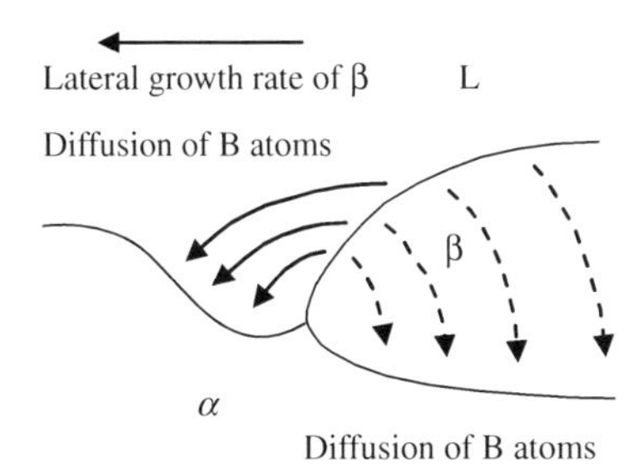

Figure 11.2a Peritectic reaction at which the secondary β phase grows along the surface of the primary α phase. The diffusion of B atoms through the β phase is coupled to the transformation of α phase into β phase.

Figure 11.2b At equilibrium, the vector sum of the surface tension forces is zero.

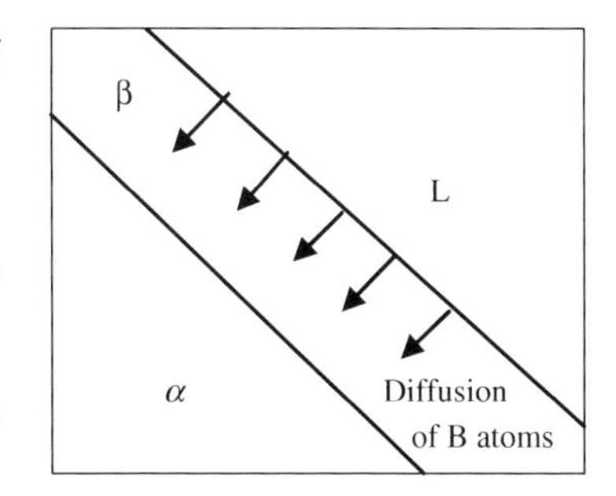

Figure 11.2c Schematic illustration of the nucleation of β phase on the α phase at the α/β interface. $L + \alpha \rightarrow \beta$.

The surface tension forces influence the lateral growth and morphology. The β phase will gradually cover the whole α phase. At equilibrium the condition

$$\boldsymbol{\sigma}_{\alpha\beta} = \boldsymbol{\sigma}_{L\alpha} + \boldsymbol{\sigma}_{L\beta} \tag{11.1}$$

is fulfilled at the three-phase point (Figure 11.2b). This topic will be further discussed on page 685.

The secondary β phase obtains a morphology similar to the primarily precipitated α phase.

As soon as the secondary β layer has been formed it separates part of the α phase and the L phase as shown in the right part of Figure 11.2a. This can be indicated schematically in the way shown in Figure 11.2c.

The thickness of the β layer after the peritectic reaction will be derived on page 686.

A peritectic reaction of type 1 is the simplest and most common peritectic reaction.

The first reaction of type 1 was observed and verified experimentally by Fredriksson and Nylén [2] (Figures 11.4a and b below).

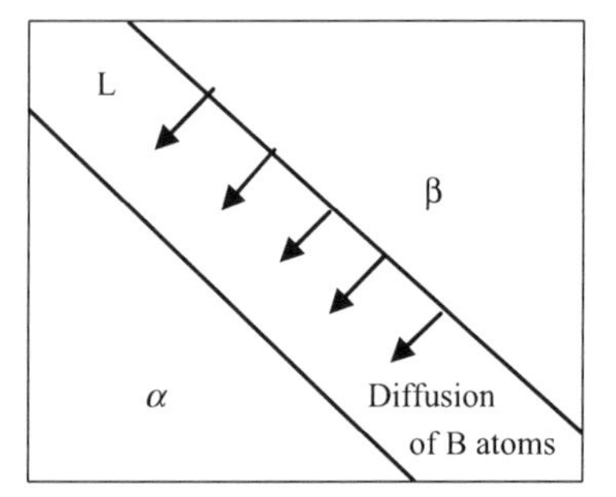

Figure 11.3 Peritectic reaction at which the melt L separates the growing secondary β phase from the shrinking primary α phase.

Type 2

During a type 2 peritectic reaction the β crystals are nucleated in the liquid. A liquid layer separates the secondary β phase from the primary α phase (Figure 11.3). After nucleation the secondary β phase grows freely in the liquid. At the same time the primary α phase dissolves partly and forms β phase, owing to diffusion of B atoms through the melt (Figure 11.3).

Type 2 reactions are not as frequent as reactions of type 1.

The second type of peritectic reaction was observed by Fredriksson and Nylén [2] for the reaction $\gamma + L \leftrightarrow \beta$ in the Al-Mn system (page 687). They reported that the secondary phase had a tendency to cover the surface of the primary phase at increasing cooling rates. Petzow and Exner [3] have reported a similar type of reaction for the Ni-Zn system.

Peritectic Reactions of Type 1

Example of a Peritectic Reaction of Type 1

The Figures 11.4a, 11.4b and 11.4c show the peritectic reaction in a unidirectionally solidified Cu-Sn alloy with 70% Sn, where the ε phase and the liquid react and give the η phase (see the phase diagram of the system Cu-Sn in Figure 11.5).

$$L + \varepsilon \rightarrow \eta$$

The temperature gradient in the figures is horizontal. The temperature in the left parts of the figures is higher than in the right parts. The temperature decreases to the right and the effects of the decreasing temperatures can be observed in the figures.

The primary ε phase consists of plate-like ε crystals, which are dark grey in the Figure 11.4 and start to precipitate at the solidification front to the left in Figure 11.4a. When the temperature decreases to and below the peritectic temperature the peritectic reaction starts and a white secondary η phase is formed, which surrounds the ε crystals. The process is shown in Figure 11.4a. This stage is shown

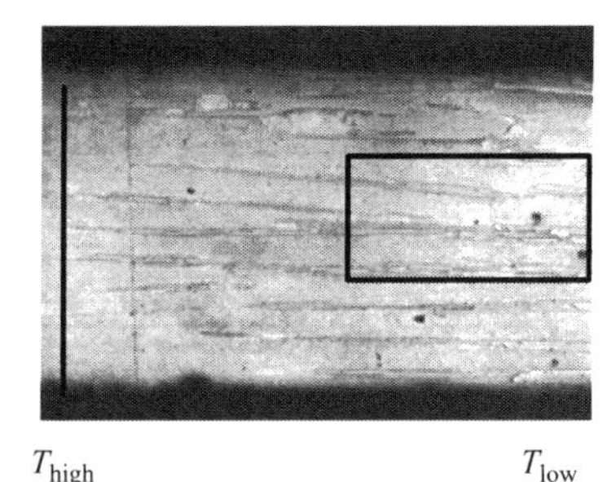

Figure 11.4a The peritectic reaction (type 1) in a unidirectionally solidified Cu–70%Sn sample. The figure shows the solidification front to the left. The rectangular area represents the peritectic reaction zone. From Mechanism of Peritectic Reactions and Transformations, H. Fredriksson, T. Nylén in Metal Science Journal, Vol. 16, The Metals Society, 1982 © Maney www.maney.co.uk/journals/mst and www.ingentaconnect.com/content/maney/msc.

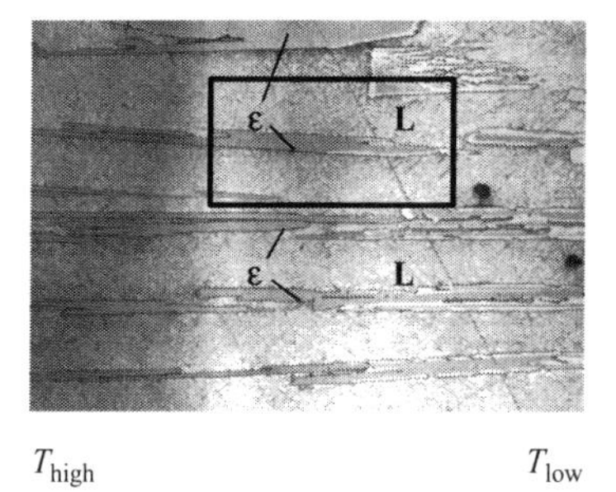

Figure 11.4b Peritectic reaction zone. Enlargement of the marked part of Figure 11.4a. (Fredriksson and Nylén). From Mechanism of Peritectic Reactions and Transformations, H. Fredriksson, T. Nylén in Metal Science Journal, Vol. 16, The Metals Society, 1982 © Maney www.maney.co.uk/journals/mst and www.ingentaconnect.com/content/maney/msc.

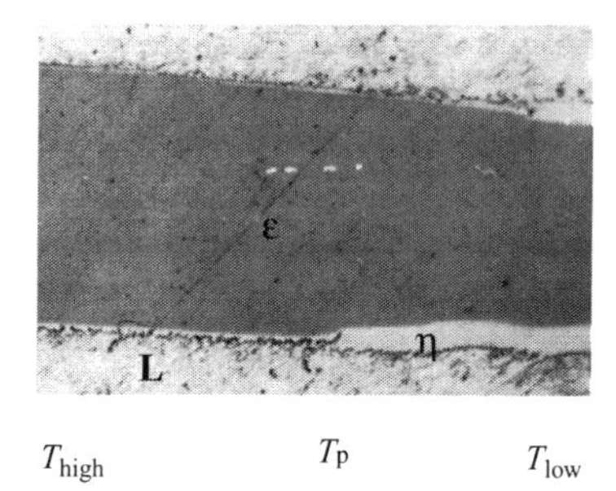

Figure 11.4c Enlargement of the marked part of Figure 11.4b. The peritectic reaction starts at the peritectic temperature T_p. From Mechanism of Peritectic Reactions and Transformations, H. Fredriksson, T. Nylén in Metal Science Journal, Vol. 16, The Metals Society, 1982 © Maney www.maney.co.uk/journals/mst and www.ingentaconnect.com/content/maney/msc.

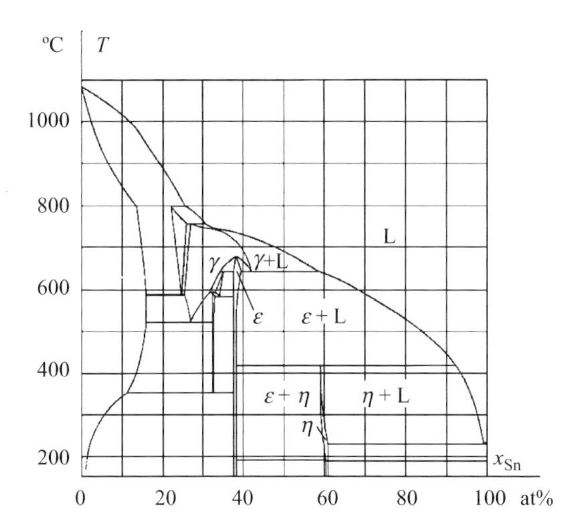

Figure 11.5 Phase diagram of the system Cu–Sn. The ε phase reacts with the melt and a secondary η phase is formed. MaxHansen © McGraw Hill 1958.

to the right in Figures 11.4a and b and more clearly in Figure 11.4c. The pale grey regions between the ε crystals consist of remaining liquid.

The phase diagram shows that the η phase, which is formed at the peritectic reaction, is richer of Sn atoms than the ε phase (earlier named α phase). The consequence is that Sn atoms diffuse through the liquid from the Sn-rich η phase (earlier named β phase) to the Sn-depleted ε phase in agreement with Figure 11.4d. The result is that some of ε phase will dissolve and the η phase grows. The larger the undercooling is and the faster the diffusion is the faster will occur the peritectic reaction.

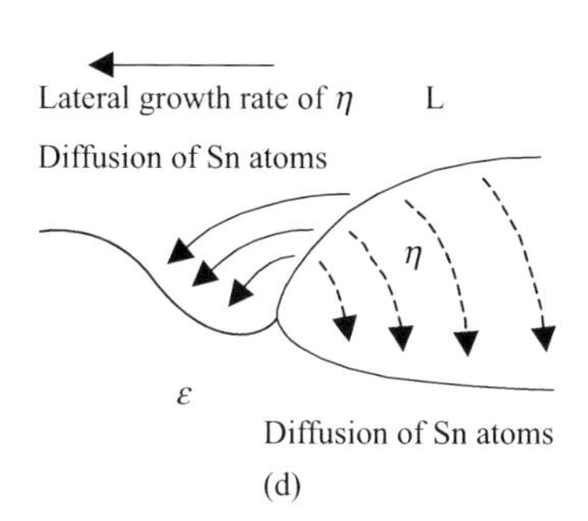

Figure 11.4d Diffusion of Sn atoms.

Growth of the β Phase in Peritectic Reactions of Type 1

Above the influence of diffusion on the rate of the peritectic reaction was pointed out. On pages 683–684 it was mentioned that surface tension also influences the rate of peritectic reactions. This topic will be discussed below.

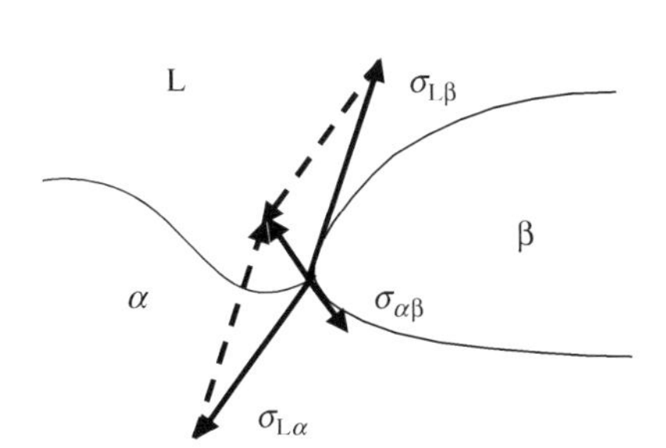

Figure 11.2b At equilibrium, the vector sum of the surface tension forces is zero.

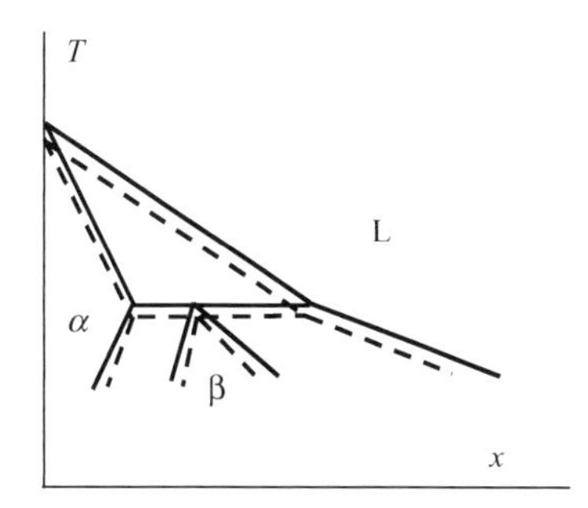

Figure 11.6 Influence of curved phases on a peritectic phase diagram.

In cases when the curvature is negative at the interface between the α phase and the liquid, the surface tension forces do not prevent the dissolution of the α phase (ε phase in the example above). As the dissolution is rapid it occurs at the same rate as the precipitation of the secondary β phase (η phase in the example above). In this case the vector sum of the surface tension forces is the dominant factor.

At the triple point the three phases are in equilibrium with each other. Because the phase boundaries are curved the equilibrium lines will change position in the phase diagram compared with the normal liquidus and solidus lines. This is shown schematically in Figure 11.6. The same phenomenon is also observed in eutectic phase diagrams.

Formation of a β Layer around the α Phase. Lateral Growth Rate

The growth laws, that control the extension of the β layer along the surface of the α phase, are analogous with other growth laws, for example the dendritic growth law (pages 514–524 in Chapter 9). As the present case concerns extension along a surface, it is best compared with the growth law of plates, that is given on page 648 [Equation (10.149) in Chapter 10]. In the present case the growth along the surface can be described by the equation

$$V_{\text{growth}} = const(\Delta T)^n \tag{11.2}$$

where

n = a constant

V_{growth} = lateral growth rate, i.e. growth rate of the β phase parallel to the α surface

ΔT = undercooling of the melt at the α/L interface.

Fredriksson and Nylén [2] have verified experimentally that the exponent n in the present case is equal to 4.

$$V_{\text{growth}} = const(\Delta T)^4 \tag{11.3}$$

V_{growth} is determined by the heat and temperature conditions, i.e. by the rate of transport of heat of solidification during formation of the solid phase β.

The influence of the surface tension is included in the constant. The β phase forces its way between the α phase and the liquid. The energy per unit area required to create one unit area of an α/β interface and one unit area of an L/β interface and remove one unit area of the original L/α interface will be

$$\sigma \text{ factor} = \sigma_{\alpha\beta} + \sigma_{L\beta} - \sigma_{L\alpha} \tag{11.4}$$

As V_{growth} is determined by external factors it can be regarded as constant in Equation (11.3). The σ factor is determined by material constants and therefore corresponds to a certain undercooling in each case. In a case with *higher* σ factor the undercooling is found to be *higher*. For this reason the σ factor in Equation (11.3) must appear in the denominator. In the simplest case Equation (11.3) can be written as

$$V_{\text{growth}} = \frac{Const}{\sigma_{\alpha\beta} + \sigma_{L\beta} - \sigma_{L\alpha}}(\Delta T)^4 \tag{11.5}$$

Thickness of the β Layer

The surface tension and the undercooling influence the thickness of the β layer. In order to calculate the thickness we will use the same principles as were used for eutectic reactions and for dendritic growth. We assumed that the growth occured at maximum rate or minimum undercooling, which gave an expression of the thickness of the β layer. Analogously the β layer, formed during a peritectic reaction, can be described by the simplified expression

$$d_\beta^0 = C\,\frac{\sigma_{\alpha\beta} + \sigma_{L\beta} - \sigma_{L\alpha}}{\Delta T} \tag{11.6}$$

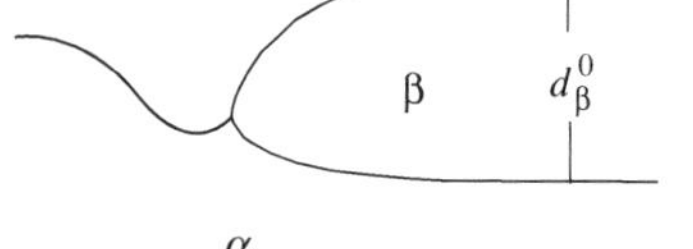

Figure 11.7 Thickness of the β layer at a peritectic reaction of type 1.

where

d_β^0 = thickness of the β layer (Figure 11.7)

C = a constant

ΔT = undercooling of the melt at the α/L interface.

Elimination of the undercooling ΔT between the Equations (11.5) and (11.6) gives the relationship

$$d_{\beta}^{0} = C \times Const^{1/4} \frac{(\sigma_{\alpha\beta} + \sigma_{L\beta} - \sigma_{L\alpha})^{3/4}}{V_{growth}{}^{1/4}} \tag{11.7}$$

Obviously the thickness of the β layer, varies with the growth rate. At constant surface tension, the thickness increases with decreasing growth rate. It is independent of time.

If the driving force (based on undercooling) for heterogeneous nucleation of the secondary β phase on the primary α phase is smaller than the driving force (based on concentration differences) of diffusion-controlled growth of β layer a number of β crystals will be formed around and cover the grains of the primary α phase.

This type of reaction has been reported by Petzow and Exner [3] in the Cd-Cu system and is shown in Figure 11.8.

The primary Cu_5 Cd_8 crystals in Figure 11.8 are bright grey. The dark matrix is Cd and the peritectically formed $CuCd_3$ crystals that surround the Cu_5 Cd_8 crystals are grey.

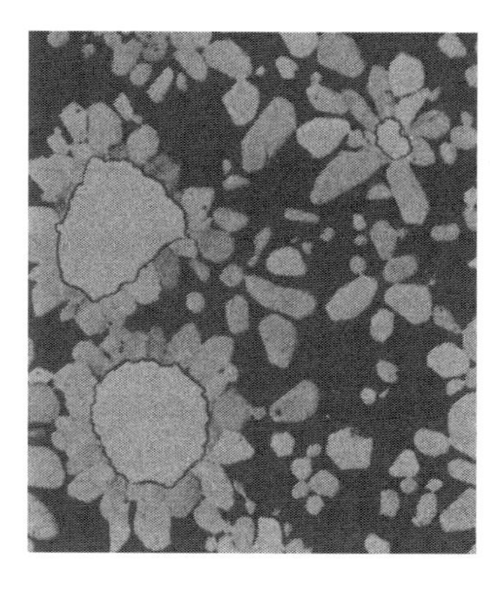

Figure 11.8 The microstructure of a Cd–10%Cu sample, in which a peritectic reaction has occurred (after Petzow and Exner). © RHI AG 1967.

Peritectic Reactions of Type 2

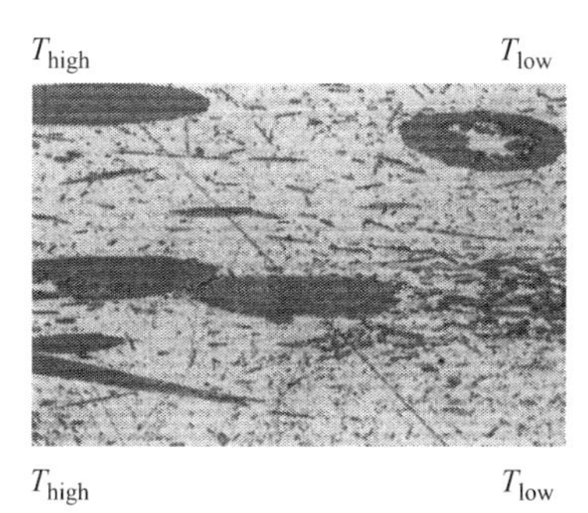

Figure 11.9a Primary precipitated γ crystals in an Al–11 wt%Mn alloy. The hollow in the crystal on the right-hand side is filled with melt. Speed = 1 mm/min.

Temperature gradient = 5 K/mm. From Mechanism of Peritectic Reactions and Transformations, H. Fredriksson, T. Nylén in Metal Science Journal, Vol. 16, The Metals Society, 1982 © Maney www.maney.co.uk/journals/mst and www.ingentaconnect.com/content/maney/msc.

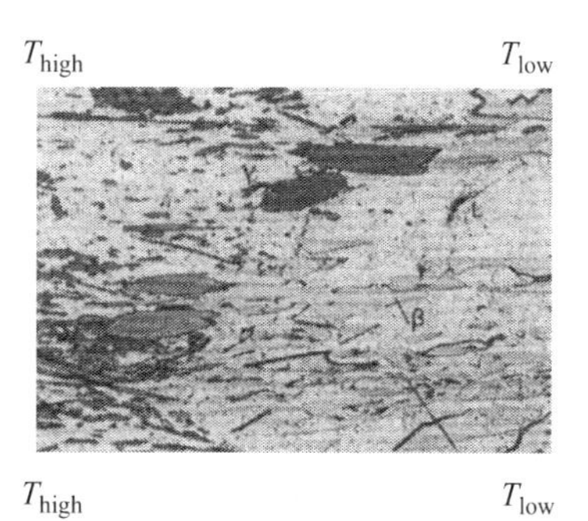

Figure 11.9b Structure of the alloy in Figure 11.9a at the start of the peritectic reaction. Speed = 1 mm/min.

Temperature gradient = 5 K/mm. From Mechanism of Peritectic Reactions and Transformations, H. Fredriksson, T. Nylén in Metal Science Journal, Vol. 16, The Metals Society, 1982 © Maney www.maney.co.uk/journals/mst and www.ingentaconnect.com/content/maney/msc.

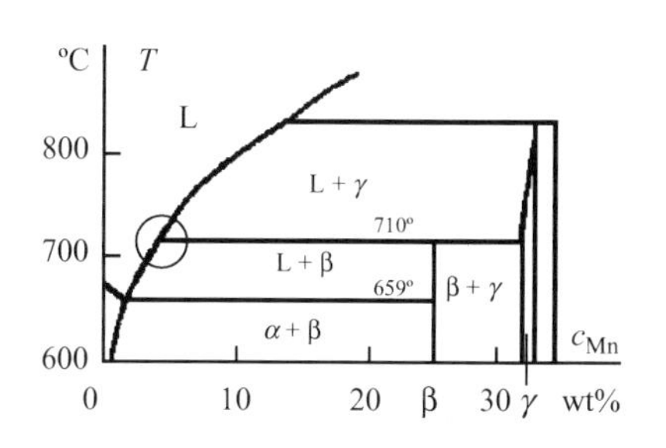

Figure 11.9c Part of the phase diagram of the binary system Al–Mn. The primary phase is the γ phase. The secondary phase, formed at the peritectic reaction, is the β phase, which is very narrow.

The Figures 11.9a and b show the peritectic reaction in a unidirectionally solidified Al-Mn alloy with 11% Mn, where the γ phase and the liquid react and give the β phase (see the phase diagram of the system Al-Mn in Figure 11.9c).

$$L + \gamma \rightarrow \beta$$

The temperature gradient in the Figures 11.9a and b is horizontal. The temperature in the left parts of the figures is higher than in the right parts. The temperature decreases to the right and the effect of the decreasing temperature can be observed in the figures.

Figure 11.9b shows that the β phase does not tend to grow around the γ phase. Figure 11.10 shows this type 2 reaction but there is a difference, compared with the type 2 reaction in Figure 11.3 on page 684.

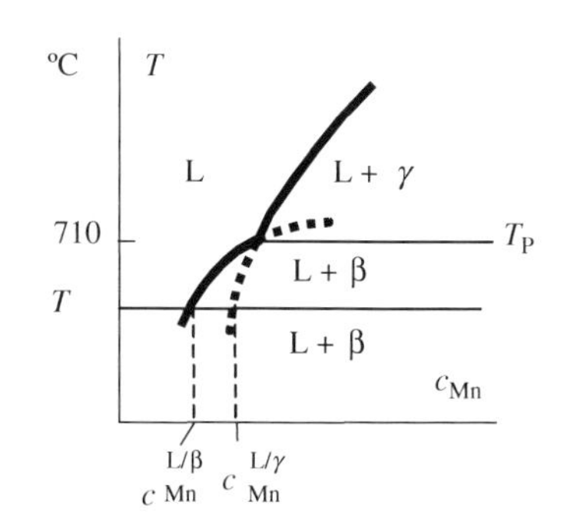

Figure 11.9d Enlargement of part of the phase diagram in Figure 11.9c.

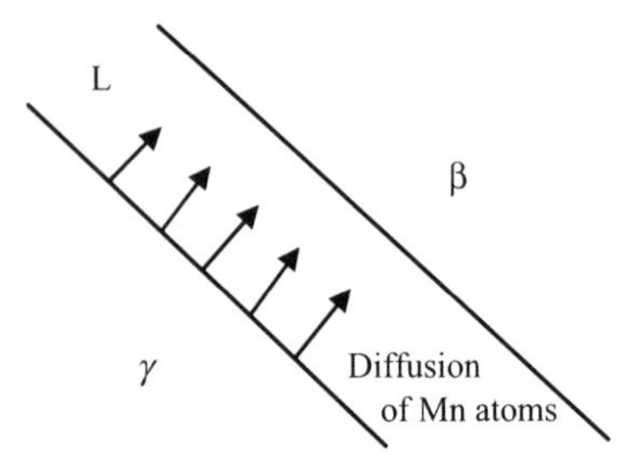

Figure 11.10 Diffusion of Mn atoms through the liquid between the γ and β phases.

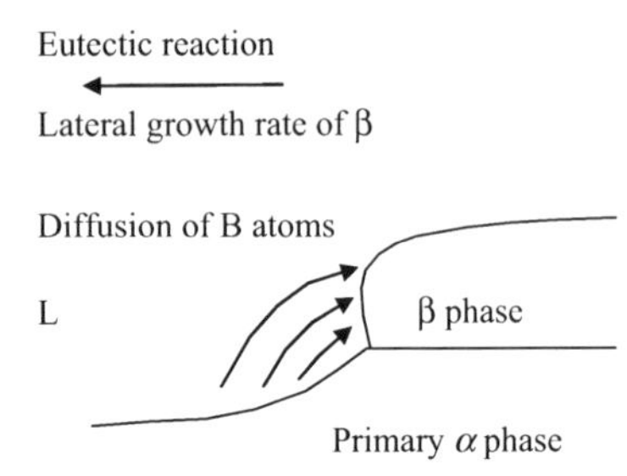

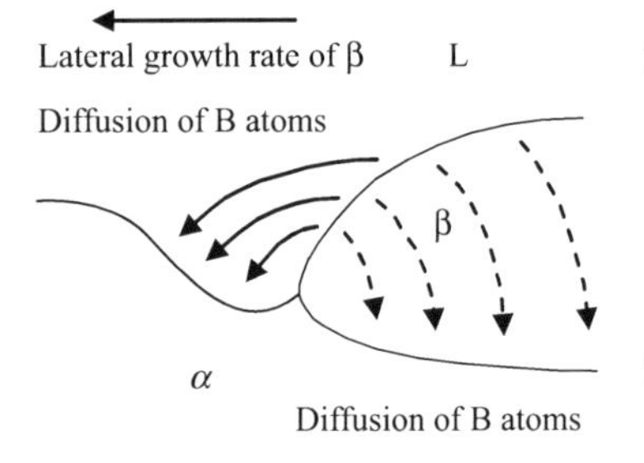

It can be observed that the diffusion of the Mn atoms in Figure 11.10 occurs in the opposite direction, compared with the diffusion of the solute atoms in Figure 11.3. This fact can easily be explained with the aid of the phase diagram in Figure 11.9d.

The direction of the diffusion of the Mn atoms through the liquid is determined by the Mn concentrations at the interfaces between the liquid L and the phases γ and β, respectively.

The Mn atoms diffuse from the Mn-rich γ phase to the Mn-depleted β phase. The phase diagrams in Figures 11.9c and 11.9d are mirror-inverted, compared to the phase diagram in Figure 11.1 on page 682. The result is that γ phase shrinks and the β phase grows.

Example 11.1

The solidification process of most alloys start with a primary precipitation of dendrites, followed by a eutectic or peritectic reaction. Both these reactions start in most cases with formation of a layer of a secondary phase, which grows around the primary phase. The diffusion and precipitation processes are completely different in the eutectic and peritectic cases, however.

Give an account of the reaction processes in the two cases.

Solution and Answer:

The solidification process consists of three subprocesses: Primary precipitation of α phase, precipitation of β phase by a eutectic/peritectic reaction and growth of the β phase.

Eutectic Solidification

1. Primary Precipitation: $L \rightarrow \alpha$
2. Eutectic Reaction: $L \rightarrow \alpha + \beta$

   The α and β phases are precipitated simultaneously from the melt. Both of them grow owing to diffusion through the melt. B atoms diffuse through the melt to the β phase. The diffusion process is shown in the margin figure.

   Diffusion of A and B atoms in the melt in front of the α and β phases results in an equilibrium with coupled growth between the α and β phases (Figure 10.6 in Chapter 10, page 590).
3. Further Growth of the β phase

   Further growth of the secondary β layer is blocked because there is no diffusion through the β layer. The solidification continuous by precipitation of the two phases directly in the liquid.

Peritectic Solidification

1. Primary Precipitation $L \rightarrow \alpha$
2. Peritectic Reaction $L + \alpha \rightarrow \beta$

   The α phase reacts with the melt and a secondary β phase is formed. The β phase grows and the α phase shrinks, i.e. is consumed owing to diffusion of B atoms through the melt. The diffusion process is shown in the margin figure.
3. Further Growth of the β phase

   Further growth of the secondary β layer occurs either by diffusion of B atoms through the β layer to the α phase or by direct precipitation (growth) of β phase in the liquid.

   The diffusion of B atoms results in the reaction α + B atoms $\rightarrow$ β i.e. results in a peritectic transformation (Section 11.2.1 below).

11.2.2 Peritectic Transformations

Every peritectic reaction is followed by a peritectic transformation when the primarily precipitated α phase will dissolve completely or partly and be replaced by β phase. Solute B atoms, which participate in the transformation, diffuse through the secondary β phase (Figure 11.2c on page 683) to the primary α-phase boundary. During a peritectic reaction the remelting of the primary α-phase occurs at the same rate as the precipitation of the β-phase. The process is controlled by the growth rate of the β-phase because there are no forces preventing dissolution. After the peritectic reaction the thickness of the β phase is d_{β}^{0} (page 686).

The growth in thickness of the secondary β layer during the peritectic transformation is only influenced by the *diffusion rate* of B atoms through the β layer.

A *peritectic transformation* of α phase into β phase occurs by long range diffusion of B atoms through the secondary β phase between the liquid and the α phase. The thickness of the β layer will normally increase during the subsequent cooling after the solidification process. The reasons for this increase in thickness are mainly three:

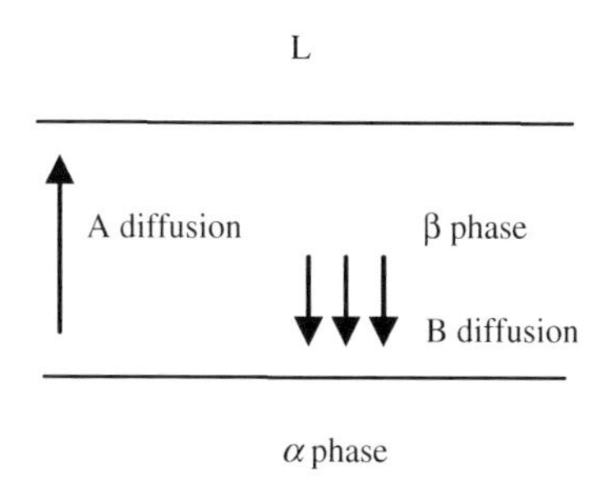

Figure 11.11 Peritectic transformation. The secondary β phase keeps the melt and the primary α phase apart (compare Figure 11.2c on page 683).

1. Precipitation of β phase directly from the liquid on the β layer.
2. Precipitation of β phase directly from the α phase owing to dissolution of the α phase, which occurs if the phase diagram has such a shape that dissolution is a spontaneous process.
3. Diffusion of B atoms through the β layer (Figure 11.11).

The precipitation of β phase directly from the liquid or the α phase depends on the shape of the phase diagram and the cooling rate. These topics will be discussed and explained on page 702–703.

The growth of the α-phase may occur if the shape of the phase diagram favours precipitation of this phase. This will be further discussed on pages 690–691.

The third process occurs because the concentration of B atoms is higher close to the liquid than close to the α phase. The diffusion process through the β layer depends on the diffusion rate, the cooling rate and the shape of the phase diagram.

Simple Theory of Peritectic Transformation

St John and Hogan [4] and also Hillert [5] have among others contributed to the theory of peritectic transformation. The theory describes the thickness of the β layer as a function of time and other quantities. In order to penetrate the transformation process we will start with an analysis of the isothermal case, i.e. the temperature is kept constant.

We assume that the growth of the β layer is controlled by the diffusion through it just below the peritectic temperature. To make the calculations somewhat simpler we will analyze a one-dimensional diffusion pattern in a plate-like structure. The diffusion process, the concentration profile and its relationship to the phase diagram are shown in Figure 11.12. Two mass balances can be set up:

Growth at the β/α interface:

$$AD_{\beta}\frac{x^{\beta/L}-x^{\beta/\alpha}}{d_{\beta}} = A\frac{\mathrm{d}d_{\beta/\alpha}}{\mathrm{d}t}\left(x^{\beta/\alpha}-x^{\alpha/\beta}\right) \tag{11.8}$$

Amount of B atoms, that iffuse through the β layer per unit time.

Added amount of B atoms per unit time in the new volume $A\ \mathrm{d}d_{\beta/\alpha}$.

Growth at the β/L interface:

$$AD_{\beta}\frac{x^{\beta/L}-x^{\beta/\alpha}}{d_{\beta}} = A\frac{\mathrm{d}d_{\beta/L}}{\mathrm{d}t}\left(x^{L/\beta}-x^{\beta/L}\right) \tag{11.9}$$

Amount of B atoms, that diffuse through the β layer per unit time

Added amount of B atoms per unit time in the new volume $A\ \mathrm{d}d_{\beta/L}$

where

A = cross-section area
d_β = thickness of the β phase
t = time
D_β = diffusion constant of B atoms in the β phase
$\mathrm{d}d_{\beta/\alpha}/\mathrm{d}t$ = growth rate of the β phase at the β/α interface
$\mathrm{d}d_{\beta/L}/\mathrm{d}t$ = growth rate of the β phase at the β/L interface
$x^{i/j}$ = concentrations defined in Figure 11.12.

The total growth rate of the β layer is the sum of the two growth rates in Equations (11.8) and (11.9).

$$\frac{\mathrm{d}d_\beta}{\mathrm{d}t} = \frac{\mathrm{d}d_{\beta/\alpha}}{\mathrm{d}t} + \frac{\mathrm{d}d_{\beta/\mathrm{L}}}{\mathrm{d}t} = \frac{D_\beta}{d_\beta}\left(x^{\beta/\mathrm{L}} - x^{\beta/\alpha}\right)\left(\frac{1}{x^{\beta/\alpha} - x^{\alpha/\beta}} + \frac{1}{x^{\mathrm{L}/\beta} - x^{\beta/\mathrm{L}}}\right) \tag{11.10}$$

Provided that the temperature T is *constant* all the concentration differences will also be constant. If and only if this condition is fulfilled Equation (11.10) on page 690 can be separated and integrated. If we replace the lower integration limit d_β^0 by zero we obtain

$$\frac{(d_\beta)^2}{2} = D_\beta\left(x^{\beta/\mathrm{L}} - x^{\beta/\alpha}\right)\left(\frac{1}{x^{\beta/\alpha} - x^{\alpha/\beta}} + \frac{1}{x^{\mathrm{L}/\beta} - x^{\beta/\mathrm{L}}}\right)t \tag{11.11}$$

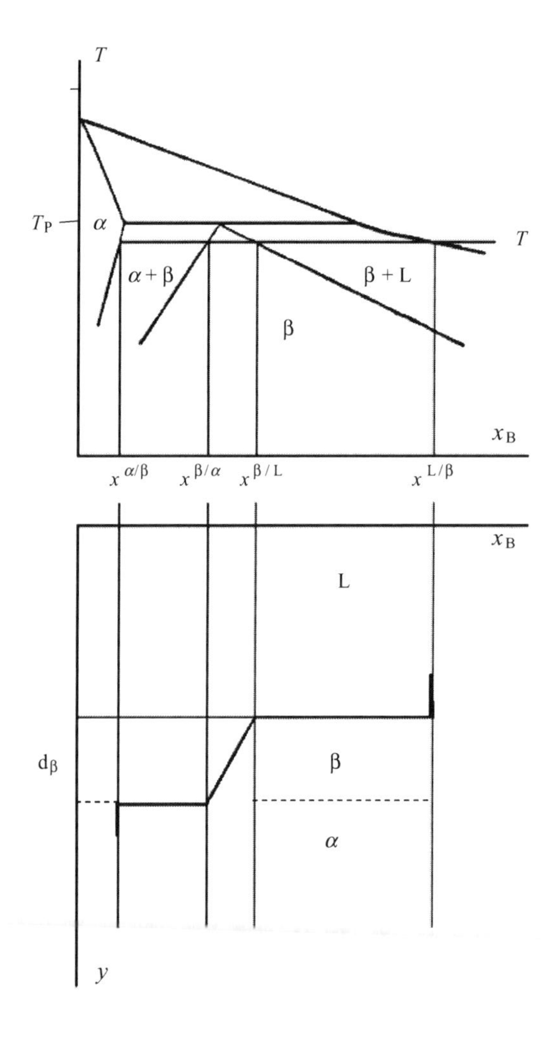

Figure 11.12 Concentration distribution during a peritectic transformation in a system with a high diffusion rate through the β phase.
Upper figure: Phase diagram.
Lower figure: Concentration distribution.

The thickness of the β layer after the peritectic reaction (pages 684–687) can normally be neglected in comparison with d_β and the total thickness will be approximately equal to d_β. This value for the total thickness will be used below.

Equation (11.11) represents the thickness of the β layer as a function of time. It is approximate, as we have neglected volume changes and the complex geometry of the real primary phase. A more accurate theory is given in Section 11.3 on page 693.

11.2.3 Influence of Some Parameters on Peritectic Transformations

Influence of Various Parameters on the Growth Rate of the β Layer at Peritectic Transformations

Equation (11.10) shows that the growth rate of the β layer depends on

- the diffusion constant of the B atoms in the β layer
- the shape of the phase diagram

Indirectly, via the concentrations of the B atoms, the growth rate is also influenced by

- the transformation temperature
- the cooling rate.

Influence of Transformation Temperature

From the phase diagram in Figure 11.11 on page 689 we can conclude that the difference $(x^{\beta/L} - x^{\beta/\alpha})$ is proportional to the undercooling $(T_P - T)$. The expressions $(x^{\beta/\alpha} - x^{\alpha/\beta})$ and $(x^{L/\beta} - x^{\beta/L})$ may either increase or decrease with increasing undercooling and decreasing temperature, depending on the slopes of the lines, which will be further discussed on pages 699–701. The correct alternative is determined with the aid of the shape of the phase diagram for the alloy in question.

Influence of Diffusion Constant

Equation (11.11) on page 690 shows that the growth rate is proportional to the diffusion constant.

For substitutionally dissolved alloying elements, as in FCC metals, the diffusion constant in the β layer is of the order 10^{-13} m^2/s or less at a temperature close to the melting point. In such a case the growth rate will be very low. The time for the peritectic transformation will consequently be extremely long and the transformation will be negligible in comparison with the precipitation of β phase from the liquid.

In other cases, such as BCC metals and metals with interstitially solved alloying elements, the diffusion rates are much higher than those for FCC metals. The diffusion processes have a large influence on the precipitation process in these cases.

- The growth rate is proportional to the diffusion constant.

Influence of Shape of the Phase Diagram

The phase diagram of a binary alloy, which is based on the material constants of the two metals, determines the concentrations of alloying element in the various phases. It is evident from Equation (11.10) that the growth rate of the β layer depends strongly on the shape of the phase diagram.

The slope of the boundary line between the $(\alpha + \beta)$ and β phases has a decisive influence on the stability of the α phase. If the slope is *positive*, as in Figure 11.12 on page 690, the α phase will dissolve and be transformed into β phase when the temperature decreases.

On the other hand, if the boundary line has a *negative* slope the α phase will stabilize at the expense of the β phase when the temperature decreases further discussed on pages 699–701.

Influence of Cooling Rate

Consider a continuously cooled sample of an alloy. The concentrations of the alloying element at the phase boundaries (liquidus and solidus lines) are functions of the temperature.

The growth rate of the β layer can be written as

$$V^{\beta}_{\text{growth}} = \frac{\mathrm{d}d_{\beta}}{\mathrm{d}t} = \left(-\frac{\mathrm{d}d_{\beta}}{\mathrm{d}T}\right)\left(-\frac{\mathrm{d}T}{\mathrm{d}t}\right) \tag{11.12}$$

where

$-\dfrac{\mathrm{d}T}{\mathrm{d}t}$ = cooling rate.

The quantity $\mathrm{d}d_{\beta}/\mathrm{d}T$ is negative because the thickness of the β layer increases when the temperature decreases. If $\mathrm{d}d_{\beta}/\mathrm{d}T$ is constant, Equation (11.12) shows that

- The growth rate is proportional to the cooling rate.

At a *unidirectional solidification experiment* on a 'peritectic' alloy the specimen is moved downwards from higher towards lower temperatures with the constant pulling rate V_{pull}. It moves in a temperature field. The temperature gradient G is kept constant and directed upwards. It is equal to

$$G = \mathrm{d}T/\mathrm{d}z \tag{11.13}$$

which is a positive quantity as T increases in the positive z direction. In this case the following relationship for the growth rate can be written as

$$V^{\beta}_{\text{growth}} = \frac{\mathrm{d}d_{\beta}}{\mathrm{d}t} = \frac{\mathrm{d}d_{\beta}}{\mathrm{d}T}\frac{\mathrm{d}T}{\mathrm{d}z}\frac{\mathrm{d}z}{\mathrm{d}t} = \left(-\frac{\mathrm{d}d_{\beta}}{\mathrm{d}T}\right)G(-V_{\text{pull}}) \tag{11.14}$$

The growth rate is always a positive quantity while V_{pull} is negative.

Thickness of the β Layer as a Function of Temperature

In the general case the thickness of the β layer can be derived with the aid of Equation (11.10) on page 690.

$$\int_0^{d_{\beta}} d_{\beta}\mathrm{d}d_{\beta} = -\int_{T_{\mathrm{P}}}^{T} D_{\beta}\left(x^{\beta/\mathrm{L}} - x^{\beta/\alpha}\right)\left(\frac{1}{x^{\beta/\alpha} - x^{\alpha/\beta}} + \frac{1}{x^{\mathrm{L}/\beta} - x^{\beta/\mathrm{L}}}\right)\left(-\frac{\mathrm{d}t}{\mathrm{d}T}\right)\mathrm{d}T \tag{11.15}$$

Equation (11.11) on page 690 is not valid as the concentrations normally vary with the temperature. The cooling rate is supposed to be known. If the temperature dependences of the concentrations are taken from the phase diagram and introduced into Equation (11.15) the integral can be solved.

Thickness of β Layer at Unidirectional Solidification

We want to find the thickness of the β layer in the unidirectional experiment as a function of temperature T. To achieve this we have to integrate Equation (11.14) in combination with Equation (11.10) on page 690.

To make this operation possible, all concentration differences in Equation (11.10) have to be transformed into temperature differences with the aid of the slopes of the different equilibrium lines in the phase diagram.

These topics will be discussed for the FeC-system in Section 11.3.1 on pages 694–696 and for alloys with substutionally dissolved alloying elements in Section 11.3.2 on pages 700–701.

11.2.4 Cascades of Peritectic Reactions

The theoretical analysis in Section 11.2.1 shows that the rate of a peritectic transformation is influenced by the diffusion rate and by the extension of the β phase region in the phase diagram. If the diffusion rate is small the peritectic transformation will be negligible compared to the peritectic reaction. The total volume of the β phase envelope surrounding the α phase after the peritectic reaction V_{total} equals the total area A times the thickness of the β layer at that time. During the further cooling an increase in thickness to $\sim d_{\beta}$ occurs, owing to precipitation of β phase directly from the liquid.

In many systems one peritectic reaction at a high temperature is followed by one or several peritectic reactions at lower temperatures. If the diffusion rate is low in the primary peritectic layer, a second peritectic layer can be formed when the second peritectic temperature is achieved. This type of a series of peritectic reactions has been studied by Petzow and Exner [3].

Figure 11.13 gives as an example the phase diagram of Al-Mn, that contains a long series of peritectic reactions. Petzow and Exner called such an array a *cascade of peritectic reactions.*

Figure 11.14b shows the microstructure formed in a Cd-24%Ni alloy resulting from two peritectic reactions at 690 °C and 495 °C (see Figure 11.14a) followed by peritectic transformations. The primary Ni crystals are covered by two 'envelopes' of secondary phases, formed at the consecutive peritectic reactions.

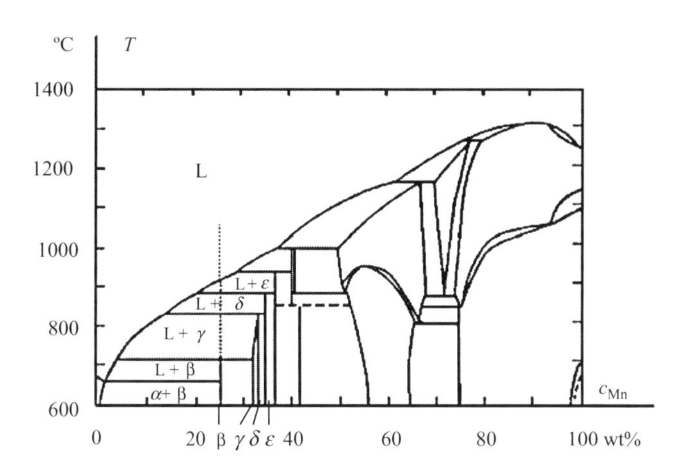

Figure 11.13 Phase diagram of the system Al-Mn including 7 peritectic reactions. Three of them occur in the alloy Al–24 wt%Mn. The reactions in the melt when it cools are in order:

L → ε (primary precipitation) $T = 900$ °C;
L + ε → δ (peritectic reaction) $T = 881$ °C;
L + δ → γ (peritectic reaction) $T = 822$ °C;
L + γ → β (peritectic reaction) $T = 710$ °C;
L → α + β (eutectic reaction) $T = 659$ °C.

From Binary Alloy Phase Diagrams Vol. 1, Ed. Thaddeus B. Massalski, American Society for Metals 1986. Reprinted with permission of ASM International. www.asminternational.org.

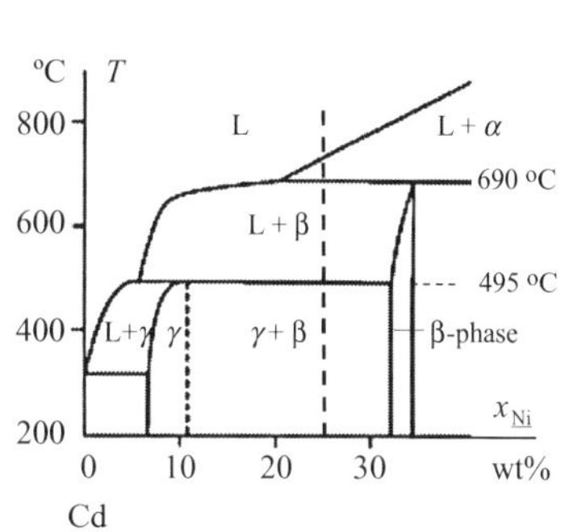

Figure 11.14a Phase diagram of Cd–Ni. From Binary Alloy Phase Diagrams Vol. 1, Ed. Thaddeus B. Massalski, American Society for Metals 1986. Reprinted with permission of ASM International. www.asminternational.org.

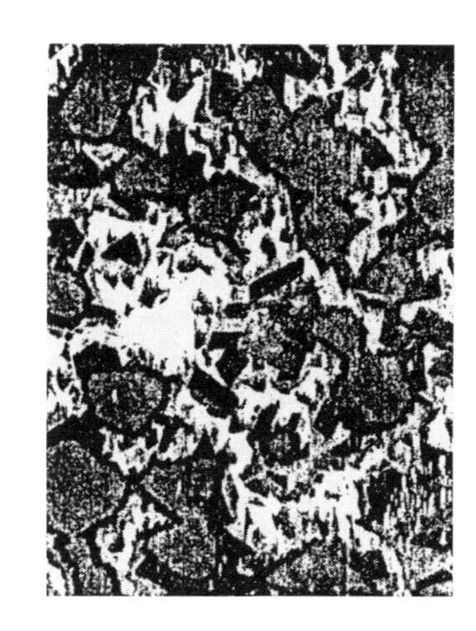

Figure 11.14b Microstructure with two peritectic envelopes in the alloy Cd–25 wt%Ni. The α phase (Ni crystals) is dark grey. The Cd matrix is white. The black envelope consists of a β layer (Cd Ni). The pale grey envelope outside the β layer consists of γ phase (Cd_5 Ni).

11.3 Peritectic Reactions and Transformations in Iron-Base Alloys

Iron-base alloys are technically and commercially the most important alloys with a manifold of applications. Peritectic reactions and transformations are common in these alloys and influence their properties strongly. For these reasons, these alloys will be extensively discussed in this section.

11.3.1 Reactions and Transformations in the Fe–C System

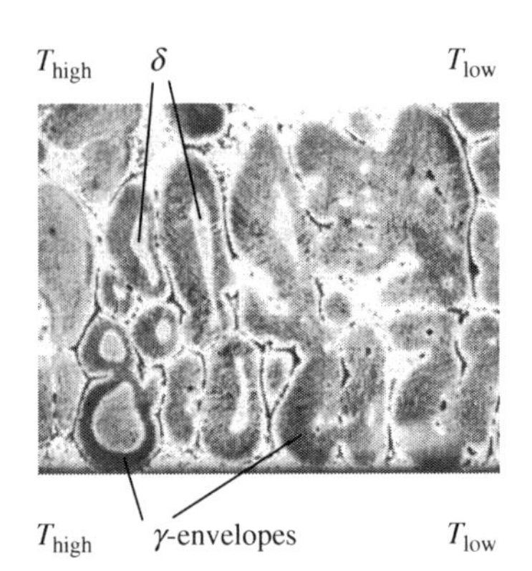

Figure 11.15 The peritectic reaction in a low carbon steel $L + \delta \rightarrow \gamma$ can be followed in the cross-section of a quenched unidirectionally solidified sample.
δ = ferrite;
γ = austenite.
The δ regions (ferrite) change colour from dark to bright at decreasing temperature, due to changed composition, which influences the etching required to produce the picture. Max Hansen © McGraw Hill 1958.

Very few experimental studies on peritectic reactions in pure Fe–C alloys have been published. Careful experiments are very difficult to perform because of the high transfomation rate. Stjerndahl [6] has performed such a study, which will be discussed here. He studied a peritectic transformation in a unidirectionally solidified sample of Fe-0.3%C with traces of phosphorous.

Part of a sample in Stjerndahl's study is shown in Figure 11.15. To the left in the figure, the structure consists of dendrites of ferrite (grey or pale grey) and former interdendritic liquid (pale with fine dendrites).

The peritectic reaction starts by formation of a layer of austenite (dark grey) around the ferrite dendrites. This layer of austenite will increase in thickness due to the transformation of ferrite and liquid into austenite.

In the right half of Figure 11.15 there is no longer any ferrite left. The peritectic transformation is very fast in this sample. The extension of the peritectic process can be measured as a length in the quenched sample.

By multiplying this length with the temperature gradient, the temperature interval within which the peritectic process occurs can be obtained. In this way the peritectic reaction and transformation in the present case was found to be performed within a temperature interval of the magnitude 1–10 K (compare Example 11.2 on page 697).

Theory of Peritectic Transformations in the Fe–C System

The simple analytical model, derived on pages 689–690, can be used to calculate the rate of the peritectic transformation in the Fe–C system.

At the start of the peritectic transformation the ferrite is surrounded by a layer of austenite and insulated from the melt. We make the following assumptions:

1. The growth of the austenite layer is controlled by the diffusion of carbon through it.
2. The carbon concentration in the austenite decreases linearly from the austenite/liquid boundary to the austenite/ferrite phase boundary.
3. The concentrations of carbon in the ferrite and in the liquid are uniform and at equilibrium. The values are found with the aid of the Fe–C phase diagram.

We apply Equation (11.10) on page 690 on the Fe–C system and obtain the following expression for the total growth rate of the γ layer

$$\frac{\mathrm{d}d_\gamma}{\mathrm{d}t} = \frac{\mathrm{d}d_{\gamma/\delta}}{\mathrm{d}t} + \frac{\mathrm{d}d_{\gamma/\mathrm{L}}}{\mathrm{d}t} = \frac{D_{\underline{\mathrm{C}}}^{\gamma}}{d_\gamma}\frac{x_{\underline{\mathrm{C}}}^{\gamma/\mathrm{L}} - x_{\underline{\mathrm{C}}}^{\gamma/\delta}}{V_{\mathrm{m}}^{\gamma}}\left(\frac{V_{\mathrm{m}}^{\delta}}{x_{\underline{\mathrm{C}}}^{\gamma/\delta} - x_{\underline{\mathrm{C}}}^{\delta/\gamma}} + \frac{V_m^L}{x_{\underline{\mathrm{C}}}^{\mathrm{L}/\gamma} - x_{\underline{\mathrm{C}}}^{\gamma/\mathrm{L}}}\right) \tag{11.16}$$

where

$\frac{\mathrm{d}d_{\gamma/\delta}}{\mathrm{d}t}$ = growth rate of γ phase at the γ/δ interface

$\frac{\mathrm{d}d_{\gamma/\mathrm{L}}}{\mathrm{d}t}$ = growth rate of γ phase at the γ/L interface

$D_{\underline{\mathrm{C}}}^{\gamma}$ = diffusion constant of carbon in the γ phase

d_γ = thickness of the γ phase

$x_{\underline{\mathrm{C}}}^{\mathrm{i/j}}$ = concentrations defined in Figure 11.16

$V_{\mathrm{m}}^{\gamma,\delta,\mathrm{L}}$ = molar volume of austenite, ferrite and melt, respectively.

The different densities of the phases have been taken into consideration by introduction of the molar volumes into Equation (11.16).

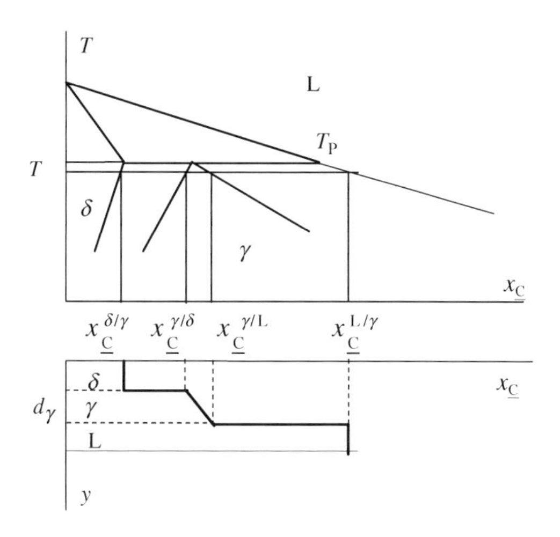

Figure 11.16 The Fe–C phase diagram and the related concentration profiles. Concentration distribution of carbon during a peritectic reaction in Fe–C alloys.

At the peritectic temperature the expression $(x_{\underline{C}}^{\gamma/L} - x_{\underline{C}}^{\gamma/\delta})$ is zero (Figure 11.16) and it increases with increasing undercooling and decreasing temperature. The two expressions $(x_{\underline{C}}^{\gamma/\delta} - x_{\underline{C}}^{\delta/\gamma})$ and $(x_{\underline{C}}^{L/\gamma} - x_{\underline{C}}^{\gamma/L})$ can either decrease or increase with increasing undercooling and decreasing temperature. No general conclusion can be drawn from Equation (11.16) concerning the growth rate. It may increase or decrease with increasing undercooling and decreasing temperature. The shape of the phase diagram of the alloy in question determines the correct alternative.

Equation (11.16) shows that the growth rate depends on the diffusion constant of the B atoms (carbon atoms or C) in the γ phase. For interstitially dissolved alloying elements such as carbon, the diffusion rates are very high and often of the order 10^{-10} m^2/s. Hence, the diffusion process has a large influence on the precipitation process in these cases.

- Equation (11.16) describes the growth rate of austenite at a constant temperature T just below the peritectic temperature T_P.

For the case that the diffusion rate in the γ phase is *high*, which is the case for carbon, Equation (11.16) can be used also to describe the reaction during the cooling process, for example at a unidirectional experiment. We will discuss this case and make three assumptions:

1. The peritectic transformation is considered at a constant cooling rate.
2. The total growth rate $\mathrm{d}d_\gamma/\mathrm{d}t$ can be written as

$$\frac{\mathrm{d}d_\gamma}{\mathrm{d}t} = \frac{\mathrm{d}d_\gamma}{\mathrm{d}T}\frac{\mathrm{d}T}{\mathrm{d}t} \tag{11.17}$$

where $(-\mathrm{d}T/\mathrm{d}t)$ is the cooling rate of the sample. Both factors on the right-hand side of Equation (11.17) are negative.

Over a small temperature interval the cooling rate is normally constant and can be written as

$$-\frac{\mathrm{d}T}{\mathrm{d}t} = -\frac{\mathrm{d}T}{\mathrm{d}z}\frac{\mathrm{d}z}{\mathrm{d}t} = -GV_{\mathrm{pull}} \tag{11.18}$$

where G is the temperature gradient $\mathrm{d}T/\mathrm{d}z$ and V_{pull} is the pulling rate during the unidirectional experiment. The specimen is moved downwards towards lower temperatures. Hence, V_{pull} is negative. G is directed upwards and positive.

3. The slopes of the lines in the phase diagram are assumed to be constant. The difference in carbon concentration can then be written (see the calculations in the box)

$$x_{\underline{C}}^{\gamma/L} - x_{\underline{C}}^{\gamma/\delta} = \frac{1}{m_{\mathrm{eff}}}(T_P - T) \tag{11.19a}$$

where

$$\frac{1}{m_{\mathrm{eff}}} = \frac{1}{|m_1|} + \frac{1}{|m_2|} \tag{11.19b}$$

The relationship (11.19b) is derived in the box below.

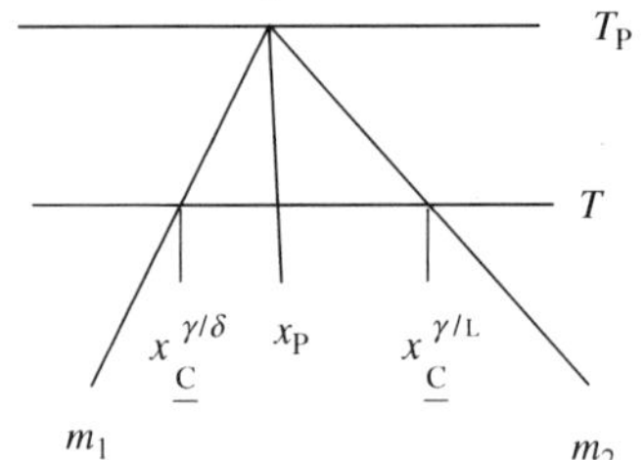

Calculation of m_{eff} as a Function of m_1 and m_2

The figure in the margin is an enlarged part of the phase diagram on page 694. Provided that the boundary lines between the phases are straight lines, the concentration difference $(x_{\underline{C}}^{\gamma/L} - x_{\underline{C}}^{\gamma/\delta})$ is proportional to the temperature difference $(T_P - T)$. This can be expressed with the aid of two equations:

$$(x_P - x_{\underline{C}}^{\gamma/\delta}) \times m_1 = T_P - T \qquad m_1 > 0 \tag{1'}$$

$$(x_P - x_{\underline{C}}^{\gamma/L}) \times m_2 = T_P - T \qquad m_2 < 0 \tag{2'}$$

Division with m_1 and m_2, respectively and subtraction of Equation (2′) from Equation (1′) give

$$x_{\underline{C}}^{\gamma/L} - x_{\underline{C}}^{\gamma/\delta} = \frac{1}{m_1}(T_P - T) - \frac{1}{m_2}(T_P - T) \tag{3'}$$

or

$$x_{\underline{C}}^{\gamma/L} - x_{\underline{C}}^{\gamma/\delta} = \left(\frac{1}{m_1} - \frac{1}{m_2}\right) \times (T_P - T) \tag{4'}$$

or, according to Equation (11.19b),

$$x_{\underline{C}}^{\gamma/L} - x_{\underline{C}}^{\gamma/\delta} = \left(\frac{1}{|m_1|} + \frac{1}{|m_2|}\right) \times (T_P - T) = \frac{1}{m_{\mathrm{eff}}} \times (T_P - T) \tag{5'}$$

By inserting the expressions in Equations (11.17), (11.18) and (11.19b) into Equation (11.16) we obtains

$$d_\gamma \mathrm{d}d_\gamma = \frac{D_{\underline{C}}^{\gamma}}{V_m^{\gamma} G(-V_{\mathrm{pull}})} \left(\frac{V_m^{\delta}}{x_{\underline{C}}^{\gamma/\delta} - x_{\underline{C}}^{\delta/\gamma}} + \frac{V_m^{L}}{x_{\underline{C}}^{L/\gamma} - x_{\underline{C}}^{\gamma/L}}\right) \frac{1}{m_{\mathrm{eff}}}(T_P - T)\mathrm{d}T \tag{11.20}$$

If the differences $(x_{\underline{C}}^{\gamma/\delta} - x_{\underline{C}}^{\delta/\gamma})$ and $(x_{\underline{C}}^{L/\gamma} - x_{\underline{C}}^{\gamma/L})$ and the cooling rate are approximately constant (small undercooling) Equation (11.20) can be integrated. By integration of the left-hand side from 0 to d_γ and the right-hand side from T_p to T and simplifying the result we obtain the thickness of the γ layer as a function of temperature:

$$(d_\gamma)^2 = \frac{D_{\underline{C}}^{\gamma}}{V_m^{\gamma} G(-V_{\mathrm{pull}})} \left(\frac{V_m^{\delta}}{x_{\underline{C}}^{\gamma/\delta} - x_{\underline{C}}^{\delta/\gamma}} + \frac{V_m^{L}}{x_{\underline{C}}^{L/\gamma} - x_{\underline{C}}^{\gamma/L}}\right) \frac{1}{m_{\mathrm{eff}}}(T_p - T)^2 \tag{11.21}$$

Equation (11.21) can be used to calculate the *peritectic transformation distance* L_γ, which is defined as the thickness of the γ layer when the peritectic transformation is finished.

The corresponding temperature decrease is called the *peritectic temperature range*. It is defined as the temperature interval during which the peritectic transformation lasts.

$$L_\gamma = \sqrt{\frac{D_{\underline{C}}^\gamma}{V_m^\gamma G}\frac{1}{m_{eff}}\left(\frac{V_m^\delta}{x_{\underline{C}}^{\gamma/\delta} - x_{\underline{C}}^{\delta/\gamma}} + \frac{V_m^L}{x_{\underline{C}}^{L/\gamma} - x_{\underline{C}}^{\gamma/L}}\right)\frac{T_P - T_{final}}{\sqrt{-V_{pull}}}} \tag{11.22}$$

where

L_γ = peritectic transformation distance

$T_P - T_{final}$ = peritectic temperature range.

The peritectic transformation distance L_γ is equal to the dendrite arm distance λ times f_γ, the fraction of γ phase during of the peritectic transformation.

$$L_\gamma = f_\gamma \lambda_{den} \tag{11.23}$$

f_γ can be derived from the phase diagram with the aid of the lever rule at the peritectic temperature.

Example 11.2

a) Derive an expression for the peritectic transformation distance for an Fe–C alloy as a function of the peritectic temperature range during which the peritectic transformation lasts. The pulling rate is V_{pull}. The alloy has the following material constants:

$$T_p = 1493°\text{C};\ V_m^\gamma = 7.44 \times 10^{-3}\ \text{m}^3/\text{kmol};\ D_{\underline{C}}^\gamma \text{ at } 1493°\text{C} = 1 \times 10^{-9}\ \text{m}^2/\text{s};$$
$$V_m^\delta = 7.64 \times 10^{-3}\ \text{m}^3/\text{kmol};\ G = 6.0 \times 10^3\ \text{K/m};\ V_m^L = 7.97 \times 10^{-3}\ \text{m}^3/\text{kmol}$$

Required concentrations and slopes of the liquidus lines can be read or derived from the phase diagram on next page.

b) Calculate the peritectic temperature range of an alloy with composition 0.74 at% C for the two cases when the pulling rate is 10.0 mm/min respectively 1.0 mm/min. The corresponding dendrite arm spacings have been determined experimentally to 0.21 mm and 0.060 mm, respectively.

The fraction of γ phase at the peritectic temperature when the peritectic reaction and transformation are over can be derived from the phase diagram on next page.

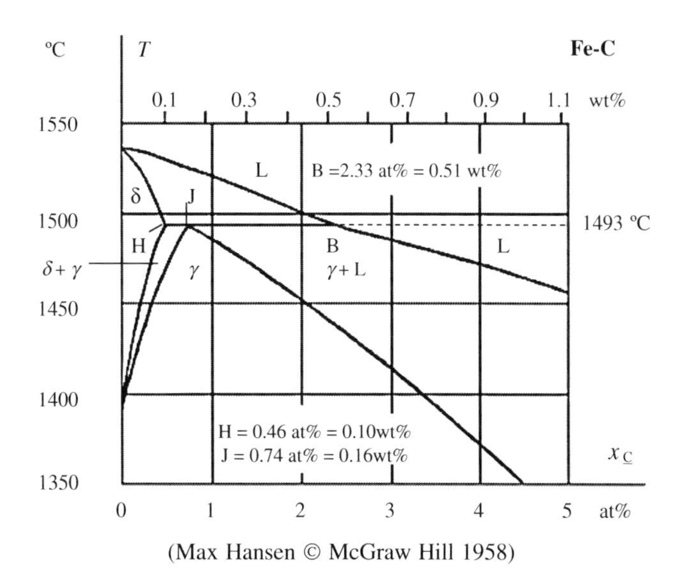

(Max Hansen © McGraw Hill 1958)

Solution:

Derivation of Required Quantities
Calculation of m_{eff}
From the phase diagram we can read the mole fractions

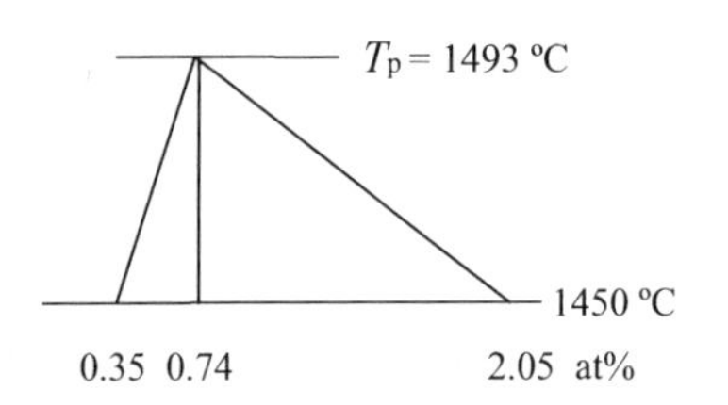

Readings from the phase diagram:

$$x_{\underline{C}}^{\gamma/\delta} - x_{\underline{C}}^{\delta/\gamma} = \text{JH} = 0.0074 - 0.0046 = 0.0028 \quad \text{(mole fraction)}$$

$$x_{\underline{C}}^{L/\gamma} - x_{\underline{C}}^{\gamma/L} = \text{BJ} = 0.0233 - 0.0074 = 0.0159 \quad \text{(mole fraction)}$$

With the aid of the figure in the margin we obtain the slopes

$$\frac{1}{m_1} = \frac{0.0074 - 0.0035}{1493 - 1450} = \frac{0.0039}{43} \text{ K}^{-1}$$

$$\frac{1}{|m_2|} = \frac{0.0205 - 0.0074}{1493 - 1450} = \frac{0.0131}{43} \text{ K}^{-1}$$

Inserting these values into Equation (11.19b) gives

$$\frac{1}{m_{\text{eff}}} = \frac{1}{|m_1|} + \frac{1}{|m_2|} = \frac{0.0039 + 0.0131}{43} = 0.000395 \text{ K}^{-1} \tag{1'}$$

Calculation of mole fractions
At $\underline{C}$-concentrations $x < 0.74$ at% the melt L solidifies as δ phase. The fraction f_δ of the δ phase at the peritectic temperature, can be calculated with the aid of the lever rule

$$f_\delta = \frac{x_B - x}{x_B - x_H} \tag{2'}$$

At the $\underline{C}$ concentration $x = 0.74$ at% and at the peritectic temperature but *before* the start of the peritectic reaction the fraction f_δ of the δ phase will be

$$f_\delta = \frac{\text{BJ}}{\text{BH}} = \frac{\text{BJ}}{\text{JH} + \text{BJ}} = \frac{0.0159}{0.0028 + 0.0159} = 0.85$$

After the peritectic reaction and transformation all the δ phase (ferrite) has been transformed into γ phase (austenite). When the solidification is completed, the structure consists of 100% austenite. In this case the fraction f_γ of the γ phase will be: $f_\gamma = \dfrac{\text{BJ}}{\text{BJ}} = 1$

a)
The given values are inserted into Equation (11.22):

$$L_\gamma = \sqrt{\frac{D_{\underline{C}}^{\gamma}}{V_m^{\gamma}} \frac{1}{G\, m_{\text{eff}}} \left(\frac{V_m^{\delta}}{x_{\underline{C}}^{\gamma/\delta} - x_{\underline{C}}^{\delta/\gamma}} + \frac{V_m^{L}}{x_{\underline{C}}^{L/\gamma} - x_{\underline{C}}^{\gamma/L}} \right) \frac{T_p - T_{\text{final}}}{\sqrt{-V_{\text{pull}}}}} \tag{3'}$$

or

$$L_\gamma = \sqrt{\frac{1 \times 10^{-9} \times 0.00395}{7.44 \times 10^{-3} \times 6.0 \times 10^3} \times \left(\frac{7.64 \times 10^{-3}}{0.0028} + \frac{7.97 \times 10^{-3}}{0.0159}\right)} \times \frac{T_p - T_{final}}{\sqrt{-V_{pull}}}$$

which gives $L_\gamma = 1.69 \times 10^{-7} \times \dfrac{T_p - T_{final}}{\sqrt{-V_{pull}}}$

b)
When the peritectic transformation is completed L_γ equals the dendrite arm spacing λ_{den} times f_γ and $f_\gamma = 1$.

$$T_p - T_{final} = \frac{L_\gamma \sqrt{-V_{pull}}}{1.69 \times 10^{-7}} = \frac{\lambda_{den} \sqrt{-V_{pull}}}{1.69 \times 10^{-7}} \tag{4'}$$

The values for case 1 given above are inserted into Equation (4′) and we obtain

$$T_p - T_{final} = \frac{\lambda_1 \sqrt{-V_1}}{1.69 \times 10^{-7}} = \frac{0.21 \times 10^{-3} \sqrt{\dfrac{0.010}{60}}}{1.69 \times 10^{-7}} = 16\ \text{K}$$

For case 2 we obtain:

$$T_p - T_{final} = \frac{\lambda_2 \sqrt{-V_2}}{1.69 \times 10^{-7}} = \frac{0.060 \times 10^{-3} \sqrt{\dfrac{0.0010}{60}}}{1.69 \times 10^{-7}} = 1.4\ \text{K}$$

Answer:

a) The required relationship is $L_\gamma = 1.7 \times 10^{-7} \times \dfrac{T_p - T_{final}}{\sqrt{-V_{pull}}}$.

b) The peritectic temperature ranges are 16 K and 1.4 K, respectively.

The peritectic temperature interval can be calculated in the same way as in Example 11.2b above with experimentally measured values of the primary dendrite arm spacing at different carbon concentrations. The result of such calculations for different C concentrations and two different pulling rates has been plotted in Figure 11.17.

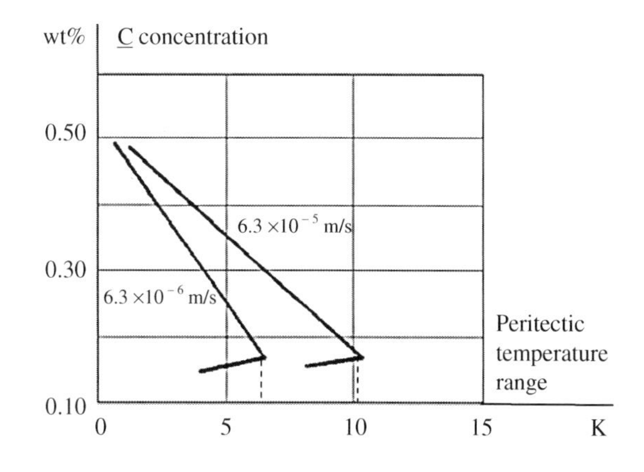

Figure 11.17 Peritectic temperature range in Fe–C alloys as a function of the carbon concentration C with the pulling rate as a parameter. The temperature gradient is 6.0×10^3 K/m.

Figure 11.17 shows that the peritectic transformations in Fe–C alloys are very rapid and completed at maximally 6–7 K and 10–11 K, respectively, below the peritectic temperature, depending on cooling rate. A high cooling rate prolongs the peritectic transformation in agreement with Equation (3′) in Example 11.2. The shape of the curves in Figure 11.17 is a result of the change in volume fraction of primarily precipitated ferrite with the initial carbon content.

11.3.2 Reactions and Transformations in Fe–M Systems

In the past section we have discussed peritectic reactions and transformations in the Fe–C system. Carbon is dissolved interstitially in the steel. In the present section an analogous discussion will be performed for systems with substitutionally dissolved alloying elements such as for example the Fe-Ni and Fe-Mn systems.

There is little information available of peritectic reactions in this type of binary systems. However, experiments made on stainless steels and on Fe-Ni-S alloys indicate that primary ferrite dendrites are surrounded by a layer of austenite. The austenite layer becomes thicker with decreasing temperature because the it grows both at the melt side and close to the primary ferrite dendrites.

Unlike $\underline{C}$, which is dissolved interstitially, the diffusion rates of substitutionally dissolved elements in austenite are *very low*. Only very small amounts of alloying elements from the liquid can pass through the austenite layer to the ferrite. Hence, the model used in Section 11.3.1 cannot be used. We must develop a new theoretical model for the present case.

The amount of ferrite precipitated before the start of the peritec-tic reaction can roughly be estimated by the lever rule as a result of the high diffusivity of carbon in ferrite. At the start of the peritectic reaction, an austenite layer is formed at the interface between ferrite and liquid. It will completely surround the ferrite and insulate it from the liquid. During the subsequent transformation, the austenite will grow at both boundaries of the γ layer.

The growth rate depends strongly on the diffusion coefficient. For substititionally dissolved alloying elements in austenite (FCC) such as Ni and Mn, the diffusion coefficient is normally of the order $10^{-13}\,\text{m}^2/\text{s}$ at a temperature close to the melting point. In these cases, the growth rate will be very low and the time for the peritectic transformation will be extremely long. The transformation can therefore be negligible in comparison with the precipitation of γ phase directly from the liquid.

In addition to the effect of the growth of the γ layer the influence of the phase diagram must be taken into account. The influence of the phase diagram is most striking when the diffusion rate of the metal atoms through the γ layer is low. This effect will be discussed more in detail below.

Theory of Peritectic Transformations in the Fe-M Systems

The Figures 11.18 and 11.19 show two types of phase diagrams in which the slopes of the equilibrium lines between the δ and γ phases are different. We will study the concentration profile during the peritectic transformation at a continuous cooling process. To emphasize the effect of the phase diagrams in the two cases we assume that the diffusion rates in the δ phase and the liquid are very high, which gives no concentration gradients in these phases. The low diffusion rate in the γ phase causes a concentration gradient in the γ region in both cases.

The concentration profiles are drawn for three different temperatures in either case: before the start of the peritectic reaction, at temperature T_P and at two lower temperatures T_1 and T_2. In the first case (Figure 11.18) the thickness of the γ layer will increase at both interfaces δ/γ and γ/L. In the second case (Figure 11.19) the γ layer will increase at the interface γ/L and decrease at the δ/γ interface. Below we will derive a quantitative expression for the growth rate of the γ layer.

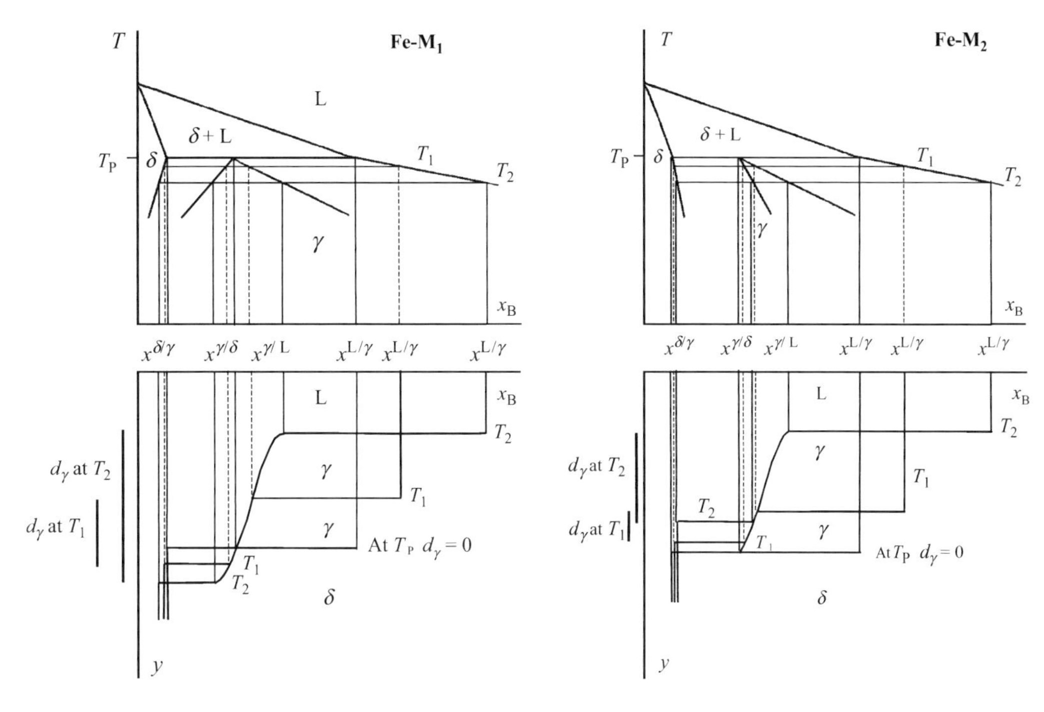

Figure 11.18 Peritectic transformation during continuous cooling in a system with a low diffusion rate in the γ phase. The volume fraction (thickness) of the γ phase increases with decreasing temperature.
Upper figure: Phase diagram.
Lower figure: Concentration distribution at three different temperatures.

Figure 11.19 Peritectic transformation during continuous cooling in a system with a low diffusion rate in the γ phase. The volume fraction (thickness) of the γ phase increases with decreasing temperature but it is smaller in Figure 11.19 than in Figure 11.18.
Upper figure: Phase diagram.
Lower figure: Concentration distribution at three different temperatures.

In order to find a more accurate expression for the total growth rate during a normal cooling process than Equation (11.10) on page 689 we have to rewrite Equations (11.8) and (11.9) on page 689:

$$A\frac{\mathrm{d}d_{\gamma/\delta}}{V_{\mathrm{m}}^{\gamma}\mathrm{d}t}\left(x^{\gamma/\delta}-x^{\delta/\gamma}\right)=A\frac{D_{\gamma}}{V_{\mathrm{m}}^{\gamma}}\left(\frac{\mathrm{d}x^{\gamma/\delta}}{\mathrm{d}y}\right)_{y=d_{\delta/\gamma}}+A\frac{D_{\gamma}}{V_{\mathrm{m}}^{\gamma}}\left(\frac{\mathrm{d}x^{\delta/\gamma}}{\mathrm{d}y}\right)_{y=d_{\delta/\gamma}}\tag{11.24}$$

Added amount of B atoms in the new volume $Add\,\gamma/\delta$, which have diffused from the γ phase into the interface γ/δ per unit time.

Amount of B atoms which diffuse from the interface γ/δ into the δ phase per unit time.

$$A\frac{\mathrm{d}d_{\gamma/\mathrm{L}}}{V_{\mathrm{m}}^{\gamma}\cdot\mathrm{d}t}\left(x^{\mathrm{L}/\gamma}-x^{\gamma/\mathrm{L}}\right)=A\frac{D_{\gamma}}{V_{\mathrm{m}}^{\gamma}}\left(\frac{\mathrm{d}x^{\gamma/\mathrm{L}}}{\mathrm{d}y}\right)_{y=d_{\mathrm{L}/\gamma}}+A\frac{D_{\gamma}}{V_{\mathrm{m}}^{\gamma}}\left(\frac{\mathrm{d}x^{\mathrm{L}/\gamma}}{\mathrm{d}y}\right)_{y=d_{\mathrm{L}/\gamma}}\tag{11.25}$$

Added amount of B atoms in the new volume $Add_{\gamma/\mathrm{L}}$, which have diffused from the interphase L/γ into the γ phase per unit time.

Amount of B atoms which diffuse from the interface L/γ into the liquid phase per unit time

The total growth rate of the γ layer is obtained if we solve the growth rates in Equations (11.24) and (11.25) and add them:

$$\begin{aligned}\frac{\mathrm{d}d_{\gamma}}{\mathrm{d}t}=\frac{\mathrm{d}d_{\gamma/\delta}}{\mathrm{d}t}+\frac{\mathrm{d}d_{\gamma/\mathrm{L}}}{\mathrm{d}t}&=\frac{D_{\gamma}}{x^{\gamma/\delta}-x^{\delta/\gamma}}\left(\frac{\mathrm{d}x^{\gamma/\delta}}{\mathrm{d}y}\right)_{y=d_{\delta/\gamma}}+\frac{D_{\gamma}}{x^{\gamma/\delta}-x^{\delta/\gamma}}\left(\frac{\mathrm{d}x^{\delta/\gamma}}{\mathrm{d}y}\right)_{y=d_{\delta/\gamma}}\\&+\frac{D_{\gamma}}{x^{\mathrm{L}/\gamma}-x^{\gamma/\mathrm{L}}}\left(\frac{\mathrm{d}x^{\gamma/\mathrm{L}}}{\mathrm{d}y}\right)_{y=d_{\mathrm{L}/\gamma}}+\frac{D_{\gamma}}{x^{\mathrm{L}/\gamma}-x^{\gamma/\mathrm{L}}}\left(\frac{\mathrm{d}x^{\mathrm{L}/\gamma}}{\mathrm{d}y}\right)_{y=d_{\mathrm{L}/\gamma}}\end{aligned}\tag{11.26}$$

Many times it is quite difficult to find exact solutions of Equation (11.26) but with the aid of numerical methods and Ficks's second law it can be solved.

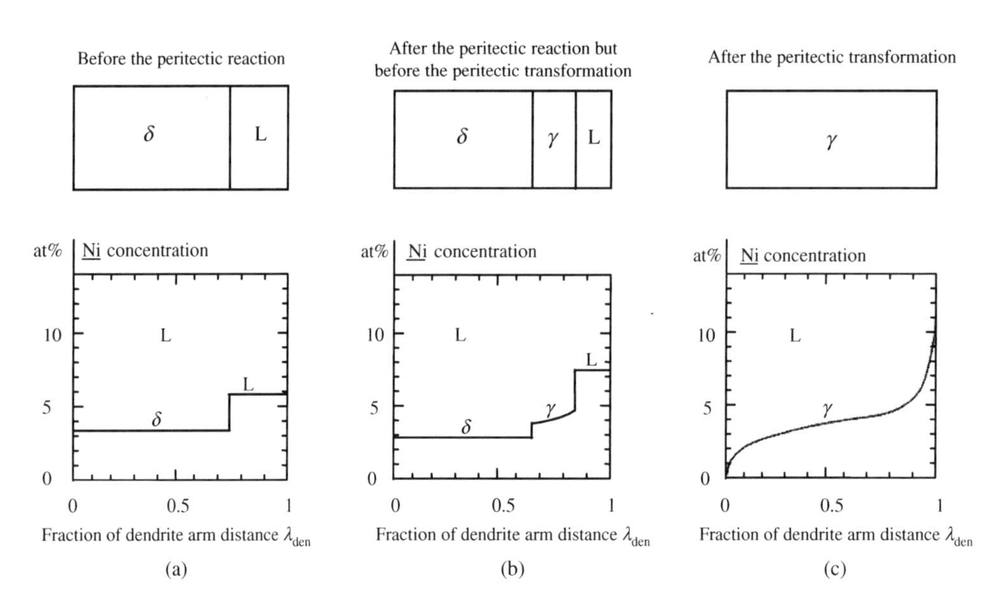

Figure 11.20 Ni distribution in an Fe–4 at %Ni alloy. The temperature gradient of 6×10^3 K/m was used. The solidification rate was 0.01 cm/min. The figures show the Ni distribution
a) before the peritectic reaction has started
b) immediately after the peritectic reaction but before the transformation
c) after the peritectic transformation.

Fredriksson and Stjerndahl [7] have performed calculations for an alloy with 4 wt% Ni at a given cooling rate. They introduced a simplification of Equation (11.26) and solved it by assuming that the diffusion in γ phase can be described in the same way as previously has been done in Chapter 9 for the segregation behaviour during the primary precipitation of austenite. The concentration of Ni in the austenite can, according to this approximation, be calculated with the aid of the Scheil' s modified equation (Equation (9.86) on page 538 in Chapter 9).

The result of the calculations is shown in Figure 11.20.

Experiments and calculations made on the Fe-Ni-Cr system show a similar curve to that in Figure 11.20. This is in accordance with the fact that the solubility of nickel in austenite is higher than that in ferrite and lower than the solubility in the liquid.

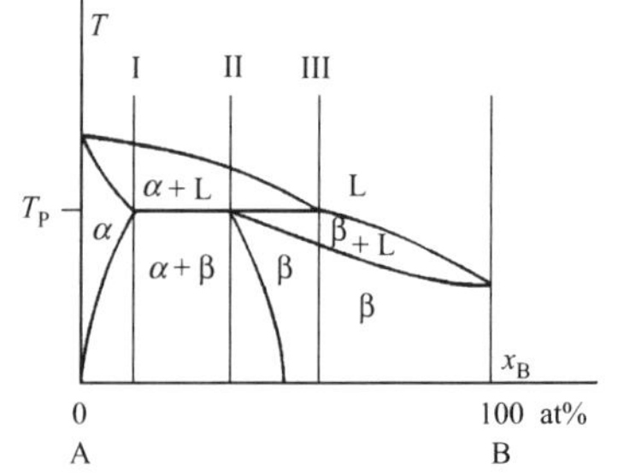

Figure 11.1 Principle sketch of a peritectic reaction in the phase diagram of a binary alloy. In iron-based alloys α is ferrite (δ iron) and β is austenite (γ iron).

11.4 Metastable Reactions in Iron-Base Alloys

Metastable Precipitation of β Phase at Peritectic Reactions

In many investigations of peritectic iron-base systems one has observed that in alloys with a composition to the left of line III in Figure 11.1, the γ phase is precipitated as a *primary* phase. This is the case for many iron-base alloys, for example stainless steel alloys and in alloyed carbon steels. The change from primary ferrite to primary metastable austenite can be achieved simply by changing the cooling rate. We will choose the Fe-Ni system as an example.

Below we will analyze the possibility of forming a primary γ phase that is not stable at composition and temperature, according to the phase diagram.

In Figure 11.21 the metastable extensions of the liquidus line L/γ and the solidus line γ/L are represented by broken lines. The melting point of γ iron is estimated to ~11 K below that of δ iron.

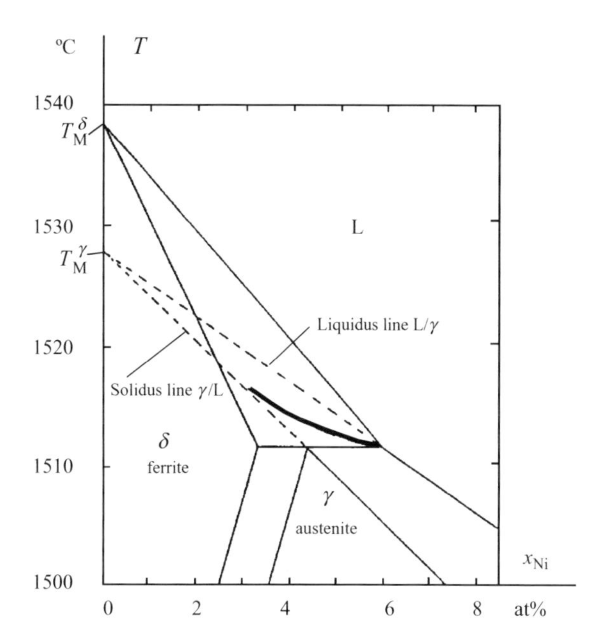

Figure 11.21 Part of the phase diagram of the binary Fe–Ni system. The thick bent line in the figure shows the temperature at which ferrite and austenite grow with the same rate as a function of the Ni concentration of the alloy. The partition coefficients can be derived from the figure: $k_{Ni}^{\delta} \approx 0.56$ $k_{Ni}^{\gamma} \approx 0.75$.

Figure 11.21 shows that there is a considerable difference in the partition coefficient $k = x^s/x^L$ between ferrite and liquid on one hand and austenite and liquid on the other.

The growth rate of austenite dendrites was calculated and compared with the growth rate of ferrite dendrites. The growth rates were calculated for an isothermal system. The alloy composition was chosen to the left of the peritectic composition of the melt (line III in Figure 11.43) and calculations of the growth rates were carried out at different undercoolings. The calculations are omitted here but the results will be reported.

Fredriksson [8] has found that only ferrite can form at low undercoolings. Austenite cannot form until the temperature of the liquid is lower than the metastable equilibrium extension of its metastable liquidus line at the given Ni concentration. Ferrite always forms at temperatures above the liquidus line of austenite – the driving force for precipitation of ferrite is larger than that of austenite. Hence, ferrite dominates the solidification process at these temperatures. As soon as the temperature of the liquid is lower than that, indicated by the liquidus line of austenite, austenite starts to precipitate.

Below a certain critical line in the phase diagram austenite has the highest growth rate and it may dominate although the temperature still is above the peritectic temperature. The reason for this is the fact that the partition coefficient k_γ of austenite is larger than that of ferrite. As $(x_{Ni}^{L/\delta} - x_{Ni}^{\delta/L}) > (x_{Ni}^{L/\gamma} - x_{Ni}^{\gamma/L})$ (Figure 11.22) a smaller amount of Ni is carried away from the solidification front when the melt solidifies as austenite instead of ferrite. For this reason the growth rate of austenite can be equal to or even larger than that of ferrite below a certain critical temperature. The critical temperature where the growth rates of austenite and ferrite are equal is indicated as a function of composition by the bent fat line in Figure 11.21.

It should be emphasized here that the partition coefficients of austenite and ferrite very well can have opposite relationships for different alloying elements in some ternary alloys. Precipitation of primary ferrite will then be promoted instead of primary austenite. Primary ferrite will be favoured by a high cooling rate owing to change of partition coefficient, caused by other alloying elements.

In addition, it should also be mentioned that an aligned eutectic lamellar-like structure can be formed in peritectic systems. This type of structure is formed when

1. primary ferrite and austenite have the same growth rates
2. the sum of the volume fractions of the two solid phases equals one.

The growth process will then occur as a growth of primary ferrite and austenite side by side in this special case.

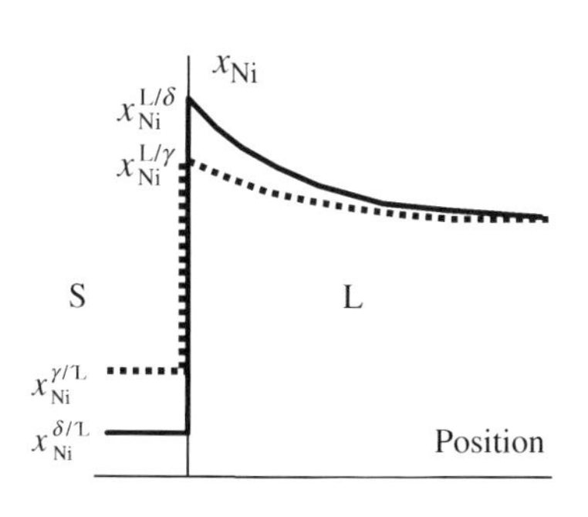

Figure 11.22 Concentration profile at the solidification front.

Transformation of δ Iron into γ Iron without a Diffusion-Controlled Peritectic Transformation

As an example of a metastable precipitation of austenite instead of formation of ferrite by a peritectic reaction followed by a transformation into austenite, we will discuss and interpret some results obtained in connection with a study of the peritectic reaction of a commercial multicomponent alloy Sanbar 64, developed at the Sandvik Corporation, Sweden.

The composition (wt%) of the alloy is as follows Fe,0.22%C,0.25%Si,0.65%Mn, 1.3%Cr,2.6% Ni,0.25%Mo and less than 0.025% P and S. The phase diagram of this alloy is not known. To interpret the mechanism of the peritectic reaction in the alloy we will use the phase diagram in Figure 11.21 as an approximation because Ni is the main alloying element.

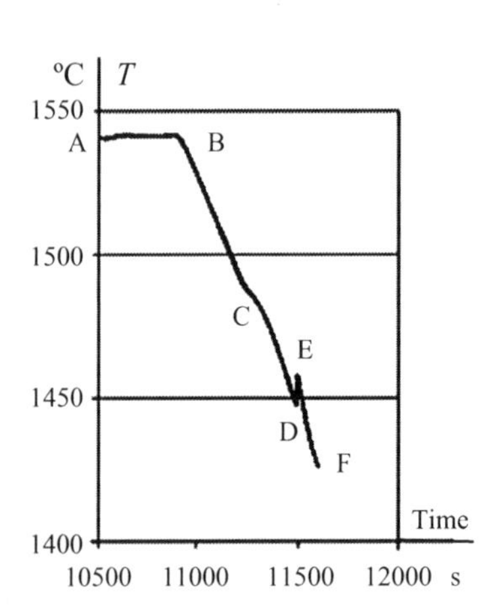

Figure 11.23 Typical cooling curve obtained for Sanbar 64 steel with the aid of DTA. Reproduced from Springer © 2004.

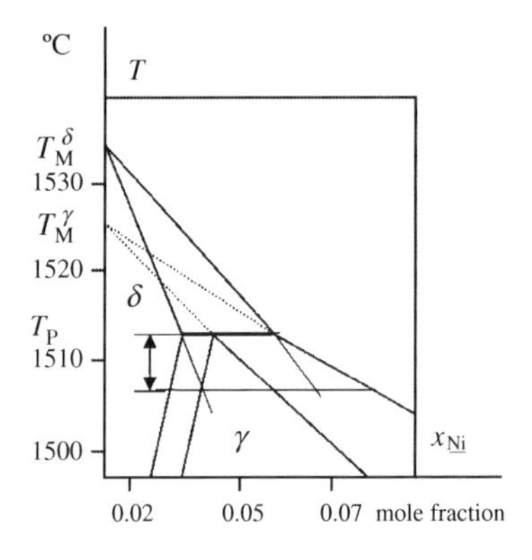

Figure 11.24 Part of the phase diagram of the system Fe–Ni.
The peritectic temperature $T_P = 1513\,^\circ C$. The undercooling is marked in the figure.

Interpretation of Experimental Evidence of Isothermal Cooling Experiments with Sanbar 64

Figure 11.23 shows an accurate cooling curves which was obtained at a cooling experiment with the aid of DTA, (differential thermal analysis, page 395 in Chapter 7), when the steel cools from the liquidus temperature down to total solidification. Based on this curve the following interpretation of the mechanism of the peritectic reaction can be made.

During the time AB the melt is kept at a temperature above the liquidus temperature. Then the melt is cooled to the liquidus temperature 1490 °C. This temperature is achieved at time C, when the precipitation of primary δ-Fe starts. It is observed as a slight temperature change.

The precipitation of δ phase goes on during the time CD when the peritectic reaction temperature 1448 °C is reached at time D. The peritectic reaction starts at time D and lasts during a very short time DE. The δ-Fe is partly transformed into γ-Fe. This process is accompanied by the observed sudden and rapid temperature rise from 1448 °C to 1458 °C. It will be discussed below.

During the time EF the specimen is further cooled and the remaining melt solidifies as austenite. The solidification is complete at time F.

The phase diagram of the alloy Sanbar 64 is unknown but we can interpret the temperature changes with the aid of the binary phase diagram of Fe-Ni.

According to the phase diagram in Figure 11.24 the peritectic reaction occurs at 1513 °C in the system Fe-Ni. The temperature rise corresponds to the undercooling, which precedes the primary precipitation of δ-Fe. The undercooling in the Fe-Ni system is marked in Figure 11.24.

In the experiment with the alloy Sanbar 64 the corresponding sudden temperature rise occurred at 1448 °C and the corresponding undercooling was 10 °C.

The temperature rise of 10 °C can be interpreted as a fast and diffusionless transformation of δ-Fe into γ-Fe, different from the usual transformation of δ-Fe into γ-Fe, which follows a normal peritectic reaction.

When the undercooling is gone, owing to the temperature increase, the direct transformation of δ-Fe into γ-Fe stops and the normal peritectic transformation controls the further transformation of δ-Fe into γ-Fe during the cooling process.

11.5 Metatectic Reactions and Transformations

A metatectic reaction can be regarded as a mixture of a eutectic and a peritectic reaction. It is a common reaction but not very well studied. It occurs in iron-base alloys with high sulfur and/or oxygen concentrations and is also common in other alloys, such as Cu-Sn and Ag-Sn alloys.

Metatectic reaction = a reaction where a primarily precipitated phase decomposes into a new solid phase and a liquid phase. The reaction occurs at or below the metatectic temperature T_{meta}.

$$\alpha \rightarrow \gamma + L$$

In Figure 11.25 a schematic phase diagram with a metatectic reaction is shown. A comparison between Figure 11.25 and nearby Figure 11.1 shows that the phase diagram of a metatectic reaction, includes liquidus lines, which look like the peritectic lines.

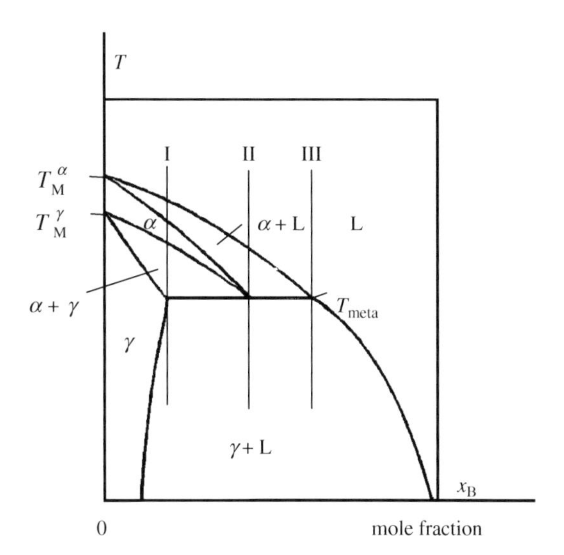

Figure 11.25 Schematic phase diagram of a binary alloy with a metatectic reaction. Depending on the appearance of the phase diagram the metatectic reaction can alternatively be mirror-inverted.

However, at a metatectic reaction the melt, the primary α phase and the two two-phase regions $\alpha + \gamma$ and $\alpha + \mathrm{L}$ form a characteristic bill, located *above* the metatectic temperature. At a peritectic reaction the secondary β phase occurs entirely at temperatures *below* the peritectic temperature.

Like the peritectic phase diagram in Figure 11.1, the metatectic phase diagram in Figure 11.25 can be divided into four regions by three lines. Under equilibrium conditions the following transformations will occur.

When an alloy with a composition to the left of line I cools, a primary precipitation of α phase occurs. At lower temperature a transformation of α phase into γ phase occurs, which goes on until the α phase is finished and only γ phase remains.

An alloy with a composition to the right of line III solidifies as γ crystals when the temperature is gradually lowered.

Alloys with compositions between lines II and III will primarily solidify as α crystals until the temperature has decreased to the metatectic temperature T_{meta}. At this temperature a metatectic reaction occurs and the α phase forms γ phase and liquid. When all α phase is finished the temperature decreases again and the liquid precipitates as γ phase until it has solidified completely.

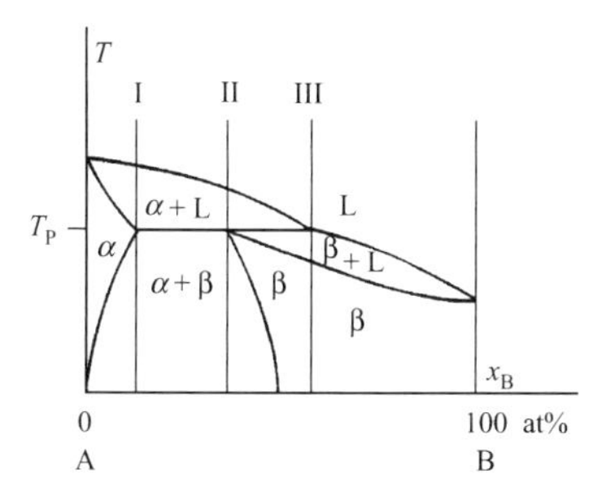

Figure 11.1 Phase diagram including a peritectic reaction.

11.5.1 Metatectic Reaction in the Cu–Sn System

The system Cu-Sn contains a metatectic reaction at 640 °C (Figure 11.26a). Nylén [9] performed a series of unidirectional solidification experiments with the Cu-45 wt%Sn alloy. Figure 11.26b shows one of his results. The solidification process was interrupted by quenching the sample.

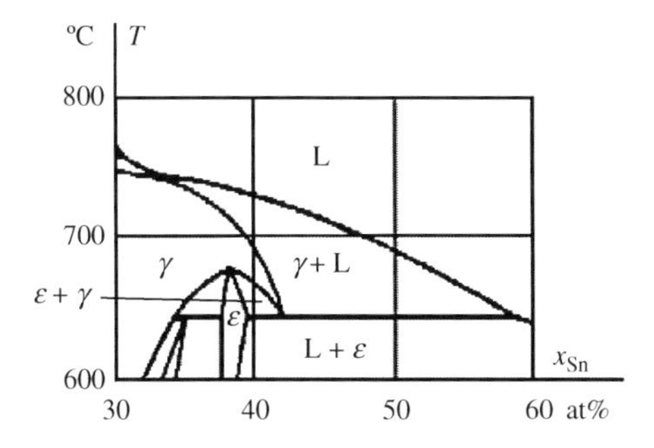

Figure 11.26a Phase diagram of the Cu–Sn system. Max Hansen © McGraw Hill 1958.

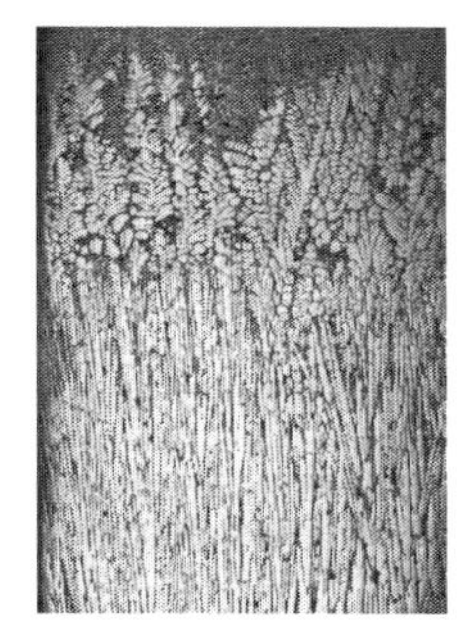

Figure 11.26b The upper part of the figure shows γ dendrites growing upwards. The metatectic reaction in the central part of the figure results in plate-like ε crystals in the lower part of the figure.

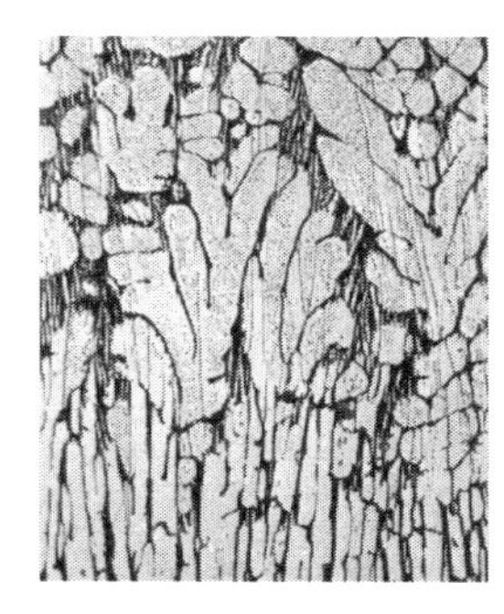

Figure 11.26c Enlargement of the central part of Figure 11.26b. The figure shows primary dendrites in the upper part of the figure and some of the secondary ε crystals in the lower part of the figure.

The dark upper part in Figure 11.26b is quenched liquid. The solidification front consists of a solid phase of γ dendrites. In the middle of the figure, the structure formed by a metatectic reaction is shown. An enlargement of this part of Figure 11.26b is shown in Figure 11.26c.

New plate-like ε crystals grow into the liquid and into the γ crystals. Between the γ crystals and the ε crystals a liquid layer is formed. The final structure shows no signs of the primarily formed γ crystals. One can conclude that the ε crystals in the lower part of Figure 11.26b are formed as a primary phase.

11.5.2 Metatectic Reactions in the Fe–O and Fe–S Systems

The most common systems known to contain a metatectic reaction are the Fe-O and Fe-S systems.

Both dissolved sulfur and oxygen are impurity elements in most steel alloys. Metatectic three-phase reactions are therefore very common in the case of precipitation of oxide and sulphide inclusions in steels.

11.5.3 Metatectic Reaction in the Fe–S System

Fredriksson and Stjerndahl [10] analyzed the metatectic reaction in the Fe-S system in a series of unidirectional solidification experiments on iron-base alloys containing sulfur. The processes were interruped by quenching the specimens.

Part of the binary phase diagram of Fe-S is given in Figure 11.27a. One of their ternary Fe-Ni-S alloys had a sulfur concentration x_0 equal to 0.07 at% S, marked in the tiny part of the binary phase diagram in Figure 11.27b. In addition the alloy contained 3 wt% Ni.

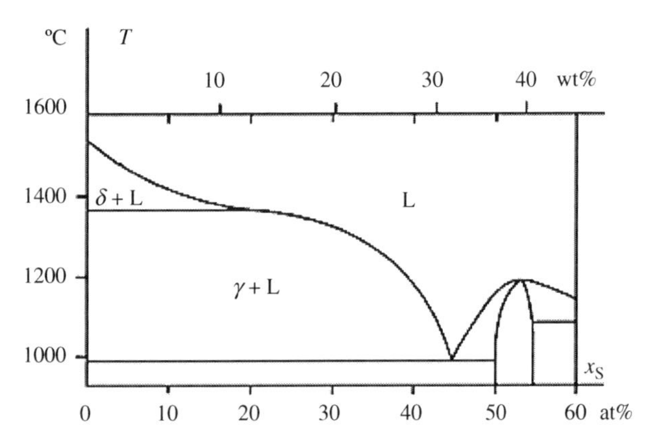

Figure 11.27a Part of the phase diagram of the Fe–S system. The details in Figure 11.27b can not be seen in Figure 11.27a, owing to difference of scales. From Binary Alloy Phase Diagrams Vol. 1, Ed. Thaddeus B. Massalski, American Society for Metals 1986. Reprinted with permission of ASM International. All rights reserved. www.asminternational.org.

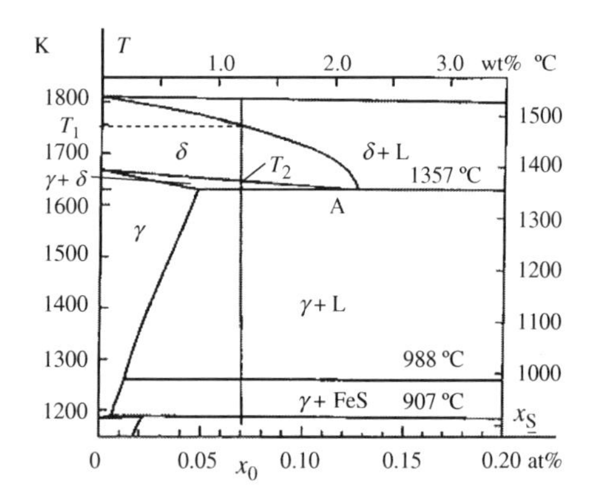

Figure 11.27b Enlargement of the phase diagram of the system Fe–S for very small concentrations of sulfur. Max Hansen © McGraw Hill 1958.

The metatectic reaction can be described as follows.

The solidification process starts with a primary precipitation of ferrite at temperature T_1. At the metatectic temperature T_{meta} (1357 °C) the reaction

$$\delta \rightarrow \gamma + L$$

occurs and results in partial remelting. The phase diagram in Figure 11.27b shows that the new liquid phase is sulfur-rich compared with the solid phases.

The structure, which was formed at the solidification experiment, is shown in Figure 11.28a. The specimen was moved in a constant temperature gradient field with the speed V_{pull}. The figure shows how the dark dendrites of ferrite grew into the white quenched melt. The specimen solidified completely as ferrite as is shown at the right-hand side of Figure 11.28a. All the sulfur was dissolved as there are no sulphide particles.

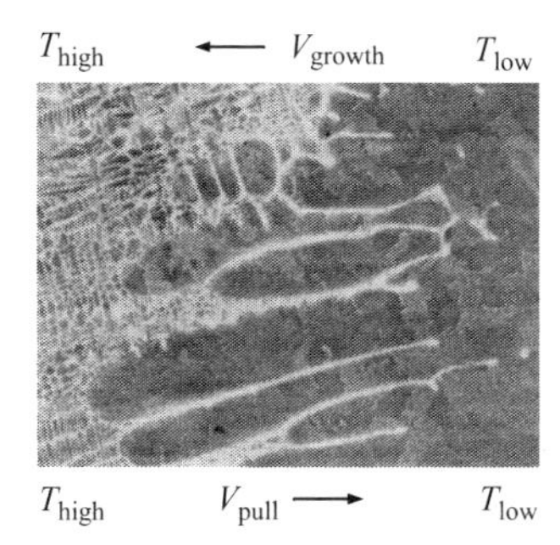

Figure 11.28a Dark ferrite dendrites, which grow into the white, quenched melt. The specimen is moved with the speed V_{pull}.

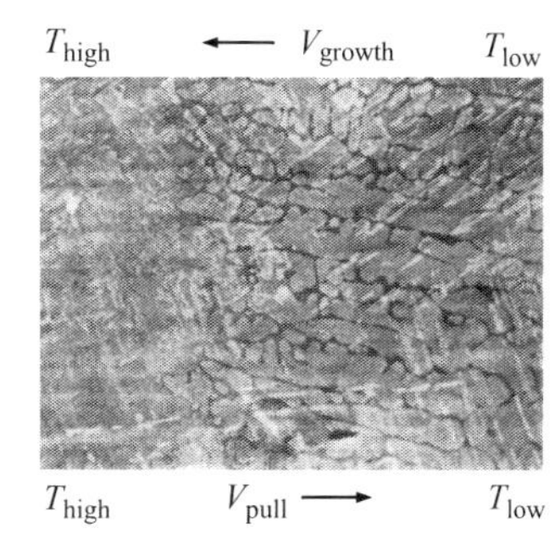

Figure 11.28b Later transformation of ferrite into austenite occurred. The figure shows ferrite dendrites on the left-hand side of the figure and dendrite-like austenite crystals on the right-hand side, that grew from right to left.

Figure 11.28b shows how the ferrite later was transformed into austenite. Only thin strings of dark ferrite remained at the right-hand side of the microphotograph at the time of quenching.

The solubility of sulfur is lower in austenite than in ferrite. For this reason undesired thin films of FeS are formed at the boundaries between ferrite and austenite.

Concentration Profile of Sulphur in the Phases before and after the Metatectic Reaction

The presence of sulfur in iron-base alloys is a severe practical problem. Above we mentioned that thin films of FeS are formed at the phase boundaries. Below we will discuss the reason for this and find the sulfur concentration distribution in the phases before and after the metatectic reaction.

The Figures 11.29 show the schematic phase diagram of the system Fe-S. The melting points T_M^{δ} and T_M^{γ} can be read to 1536 °C and 1392 °C, respectively, in Figure 11.27b.

An alloy with a sulfur concentration between lines II and III, which corresponds to a $\underline{S}$ concentration higher than that marked A in Figure 11.27b on page 706, has not solidified completely when the metatectic temperature is reached. The metatectic reaction shows many similarities with the better known peritectic reaction and transformation. The metatectic reaction occurs by nucleation and lateral growth of a secondary phase γ around a primary phase δ and by nucleation of a melt L at the interface between the solid primary phase δ and the secondary phase γ. The growing sulfur-rich liquid film is moving with the phase boundaries while the reaction proceeds.

In order to analyze the process more carefully we consider an iron-base alloy with a sulfur concentration in the region between the lines II and III in Figure 11.29a. When the alloy cools a primary δ phase is precipitated. The precipitation goes on until the metatectic temperature T_{meta} is reached. Figure 11.29a shows the sulfur distribution when the temperature equals the metatectic temperature but *before* the metatectic reaction has started. So far there is no secondary γ phase and no remelting.

The metatectic reaction starts at temperature T_{meta}. To obtain a reasonable rate of the metatectic reaction, the temperature is lowered to T, which gives a minor undercooling $(T_{meta} - T)$. Then γ phase precipitates. The concentration of $\underline{S}$ in the δ layer increases along the bent curve and stops at the value determined by the temperature T (Figure 11.29b). At temperature T a thin liquid layer is formed between the δ and γ phases.

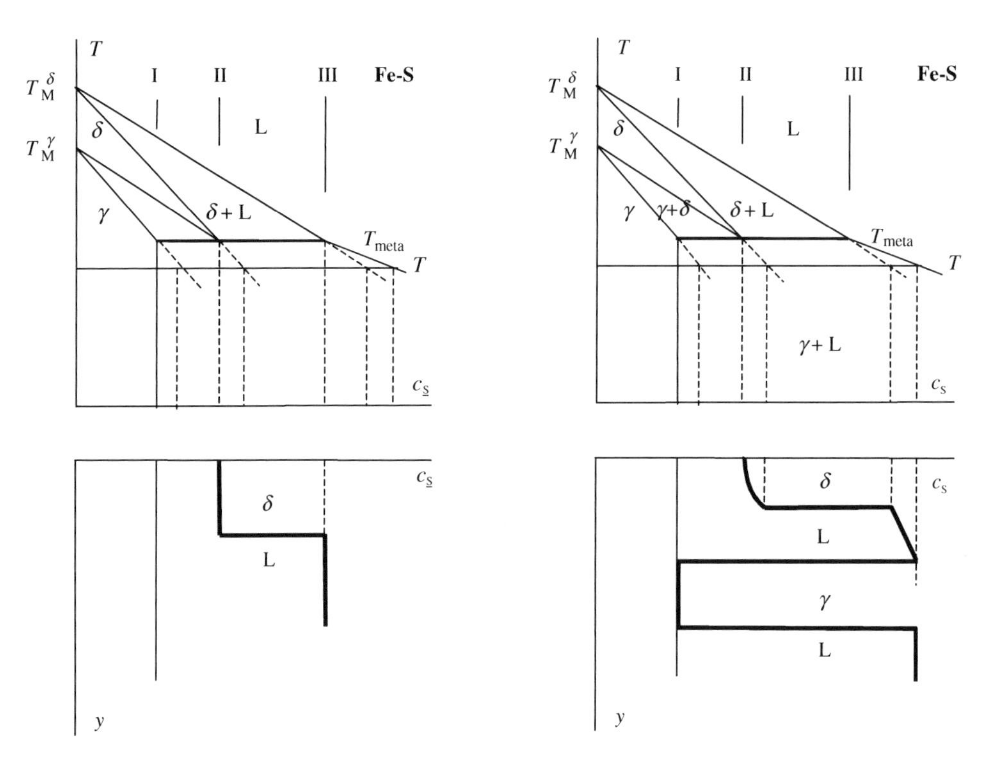

Figure 11.29a Upper figure: Schematic phase diagram of the Fe–S system. Lower figure: Concentration distribution of S in the δ phase and in the melt *before* the metatectic reaction at the temperature T_{meta}.

Figure 11.29b Transformation of ferrite into austenite *during* a metatectic reaction in the system Fe–S.
Upper figure: Schematic phase diagram of the Fe–S system.
Lower figure: Concentration distribution of S *during* the metatectic reaction, from temperature T_{meta} to the constant temperature T.

The metatectic reaction occurs mainly at temperature T. Figure 11.29b shows the concentration distribution of S at temperature T when the metatectic reaction $\delta \rightarrow \gamma + \text{L}$ occurs. It goes on until all δ phase is finished and a layer of γ phase is formed. The figure shows that the S concentration in the liquid, formed during the metatectic reaction, is considerably higher than the S concentration in the γ phase.

The primary δ phase contains less than the equilibrium value of the S concentration. The difference in S concentration between the melt and the δ phase results in diffusion of S atoms from and through the melt to the shrinking δ phase. The γ phase has a low concentration of S when it grows at the expence of the δ phase. The volume fraction of L between δ and γ will increase.

When all δ phase is dissolved a small drop of L may be found in the central parts of the dendrites and we obtain liquid on both sides the γ phase.

11.6 Microsegregation in Iron-Base Alloys

The solidification of iron-base alloys has been of research interest since the production of steel ingots started at the middle of the nineteenth century. Since then much attention has been paid to fundamental understanding of the formation and coarseness of dendrite structures and the related microsegregation phenomena. This topic was treated in Chapter 9.

Understanding of the formation of microsegregation in iron-base alloys is very important because they influence not only the final properties of the materials but also macrosegregations in ingots. Here we will discuss the influence of various alloying elements on peritectic reactions in iron-base alloys and their microsegregation behaviour.

11.6.1 *Laws and Concepts of Microsegregation*

Scheil's Equation

Microsegregation has been treated briefly in this book. On page 441 in Chapter 8 and page 537 in Chapter 9 we discussed Scheil's equation, which gives the concentration x^s of the alloying element in the solid as a function of the solidification fraction f

$$x^s = kx_0^L(1-f)^{-(1-k)} \tag{11.27}$$

Scheil's equation is valid when the diffusion in the solid phase is zero or very small.

Scheil's Modified Equation

Scheil's equation indicates that the concentration approaches infinity at the end of the solidification process ($f \approx 1$). The concentration gradient becomes very high and the driving force for back diffusion will increase. In this case the back diffusion in the solid phase cannot be neglected.

When back diffusion (page 538 in Chapter 9) is taken into consideration it is necessary to modify Scheil's equation. Scheil's modified equation for dendritic growth can be written as

$$x^s = kx_0^L\left(1-\frac{f}{1+D_s\dfrac{4\theta}{\lambda_{den}^2}}\right)^{-(1-k)} \tag{11.28}$$

Segregation Ratio

It is often useful to characterize the segregation tendency of an alloying element by a single quantity. Many methods to do so have been reported in literature. One such quantity used to describe segregation behaviour of an alloying element is the *segregation ratio S* or *degree of segregation.*

S = the ratio of the highest and the lowest values of the concentration of the alloying element in a dendrite crystal aggregate.

$$S = \frac{x_{max}^s}{x_{min}^s} \tag{11.29}$$

In the absence of microsegreration the maximum and minimum values are equal and S is equal to 1. The more pronounced the microsegregation is the higher will be the value of S.

Segregation Ratio in a Dendrite Arm as a Function of the Solidification and Homogenization Time

If the alloying elements are interstitially dissolved or dissolved in structures where the atoms are more loosely packed than in the FCC or HCP structures the diffusion rates may be so high that neither Scheil's equation nor Scheil's modified equation are valid. The reason is that the concentration in the central parts of a dendrite arm changes gradually during the solidification process. We want a quantitative relationship between the segregation ratio and the solidification time for this case.

Calculation of S

For this purpose we will study a number of growing parallel dendrite arms in an alloy melt at the constant distance λ_{den} from each other. The arms, which can be approximated by parallel plates, start to solidify at time $t = 0$. The solidification is completed at time $t = \theta$. The width of each dendrite arm is then λ_{den} and all the melt has solidified.

Owing to microsegregation the solute distribution varies within the dendrite arms. The solute concentration is a function of both time and position. If we introduce a coordinate system with $y = 0$

at the centre of a dendrite arm and apply the general Equation (9.85) on page 537 in Chapter 9, the solute concentration can be written as

$$x^{s}(y,t) = x^{s/L} - (x^{s/L} - x_{min})e^{-\frac{\pi^2 D_s}{f_0^2 \lambda_{den}^2} t} \cos\frac{\pi y}{f_0 \lambda_{den}} \qquad (11.30)$$

where f_0 is the fraction solidified at time t and $-\lambda_{den}/2 \leq y \leq \lambda_{den}/2$.

The solute concentration has maxima at the last solidified parts in the interdendritic regions, and minima at the centres of the dendrite arms, i.e. at the first solidified parts. The solute concentration in the solid at the liquid/solid interface is identical with the maximum concentration x^{s}_{max} in Equation (11.29).

Calculation of x^{s}_{max}

Homogenization of the dendrite arms occurs during the whole solidification process. Hence, the homogenization time t_{hom} is equal to the solidification time θ.

As solidification and homogenization merely involves replacements of solute atoms between the dendrites and the melt within the dendrites, *the average solute concentration in a completely solidified dendrite must be equal to the initial solute concentration in the melt* x_0^L (Figure 11.30). This condition can be described mathematically by forming the average solute concentration by inserting $t = t_{hom}$, $f_0 = 1$, $y = \lambda_{den}/2$ and $x_{min} = kx_0^L$ into Equation (11.30) and set it equal to x_0^L.

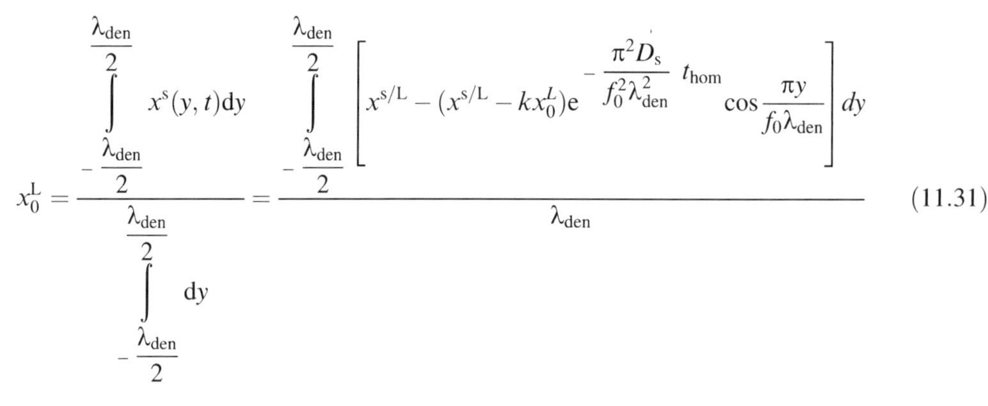

$$x_0^L = \frac{\int_{-\frac{\lambda_{den}}{2}}^{\frac{\lambda_{den}}{2}} x^{s}(y,t)dy}{\int_{-\frac{\lambda_{den}}{2}}^{\frac{\lambda_{den}}{2}} dy} = \frac{\int_{-\frac{\lambda_{den}}{2}}^{\frac{\lambda_{den}}{2}} \left[x^{s/L} - (x^{s/L} - kx_0^L)e^{-\frac{\pi^2 D_s}{f_0^2 \lambda_{den}^2} t_{hom}} \cos\frac{\pi y}{f_0 \lambda_{den}} \right] dy}{\lambda_{den}} \qquad (11.31)$$

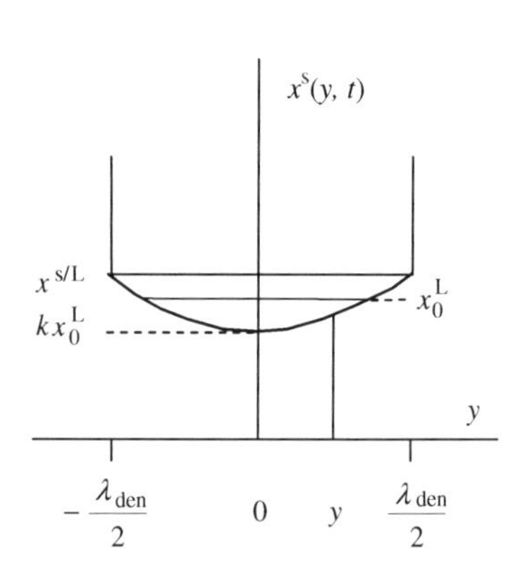

Figure 11.30 Solute distribution $x^{s \to L}(y, t)$ as a function of position in a solidified dendrite arm.

where

t_{hom} = homogenization time = local solidification time
y = distance from $y = 0$
λ_{den} = distance between primary plates (dendrites)
$x^{s/L}$ = solute concentration in the solid phase at the interface
x_0^L = initial concentration of solute in the melt
$k = x^{s}/x^{L}$ = partition coefficient of the solute between the solid and the liquid
D_s = diffusion coefficient of the solute in the solid.

Integration of Equation (11.31) gives

$$x_0^L \lambda_{den} = x^{s/L}\lambda_{den} - (x^{s/L} - kx_0^L)e^{-\frac{\pi^2 D_s}{\lambda_{den}^2} t_{hom}} \left[\sin\frac{\pi y}{\lambda_{den}} \right]_{-\lambda_{den}/2}^{\lambda_{den}/2} \frac{\lambda_{den}}{\pi} \qquad (11.32)$$

We solve $x^{s/L}$ from Equation (11.32) and obtain

$$x^{s}_{max} = x^{s/L} = x_0^L \frac{1 - \frac{2k}{\pi} e^{-\frac{\pi^2 D_s}{\lambda_{den}^2} t_{hom}}}{1 - \frac{2}{\pi} e^{-\frac{\pi^2 D_s}{\lambda_{den}^2} t_{hom}}} \qquad (11.33)$$

The concentration distribution of the solute in a dendrite during the solidification is treated on pages X101-116 in Chapter 9. The solute distribution at the beginning and near end of the solidification process is shown in Figure 11.31.

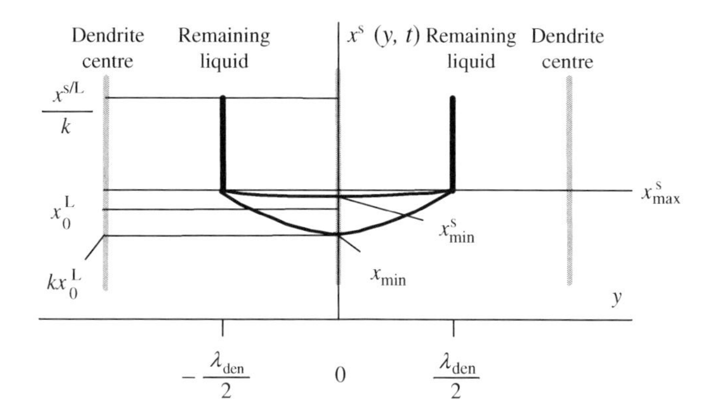

Figure 11.31 Distribution of the alloying element in a dendrite at the end of the solidification process. The solute concentration at the centre of the solidifying dendrite increases gradually with increasing solidification fraction.

Calculation of x_{min}^s

The *minimum value* x_{min}^s is obtained if we introduce $t = t_{hom}$, $f_0 = 1$, $y = 0$. and $x_{min} = kx_0^L$ into Equation (11.30).

$$x_{min}^s = x^{s/L} - (x^{s/L} - kx_0^L)e^{-\frac{\pi^2 D_s}{\lambda_{den}^2} t_{hom}} \cos 0 \tag{11.34}$$

Calculation of S as a Function of the Homogenization Time

The expression for $x^{s/L}$ in Equation (11.33), that is equal to x_{max}^s, is introduced into Equation (11.34). and the segregation ratio x_{max}^s / x_{min}^s can be solved. After reduction we obtain

$$S = \frac{1}{1 - \left(1 - \dfrac{k\left(1 - \dfrac{2}{\pi} e^{-\frac{\pi^2 D_s}{\lambda_{den}^2} t_{hom}}\right)}{1 - \dfrac{2k}{\pi} e^{-\frac{\pi^2 D_s}{\lambda_{den}^2} t_{hom}}}\right) e^{-\frac{\pi^2 D_s}{\lambda_{den}^2} t_{hom}}} \tag{11.35}$$

Segregation Ratio in a Dendrite Arm as a Function of the Solidification Interval and the Cooling Rate

Calculation of the segregation ratio S with the aid of Equation (11.35) requires knowledge of the local solidification time θ, which is equal to the homogenization time, and the primary plate spacing λ_{den}. They are related to the cooling rate ($-\mathrm{d}T/\mathrm{d}t$) by the following relationships:

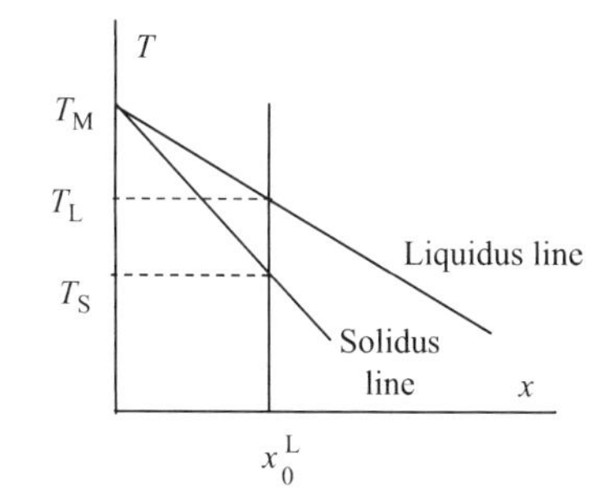

Figure 11.32 Solidification interval $T_L - T_s$.

$$t = \theta = \frac{\Delta T}{-\dfrac{\mathrm{d}T}{\mathrm{d}t}} \tag{11.36}$$

$$\lambda_{den} = A\left(-\frac{\mathrm{d}T}{\mathrm{d}t}\right)^n \tag{11.37}$$

ΔT is the solidification interval of the alloy. It is defined as (Figure 11.32)

$$\Delta T = T_L - T_S \tag{11.38}$$

where
T_L = the temperature when the dendrite starts to solidify
T_S = the temperature when the solidification is completed.

A and n in Equation (11.37) are experimentally determined constants.

If the expressions in Equations (11.36) and (11.37) are introduced into Equation (11.35) the final expression of S is found, which depends on the *phase diagram* (k) (see Table 11.1), the *material constants* (especially D_s) and the *cooling rate*:

Table 11.1 Partition coefficients of some common alloying elements in iron-base alloys

Element	k_δ	k_γ
Carbon	0.17	0.41
Chromium	~ 0.86	~ 0.7
Molybdenum	~ 0.68	~ 0.52
Nickel	~ 0.56	~ 0.75
Niobium	~ 0.18	~ 0.12
Phosphorus	~ 0.15	~ 0.09
Vanadium	~ 0.56	~ 0.50

$$S = \frac{1}{1 - \left(1 - \dfrac{k\left(1 - \dfrac{2}{\pi} e^{-\dfrac{\pi^2 D_s \Delta T}{A^2\left(-\dfrac{dT}{dt}\right)^{1+2n}}}\right)}{1 - \dfrac{2k}{\pi} e^{-\dfrac{\pi^2 D_s \Delta T}{A^2 \cdot \left(-\dfrac{dT}{dt}\right)^{1+2n}}}}\right) e^{-\dfrac{\pi^2 D_s \Delta T}{A^2\left(-\dfrac{dT}{dt}\right)^{1+2n}}}} \tag{11.39}$$

With the aid of Equation (11.39) the segregation ratio S can be calculated for different alloying elements in iron-base alloys. These topics will be discussed below.

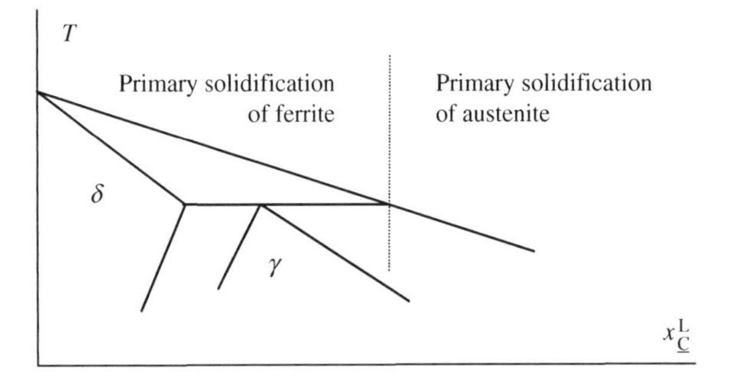

Figure 11.33 Simplified phase diagram of the system Fe–C. Solidification regions for precipitation of ferrite and austenite.

Iron-base alloys solidify either as ferrite (δ-iron) or austenite (γ-iron). Figure 11.33 shows that the carbon concentration controls the primary solidification mode.

11.6.2 Segregation Behaviour of Alloying Elements during Primary Precipitation of Ferrite

In many alloys with primary ferrite solidification the segregation ratio S is found to be equal or very close to 1. This means that there is no or very little microsegregation in the structure. This may depend on either of two reasons

1. The partition coefficient between liquid and solid for the alloying element is equal to one.
2. The diffusion rates of the alloying elements in ferrite are fairly high, which causes back diffusion in the solid phase.

Phase diagrams of iron-base alloys show, however, that their partition coefficients very seldom are equal to 1. Results from quenching experiments indicate that there always exists some segregation of alloying elements in the dendrite structure during the solidification process. We also know that diffusion rates of both substitutionally dissolved atoms and especially interstitially dissolved atoms in ferrite (BCC structure) are quite high.

These circumstances and the fact that the segregation ratio S increases with increasing cooling rate, indicate that the back diffusion of alloying elements during the solidification process is of major importance for the microsegregation in primary precipitated ferrite dendrites.

Segregation Ratio of Ferrite with Different Alloying Elements

With the aid of Equation (11.39) above the segregation ratio S of three different systems, Fe–Cr, Fe-Nb and Fe–C, have been calculated and the result is shown in Figure 11.34.

The k values of C, Cr and Nb were calculated from the phase diagrams of the three systems Fe–C, Fe–Cr and Fe-Nb. They are given in Table 11.1 on page 712.

Fe–Cr and Fe-Nb are chosen to show the influence of the phase diagram, i.e. the partition coefficient k. Cr and Nb have nearly the same diffusion rates but different k values. Table 11.1 shows that Cr has a much higher k value than Nb. Figure 11.34 shows that

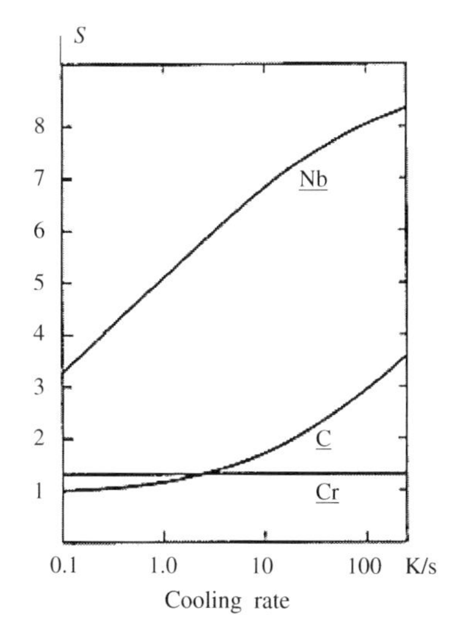

Figure 11.34 Segregation ratio S of Nb, C and Cr as functions of the solidification rate at ferrite solidification. The temperature gradient was chosen as 6000 K/m in the calculations.

- The smaller the k value is, the larger will be the segregation ratio S and vice versa.

Nb and C have roughly the same k values but there is a large difference between their diffusion rates. The Nb atoms, that are substitutionally dissolved, have a low diffusion coefficient. The small C atoms are interstitially dissolved and have a high diffusion coefficient. Hence, C has a much higher D value than Nb. Figure 11.34 shows that

- The smaller the D_s value is the larger will be the segregation ratio S and vice versa.

Figure 11.34 shows that the segregation ratio S is equal or very close to 1 for C and Cr unless the cooling rate is very high. There is always a considerable segregation for Nb. However, the results of the calculations describe the segregated structure at the very end of the solidification process. During subsequent cooling, the alloy will homogenize further and there will probably be very little segregation left at room temperature.

Figure 11.34 also confirms the assumption that the lever rule gives a fairly good approximation of the concentration of the alloying elements in a ferrite structure during the solidification process at low cooling rates. The lever rule is optimal for interstitially dissolved elements such as C, N and H.

- The larger the cooling rate is the larger will be the segregation ratio and vice versa.

11.6.3 Segregation Behaviour of Alloying Elements during Primary Precipitation of Austenite

Most alloys with an austenite solidification pattern show a type of microsegregation quite different from alloys with ferrite solidification. The main reason for this is the much lower diffusion rates in austenite structures (FCC) compared to those in the ferrite structures (BCC). For this reason higher segregation ratios can be expected for austenite than for ferrite structures.

Owing to the low diffusion rate, it is not necessary to take the full effect from the back diffusion of alloying elements in the melt into the previously solidified material, into account. A segregation model, which describes the segregation pattern during the solidification process quite well, is the modified Scheil's segregation Equation (11.28) on page 709.

Equation (11.28) is often written as

$$x^{L} = x_0^{L}\left(1 - \frac{f}{1+B}\right)^{-(1-k)} \tag{11.40}$$

where

$$B = \frac{4kD_s\theta}{\lambda_{den}^2} \tag{11.41}$$

The parameter B is a correction term to the basic Sheil's segregation equation. Its origin is the back diffusion.

Equation (11.40) shows that as long as B is much smaller than 1 the distribution of alloying elements can fairly well be described by the original Scheil's segregation equation. When the expression $(1-f)$ decreases towards the end of the solidification process, the influence of the back diffusion increases. This is an effect of the very steep concentration gradients at the end of the solidification process between the previously solidified material and the remaining liquid.

Equation (11.40) can be used to estimate the segregation ratio of alloying elements in an austenitic structure. By use of the definition of the segregation ratio [Equation (11.29) page 709] and application of $x^s = kx^L$ at $f = 1$ (maximum segregation) and at $f = 0$ (minimum segregation) we obtain the following relationship

$$S = \left(\frac{B}{1+B}\right)^{-(1-k)} \tag{11.42}$$

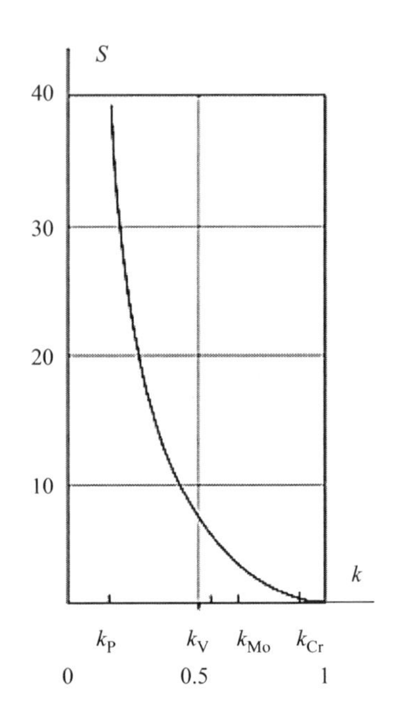

Figure 11.35 The segregation ratio as a function of the partition coefficient. Theoretical calculations of the segregation ratio of steel alloys as a function of the partition constant k.

The partition coefficient between austenite and liquid varies often very much for different elements. From Equation (11.42) and use of the definition of the parameter B [Equation (11.41)] it is possible to calculate the segregation ratio S as a function of the partition coefficient k. Figure 11.35 shows the result of such calculations. As expected S increases very much with decreasing k.

It is also worth while to note that there is practically no segregation for k values between 0.90 and 1.00. In Figure 11.35 the k values of some common alloying elements in steel are marked. As expected, phosphorus shows the largest segregation ratio.

By inserting the definition of the parameter B [Equation (11.41) on page 714] and the relationships for the solidification time and dendrite distance [Equations (11.37) and (11.38) on page 711] into Equation (11.42), we obtain another expression, which describes the influence of the phase diagram (k) and the cooling rate on the segregation ratio in an austenitic structure.

$$S = \left(\frac{1 + \dfrac{4kD_s\Delta T}{A^2\left(-\dfrac{dT}{dt}\right)^{1+2n}}}{\dfrac{4kD_s\Delta T}{A^2\left(-\dfrac{dT}{dt}\right)^{1+2n}}}\right)^{(1-k)} \tag{11.43}$$

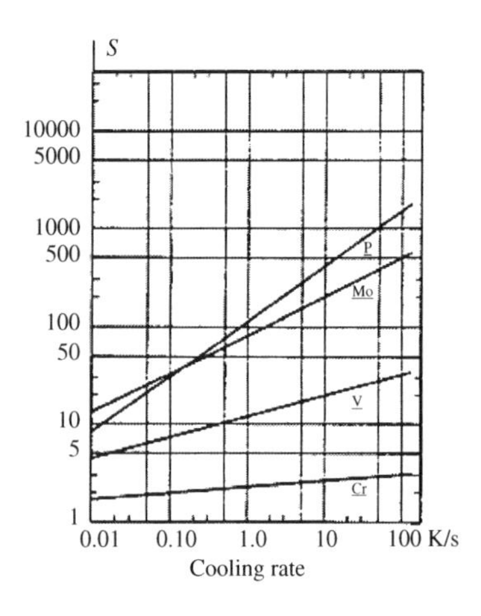

Figure 11.36 Segregation ratios of P, Mo, V and Cr as a function of solidification rate at austenite solidification. The temperature gradient was chosen as 6000 K/m in the calculations. At the calculations the same values of the constants n and A (page 712) as in Figure (11.34) have been used. The diffusion rate of Mo is much smaller than that of V.

Equation (11.43) can be used to describe the effect of the partition coefficient and also the cooling rate when n and A are constants. The segregation ratios of four different alloying systems Fe-P, Fe-Mo, Fe-V and Fe–Cr have been calculated. The same A and n values (page 713) as in Figure 11.34 were used. The partition coefficients are given in Table 11.1 on page 712.

Segregation Ratio of Austenite with Different Alloying Elements

The results of the calculations are shown in Figure 11.36 as a function of the cooling rate. The figure shows that

- The smaller the k value is the larger will be the segregation ratio S and vice versa.
- The smaller the D_s value is the larger will be the segregation ratio S and vice versa.
- The larger the cooling rate is the larger will be the segregation ratio and vice versa.

The discussion above and other experimental evidence show that the same statements concerning the segregation behaviour of alloying elements during precipitation are valid for both ferrite and austenite. The only difference is that the diffusion constant D_s for substitutionally dissolved alloying elements is about 100 times smaller in austenite than that in ferrite. The consequence is that

- The segregation ratio increases strikingly with increasing cooling rate in austenite much more than in ferrite.

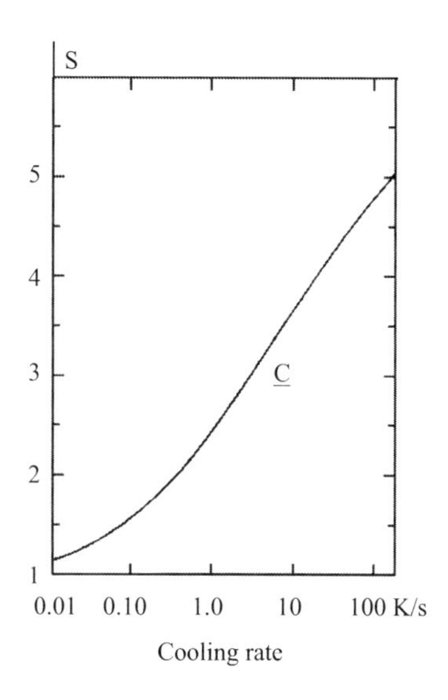

Figure 11.37 Segregation ratio of $\underline{C}$ as a function of the cooling rate at austenite solidification. The temperature gradient was 6000 K/m. In the calculations the same values of the constants n and A (page 713) as in Figure 11.34 have been used. Reproduced with permission from Scripta Metallurgica, 3, 4, 249-252 Copyright (1969) Elsevier.

Carbon Segregation in Austenite

The segregation of carbon during primary precipitation of ausenite is much smaller than that of other alloying elements owing to the higher diffusion rate of interstitially dissolved elements compared to substitutionals.

Owing to the high diffusion rate of carbon, Equation (11.43) can *not* be used to calculate the segregation ratio because the carbon concentration at the centre of a dendrite arm is not constant in time during the solidification process. The segregation ratio of carbon can, however, as before be estimated with the aid of Equation (11.39) on page 711. The result of these calculations is shown in Figure 11.37.

Figure 11.37 shows that the lever rule, in this case too, is a fairly good approximation of the concentration of interstitially dissolved alloying elements, such as $\underline{C}$, $\underline{N}$ and $\underline{H}$, in an austenitic structure during a solidification process at low cooling rate.

11.6.4 *Influence of Carbon on Segregation Behaviour in the Fe–Cr–C Systems*

Very few investigations have been done on how the interaction between different alloying elements influences the segregation behaviour in multicomponent systems. Addition of a third element to a binary alloy often affects the segregation behaviour of the original solute significantly.

An interesting and commercially important three-component iron-base alloy is the Fe-Cr-C system. The segregation ratio of chromium as a function of the carbon concentration has been determined experimentally in an Fe-1.5%Cr steel and some results are shown in Figure 11.38. A binary Fe–Cr alloy does not segregate at all (ferrite solidification). Additions of carbon increase the segregation ratio up to as much as five times at 1.5% $\underline{C}$. The figure shows that a further increase of the carbon concentration causes a decrease of the segregation ratio.

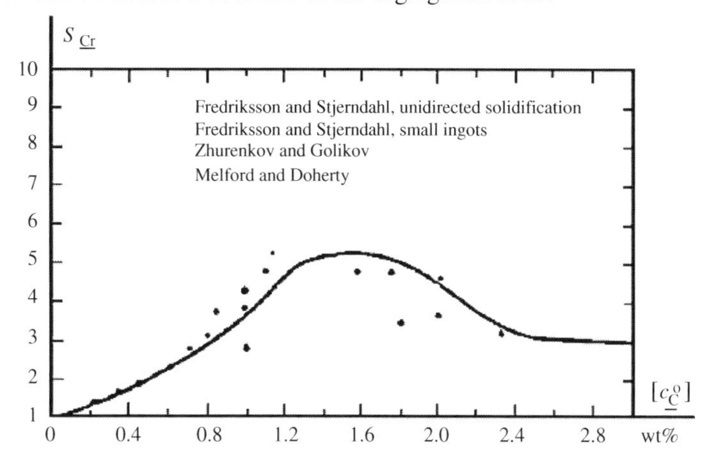

Figure 11.38 Segregation ratio S of $\underline{Cr}$ as a function of the carbon concentration. Several independent measurements are included in the figure. Max Hansen © McGraw Hill 1958.

The reasons for the influence of carbon on the segregation ratio according to Figure 11.38 have been discussed thoroughly in the literature. The increase of segregation ratio at low carbon contents can be explained by a decrease of the partition coefficient of chromium owing to a thermodynamic interaction between chromium and carbon in the solid (austenite) and liquid phases.

The decrease of the segregation ratio $S_{\underline{Cr}}$ at higher carbon concentrations than 2.5 wt% can be explained with the aid of the ternary phase diagram of Fe–C-Cr. One of the most detailed studies of this system has been performed by Bungart and his co-workers [11]. Their phase diagram is given in Figure 11.39 on the next page.

The line E_1 E_2 E_3 in Figure 11.39 represents eutectic valleys (Section 10.5 in Chapter 10) where eutectic surfaces intersect. The valleys have lower temperatures than the surrounding eutectic surfaces and correspond to equilibrium (lowest possible energy).

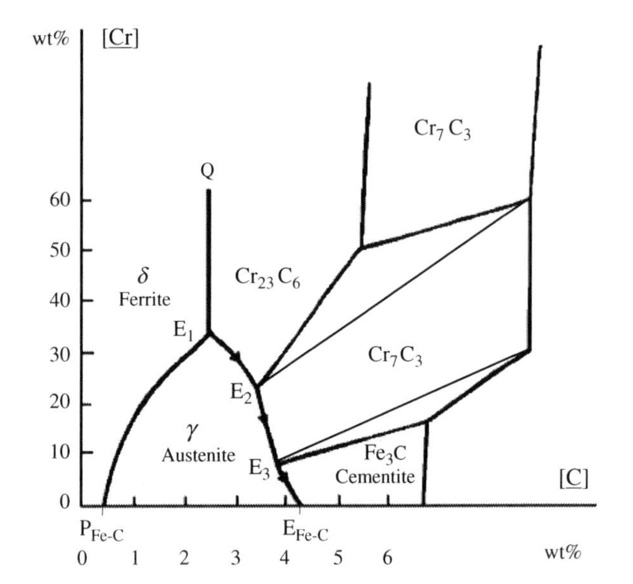

Figure 11.39 Simplified corner of the ternary phase diagram of the system Fe–Cr–C according to Bungart. Redrawn from Bundardt, Karl von et al.: Untersuchungen über den Aufbau des Systems Fe-Cr-C. Archiv für das Eisenhüttenwesen 29, 1958, 193–203.

The phase diagram shows that the Cr concentration is lower in a C-rich alloy than in a C-depleted alloy with the same Cr concentration. The reason is the negative slope of the line E_1 E_2 E_3 E_{Fe-C}.

11.7 Transitions between Peritectic and Eutectic Reactions in Iron-Base Alloys

It is well known that pure iron exists in two allotropic crystal forms: ferrite or δ-Fe (BCC structure) and austenite or γ-iron (FCC structure). Various alloying elements influence the extensions of these two crystal forms.

Some elements, for instance C, Ni and Mn, widen the temperature range for stable austenite. Such elements are called *austenite stabilizers.* Other elements, for example Cr, Si and Mg, narrow the temperature range for austenite. They are called *ferrite stabilizers.*

The property, which decides the stabilizing character of the elements, is the heat of mixing of the element in the two phases. An element with a positive heat of mixing in ferrite and a negative heat of mixing in austenite is an austenite stabilizer.

Austenite-stabilizing elements have in most cases phase diagrams, which are characterized by one or several peritectic reactions. A typical example of such an element is nickel. The phase diagram in Figure 11.40 shows that the temperature region of austenite is large.

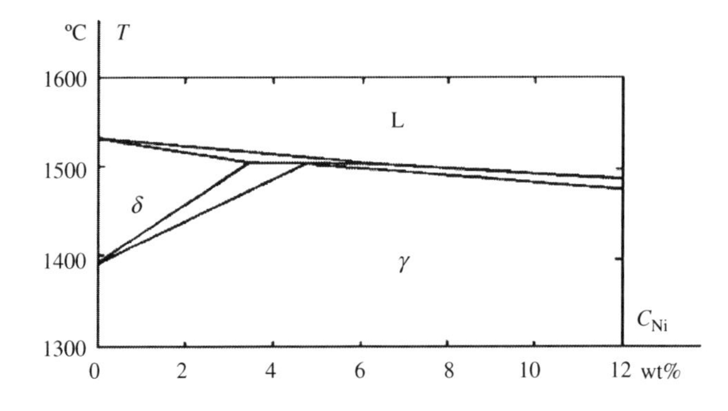

Figure 11.40 Part of the phase diagram of the system Fe–Ni.

Ferrite-stabilizing elements in steel give rise to a so-called γ loop in their phase diagrams. A typical example of such an element is chromium (Figure 11.41). The phase diagram shows that the γ region is surpressed to a comparatively small part of the phase diagram.

Many alloys contain a mixture of austenite stabilizing and ferrite stabilizing elements in order to satisfy special demands on the alloys. Stainless steel alloys are examples of this combination

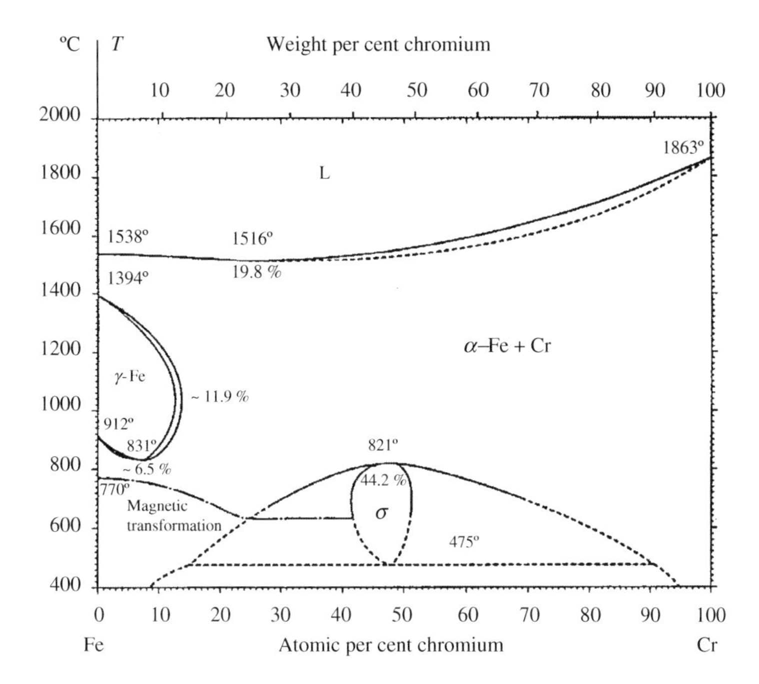

Figure 11.41 Phase diagram of the Fe–Cr system.

of alloying elements. The main alloying elements in stainless steel are in most cases nickel or chromium. The effect of a mixture of these alloying elements will be discussed below.

11.7.1 Transition from Peritectic to Eutectic Reaction in Stainless Steel

Figures 11.40 and 11.41 show that the phase diagram of the binary Fe-Ni system contains a peritectic reaction while the phase diagram of the binary Fe–Cr system contains a so-called γ loop.

There is a considerable difference in segregation behaviour between alloys, which solidify with a peritectic reaction and alloys, which solidify with a eutectic reaction. The microsegregation is less pronounced during a eutectic reaction than during a peritectic reaction.

In this section we will discuss the possibility that the peritectic reaction may change into a eutectic reaction with an increasing concentration of chromium.

When Cr is added to a binary Fe-Ni alloy the peritectic three-phase line is changed into a three-phase triangle. The higher the Cr concentration is the lower will be the required temperature for the three-phase reaction. The liquid composition as well as the two solid phases δ and γ change their compositions.

This is shown in Figures 11.42a and b. The figures show two isothermal sections of the ternary Fe-Ni-Cr phase diagrams, one at 1480 °C and the other at 1440 °C. The liquidus and solidus lines are drawn together with the three-phase tie lines. The tie lines form two triangles ABC and A′B′C′,

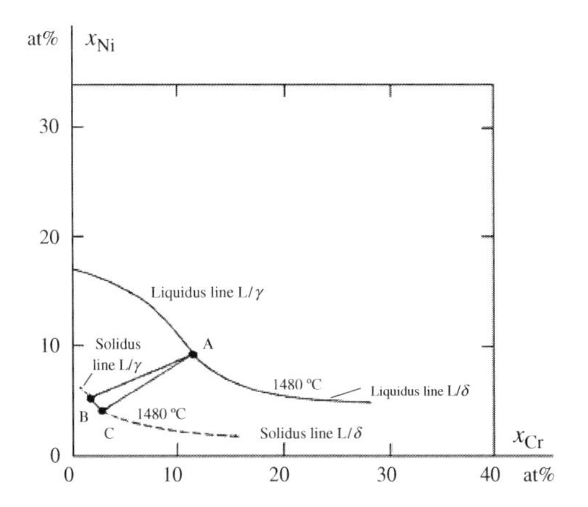

Figure 11.42a Isothermal cross-section of the phase diagram Fe–Cr-C at 1480 °C.

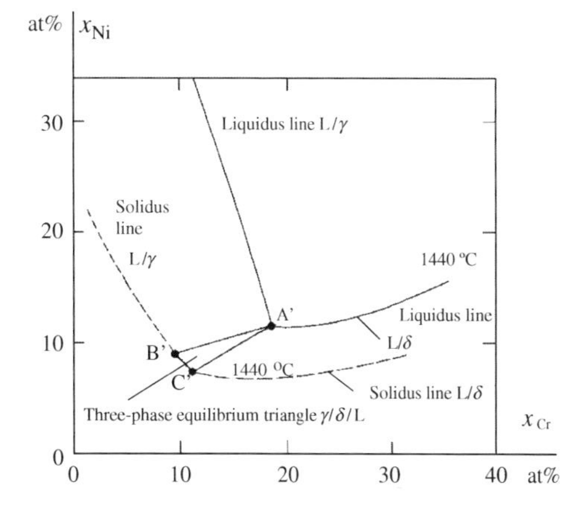

Figure 11.42b Isothermal cross-section of the phase diagram Fe–Cr-C at 1440 °C.

respectively, in the two isothermal cross-sections. The points A and A′ show the composition of the liquid in equilibrium with ferrite and austenite. The composition of these two phases, in equilibrium with the liquid, are given by the points B, B′ and C, C′, respectively.

If we gather all isothermal sections and project the points ABC into one projection, we obtain lines, which show the extension of the three-phase composition. Figure 11.42c shows such a projection. The figure shows that the three-phase line of the liquid composition moves more and more towards higher $\underline{\text{Ni}}$ concentrations with increasing $\underline{\text{Cr}}$ concentration.

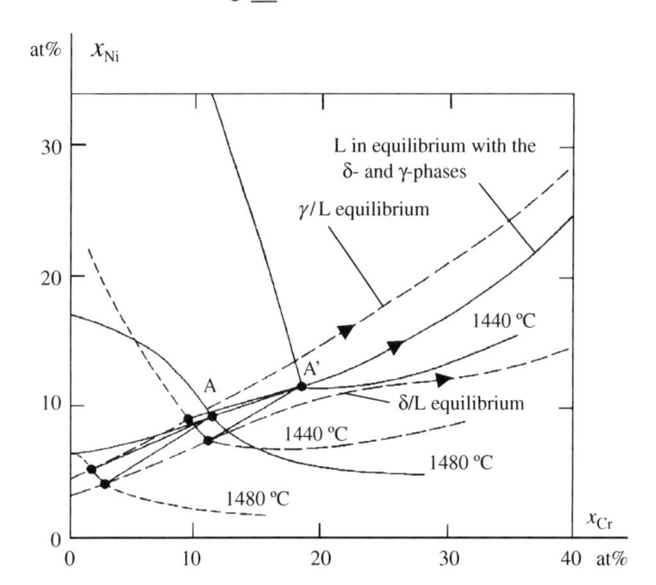

Figure 11.42c Projection of the ternary phase diagram of Fe–Cr–Ni. The temperature axis is perpendicular to the plane of the paper.

The directions of the tie lines AB and AC shows the direction in which the liquid composition changes during the solidification process. Now we can analyze more carefully what happens with a liquid of composition A in Figures 11.42a and 11.42c.

If we lower the temperature and assume that both δ and γ are formed, we can conclude from the phase diagram that the liquid composition moves towards the austenite liquidus surface. This will result in a pure peritectic reaction and transformation.

If we perform the same analysis for a liquid with the composition A′ and assume that δ phase is precipitated primarily, we find that the liquid composition moves towards the austenite liquidus surface. The result is that austenite starts to precipitate. In that case, the liquid composition moves towards the ferrite surface and ferrite starts to precipitate. We obtain a simultaneous precipitation of δ and γ when the temperature decreases. This phenomenon can be characterized as a eutectic reaction.

Three-Phase Fe-Ni-Cr Line

The phase diagram in Figure 11.42c shows that the three-phase reaction changes from a pure peritectic reaction to a pure eutectic reaction owing to the influence of a second alloying element (compare Section 11.6.4 on page 715). The concentration changes of Ni and Cr in the liquid during the peritectic reaction can be calculated. These changes can be found by taking the derivative of Scheil's Equation (11.27) on page 709 with respect to f_s twice, once for Ni and once for Cr. The calculations have been performed in the box and the resulting expression is

$$\frac{\mathrm{d}x^{L}_{\underline{\text{Ni}}}}{\mathrm{d}x^{L}_{\underline{\text{Cr}}}} = \frac{x^{L}_{\underline{\text{Ni}}}\left(1 - k^{\gamma/L}_{\underline{\text{Ni}}}\right)}{x^{L}_{\underline{\text{Cr}}}\left(1 - k^{\gamma/L}_{\underline{\text{Cr}}}\right)} \tag{11.44}$$

where

$x^{L}_{\underline{\text{Ni}},\underline{\text{Cr}}}$ = mole fraction of Ni and Cr, respectively, in the liquid

$k^{\gamma/L}_{\underline{\text{Ni}},\underline{\text{Cr}}}$ = partition coefficient of Ni and Cr, respectively, between the solid γ phase and the liquid.

Slope of the Three-Phase Fe-Ni-Cr Line

Scheil's Equation for $\underline{\text{Ni}}$ can be written as

$$x^{\text{s}}_{\underline{\text{Ni}}} = k^{\text{L}/\gamma}_{\underline{\text{Ni}}} x_0 (1 - f_{\text{s}})^{-(1 - k^{\text{L}/\gamma}_{\underline{\text{Ni}}})} \quad (1')$$

The e-logarithm of Equation (1′) is:

$$\ln x^{\text{s}}_{\underline{\text{Ni}}} = \ln k^{\text{L}/\gamma}_{\underline{\text{Ni}}} x_0 - (1 - k^{\text{L}/\gamma}_{\underline{\text{Ni}}}) \ln(1 - f_s) \quad (2')$$

Equation (2′) is differentiated:

$$\frac{\text{d}x^{\text{s}}_{\underline{\text{Ni}}}}{x^{\text{s}}_{\underline{\text{Ni}}}} = \frac{-(1 - k^{\text{L}/\gamma}_{\underline{\text{Ni}}})\text{d}(1 - f_{\text{s}})}{1 - f_{\text{s}}} \quad (3')$$

A completely analogous equation is obtained for $\underline{\text{Cr}}$:

$$\frac{\text{d}x^{\text{s}}_{\underline{\text{Cr}}}}{x^{\text{s}}_{\underline{\text{Cr}}}} = \frac{-(1 - k^{\text{L}/\gamma}_{\underline{\text{Cr}}})\text{d}(1 - f_{\text{s}})}{1 - f_{\text{s}}} \quad (4')$$

The solidification fraction is the same in both cases. Equation (3′) is divided by Equation (4′). The factors $\text{d}(1 - f_{\text{s}})$ and $(1 - f_{\text{s}})$ cancel and we obtain

$$\frac{\text{d}x^{\text{L}}_{\underline{\text{Ni}}}}{\text{d}x^{\text{L}}_{\underline{\text{Cr}}}} = \frac{x^{\text{L}}_{\underline{\text{Ni}}}(1 - k^{\gamma/\text{L}}_{\underline{\text{Ni}}})}{x^{\text{L}}_{\underline{\text{Cr}}}(1 - k^{\gamma/\text{L}}_{\underline{\text{Cr}}})} \quad (5')$$

The ratio of $\text{d}x^{\text{L}}_{\underline{\text{Ni}}}$ and $\text{d}x^{\text{L}}_{\underline{\text{Cr}}}$ corresponds to the same direction as the slope of the tie line between austenite and liquid. If the slope of the tie line is *smaller* than the slope of the three-phase line, then ferrite must precipitate at the same time as austenite. This occurs in the phase diagram in Figure 11. NaNc when the concentrations are roughly 12 at% Ni and 19 at% Cr, i.e. at point A′.

This point is the transition point from a peritectic reaction at lower $\underline{\text{Cr}}$ concentration to a eutectic reaction at higher $\underline{\text{Cr}}$ concentration.

Summary

■ Peritectic Reactions and Transformations

Peritectic Reaction

A peritectic reaction is a reaction where a liquid phase L reacts with a primarily precipitated phase α and forms a new secondary phase β.

$\text{L} + \alpha \rightarrow \beta$

Peritectic Transformation

A peritectic transformation is the secondary process, which follows after a peritectic reaction when the α phase is transformed into β phase.

The transformation requires B atoms as the β phase has a higher concentration of solute atoms than the α phase. The B atoms diffuse through the β phase to the α phase.

$\alpha + \text{B atoms} \rightarrow \beta$

Peritectic Reactions

Peritectic Reaction of Type 1

Nucleation and growth of β crystals when the primary α phase, the liquid L and the β phase stay in contact with each other.

Growth rate: $$V_{\text{growth}} = \frac{Const}{\sigma_{\alpha\beta} + \sigma_{L\beta} - \sigma_{L\alpha}} (\Delta T)^4$$

The thickness of β layer depends on the surface tensions and the undercooling.

$$d_\beta^0 = C \frac{\sigma_{\alpha\beta} + \sigma_{L\beta} - \sigma_{L\alpha}}{\Delta T_{\text{surface}}}$$

Peritectic Reaction of Type 2

The β crystals are nucleated in the liquid. A liquid layer separates the growing secondary β phase from the shrinking primary α phase.

After nucleation the secondary β phase grows freely in the liquid. At the same time the primary α phase dissolves partly and forms β phase, owing to diffusion of B atoms through the melt.

Type 1 reactions are more common than reactions of type 2.

Peritectic reactions often occur in a cascade.

Peritectic Transformations

Every peritectic reaction is followed by a peritectic transformation when the primarily precipitated α phase will completely or partly be dissolved and replaced by β phase.

After the peritectic reaction the thickness of the β phase is d_β^0. During the peritectic transformation the β layer grows with time.

The growth of the β layer occurs with the aid of

- precipitation of β phase directly from the liquid on the β layer
- precipitation of β phase directly from the α phase owing to dissolution of the α phase, which occurs if the phase diagram has such a shape that dissolution is a spontaneous process (depends of the shape of the phase diagram)
- diffusion of B atoms through the β phase to the α phase, which results in the reaction

α + B atoms → β.

Thickness of the β Layer

Provided that the differences in densities between the phases are neglected (the molar volumes are approximately equal) V_{growth} can be written as

$$\frac{dd_\beta}{dt} = \frac{dd_{\beta/\alpha}}{dt} + \frac{dd_{\beta/L}}{dt} = \frac{D_\beta}{d_\beta}\left(x^{\beta/L} - x^{\beta/\alpha}\right)\left(\frac{1}{x^{\beta/\alpha} - x^{\alpha/\beta}} + \frac{1}{x^{L/\beta} - x^{\beta/L}}\right)$$

In addition to influence of the diffusion constant and the transformation temperature, which determines the concentrations, V_{growth} depends on the cooling rate and the shape of the phase diagram.

The growth rate is proportional to the cooling rate.

The slope of the boundary line between the (α + β) and β phases has a decisive influence on the stability of the α phase.

If the slope is *positive* the α phase will dissolve and be transformed into β phase when the temperature decreases.

On the other hand, if the boundary line has a *negative* slope the α phase will stabilize at the expense of the β phase when the temperature decreases.

In both cases, at constant temperature, the growth equation can be integrated and gives the thickness of the β layer, formed during the transformation

$$\frac{(d_\beta)^2}{2} = D_\beta\left(x^{\beta/\mathrm{L}} - x^{\beta/\alpha}\right)\left(\frac{1}{x^{\beta/\alpha} - x^{\alpha/\beta}} + \frac{1}{x^{\mathrm{L}/\beta} - x^{\beta/\mathrm{L}}}\right)t$$

Growth Rate of the β Layer

$$V_{\mathrm{growth}}^{\beta} = \frac{1}{G}\left(-\frac{\mathrm{d}T}{\mathrm{d}t}\right) \quad \text{where} \quad G = \frac{\mathrm{d}T}{\mathrm{d}z}$$

At unidirectional solidification:

$$V_{\mathrm{growth}}^{\beta} = \frac{\mathrm{d}d_\beta}{\mathrm{d}t} = \frac{\mathrm{d}d_\beta}{\mathrm{d}T}\frac{\mathrm{d}T}{\mathrm{d}z}\frac{\mathrm{d}z}{\mathrm{d}t} = \left(-\frac{\mathrm{d}d_\beta}{\mathrm{d}T}\right)G(-V_{\mathrm{pull}})$$

Peritectic reactions and transformations in the Fe–C and Fe-M systems are extensively discussed in the text.

■ Reactions and Transformations in the Fe–C System

At unidirectional solidification of low-carbon steel the peritectic transformation of primary δ phase into γ phase has been studied.

Thickness of the γ layer as a function of temperature:

$$(d_\gamma)^2 = \frac{D_{\underline{\mathrm{C}}}^{\gamma}}{V_{\mathrm{m}}^{\gamma}G(-V_{\mathrm{pull}})}\frac{1}{m_{\mathrm{eff}}}\left(\frac{V_{\mathrm{m}}^{\delta}}{x_{\underline{\mathrm{C}}}^{\gamma/\delta} - x_{\underline{\mathrm{C}}}^{\delta/\gamma}} + \frac{V_m^L}{x_{\underline{\mathrm{C}}}^{\mathrm{L}/\gamma} - x_{\underline{\mathrm{C}}}^{\gamma/\mathrm{L}}}\right)(T_{\mathrm{p}} - T)^2$$

Peritectic temperature range $= T_{\mathrm{P}} - T_{\mathrm{final}}$

Peritectic transformation distance:

$$L_\gamma = \sqrt{\frac{D_{\underline{\mathrm{C}}}^{\gamma}}{V_{\mathrm{m}}^{\gamma}G}\frac{1}{m_{\mathrm{eff}}}\left(\frac{V_{\mathrm{m}}^{\delta}}{x_{\underline{\mathrm{C}}}^{\gamma/\delta} - x_{\underline{\mathrm{C}}}^{\delta/\gamma}} + \frac{V_m^L}{x_{\underline{\mathrm{C}}}^{\mathrm{L}/\gamma} - x_{\underline{\mathrm{C}}}^{\gamma/\mathrm{L}}}\right)}\frac{T_{\mathrm{P}} - T_{\mathrm{final}}}{\sqrt{-V_{\mathrm{pull}}}}$$

■ Metastable Reactions in Iron-Base Alloys

In iron-base alloys the primarily precipitated phase is α phase or *ferrite* (δ) if the alloy has a solute composition to the left of line III. At a temperature below the peritectic temperature a peritectic reaction occurs and the α phase is transformed into β phase or *austenite* (γ). During the following peritectic transformation the ferrite dissolves and the final product is pure austenite.

However, for many iron-base alloys a primary precipitation of austenite at temperatures $> T_{\mathrm{P}}$ has been observed in alloys with compositions to the left of line III.

This contradicts the phase diagram and can be explained by the formation of a metastable γ phase, instead of the normal δ phase, at temperatures above the peritectic temperature.

Formation of a metastable γ phase is a result of the relative sizes of the partition coefficients and the concentration profile at the solidification front. Below the peritectic temperature the normal peritectic reaction and trans-formation of ferrite into austenite occurs.

■ Metatectic Reactions and Transformations

Metatectic Reaction

A metatectic reaction is a reaction where a primarily precipitated phase decomposes into a liquid phase and a new solid phase.

The reaction occurs at or above the metatectic temperature T_{meta}.

The phase diagram of a system with a metatectic reaction has a very characteristic appearance ('bird's beak'). Metatectic reactions in Cu-Sn and Fe-S are discussed in the text.

■ Microsegregation in Iron-Base Alloys

As a measure of the microsegregation the *segregation ratio* can be used.

Segregation Ratio

The ratio of the highest and the lowest values of the concentration of the alloying element in a dendrite crystal aggregate.

$$S = \frac{x^{s}_{max}}{x^{s}_{min}}$$

If there is no microsegregation $S = 1$, otherwise $S > 1$.

Calculation of the Segregation Ratio in a Dendrtie Arm

The segregation ratio in a dendrite arm can be derived with the aid of the solute concentration in a dendrite arm as a function of position and time. The back diffusion of the solute is taken into consideration.

$$S = \frac{1}{1 - \left(1 - \frac{k\left(1 - \frac{2}{\pi} e^{-\frac{\pi^2 D_s}{\lambda_{den}^2} t_{hom}}\right)}{1 - \frac{2k}{\pi} e^{-\frac{\pi^2 D_s}{\lambda_{den}^2} t_{hom}}}\right) e^{-\frac{\pi^2 D_s}{\lambda_{den}^2} t_{hom}}}$$

The homogenization is assumed to be complete when the solidification process is ended, i.e. $t_{hom} = \theta$.

The value of S depends strongly on the partition coefficient k, the diffusion constant D_s and the cooling rate or, in terms of the solidification interval and the cooling rate

$$S = \frac{1}{1 - \left(1 - \frac{k\left(1 - \frac{2}{\pi} e^{-\frac{\pi^2 D_s \Delta T}{A^2\left(-\frac{dT}{dt}\right)^{1+2n}}}\right)}{1 - \frac{2k}{\pi} e^{-\frac{\pi^2 D_s \Delta T}{A^2 \cdot \left(-\frac{dT}{dt}\right)^{1+2n}}}}\right) e^{-\frac{\pi^2 D_s \Delta T}{A^2\left(-\frac{dT}{dt}\right)^{1+2n}}}}$$

■ Segregation Behaviour of Alloying Elements during Primary Precipitation of Ferrite

In many alloys with primary ferrite solidification the segregation ratio S is found to be equal or very close to 1.

The diffusion rates of the alloying elements are quite high both of substitutionally dissolved atoms and especially of interstitially dissolved atoms in ferrite (BCC structure).

Influence of k, D_s and Cooling Rate on the Segregation Ratio

- The smaller the k value is, the larger will be the segregation ratio S and vice versa.
- The smaller the D_s value is, the larger will be the segregation ratio S and vice versa.
- The larger the cooling rate is, the larger will be the segregation ratio and vice versa.

■ Segregation Behaviour of Alloying Elements during Primary Precipitation of Austenite

Most alloys with an austenite solidification pattern show a microsegregation quite different from that in ferrite. The main reason for this is the much lower diffusion rate in the austenite structure (FCC) compared to the ferrite structure (BCC), which gives higher segregation ratios in austenite than ferrite structures.

Owing to the low diffusion rate of alloying elements in the previously solidified material it is not necessary to take the full effect from the back diffusion, into account. A segregation model, which quite well describes the segregation pattern in austenite, during the solidification process is the modified Scheils segregation equation

$$x^s = kx_0^L \left(1 - \frac{f}{1 + D_s \dfrac{4\theta}{\lambda_{den}^2}} \right)^{-(1-k)}$$

Influence of k, D_s and Cooling Rate on the Segregation Ratio

- The smaller the k value is, the larger will be the segregation ratio S and vice versa.
- The smaller the D_s value is, the larger will be the segregation ratio S and vice versa.
- The larger the cooling rate is, the larger will be the segregation ratio and vice versa.

These statements are valid for both ferrite and austenite. The only difference is that the diffusion constant D_s for substitutionally dissolved alloying elements is about 100 times smaller in austenite than in ferrite. The consequence is that

- The segregation ratio increases strikingly with increasing cooling rate in austenite, much more than in ferrite.

Carbon Segregation in Austenite

The segregation of carbon during primary precipitation of austenite is much smaller than that of other alloying elements.

The reason is that carbon is interstitially dissolved and has a much higher diffusion rate in the solid phase than substitutionally dissolver elements.

In the case of carbon the back diffusion has to be considered and Scheil's modified segregation equation is not valid. Instead the segregation relationship, valid for alloying elements in ferrite, has to be used.

■ Influence of Additional Elements on Segregation in Iron-Base Alloys

Addition of a third element to a binary alloy often affects the segregation behaviour of the original solute significantly.

The influence of carbon on the segregation of chromium in the system Fe–Cr-C is discussed as an example.

■ Transitions between Peritectic and Eutectic Reactions in Iron-Base Alloys

Steel alloys solidify either as austenite (γ) or ferrite (δ). The ferrite or austenite structure is a consequence of the influence of the alloying elements. They can be divided into two groups, austenite-stabilizing and ferrite-stabilizing elements.

Austenite-stabilizing elements in steel have phase diagrams, which in most cases contain a peritectic reaction. A typical example of such an element is nickel.

Ferrite-stabilizing elements in steel contain a so-called γ loop in their phase diagrams. A typical example of such an element is chromium.

Many alloys contain a mixture of these two types of alloying elements to satisfy other demands on the alloys. Stainless steel alloys are examples of this combination of alloying elements.

The main alloying elements in stainless steel are in most cases nickel or chromium. The effect of a mixture of these alloying elements is discussed in the text.

Exercises

11.1 Describe the difference between peritectic reactions and peritectic transformations. Draw figures that describe the diffusion field connected with the process.

11.2 (a) Describe the diffusion profile during a eutectic reaction and draw a figure.

(b) The solidification process of most alloys starts with a primary precipitation of dendrites, followed by a eutectic or a peritectic reaction. Both reactions start in most cases with formation of a frame of a secondary phase, which grows around of the primary phase. However, the diffusion and the precipitation processes are entirely different in the two cases.

Explain and illustrate the difference at formation of the frame between the eutectic and peritectic processes, respectively.

11.3 The phase diagram of the system Cu-Cd on next page shows that there is a peritectic reaction at 397 °C.

$$\delta + \mathrm{L} \rightarrow \varepsilon$$

The peritectic reaction is followed by a peritectic transformation where a secondary phase forms and covers the primary phase. B atoms diffuse through the growing layer of the secondary phase.

At an experiment a number of samples was isothermally treated at 380 °C in order to achieve equilibrium. The thickness of the ε layer was measured as a function of time. The relationship was found to be

$$d = 1.74 \times \sqrt{t}$$

where the thickness of the ε layer was measured in μm and the time in minutes. Figure 11.3–2 on next page shows their result graphically.

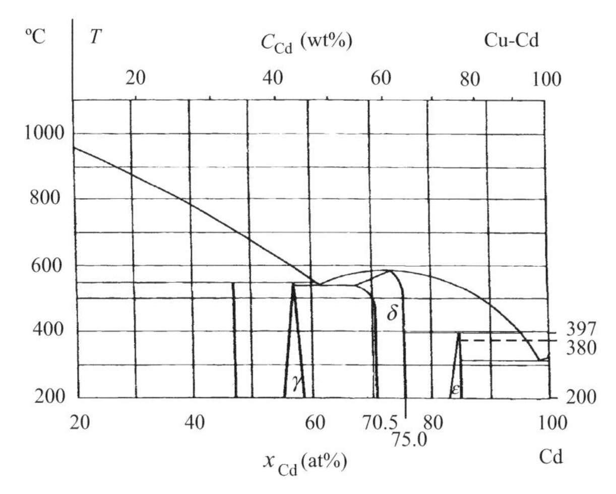

Figure 11.3–1 Part of the phase-diagram of the system Cu-Cd.

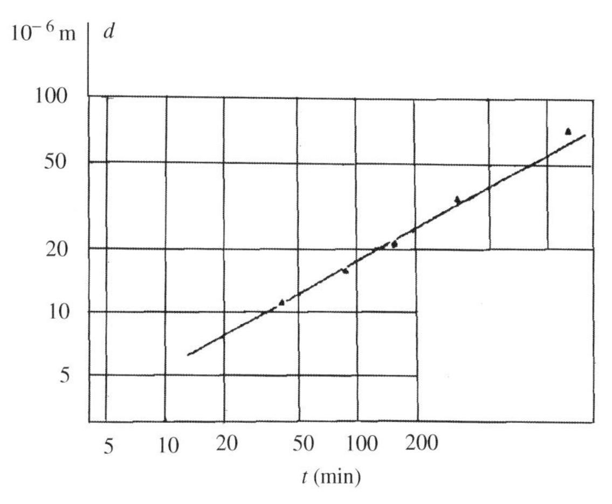

Figure 11.3–2 Relationship between the thickness d of the ε layer and the square root of the time t.

a) Verify the value of the *constant* with the aid of pairs of d_ε and t values in Figure 11.3–2 and recalculate it into SI units.

b) Derive the value of the diffusion constant D_ε measured in m^2/s of the atoms which diffuse through the solid ε layer during the peritectic transformation. Compare the result with the self diffusion coefficient of Cd, that is of the magnitude $10^{-12}\ m^2/s$.

11.4 Emi et al. has studied peritectic reactions in FeNi alloys with the aid of optical laser microscopy. The result of an experiment with an Fe4.86 at%Ni alloy is shown in Figure 11.4–1 below.

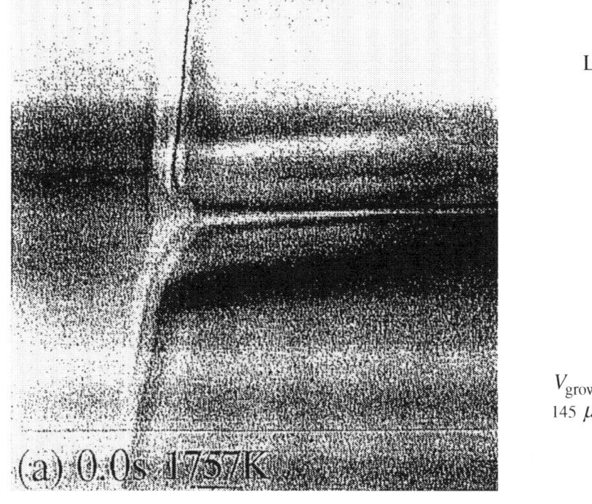

Figure 11.4–1

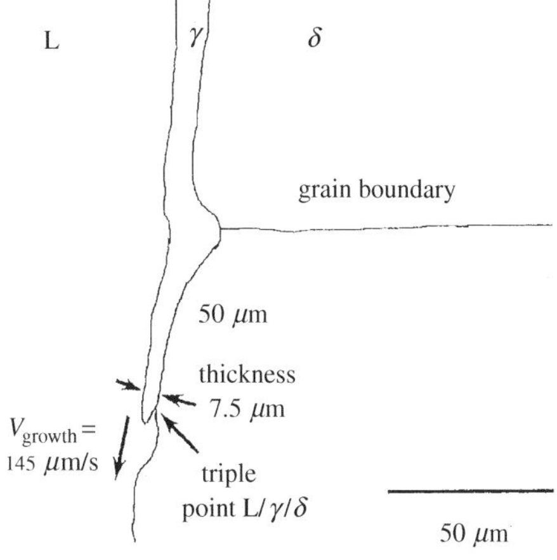

Figure 11.4–2

Figure 11.4–1 shows that the γ phase (austenite) forms a deep valley between the melt and the δ phase. Figure 11.4–2 shows the surface tension balance at the bottom of the γ phase, i.e. at the triple point L/γ/δ.

The balance between the three surface tension forces is shown in Figure 11.4–3.

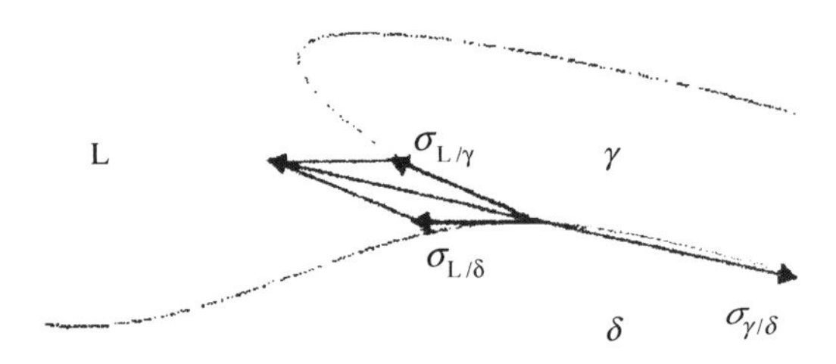

Figure 11.4–3

This type of surface balance is discussed briefly on pages 683 and 685. The surface tension between two solid phases is much smaller than that between a liquid and a solid phase (see Table 3.3 on page 145 in Chapter 3) at equilibrium conditions. The surface tension between two solid phases (coherent or semicoherent surfaces) is of the magnitude 0.2 J/m^2 and of the magnitude 1 J/m^2 between a metal melt and a solid phase (completely incoherent surfaces).

A suggested explanation of the experimental observations is that the density difference or rather volume difference between the γ and δ phases causes strain at the interphase between the two phases during the solidification process. This strain increases strongly the surface tension between the γ and δ phases during the growth.

Estimate the strain energy, the surface tension $\sigma_{\gamma/\delta}$ during the solidification process and the surface energy coupled to the interatomic interaction at the surface. In order to simplify the problem you may assume that the surface tensions $\sigma_{L/\gamma}$ and $\sigma_{L/\delta}$ are equal.

Material constants:

$\sigma_{L/\gamma} = 1\ \text{J/m}^2$
thickness of interphase $= 5 \times 10^{-9}$ m
Young's modulus of elasticity $= 70$ MPa

$$\frac{-\Delta\rho_{\delta-\gamma}}{\rho} = \frac{\Delta V}{V} = 2.5 \text{ volume \%}.$$

11.5 In a unidirectional experiment with an Al alloy with 0.6 wt% Ti the melt is heated to 900 °C and is left to cool. After some time embryos are formed in the melt and some of them survive and grow as β-crystals in the melt.

Figure 1 shows a cross-section of three consecutive β-crystals where the distance between them is filled of melt. When time passes the temperature goes down, the nuclei grow and two phases, primary β and later α precipitate.

Interpret the solidification process in detail with the aid of the phase diagram sketched in Figure 2. Explain the appearance of Figure 1, i.e. the temperatures when α-phases are formed, which components the phases contain and concentration changes as a function of time. Comment also the difference in solidification of α phase between the β-crystals on both sides of the α-phase.

11.6 In a series of unidirectional solidification experiments on peritectic alloys in the Cu-Sn system an unusual observation was made. A eutectic-like structure of the solid phase was obtained for one of the alloys with a special composition when it solidified within a short range of solidification rates. The alloys analyzed contained 10–15 at% Sn.

Normally the peritectic reaction starts with a primary precipitation of α phase, followed by a peritectic reaction $L + \alpha \rightarrow \beta$.

Consider the possibility to obtain a eutectic-like reaction in a peritectic system. Use the phase diagram of the Cu-Sn system for this purpose.

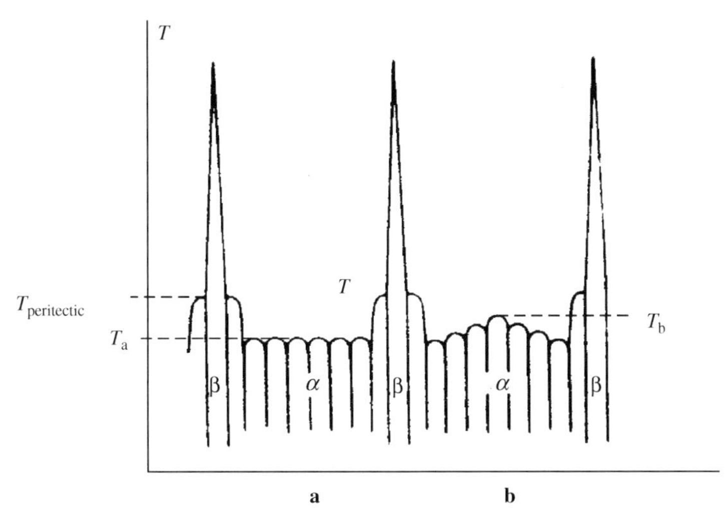

Figure 11.5–1 Cross-section of growing crystal as a function of temperature.

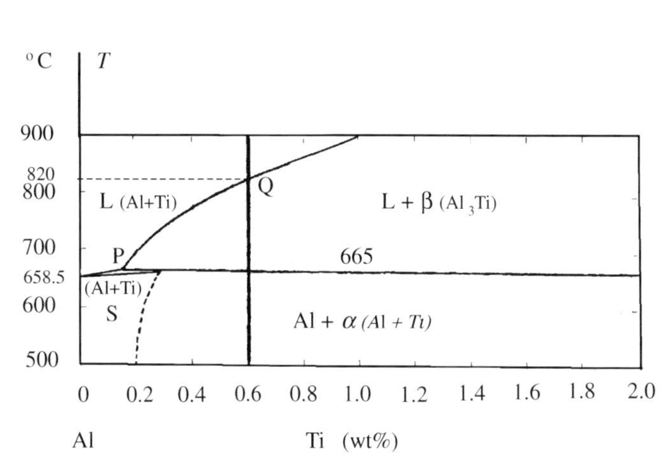

Figure 11.5–2 Sketch of part of the phase diagram of the system Al-Ti.

The two phases follow the same quadratic growth law. For the α phase it can be written as

$$V^{\alpha}_{\text{growth}} = C \times \frac{(x^{L/\alpha} - x_0^L)^2}{x^{L/\alpha} - x^{\alpha/L}}$$

Hint: Assume that both the α and β phases are precipitated simultaneously, i.e. they are both primary phases. See also pages in Chapter 11.

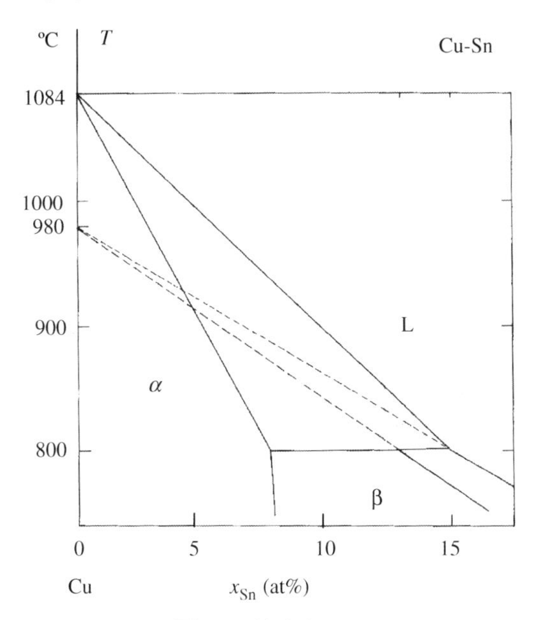

Figure 11.6–1

References

1. H. W. Kerr et al., *Acta Metallurgica*, **22**, 1974, 677.
2. H. Fredriksson, T. Nylén, On the mechanism of peritectic reactions and transformations, *Metals Science*, **16**, 1982, 283–294.
3. G. Petzow, H. E. Exner, Zur Kenntnis Peritektisher Umvandlungen, *Radex-X20 Rundschau*, **3/4**, 1967, 534–539.

4. D. H. St John, L. M. Hogan, The peritectic transformation, *Acta Metallurgica*, **25**, 1977, 77–81.
5. M. Hillert, Eutectic and peritectic reactions, in *Solidification and Casting of Metals*, The Metals Society, London **192**, 1979, 81–87.
6. J. Stjerndahl,The solidification process of iron-base alloys, Doctoral Thesis, KTH, Stockholm, 1978.
7. H Fredriksson, J. Stjerndahl, Solidification of iron-base Alloys, *Metal Science*, **16**, 1982, 575–585.
8. H Fredriksson, Segregation phenomena in iron-base alloys, *Scandinavian Journal of Metallurgy*, **5**, 1976, 27–32.
9. T. Nylen,Investigation of the peritectic/metatectic reaction by controlled solidification experiments, Master Thesis, KTH, Stockholm, 1976.
10. H Fredriksson, J. Stjerndahl, On the formation of a liquid phase during cooling of steel, Metallurgical Transactions, **6B**, 1975, 661–665.
11. C. von Bungart, Untersuchungen űber den Aufbau des Systems Fe–Cr-C, *Archiv für das Eisenhuttenwesen*, **29**, 1958, 163–167.

12

Metallic Glasses and Amorphous Alloy Melts

Solidification and Crystallization Processing in Metals and Alloys, First Edition. Hasse Fredriksson and Ulla Åkerlind.

12.1 Introduction

In Chapters 9, 10 and 11 we have discussed the crystalline solidification structures of metals and alloys in faceted, dendritic, eutectic, peritectic and metatectic growth. The solidified alloys and pure metals have regular structures with a three-dimensional periodicity of the atoms.

In this last chapter, so-called amorphous alloys with no crystalline structure at all will be discussed. In amorphous alloys the atoms are distributed at random, just as in a liquid, but with the difference that the mobilities of the atoms in amorphous alloys are much lower than those in liquids. Such alloys can be regarded as very special liquids with extremely high viscosities and very strong forces between the atoms. Their metallurgical and mechanical properties are excellent, which makes the alloys very useful for practical applications. They are all in a metastable state. Production methods, properties, applications, theory and structures of metallic glasses will be discussed below.

12.1.1 *History of Development of Amorphous Alloys*

The art of making transparent glass has been known for more than 2000 years. This innovation had little influence on the daily life of man during the ancient and medieval times. The construction of optical instruments in Holland in the seventeenth century enabled Galilei to focus his simple telescope towards the night sky, which resulted in a dramatic change of the idea of the universe. Glass materials are transparent and are made of silicates (sand) molten together with various of oxides and fluorides of common metals such as Na, K, Ca and Pb.

The discovery of the very first amorphous alloy was made in 1960 when Duwez and his coworkers [1] in the US unexpectedly discovered an amorphous alloy during rapid solidification of small amounts of the alloy $Au_{75}Si_{25}$ (Au–25 at%Si). The sensational discovery created a great interest in amorphous alloys among metallurgists and material scientists all over the world and numerous production methods were developed during the following decades. The new materials with their remarkable properties were named *metallic glasses* with a common name, in analogy with ordinary glass.

A serious disadvantage with the first generation of metallic glasses was the necessary very high cooling rate and the thin dimensions of the metallic glass products. The early metallic glasses required a cooling rate of at least the magnitude 10^5 K/s and resulted in samples of thickness of the order of 1–100 μm.

A great break-through in the development of metallic glasses occurred when Kui et al. [2] reported their preparation of a 10-mm diameter ingot of the fluxed $Pd_{40}Ni_{40}P_{20}$ glass in 1984. They discovered that the cooling rate could be strongly reduced and that the thickness of the metallic glass increased simultaneously. By adding copper to the Pd–Ni–P alloy and varying the Ni and Cu concentrations Inoue, et al. [3] reached an optimal metallic glass diameter of 72 mm in 1997. Metallic glasses with ingot diameters > 1 mm are called *bulk metallic glasses*.

The development during the 1980s, 1990s and the first decade of the new millennium resulted in an intense and systematic search for new combinations of atoms in order to find bulk metal glasses with thicker and thicker ingot diameters, produced at lower and lower cooling rates.

The second break-through came in the first decade of the twentieth century. The experimental development was ahead of the theoretical research but a new promising theory, the ECP model (efficient cluster packing), was developed. It could explain some features of metallic glasses but much work remains to be done.

Metallic glasses represent a very dynamic research field in material science with wide applications. The development advances rapidly, is escalating and will probably continue for a long time. At present (around the year 2012) more than 300 different metallic glasses are known and manufactured commercially.

12.2 Basic Concepts and Definitions

12.2.1 Glass-Transition Temperature. CCT Diagrams. Critical Cooling Rate for Formation of Metallic Glasses. Crystallization Temperature

The new metallic, non-crystalline materials, i.e. the metallic glasses, were primarily obtained when their melts were exposed to high cooling rates. The so-called CCT diagram (*continuous cooling transition*) in Figure 12.1 shows how metallic glasses are formed.

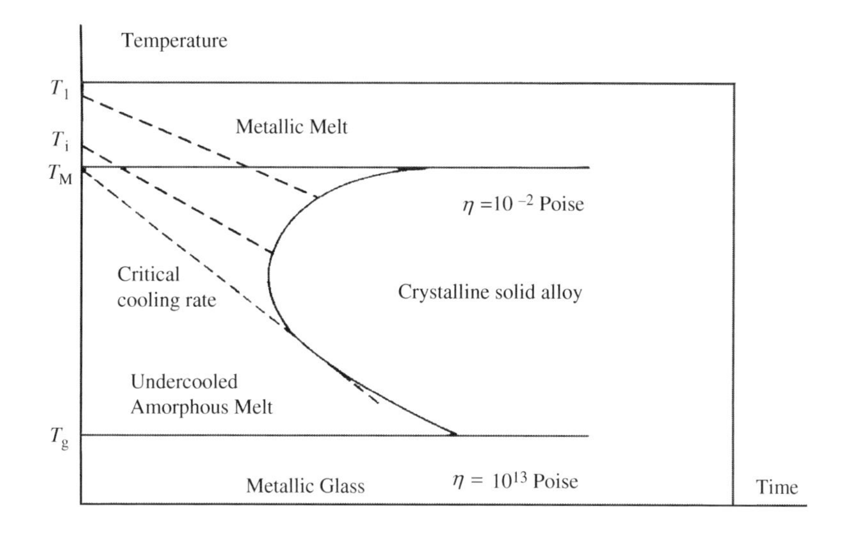

Figure 12.1 CCT diagram of an amorphous alloy.

Each metallic glass has its own characteristic CCT curve. Such curves are based on experimental data and can be constructed in the following way.

Consider an alloy melt with the temperature T_l ($T_l > T_M$, the melting-point temperature of the alloy). At the time $t = 0$ the alloy melt starts to cool with a given cooling rate, indicated in Figure 12.1 by a line with the slope equal to the cooling rate and becomes undercooled. The nucleation and the growth rate of the nuclei start but are initially low and the fraction of crystals formed is small. At the time $t = t_{cr}^1$ a small fraction, say $f = 1\%$, of visible crystals can be observed. and the corresponding temperature T_{cr}^1 is measured. The point $\left(t_{cr}^1, T_{cr}^1\right)$ represents one point on the boundary curve between the undercooled melt and the crystallized region. The procedure is repeated for a number of other values of the temperature of the melt at $t = 0$ and other cooling rates. All the values $\left(t_{cr}^1, T_{cr}^1\right)$ are plotted in the CCT diagram and the curve can be drawn.

The boundary curve approaches asymptotically the line $T = T_M$ because the nucleation and growth rate of the nuclei approaches zero when the undercooling becomes very small.

The curve shows a minimum time at the 'nose' of the curve and increases again at lower temperatures. The reason is that the mobility or diffusion rates of the atoms in the melt decreases exponentially with temperature and the growth of the nuclei decreases. This part of the curve is often dashed because it can not be constructed from experimental data.

It is difficult to define the *critical cooling rate* strictly because it depends on the temperature of the initial melt and can not be equal to the melting-point in practice. In spite of this the common and most practical definition is:

- The critical cooling rate is the one that forms a tangent to the boundary curve from the melting-point at $t = 0$.
  The boundary curve is based on the crystallization fraction $f = 0.01$.

This line is drawn in Figure 12.1. Each alloy has its own characteristic value of the critical cooling rate. If the cooling rate exceeds the critical cooling rate the melt becomes more and more undercooled and no crystallization occurs. At a certain temperature, its *glass-transition temperature* T_g, the melt is transformed into a metallic glass. It has no crystalline structure and the viscosity of the undercooled melt increases rapidly when the temperature approaches T_g. The viscosity of the metallic glass is a material constant, specific to the alloy in question, of the order of 10^{13} Poise. As a comparison, it can be mentioned that the viscosity of alloys at their melting point is of the magnitude 10^{-2} Poise.

At temperatures equal to and below T_g the alloy is *metastable*. At temperatures below the glass-transition temperature the metallic glass remains stable. At temperatures above T_g the alloy becomes softer as the viscosity decreases and it can crystallize.

If the temperature of a bulk metallic glass is successively increased it becomes transformed into an amorphous-like melt with a lower viscosity than the metallic glasses. The *crystallization temperature* is called T_x. The difference $\Delta T_x = T_x - T_g$ can also be used to characterize metallic glasses.

The transition from a viscous metastable alloy melt into a metallic glass occurs at cooling rates larger than the critical cooling rate at the transition temperature T_g. The undercooling is very high and equals $(T_M - T_g)$. Hence, the transition occurs far from equilibrium.

12.2.2 Bulk Metallic Glasses. Glass-Forming Ability. Glass-Temperature Ratio

As was mentioned on page 730, *bulk metallic glasses* can be produced in dimensions of mm size or more. This concept is so frequently used that a special abbreviation, BMG, has been introduced and is generally accepted.

- BMG = the common name of metallic glasses that can be produced with a thickness > 1 mm. The critical sample thickness of a BMG is called d_{cr} in this book.
  Another concept, that is often used, is *glass-forming ability*, or briefly GFA. GFA is a measure of how easily an alloy melt can be transformed into a metallic glass. It can be defined as
- GFA = the minimum dimension of the largest sample that can be made fully glassy.

The so-called *glass-temperature ratio* T_{gr} can be used as a useful criterion and also as a measure of the GFA. The glass-temperature ratio is defined by the relationship

$$T_{gr} = \frac{T_g}{T_M} \tag{12.1}$$

The ratio of the absolute glass-transition temperature to the absolute melting-point temperature is found to be in the range 0.45–0.65 for most BMGs.

- The higher the value of T_{gr} is for the alloy, the easier will be the transformation into a metallic glass.

So far it has not been possible to transform pure metals into metallic glasses. Their GFAs are zero. Calculations based on homogeneous nucleation show that the critical cooling rate in this case must be of the order of 10^{11}–10^{13} K/s. It has not been possible to achieve such extremely high cooling rates in practice.

The BMGs usually consist of three or more kinds of atoms. The composition is given in at% for each component. The percentages are written as subscripts and start with the solvent. Hence, the sum of the subscripts is always equals to 100.

For the sake of clarity the definitions and the nomenclature above are listed in Table 12.1.

Table 12.1 Some definitions and nomenclature of metallic glasses

T_M	= melting-point temperature at standard pressure (1 atm)
R_{cool}^{cr}	= the critical cooling rate for glass transformation i.e. the smallest cooling rate at which an undercooled alloy melt can be transformed into a metallic glass
T_g	= glass transition temperature i.e. the temperature at which a metal melt can be transformed into a metallic glass
T_{gr}	= glass temperature ratio T_g/T_M
T_x	= the temperature at which a metallic glass starts to crystallize
ΔT_x	= $T_x - T_g$.
BMG	= bulk metallic glass where a metallic glass is an amorphous solid alloy with a thickness > 1 mm
GFA	= glass forming ability i.e. the minimum dimension d_{cr} of the largest sample that can be made fully glassy
d_{cr}	= critical sample thickness of a metallic glass, also called d_{max}.

12.2.3 *Experimental Determination of Glass-Transition Temperature and Critical Cooling Rate*

Glass-Transition Temperature T_g

The glass-transition temperature is a material constant – each alloy has its own characteristic value of T_g.

The glass-transition temperature T_g can be determined by heating a sample of the alloy slowly from room temperature in a DTA equipment (see page 395 in Chapter 7). The temperatures of the sample and a reference sample are recorded as functions of time during heating and a curve that shows the temperature difference $\Delta T = T_{sample} - T_{ref}$ can be plotted as a function of T_{sample}. Figure 12.2 shows such a curve. No phase transformations occur in the reference sample. A phase transformation in the sample is recorded as a sudden change of ΔT on the vertical axis. The transition temperature of the phase transformation can be read directly on the horizontal axis.

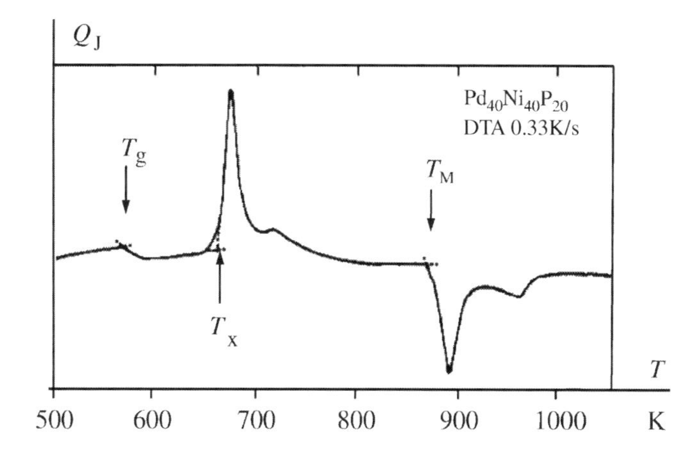

Figure 12.2 DTA curve of the metallic glass $Pd_{40}Ni_{40}P_{20}$. It is interpreted in the text.

In reality, there are two different metallic glasses, a major and a minor one. We discuss only the one with strong maxima and minima. Reproduced with permission from Trans Tech © 1999.

Figure 12.2 gives an example of such a curve. It shows a recording of a sample of the alloy $Pd_{40}Ni_{40}P_{20}$, Below 500 K the alloy is a metastable metallic glass. At the transition temperature. T_g, it becomes transformed into an undercooled amorphous melt. Energy is consumed that corresponds to the minor decrease of ΔT at the temperature marked T_g. At the temperature T_x the curve rises suddenly, when the amorphous alloy starts to crystallize and a large amount of heat is released when the atomic structure is drastically changed. On further heating a deep minimum occurs in the curve. It corresponds to the phase transformation from solid crystal to melt at the melting-point temperature of a metallic glass.

In this way, the glass-transition temperature, the crystallization temperature and the melting point can easily be determined simultaneously for metallic glasses.

Critical Cooling Rate for Glass Transformation R_{cool}^{cr}

The critical cooling rate can be determined with the aid of cooling curves recorded by a similar DTA technique as we discussed on page 733 for heating.

The method to determine R_{cool}^{cr} is demonstrated in the example below.

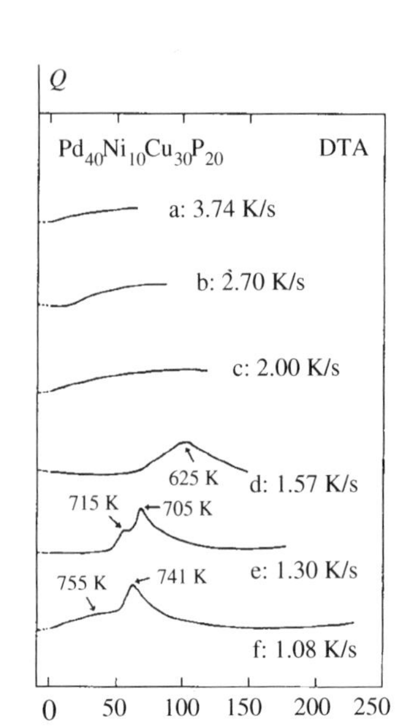

DTA (differential thermal analysis) cooling curves of ingots with a mass of 20 mg of the alloy $Pd_{40}Cu_{30}Ni_{10}P_{20}$ which were cooled at various cooling rates. The molten alloy was heated for 900 s at 1004 K before cooling. Reproduced with permission from Trans Tech © 1999.

Example 12.1

The figure in the margin shows a number of DTA cooling curves of ingots with a mass of 20 mg of the alloy $Pd_{40}Cu_{30}Ni_{10}P_{20}$.. The melting-point of the alloy is 804 K. The glass-transition temperature of the alloy is 572 K.

Interpret the shapes of the curves and derive the critical cooling rate for glass transformation. The curves e and f show double maxima due to precipitation of two types of crystals. Disregard the minor peaks and concentrate on the major type of crystals.

Solution:

The cooling curves can be used to identify the region of crystallization in a CCT diagram that will be drawn.

When the melt starts to solidify as crystals, heat is released and the temperature increases slightly before the additional heat is removed to the surroundings by cooling.

Obviously no crystallization occurs at the cooling rates, which correspond to the curves a, b and c.

Curve d shows a temperature increase at the cooling rate 1.57 K/s. At this cooling rate crystallization starts. The time t is read to 70 s.

In a CCT diagram we start at the point (0; 804), i.e. at the melting-point of the alloy. After about 70 s the temperature has decreased to $804 - 1.57 \times 70 = 694$ K. This point lies on the boundary curve for crystallization and is in the table below. The tangent point is (70; 694).

The critical cooling rate must be larger than 1.57 K/s. In the absence of better information we assume that $R_{cool}^{cr} = 1.58\ \text{K/s}$.

The curves e and f show temperature increases, which are also related to crystallization. With the aid of the given cooling rates and the times, the corresponding temperatures at the points of beginning temperature increase can be calculated in the same way as above. In addition, the corresponding times are read from the curves. The resulting values are given in the table below.

t (s)	R_{cool} (K/s)	$T = 804 - R_{cool}t$ (K)	Estimated values of T	Curve
~ 70	1.57	694	not readable	d
~ 50	1.30	739	725 (interpolation)	e
~ 55	1.08	745	755	f

The accuracy is not so good but the table indicates that the calculated and the read temperatures show reasonable agreement, which supports the interpretation of the curves.

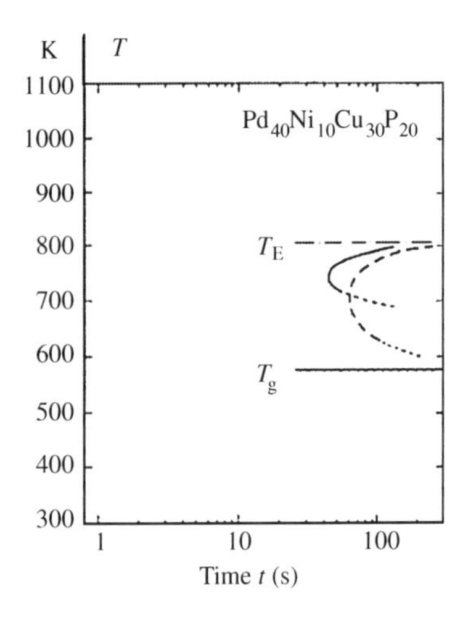

CCT diagram of the alloy $Pd_{40}Cu_{30}Ni_{10}P_{20}$. The crystallization region and the tangent point, related to the critical cooling rate, are shown in the figure. The latter can not be drawn because the scale on the t axis is logarithmic. Reproduced with permission from Trans Tech © 1999.

Answer:

The initial temperature increase of the cooling curves corresponds to heat release due to solidification. If the values are plotted in a time–temperature diagram the outlines of the crystallization region are shown in the figure.

The critical cooling rate that gives crystallization is 1.57 K/s. The critical cooling rate for glass transition may be set to 1.58 K/s. At higher cooling rates (cooling curves a, b and c in the figure in the text) no crystallization is possible.

12.3 Production of Metallic Glasses

12.3.1 Production of Thin Metallic Glasses

Production Methods

It is well known from former structure chapters that the cooling rate during solidification of an alloy influences its crystalline structure very much. The faster the alloy solidifies the smaller will be the grain size in its structure.

The condition for production of metastable amorphous alloys is *very rapid solidification of the alloy*. Many alternative methods are used to achieve the necessary condition for formation of amorphous alloys. Only a few of them will be mentioned here.

1. splat quenching
2. melt spinning, melt extraction and planar flow casting
3. atomization processes.

Splat Quenching

The oldest technique is the gun technique. With the aid of a shock wave liquid droplets of a hot liquid alloy are thrown at high speed towards the surface of a strongly chilled substrate. When they hit the surface they spread out and form a thin layer of liquid. The layer solidifies practically instantly due to the effective cooling, which gives a very high cooling rate in the thin layer or foil.

Cooling rates from 10^4 K/s up to 10^{10} K/s have been measured with this technique. If the cooling rate is high enough for the alloy in question the layer does not crystallize at all. Instead, a thin layer of metastable amorphous alloy is formed. The cooling rates depend on the metal in the substrate and on its roughness. Silver gives the highest cooling rate but the substrate is normally made of copper. A smooth polished surface increases the cooling rate.

The disadvantages of the method are that only small quantities ($\leq$ 100 mg) of thin amorphous materials (thicknesses ~1–100 μm) can be produced in this way. The methods 2 and 3 above allow more substantial amounts of rapidly solidified alloys.

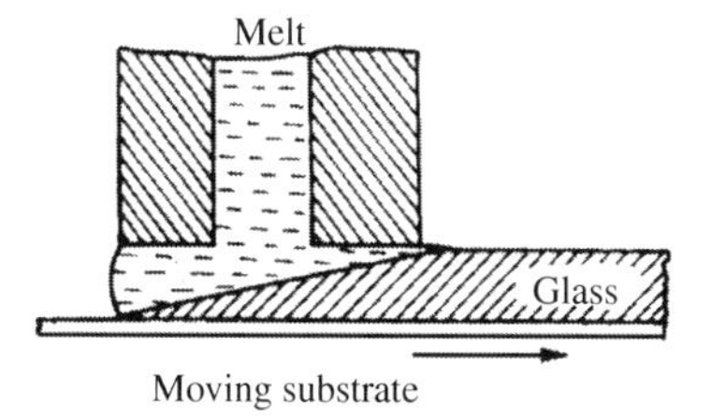

Figure 12.3 Planar flow casting process for production of ribbons. Reproduced with permission from Trans Tech © 1987.

Melt Spinning, Melt Extraction and Planar Flow Casting

A number of methods with a rotating wheel have been developed in order to obtain larger amounts of rapidly solidified materials. The advantage is that they allow continuous production. The principle is the same for them all. Figure 12.3 shows a typical equipment for ribbon production.

A stream of liquid alloy is poured as a thin layer on a moving substrate, which is strongly cooled. If the cooling rate on the substrate exceeds a certain critical value amorphous materials will form. Depending on the design of the substrate the equipment produces continuously stripes, wires or sheets of amorphous material.

The cooling rates are of the order of ~10^5–10^7 K/s and depend strongly on the roughness of the substrate surface. The smoother the surface is the higher will be the cooling rate.

With the aid of these methods strips or sheets of the thicknesses ~50–500 μm can be obtained. Ribbons with a width of 10 cm or more can be produced. This type of material is used in transformers.

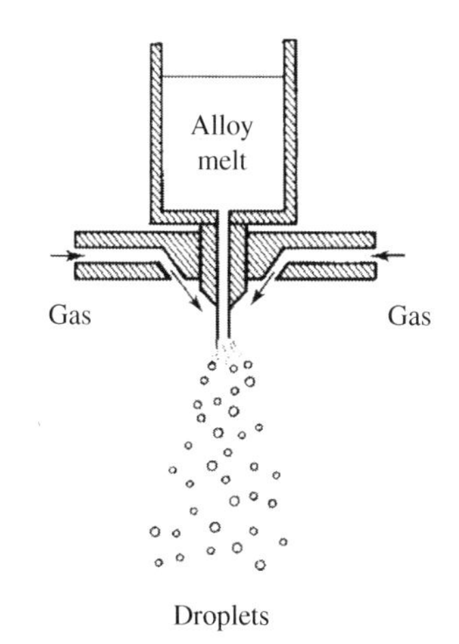

Figure 12.4 Atomizing process for production of amorphous powder of alloys. Reproduced from Cambridge University Press © 1997.

Atomization Processes

Atomization processes are used to produce powder of amorphous materials.

Alloy melt in a crucible is pressed through a nozzle in the bottom of the crucible. The melt is disintegrated into small droplets by a gas, blown into the stream through several nozzles. A mixture of small droplets in a high speed gas stream is formed (Figure 12.4). The high speed gas cools the droplets strongly. The speed of the gas determines the sizes of the droplets and their cooling rates.

If the cooling rates are large enough rapid solidification occurs and a fine-grained powder of amorphous alloy is formed. Cooling rates of 10^3 K/s may be achieved.

The liquid alloy is cooled so rapidly that a high undercooling arises. The solidification occurs far from equilibrium and the alloy ends up in a metastable state. The solidification time is of the order of milliseconds.

Droplets of amorphous alloys with diameters of the magnitude 1 μm–1 mm can be produced with this method. The largest droplet may be only partially amorphous.

Fluxing

The alloy melt used for production of metallic glasses must have a high purity.

It is especially important to keep the concentrations of oxygen and hydrogen as low as possible. Hydrogen makes the amorphous alloys brittle. Oxygen often gives unwanted oxide inclusions, that may act as nucleants for crystals. If the melt contains alloying elements with high affinity to nitrogen, nitrides may form, unless the nitrogen concentration is kept low.

Alloys used in the production of metallic glasses are often produced by high-vacuum melting processes in order to reduce the concentrations of oxygen and hydrogen. Another method is arc-melting. An electrode is made by compacting powder of the alloying elements in the proportions of the wanted alloy. The electrode is arc-melted and cast in a water-cooled copper mould. The ingot is remelted several times in order to suppress the impurities and the gas concentration as much as possible.

Impurities or heterogeneities in the melt act as nucleation sites for crystals. Crystals may also nucleate at the surface between the surrounding atmosphere during the production process. In order to reduce the concentration of heterogeneities in the melt and also purify the surface and prevent the melt from reacting with the atmosphere, the melt is often covered with a molten slag such as boron oxide B_2O_3 or other oxides, fluorides or chlorides.

This process is called *fluxing* of the melt. It was for instance used by Greer et al. [2] in the development of the first bulk metallic glass.

Fluxing is often combined with arc-melting (page 739).

12.3.2 Production of Bulk Metallic Glasses

The disadvantages with the early metallic glasses in the 1960s and the 1970s were the high cooling rates required and the small thicknesses of the amorphous alloy products.

The first metallic glass (an Au–Si alloy with a critical cooling rate of the order of 10^5 K/s), was followed among others by Pt–Ni–P and Pd–Ni–P amorphous alloys of various compositions and critical cooling rates of the magnitude 10^2 K/s. These alloys were among the first of the bulk metallic glasses (BMGs). The prefix 'bulk' indicates that the *maximum sample thicknesses* d_{max} of the new metallic glasses are drastically increased compared to the first ones.

Development of New BMGs

In his book 'Bulk Amorphous Alloys' Inoue [4] has given an extensive survey of preparation and characteristics of BMGs.

Since the 1990s a systematic and intense search is going on for finding multiple-component amorphous glasses with successively better properties. An example of this type of research is given in the box. A great number of scientific and engineering data for metallic glasses is available nowadays. Low R_{cool}^{cr} and large d_{max} occur simultaneously. The best BMGs today (in the year 2010) are almost compatible with ordinary glass.

Pinpoint Approach to Select the Best BMG of the Quarternary System Mg–Cu–Ag–Y

After Li, et al. [5].

The best BMG means the one that has the optimal GFA, i.e. that has the largest diameter D_{cr} when cast in a cylindrical Cu mould. In order to simplify the problem the quarternary system can be regarded as the pseudoternary system Mg–(Cu, Ag)–Y, with reference to the ternary system Mg–Cu–Y.

Methods are known for ternary systems to find the optimal composition that gives the largest diameter. For the 3-dimensional ternary alloys the best BMG corresponds to a point in the bottom triangle in Figure 1.

The process of finding the best BMG in the quarternary system can be visualized in the 3-dimensional diagram, shown in the figure. The term pseudoternary means that the prism-shaped space in Figure 1 is cut by horizontal planes, each with a fixed Ag to Cu ratio $Cu_{1-x}Ag_x$. Together Ag and Cu correspond to the third component. In Figure 1 the three planes corresponding to $z = 0.1$, 0.2 and 0.3 are drawn.

For each of these planes the best BMG is derived and plotted as points in Figure 1. Including the bottom plane there are four points known, which forms a pathway that includes the optimal composition. It is marked in Figure 1. The highest GFA value along the pathway is the best BMG of the quarternary system. It is found to be $D_{cr} = 16$ mm for $x = 0.2$. The composition of the alloy is $Mg_{54}Cu_{28}Ag_7Y_{11}$.

This power of the pinpoint 3-dimensinal approach is that it localizes the best glass-forming composition in a given system. Figure 2 shows that a simple Ag substitution (vertical arrow) misses the best BMG former. The vertical arrow points in a direction very different from that of the vertical arrow.

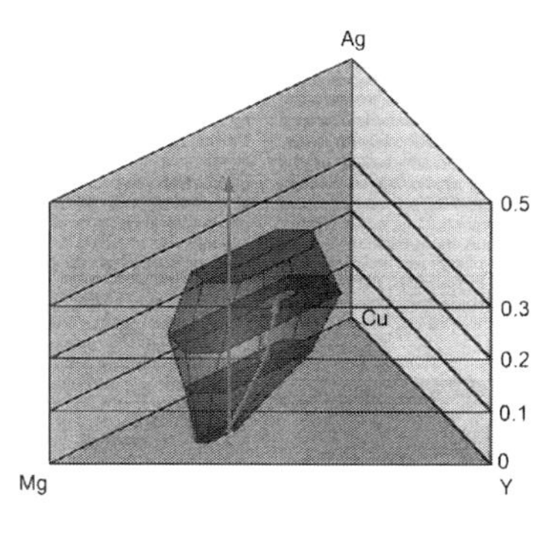

Figure 1 Phase diagram of Mg–Cu–Ag–Y.

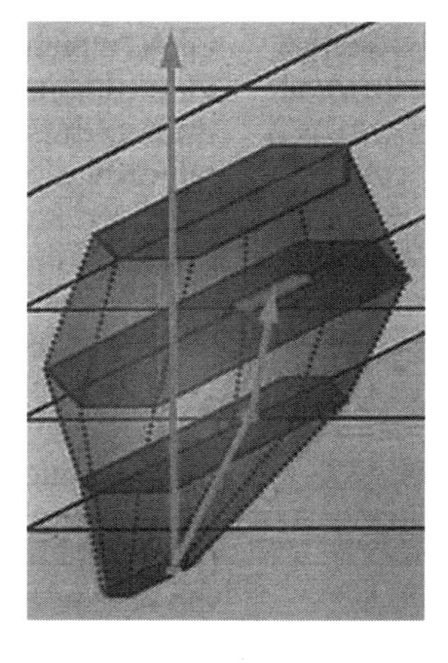

Figure 2 Magnification of the phase diagram Mg–Cu–Ag–Y.

Thumb Rules for BMGs with High GFA

In order to facilitate glass forming by cooling it is desirable to suppress crystallization during cooling as much as possible. The wide experience of finding new amorphous alloys with high glass-forming ability, GFA, can be summarized in the following three 'thumb rules'.

High glass-forming ability is obtained

1. if the bulk metallic glass consists of a multi-component system with three or more elements.
2. if there is a significant difference of at least 12% in atomic size ratios among the main constituent elements.
3. if the heat of mixing of the components is negative.

The validity of point 1 is shown by comparison of Figures 12.5 and 12.6 on next page.

Figure 12.5 illustrates the *first* condition. It shows how drastically the crystallization process is suppressed when a fourth element is included in the alloy. The crystallization is moved strongly to the right, the crystallization temperature T_x is lowered and reduces the critical cooling rate for glass transformation very much. In addition, the critical thickness d_{cr} is strongly increased. Figure 12.6 shows a similar effect. In this case, even more elements are added and the effect is even better pronounced than in Figure 12.5.

The final critical cooling rate is very low indeed and corresponds to a large maximum thickness of the BMG of the alloy $Pd_{40}Cu_{30}Ni_{10}Ni_{10}P_{20}P_{20}$. In 1997 Inoue and his group produced a sample with a thickness of 72 mm [3].

The *second* condition facilitates cluster formation (groups of solvent atoms around a solute atom) in the melt which suppresses crystallization.

The *third* condition (negative heat of mixing) refers to the heat release emitted when one element is added to a melt with predetermined composition. Negative heat of mixing means strong attractive forces between the unlike atoms in the melt. This indicates cluster formation in the melt. The clusters are held together by strong forces, that are hard to break. This makes a crystallization process in the melt unlikely.

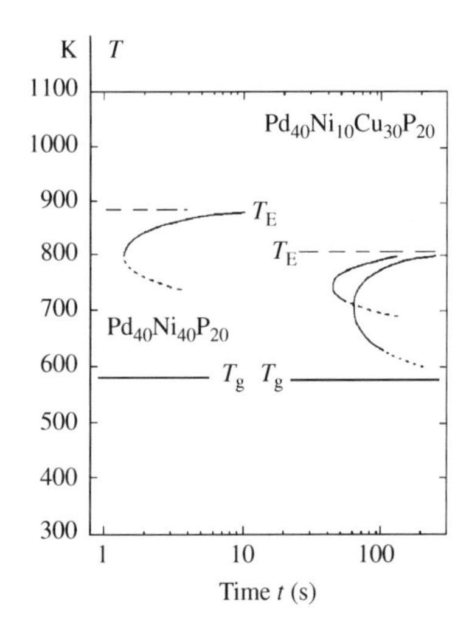

Figure 12.5 CCT diagram of the BMG alloys $Pd_{40}Ni_{40}P_{20}$ and $Pd_{40}Cu_{30}Ni_{10}P_{20}$. The addition of a fourth component Cu moves the crystallization region strongly towards the left.
T_E = the eutectic temperature. Reproduced with permission from Trans Tech © 1999.

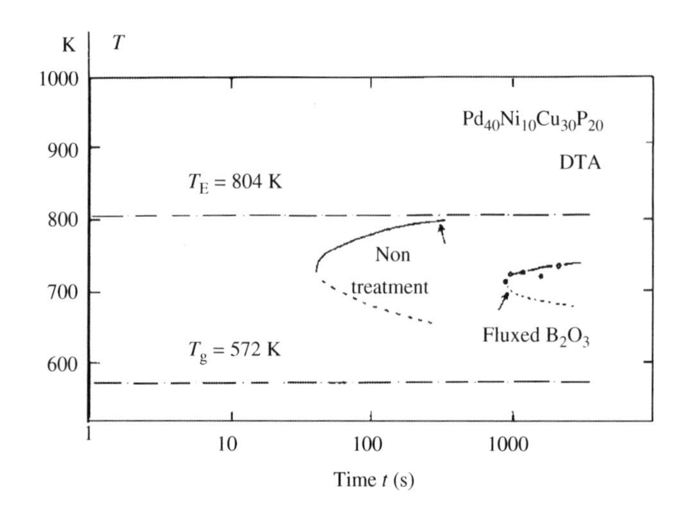

Figure 12.6 CCT diagram of a 2 g sample of the BMG alloy $Pd_{40}Cu_{30}Ni_{10}P_{20}$ with and without flux treatment of the melt with B_2O_3 before the DTA cooling. The identification of the crystallization regions are based on cooling curves of the same type as those shown in Example 12.1 within the range 0.033–3.74 K/s. Reproduced with permission from Trans Tech © 1999.

Table 12.2 below gives a review of some important BMGs developed during the last decades. Greer [6] has collected the table, which clearly reflects the way to metallic glasses with successively thicker sheets and cylinders. Today, the maximum thickness is of the magnitude 1 cm, on its way to the order of 10 cm.

Table 12.2 Bulk metallic glasses with critical cylinder diameter ≥10 mm

System	Composition	Critical diameter (at%)	Production Method (mm)	Year
Pd-based	$Pd_{40}Ni_{40}P_{20}$	10	Fluxing	1984
	$Pd_{40}Cu_{30}Ni_{10}P_{20}$	72	Water quenching	1997
Zr-based	$Zr_{65}Al_{7.5}Ni_{10}Cu_{17.5}$	16	Water quenching	1993
	$Zr_{41.2}Ti_{13.8}Cu_{12.5}Ni_{10}Be_{22.5}$	25	Copper mould casting	1996
Cu-based	$Cu_{46}Zr_{42}Al_7Y_5$	10	Copper mould casting	2004
	$Cu_{49}Hf_{42}Al_9$	10	Copper mould casting	2006
RE-based	$Y_{36}Sc_{20}Al_{24}Co_{20}$	25	Water quenching	2003
	$La_{62}Al_{15.7}Cu_{11.15}Ni_{11.15}$	11	Copper mould casting	2003
Mg-based	$Mg_{54}Cu_{26.5}Ag_{8.5}Gd_{11}$	25	Copper mould casting	2005
	$Mg_{65}Cu_{7.5}Ni_{7.5}Zn_5Ag_5Y_5Gd_5$	14	Copper mould casting	2005
Fe-based	$Fe_{48}Cr_{15}Mo_{14}Er_2C_{15}B_6$	12	Copper mould casting	2004
	$(Fe_{44.3}Cr_5Co_5Mo_{12.8}Mn_{11.2}C_{15.8}B_{5.9})_{98.5}Y_{1.5}$	12	Copper mould casting	2004
	$Fe_{41}Co_7Cr_{15}Mo_{14}C_{15}B_6Y_2$	16	Copper mould casting	2005
Co-based	$Co_{48}Cr_{15}Mo_{14}C_{15}B_6Er_2$	10	Copper mould casting	2006
Ti-based	$Ti_{40}Zr_{25}Cu_{12}Ni_3Be_{20}$	14	Copper mould casting	2005
Ca-based	$Ca_{65}Mn_{15}Zn_{20}$	15	Copper mould casting	2004
Pt-based	$Pt_{42.5}Cu_{27}Ni_{9.5}P_{21}$	20	Water quenching	2004

Table 12.2 confirms the rules of thumb on page 737. However, it is a very puzzling fact that even a minor change in composition may cause a drastic increase of the GFA.

Production Methods of BMGs

Thanks to the remarkable development during the last three decades from thin metal glasses to bulk metal glasses, ordinary casting methods can be used for production of BMGs. The castings are the basis for making a manifold of products for a variety of applications.

Examples of the most common casting methods used for BMGs are briefly described below. More information about the experimental conditions on casting of metallic glasses is given in Section 12.7.

Arc-Melting

Bulk metallic glasses have low melting points, much lower than the metal components of the alloys. Many of the BMGs also consist of metals, which have high affinity to oxygen. In such cases special methods have to be used to prepare the alloy melt before casting. The most common method is an arc-melting process in a water-cooled copper crucible.

The components of the sample are placed in the water-cooled Cu crucible included in a vacuum-tight chamber with an argon atmosphere, free from oxygen. The sample is melted by the arc at high temperature. In order to obtain the alloy as pure as possible and free from heterogeneities the alloy is repeatedly remelted up to 10 times. The melt is then cast for example with the aid of one of the methods described below.

Arc-Melting of Zr–Al–TM Alloys

Quarternary and penternary alloys of the type Zr–Al–TM, where TM = Co, Ni Cu, have comparatively high melting-points. For this reason, a high-temperature source is required to melt these alloys. In such cases an electric arc is a suitable source.

Arc-melting has the disadvantage, though, that it is very hard to suppress the crystalline phase of the compound Zr_2TM completely during cooling. The reason is heterogeneous nucleation caused by incomplete melting at the bottom side in contact with the copper hearth.

An amorphous phase of the alloy is formed in the inner region where the heat flux disappears and the cooling power is reduced. However, the low cooling rate obtained by arc-melting is high enough to form an amorphous phase in the Zr-Al-Co-Ni-Cu system.

High-Pressure Die Casting

Ternary alloys of the types Mg–Ln–TM and Ln–Al–TM have wide supercooled liquid regions and high GFA (glass-forming ability). They can be produced by the high-pressure die casting method in the shape of cylinders or sheets up to a maximum thickness of 10 mm (Figure 12.7).

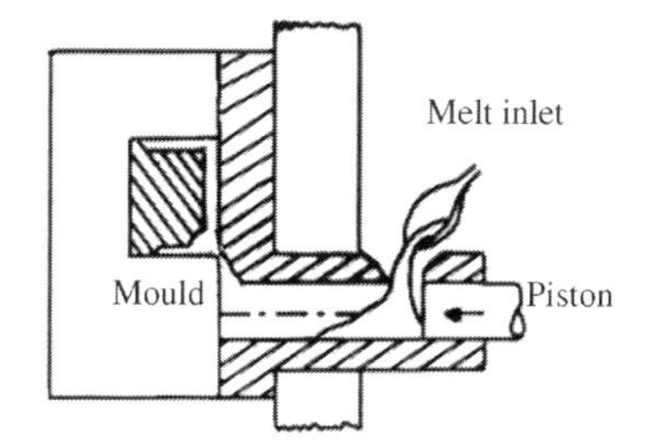

Figure 12.7 High-pressure die casting machine.

During casting the supercooled metal melt is transferred into the shot cylinder. The piston is then pushed inwards and forces the melt into the mould. The method is described on page 8 in [7].

An example is the shiny bulk metal glass $Mg_{65}Cu_{25}Y_{10}$ which has a tensile strength that is three times larger than that of ordinary Mg-based cast alloys.

Water Quenching

Water quenching of a BMG melt means that the molten alloy, encased in a quartz tube, is transformed into an amorphous solid alloy simply by quenching the tube into water.

The method can be applied on BMGs that have critical cooling rates of the order of 1–10 K/s, that is compatible with those for ordinary glass. As an example we chose $Pd_{40}Cu_{30}\,Ni_{10}\,P_{20}$, the alloy in Example 12.1 on page 734, that has a critical cooling rate of 1.6 K/s.

Increase of the number of components in a ternary alloy from 3 to 4 or 5 is a well-known method to increase the glass-forming ability GTA of an alloy. In most cases, the ratio T_g/T_M does not change very much on the change from a ternary to a quaternary or penternary system. On the other hand, there is a considerable increase of the difference $\Delta T_x = T_x - T_g$ (Figure 12.1 on page 731). For the amorphous alloy $Pd_{40}Cu_{30}\,Ni_{10}\,P_{20}$ $\Delta T_x = 98$ K and is nearly twice as large as that of the ternary alloy $Pd_{40}\,Ni_{40}\,P_{20}$, which was found in the 1970s.

In addition, T_g, ΔT_x, T_x and $(-\Delta H)$ are the same for a bulk sample and a thin sample of the amorphous alloy $Pd_{40}Cu_{30}Ni_{10}P_{20}$. The conclusion is that the cooling rate does not influence the structure of a BMG.

Figure 12.8 Shiny cylinder of the amorphous $Pd_{40}Cu_{30}Ni_{10}P_{20}$ alloy with radius = 20 mm and the length = 75 mm.

Figure 12.8 shows the appearance of a cylindrical BMG of the alloy $Pd_{40}Cu_{30}Ni_{10}P_{20}$, cast by water quenching.

The penternary amorphous alloy $Zr_{60}Al_{10}Ni_{10}Cu_{15}Pd_5$ shows similar properties as the BMG alloy $Pd_{40}Cu_{30}Ni_{10}P_{20}$ and can also be cast by water quenching.

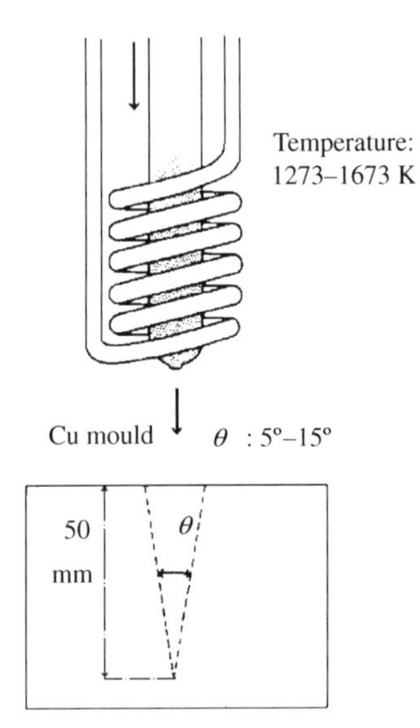

Figure 12.9 Casting of a molten alloy into a wedge-shaped copper mould. Registration of continuous cooling curves was made at several points on the vertical axis. Reproduced with permission from Trans Tech © 1999.

Copper Mould Casting

Copper mould casting means simply that melt is ejected into a cooled copper mould for solidification. The mould is made of copper because of its high thermal conductivity. As an example of the method we will study a wedge-shaped copper mould in order to illustrate the complications when the glass transition is partial and not complete. Inoue has reported this experiment on pages 29–32 in his book 'Bulk Amorphous Alloys' [4].

Figure 12.9 shows a sketch of such an equipment, The design of a series of copper moulds is shown at the bottom of the figure. Their heights are 50 mm and the angle θ can be varied from 5° to 15°. The ejection pressure is kept constant and equal to 0.05 MPa. The temperature of the melt at ejection can be varied from 1273 K to 1573 K.

The cooling rates of the ejected melt are measured by a thermocouple at five different sites along the central vertical axis. The solidification structure of the wedge-shaped specimens is examined by X-ray examination and by optical and transmission electron microscopy.

The cooling rate along the wedge depends primarily on width of the wedge, which increases with increasing height. The cooling rates decrease with increasing height, Close to the bottom of the wedge the cooling rate is rapid and the melt solidifies there as a metallic glass. In the upper regions the cooling rate is lower and may be slow enough for incomplete glass transition and partial crystallization. This occurs above a certain critical height (see Figure 12.11).

Determination of Critical Cooling Rate and Critical Thickness

Solidification tests with the wedge method offers a practical method for determination of the critical cooling rate and the critical thickness for metallic glasses. The method is analogous with the one used for cast iron to avoid undesired solidification of white iron instead of grey iron (Example 6.9 on page 181 in [7].

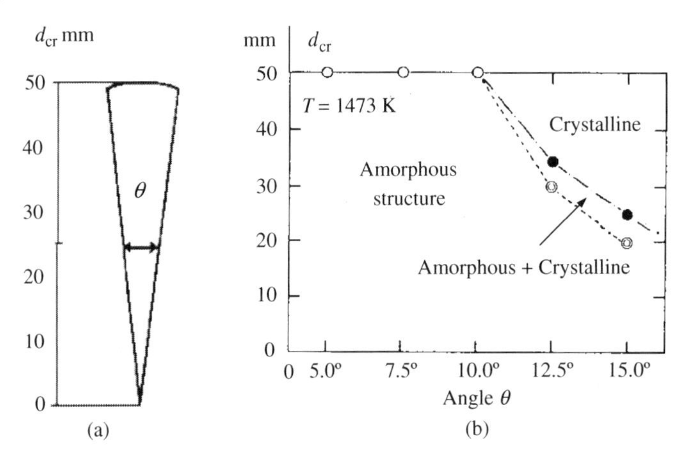

Figure 12.10 Relationship between the depth d of the cavity and the angle θ. Reproduced with permission from Trans Tech © 1999.

Inoue used the method described in Figure 12.9 for solidification analyzes of the metallic glass $Zr_{60}Al_{10}Ni_{10}Cu_{15}Pd_5$ mentioned above.

Figure 12.10 shows the relationship between the depth d and the angle θ for formation of an amorphous phase at a melt temperature of 1473 K. If $\theta < 10°$ an amorphous phase is obtained for all d values in the range 1–50 mm. If $\theta > 10°$, no single amorphous phase is formed for higher θ values. The curve shows that the maximum depth for formation of an amorphous phase is 30 mm for $\theta = 12.5°$ and only 20 mm at $\theta = 15°$.

Figure 12.11 shows that the region C has an amorphous structure while the upper region consists of a two-phase region B with an amorphous and a crystalline phase and a region A with a pure crystalline phase.

The formation of an amorphous phase is not complete and the material is wasted and can not be used as BMG.

With the aid of Figures 12.12 and 12.13 we can analyze the reasons for this failure.

Figure 12.12 shows the temperature–time curves measured at different heights in the wedge-shaped casting. The temperature–time curves a and b show that the temperature decreases continuously, down

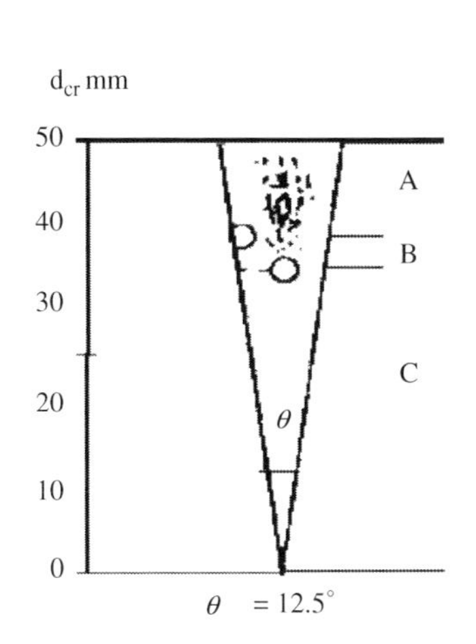

Figure 12.11 Wedge with three regions of different structures. Reproduced with permission from Trans Tech © 1999.

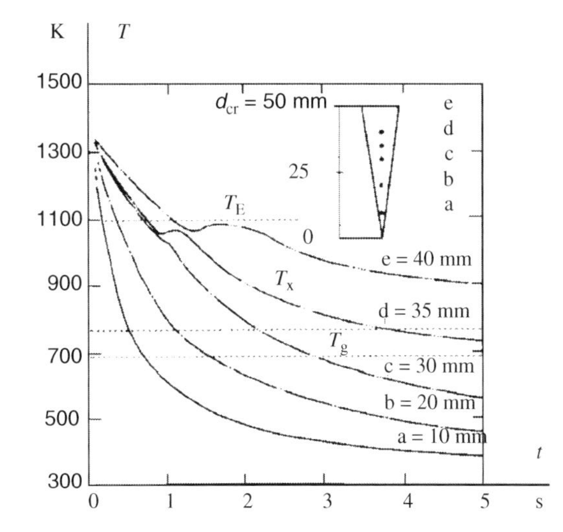

Figure 12.12 Temperature–time curves at different sites a – e in the transverse cross-section of the cast $Zr_{60}Al_{10}Ni_{10}Cu_{15}Pd_5$ alloy with the wedge shape $\theta = 12.5°$. T_g, T_x and the eutectic temperature T_E are marked in the figure. Reproduced with permission from Trans Tech © 1999.

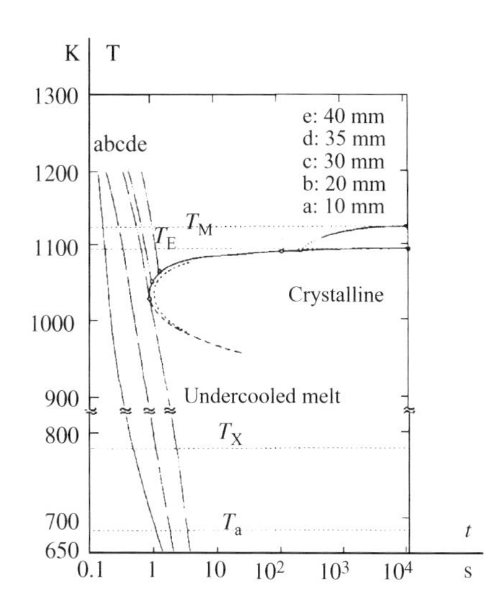

Figure 12.13 CCT curves for the transformation from supercooled melt to crystalline phases for a $Zr_{60}Al_{10}Ni_{10}Cu_{15}Pd_5$ alloy. Reproduced with permission from Trans Tech © 1999.

to and below the glass-transition temperature T_g. The curves c, d and e show inflexion points and a short region of nearly constant temperature at temperatures below the melting point T_M.

The appearances of the c, d and e curves can be interpreted as a result of a crystallization process in each of the three sites. During the crystallization process heat is released, which affects the temperature and interrupts the continuous temperature drop. The inflexion points can be interpreted as the start temperature for formation of crystals. The CCT diagrams in Figure 12.13 have been constructed with the aid of these temperatures and the cooling curves.

Characteristics of Genuine BMGs

The discussion on pages 739–740 shows the need for testing the quality of BMGs cast in copper moulds.

The castings have to be examined carefully in order to find out whether they are genuine BMG with no regions of crystallized regions or not. The following characteristics of 100% BMGs can be used.

1. The shiny appearance of the surface indicates that the ingots show no traces of crystallized material.
2. X-ray diffraction patterns from longitudinal and transverse cross-sections show only patterns of a wide primary ring with a strong maximum. The pattern resembles the X-ray patterns of a liquid, which have the same type of disordered structure as solid amorphous alloys. Such structures will be discussed in Section 12.4. The absence of narrow X-ray lines exclude the presence of crystalline structures.
3. The shrinkage cavities that are often formed inside cast crystalline materials, are missing completely. This confirms the absence of crystalline phases.
4. Measures of the tensile stress showed normal values for BMGs, around 1600–1620 MPa.
5. The DSC (differential scanning calorimetry, page 404 in Chapter 7) or DTA curve of a sample, taken from the central region shows a distinct endothermic reaction, owing to the glass transition, followed by an undercooled amorphous melt region and finally an exothermic reaction due to crystallization.

12.4 Experimental Methods for Structure Determination of Metallic Glasses and Amorphous Alloy Melts

The experimental methods to study the structures of amorphous melts and metallic glasses are the same as those used for crystals and ordinary liquids and melts. The dominant method is X-ray examination, but also electron diffraction and neutron scattering can be used. The experimental methods of X-ray examination of crystals and liquids and the interpretation of the results are described in Section 1.2 in [8] or in any other equivalent book.

12.4.1 X-ray Examination of Crystals and Melts

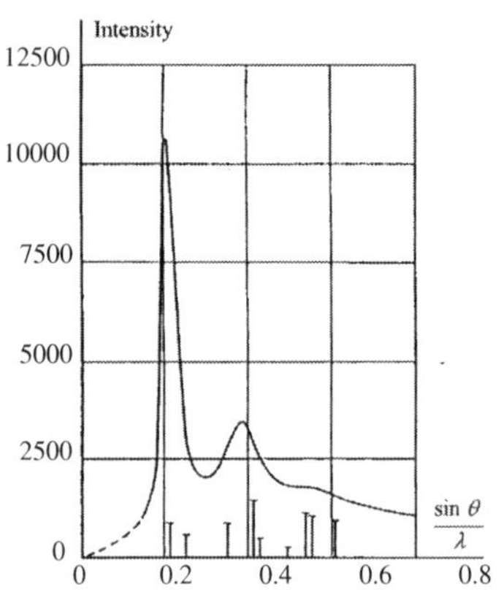

Figure 12.14 The intensity of scattered X-ray radiation as a function of sin θ/λ (see page 5 in [8]. The curve originates from scattering in liquid gold at 1100 °C.

Parallel X-rays hit a specimen and the intensity of the scattered radiation is recorded as a function the deviation angle. It is well known that scattering in crystals gives a few narrow sharp lines. The scattered radiation from liquids and melts result in a few wide peaks separated by minima. Examples of such X-ray diagrams are given on page 5 in [8] for two monoatomic metal melts, liquid gold and liquid zinc. The former one is shown in Figure 12.14.

The different appearance between the X-ray diagrams for crystals and melts depend entirely on their structures.

Interpretation of X-ray Diagrams for Metal Melts. Structures of Metal Melts

The atoms in liquids and melts are in incessant motion and have no permanent average positions. The distribution of the atoms can only be treated with the aid of statistical methods. For this purpose we will introduce some new concepts.

From X-ray Plots to Atomic Distribution Diagrams

Consider the distribution of atoms in a liquid around an arbitrary atom and choose this atom as the origin. The probability dW_r of finding another atom within the volume element dV at a distance r from the origin can be written as

$$dW_r = w_r dV = w_r 4\pi r^2 dr \tag{12.2}$$

or

$$W_r = \int_{r_1}^{r_2} w_r 4\pi r^2 dr \tag{12.3}$$

where

r = distance from the origin
dW_r = the probability of finding another atom at distance r from each other
w_r = probability of finding another atom in a unit volume at distance r from the origin
$4\pi r^2 dr$ = volume of a spherical shell with radius r and thickness dr at distance r from the origin
W_r = the probability of finding two atoms at a distance r from each other where $r_1 \leq r \leq r_2$. r_1 and r_2 are the intersections between the curve and the r-axis.

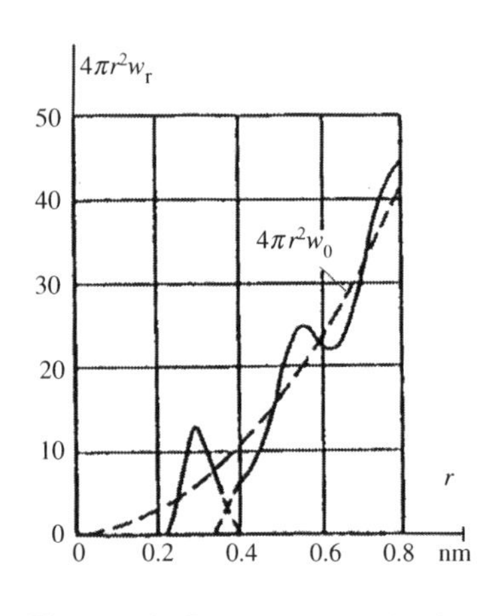

Figure 12.15 Atomic distribution diagram of liquid gold at 1100 °C. The dashed line corresponds the average atomic distribution w_0 of atoms (no coordination shells).

It is possible to derive the probability w_r as a function of the X-ray intensity theoretically. When this function is known it is possible to draw so-called atomic distribution diagrams on the basis of the X-ray intensity diagrams.

By plotting $4\pi r^2 w_r dr$ as a function of the radius r we obtain the atomic distribution diagram of the liquid in question. Figure 12.15 shows the atomic distribution diagram of liquid gold (Figure 12.14) as an example.

From the atomic distribution diagrams several important quantities can be derived:

- the minimum distance r_0 between two gold atoms (can be read from the diagram)
- the nearest-neighbour distance between two atoms i.e. the r value of the maximum of the first peak
- the coordination shells, i.e. the peaks
- the coordination numbers in the first shell, i.e. i.e. the area under the first peak.

12.4.2 Bragg Scattering in Crystals, Metallic Glasses and Amorphous Alloy Melts by Electron Diffraction and Neutron Scattering

High-energy electrons and neutrons can be regarded as particles, which move with high speed but they also show wave properties. Each particle has a wavelength that is a function of its energy and mass. When a particle beam hits a target of the sample to be examined, a diffraction pattern that can be recorded or appear on a photographic plate. Figures 12.16, 12.17 and 12.18 show two types of spectrometers.

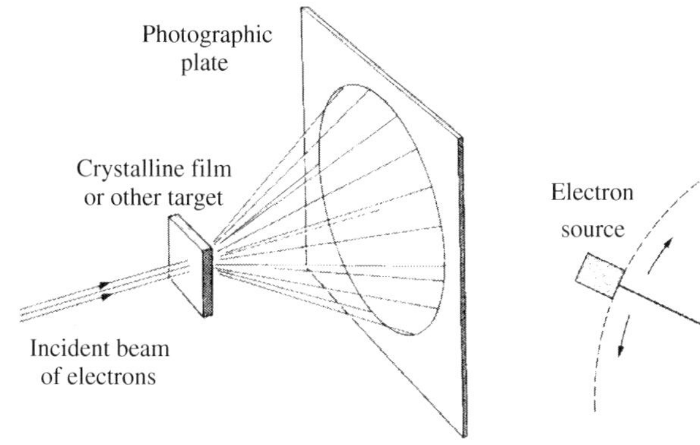

Figure 12.16 Equipment for observation of electron diffraction through a crystalline or liquid material.

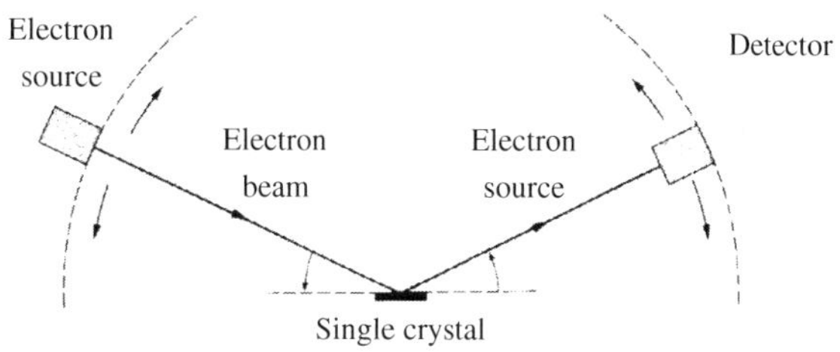

Electron source
Electron beam
Electron source
Detector
Single crystal

Figure 12.17 Bragg scattering of electrons.

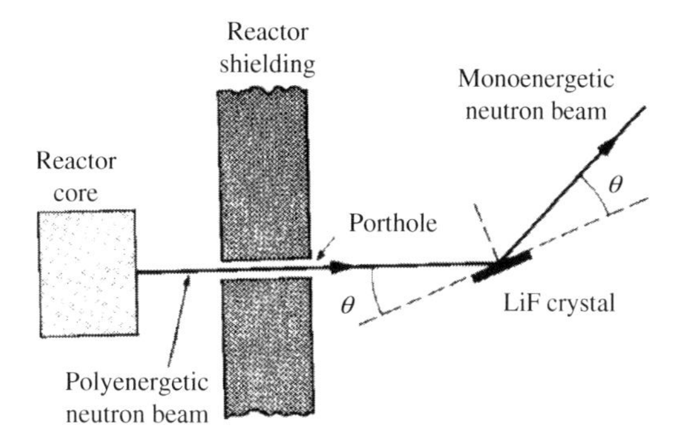

Figure 12.18 Neutron diffraction spectrometer.

The basic equation, valid for X-rays, electrons and neutrons, is Bragg's law:

$$2d \sin \theta = p\lambda \tag{12.4}$$

where λ is the wavelength of the radiation.

As the wavelength λ and the diffraction angle θ are known the distance d can be measured and the structure of the specimen can be interpreted.

Applications

Figures 12.19 and 12.20 show two examples. Figure 12.19 shows the diffraction pattern of electrons reflected at a crystal powder target. Similar patterns have also been observed for X-rays and neutron beams. Figure 12.20 shows that metallic glasses also give ring systems of the diffuse type found in liquids.

When the electron beam passes a crystalline specimen another type of pattern, called a von Laue spot pattern or simply a von Laue diagram, is observed.

Figure 12.21 shows a von Laue pattern for graphite. The number and positions of the spots supply information of the symmetry of the examined type of crystal.

Figure 12.22 shows a Laue diagram of a quasi-crystalline solid. Quasicrystals are formed as a metastable solid at high cooling rates in some systems instead of an amorphous phase. Quasicrystals will be discussed in Section 12.5.2 on page 750.

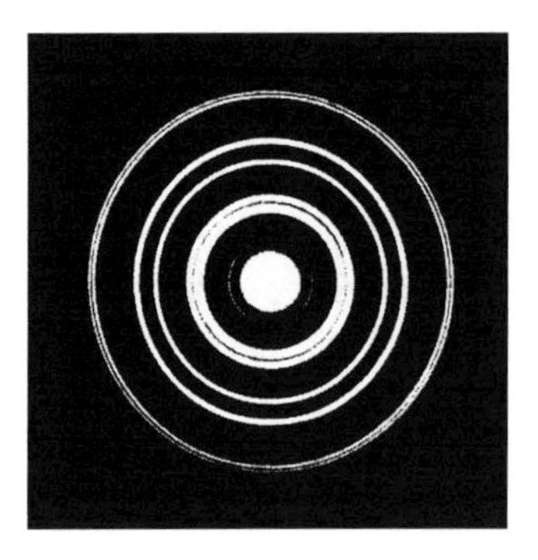

Figure 12.19 Diffraction pattern of electrons scattered by crystal powder.

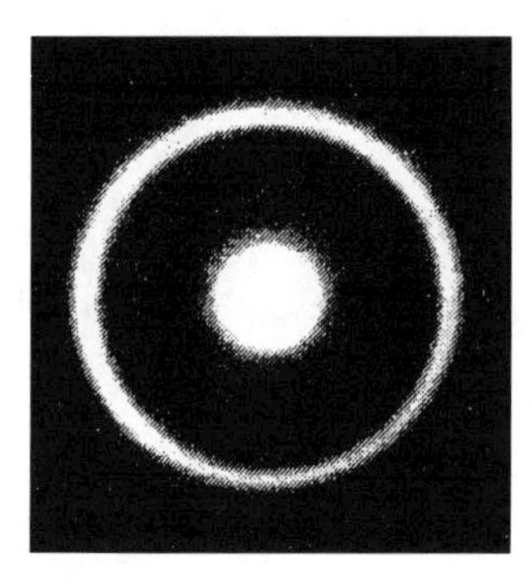

Figure 12.20 Diffraction pattern of electrons scattered by the metallic glass Co_{78} Ti_{22}.

Figure 12.21 Laue diagram of a crystalline material. Diffraction of 80 kV electrons through graphite (carbon).

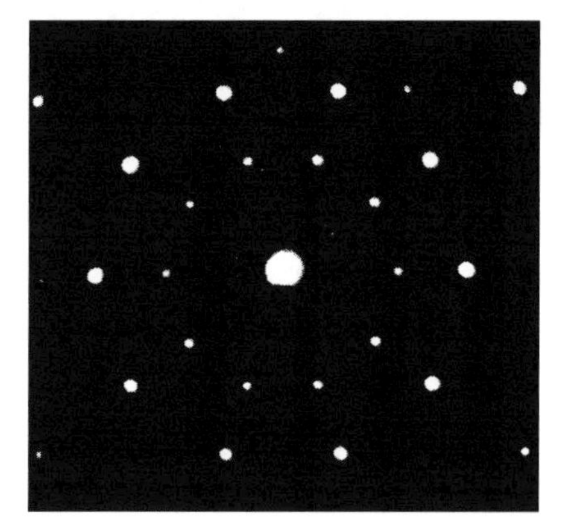

Figure 12.22 Laue diagram of a quasicrystalline material. High-resolution diffraction pattern of $Al_{86}Mn_{14}$ quasicrystals obtained by transmission electron microscopy. Reproduced with permission from Trans Tech © 1987.

12.5 Structures of Metallic Glasses

12.5.1 Structure Models of Amorphous Alloys

Partial Distribution Functions

Metallic glasses show the same type of X-ray diagrams as liquid metals and alloys and have the same type of structures. For this reason, it is necessary to use statistical methods to describe the distribution of atoms in amorphous alloys. The main concepts are listed below.

The radial distribution function:

- $g(r) =$ the probability of finding two atoms at the distance r from each other.
  The radial density function:
- $\rho\ (r) =$ the average number of atoms per unit volume at distance r from another atom.
  The function
- $N(r) =$ the average total number of atoms within a distance r from another atom.

Figure 12.23 shows the radial distribution function $g(r)$ as a function of r. The dashed curve represents the average distribution function obtained for a constant value of the radial density function. In the absence of interaction forces between the atoms the experiments would show this result. The true experimental curves do show maxima and minima that are interpreted as a shell structure. Obviously, there exists an interaction between the atoms, which is called *short-range order interaction*.

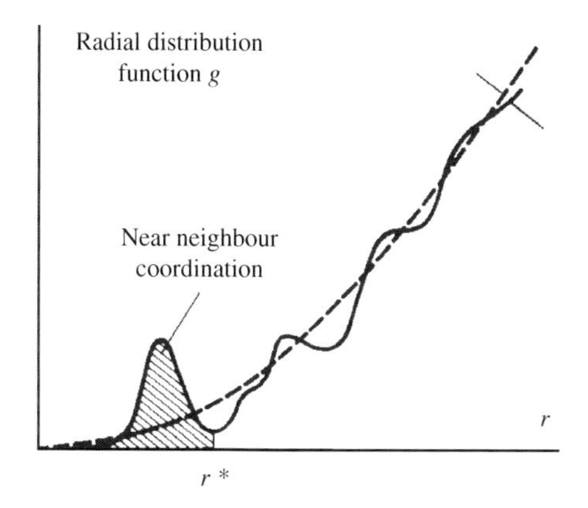

Figure 12.23 Radial distribution function $g(r)$ in an amorphous alloy. Reproduced from Cambridge University Press © 1997.

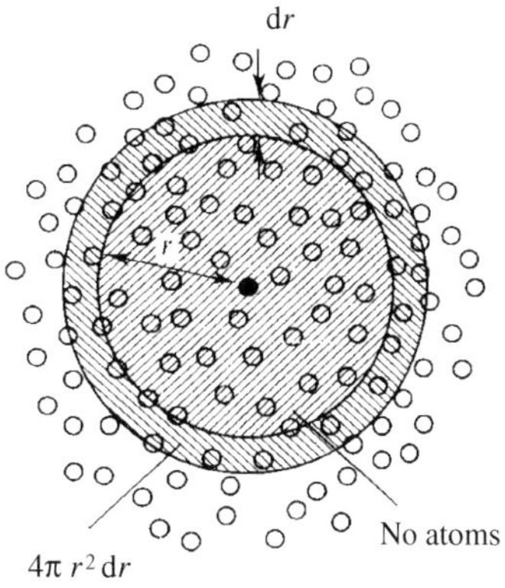

Figure 12.24 Relationship between the radial density function g and the radial density function ρ. Reproduced from Cambridge University Press © 1997.

The average number of atoms dN in the shell between r and $r + dr$ is given by the radial distribution function multiplied by the shell thickness dr. It is also given by the radial density function multiplied by the volume of the shell (Figure 12.24):

$$g = 4\pi r^2 \rho dr = \frac{dN}{dr} \tag{12.5}$$

which gives

$$N = \int_0^r g dr = \int_0^r 4\pi r^2 \rho dr \tag{12.6}$$

The *coordination number* Z can be calculated from the function

$$Z = \int_0^{r^*} g dr \tag{12.7}$$

where r^* is the r value of the minimum after the first peak.

However, amorphous alloy are much more complicated than the melt of a pure metal. For this reason the theory must be generalized and new concepts have to be introduced. Most metallic glasses consist of three or more atoms of different kinds, number and sizes. The probability g is different for different kinds of atoms and this fact must be taken into consideration. Partial probabilities must replace g. For the simplest case, a binary amorphous alloy with A and B atoms there are three different g functions.

$$g = \sum_{i,j=A,B} n_{ij} g_{ij} \tag{12.8}$$

where

$n_{i,j}$ = the fractional number of i, j pairs of atoms
$g_{i,j}$ = the partial radial function for the i, j atom pairs
$n_{AA} = x_A^2$
$n_{AB} = 2x_A x_B$
$n_{BB} = x_B^2$
x_A = atomic fraction of A
x_B = atomic fraction of B.

Metallic glasses are normally ternary, quaternary or even penternary. It is easy to realize that difficulties of theoretical calculations increase strongly with the number of components. Our ambition in this book is merely to indicate the basic ideas behind the theoretical methods and describe the development and recent advances of the theoretical research in this field.

Bernal's Hard-Sphere Model. Voronoi Cells

In the early 1960s Bernal made an attempt to interpret the structure of a melt. His approach was generally in good agreement with the results of X-ray examination of metal melts. Bernal and his coworkers found that the structure of a melt can be successfully described by the following geometrical model.

The atoms are represented by hard spheres situated at the corners of five different types of rigid polyhedra with the following properties:

1. The edges of the polyhedra must all be of approximately the same lengths.
2. It is impossible to introduce one more sphere into the centre of the melt without stretching the distances between the atoms.

The five types of polyhedra, the so-called *canonical holes* or *Bernal polyhedra*, are tetrahedra (73%), half-octahedra (20%), tetragonal dodecahedra (3%), trigonal prisms (3%) and Archimedean anti-prisms (1%), The percentages represent the fractions present in a pure metal melt.

Bernal's liquid model is obviously a hybrid between a close-packed random heap of atoms and atoms arranged in an ordered state. The model is discussed on pages 11–12 in [8].

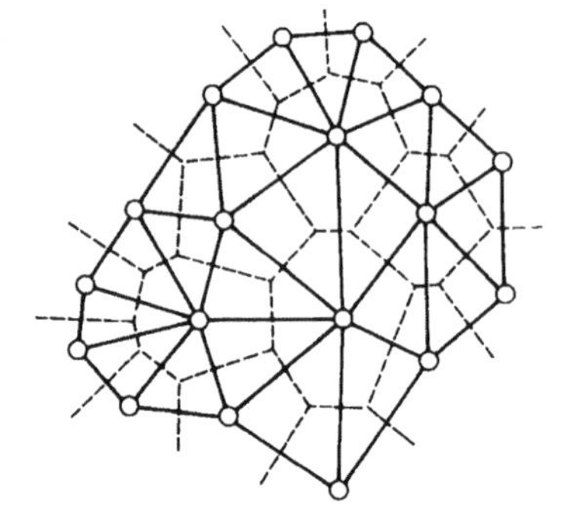

Figure 12.25 Bernal holes and Voronoi cells. Reproduced from Cambridge University Press © 1997.

Voronoi Cells

Owing to the polyhedra, so-called Bernal holes appear everywhere in the melt. In Figure 12.25 the Bernal holes (in two dimensions) are marked by full lines between the atoms. The dashed lines enclose so-called *Voronoi cells.*

- Voronoi cell = the space that is closer to an atom than to all other atoms.

The dashed lines in Figure 12.25 correspond to normal planes, half-way between two atoms.

The Voronoi cells correspond to Wigner–Seitz cells in crystals (page 137 in [8]) but with the difference that they are all different. The collection of Voronoi cells fill the space of the melt completely.

Models of Metallic Glass Structures

Partial Coordination Numbers in Metallic Glasses

We have seen on pages 744–745 that the coordination number Z of the atoms can be derived from the atomic-distribution diagram for melts of pure metals.

With the aid of the extended theory described on page 745, the partial coordination numbers Z_{AA}, Z_{AB}, and Z_{BB}, of a binary alloy AB can be determined separately from three X-ray scattering experiments on three alloys, e.g. with different compositions (x_A and $x_B = 1 - x_A$). The X-ray intensity diagrams can be transformed into partial atomic-distribution diagrams, one for each alloy. The three $g_{i,j}$ functions are plotted as functions of r for each alloy. From these diagrams and Equations (12.5)–(12.8) on page 745) the partial coordination numbers Z_{AA}, Z_{AB}, and Z_{BB} can be determined.

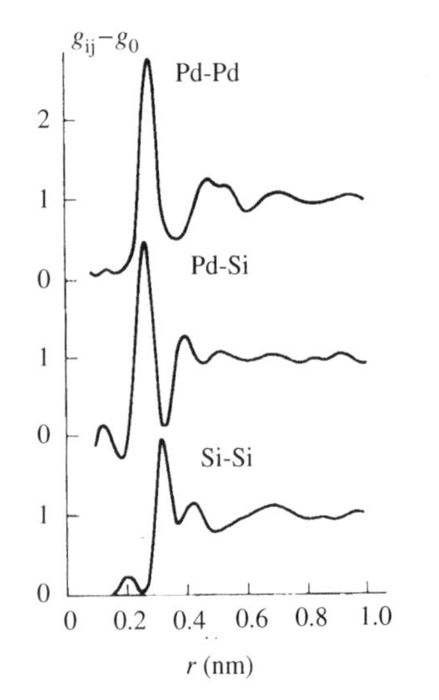

Figure 12.26 Partial radial distribution functions $g_{i,j}$ for amorphous Pd–Si. g_0 represents an *even*, average distribution everywhere (see Figure 12.23 on page 745). g_{ij} represents the *deviation* from the average distribution. Reproduced from Cambridge University Press © 1997.

Figure 12.26 shows a simple example. The simplest possible binary 'alloys' of a mixture of Pd and Si are Pd–Pd (100 at% Pd + 0 at% Si), Pd–Si (50 at% Pd + 50 at% Si) and Si–Si (0 at% Pd + 100 at% Si).

The fractions of ij bonds can be derived from the partial radial distribution functions. This method has been used determine the partial coordination numbers of the components in the metallic glasses. Normally, there are more than two components. Some examples are given in Table 12.3.

Table 12.3 Measured solvent atom fraction around solute atoms B and random distribution of solvent atoms A around solute atoms B in some binary metallic glasses

Alloy A-B	Fraction of A neighbours of B (at%)	Fraction of A neighbours of B if there were no ordering at all (at%)
$Fe_{80}B_{20}$	100	80
$Ni_{81}B_{19}$	100	81
$Ni_{64}B_{36}$	88.8	64
$Cu_{57}Zr_{43}$	56.2	57
$Zr_{65}Ni_{35}$	70.1	65
$Zr_{57}Be_{43}$	70	57

Table 12.3 shows that the B (boron) atoms are surrounded by Fe atoms to 100% for the alloy Fe_{80} B_{20}. In the alloy Ni_{81} B_{19} the B (boron) atoms are surrounded only by Ni atoms. These amorphous alloys are said to be *solute centred* and have no boron–boron bonds at all.

Each Zr atom in the alloy Cu_{57} Zr_{43} seems to be surrounded by ~56% Cu atoms, which almost corresponds to a random distribution. In the last two alloys the number of Zr–Ni and Zr–Be bonds, respectively, is larger than expected from a random distribution.

Similar experimental examinations of many other metallic glasses have confirmed the generally accepted model of the structure of metallic glasses:

1. A metallic glass can be regarded as a liquid with very high viscosity.
2. The atoms are mainly arranged in clusters with strong forces between the atoms.
3. The interactions between the clusters are weaker than that between the cluster atoms.

Bernal's Hard-Sphere Model and Metallic Glass Clusters

Bernal's hard-sphere model for pure metal melts presupposes that the melt is built up from hard spheres of equal sizes.

Metallic glasses consist of two, three or more different elements with different radii. They have high, negative heats of mixing, which agrees well with the assumption that metallic glasses are built up from clusters with a central atom of one of the elements surrounded by a 'shell' of nearest neighbours of the other elements. Both solute-centred and solvent-centred clusters occur.

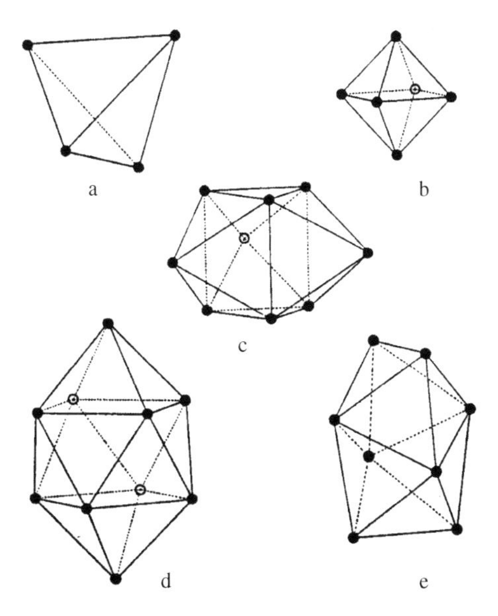

Figure 12.27 Unit blocks in Bernal's liquid model. The five canonical holes of random close-packing are: a) tetrahedron b) octahedron c) dodecahedron d) trigonal prism e) Archimedean antiprism.

Table 12.4 Number of nearest neighbours in 'Bernal clusters'.

Polyhedron	Numbers
Tetrahedron	4
Octahedron	6
Dodecahedron	9
Trigonal prism	10
Archemedean antiprism	8

The Bernal hard-sphere model applied to metallic glasses can be described as follows. The five Bernal holes can be regarded as the skeletons of the clusters. In a solute-centred metallic glass a solute atom is placed inside a Bernal hole, while the nearest neighbours are located at its corners.

Figure 12.27 on page 747 shows the five Bernal's holes. It is easy to count the numbers of nearest neighbours in each case as they agree with the number of corners. The numbers are listed in Table 12.4.

In metallic glasses other types of clusters with 7, 11, 12 or more atoms around the central atoms may form. If the atoms are assumed to be hard spheres with different radii for different types of atoms, the sizes of the clusters can be calculated as they depend on the ratio of the atomic radii. The ratios for the hard spheres have been calculated and the result is presented in Figure 12.28.

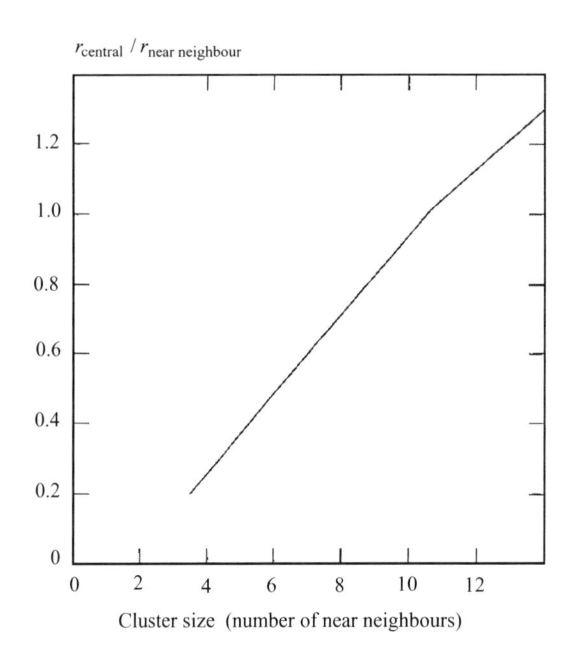

Figure 12.28 Relationship between cluster size and the ratio of the radii of the central atom B and the surrounding atoms A.

The hard-sphere model indicates what types of structures that can exist in the metallic glasses for space reasons. Other types of holes than the Bernal holes will stretch or compress the distances between the atoms. The strain related to such changes will increase the potential energy of the system. The Bernal holes correspond to equilibrium, i.e. a minimum of the potential energy.

The Structure of the Fe–B and Ni–B Metallic Glasses

In order to emphasize the relationship between the cluster type and the ratio of the radii of the central atom and its nearest neighbours we will discuss the structures and clusters of the metallic glass alloys Fe–B and Ni–B (see Table 12.3 on page 747). As a guide and inspiration we start with the structure of the Si–O system, i.e. of quartz, SiO_2, given in the box on next page.

Even if the nature of the forces between the atoms are very different in crystalline silica and the two metallic glasses Fe–B and Ni–B, respectively, there are analogies.

The ratios of the atomic radii r_B/r_{Fe} and r_B/r_{Ni} are both ~ 0.3. A check in Figure 12.28 shows that the most likely building unit is a tetrahedron. A comparison with the upper-left figure in the box on the next page shows that it is possible to place the boron atom in the centre of a tetrahedron.

Hence, the metallic glasses Fe–B and Ni–B are likely to be built up by tetrahedra of Fe_4B and Ni_4B clusters, respectively. The bonds between the clusters are very weak in both cases. The Fe atoms and Ni atoms are not shared among the clusters as in the case of quartz.

Structures of Crystalline and Molten Silica

From X-ray measurements of crystalline and molten quartz one has built a two-dimensional model of their structures. The figure below shows the result.

The X-ray diffraction studies have shown that the building units of crystalline quartz are SiO_4 tetrahedra, shown in the upper part of the figure, where the proportions between the black O atoms and the white Si atoms are drawn on the proper scale.

The lower left figure shows the tetrahedra of crystalline SiO_4, where the centres of the atoms are marked in the figure.

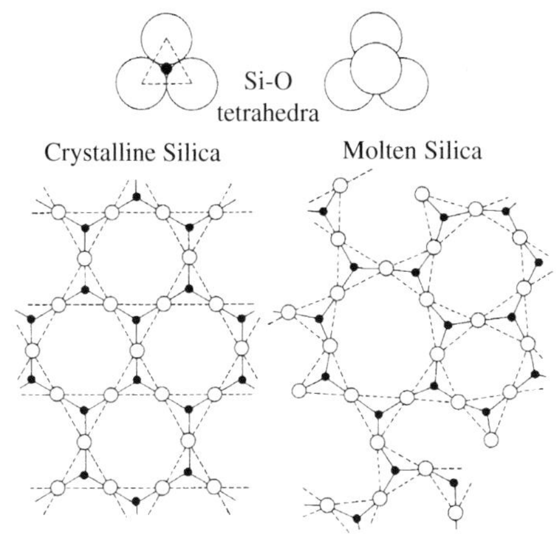

(Reproduced from Elsevier © 1974.)

The lower-right figure shows that the tetrahedron structure persists partly in the melt, even if some of them are deformed. The melt has a high viscosity.

The small O atom is in the centre of the tetrahedral as expected from the hard-sphere model.

Example 12.2

Consider the metallic glass $Cu_{57}Zr_{43}$ in Table 12.3 on page 747. The ratio of the atomic radii of Cu and Zr is ~ 0.8.

a) What type of cluster is likely to form when copper and zirconium in the proper proportions are melted together?
b) What happens to the atoms that are not involved in the clusters?

Solution:

a)
With the aid of Figure 12.28 on page 748 we can read that the ratio of atomic radii equal to 0.8 corresponds to a cluster size of 8 atoms, located at the corners of an Archimedean prism. The cluster consists of one Cu atom inside the prism with 8 Zr atoms at the corners of the prism.
b)
Each metallic glass unit can form maximum 5 clusters and leave at least 3 Zr atoms and 52 Cu atoms. The single atoms are free in the sense that they can move everywhere within the melt but can not leave it. The free atoms show a random distribution in the melt.

Answer:

a) The clusters consist of 8 Zr atoms in the corners of an Archimedean prism that surround one central Cu atom.
b) Only a small fraction of such clusters can form. In addition, the melt contains a random distribution of a minor fraction of free Zr atoms and an excess of free Cu atoms.

Figure 12.29 Model of a simple binary metallic glass. The quasiequivalent clusters share faces, edges or vertices in the packing configuration of a Zr–Pt metallic glass. The black atoms are the solute (Pt) atoms. Some of them are hidden behind solvent atoms. Reproduced from Nature © 2006.

Structure of Metallic Glasses

Knowledge of the partial coordination numbers is helpful for determination of the structure of metallic glasses. As an example (given by the guest editors Greer and Ma on page 612 in [9]) we will discuss the structure of an amorphous Zr–Pt alloy, which have relatively few solute atoms Pt. In analogy with the first alloys in Table 12.3 on page 747 there are no Pt–Pt bonds. The Pt atoms are surrounded by 100% solvent atoms Zr. Recent research has shown that low solute contents is likely to result in a dense-packed structure of quasiequivalent clusters. Each cluster consists of a Pt atom surrounded by Zr atoms. Figure 12.29 shows a model of such a structure.

There is a strong tendency of formation of as many bonds as possible between unlike atoms because of the negative heat of mixing, which favours glass forming. In each amorphous alloy the atoms are arranged in such a way that the total energy is minimized. The cluster size and type depends on the relative sizes of solute and solvent. The clusters overlap but the cluster–cluster distances will be larger than the solute–solvent distances. This leads to so-called *medium-range order* because of interaction between solvent atoms and a dense packing structure in space.

If the solute Pt atoms are replaced by much smaller Be atoms, the structure would be considerably changed because the small Be atoms can not accommodate the same number of Zr atoms as the Pt atoms. Therefore, the solute concentration must be increased and the cluster structure will be radically different. The structure and properties of metallic glasses obviously varies strongly with the composition of the alloy.

12.5.2 Recent Progress in Structure Research of Metallic Glasses. The Efficient Cluster Packing Model. Quasicrystals

The discovery of bulk metal glasses was a great step forward from the experimental point of view. The development is illustrated in Table 12.2 on page 738. The properties of the new metallic glasses were examined with the aid of available experimental methods and models were set up to explain their behavior. The atom distribution showed lack of symmetry, but was not completely random as both short-range order and medium-range order were found. The struggle to find a model that could explain the characteristics of the bulk metallic glasses was not successful during the 1990s. None of the existing models could, for example, explain why the most stable metallic glasses have three or more components, why the solute and solvent atoms preferably differ in sizes. It was also hard to find a general model that could explain the medium-range order satisfactorily. Doubts were raised that it would be impossible to find an acceptable model that could describe the many features of the metallic glasses.

The Efficient Cluster Packing Model ECP

The situation changed at the beginning of the new millennium. It was known that the properties of the metallic glasses were associated with their dense packing structure. Efficient filling of space was an important condition to find an adequate model. Attempts were made with the aid of computers to build efficient cluster packing models that describe a new way to achieve dense atomic packing.

Simultaneous new sophisticated experimental methods were developed, such as wide-angle and anomalous X-ray diffraction, fluctuation electron microscopy, high-resolution transmission electron microscopy and positron annihilation spectroscopy.

Quantitative computer calculations with the aid of the ECP model have confirmed many characteristic of metallic glasses, for example:

- Agreement between predicted and observed coordination numbers within the experimental error.
- Calculated solute–solute medium-range order show good agreement with experimental measurements up to the limit ~ 1 nm.
- Composition research of new metallic glasses is facilitated as the calculations serve as a useful guide.

- Calculated and measured densities show acceptable agreement.
- Defects are compatible with the ECP model.
- The ECP model predicts no more than 4 different atomic sizes.

The ECP model is described by Miracle and his coworkers in [10].

Short-Range Atomic Order

Short-range order depends on the electronic structure and the topology of the metallic glass. It is often extremely sensitive to even small changes in composition. Short-range order controls the metallic glass formation and the stability of the glass. On decomposition, phases with a similar structure is favoured and unlike structures are obstructed.

Glasses are supercooled liquids with extremely high viscosities and dense random-packing structures. To obtain highest possible atomic density in metallic glasses with a small metalloid component, e.g. phosphorous, it was earlier assumed that the P atoms could fill the empty holes. However, the holes turned out to be too few and too small to be able to accommodate the P atoms. Instead, a strong preference for icosahedral short-range order has been predicted in liquids and metallic glasses and is obtained in many experiments and computer simulations. A metallic glass has the structure that corresponds to the lowest possible total energy.

The icosahedral short-range order (Figure 12.30) is confirmed by X-ray and electron diffraction. It is much more common than used to be expected. The ECP model predicts that the dominant clusters have coordination numbers within the interval 8–20, depending on the relative solute size. However, the normal short-range order in the ECP model is solute-centred clusters, while icosahedral short-range order is associated to solvent-centred clusters. These facts need further research work.

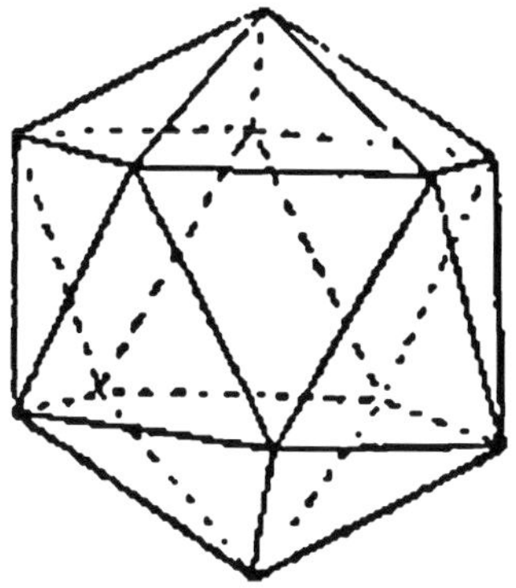

Figure 12.30 Icosahedron. An icosahedron is a polyhedron with 12 corners, 20 triangle faces and 30 edges. The figure shows that the icosahedron has a 5-fold symmetry.

Medium-Range Atomic Order

A combination of experimental and computer methods turned out to be very fruitful. Both solute-centred and solvent-centred clusters were studied, i.e. a central solute atom, surrounded by mainly solvent atoms and vice versa. The short-range order observations were found to be in agreement with the well-known large negative heat of mixing. The principle of no solute–solute bonds offered a simple explanation of the medium-range order of solute atoms and a nearly complete disorder of the solvent atoms, which resulted in a close-packed orientation of the clusters in space. Fluctuation electron microscopy confirmed the theoretical results concerning three- and four-body interaction and medium-range order.

Quasicrystals

There is an empirical rule that says that no crystal structure can possess a fivefold rotational symmetry. No crystals that violate this rule have ever been observed. There is also a fundamental rule, which is taken for granted, that a Bragg diffraction pattern, e.g. a point diffraction (von Laue diffraction) pattern can only appear if the atoms are arranged in a periodic translation order.

For this reason it was a scientific sensation when Shechtnan and his coworkers in 1984 [11] discovered a fivefold point diffraction diagram when they examined a micrometre grain of a rapidly cooled alloy. Later, their discovery has been confirmed many times: *Metallic glasses may show point diffraction patterns* in spite of their total absence of regular order of their atoms. They violate both rules mentioned above. Materials with these properties are called *quasicrystals.*

Figure 12.22 on page 744 is an excellent example of a point diffraction pattern of a metallic glass that shows fivefold symmetry with 2×10 points in the rings.

Another example of fivefold symmetry is given in Figure 12.31 that shows a SEM (scanning electron microscope) picture of the surface of a rapidly quenched quasicrystal.

It is now understandable that icosahedrons with their fivefold symmetry are involved in the theoretical calculations on metallic glasses.

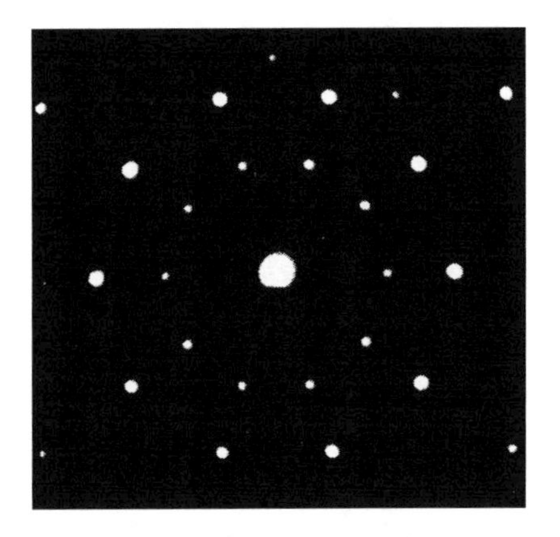

Figure 12.22 Laue diagram of a quasicrystalline material. High-resolution diffraction pattern of $Al_{86}Mn_{14}$ quasicrystals obtained by transmission electron microscopy.

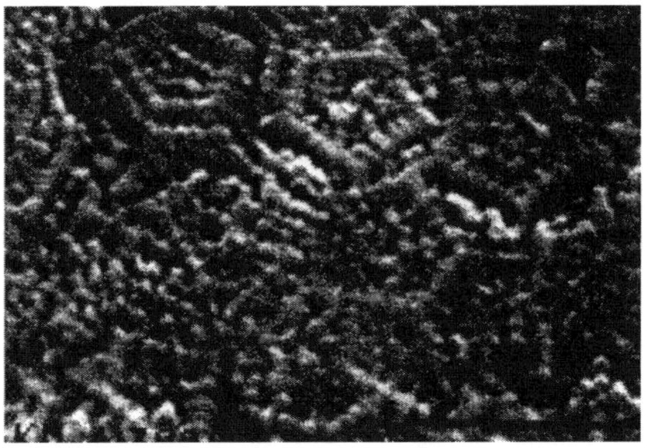

Figure 12.31 Topology of the surface of a rapidly quenched quasicrystal. The 5-fold symmetry is present also in the metallic glass.

Free Volume

Free volume is defined as the volume in excess of the ideal densely packed amorphous structure.

Density fluctuations always exist in metallic glasses and are an important feature that influence their properties. They are caused by holes but not holes in the sense of a vacancy of a single atom. The holes in metallic glasses consist of 'interstitial void clusters' between the clusters that form the glass.

The theory predicts two kinds of voids, intrinsic *small voids* with 1–9 missing atoms and *large voids* with > 10 missing atoms. The larger voids are assumed to be strongly non-spherical and can be removed by annealing. The small voids do not disappear at annealing. The large voids are essential for atomic transport within the metallic glass, which is not the case for the small voids. We will come back to the influence of the free volume in connection with the properties of metal glasses.

In his doctoral dissertation 'Phase Transformations in Computer Simulated Icosahedrally Ordered Phases' in 2003 Fredrik Zetterling [12] performed calculations on large clusters of icosahedra. He showed that holes appear automatically as a consequence of the simulation calculations. Some of his illustrations are shown below.

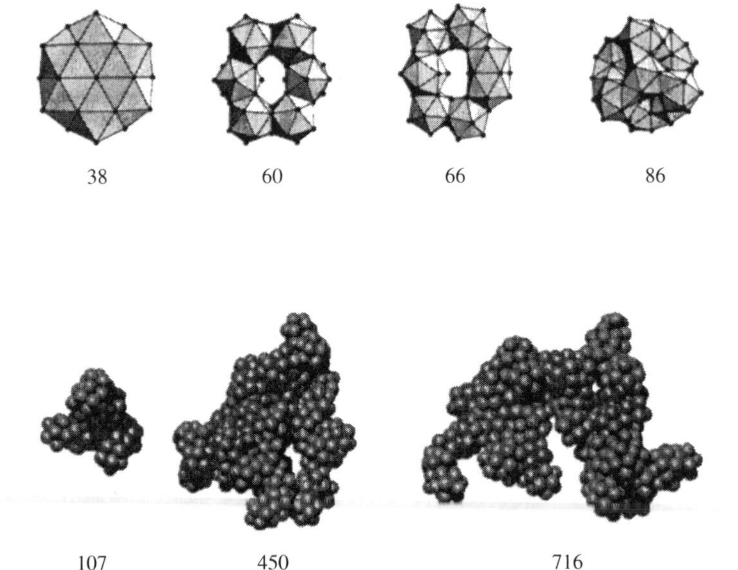

Figure 12.32 Representative icosahedron structures. The numbers represent the number of atoms in the clusters. Reproduced from Royal Society of Chemistry © 2001.

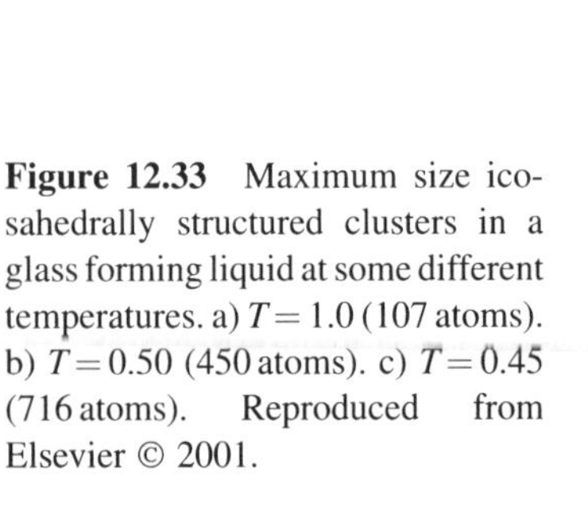

Figure 12.33 Maximum size icosahedrally structured clusters in a glass forming liquid at some different temperatures. a) $T = 1.0$ (107 atoms). b) $T = 0.50$ (450 atoms). c) $T = 0.45$ (716 atoms). Reproduced from Elsevier © 2001.

Structural Dynamics

Finally, it will be mentioned that ECP is a powerful but *static* model of metallic glasses, valid at a constant temperature. During the first decade of the new millennium the so-called *structural dynamics* has been developed, i.e. the structures of metallic glasses are studied as a function of time and temperature.

12.6 Comparison of the Structures of Metallic Glasses and Amorphous Alloy Melts. Rough Models of Metallic Glasses and Amorphous Alloy Melts

12.6.1 Rough Model of Amorphous Alloy Melts

Intense efforts have been made to understand the structures of metal glasses, but much work still remains. A short review of the present knowledge of the structures of metallic glasses is sketched below. It will be used for a comparison between the structures of the metallic glasses and the structures of the far more unknown amorphous alloy melts. In this way one may find a tentative model for the amorphous alloy melts, which is compatible with the present ideas of the structures of the metallic glasses.

12.6.2 Structures of Metallic Glasses

Based on experimental evidence (see Section 12.4) and development of the theory (Miracle et al [10]) the following rough model of metallic glasses can be sketched.

The metallic glasses always have the structure that results in lowest possible total energy. This is achieved by the densest possible packing of atoms and groups of atoms, so-called clusters. The relative sizes of the atoms are important. The coordination numbers vary between 8 and 20.

Two types of clusters seem to occur:

- a solute atom surrounded by solvent atoms;
- a solvent atoms surrounded by solute atoms.

The clusters vary in size depending on the sizes and the coordination numbers of the atoms. This influences the possibility of an efficient packing structure. If the central atom is large and the surrounding atoms are small the number of atoms in the cluster can be large. If the central atom is small and the surrounding atoms are large there is only enough space for a few surrounding atoms. Such clusters will be small and contain only a few atoms. The solute-centred type of cluster is more common than the solvent-centred one.

The clusters are assumed to be stable units even if they interact with each other and the outer atoms are added and disappear incessantly. The forces between the atoms within the clusters are responsible for the short-range order. The medium-range order is associated to the interaction between the clusters.

The so-called ECP theory (efficient cluster packing model) focuses on dense packing and is a powerful tool for calculations. It predicts the existence of voids of various sizes, which seems to play an important role in the structures.

12.6.3 Structures of Amorphous Alloy Melts

At the glass-transition temperature T_g an undercooled alloy melt can be transformed into a metallic glass, provided that the cooling rate exceeds the critical cooling rate (Figure 12.1 on page 731 and Table 12.1 on page 733).

Below the temperature T_g the alloy is an amorphous solid, i.e. a metallic glass. Above T_g the alloy is an amorphous melt if its temperature is low enough to avoid crystallization.

In order to find a model for the structure of amorphous alloy melts it is reasonable to start with the properties of such undercooled melts and compare them with the corresponding metallic glasses.

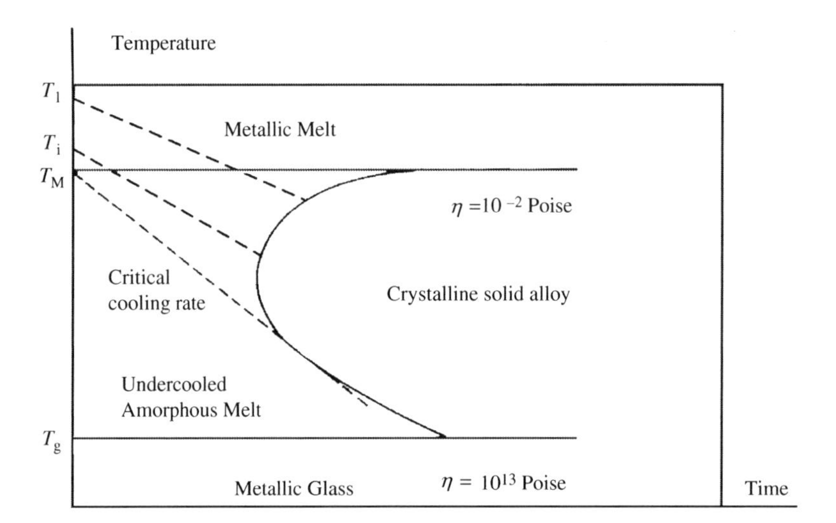

Figure 12.1 CCT diagram of an amorphous alloy.

12.6.4 Comparison between Some Properties of Metallic Glasses and Amorphous Alloy Melts

Viscosity

Amorphous alloy melts have viscosities of the order of 10^{-2} Poise, which vary strongly with the temperature.

Metallic glasses have extremely high viscosities of the order of 10^{13} Poise.

Diffraction Patterns

Very few X-ray and neutron-scattering measurements on amorphous alloy melts and metallic glasses with the same composition have been performed and reported in the scientific literature. One of the few publications is Fukunaga's and Suzuki's examination of the binary $Pd_{80}\ Si_{20}$ alloy [13].

Fukunaga and Suzuki performed neutron-scattering experiments on the metallic glass $Pd_{80}\ Si_{20}$ at room temperature and the same amorphous alloy at 980 °C as a melt. From the measurements they derived the radial distribution function $g(r)$ (page 745) for the metallic glass and the melt and plotted the deviation from an even distribution $g(r) - g_0$ versus r. These functions are shown in Figure 12.34.

Figure 12.34 shows that the difference between the curves for the metallic glass and the molten amorphous alloy is very small. The melt shows a short-range order similar to that of the metallic glass but the short-range order is somewhat more pronounced in the melt.

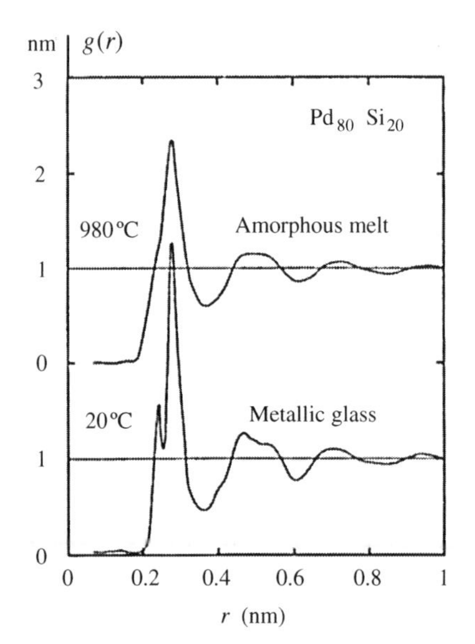

Figure 12.34 The pair potential function $g(r)$ divided by the even distribution function g_0 plotted versus r for the metallic glass and the amorphous alloy melt of $Pd_{80}\ Si_{20}$.

The conclusion is that the amorphous alloy melt also has a cluster structure that is essentially the same in the molten and glassy alloy. Convection occurs in the molten amorphous melt, which counteracts and reduces the viscosity. Convection is absent in the metallic glass because of the extremely high viscosity. This explains the startling difference in viscosity of the molten and the glassy amorphous alloy.

Internal Forces

When a metallic glass is heated and the temperature T_g is reached a transition occurs when the metallic glass is transformed into an undercooled amorphous melt. The temperature decreases because energy is required to break the bonds between the particles in the metallic glass. This energy is taken from the thermal energy of the surroundings. Figure 12.2 on page 733 shows that the potential-energy difference between the amorphous melt and the metallic glass is very small. The conclusion is that the internal forces are approximately the same in the melt and in the glass.

Thermal Expansion

The thermal expansion and the volume change that has been observed at the transformation of an amorphous alloy melt into a metallic glass indicates a change of the potential energy of the system at the phase transformation. Very few measurements on thermal expansion have been published in the scientific literature. One of the few is the study by Kimura et al. [14] on Pd–Ni–Cu–P alloys of various compositions.

They showed that the thermal volume-expansion coefficient, which is a material constant, varied from $8 \times 10^{-6}\,K^{-1}$ to $18 \times 10^{-6}\,K^{-1}$ for the metallic glasses and from $3 \times 10^{-2}\,K^{-1}$ to $3.4 \times 10^{-2}\,K^{-1}$ for the corresponding amorphous alloy melts.

The thermal expansion of the metallic glasses is obviously very small compared to that of the amorphous alloy melts. The difference is nearly a factor 10^{-4}.

Both metallic glasses and amorphous alloy melts are liquids and there is no abrupt change but a smooth volume change when the system changes from an amorphous alloy melt into a metallic glass and vice versa. When crystalline materials change from a liquid into a solid state, the transformation is abrupt.

The thermal expansion coefficient for common metals is $\sim$ 2–5%, i.e. very much larger than those of metallic glasses and amorphous melts.

Molar Heat Capacity

The molar heat capacity at constant pressure C_p, which is a material constant, has been measured (Inoue [4]) for the metallic glasses Zr–Al–Cu of various compositions. The measurements were performed after annealing the samples for 12 h at different temperatures from 400 K up to the glass-transition temperature and at one temperature above the glass transition region 630–675 K. The samples were rods, cast in a copper mould. The result of one measurement is shown in Figure 12.35.

Figure 12.35 shows that the molar heat capacities initially decrease with increasing temperature, pass a minimum value and then increase rapidly. The more the temperature approaches the glass-transition temperature ($\sim$680 K) the steeper will be the curves.

The decrease of the C_p values is a result of an exothermic reaction during the heating that reduces the vibration contribution to C_p. The reaction probably reflects a minor rearrangement of the atoms inside the clusters or of the free atoms between the clusters. This increases the short-range order and a relaxation process starts spontaneously. This is related to the negative heat of mixing to form a more densely packed cluster, which corresponds to lowest possible potential energy of the system.

The steep increase of the curves in the vicinity of the glass-transition temperature T_g is the result of an order–disorder transformation and proves that a rearrangement of the structure occurs at the metallic glass temperature. Such processes have been discussed extensively on pages 316–317 in [8]

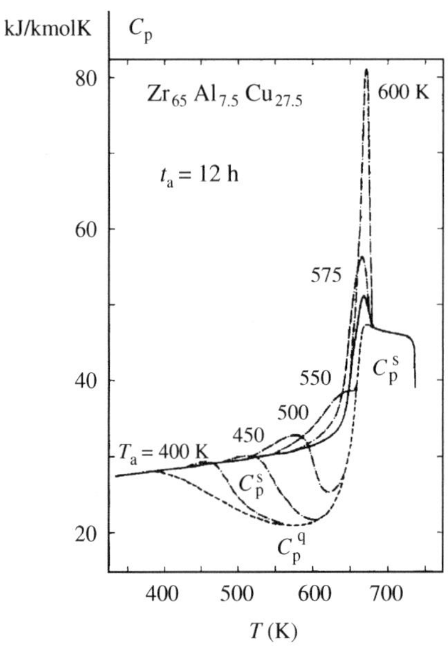

Figure 12.35 The molar heat capacity C_p as a function of temperature for the alloy $Zr_{65}Al_{7.5}Cu_{27.5}$ annealed for 12 hours at temperatures from 400 K to 620 K. The solid line represents the curve for a sample heated to 690 K. t_a = annealing time, T_a = annealing temperature, dT/dt = heating rate. Reproduced with permission from Trans Tech © 1999.

for β brass and on pages 329–330 in [8] for ferromagnetic materials close to the Curie point. The order–disorder transformation in β brass (CuZn) is decribed in the box below.

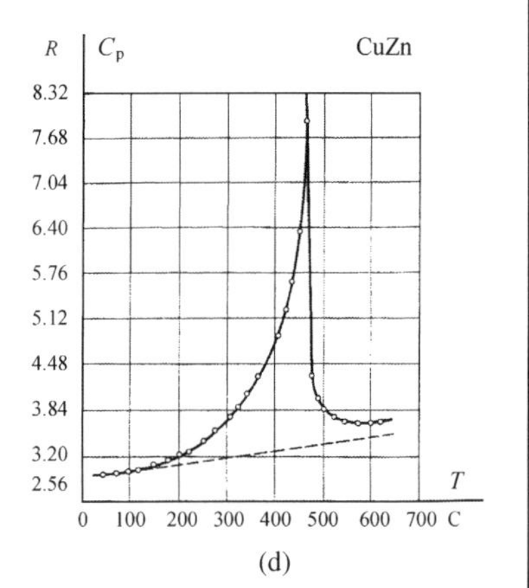

(d)

Order–Disorder Transformation in β brass

As the properties of solids depend strongly on their structure a sudden transformation from one structure to another will result in abrupt changes of the properties of the solid, for example its heat capacity and its thermal and electrical conductivities. As a concrete example we will discuss the alloy β brass, which consists of 46–50% Zn dissolved in Cu. Its chemical composition is close to CuZn.

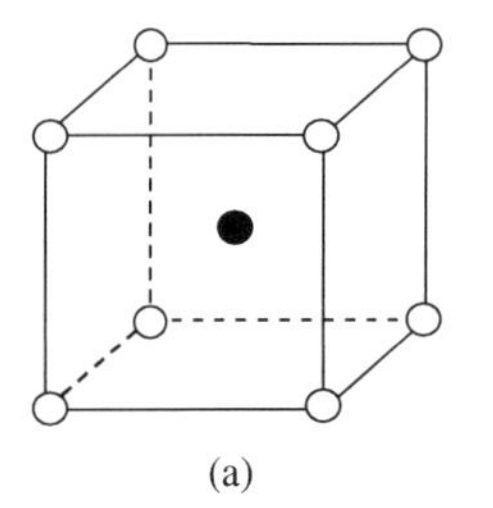
(a)

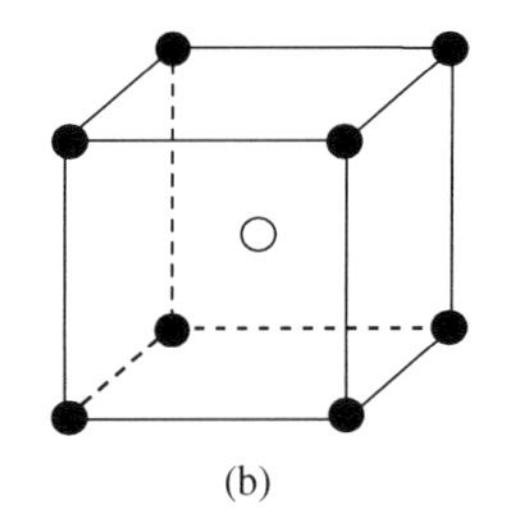
(b)

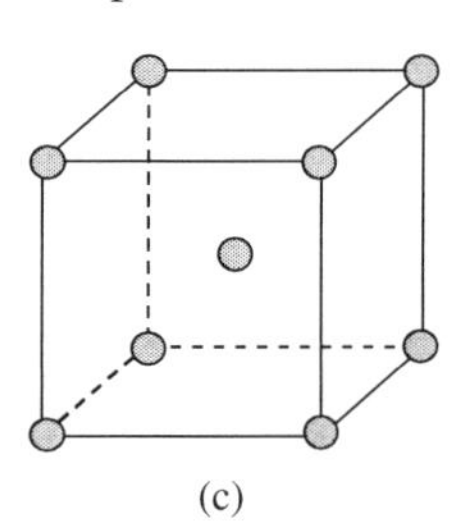
(c)

At room temperature β brass is an ordered alloy. The unit cell corners are occupied by Cu atoms and the centre by a Zn atom (Figure a) or reversed (Figure b). The structure can be described as two simple cubic structures of either type of atoms. The structures are displaced relative to each other by $a/2$, where a is the lattice constant and side of the cubes.

The order is maintained at constant temperature. When the temperature changes and the properties, for example the heat capacity, gradually change until the temperature reaches a critical value T_{cr}. At the critical temperature the kinetic motion is violent enough to let the atoms change sites rather freely and the structure becomes disordered. It can be described as a BCC phase with a completely random distribution of the Cu and Zn atoms (Figure c).

Figure d shows the molar heat capacity as a function of temperature. The critical temperature T_{cr} at which the order–disorder transformation occurs, is approximately 740 K or 467 °C, read from Figure d, which is identical with Figure 6.23 on page 316 in [8]. Energy is required to change the system from an ordered to a disordered state. The increased need of energy appears as an abrupt increase of C_p in the vicinity of the critical temperature. When the transformation is carried through the C_p value decreases rapidly to the 'normal' value of the disordered state.

Electronic Structures

The increase of the molar heat capacity with increasing temperature at the glass-transition temperature proves that the transition from a metallic glass to an amorphous alloy melt is an order–disorder transformation via an intermediate phase, analogous to β brass in the box above.

This implies that there must be some ordering in the structures of amorphous alloy melts and metallic glasses independent of the type of atoms. Apart from the formation of larger atomic units, i.e. clusters their positions are distributed at random.

Instead, it is the order related to the *electronic structure* of the amorphous alloy melt and the metallic glass and *the bonds* between the clusters that changes at temperature T_g.

In metallic intermediate phases with an ordered structure, such as β brass, the forces between the atoms are partly ionic and partly metallic (see Sections 3.2, 3.3, 3.4 in [8] or other comparable literature) concerning ionic and metallic bonds.

In a binary intermediate phase of AB type the A atoms are either positive or negative ions and the B atoms have opposite charges. In a metallic glass one type of cluster or atom may be positive or negative ions, while the other type of cluster or atom are ions with the opposite sign.

Hence, all the excess electrons and holes (missing electrons are equivalent with positive charges) are bound to a cluster or an atom. Therefore, the electrons and holes can not move freely from one cluster or atom to another.

Exchange of the positions of the ions is necessary for deformation or a flow of the material. Hence, the electrical structure offers considerable obstacles for position changes and contributes to the high viscosity of metallic glasses.

12.6.5 Rough Compatible Models of Metallic Glasses and Amorphous Alloy Melts

The model of an amorphous melt must be able to explain the differences between an amorphous melt and a dense-packed solid metallic glass. It must also be compatible with the model of metallic glasses that is far better known than that of amorphous melts.

The fact that a 5-fold symmetry (page 751) appears both is amorphous melts and metallic glasses can not be random. It gives associations to Bernal's hard-sphere model of a liquid (pages 11–12 in [8]). He considered a liquid as a mixture of polyhedra with atoms in the corners.

That the internal forces are of the same magnitude and that the viscosities differ by a factor 10^{15} are two facts that are hard to combine.

A Tentative Model

One possibility to combine the contradicting experimental facts could be to assume that

1. Both amorphous alloy melts and metallic glasses consist of *clusters* and *free atoms*.
   In the amorphous *melt* the single atoms are free in the sense that they are not bound to any special cluster and can move easily within the whole melt.
   In the *metallic glass* the single atoms are bound and *not* free to move within the glass.
2. The clusters vary in size and are arranged in *superclusters* in the shape of polyhedra of different sizes and types with clusters instead of atoms in the corners. If so, it is not possible to obtain a complete filling of the space. Voids of various sizes appear in the structures of both amorphous alloy melts and metallic glasses. They are stable, i.e. do not disappear.
3. The free atoms serve as a lubricant between the heavy and large clusters. The free atoms are probably comparatively small and can easily move into and out of the clusters and the space between the clusters in the amorphous alloy melts.

Can the Model Explain the Experimental Observations?

A minimum demand on an acceptable model is that it can explain all experimental observations including the ones in Section 12.6.4.

Viscosity

The difference in viscosity between a metallic glass and an amorphous alloy melt is very large. This difference is hard but not impossible to explain. The amorphous alloy melt consists of clusters and free or single atoms. The viscosity of the amorphous alloy melts can be explained by the high mobility of the free atoms. The large and heavy superclusters in the amorphous alloy melt have a very low mobility and the free atoms serve as lubricant and enables layers of superclusters to slide relative to each other.

It is more difficult to explain the dramatic change of the viscosity at and below the glass-transition temperature. If the free atoms disappear, only the superclusters of different sizes and orientations would remain in the metallic glass. It is easy to understand that this would result in a high viscosity as the lubricant is no longer available.

Of course the free atoms can not disappear but they may be trapped inside the clusters. This assumption is supported by the fact that the free atoms often consist of small metalloid atoms such as B, C and P atoms, which can be accommodated inside the clusters in the metallic glasses. It is well known that most good metallic glasses contain metalloid atoms.

The reason for trapping the free atoms inside the clusters would be that such a structure would result in a more close-packed structure than other structures and consequently correspond to the lowest possible potential energy.

Diffraction Patterns

The assumption of superclusters can explain the presence of quasicrystals (icosahedra) in the amorphous melt and in the metallic glasses (pages 751–752).

Internal Forces

The clusters, which consist of a central solute atom surrounded by solvent atoms or vice versa, must be stable with strong forces between the atoms. The 5-fold symmetry in the shape of quasicrystals (page 751) occurs both in the undercooled amorphous melt and in the metallic glass. The minor difference in potential energy (enthalpy) indicates that neither the forces between the atoms in the clusters nor the interaction between the clusters in the superclusters are changed at the phase transition and gives no contribution to the change in potential energy. Instead a weak interaction between the superclusters may be responsible for the energy change and change of the temperature at the phase transformation (pages 740 and 733).

Electrical Conductivity

Measurements of the electronic conductivity, alternatively the resistivity, have been performed and analyzed for some metallic glasses and their amorphous alloy melts. One of them is the study by Johnson [18] on the bulk metallic glass $Pd_{44}Ni_9Cu_{27}P_{20}$ in 1996. His results are shown in Figure 12.36.

The figure shows that the resistivity decreases with increasing temperature. After the first jerk at T_g comes another two jerks called I and II in the undercooled region and the resistivity falls rapidly when the material crystallizes.

The decrease of the resistivity in the metallic glass is opposite to what is observed in crystalline materials. The decrease with increasing temperature may be the result of a relaxation process of the atomic structure in the metallic glass with a higher mobility of the electrons as a result. Some of the electrons are no longer bound or missing as ions but appear as free electrons, which can move easily.

The decrease of the resistivity or increase of the electrical conductivity at temperature T_g when the metallic glass transforms to an amorphous alloy melt is a result of more free electrons in the conduction bands when the ionic bonds in the metallic glass are broken.

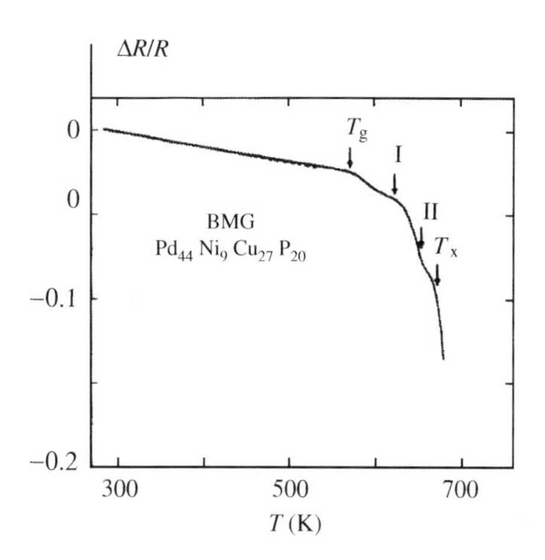

Figure 12.36 The relative resistivity $\Delta R/R$ as a function of temperature for a cast $Pd_{44}Ni_9Cu_{27}P_{20}$ amorphous sheet with the thickness 1 mm and a width of 5 mm. Reproduced with permission from Trans Tech © 1999.

The resistivity decreases significantly at the temperatures marked by I and II. This two-stage decrease of the resistivity corresponds to relaxation processes of the atomic structure in the amorphous alloy melt with a higher mobility of the electrons as a result.

The sudden and large decrease of the resistivity at temperature T_x is due to crystallization of the undercooled amorphous alloy melt. A relaxation process occurs, the atoms form a lattice and a large number of electrons become free instead of bound to clusters or single atoms. The bonds in the sample become metallic to 100%.

12.7 Casting of Metallic Glasses. Crystallization Processes in Amorphous Alloy Melts

A CCT diagram is a tool to judge the possibilities to produce a metallic glass. The CCT diagrams are often determined by experimental methods (pages 733–734). CCT diagrams can alternatively be drawn with the aid of either analytical or numerical theoretical models. It is hard to find all the physical data needed to obtain a correct theoretical description of the process. However, a theoretical model may help to design a process or a particular material.

The model must include

- the heat flow
- the volume fraction
- the interphase kinetics.

These three parts have been discussed earlier in Chapters 4, 6 and 7. We will refer to them in the following discussion.

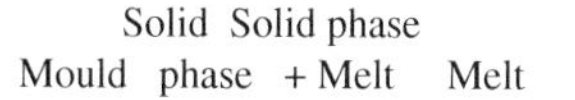

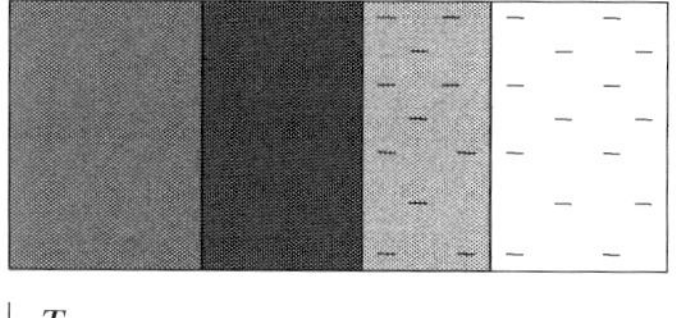

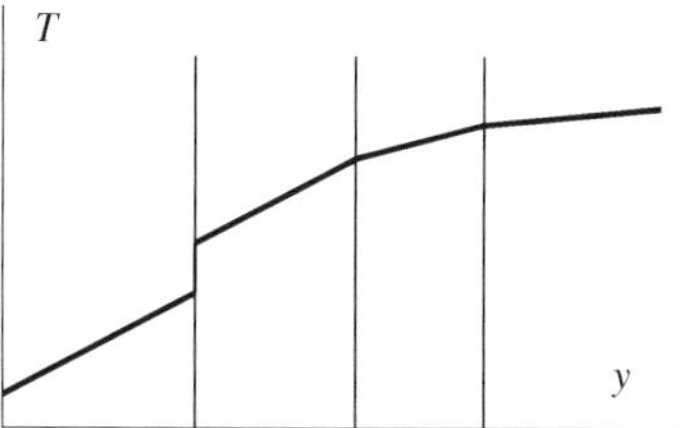

Figure 12.37 Temperature distribution in a cast, solidifying metal melt.

12.7.1 Cooling Rate

In Chapter 7 the heat transfer during the solidification process was briefly discussed. A more extensive analysis is given in the book 'Material Processing during Casting'. The heat transport during a casting process is schematically described in Figure 12.37 (Figure 4.1 in [7]). The heat transport is divided into several steps, heat transport through the melt to the solidification front and from the solidification front:through the solidified layer to the mould. Between the solidified layer and the mould there is an air gap, which is a considerable restriction for heat transport.

In casting operations of metallic glasses the thickness of the product is small compared to other casting processes. In most cases a Cu mould is used. In this case, the temperature profile will be the one described in Figure 12.38 (Figure 4.27 in [7]).

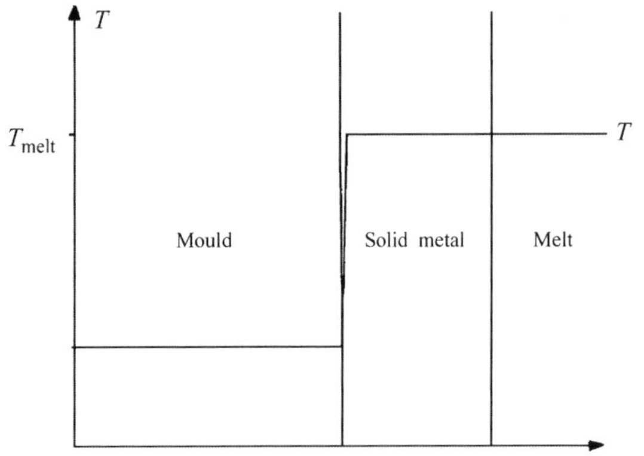

Figure 12.38 Temperature distribution in Cu mould and metallic glass.

The heat transport is controlled by the relationship

$$\frac{dQ}{dt} = hA(T_m - T_{mould}) \tag{12.9}$$

where
h = heat transfer coefficient
A = cooling area
T_m = temperature of the metallic glass
T_{mould} = temperature of the mould.

The temperature of the mould decreases with time. The cooling rate can be derived from the relationship

$$\frac{dQ}{dt} = hA(T_m - T_{mould}) = V\rho c_p \frac{dT_m}{dt} \tag{12.10}$$

where
V = volume of the sample
ρ = density of the metallic glass
c_p = heat capacitivity of the metallic glass
$\frac{dT_m}{dt}$ = cooling rate of the metallic glass
t = time.

By integration of Equation (12.10), the cooling curve can be calculated, i.e. the temperature as a function of time. The curve is analogous to the CCT diagram in Figure 12.1 on page 731.

12.7.2 *Volume Fraction*

During the cooling process small crystals are formed when the temperature becomes lower than the melting point. The crystals start to grow at this temperature and heat of fusion is released. This will affect the temperature curve.

The volume fraction of solid formed per unit time can be derived with the aid of the energy law, i.e. an equation of the type (12.10) with consideration to the heat of fusion:

$$\frac{dQ}{dt} = hA(T_m - T_{mould}) = V\rho c_p \frac{dT_m}{dt} + V\rho(-\Delta H)\frac{df}{dt} \tag{12.11}$$

where
$-\Delta H$ = heat of fusion
f = fraction solid.

The volume fraction of solid depends on

- the number of growing crystals
- the rate of crystal growth
- the size of the crystals.

The analysis will be simplified if we use a simple model of spherical growth.

Consider N crystals that grow in a spherical way. As the fraction of solid is small, crystal collisions can be disregarded. In this case the fraction of crystals can be written as

$$f = N \times \frac{4\pi r^3}{3} \tag{12.12}$$

where
N = number of crystals per unit volume
r = radius of one of the spherical crystals.

Equation (12.12) is derived with respect to time and the result is inserted into Equation (12.11). The result is

$$\frac{dQ}{dt} = hA(T_m - T_{mould}) = V\rho c_p \frac{dT_m}{dt} + VN\rho(-\Delta H) \times 4\pi r^2 \frac{dr}{dt} \tag{12.13}$$

Obviously, both the number of crystals per unit volume and their growth rate are important quantities. It is inevitable that crystals form (~1%) when the amorphous alloy melt cools but it is important to suppress crystallization as much as possible. Therefore, the number of crystals per unit volume shall be kept as low as possible in order to minimize the growth rate.

12.7.3 The Kinetic Process

In most casting processes the nucleation of crystals occurs when the temperature of the melt falls below the liquidus temperature. This is also the case when amorphous alloy melts are cast.

The interface kinetics during crystal growth has been extensively discussed in Chapter 6, primarily for pure metal melts and secondly for binary alloys.

Metallic glasses are very complex and are often built up of three or more components. In addition, the amorphous alloy melt has a cluster structure. The crystals that form have in most cases a composition other than that of the melt. The structure of the crystals very seldom corresponds to the structure of the clusters in the melt. The cluster structure of the melt often breaks up during the crystal growth. The splitting of the clusters may occur at the interface between the melt and the solid.

When the clusters break the atoms have to find new positions, now in the crystal lattice. At the same time an extensive and complex diffusion of atoms occurs across the interphase in order to obtain crystals with lowest possible energy and a matching composition. Several steps are built up simultaneously. The steps are coupled in a series and the slowest step controls the whole solidification process.

In Chapter 6 we regarded the interface kinetic process as a thermally activated process shown in Figure 12.39 (Figure 6.24 in Chapter 6). Each step can be described as follows. Each step is characterized by an activation energy and a driving force. The sum of all the driving forces gives the total driving force. The step with the largest activation energy will be the slowest one. This step will control the kinetic interphase process. We also discussed the relaxation process in the crystals in Chapter 6. If the interphase kinetics is slow the precipitated crystals are close to their equilibrium states during the growth.

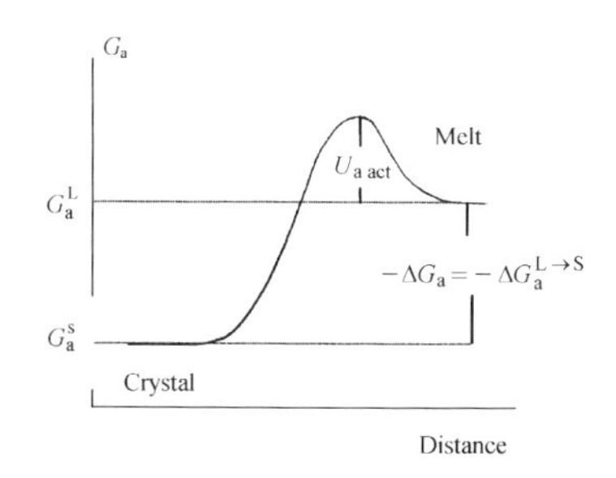

Figure 12.39 Free-energy diagram of the interphase between a crystal with a rough surface and its liquid phase as a function of position.

The kinetic interphase process is essentially the same for metals and metallic glasses. Therefore, we restrict the discussion here to amorphous alloy melts with strong clusters.

The kinetic process in these melts can be described by an equation similar to Equation (6.34) on page 293. Consider a case with a given thickness of the interphase between the metallic glass and the melt. The clusters enter the interphase from the melt. By a thermally activated process they split up into smaller units or free atoms. The units in the interphase rearrange and combine into a crystal lattice, often as a compound with different composition and atomic structure than the clusters. This can also be described as a thermally activated process. The whole process occurs in the interphase. In analogy with the equations in Chapter 6 the growth rate of the crystals can be written as

$$V_{\text{growth}} = \frac{\mathrm{d}r}{\mathrm{d}t} = \frac{A}{\delta}\left(\nu_0^{\text{cluster}} \exp\frac{-Q^{\text{cluster}}}{RT} - \nu_0^{\text{crystal}} \exp\frac{-Q^{\text{crystal}} - \Delta G^{\text{L}\to\text{s}}}{RT}\right) \quad (12.14)$$

where

V_{growth} = growth rate of the crystals
$-\Delta G^{\text{L}\to\text{s}}$ = driving force of crystallization
δ = thickness of the interphase
A = a constant (m^2)
ν_0^{i} = atomic vibration frequency, number of opportunities per second to form a crystal or split a cluster, respectively
Q^{i} = activation energy to transfer atoms from the interphase into the crystal or to transfer clusters into the interphase and split them, respectively.

The crystals that form are often chemical compounds. For simplicity we assume that the activation energy to split clusters and form crystal compounds are equal and that the two vibrational frequencies also are the same. These approximations simplify Equation (12.14) to

$$V_{\text{growth}} = \frac{\mathrm{d}r}{\mathrm{d}t} = \frac{A\nu_0}{\delta} \exp\frac{-Q}{RT}\left(1 - \exp\frac{-\Delta G^{\text{L}\to\text{s}}}{RT}\right) \quad (12.15)$$

The expression for dr/dt can be inserted into Equation (12.13), which gives

$$\frac{\mathrm{d}Q}{\mathrm{d}t} = hA(T_{\text{m}} - T_{\text{mould}})$$

$$= V\rho c_{\text{p}}\frac{\mathrm{d}T_{\text{m}}}{\mathrm{d}t} + VN\rho(-\Delta H) \times 4\pi r^2 \frac{A\nu_0}{\delta} \exp\frac{-Q}{RT}\left(1 - \exp\frac{-\Delta G^{\text{L}\to\text{s}}}{RT}\right) \quad (12.16)$$

The radius of the crystals as a function of time can be calculated with the aid of a numerical method. The temperature of the amorphous alloy melt is calculated stepwise for different values of the time t. The calculations show that the crystal fraction that forms is strongly influenced by a *decrease of the heat of fusion* and an *increase of the activation energy*, especially the latter.

An analysis of the function (12.15) shows that the growth rate initially increases with decreasing temperature, owing to an increase of the driving force ($-\Delta G^{\text{L}\to\text{s}}$). The growth rate reaches a maximum value at a certain temperature and then decreases and approaches zero. This behaviour is caused by the exponential term exp ($-Q/RT$). The higher the Q value is, the lower will be the maximum value of the growth value and the higher will be the temperature at which the growth stops.

It is expected that the activation energy is very high when the melt is built up by clusters.

- The more stable the clusters are, the higher will be the activation energy Q.
- The more negative the heat of mixing is, the more stable will be the clusters.

In amorphous alloy melts with very stable clusters, the clusters may not split up. In this case, they crystallize as clusters. This may be the explanation for formation of quasicrystals. In this case the activation energy is low and comparable with the activation energy for diffusion in melts.

Nucleation of Crystals

Metallic glasses are produced at metallurgical processes under extremely careful conditions. The amorphous alloy melt is in most cases remelted several times in vacuum or in an atmosphere free from oxygen and nitrogen. This type of process is performed in order to reduce the number of heterogeneities, which can act as nucleating agents in the melt. Equation (12.13) shows that the number of growing crystals has a large influence on the fraction of solid that forms.

- The number of heterogeneities per unit volume must be kept as low as possible.

  If the melt is pure enough, no crystals are formed when the temperature of the melt reaches the melting point in the casting process. However, crystals may form by homogeneous nucleation (see Chapter 4). The activation energy for nucleation is the factor in his equation, that has the largest influence on the nucleation temperature, for amorphous melts.

  In order to suppress the crystallization process:

- The physical parameters shall be selected in such a way that the nucleation temperature is lower than the maximum crystal growth temperature.

  The nucleation temperature is influenced by the free energy of the surface and by the heat of solidification.

- The surface free energy shall be as high as possible and the heat of fusion shall be as low as possible to obtain lowest possible nucleation temperature.

12.8 Classification of Metallic Glasses

12.8.1 Composition Classification

The metallic glasses can be classified as two main groups after composition:

- metal–metalloid glasses;
- metal–metal glasses.

Metal–Metalloid Glasses

The *metal–metalloid* glasses contain one or several metals, mainly light or heavy transition elements or rare-earth metals in the periodic table of elements such as Ca, Ba, Mg, Al and Fe, Ni, Cu, Zn, Zr, Pd, alloyed with one or several metalloid elements such as B, C, Si and P. The light and small metalloid elements contribute to the hard packing structure, which characterizes the metallic glasses.

Binary alloy systems such as Fe–B, and Ni–B and Pd–Si can be transformed into metallic glasses fairly easy. Their glass forming abilities GFA are often increased when a second type of metalloid is added. Examples of such ternary alloys are Fe–P–C, Ni–P–B and Ni–Si–B. Their thermal stability is also improved by addition of a second metalloid element or a second transition metal.

Examples of BMGs of the metal–metalloid type are given in Table 12.2 on page 738.

Metal–Metal Glasses

The *metal–metal glasses* cannot match the metal–metalloid glasses neither concerning the range of properties nor in applications so far. They have no strictly defined composition requirements but two or more metals are normally involved. The metals can either be light or heavy transition elements *or* metals such as Ca, Ba, Mg or Al from the main groups in the periodic table *or* rare-earth metals such as La or Gd. Common examples are the following pairs: ($Cu_{65-40}Zr_{35-60}$), ($Ni_{65-40}Ta_{35-60}$), ($Mg_{70}Zn_{30}$), and ($La_{82-74}Au_{18-26}$).

12.8.2 Other Classification

In the following section the metallic glasses have been classified according to *general properties*, common for all metallic glasses, and *special properties*, developed for specific purposes, both with applications.

It is possible to 'design' BMGs with properties for special purposes, to enhance for example some specific mechanical, thermal, electrical or magnetic property. Such research is going on currently and further progress can be expected in the future.

12.9 Properties and Applications of Metallic Glasses

Before we discuss different properties of metallic glasses, a short comparison between properties of traditional metals, traditional glasses and metallic glasses is given below.

We also summarize the recent development of metallic glass research, which defines the tools and possibilities available today for design of new BMGs.

12.9.1 Comparison of Properties between Metals, Common Glasses and Metallic Glasses

Common glass is an undercooled liquid, made of a mixture of sand (silicon and oxygen) and metal oxides (sodium and for example potassium and lead) Metallic glasses resemble in many respects of traditional glass but there are also important differences.

Both types of glasses are non-crystalline with high viscosities. The types of interatomic bonds in metallic glasses are totally different from those in traditional glasses, which give diverging thermal, optical, electrical and magnetic properties. Thermal and electrical properties have been discussed in Section 12.6 on pages 753 and 758–759, respectively.

Common glasses have covalent bonds (Section 3.3.3 in [8]) while the bonds of metallic glasses are much more complicated, as indicated in Sections 12.5 and 12.6. Table 12.5 gives a survey of some properties of traditional glasses and the new metallic glasses.

Table 12.5 Comparison of the properties of traditional metals, traditional glasses and metallic glasses (after Anantharaman and Suryanarayana)

Characteristic/Property	Traditional Metal	Traditional Glass	Metallic Glass
Structure	Crystalline	Amorphous	Amorphous
Bonding	Metallic	Covalent	Metallic and Ionic
Yield stress	Not ideal	Almost ideal	Almost ideal
Forkability	Good, Ductile	Poor, Brittle	Good, Ductile
Hardness	Low – High	Very High	Very High
Ultimate stress	Low – High	Low	High – Very High
Optical character	Opaque	Transparent	Opaque
Thermal/electrical conductivity	Very Good– Good	Bad	Very Good
Corrosion resistance	Poor – Good	Very Good	Very Good
Magnetic properties	Various, mainly para- and ferro-magnetic	Non-existent	Various, mainly para- and ferro-magnetic

Table 12.5 shows clearly that metallic glasses in many respects have desirable properties, which are very useful in practical applications.

12.9.2 Tools and Possibilities for Design of New Bulk Metallic Glasses

1. BMGs have been developed to dimensions 1–25 (maximum 72) mm based on common engineering metals such as Mg, Al, Ti, Cu and Fe, which replace among others Pd (expensive) and Be toxic). Methods for systematic search for new BMGs have been developed.
2. The high GFA provide processing parameters and composition required for reproducibility in industrial mass production.
3. The macroscopic sizes extend the limits for possible applications.

12.9.3 Mechanical Properties and Applications of Metallic Glasses

All metallic glasses have very high viscosities and excellent mechanical properties. The viscosities of BMGs and amorphous melts have been discussed earlier in this chapter and will be omitted here.

General Mechanical Properties of Metallic Glasses

Below we will discuss the general mechanical properties of metallic glasses briefly and give some examples of applications.

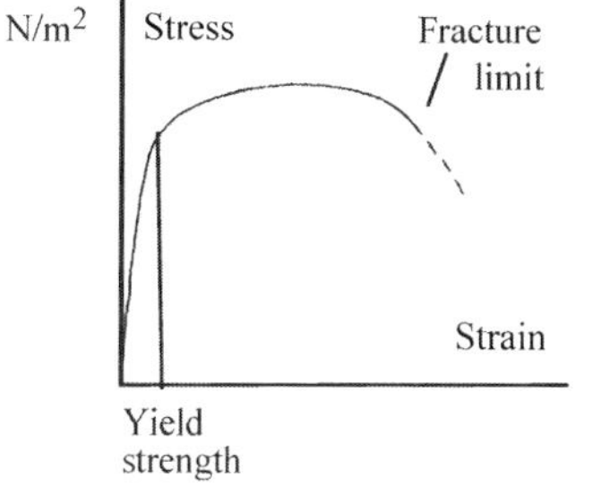

Figure 12.40 Yield strength and fracture limit.

Elasticity

Crystalline materials have a maximum elastic strain range $\Delta l/l$ of the *magnitude* 0.2%. Within this interval Hooke's law is valid.

$$\sigma = Y\varepsilon \tag{12.17}$$

where

σ = stress F/A caused by tensile or compressive forces
Y = modulus of elasticity of the material
ε = elastic strain $\Delta l/l$.\

The yield strength, i.e. the upper limit of the elastic strain range, of the metallic glasses is of the order of 2%, about 10 times as high as for crystalline materials (Figure 12.40).

Table 12.6 Room temperature compressive fracture strengths of some BMGs

Type of BMG	Fracture strength σ_f (GPA)
Co-Fe-Ta-B	5.2
Co-Fe-Ta-B-Mo	5.5
Fe-based	~4
Ni.based	~3
Zr-based	~2
Cu-based	~2
Steel	~1.6

Mechanical Strength

Metallic glasses have no slip systems (atomic planes where the atoms can move easily), neither lattice dislocations nor grain boundaries. They show high yield strengths and high values of fracture limits. Table 12.6 gives some examples.

Their hardnesses and fracture strengths σ_f are correlated with Young's modulus Y and also with the glass-forming temperature T_g. For the hardest BMGs the correlation between σ_f and T_g is particularly evident. When T_g changes from 700 K to 900 K the corresponding σ_f value increases from less than 2 GPA to more than 5 GPA.

Specific Mechanical Properties of Metallic Glasses

Heterogeneous Deformation

In the absence of dislocations and slip deformation some BMGs deform by formation of shear bands. In the deformation zones a sudden drop of the viscosity occurs in the deformation zones up to liquid levels. This facilitates bending and other deformations strongly.

Most BMGs fail but some undergo high plastic deformation under high confining pressure. This phenomenon has been observed for Pt-, Cu-, Pd-, Ti-, and Zr-based BMGs. These glassy alloys show extensive deformation. Spaepen [16] has successfully been able to find a mechanism for shear softening, based on the theory of the free excess volume.

However, it is still hard to understand why individual shear bands cease to operate in these BMGs instead of resulting in a predicted and frequently observed disastrous fracture by shear softening on one or a few bands.

The answers seem to be related to the ordered ionic character of the metallic glasses. A comparison can be made with common ordered metallic phases, which also show the same type of behavior. The reason is that the ionic structure does not allow atoms with the same type of charge to come close to each other and the slip theory can not be applied. If the material slips along one plane, atoms with the same charge may be neighbours now and then, which is forbidden. The deformation can only occur when electrons are excited up to the conduction band. A large amount of extra energy must be supplied to the material.

The additional energy is supplied in certain regions with high stress concentrations, such as vacancies and inclusions. The electrons of a group of atoms become transferred into the conduction band. This excitation process spreads along bands with high stress concentration and a deformation band is formed. An example is given in Figure 12.41.

Figure 12.41 a) The line AB shows the plane on which a slip-like correlated motion of atoms has occurred. b) Typical rearrangement of the atoms in the plastic deformation. © Research Institutes, Tohoku University 1978.

Poisson's Number

$$\varepsilon_{trans} = \nu\varepsilon$$

where

ε = strain in the length direction

ν = Poisson's ratio

ε_{trans} = strain in a transverse direction

Plasticity

As shown above, metallic glasses have many excellent mechanical properties but ductility was not initially one of them. Later, BMGs with very high deformation without fracture were developed. Schroers and Johnson [17] proposed that the large macroscopic compressive plasticity found in the BMG $Pt_{57.5}Cu_{14.7}Ni_{53}P_{22.5}$ is caused by an unusually high value of Poisson's ratio ν. It is defined as the ratio of the strain in the length direction and the strain in a perpendicular direction.

Poisson's ratio is about 0.3 for normal metals. The BMG mentioned above had a value of ν equal to 0.41. When Liu et al. [18] adjusted the composition of the metallic glass $Zr_{65}Cu_{15}Ni_{10}Al_{10}$ to give an optimal value of ν, it could stand extremely high deformations without fracture, both in compression and in bending.

Another way to improve poor uniaxial ductility of metallic glasses, which arises from unstable propagation of shear bands, is to interrupt those shear bands by introduction of a second phase in the BMG. This can be done by production of *porous* or *foamed metallic glasses*. A gas, e.g. hydrogen, dissolved in the amorphous melt precipitates during cooling. Figures 12.42a and b show two examples of such structures, given in an article on porous and foamed amorphous metals by Brothers and Dunand [19].

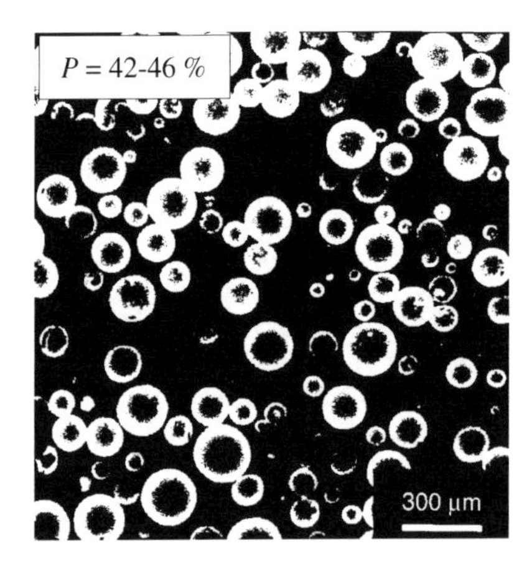

Figure 12.42a Pd-based foam (porosity $P = 42–46\%$) made by precipitation of dissolved hydrogen during cooling. T. Wada, A. Inoue in Materials Transactions © The Japan Institute of Metals 2004.

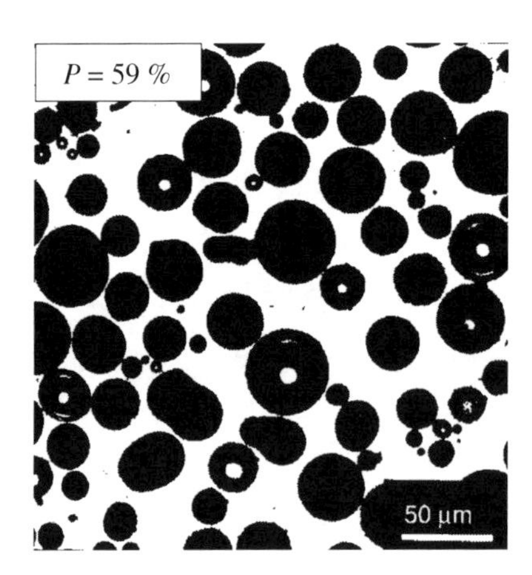

Figure 12.42b Zr-based foam, made by infiltration of a bed of hollow carbon spheres. Volume fraction of spheres in the foam (porosity) equals 59%. A. H. Brothers, D. C. Dunand, Appl. Phys. Lett. 84, 1108, (2004). Copyright 2004, American Institute of Physics.

The influence of porosity P (volume fraction of gas) on compressive stress is shown in Figure 12.43.

Pores interrupt the shear bands when they intersect. Figure 12.43 shows that even a small porosity of about 4% is enough to maintain a high ductility (1.5 GPa). The higher the porosity is, the lower will be the strength of the BMG.

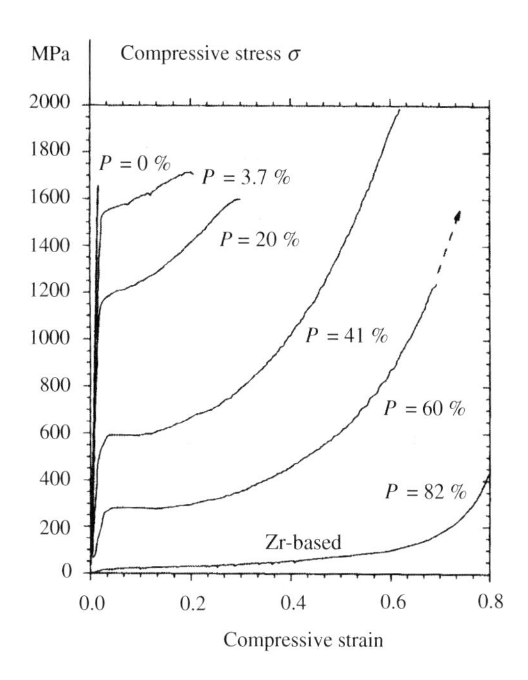

Figure 12.43 Compressive stress-strain curves for porous BMGs with the porosity as a parameter. The lowest curve refers to the BMG $Zr_{57}Nb_5Cu_{15.4}Ni_{12.6}Al_{10}$. The other curves refer to the BMG $Pd_{42.5}Cu_{30}Ni_{7.5}P_{20}$.

By controlling levels of porosity, compressive strength can be varied from near-maximum values to very low values. Simultaneously, the limit of compressive failure strain (yield strength) can be varied from 2% to about 80%.

Applications

General Applications

It is easy to understand that with more than 300 BMGs commercially available, the BMGs already appear in many practical applications. Owing to their excellent mechanical properties and hardness the BMGs are used in manifold of industrial products. We can only mention a few of them here. Table 12.7 shows some typical BMGs and their reported properties.

Table 12.7 Some typical BMGs with sufficient GFA to permit casting into fully glassy forms > 10 mm in diameter

Composition	Diameter (mm)	Reported properties	Publication year
$Zr_{41.2}Ti_{13.8}Cu_{12.5}Ni_{10.0}Be_{22.5}$			1993
$Zr_{55}Al_{10}Ni_5Cu_{30}$	30	None mentioned	1996
$Pd_{40}Ni_{40}P_{20}$	25	High fracture strength (1.70 GPa)	1996
$Pd_{40}Cu_{30}Ni_{10}P_{20}$	72	High fracture strength (1.68 GPa)	1996
$Pt_{57.5}Cu_{14.7}Ni_{5.3}P_{22.5}$	12	High fracture strength (1.47 GPa) Compressive ductility (20 %)	2004
$Pd_{35}Pt_{15}Cu_{30}P_{20}$	30	High fracture strength (1.41 GPa)	2006
$Mg_{65}Y_{10}Cu_{15}Ag_5Pd_5$	12	None mentioned	2001
$Mg_{56}Cu_{26.5}Ag_{8.5}Gd_{11}$	25	High fracture strength (1.0 GPa)	2005
$Fe_{48}Cr_{16}Mo_{14}C_{15}B_6Er_2$	12	None mentioned	2004
$Co_{48}Cr_{15}Mo_{14}C_{15}B_6Tm_2$	10	High fracture strength (4.00 GPa) None magnetic	2006
$Ni_{50}Pd_{30}P_{20}$	21	High fracture strength (1.78 GPa) Compressive ductility	2007
$Cu_{46}Zr_{42}Al_7Y_5$	10	None mentioned	2004
$Cu_{44.25}Zr_{36}Ti_5Ag_{14.75}$	10	High fracture strength (1.97 GPa)	2006
$Cu_{40}Zr_{44}Ag_8Al_8$	15	High fracture strength (1.85 GPa) Compressive ductility	2006

BMGs are used for production of

1. thin electronic casings such as mobile telephones, cameras and similar products.
2. sports wear such as golf clubs, tennis rackets, baseball and soft ball bats, skis and snowboards, bicycle parts, fishing equipment and so on.
3. medical devices and equipment for reconstructive supports such as fracture fixations, surgical blades and spinal implants.
4. minor miscellaneous products, e.g. watch casings, finger rings, fountain pens and so on.
5. precision devices for various purposes due to no shrinking during solidification, such as precision gears for micromotors, diaphragms for pressure sensors (e.g. fuel injections in cars).

Specific Applications
BMGs with high strength and high plasticity are excellent materials for springs of different kinds, such as automobile valve springs, and wires.

12.9.4 Chemical Properties and Applications of Metallic Glasses

Corrosion Resistance
One of the most striking and useful properties of BMGs, particularly the ferrous metallic glasses, is their corrosion resistance.

Metal–metal Ni-based BMGs, e.g. Ni–Nb–Ti–Zr–Co–Cu, show high tensile strength (~ 2.7 GPA) but have a limited GFA ($d_{cr} \leq 10$ mm).

Metal–metalloid Ni-based BMGs are more widely applied because of their better GFA with about the double critical diameter, compared to the metal–metal Ni-based BMGs, They show a fairly high compressive plastic strain (~1.8 GPA) and good corrosion resistance (~7.5%) One example is the metallic glass $Ni_{50}Pd_{30}P_{20}$ that can be cast as a cylinder with a diameter of 21 mm.

Cu-based BMGs show high mechanical strength (> 2 GPA) but fairly low GFA ($d_{cr} \sim 4$ mm). Their corrosion resistance is lower than that of the Ni-based BMGs but can be improved by a small addition of Nb.

Applications

Ni-based BMGs are generally a better alternative than ordinary metals in cases with a corrosive environment.

Great expectations are attached to the development of *fuel cells*, which would reduce the consumption of fossil fuels and reduce the emission of CO_2 gas into the atmosphere and limit the global warming.

In addition, fuel cells, which transform chemical energy directly into electrical energy, are more efficient than all other energy-converting engines.

The development of fuel cells has been going on for a long time and is a challenge indeed. The development of the proton exchange membrane fuel cells (PEMFC) which includes the use of the superior corrosion resistance and viscous deformability of BMGs, is promising but much work remains to be done. This type of fuel cell with high output current density and low-temperature operation, would be convenient for domestic use and for automobile applications.

Technical problems with the main constituent parts, e.g. the membrane, the catalyst and the separator are not solved, nor how to supply fuel and oxidizer to the reaction chamber, how to store the energy and reinforce the cell mechanically. It is urgent to make the system lighter and smaller and reduce the costs.

The Ni-based BMGs are of great interest as replacement materials because of their high strength, superior corrosion resistance and viscous deformability. The corrosion rates for a typical Ni-based BMG in a sulfuric acid solution is about 10 times lower than that of stainless steel. A Ni-based BMG is a promising material for next generation of separators in fuel cells. Test with prototypes show that the Ni-based BMG generates a high voltage. Durability tests at a current density of 0.5 A/m^2 show no degradation after 350 hours.

12.9.5 *Magnetic Properties and Applications of Ferrous Metallic Glasses*

Duhaj (Tjeckien) spun the first ductile glassy ribbons of a ferrous metallic glass in 1972. Ferrous metallic glasses differ considerably in composition from common steel. They have a much higher fraction of metalloids than conventional steel.

Ferromagnetic BMGs of the type (Fe, Co, Ni)(B, Si) are cast in Cu moulds with critical diameters up to 7.7 mm when fluxed with B_2O_3. They show

- high strength (2–4 GPa);
- high corrosion resistance;
- very low acoustic attenuation;
- high electrical resistivity;
- excellent magnetic properties;
- inexpensive processing into wires and ribbons.

The properties of ordinary ferromagnetic materials have been discussed on pages 323–333 in [8] or can be studied in any other comparable book. A short summary is given in the box.

Basic Magnetic Concepts. Hysteresis Loop

The easiest way to magnetize a ferromagnetic specimen is to place it in a homogeneous magnetizing *H* field inside a coil with an electric current *I*. The *H* field is proportional to the current *I* and creates magnetic field *B* in the specimen,

$$B = \mu\mu_0 H \tag{12.18}$$

where

H = magnetizing field (A/m)
μ = relative permeability of the material (magnitude 10^3)
μ_0 = permeability in vacuum
B = magnetic field, caused by the magnetizing field.

The alignment of the spins of a large number of atoms create regions called *Weiss domains* that are firmly held together. Their size is 10^{-3}–10^{-4} cm. If the *B* field is zero the resulting magnetic moments of all the domains is zero because they are orientated in arbitrary directions. In a magnetizing *H* field the domains orient themselves partially in the direction of the field and gives a resulting *B* field.

If the coil is fed with an alternating current the magnetizing field changes direction in a cyclic way and the *B* field varies with time according to the so-called hysteresis loop. The enclosed area represents the energy per mass unit, required to run through the cycle once.

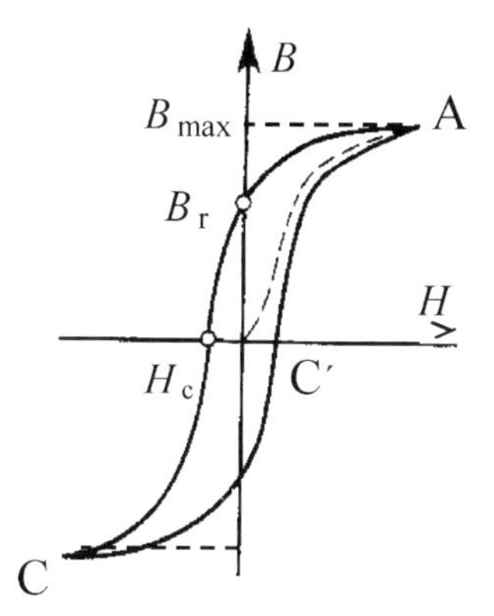

Hysteresis loop of a ferromagnetic material.
B_r = remanence
H_c = coercitivity.

Magnetic Properties

Ferrous BMGs are extremely easy to magnetize and remagnetize. A magnetizing field of 10^{-2} A/m – that is 100 times weaker than the magnetic field of the earth – is enough for magnetizing the metallic glass.

Magnetization and remagnetization occurs with the aid of changes of sizes and positions of the domains walls. Domains in or close to the direction of the magnetizing field grow at the expense of others in other directions.

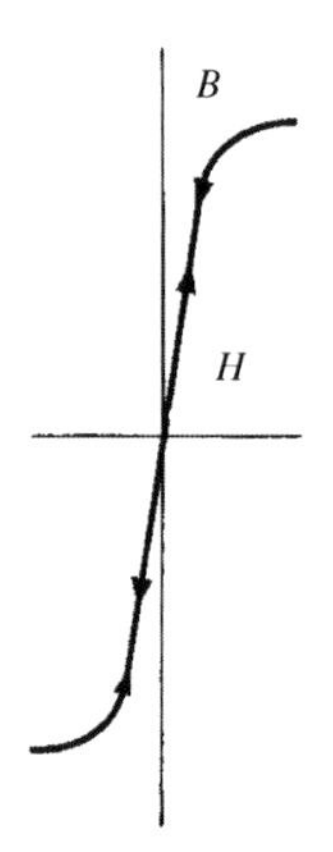

Figure 12.44 Hysteresis loop of a ferrous metallic glass, somewhat exaggerated. Reprinted with permission from H. Ma et al. Appl. Phys. Lett. 87, 181915 (2005). Copyright 2005, American Institute of Physics.

- The domain walls move very easily in BMGs.

The reason is that there are no grain boundaries in metallic glasses that can prevent the domain walls from moving.

As the domain walls can move so easily, a magnetization–remagnetization cycle in the hysteresis loop requires very little energy, the so-called hysteresis losses that are emitted as heat. Such ferromagnetic materials are characterized as *soft*. Ferrous BMGs are supersoft with an extremely narrow hysteresis loop (Figure 12.44).

In addition to the ease of moving the domain wall there is another reason for this. Because of the high electrical resistivity of the BMG the eddy currents, which always appear in the material when the magnetic field changes, are small.

Ferrous BMGs have high saturation magnetizations, low coercivities and low Curie points. For comparison we list some representative material constants for three BMGs and a soft magnetic steel (used in transformers) in Table 12.8.

Table 12.8 Material constants of some ferrous metallic glasses and a soft magnetic steel

Material constant	Some ferromagnetic glasses			Soft magnetic steel
	$Fe_{80}B_{20}$	$Fe_{67}Co_{18}B_{14}Si_1$	$Fe_{81}B_{13.5}Si_{3.5}C_2$	Fe3%Si
Saturation magnetization B_{max} (Tesla)	1.6	1.8	1.6	2.0
Coecivity H_c (A/m)	3.2	6.4	6.4	7.2
Permeability μ_{max} (T/[A/m])	4.0×10^{-7}	2.5×10^{-7}	3.8×10^{-7}	0.38×10^{-7}
Curie point (°C)	374	415	415	735
Resistivity ρ (Ωm)	1.4×10^{-6}	1.3×10^{-6}	1.2×10^{-6}	0.46×10^{-6}

Applications

Because of their soft magnetic properties, owing to easily mobile domain walls, the ferrous metallic glasses show

- high relative permeability μ_{max};
- high remanence;
- low coercivity H_c;
- very rapid magnetization and remagnetization in case of time-dependent magnetizing fields;
- high resistivity.

The ferrous BMGs are easy to produce by squeeze casting and are also produced as thin ribbons up to meter widths by melt spinning (Figure 12.3 on page 735). They are very attractive and widely used in practical applications such as:

1. Cores in transformers because of their low hysteresis losses, that can be reduced by ~75% of the losses in crystalline ferromagnetic cores. They can be used at frequencies up to 10 MHz.
   Other applications are amplifiers, switches, lamp ballasts and modern electronic devices.
2. Cores in high-frequency AC generators and AC motors up to 10 MHz. Amorphous metals can provide extremely high efficiencies. The hysteresis losses can be reduced to 90% of the losses in conventional three-phase motors. The dimensions of the motors can be reduced in size and dimensions and in addition be much more efficient.
3. Shielding of modern electronic devices such as for example lap-tops and magnetic tape cassettes. The latter need both a magnetic shield to protect the tape head from stray fields and a shield for the spring that holds the tape close to the head.

Magnetic shields made of thin bands of metallic glasses are often woven into carpet-like sheets or plates.

12.9.6 Future Development. Expectations and Desires

It is natural to hope that the present development will continue, i.e. that the understanding of the structure of metallic glasses will be better and better. As a consequence, it will be easier to predict and find new bulk metallic glasses for various purposes.

Bulk metallic glasses:

- will hopefully be made of cheap components and be easy to produce on an industrial scale, e.g. by continuous casting. This will give abundant access to metallic glasses and enable new extensive applications within widely different branches, e.g. house building in earthquake territories (mechanical strength) and establishments in the sea for extraction of wave energy and transformation into electrical energy (mechanical strength, corrosion resistance, magnetic and electrical properties).
- will contribute to saving all sorts of resources and limit the present waste of natural resources such as energy, minerals, oil and metals before it is too late. The material resources on earth are limited but access to energy is practically unlimited. The problem is not access to energy but to transport and store it in a safe and easy way. Development of light and reliable batteries is a challenge.
- are a resource in the struggle for clean air, water and ground or shortly environment care. It is urgent to get rid of fossil fuels to reduce the global heating due to formation of CO_2. The increase of the standard of living in for example the large Asian countries is of course positive, but may also be a threat against or at least a serious trial on the environment. The construction of electrical cars (fuel cells) has been discussed earlier in this chapter. It is a concrete example on the benefit of metallic glasses in connection with the fight against pollution of the world.

Summary

■ Some Basic Concepts and Definitions

Metallic Glass

Amorphous alloy with excellent mechanical properties, produced by cooling an amorphous alloy melt below the so-called glass transition temperature T_g at a cooling rate below the so-called critical cooling rate.

Critical Cooling Rate

The critical cooling rate is the one, which forms a tangent to the boundary curve from the melting-point at $t = 0$. The boundary curve is based on the crystallization fraction $f = 0.01$.

The boundary curve is a part of a cooling curve, i.e. temperature-time plot, the so-called CCT diagram (continuous cooling transition). Each metallic glass has its own characteristic CCT curve.

Methods of measuring T_g and the critical cooling rate are given in the text.

BMG

Bulk metallic glass = metallic glass with a diameter > 1 mm when cast as a cylinder.

GFA

Glass-forming ability = the minimum dimension of the largest sample that can be made fully glassy.

■ Production Methods

Thin Metallic Glasses

- splat quenching
- melt spinning
- atomization.

Bulk Metallic Glasses

- arc-melting
- high pressure die casting
- water quenching
- copper mould casting (most common).

Rules of Thumb for BMGs with High GFA

- multicomponent system with three or more elements;
- significant difference of at least 12% in atomic size ratios among the main constituent elements;
- negative heat of mixing of the components.

■ Tools for Structure Research

Experimental Methods

X-ray examination, electron diffraction, neutron scattering.

■ Atomic Distribution Diagrams

Partial Distribution Functions. Partial Coordination Numbers

Metallic glasses show the same type of X-ray diagrams as liquid metals and alloys and have the same type of structures. For this reason it is necessary to use statistical methods to describe the distribution of atoms in amorphous alloys. The main concepts are:

The radial distribution function:

- $g(r)$ = the probability of finding two atoms at the distance r from each other.

The radial density function:

- $\rho\ (r)$ = the average number of atoms per unit volume at distance r from another atom.

$$g = 4\pi r^2 \rho \, \mathrm{d}r = \frac{\mathrm{d}N}{\mathrm{d}r}$$

The function $N(r)$

- $N(r)$ = the average total number of atoms within a distance r from another atom.

$$N = \int_0^r g\mathrm{d}r = \int_0^r 4\pi r^2 \rho \, \mathrm{d}r$$

The coordination number Z can be calculated from the function

$$-Z = \int_0^{r^*} g\mathrm{d}r$$

Most metallic glasses consist of three or more components. The probability g is different for different kinds of atoms and this fact must be taken into consideration. Partial probabilities must replace g and partial coordination numbers Z_{ij} can be derived from radial function plots.

The Efficient Cluster Packing Model ECP

It is known that the properties of the metallic glasses are associated with their dense packing structure. Efficient filling of space is an important condition to find an adequate model. Successful attempts with the aid of computers have been made to build efficient cluster packing models that describe a new way to achieve dense atomic packing. This method is called the ECP model. It is supported by new experimental methods.

■ Models of Metallic Glasses

Bernal's Hard-Sphere Model

In the early 1960s Bernal made an attempt to interpret the structure of a melt. Bernal and coworkers found that the structure of a melt can be successfully described by the following geometrical model.

The atoms are represented by hard spheres situated at the corners of five different types of rigid polyhedra with the following properties:

1. The edges of the polyhedra must all be of approximately the same length.
2. It is impossible to introduce one more sphere into the centre of the melt without stretching the distance between the atoms.

The five types of polyhedra, the so-called canonical holes or Bernal polyhedra, are tetrahedra, half-octahedra, tetragonal dodecahedra, trigonal prisms and Archimedean antiprisms.

The Cluster Model

Knowledge of the partial coordination numbers is helpful for determination of the structure of metallic glasses.

- The atoms are arranged in either solute-centred or solvent-centred clusters.

Short-Range Order Interaction

Metallic glasses are supercooled liquids with extremely high viscosities and dense random-packing structures. To obtain the highest possible atomic density in metallic glasses with a small metalloid component, e.g. phosphorous, it was earlier assumed that the P atoms could fill the empty holes. However, the holes turned out to be too few and too small to be able to accommodate the P atoms.

Instead, a strong preference for icosahedral short-range order has been predicted in liquids and metallic glasses and is obtained in many experiments and computer simulations. A metallic glass has the structure that corresponds to lowest possible total energy.

The icosahedral short-range order is confirmed by X-ray and electron diffraction. It has a 5-fold symmetry.

The ECP model predicts that the dominant clusters have coordination numbers within the interval 8–20, depending on the relative solute size. However, the normal short-range order in the ECP model deals with solute-centred clusters while icosahedral short-range order is associated to solvent-centred clusters.

Medium-Range Order Interaction

There is a strong tendency of formation of as many bonds as possible between unlike atoms because of the negative heat of mixing, which favours glass forming. In each amorphous alloy the atoms are arranged in such a way that the total energy is minimized.

The cluster size and type depend on the relative sizes of the solute and solvent. The clusters overlap but the cluster–cluster distances will be larger than the solute–solvent distances. This leads to so-called *medium-range order* because of interaction between solvent atoms and a *dense packing structure in space.*

The Efficient Cluster Packing Model ECP

It was known that the properties of the metallic glasses were associated with their dense packing structure. Efficient filling of space was an important condition to find an adequate model. Attempts were made with the aid of computers to build efficient cluster-packing models that describe a new way to achieve dense atomic packing. Free volume inside the dense-packed structure occurs and seems to play an important role in the structure.

Simultaneously, new sophisticated experimental methods were developed, such as wide-angle and anomalous X-ray diffraction, fluctuation electron microscopy, high-resolution transmission electron microscopy and positron annihilation spectroscopy.

Quantitative computer calculations with the aid of the ECP model have confirmed many characteristic of metallic glasses.

Quasicrystals

Metallic glasses may show point diffraction patterns in spite of their total absence of regular order of their atoms. Materials with these properties are called quasicrystals.

Quasicrystals show 5-fold symmetry.

Casting of Metallic Glasses

It is urgent to suppress crystallization as much as possible in order to obtain good metallic glasses. It is necessary to suppress the fraction of crystals to $< 1\%$. A high activation energy counteracts crystallization.

- The more stable the clusters are, the higher will be the activation energy.
- The more negative the heat of mixing of the components is, the more stable will be the clusters.

The nucleation must be kept low. For this reason the number of heterogeneities per unit volume in the amorphous melt is reduced by

- repeated remelting;
- fluxing, i.e. refining with molten B_2O_3.

In the case of homogeneous nucleation it is urgent

- to select the physical properties of the amorphous alloy melt in such a way that the nucleation temperature is lower than the maximum crystal growth temperature.

■ Classification of Metallic Glasses

The metallic glasses can be classified as two main groups after composition:

- metal–metalloid glasses;
- metal–metal glasses.

The *metal–metalloid glasses* contain one or several metals, mainly light or heavy transition elements or rare-earth metals in the periodic table of elements such as Ca, Ba, Mg, Al and Fe, Ni, Cu, Zn, Zr, Pd, alloyed with one or several metalloid elements such as B, C, Si and P. The light and small metalloid elements contribute to the hard packing structure, which characterizes the metallic glasses. Binary alloy systems such as Fe–B, and Ni–B and Pd–Si can be transformed into metallic glasses fairly easy.

The *metal–metal glasses* cannot match the metal–metalloid glasses neither concerning the range of properties nor in applications so far. They have no strictly defined composition requirements but two or more metals are normally involved. The metals can either be light or heavy transition elements *or* metals such as Ca, Ba, Mg or Al from the main groups in the periodic table *or* rare-earth metals such as La or Gd.

■ Properties of Metallic Glasses

Mechanical Properties

Metallic glasses have excellent mechanical properties compared to common metals.

- maximum elastic strain (~10 times better than metals) and higher fracture limit;
- mechanical strength (up to 5×10^3 times better than ordinary metals and alloys).

Most BMGs fail but some undergo high plastic deformation, particularly in the shape of foam.

Chemical Properties

Some BMGs show

- excellent corrosion resistance (~10 times better than metals).

Magnetic Properties

Ferrous magnetic glasses are very easy to magnetize and remagnetize because the domain walls can move easily in the absence of lattice faults. The have

- high saturation magnetization;
- low coercivity;
- low Curie points;
- very low heat losses at magnetization and remagnetization.

Applications of Metallic Glasses

Hundreds of metallic glasses are commercially available and have a manifold of various applications.

Examples are given in the text, such as

- casings for mobile telephones, cameras, lap-tops and medical devices and instruments (mechanical strength);
- fuel cells (corrosion);
- high frequency motors and in all types of transformers (magnetic properties).

Exercises

12.1 In many textbooks phase transformations of materials are presented as a first-order or second-order transition. Schewmon explains graphically in his book [20] how some basic thermodynamic properties behave when the materials go through transitions of the first and second order, respectively. These concepts are defined in the figure below.

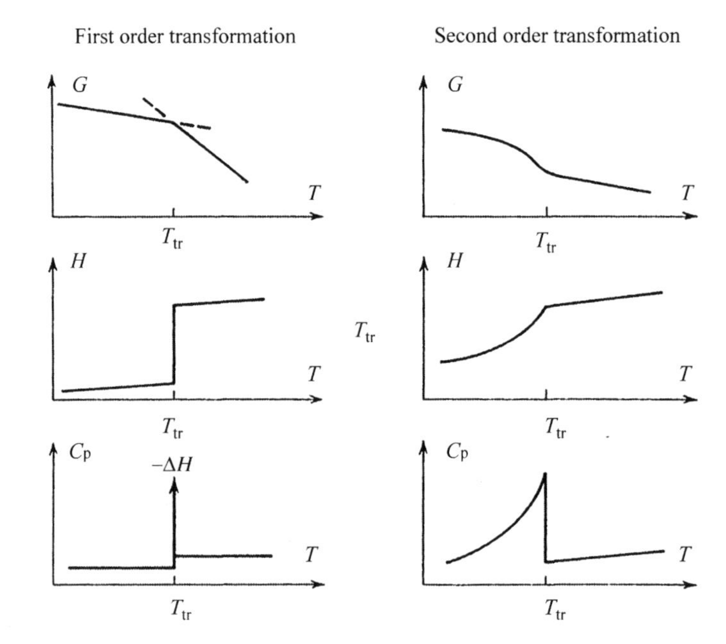

Figure 12.1–1

The left column of graphs shows how various thermodynamic properties change when the material goes through a first-order transition, for example melting.

The right column shows how the same properties change when the material goes through a second-order transition.

Analyze the glass-transformation process with respect to $(-\Delta H)$ and C_p in Chapter 12 and discuss which type of transition characterizes the glass-forming process best.

12.2

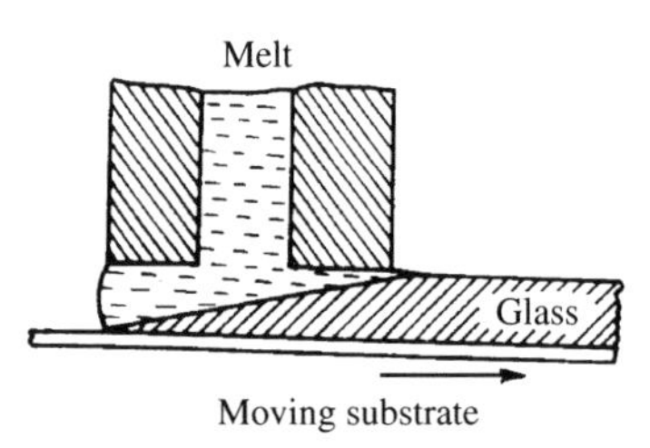

Figure 12.2–1

Formation of crystals is unavoidable at production of metallic glasses but it is urgent to suppress crystallization as much as possible.

In a melt-spinning process observations show that small crystals are formed on the surface of the ribbon. The crystals grow inwards into the ribbon. It is important to analyze the process parameters in order to minimize the thickness of the crystalline layer on the surface in favour of the metallic glass ribbon.

Perform a theoretical analysis of the melt-spinning process and calculate the thickness y of the crystal layer as a function of the temperature T of the crystals and the ribbon. The growth of the crystals follows the law

$$V_{growth} = \beta_{kin}(T_M - T)$$

where β_{kin} is the growth constant, T_M is the melting point of the crystals and T the temperature of crystals and the ribbon.

12.3 Observations have shown that metallic glasses may form within the composition range 20 at% Si $\leq x_{Si} \leq$ 30 at%Si. The diameters of the Al and Si atoms are 0.202 nm and 0.157 nm, respectively.

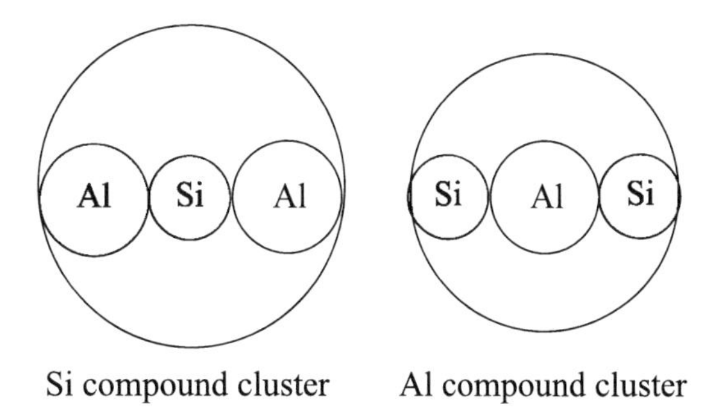

Figure 12.3–1

Al and Si atoms both form simple clusters, i.e. a central Si atom is surrounded by a number of Al atoms and vice versa. There also seems to exist a long range interaction between the clusters, which can be called 'compound clusters'. A central cluster is surrounded by clusters of the opposite kind. The clusters and compound clusters are named after their central atom.

Calculate the most likely cluster structure of the melt and estimate the composition of the metallic glass with the highest GFA.

12.4 Metallic glasses can be obtained from a metal melt, provided that it is exposed to a very high cooling rate. The growth rate of a crystal can roughly be described as a function of the undercooling with the aid of the left figure below. The nucleation rate of crystals in a metal melt as a function of the undercooling can be described with the aid of the sketched figure on the right-hand side.

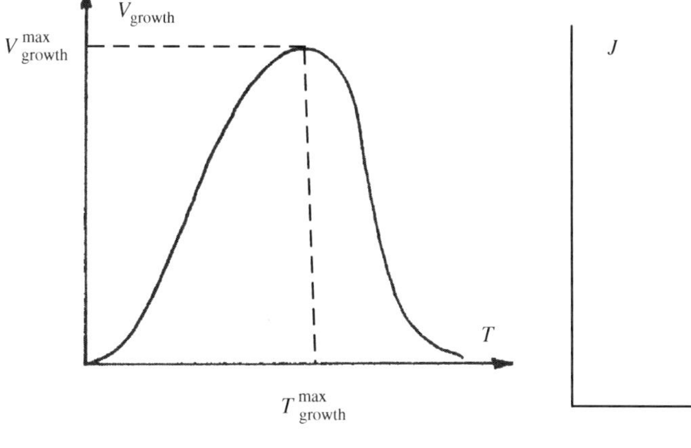

Figure 12.4–1 Growth rate as a function of temperature.

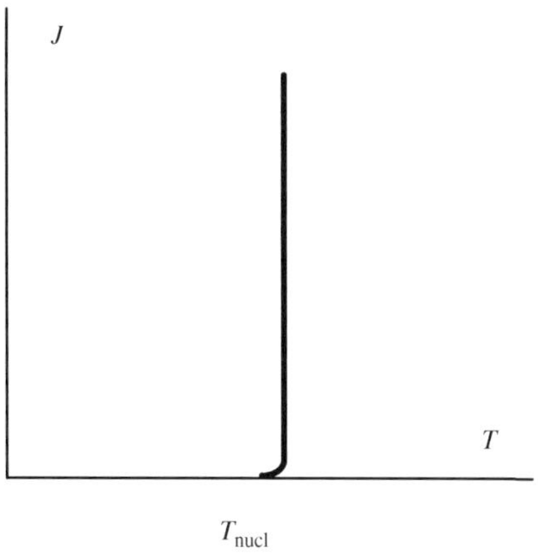

Figure 12.4–2 Nucleation rate as a function of temperature.

What is the relative positions of the two curves along the T-axis, required for obtaining a metallic glass with lowest possible cooling rate?

Derive expressions for T_{growth}^{max} and T_{nucl} as functions of the melting-point temperature T_M and discuss the relative positions of the two curves.

12.5

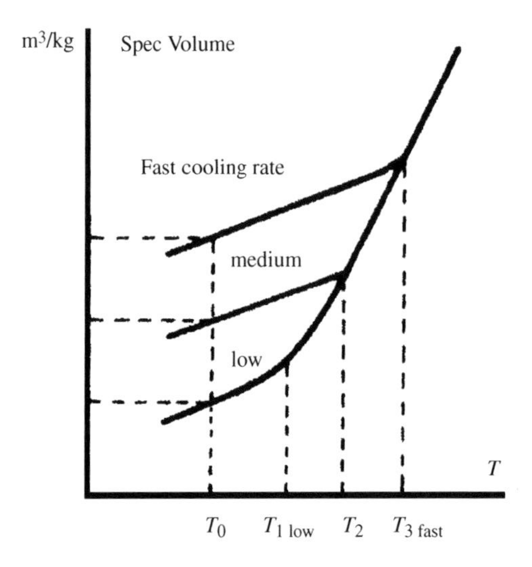

Figure 12.5–1

The figure shows the effect of volume change during cooling of a glass-forming metal. The figure shows that the cooling rate has a large influence on the specific volume of a glass-forming material.

Discuss the reason for this fact and give a likely explanation.

12.6 The viscosities of amorphous alloy melts have been extensively studied in the 2000th. The figure shows some measurements of this kind. The viscosity η (divided by a constant of the order 10^{-3}) is plotted versus the ratio T_g/T, in a diagram. The result is shown in the figure below.

The figure shows clearly that there are two separate groups of amorphous alloy melts, one with low viscosities and one with high viscosities.

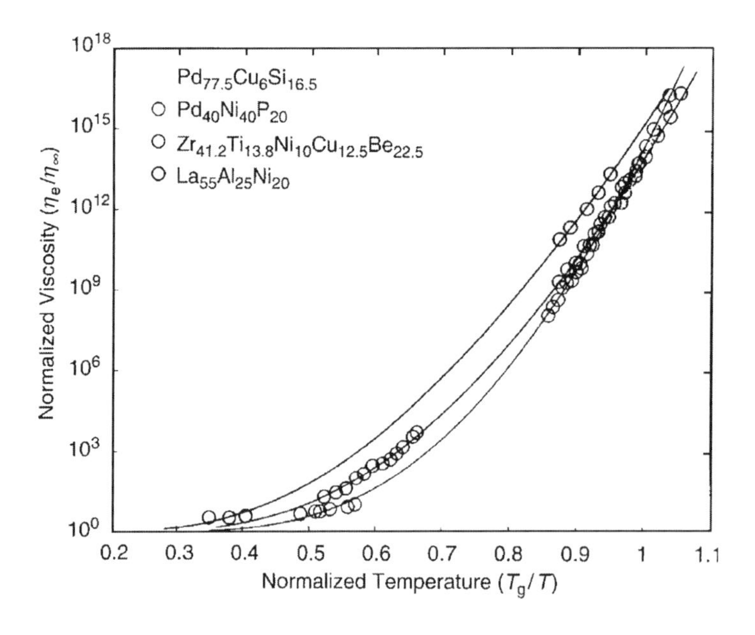

Figure 12.6–1

Analyze the reason for this difference between the two groups and give a likely explanation.

References

1. W. Klement, R.H. Willens, and P. Duwez, *Nature* **187**, 869, 1960.
2. H.-W. Kui, A. L. Greer, and D. Turnbull, *Applied Physics Letters*, **63**, 1993, 2342.
3. A. Inoue, N. Nishiyama, and H. M. Kimura, *Material Transactions*, **JIM, 38**, 1997, 179.
4. A. Inoue, *Bulk Amorphous Alloys, Preparation and Fundamental Characteristics*, Material Science Foundation 4, Zűrich, Trans. Tech Publications, 1998.
5. Y. Li, S.J. Poon, G.J. Shiflet, J. Xu, D.H. Kim and J.F. Löffler. *MRS Bulletin* **32**, 2007, pages 624–628.
6. A.L. Greer, Metallic Glasses on the Threshold, *Materials Today*, **12**, 2009, 14–22.
7. H. Fredriksson and U. Åkerlind, *Materials Processing during Casting*, John Wiley & Sons, 2005.
8. H Fredriksson and U. Åkerlind, *'Physics for Functional Materials'*, John Wiley & Sons, 2008.
9. A. L. Greer and E. Ma, Bulk Metallic Glasses; At the Cutting Edge of Metals Research, *MRS Bulletin*, **32**, 2007, 611–657.
10. D. B. Miracle, T. Egami, K. M. Flores, and K. F. Kelton, Structural Aspects of Metallic Glasses, *MRS Bulletin*, **32**, 2007, pages 629–634.
11. D. Shechtma, I. Blech, D. Gratias, and J. W. Cahn, Metallic Phases with Long-Range Orientation Order and No Translational Symmetry, *Physical Review Letters*, **53**, 1984, 1951–1953.
12. F. Zetterling, *Doctoral Dissertation*, Royal Institute of Technology, Stockholm 2003.
13. T. Fukunaga and K. Suzuki, *Science Report Research*, Institute of Tohoku University, **A26**, 1981, 153.
14. H. M. Kimura, A. Inoue, N. Nishiyama, K. Sasamori, O. Haruyama, and T. Masumoto, *Science Reports*, Research Institute Tohoku University, **A43**, 1997, 216.
15. A Inoue, *Bulk Amorphous Alloys, Preparation and Fundamental Characteristics*, Material Science Foundation 4, Zűrich, Trans. Tech Publications, 1998, page 69.
16. F. Spaepen and A. I. Taub, *Flow and fracture, in: Amorphous Metallic Alloys*. London, Butterworth, 1983, page 257.
17. J. Schroers and W. L. Johnson, *Physical Review Letters*, **93**, 2004, 255.
18. Y. H. Liu, G. Wang, R. J. Wang, D. Q. Zhau, M. X. Pan, W. H. Wang, *Science*, **315**, 2007, 315.
19. A. H. Brothers and D. C. Dunand, Porous and Foamed Metals, *MRS Bulletin*, **32**, 2007, 639–643.
20. P. G. Schewmon, *Transformations in Metals*, New York, McGraw-Hill, Exercise 12.1, 1969.

Answers to Exercises

Chapter 2

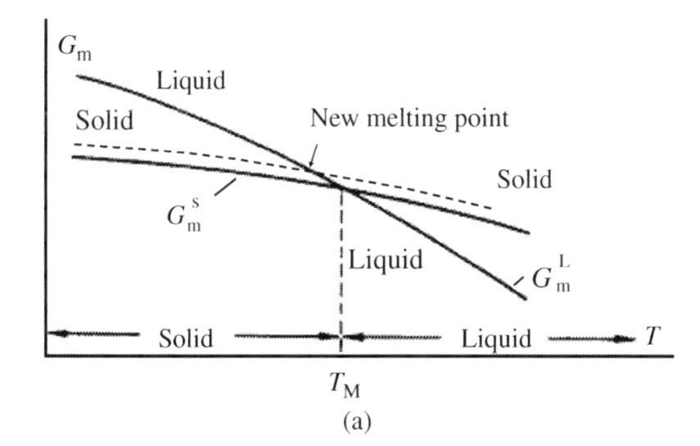

Figure 2.1–1

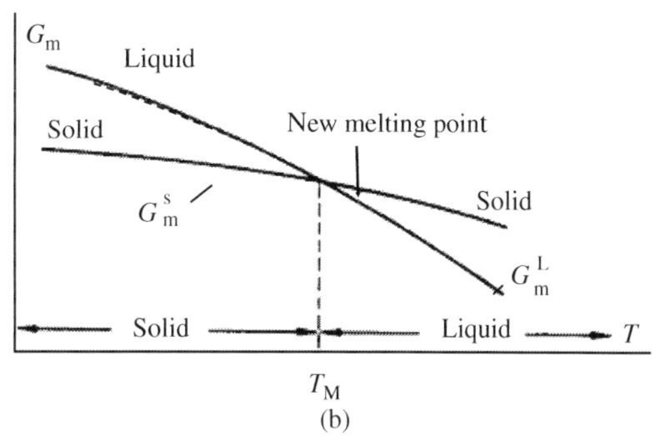

Figure 2.1–2

2.1a The left figure shows that the melting point will be lowered.

2.1b The right figure shows that the melting point will be increased.

2.1c $\Delta G_m^{L\to\alpha} = 3.3 \times 10^3$ kJ/kmol.

2.1d $-\Delta G_m^{L\to\alpha} < -\Delta G_m^{V\to\alpha}$.

2.2a 1.6×10^{-3} K.

2.2b 16 K.

2.2c 1.6×10^3 K.

2.3a The driving force of solidification of the alloy can be written as

$$-\Delta G^{L\to\alpha} = \left(1-x_B^{\alpha/L}\right)\left(\mu_A^L-\mu_A^{\alpha/L}\right)-(-x_B^{\alpha/L}\left(\mu_B^{\alpha/L}-\mu_B^L\right)$$

or

$$-\Delta G^{L\to\alpha} = \left(1-x_B^{\alpha/L}\right)\left(-\Delta G_A^{L\to\alpha}\right)+x_B^{\alpha/L}\left(-\Delta G_B^{L\to\alpha}\right)$$

The driving force of solidification is shown as the distance PQ in the Figure 2.3.

Solidification and Crystallization Processing in Metals and Alloys, First Edition. Hasse Fredriksson and Ulla Åkerlind.

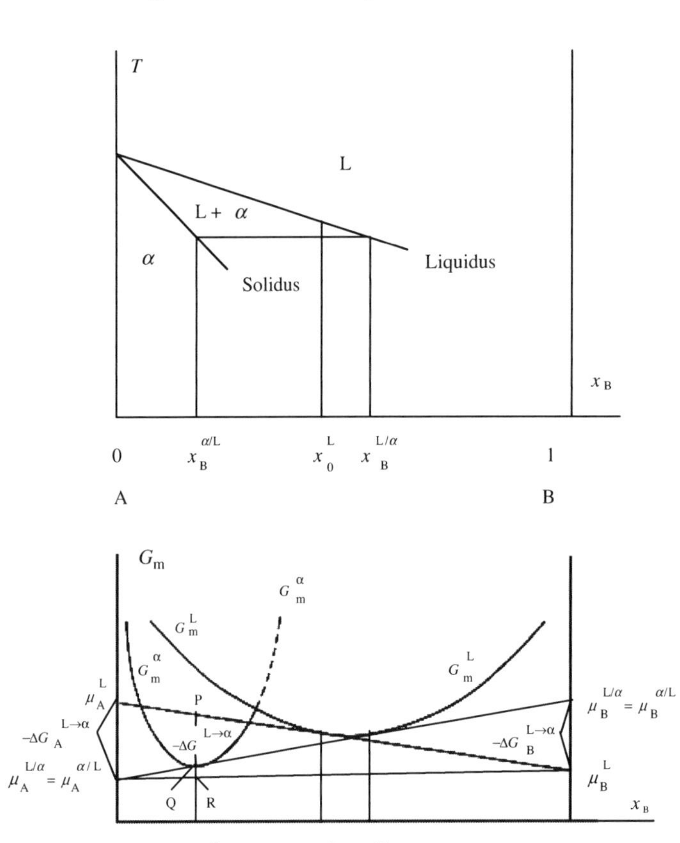

Figure 2.3a

2.3b Calculation of $-\Delta G_A^{L\to\alpha} = RT\left(x_B^{L/\alpha} - x_0^L\right)$ and $-\Delta G_B^{L\to\alpha} = -RT\frac{x_B^{L/\alpha}-x_0^L}{x_B^{L/\alpha}}$ and introduction of these expressions into the equation $-\Delta G^{L\to\alpha} = \left(1-x_B^{\alpha/L}\right)\left(-\Delta G_A^{L\to\alpha}\right)+x_B^{\alpha/L}\left(-\Delta G_B^{L\to\alpha}\right)$ gives the relationship $\frac{dr}{dt} = \frac{D}{r}\frac{V_m^\alpha}{V_m^L}\frac{1}{x_B^{L/\alpha}-x_B^{\alpha/L}}\frac{-\Delta G^{L\to\alpha}}{RT\left(1-x_B^{\alpha/L}-\frac{x_B^{\alpha/L}}{x_B^{L/\alpha}}\right)}$.

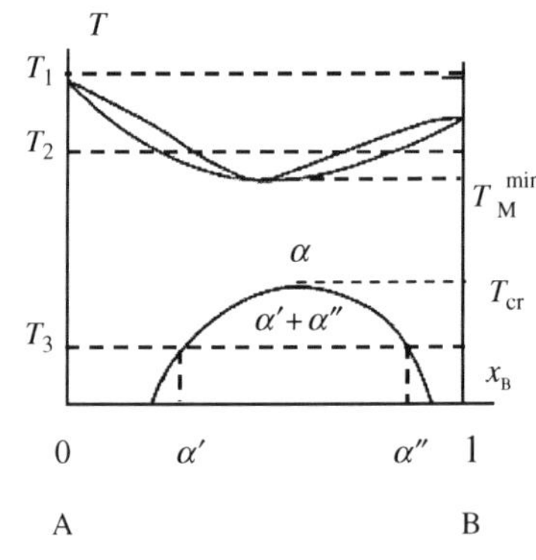

Figure 2.5a

2.4a 1670 K and 1.6 K.

2.4b 2.78×10^7 J/kmol.

2.4c 0.02 at%.

2.5a The phase diagram of the hypothetical A–B system would show two two-phase regions with a common minimum for the liquidus and solidus curves and a miscibility gap in the solid region at low temperatures.

2.5b The heat of mixing of the binary Cs–Rb system at the minimum liquidus temperature is 182 kJ/kmol. The theoretical critical temperature at the top of the miscibility gap is –229°C. Therefore, the miscibility gap is not shown in the phase diagram.

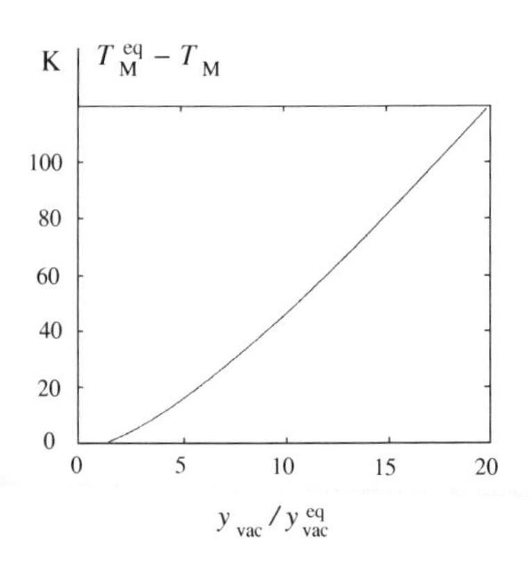

Figure 2.6–2

2.6 $T_M^{eq}-T_M = \frac{R(T_M^{eq})^2}{-\Delta H_m^{fusion}} y_{vac}^{eq} \times \frac{y_{vac}}{y_{vac}^{eq}}\ln\frac{y_{vac}}{y_{vac}^{eq}}$

The function $T_M^{eq}-T_M = 2.0 \times n\ln n = 2.0 \times \frac{y_{vac}}{y_{vac}^{eq}}\ln\frac{y_{vac}}{y_{vac}^{eq}}$ versus $\frac{y_{vac}}{y_{vac}^{eq}}$ is plotted in the figure.

2.7 The regular solution parameter between Cu atoms and vacancies in the alloy is $\approx -1.8 \times 10^6$ J/kmol. The interaction parameter $L_{Cu/vac}$ is strongly negative. The Cu atoms attract the vacancies because they are larger than the Al atoms. The concentration of vacancies around the Cu atoms results in relaxation of the crystal lattice.

Chapter 3

3.1 $\sigma_{LC1} = 0.79\ \text{J/m}^2$ and $\sigma_{LC2} = 0.75\ \text{J/m}^2$. The reason for the discrepancy between σ_{LC1} and σ_{LC2} is anisotropy owing to the crystallographic orientation.

3.2 Yes indeed. The thin channels can lift the melt 55 m.

3.3 $\sigma_{\alpha\text{Fe-P}} = 0.12\ \text{J/m}^2$ [based on $x_P^{max} \approx 1.7 \times 10^{-4}$ (mole fraction) and $\beta \approx 2.7 \times 10^2$].

3.4a $\zeta = \dfrac{2RT_{max}}{ZN_A} = 2.34 \times 10^{-21}\ \text{J/bond}.$

3.4b $n = \dfrac{3}{d^2} = \dfrac{3}{(2 \times 0.165 \times 10^{-9})^2} \approx 2.75 \times 10^{19}\ \text{bonds/m}^2.$

3.4c $\sigma = \dfrac{n}{2}(0.998^2 - 2 \times 0.998 \times 0.002)\,\zeta = 0.032\ \text{J/m}^2.$

3.5 $\dfrac{h}{r} = 2 \times \left(1 + \dfrac{RTd(1-\beta)}{V_m\sigma^0} x_0^L\right) \quad 0 < x_0^L < 1.8 \times 10^{-4}$

The function $\dfrac{h}{r} = 2 \times (1 - 1.4 \times 10^3 x_0^L)$ is plotted in the figure.

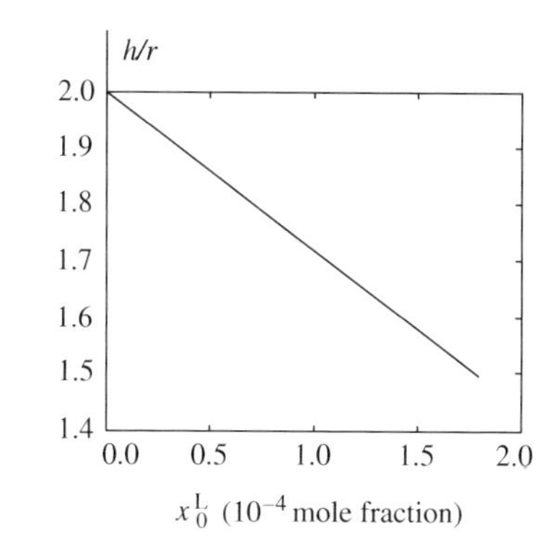

Figure 3.5–4

Return from the approximate model to the hexagonal graphite crystals: *h/a* decreases linearly with increasing x_0^L.

Table 3.5–1

x_0^L	0	1.8×10^{-4}
$\dfrac{h}{a}$	2	1.5

The sulfur concentration is in most cases higher than the mole fraction 1.8×10^{-4}. The shape of the crystals changes gradually towards a plate-like graphite morphology. At high sulfur concentrations the linearity between concentration and *h/a* is no longer valid, which is mentioned in the text.

3.6 **Table 3.6–1**

$2r$	T_M^σ	$\sigma = \dfrac{-\Delta H}{3V_m T_M} \times 2r(T_M - T_M^\sigma)$
10^{-9}m	K	J/m^2
30.0	~1300	0.43
25.0	~1290	0.95
20.0	~1270	1.71
15.0	~1240	2.35
10.0	~1190	2.75
5.0	~1080	2.61

- The solid-solid surface tension is of the order 0.5 J/m^2 between solid grains (see Table 3.3 on page 145 in Chapter 3). At large grain sizes the theoretical value approaches the normal value closely. When the grain size decreases the calculations indicate an increase of the surface tension.
- The reason can either be an increase of the interphase thickness or an increase of the lattice defects inside the grain that results in a decrease of the heat of fusion. One can conclude that the atomic structure inside the grains is affected for grain sizes less than 10 times the thickness of a grain boundary. The grain-boundary thickness is of the order of 1 nm.

3.7a Gibbs' interface is a 'mathematical' interface. Its thickness is zero. It is the junction between the liquid phase and the solid phase. Guggenheim's interphase describes the interphase as a third phase between the solid and liquid phases. It has a certain thickness. It consists of two Gibbs' interfaces and one disordered interphase with thickness d in between. Guggenheim's description of an interphase is more realistic than Gibbs' model. Gibbs' interface describes correctly the condition for twin boundaries.

3.7b $G^{l/s} - G^{s} = \dfrac{V_m \sigma}{d}$, which is identical with Equation (3.2) on page 100 in Chapter 3, is the link between the two interface models.

3.7c Jackson's and Temkin's solid/liquid interphase models are based on broken liquid/solid bonds over the interface. The calculated values of the interphase energies (see Examples 3.4 on page 137 and Example 3.5 on page 140 both in Chapter 3) are rather low compared with the experimentally measured interphase solid/liquid values.

Chapter 4

4.1 $r^* = \dfrac{8\pi\sigma^{L/s} V_m}{4\pi\left[-\Delta G^{L\to s}\right]} = \dfrac{2\sigma^{L/s} V_m}{-\Delta G^{L\to s}}$

The equation is identical with Equation (4.3) on page 168 in Chapter 4.

$$\Delta G^* = \frac{16}{3} \times \frac{\left(\sigma^{L/s}\right)^3 V_m^2}{\left(-\Delta G^{L\to s}\right)^2}.$$

The equation is identical with Equation (4.4) on page 169 in Chapter 4.

4.2 The surface energy equals 0.25 J/m^2.

4.3 The activation energy for grain nucleation of a pure metal can be written as

$$\Delta G_r^* = -\frac{4\pi r^{*3}}{3}\frac{\Delta T \Delta H_m^{fusion}}{T_M V_m} + 4\pi r^{*2}\sigma_0\left(1 - \frac{\delta}{r^*}\right)$$

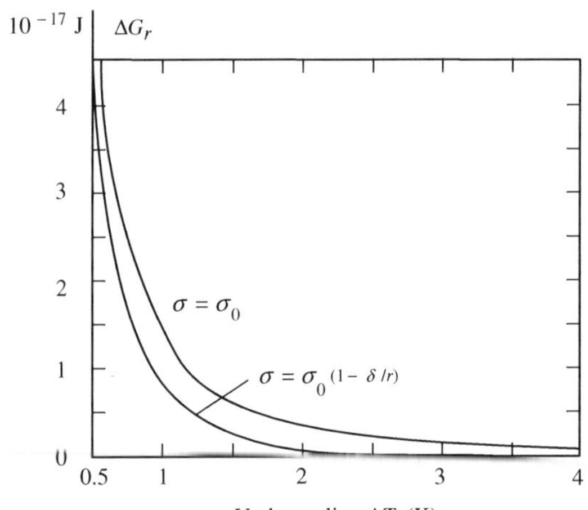

Figure 4.3–2

The activation energy for grain nucleation decreases with the undercooling, as in the classical theory of homogeneous nucleation.
The activation energy is lower in the case of a surface tension that depends on the particle size, compared to the classical theory. For small particles, the surface tension will be lower than that of a flat surface and less energy will be required for the growth of the particle.

4.4 The fact that the $FeOAl_2O_3$ oxide is easier to nucleate than the Al_2O_3 oxide means that $\frac{(\Delta G^*)_{FeOAl_2O_3}}{(\Delta G^*)_{Al_2O_3}} < 1$. It can be shown that $\frac{(\Delta G^*)_{FeOAl_2O_3}}{(\Delta G^*)_{Al_2O_3}} \approx 2 \times \left(\frac{\sigma^{FeOAl_2O_3/L}}{\sigma^{Al_2O_3/L}}\right)^3$, which gives $\sigma^{FeOAl_2O_3/L} < \sqrt[3]{0.5} \times \sigma^{Al_2O_3/L} = 0.8 \times \sigma^{Al_2O_3/L}$.
The observed nucleation of $FeOAl_2O_3$ oxide can be explained by the fact that the surface tension $\sigma^{FeOAl_2O_3/L}$ between the iron melt and the liquid $FeOAl_2O_3$ oxide is lower than the surface tension 0.8 $\sigma^{Al_2O_3/L}$ between the melt and the solid Al_2O_3 particles. The Gibbs' free energy for homogeneous nucleation of the liquid $FeOAl_2O_3$ oxides is therefore lower than that of the Al_2O_3 particles. Hence, the nucleation of $FeOAl_2O_3$ is thermodynamically favoured.

4.5 Expressions for the necessary nucleation temperature for homogeneous nucleation of graphite and cementite are derived and plotted in the Fe–C phase diagram. As the necessary nucleation temperature for homogeneous nucleation of cementite is lower than that of graphite, cementite is easier to nucleate than graphite.

Nucleation of Graphite

$\left(\ln \frac{x^L}{x^L_{eq}}\right)^2 = \frac{16\pi}{3} \frac{\sigma^3 V_m^2}{55 k_B R^2 \left(T_\alpha^{nucl}\right)^3}$. Inserting the numerical values gives

$$\ln \frac{x^L}{x^L_{eq}} = \sqrt{\frac{16\pi}{3} \times \frac{0.20^3 \times (5.4 \times 10^{-3})^2}{55 \times 1.38 \times 10^{-23} \times (8.31 \times 10^3)^2 \left(T_\alpha^{nucl}\right)^3}} = \frac{8.61 \times 10^3}{\left(T_\alpha^{nucl}\right)^{3/2}}$$

Nucleation of Cementite

$\left(\ln \frac{(1-x^L)^3 x^L}{\left(1-x^L_{eq}\right)^3 x^L_{eq}}\right)^2 = \frac{16\pi}{3} \frac{\sigma^3 V_m^2}{55 k_B R^2 \left(T_\alpha^{nucl}\right)^3}$. Inserting the numerical values gives

$$\ln \frac{(1-x^L)^3 x^L}{\left(1-x^L_{eq}\right)^3 x^L_{eq}} = \sqrt{\frac{16\pi}{3} \times \frac{0.20^3 (5.8 \times 10^{-3})^2}{55 \times (1.38 \times 10^{-23})(8.31 \times 10^3)^2 \left(T_\alpha^{nucl}\right)^3}} = \frac{9.25 \times 10^3}{\left(T_\alpha^{nucl}\right)^{3/2}}$$

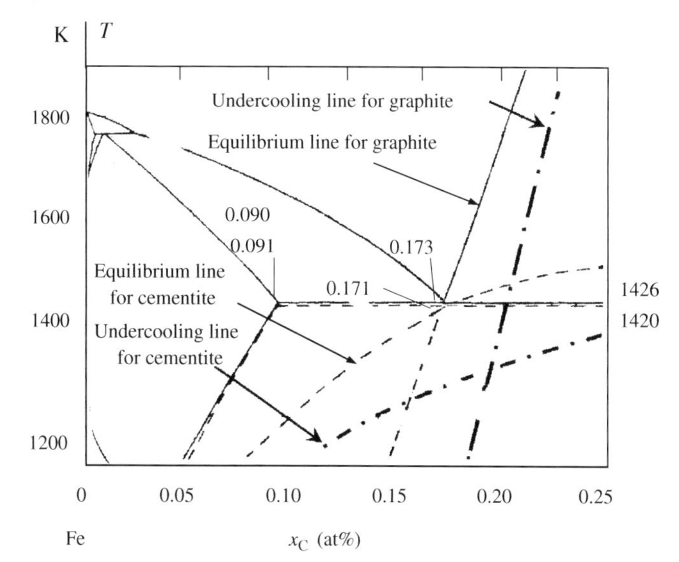

Figure 4.5–4

4.6 The maximum undercooling is 35 K.

4.7

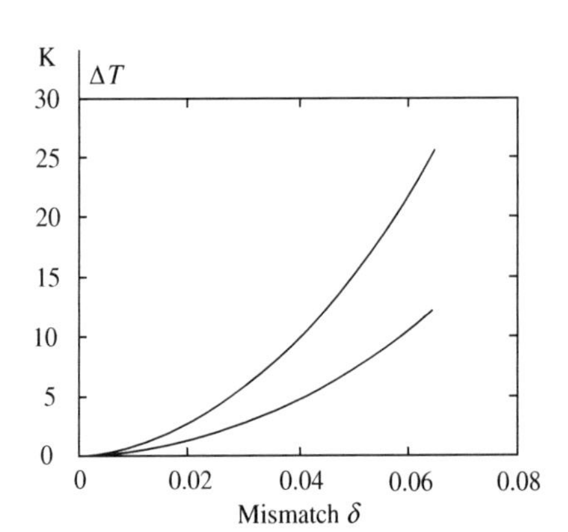

Figure 4.7–2

Upper curve: All layers fully strained. $\Delta T = \dfrac{Y\delta^2 V_m T_M}{-\Delta H_m^{fusion}} = 6.1 \times 10^3 \times \delta^2$

Lower curve: Partly strained layers. $\Delta T = 0.47 \times \dfrac{Y\delta^2 V_m T_M}{-\Delta H_m^{fusion}} \times \delta^2 = 2.9 \times 10^3 \times \delta^2$

4.8 Addition of Al3 wt%Ti to the melt exceeding a total Ti concentration higher than 0.22 wt% makes homogeneous nucleation of Al_3Ti particles possible. These particles can then act as inoculants. This is the explanation for the obtained fine-grain structure.

Chapter 5

5.1 Driving force *with no* consideration to strain

$$-\Delta G_{Zn}^{V \to s} = RT \ln p_0 - RT \ln p_T = 6947R \times \left(1 - \frac{T}{T_0}\right)$$

The function

$$-\Delta G_{Zn}^{V \to s} = 6947 \times 8.31 \times 10^3 \times \left(1 - \frac{T}{773}\right) \text{(J/kmol)}$$

is plotted in the diagram below.
The total driving force *with* consideration to strain:

$$-\Delta G_{Zn\ strain}^{V \to s} = -\Delta G_{Zn}^{V \to s} - G_{strain}$$

or

$$-\Delta G_{Zn\ strain}^{V \to s} = 6947R \times \left(1 - \frac{T}{T_0}\right) - Y_{Zn}\left(\frac{a_{Cu}[1+\alpha_{Cu}(T-298)] - a_{Zn}[1+\alpha_{Zn}(773-298)]}{a_{Zn}[1+\alpha_{Zn}(773-298)]}\right)^2 V_m^{Zn}$$

The function

$$-\Delta G_{Zn\,strain}^{V \to s} = 6947 \times 8.31 \times 10^3 \times \left(1 - \frac{T}{773}\right) - 8.65 \times 10^8 \left(\frac{1.031 \times [1+1.6 \times 10^{-5}(T-298)]}{[1+2.6 \times 10^{-5}(T-298)]} - 1\right)$$

(J/kmol) is plotted in the figure.

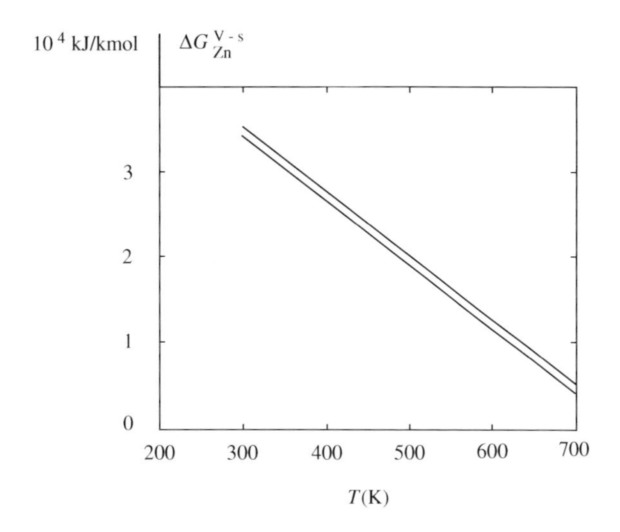

Figure 5.1–1

The lower curve represents the true driving force. The vertical difference between the curves is the strain energy. It is small, unavoidable and decreases slowly with increasing temperature.

5.2 $$\frac{V_{\text{growth}}^{\text{strain}}}{V_{\text{growth}}} = \frac{\dfrac{p}{p_{\text{eq}}} - \exp\dfrac{Y\varepsilon^2 V_{\text{m}}}{RT}}{\dfrac{p}{p_{\text{eq}}} - 1}$$ where T is the temperature of the vapour close to the film

The function $\dfrac{V_{\text{growth}}^{\text{strain}}}{V_{\text{growth}}} = \dfrac{\dfrac{p}{p_{\text{eq}}} - 1.62}{\dfrac{p}{p_{\text{eq}}} - 1}$ is plotted in the figure.

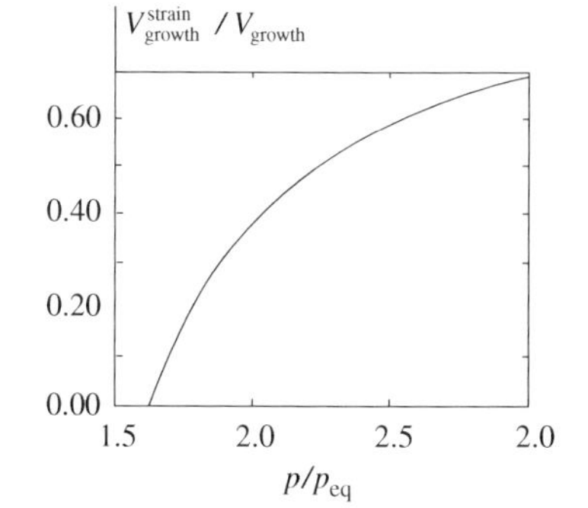

Figure 5.2–1

5.3 $$T_0 = \frac{T_{\text{comp}}^{3/2}}{T_{\text{comp}}^{1/2} - \dfrac{1}{40\,400} \times \dfrac{V_{\text{m}}}{R}\sqrt{\dfrac{16\pi\sigma^3}{3 \times 55k_{\text{B}}}}} = \frac{x^3}{x - \dfrac{1}{40\,400} \times \dfrac{V_{\text{m}}}{R}\sqrt{\dfrac{16\pi\sigma^3}{3 \times 55k_{\text{B}}}}}$$ where $x = \sqrt{T_{\text{comp}}}$.

The functions $T_0 = \dfrac{x^{3/2}}{x^{1/2} - 5.49}$ and $T_{\text{comp}} = x^2$ are plotted in the diagram below for the interval $33.4 \leq x \leq 36$.

- T_0 must exceed the melting point of gold ($1063 + 273 = 1336$ K.
- To avoid droplet precipitation T_{comp} must exceed $x_{\text{cr}}^2 = 33.4^2 = 1117$ K.
- Each value of T_0 corresponds to a value of T_{comp}, which can be read from the diagram.
- The largest temperature gap (which promotes the coating) is obtained for $x = 33.4$ but it is too close to the melting point of gold and the critical value of T_{comp} for droplet precipitation.

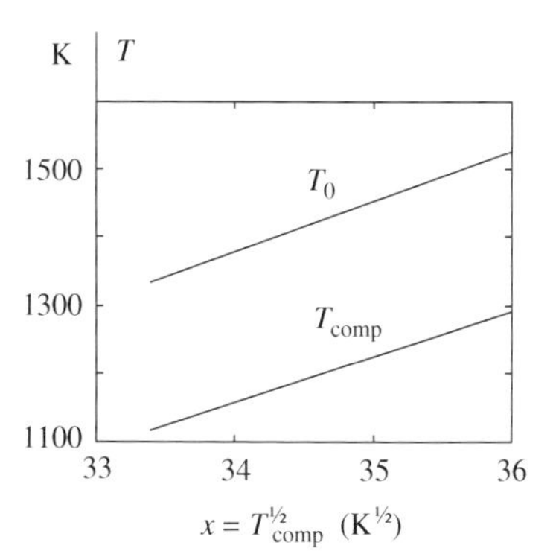

Figure 5.3–1

A convenient choice is $x = 33.5$, which corresponds to $T_0 = 1342$ K and $T_{\text{comp}} = 1122$ K. As the temperature gap decreases rather slowly with increasing x values higher values of the temperatures can be chosen for safety.

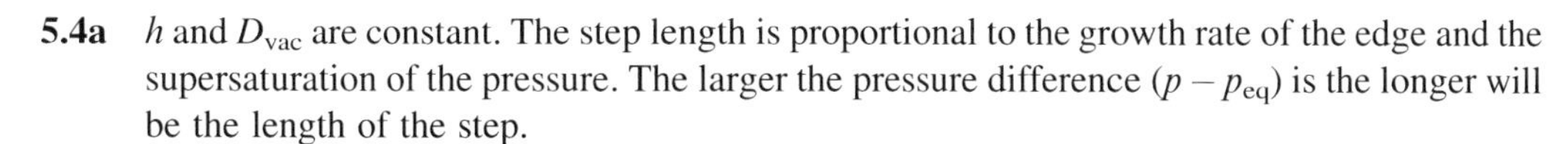

5.4a h and D_{vac} are constant. The step length is proportional to the growth rate of the edge and the supersaturation of the pressure. The larger the pressure difference ($p - p_{\text{eq}}$) is the longer will be the length of the step.

5.4b D_{vac} is constant. Both h and λ decrease with increasing supersaturation. When h and/or λ decrease/decreases to a value of the same magnitude as the thickness of the interphase, the growth mechanism will change. The steps become included into the interphase and the interphase can be considered as rough.

5.4c All sorts of defects, which break the regular lattice pattern in crystals, deteriorate the properties of crystals, such as mechanical strength and electrical conductivity, particularly for semiconductors. Lattice defects also cause crack formation and strain in the materials.

5.5 Activation energy/atom = 1.6 eV/atom. $\beta_{vapour} = 7.4 \times 10^{-17}$ m³/Ns at 700 °C provided that the pressure is measured in Pa (N/m²).

5.6 The Gibbs' free energy diagram shows that graphite is always the stable phase as the curve is positive for all temperatures. However, in a substrate the C atoms in the graphite are always strained. Therefore the graphite layer has extra free energy owing to the addition of strain energy. In this case, diamond will be more stable than graphite. Hence graphite is transformed into diamond. The higher the temperature is the more strain energy is required to transform graphite into diamond.

Chapter 6

6.1a The relaxation does not end until infinity. For practical reasons you have to set up a limit, for example 10% or 1% of the initial value and calculate the time required to reach this limit.

6.1b $t = \tau \times \ln 2.$

6.1c $t = \tau \times \ln 100 = 2\tau \times \ln 10.$

6.1d A reduction with the factor $1/e^5 = 1/2.718^5 = 0.00674$.

6.2a $$V_{growth} = \frac{D_0}{\delta} \exp \frac{-U_{act}}{RT} \times \frac{-\Delta H}{RTT_M} \times (T_M - T)$$

6.2b $$V_{growth} = \frac{D_0}{\delta} \exp \frac{-U_{act}}{RT} \times \frac{-\Delta H + Y\varepsilon^2 \dfrac{M}{\rho}}{RTT_M} \times (T_M - T)$$

6.3 $$\frac{D_{relax}^{111}}{D_{relax}^{100}} = 10^4$$

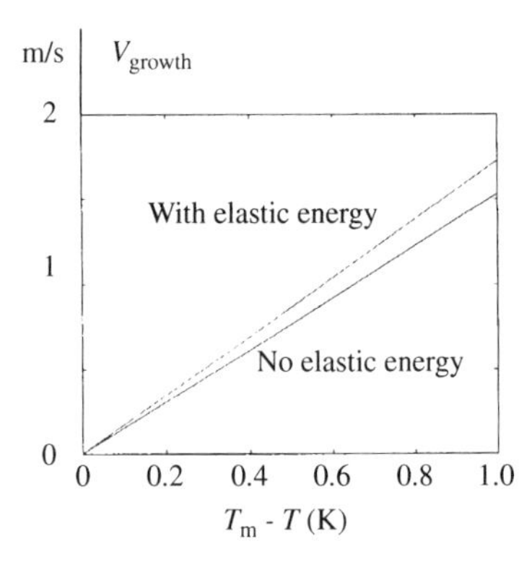

Figure 6.2–1

6.4a Consider Figure 6.24 and Equation (6.23) on page 290 in Chapter 6. The figure shows that the driving force of solidification is $-\Delta G_a$. In addition, only those atoms which have energy enough to over come the energy barrier $-\Delta U_a$ can enter the crystal, i.e. solidify. The driving force is related to the heat of fusion via the relationship.

$$-\Delta G^{total} = (-\Delta H_0) \frac{T_M - T}{T_M} \text{ (Equation (6.23) on page 290)}$$

6.4b $$-\Delta H_{strain} = -\Delta H_0 \sqrt{\frac{V_{growth}\delta}{\pi D_{relax}}} \text{ (Equation (6.21) on page 288 in Chapter 6)}$$

The energy is stored in the solid and is gradually released.

6.4c The total heat of fusion is the sum of the promptly released heat of fusion and the strain energy. The promptly released heat of fusion is the one you can measure as a function of V_{growth}.

$$-\Delta H_0 = (-\Delta H_{prompt}) + (-\Delta H_{strain})$$

or

$$-\Delta H_{prompt} = (-\Delta H_0)\left(1 - \sqrt{\frac{\delta}{\pi D_{relax}}} \times \sqrt{V_{growth}}\right)$$

$(-\Delta H_0)$ can not be measured because it corresponds to V_{growth}. The only way is to measure $(-\Delta H_{prompt})$ as a function of $\sqrt{V_{growth}}$ and extrapolate to $\sqrt{V_{growth}} = 0$.

6.4d The expected fraction of strain energy is $\approx$ 0.003 or 0.3% of the total heat of fusion. Such a small fraction is hard to observe and can be neglected.

6.5 The functions $x_{eq}^{L/\alpha} - \frac{x_0^L + 1 - x_{eq}^{L/\alpha} + 0.10 \times x_{eq}^{L/\alpha}}{2} - \sqrt{\left(\frac{x_0^L + 1 - x_{eq}^{L/\alpha} + 0.10 \times x_{eq}^{L/\alpha}}{2}\right)^2 - 0.10 \times x_0^L}$

and $(x_{eq}^{L/\alpha} - x_{eq}^{\alpha/L})$ are plotted in the figure as functions of $x_{eq}^{L/\alpha}$.

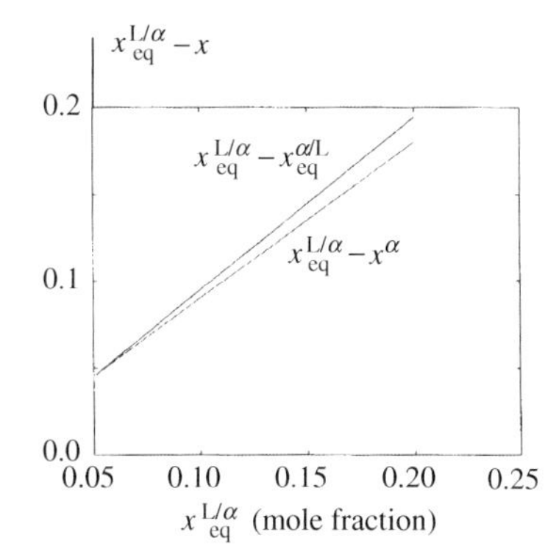

Figure 6.5–2

The lower the temperature is, the larger will be the growth rate and the difference between x^{α} and $x_{eq}^{\alpha/L}$.

6.6a Pressure = 65 kbar. Temperature = 1795 K.

6.6b 0.7×10^{-10} m/s and 4×10^2 hours.

6.7a The radius of the maximum droplet is 1.1×10^{-4} m.

6.7b The desired function is $r_{L_2} = \sqrt[3]{\frac{0.043}{N}}$ m.

6.8
$$V_{BC\ growth} = D^L \frac{\frac{x_0^{LB} - x_0^{LC}}{2} + \sqrt{\left(\frac{x_0^{LB} - x_0^{LC}}{2}\right)^2 + 10^{-3} - x_0^{LB}}}{\frac{x_0^{LB} - x_0^{LC}}{2} + \sqrt{\left(\frac{x_0^{LB} - x_0^{LC}}{2}\right)^2 + 10^{-3} - x_{eq}^{BC/L}}} \frac{V_m^{BC}}{V_m^L} \frac{1}{r_{BC}}$$

Chapter 7

7.1 The solidification time of a turbine blade is about 5 min (4 ¾ min).

7.2a The solidification time of the copper ingot is about 2.9 hours.

7.2b The solidification time of the steel ingot is about 1.5 hours. The steel ingot solidifies much more rapidly than the Cu ingot mainly because it has a much larger driving force of solidification than the Cu ingot.

7.3 The critical cooling rate is 6×10^5 K/s.

7.4 The heat capacitivity of the melt is 3.0 J/kg K.

7.5a 100 kJ/kg on earth and 130 kJ/kg in space. Lattice defects are formed during rapid solidification. The fraction of lattice defects is smaller in space than on earth.

7.5b The tabulated value of the heat of fusion differs greatly from those of the experimental values for both samples. The explanation is the rapidity of the solidification process, which corresponds to a long relaxation time and a large relaxation energy both for the reference sample on earth and for the space sample. The tabulated value of the heat of fusion corresponds to a total absence of lattice defects and an infinitely low growth rate of the solidification front.

7.6a The alloy starts to solidify when the liquidus line is reached, The phase diagram shows that this occurs at 1516 °C. The alloy cools and solidifies partly until the peritectic temperature 1493 °C. (read from the phase diagram) is reached. These two simultaneous processes last for 23 min. The total solidification time is 388 min. During the last 365 min the temperature is constant 1493 °C while the alloy solidifies completely. The indication on this is that the solid alloy starts to cool and the temperature decreases.

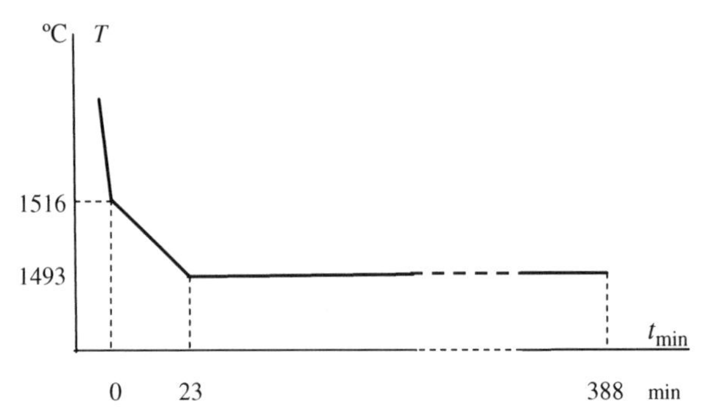

7.6b The peritectic reaction will occur at a constant temperature owing to the low diffusivity of carbon and the low cooling rate. At higher cooling rates than the one in the present case the peritectic reaction will occur within a temperature interval. For elements with a lower diffusivity than that of carbon the time for the reaction will be much shorter than for carbon. The solidification will proceed with precipitation of a second phase (peritectic phase) directly from the melt and the curve will look like a solidification curve of a process when only one phase is formed.

Chapter 8

8.1 15 cycles are required. At the beginning of the rod the impurity concentration has been reduced by a factor 0.095 of the initial impurity concentration after 15 cycles.

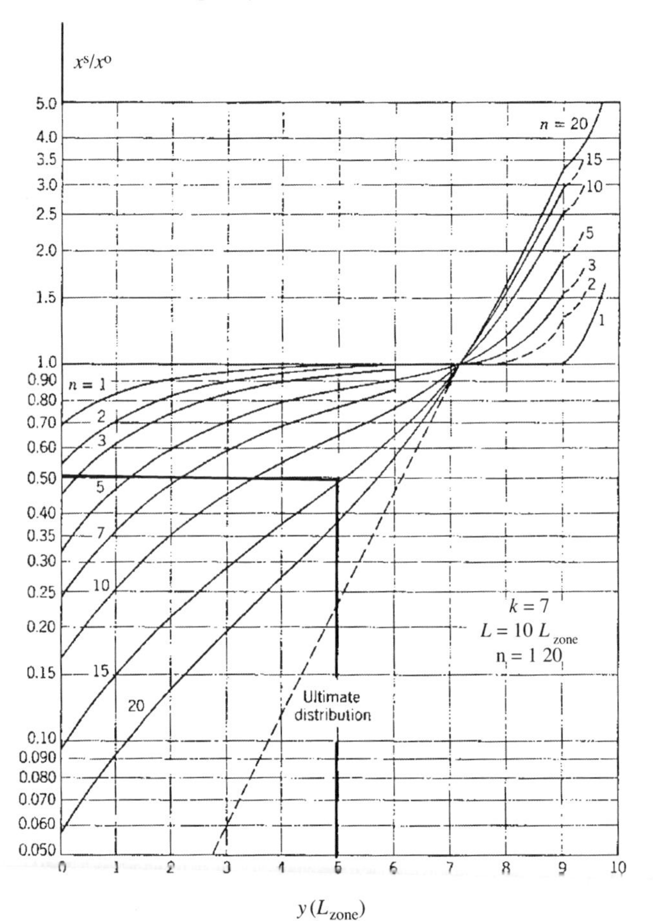

Figure 8.1–2

8.2a $x^{s} = x_{0}^{L}\left[1-(1-k_{0})\exp\frac{-k_{0}y}{L_{zone}}\right]$

The function $x^{s} = 0.0050 \times \left[1-(1-0.145)\exp\frac{-0.145y}{L_{zone}}\right]$ is plotted in the figure.

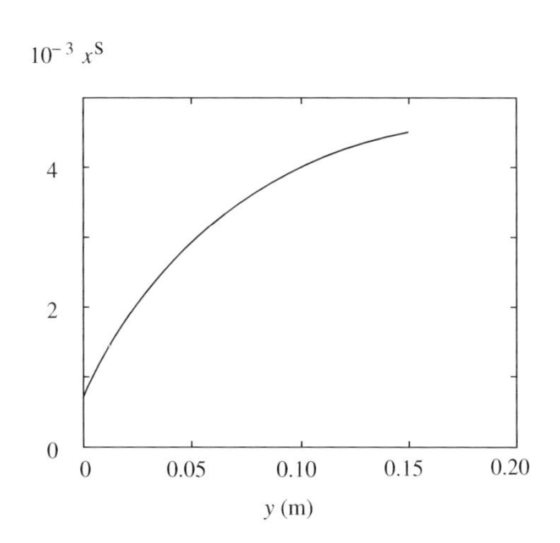

Figure 8.2–5

Slow pulling rate and strong convection promotes zone refining.

8.2b Strong convection $\rightarrow$ small δ $\rightarrow$ rapid diffusion.
Small convection $\rightarrow$ large δ $\rightarrow$ slow diffusion.
Concentration profile of the alloying element:

$$x^{L} = x^{s}\left(1+\frac{1-k_{0}}{k_{0}}\,e^{-\frac{V_{growth}}{D_{L}}y}\right) \qquad y \leq \delta$$

$$(x^{L})_{y=\delta} = x^{s}\left(1+\frac{1-k_{0}}{k_{0}}\,e^{-\frac{V_{growth}}{D_{L}}\delta}\right) \qquad y \geq \delta$$

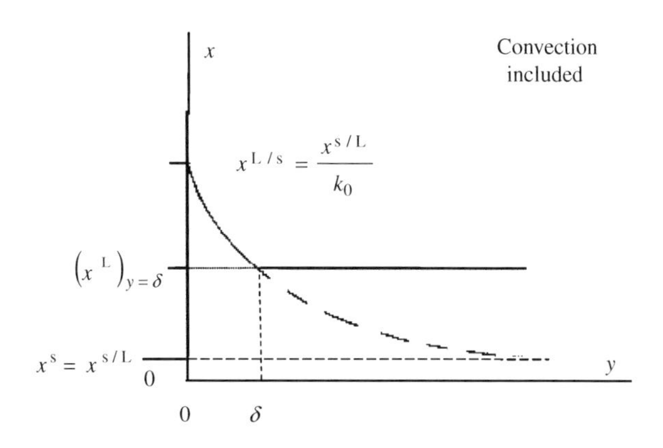

Figure 8.2–6

8.3 The impurity concentration in the rod as a function of the distance from the left end of the rod is

$$x^{s} = k\,x^{L} = x_0^{L}\left[1-(1-k)\exp-\frac{k\,V_{m}^{L}}{L_{1}^{zone}\,V_{m}^{s}}\,y\right] \qquad 0 < y < 10\,\text{cm and narrow zone width.}$$

The impurity concentration in the rod as a function of the distance from the new origin y_1 is

$$x^{s} = k\,x^{L} = x_0^{L}\left[1-\left(1-k\frac{x_1^{0}}{x_0^{L}}\right)\exp-\frac{k\,V_{m}^{L}}{L_{1}^{zone}\,V_{m}^{s}}\,\delta\right] \qquad y > 10\,\text{cm and wide zone width.}$$

The functions are sketched in the figure below.

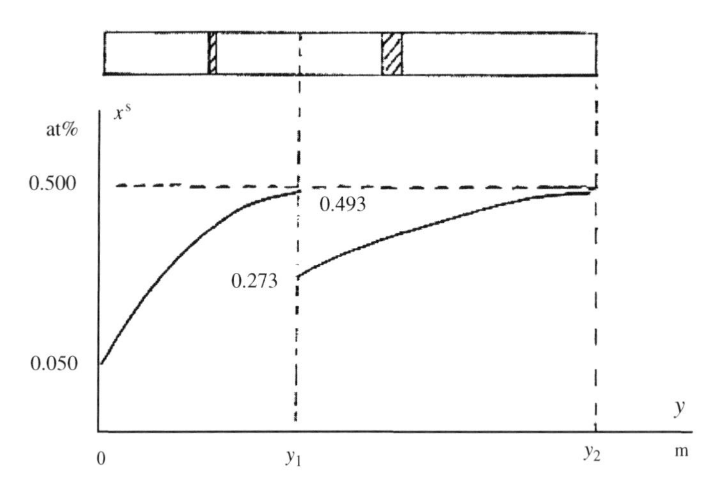

Figure 8.3–2

8.4a

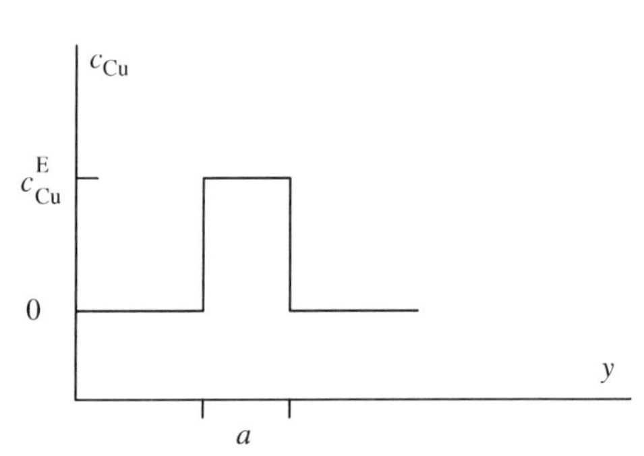

Figure 8.4–1

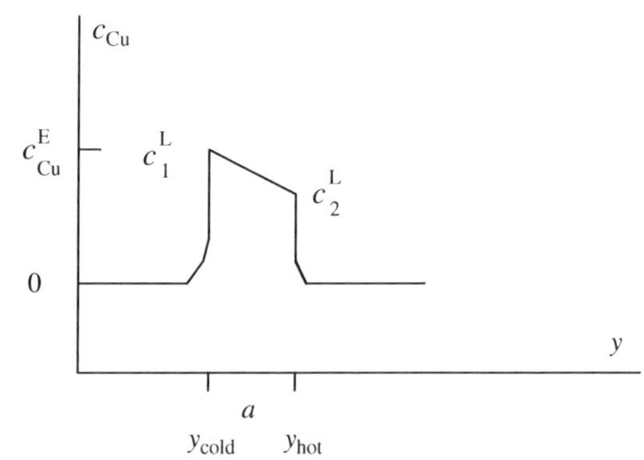

Figure 8.4–2

8.4b $V_{hot} = D_L \dfrac{-aG}{m_L c_E^L + aG} = 0.22\,\text{m/s}$ and $V_{cold} = D_L \dfrac{-aG}{m_L c_E^L (1-k)\,a} = 0.21\,\text{m/s}$

8.4c The area under the curve represents the amount of the alloying element. When the eutectic slice moves in the temperature gradient the amount of the alloying element is kept constant, i.e. $A_1 = A_2$. For this reason the dark area at the hot end must be equal to that of the cold end. Obviously the hot end will move faster than the cold end.

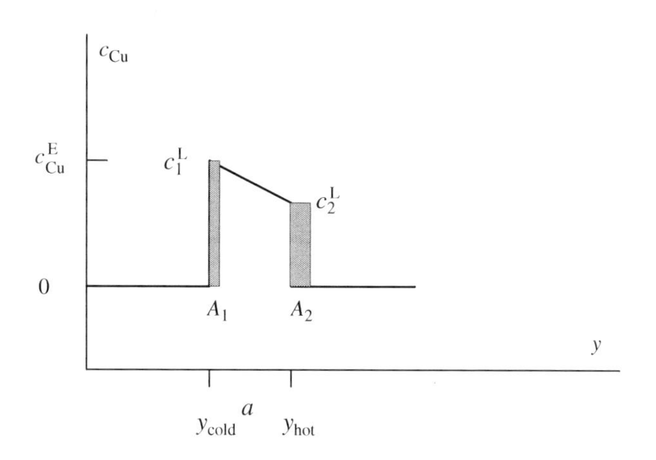

Figure 8.4–6

$V_{hot} > V_{cold}$ because of the material balance $A_1 = A_2$.

8.5 $r = \sqrt{2D_L\left(t - \frac{1-k}{0.00575}\ln\frac{1-k+0.00575\,t}{1-k}\right)}$

The figure shows the function for the time interval $0 < t < 3600\,\text{s} = 1$ hour.

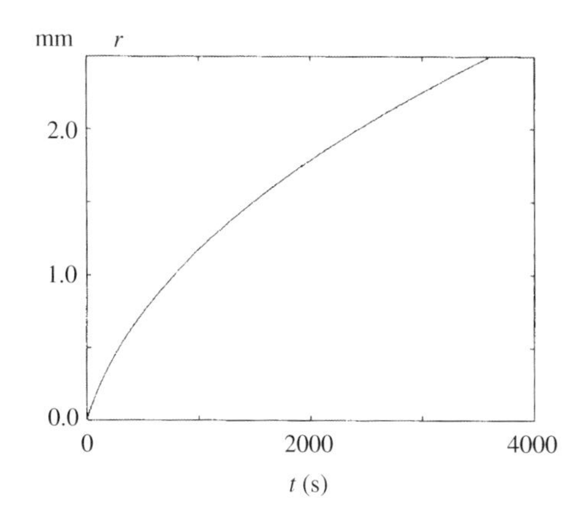

Figure 8.5–2

8.6a $c^L = c_0^L\left[1 - \left(\frac{1}{k} - 1\right)\exp\left(-\frac{V_{growth}}{D_L}y\right)\right]$

8.6b $c^L = 10 - 90\exp\left(-1.6 \times 10^4\,y\right)$ and $c^L = 10 - 90\exp\left(-8.0 \times 10^4\,y\right)$ after increase of the rate of solidification.

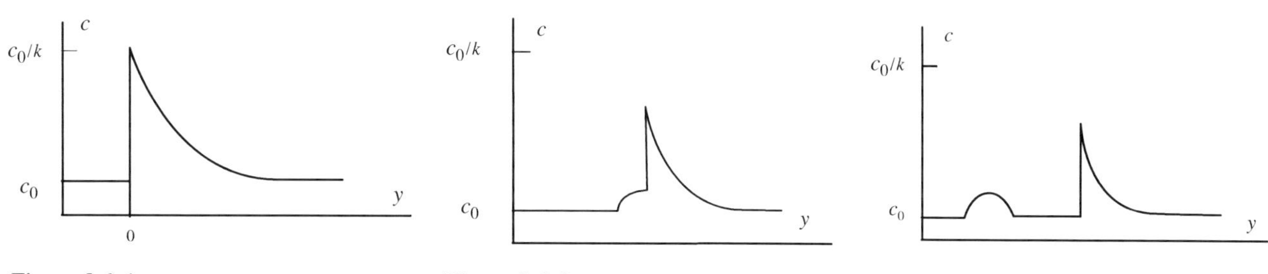

Figure 8.6–1 **Figure 8.6–2** **Figure 8.6–3**

The solidification wave carries and deposits the alloying element at solidification. The additional amount of the alloying element that earlier was present in the 'wave' (left figure) is now in the 'band' to the left in the right figure. The material balance requires that the area under the curve in the left figure is equal to the sum of the areas under the curve in the right figure. The 'band' appears as an increase of the concentration of the alloying element in a stripe parallel with the solidification front.

8.7 The magnitude of the diffusion coefficient is $8.7 \times 10^{-10}\,\mathrm{m^2/s}$.

8.8a No constitutional cooling if $G_L \leq m_L x_0^L \dfrac{1-k_0}{k_0} \dfrac{V_{growth}}{D_L}$.

8.8b $G_L \leq 9 \times 10^4\ \mathrm{K/m}$.

Chapter 9

9.1 Hint: Try to find OA and OM in terms of the edge length a of the octahedron and form the ratio OA/OM. What is the relationship between this ratio and the growth rates?

9.2a

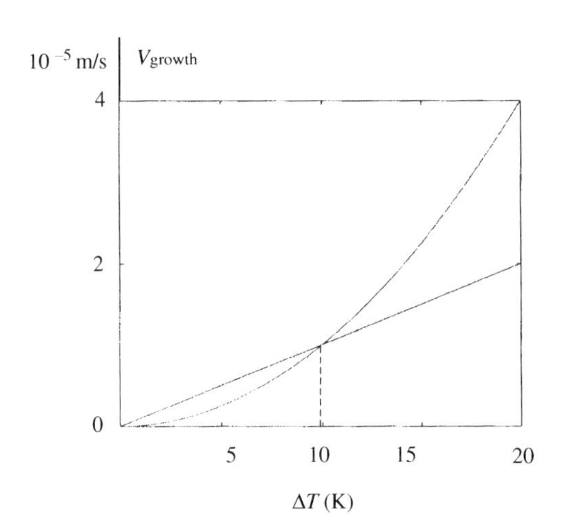

Figure 9.2–2

$$V^{a}_{growth} = 1.0 \times 10^{-6} \times \Delta T\ \mathrm{m/s} \quad \text{and} \quad V^{c}_{growth} = 0.10 \times 10^{-6} \times \Delta T^2\ \mathrm{m/s}.$$

9.2b At a lower undercooling than 10 K the base plane grows faster than the c-direction and the shape of the crystal will be compressed in the c-direction, compared with the hexagonal crystal.

At a higher undercooling than 10 K the growth in the c-direction will dominate over the growth in the base plane and the crystal will be more and more bar-like.

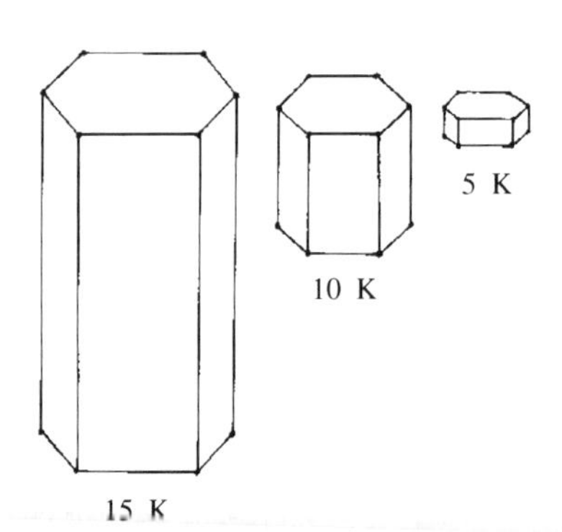

Figure 9.2–4

9.3a $V^{100}_{\text{growth}} = \dfrac{0.0616 \times 10^{-9}}{r_c}$ and $V^{111}_{\text{growth}} = 0.027 \times 10^{-3}$ m/s

The diffusion-controlled growth rate V^{100}_{growth} varies strongly with the size of the crystal, especially for small crystals, which grow very rapidly. The growth rate V^{111}_{growth} is constant and is controlled by the kinetics.

9.3b When $V^{100}_{\text{growth}} = V^{111}_{\text{growth}}$ the radius $r^{100}_c \approx 2\,\mu$m and r^{111}_c is practically zero (magnitude $10^{-5}\,\mu$m). This fact agrees well with the observation that the facets with the slowest growth rate always win.

9.4 $\Delta T_{\text{kin}} = 2.4$ K $\Delta T_{\text{mass}} = 2.6$ K $\beta_{\text{kin}} = 7$ m/s.

9.5 The degrees of segregation of Ni, Cr and Mo in the solid steel are 1.5, 1.6 and 3.1, respectively.

9.6 Scheil's equation and Scheil's modified equation can not be used to study the micro-segregation in the centres of the dendrites. Instead the function

$$x(0,t) = x^{s/L} - \left(x^{s/L} - x_{\min}\right) e^{-\frac{4\pi^2 D_s}{\lambda^2} t}$$

is used. The calculated values for the Cu mole fractions in the dendrite centres with consideration to backdiffusion is calculated to 0.011 and 0.013, respectively, which can be compared to the experimental values 0.013 and 0.015, respectively. They differ very much from the experimental values. Obviously back diffusion can not explain the deviation from the experimental values.

As is shown from the phase diagram the alloy with the initial Cu concentration 0.025 starts to solidify at a temperature at about 640 °C (above the line $T = 900$ K). This temperature is very much higher than the equilibrium temperature 548 °C. The most likely reason is that the theory is not valid far from equilibrium.

9.7 $C = 7.4 \times 10^2$ m^3/s.
The experimental value of the constant C is approximately 1000 times larger than the value expected for a surface-tension-driven process. A reasonable explanation could be that the coarsening process is a result of strong deviation from equilibrium. Coarsening is the way for the material to achieve equilibrium.

9.8 **Table 9.8a**

T_{quench} °C	$f_{\text{experimental}}$ wt%	f_{Scheil} wt%
630	0.70	0.33
620	0.82	0.23

Table 9.8b

T_{quench} °C	$c^{s}_{\text{experimental}}$ wt%	c^{s}_{Scheil} wt%
630	0.52	0.38
620	0.95	0.29

The experimental values and the figure in the text show that stirring results in a higher Cu concentration than no stirring. The volume fraction of solid was also larger with than without stirring.

The theoretical calculations deviate strongly from the experimental values for the fraction of solid. The reason is the same as in Exercise 7.6. The theory implies equilibrium and at quench temperatures of 620–630 °C we are very far from equilibrium temperature (548 °C).

Chapter 10

10.1 According to the theory in Chapter 10 [Equation (10. 54) on page 604]

$$\Delta T_E = \frac{1}{2}\left(m^L_\alpha + m^L_\beta\right)\left(x^L_E - x^L_0\right) + \sqrt{\frac{\sigma_{\alpha\beta} A_{\text{lamella}} B_{\text{lamella}} C_{0\ \text{lamella}}}{RTD_L}}\sqrt{V_{\text{growth}}}$$

the slope of the lines in the figure in the text is identical with the expression

$$\text{Slope} = \sqrt{\frac{\sigma_{\alpha\beta} A_{\text{lamella}} B_{\text{lamella}} C_{0\ \text{lamella}}}{RTD_{\text{L}}}} \quad \text{where } A_{\text{lamella}} = \frac{m_{\beta}^{\text{L}} V_{\text{m}}^{\beta}}{f_{\beta}} - \frac{m_{\alpha}^{\text{L}} V_{\text{m}}^{\alpha}}{f_{\alpha}}$$

$$B_{\text{lamella}} = \frac{m_{\beta}^{\text{L}}}{V_{\text{m}}^{\beta}} - \frac{m_{\alpha}^{\text{L}}}{V_{\text{m}}^{\alpha}} \quad \text{and} \quad C_{0\ \text{lamella}} = f_{\alpha} f_{\beta} V_{\text{m}}^{\text{L}} (x_{\text{E}}^{\beta/\text{L}} - x_{\text{E}}^{\alpha/\text{L}})$$

The experimental value of the constant is of the magnitude $1.0 \times 10^2\ \text{K}/\sqrt{\text{m/s}}$ to be compared with the theoretical value $1.4 \times 10^2\ \text{K}/\sqrt{\text{m/s}}$. The discrepancy can be explained by the uncertainties in the selected values of the surface tension σ and especially the diffusion constant D_{L}. Hence, the agreement is acceptable.

10.2 The theoretical value of the constant C_{rod} equals $2.8 \times 10^{-14}\ \text{m}^3/\text{s}$. The experimental value is $2.0 \times 10^{-14}\ \text{m}^3/\text{s}$.

The discrepancy can be explained by the uncertainties in the selected values of the surface tension σ and especially the diffusion constant D_{L}. Hence, the agreement is acceptable.

10.3a $V_{\text{growth}} = \dfrac{0.72 \times 10^{-4}}{s}$

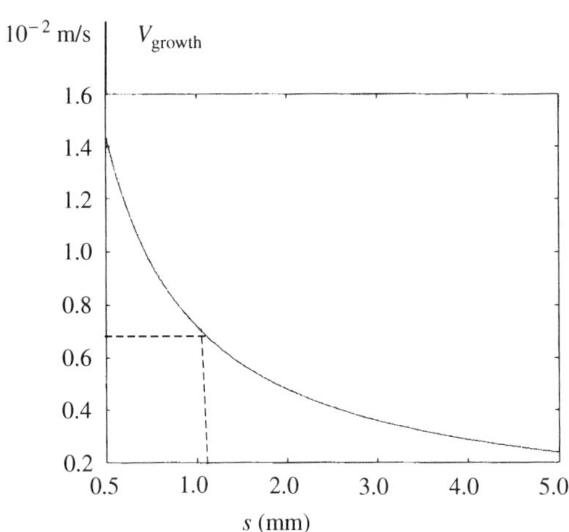

Figure 10.3a

10.3b $V_{\text{growth}}^{\text{cr}} = 0.64 \times 10^{-2}$ m/s.

Reading in the Figure 10.3a gives $s_{\text{cr}} = 11$ mm for $V_{\text{growth}}^{\text{cr}} = 0.64 \times 10^{-2}$ m/s.

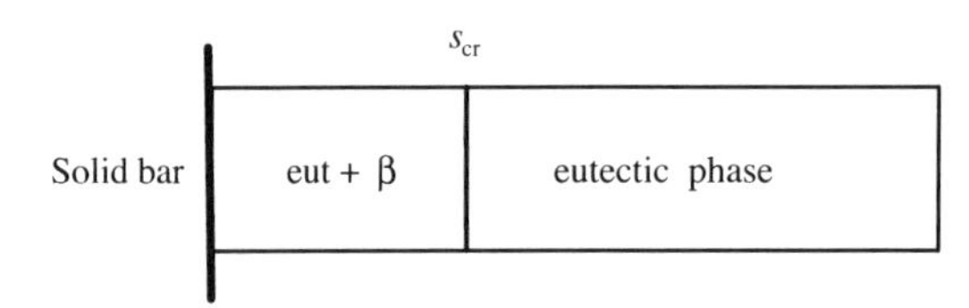

Figure 10.3–1

Eutectic phase consists of α and β phases at constant proportions.

10.4a

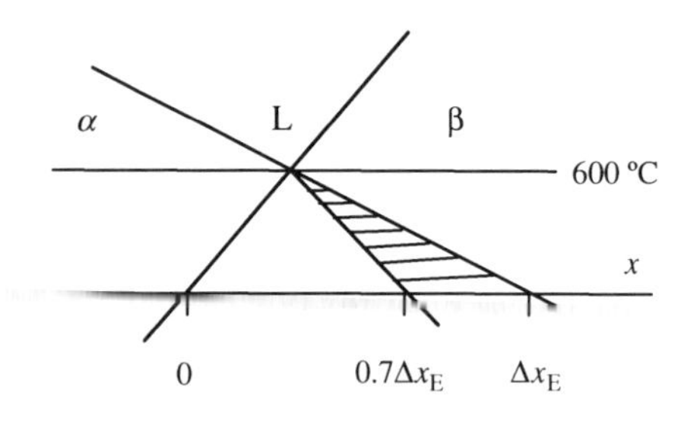

Figure 10.4–3

Hint: Use the standard method, described in Chapter 10 on pages 628–633, to find the boundary line between the α phase and the coupled zone and the β phase and the coupled zone, respectively.

In the present case the coupled zone touches the prolonged liquidus (α) line because $\Delta x_\beta = \Delta x_E$.

10.4b $\mu_E \geq 5.8 \times 10^{-6}$ m/s. The coupled zone disappears if $\Delta x_\alpha + \Delta x_\beta - \Delta x_E \leq 0$.

10.5 Scheil's equation for a binary eutectic reaction can be written as

$$x_P^{L\ bin} = x_P^{0\ L\ bin}(1 - f_{\alpha+\beta})^{-\left[(1-k_P^{\alpha/L})f_\alpha + (1-k_P^{\beta/L})f_\beta\right]}$$

The desired function is

$$f_{\alpha+\beta} = 1 - \left(\frac{x_P^0}{0.101}\right)^{1/0.818} = 1 - 0.101^{-1.22} \times (x_P^0)^{1.22}$$

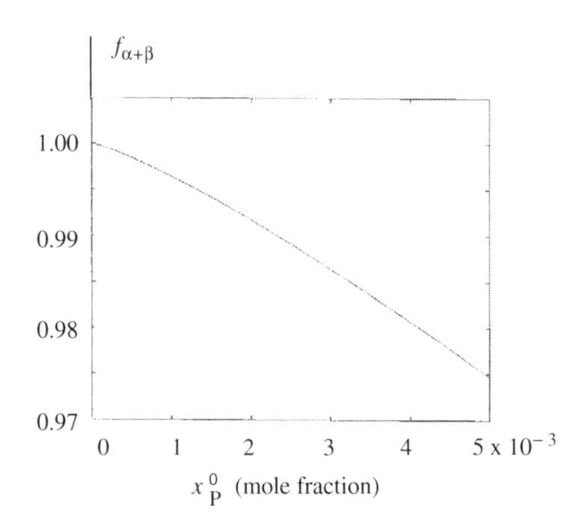

Figure 10.5–3

10.6 – Anisotropy of the crystals, which influences the surface tension between the two solid phases. This topic will be discussed briefly on page 668 in terms of surface energy.
– The cooling conditions influence the type of structures, as is shown in the figure in the text.
– Small fractions of impurities are also known to be able to change the type of structure.

A very common explanation is that lamellar eutectic structure is promoted by favourable surface-tension conditions between the two growing phases α and β. These conditions are probably no longer valid at higher growth rates, which gives the sudden transition shown in the figure in the text.

Chapter 11

11.1 **Peritectic reaction:** A reaction where a liquid phase L reacts with a primarily precipitated phase α and forms a new secondary phase β.

$$L + \alpha \rightarrow \beta$$

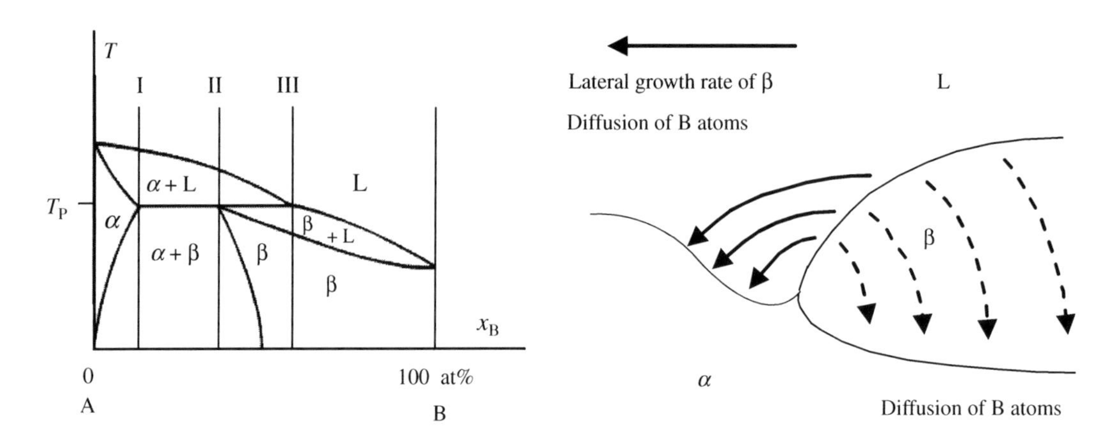

Figure 11.1–1 **Figure 11.1–2**

Peritectic transformation: The secondary process, which follows after a peritectic reaction, when the α phase is transformed into β phase.

$$\alpha + \text{B atoms} \rightarrow \beta$$

The β phase grows around the α phase and cover it. B atoms diffuse through the β layer and react with the α phase and form β phase. The β phase becomes thicker at the expense of the α phase. The α phase dissolves at the interphase.

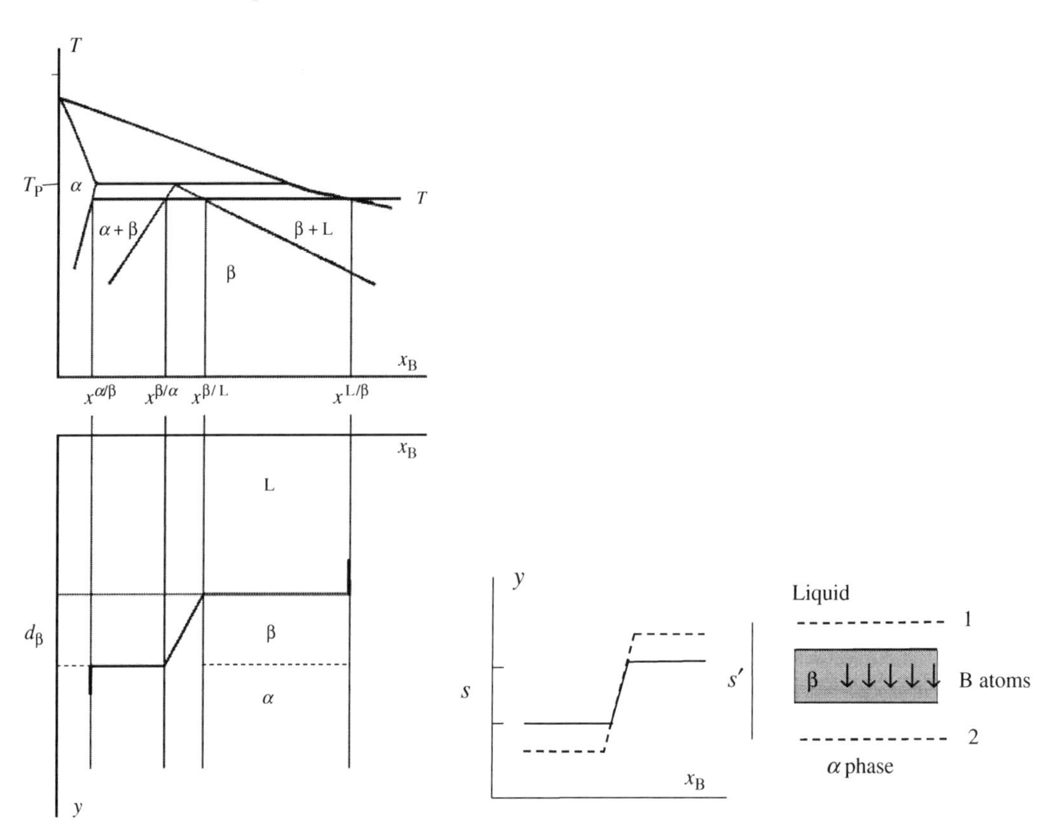

Figure 11.1c **Figure 11.1d**

At interphase 1 the β phase will precipitate from the melt. B atoms will diffuse through the β layer and increase the B concentration in the α phase, which gives a dissolution of the α phase at interphase 2.

11.2a Eutectic solidification of a liquid is defined as a simultaneous precipitation of two or more phases via a eutectic reaction L $\rightarrow \alpha + \beta$.

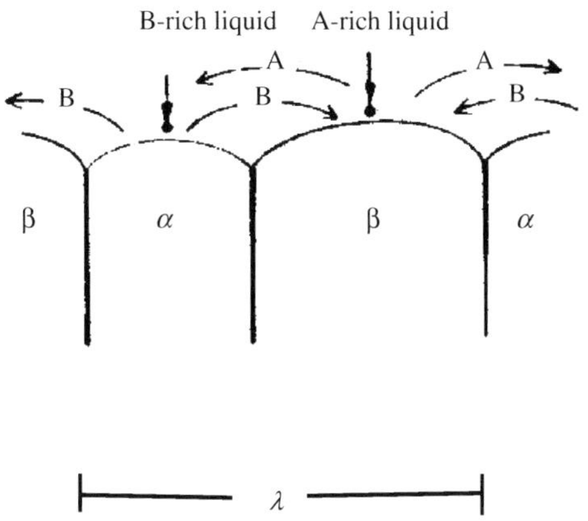

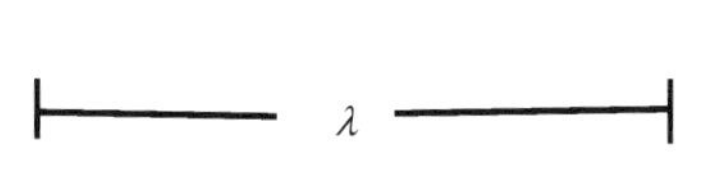

Figure 11.2a (left)

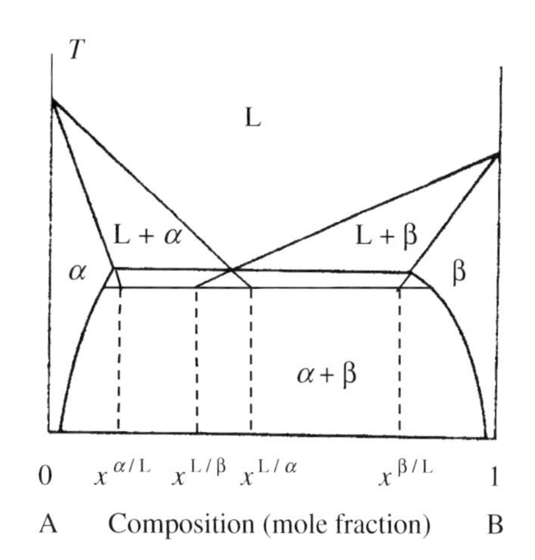

Figure 11.2a (right)

11.2b The peritectic phase diffusion in the melt during the transformation has been described in Exercise Figure 11.1. The eutectic diffusion in the melt close to the lamellas has been discussed in Exercise Figure 11.2a. A comparison between Exercise Figure 11.1b and Exercise Figure 11.2a gives the desired comparison between the diffusion in the melt in the two cases.

Peritectic reaction

Diffusion in the melt:

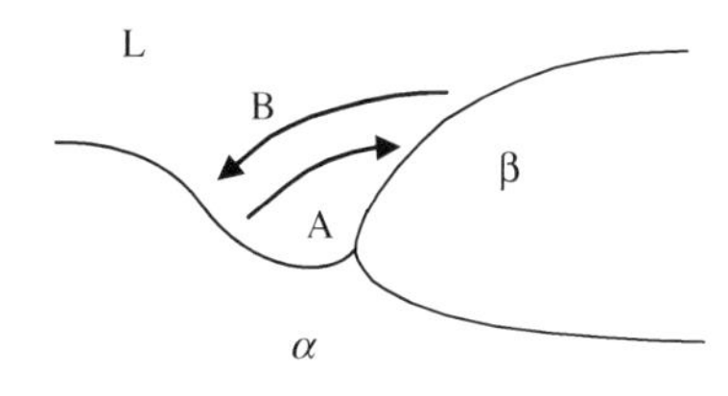

$L + \alpha \rightarrow \beta$

Peritectic transformation goes on until only β phase remains.

Eutectic reaction

Diffusion in the melt

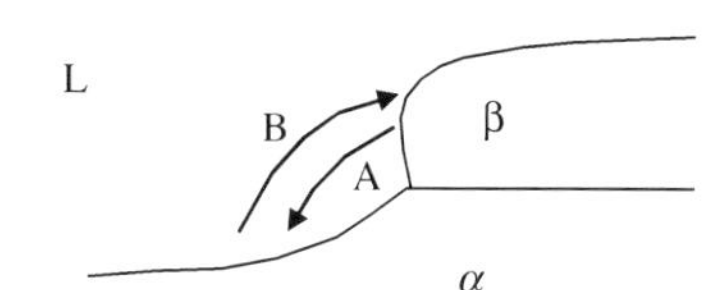

$L \rightarrow \alpha + \beta$

A mixture of $\alpha + \beta$ phases at constant proportions is formed.

11.3a $C = \dfrac{d}{\sqrt{t}} = 0.224 \times 10^{-6}\ \text{m s}^{-0.5}$.

11.3b $D_\varepsilon = \dfrac{C^2}{2} \times \dfrac{1}{\left(x^{\varepsilon/L} - x^{\varepsilon/\delta}\right) \times \left(\dfrac{1}{x^{\varepsilon/\delta} - x^{\delta/\varepsilon}} + \dfrac{1}{x^{L/\varepsilon} - x^{\varepsilon/L}}\right)} \approx 1 \times 10^{-12}\ \text{m}^2/\text{s}.$

D_ε is of the same magnitude as D_{Cd}. The Cd atoms diffuse through an intermediate phase ε, which has a complicated structure and is harder to pass for the Cd atoms than a pure Cd melt. The passage of the intermediate phase may reduce the diffusion constant strongly. In the present case the distance through the intermediate phase is short and no reduction is observed.

11.4 $\sigma^{\delta/\gamma} = 1.9\ \text{J/m}^2\ \sigma_{el} = 0.2\ \text{J/m}^2\ \sigma_{ch} = 1.7\ \text{J/m}^2$. The experimental result in the present case violates strongly the experience that solid–solid maximum surface energies are of the magnitude of 0.2 J/m^2. The solidification process occurs far from equilibrium.

Resulting Surface Tension at Equilibrium

At equilibrium the angle θ is 75° instead of 15°. This case, which is **not** valid here, corresponds to the figures below. It agrees with the normal magnitudes of surface tensions.

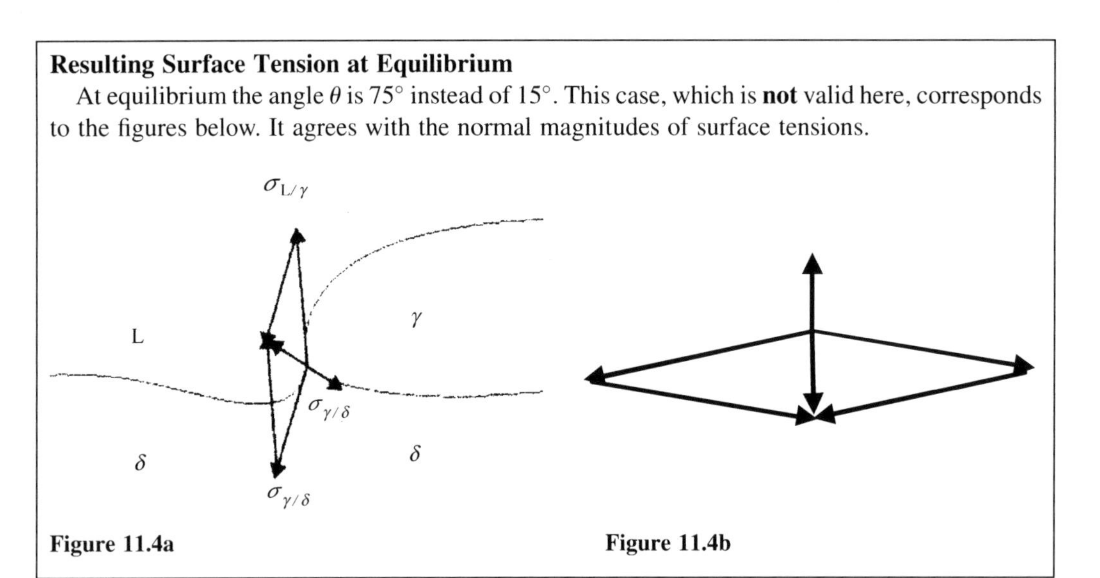

Figure 11.4a **Figure 11.4b**

11.5 *Temperature interval* 900–665 °C:
Cooling until β phase (Al_3Ti) starts to precipitate at 820 °C. The compositions of the precipitated β phase and the melt change gradually and follow the bent curve down to the point (0.15 wt%, 665 °C).

Temperature interval 665–658.5 °C:
At 665 °C a peritectic reaction occurs.

$$L + \beta \rightarrow \alpha$$

α phase precipitates outside of the β phase as is shown in Figure 11.4a. The α phase consists of solid Al + 0.15 wt% Ti.

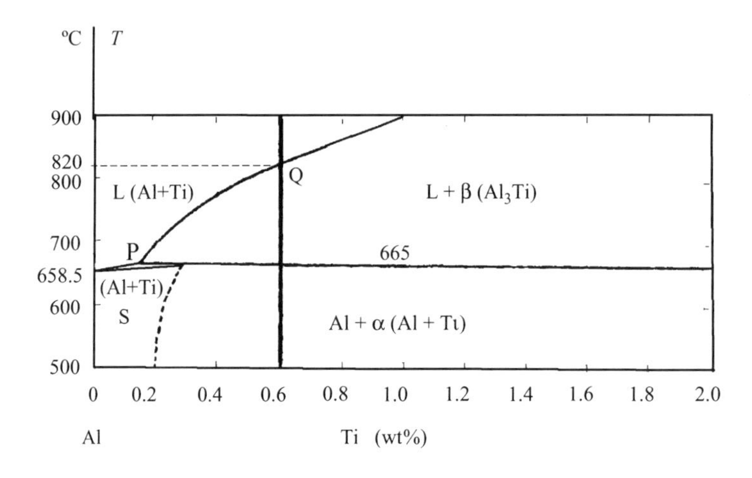

Figure 11.5–2

The layer of α phase grows successively until the temperature has decreased down to 658.5 °C, the melting point of pure aluminium. The composition of the melt and the precipitated α phase decreases and follows the curve PM. At point M the temperature is T_M: and the sample is almost solid. The last solidified melt consists of pure Al phase.

Solidification patterns a *and* b:
Local variations of the Ti concentration in the melt during the temperature interval 665–658.5 °C.

11.6 Two criteria must be fulfilled if a eutectic structure is to arise:

$$V^{\alpha}_{growth} = V^{\beta}_{growth} \text{ that corresponds to } \frac{(x^{L/\alpha}-x^{L}_{0})^2}{x^{L/\alpha}-x^{\alpha/L}} = \frac{(x^{L/\beta}-x^{L}_{0})^2}{x^{L/\beta}-x^{\beta/L}}$$

$$f_{\alpha}+f_{\beta} = 1 \text{ that corresponds to } \frac{x^{L/\alpha}-x^{L}_{0}}{x^{L/\alpha}(1-k^{\alpha/L})} + \frac{x^{L/\beta}-x^{L}_{0}}{x^{L/\beta}(1-k^{\beta/L})} = 1$$

These relationships are plotted in the phase diagram. The curves intersect at point E where both relationships are exactly valid. Eutectic-like structure is precipitated there and in the close vicinity.

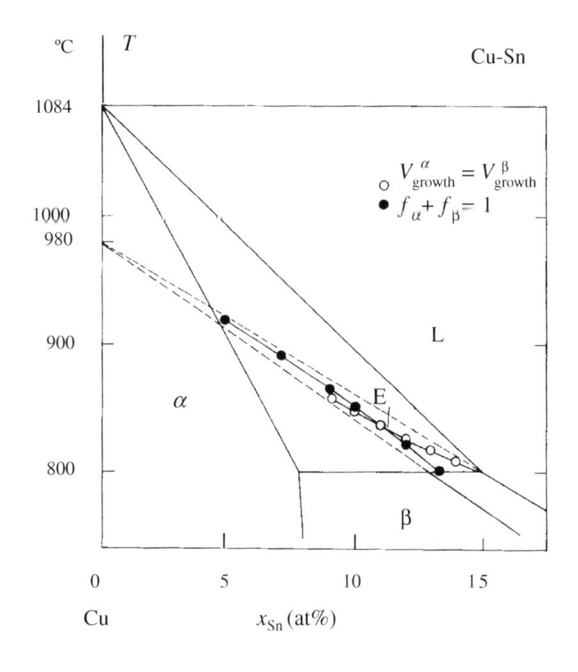

Figure 11.6–4

At point E both criterion are fulfilled. In the vicinity of point E the criteria are almost fulfilled but outside this region the conditions are not fulfilled and the structures lack the eutectic-like appearance.

Chapter 12

12.1 Comparison between Figures 12.2 on page 733 and Figure 12.35 on page 756 both in Chapter 12 and the text figure show that a glass-transition process is a second-order transition.

12.2 $y = a(T_M-b)\ln\dfrac{T_M-b}{T_g-b} - a(T_M-T_g)$

where

T_g = glass transition temperature
T_M = melting point temperature

$$a = \frac{s\rho c_p \beta_{kin}}{h+\rho(-\Delta H)\beta_{kin}} \quad \text{and} \quad b = \frac{hT_{mould}+\rho(-\Delta H)\beta_{kin}T_M}{h+\rho(-\Delta H)\beta_{kin}}$$

12.3 With the aid of readings in the diagram you obtain:
1 Si atom can accommodate ≈ 8 surrounding Al atoms.
1 Al atom can accommodate ≈ 15 surrounding Si atoms.
1 Al cluster (1 Al + 15 Si) can accommodate ≈ 12 Si clusters (1 Si + 8 Al).

The composition of a melt of Al compound clusters will consist of ≈22 at% Si atoms and ≈78 at% Al, i.e. $x_{Si} \approx 0.22$ and $x_{Al} \approx 0.78$.

12.4 $T_{max}^{growth} = \dfrac{QT_M}{Q + RT_M}$ and $T_{nucl} = T_M - Const \times \dfrac{\sigma^{3/2}}{-\Delta H}$

The growth condition depends strongly on the activation energy Q. A melt with clusters where the interaction between the atoms is large has a large Q value. T_{max}^{growth} is close to but always lower than T_M. Melts with strong interaction between their atoms have large negative values of their heat of mixing.

The undercooling for nucleation increases with increasing surface tension and decreasing heat of fusion.

12.5

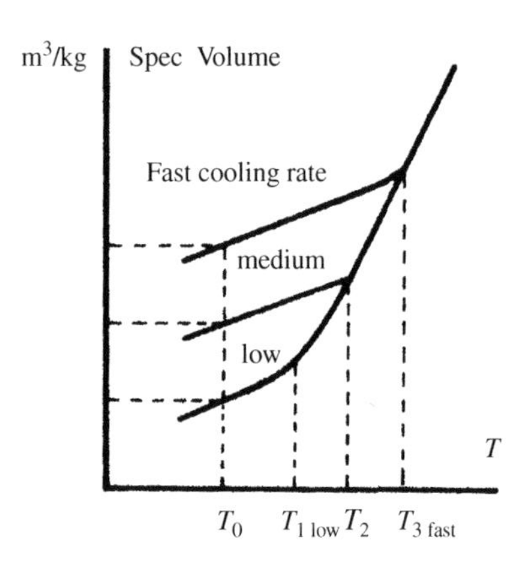

Figure 12.5–1

A melt consists of clusters and free atoms. The number of clusters increases with decreasing temperature. Time is required to achieve equilibrium cluster structure. At high cooling rates the equilibrium structure will not be reached until the glass transition occurs.

The glass material formed at high cooling rates contains a large fraction of 'lattice defects', which results in a larger volume than the volume formed at low cooling rates. In the latter case the cluster structure in the melt has been fully developed before the glass transition occurs.

12.6 At low temperatures the number of clusters is large and their mobilities are low.
At high temperatures the number of clusters is low and the mobilities of the clusters and free atoms are high. The viscosities at low and high temperatures reflect these facts.

Index

In addition to the current list of specimens you may look under the letters related to the particular chapters listed below.

Chapter	Title	Letter
1	Thermodynamic Concepts and Relationships	T
2	Thermodynamic Analysis of Solidification Processes in Metals and Alloys	T
3	Properties of Interfaces	P
4	Nucleation	N
5	Crystal Growth in Vapours	C
6	Crystal Growth in Liquids and Melts	C
7	Heat Transport during Solidification Processes	H
7	Thermal Analysis	T
8	Crystal Growth Controlled by Heat and Mass Transport	C
9	Dendritic Solidification Structures	D
9	Faceted Solidification Structures	F
10	Eutectic Solidification Structures	E
11	Peritectic Solidification Structures	P
12	Metallic Glasses and Amorphous Alloy Structures	M

Solidification and Crystallization Processing in Metals and Alloys, First Edition. Hasse Fredriksson and Ulla Åkerlind.